THE *PRINCIPIA*

The publication of this work has been made possible in part by a grant from the National Endowment for the Humanities, an independent federal agency. The publisher also gratefully acknowledges the generous contribution to this book provided by the General Endowment Fund of the Associates of the University of California Press.

Portrait of Isaac Newton at about the age of sixty, a drawing presented by Newton to David Gregory. For details see the following page.

Portrait of Isaac Newton at about the age of sixty, presented by Newton to David Gregory (1661–1708). This small oval drawing (roughly 3¾ in. from top to bottom and 3¼ in. from left to right) is closely related to the large oval portrait in oils made by Kneller in 1702, which is considered to be the second authentic portrait made of Newton. The kinship between this drawing and the oil painting can be seen in the pose, the expression, and such unmistakable details as the slight cast in the left eye and the button on the shirt. Newton is shown in both this drawing and the painting of 1702 in his academic robe and wearing a luxurious wig, whereas in the previous portrait by Kneller (now in the National Portrait Gallery in London), painted in 1689, two years after the publication of the *Principia*, Newton is similarly attired but is shown with his own shoulder-length hair.

This drawing was almost certainly made after the painting, since Kneller's preliminary drawings for his paintings are usually larger than this one and tend to concentrate on the face rather than on the details of the attire of the subject. The fact that this drawing shows every detail of the finished oil painting is thus evidence that it was copied from the finished portrait. Since Gregory died in 1708, the drawing can readily be dated to between 1702 and 1708. In those days miniature portraits were commonly used in the way that we today would use portrait photographs. The small size of the drawing indicates that it was not a copy made in preparation for an engraved portrait but was rather made to be used by Newton as a gift.

The drawing captures Kneller's powerful representation of Newton, showing him as a person with a forceful personality, poised to conquer new worlds in his recently gained position of power in London. This high level of artistic representation and the quality of the drawing indicate that the artist responsible for it was a person of real talent and skill.

The drawing is mounted in a frame, on the back of which there is a longhand note reading: "This original drawing of Sir Isaac Newton, belonged formerly to Professor Gregory of Oxford; by him it was bequeathed to his youngest son (Sir Isaac's godson) who was later Secretary of Sion College; & by him left by Will to the Revd. Mr. Mence, who had the Goodness to give it to Dr. Douglas; March 8th 1870."

David Gregory first made contact with Newton in the early 1690s, and although their relations got off to a bad start, Newton did recommend Gregory for the Savilian Professorship of Astronomy at Oxford, a post which he occupied until his death in 1708. As will be evident to readers of the Guide, Gregory is one of our chief sources of information concerning Newton's intellectual activities during the 1690s and the early years of the eighteenth century, the period when Newton was engaged in revising and planning a reconstruction of his *Principia*. Gregory recorded many conversations with Newton in which Newton discussed his proposed revisions of the *Principia* and other projects and revealed some of his most intimate and fundamental thoughts about science, religion, and philosophy. So far as is known, the note on the back of the portrait is the only record that Newton stood godfather to Gregory's youngest son.

ISAAC NEWTON

THE *PRINCIPIA*

Mathematical Principles of Natural Philosophy

A New Translation
by I. Bernard Cohen and Anne Whitman
assisted by Julia Budenz

Preceded by

A GUIDE TO NEWTON'S *PRINCIPIA*
by I. Bernard Cohen

UNIVERSITY OF CALIFORNIA PRESS
Berkeley Los Angeles London

University of California Press
Berkeley and Los Angeles, California

University of California Press
London, England

Library of Congress Cataloging-in-Publication Data

Newton, Isaac, Sir, 1642–1727.
[Principia. English]
The Principia : mathematical principles of natural philosophy / Isaac Newton ; a new translation by I. Bernard Cohen and Anne Whitman assisted by Julia Budenz ; preceded by a guide to Newton's Principia by I. Bernard Cohen.
p. cm.
Includes bibliographical references and index.
ISBN 0-520-08816-6 (alk. paper).—ISBN 0-520-08817-4 (alk. paper)
1. Newton, Isaac, Sir, 1642–1727. Principia. 2. Mechanics—Early works to 1800. 3. Celestial mechanics—Early works to 1800. I. Cohen, I. Bernard, 1914– . II. Whitman, Anne Miller, 1937–1984. III. Title.
QA803.N413 1999
531—dc21 99-10278
CIP

Printed in the United States of America

9 8 7 6 5 4 3 2 1

The paper used in this publication meets the minimum requirements of American National Standard for Information Sciences—Permanence of Paper for Printed Library Materials, ANSI Z39.48-1984

This translation is
dedicated to

D. T. WHITESIDE

with respect and affection

Contents

Preface *xi*

A GUIDE TO NEWTON'S *PRINCIPIA* 1

Contents of the Guide *3*

Abbreviations *9*

CHAPTER ONE: A Brief History of the *Principia* 11

CHAPTER TWO: Translating the *Principia* 26

CHAPTER THREE: Some General Aspects of the *Principia* 43

CHAPTER FOUR: Some Fundamental Concepts of the *Principia* 85

CHAPTER FIVE: Axioms, or the Laws of Motion 109

CHAPTER SIX: The Structure of Book 1 128

CHAPTER SEVEN: The Structure of Book 2 161

CHAPTER EIGHT: The Structure of Book 3 195

CHAPTER NINE: The Concluding General Scholium 274

CHAPTER TEN: How to Read the *Principia* 293

CHAPTER ELEVEN: Conclusion 369

THE *PRINCIPIA* (Mathematical Principles of Natural Philosophy) 371

Halley's Ode to Newton *379*

Newton's Preface to the First Edition *381*

Newton's Preface to the Second Edition *384*

Cotes's Preface to the Second Edition *385*

Newton's Preface to the Third Edition *400*

Definitions 403

Axioms, or the Laws of Motion 416

BOOK 1: THE MOTION OF BODIES 431

BOOK 2: THE MOTION OF BODIES 631

BOOK 3: THE SYSTEM OF THE WORLD 791

General Scholium *939*

Contents of the Principia *945*

Index of Names *973*

Preface

ALTHOUGH NEWTON'S *PRINCIPIA* has been translated into many languages, the last complete translation into English (indeed, the only such complete translation) was produced by Andrew Motte and published in London more than two and a half centuries ago. This translation was printed again and again in the nineteenth century and in the 1930s was modernized and revised as a result of the efforts of Florian Cajori. This latter version, with its partial modernization and partial revision, has become the standard English text of the *Principia*.

Motte's version is often almost as opaque to the modern reader as Newton's Latin original, since Motte used such older and unfamiliar expressions as "sub-sesquialterate" ratio. Additionally, there are statements in which the terms are no longer immediately comprehensible today, such as book 3, prop. 8, corol. 3, in which Motte writes that "the densities of dissimilar spheres are as those weights applied to the diameters of the spheres," a statement unaltered in the Motte-Cajori version. Of course, a little thought reveals that Newton was writing about the densities of nonhomogeneous spheres and was concluding with a reference to the weights divided by the diameters. The Motte-Cajori version, as explained in §2.3 of the Guide to the present translation, is also not satisfactory because it too is frequently difficult to read and, what is more important, does not always present an authentic rendition of Newton's original. The discovery of certain extraordinary examples in which scholars have been misled in this regard was a chief factor in our decision to produce the present translation.

When we completed our Latin edition, somewhat awed by the prospect of undertaking a wholly new translation, we thought of producing a new edition of Motte's English version, with notes that either would give the reader a modern equivalent of difficult passages in Motte's English prose or would contain some aids to help the reader with certain archaic mathematical expressions. That is, since Motte's text had been a chief means of disseminating Newton's science for over two centuries, we considered treating it as an important historical document in its own right. Such a plan was announced in our Latin edition, and we even prepared a special interleaved copy of the facsimile of the 1729 edition to serve as our working text.*

After the Latin edition appeared, however, many colleagues and some reviewers of that edition insisted that it was now our obligation to produce a completely new translation of the *Principia*, rather than confine our attentions to Motte's older pioneering work. We were at first reluctant to accept this assignment, not only because of the difficulty and enormous labor involved, but also because of our awareness that we ourselves would thereby become responsible for interpretations of Newton's thought for a long period of time.

Goaded by our colleagues and friendly critics, Anne Whitman and I finally agreed to produce a wholly new version of the *Principia*. We were fortunate in obtaining a grant from the National Science Foundation to support our efforts. Many scholars offered good advice, chief among them our good friends D. T. Whiteside and R. S. Westfall. In particular, Whiteside stressed for us that we should pay no attention to any existing translation, not even consulting any other version on occasions when we might be puzzled, until after our own assignment had been fully completed. Anyone who has had to translate a technical text will appreciate the importance of this advice, since it is all too easy to be influenced by other translations, even to the extent of unconsciously repeating their errors. Accordingly, during the first two or three rounds of translation and revision, we recorded puzzling or doubtful passages, and passages for which we hoped to produce a final version that would be less awkward than our preliminary efforts, reserving for some later time a possible comparison of our version with others. It should be noted that in the final two rounds of our revision, while checking some difficult passages and comparing some of our renditions with others, the most useful works for such purpose were Whiteside's own translation of an early draft of what corresponds to most of book 1 of the *Principia* and the French translation made in the

*An Index Verborum of the Latin edition of the *Principia* has been produced by Anne Whitman and I. Bernard Cohen in association with Owen Gingerich and Barbara Welther. This index includes the complete text of the third edition (1726) and also the variant readings as given in the Latin edition of the *Principia* (edited by Alexandre Koyré, I. Bernard Cohen, and Anne Whitman), published in 1972 by the Harvard University Press and the Cambridge University Press. Thus the Index includes the complete text of the three authorized Latin editions (1687, 1713, 1726) as well as the MS annotations in both Newton's "annotated" and "interleaved" personal copies of the first and second editions. The Index is on deposit in the Burndy Library of the Dibner Institute (Cambridge, Mass.), where it may be consulted. Microfilm copies can be purchased.

Very useful tools for scholars and students are the planned Octavo editions of the first and third Latin editions of Newton's *Principia*; the latter will include this English translation. The high-resolution facsimiles on CD-ROM allow readers to view the original book and search the complete Latin texts and translation. For publication information, see the Octavo web site: www.octavo.com.

mid-eighteenth century by the marquise du Châtelet. On some difficult points, we also profited from the exegeses and explanations in the Le Seur and Jacquier Latin edition and the Krylov Russian edition. While neither Anne Whitman nor I could read Russian, we did have the good fortune to have two former students, Richard Kotz and Dennis Brezina, able and willing to translate a number of Krylov's notes for us.

The translation presented here is a rendition of the third and final edition of Newton's *Principia* into present-day English with two major aims: to make Newton's text understandable to today's reader and yet to preserve Newton's form of mathematical expression. We have thus resisted any temptation to rewrite Newton's text by introducing equations where he expressed himself in words. We have, however, generally transmuted such expressions as "subsesquiplicate ratio" into more simply understandable terms. These matters are explained at length in §§10.3–10.5 of the Guide.

After we had completed our translation and had checked it against Newton's Latin original several times, we compared our version with Motte's and found many of our phrases to be almost identical, except for Motte's antique mathematical expressions. This was especially the case in the mathematical portions of books 1 and 2 and the early part of book 3. After all, there are not many ways of saying that a quantity A is proportional to another quantity B. Taking into account that Motte's phrasing represents the prose of Newton's own day (his translation was published in 1729) and that in various forms his rendition has been the standard for the English-reading world for almost three centuries, we decided that we would maintain some continuity with this tradition by making our phrasing conform to some degree to Motte's. This comparison of texts did show, however, that Motte had often taken liberties with Newton's text and had even expanded Newton's expressions by adding his own explanations—a result that confirmed the soundness of the advice that we not look at Motte's translation until *after* we had completed our own text.

This translation was undertaken in order to provide a readable text for students of Newton's thought who are unable to penetrate the barrier of Newton's Latin. Following the advice of scholarly friends and counselors, we have not overloaded the translation with extensive notes and comments of the sort intended for specialists, rather allowing the text to speak for itself. Much of the kind of editorial comment and explanation that would normally appear in such notes may be found in the Guide. Similarly, information concerning certain important changes in the text from edition to edition is given in the Guide, as well as in occasional textual notes. The table of contents for the Guide, found on pages 3–7, will direct the reader to specific sections of the *Principia*, or even to particular propositions.

The Guide to the present translation is intended to be just that—a kind of extended road map through the sometimes labyrinthine pathways of the *Principia*. Some propositions, methods, and concepts are analyzed at length, and in some instances critical details of Newton's argument are presented and some indications are given of the alterations produced by Newton from one edition to the next. Sometimes reference is made to secondary works where particular topics are discussed, but no attempt has been made to indicate the vast range of scholarly information relating to this or that point. That is, I have tended to cite, in the text and in the footnote references, primarily those works that either have been of special importance for my understanding of some particular point or some sections of the *Principia* or that may be of help to the reader who wishes to gain a more extensive knowledge of some topic. As a result, I have not had occasion in the text to make public acknowledgment of all the works that have been important influences on my own thinking about the *Principia* and about the Newtonian problems associated with that work. In this rubric I would include, among others, the important articles by J. E. McGuire, the extremely valuable monographs on many significant aspects of Newton's science and philosophic background by Maurizio Mamiani (which have not been fully appreciated by the scholarly world because they are written in Italian), the two histories of mechanics by René Dugas and the antecedent documentary history by Léon Jouguet, the analysis of Newton's concepts and methods by Pierre Duhem and Ernst Mach, and monographic studies by Michel Blay, G. Bathélemy, Pierre Costabel, and A. Rupert Hall, and by François de Gandt.

I also fear that in the Guide I may not have sufficiently stressed how greatly my understanding of the *Principia* has profited from the researches of D. T. Whiteside and Curtis Wilson and from the earlier commentaries of David Gregory, of Thomas Le Seur and François Jacquier, and of Alexis Clairaut. The reader will find, as I have done, that R. S. Westfall's *Never at Rest* not only provides an admirable guide to the chronology of Newton's life and the development of his thought in general, but also analyzes the whole range of Newton's science and presents almost every aspect of the *Principia* in historical perspective.

All students of the *Principia* find a guiding beacon in D. T. Whiteside's essays and his texts and commentaries in his edition of Newton's *Mathematical Papers*, esp. vols. 6 and 8 (cited on p. 9 below). On Newton's astronomy, the concise analysis by Curtis Wilson (cited on p. 10 below) has been of enormous value. Many of the texts quoted in the Guide have been translated into English. It did not seem necessary to mention this fact again and again.

From the very start of this endeavor, Anne Whitman and I were continuously aware of the awesome responsibility that was placed on our shoulders, having in mind all too well the ways in which even scholars of the highest distinction have been misled by inaccuracies and real faults in the current twentieth-century English version. We recognized that no translator or editor could boast of having perfectly understood Newton's text and of having found the proper meaning of every proof and construction. We have ever been aware that a translation of a work as difficult as Newton's *Principia* will certainly contain some serious blunders or errors of interpretation. We were not so vain that we were always sure that we fully understood every level of Newton's meaning. We took comfort in noting that even Halley, who probably read the original *Principia* as carefully as anyone could, did not always fully understand the mathematical significance of Newton's text. We therefore, in close paraphrase of Newton's own preface to the first edition, earnestly ask that everything be read with an open mind and that the defects in a subject so difficult may be not so much reprehended as kindly corrected and improved by the endeavors of our readers. Readers are requested to send to our publisher any suggestions for the improvement of our text as well as corrections of errors of fact or interpretation in the Guide.

I. B. C.

Some Acknowledgments

Anne Whitman died in 1984, when our complete text was all but ready for publication, being our fourth (and in some cases fifth and even sixth) version. It was her wish, as well as mine, that this translation be dedicated to the scholar whose knowledge of almost every aspect of Newton's mathematics, science, and life is unmatched in our time and whose own contributions to knowledge have raised the level of Newtonian scholarship to new heights.

We are fortunate that Julia Budenz has been able to help us with various aspects of producing our translation and especially in the final stages of preparing this work for publication.

It has been a continual joy to work with the University of California Press. I am especially grateful to Elizabeth Knoll, for her thoughtfulness with regard to every aspect of converting our work into a printed book, and to Rose Vekony, for the care and wisdom she has exercised in seeing this complex work through production, completing the assignment so skillfully begun by Rebecca Frazier. I have profited greatly from the many wise suggestions made by Nicholas Goodhue, whose command of Latin has made notable improvements in both the Guide and the translation. One of the fortunate aspects of having the translation published by the University of California Press is that we have been able to use the diagrams (some with corrections) of the older version.

I gladly acknowledge and record some truly extraordinary acts of scholarly friendship. Three colleagues—George E. Smith, Richard S. Westfall, and Curtis Wilson—not only gave my Guide a careful reading, sending me detailed commentaries for its improvement; these three colleagues also checked our translation and sent me many pages of detailed criticisms and useful suggestions for its improvement. I am particularly indebted to George Smith of Tufts University for having allowed me to make use of his as yet unpublished *Companion to Newton's "Principia,"* a detailed analysis of the *Principia* proposition by proposition. Smith used the text of our translation in his seminar on the *Principia* at Tufts during the academic year 1993/94 and again during 1997/98. Our final version has profited greatly from the suggestions of the students, who were required to study the actual text of the *Principia* from beginning to end. I am happy to be able to include in the Guide a general presentation he has written (in his dual capacity as a philosopher of science and a specialist in fluid mechanics) on the contents of book 2 and also two longish notes, one on planetary perturbations, the other on the motion of the lunar apsis. I have also included a note by Prof. Michael Nauenberg of the University of California, Santa Cruz, on his current research into the origins of some of Newton's methods.

I am grateful to the University Library, Cambridge, for permission to quote extracts and translations of various Newton MSS. I gladly record here my deep gratitude to the staff and officers of the UL for their generosity, courtesy, kindness, and helpfulness over many years.

I am especially grateful to Robert S. Pirie for permission to reproduce the miniature portrait which serves as frontispiece to this work. The following illustrations are reproduced, with permission, from books in the Grace K. Babson Collection of the Works of Sir Isaac Newton, Burndy Library, Dibner Institute for the History of Science and Technology: the title pages of the first and second editions of the *Principia*; the half title, title page, and dedication of the third edition; and the diagrams for book 2, prop. 10, in the Jacquier and Le Seur edition of the *Principia*.

I gratefully record the continued and generous support of this project by the National Science Foundation, which also supported the prior production of our Latin edition with variant readings. Without such aid this translation and Guide would never have come into being.

Addendum

I am particularly grateful to four colleagues who helped me read the proofs. Bruce Brackenridge checked the proofs of the Guide and shared with me many of his insights into the methods used by Newton in the *Principia*, while George Smith worked through the proofs of each section of the Guide and also helped me check the translation. Michael Nauenberg and William Harper helped me find errors in the Guide. A student, Luis Campos, gave me the benefit of his skill at proofreading.

Readers' attention is called to three collections of studies that are either in process of publication or appeared too late to be used in preparing the Guide: *Planetary Astronomy from the Renaissance to the Rise of Astrophysics, Part B: The Eighteenth and Nineteenth Centuries*, ed. René Taton and Curtis Wilson (Cambridge: Cambridge University Press, 1995); *Isaac Newton's Natural Philosophy*, ed. Jed Buchwald and I. B. Cohen (Cambridge: MIT Press, forthcoming); and *The Foundations of Newtonian Scholarship: Proceedings of the 1997 Symposium at the Royal Society*, ed. R. Dalitz and M. Nauenberg (Singapore: World Scientific, forthcoming). Some of the chapters in these collections, notably those by Michael Nauenberg, either suggest revisions of the interpretations set forth in the Guide or offer alternative interpretations. Other contributions of Nauenberg are cited in the notes to the Guide.

A GUIDE TO NEWTON'S *PRINCIPIA*

By I. Bernard Cohen
with contributions by Michael Nauenberg (§3.9)
and George E. Smith (§§7.10, 8.8, 8.15, 8.16, and 10.19)

Contents of the Guide

Abbreviations 9

CHAPTER ONE: A Brief History of the *Principia* 11

1.1 *The Origins of the* Principia *11*

1.2 *Steps Leading to the Composition and Publication of the* Principia *13*

1.3 *Revisions and Later Editions* *22*

CHAPTER TWO: Translating the *Principia* 26

2.1 *Translations of the* Principia: *English Versions by Andrew Motte (1729), Henry Pemberton (1729–?), and Thomas Thorp (1777)* *26*

2.2 *Motte's English Translation* *28*

2.3 *The Motte-Cajori Version; The Need for a New Translation* *29*

2.4 *A Problem in Translation: Passive or Active Voice in Book 1, Secs. 2 and 3; The Sense of "Acta" in Book 1, Prop. 1* *37*

2.5 *"Continuous" versus "Continual"; The Problem of "Infinite"* *41*

CHAPTER THREE: Some General Aspects of the *Principia* 43

3.1 *The Title of Newton's* Principia; *The Influence of Descartes* *43*

3.2 *Newton's Goals: An Unpublished Preface to the* Principia *49*

3.3 *Varieties of Newton's Concepts of Force in the* Principia *54*

3.4 *The Reorientation of Newton's Philosophy of Nature; Alchemy and the* Principia *56*

3.5 *The Reality of Forces of Attraction; The Newtonian Style* *60*

3.6 *Did Newton Make the Famous Moon Test of the Theory of Gravity in the 1660s?* *64*

3.7 *The Continuity in Newton's Methods of Studying Orbital Motion: The Three Approximations (Polygonal, Parabolic, Circular)* *70*

3.8 *Hooke's Contribution to Newton's Thinking about Orbital Motion* 76
3.9 *Newton's Curvature Measure of Force: New Findings (by Michael Nauenberg)* 78
3.10 *Newton's Use of "Centrifugal Force" in the* Principia 82

CHAPTER FOUR: Some Fundamental Concepts of the *Principia* 85
4.1 *Newton's Definitions* 85
4.2 *Quantity of Matter: Mass (Def. 1)* 86
4.3 *Is Newton's Definition of Mass Circular?* 89
4.4 *Newton's Determination of Masses* 93
4.5 *Newton's Concept of Density* 94
4.6 *Quantity of Motion: Momentum (Def. 2)* 95
4.7 *"Vis Insita": Inherent Force and the Force of Inertia (Def. 3)* 96
4.8 *Newton's Acquaintance with "Vis Insita" and "Inertia"* 99
4.9 *Impressed Force; Anticipations of Laws of Motion (Def. 4)* 102
4.10 *Centripetal Force and Its Three Measures (Defs. 5, 6–8)* 102
4.11 *Time and Space in the* Principia; *Newton's Concept of Absolute and Relative Space and Time and of Absolute Motion; The Rotating-Bucket Experiment* 106

CHAPTER FIVE: Axioms, or the Laws of Motion 109
5.1 *Newton's Laws of Motion or Laws of Nature* 109
5.2 *The First Law; Why Both a First Law and a Second Law?* 109
5.3 *The Second Law: Force and Change in Motion* 111
5.4 *From Impulsive Forces to Continually Acting Forces: Book 1, Prop. 1; Is the* Principia *Written in the Manner of Greek Geometry?* 114
5.5 *The Third Law: Action and Reaction* 117
5.6 *Corollaries and Scholium: The Parallelogram Rule, Simple Machines, Elastic and Inelastic Impact* 118
5.7 *Does the Concept of Energy and Its Conservation Appear in the* Principia? 119
5.8 *Methods of Proof versus Method of Discovery; The "New Analysis" and Newton's Allegations about How the* Principia *Was Produced; The Use of Fluxions in the* Principia 122

CHAPTER SIX: The Structure of Book 1 128
6.1 *The General Structure of the* Principia 128
6.2 *Sec. 1: First and Ultimate Ratios* 129

6.3 *Secs. 2–3: The Law of Areas, Circular Motion, Newton's Dynamical Measure of Force; The Centripetal Force in Motion on an Ellipse, on a Hyperbola, and on a Parabola* 131
6.4 *The Inverse Problem: The Orbit Produced by an Inverse-Square Centripetal Force* 135
6.5 *Secs. 4–5: The Geometry of Conics* 136
6.6 *Problems of Elliptical Motion (Sec. 6); Kepler's Problem* 138
6.7 *The Rectilinear Ascent and Descent of Bodies (Sec. 7)* 140
6.8 *Motion under the Action of Any Centripetal Forces (Sec. 8); Prop. 41* 141
6.9 *Specifying the Initial Velocity in Prop. 41; The Galileo-Plato Problem* 143
6.10 *Moving Orbits (Sec. 9) and Motions on Smooth Planes and Surfaces (Sec. 10)* 147
6.11 *The Mutual Attractions of Bodies (Sec. 11) and the Newtonian Style* 148
6.12 *Sec. 11: The Theory of Perturbations (Prop. 66 and Its Twenty-Two Corollaries)* 155
6.13 *Sec. 12 (Some Aspects of the Dynamics of Spherical Bodies); Sec. 13 (The Attraction of Nonspherical Bodies); Sec. 14 (The Motion of "Minimally Small" Bodies)* 159

CHAPTER SEVEN: The Structure of Book 2 161
7.1 *Some Aspects of Book 2* 161
7.2 *Secs. 1, 2, and 3 of Book 2: Motion under Various Conditions of Resistance* 165
7.3 *Problems with Prop. 10* 167
7.4 *Problems with the Diagram for the Scholium to Prop. 10* 171
7.5 *Secs. 4 and 5: Definition of a Fluid; Newton on Boyle's Law; The Definition of "Simple Pendulum"* 177
7.6 *Secs. 6 and 7: The Motion of Pendulums and the Resistance of Fluids to the Motions of Pendulums and Projectiles; A General Scholium (Experiments on Resistance to Motion)* 180
7.7 *The Solid of Least Resistance; The Design of Ships; The Efflux of Water* 181
7.8 *Sec. 8: Wave Motion and the Motion of Sound* 184
7.9 *The Physics of Vortices (Sec. 9, Props. 51–53); Vortices Shown to Be Inconsistent with Keplerian Planetary Motion* 187

7.10 *Another Way of Considering Book 2: Some Achievements of Book 2 (by George E. Smith) 188*

CHAPTER EIGHT: The Structure of Book 3; The System of the World 195

8.1 *The Structure of Book 3 195*

8.2 *From Hypotheses to Rules and Phenomena 198*

8.3 *Newton's "Rules for Natural Philosophy" 199*

8.4 *Newton's "Phenomena" 200*

8.5 *Newton's Hyp. 3 201*

8.6 *Props. 1–5: The Principles of Motion of Planets and of Their Satellites; The First Moon Test 204*

8.7 *Jupiter's Perturbation of Saturn as Evidence for Universal Gravity 206*

8.8 *Planetary Perturbations: The Interaction of Jupiter and Saturn (by George E. Smith) 211*

8.9 *Props. 6 and 7: Mass and Weight 217*

8.10 *Prop. 8 and Its Corollaries (The Masses and Densities of the Sun, Jupiter, Saturn, and the Earth) 218*

8.11 *Props. 9–20: The Force of Gravity within Planets, the Duration of the Solar System, the Effects of the Inverse-Square Force of Gravity (Further Aspects of Gravity and the System of the World): Newton's Hyp. 1 231*

8.12 *Prop. 24: Theory of the Tides; The First Enunciation of the Principle of Interference 238*

8.13 *Props. 36–38: The Tidal Force of the Sun and the Moon; The Mass and Density (and Dimensions) of the Moon 242*

8.14 *Props. 22, 25–35: The Motion of the Moon 246*

8.15 *Newton and the Problem of the Moon's Motion (by George E. Smith) 252*

8.16 *The Motion of the Lunar Apsis (by George E. Smith) 257*

8.17 *Prop. 39: The Shape of the Earth and the Precession of the Equinoxes 265*

8.18 *Comets (The Concluding Portion of Book 3) 268*

CHAPTER NINE: The Concluding General Scholium 274

9.1 *The General Scholium: "Hypotheses non fingo" 274*

9.2 *"Satis est": Is It Enough? 277*

9.3 *Newton's "Electric and Elastic" Spirit 280*

9.4 *A Gloss on Newton's "Electric and Elastic" Spirit: An Electrical Conclusion to the* Principia *283*

CHAPTER TEN: How to Read the *Principia* 293
10.1 Some Useful Commentaries and Reference Works; Newton's Directions for Reading the Principia *293*
10.2 Some Features of Our Translation; A Note on the Diagrams 298
10.3 Some Technical Terms and Special Translations (including Newton's Use of "Rectangle" and "Solid") 302
10.4 Some Trigonometric Terms ("Sine," "Cosine," "Tangent," "Versed Sine" and "Sagitta," "Subtense," "Subtangent"); "Ratio" versus "Proportion"; "Q.E.D.," "Q.E.F.," "Q.E.I.," and "Q.E.O." 305
10.5 Newton's Way of Expressing Ratios and Proportions 310
10.6 The Classic Ratios (as in Euclid's Elements, *Book 5) 313*
10.7 Newton's Proofs; Limits and Quadratures; More on Fluxions in the Principia *316*
10.8 Example No. 1: Book 1, Prop. 6 (Newton's Dynamical Measure of Force), with Notes on Prop. 1 (A Centrally Directed Force Acting on a Body with Uniform Linear Motion Will Produce Motion according to the Law of Areas) 317
10.9 Example No. 2: Book 1, Prop. 11 (The Direct Problem: Given an Ellipse, to Find the Force Directed toward a Focus) 324
10.10 Example No. 3: A Theorem on Ellipses from the Theory of Conics, Needed for Book 1, Props. 10 and 11 330
10.11 Example No. 4: Book 1, Prop. 32 (How Far a Body Will Fall under the Action of an Inverse-Square Force) 333
10.12 Example No. 5: Book 1, Prop. 41 (To Find the Orbit in Which a Body Moves When Acted On by a Given Centripetal Force and Then to Find the Orbital Position at Any Specified Time) 334
10.13 Example No. 6: Book 1, Lem. 29 (An Example of the Calculus or of Fluxions in Geometric Form) 345
10.14 Example No. 7: Book 3, Prop. 19 (The Shape of the Earth) 347
10.15 Example No. 8: A Theorem on the Variation of Weight with Latitude (from Book 3, Prop. 20) 350
10.16 Example No. 9: Book 1, Prop. 66, Corol. 14, Needed for Book 3, Prop. 37, Corol. 3 355
10.17 Newton's Numbers: The "Fudge Factor" 360
10.18 Newton's Measures 362
10.19 A Puzzle in Book 1, Prop. 66, Corol. 14 (by George E. Smith) 365

CHAPTER ELEVEN: Conclusion 369

Abbreviations

Throughout the notes of the Guide, ten works are referred to constantly under the following standardized abbreviations:

Anal. View — Henry Lord Brougham and E. J. Routh, *Analytical View of Sir Isaac Newton's "Principia"* (London: Longman, Brown, Green, and Longmans, 1855; reprint, with an introd. by I. B. Cohen, New York and London: Johnson Reprint Corp., 1972).

Background — John Herivel, *The Background to Newton's "Principia": A Study of Newton's Dynamical Researches in the Years 1664–84* (Oxford: Clarendon Press, 1965).

Corresp. — The Royal Society's edition of *The Correspondence of Isaac Newton*, 7 vols., ed. H. W. Turnbull (vols. 1–3), J. F. Scott (vol. 4), A. Rupert Hall and Laura Tilling (vols. 5–7) (Cambridge: Cambridge University Press, 1959–1977).

Essay — W. W. Rouse Ball, *An Essay on Newton's "Principia"* (London: Macmillan and Co., 1893; reprint, with an introd. by I. B. Cohen, New York and London: Johnson Reprint Corp., 1972).

Introduction — I. B. Cohen, *Introduction to Newton's "Principia"* (Cambridge, Mass.: Harvard University Press; Cambridge: Cambridge University Press, 1971).

Math. Papers — *The Mathematical Papers of Isaac Newton*, 8 vols., ed. D. T. Whiteside (Cambridge: Cambridge University Press, 1967–1981).

Never at Rest	R. S. Westfall, *Never at Rest: A Biography of Isaac Newton* (Cambridge, London, New York: Cambridge University Press, 1980).
"Newt. Achievement"	Curtis Wilson, "The Newtonian Achievement in Astronomy," in *Planetary Astronomy from the Renaissance to the Rise of Astrophysics, Part A: Tycho Brahe to Newton*, ed. René Taton and Curtis Wilson, vol. 2A of *The General History of Astronomy* (Cambridge and New York: Cambridge University Press, 1989), pp. 233–274.
Newt. Revolution	I. B. Cohen, *The Newtonian Revolution* (Cambridge, London, New York: Cambridge University Press, 1980).
Unpubl. Sci. Papers	*Unpublished Scientific Papers of Isaac Newton*, ed. A. Rupert Hall and Marie Boas Hall (Cambridge: Cambridge University Press, 1962).

A useful bibliographical guide to Newton's writings and the vast secondary literature concerning Newton's life, thought, and influence is Peter Wallis and Ruth Wallis, *Newton and Newtoniana, 1672–1975: A Bibliography* (Folkestone: Dawson, 1977). A generally helpful reference work is Derek Gjertsen, *The Newton Handbook* (London and New York: Routledge and Kegan Paul, 1986).

The abbreviation "ULC" in such references as "ULC MS Add. 3968" refers to the University Library of Cambridge University.

CHAPTER ONE

A Brief History of the *Principia*

The Origins of the Principia — 1.1

Isaac Newton's *Principia* was published in 1687. The full title is *Philosophiae Naturalis Principia Mathematica,* or *Mathematical Principles of Natural Philosophy*. A revised edition appeared in 1713, followed by a third edition in 1726, just one year before the author's death in 1727. The subject of this work, to use the name assigned by Newton in the first preface, is "rational mechanics." Later on, Leibniz introduced the name "dynamics." Although Newton objected to this name,[1] "dynamics" provides an appropriate designation of the subject matter of the *Principia*, since "force" is a primary concept of that work. Indeed, the *Principia* can quite properly be described as a study of a variety of forces and the different kinds of motions they produce. Newton's eventual goal, achieved in the third of the three "books" of which the *Principia* is composed, was to apply the results of the prior study to the system of the world, to the motions of the heavenly bodies. This subject is generally known today by the name used a century or so later by Laplace, "celestial mechanics."

1. Newton's objections were not based on the name in terms of its Greek roots or its adequacy or inadequacy to describe the subject matter. Rather, he took umbrage at Leibniz's having devised a name as if he had been the inventor of the subject, whereas Newton believed that he himself had been the primary creator. In a private memorandum (my *Introduction*, p. 296, §6), Newton wrote that "Galileo began to consider the effect of Gravity upon Projectiles. Mr. Newton in his *Principia Philosophiae* improved that consideration into a large science. Mr. Leibnitz christened the child by [a] new name as if it had been his own, calling it *Dynamica*." In another such memorandum (ibid., p. 297), he declared that Leibniz "changed the name of vis centripeta used by Newton into that of sollicitatio paracentrica, not because it is a fitter name, but to avoid being thought to build upon Mr. Newton's foundation." He also held that Leibniz "has set his mark upon this whole science of forces calling it Dynamick, as if he had invented it himself & is frequently setting his mark upon things by new names & new Notations."

The history of how the *Principia* came into being has been told and retold.[2] In the summer of 1684, the astronomer Edmond Halley visited Newton in order to find out whether he could solve a problem that had baffled Christopher Wren, Robert Hooke, and himself: to find the planetary orbit that would be produced by an inverse-square central force. Newton knew the answer to be an ellipse.[3] He had solved the problem of elliptical orbits earlier, apparently in the period 1679–1680 during the course of an exchange of letters with Hooke. When Halley heard Newton's reply, he urged him to write up his results. With Halley's prodding and encouragement, Newton produced a short tract which exists in several versions and will be referred to as *De Motu*[4] (On Motion), the common beginning of all the titles Newton gave to the several versions. Once started, Newton could not restrain the creative force of his genius, and the end product was the *Principia*. In his progress from the early versions of *De Motu* to the *Principia*, Newton's conception of what could be achieved by an empirically based mathematical science had become enlarged by several orders of magnitude.

2. E.g., my *Introduction*, Westfall's *Never at Rest*, Whiteside's introduction in *Math. Papers* (vol. 6), Herivel's *Background*, and more recently A. Rupert Hall, *Isaac Newton: Adventurer in Thought* (Oxford and Cambridge, Mass.: Blackwell, 1992).

3. Our source for this anecdote may be found in the notes accumulated by John Conduitt, husband of Newton's niece and Newton's successor at the Mint, for a proposed biography of Newton. Conduitt got the story from the mathematician Abraham de Moivre. The main lines of the story are undoubtedly correct, but we may doubt the accuracy of the details, since this is a secondhand record of an event that had happened about half a century earlier. What was the exact question that Halley would have asked Newton?

The question recorded by Conduitt has puzzled critical historians, because it does not have a simple answer. There has even been some speculation whether Halley may have asked Newton for the force acting in the case of an elliptical orbit rather than for the orbit produced by an inverse-square force. It is doubtful whether Conduitt knew enough mathematics to see the difference between the two. But, in fact, there is a real difference. As Newton shows in the *Principia*, in prop. 11, and as he proved in the drafts of *De Motu*, an elliptical orbit does imply an inverse-square force. Yet, as readers of the *Principia* would have been aware, an inverse-square force does not necessarily imply an elliptic orbit, rather a conic section (which can be an ellipse, a parabola, or a hyperbola).

Of course, Halley's question may have implied (or have been thought by Newton to have implied) an orbit of a planet or possibly a planetary satellite. Since such an orbit is a closed curve, and therefore not a parabola or a hyperbola, Halley's question to Newton would then have been, in effect, What is the planetary orbit (or closed orbit) produced by an inverse-square force? In this case, the answer would legitimately be the one recorded by Conduitt.

4. The several versions of *De Motu* may be found (with translations and commentary) in Whiteside's edition of *Math. Papers* 6:30–80; the Halls' *Unpublished Sci. Papers*, pp. 237–239, 243–292; Herivel's *Background*, pp. 256–303; and, earlier, in Rouse Ball's *Essay*, pp. 31–56, and in Stephen P. Rigaud, *Historical Essay on the First Publication of Sir Isaac Newton's "Principia"* (Oxford: Oxford University Press, 1838; reprint, with an introd. by I. B. Cohen, New York and London: Johnson Reprint Corp., 1972), appendix, no. 1, pp. 1–19. For a facsimile reprint of the MSS of *De Motu*, see n. 5 below.

As first conceived, the *Principia* consisted of two "books" and bore the simple title *De Motu Corporum* (On the Motion of Bodies).[5] This manuscript begins, as does the *Principia*, with a series of Definitions and Laws of Motion, followed by a book 1 whose subject matter more or less corresponds to book 1 of the *Principia*.[6] The subject matter of book 2 of this early draft is much the same as that of book 3 of the *Principia*. In revising this text for the *Principia*, Newton limited book 1 to the subject of forces and motion in free spaces, that is, in spaces devoid of any resistance. Book 2 of the *Principia* contains an expanded version of the analysis of motion in resisting mediums, plus discussions of pendulums,[7] of wave motion, and of the physics of vortices. In the *Principia*, the system of the world became the subject of what is there book 3, incorporating much that had been in the older book 2 but generally recast in a new form. As Newton explained in the final *Principia*, while introducing book 3, he had originally presented this subject in a popular manner, but then decided to recast it in a more mathematical form so that it would not be read by anyone who had not first mastered the principles of rational mechanics. Even so, whole paragraphs of the new book 3 were copied word for word from the old book 2.[8]

Steps Leading to the Composition and Publication of the Principia 1.2

The history of the development of Newton's ideas concerning mechanics, more specifically dynamics, has been explored by many scholars and is still the subject of active research and study.[9] The details of the early development of Newton's

5. On this first draft of book 1, see my *Introduction*, chap. 4 and suppl. 3, where it is referred to as Newton's Lucasian Lectures (LL) because Newton later deposited this MS in the University Library as if it were the text of his university lectures for 1684 and 1685. This text has been edited and translated by D. T. Whiteside in vol. 6 of *Math. Papers*, and Whiteside has also prepared a facsimile edition of the whole MS, together with the drafts of *De Motu*, under the general title *The Preliminary Manuscripts for Isaac Newton's 1687 "Principia," 1684–1685* (Cambridge and New York: Cambridge University Press, 1989).

6. This early book 1 concluded (as did *De Motu*) with a brief presentation of motion in resisting fluids, which was later considerably expanded so as to become the first sections of book 2 of the *Principia*.

7. Pendulums are also discussed in book 1.

8. A new translation of this early version of book 3, by Anne Whitman and I. Bernard Cohen, is scheduled for publication by the University of California Press. In order to distinguish this work from book 3 of the *Principia* (with its subtitle "De Systemate Mundi"), we have called this early version *Essay on the System of the World*. A list of the paragraphs that are the same in both versions may be found in a supplement to our edition of the *Principia* with variant readings, cited in n. 45 below.

9. The books and articles devoted to this topic are so numerous, and continue to appear at so rapid a rate, that it would hardly be practical to cite them all. The most accessible and authoritative presentations are to be found in Curtis Wilson's "Newt. Achievement" and especially in D. T. Whiteside, "Before the *Principia*: The Maturing of Newton's Thoughts on Dynamical Astronomy, 1664–84," *Journal for the History of Astronomy* 1 (1970): 5–19; "The Mathematical Principles Underlying Newton's *Principia*," ibid.,

ideas about force and motion, however interesting in their own right, are not directly related to the present assignment, which is to provide a reader's guide to the *Principia*. Nevertheless, some aspects of this prehistory should be of interest to every prospective reader of the *Principia*. In the scholium to book 1, prop. 4, Newton refers to his independent discovery (in the 1660s) of the $\frac{v^2}{r}$ rule for the force in uniform circular motion (at speed v along a circle of radius r), a discovery usually attributed to Christiaan Huygens, who formally announced it to the world in his *Horologium Oscillatorium* of 1673.[10] It requires only the minimum skill in algebraic manipulation to combine the $\frac{v^2}{r}$ rule with Kepler's third law in order to determine that in a system of bodies in uniform circular motion the force is proportional to $\frac{1}{r^2}$ or is inversely proportional to the square of the distance. Of course, this computation does not of itself specify anything about the nature of the force, whether it is a centripetal or a centrifugal force or whether it is a force in the sense of the later Newtonian dynamics or merely a Cartesian "conatus," or endeavor. In a manuscript note Newton later claimed that at an early date, in the 1660s, he had actually applied the $\frac{v^2}{r}$ rule to the moon's motion, much as he does later on in book 3, prop. 4, of the *Principia*, in order to confirm his idea of the force of "gravity extending to the Moon."[11] In this way he could counter Hooke's allegation that he had learned the concept of an inverse-square force of gravity from Hooke.

A careful reading of the documents in question shows that sometime in the 1660s, Newton made a series of computations, one of which was aimed at proving that what was later known as the outward or centrifugal force arising from the

116–138. See also my *Newt. Revolution*, chaps. 4 and 5; R. S. Westfall's *Never at Rest* and his earlier *Force in Newton's Physics* (London: Macdonald; New York: American Elsevier, 1971), chaps. 7 and 8; Herivel's *Background*. A splendid review of this topic is available in Hall, *Isaac Newton: Adventurer in Thought*, pp. 55–64. A list of other scholars who have made contributions to this subject would include, among others, Bruce Brackenridge, Herman Ehrlichson, J. E. McGuire, and Michael M. Nauenberg.

10. The text of Newton's early discovery of the $\frac{v^2}{r}$ rule is published (from Newton's "Waste Book") in *Background*, pp. 130–131.

11. This celebrated autobiographical document was first printed in *A Catalogue of the Portsmouth Collection of Books and Papers Written by or Belonging to Sir Isaac Newton, the Scientific Portion of Which Has Been Presented by the Earl of Portsmouth to the University of Cambridge*, ed. H. R. Luard et al. (Cambridge: Cambridge University Press, 1888), and has been reprinted many times since. A corrected version, taken from the manuscript in the Cambridge University Library (ULC MS Add. 3968, §41, fol. 85) may be found in my *Introduction*, pp. 290–292.

earth's rotation is less than the earth's gravity, as it must be for the Copernican system to be possible. He then computed a series of forces. Cartesian vortical endeavors are not the kind of forces that, in the *Principia*, are exerted by the sun on the planets to keep them in a curved path or the similar force exerted by the earth on the moon. At this time, and for some years to come, Newton was deeply enmeshed in the Cartesian doctrine of vortices. He had no concept of a "force of gravity" acting on the moon in anything like the later sense of the dynamics of the *Principia*. These Cartesian "endeavors" (Newton used Descartes's own technical term, "conatus") are the magnitude of the planets' endeavors to fly out of their orbits. Newton concludes that since the cubes of the distances of the planets from the sun are "reciprocally as the squared numbers of their revolutions in a given time," their "*conatus* to recede from the Sun will be reciprocally as the squares of their distances from the Sun."[12]

Newton also made computations to show that the endeavor or "conatus" of receding from the earth's surface (caused by the earth's daily rotation) is 12½ times greater than the orbital endeavor of the moon to recede from the earth. He concludes that the force of receding at the earth's surface is "4000 and more times greater than the endeavor of the Moon to recede from the Earth."

In other words, "Newton had discovered an interesting mathematical correlation within the solar vortex,"[13] but he plainly had not as yet invented the radically new concept of a centripetal dynamical force, an attraction that draws the planets toward the sun and the moon toward the earth.[14] There was no "twenty years' delay" (from the mid-1660s to the mid-1680s) in Newton's publication of the theory of universal gravity, as was alleged by Florian Cajori.[15]

In 1679/80, Hooke initiated an exchange of correspondence with Newton on scientific topics. In the course of this epistolary interchange, Hooke suggested to Newton a "hypothesis" of his own devising which would account for curved orbital motion by a combination of two motions: an inertial or uniform linear component

12. *Corresp.* 1:300; see A. Rupert Hall, "Newton on the Calculation of Central Forces," *Annals of Science* 13 (1957): 62–71.

13. Hall, *Isaac Newton: Adventurer in Thought*, p. 62. This work gives an excellent critical summary of Newton's thoughts about celestial motions during the 1660s.

14. For the documents and an analysis, see Hall, "Newton on the Calculation of Central Forces," pp. 62–71; also *Background*, pp. 192–198, 68–69; and esp. *Never at Rest*, pp. 151–152. See, further, *Newt. Revolution*, esp. pp. 238–240. A splendid review of this subject is available in D. T. Whiteside, "The Prehistory of the *Principia* from 1664 to 1686," *Notes and Records of the Royal Society of London* 45 (1991): 11–61, esp. 18–22.

15. Florian Cajori, "Newton's Twenty Years' Delay in Announcing the Law of Gravitation," in *Sir Isaac Newton, 1727–1927: A Bicentenary Evaluation of His Work*, ed. Frederic E. Brasch (Baltimore: Williams and Wilkins, 1928), pp. 127–188.

along the tangent to the curve and a motion of falling inward toward a center. Newton told Hooke that he had never heard of this "hypothesis."[16] In the course of their letters, Hooke urged Newton to explore the consequences of his hypothesis, advancing the opinion or guess that in combination with the supposition of an inverse-square law of solar-planetary force, it would lead to the true planetary motions.[17] Hooke also wrote that the inverse-square law would lead to a rule for orbital speed being inversely proportional to the distance of a planet from the sun.[18] Stimulated by Hooke, Newton apparently then proved that the solar-planetary force is as the inverse square of the distance, a first step toward the eventual *Principia*.

We cannot be absolutely certain of exactly how Newton proceeded to solve the problem of motion in elliptical orbits, but most scholars agree that he more or less followed the path set forth in the tract *De Motu* which he wrote after Halley's visit a few years later in 1684.[19] Essentially, this is the path from props. 1 and 2 of book 1 to prop. 4, through prop. 6, to props. 10 and 11. Being secretive by nature, Newton didn't tell Hooke of his achievement. In any event, he would hardly have announced so major a discovery to a jealous professional rival, nor in a private letter. What may seem astonishing, in retrospect, is not that Newton did not reveal his discovery to Hooke, but that Newton was not at once galvanized into expanding his discovery into the eventual *Principia*.

Several aspects of the Hooke-Newton exchange deserve to be noted. First, Hooke was unable to solve the problem that arose from his guess or his intuition; he simply did not have sufficient skill in mathematics to be able to find the orbit produced by an inverse-square force. A few years later, Wren and Halley were equally baffled by this problem. Newton's solution was, as Westfall has noted, to invert the problem, to assume the path to be an ellipse and find the force rather

16. The Newton-Hooke correspondence during 1679/80 is to be found in *Corresp.*, vol. 2. See, in this regard, Alexandre Koyré, "An Unpublished Letter of Robert Hooke to Isaac Newton," *Isis* 43 (1952): 312–337, reprinted in Koyré's *Newtonian Studies* (Cambridge, Mass.: Harvard University Press, 1965), pp. 221–260. Also J. A. Lohne, "Hooke versus Newton: An Analysis of the Documents in the Case of Free Fall and Planetary Motion," *Centaurus* 7 (1960): 6–52.

17. Later, Newton quite correctly insisted that Hooke could not prove this assertion. In any event he himself had already been thinking of an inverse-square force.

18. Newton was to prove that this particular conclusion or guess of Hooke's was wrong. The force on a planet at a point P (see book 1, prop. 1, corol. 1) is inversely proportional to the perpendicular distance from the sun to the tangent to the curve at P. We shall take note, below, that Hooke's rule, previously stated by Kepler, is true only at the apsides.

19. There has, however, been some consideration given to the possibility that what Newton wrote up at this time was a prototype of the paper he later sent to John Locke.

This work is available, with a commentary by D. T. Whiteside, in *Math. Papers*, vol. 6, and in Herivel's *Background* and the Halls' edition of *Unpubl. Sci. Papers*.

than "investigating the path in an inverse-square force field."[20] Second, there is no certainty that the tract *De Motu* actually represents the line of Newton's thought after corresponding with Hooke; Westfall, for one, has argued that a better candidate would be an essay in English which Newton sent later to John Locke, a position he maintains in his biography of Newton.[21] A third point is that Newton was quite frank in admitting (in private memoranda) that the correspondence with Hooke provided the occasion for his investigations of orbital motion that eventually led to the *Principia*.[22] Fourth, as we shall see in §3.4 below, the encounter with Hooke was associated with a radical reorientation of Newton's philosophy of nature that is indissolubly linked with the *Principia*. Fifth, despite Newton's success in proving that an elliptical orbit implies an inverse-square force, he was not at that time stimulated—as he would be some four years later—to move ahead and to create modern rational mechanics. Sixth, Newton's solution of the problems of planetary force depended on both his own new concept of a dynamical measure of force (as in book 1, prop. 6) and his recognition of the importance of Kepler's law of areas.[23] A final point to be made is that most scholarly analyses of Newton's thoughts during this crucial period concentrate on conceptual formulations and analytical solutions, whereas we know that both Hooke[24] and Newton made

20. *Never at Rest*, p. 387. I have discussed this matter in my *Introduction*, pp. 49–52, in relation to the question of what Halley asked Newton on the famous visit in the summer of 1684 and what Newton would have replied.

21. Most scholars date the Locke paper after the *Principia*. An earlier dating was suggested by Herivel in 1961 and reaffirmed in his *Background*, pp. 108–117. This assigned date was then challenged by the Halls in 1963, and supported by Westfall in 1969, whose arguments were refuted by Whiteside in 1970. See the summary in Westfall's *Never at Rest*, pp. 387–388 n. 145.

An admirable discussion of the various attempts to date this work is given in Bruce Brackenridge, "The Critical Role of Curvature in Newton's Developing Dynamics," in *The Investigation of Difficult Things: Essays on Newton and the History of the Exact Sciences*, ed. P. M. Harman and Alan E. Shapiro (Cambridge: Cambridge University Press, 1992), pp. 231–260, esp. 241–242 and n. 35. Brackenridge concludes by agreeing with Whiteside that the date of a "prototype manuscript" on which this tract is based should be fixed at August 1684, shortly after Halley's visit.

22. Newton to Halley, 27 July 1686, *Corresp.* 2:447; my *Introduction*, suppl. 1. My own awareness of the significance of the Hooke-Newton correspondence (in suggesting a fruitful way to analyze celestial orbital motions) derives from a pioneering study by R. S. Westfall, "Hooke and the Law of Universal Gravitation," *The British Journal for the History of Science* 3 (1967): 245–261.

23. On the problems of using Kepler's law of areas and the various approximations used by seventeenth-century astronomers in place of this law, see Curtis Wilson, "From Kepler's Laws, So-Called, to Universal Gravitation: Empirical Factors," *Archive for History of Exact Sciences* 6 (1970): 89–170; and my *Newt. Revolution*, pp. 224–229.

24. Patri Pugliese, "Robert Hooke and the Dynamics of Motion in the Curved Path," in *Robert Hooke: New Studies*, ed. Michael Hunter and Simon Schaffer (London: Boydell Press, 1989), pp. 181–205. See, further, Michael Nauenberg, "Hooke, Orbital Motion, and Newton's *Principia*," *American Journal of Physics* 62 (1994): 331–350.

important use of graphical methods, a point rightly stressed by Curtis Wilson.[25] Newton, in fact, in an early letter to Hooke, wrote of Hooke's "acute Letter having put me upon considering . . . the species of this curve," saying he might go on to "add something about its description quam proxime,"[26] or by graphic methods. The final proposition in the *Principia* (book 3, prop. 42) declared its subject (in the first edition) to be: "To correct a comet's trajectory found graphically." In the second and third editions, the text of the demonstration was not appreciably altered, but the statement of the proposition now reads: "To correct a comet's trajectory that has been found."

When Newton wrote up his results for Halley (in the tract *De Motu*) and proved (in the equivalent of book 1, prop. 11) that an elliptical orbit implies an inverse-square central force, he included in his text the joyous conclusion: "Therefore the major planets revolve in ellipses having a focus in the center of the sun; and the radii to the sun describe areas proportional to the times, exactly ["omnino"] as Kepler supposed."[27] But after some reflection, Newton recognized that he had been considering a rather artificial situation in which a body moves about a mathematical center of force. In nature, bodies move about other bodies, not about mathematical points. When he began to consider such a two-body system, he came to recognize that in this case each body must act on the other. If this is true for one such pair of bodies, as for the sun-earth system, then it must be so in all such systems. In this way he concluded that the sun (like all the planets) is a body on which the force acts and also a body that gives rise to the force. It follows at once that each planet must exert a perturbing force on every other planet in the solar system. The consequence must be, as Newton recognized almost at once, that "the displacement of the sun from the center of gravity" may have the effect that "the centripetal force does not always tend to" an "immobile center" and that "the planets neither revolve exactly in ellipses nor revolve twice in the same orbit." In other words, "Each time a planet revolves it traces a fresh orbit, as happens also with the motion of the moon, and each orbit is dependent upon the combined motions of all the planets, not to mention their actions upon each other."[28] This led him to the melancholy conclusion: "Unless I am much mistaken, it would exceed the force of human wit to consider so many causes of motion at the same time, and to define the motions by exact laws which would allow of any easy calculation."

25. "Newt. Achievement," pp. 242–243.

26. Newton to Hooke, 13 December 1679, *Corresp.* 2:308. Wilson's suggested reconstruction occurs in "Newt. Achievement," p. 243.

27. *Unpubl. Sci. Papers*, pp. 253, 277.

28. Ibid., pp. 256, 281.

We don't know exactly how Newton reached this conclusion, but a major factor may have been the recognition of the need to take account of the third law, that to every action there must be an equal and opposite reaction. Yet, in the texts of *De Motu*, the third law does not appear explicitly among the "laws" or "hypotheses." We do have good evidence, however, that Newton was aware of the third law long before writing *De Motu*.[29] In any event, the recognition that there must be interplanetary perturbations was clearly an essential step on the road to universal gravity and the *Principia*.[30]

In reviewing this pre-*Principia* development of Newton's dynamics, we should take note that by and large, Newton has been considering exclusively the motion of a particle, of unit mass. Indeed, a careful reading of the *Principia* will show that even though mass is the subject of the first definition at the beginning of the *Principia*, mass is not a primary variable in Newton's mode of developing his dynamics in book 1. In fact, most of book 1 deals exclusively with particles. Physical bodies with significant dimensions or shapes do not appear until sec. 12, "The attractive forces of spherical bodies."

Newton's concept of mass is one of the most original concepts of the *Principia*. Newton began thinking about mass some years before Halley's visit. Yet, in a series of definitions which he wrote out some time after *De Motu* and before composing the *Principia*, mass does not appear as a primary entry. We do not have documents that allow us to trace the development of Newton's concept of mass with any precision. We know, however, that two events must have been important, even though we cannot tell whether they initiated Newton's thinking about mass or reinforced ideas that were being developed by Newton. One of these was the report of the Richer expedition, with evidence that indicated that weight is a variable quantity, depending on the terrestrial latitude. Hence weight is a "local" property and cannot be used as a universal measure of a body's quantity of matter. Another was Newton's study of the comet of 1680. After he recognized that the comet turned around the sun and after he concluded that the sun's action

29. See D. T. Whiteside's notes in *Math. Papers* 5:148–149 n. 152; 6:98–99 n. 16.

30. A quite different reconstruction of Newton's path to universal gravity has been proposed by George Smith. He suggests: "The 'one-body' solutions of the tract 'De Motu' expressly entail that the a^3/T^2 value associated with each celestial central body is a measure of the centripetal tendency toward it. The known values for the Sun, Jupiter, Saturn, and the Earth can then be used, in conjunction with the principle that the center of gravity of the system remains unaffected (corollary 4 of the Laws of Motion), first to conclude independently of any explicit reference to mass that the Copernican system is basically correct (as in the 'Copernican scholium' of the revised version of 'De Motu'), and *then* to infer that the gravitational force acting celestially is proportional to the masses of the central and orbiting bodies. The final step to universal gravitation then follows along the lines of Propositions 8 and 9 of Book III of the *Principia*." See, also, Wilson's "From Kepler's Laws" (n. 23 above).

on the comet cannot be magnetic, he came to believe that Jupiter must also exert an influence on the comet. Clearly, this influence must derive from the matter in Jupiter, Jupiter acting on the comet just as it does on its satellites.

Once Newton had concluded that planets are centers of force because of their matter or mass, he sought some kind of empirical confirmation of so bold a concept. Since Jupiter is by far the most massive of all the planets, it was obvious that evidence of a planetary force would be most manifest in relation to the action of Jupiter on a neighboring planet. In happened that in 1684/85 the orbital motions of Jupiter and Saturn were bringing these two planets to conjunction. If Newton's conclusion were correct, then the interactions of these two giant planets should show the observable effects of an interplanetary force. Newton wrote to the astronomer John Flamsteed at the Royal Observatory at Greenwich for information on this point. Flamsteed reported that Saturn's orbital speed in the vicinity of Jupiter did not exactly follow the expected path, but he could not detect the kind of effect or perturbation that Newton had predicted.[31] As we shall see, the effect predicted by Newton does occur, but its magnitude is so tiny that Flamsteed could never have observed it. Newton needed other evidence to establish the validity of his force of universal gravity.

Newton's discovery of interplanetary forces as a special instance of universal gravity enables us to specify two primary goals of the *Principia*. The first is to show the conditions under which Kepler's laws of planetary motion are exactly or accurately true; the second is to explore how these laws must be modified in the world of observed nature by perturbations, to show the effects of perturbations on the motions of planets and their moons.[32]

It is well known that after the *Principia* was presented to the Royal Society, Hooke claimed that he should be given credit for having suggested to Newton the idea of universal gravity. We have seen that Hooke did suggest to Newton that the sun exerts an inverse-square force on the planets, but Newton insisted that he didn't need Hooke to suggest to him that there is an inverse-square relation. Furthermore, Newton said that this was but one of Hooke's guesses. Newton again and again asserted that Hooke didn't know enough mathematics to substantiate his guess, and he was right. As the mathematical astronomer Alexis Clairaut said

31. *Corresp.* 2:419–420.

32. These two goals are discussed in my "Newton's Theory vs. Kepler's Theory and Galileo's Theory: An Example of a Difference between a Philosophical and a Historical Analysis of Science," in *The Interaction between Science and Philosophy*, ed. Yehuda Elkana (Atlantic Highlands, N.J.: Humanities Press, 1974), pp. 299–388.

of Hooke's claim, a generation later, it serves "to show what a distance there is between a truth that is glimpsed and a truth that is demonstrated."[33]

In explaining his position with respect to Hooke's guess, Newton compared his own work with that of Hooke and Kepler. Newton evidently believed that he himself had "as great a right" to the inverse-square law "as to the ellipsis." For, just "as Kepler knew the orb to be not circular but oval, and guessed it to be elliptical, so Mr. Hooke, without knowing what I have found out since his letters to me,"[34] knew only "that the proportion was duplicata quam proxime at great distances from the centre," and "guessed it to be so accurately, and guessed amiss in extending that proportion down to the very centre." But, unlike Hooke, "Kepler guessed right at the ellipsis," so that "Mr. Hooke found less of the proportion than Kepler of the ellipsis."[35] Newton believed that he himself deserved credit for the law of elliptical orbits, as well as the law of the inverse square, on the grounds that he had proved both in their generality.[36] In the *Principia* (e.g., in the "Phenomena" in book 3), Newton gave Kepler credit only for the third or harmonic law. At the time that Newton was writing his *Principia*, there were alternatives to the area law that were in use in making tables of planetary motion. Newton proposed using the eclipses of Jupiter's satellites (and later of those of Saturn) to show that this law holds to a high degree. But the law of elliptical orbits was of a different sort because there was no observational evidence that would distinguish between an ellipse and other ovals. Thus there may have been very different reasons for not giving Kepler credit for these two laws.

At one point during the exchange of letters with Halley on Hooke's claims to recognition, Newton—in a fit of pique—threatened to withdraw book 3 altogether.[37] We do not know how serious this threat was, but Halley was able to explain matters and to calm Newton's rage. Halley deserves much praise for his services as midwife to Newton's brainchild. Not only was he responsible for goading Newton into writing up his preliminary results; he encouraged Newton to produce the *Principia*. At an early stage of composition of the *Principia*, as I discovered while preparing the Latin edition with variant readings, Halley even

33. "Exposition abregée du système du monde, et explication des principaux phénomènes astronomiques tirée des *Principes* de M. Newton," suppl. to the marquise du Châtelet's translation of the *Principia* (Paris: chez Desaint & Saillant [&] Lambert, 1756), 2:6.

34. Newton was referring to the problem of the gravitational action of a homogeneous sphere on an external particle; see Whiteside's note in *Math. Papers* 6:19 n. 59.

35. Newton to Halley, 20 June 1686, *Corresp.*, vol. 2.

36. Ibid. "I do pretend [i.e., claim]," Newton wrote, "to have done as much for the proportion [of the inverse square] as for the ellipsis, and to have as much right to the one from Mr. Hooke and all men, as to the other from Kepler."

37. See §3.1 below.

helped Newton by making suggestive comments on an early draft of book 1, the manuscript of which no longer exists.[38]

Although publication of the *Principia* was sponsored by the Royal Society, there were no funds available for the costs of printing, and so Halley had to assume those expenses.[39] Additionally, he edited the book for the printer, saw to the making of the woodcuts of the diagrams, and read the proofs. He wrote a flattering ode to Newton that introduces the *Principia* in all three editions,[40] and he also wrote a book review that was published in the Royal Society's *Philosophical Transactions*.[41]

1.3 *Revisions and Later Editions*

Within a decade of publication of the *Principia*, Newton was busy with a number of radical revisions, including an extensive restructuring of the opening sections.[42] He planned to remove secs. 4 and 5, which are purely geometrical and not necessary to the rest of the text, and to publish them separately.[43] He also developed plans to include a mathematical supplement on his methods of the calculus, his treatise *De Quadratura*. Many of the proposed revisions and restructurings of the 1690s are recorded in Newton's manuscripts; others were reported in some detail by David Gregory.[44] When Newton began to produce a second edition, however, with the aid of Roger Cotes, the revisions were of a quite different sort. Some of the major or most interesting alterations are given in the notes to the present translation. The rest are to be found in the *apparatus criticus* of our Latin edition of the *Principia* with variant readings.[45]

There were a number of truly major emendations that appeared in the second edition, some of which involved a complete replacement of the original text. One

38. See my *Introduction*, suppl. 7.

39. A. N. L. Munby estimated the size of the first edition at some 300 or 400 copies, but this number has recently been increased to perhaps 500. See Whiteside, "The Prehistory of the *Principia*" (n. 14 above), esp. p. 34. Whiteside reckons that, granting this larger size of the edition, Halley would not have suffered financially by paying the printing costs of the *Principia* and would even have made not "less than £10 in pocket for all his time and trouble."

40. On the alterations in the poem in successive editions of the *Principia*, see our Latin edition, cited in n. 45 below.

41. *Philosophical Transactions* 16, no. 186 (Jan.-Feb.-March 1687): 291–297, reprinted in *Isaac Newton's Papers and Letters on Natural Philosophy*, ed. I. B. Cohen and Robert E. Schofield, 2d ed. (Cambridge, Mass.: Harvard University Press, 1978), pp. 405–411.

42. See my *Introduction*, chap. 7, and esp. *Math. Papers*, vol. 6.

43. *Introduction*, p. 193.

44. Ibid., pp. 188–198.

45. *Isaac Newton's "Philosophiae Naturalis Principia Mathematica": The Third Edition (1726) with Variant Readings*, assembled and edited by Alexandre Koyré, I. Bernard Cohen, and Anne Whitman, 2 vols. (Cambridge: Cambridge University Press; Cambridge, Mass.: Harvard University Press, 1972).

of these was the wholly new proof of book 2, prop. 10, a last-minute alteration in response to a criticism made by Johann Bernoulli.[46] Another occurred in book 2, sec. 7, on the motion of fluids and the resistance encountered by projectiles, where most of the propositions and their proofs are entirely different in the second edition from those of the first edition. That is, the whole set of props. 34–40 of the first edition were cast out and replaced.[47] This complete revision of sec. 7 made it more appropriate to remove to the end of sec. 6 the General Scholium on pendulum experiments which originally had been at the end of sec. 7. This was a more thorough revision of the text than occurred in any other part of the *Principia*.

Another significant novelty of the second edition was the introduction of a conclusion to the great work, the celebrated General Scholium that appears at the conclusion of book 3. The original edition ended rather abruptly with a discussion of the orbits of comets, a topic making up about a third of book 3. Newton had at first essayed a conclusion, but later changed his mind. His intentions were revealed in 1962 by A. Rupert Hall and Marie Boas Hall, who published the original drafts. In these texts, Newton shows that he intended to conclude the *Principia* with a discussion of the forces between the particles of matter, but then thought better of introducing so controversial a topic. While preparing the second edition, Newton thought once again of an essay on "the attraction of the small particles of bodies," but on "second thought" he chose "rather to add but one short Paragraph about that part of Philosophy."[48] The conclusion he finally produced is the celebrated General Scholium, with its oft-quoted slogan "Hypotheses non fingo." This General Scholium ends with a paragraph about a "spirit" which has certain physical properties, but whose laws have not as yet been determined by experiment. Again thanks to the researches of A. Rupert Hall and Marie Boas Hall, we now know that while composing this paragraph, Newton was thinking about the new phenomena of electricity.[49]

Another change that occurs in the second edition is in the beginning of book 3. In the first edition, book 3 opened with a preliminary set of *Hypotheses*.[50] Perhaps

46. See D. T. Whiteside's magisterial discussion of this episode, together with all the relevant documents concerning the stages of alteration of book 2, prop. 10, in *Math. Papers* 8:50–53, esp. nn. 175, 180, and esp. §6, appendix 2.1.52 in that same volume. See also §7.3 below and my *Introduction*, §9.4.

47. These props. 34–40 of the first edition (translated by I. Bernard Cohen and Anne Whitman) will be published, together with a commentary by George Smith, in *Newton's Natural Philosophy*, ed. Jed Buchwald and I. Bernard Cohen (Cambridge: MIT Press, forthcoming).

48. *Unpubl. Sci. Papers*, pp. 320–347 (see §9.3 below); Newton to Cotes, 2 Mar. 1712/13. On the production of the second edition, see the texts, notes, and commentaries in *Correspondence of Sir Isaac Newton and Professor Cotes*, ed. J. Edleston (London: John W. Parker; Cambridge: John Deighton, 1850).

49. See §9.3 below.

50. See §8.2 below.

in reply to the criticism in the *Journal des Sçavans*,[51] Newton now renamed the "hypotheses" and divided them into several classes. Some became *Regulae Philosophandi*, or "Rules for Natural Philosophy," with a new rule (no. 3). Others became "Phenomena," with new numerical data. Yet another was transferred to a later place in book 3, where it became "hypothesis 1."

Newton also made a slight modification in the scholium following lem. 2 (book 2, sec. 2), in reference to Leibniz's method of the calculus. He had originally written that Leibniz's method "hardly differed from mine except in the forms of words and notations." In the second edition Newton altered this statement by adding that there was another difference between the two methods, namely, in "the generation of quantities." This scholium and its successive alterations attracted attention because of the controversy over priority in the invention of the calculus. In the third edition, Newton eliminated any direct reference to Leibniz.

Critical readers of the *Principia* paid close attention to the alteration in the scholium following book 3, prop. 35. In the second edition, the original short text was replaced by a long discussion of Newton's attempts to apply the theory of gravity to some inequalities of the moon's motion.[52] Much of the text of this scholium had been published separately by David Gregory.[53]

Many of Newton's plans for the actual revisions of the first edition, in order to produce a second edition, were entered in two personal copies of the *Principia*. One of these was specially bound and interleaved. Once the second edition had been published, Newton again prepared an interleaved copy and kept track of proposed alterations or emendations in his interleaved copy and in an annotated copy. These four special copies of the *Principia* have been preserved among Newton's books, and their contents have been noted in our Latin edition with variant readings.[54]

Soon after the appearance of the second edition, Newton began planning for yet another revision. The preface which he wrote for this planned edition of the late 1710s is of great interest in that it tells us in Newton's own words about some of the features of the *Principia* he believed to be most significant. It is printed below in §3.2. Newton at this time once again planned to have a treatise on the calculus published together with the *Principia*. In the end he abandoned this effort. Later on, when he was in his eighties, he finally decided to produce a new edition. He

51. See my *Introduction*, chap. 6, sec. 6.

52. See §8.14 below.

53. For details see *Isaac Newton's Theory of the Moon's Motion (1702)*, introd. I. Bernard Cohen (Folkestone: Dawson, 1975).

54. These four special copies are described in my *Introduction*; Newton's MS notes appear in our edition with variant readings (n. 45 above).

chose as editor Dr. Henry Pemberton, a medical doctor and authority on pharmacy and an amateur mathematician.

The revisions in the third edition were not quite as extensive as those in the second edition.[55] A new rule 4 was added on the subject of induction, and there were other alterations, some of which may be found in the notes to the present translation. An important change was made in the "Leibniz Scholium" in book 2, sec. 2. The old scholium was replaced by a wholly different one. Newton now boldly asserted his own claims to be the primary inventor of the calculus, referring to some correspondence to prove the point.[56] Even though Leibniz had been dead for almost a decade, Newton still pursued his rival with dogged obstinacy. Another innovation in the third edition appeared in book 3, where Newton inserted (following prop. 33) two propositions by John Machin, astronomy professor at Gresham College, whose academic title would later lead to the invention of a fictitious scientist in the Motte-Cajori edition.[57]

By the time of the third edition, Newton seems to have abandoned his earlier attempts to explain the action of gravity by reference to electrical phenomena and had come rather to hope that an explanation might be found in the actions of an "aethereal medium" of varying density.[58] In his personal copy of the *Principia*, in which he recorded his proposed emendations and revisions, he at first had entered an addition to specify that the "spirit" to which he had referred in the final paragraph of the General Scholium was "electric and elastic."[59] Later on, he apparently decided that since he no longer believed in the supreme importance of the electrical theory, he would cancel the whole paragraph. Accordingly, he drew a line through the text, indicating that this paragraph should be omitted. It is one of the oddities of history that Andrew Motte should have learned of Newton's planned insertion of the modifier "electricus et elasticus" but not of Newton's proposed elimination of the paragraph. Without comment, Motte entered "electric and elastic" into his English version of 1729. These words were in due course preserved in the Motte-Cajori version and have been quoted in the English-speaking world ever since.

55. See my *Introduction*, chap. 11.

56. See A. Rupert Hall, *Philosophers at War: The Quarrel between Newton and Leibniz* (Cambridge, London, New York: Cambridge University Press, 1980), and especially *Math. Papers*, vol. 8.

57. See §2.3 below.

58. See the later Queries of the *Opticks* and the discussion by Betty Jo Dobbs, *Janus Faces* (§3.1, n. 10 below).

59. See §9.3 below.

CHAPTER TWO

Translating the *Principia*

2.1 *Translations of the* Principia: *English Versions by Andrew Motte (1729), Henry Pemberton (1729–?), and Thomas Thorp (1777)*

Newton's *Principia* has been translated in full, in large part, or in close paraphrase into many languages, including Chinese, Dutch, English, French, German, Italian, Japanese, Mongolian, Portuguese, Romanian, Russian, Spanish, and Swedish.[1] The whole treatise has been more or less continuously available in an English version, from 1729 to the present, in a translation made by Andrew Motte and modernized more than a half-century ago. In addition, a new version of book 1 by Thomas Thorp was published in 1777.[2]

Motte's translation of 1729 was reprinted in London in 1803 "carefully revised and corrected by W. Davis" and with "a short comment on, and defence of, the Principia, by W. Emerson."[3] This edition was reissued in London in 1819. Motte's text also served as the basis for the "first American edition, carefully revised and corrected, with a life of the author, by N. W. Chittenden," apparently published in New York in 1848, with a copyright date of 1846. There were a number of further New York printings.[4]

1. Most of these versions are listed in our Latin edition with variant readings (§1.3, n. 45 above), appendix 8.

2. Isaac Newton, *Mathematical Principles of Natural Philosophy*, trans. Robert Thorp, vol. 1 (London: W. Strahan and T. Cadell, 1777; reprint, with introd. by I. Bernard Cohen, London: Dawsons of Pall Mall, 1969).

3. See our Latin edition with variant readings (§1.3, n. 45 above), 2:863.

4. See P. J. Wallis, "The Popular American Editions of Newton's *Principia*," *Harvard Library Bulletin* 26 (1978): 355–360.

We do not know much about Andrew Motte. He was the brother of Benjamin Motte, a London printer who was the publisher of Andrew's English version.[5] Andrew was the author of a modest work entitled *A Treatise of the Mechanical Powers, wherein the Laws of Motion and the Properties of those Powers are explained and demonstrated in an easy and familiar Method*, published by his brother in London in 1727. This book, which informs us that Andrew had been a lecturer at Gresham College, displays an acquaintance with the basics of Newtonian physics, providing a useful and correct discussion of the three laws of motion of Newton's *Principia*. Andrew Motte was also a skilled draftsman and engraver. Each of the two volumes of his translation contains an allegorical frontispiece which he had drawn and engraved ("A. Motte invenit & fecit"), and a vignette engraved by him introduced each of the three books.

The new translation of book 1 published in 1777 was the work of Robert Thorp. A second edition was published in 1802.[6] Thorp had excelled in mathematics as an undergraduate at Cambridge, where he was graduated B.A. in 1758 as "Senior Wrangler." He taught Newtonian science as a Fellow of Peterhouse, receiving his M.A. in 1761. In 1765 he was one of three annotators of a Latin volume of excerpta from the *Principia*. He later followed an ecclesiastical career, eventually becoming archdeacon of Durham.

Thorp's translation was accompanied by a commentary that is perhaps more to be valued than his translation. His goal was to make Newton's "sublime discoveries" and the "reasonings by which they are established" available to a reading public with a knowledge only of elementary mathematics. He boasted that he had made explicit "every intermediate step omitted by the author." A sampling of his commentary would seem to indicate that this latter boast was an exaggeration. I have not attempted to check the accuracy of this version, but a single example, discussed in detail in §2.3 below, shows so gross a distortion of Newton's straightforward Latin text that one would hardly consider it a reliable rendition of the *Principia*.[7]

5. Isaac Newton, *The Mathematical Principles of Natural Philosophy*, trans. Andrew Motte, 2 vols. (London: Benjamin Motte, 1729; reprint, with introd. by I. Bernard Cohen, London: Dawsons of Pall Mall, 1968).

6. See I. Bernard Cohen, introduction to reprint of first edition (cited in n. 2 above), p. i; cf. Peter Wallis and Ruth Wallis, *Newton and Newtoniana, 1672–1975: A Bibliography* (Folkestone: Dawson, 1977), nos. 28, 28.2, 29.

7. Thorp never published his version of books 2 and 3, although these were announced as forthcoming in 1776.

Information about Robert Thorp (1736–1812) can be found in the *Dictionary of National Biography* at the beginning of the entry on his son, Charles Thorp (1783–1862).

Soon after Newton died, Henry Pemberton made a series of public announcements that he had all but ready for the press an English translation of the *Principia*, together with an explanatory commentary.[8] Neither of these works was ever published, and the manuscripts of both have disappeared.

2.2 *Motte's English Translation*

The Motte version was the result of heroic enterprise, the full measure of which can be fully appreciated only by someone who has gone through the same effort. And yet Motte's text presents great hurdles to today's reader, for whom the style will seem archaic and at times very hard to follow. Motte uses such unfamiliar language for ratios as "subduplicate," "sesquiplicate," and "semialterate." He also sometimes, but not always, translates into English certain words of Latin or Greek-Latin origin that Newton introduced into his text but that had not yet become standard. An example of this double usage is "phaenomena," occurring in Motte's text almost side by side with the English translation "appearances of things." Another term which Motte sometimes translates and sometimes keeps in Latin is "inertia," which—like "phaenomena"—appears to have been not yet common either in treatises on mechanics or in ordinary English. Motte introduces "inertia" in def. 3, referring first to "the inactivity of the Mass," then to "the inactivity of matter," and finally to "*Vis Inertiae*, or Force of Inactivity."

Motte's translation has additional problems. Some phrases and passages are based on the second edition and do not therefore represent the final state of the third and ultimate authorized edition.[9] Motte introduced phrases that are not in Newton's text at all. Some of these intrusive phrases consist of explanatory insertions to help the reader through an otherwise difficult passage. In the final sentence of the concluding General Scholium there is an insertion derived from Newton's proposed manuscript emendation.[10]

Not only do many of Motte's technical mathematical terms seem unintelligible today, but his language abounds in archaic phrases that seem awkward and difficult to understand. Motte's style of punctuation adds yet an additional hurdle for the would-be reader. (On the relation between our translation and Motte's, see §10.2 below.)

8. See I. Bernard Cohen, "Pemberton's Translation of Newton's *Principia*, with Notes on Motte's Translation," *Isis* 54 (1963): 319–351.

9. These are listed in the article cited in n. 8 above.

10. On Motte's insertion of the words "electric and elastic," see §2.3 and §9.3 below.

The Motte-Cajori Version; The Need for a New Translation 2.3

In 1934 the University of California Press published a handsome English version of the *Principia*. The trade edition was presented in a half-leather binding, in a large-size format with an elegant typeface and a title page printed in two colors. This version was presented as a revision of Motte's translation, which, the title page declared, was "revised, and supplied with an historical and explanatory appendix, by Florian Cajori." The Motte-Cajori version has become widely used as a result of several hardback reprints and a great many further reprints in a two-volume paperback edition. This text has become "enshrined" in the various editions and reprints of the *Encyclopaedia Britannica*'s "Great Books of the Western World."

Florian Cajori had a long and distinguished career as a historian of mathematics.[11] When he died in August 1930,[12] he left among his papers two partially completed works: the edition of the *Principia* that bears his name and an edition of Newton's *Opticks* that was never completed for publication.[13]

In a brief editor's note to the printed Motte-Cajori revision, R. T. Crawford states that he had been "invited by the University of California Press to edit this work," but he gives no information concerning the degree of such editing that had been required. There is no way of telling how much of the actual revision had been completed by Cajori at the time of his death.

Cajori's expressed goal was to eliminate "certain mathematical expressions [and notations] which are no longer used in mathematics and are therefore not immediately understood by a reader familiar only with modern phraseology." Accordingly, he "altered the translation by substituting for the old, corresponding modern terminology." As he explained:

> Most frequent of the obsolete terms are "duplicate ratio," "subduplicate ratio," "triplicate ratio," "subtriplicate ratio," "sesquiplicate ratio," "subsesquiplicate ratio," "sesquialteral ratio." For these I have used, respectively, the terms "square of the ratio," "square root of the ratio," "cube of the ratio," "cube root of the ratio," "3/2th power of the ratio," "2/3th

11. His major scholarly works include *A History of Mathematical Notations*, 2 vols. (Chicago: Open Court Publishing Co., 1928–1929); *A History of the Conceptions of Limits and Fluxions in Great Britain from Newton to Woodhouse* (Chicago: Open Court Publishing Co., 1919); *A History of Mathematics*, 2d ed., rev. and enl. (New York: Macmillan Co., 1911).

12. See the obituary by Raymond Clare Archibald, "Florian Cajori (1859–1930)," *Isis* 17 (1935): 384–407, with a bibliography of Cajori's writings.

13. Ibid.

> power of the ratio," "ratio of 3 to 2." In a few rare occurrences, the old usage of the term "proportion" corresponds to the modern "ratio."

Cajori also "discarded the vinculum as the sign of aggregation" and "introduced round parentheses instead." In some places, additionally, "where the reasoning is unusually involved," he "introduced the modern notation for proportion, $a : b = c : d$, in place of the rhetorical form used in the *Principia*."[14] The Motte-Cajori version introduced other changes in Motte's translation.[15] Some of these were intended to make Motte's prose easier to read for a twentieth-century student, that is, to remove archaic or unfamiliar modes of expression. But others were introduced with the goal of improving Newton's own text, to make it become more like a work of the twentieth than of the seventeenth century.

In the modernizations of Motte's prose, some real errors were introduced. These arose from the way in which the alterations were made: largely by considering Motte's version as if it were an author's original manuscript that needed revision, rephrasing, and rewriting. In this process of editorial recasting of Motte's prose, little if any attention was paid to Newton's original Latin text on which Motte's translation was based. In some cases, there was a failure to consult not only the original Latin of 1726, but even the original Motte version of 1729.

In the Motte-Cajori version of book 1, prop. 94, reference is made to a certain curve which "will be a parabola, whose property is that of a rectangle under its given latus rectum, and the line IM equal to the square of HM." Clearly it makes no sense to say that a "property" of the parabola "is that of a rectangle under its given latus rectum," nor is it geometrically (nor, for that matter, dimensionally) possible to have a "line IM equal to the square of HM." In the original Latin text, it is said that the curve "erit . . . parabola, cujus haec est proprietas, ut rectangulum sub dato latere recto & linea IM aequale sit HM quadrato." That is, the property of the parabola is simply that the square of HM equals the product (or "rectangle") of IM and the given latus rectum. Motte's original (1729) text contains a faithful[16] rendition: "a parabola, whose property is, that a rectangle under its given latus rectum and the line IM is equal to the square of HM" (1:312). In the 1803 edition

14. *Sir Isaac Newton's Mathematical Principles of Natural Philosophy and His System of the World*, trans. Andrew Motte, rev. Florian Cajori (Berkeley: University of California Press, 1934), pp. 645–646 n. 16.

15. This kind of revision, however, was not done as carefully as one might have wished. For example, in book 2, prop. 10, we find "oo" unchanged to "o^2" in the same line as "o^3." Also "QQ," "aaoo," "nnoo," and "2 dnbb."

16. In spite of a misprint that yields "be" for "will be." It may also be noted in passing that Thorp's version of 1777, although employing what is now antiquated punctuation, is completely accurate and literal: "will be a parabola, of which this is a property, that the rectangle under its given *latus rectum*, and the line IM is equal to the square of HM" (p. 352).

of Motte's translation, the then old-fashioned comma following "is" was eliminated and an incorrect and misleading "of" was inserted so as to make the sentence read: "a parabola, whose property is that of a rectangle under its given latus rectum and the line IM is equal to the square of HM" (1:203). In the next edition (1819), somebody evidently thought that clarification might result from adding a new comma following "latus rectum" so as to give the rendition, "a parabola, whose property is that of a rectangle under its given latus rectum, and the line IM is equal to the square of HM" (1:205). The final distortion, in the Motte-Cajori version, was to keep the new comma of the 1819 text but to eliminate the "is" before "equal to." This example suggests that the text which was being revised was not, as declared on the title page, the original (1729) edition but either the second revision (1819) or perhaps one of the mid-nineteenth-century versions based upon it.

Evidence that the revision of Motte's text was not made with reference to the Latin third edition may be seen in the simple fact that on the many occasions where Motte's text has kept a word or phrase of the second edition, these pass on to the Motte-Cajori version. I have published elsewhere a somewhat extended set of examples of Motte's dependence on the second, rather than the third, edition.[17] In every such instance, I found that the Motte-Cajori version follows Motte's text. Had the revision been made with reference to the Latin text, these variants would have been discovered and updated to Newton's final version.[18] Because the Motte text was revised without reference to the Latin original, the Motte-Cajori version is at times misleading as a result of the words and phrases introduced by Motte and not found in Newton's original. Most of these are not of tremendous significance.[19] But

17. Cohen, "Pemberton's Translation of Newton's *Principia*" (n. 8 above), esp. appendixes 2 and 3 on the Motte-Cajori text; also "Newton's Use of 'Force,' or, Cajori versus Newton: A Note on Translations of the *Principia*," *Isis* 58 (1967): 226–230.

18. In the revision of the very beginning parts of the *Principia*, however, some attention was paid to the Latin texts of the several editions. Thus n. 3 (on p. 628) calls attention to the fact that in the first edition the preface bears no date and no author's name, that the "signature 'Is. Newton' and the date 'Dabam Cantabrigiae, e Collegio S. Trinitatis, Maii 8. 1686' first appear in the second edition, 1713." At the end of this note, reference is made to ten other notes, nn. 3, 19, 24, 26, 27, 29, 30, 39, 42, 45, in which certain "changes in the second edition of the *Principia* are mentioned." A careful reading of these ten notes, however, reveals that only one even mentions the second edition. Evidently, at some stage of preparation of the work for the printer, the notes were renumbered but the numbers in the cross-references were not changed. Cajori himself never completed the composition of all his planned annotations.

It should be observed that in n. 30 (pp. 653–654), Cajori gives extracts both in Latin and in English translation from book 2, lem. 2, to illustrate Newton's use (in the first edition) of "infinitely small quantities (that is, fixed infinitesimals)." In n. 14, Cajori displayed the Latin texts of the laws of motion, showing how the first law was very slightly altered in the third edition.

19. Two examples will show how phrases inserted by Motte and not found in Newton's text were kept in the Motte-Cajori version. In the last sentence of the discussion of def. 1, Motte inserted the words "fine dust." Newton wrote, "Idem intellige de nive & pulveribus," which was rendered by Motte as "The

in the case of the qualifying words "electric and elastic" in the General Scholium, some scholars have been misled into believing these words to be part of Newton's published text.[20] In another example, discussed below, scholars have supposed that Motte, rather than the reviser for the Motte-Cajori version, was responsible for altering Newton's "refraction of the heavens" into "crumbling of the heavens."

An example that shows how the Motte-Cajori version did not consistently pay attention to the Latin original would seem to suggest that Cajori himself may not have been responsible for the revisions. At issue is a minor point, a reference (in the scholium following lem. 11 in book 1, sec. 1) to "perplexed demonstrations," a literal translation by Motte of Newton's "perplexas demonstrationes." In the Motte-Cajori version this is altered to "involved demonstrations." Had Newton's text been consulted, the reviser would have quickly discovered that in the third edition this phrase was changed to "longas demonstrationes," which may be translated as "lengthy demonstrations." The reason why this example is of interest is that only a dozen years earlier, in his history of fluxions, Cajori had published this very passage both in the original Latin and in English translation. He had obviously made a careful collation of the several editions of the *Principia*, and he explicitly noted that in "the third edition 'longas' takes the place of 'perplexas.' "[21]

A fault of a much more serious kind is the attempt to modernize Newton's thought and not just Motte's translation. This appears in the more or less continual elimination of the word "force" ("vis") in the many references by Newton to "force of inertia" ("vis inertiae"). In each such case, it is true, the resulting sentence seems to make good sense to a twentieth-century reader and does so to such a degree as to raise an important question of why Newton used "force" in this case.[22] But this is a wholly separate question from translating what Newton wrote and presenting Newton's thought. The same tactic of making Newton's physics seem more correct may be seen in the Motte-Cajori substitution of the verb "continues" for Motte's more volitional "perseveres" in the statement of the first law of motion,

---

same thing is to be understood of snow, and fine dust or powders." In the discussion of def. 5, Motte makes Newton say of the moon that it "may be perpetually drawn aside towards the earth, out of the rectilinear way, which by its innate force it would pursue; and would be made to revolve in the orbit which it now describes." The reviser altered "perpetually" to "continually," but did not know, or did not indicate, that two clauses—"which by its innate force it would pursue" and "which it now describes"—are not to be found in Newton's Latin text, which reads simply "retrahi semper a cursu rectilineo terram versus, & in orbem suum flecti."

20. See §2.2 above and esp. §9.3 below. On this phrase see A. Rupert Hall and Marie Boas Hall, "Newton's Electric Spirit: Four Oddities," *Isis* 50 (1959): 473–476; A. Koyré and I. B. Cohen, "Newton's 'Electric and Elastic Spirit,' " *Isis* 51 (1960): 337.

21. Cajori, *Limits and Fluxions*, p. 5.

22. See the discussion of "force of inertia" in §3.3 below.

where "Every body perseveres in its state . . . " is modernized to "Every body continues in its state . . . " Newton's Latin version, in all three editions, uses the verb "perseverare."

Rather than make a parade of the faults of the Motte-Cajori edition, let me call attention to some specific errors which are foremost among those that impelled us to undertake a new translation. Let me note, however, that it is easier to find errors in someone else's translation than to make a new one.

The first of these faults occurs in the Motte-Cajori version of Newton's discussion of comets in book 3, prop. 41. Here Newton notes that one of the opinions which exist concerning the tails of comets is that they "arise from the refraction of light in its progress from the head of the comet to the earth." The reviser, following Motte, here correctly renders "refractio" as "refraction." In the paragraph as a whole, there are seven instances of "refractio," consistently translated by Motte as "refraction."

In two instances, however, "refractio" is transformed by the Motte-Cajori version into "crumbling," with the result that "si cauda oriretur ex refractione materiae coelestis" becomes "if the tail was due to the crumbling of the celestial matter." Similarly, "igitur repudiata coelorum refractione" turns into "the crumbling, therefore, of the heavens being thus disproved." Unfortunately this rendering has misled some scholars who supposed that this was Motte's translation rather than an unwarranted alteration made by the twentieth-century reviser.

Among the faults in the Motte-Cajori version, one (occurring on page 3) is astonishing. It appears in the midst of a discussion of def. 5, in a statement about the motion of a projectile near the surface of the earth. Newton is considering what would happen in the case of an absence of air resistance. The Motte-Cajori version would have Newton say: "The less its gravity is, or the quantity of its matter, or the greater the velocity with which it is projected, the less will it deviate from a rectilinear course, and the farther it will go." Newton is thus presented as believing that the distance through which a forward-moving body falls, which is related to its acceleration in free fall, depends on its weight ("its gravity") or its mass ("or the quantity of its matter"). Newton would appear to be ignorant of the experiments of Galileo and others and to have made a gross error in the elementary physics of motion. Newton, of course, knew better, as may be seen elsewhere in the *Principia*.

The error is not Newton's, but was introduced in the Motte-Cajori version. A more exact rendering of the Latin text reads: "The less its gravity in proportion to its quantity of matter, or the greater the velocity with which it is projected, the less it will deviate from a rectilinear course and the farther it will go." That is, if gravity or weight is not proportional to the mass but is in some lesser proportion, then in any given time the projected body will fall through a smaller distance and so "the less it will deviate from a rectilinear course and the farther it will go."

The words used by Motte give a correct statement, although the text seems awkward (and there is a comma that troubles a modern reader): "The less its gravity is, for the quantity of its matter . . . " That is, in a modern version, "The less the gravity or weight is in proportion to the mass . . . " Most likely, some twentieth-century editor or proofreader may have been confused by the fact that Motte's eighteenth-century style of punctuation is so different from our own, especially his seemingly excessive use of commas, notably the comma after "its gravity is" to indicate a slight pause in reading. Very probably someone, without paying attention to the sense of the text, thought the phrase between commas was appositive or parenthetical and so altered Motte's "for" to "or" and made nonsense of the text. Another possibility is that this substitution resulted from a misprint in which the "f" of "for" was dropped.

Newton was not thinking in terms of a purely hypothetical situation, but rather of the physical condition of bodies immersed in a fluid or medium. If the "proportion" of a body's weight to mass is determined in a vacuum, and the body is then weighed in air, the weight of the body and hence the proportion of the body's weight to mass will be diminished by the buoyancy of air. The effect of buoyancy on weight enters Newton's considerations in a number of problems in the *Principia*.

Let me turn now to an error which is so gross that I find it difficult to believe it can be attributed to a scholar of Cajori's stature. Cajori had actually made a careful study of the Latin text of the *Principia* a decade before his death. In his *History of the Conceptions of Limits and Fluxions*, he included extracts from the *Principia* in the original Latin and in English translation.[23] In each case, he gave an extract from the Latin first edition, carefully noting the alterations made by Newton in subsequent editions. He even took pedantic note of such fine points of variation as that (in the scholium following book 2, lem. 2), " 'pervenientis' takes the place of 'pergentis,' " or that " 'intelligi eam' takes the place of 'intelligieam,' " or that " 'at' takes the place of 'et.' "[24] Similar textual considerations appear in other parts of Cajori's history of fluxions.[25] It seems hard to believe that a scholar who had paid such particular care in examining Newtonian texts could have been responsible

23. Cajori, *Limits and Fluxions*, chap. 1.

24. With respect to book 2, sec. 2, lem. 2, of the *Principia*, Cajori recorded such minor details as the substitution of "eorum" for "earum" along with "Lateris" for "Termini," as well as major changes.

25. For example, on p. 51, in presenting E. Stone's translation (London, 1730) of the preface to the marquis de l'Hôpital's textbook, Cajori used square brackets to indicate words "interpolated by Stone."

for such travesties of Newton's thought and for such absurdities as occur in the following two examples from the Motte-Cajori edition of the *Principia*.

The scholium to the definitions (as in the Motte-Cajori version) discusses "quantities themselves" and our "sensible measures" of them, making an important distinction between "relative quantities" and the quantities whose names they carry. The measures are "relative quantities" and are commonly used "instead of the quantities themselves." Then, arguing that "the meaning of words is to be determined by use," Newton asserts that "by the names time, space, place, and motion," their sensible "measures are to be understood" except in "purely mathematical" expressions. Then Newton declares, in the Motte-Cajori version:

> On this account, those violate the accuracy of language, which ought to be kept precise, who interpret these words for the measured quantities.

Nor do those "who confound real quantities with their relations and sensible measures" less "defile the purity of mathematical and philosophical truths."

The philosophical position here expressed is of real significance in relation to Newton's thought on such important issues as absolute and relative time and space. No reader of these lines can fail to be moved by the majestic rhetoric and the stress on "the accuracy of language, which ought to be kept precise." The only problem is that this particular expression is not Newton's. Nor, for that matter, is it Motte's. Newton's text is rather simple and straightforward:

> Proinde vim inferunt sacris literis, qui voces hasce de quantitatibus mensuratis ibi interpretantur. Neque minus contaminant mathesin & philosophiam, qui quantitates veras cum ipsarum relationibus & vulgaribus mensuris confundunt.

We have rendered this in English as "those who there interpret these words . . . do violence to the Scriptures." Here "sacris literis" refers to sacred writings, specifically Holy Writ or Scripture, as in Motte's version, which accurately renders Newton's text by "they do strain the Sacred Writings, who there interpret those words . . . "

How could the Motte-Cajori version have gone so far astray? The answer is that in the Motte-Cajori text, this sentence was based on the translation by Thorp, discarding Motte's version altogether. Thorp rendered "sacris literis" not as "sacred writings" but as "language, which ought to be held sacred," with an "accuracy" thrown in for good measure. The Motte-Cajori version adopts Thorp's "violate the accuracy of language" but then alters "which ought to be held sacred" to become "which ought to be kept precise."[26] The result is that students of Newton are led astray by believing that this pretty sentiment about the accuracy of language was written by Newton.

A final example is so ridiculous that it is difficult to believe it may be found in the *Principia*. This bizarrerie occurs in the third book where, in the third edition, Newton introduces two propositions by John Machin on the motion of the nodes of the moon's orbit. In an introductory scholium, Newton wrote:

> Alia ratione motum nodorum *J. Machin Astron. Prof. Gresham. & Hen. Pemberton* M.D. seorsum invenerunt.

Motte translated this unambiguous sentence correctly, but he abbreviated Machin's title, Gresham Professor of Astronomy (or Professor of Astronomy in Gresham College), so as to read "Astron. Prof. Gresh." His text reads as follows:

> Mr. *Machin* Astron. Prof. Gresh. and Dr. *Henry Pemberton* separately found out the motion of the nodes by a different method. (2:288)

The names Machin and Henry Pemberton were printed in italics.

Someone evidently decided that Motte's "Prof. Gresh." must refer to a third astronomer, not recognizing that these words were simply an abbreviated form of Machin's title. Assuming that Motte or the printer had forgotten to put this astronomer's name in italics, the reviser remedied the supposed error, with the result that—for more than half a century—this statement has been made to read:

> Mr. *Machin*, Professor *Gresham*, and Dr. *Henry Pemberton*, separately found out the motion of the moon by a different method. (P. 463)

These examples leave no doubt that the Motte-Cajori version cannot be accepted as an authentic presentation of Newton's *Principia*. It was the discovery

26. In a preliminary version of this passage, in the long series of definitions that Newton wrote out after *De Motu* and just before composing the *Principia*, there is a final paragraph mentioning this very issue. Newton's statement of his position may be translated as follows: "Moreover, it has been necessary to distinguish absolute and relative quantities carefully from each other because all phenomena may depend on absolute quantities, but ordinary people who do not know how to abstract their thoughts from the senses always speak of relative quantities, to such an extent that it would be absurd for either scholars or even Prophets to speak otherwise in relation to them. Thus both Sacred Scripture and the writings of Theologians must always be understood as referring to relative quantities, and a person would be laboring under a crass prejudice if on this basis he stirred up arguments about absolute [*changed to* philosophical] notions of natural things," that is, "notions taken in a philosophical sense." The context here (Prophets, Theologians) should remove any doubt, if there can be any, that in the *Principia*, Newton was referring to the interpretation of Scripture. The Latin text is published in *Math. Papers* 6:102 and in *Background*, p. 307.

of the last two—the discussion of the accuracy of language and the reference to Professor Gresham—that provided the final impetus to our decision to produce a wholly new translation.

A Problem in Translation: Passive or Active Voice in Book 1, Secs. 2 and 3; The Sense of "Acta" in Book 1, Prop. 1 2.4

In the opening portions of the *Principia*, Newton has used the passive rather than the active voice, notably in secs. 2 and 3 of book 1. Thus he writes, "Corpus omne, quod movetur in linea aliqua curva . . . " (prop. 2); "Si corpus . . . in orbe quocunque revolvatur . . . " (prop. 6); "Gyretur corpus in circumferentia circuli . . . " (prop. 7). In some English versions, the verbs in question are translated in the passive (e.g., "Every Body which is moved" or "Let a body be revolved"). But the reason why the verbs "movere" and "revolvere" appear in the passive is that in classical Latin, both were considered to be transitive verbs and so could not properly be in the active voice without a direct object. Thus the verb "revolvere" appears in the *Oxford Latin Dictionary* only as a transitive verb; no example appears to have been found of the use of this verb intransitively in the classical period. The verb "movere," however, does appear in this dictionary as "tr. (intr.)," but all major examples are transitive and the examples given of ordinary motion (a shift of position) or impelling are in every case transitive.[27]

With regard to "gyrare," the situation is somewhat different. The verb "gyrare" does not appear in the *Oxford Latin Dictionary*, which is based on texts not later than the third century, although both the noun "gyrus" and the adjective "gyratus" are listed. This would accord with the fact that in the entry for the verb "gyrare" in Lewis and Short, all the examples are taken from later Latin; in most cases the verb is used intransitively. As will be seen below, Newton seems to have preferred the passive form of "gyrare" so as to give it a form in harmony with the two other verbs, "movere" and "revolvere."

There is ample evidence that at the time of the *Principia* and later, Newton—when writing in English—made use of the active and not the passive voice to specify bodies revolving in orbit. For example, shortly after composing the *Principia*, Newton wrote out in English a series of "Phaenomena" (ULC MS Add.

27. On this score, Gildersleeve and Lodge (§213*a*) declare that transitive verbs "are often used intransitively, in which case they serve simply to characterize the agent," a feature that "is true especially of verbs of *movement*." This particular usage is said to be "found at all periods." Although the examples given include the verbs "movere" and "vertere" (but not "revolvere"), this particular usage ("to characterize the agent") clearly does not occur in Newton's propositions concerning motion, since their goal is to characterize the motion produced and not the agent producing the motion.

4005, fols. 45–49), first published by A. Rupert Hall and Marie Boas Hall. Here we find "Phaenom. 7": "The Planets Mercury & Venus revolve about the Sun according to the order of the twelve signes." Similarly, "Phaenom. 8," "Phaenom. 9," and "Phaenom. 10" declare, respectively, that each of the superior planets—Mars, Jupiter, and Saturn—"revolves about the Sun."[28]

A second document in English that uses the active rather than the passive voice is Newton's essay on planetary motion in ellipses, sent to the philosopher John Locke a few years after the publication of the first edition of the *Principia*.[29] In prop. 2, a body is said to "revolve in the circumference of the Ellipsis," while in prop. 3 a body is said "to revolve in the Perimeter of the Ellipsis."

In the 1717/18 edition of the *Opticks*, Newton, writing in English, notes in query 28 that "Planets move . . . in Orbs concentrick" and "Comets move . . . in Orbs very excentrick." Similarly, in query 31, he writes that "Planets move . . . in Orbs concentrick" and "Comets move in very excentric Orbs."[30] When the *Opticks* was translated into Latin, these references to motion in orbit were rendered in the passive voice, "moventur" and not "movent."[31]

An early version of the propositions in book 1, secs. 2 and 3, of the *Principia* appears in the tract *De Motu*, composed in the autumn of 1684. In this preliminary text, both the verbs "movere" and "gyrare" do appear in the active voice. Thus prob. 1 and prob. 2 begin: "Gyrat corpus in . . ."[32] In prob. 4 (which was to become prop. 17 of the *Principia*), Newton used the active form of "movere," writing ". . . ut corpus moveat in Parabola vel Hyperbola."

During the winter of 1684/85, when Newton composed the draft of what was to become book 1 of the *Principia* (the text which I have called the Lucasian Lectures),[33] he rewrote the propositions of *De Motu* and added others, now using both the verbs "gyrare" and "movere." In doing so he revised the statement, just

28. *Unpubl. Sci. Papers*, pp. 378–385.

29. Ibid., pp. 293–301. A study (by D. T. Whiteside and supported by Bruce Brackenridge) would assign the date of 1684 to a prototype, now lost, of this essay.

30. Query 28, ult. par. (Isaac Newton, *Opticks*, based on the 4th ed., London, 1730 [New York: Dover Publications, 1952], p. 368); query 31 (ibid., p. 402).

31. For query 28 the case is a little different because the text of what became query 28 in the later English editions was revised between its first publication in Latin (1706) and the later printing in English (1717/18). That is, the specific reference to the way the planets and comets "move" does not appear in the Latin text of 1706. But in the later Latin edition (London, 1719), revised so as to conform to the second English edition (1717/18), the word "move" is rendered by "moventur" and not by "movent."

The Latin version was an "authorized" one, made by Samuel Clarke, a close associate of Newton's and Newton's spokesman in the Leibniz-Clarke correspondence.

32. The text of *De Motu* has been often printed: by Rigaud, by Rouse Ball, by the Halls, by Herivel, and most recently, by Whiteside in *Math. Papers*, vol. 6.

33. *Introduction*, §4.2.

quoted, of prob. 4, altering "moveat" so that it became what Whiteside has called "the more natural 'moveatur.'" This change reflects Newton's training in Latin in which the verb "movere" was to be used transitively (or primarily transitively) or reflexively.[34] In the Lucasian Lectures, however, he continued to use the active form of "gyrare" (e.g., prop. 8, "Gyrat corpus in spirali PQS..."; props. 9, 10, "Gyrat corpus in Ellipsi..."). As we have seen, "gyrare" is usually an intransitive verb.

The actual text of the Lucasian Lectures provides firm evidence that for Newton the passive form of the Latin verb "movere" is the equivalent of the English active "move" (and not "is moved"). Prop. 11 ("Movetur corpus in Hyperbola...") and prop. 12 ("Movetur corpus in perimetro Parabolae...") set forth the same concept of orbital motion under the action of a central force as the previous props. 8, 9, and 10. The only distinction between the two groups is the choice of verb. To express the same idea, one verb ("movere") requires the use of the passive, while the other ("gyrare") does not.

When the final manuscript for the printer was produced from the Lucasian Lectures text, however, Newton evidently decided that he would adopt a uniform style, and so he expressed all the verbs relating to orbital motion in the passive voice, though retaining the use of the present active participle in a reflexive sense in accord with regular Latin usage.[35] He also shifted the mood from the indicative to the jussive subjunctive, thus replacing "movetur" by "moveatur." In this revision, he also changed the verb in some propositions. For example, he rewrote prop. 7 with a new verb ("gyrare" in place of "movere"), using the passive voice ("Gyretur corpus in circumferentia circuli..."). Props. 10 and 11 (renumbered from props. 8 and 9) became "Gyretur corpus in Ellipsi..." Prop. 11 (formerly prop. 10) was given a new verb so as to become "Revolvatur corpus in Ellipsi..." in place of "Gyrat corpus in Ellipsi..." As has been noted, the verb "revolvere" occurs in classical Latin as a transitive verb and so appears here in the passive, in a form similar to "moveatur." Here is further evidence that at the time of composing the *Principia*, Newton's preferred style when writing in Latin about orbital motion was to use the passive of the verb "gyrare" as well as "revolvere" and "movere," but the sense is that of the active equivalents in English.

Prop. 1 of book 1 of the *Principia* differs a little from the others in using the past participle "acta." In the earlier *De Motu*, this proposition reads: "Gyrantia

34. But he did not change all the occurrences of the active voice into the passive, retaining the active "moveat" in prop. 7, "Moveat corpus in circulo PQA..." (ULC MS Dd.9.46); see *Math. Papers*, vol. 6.

35. On the use of the present active participle in a reflexive sense see Kühner-Stegmann, 1:108–110 (§28.4); A. Ernout and F. Thomas, *Syntaxe latine*, 2d ed., §224; G. M. Lane, *Latin Grammar*, §1482; Hale and Buck, *Latin Grammar*, §288.3.*a*.

omnia radiis ad centrum ductis areas temporibus proportionales describere" ("All rotating [bodies] by radii drawn to the center describe areas proportional to the times"). In the Lucasian Lectures, this statement was considerably expanded in several stages, during which the reference to "radiis" was dropped, apparently inadvertently. At this time, Newton altered the beginning to "Areas, quas corpora in gyros ad immobile centrum virium ductis describunt . . . " Then, in the *Principia*, he restored "radiis," needed for the statement to make sense, and added the participle "acta." Thus, in this final version, which appears unchanged in the second and third editions of the *Principia*, the text reads:

> Areas, quas corpora in gyros acta radiis ad immobile centrum virium ductis describunt, & in planis immobilibus consistere, & esse temporibus proportionales.

Except for the phrase "in gyros acta" the meaning is quite plain. "The areas which bodies . . . describe by radii drawn to an unmoving center of forces" are said to have two properties: they lie "in unmoving planes and are proportional to the times."

The phrase "in gyros acta," however, is puzzling. "Acta" is a perfect passive participle, and the phrase literally means "having been impelled (driven) into orbits." So understood, the phrase seems to be incomplete according to Newtonian dynamics; that is, it would apparently be referring only to the initial component of linear inertial motion that first put the body into orbit, with no mention of the continually acting centripetal force necessary to maintain the body in orbit. It seems probable, therefore, that "acta" should be understood in a present sense, "in gyros acta" being taken as equivalent to the clause "quae in gyrum aguntur" (*lit.* "that are [continually] being impelled into orbit") which Newton uses in the fourth sentence of his discussion of def. 5 but which he could not have used here without the ungainliness of a relative clause within another relative clause (that is, prop. 1 would have begun: "Areas, quas corpora quae in gyrum [or gyros] aguntur . . . "). Compare the use of "agitata" in the title of book 1, sec. 8: "De inventione orbium in quibus corpora viribus quibuscunque centripetis agitata revolvuntur," which we have translated as "To find the orbits in which bodies revolve when acted upon by any centripetal forces." The exact sense of the participle "agitata" was indicated in Andrew Motte's translation by the phrase "being acted upon."[36] Of course, Newton could have avoided any such construction by using "gyrantia omnia," which he

36. On the use of the perfect participle in a present sense see Kühner-Stegmann, 1:757–758 (§136.4.*b*.α); A. Ernout and F. Thomas, *Syntaxe latine*, 2d ed., §289; Hale and Buck, *Latin Grammar*, §601.2; Woodcock, *A New Latin Syntax* (1959), p. 82, §103 (the cited examples include "acta"). Lewis and Short list "agito" and "ago" among the synonyms of "movere."

had used earlier, or something like "corpora in orbibus revolventia" (cf. "planetarum in orbibus . . . revolventium" near the end of prop. 41 of book 3).

In order to convey the special sense of "acta" and to keep this sentence in harmony with the principles of Newtonian physics, we have rendered Newton's discussion of def. 5 by: "And the same applies to all bodies that are made to move in orbits." In our translation of prop. 1, we have used a similar version: "The areas which bodies made to move in orbits describe . . . "

"Continuous" versus "Continual"; The Problem of "Infinite" 2.5

A real problem arises with respect to the use of "continuous" (and "continuously") versus "continual" (and "continually"). According to Eric Partridge's *Usage and Abusage*, these terms represent different concepts and "must" not be confused. They are not synonyms because "continuous" is defined as "connected, unbroken; uninterrupted in time or sequence," whereas "continual" is defined as "always going on." Fowler's *Modern English Usage* stresses the difference at greater length. Something is "continual" if it "either is always going on or recurs at short intervals and never comes (or is regarded as never coming) to an end." By contrast, "continuous" means that "no break occurs between the beginning and the (not necessarily or even presumably long-deferred) end." Neither Partridge nor Fowler, however, was concerned with the special sense in which "continuous" is used in current mathematics. To prevent any mathematical misinterpretation, we have tended to use "continual" rather than "continuous."

Another problem of a similar kind arose with regard to Newton's use of the Latin phrase "in infinitum" (as in the proof of book 1, prop. 1) and the Latin adverb "infinite" (as in book 1, prop. 4, schol.). Some translators, such as Motte, have kept the phrase "in infinitum" in the original Latin, and they have used the same phrase for the adverb "infinite." In order to avoid philosophical discussions of the meaning of "infinite" and "infinitely," we have translated these words by "indefinite" and "indefinitely," in the sense of being "without limit," which was the practice (as we later found) adopted by the late-nineteenth-century editors, such as John H. Evans and P. T. Main, who prepared texts of parts of the *Principia* for the use of English college students. Thus, our choice accords with modern mathematical practice.

From an etymological point of view, there is no real difference between "infinite," "indefinite," and "without limit." "Infinite" comes from a joining of the Latin prefix "in-" (meaning "not") and "finitus," the past participle of "finire" ("to limit"). "Indefinite" comes from "in-" plus "definite" (which means literally "having distinct limits"), which comes from the Latin "definitus," the past participle of "definire" ("to set definite or precise limits"). The word "infinite," however, has philosophical and mathematical overtones which we hoped to avoid by using "indefinite" or "without limit," a course which also enabled us to avoid any problems arising from the special senses of "infinite" in mathematics.[37]

In the discussion of book 1, prop. 1, Evans and Main would have a body being "perpetually drawn off from the tangent" by a force which "will act incessantly." Percival Frost similarly avoids "infinitely," and he adopts "indefinitely." For him, however, the limiting force "will act continuously." The problem for the translator is well exemplified by this proposition, which contains Newton's polygonal model. In the discussions of book 1, prop. 1, there is apt to be agreement that the limit of the polygonal trajectory is a curve, and therefore "continuous," whereas—as we shall see below—there is considerable discussion concerning whether the force can similarly be taken to act in the limit in a "continuous" manner. Accordingly, we have generally chosen the adjective "continual" and the associated adverb "continually," in order to avoid the confusion that would arise because "continuous" and its adverb "continuously" have a special technical sense in today's mathematics. We are, however, mindful of the warnings of such authors as Partridge and Fowler. We were strengthened in our decision to use a seemingly incorrect use of "continual" and "continually" when we discovered that Newton himself, when writing in English, tended to use "continual" rather than "continuous." In the essay on elliptical motion which he sent to John Locke in 1690, shortly after publishing the *Principia*, he favored "continuall" ("the attraction may become continuall") and "continually" ("the attraction acts not continually but by intervals") rather than "continuous" and "continuously."[38]

37. There is, of course, the even more vexing problem of the associated terms "infinitely small," "infinitesimal," and "infinitesimally small."

38. For illumination on this topic, the reader should consult the discussions by D. T. Whiteside and Eric Aiton, and especially the writings of Michel Blay. It has been argued that in book 1, props. 1 and 2, and therefore elsewhere, in the limit the force may become "continuous" while the polygon does not become a "continuous" curve. Needless to say, it is to avoid just such a controversy that we have chosen the more neutral term "continual." On this topic see, notably, Eric J. Aiton, "Parabolas and Polygons: Some Problems concerning the Dynamics of Planetary Orbits," *Centaurus* 31 (1989): 207–221; D. T. Whiteside, "Newtonian Dynamics: An Essay Review of John Herivel's *The Background to Newton's 'Principia,'*" *History of Science* 5 (1966): 104–117, and also "The Mathematical Principles Underlying Newton's *Principia Mathematica*" (§1.2, n. 9 above), and the introduction to Whiteside, *Preliminary Manuscripts for Isaac Newton's 1687 "Principia"* (§1.1, n. 5 above); also Herman Ehrlichson, "Newton's Polygonal Model and the Second Order Fallacy," *Centaurus* 35 (1992): 243–258.

CHAPTER THREE

Some General Aspects of the *Principia*

The Title of Newton's Principia; *The Influence of Descartes* 3.1

When Newton enlarged the *Principia* from two books to three, he changed the title from *De Motu Corporum* to the more grandiose title *Philosophiae Naturalis Principia Mathematica*. In the first edition, the title page displays the words *Philosophiae* and *Principia* in large capital letters, stressing these two nouns at the expense of the modifiers *Naturalis* and *Mathematica* (see figs. 3.1, 3.2). We don't know for certain whether this emphasis was initially the result of a conscious decision by Halley or whether it came about through the printer's design, merely stressing the nouns at the expense of the adjectives. But there can be no doubt that Newton himself favored this emphasis since the same stress on *Philosophiae* and *Principia* was a feature of the title page of the second edition (1713). In the third edition, the stressed pair of words were printed not only in larger type but in red ink, so as to set them apart from the modifiers. The third and final approved edition displayed a half title reading

NEWTONI P R I N C I P I A *PHILOSOPHIÆ*

or "Newton's Principles of Philosophy."

Even without such emphasis, however, most readers would be aware that Newton's title was a simple transformation of Descartes's *Principia Philosophiae*.[1] Thus the title boldly declared that Newton had not written a work on general

1. D. T. Whiteside holds a different opinion, namely, that the "book's title of Philosophiae naturalis principia mathematica was a catchpenny one contrived to 'help ye sale of ye book' (rather than a hint at Newton's immense debt to Descartes's own *Principia philosophiae* in forming many of his basic ideas on motion), as is well known." See his "The Prehistory of the *Principia*" (§1.2, n. 14 above), esp. p. 34.

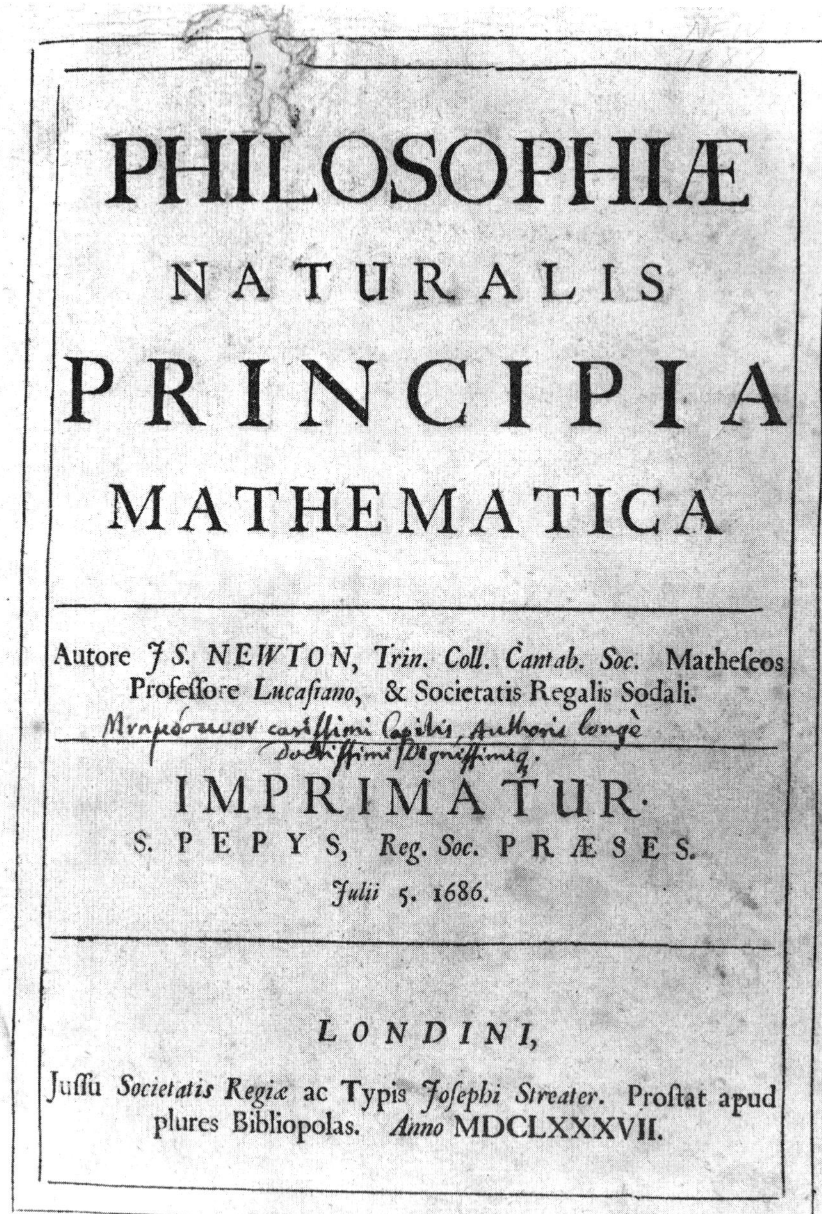

PHILOSOPHIÆ
NATURALIS
PRINCIPIA
MATHEMATICA

Autore *JS. NEWTON*, *Trin. Coll. Cantab. Soc.* Matheseos Professore *Lucasiano*, & Societatis Regalis Sodali.

Μνημόσυνον carissimi Capitis, Authoris longè Doctissimi Dignissimiq.

IMPRIMATUR.
S. PEPYS, *Reg. Soc.* PRÆSES.
Julii 5. 1686.

LONDINI,
Jussu *Societatis Regiæ* ac Typis *Josephi Streater*. Prostat apud plures Bibliopolas. *Anno* MDCLXXXVII.

Figure 3.1. Title page of the first edition of the *Principia* (London, 1687). This copy, in the Grace K. Babson Newton Collection in the Burndy Library, belonged to Edmond Halley and contains manuscript corrections by Halley and by Newton.

Some copies of the first edition have a variant title page, which is a cancel, pasted onto the stub of the original. This variant form of the title page differs only in the imprint, "*LONDINI*, Jussu *Societatis Regiae* ac Typis *Josephi Streater*. Prostant Venales apud *Sam. Smith* ad insignia Principis *Walliae* in Coemeterio D. *Pauli*, aliosq; nonnullos Bibliopolas. *Anno* MDCLXXXVII."

PHILOSOPHIÆ
NATURALIS
PRINCIPIA
MATHEMATICA.

AUCTORE
ISAACO NEWTONO,
EQUITE AURATO.

EDITIO SECUNDA AUCTIOR ET EMENDATIOR.

CANTABRIGIÆ, MDCCXIII.

Figure 3.2. Title page of the second edition (Cambridge, 1713), from the copy in the Grace K. Babson Newton Collection in the Burndy Library. Note that, as in the title page of the first and third editions, the words "PHILOSOPHIAE" and "PRINCIPIA" are emphasized.

philosophy, as Descartes had done, but had limited the subject matter to natural philosophy. Additionally, Newton's title declared that his principles were mathematical.

Readers of Newton's *Principia* could not help but be aware of its anti-Cartesian bias, its goal of replacing the earlier *Principia* and its faulty system of vortices. Book 2 of Newton's *Principia* concludes by showing that the Cartesian system cannot be reconciled with Kepler's laws,[2] which was probably the reason why there arose a belief that the whole purpose of book 2 was to confute Descartes.[3] Thus Newton's text would openly reinforce the anti-Cartesian sense of the title.

Despite the hostile attitude to Cartesian science, Newton's debt to Descartes appears again and again in his *Principia*. To the cognoscenti the presentation of Newton's laws as "Axiomata, sive Leges Motus" (Axioms, or the Laws of Motion) would have had rather obvious overtones of Descartes's *Principia*, where the laws of motion had been called "Regulae quaedam sive Leges Naturae" (Certain Rules or Laws of Nature). It was from Descartes's *Principia* that Newton learned the law of inertia, which in fact appears in law 1 in both the *Principia* of Descartes and the *Principia* of Newton.[4] In Descartes's presentation of the law, Newton gained the important concept of motion as a "state" ("status"). Newton's presentation (in def. 3 and law 1) of inertia uses the same language as Descartes's, even to the expression "quantum in se est."[5] In the early part of Newton's *Principia*, the repeated expression "conatus recedendi a centro," or "endeavor to recede from the center," is a direct quotation of a phrase used by Descartes, who was more concerned with "conatus," or "endeavor," than with the action of forces as such.[6]

It was from Descartes that Newton learned of the action of an impulsive force on a body in uniform rectilinear motion, the principle used in the first proposition of the first book of the *Principia*. Here[7] Newton begins with a body in uniform rectilinear motion; he supposes that an impulsive force or a blow is directed toward some point S, and he then determines the new rectilinear path. After an interval he repeats the process, and does so again and again, producing a trajectory in the form of an extended polygon. Although this procedure is often said to have been derived by Newton from Galileo's *Two New Sciences*, it rather comes directly from

2. Book 2, sec. 9, schol. foll. prop. 52, and esp. schol. foll. prop. 53; see §7.9 below.

3. See §7.1 below.

4. See my "Newton and Descartes," in *Descartes: Il metodo e i saggi*, ed. Giulia Belgioioso et al. (Rome: Istituto della Enciclopedia Italiana, 1990), pp. 607–634. Newton encountered the Cartesian concept of inertia before reading Descartes's *Principia*; see §4.8, below.

5. See my "'Quantum in se est': Newton's Concept of Inertia in Relation to Descartes and Lucretius," *Notes and Records of the Royal Society* 19 (1964): 131–156.

6. See my "Newton and Descartes" (n. 4 above).

7. See §§5.3 and 10.8 below.

Descartes's *Dioptrique*, one of the three treatises published by Descartes in 1637 as examples of the application of his "method." Here Descartes considers a moving object with constant speed in some medium such as air. He supposes that when the moving object reaches the interface of that medium and some other medium such as water, it receives a blow or impulsive force, as from a tennis racket, which changes both the direction and magnitude of the motion.[8] Although Descartes's goal was to establish the optical law of refraction, he invoked the analogy of the motion of a real object, that is, a moving tennis ball receiving a blow from a racket.

The researches of Betty Jo Dobbs have, among other things, revised the chronology of some of Newton's writings, making us aware that Descartes's ideas were being actively studied and criticized by Newton on the very eve of the *Principia*, between the composition of *De Motu* and the writing of the *Principia*. Her revision of the chronology shows not only that Newton's untitled essay that begins "De Gravitatione et Aequipondio fluidorum," referred to hereafter as Newton's essay "De Gravitatione,"[9] was composed at this late date but also that it was in this essay "that Newton really broke with Descartes."[10] This new chronology helps to explain why Newton's *Principia* exhibits so overt an anti-Cartesian bias since during the years just before the *Principia*, Newton was developing his own concepts in a matrix of active hostility to Cartesian philosophy. Although there are specific references to Descartes and his *Principia* in "De Gravitatione," neither one is mentioned by name in the *Principia* in the context of dynamics or celestial motion. Cotes, however, indicated in the index he prepared for the second edition that it was the followers of Descartes whom Newton had in mind in his reference (in the conclusion of the General Scholium at the end of book 2, sec. 6) to a "subtle matter" which had to be rejected.

In the *Principia* Newton refers to Descartes by name four times. In book 1, prop. 96, schol., Newton says that the law of refraction was "discovered" (in one form) by Snel and then "set forth" or "published" (in another form) by Descartes. In book 1, prop. 97, corol. 1, Newton writes of certain curves "which Descartes exhibited with respect to refractions in his treatises on optics and geometry," adding that Descartes "concealed" the methods whereby he had found these curves. A fourth explicit mention of Descartes occurs in book 3, prop. 6, corol. 2, in the second and third editions. This corollary is discussed at length in §8.5 below.

8. On the indebtedness of Newton to Descartes in relation to optics, see Maurizio Mamiani, "Newton e Descartes: La *Questio* manoscritta *Of Colours* e *La Dioptrice*," in *Descartes: Il metodo e i saggi* (n. 4 above), pp. 335–340.

9. Published in *Unpubl. Sci. Papers*, pp. 89–156.

10. Betty Jo Teeter Dobbs, *The Janus Faces of Genius: The Role of Alchemy in Newton's Thought* (Cambridge: Cambridge University Press, 1992), p. 148.

The essay "De Gravitatione" records how Newton pored through his copy of Descartes's *Principia* in order "to dispose of [Descartes's] fictions."[11] Since Newton was deep in a critical study of Descartes's *Principia* on the eve of composing his own work, we may all the more readily appreciate that his goal was to replace Descartes's *Principia* by a new kind of *Principia*. Newton would thus seem to have produced a title for his own opus by transforming Descartes's *Principia Philosophiae* into *Philosophiae Naturalis Principia Mathematica*.

Newton himself often referred to his work by the title of Descartes's treatise, as if he meant his own *Principia Philosophiae* as opposed to Descartes's, and so did many of his followers.[12] On at least one occasion in print, Newton referred to his book as *his* "Principles of Philosophy"; this was in the anonymous book review of the *Commercium Epistolicum*, the report of the Royal Society on the question of priority in the invention of the calculus. In this book review, in which Newton writes about himself in the third person, it is said that "by the help of the new *Analysis* Mr. *Newton* found out most of the Propositions in his *Principia Philosophiae*."[13] John Clarke, brother of Samuel (the Clarke in the Leibniz-Clarke correspondence), in his presentation of "Sir Isaac Newton's Principles of Natural Philosophy," explained how Newton's "first two Books . . . of the Principles of Philosophy" treat the subject "in a strict mathematical way."[14] On opening a copy of the third and final edition of the *Principia*, the reader would find a half title preceding the license of publication and the portrait of Newton. This half title, as we have seen, boldly proclaims that the treatise is "Newtoni Principia Philosophiae," that is, "Newton's Principles of Philosophy."

In 1686, Newton was so angered by Hooke's demanding credit for the inverse-square law that he told Halley he intended to suppress book 3. He was concerned, however, whether the title would still be appropriate. Writing to Halley of his decision to eliminate book 3, he said that the "first two books without the third" will not "so well bear the title of Philosophiae naturalis Principia Mathematica," and so he had "altered it to this, De Motu corporum libri duo." Then he had "second thoughts," he continued, and decided to "retain the former title," which

11. *Unpubl. Sci. Papers*, pp. 113, 146–147; see Dobbs, *Janus Faces*, p. 145.

12. Newton (and others) also called the book "Principia," "Principia Mathematica," "the Book of Principle," "Mathematical Principles of Philosophy," "the Book of Mathematical Principles of Philosophy," "the Book of Principles," and "Mathematical Principles."

13. *Philosophical Transactions* 29, no. 342 (Jan.-Feb. 1715): 173–224, esp. 206. See §5.8, n. 25 below. This document is reprinted in facsimile in Hall, *Philosophers at War* (§1.3, n. 56 above).

14. John Clarke, *A Demonstration of Some of the Principal Sections of Sir Isaac Newton's "Principles of Natural Philosophy"* (London: printed for James and John Knapton, 1730; facsimile reprint, with introd. by I. Bernard Cohen, New York and London: Johnson Reprint Corp., 1972), p. iv.

"will help the sale of the book which I ought not to diminish now tis yours."[15] Newton did change his mind, and book 3 was not eliminated from the *Principia*.

Newton's Goals: An Unpublished Preface to the Principia 3.2

In addition to the published prefaces to the three editions of the *Principia*, Newton wrote a number of draft versions of prefaces which he never completed for publication. Many of these were written at a time (mid and late 1710s) when Newton was planning a new edition which would include a tract on his method of fluxions. One purpose of these proposed prefaces was to insist that fluxions (or the calculus) had actually been used in the preliminary stages of writing the *Principia*; another was to assert his priority over Leibniz in the first invention of the calculus.[16] Although Newton often alleged that he had originally found many propositions and had produced their demonstrations by one method (analysis) and had then rewritten them according to another method (synthesis), there is no evidence whatever to support this claim and much circumstantial evidence to refute it.[17]

Among the draft prefaces, mostly dating from the years after the second edition of the *Principia* (1713), one is of special interest because it sets forth clearly what Newton considered to be the goals and achievements of the *Principia*. This draft also discusses Newton's mathematical methods with regard to specific propositions. Newton has here given us a valuable summary of what he considered to be some of the principal features of his treatise.

The following translation is based on Newton's Latin original.[18]

> *Unpublished Preface to the* Principia[19]
>
> The ancient geometers investigated by analysis what was sought [i.e., found their solutions to problems by the method of analysis], demonstrated by synthesis what had been found, and published what had

15. Newton to Halley, 20 June 1686, *Corresp.* 2:437.

16. Some of the most important of these drafts have been published, together with English translations, in *Math. Papers*, vol. 8.

17. This topic is discussed in §5.8 below.

18. ULC MS Add. 3968, fol. 109. D. T. Whiteside has given a Latin transcription (and a translation of the opening paragraphs), together with an extensive commentary, in *Math. Papers* 8:442–459. Whiteside's Latin text contains some important passages—based "on original drafts now in private possession"—which supplement and complete the version in the University Library.

19. Our translation is based on Newton's manuscript in the University Library. We have, however, added some additional passages taken from Newton's preliminary draft version as published by Whiteside.

been demonstrated so that it might be received into geometry. What was resolved was not immediately received into geometry; a solution by means of the composition of demonstrations was required.[20] For all the power and glory of geometry consisted in certainty of things, and certainty consisted in demonstrations clearly composed [i.e., demonstrations according to the method of synthesis, or composition]. In this science, what counts is not so much brevity as certainty. And accordingly, in the following treatise I have demonstrated by synthesis the propositions found by analysis.[21]

The geometry of the ancients was indeed concerned with magnitudes; but propositions concerning magnitudes were sometimes demonstrated by means of local motion, as when the equality of two triangles in prop. 4 of book 1 of Euclid's *Elements* was demonstrated by transferring either triangle into the place of the other. But also the generation of magnitudes by continual motion was accepted in geometry, as when a straight line was multiplied by a straight line to generate an area, and an area was multiplied by a straight line to generate a solid. If a straight line which is multiplied by another one is of a given length, a parallelogram area will be generated. If its length is continually changed according to some fixed law, a curvilinear area will be generated. If the magnitude of an area which is multiplied by a straight line is continually changed, a solid terminated by a curved surface will be generated. If the times, forces, motions, and velocities of motions are represented by lines, areas, solids, or angles, these quantities can also be dealt with in geometry.

Quantities increasing by a continual flux we call fluents, and the velocities of increasing we call fluxions, and the momentaneous increments we call moments, and the method by which we deal with this sort of quantities we call the method of fluxions and moments; and this method uses either synthesis or analysis [i.e., is either synthetic or analytic].

The synthetic method for fluxions and moments occurs here and there in the following treatise, and I have put its elements in the first eleven lemmas of book 1 and in lem. 2, book 2. [*The preliminary draft contains the following two sentences:* We found propositions which follow in this treatise by the analytical method, but we demonstrated them

20. "Resolutio" and "compositio" are the Latin equivalents of "analysis" and "synthesis." In the first sentence Newton uses the Graeco-Latin words "analysis" and "synthesis," but in this second sentence he shifts to the Latin equivalents.

21. See §5.8 below.

synthetically so that the [natural] philosophy of the heavens, set forth in the third book, should be founded on propositions demonstrated geometrically. And he who properly understood the method of composition cannot be ignorant of the method of resolution.]

Examples of the method of analysis occur in prop. 45 and the scholium to prop. 92, book 1, and in props. 10 and 14, book 2.[22] But the analysis by which the propositions were found can additionally be learned by inverting the compositions to demonstrations. {I have appended to the *Principia* a treatise concerning this analysis, taken from pages previously published.}[23]

The aim of the Book of Principles was not to give detailed explanations of the mathematical methods, nor to provide exhaustive solutions to all the difficulties therein relating to magnitudes, motions, and forces, but to deal only with those things which relate to natural philosophy and especially to the motions of the heavens; and thus what contributed little toward this end I have either entirely omitted or only lightly touched on, omitting the demonstrations.

[*In the preliminary draft, a canceled paragraph reads:* It has been objected that the corollary of prop. 13 of book 1 in the first edition was not demonstrated by me.[24] For they assert that a body P, going off along a straight line given in position, at a given speed and from a given place under a centripetal force whose law is given (i.e., reciprocally proportional to the square of the distance from the center of force), can describe a great many curves. But they are deluded. If either the position of the straight line or the speed of the body be changed, then the curve to be described can also be changed and from a circle become an ellipse, from an ellipse a parabola or hyperbola. But where the position of the straight line, the body's speed, and the law of central force stay the same, differing curves cannot be described. And in consequence, if from a given curve there be determined the (generating) centripetal force, conversely the curve will be determined from the central force given. In the second edition I touched on this in a few words merely; but in each [edition] I displayed a construction

22. In the preliminary draft, Newton has an additional phrase, "case 3, where I called second moments the difference of moments and the moment of a difference." He also refers here to "book 2, problem 3 [prop. 5]," noting that this is "where I called the fluxion of the fluxion of curvature the variation of its variation." On the claim to have used a "second difference," see Whiteside's lengthy discussion in *Math. Papers* 8:456–457 n. 44.

23. Newton has bracketed this sentence to indicate that it is to be deleted.

24. See n. 26 below.

of this corollary in prop. 17 whereby its truth would adequately come to be apparent, and exhibited the problem generally solved in book 1, prop. 41.]

In the first two books I dealt with forces in general,[25] and if they tend toward some center, whether unmoving or moving, I called them centripetal (by a general name), not inquiring into the causes or species of the forces, but considering only their quantities, directions [*lit.* determinations], and effects. In the third book I began to deal with gravity as the force by which the heavenly bodies are kept in their orbits. I learned that the forces by which the planets are kept in their orbits, in receding from the planets toward whose centers those forces tend, decrease in the doubled ratio of the distances [i.e., vary inversely as the square of the distances] from the centers, and that the force by which the moon is kept in its orbit around the earth, in descending to the surface of the earth, comes out equal to the force of our gravity, and thus either is gravity or duplicates the force of gravity.

[*The preliminary draft contains the following canceled passage:* The demonstration of the first corollary of book 1, prop. 13, I omitted in the first edition as being abundantly obvious enough; in the second, at the demand of a Cambridge friend, I expressed it in a few words; while in prop. 17, book 1, I constructed all the cases of this corollary. And furthermore, in prop. 41 of book 1, I set forth with demonstration a general rule which embraces this corollary as a particular case.[26] And the purpose of this third book is to expound the properties, forces, directions, and effects of gravity.]

[*The uncanceled text of the preliminary draft continues:* And after explaining in addition that the cause of gravity, whatever it may be in the end, acts on all bodies and on all parts of all bodies in proportion to the quantity of matter in each and penetrates with equal force and efficacy right to the center of all bodies, I treated gravity as the universal force by which all the planets act on one another; and the sun, on account of its size, acts with the greatest force on all the other planets and brings it about that when tending to the sun they [continually *del.*] describe orbits curved toward it just as a cast stone, in tending toward

25. In the preliminary draft, Newton introduces a qualifying clause: "not defining of what kind they are except in cases where there is a question of pendulums and oscillating bodies."

26. On this corollary, see §6.4 below; also my *Introduction*, pp. 293–294, and Whiteside's commentary in *Math. Papers* 6:147 n. 124, and vol. 8, §2, appendix 6. In the preliminary draft, the penultimate sentence of the preceding paragraph is followed by: "I began to treat gravity as the force by which the heavenly bodies are retained in their orbits."

the earth while it moves through the air, describes a line curved toward the earth.]

The Chaldeans long ago believed that the planets revolve in nearly concentric orbits[27] around the sun and that the comets do so in extremely eccentric orbits, and the Pythagoreans introduced this philosophy into Greece. But it was also known to the ancients that the moon is heavy toward the earth, and that the stars are heavy toward one another, and that all bodies in a vacuum fall to the earth with equal velocity and thus are heavy in proportion to the quantity of matter in each of them. Because of a lack of demonstrations, this philosophy fell into disuse,[28] and I did not invent it but have only tried to use the force of demonstrations to revive it. But it has also been shown in the *Principia* that the precession of the equinoxes and the ebb and flow of the sea and the unequal motions of the moon and the orbits of comets and the perturbation of the orbit of Saturn by its gravity[29] toward Jupiter[30] follow from the same principles and that what follows from these principles plainly agrees with the phenomena.[31] I have not yet learned the cause of gravity from the phenomena.

He who investigates the laws and effects of electric forces with the same success and certainty will greatly promote philosophy [i.e., natural philosophy], even if perhaps he does not know the cause of these forces. First, the phenomena should be observed, then their proximate causes—and afterward the causes of the causes—should be investigated, and finally it will be possible to come down from the causes of the causes (established by phenomena) to their effects, by arguing a priori. [And the actions of the mind of which we are conscious[32] should be numbered among the phenomena *del.*] Natural philosophy should be founded not

27. Newton here uses "orbibus" (*lit.* "orbs," "circles," "orbits"), which for ancient astronomy could be orbs or spheres (as in "celestial spheres"). In the preliminary draft, Newton referred to "the ancients" rather than specifying the Chaldeans.

28. In the preliminary draft, this clause reads: "this philosophy was not passed down to us and gave place to the popular opinion concerning solid orbs."

29. In the preliminary draft, Newton first wrote "gravitation" ("gravitationem"), which he then replaced by "gravity" or weight ("gravitatem").

30. In the preliminary draft Newton initially wrote: "that in like manner Saturn and Jupiter by mutual gravity attract each other near conjunction."

31. In the draft Newton initially wrote: "are determined by the same theory and [agree] very well with observations accurately made."

32. Newton first wrote "quae nobis innotescunt" ("which come to our notice"; "which become known to us"; "which we learn from experience"), and then changed it to "quarum conscii sumus" ("of which we are conscious").

on metaphysical opinions, but on its own principles; and . . . [MS breaks off].[33]

[*The preliminary draft contains an additional paragraph reading:* What is taught in metaphysics, if it is derived from divine revelation, is religion; if it is derived from phenomena through the five external senses, it pertains to physics ["ad Physicam pertinet"]; if it is derived from knowledge of the internal actions of our mind through the sense of reflection, it is only philosophy about the human mind and its ideas as internal phenomena likewise pertain to physics. To dispute about the objects of ideas except insofar as they are phenomena is dreaming. In all philosophy we must begin from phenomena and admit no principles of things, no causes, no explanations, except those which are established through phenomena. And although the whole of philosophy is not immediately evident, still it is better to add something to our knowledge day by day than to fill up men's minds in advance with the preconceptions of hypotheses.]

3.3 *Varieties of Newton's Concepts of Force in the* Principia

Newton believed in and made use of many varieties of force in the course of his scientific thinking.[34] These include—among others—"active" and "passive" forces,[35] chemical and other short-range or interparticle forces, and various forces associated with (or arising from) various concepts of aether.[36] Although the main subject of the *Principia* is the dynamics of macro-bodies, Newton referred, in the preface to the first edition, to his hope that a similar analysis could be made for the forces between the particles of matter. In sec. 14 of book 1 he explored the action of a gravity-like force field on particles that produces effects analogous to those observed in rays of light. In book 2, prop. 23, he discussed a force of repulsion between particles that would produce an analogue of the kind of "elastic fluid"

33. Whiteside suggests, in conclusion: "and principles must be derived from phenomena." In the preliminary draft there is a canceled sentence reading: "and among the phenomena must be counted the actions of the mind of which we are conscious."

34. R. S. Westfall, *Force in Newton's Physics* (London: Macdonald; New York: American Elsevier, 1971); see also Max Jammer, *Concepts of Force* (Cambridge, Mass.: Harvard University Press, 1957), chap. 7, "The Newtonian Concept of Force."

35. On this topic see Dobbs, *Janus Faces* (n. 10 above). On "active" and "passive" forces, see also James E. McGuire, "Force, Active Principles, and Newton's Invisible Realm," *Ambix* 15 (1968): 154–208; and "Neoplatonism and Active Principles," in *Hermeticism and the Scientific Revolution*, ed. Robert S. Westman and J. E. McGuire (Los Angeles: Clark Memorial Library, 1977), pp. 93–142.

36. On the development of the varieties of aether in Newton's thought, see Dobbs, *Janus Faces* (n. 10 above).

or compressible gas observed in nature, to which Boyle's law applies.[37] In book 1, sec. 12, and in book 3, props. 8, 19, and 20, he examined mathematically the sum of the gravitational forces of the particles composing a body in relation to the gravitational force of the body as a whole.

In the *Principia*, Newton classified forces according to several different kinds of properties. For example, there were internal forces of bodies, primarily the "vis insita" or "inherent force" (discussed in §4.7 below) and "vis inertiae" or "force of inertia." Then there were external forces, among them gravity, electricity, magnetism,[38] pressure, impact, elastic force, and resistance. A primary dichotomy that separated the external from the internal forces was, for Newton, that external forces can change a body's state of motion or of rest while others that act internally resist any such change in state. Newton mentions forces of cohesion which cause the particles composing bodies to stick together.[39] He also makes use of static forces, as in the analysis of the classic machines in corollaries to the laws of motion.

In the early parts of the *Principia*, the "force of gravity" is presented as the force producing terrestrial weight, whatever might be its cause, without any specification of whether it is the result of a pull or a push or even of some kind of aether or a shower of particles. He admits (in the concluding scholium to book 1, sec. 11) that most likely this kind of force arises from impulsion.[40] This "force of gravity" is not as yet identified with the "centripetal force" that produces the falling of the moon or the inverse-square centripetal force of the sun on planets, as developed in the tract *De Motu*. In the *Principia*, this force is introduced (def. 5) as "that force, whatever it may be, by which the planets are . . . compelled to revolve in curved lines," a force which he will later (in the introduction to book 1, sec. 11, of the *Principia*) simply call "attraction."[41] It is this "force of gravity," in the sense of terrestrial gravity, that appears in the essay "De Gravitatione," which we now know was written roughly at the same time as *De Motu*. Eventually Newton will prove (*Principia*, book 3, prop. 4) that terrestrial gravity extends to the moon and is the force responsible for the moon's orbital motion,[42] and he will advance good arguments that solar-planetary forces are instances of that same gravity, so that gravity leaps from a local terrestrial force to become universal gravity. For the

37. See §7.5 below.

38. The magnetic force appears in book 3, prop. 6, corol. 5.

39. He does not explore the nature of such forces or their quantitative expression; see his examples to illustrate law 1.

40. The discussions arising from Newton's expressed preference for impulsion are mentioned in §6.11 below.

41. The problems arising from the use of the words "attraction" are discussed in §6.11 below.

42. See Newton's presentation of this result in the draft preface published in §3.2 above.

modern reader, the two are quite distinct and have different dimensions; that is, one is an impulse (F × *dt*), the other a force (F). As we shall see below, Newton avoided the problem of dimensionality because he was dealing with ratios rather than equations.

Whoever reads the definitions and laws will at once become aware that Newton introduces both instantaneous forces of impact and continuous forces of pressure. As we shall see (§5.4 below), Newton makes an easy mathematical transition from instantaneous to continuous forces. Law 1 is stated for continuous forces, as may be seen in the illustrative examples, whereas law 2 is stated for impulsive or instantaneous forces, as may be seen in the application of law 2 in corol. 1 to the laws.[43] Because of the Newtonian mode of mathematical transition from instantaneous to continually acting forces, the distinction between the two is not a fundamental one in the *Principia*.

The primary distinction made in the definitions in the *Principia* is between those "forces" that preserve a body's state of motion or of rest and those that change the body's state. Today's reader will be puzzled by def. 3, in which Newton introduces a "force" of inertia, using "force" in a sense very different from later usage. No doubt this was a legacy from the traditional (pre-inertial) natural philosophy which held that there can be no motion without a mover.[44] In the Motte-Cajori version of the *Principia*, as has been mentioned, the word "force" in the context of "force of inertia" is generally removed from Newton's text. Although this procedure does violence to Newton's expression and concept, the result makes perfect sense to a post-Newtonian reader. In assigning the name of "vis inertiae" (in def. 3) to the then-current "vis insita," Newton formally introduced "inertia" into the common language of rational mechanics.

3.4 *The Reorientation of Newton's Philosophy of Nature; Alchemy and the* Principia

The epistolary exchange with Hooke in 1679/80, which gave Newton a powerful new way of analyzing orbital motion, took place at a time of profound alteration in Newton's philosophy of nature, one so great that R. S. Westfall has referred to it as a "profound conversion." This event, which in retrospect seems a necessary precondition for Newton's having been able to produce the system of the *Principia*, is attributed by Westfall to the combined influence of "two unlikely partners," which he identified as "alchemy and the cosmic problem of orbital mechanics." Until now, the stages of this development have been difficult to determine

43. See the discussion of this topic in §§5.2, 5.3 below.

44. See §4.7 below.

with any exactness because of the problems in fixing the dates of certain writings that would document the successive stages of development of Newton's thought between 1679 (when he wrote a famous letter to Boyle about his philosophy of nature) and the papers that are definitely of the period 1686/87. By 1686/87, when Newton was composing a draft "Conclusio" to the *Principia* and also discussing similar problems in a proposed preface, he had evidently embraced the concept of short-range interparticle forces. Westfall has particularly stressed the fact that when Hooke "posed the problem of orbital motion in terms of an attraction to the center," Newton apparently "accepted the concept of attraction without even blinking."[45] Newton was still "a mechanical philosopher in some sense," but not any longer in the strict sense in which that designation was usually understood. Whereas a strict mechanical philosopher sought the explanation of all phenomena in terms of what Boyle once called those "two grand and most catholick principles of bodies, matter and motion,"[46] Newton came to believe that "the ultimate agent of nature would be . . . a force acting between particles rather than a moving particle itself."[47]

Newton's unitary philosophy of nature was expressed in the unpublished "Conclusio," where he stated that "Nature is exceedingly simple and conformable to herself." Thus, whatever "holds for greater motions, should hold for lesser ones as well." Since gross bodies "act upon one another" by "various kinds of natural forces," such as "the forces of gravity, of magnetism, and of electricity," Newton was led to "suspect" that there should be "lesser forces, as yet unobserved, of insensible particles."[48] The evidence Newton adduced to support his position grew to be—as Westfall has observed—"an early draft of what later became familiar as Query 31 of the *Opticks*"; his text came to include the various phenomena that had "played roles connected with aethereal mechanisms in the 'Hypothesis of Light' " which he had presented to the Royal Society in the 1670s. In arguing for interparticle forces, Westfall concluded, Newton drew heavily on chemical phenomena which, "without exception," had "appeared in his alchemical papers, some among his own experiments, some among other papers, most of them in both."[49] Thus Newton's

45. *Never at Rest*, p. 388.

46. *The Works of the Honourable Robert Boyle*, ed. Thomas Birch (London: printed for J. and F. Rivington . . . , 1772), vol. 3, "Origins of Forms and Qualities," p. 14.

47. *Never at Rest*, p. 390. Westfall remarks that Newton now adhered to "a dynamic mechanical philosophy in contrast to a kinetic."

48. See *Unpubl. Sci. Papers*, pp. 183–213.

49. In particular, Westfall (*Never at Rest*, p. 389) notes that "certain alchemical themes, such as the role of fermentation and vegetation in altering substances, the peculiar activity of sulfur, and the combination of active with passive, also made their way into the new exposition of the nature of things." His "alchemical compound, the net, even supplied its name to his new conception of matter."

acceptance of forces as fundamental entities was conditioned to a significant degree by his studies of alchemy.

Reference has already been made to the work of Betty Jo Dobbs, who has altered many aspects of our beliefs about Newton's ideas and the sequence of their development, notably through her intensive studies of Newton's alchemical and theological writings and her new dating of two important tracts of Newton's, his "De Aere et Aethere" and his essay "De Gravitatione." She has shown that both should be assigned to years after the correspondence with Hooke, to the period just before the writing of the *Principia*.[50]

This new chronology may now be applied to explain a number of problems relating to the concepts of the *Principia*. For example, we shall see that Newton makes use of the term "vis insita" some time after the tract *De Motu*, in some preliminaries to the *Principia*. A somewhat similar concept of an inner or inherent force appears as a "vis indita" in "De Gravitatione."[51] With the new dating, we can now understand how Newton—in the period just before the *Principia*—made an easy transition from "vis indita" to "vis insita" and then to "vis inertiae." But if this tract had been written many years earlier, we should have to explain the gap of twenty years between Newton's consideration of such an inner force in "De Gravitatione" and his return to his subject many years later, on the eve of the *Principia*.

Betty Jo Dobbs's work is not, of course, primarily devoted to questions of chronology. Rather, her goal is to show that Newton's alchemy cannot be separated from the rest of his scientific thought. Her extended argument draws on an extensive examination of a large variety of different manuscript writings of Newton. There can no longer be any doubt that Newton's thinking in the area of mechanics or cosmic physics was closely related to his very deep concern with alchemy and the structure and interactions of matter. She concludes that "all issues of passivity and activity, of mechanical and nonmechanical forces, were enmeshed for Newton in a philosophical/religious complex," which she has set out to understand. Although she notes that "Newton's first encounter with attractive and 'active' principles may well have been in his alchemical study," she now finds that "his application of such ideas to the force of gravity was almost certainly mediated by several other considerations." She stresses Newton's considerations of a universal "vegetative principle operating in the natural world, a principle that he understood to be the secret, universal, animating spirit of which the alchemists spoke."

50. Dobbs, *Janus Faces* (n. 10 above).

51. See §4.7 below; Dobbs accepts Herivel's misreading of "vis insita" for "vis indita" in "De Gravitatione."

Betty Jo Dobbs has shown in detail how certain concepts which appear in the *Principia* are closely intertwined with Newton's general concerns in a domain that includes his religious beliefs, his ideas about space and matter, his concern with traditions of ancient wisdom, his explorations of alchemy, his considerations of active principles in matter as well as passive principles, and his interest in general (e.g., "vegetative") "spirits." Accordingly, if our goal is to understand the thought of Isaac Newton or the set of his mind at the time that he was writing the *Principia*, we cannot separate his mathematically developed rational mechanics from the rest of his thought.

For example, consider Newton's conclusions about the void, about a space free of any kind of Cartesian matter or aether that would be coarse enough to produce a force which could affect the motion of planets. These conclusions were founded on astronomical results (the fact that there was no observed long-term departure from Kepler's laws), but also were intimately related to Newton's views about space and place and the nature of God.

The mature Newton put forward and maintained a sharp distinction between "experimental philosophy" and areas of philosophy that reached significantly beyond that which could be established empirically. This sharp distinction and his continual efforts to maintain it may possibly rival any other contribution he made to science. Accordingly, even though we must grant the continuity of Newton's overall philosophic thought, we too should make a distinction between experimental philosophy and the other areas of philosophy.

It is a fact that the *Principia* can be read as a work in physics, as has been done successfully for three centuries, without taking cognizance of what—to modern eyes—will seem to be extrascientific considerations. Even Newton's discussion of absolute space, in the scholium following the Definitions, can be passed over without preventing an understanding of Newton's exposition of rational mechanics and the system of the world—all the more so since throughout most of the *Principia*, Newton makes use of relative rather than absolute space. In the first edition, there was a possibly alchemical overtone to hypothesis 3 at the beginning of book 3, in which Newton mentioned the transformation of matter into matter of any other kind. But there was only one single reference to this "hypothesis" in the *Principia*, and by the time of the second edition this "hypothesis" and the reference to it had been eliminated.[52] Of course, the concluding General Scholium, first printed in the second edition, did introduce questions about the being and

52. It is curious that in "hypothesis 3," writing as an alchemist, Newton referred to matter being "transformed," while in secs. 4 and 5 of book 1, writing as a mathematician, he referred to the transformation of geometric figures (as by projection) in terms of "transmutation."

attributes of God, along with some aspects of an "electric and elastic spirit," but this was clearly a theologico-philosophical appendix and was never considered to be an integral part of the *Principia*.

Newton wrote the *Principia* in a way that excluded any insight on the part of the reader (save perhaps for the lone "hypothesis 3" in the first edition) that for Newton the system of rational mechanics (books 1 and 2) and the dynamics of the system of the world (book 3) were not ends in themselves but only parts of a unified general ordering of the universe that encompassed alchemy, the theory of matter, "vegetative" and other "spirits," prophecy, the wisdom of the ancients, and God's providence. It is, furthermore, a fact of record and of notable interest that this "holistic" view of Newton's thought would never have been discovered by merely reading Newton's text of the *Principia* and had to wait for some three hundred years for scholars—notably Betty Jo Dobbs, R. S. Westfall, J. E. McGuire, A. Rupert Hall and Marie Boas Hall, myself, and others—to reveal it. From time to time, as these scholars have shown, Newton planned to introduce into the *Principia* some extracts from ancient sages and philosophers or other hints of his fundamental and all-encompassing concerns. Yet, in the end, he resisted the temptation to "show his hand," and the *Principia* remained an austere presentation of mathematical principles and their applications to natural philosophy.

In retrospect, one of the most extraordinary things about the *Principia* is that Newton could so thoroughly eliminate the traces of his concern for topics which today we would consider nonscientific. After all, it has taken us three centuries of intellectual archaeology to get behind the facade of the *Principia* and to reveal the full context of his thought. How could Newton have kept these considerations out of his *Principia* to such a degree that readers have had only the barest hint of their existence until our own times? Some years ago, I put this question to Frances Yates, the doyenne of studies in this area. She replied simply but very profoundly that, after all, "Newton was a genius." Who would not agree! We cannot expect a person of such creative genius to behave in an ordinary fashion.

3.5 *The Reality of Forces of Attraction; The Newtonian Style*

In the *Principia*, Newton adopted a mode of presentation that enabled him to put aside, for the moment at least, any considerations other than those directly related to mathematics and mathematical conceptions of physics. I have called this manner of composing the *Principia* the "Newtonian style," and have shown how it describes Newton's procedure in developing the propositions of the *Principia* and then applying them to the world of experiment and observation. It should be noted, however, that in doing so I am merely systematizing Newton's own procedure in the *Principia* and restating what Newton himself has said. This style enabled Newton to discuss such concepts as matter (or mass) and force and even space (in relation to displacements, velocities, accelerations, and orbits) without divulging any of his fundamental beliefs about God and nature or about the world in general. In this way he could even consider forces acting over vast distances without being (at least in the first stages) inhibited by concerns about the way in which such forces can possibly act. In private documents, however, and in attempts to explain how gravity can act, he would resort to all the interrelated realms of his belief. Yet the Newtonian style enabled him to deal with forces of attraction in the *Principia* without having to consider any larger questions of a philosophical kind. This separation, as is indicated by the final paragraph of the draft preface quoted earlier, represents a considerable intellectual achievement. In particular, we should take note that this attitude enabled Newton to explore the conjectured consequences of philosophic questions as a form of "dreaming," without thereby necessarily undermining in any way the results of the *Principia*, without thereby producing a "philosophical romance" in the way that Descartes was said to have done. I repeat what Newton said in the last paragraph of that preface: "And although the whole of philosophy is not immediately evident, still it is better to add something to our knowledge day by day than to fill up men's minds in advance with the preconceptions of hypotheses." Certain fundamental truths, such as the universality of the force of gravity acting according to the inverse-square law, were derived directly from mathematics; but in Newton's mind even such a law—once found—had to be fitted into his general scheme of thought, and he came to believe that certain aspects of this law had been known to the ancient sages.

Following the reorientation of Newton's philosophy of nature, he came to believe that interparticle forces of attraction and repulsion exist. Such forces, according to Newton, are sufficiently short-range in their action (as he makes quite explicit in query 31 of the *Opticks*) that they do not raise a major problem of understanding their mode of action. They do not, in other words, fall into the category of the forces acting at a distance. His studies of matter, and in particular of alchemy, had made the existence of these forces seem reasonable. But does the reasonableness of such short-range forces provide a warrant for belief in the existence of long-range forces acting over huge distances? Consider the gravitational force between the sun and the earth: this force must act through a distance of about one hundred million miles. Even worse from the conceptual point of view is the force between the sun and Saturn, some thousands of millions of miles. Eventually Newton was to conclude that comets are a sort of planet, with the result that the solar gravitational force must extend way out beyond the limits of the visible solar system, and still be powerful enough to turn a comet around in its path so that it will return. As Newton said in the General Scholium, gravity "really exists," but how is one to "assign a cause"?

Shortly after completing the *Principia*, Newton wrote a strong letter to Richard Bentley declaring that no person in his right mind could believe that a body can "act where it is not." We know that after the *Principia* had been published, Newton tried one means after another to account for gravity, including a kind of aethereal explanation suggested by Fatio de Duillier, and electricity. He finally adopted a kind of "aethereal medium," discussed in the Queries of the later editions of the *Opticks*, which varies in density according to the matter of bodies located in it.

Although we must, I believe, agree with R. S. Westfall and Betty Jo Dobbs that alchemy conditioned Newton to accept the idea of force as a primary element of nature's operations, we must also recognize that alchemy could not fully justify the kind of long-range action at a distance required of a force of universal gravity. Alchemy seems to have given Newton an overall philosophical or conceptual framework that allowed forces to be primary elements of nature's operations. It is even possible that without such a framework, he might have been reluctant to develop a system of physics based on forces, one in which a central place would be given to the elaboration of the properties and effects of a force of gravity. But these are aspects of the mindset or basic philosophy of Newton the individual and are not features of the *Principia*. The only real sign of any such view in the *Principia* itself is the remark in the preface to the first edition, where Newton expressed the hope of deriving "the other phenomena of nature" from "mechanical principles" in a manner similar to the one he had employed for gross bodies, and the "hypothesis 3" of the first edition (eliminated in the later editions) concerning the transformations of matter. In the remarks in the preface to the first edition, however, Newton is meticulous in treating the broader philosophical picture of fundamental forces as a conjecture which may or may not gain support later, depending on whether the method he has put on display in the *Principia* leads to comparable results in other domains of phenomena. In this sense, alchemy may have been in part behind the remarks in the preface, but this in no way lessens the integrity of the method he is presenting. The very tone of the hopes expressed concerning "other phenomena of nature" declares the fundamental difference between any such philosophical conjecture and experimental philosophy.

This insistence on maintaining a sharp distinction between empirical science and philosophy is a basic element in his disagreement with the point of view of Descartes, who makes no such separation. In this perspective, the Newtonian style of the *Principia* appears to be in harmony with the views Newton expressed in the draft preface. Here he says explicitly that it is no more than "dreaming" to "dispute about the objects of ideas except insofar as they are phenomena." In natural philosophy we must, he declares, "begin from phenomena and admit no principles of things, no causes, no explanations, except those which are established through phenomena." The aim of the *Principia*, he insists, is "to deal only with those things which relate to natural philosophy" and natural philosophy should not "be founded ... on metaphysical opinions."

Similarly, in his "recensio libri" or review of the *Commercium Epistolicum*, which he wrote and published in the *Philosophical Transactions* in 1715, he insisted that the philosophy which he has pursued in both the *Principia* and the *Opticks* "is Experimental." This "Experimental Philosophy" teaches "the Causes of things" only to the degree that "they can be proved by Experiments." Accordingly, we should "not fill this Philosophy with Opinions which cannot be proved by Phaenomena."

Of course, there must be some continuity between the position with respect to forces as expressed in the *Principia* and the rest of Newton's philosophy, including his beliefs arising from his studies of alchemy. The transformation Newton went through in relaxing the strictures of the mechanical philosophy may have been motivated in part by alchemy, but it also required a new method that allowed him to continue maintaining the sharp distinction between experimental philosophy and the rest of philosophy, while nevertheless delving into the realm of unseen forces acting at a distance. I have called this new method the Newtonian style, developed to a high degree in the *Principia*. The elaboration of this style was a fundamental part of Newton's basic philosophy that emerged in the early 1680s.

From this perspective we can understand how Newton faced the problem of dealing with forces. He was able to make use of Hooke's suggestion of a central force and later to develop the mathematical properties of such forces in *De Motu*—and ultimately to produce the *Principia*—without being deterred by the disdain he continued to feel for long-range forces of attraction. How could he have done so? I believe that the answer has two parts. No doubt his general belief that there "really" are forces in nature, and the corollary that forces are fundamental concepts in natural philosophy, received strong support from (and even may have originated in) his studies of alchemy. But an analysis of the actual structure of the *Principia* shows also that Newton developed a "style" which is plainly set forth in that work—both in precept and in example—that enabled him to explore freely the properties of a kind of force which his contemporaries banned from natural philosophy. This force was of a kind that even Newton could not fully approve without assigning some underlying "cause" or mode of action. The Newtonian style, as we shall see,[53] is independent of Newton's own strong belief in the reality of forces of attraction—in

53. See §6.11 below for a further analysis of the Newtonian style and its consequences.

particular, the force of universal gravity. The Newtonian style constituted a mode of discourse which could enable him to develop the properties of such a force without need to discuss either whether forces of that sort are part of good science or how these forces act to produce their effects.

In the General Scholium, Newton explains why the cause of gravity cannot be "mechanical," by which he meant attributable to mechanical causes, that is, matter and motion. The reason is not hard to find. Suppose that the force pushing a body gravitationally is "mechanical," say the result of a continual bombardment of particles in motion. In this case, the magnitude of the force on any given body would be proportional to the number of impacts per unit time, that is, proportional to the surface area exposed to the particles. But gravity is not proportional to the surface area but rather to the mass of a body. Fatio de Duillier wrote that Newton once allowed him to transcribe a note from one of his personal copies of the *Principia*, in which—according to Fatio—Newton had said that the only "possible Mechanical cause of Gravity" was the one that Fatio "had found out."[54] Fatio added, however, that Newton "would often seem to incline to think that Gravity had its Foundation only in the arbitrary will of God,"[55] a belief we have seen to have also been recorded by David Gregory.

3.6 *Did Newton Make the Famous Moon Test of the Theory of Gravity in the 1660s?*

One of the great climaxes of the *Principia* occurs in book 3, prop. 4, when Newton shows that if gravity extends from the earth as far out as the moon and does so according to the law of the inverse square, the "falling of the moon" would be exactly as observed. Here is convincing evidence that the moon moves in its orbit under the inverse-square force of gravity. In an autobiographical sketch, often reprinted, Newton claimed that he had made such a moon test in the 1660s, attempting to find out whether gravity extends as far as the moon. In this sketch, he describes his invention of the differential and integral calculus and his discoveries about light and color; he then refers to his thoughts concerning "gravity extending to the orb of the Moon." He had already found, he writes, how to "estimate the force with which [a] globe revolving within a sphere presses the surface of the sphere," and he was acquainted with Kepler's third or harmonic law of planetary motion. Combining these two rules, he "deduced that the forces which keep the Planets in their Orbs must [be] reciprocally as the squares of their distances from

54. Nicolas Fatio de Duillier, "La cause de la pesanteur: Mémoire présenté à la Royal Society le 26 février 1690," ed. Bernard Gagnebin, *Notes and Records of the Royal Society of London* 6 (1949): 105–160. See my *Introduction*, p. 185.

55. *Introduction*, p. 185.

the centers about which they revolve." That is, he combined his own discovery that the "force" in uniform circular motion is measured by $\frac{v^2}{r}$ with Kepler's law that $r^3 \propto T^2$ to obtain the result that the force keeping the planets in their orbits must be proportional to $\frac{1}{r^2}$. He claimed that he "thereby compared the force requisite to keep the Moon in her Orb with the force of gravity at the surface of the earth" and "found them to agree pretty nearly."[56] This statement has given rise to exaggerated claims concerning the state of Newton's knowledge in the 1660s, including the supposition that he knew the "law of gravitation" at this early date and that, accordingly, there was a "twenty years' delay" between the time of discovery and the date of publication (in 1687, in the *Principia*).[57]

There are several documents that shed light on Newton's actual research into problems of gravity and planetary and lunar motion in the 1660s. In these, however, Newton shows that he was still thinking about curved or orbital motion in Cartesian terms and hence based his studies of motion on a centrifugal endeavor, or "conatus," rather than on a centripetal force, as he was to do after 1679/80. This statement refers to Newton's basic conceptual framework and not to the mathematical results he achieved. Indeed, in such an example we may gain an insight into the gulf that separates our present viewpoint (essentially Newton's mature viewpoint) from the mid-seventeenth-century manner of conceiving dynamical problems (essentially that of Newton's young manhood).

The calculations to which Newton referred in his autobiographical memorandum are found in two documents. One of them, called the Waste Book, is a large bound volume belonging to his stepfather, Barnabas Smith, who had been compiling a kind of theological glossary. Newton used the blank spaces on the pages for notes and calculations. Here, indeed, on the first page, Newton—as he said—estimated the force with which a globe "revolving within a sphere presses the surface of the sphere." He recorded his discovery (independently of Huygens) that in uniform motion at speed v in a circle of radius r, the centrifugal endeavor is measured by $\frac{v^2}{r}$.[58]

56. First published in the *Catalogue of the Portsmouth Collection* (§1.2, n. 11 above). A more complete and accurate version is available, along with similar documents, in my *Introduction*, suppl. 1.

57. Cajori, "Newton's Twenty Years' Delay in Announcing the Law of Gravitation" (§1.2, n. 15 above).

58. Herivel's *Background* contains an edited version of this extract from the Waste Book, along with other entries of Newton's relating to dynamics. The Waste Book is in the University Library, Cambridge, where it is catalogued as ULC MS Add. 4004.

In one of the calculations in a manuscript that has become known as the Vellum Manuscript,[59] Newton sought to answer a Copernican problem investigated by Galileo in his *Two New Sciences*: how it is that bodies on the surface of the earth are not flung into space by the earth's rotation and annual revolution. Newton compared the centrifugal endeavor at the surface of a rotating earth with the force of gravity, determining the latter by means of an experiment with a conical pendulum 81 inches long, at an incline of 45 degrees to the vertical.[60] He showed that the force of gravity is so far greater than the centrifugal endeavor produced by the earth's rotation that bodies on earth are not flung off as a result of the rotation. He made a similar calculation for the effects of the earth's annual motion.

Shortly afterward, in another document, Newton recalculated the ratio of the earth's surface centrifugal endeavor to gravity by a somewhat different method.[61] This time, he wrote, he went a step further and compared the "endeavor of the moon to recede from the center of the earth" with the force of gravity at the earth's surface. The force of gravity proved to be a little more than 4,000 times as great. It is in this second document[62] that Newton combines his rule for centrifugal endeavor with Kepler's third law to conclude that for the planets the "endeavors to recede from the sun" are "reciprocally as the squares of the distances from the sun."

This document contains a lengthy passage in which Newton compares the centrifugal endeavor of the revolving moon with the centrifugal endeavor of a body on a rotating earth. Earlier, he had compared the centrifugal endeavor at the earth's surface to gravity, but now he compares the centrifugal endeavor at the earth's surface with the centrifugal "endeavor of the moon to recede from the center of the earth." We may take note that Newton is writing specifically about the moon's "endeavor . . . to recede" ("conatus Lunae recedendi a centro terrae") and not in any sense about a centripetal force. He now makes use of the rule he has found, according to which the endeavors from the center of different circles are as the diameters divided by the squares of the periodic times. He finds that the centrifugal endeavor at the earth's surface is about 12½ times greater than the moon's orbital centrifugal endeavor and concludes, without comment, that the "force of gravity" ("vis gravitatis") at the earth's surface is somewhat more than

59. ULC MS Add. 3958, fol. 45; Herivel, *Background*, pp. 183–191; *Corresp.* 3:46–54. This manuscript, as described by Herivel, is "a torn legal parchment on which a lease is engrossed"; Newton has entered calculations on the back side.

60. For details, a convenient and accurate summary is available in Westfall, *Never at Rest*, p. 150; the relevant documents may be found in Herivel's *Background*.

61. *Never at Rest*, p. 151.

62. ULC MS Add. 3958.5, fol. 87; *Background*, pp. 195–197; *Corresp.* 1:297–303; cf. Hall, "Newton on the Calculation of Central Forces" (§1.2, n. 12 above); Ole Knudsen, "A Note on Newton's Concept of Force," *Centaurus* 9 (1963–1964): 266–271.

4,000 times greater than the moon's orbital endeavor to recede from the center of the earth. He does not comment on the significance of this result.

Newton next introduces another computation in which he finds the ratio of the moon's orbital centrifugal endeavor (or the moon's endeavor to recede from the earth) to the earth's orbital centrifugal endeavor (or endeavor to recede from the sun) to be about 5 to 4. This ratio, combined with the previous one, enables him to find the ratio of the earth's orbital endeavor to the force of terrestrial gravity. He finds that the earth's gravity is 5,000 times greater than the earth's endeavor to recede from the sun.

The next paragraph discloses a possible significance of the number 4,000 which Newton has previously found to be the ratio of the earth's gravity to the moon's orbital endeavor to recede. This paragraph begins with a statement of Kepler's third law, but without any reference to Kepler. He states that from this law, assuming that the orbits of the planets can be considered to be circles, the orbital endeavors of the planets are inversely as their distances from the sun. In the autobiographical sketch, Newton says that he combined his $\frac{v^2}{r}$ rule with Kepler's third law to obtain the result that the endeavor to recede is inversely proportional to the square of the distance. It is quite easy to see how this could be done.

Let the planets be considered to move in circular orbits (radius $= r$, period $=$ T). Then one has only to use simplest algebra in order to combine the $\frac{v^2}{r}$ rule discovered by Newton with Kepler's law that r^3 is proportional to T^2. Let E be the centrifugal endeavor.

$$E \propto \frac{v^2}{r} = \frac{1}{r} \times \left(\frac{2\pi r}{T}\right)^2 = \frac{4\pi^2 r^2}{rT^2} \propto \frac{r^2}{rT^2}.$$

Since Kepler's third law says that $T^2 \propto r^3$,

$$E \propto \frac{r^2}{rT^2} \propto \frac{r^2}{r^4} = \frac{1}{r^2}.$$

Newton thus concludes that the centrifugal endeavors of the planets are inversely proportional to the squares of their distances from the sun.

Newton's later statement that he compared the moon's motion with gravity extended to the moon can be interpreted as follows. Assume that terrestrial gravity extends to the moon and that it is diminished with distance in the same proportion as the planets (for which Newton has computed the $E \propto \frac{1}{r^2}$ rule). Then, since

Newton—in another document—takes the moon to be 59 or 60 earth radii from the earth's center, the force of the earth's gravity at the moon's orbit should be $\frac{1}{60^2}$ of what it is at the earth's surface or at a distance of one earth radius from the earth's center. Thus there is a theoretical conclusion that the force of gravity at the earth's surface should be about 3,600 times as great as the moon's orbital centrifugal endeavor, which Newton has computed to be 4,000. As Herivel has pointed out, this discrepancy arises from Newton's adoption of an incorrect figure for the radius of the earth.[63]

The difference between 4,000 and 3,600 is 400 parts in 4,000, or one part in ten, or 10 percent. When, therefore, Newton later wrote of the agreement between theory and observation in the words "pretty nearly," he had either forgotten the actual numbers or was exaggerating the accuracy of the comparison. Possibly, however, the words "pretty nearly" were intended to mean "not quite."[64]

Another version is reported by Henry Pemberton, who was the editor of the third edition of the *Principia* under Newton's direction. According to Pemberton, Newton's "computation did not answer expectation," with the result that Newton "concluded that some other cause must at least join with the action of the power of gravity on the moon."[65] A somewhat similar, but more detailed, version is given by William Whiston, Newton's successor as Lucasian Professor at Cambridge. According to Whiston, Newton was "somewhat disappointed" to find that "the Power that restrained the Moon in her Orbit" was "not quite the same that was to be expected, had it been the Power of Gravity alone, by which the Moon was there influenced." This "Disappointment," Whiston added, "made Sir Isaac suspect that this Power was partly that of Gravity, and partly that of Cartesius's Vortices."[66] Both Pemberton and Whiston were aware that Newton's error arose from an imperfect value for the radius of the earth.

There are a number of puzzling aspects of this affair. The first must certainly be the curious fact that Newton did not submit the second of these documents when the question arose of whether Newton learned of the inverse-square planetary force from Hooke or had known it earlier. Instead, Newton referred Halley to his "hypothesis" on the nature of light, submitted to the Royal Society in 1675, which made out a much weaker case for Newton than the earlier document on

63. *Background*, pp. 68, 198 n. 8.

64. This is the suggestion of Herivel, *Background*, p. 68.

65. Henry Pemberton, *A View of Sir Isaac Newton's Philosophy* (London: S. Palmer, 1728), preface.

66. William Whiston, *Memoirs of the Life and Writings of M. William Whiston, Containing Memoirs of Several of his Friends also, Written by Himself*, 2d ed. (London: J. Whiston and B. White, 1753), 1:35–36.

motion. We know that Newton had the manuscript with his calculations readily available shortly after the publication of the *Principia* because, as Westfall has pointed out,[67] this must be the document to which Gregory referred when he wrote in 1694, some two decades before the date of Newton's autobiographical memorandum, that Newton showed him a paper written before 1669 that contained "all the foundations of his philosophy . . . , namely the Gravity of the Moon to the Earth, and of the planets to the Sun, and in fact all this even then is subjected to calculation."[68]

But even though it seems that Newton did make some kind of moon test in the 1660s, what kind of test was it? What was he testing? Can Newton's calculations be considered a moon test of the theory of gravitation, as has been alleged? The answer to this third question is decidedly in the negative. First of all, the theory of gravitation is based on the central concept of a universal force. In the 1660s, Newton was considering something quite different, namely, a Cartesian endeavor, or "conatus," rather than a force, and certainly not a universal force. This endeavor was centrifugal,[69] or directed outward, tending to produce an outward displacement, one having a direction away from the central body. Universal gravity is centripetal, directed inward, tending to produce an inward displacement. Furthermore, in Newton's mature physics, gravity produces a centrally directed acceleration, so that the orbiting body is continually falling inward because there is an unbalanced centripetal force. In the 1660s, Newton was still thinking of a centrifugal endeavor that had to be balanced so as to keep the body in orbit, in a kind of equilibrium.[70] Finally, in comparing the gravity on the surface of the earth with the force gravity would have at the moon, Newton assumed that the measure of the force was the inverse square of the distance from the center of the earth. Anyone who has read sec. 12 of book 1 of the *Principia* knows how difficult this was to prove.

In short, Newton evidently did make a moon test of sorts in the 1660s. He evidently did conceive of gravity extending to the moon, but he certainly did not compare two "forces," one a "force requisite to keep the Moon in her Orb," the other the force of gravity. At this time, he did not as yet have the concept of such a "force," which he developed only a decade or so later, in the years before the

67. *Never at Rest*, p. 152 n. 36.

68. *Corresp.* 1:301.

69. Descartes did not use "centrifugal" in relation to his concept of "conatus," or endeavor; that term was introduced by Christiaan Huygens.

70. For a contrast between Newton's Cartesian "conatus" of the 1660s and his later concept of force, see *Never at Rest*, pp. 153–155; for more details, see Westfall's *Force in Newton's Physics* (n. 34 above).

writing of the *Principia*. Then, at last, he came to understand the way in which forces produce changes in motion, and he invented the name "centripetal force."

3.7 *The Continuity in Newton's Methods of Studying Orbital Motion: The Three Approximations (Polygonal, Parabolic, Circular)*

In the 1660s, when Newton was still a young man in his twenties, he set forth three modes of dealing with curved or orbital motion, each of which appears decades later in the *Principia*. These have been named by Bruce Brackenridge the polygonal, the parabolic, and the circular approximations.[71] The only major difference between the early and the later applications of these approximations is that in the 1660s (as we have seen in §3.6) Newton was still thinking in the Cartesian terms of centrifugal "conatus," or endeavor; he had not as yet developed the concept of centripetal force. The difference between this early view and the later one is not in the mode of computing the value of the endeavor or force, but in the intellectual framework. In one case, curved motion is produced by a balance, a kind of statics; in the other, curved motion results from an unbalanced (acceleration-producing) force and a tangential inertial component, an example of dynamics.

The polygonal approximation appears in the Waste Book[72] in an entry of the mid-1660s. Newton is determining the measure of centrifugal force in circular motion. His method (as has been mentioned in §2.5 above) is to start with a trajectory consisting of a square inscribed in the circle. At each vertex, the ball will strike the circle with an outward force and be reflected. Newton considers each such collision between the ball and the curve to be perfectly elastic so that there is a reflection in which the angle of incidence is equal to the angle of reflection and in which there is a change in the direction but not in the magnitude of the motion. Newton considers what we would call an impulse or impulsive force (the product of the force and the time-interval in which the force acts) rather than the force itself. The measure is the difference between where the body would have gone in the absence of the impulsive force and where it went under its action. He increases the number of sides so as to form an n-sided polygon and finally allows the number of sides to become infinite. In the limit, the polygon becomes a circle and the force

71. Bruce Brackenridge, *The Key to Newton's Dynamics: The Kepler Problem and the "Principia,"* with English translations by Mary Ann Rossi (Berkeley, Los Angeles, London: University of California Press, 1996). My discussion of Newton's three modes of analysis depends highly on Brackenridge's presentations.

72. ULC MS Add. 4004. See *Background*, pp. 128–131.

acts continually.[73] In this early document, Newton does not demonstrate that the force is as $\frac{v^2}{r}$ but rather derives another property of the motion, one that can be reduced to the $\frac{v^2}{r}$ rule. Essentially this same method, including the transition from a polygon to a smooth curve and a force that acts continually, is found in the *Principia*, book 1, props. 1 and 2, and also in their antecedents in the preliminary tract *De Motu*. In a scholium to book 1, prop. 4, of the *Principia*, Newton mentions this mode of analysis, developed many years earlier. In this later document he does demonstrate the force to be as $\frac{v^2}{r}$.

The parabolic approximation appears in a document on circular motion written in 1669.[74] In this analysis, a body moves along a circle under the action of some force F, directed from some central point C toward any point P on the curve. Newton considers the action of the force in a very small (actually infinitesimal) region around the point P. Because the region is so small, the force may be considered to be constant. The body has two components of motion. One, uniform and linear, along the tangent, has been given to the body at the start of the motion. The other is an inward accelerated motion produced by the force F. Newton knew, from Galileo's work, that the effect of a constant force is to produce a constant acceleration. Thus, in the region under consideration, the situation is formally identical with the conditions of projectile motion studied by Galileo. Hence the body's trajectory in the neighborhood of P will be in the form of a parabola. This mode of analysis readily yields the force in circular motion and appears in the *Principia* in book 1, prop. 4, and the previous version in *De Motu*. This method also appears in the *Principia* in book 1, prop. 6, and in props. 10 and 11, and elsewhere, and also in their anticipations in *De Motu*. In these later examples, the method is extended from considerations of motion in a circle to motion along any curved path.

73. The issue of the mathematical rigor in Newton's analysis has been, and still remains, a subject of debate among scholars. It has been argued, for example (by Michel Blay among others), that even if the limit of the polygon can legitimately be taken to be a continuous curve, the limit of a series of impulsive blows cannot be a continually acting force. On the question of mathematical rigor in Newton's argument, see D. T. Whiteside's discussion in *Math. Papers* 6:6 n. 12; 35–39 n. 19; 41 n. 29; Aiton, "Parabolas and Polygons" (§2.5, n. 38 above). See, further, my "Newton's Second Law and the Concept of Force in the *Principia*," in *The Annus Mirabilis of Sir Isaac Newton, 1666–1966*, ed. Robert Palter (Cambridge, Mass., and London: MIT Press, 1970), pp. 143–185.

74. This document and the method of parabolic analysis were first published by Hall, "Newton on the Calculation of Central Forces." A lengthy discussion of this document and its significance is given by Brackenridge in his *Key to Newton's Dynamics*; see also Herivel, *Background*, pp. 192–198. An important analysis by Whiteside appears in *Math. Papers* 1:297–303.

The third or circular approximation (or, more properly, circular measure) presents problems that are not present in the other two because in this case we have no example in which the method is applied in the 1660s. We do, however, have an unambiguous statement by Newton, written in the 1660s in the Waste Book,[75] in which he says that he can solve the problem of finding the force producing elliptic motion by means of a tangent circle, that is, a circle of curvature. The statement actually reads as follows:

> If the body is moved in an Ellipsis then its force in each point (if its motion in that point bee given) may bee found by a tangent circle of equall crookednesse with that point on the Ellipsis.

Newton's mathematical manuscripts make it possible, as Whiteside has shown, to document Newton's development of a method of measuring the "crookednesse" or degree of bending of curves by means of a tangent circle.[76] This is the circle that at the point under study most closely approximates the curve at that point; it is known today as the osculating circle. Newton was very early aware that the curvature of a circle is constant, that the curvature of a logarithmic or equiangular spiral changes uniformly, and that the curvature of a conic varies nonuniformly. In today's language of mathematics, the circle that most closely approximates a curve at some given point, the osculating circle, has the same first and second derivatives at that point as the curve whose "crookednesse" is being measured. Newton's manuscripts show that he knew the main steps in finding the curvature of a conic as early as December 1664.[77]

Although Newton's early mathematical work on curvature is well documented,[78] there is—as has been mentioned—no similar set of documents in which we may see Newton applying the method of curvature to problems of force and motion. There is, however, an example of its use in the tract *De Motu* of 1684/85 and in a canceled passage of the draft of the first edition of the *Principia* (1687). Cognoscenti will recognize that in that first edition the concept of curvature "is basic to lem. 11" as it is in all editions. In the first edition, however, "this lemma

75. Also published by Herivel (*Background*, p. 60). The original is found on fol. 1, right under the analysis of circular motion by the polygonal method. Herivel was uncertain concerning the reading of the verb toward the end of the passage and suggested "would," but Brackenridge has found that it should be "may," so that Newton is saying that the "force . . . may be found by a tangent circle."

76. See, notably, Whiteside's discussion in *Math. Papers* 1:546 n. 3.

77. Ibid.

78. This has been done by Whiteside in his edition of Newton's *Math. Papers*.

is applied only to the problems of circular and spiral motions"[79] (props. 4 and 9). There is, however, one example in the first edition in which Newton does make use of the method of curvature, namely, book 3, prop. 28. This may be taken as evidence that at least by 1687 Newton had fully developed this way of dealing with dynamical problems. A document displaying in full the use of this method in the context of dynamics is the tract that Newton sent to John Locke in 1690, a few years after the *Principia* had been written and published.[80] It is possible that in some form the ideas in this tract can be dated even earlier, perhaps even as early as 1684.[81]

A few years after the publication of the first edition of the *Principia*, in the 1690s, David Gregory learned from Newton about some proposed revisions. His recorded summary reveals Newton's plan for the new edition that closely "echoes the cryptic statement on curvature in the Waste Book made some thirty-five years earlier."[82] Whiteside identified and published the extant pages of manuscript relating to such a revision,[83] showing how Newton—among other novelties—planned to introduce the method of curvature in his proposed recasting of the first part of the *Principia*.

In the second edition of the *Principia*, this method does appear, although not always in an obvious way. In this edition, Newton added new corollaries (3, 4, and 5) to book 1, prop. 6, revised the proof of prop. 7, and added alternate proofs ("idem aliter") to props. 9 and 10. These new corollaries to prop. 6 introduce the circle of curvature. For example, corol. 3 refers specifically to the osculating circle, the case of an orbit which "makes with the circle an angle of contact . . . which is the least possible." This concept of an angle of contact "which is the least possible" is discussed in the earlier lem. 11 and its associated scholium, although there is no explicit reference there to the radius of curvature. The revisions of the proof of prop. 7 and the alternate proofs added to props. 9 and 10 also make use of the method of curvature, of constructing the circle that most closely approximates the curve at a given point. The actual way in which curvature is being used is far from being immediately obvious, but the concerned reader will find a thorough

79. *Math. Papers* 1:27 n. 43.

80. Published in *Unpubl. Sci. Papers*.

81. Brackenridge, "The Critical Role of Curvature in Newton's Developing Dynamics" (§1.2, n. 21 above).

82. Bruce Brackenridge, "Newton's Mature Dynamics: A Crooked Path Made Straight" (paper read to the symposium on "Isaac Newton's Natural Philosophy," held at the Dibner Institute, Cambridge, Mass., in November 1995).

83. These are published and analyzed by Whiteside in *Math. Papers*, vol. 6.

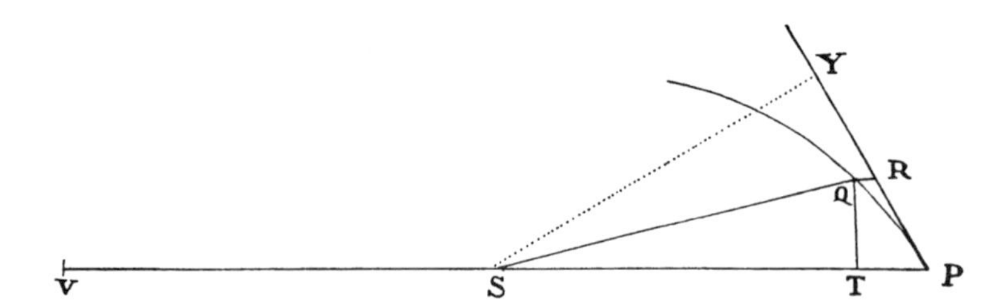

Figure 3.3. The centripetal force in orbital motion in a spiral (the diagram for book 1, prop. 9; third edition). Note that the line PS is extended to the point V, which appears only in the alternative proof added in the second and third editions.

and useful expansion of these somewhat cryptic statements in Bruce Brackenridge's articles and recent book.[84]

Although Newton does not refer in the second edition to the circle of curvature by this name, the diagrams contain markings that uniquely refer to such a circle. Thus, in the diagram for book 1, prop. 9, the line PS (see fig. 3.3) is extended to the point V, where PV is a chord of curvature (or a chord of the circle of curvature) through the point S. In prop. 6, the line of force from the point P on a curved orbit to the center of force S is extended out to a point V, used as the center of a circle of curvature in corol. 3 in the later editions. (See §10.8 below and fig. 10.6b.) In prop. 10, a point V is added (see fig. 3.4) on the diameter PCG of the ellipse, to indicate a point of terminus for the chord of curvature through the center C of the ellipse. In prop. 11, however, the alternate proof does not involve the circle of curvature in the same direct way as in the preceding prop. 10, and so in the diagram for the revised prop. 11 there is no need for the point V. It is amusing to note that, because of the desire to save money by using the same diagram in different propositions, the diagram that appears in book 1, prop. 10, in the second edition, is used again for prop. 11, even though the point V has no role in prop. 11. In the third edition, a different diagram occurs in these two propositions.

It has been suggested by Michael Nauenberg that Newton may have approached the problem of curved orbits long before the *Principia* and that he did so by using the method of curvature in combination with graphical and numerical analysis. Part of his argument is based on the curve appearing in Newton's letter to Hooke of 13 December 1679, a representation of the orbit of a body subjected

84. *The Key to Newton's Dynamics* (n. 71 above). In the second and third editions, Newton also derives a second measure of force ($SY^2 \times PV$) which clearly depends on curvature, because PV is the chord of curvature at the point P. This curvature measure is employed in the alternate solutions given for props. 7, 9, and 10.

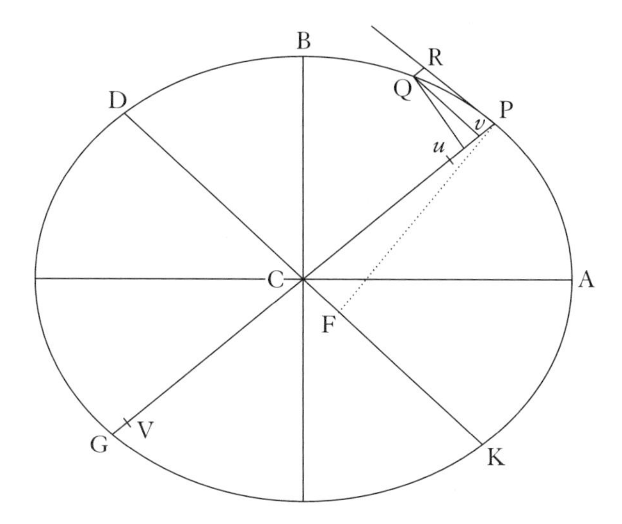

Figure 3.4. Motion in an elliptical orbit under the action of a force directed toward the center (the diagram, with some portions removed, for book 1, prop. 10; third edition). Note the point V on the diameter PCG of the ellipse.

to a centrally directed force of constant value, which Nauenberg shows could have been generated by an iterative numerical method.[85] Another part of Nauenberg's reconstruction is to assume that Newton could have devised an analytical measure of the force producing a curved orbit by using the methods he had developed in his *Methods of Series and Fluxions* of 1671. To support this claim, Nauenberg points to the above-mentioned letter of December 1679 to Hooke, in which Newton not only mentions the curve produced by a constant central force but also claims that one could suppose a centrally directed force that varied inversely as some (unspecified) power of the distance, in which case a body might "by an infinite number of spiral revolutions descend continually till it cross the center." Nauenberg takes note that this shows that Newton "must" have known that an inverse-cube force

85. Michael Nauenberg, "Newton's Early Computational Method for Dynamics," *Archive for History of Exact Sciences* 46 (1994): 221–252.

Newton's diagram in this letter has been the subject of considerable scholarly discussion. Nauenberg argues that Newton was using a correct method but that he made a blunder when drawing the figure. He believes that Newton's method consisted basically of approximating the curve by using discrete arcs of the circle of curvature. He then proceeds (pp. 235–239) to use that method to construct what he believes to have been the correct figure intended by Newton. This line of argument, it should be noted, has no basis in documentary evidence.

Professor Nauenberg's research of reconstruction and analysis has produced a new interpretation of the development of Newton's methods of dynamics in the years before the *Principia*. A summary of his evidence and results is given in §3.9 below. This reconstruction explains a number of aspects of the development of Newton's thought but does have some gaps. For example, Nauenberg must assume that documents some years apart refer to the identical methods.

produces an equiangular or logarithmic spiral. How else, he argues, could Newton have known that after "an infinite number of spiral revolutions," the curve would pass through the center of force or the pole of the spiral? A numerical solution would yield only a finite number of revolutions, obviously tending toward the central region, but would hardly provide a basis to justify Newton's bold and specific reference "till it cross the center."[86] Nauenberg reminds the reader that book 1, prop. 9, of the *Principia* is devoted to the problem of finding the force that produces an equiangular spiral; the inverse problem appears in book 1, prop. 41, corol. 3.

In retrospect, what may be most astonishing about Newton's three modes of analysis of the actions of forces is their constancy. Appearing first in the youthful exercises toward a science of dynamics, they survived and became a feature of the mature dynamics developed in the *Principia*. In the meantime he had undergone many changes in his fundamental philosophy of nature, not only shifting—as we have seen—from a Cartesian "conatus" to a dynamical force, but undergoing a variety of changes in belief about an "aether." In all these decades of changes in thought about the fundamentals of force and motion and the nature of space, however, he never lost sight of his original three modes of analysis. This may be taken as a measure of the independence of his mathematical methods and analyses from any philosophical or theological matrix in which they were embedded.

3.8 *Hooke's Contribution to Newton's Thinking about Orbital Motion*

An awareness of these three early approximations and their continued use in Newton's later dynamics helps to specify exactly what Newton could have learned from Hooke at the time of their epistolary exchange during 1679 and 1680. Newton already knew about curved paths being produced by two components of motion when one is uniform (or inertial) straight forward and the other directed downward and accelerated. He had learned from his reading that this form of conceiving the motion of a projectile had enabled Galileo to show that in the absence of any air resistance, the path would be a parabola. With regard to circular or general curved orbital motion, however, Newton was still under the spell of Cartesian physics and was thinking about centrifugal rather than centripetal forces, or rather endeavors. Thus, in the case of orbital motion, he was contemplating a radial displacement that was tangentially directed outward, away from the center, rather than the

86. Nauenberg has made an attractive reconstruction, based on two major assumptions for which there is no direct evidence. One is that Newton knew from his analysis of the logarithmic spiral that the curve would eventually reach the center. The other is that he purposely concealed or did not specify the power of the distance in the expression for the force producing the spiral.

more fruitful concept of a radial displacement inward. He had, apparently, already advanced to the study of curved orbits by the method of curvature.

What Hooke did for Newton, therefore, was not to tell him how to analyze curved motion into components, but rather to reverse the direction of his concept of displacement in orbital motion, to shift from an outward to an inward displacement. Once Newton began to think of centrally directed displacements and their cause in the action of forces, he recognized, as Hooke did not, that the key to a quantitative specification of an inward displacement lay in Kepler's law of areas. He would then have advanced along essentially the sequence he later exhibited in *De Motu* and elaborated in the *Principia*, of which the key propositions are book 1, props. 1, 2, 4, 6, 10, and 11. If, in 1684, he composed a prototype of the tract he sent to Locke in 1690, then he would have also proceeded by using the method of curvature. (For a complementary view of Hooke's contribution to the development of Newton's thinking about celestial dynamics, see §3.9 below.)

Looking at these events in the reverse order of time, it may seem puzzling that prior to Hooke's suggestion in 1689/90, Newton simply did not apply to planetary motion the principles of analysis that Galileo had worked out for projectile motion. In retrospect, these may seem to be two examples of a single principle or mode of analysis. There was, however, a tremendous conceptual gap between the downward accelerating force of gravity in the case of the parabolic path of a projectile and the apparently centrifugal endeavor of an orbiting body, a gap that required a gigantic intellectual leap forward, a complete reorientation of the framework of Newton's thinking. He needed just the kind of impetus in this regard that Hooke was able to give him. What is perhaps most interesting in retrospect is that Newton's three modes of analysis, developed in the 1660s, could be then conceptually inverted by reversing the direction of the endeavor and displacement and converting the endeavor into a force and that this could be done without producing any fundamental alteration in the mathematical analysis he had developed. Of course, for a body to stay in an orbital path there must be an opposing or centrally directed endeavor or pressure to counteract the centrifugal endeavor. It must also be noted that in the *Principia*, there are some references to a "vis centrifuga," notably (just as one would expect) in book 3, where Newton explores the effect of the earth's rotation on a body's weight. (See, further, §3.10 below.)

Obviously, this contribution of Hooke was a deciding factor in setting Newton on the path that would lead him to the *Principia*. Hooke, as is well known, did not claim that this was his contribution. Rather, he told Halley and others that Newton had learned of the inverse-square force from him and that he was therefore primarily the author of Newton's theory of gravity. In its most exaggerated form, this claim was stated by John Aubrey in his *Brief Lives*. Aubrey wrote that Hooke corrected Newton's error, teaching Newton that "the gravitation was reciprocall to the square of the distance: which is the whole coelastiall theory, concerning which Mr. Newton has made a demonstration, not at all owning he receiv'd the first Intimation of it from Mr. Hooke." Aubrey concluded that Newton took from Hooke, "without acknowledging" the source, "the greatest Discovery in Nature that ever was since the World's Creation."[87]

Newton's rage, on hearing about Hooke's claim, took the form of threatening to withdraw book 3 of the *Principia*. He calmed down, however, and took satisfaction in sending Halley some proofs that he had known of the inverse-square relation before Hooke suggested it to him. As the documents discussed in §§3.6 and 3.7 make clear, Newton already knew the inverse-square relation when he received Hooke's letter. He was alert to the fact that Hooke's mathematics was faulty, since Hooke had written that a consequence of an inverse-square attraction would be an orbital velocity "reciprocall to the Distance" from the center of attraction. Newton proved that the velocity should rather be inversely proportional to the perpendicular distance from the center to the tangent. The two are equal only in the apsides. In any event, as Curtis Wilson has concluded, in stating his ideas to Newton, "Hooke did not know whereof he wrote"; the "dynamics of the passage is hopelessly wrong, the mathematics unsalvageable."[88] A mere shift in the direction of the force (from centrifugal to centripetal) would not automatically suggest either (1) that in an ellipse the force is directed toward a focus rather than the center, or (2) that the key to such motion is the law of areas. Michael Nauenberg's analysis (for which see §3.9 below) and its elaboration by Bruce Brackenridge set forth a plausible path by which Newton, following the stimulus of Hooke's suggestion, discovered the solution to the problem of orbital motion. This new research not only indicates the stages of Newton's tremendous creative leap forward but makes precise the effect of the encounter with Hooke.

3.9 *Newton's Curvature Measure of Force: New Findings* (BY MICHAEL NAUENBERG)

An analysis of Newton's letter to Hooke of 13 December 1679 indicates that Newton had developed the dynamics of orbital motion far beyond what has been generally understood.[89] In particular, this letter shows that Newton had developed a method by "points *quam proxime*" to compute the geometrical shape of orbits

87. Oliver Lawson Dick, *Aubrey's Brief Lives, Edited from the Original Manuscripts* (Ann Arbor: University of Michigan Press, 1962).

88. Curtis Wilson, "Newton's Orbit Problem: A Historian's Response," *The College Mathematics Journal* 25 (1994): 193–200, esp. 195.

89. Nauenberg, "Newton's Early Computational Method" (n. 85 above).

for general central forces, although by his own later accounts he had not as yet discovered that central forces imply the law of areas. A diagram drawn in a corner of this letter shows an orbit for the case of a constant central force, one which shows the symmetries expected from energy conservation and the reversal of the direction of time. Moreover, the geometrical shape of the orbit, a rotating oval figure, is drawn correctly, although the angle between the successive apsides is somewhat off the mark—an error in the drawing and not in the computation of the orbit.[90]

In the text of the letter to Hooke, Newton discusses in a qualitative manner the changes in the orbit which occur when the force is increasing toward the center, and he indicates the existence of a force law which leads to a spiral orbit. Newton refers to an orbit with "an infinite number of spiral revolutions." Because such an orbit cannot be found by using only numerical and graphical approximations, this statement provides important evidence indicating that Newton seems to have found an analytic solution. In his letter to Hooke, Newton did not mention that the spiral orbit was produced by a force depending on the inverse cube of the radius. This radial dependence does appear in a scholium to the tract *De Motu* of 1684, but without any proof.[91]

There is additional evidence that may indicate that at the time of his correspondence with Hooke, Newton knew the law of the inverse cube for the force producing a spiral. This occurs in a scholium in the draft of book 1 of the *Principia*, deposited in the University Library as the Lucasian Lectures.[92] This draft, dated 1684, contains a scholium in which Newton describes the same approximate orbits as in his correspondence with Hooke. This time, however, he includes the radial dependence of the force and also the corresponding numerical values for the angles between successive apsides; these are in close agreement with the exact values.

This scholium was crossed out and not included in the first edition of the *Principia*. The reason is that Newton had developed a more powerful analytic method for evaluating these angles, a method that appears in sec. 9 of the *Principia* under the title "mobile orbits." This method would appear to have evolved from Newton's earlier graphical and numerical method, which had shown that for small eccentricities orbital motion for central forces is well approximated by a rotating oval figure.

Newton did not reveal his early computational method to Hooke, but we can be certain that it could not have been based entirely on any of the measures of force presented in book 1, prop. 6, of the *Principia*. The reason is that these measures

90. For details see the article cited in the preceding note.

91. *Background*, p. 281; *Math. Papers* 6:43.

92. Whiteside, *Preliminary Manuscripts* (§1.1, n. 5 above), pp. 89–91.

all depend on measuring time by the area swept out by a radius, that is, on the law of areas. There is, however, an important clue to Newton's early method in the cryptic remark (1664/65) in the Waste Book. Here Newton says that the force acting on a body moving in an elliptic orbit is determined at each point if the velocity and the crookedness (that is, the curvature) of the ellipse are given. This remark indicates that Newton had considered the extension to elliptic orbits of the rule that he (and also Huygens) had found for circular motion.

Moreover, Newton's mathematical manuscripts from 1664 to 1671 show that he had developed also the same "method of evolution" as Huygens, who had developed it somewhat earlier and who gave it its name in his *Horologium Oscillatorium* (Leiden, 1673). In this method, each short segment of a curve is approximated by the arc of a circle with its radius and center adjusted for an optimal fit, which is obtained by requiring that three intersections of the circle with a segment of the curve coincide in the limit.[93] One of Newton's mathematical definitions of curvature is to equate curvature with the inverse of the radius of this circle. The method of evolution also shows how a curve can be obtained (if its curvature is known) by approximating a segment of the curve at each point by an arc of the circle of curvature.

For an orbital curve, Newton's cryptic remark about crookedness implies that the curvature can be determined if the force and the velocity are known as functions of position. In the *Principia*, Newton determines the velocity by the law of areas for central forces. Since he was unaware of this law or of its significance at the time of his correspondence with Hooke, he needed an alternative way of obtaining the velocity. One straightforward way of approximating the velocity is to equate the change in velocity along an arc of the curve with the tangential component of the force multiplied by the elapsed time interval. (This time interval is equal to the length of the arc divided by the initial velocity along the arc.) In the limit as the arc-length vanishes, this generalization of Galileo's law for uniform acceleration gives rise to what we now call the principle of conservation of energy, which Newton introduces in book 1, props. 39 and 40. Indeed, Newton applied this principle in 1671 to determine the velocity of a cycloidal pendulum weight as a function of position and obtained its period by integration, much as Huygens had done in 1659.[94]

Thus Newton could have developed a curvature approach to orbital dynamics based on the decomposition of the force into a component normal to the orbit

93. Leibniz's original definition of an osculating circle requires the coincidence of four intersections, which is incorrect except at special points of maximum and minimum curvature.

94. M. Nauenberg, "Huygens, Newton on Curvature and Its Application to Orbital Dynamics," *De Zeventiende eeuw: Cultuur in de Nederlanden in interdisciplinair perspectief* 12 (1996): 215–234.

related to curvature and velocity and a tangential component that determines the time rate of change of the velocity. It can be verified that all of the approximate results in Newton's letter to Hooke can be reproduced by this approach.[95] In particular, the special role of the spiral orbit that appears in this letter now becomes apparent because this spiral (identified in book 1, prop. 9, as an equiangular spiral) has the property, previously derived by Newton in 1671, that its curvature is simply inversely proportional to the radial distance. For this case, it can be shown in a few lines, by using the curvature approach, that the central force varies inversely as the cube of the distance from the center to the spiral.[96]

In his letter of 24 November 1679, Hooke suggested to Newton that the motion of the planets around the sun be considered by "compounding" the motion (that is, the velocity) along the tangent to the orbit with an attractive motion (that is, change of velocity) toward the sun. Hooke was apparently unable to give mathematical expression to his proposal, which can be done if the central force is approximated by a sequence of centrally directed impulses.[97] Newton had applied this idea earlier to uniform circular motion. In the limit, as the times between successive impulses vanish, the central force acts continually, leading to the well-known relation $f \propto \frac{v^2}{r}$ between centrifugal force f, velocity v, and radius r (for which see the scholium to the *Principia*, book 1, prop. 4). For noncircular motion, such a sequence of impulses leads immediately to the law of areas, as Newton had demonstrated earlier, in book 1, prop. 1, of the *Principia*. Since Newton discovered the connection of this law with the action of central forces shortly after his correspondence with Hooke, it would appear that he first implemented the impulse approximation at this time. In the curvature approach, the law of areas is valid only in the analytic limit, but it remains hidden and must be derived from curvature relations.

This line of reasoning leads to the conclusion that Hooke's suggestion to Newton in 1679 led Newton to an approach to orbital dynamics complementary to his earlier curvature method and revealed the significance of the law of areas for central forces. Newton then applied the law to evaluate the velocity which appears in his curvature method, thereby obtaining a new curvature measure of force. This measure, however, was not introduced in book 1 of the first edition

95. See Nauenberg, "Newton's Early Computational Method" (n. 85 above).

96. Ibid.

97. In a manuscript dated September 1685, however, Hooke implemented this method graphically for a central force which varies linearly with distance, and he demonstrated that the orbit is an ellipse. See M. Nauenberg, "Hooke, Orbital Motion and Newton's *Principia*," *American Journal of Physics* 52 (1994): 331–350 and idem, "On Hooke's 1685 Manuscript on Orbital Motion," *Historia Mathematica* 25 (1998): 89–93.

of the *Principia*. In the early 1690s, Newton conceived plans for radical revision of the first part of the *Principia* that would be based on the method of curvature, but this plan was abandoned. The method of curvature appears in book 1 of the *Principia* for the first time in the second edition (1713) in a new corol. 3 to book 1, prop. 6. But in the first edition, in book 3, prop. 28, this measure is applied to the perturbation of the lunar motion by the gravitational force of the sun.

Apparently, Newton was reluctant to reveal his early computational method of curvature, even while writing the text of the first edition of the *Principia*. Accordingly, it is perhaps not surprising that it has been difficult to elucidate the nature of his method by points *quam proxime*, to which he alluded in his correspondence with Hooke.

3.10 *Newton's Use of "Centrifugal Force" in the* Principia

In the *Principia* the word "centrifugal" appears only a few times, and in several different senses. In book 1, prop. 4, schol., Newton mentions Huygens, who believed in a "vis centrifuga" and who gave it this name. In all three editions, Newton introduced a summary of an early study of the outward ("centrifugal") force exerted on a circle by a body moving on a polygon inscribed in the circle.

At the end of book 1, sec. 2, Newton notes that if an orbital path is turned into a hyperbola, the "centripetal" force becomes "centrifugal." Similarly (prop. 12, alt. proof), a change from a centripetal "into a centrifugal force" will cause a body to "move in the opposite branch of the hyperbola." In prop. 41, corol. 1, Newton explores the result of changing a centripetal force into a centrifugal force. In book 1, prop. 66, corol. 20, Newton denies that water could be "sustained" in a special condition "by its own centrifugal force." These are mathematical problems, none of which deals with the kind of motion that occurs in the orbits of planets.

In book 2, prop. 23, "centrifugal force" is used for the forces of repulsion between particles of "elastic" fluids, or gases.

In book 3, "centrifugal force" occurs twelve times in prop. 19, in explaining how weight or gravity on the surface of a rotating planet is reduced by the "force" arising from rotation. In prop. 36, corol., and prop. 37, corol. 7, "centrifugal" is used in the same way with respect to the earth's rotation. We should note, in this connection, that many engineers and teachers of physics still use the idea of a centrifugal force in explaining how the earth's rotation affects the weight of a body.

A somewhat different use of "centrifugal" appears in book 3, prop. 4, schol., written for the third edition, on the orbital motion of a hypothetical terrestrial moon. Newton says that if this small moon "were deprived of all the motion with which it proceeds in its orbit," it would "descend to the earth" because of "the absence of the centrifugal force with which it had remained in its orbit." In Newton's day and in succeeding decades, Newtonians tended to use "centrifugal force" (and also "projectile force") to denote the inertial (tangential) component of curved motion.

This usage appears in the works of two staunch Newtonians, writing about the science of Newton's *Principia* in the years after the first edition (1687). William Whiston was the person chosen by Newton to be his deputy when he left Cambridge for London; he later became Newton's successor as Lucasian Professor. Whiston's published lectures on the science of the *Principia* were quoted at length in John Harris's *Lexicon Technicum* (London, 1704).

Whiston explained the motion of the planets in terms of Newtonian rational mechanics and celestial dynamics, as follows:

> [The] true meaning . . . of *Attempt* or *Endeavour* to get farther off the Centre of Motion, is only this, That all Bodies being purely Passive, and so incapable of altering their uniform Motion along those straight Lines or Tangents, to their Curves, in which they are every Moment; do still tend inwards in the same Lines, and retain their Propension or Effort towards that Rectilinear Motion all the Time they are obliged to move in Curves; and consequently at every point of their Course, endeavour to fly off by their Tangents.

In short, in curved or orbital motion, the endeavor to move out along the tangent tends to take the body away from the center, or "to recede from that Centre," and so the "*Conatus recedendi a centro*" arises "from no other Affection than that of Inactivity [i.e., inertia], or of persevering in a Rectilinear Motion." Accordingly, he concludes,

> tho' the *vis centripeta*, or Power of Gravitation, be an Active and Positive Force, continually renew'd and impress'd on Bodies; yet the *vis centrifuga*, or conatus recedendi a centro motus is not so, but the meer Consequence and Result of their Inactivity [i.e., inertia].

John Harris was also a thoroughgoing Newtonian. His *Lexicon Technicum* (vol. 1, 1704; vol. 2, 1710) is a primer of the Newtonian natural philosophy and mathematical system of the world, with copious extracts from Newton's writings. Newton entrusted to Harris the first publication of his essay "De natura acidorum." In volume 1, Harris defines "CENTRIFUGAL *Force*" simply and straightforwardly as

> that Force by which all Bodies which move round any other Body in a Circle or an *Ellipsis* do endeavour to fly off from the Axis of their Motion in a Tangent to the Periphery of it.

In volume 2, Harris explains "CENTRIFUGAL *Force*" at greater length, as follows:

> All moving Bodies endeavour, after a rectilinear Motion. . . . Whenever therefore they move in any Curve, there must be something that draws them from their rectilinear Motion, and detains them in the *Orbit*. Whenever this Force, which is called *Centripetal*, ceases, which attracts them toward a Centre, the moving Body would strait go off in a Tangent to the Curve in that very Point and so wou'd get farther and farther from the Centre or Focus of its former Curvilinear Motion: And that Endeavour to fly off in the Tangent is the *Centrifugal Force*.

These extracts leave no doubt that Newtonians used the term "centrifugal force" for what we would call the tangential component of curved or orbital motion, or the inertial component. This "force" is, therefore, as Whiston noted, nothing other than the result of a body's "inertia."

This use of "centrifugal force" is doubly puzzling to today's readers. For there is not only the seemingly odd use of the adjective "centrifugal," but also the designation "force." In the *Principia*, however, Newton generally writes of a body's inertial tendency in terms of a force, one that he calls a "vis inertiae" or "force of inertia," and which he says is the same as a "vis insita" by another name. This is not the kind of force that causes a body to undergo a change in its state of motion but, as Newton makes clear in def. 4, rather one that keeps a body in its state of motion.

CHAPTER FOUR

Some Fundamental Concepts of the *Principia*

Newton's Definitions 4.1

In a scholium to the Definitions (discussed in §4.11), Newton says that he has undertaken "to explain the senses in which less familiar words are to be taken in this treatise." These do not include "time, space, place, and motion," which are "familiar to everyone." It is important, however, he notes, to take into account that "these quantities are popularly conceived solely with reference to the objects of sense perception." Newton says that he will "distinguish these quantities into absolute and relative, true and apparent, mathematical and common."

Preliminary to any such discussion is the set of eight definitions of "less familiar" words. By "less familiar" Newton does not suppose that the words themselves are not well known; rather, he means that he is using these words in an unfamiliar sense to express concepts that are new. The first of these is "quantity of matter" (def. 1) as a different measure of matter from the ones then in use, a measure that he also calls "body" or "mass." The second (def. 2), "quantity of motion," is a new concept of the measure of motion to the degree that it is based on the previous measure of matter. Then Newton gives the meaning he assigns to the current concept of "vis insita," or inherent force (def. 3), introducing a new concept and name of "vis inertiae," or "force of inertia," that arises from the "inertia materiae," or "inertia of matter," thus formally introducing "inertia" into the standard language of rational mechanics. Next (def. 4), Newton gives his definition of "impressed force." He then (def. 5) introduces a general concept of a centrally directed force, using the name "centripetal force," which he has invented. The final three definitions (defs. 6, 7, 8) are devoted to a set of three measures of centripetal force: the "absolute," the "accelerative," and the "motive" measures.

There are a number of curious features of these definitions. First of all, as we shall see, defs. 1 and 2 differ in form from defs. 3–8. Second, in presenting these definitions, Newton actually anticipates the first of the "axioms, or laws of motion," which are supposedly based on the definitions. Third, def. 1 proves not to be a definition in the usual sense, but rather states a rule or relation between quantities.

4.2 *Quantity of Matter: Mass (Def. 1)*

The subject of the opening definition (def. 1) of the *Principia* is "quantity of matter." Newton defines quantity of matter as a measure of matter; in particular, he writes, it "arises from" (that is, originates in) "its density and volume jointly," that is, from the product of the density and volume. In choosing this measure, Newton was aware that there were other measures of matter in current use, such as Descartes's extension, or space occupied, which is somewhat similar to Kepler's "moles," or bulk. Another such measure is weight, favored by Galileo.

Newton produced the final refinement of his own concept only when he was in the last stages of writing the *Principia*. Mass (or quantity of matter) does not appear as one of the definienda in the tract *De Motu*. Since in most of that tract, Newton was dealing with essentially mass points of unit mass, he had no need of this concept. Yet, in the concluding portion of *De Motu* dealing with resistance forces, the lack of the notion of mass produces a severe limitation. Because there is no concept of mass, the method of measuring resistance forces proposed in the final scholium yields only an accelerative measure, and hence a different measure for bodies of the same shape but different weight. This is strong evidence that he had not formed either the concept of mass or the associated concept of motive force at the time of composing *De Motu*.

The definition of quantity of matter seems to have first arisen as a problem for Newton when he proposed a set of some eighteen definitions, together with a set of laws, in an unfinished manuscript[1] of book 1 of the *Principia*, written some time after *De Motu* and before he had composed the first full draft of book 1 of the *Principia* (the text of which he later deposited in the University Library as his Lucasian Lectures).[2] The sixth of this preliminary set of definitions is "density of a

1. The Latin text of this manuscript is published by Whiteside in *Math. Papers* 6:188–192. An earlier edited transcript, with English translation, is given in Herivel, *Background*, pp. 304–312 (supplemented by my *Introduction*, pp. 92–96).

2. This text, made of up several parts, representing different stages of composition and revision, is the basis of Whiteside's edition (with translation and commentary) in vol. 6 of *Math. Papers*. For details, see my *Introduction*, pp. 310–321, where this manuscript is called "LL" ("Lucasian Lectures"). A facsimile

body." In the discussion, Newton makes reference to "quantitas seu copia materiae," or "quantity or amount of matter," which is not explicitly defined, but is considered with respect to "the quantity of space occupied." The following definition (def. 7) in this preliminary set declares the sense in which Newton understands "pondus," or "weight." Newton's meaning is not the simple and conventional one. By "weight," he writes, "I mean the quantity or amount of matter being moved, apart from considerations of gravity, so long as there is no question of gravitating bodies." That is, he is groping for a measure of the quantity of matter. He uses the customary term "weight," for which as yet he has found no substitute. Nevertheless, he does not want "weight" to be interpreted in the usual sense. Rather, he has in mind what he will later conceive to be the inertial property of mass, as is evident in his reference to motion and not to a downward force, the way in which weight is usually conceived. He cannot use "weight" in this ordinary sense because to do so would imply that his measure, or "quantity of matter," in any given sample is a variable and not a fixed property. That is, he is aware of the experiences of astronomers who carried clocks or seconds pendulums from centers such as London or Paris to some distant place. He knows weight varies from one place to another.[3] It is weight, in this ordinary sense, of heavy (literally, gravitating) bodies which Newton says he has found by experiment to be "proportional to their quantity of matter." By "analogy," he concludes, the "quantity of matter" in bodies "can be represented or designated." Newton here refers to, but does not describe in detail, the experiments with pendulums from which this "analogy" can be inferred. The materials used in these experiments are said to be gold, silver, lead, glass, sand, common salt, water, wood, and wheat.[4]

A final sentence, which is later canceled, then declares that on the basis of "this analogy," and "because of the want of a suitable word," he will "represent and designate quantity of matter by weight," even though he is aware that this usage is not appropriate for bodies in which there is no consideration of gravitating or of "heaviness." The latter category could include planetary satellites being "heavy" toward their parent planet. Here, and in a canceled earlier version, we may see how Newton is working his way toward a new concept for which he has not as

---

reproduction of this manuscript has been published, with an introduction, by Whiteside in *Preliminary Manuscripts for Isaac Newton's 1687 "Principia"* (§1.1, n. 5 above).

3. This important property of weight was discovered when astronomers on expeditions, notably Jean Richer and Edmond Halley, found that when they traveled from Paris or London to another and distant latitude, their pendulum clocks no longer maintained the same period as before. Newton (in book 3 of the *Principia*) explained this phenomenon of a variation in period as a variation in the acceleration of free fall or of gravity, that is, a variation in the weight of the pendulum bob.

4. These experiments are discussed in §8.9 below.

yet found a name. The "quantity of a body," he suggests, can be estimated from the "amount of corporeal matter," which "tends to be proportional to the weight." In the *Principia*, in def. 1, Newton will repeat that quantity of matter can be determined by weighing.

Before very long, by the winter or early spring of 1684/85, Newton had advanced from this preliminary "want of a suitable word" to the choice of "mass." In the first draft of book 1 of the *Principia*,[5] the first five definitions are essentially the same as in the later published version. In this draft, however, the last three definitions (defs. 6-7-8) of the printed version are included as part of the discussion of def. 5, but without being considered separate subjects of numbered definitions. The first definition of this draft is the same as in the later printed text. He now uses the name "quantity of matter" ("quantitas materiae"). He says, in his discussion of def. 1, that he will designate "quantity of matter" by two names, "body" and "mass," thus introducing—in a new and specific sense—a concept and a name that have been at the foundations of physical science ever since.

In the discussion of def. 1 in the *Principia*, Newton repeats his earlier statement that "mass" (or "quantity of matter") "can always be known from a body's weight." He now explains that "very accurate experiments with pendulums" have shown that mass is "proportional to the weight."[6] It is important to note that the only way that Newton says that mass can be determined is by weighing, by finding gravitational forces. As we shall see in detail below, nowhere in the *Principia* does Newton in fact begin by determining a body's density and volume and then computing the mass.[7] Rather, he uses a combination of dynamical and gravitational considerations to determine a mass which then yields a density.

The mathematical underpinnings of these experiments is the subject of book 2, prop. 24, and the details of the performance of the experiments are introduced in book 3, prop. 6. Newton made use of a pair of pendulums "exactly like each other with respect to their weight, shape, and air resistance," suspended "by equal eleven-foot cords." He compared the periods when equal weights of different substances were introduced, the same substances he had mentioned earlier. The pendulum experiments showed to a high degree of accuracy that at any given location, the quantity of matter or mass of different substances is proportional to their weight.

5. See n. 2 above.

6. Newton evidently did not consider it necessary to explain that although the weight of any object on the earth will vary from one latitude to another, a weighing device (in which weights are determined on a balance, but not by a spring) will always give the same numerical weight to an object, no matter what its geographical location. The reason is, of course, that the object being weighed and the standard weight or weights with which it is being compared will both vary by the same factor.

7. See §4.3 below.

By induction, this result may be generalized to all substances. Albert Einstein once remarked that Newton's concern for this problem doubly exhibited the signs of Newton's genius: first, for recognizing that proof is needed that mass is proportional to weight, and second, for equally recognizing that in classical mechanics the only possible proof is by experiment.[8]

Is Newton's Definition of Mass Circular? 4.3

Newton's definition of quantity of matter, or mass, has been subjected to intense scrutiny by many scientists and philosophers. One of the charges leveled against the definition, notably by the eminent philosopher of science Ernst Mach,[9] is its alleged circularity. In the post-Newtonian world, density is considered to be mass per unit volume; accordingly, the definition of mass is circular if mass is also defined as the product of density and volume or as proportional to that product. In order to understand the sense of def. 1 and its role in the *Principia*, however, we need a historical as well as a logical analysis. We should begin by taking note that Newton is defining a new and invariable measure of matter. This measure, as has been mentioned, has the property that it does not change if a sample of matter is heated, bent, stretched, squeezed or compressed, or transported from one place on the earth to another or even to a position out in space—on the moon, on Jupiter, or way out in the heavens. Newton's mass is one of those qualities which (as Newton was to say in rule 3) "cannot be intended and remitted."

In discussing quantity of matter, in the opening of the *Principia*, Newton refers specifically to the experimental phenomena of the compression of gases, as in Boyle's experiments. If a gas is compressed in a sealed apparatus, the act of compressing the gas does not alter the quantity of matter in the gas so long as none of the gas escapes during the compression. And it is the same for the expansion of a confined gas. If a given quantity of gas is compressed into half the original volume, the matter is more compact or more dense by a factor of 2. Similarly, if the gas is allowed to expand until it fills twice the original volume, there will be exactly half as much matter in the original space as at the start; the matter will be half as compact or half as dense. In this line of reasoning, we may see how the quantity of matter does, as Newton says, arise or originate from two input factors,

8. In the conceptual framework of post-Einsteinian physics, what Newton recognized amounts to the proportionality of gravitational and inertial mass.

9. Ernst Mach, *The Science of Mechanics: A Critical and Historical Account of Its Development*, trans. Thomas J. McCormack, 6th ed., with revisions from the 9th German ed. (La Salle, Ill.: Open Court Publishing Co., 1960), chap. 2, §7: "As we can only define density as the mass of unit volume, the circle is manifest."

the space occupied and the degree of compression or expansion, the two factors mentioned by Newton in the sentence containing the formal definition of quantity of matter.

It is to be especially noted that in the definitions set forth at the beginning of the *Principia*, Newton does not define density; nor is density discussed in the scholium following the definitions. Newton apparently assumes that readers will know what is meant by this term, will understand that density is some kind of thus far unspecified measure of the degree to which matter is compacted into a given space.

In Newton's day, we must remember, there was not as yet a system of units for physical quantities; in particular, there was as yet no unit of mass. Force was measured in pounds (as in weight) or some equivalent, and volume was often given in terms of liquid measures; but there was no unit for density. As a result, density was expressed as a relative quantity. That scientists tended to use density in a general sense without a specific definition may be seen in the treatment of density in John Harris's *Lexicon Technicum* (London, 1704), a very Newtonian dictionary of the sciences. Here we find two definitions relating to density. The first is "dense." A body, according to Harris, is "said to be Dense or Thick when it hath more of Matter in proportion to the Space or Room it takes up, than other Bodies have." This leads to the next definition, in which it is said that "the being under these Circumstances is called the DENSITY *of Bodies*," just as "that which produces it is called *Condensation*." Harris then gives the densities of certain substances as ratios. For example, "The Density of Water to Air is as Mr. *Is. Newton* states it, as 800, or 850 to 1." Or, "The Density of Quick-silver to Water is as 13½ to 1." Because in these examples, and the others he gives, the densities are relative numbers or ratios, they are rather what we would think of as specific gravities.[10] As we shall see below, Newton himself considered densities to be the same as specific gravities.

It has been mentioned that the actual wording of def. 1 and of def. 2, which depends on def. 1, differs from that of all the other definitions in the *Principia*. The others begin by stating that the definiendum is something or other. For example, "Inherent force of matter is . . . ," "The accelerative quantity of centripetal force is . . . ," or "A fluid is . . . "[11] That is, they all begin by stating the definiendum and then use the verb "esse," "to be." This is the standard form, as found in Euclid: "Punctum est . . . ," "Linea est . . . ," or "Superficies est . . . ," the definitions of a

10. Later on in the *Lexicon Technicum*, there is a very long entry on specific gravity, extending over several pages. On density and specific gravity, see further §4.5 below.

11. Defs. 3, 7; book 2, sec. 5.

point, line, or surface. Accordingly, def. 1 is stated in two more or less separate parts, the first of which is in traditional form and reads, "Quantitas materiae est mensura ejusdem," or "Quantity of matter is a measure," that is, is a (or the) measure of matter. The second part would then refer to this measure as "orta ex illius densitate et magnitudine conjunctim." Note that in this portion of the definition, Newton is using the past participle of the deponent verb "oriri," "to arise." He is stating formally the property of mass that we have just been discussing. By his choice of language, Newton indicates that this second part of the definition is not the primary part of the definition; it is clearly not stated in the primary form in which definitions are usually presented and in which all the other definitions of the *Principia* are given. It would seem as if this part of the definition is rather a rule, stating a relation between the new concept of mass (as a measure of matter) and the concepts of volume and the intuitively known density.

This sense of def. 1 was carefully maintained by good Newtonians, such as John Harris. In the *Lexicon Technicum* (London, 1704), s.v. "MATTER," Harris plainly states that the "Quantity of Matter in any Body, is its Measure, arising from the joint Consideration of the *Magnitude* and *Density* of that Body." Harris does not say that quantity of matter is defined in terms of, or is defined in proportion to, the product of density and volume.

In the light of the stages of evolution of the definitions and Newton's care in revising and rewriting them, we may be certain that the actual form of def. 1, radically differing from defs. 3–8, was the result of intent and not merely an accident of style. This interpretation of def. 1 is further buttressed by two kinds of evidence: Newton's successive alterations of the numerical examples he used to illustrate the definition and his treatment of the masses and densities of the sun, the moon, and the planets in book 3. As may be seen in the variants of our Latin edition, and in the note to def. 1 in the present translation, once Newton had published the *Principia*, he spent a considerable amount of time and energy with the numerical examples that explain the meaning of def. 1. At first glance, it will even seem inconceivable that Newton had to take so much time and energy in order to explain what might have seemed to be a simple and straightforward ratio of mass to the product of density and volume. But Newton, as we have seen, very carefully chose not to define mass by saying that mass is the product of density and volume and even eschewed saying that mass is proportional to the product of density and volume. Accordingly, there was a need to explain what he had written by giving numerical examples. In the first edition, Newton gave a single numerical example to explain how quantity of matter "arises" from density and volume. Once the *Principia* had been published, however, he went over his text a number of times, fiddling with the words of explanation and devising additional numerical explanatory examples to make the sense of his definition clear to readers. Unless cognizance is taken of the reason why, it will seem absurd that Newton had to go to such great lengths to explain what would otherwise have been so simple.

In the centuries following the publication of the *Principia*, the fundamental units of classical dynamics have been mass, length, and time, associated with dimensional analysis (based on M, L, T). In the *Principia*, however, Newton is generally not concerned with units or with dimensionality. The reason is that, except for such quantities as distances fallen as measures of gravity, lengths of seconds pendulums at various places, and the difference in size of the earth's equatorial and polar axes, Newton tends to be concerned with ratios of quantities rather than with quantities themselves. Thus he does not establish a unit of mass and then compute individual masses. As has been mentioned, although he has a unit of force (English pound of weight, and its French equivalent) he does not compute the magnitude of specific gravitational or dynamical forces. Rather, he compares masses with other masses, and forces with other forces; for example, the force of the moon to move the sea compared with the force of the sun to move the sea, the force of the earth on the moon compared with the force of gravity on the earth's surface, or the mass of Jupiter or of Saturn compared with the mass of the sun. Since he has no unit for density, he does not compute densities, but rather limits his calculations to relative densities. It is, in fact, because the *Principia* sets forth a dimensionless physics that Newton can make a transition from a law of impulsive force $F \propto d(mV)$ to $F \propto \frac{d(mV)}{dt}$, basing the shift on dt being a constant.[12] The modern reader would be troubled by this example because the constant of proportionality in the two cases must be of a different dimensionality.

Newton's primitive or primary quantities are not the mass, length, and time of post-Newtonian classical physics. Length (as space) and time are discussed in the scholium to the definitions, but the actual definienda of the *Principia* are mass (quantity of matter), momentum (quantity of motion), and various types of forces and their measures. Momentum (def. 2) is a derived or secondary quantity, arising from mass and velocity, while velocity (or speed) and acceleration are also secondary quantities derived from space and time. Newton's primary quantities in the definitions of the *Principia* and the associated scholium are thus mass, momentum, force, space, and time; they include the intuitive concepts of density, velocity, and acceleration, which are not specifically defined.

12. See §5.4 below.

Newton's Determination of Masses 4.4

In the *Principia*, mass is related to force in two ways: (1) dynamically, through the second law of motion; (2) gravitationally, through the law of gravity. From this point of view, the argument about circularity is wrongly conceived, being based on the supposition that in def. 1, mass is a secondary quantity defined in terms of the primary quantities, density and volume. But in the *Principia*, mass is a primary and not a secondary property, and it is not explicitly or properly defined in terms of other quantities which are primary. Mass is defined only by implication. It has the property of inertia (def. 3). It appears (via momentum, or "quantity of motion") in the second law as the measure of a body's resistance to a change in motion (for impulsive forces) or as the measure of a body's resistance to acceleration (for continuous forces).

In the *Principia*, Newton does not in general determine masses by finding densities and volumes, but by dynamical (inertial and gravitational) considerations. Indeed, a careful reading of def. 1 and its discussion shows that Newton does not say that he will determine masses by finding densities and volumes. Rather he tells the reader that masses are known from weighing.

This view of def. 1 may find some support in John Clarke's *Demonstration of Some of the Principal Sections of Sir Isaac Newton's Principles of Natural Philosophy* (London, 1730). John Clarke, a sound interpreter of Newton, was the brother of Samuel Clarke, who translated Newton's *Opticks* into Latin at Newton's request and who was Newton's spokesman in the Leibniz-Clarke correspondence. In this work, Clarke does not define quantity of matter in any usual sense; he merely states how it is to be measured. That is, he says: "The Quantity of Matter is to be measured by the Density and Magnitude together, or by the Density multiplyed into the Magnitude."

On at least two occasions, Newton himself gave a simple statement of the relationship between mass, density, and volume. In a discussion of the particles of matter in book 3, prop. 6, corol. 4, he says that "particles have the same density when their respective forces of inertia [or inertial masses] are as their sizes." This is very similar to a definition occurring in the essay "De Gravitatione," which seems to have been written shortly before Newton composed the *Principia*. In this essay, there is no definition of mass or quantity of matter, but there is a definition of density (def. 15), which reads (in translation): "Bodies are denser when their inertia is more intense, and rarer when it is more remiss."[13] At one time, soon after the *Principia* had been published, Newton thought of adding in some later edition

13. *Unpubl. Sci. Papers*, pp. 89–157.

an amplification of his explanation of def. 1. He would now make an explicit statement of the proportionality of mass and the product of density and volume. In a list of manuscript errata and emendata, which he wrote out in one of his personal copies of the *Principia*, he would amplify his discussion by saying: "For this quantity, if the density is given, is as the volume and, if the volume is given, is as the density; and, therefore, if neither is given, is as the product of both." In the end, he rejected this emendation, deciding not to make a statement that would seem to interpret the definition as a proportion. He evidently wanted the definition itself to stand as he had first published it, not as a definition in the usual sense, and without an express statement of the ratio which the rule in the second part of the definition implied.

4.5 *Newton's Concept of Density*

In Newton's day, as has been mentioned, density was not given in terms of some particular set of units, but rather as a relative number, in comparison with the density of water or of some other substance or body. Thus, in the *Principia*, Newton compared the density of the moon with the density of the earth, but did not evaluate either density on some numerical scale. Since densities were given as ratios, they were indistinguishable in practice from specific gravities. Indeed, density could even be considered as a quantitative property of substances (or of bodies) found directly by experimental determinations of specific gravity.

We may see an example of Newton's conflation of density and specific gravity by turning to the *Opticks*. Here (book 2, pt. 3, prop. 10), Newton writes of "the Densities of . . . Bodies" which are to be "estimated by their Specifick Gravities." In a table (book 2, pt. 3), Newton has a column marked, "The density and specifick gravity of the Body."[14]

14. In the essay "De Gravitatione," to be dated at some time just before the writing of the *Principia*, Newton introduces a quantity which is determined by the factors of specific gravity and bulk. Here, however, he is not defining quantity of matter, but rather weight. He declares that, "speaking absolutely" ("absolute loquendo"), gravity ("gravitas") is that which results from the "specific gravity and the bulk of the gravitating body" ("ex gravitate specifica et mole corporis gravitantis").

In this definition, Newton takes specific gravity as weight per unit volume, rather than the relative weight compared with some standard. Specific gravity was defined in Newton's day (as in ours) as the weight of a volume of any substance compared with the weight of an equal volume of water. This definition of specific gravity is manifest in the long discussion of specific gravity in Harris's *Lexicon Technicum* and in the tables of specific gravities such as those published in the *Philosophical Transactions* (no. 199) reprinted by Harris.

It is to be noted that in this definition, Newton is using "absolute" weight in a sense that is not

Henry Crew has suggested[15] an interpretation of density that does not depend on mass and that would accordingly eliminate the seeming circularity in Newton's definition of mass in the *Principia*. In this reconstruction, there are independent definitions of quantity of matter, or mass, and density. Newton was a convinced atomist; he firmly believed that all matter is composed of some sort of ultimate or fundamental particles. Accordingly, Newton could very well have had in mind that mass is a measure of the number of fundamental corpuscles or particles of matter in a given sample (what he had earlier called "the amount of corporeal matter"). In this case density would be a measure of the closeness with which the component fundamental particles are packed together.

Although Newton did not formally define density anywhere in the *Principia*, he did do so in the set of definitions which he drew up just before composing the *Principia*.[16] Here (in def. 6), he wrote that the density of a body is "the quantity or amount of matter" ("quantitas seu copia materiae") "compared with the quantity of occupied space" ("collata cum quantitate occupati spatii"). In this same document, quantity of matter is defined (def. 1) as "that which arises from its density and volume jointly," which is similar to the text of the *Principia*, except that the latter states additionally that Newton is defining a "measure" of matter. But the definition of density was abandoned almost at once, presumably because Newton recognized the circularity. Accordingly, this definition of density does not appear in either of two drafts of book 1 of the *Principia*.[17]

Quantity of Motion: Momentum (Def. 2) 4.6

Def. 2 states that the measure of motion adopted by Newton is the one arising from the mass and velocity, our momentum. Although Newton does not say so specifically, as he had done in the case of quantity of matter, it is this quantity of motion that he means when, as is often the case, he writes simply of "motion."

---

intuitively obvious, since the result is a relative weight. In any event, as we have seen, in the *Principia*, Newton generally computed relative weights.

Perhaps the most interesting part of this definition is the use of the word "moles" or bulk rather than magnitude or volume. This was the term used more than a half-century earlier by Johannes Kepler in his *Astronomia Nova* (1609).

This particular definition did not appear in *De Motu*, nor in the succeeding set of definitions, nor in the first and second drafts of the *Principia*.

15. Henry Crew, *The Rise of Modern Physics*, rev. ed. (Baltimore: Williams and Wilkins, 1935), pp. 127–128.

16. See my *Introduction*, §4.3.

17. That is, this is not a subject of definition in either the "Lucasian Lectures" or the draft of the *Principia* sent to the printer. See, on this topic, Herivel's *Background*, pp. 25–26.

For example, in law 2, Newton writes that a "change in motion" is "proportional to the motive force." Here he means "change in the quantity of motion" or, in our terminology, change in momentum.

4.7 *"Vis Insita": Inherent Force and the Force of Inertia (Def. 3)*

Def. 3 is, in many ways, the most puzzling of all the definitions in the *Principia*. Its subject is what Newton calls a "vis insita" (often translated as "innate force"), which we have called "inherent force," said by Newton to be another name for "vis inertiae," or "force of inertia." This "force," according to Newton, is a "power" that is in "every body" and that, subject to one limitation, causes the body to persevere in its "state either of resting or of moving uniformly straight forward," that is, moving ahead in a straight line at constant speed. The limitation is expressed by Newton in the phrase "quantum in se est," often translated by "as much as in it lies," which has the sense of "so far as it is able" and implies that there may be circumstances that prevent the body from so persevering. Examples might be contact with another body or collision with another body, the action of an outside force, immersion in a resisting medium, and so on. This definition is of notable interest because it implies the first law of motion, which does not appear in the *Principia* until a little later on. Today's reader will also be struck by the fact that Newton uses the word "force" in relation to "inertia" ("vis inertiae"), although—as Newton is at pains to explain—this is an internal force and not the kind of force which (according to the second law) acts externally to change a body's state of rest or of motion. Unless we follow Newton's instructions and make a sharp cleavage between such an internal "force" and external forces, we shall fail to grasp fully the Newtonian formulation of the science of dynamics.

Much has been written about the meaning and significance of Newton's concept of "vis insita." For reading and understanding Newton's *Principia*, however, there is no need to explore the many possible philosophical implications of this term. The name "vis insita" was not an invention of Newton's, in the sense that we have seen that "mass" was. This term had already been used in writings on the physical sciences, notably by Kepler, and could be found in dictionaries that were standard reference works at the time when Newton was writing the *Principia*.[18]

In the draft of "Definitiones" and "Leges Motus," dating from just before the writing of the *Principia*,[19] Newton essayed a def. 12 of "Corporis vis insita innata et essentialis." A few years later, in a letter to Richard Bentley, Newton wrote (in

18. See, further, §4.8 below.

19. See n. 2 above.

English) of an internal force as "innate, inherent, and essential,"[20] which I believe gives sufficient warrant for translating Newton's "vis insita" by "inherent force," rather than—as is customary—by "innate force," which has decidedly Cartesian overtones and raises many philosophical questions which we need not explore here.[21]

Further support for this rendering of "vis insita" is given in John Clarke's *Demonstration of Some of the Principal Sections of Sir Isaac Newton's Principles of Natural Philosophy*. In his translation of Newton's def. 3, Clarke wrote of Newton's "vis insita" as "the inherent Force of Matter."

Newton's definition of inherent force has an aspect in common with the prior definitions of quantity of matter and quantity of motion in that Newton is not inventing the concept but rather giving his own interpretation of a concept then in use. In the case of "vis insita" in def. 3, Newton says, this "force" does not "differ in any way" from "the inertia" of a body's "mass" except that it usually has a different conceptual basis. He then declares that because of the "inertia of matter," all bodies resist any change in state and so only with difficulty are made to change their state—whether of resting or of moving straight forward. Because of this resistance arising from the "inertia of matter," Newton says that he will introduce a new name for the "vis insita" and will call it a "vis inertiae," or force of inertia. He then explains how this "force" is exerted by a body in order to resist a change in state.

In the Lucasian Lectures, the preliminary version of the *Principia*, Newton first wrote, "3. *Materiae vis insita* est potentia resistendi . . . ," then inserted "inertia sive," so as to make the sentence read: "*Materiae vis insita* est inertia sive potentia resistendi . . . " That is, he would have def. 3. begin, "Inherent force of matter is inertia or the power of resisting . . . " He later changed his mind and canceled this insertion. In the final text of the *Principia*, this same idea is expressed in both the first and the third sentence of the discussion of def. 3, where "vis insita" is identified as "vis inertiae." Later on, in book 3, rule 3, introduced in the second edition of the *Principia*, Newton affirmed once again, "Per vim insitam intelligo solam vim inertiae," that is, "By inherent force I mean only the force of inertia."

Since Newton has said that he will use the name "vis inertiae" for "vis insita," we should not expect to find many occurrences of the latter in the main body of the *Principia*. And indeed this expression does not appear very often, there being only

20. Newton to Bentley, 25 February 1692/93, *Corresp.* 3:254.

21. Of course, the rendering "inherent force" has problems of its own, since the Latin word for "inherent" is "inhaerens"; in the *Principia*, however, Newton never uses the adjective "inhaerens." In Cicero's *Tusculan Disputations* (4.11.26), reference is made to an "Opinatio inhaerens et penitus insita," that is, a conjecture or opinion that is "inherent and deeply implanted."

fifteen such occurrences in the *Principia*. In each case, it is plain that "vis insita" or some variant form means no more than "vis inertiae" and is used synonymously with that latter expression.[22]

Since this "force of inertia" is passive, it cannot (of and by itself) alter a body's state of motion or of rest. This is merely another way of saying that because of the inertness of matter, a body cannot (of and by itself) change its own state of motion or of rest. But whenever an external force has acted to change a body's state of motion or of rest, the body (according to def. 4) "perseveres in any new state solely by the force of inertia."[23]

Reams of paper have been expended on elucidating Newton's point of view and expression with regard to both "vis insita" and "vis inertiae." We need not take account of these extensive discussions of Newton's "vis insita." Following Newton's directions, we have license to replace "inherent force" by "force of inertia." Thus the only question at issue is the one that has been mentioned in the preceding chapter: that by introducing the "force" of inertia rather than simply "inertia" as a property of matter,[24] Newton (if only on an unconscious or psychological level) has not fully abandoned the ancient notion that every motion must require a "mover" or some kind of moving force, even if a very special kind of internal force. We must not, however, confuse this force with other and more traditional forces so as, for example, to use the parallelogram law to find the resultant force of the combination of a force of inertia and a centripetal force.[25] Nor should we suppose that this force has any properties other than those specified by Newton.

22. There is an occurrence of "vis insita" in the scholium to the laws and only three others in the whole of book 1. One of these is in the opening sentence of the proof of prop. 4 as given in the first edition; this was eliminated when a new proof was introduced in the second edition. The other two (in prop. 1 and prop. 66, corol. 20) could just as easily have been replaced by "force of inertia," since both merely refer to a body's motion "according to its inherent force" ("vi insita"; "per vim insitam").

In the two final editions, in addition to these five occurrences (twice in def. 3, once in the scholium to the laws, and once in prop. 1 and in corol. 20 to prop. 66), the term appears ten more times: nine times in book 2 and once in book 3, rule 3, where "vis insita" is said to be merely "vis inertiae."

In book 2, "vis insita" is used in the opening sentence of the corollary to prop. 1, in the scholium to lem. 3, twice in the general scholium at the end of sec. 6, in the proof of prop. 53, and in the statement of props. 2, 5, 6, and 11. Of these nine occurrences of "vis insita," all but two are of the form "sola vi insita" or some variant of it, and all are used synonymously with some form of "vis inertiae."

23. In the first draft of the *Principia*, as in the printed version, law 1 does not mention "vis insita," which had been relegated to the definitions.

24. On this point, see the important study by W. A. Gabbey, "Force and Inertia in Seventeenth-Century Dynamics," *Studies in History and Philosophy of Science* 2 (1971): 32–50.

25. Newton combines forces by using a vector addition of the displacements they produce in some given time. This topic is presented in §5.3 below. Newton, therefore, does not combine a vis insita and an accelerating force, although he may combine the motion in some given time that is associated with each of these forces.

Numerous examples come to mind to illustrate what Newton means when he says that this "force" is exerted by a body only when the body resists a change in state. Consider a heavy truck moving on horizontal rails (so that friction is minimized). A great effort is required in order to get the truck moving from rest (as if one is overcoming a force of resistance to a change in state), but it takes relatively little effort to keep the truck moving. Again, to bring the moving truck to a halt requires an effort that makes it seem as if there is a force to be overcome. A similar example[26] would be a forward-moving projectile that hits a wall, at which point a force is exerted which may be great enough to make a hole in the wall.

Newton's Acquaintance with "Vis Insita" and "Inertia" 4.8

It has been mentioned that in def. 3, Newton was not introducing a new concept, as some interpreters have assumed, but was rather giving his own definition of a term then in use, just as he had given his own definition of both "quantity of matter" and "quantity of motion." Although by no means a common expression, "vis insita" was in fairly general usage in Newton's day and appears in a number of books with which he was familiar. For example, this term occurs in Henry More's *The Immortality of the Soul* (London, 1679), a work that Newton read carefully, making notes as he went along, referring to him as the "excellent Dr Moore [i.e., More] in his booke of the soules immortality."[27] In this work there is a reference to an "innate force or quality (which is called heaviness) implanted in earthly bodies." This is rendered in a later Latin version as "innatam quandam vim vel qualitatem (quae Gravitas dicitur) corporibus terrestribus insitam." This literal sense of "implanted" is not the only one in use; traditionally, "insitus" was used in a general fashion for a quality that is "inherent" or "natural." This usage goes back at least to Cicero, in whose *De Natura Deorum*[28] and other works it appears frequently. The phrase "vis insita" also occurs in an ode of Horace.

Another author in whose writings Newton encountered the expression "vis insita" is Johann Magirus, who wrote a handbook of Aristotelian philosophy which

26. Suggested to me by George Smith.

27. Henry More, *The Immortality of the Soul, So Farre Forth as it is Demonstrable from the Knowledge of Nature and the Light of Reason* (London: printed by J. Flesher for William Morden, 1659), p. 192. More's book was translated into Latin as *Enchiridion Metaphysicum; sive, De Rebus Incorporeis Succincta & Luculenta Dissertatio*, in vol. 1 of his *Opera Omnia* (London: typis impressa J. Macock, 1679). Newton's college notebook is marked "Isaac Newton/Trin: Coll Cant/1661" and is listed in the University Library, Cambridge, as MS Add. 3996 (see fol. 89r).

28. There is a long and detailed note on this subject in Arthur Stanley Pease's edition of Cicero's *De Natura Deorum*, vol. 1 (Cambridge: Harvard University Press, 1955), pp. 289ff.

was studied by Newton as a Cambridge undergraduate. According to Magirus, "Motion is *per se* or proper when a movable body moves by its own power [*sua virtute*]: thus man is said to move *per se* because he moves wholly and by his innate [or inherent] force [*insita vi sua*]."[29] There are other occurrences of "insitus" or "insita" in Magirus's text, in association sometimes with "vis" and at other times with "virtus." This word also appears in the Latin translations of Aristotle's *Ethics*, a work studied by Newton as an undergraduate.[30]

The term "vis insita" occurs in the popular philosophical dictionary by Rudolph Goclenius, as part of his discussion of the varieties of "vis." Here he writes that "Vis insita est, vel violenta. Insita, ut naturalis potestas."[31] That is, "Force is inherent [or innate] or violent. Inherent, as a natural power."

This term occurs frequently in Kepler's *Astronomia Nova* (1609) and in his *Epitome Astronomiae Copernicanae* (1618–1621). While there is no evidence as to whether—and if so, to what degree—Newton had read either of these two works, he might well have encountered this notion in the literature.[32]

The source of Newton's term "inertia" has always been something of a puzzle because of the date traditionally assigned to Newton's essay "De Gravitatione." In the course of my research I had found in this essay a clue to Newton's encounter with the term "inertia." Since we now know that Newton wrote this essay just before composing the *Principia*, we may understand why he introduced the term "inertia" only at the stages of actually writing the *Principia*, and not in prior statements about force and motion. It was difficult to imagine how he would at this last moment of creation have turned back to a subject explored some two decades earlier. Newton's "De Gravitatione" is primarily a critique of Descartes's *Principia*, but it also contains a reference by Newton to Descartes's published correspondence, in particular to an exchange of letters between Descartes and Marin Mersenne about the physics of motion. Following this clue I discovered that one of the topics of this epistolary exchange was "inertia" and "natural inertia," an "inertness" of matter that would tend to make bodies come to rest whenever the moving force ceases to act. Presumably Mersenne was asking Descartes about Kepler's use of the concept "inertia" although neither he nor Descartes refers to Kepler by name. Although the Cartesian letters took a critical position with respect to this concept, here was

29. Johann Magirus, *Physiologiae Peripateticae Libri Sex, cum Commentariis* . . . (Cambridge: ex officina R. Daniels, 1642), book 1, chap. 4, §28. Newton actually copied out the first part of this quotation in his notebook.

30. Book 2, chap. 1, 1103a. Specifically, this word occurs in Magirus's version of Aristotle's *Ethics*.

31. Rudolph Goclenius, *Lexicon Philosophicum, Quo Tanquam Clave Philosophiae Fores Aperiuntur* (Frankfurt: typis viduae Matthiae Beckeri, 1613), p. 321.

32. On the history and diffusion of the concept of "vis insita," see my *Newt. Revolution*.

an obvious source of Newton's acquaintance with the term "inertia" in a physical context, specifically the context of motion. The new dating of this essay by Betty Jo Dobbs[33] implies that Newton encountered the term "inertia" just as he was getting ready to produce the *Principia*. Since, in "De Gravitatione," he was taking a strong anti-Cartesian position, he might even have taken pleasure in adopting a concept that Descartes had spurned.[34]

Descartes did not, so far as I have been able to determine, use the term "inertia" or "natural inertia" in his writings on motion (except in his correspondence); he certainly did not refer to "inertia" or "natural inertia" in his presentation of the laws of motion in his *Principia Philosophiae*. Newton apparently liked the idea of the "inertness" of gross matter and took it over for his own. Newton's concept of inertia introduced an important alteration that made a significant transformation of the concept discussed by Descartes. "Inertia" in Descartes's usage was a property of "inertness," whereby matter could not of and by itself keep moving when the moving force ceases to act. Newton's concept ("inertia materiae") implied a "force" of maintaining whatever state a body happens to be in, whether a state of rest or of uniform rectilinear motion. With the elimination of the "force," Newton's concept became the standard of classical physics.

When Newton introduced the term "inertia" he did not know who had introduced this term into the study of motion, presumably because neither Mersenne nor Descartes had mentioned Kepler in this context. Some years after the publication of the *Principia*, however, Newton was criticized by Leibniz (in the latter's *Théodicée*) for having taken this concept from Kepler,[35] who had indeed used the term "inertia" in relation to physics. But Kepler argued for the position disdained by Descartes, that because of the "inertness" of matter, a body would come to rest wherever it happened to be when the moving force ceases to act. When Newton read Leibniz's criticism, he folded the corner of the page of his personal copy of the *Théodicée* to mark the lines in question. He also entered into his own copy of the *Principia* a proposed emendation of def. 3. He would now add a sentence reading: "I do not mean Kepler's force of inertia by which bodies tend toward rest"; rather, he meant "the force of remaining in the same state, whether of resting or of moving." In the event, he decided that this alteration was unnecessary.[36]

33. *Janus Faces* (§3.1, n. 10 above).

34. I am happy to be able to advance this additional argument in support of Prof. Dobbs's new dating of "De Gravitatione."

35. I. B. Cohen, "Newton's Copy of Leibniz's *Théodicée*, with Some Remarks on the Turned-Down Pages of Books in Newton's Library," *Isis* 73 (1982): 410–414.

36. Ibid.

4.9 *Impressed Force; Anticipations of Laws of Motion (Def. 4)*

In the presentation of his concept of "impressed force," in def. 4, Newton once again makes use of a traditional (this time a late medieval) concept, but again in a wholly novel way. Impressed force is said to be the "action"—and only the action—that is "exerted on a body to change its state," whether a state of rest or of motion. It is here, after this partial reference to the second law (insofar as forces act to change a body's state of motion or of rest), that Newton anticipates the first law by stating that "a body perseveres in any new state solely by the force of inertia." Newton then says that impressed force has various sources, among them percussion (or impetus), pressure, and centripetal force.

In the remaining parts of the definitions, Newton discusses only centripetal force, which was a new concept, bearing a name that he had invented and that would be unfamiliar to his readers. He assumed that there was no need to say anything about percussion and pressure.

4.10 *Centripetal Force and Its Three Measures (Defs. 5, 6–8)*

In def. 5 of the *Principia*, Newton introduces the term "centripetal force," a name that he invented for the kind of force by which "bodies are drawn from all sides, are impelled, or in any way tend, toward some point as to a center." He gives three examples: terrestrial gravity ("by which bodies tend toward the center of the earth"), magnetic force, and "that force, whatever it may be, by which the planets are continually drawn back from rectilinear motions and compelled to revolve in curved lines." Newton then gives a variety of examples, including a stone whirled in a sling, the motion of a projectile, an artificial satellite, and the moon. The newness of this concept may be seen in the very fact that so many examples are needed to illustrate it.

Although Newton gives a number of physical examples, he also indicates that—at least at this stage of his presentation—his concern is primarily mathematical and not physical; that is, he is developing the mathematical properties of centripetal forces rather than investigating their physical nature, mode of action, or physical properties. Accordingly, Newton says that his assignment in the *Principia* is the task of mathematicians, namely, to solve two mathematical problems: to find the mathematical law of force that keeps a body "exactly in any given orbit with a given velocity" and to find "the curvilinear path into which a body leaving any given place with a given velocity is deflected by a given force."

In the final three definitions (defs. 6, 7, 8), Newton introduces the three measures he will use in considering centripetal forces. He refers to the "quantity" of such a force rather than measure, using "quantity" in the sense of defs. 1 and 2, where "quantity" of matter was its "measure" (or mass) and "quantity" of motion was its "measure" (momentum).

The accelerative quantity (def. 7) of centripetal force is that measure "proportional to the velocity which it generates in a given time." Here Newton has to introduce "in a given time" because he is dealing with a continuous rather than an impulsive force. Since this measure is proportional to the velocity generated in a given time, it is proportional to the acceleration produced, as Newton explicitly notes. Basically, the accelerative measure is the effect of a given force that would act on a unit mass, the quantity we now refer to as the field. It should be noted that in this definition, Newton is subsuming the second law of motion for continuous forces, even though he will not formally announce the second law until the following section of the *Principia*. An example of this measure, according to Newton, is gravity, which "equally accelerates all falling bodies" (at any given place) save for air resistance. This is the measure which Newton generally has in mind in book 1, up to sec. 11, whenever he refers to force.

A second such measure (def. 8) is the "motive quantity" of a force. This measure is proportional to the quantity of motion generated "in a given time," that is, the momentum generated in a given time. An example is weight, which Newton, writing as a mathematician, calls "centripetency, or propensity toward a center." Since quantity of motion, or momentum, is proportional to mass, so the motive quantity of a force must be proportional to mass. Once again, we note the anticipation of the second law of motion in these definitions.

Newton has a third measure, the "absolute quantity," said to be that measure which is proportional to "the efficacy of the cause propagating it from a center through the surrounding regions." In def. 6, Newton elucidates this concept in terms of the magnetic force exerted by a lodestone. The magnitude of this force differs from one lodestone to another and depends on size (a large lodestone generally being more powerful than a small one) and on the intensity of the magnetic force (which can differ in lodestones of the same size).

Throughout the *Principia*, Newton will refer to these measures, "for the sake of brevity," as "motive, accelerative, and absolute." Newton insists once again that his concept of force "is purely mathematical," that he is "not now considering the physical causes and sites of forces."

The sense of these three measures of force can be gathered by considering the measures in relation to the gravitational force. The "accelerative quantity" is the measure of the gravitational force "proportional to the velocity which it generates in a given time." In modern terms, this is $\frac{dv}{dt}$, or the acceleration. On earth, this measure is the familiar g, often called the acceleration of gravity. By Galileo's experiments and Newton's dynamical analysis, at any given point on the earth and in the absence of air resistance, this is the acceleration of free fall for all bodies, irrespective of their difference in mass. On the surface of the moon, or of any of the planets, or of the sun, this acceleration of gravity or of free fall has a quite different value.

This measure is the one used most frequently in the *Principia*, usually without specific indication that it is this measure of force which Newton intends. For example, in the corollaries to book 3, prop. 8, Newton computes the "weight" or "gravity" (or gravitational force) exerted by the sun, Jupiter, Saturn, and the earth on "equal bodies," that is, bodies of equal masses, at the distance respectively of Venus, a planetary satellite, or the moon from the centers of their orbits. He also computes this "weight" on the surfaces of the sun, Jupiter, Saturn, and the earth as well as at equal distances from the centers of the sun, Jupiter, Saturn, and the earth. Since Newton is dealing with relative rather than absolute quantities, this weight of equal bodies is equivalent to the weight per unit mass. According to Newton's second law $F = mA$, the accelerative measure of the force is $\frac{F}{m} = A$ or, in the case of the weight force, $W = mg$, yielding $\frac{W}{m} = g$. For constant masses m, or for a unit mass, the measure of F is A and the measure of W is g.

In the case of gravitation, the "motive quantity" is by definition the measure of the gravitational force "proportional to the [quantity of] motion which it generates in a given time." In modern terms, this is $\frac{d(mv)}{dt}$, or the force itself that acts on a particular body, that is, its weight. Since m is a constant for any given body in classical Newtonian physics, this measure of a force $\frac{d(mv)}{dt}$ is equal to $\frac{mdv}{dt}$, the expression for a force according to the usual form of the second law. Here we may see vividly how Newton anticipates the axioms or laws of motion in the prior definitions.

In the case of weight or gravitational force, then, the accelerative quantity or measure is the acceleration of free fall. Although this is the same for all bodies placed in some given situation with regard to the earth or other central body, it differs from one latitude on earth to another. It also differs on the surface of the earth, of the moon, of a planet, or of the sun, or at varying distances from the center of such a central body. The motive quantity or measure of the gravitational force turns out to be its actual weight toward the central body, whether the earth, the moon, the sun, the planets, and so on.

The absolute quantity or measure is a little harder to understand. In terms of the gravitational force, however, we can see how this concept might arise. The weight force experienced by a body of given mass is different on the surface of the earth, the moon, the sun, or any of the planets. In the previous example, the corollaries to book 3, prop. 8, Newton computes such different values of the gravitational force as it acts on bodies of equal mass on those surfaces. He also computes the values of the gravitational forces at equal distances from the centers of the sun, Jupiter, Saturn, and the earth. These values differ greatly among themselves, the one for the sun being several orders of magnitude greater than the others. One way of describing this situation is to say that the sun is more efficacious in production of the gravitational force, or has a greater gravitational "potency," than the other three. In Newton's language, the absolute quantity or measure of the gravitating force of the sun is greater than the similar force of the other three.

In the particular example of gravitation, the factor causing the difference in absolute quantity is the mass. And, indeed, in the corollaries to prop. 8, Newton uses this fact in order to compute the relative masses of the sun, Jupiter, Saturn, and the earth. But for other types of force, this variation in potency is not caused by a difference in mass. The example Newton gives is that of magnetic force, which varies from one lodestone to another. Lodestones differ greatly in their "potency," that is, in their ability to produce magnetic force, but this variation is not a direct function of mass.

Throughout most of the *Principia*, especially the first sections of book 1, Newton makes use of the accelerative measure, essentially dealing with unit masses and single force fields. Generally speaking, whenever Newton introduces "force" without any qualification, he intends accelerative measure. In some cases, however, he does introduce absolute forces. An example is book 1, prop. 69, where Newton postulates that a body A attracts several others (B, C, D, . . .) "by accelerative forces that are inversely as the squares of the distances," while some other body (B) also attracts the rest of the bodies (A, C, D, . . .) according to the same law of force. In this case, according to prop. 69, "the absolute forces of the attracting bodies A and B will be to each other in the same ratio as the bodies [i.e., the masses of the bodies A and B]." Absolute forces are also introduced in book 1, prop. 65, case 1; book 1, prop. 66, corol. 14, and elsewhere. Masses are introduced explicitly in book 1, sec. 11, in prop. 57 and elsewhere. Thus the level of physical discourse undergoes a major shift in sec. 11, going from "accelerative" to "motive" forces and thereby inaugurating the science of dynamics. From a strict point of view, the subject of book 1, secs. 1–10, is largely kinematics despite the use of the term "force."

Newton concludes with a paragraph stressing once again that when he uses "words signifying attraction, impulse, or any sort of propensity toward a center," he is considering such forces "not from a physical but only from a mathematical point of view." This is the sense in which he will "call attractions and impulses accelerative and motive." The reader is warned not to think that by using words of this kind Newton is "anywhere defining a species or mode of action or a physical cause or reason" or "attributing forces in a true and physical sense to centers (which are mathematical points)" if he should "happen to say that centers attract or that centers have forces."

4.11 *Time and Space in the* Principia; *Newton's Concept of Absolute and Relative Space and Time and of Absolute Motion; The Rotating-Bucket Experiment*

The section of definitions concludes with a scholium on space and time. The actual subjects discussed are time, space, place, and motion. Newton begins by distinguishing "these quantities into absolute and relative, true and apparent, mathematical and common." In the ensuing discussion, Newton deals with "absolute, true, and mathematical time," as compared with "relative, apparent, and common time"; but for space, place, and motion, the distinction is between "absolute" and "relative." In arguing for an absolute time, Newton cites the example of astronomical time in which there is a "true" time, absolutely moving forward at a uniform rate; this is called mean time, a mathematical concept. In practice, astronomers observe local or apparent time and compute an "equation of time" which yields the difference between the two at any given moment. As we shall see below, in discussing the laws of motion in §5.4, the concept of time in Newton's formulation of his method of fluxions is the same as the one employed here.

Newton's pure mathematics was apt to be expressed in the language of motion, including such kinematical principles as the parallelogram law. To such an extent is this the case that the reader must constantly exercise the greatest caution in order not to misread a purely mathematical text as if it were an exercise in physical kinematics.[37] The fundamental independent variable in this mathematics is time, a uniformly flowing quantity, with properties like the absolute time set forth in the scholium. It is with regard to this time, in mathematics, that Newtonian fluxions such as x and y are determined; the $\frac{dx}{dt}$ or $\frac{dy}{dt}$ of the Leibnizian calculus are for Newton the fluxions $\dot{x}$ and $\dot{y}$, to use Newton's later notation.

Newton believed in an absolute space, just as he believed in an absolute time, but he was fully aware that in common practice, as in physical science, we make use of relative rather than absolute space. Indeed, toward the end of the scholium, Newton returns to common usage and its problems. It is here that he distinguishes

37. For details see the early volumes of Whiteside's edition of *Math. Papers* and my *Newt. Revolution*, pp. 55–56.

between actual "quantities being measured" (or quantities themselves) and their "sensible measures." It was in this context that we have seen (§2.3 above) how Newton's comparison with Scriptures was converted into "accuracy of language."

In the course of this presentation, Newton presents two experiments, often wrongly believed to provide the basis of his argument rather than to serve as an illustration of it. These two experiments both deal with the centrifugal effects of rotation and have often been treated as different aspects of the same point in Newton's argument, even though there is a lengthy discussion that separates them. The second of these, occurring in the final paragraph of the scholium, is apparently a "thought experiment," one that he does not say he has performed or witnessed. In the experiment, two balls, "at a given distance from each other with a cord connecting them," are to be whirled about their common center of gravity. In this case, Newton writes, "the endeavor of the balls to recede from the axis of motion could be known from the tension of the cord," thereby yielding "the quantity of circular motion."

The other one, appearing earlier in the scholium, is the celebrated rotating-bucket experiment, in which a bucket partly filled with water is suspended by a cord. The experimenter turns the bucket around until "the cord becomes twisted tight." When the experimenter releases his hold and allows the string to unwind, the water in the rotating bucket "will gradually recede from the middle and rise up the sides of the vessel, assuming a concave shape." As the bucket rotates faster, the water at the sides will "rise further and further," and the more concave the surface of the water will become. The "rise of the water" up the side of the bucket, according to Newton, "reveals its endeavor to recede from the axis of motion," so that "such an endeavor" enables one to "find out and measure the true and absolute circular motion of the water." In reporting the results of this experiment, Newton uses the phrase "as experience has shown me," indicating that he has actually performed this experiment.

One very interesting aspect of these experiments is that in discussing them, Newton makes use of Cartesian language. Newton's reference to a "conatus recedendi ab axe motus," or "endeavor to recede from the axis of motion," is taken directly from Descartes's *Principia*.[38] This Cartesian phrase occurs in the description of both experiments.

Newton concludes this scholium with a notice that in the text which follows he will explain more fully "how to determine true motions from their causes, effects, and apparent differences." Conversely, he will explain "how to determine

38. See my "Newton and Descartes" (§3.1, n. 4 above).

from motions, whether true or apparent, their causes and effects." This was the purpose, Newton concludes, for which he had "composed the following treatise."

A recent analysis by Robert Rynasiewicz[39] shows that Newton's strategy of argument in this scholium is often misunderstood, it being supposed that Newton sought to establish "the existence of absolute motion" on the basis of "the centrifugal effects manifested in such instances as the rotating bucket experiment and the example of the globes." Rynasiewicz shows that the introduction of the rotating-bucket experiment comes at the end of "a sequence of five [arguments], all of which seek to show that the true rest or motion of a body cannot be defined as some preferred type of motion relative to other bodies as had been proposed, notably, by Descartes."[40] The "cumulative purpose" of these five arguments, Rynasiewicz concludes, is to convince the reader that "true motion and rest can be adequately understood only with reference to motionless places, and hence to absolute space as characterized in the scholium." The discussion of the whirling balls comes only after the main argument "has been brought to a close." Its purpose is to illustrate that there are cases in which one "can acquire evidence concerning the true motion of individual bodies, given that the absolute space in which they move cannot be perceived."

39. Robert Rynasiewicz, "By Their Properties, Causes and Effects: Newton's Scholium on Time, Space, Place and Motion," *Studies in History and Philosophy of Science* 26 (1995): 133–153, 295–321.

40. This may explain why Newton, in countering Descartes's arguments, tended to use Cartesian language and concepts.

CHAPTER FIVE

Axioms, or the Laws of Motion

Newton's Laws of Motion or Laws of Nature 5.1

Newton's three laws of motion were set forth under the general heading "Axiomata, sive Leges Motus," or the "Axioms, or the Laws of Motion." As mentioned (§3.1 above), it is difficult to believe he was not (even if unconsciously) making a direct improvement on the laws announced by Descartes in his *Principia* as "Laws of Nature" or "Regulae quaedam sive Leges Naturae." On at least one occasion, in a letter to Roger Cotes during the preparation of the second edition of the *Principia*, Newton inadvertently called one of his laws of motion by Descartes's name, "Law of nature."

Newtonians generally called Newton's laws of motion the "laws of nature." For example, the major textbook of Newtonian physics, W. J. 'sGravesande's *Introduction to Sir Isaac Newton's Philosophy*, presented the laws of motion in a chapter called "Concerning the Laws of Nature."[1] In the Newtonian *Lexicon Technicum* of John Harris (London, 1704), we find—under "MOTION, *Its Laws*"—the three laws, introduced by the statement: "The Incomparable Mr. *Isaac Newton* gives but these Three Laws of Motion, which may be truly called *Laws of Nature*."[2]

The First Law; Why Both a First Law and a Second Law? 5.2

The first law states the law or principle of inertia which Newton had learned from Descartes's *Principia*, where it is also a first law. For Descartes, as has been

1. W. J. 'sGravesande, *Mathematical Elements of Physicks, prov'd by Experiments, being an Introduction to Sir Isaac Newton's Philosophy*, revised and corrected by John Keill (London: printed for G. Strahan . . . , 1720), vol. 1, book 1, pt. 1, chap. 16.

2. Jane Ruby, "The Origins of Scientific 'Law,'" *Journal of the History of Ideas* 47 (1986): 341–360.

mentioned, the statement of the principle of inertia requires two laws rather than one, a feature which indicates that Descartes's "laws of nature" are not formulated in a way that would enable him to infer forces from motions, as Newton does in his *Principia*. Often the question has been raised why there are both a first and a second law since, as Rouse Ball put it, the first law "seems to be a consequence of the second law," so that "it is not clear why it was enunciated as a separate law."[3] That is, if the second law is that $F = mA$ or $F = \frac{mdV}{dt} = \frac{d(mV)}{dt}$, then it follows that if $F = 0$, $A = \frac{dV}{dt} = 0$. The only trouble with this line of thought is that $F = mA = \frac{mdV}{dt}$ is the second law for a continually acting force F, whereas Newton's second law (as stated in the *Principia*) is expressed in terms of an impulsive force. For such a force, the "force" is proportional to the change in momentum and not the rate of change in momentum. Thus a possible clue to Newton's thinking is found in the examples used in the discussion of the first law, analyzed below, each one of which is (unlike the forces in law 2) a continually acting force. Accordingly, we may conclude that law 1 is not a special case of law 2 since law 1 is concerned with a different kind of force.

A more interesting question to be raised, however, is why Newton believed he needed a first law since the principle of inertia had already been anticipated in def. 3 and also in def. 4. It would seem that Newton's first law was not so much intended as a simple restatement of the principle previously embodied in def. 3 as it was a condition for the existence of certain insensible forces, not otherwise known to us. The most significant such force for Newton was the centripetal force acting on the moon and planets, which he eventually would identify with gravity; we are aware of such a force only because the moon and planets do not exhibit the uniform rectilinear motion that occurs when there is no external force. That is, our awareness of such a force is based on the first law and the observed fact that the planets do not follow a uniform rectilinear path.

No doubt, another factor in Newton's decision to have a separate law 1 and law 2 was the model he found in Huygens's *Horologium Oscillatorium* of 1673, a work he knew well. In part 2 of that work, Huygens began the analysis of gravitational dynamics with three laws which he labeled "hypotheses," in the same way that Newton—a decade later—in *De Motu* would set forth the laws of motion as "hypotheses." Huygens's first one was somewhat similar to Newton's first law, stating that if "gravity did not exist, and no resistance of air opposed the motion of a body," a body once set in motion "would continue its motion with a uniform

3. *Essay*, p. 77.

velocity in a straight line." Huygens's second "hypothesis" introduced the action of gravity (whatever may be its cause), as a result of which "the body's motion will be compounded of the original uniform motion and the falling motion produced by gravity." Huygens's separation of projectile motion into two parts—the first a uniform motion in the absence of any external forces and the second a nonuniform motion added to it[4]—may very well have been a factor in Newton's decision to formulate his own first two laws in the way he did. It is to be noted, however, that the conditions of Huygens's two hypotheses are somewhat different from those of Newton's first two laws.[5]

After stating the first law, Newton gives three examples of the longtime persistence of linear motion: a projectile, a spinning hoop or top, and the orbital motion of bodies. On first encounter, these examples may seem confusing since each involves curved paths and yet the subject of the first law is uniform linear (or rectilinear) motion. But Newton shows, in the examples of projectiles and spinning hoops (or tops), that it is only the tangential or linear component that is inertial, not the curved motion. A spinning object keeps spinning because the particles of which it is composed act "by their cohesion" and thus "continually draw one another back from rectilinear motions." In this explanation, Newton is assuming the analysis of curved motion into a tangential inertial component and a component of falling inward. We may take note here that Newton refers to the force of cohesion among particles as the accelerative force.

The Second Law: Force and Change in Motion 5.3

Newton's second law, as stated in the *Principia*, sets forth a proportionality between a "force" and the resulting "change in motion," by which Newton means change in quantity of motion or change in momentum. Since this is not the more familiar version of the second law, in which a force produces an acceleration or a change in momentum in a given (or unit) time, some writers have seen a need to introduce a correction to Newton's statement of the law.[6]

Newton's explanation should leave no reasonable doubts that in the statement of this law, "force" means an impulsive force or impulse. Thus, he explains, if a

4. Christiaan Huygens, *The Pendulum Clock; or, Geometric Demonstrations concerning the Motion of Pendula as Applied to Clocks*, trans. Richard J. Blackwell (Ames: Iowa State University Press, 1986).

5. That is, in Huygens's hypothesis 1, the only force under consideration arises from the resistance of the air, while in his hypothesis 2 the only force is gravity.

6. Thus Rouse Ball, without comment, states the law as follows: "*Law 2.* The change of momentum [per unit of time] is always proportional to the moving force impressed, and takes place in the direction in which the force is impressed." It apparently never occurred to him to try to find out what Newton meant, rather than to introduce "per unit of time."

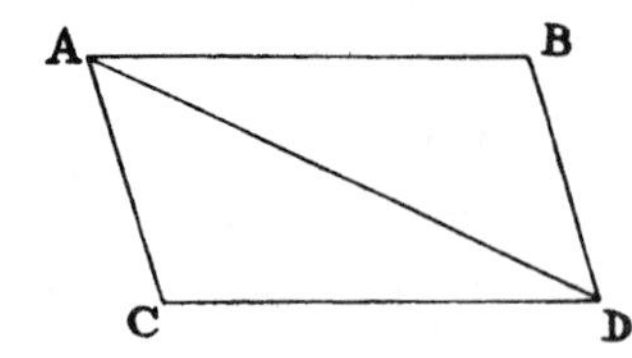

Figure 5.1. The parallelogram rule for the motion produced by an impulsive force (corol. 1 to the Laws of Motion).

force F generates a momentum mV, then a force 2F will generate a momentum 2mV, a force 3F will generate a momentum 3mV, and so on, all of which is true for impulsive forces. And he takes note that it does not matter whether that force be impressed "simul & semel" or "gradatim & successive," that is, "all at once" ("instantaneously") or "successively" ("by degrees"). In other words, the same momentum 3mV will be generated whether a force 3F is applied as a single impulse all at once or as three successive impulses F, or in two successive impulses F and 2F, and so on. A source of confusion to today's reader may be Newton's use here of the adverb "gradatim," which means "gradually" or "by degrees." Today, the English adverb "gradually" has lost its original sense of "step by step," but in Newton's day, "gradatim" or "gradually" still meant "by degrees," by stages or by steps.

That an impulse is the primary sense of force in the second law (as stated) is also made clear in the first application of the law in the succeeding corol. 1. Here (see fig. 5.1) Newton lets a body be acted on by a force M at a point A and then move with uniform motion from A to B; similarly, a force N would cause a uniform motion from A to C. Let both forces be applied simultaneously at A and the body will move along AD, the diagonal of the parallelogram ABDC with sides AB and AC. In all these cases, the force does not act continually during the motion, but only at the point A where it is impressed. Thus Newton specifies that the motion produced by the force is "uniform." He thus invokes def. 4, where he has said that a body maintains, by its "force of inertia," any motion it has acquired.

Of course, Newton knew the second law in the form in which we commonly use it today, that is, as a law for continually acting forces, and used it as such in the *Principia*. Thus, in the proof of book 2, prop. 24, Newton writes that "the velocity that a given force can generate in a given time in a given quantity of matter is as the force and the time directly and the matter inversely." Observe that here Newton adds the phrase "in a given time," which is required for a continually acting force. He then states that what he has been saying is "manifest from the second law of motion." He thus implies that the law for continually acting forces follows from the impulsive version in which the second law has been stated as an axiom.

The primacy of impulsive forces appears notably in the introduction to sec. 11 (book 1), where Newton suggests that the forces of "attraction" should more properly "be called impulses." We may hazard a guess why Newton chose to make impulses primary and continually acting forces secondary or derivative. His ultimate goal in the *Principia* was to explore the actions of universal gravity in various aspects of the system of the world. As has been mentioned, out in space, the existence of such a force of attraction is not manifested in relation to a directly experienced phenomenon, but is only inferred by logic and a theory of rational mechanics, on the grounds that the planets do not move uniformly straight forward. In fact, as Newton knew full well and as events were to prove, many of Newton's contemporaries abhorred the very concept of such a force of attraction. But no one in Newton's day could doubt that a force had been applied in the action of an impulsive force, where almost always a plainly discernible physical event is associated with an observable change in momentum. Thus Newton, by giving primacy to impulsive forces, would have been basing his system of rational mechanics on a concept easily acceptable to his contemporaries, rather than one which implied action at a distance and was of a kind that was supposed by many to have been banished from sound natural philosophy.

Newton never actually made a formal statement of the second law in the algorithm of fluxions or the calculus. The first person to do so seems to have been Jacob Hermann in his *Phoronomia* (1716), in which he writes

$$G = M dV : dT$$

where, he says, "G signifies weight or gravity applied to a variable mass M."[7]

In the scholium to the laws, in a portion added in the second edition, Newton attributed the second law (as well as the first law) to Galileo, even alleging that it was by use of the second law that Galileo had discovered the law of falling bodies. Here Newton showed that he was not aware of how Galileo had presented his discovery.[8] Galileo certainly did not know Newton's first law. As to the second law, Galileo would not have known the part about change in momentum in the Newtonian sense, since this concept depends on the concept of mass which was invented by Newton and first made public in the *Principia*.

7. Jacob Hermann, *Phoronomia; sive, De Corporum Solidorum et Fluidorum* (Amsterdam: apud Rod. & Gerh. Wetstenios, 1716), p. 57; see, further, my *Newt. Revolution*, pp. 143–146.

8. See I. B. Cohen, "Newton's Attribution of the First Two Laws of Motion to Galileo," in *Atti del Symposium Internazionale di Storia, Metodologia, Logica e Filosofia della Scienza: Galileo nella storia e nella filosofia della scienza*, Collection des travaux de l'Académie internationale d'histoire des sciences, no. 16 (Vinci and Florence: Gruppo Italiano di Storia delle Scienze, 1967), pp. xxv–xliv.

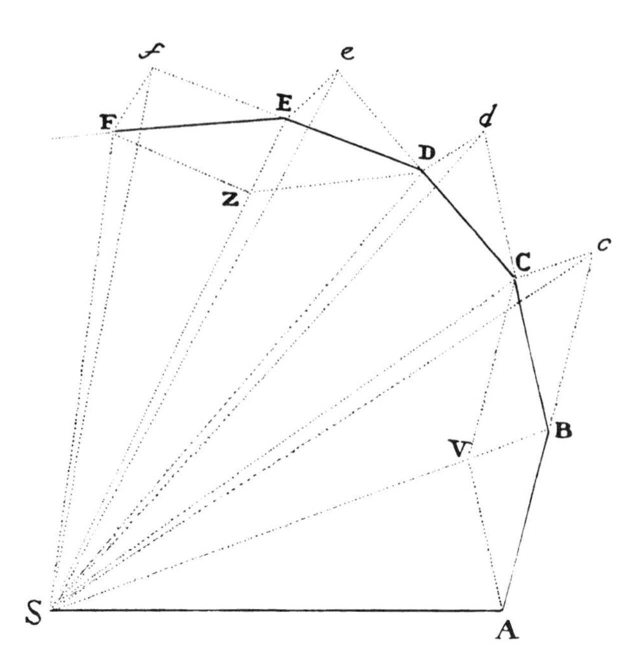

Figure 5.2. The polygonal path produced by a succession of impulsive forces (book 1, prop. 1).

5.4 *From Impulsive Forces to Continually Acting Forces: Book 1, Prop. 1; Is the* Principia *Written in the Manner of Greek Geometry?*

Newton's method of advancing from impulsive to continually acting forces appears in book 1, prop. 1. Newton begins by considering (see fig. 5.2) a body moving with uniform motion along a straight line from A to B in some time *t*. When the body reaches B, it is given a thrust or impulsive force directed toward some point S; this causes an instantaneous change in the body's "quantity of motion," or momentum. That is, according to the second law, there will be a change in both the magnitude and the direction of motion so that the body will now move ahead along the straight line BC with its new speed. In prop. 1, after the passage of another time *t*, when the body reaches C, it once again receives a thrust toward S and now moves along the line CD. In this way, a succession of regularly timed impulses produces a polygonal path ABCD . . . and a series of triangles ASB, BSC, CSD . . . which have equal areas. (For details, including the proof that the triangles have equal areas, see §10.8 below.) Newton then considers the limit as the number of triangles is increased indefinitely and their width diminished indefinitely, whereupon their "ultimate perimeter" will be "a curved line" and the "centripetal force" will "act uninterruptedly." In the limit, any "areas described" will be proportional to "the times of description." Here we see plainly the manner in which Newton proceeds from a succession of instantaneous or impulsive forces to a force which acts continually.

It should be noted that in this first proposition of the first book, Newton has made a statement whose implication is of the greatest importance. He has indicated in the first numbered proposition (as he has done in the antecedent sec. 1) that the *Principia* depends on the theory of limits, that this is not a book written in the style of classical or Euclidean geometry. The reader, presumably, has already gathered from the antecedent sec. 1 that it is only by a superficial examination that the *Principia* may appear to be written like a text in Greek geometry. Be that as it may, the limiting process introduced in prop. 1 of book 1 boldly declares the mathematical character of the *Principia*.

The proof of prop. 1 appears deceptively simple, but a full analysis reveals certain hidden assumptions. These have been brought to light through the research and thought of D. T. Whiteside, first in 1966,[9] then again in 1974,[10] and once more in 1991.[11] Discussing Newton's presentation of the transformation of the polygonal path ABCDEF into a curve ABCDEF, Whiteside pinpoints the problem of how "the aggregate of the infinity of discrete infinitesimal impulses of force acting 'instantly' towards S at 'every' point of the arc BF becomes a force acting 'without break' over the arc BF 'ever' towards the centre S."[12] His critical study of Newton's mode of address to this problem can be warmly recommended to every reader. The difficulties are epitomized in Whiteside's remark that even Halley "did not appreciate that in Newton's argument the chain-polygon ABCD . . . ceases to be angular only in the limit as its component linelets come each to be 'infinitely' small."

In the third paragraph of the scholium following book 1, prop. 4, Newton refers to an earlier proof of his (taken from the Waste Book),[13] in which he had found that for circular motion at uniform speed v in a circle of radius r, the "centrifugal force" is proportional to $\frac{v^2}{r}$, which must be equal to the contrary force.

9. Whiteside, "Newtonian Dynamics" (§2.5, n. 38 above), an essay-review of Herivel's *Background*. Here Whiteside points out that "the assumptions that the continuous action of the centrifugal 'endeavour' may be determined through the combined effect of an infinite number of instantaneous 'pressions,' and that a circle of defined curvature may be approximated by a polygon of infinitely many infinitely small sides, each require that the deviations cC, dD, . . . and the polygon's sides bc, cd, . . . be of a second infinitesimal order."

10. *Math. Papers* 6:35–37 n. 19.

11. Whiteside, "The Prehistory of the *Principia* from 1664 to 1686" (§1.2, n. 14 above), esp. p. 30.

12. Ibid. On p. 31, Whiteside notes that in "Theorem 1 even the smooth arc BF engendered when the force acts 'continuously,' via infinitesimal impulses 'instantly' applied at successive points, must itself be put to be infinitely small."

13. *Background*, pp. 128–130.

In this early proof of the 1660s, Newton first considers a sequence of impulses producing a polygonal path and then proceeds to the limit (having the sides of the polygon be "diminished indefinitely"), so that the sequence of impulses becomes a continually acting force. Here, once again, we see Newton's use of an intuitive limit-process to make a transition from primary impulses to secondary forces that act continually.

Newton's transition from impulsive (instantaneous) to continually acting forces essentially bids us conceive of these forces as a sequence of infinitesimal impulses. In Newton's theory of mathematical time, this quantity flows uniformly and continuously but is itself composed of infinitesimal "particles" of time (essentially "dt"), as Newton says again and again in the *Principia*. This doctrine of time is also of fundamental importance in Newton's theory of the calculus, where Newton indicates that "fluxions" are the "velocities" with which quantities increase or decrease.[14] Because of the uniform flow of mathematical time, and its quasi-discrete character (dt being constant), a number of forms of the second law become equivalent, among them

$$(1) \quad F \propto dV$$

$$(2) \quad F \propto \frac{dV}{dt}$$

Statements (1) and (2) may also be written as

$$(1a) \quad F = k_1 d(mV)$$

$$(2a) \quad F = k_2 \frac{d(mV)}{dt}$$

where k_1 and k_2 are constants of proportionality. Clearly the constants k_1 and k_2 will have different dimensionality, since dt is absorbed into k_1 but not into k_2. We today make a clear-cut distinction between these two forces and would not, therefore, normally use the same letter "F" in both equations 1 and 2; that is, in eq. 1 the force is impulsive, whereas in eq. 2 the force is continually acting.

In the case of the force itself being a variable, "F" must be the average value during time dt. It must be kept in mind that the dimensions of the constant of proportionality would not have posed a problem for Newton, because he expressed

14. See *Math. Papers* 3:17, 73, and esp. 71–72 nn. 82, 84.

the principles of motion in proportions in a rhetorical style and not in equations.[15] He thus had no need to consider the dimensions of constants.

In this context, it should be noted that, somewhat paradoxically, time does flow uniformly in Newton's physics. Accordingly, it would be perhaps more accurate to say that Newtonian time is the measure of the flux that can be divided into infinitesimals of equal "length," rather than to insist that Newtonian time is itself made up of some sort of discrete infinitesimal units.[16] On this score, note that in *De Quadratura*, Newton unambiguously referred to fluxions as "the Augments of the Fluents generated in equal but very small Particles of Time."[17]

The Third Law: Action and Reaction 5.5

Newton's third law states the equality of "action and reaction." This law has been cited as "the most important achievement of Newton's with respect to the principles [of dynamics]."[18] It is the only one of the laws that Newton himself did not assign to his illustrious predecessor, Galileo. Newton seems to have come upon this law while contemplating the varieties of elastic and inelastic collisions which are introduced in the scholium to the laws.[19] The search for the laws of collisions had been a central problem in the seventeenth century, having been the subject of an almost wholly erroneous set of rules given in Descartes's *Principia*.[20]

Newton's third law has often been a source of confusion. It states that if a body A exerts a force F on a second body B, then that body B will exert on body A an oppositely directed force of magnitude equal to F. The common error is to assume that these two forces, equal in magnitude and opposite in direction, can produce a condition of equilibrium. Thus, in arguing for a bicameral legislature at the end

15. See, further, my "Newton's Second Law and the Concept of Force in the *Principia*" (§3.7, n. 73 above), esp. appendix 1 ("Continuous and Impulsive Forces and Newton's Concept of Time") and appendix 2 ("Finite or Infinitesimal Impulses, 'Particles' of Time, Forces, and Increments of 'Quantity of Motion' ").

16. In analyzing some proposed revisions of the second law, I was led at first into error by considering that a certain force ("vis motrix impressa") acts over a finite time interval, rather than over a first-order infinitesimal time *dt*. See the preliminary printing of my article "Newton's Second Law" (referred to in n. 15 above) in *The Texas Quarterly* 10, no. 3 (autumn 1967): 127–157. For this important aspect of Newton's notes on the second law I am indebted to D. T. Whiteside.

17. See also J. M. F. Wright's *Commentary on Newton's "Principia"* (London: printed for T. T. & J. Tegg, 1833), vol. 1, esp. §43, where the significance of "considering *dt* constant" is introduced, along with the warning that the various forms of the second law require "properly adjusting the units of force."

18. E.g., Mach, *The Science of Mechanics* (§4.3, n. 9 above), p. 243.

19. See §1.2, n. 29 above, on Newton's early knowledge of the third law.

20. See William R. Shea, *The Magic of Numbers and Motion: The Scientific Career of René Descartes* (Canton, Mass.: Science History Publications, 1991), pp. 296–299.

of the eighteenth century, John Adams drew on his faulty recollections of college physics to declare that his opponent, Benjamin Franklin, had forgotten "one of Sir Isaac Newton's laws of motion, namely,—'that reaction must always be equal or contrary to action,' or there can never be any *rest*." Adams believed that for the American Congress to attain the kind of equilibrium needed for stability, there must be a balance between an elected House of Representatives and an appointed Senate, and he quite wrongly supposed that his position was supported by analogy with Newton's third law.[21] He had forgotten that the forces in the third law act on different bodies and so cannot produce a balance or an equilibrium.

5.6 *Corollaries and Scholium: The Parallelogram Rule, Simple Machines, Elastic and Inelastic Impact*

The laws are followed by a series of corollaries, of which corol. 1 presents the parallelogram law for impulsive forces, that is, for the velocities that such forces produce. Corol. 2 shows how static forces may be resolved into components and compounded into resultants. This principle is then applied to explain the action of forces in the classic simple machines. The remaining corollaries deal with systems of bodies. The subject of corol. 3 is the total momentum of a system of interacting bodies, which is not changed by the action of these bodies on one another. Corol. 4 states that the "common center of gravity" of a system of "two or more bodies" does not change its "state" of motion or rest "as a result of the actions of the bodies upon one another." This pair of corollaries enables Newton to consider the dynamics of the solar system (as in the early parts of book 3) or of any isolated system of interacting bodies independently of whether the system as a whole is itself at rest or moves uniformly in a straight line with respect to absolute space.

In a scholium, Newton presents the results of experiments with colliding pendulums. His discussion is notable for his distinction between elastic and nonelastic impacts. It is the line of thought developed here that suggests the likelihood of Newton's discovery of the third law through studies of such impacts.

The analysis of machines is said by Newton to have been introduced "only to show the wide range and the certainty of the third law of motion," since his "purpose here is not to write a treatise on [practical] mechanics." This stress on the third law, including an extensive discussion of a thought experiment in order to show that the third law "is valid also for attraction," bolsters the argument for

21. See I. B. Cohen, *Science and the Founding Fathers* (New York: W. W. Norton and Co., 1995), chap. 4.

the importance of the third law in Newton's crucial step toward universal gravity during the course of his revision of *De Motu* in 1684/85.

5.7 *Does the Concept of Energy and Its Conservation Appear in the* Principia?

The British physicist Peter Guthrie Tait claimed that in the *Principia*, in the scholium concluding the section of the laws of motion, Newton "stated, so far as the development of experimental science in his time permitted, the great law of Conservation of Energy." Tait decided that the context of the *Principia* indicated that Newton meant by "actio" the quantity which "is now called *rate of doing work* or *horsepower*." Similarly, "the *reactio*, as far as acceleration is concerned, is precisely what is now known as *rate of increase of kinetic energy*." Tait also completely rewrote a sentence from the scholium, introducing such phrases as "work against friction," work against "molecular forces," and "work . . . expended in overcoming resistance to acceleration." His version, taken from his *Sketch of Thermodynamics* (Edinburgh: David Douglas, 1877), chap. 2, is printed side by side with our translation.

Our translation	*Tait's version*
For if the action of an agent is reckoned by its force and velocity jointly, and if, similarly, the reaction of a resistant is reckoned jointly by the velocities of its individual parts and the forces of resistance arising from their friction, cohesion, weight, and acceleration, the action and reaction will always be equal to each other in all examples of using devices or machines.	Work done on any system of bodies has its equivalent in the form of work done against friction, molecular forces or gravity, if there be no acceleration; but if there be acceleration, part of the work is expended in overcoming resistance to acceleration, and the additional kinetic energy developed is equivalent to the work so spent.

It is certainly true that Newton's "force and velocity jointly" can be translated out of context into its post-Newtonian equivalent of "rate of doing work" since $\mathrm{F}\frac{ds}{dt} = \frac{d(\mathrm{F} \times s)}{dt}$ for constant F. Nevertheless, the concept of work as a useful physical entity for rational mechanics was developed only long after the *Principia* was published, and the reckoning of "action" by "force and velocity jointly" is accordingly not a feature of the *Principia* as Newton wrote it. In the discussion of machines, however, in the scholium at the end of the laws of motion, Newton does recognize that the functioning of machines is based on the principle of "our being able to increase the force by decreasing the velocity, and vice versa." This fact, however interesting and important, would have been obvious to anyone with

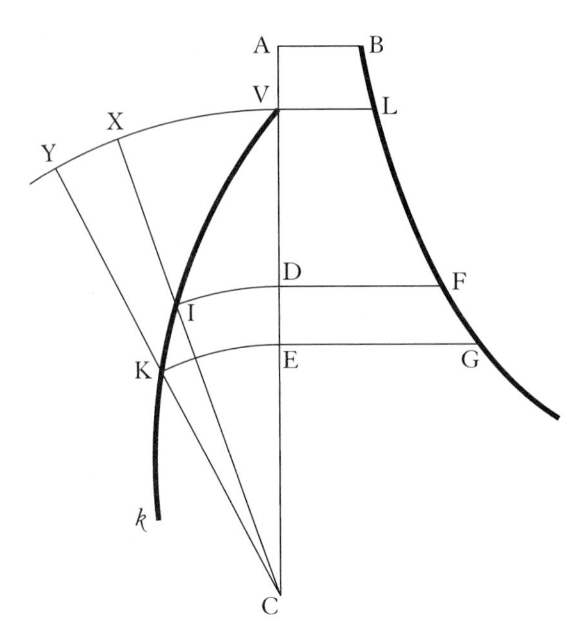

Figure 5.3. Determining the velocity of a body falling under the action of a force. In this simplified version of the diagram accompanying book 1, prop. 41, VIK*k* represents the path of a body let fall from V under the action of a centripetal force directed toward C. Another body is let fall from A and at some point D has acquired the same velocity that the first body has at I, where DI is the arc of a circle centered at C.

the slightest acquaintance with machines such as a pulley system or a crowbar or a steelyard.

A comparison of the two texts will show that Tait has not only "modernized" Newton's text by the introduction of "kinetic energy" and "work" but has also transformed Newton's "cohesion" into "molecular forces." Each reader will have to decide whether Tait's radical rephrasing or translation of Newton's sentence accurately conveys Newton's intent.

It should be noted that in book 1, prop. 40, Newton is concerned with evaluating a quantity that we would recognize in terms of the concepts of work and energy. (In fig. 5.3 I have presented a simplified version of Newton's diagram and in fig. 5.4 a portion of it, rotated through 90 degrees.) In this proposition, a body is allowed to fall (starting from rest) from the point A under the action of a centripetal force directed toward C. The abscissas (AD, AE) represent the displacement s of the falling body from the starting point A. The curve represents the force F; that is, the ordinates FD, GE are proportional to the magnitude of the force. Hence the area under the curve ABFD is $\int F ds$. As noted by Whiteside (*Math. Papers* 6:338 n. 191), Newton shows that this area has the value $\frac{1}{2}v^2$, where

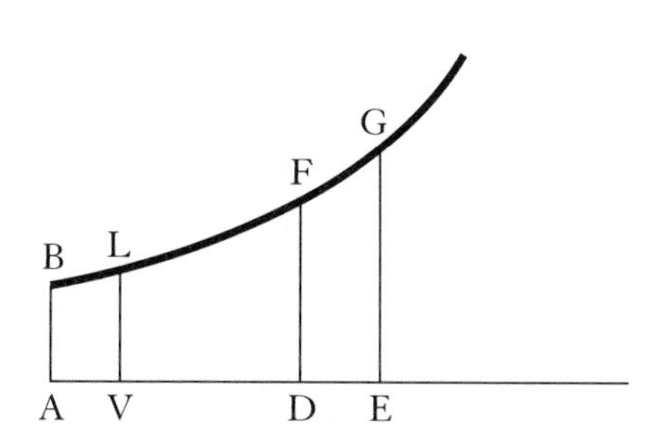

Figure 5.4. A portion of fig. 5.3 rotated through 90 degrees. The ordinates FD and GE are proportional to the magnitude of the force F, and the abscissas VD and VE represent the displacement s of the falling body. Hence, the curve BLFG is a graph of the force F as a function of the distance s fallen, and the area under the curve $\int Fds$ represents the gain in kinetic energy; it has the value $\frac{1}{2}v^2$ where v is velocity at the point D.

v is the velocity at the point D. This integral represents the product of a force and a displacement—that is, what we know now as work or energy. If we substitute a mass m for Newton's unit mass, $\frac{1}{2}mv^2$ is the gain in kinetic energy acquired in falling.

Furthermore, as Aiton and others have pointed out,[22] Newton's ratios can be readily translated into the language of the Leibnizian calculus so that we have not only $v^2 = 2\int Fds$ but also $t = \int \frac{ds}{\sqrt{(2\int Fds)}}$. In prop. 41 (see §10.12 below), Newton analyzes the motion of a body along a curved path, again under the action of a centripetal force F. Here he derives the equivalent of the following equation for v (the velocity at any point along the curved trajectory), where v_0 is an initial velocity at point V, $r = \mathrm{CP}$, $a = \mathrm{CV}$, and F is the centripetal force:

$$v^2 = v_0^2 + 2\int Fdr.$$

As Aiton notes, today's reader will recognize that this equation "expresses the invariance of the sum of the kinetic and gravitational potential energies in an orbit." He concludes, "Evidently Newton understood, without having an explicit terminology, the relation between work and kinetic energy."

It should be noted, however, that—as Whiteside pointed out and as Guicciardini has demonstrated in detail[23]—Newton's contemporaries did not at once grasp

22. Eric J. Aiton, "The Inverse Problem of Central Forces," *Annals of Science* 20 (1964): 81–99, and the updated version published as "The Contributions of Isaac Newton, Johann Bernoulli and Jakob Hermann to the Inverse Problem of Central Forces," in *Der Ausbau des Calculus durch Leibniz und die Brüder Bernoulli: Symposion der Leibniz-Gesellschaft und der Bernoulli-Edition der Naturforschenden Gesellschaft in Basel, 15. bis 17. Juni 1987*, ed. Heinz-Jürgen Hess and Fritz Nagel, *Studia Leibnitiana*, Sonderheft 17 (Stuttgart: Franz Steiner Verlag Wiesbaden, 1989), pp. 48–58.

23. Whiteside's elaborate discussion of this issue appears in *Math. Papers* 8:348–351 in the form of an extended commentary on book 1, prop. 41; cf. Niccolò Guicciardini, "Johann Bernoulli, John Keill and the Inverse Problem of Central Forces," *Annals of Science* 52 (1995): 537–575.

the power of Newton's mode of analysis and in effect had to rediscover in the language of the calculus what Newton had already equivalently accomplished.

5.8 *Methods of Proof versus Method of Discovery; The "New Analysis" and Newton's Allegations about How the* Principia *Was Produced; The Use of Fluxions in the* Principia

On a number of occasions, in print and in manuscript, Newton alleged that he had discovered and proved the principal propositions of the *Principia* in one mode and had then recast the demonstrations so as to present his results in a quite different mode. We have seen one example of such a declaration in the unpublished preface (§3.2 above). The most blatant assertion to that effect that Newton made in print occurs in the book review he wrote of the *Commercium Epistolicum* (London, 1722), the account of the rival claims of Leibniz and Newton to priority in the invention of the calculus.[24] Here, writing about himself in the third person, Newton said:

> By the help of the new *Analysis* Mr. Newton found out most of the Propositions in his *Principia Philosophiae*: but because the Ancients for making things certain admitted nothing into Geometry before it was demonstrated synthetically, he demonstrated the Propositions synthetically, that the System of the Heavens might be founded upon good Geometry. And this makes it now difficult for unskilful men to see the Analysis by which those Propositions were found out.[25]

In a manuscript autobiographical statement, intended as a draft of a preface to a later edition of the *Principia*, Newton was more explicit, stating:

> The Propositions in the following book were invented by Analysis. But considering that the Ancients (so far as I can find) admitted nothing into Geometry before it was demonstrated by Composition I composed what I invented by Analysis to make it Geometrically authentic & fit for the publick. And this is the reason why this Book was written in words at length after the manner of the Ancients without Analytical

24. See Hall, *Philosophers at War* (§1.3, n. 56 above); also *Math. Papers*, vol. 8.

25. Isaac Newton, "[Recensio Libri =] An Account of the Book entitled *Commercium Epistolicum Collinii & Aliorum, De Analysi Promota*, published by order of the Royal Society, in relation to the dispute between Mr. Leibnits and Dr. Keill, about the right of invention of the new geometry of fluxions, otherwise call'd the differential method," *Philosophical Transactions* 29 (1715): 173–224. See §3.1, n. 13 above. The title of the book given in the *Philosophical Transactions* is not the exact title of the *Commercium Epistolicum* (first edition) itself.

> calculations. But if any man who understands Analysis will reduce the Demonstrations of the Propositions from their composition back into Analysis which is very easy to be done; he will see by what method of Analysis they were invented. And by this means the Marquess de l'Hospital was able to affirm that this Book was [presque tout de ce Calcul] almost wholy of the infinitesimal Analysis.[26]

Perhaps the most extravagant such claim is one that begins:

> By the inverse Method of fluxions I found in the year 1677 the demonstration of Keplers Astronomical Proposition vizt. that the Planets move in Ellipses, which is the eleventh Proposition of the first book of Principles.[27]

Here Newton would have a reader believe he had used the methods of the calculus in producing and proving the propositions of the *Principia* and also that he had solved the problem of the forces producing elliptical orbits a little more than two years before the exchange of letters with Hooke that set him on the right path.

In order to understand why Newton made such claims, we must keep in mind that they were made in the mid-1710s, when Newton was deep in the bitter controversy with Leibniz about priority in the invention of the calculus; he wanted to show that he understood and was using the calculus long before Leibniz. There is, however, not a shred of evidence that the *Principia* had been produced according to a secret mathematical agenda that differed from the one used in presenting his results. There is no letter, no draft of a proposition, no text of any kind—not even a lone scrap of paper—that would indicate a private mode of composition other than the public one we know in the *Principia*.

I know of no document to contradict D. T. Whiteside's judgment of 1970, reaffirmed in 1981, that the "published state of the *Principia* . . . is *exactly* that in which it was written."[28] Although Newton wrote that it "is easy" for "any man who understands Analysis" to reduce the demonstrations "from their composition back into Analysis," he "must have been dejected to see in just how relatively few

26. *Introduction*, p. 294 (from ULC MS Add. 3968, fol. 101). The expression "presque tout de ce Calcul" was enclosed within square brackets (indicating a quotation) by Newton himself. The manuscript goes on to discuss aspects of the controversy with Leibniz. The reference is to the preface of the marquis de l'Hôpital's *Analyse des infiniment petits* . . . (Paris, 1696).

27. *Introduction*, p. 295 (from ULC MS Add. 3968, fol. 402).

28. D. T. Whiteside, "The Mathematical Principles Underlying Newton's *Principia Mathematica*," *Journal for the History of Astronomy* 1 (1970): 116–138; *Math. Papers* 8:442–443 n. 1.

of the *Principia*'s 'composed' propositions—if indeed any at all!—he could himself trace any variant preliminary analysis."[29]

In a certain sense, the external form of the *Principia* does seem to have been cast in the style of Greek geometry, since much of the text deals with geometric figures and presents proofs in single successive steps to a conclusion, either "Q.E.D." ("Quod erat demonstrandum"), "Which was to be demonstrated," or "Q.E.I." ("Quod erat inveniendum"), "Which was to be found," or "Q.E.F." ("Quod erat faciendum"), "Which was to be done," or "Q.E.O." ("Quod erat ostendendum"), "Which was to be shown." For the most part, as Newton said, the *Principia* is "written in words at length after the manner of the Ancients" and does not generally proceed by series of interlocking algebraic or differential equations. Even the ratios or proportions tend to be written out in "rhetorical form," in words and sentences within paragraphs of prose, rather than set out in algebraic style or converted into equations. To this extent, the *Principia* certainly has the external look of Greek geometry.

But anyone who goes beyond the mere appearance of the *Principia* to read the proofs will at once recognize that for the most part, proofs proceed in two steps. First, Newton sets up a series of geometric relations, usually expressed as a set of ratios or proportions. In some cases, as in book 1, prop. 11 (see §10.9 below), he will be able to combine these so as, with suitable cancellation of terms, to obtain a single or simpler statement of proportionality. The second step is to introduce a limit as some quantity tends toward zero or as some other increases without limit. In book 1, prop. 1, Newton lets "the number of triangles be increased and their width decreased indefinitely," and the result is the "ultimate" condition. In book 1, prop. 11, he is concerned with "the ratio of Qv^2 to Qx^2" under the limiting condition of "the points Q and P coming together." In book 1, prop. 6, the very statement of the proposition indicates its infinitesimal quality, since a body is said to describe "any just-nascent arc in a minimally small time." Again and again, Newton refers to a "minimally small time," a "line-element" ("lineola" or, in older English, "linelet"), an "area-element," or a minimally small arc, indicating that if the infinitesimal calculus is not used explicitly, there is an infinitesimal level of discourse that marks the *Principia* as a work of the seventeenth century and certainly not an example of Greek geometry. This distinction is apparent in propositions in which, as in book 1, prop. 41, an instantaneous velocity is expressed as the quotient of an infinitesimal increment of arc and an infinitesimal element of time.

29. *Math. Papers* 8:442 n. 1. According to Whiteside, Newton tried to "bolster his case" by "laying emphasis on the handful of lemmas, theorems and problems where (so he claimed) he had left his 'preceding' analysis without recasting his argument into equivalent synthetically demonstrated form."

In book 1, prop. 45, Newton uses the beginning terms of the series expansion of the binomial $(T - X)^n$, in what he calls "our method of converging series," in order to produce a solution of the problem of determining the motion of the apsides in certain perturbed elliptical orbits. Similarly,[30] in the scholium to book 1, prop. 93, he introduces a converging series expansion of $(A + o)^{\frac{m}{n}}$. In a list entitled "On the Method of Fluxions," which Newton inserted into his interleaved copy of the second edition,[31] he indicated that in this example he was using a combination of his methods of fluxions ("moments") and infinite series, noting that "the method of solving Problems by series and moments together is set forth." In this list, the other examples of the use of fluxions and infinite series were book 1, sec. 1, on "first and ultimate ratios," and lem. 2 and prop. 14 of book 2.[32] This prop. 14 makes explicit use of "moments" as well as limits. For example, Newton has

$$AK = \frac{AP^2 + 2BA \times AP}{Z}$$

and determines the "moment" KL of AK. That is, he differentiates AK, so as to get

$$KL = \frac{2AP \times PQ + 2BA \times PQ}{Z}$$

where PQ is the differential of AP. Infinite series also appear elsewhere, as in book 2, prop. 18, prop. 21, prop. 22 and its scholium.

In addition to the foregoing features, the *Principia* also makes explicit use of the calculus, of Newton's method of fluxions. For example, in the statement of book 1, prop. 41, and elsewhere, there occurs the phrase "granting the quadratures of curvilinear figures," by which Newton means that the proof requires the ability to integrate certain specified functions, to find the area under certain curves. We may thus understand why Newton planned to include, in some later proposed editions of the *Principia*, his tract *De Quadratura*,[33] initially published as a supplement to the

30. See the discussion in *Math. Papers* 8:446 n. 13.

31. See §1.3 above; this volume is in the University Library, Cambridge, Adv.b. 39.2. All of the notes made by Newton are included in our Latin edition with variant readings.

32. These are discussed below in §6.2.

33. See Newton's projected preface to a joint edition of the *Principia* and the *De Quadratura* in *Math. Papers* 8:625–632. There is also an English draft of a preface to the *Principia* (ca. 1719), in *Math. Papers* 8:647–655, in which many of the claims about the composition of the *Principia* are repeated; as also in some preliminary drafts of a preface to the *De Quadratura* (also ca. 1719), planned for inclusion in a new edition of the *Principia*, ibid., pp. 656–669 (plus a Latin version, ibid., pp. 670–675).

Opticks (London, 1704), and actually Newton's first full publication of his method of fluxions.

Fluxions also appear in book 2, prop. 11, as is made evident in the second paragraph of the proof, where Newton introduces the condition that "the area-element DE*ed* be a minimally small given increment of time." Then, Newton writes, "the decrement of $\frac{1}{\mathrm{GD}}$, which (by book 2, lem. 2) is $\frac{\mathrm{D}d}{\mathrm{GD}^2}$, will be as . . ." In context, it is obvious that by "decrement" he means "velocity of decrement" or fluxion. A third reference occurs in the conclusion to book 2, sec. 4, following prop. 18, where Newton says: "I have presented the method of dealing with these problems in book 2, prop. 10 and lem. 2, and I do not wish to detain the reader any longer in complex inquiries of this sort."

Lem. 2 of book 2 is devoted to a brief exposition of Newton's method of fluxions. This is, in fact, Newton's first presentation in print on this subject. Here he shows how to differentiate products such as AB, yielding $b\mathrm{A} + a\mathrm{B}$; powers such as A^2, yielding $2a\mathrm{A}$; triple products such as ABC, yielding $c\mathrm{AB} + b\mathrm{CA} + a\mathrm{BC}$; and A^n for positive integral values of n, yielding $n\mathrm{A}^{n-1}$. He then goes on to more general examples such as $\mathrm{A}^n\mathrm{B}^m$ where n and m can be any rational power of A and B, positive or negative. This leads him to the case of polynomials with rational coefficients.[34] Although Newton uses the expression "moments" in lem. 2, he does explain in that lemma that in place of "moments" one can use "velocities of increments and decrements" which can also be called "motions" and "fluxions of quantities." In the scholium following lem. 2, considerably recast for the second edition, Newton—stating his claims to priority in the invention of the calculus—refers to his "general method," of which "the foundation . . . is contained in the preceding lemma."[35] (See, further, p. 166 below.)

Fluxions actually appear explicitly in book 3, lem. 2 (just before prop. 39), toward the end of the demonstration. Here Newton unabashedly, and without any explanation or apology to the reader, writes of the ratio of "the total fluent quantity whose fluxion is $\mathrm{AC}^4 - 4\mathrm{AC}^2 \times \mathrm{CX}^2 + 3\mathrm{CX}^4$" to "the total fluent quantity whose fluxion is $\mathrm{AC}^4 - \mathrm{AC}^2 \times \mathrm{CX}^2$." This is, according to "the method of fluxions," as

34. In the succeeding parts of book 2 and in book 3, there are only three direct references to lem. 2, although the method is used elsewhere. In the third paragraph of the proof of book 2, prop. 29, Newton says that since the resistance (R) is as the square of the velocity (V^2), the "increment of the resistance" will be, by lem. 2, "as the velocity and the increment of the velocity jointly." That is, since $\mathrm{R} \propto \mathrm{V}^2$ or $\mathrm{R} = k\mathrm{V}^2$, $d\mathrm{R} = 2k\mathrm{V}d\mathrm{V}$ or $d\mathrm{R} \propto \mathrm{V}d\mathrm{V}$.

35. On this scholium, see Whiteside's notes in *Math. Papers* 8:628–632, and esp. pp. 633–646, on "The 'Mind' of the *Principia*'s Fluxions Scholium," Newton's "Enarratio Plenior Scholii Praecedentis" or "Mens Scholii Praecedentis."

$AC^4 \times CX - \frac{4}{3}AC^2 \times CX^3 + \frac{3}{5}CX^5$ to $AC^4 \times CX - \frac{1}{3}AC^2 \times CX^3$.[36] The next result is obtained, he writes, "by the method of fluxions." So much for those who say the *Principia* does not use the calculus and is written in the style of Greek geometry! We may well understand why Newton was so pleased with the statement quoted earlier, in which the marquis de l'Hôpital declared that the *Principia* was "presque tout de ce Calcul," "almost wholy of the infinitesimal Analysis." At first Newton began to write "infinitesimal calculus" but then deleted the word "calculus" in favor of "analysis."[37]

36. Replace Newton's AC by A and CX by X. Then he is saying that if

$$Z = A^4X - \tfrac{4}{3}A^2X^3 + \tfrac{3}{5}X^5$$

and

$$Y = A^4X - \tfrac{1}{3}A^2X^3$$

then the desired fluxions are

$$\frac{dZ}{dX} = A^4 - 4A^2X^2 + 3X^4$$

and

$$\frac{dY}{dX} = A^4 - A^2X^2.$$

37. *Introduction*, p. 294.

CHAPTER SIX

The Structure of Book 1

6.1 *The General Structure of the* Principia

The *Principia*, in its final form, consists of four prefaces, a set of "Definitions," "Axioms, or the Laws of Motion," books 1 and 2 on "The Motion of Bodies," book 3 on "The System of the World," and a concluding "General Scholium." The subject of book 1 is motion in free spaces, that is, in spaces devoid of resistance, while book 2 deals with motion under several different kinds of resistance. Here Newton's subject is extended to include many other topics of natural philosophy, such as the principles of wave motion, along with some aspects of the general theory of fluids.

Book 1 has a number of distinct parts. The opening sec. 1, "The method of first and ultimate ratios," is intended to set forth a mathematical framework of the proofs, constructions, and exercises that are to be used throughout books 1 and 2 and that accordingly are the ultimate foundation for book 3. In our modern terminology, sec. 1 sets forth the theory of limits and its applications.

Secs. 2 and 3 develop the laws of centripetal or centrally directed forces, including the proof (props. 1, 2, 3) that Kepler's law of areas is both a necessary and a sufficient condition for a centrally directed force acting on a body with a component of inertial motion.

Following a presentation of Newton's measure of centripetal force (prop. 6), there are proofs (props. 11, 12, 13) that elliptical, hyperbolic, and parabolic orbital motions imply a focally directed force inversely proportional to the square of the distance. Prop. 14 deals with Kepler's third (or harmonic) law and prop. 15 with orbital velocity at any point on an orbit. This marks the boundary of how much of book 1 Newton recommended that even those with good mathematical training should read.

Sec. 1: First and Ultimate Ratios 6.2

The title of book 1, sec. 1, is "The method of first and ultimate ratios," to which Newton added the subtitle "for use in demonstrating what follows" ("cujus ope sequentia demonstrantur," *lit.* "with the help of which the following things are demonstrated"). It is thus made to appear that Newton is setting up a rigorous theory of limits, expressed in a set of eleven lemmas, and will then refer back to them in using the method of limits in the proofs that follow. In fact, however, these lemmas do not "everywhere play the central auxiliary role which Newton here foresees."[1] That is, "It would appear that his initial vision of presenting a logically tight exposition of the principles of motion under accelerative forces faded more and more when he came in detail to cast his arguments, and that he was happy after a while to lapse into the less rigorously justified mode of presentation which he largely exhibits in his published *Principia*."[2] Of course Newton frequently made use of the result (as in lem. 11, corol. 1) that in the limit, tangents, arcs, and chords (Newton here sometimes calls them "sines") "become ultimately equal." In considering sec. 1, furthermore, we must be careful lest we overrate the novelty of the lemmas since "in seeking to formulate general lemmatical theorems to serve as a basis for the many particular geometrical quadratures, rectifications and limit-equalities he will thereafter make, Newton is solidly in a contemporary tradition comprehending the varied researches and systematisations of such men as Fermat, Blaise Pascal, Huygens, James Gregory and his lately deceased Cambridge colleague Isaac Barrow."[3]

Lem. 1 sets forth the basic concept of "quantities" and "the ratios of quantities" which ("in any finite time") "constantly tend to equality" and become "ultimately equal." That is, they approach each other so closely that their difference becomes "less than any given quantity." Here Newton introduces the essential concept of limit. The quantities or ratios of quantities are said by Newton to approach each other so closely in "any finite time" that "before the end of that time" ("ante finem temporis"), however small the difference between them may be, there will be a still smaller difference as one proceeds ever nearer to the ultimate state or limit.

In the scholium at the conclusion of sec. 1, Newton indicates the important difference between the "ultimate ratio of vanishing quantities" or "the ratio with which they vanish" and the ratio of those quantities "before they vanish or after they have vanished." In this sense, in the scholium, Newton explains that those

1. *Math. Papers* 6:108 n. 40.
2. Ibid.
3. Ibid. See, further, D. T. Whiteside, "Patterns of Mathematical Thought in the Later Seventeenth Century," *Archive for History of Exact Sciences* 1 (1961): 179–388, esp. chaps. 9, 10, pp. 331–355.

"ultimate ratios with which quantities vanish are not actually ratios of [the separate and individual] ultimate quantities." Rather, the "ultimate ratios" of constantly decreasing quantities are "limits which the ratios of quantities decreasing without limit are continually approaching." These limits toward which the ratios continually converge are (as has previously been set forth in lem. 1) specified by the property that the ratios "approach so closely" to these limits that the difference between them "is less than any given quantity"; the ratios, however, "can never exceed and can never reach" such limits "before the quantities are decreased indefinitely."

Lems. 2, 3, and 4 are devoted to methods of quadrature of curves, finding the area under a curve by considering an inscribed and circumscribed set of parallelograms, and then considering "the width of these parallelograms" to be "diminished and their number increased indefinitely." In the statement of lems. 2, 3, and 4 Newton refers to "parallelograms," but in the proof of lem. 2, he seems to have restricted the argument to rectangles (as also would seem to be the case in the diagram). The proof can be equally valid for parallelograms, however, and many expositors (e.g., Frost) merely substituted "parallelograms" for Newton's "rectangles" and skewed the diagram so that AK, BL, CM, D*d*, E*o* are no longer perpendicular to AE.

Lems. 6, 7, and 8 declare that an evanescent arc and its chord and tangent are ultimately equal. In lem. 7, Newton is using "tangent" in a special sense to indicate the segment of an indefinite tangent line intercepted between one end A of an arc (the point of tangency) and a line drawn from the other end B of the arc to the indefinite tangent, at some fixed finite angle to the chord. These results are used frequently throughout the *Principia*. Newton also makes much use of lem. 10, in which he considers the initial displacement produced by a finite force acting on a body, whether the force is constant or uniformly increasing or decreasing. Essentially, he proves that during an indefinitely small time, the action of the force will cause a displacement proportional to the square of the time.

Lem. 11 sets forth Newton's measure of curvature. Here he uses "subtense" in two very different senses in the opening of the proof. In one sense it denotes a line drawn from one end of an arc (B) to a line which is tangent to the curve at the other end of the arc (A), which is the subtense (or subtangent) of the angle of contact. But Newton also uses "subtense" to denote the chord AB.

The concluding scholium compares and contrasts the method of limits with the method of exhaustion (and *reductio ad absurdum*) and the method of indivisibles. Here Newton explains that "what has been demonstrated concerning curved lines and the [plane] surfaces comprehended by them" can be easily extended to "curved surfaces and their solid contents." He concludes by alerting the reader to the sense in which he will (in order "to make things easier to comprehend") use the expressions "minimally small" or "vanishing" or "ultimate." These are not to be understood as "quantities that are determinate in magnitude," but rather as "quantities that are to be decreased without limit."

6.3 *Secs. 2–3: The Law of Areas, Circular Motion, Newton's Dynamical Measure of Force; The Centripetal Force in Motion on an Ellipse, on a Hyperbola, and on a Parabola*

Secs. 2 and 3, on the laws and effects of centrally directed forces, complete the part of book 1 that Newton recommended to general readers. In props. 1–3 (sec. 2), Newton shows how the law of areas is related to the principle of inertia. That is, he proves that the law of areas is both a necessary and a sufficient condition for a centrally directed force acting on a body with an initial component of inertial motion.[4] The method pursued in prop. 1 has already been discussed in §5.4. In the second edition, Newton introduced a new corollary (corol. 1) stating that the velocity at any point on a curved orbit is inversely proportional to the perpendicular dropped from the center of force to the tangent at that point. Thus, in the first corollary to the first proposition of the first book of the *Principia*, Newton corrected Hooke, who, in a letter during the 1679/80 correspondence, stated that the velocity would be inversely proportional to the distance from the center of force to the point in question.

Newton does not mention Kepler's name in relation to the area law, nor will he do so in relation to the law of elliptical orbits (prop. 11) or the harmonic law (prop. 15).[5] The subject of prop. 4 is the force in uniform circular motion. Newton's analysis is based on considering this to be a case of accelerated motion;

4. See Bruce Pourciau, "Newton's Solution of the One-Body Problem," *Archive for History of Exact Sciences* 44 (1992): 125–146. On Newton's early ideas on the dynamics of circular motion and the stages of development leading up to the final elaboration in the *Principia*, see Brackenridge, *The Key to Newton's Dynamics* (§3.7, n. 71 above).

5. In fact, nowhere in book 1 or in book 2 does Newton refer to Kepler. In book 3, "On the System of the World," however, Kepler does appear in the introductory section of "Phenomena" as the acknowledged discoverer of the third law of planetary motion, but not of the first two laws. In book 3, Kepler's name also appears frequently in the discussion of the positions and structure of comets. On Newton and Kepler, see my "Kepler's Century, Prelude to Newton's," *Vistas in Astronomy* 18 (1975): 3–36.

In the earlier tract *De Motu*, however, written just before the *Principia*, Newton referred to elliptical orbits of planets as being "entirely as Kepler supposed." In later manuscript notes about the history of his ideas, Newton wrote of how he found that the orbits of planets "would be such ellipses as Kepler had described." He also wrote of how he found "the demonstration of Keplers Astronomical Proposition . . . that the Planets move in Ellipses."

Newton's nephew-in-law, John Conduitt, recorded in his notes for a proposed biography of Newton that the "proposition that the areas described in equal times were equal" was only "assumed by Kepler," but "was not by him demonstrated, of which demonstration the first glory is due to Sir Isaac." For details see my *Newt. Revolution*, chap. 5, esp. p. 252; also *Introduction*, suppl. 1.

Newton's views concerning the discovery of elliptical planetary orbits are found in his response to a

since in equal times equal areas are swept out with regard to the center, there must be a centrally directed force. The magnitude of the force is shown to be proportional to $\frac{v^2}{r}$ where v is the orbital speed and r is the radius of the circle. One of the corollaries shows that the Huygens-Newton $\frac{v^2}{r}$ rule, combined with Kepler's harmonic law, yields an inverse-square centripetal force. (The way in which the combination of this rule and law yields an inverse-square law has been discussed in §3.6 above.) In a scholium, Newton introduced a reference to the work he had done on this topic in the 1660s. The purpose of this scholium was to bolster Newton's claims to priority over Hooke in the matter of the inverse-square law. The proof of prop. 4 was considerably revised for the second edition. In corol. 6, Newton proves that if the periodic time is as the $\frac{3}{2}$ power of the radius, the force will be inversely as the square of the radius, and conversely. In corol. 7, he generalized corol. 6 to the case of the time being as any power of the radius.

Prop. 6 discloses Newton's dynamical measure of a centripetal force. This is the key to his solution of finding the force in motion along an elliptical path. The derivation of this measure is set forth in §10.8 below. In the second edition, the presentation of this measure of force is reworked and expanded, and Newton introduces an alternative demonstration. In the second edition, furthermore, as we have seen (§3.7 above), Newton also introduced alternative proofs for props. 7, 9, 10, and 11.

In prop. 8, Newton seeks the force on a body moving in a semicircle when the center of the force is very remote. The subject of prop. 9 is the equiangular spiral; the force is shown to be as the inverse cube of the distance. In prop. 10, the body moves along an ellipse under the action of a force directed to the center of the ellipse. A scholium takes note that when the center moves infinitely far away, the curve becomes a parabola and the force will become constant. Newton remarks that this is the theorem discovered by Galileo for the motion of projectiles. This ends sec. 2.

Sec. 3 opens with prop. 11. Newton draws on the previous propositions (notably props. 6 and 10) to prove that motion along an elliptical orbit implies that the centrally directed force will be inversely as the square of the distance. His proof is

---

letter from Halley informing him of Hooke's claim to have discovered the inverse-square law of gravitational force. In reply (29 June 1686), Newton invoked the analogy of the discovery of elliptical orbits, writing that Kepler had only "guessed" the ellipticity of planetary orbits, just as Hooke had only guessed the law of force. Newton concluded that he himself was entitled to credit for both "the Ellipsis" and the inverse-square law.

set forth in detail, step by step, in §10.9. It should be noted that Newton assumes that in such elliptical motion there actually is a force directed toward one of the foci. In fact, as we have seen, this is equivalent (by the preceding props. 1 and 2) to having equal areas swept out in any equal times by a focal radius. Props. 12 and 13 then show that for hyperbolic and parabolic orbits, the force must also be as the inverse square. Accordingly, Newton has solved what is known as the direct problem; he has proved that equal-area orbital motion along any conic section implies an inverse-square force. In the second edition, as has been mentioned earlier, Newton introduces alternative proofs for props. 11 and 12.

The reader who is unacquainted with Newton's procedure in the *Principia* would naturally suppose that after solving the direct problem in props. 11-12-13, Newton would turn to the indirect problem. But Newton does not devote a whole proposition to the determination of the trajectory produced by a centrally directed inverse-square force. Rather, he relegates this important assignment to a corollary, corol. 1 to prop. 13. In the first edition, as we shall see below, he did not even bother to include a proof. This procedure must seem all the more astonishing when we consider the importance of the inverse problem in providing the basis for Newton's development of the system of the world in book 3.

In prop. 14, several (noninteracting) bodies are supposed to move about a common center, each subject to an inverse-square centripetal force; it is proved that the latera recta are as the squares of the areas described by focal radii in any given (same) time. In the following prop. 15, under the same conditions, the periodic times in the resultant elliptical orbits are shown to be as the $3/2$ power of the major axes, which is Kepler's third law. Prop. 16 then shows how to find the velocity of a body moving on a conic section.

In prop. 17, Newton supposes an inverse-square force of known "absolute" quantity that is directed toward a given point. The proposition states that it is required to find the "line" (i.e., the curve) which a body will describe when let go from some given point in a given direction with a given velocity along a straight line. In fact, in the demonstration Newton assumes that the orbit will be a conic and he proceeds to find the conic in question, taking the center of force as the principal focus (S) and determining the latus rectum and the position of the other focus (H). He shows which conditions yield the different species of conic.

Prop. 17 is easily misread, leading to the conclusion that Newton here proves what he has proposed in the statement of the proposition, that is, to find the trajectory produced by the action of a centrally directed inverse-square force on a body "with a given velocity along a given straight line." This is actually the subject of the previous prop. 13, corol. 1, discussed below. What Newton is proving in prop. 17, however, is not quite the inverse problem of forces and orbits. As has just been mentioned, in prop. 17 he assumes (although he does not say so) that under the given conditions (an inverse-square force directed toward a fixed center and an initial component of inertial motion) the orbit *is* a conic and he then seeks to determine the particular conic under various conditions. Hence, in the statement of the proposition, as Whiteside has noted, the word "line" or "curve" should more restrictively read "conic, namely, with a focus at the force-center." As a matter of fact, in the preliminary statement of this proposition in the earlier *De Motu*, Newton did not propose to find a "line" or curve in general. Rather, under the same initial conditions, he seeks "the ellipse" which the body will traverse. Sec. 3 concludes with prop. 17.[6]

In the first and later editions of the *Principia*, there are four corollaries, which are more or less the same in all three editions. Of these, corols. 3 and 4 to prop. 17 were not part of the original text of the preliminary version of book 1 (the "Lucasian Lectures"), where this proposition is numbered 16 rather than 17; nor are they present in the earlier version of this proposition in the tract *De Motu*. Corols. 3 and 4 were, however, added later on the blank verso side of fol. 56 of the Lucasian Lectures and became part of the manuscript (now in the library of the Royal Society) from which the first edition was composed and printed. These two corollaries show that Newton had a method of dealing with orbits with perturbed motion. In corol. 4, Newton introduces a method for determining the trajectory "very nearly" when a body moving in orbit is "continually perturbed by some force impressed from outside." Newton does not here develop the method in detail, but he explains that it consists of determining "the changes which the [impressed] force introduces at certain points" and then estimating "from the order of the sequence the continual changes at intermediate places." This may be paraphrased in a more readily understandable way as a method of "estimating, by interpolation, the changes continually made at intermediary points." Michael Nauenberg has found in corols. 3 and 4 a recognition by Newton, just as he was composing the *Principia*, that prop. 17 "gave him the basis for a new perturbation method," a method identified by Nauenberg (in a paper presented at the Symposium on the Foundation of Newtonian Scholarship held in March 1997 at the Royal Society, London) as the one outlined by Newton in the manuscript published as an appendix to the introduction to the *Catalogue of the Portsmouth Collection*.

6. Robert Weinstock, *The College Mathematics Journal* 25 (1994): 222, claims to have found an error in prop. 17, that an inverse-square force produces an orbit in the shape of a conic. He observes that this error has been made by "not only Newton, but also L. Euler and I. B. Cohen."

The Inverse Problem: The Orbit Produced by an Inverse-Square Centripetal Force 6.4

Props. 11, 12, and 14 lead to the statement—in corol. 1 of prop. 13—of the general converse: that an inverse-square force, applied to a body with a component of linear motion, will produce a curved orbit of the second degree, that is, a conic section. In the first edition, Newton merely stated this result without proof,[7] but in response to criticism he sketched out a proof in the second edition, which appears with one correction in the third.[8]

Readers may be puzzled as to why this important result was relegated to a mere corollary to prop. 13 rather than prominently displayed as a proposition in its own right. Perhaps Newton believed that this result was so obvious that it did not merit more than a corollary. In this regard, we may note that Brougham and Routh maintained that one can easily make of Newton's separate cases (ellipse, hyperbola, parabola) a "compendious theorem" including all three and then "work backwards" to obtain the value of a given perpendicular (SY) and a radius of curvature (R). "But no curves can have the same value of SY and R except the conic sections," they observe, "because there are no other curves of the second order, and those values give quadratic equations between the co-ordinates."[9]

In recent years this issue has been given prominence by two charges[10] raised against Newton. The first is that Newton did not initially believe he needed a proof; the second, that the eventual proof is not a proof at all. I find it difficult to believe that Newton might have thought that the statement of an inverse-square force implying a conic orbit would follow without proof from his individual proofs that a circle, ellipse, hyperbola, and parabola require an inverse-square centripetal force directed toward a focus (or a center, in the case of a circle). Were this so, Newton

7. In his unpublished commentary on the *Principia* (based on the first edition), David Gregory observed that corol. 1 is "the converse of the three preceding propositions." He could not help expressing the "wish that the author had demonstrated that a body cannot be carried by this law of centripetal force in a curve other than a conic section." This is "likely," he noted, since "no curve of the same [i.e., second] degree exists other than conic sections." Quoted, in translation, from Gregory's "Notae in Newtoni Principia Mathematica Philosophiae Naturalis." There are four MS versions of this text, in Aberdeen, Edinburgh, Oxford (Christ Church), and London (Royal Society, in Gregory's hand); see my *Introduction*, pp. 189–191, and W. P. D. Wightman, "David Gregory's Commentary on Newton's 'Principia,'" *Nature* 179 (1957): 393–394.

8. For details, see *Math. Papers* 6:146–149 n. 124.

9. *Anal. View*, pp. 55–57.

10. Raised initially by Robert Weinstock, "Dismantling a Centuries-Old Myth: Newton's *Principia* and Inverse-Square Orbits," *American Journal of Physics* 50 (1982): 610–617; also "Long-Buried Dismantling of a Centuries-Old Myth: Newton's *Principia* and Inverse-Square Orbits," *American Journal of Physics* 57 (1989): 846–849.

would have been guilty of so elementary an error in the logic of mathematics as to belie his competence to have written the *Principia*. Essentially this error would have been to assume that a proof that "A implies B" proves of and by itself that "B implies A." The very fact that there is both a prop. 1 and a prop. 2 in sec. 2 shows that Newton was perfectly aware that the converse of any proposition needs a separate proof. Prop. 1 proves that a centripetal force, applied to a body with uniform motion, implies the law of areas; prop. 2 proves, conversely, that the law of areas implies a centripetal force.

Let me now turn to the second part of the charge against Newton, that the proof which he sketched out was no proof at all. A number of articles have been generated on this topic. A careful and very thorough mathematical examination of "the logical structure . . . as well as the details" of Newton's proof has been made by Bruce H. Pourciau, who concludes that, while Newton's "argument does indeed contain a flaw, it is a minor omission rather than a serious logical error." By rectifying this omission, Pourciau shows "how Newton's outline expands into a convincing proof that inverse-square orbits must be conics."[11]

Newton discussed this subject in a private memorandum, in which he wrote: "The Demonstration of the first Corollary of the 11th, 12th & 13th Propositions being very obvious, I omitted it in the first edition & contented my self with adding the 17th Proposition whereby it is proved that a body going from any place with any velocity will in all cases describe a conic Section: which is that very Corollary."[12] As we have seen, however, this is in no way a proper statement of the contents of prop. 17.

6.5 *Secs. 4–5: The Geometry of Conics*

Secs. 4 and 5 (book 1) are devoted to various propositions and lemmas relating to the geometry of conics. Their presence is justified by Newton's insertion of the word "orbit." In fact, however, Newton had composed these two sections as separate tracts well before writing the *Principia*.[13] Having them in more or less final form, he decided that the *Principia* would be a convenient place to publish them.

11. Bruce H. Pourciau, "On Newton's Proof That Inverse-Square Orbits Must Be Conics," *Annals of Science* 48 (1991): 159–172; see also Eric J. Aiton, "The Solution of the Inverse Problem of Central Forces in Newton's *Principia*," *Archives internationales d'histoire des sciences* 38 (1988): 271–276. See also the extended commentary by D. T. Whiteside in *Math. Papers*, vol. 6.

12. *Introduction*, suppl. 1, §3, p. 294 (from ULC MS Add. 3968, fol. 101). There are a number of similar statements by Newton in various manuscripts (*Introduction*, suppls. 1, 8).

13. See *Introduction*, chap. 4, §4.

In book 3, Newton explains why he included them. When he composed book 1 and sent the completed text to Halley for the printer (as we know from Newton's correspondence with Halley), Newton had not as yet developed an adequate method for determining the parabolic orbits of comets. His plan at that time was to use some of these propositions (as he explains in book 3, prop. 41) in solving "this exceedingly difficult problem." It was only "later on," he says, that he conceived the "slightly simpler solution" (using approximations) displayed in prop. 41. By that time, however, book 1 had been printed, and it was too late to remove secs. 4 and 5.

Newton was aware that these two sections, however important they might be for the subject of geometry, were really out of place in the *Principia*, that they distracted the attention of a reader concerned with dynamics, whose course of study would lead directly from sec. 3 to sec. 6. Accordingly, in the 1690s, while planning a second edition, he proposed that these two sections be eliminated from the main text and, as David Gregory recorded, "become parts of a separate treatise," a two-part supplement, of which one part would embody these two sections and the other would "contain his Method of Quadratures," or his methods of the calculus.[14]

In secs. 4 and 5, Newton gives original and "elegant constructions and proofs" for "describing a conic to satisfy five conditions of the type either to pass through a given point or else to touch a given line: and he deals with all the six cases so arising."[15] He elaborated the geometry of conics as "the locus of a point, the product of whose distances from two given lines is proportional to the distance from a third line, or else to the product of its distances from a third and a fourth given line."[16] In the course of this presentation, Newton made use of projective transformations. In book 1, sec. 5, lem. 22, for example, the problem is "to change" ("mutare") figures "into other figures of the same class." The text begins with the requirement "to transmute" ("transmutare") a given figure. The projection is described as a "continual motion" whereby "each of the points in the first figure will yield a corresponding point in the new figure." An elaborate and enlightening historical and critical commentary on secs. 4 and 5 has been given by D. T. Whiteside.[17]

14. *Introduction*, pp. 193–194; Gregory's memorandum is published in *Corresp.* 3:384. On Newton's plans for these two sections, see *Math. Papers* 6:229–298, and esp. 229 n. 1.

15. H. W. Turnbull, *The Mathematical Discoveries of Newton* (London and Glasgow: Blackie and Sons, 1945), p. 54.

16. Ibid., p. 55.

17. *Math. Papers* 6:229–299.

6.6 *Problems of Elliptical Motion (Sec. 6); Kepler's Problem*

The topic of sec. 6 is how to find the position of an orbiting body at any given time. Here Newton is concerned with orbits found in nature, that is, parabolas and ellipses. In lem. 28, and the following prop. 31 and scholium, Newton addresses the generalities of what is known as Kepler's problem. Kepler was unable to find a "rational" solution to the problem of finding a focal sector of an ellipse having a given simple ratio to another focal sector. Hence there was no way to use Kepler's area law as a means of exact determination of planetary positions. The astronomers in the post-Keplerian era had recourse to various approximations in place of the area law, centering on the uniform rotation of a radius vector about the empty focus of the ellipse plus some suitable correction factor.[18] Lem. 28 boldly states: "No oval figure exists whose area, cut off by straight lines at will, can in general be found by means of equations finite in the number of their terms and dimensions."

Newton's lem. 28 encountered problems from the very start. As Whiteside's commentary shows (*Math. Papers* 6:302–309 nn. 119–127), Newton's first enunciation in the manuscript version was too general and had to be emended for the first printed edition. Then Leibniz detected a slip, but—according to Whiteside—Johann Bernoulli "accepted its truth unhesitatingly" and incorporated it into a paper of 1691. Leibniz claimed to have found a flaw in Newton's proof and produced an example purporting to show that lem. 28 cannot stand in all its generality as stated. Yet, as Whiteside observes, "Leibniz's Bernoullian lemniscate has a double loop, crossing itself at the origin." Huygens also produced what he considered to be a counterexample, but this was a "double parabola" which—to quote Whiteside once again—has "two pairs of conjugate infinite branches," a feature to which "Newton was later . . . to take explicit exception." Whiteside concludes that both Leibniz's lemniscate and Huygens's double parabola are not "true closed 'ovals' in the simple sense of the term." In any event, in the second edition of the *Principia*, Newton added a final sentence to the demonstration of lem. 28 in which he made it clear that he was "speaking of ovals that are not touched by conjugate figures extending out to infinity."

Rouse Ball found fault with Newton's proof on the grounds that "the exact quadrature" of certain ovals "is possible."[19] Here he was merely echoing the

18. On the problems of using Kepler's law of areas and the various approximations used by seventeenth-century astronomers in place of this law, see Wilson, "From Kepler's Laws, So-Called, to Universal Gravitation" (§1.2, n. 23 above); and my *Newt. Revolution*, pp. 224–229.

19. *Essay*, p. 83. For a critique of another discussion of this lemma by Rouse Ball, see *Math. Papers* 6:307 n. 126.

judgment of Brougham and Routh.[20] In his commentary on lem. 28, Whiteside announced (*Math. Papers* 6:302–303 n. 121) his finding that Newton's proof contains a fundamental flaw, a claim he was able to bolster by producing "a counter-example to Newton's assertion" (p. 207 n. 26).

More recently, lem. 28 has been studied by V. I. Arnol'd, who was tremendously impressed by "two purely mathematical pages" of Newton's presentation. Arnol'd found these to be remarkable because they contain "an astonishingly modern topological theorem on the transcendence of Abelian integrals." Arnol'd essentially restated the theorem in the following (more correct) form: "Every algebraically integrable oval has singular points; all smooth ovals are algebraically non-integrable."[21] In this example we may see an admirable illustration of Clifford Truesdell's dictum that a critical reading of mathematical or mathematically based works of the past tends to involve a judgment concerning whether a proof or other presentation is true or false.

Newton addresses the solution of Kepler's problem[22] directly in the succeeding prop. 31. He introduces an approximate solution using a curve (identified by Whiteside as a prolate cycloid) whose "description . . . is difficult," and so in a scholium he resorts to the use of "a solution that is approximately true." We may note here the introduction of what is generally known today as the Newton-Raphson method of approximation.[23]

In the first edition, the scholium began with what has been called an "ingenious but unpractical geometrical mode of construction,"[24] which was replaced in the second and third editions by Newton's "computationally superior analytical methods, making use of iterations and expansion into series." The opening sentence

20. *Anal. View*, p. 72: "Newton . . . begins his investigation by a lemma [lem. 28], in which he endeavours to demonstrate that no figure of an oval form, no curve returned into itself and without touching any infinite arch, is capable of definite quadrature." Their further comment reads: "It is rarely, indeed, that the expression 'endeavour' can be applied to Sir Isaac Newton. But some have questioned the conclusiveness of his reasoning in this instance."

21. V. I. Arnol'd, *Huygens and Barrow, Newton and Hooke: Pioneers in Mathematical Analysis and Catastrophe Theory from Evolvents to Quasicrystals* (Basel, Boston, Berlin: Birkhäuser Verlag, 1990), appendix 2, "Lemma XVIII of Newton's *Principia*"; S. Chandrasekhar, *Newton's "Principia" for the Common Reader* (Oxford: Clarendon Press, 1995), p. 133. Chandrasekhar (citing the authority of Arnol'd) held that lem. 28 "is a striking manifestation of Newton's mathematical insight," enabling him "to surpass the level of scientific understanding of his time by two hundred years." Chandrasekhar did not take cognizance of Whiteside's critique of Newton's demonstration.

22. On this subject see J. C. Adams, "On Newton's Solution of Kepler's Problem," *Monthly Notices of the Royal Astronomical Society* 43 (1882): 43–49. The degree to which Kepler's problem has attracted the attention of astronomers is shown by the fact that a bibliography of 1900 lists 123 papers on this subject.

23. On this usage, and the prehistory of this method of approximation, see *Math. Papers* 6:317 n. 148, and—for a mathematical commentary on prop. 31 and its scholium—ibid., pp. 308–323.

24. *Math. Papers* 6:314 n. 114.

of the scholium was altered in the second edition to its present reading. Originally, in the first edition, Newton wrote: "Because of the difficulty in describing that curve, it is preferable to use constructions in working practice that are approximately true." That is, in the first edition Newton found it preferable to use graphic "constructions" ("constructiones") whereas in the second (and third) he produced a "solution" ("solutionem"). He also dropped the condition of "in praxi mechanica" ("in working practice").

6.7 *The Rectilinear Ascent and Descent of Bodies (Sec. 7)*

Thus far (secs. 2, 3) Newton has discussed orbits that are conics and (secs. 4, 5) the geometry of conics, plus (sec. 6) motions in given conics. In sec. 7, Newton shifts from curved motions to the rectilinear "ascent and descent" of bodies under various conditions of force. The first proposition (prop. 32) displays Newton's method of proceeding to a limit and also shows how his geometric constructions of conics can be made to serve problems of linear descent. In retrospect, however, while prop. 32 is a kind of *tour de force* of Newton's geometric method, it also appears a cumbersome and unnecessarily complex means of solving a problem that can be reduced to far simpler analytic procedures. This proposition, or its equivalent, appears in the original *De Motu*.

Since prop. 32 is analyzed in §10.11 below, where each step is set out separately for ease of study, we need only mention here that Newton's construction depends on consideration of motion along an ellipse under the action of an inverse-square force directed toward a focus S. In the limit, as the minor axis tends to zero, the orbit tends to a rectilinear path (the point S tending to the lower apsis), which is the solution to the problem.

The final proposition of sec. 7, prop. 39, differs in one important respect from all other propositions thus far. The statement of the proposition includes the specific condition of granting "the quadratures of curvilinear figures," that is, of performing integration or of finding the area under certain curves. In other words, the proof requires certain specified integrations. This proposition was a late addition to the manuscript. In fact, with respect to the nature of its subject matter as well as the condition of granting a quadrature, prop. 39 could be more logically grouped with props. 40 and 41 and thus could well have been the opening proposition of sec. 8. Like props. 40 and 41, prop. 39 deals with motion under the action of a centripetal force of some unspecified kind. The primary result of prop. 39 is to show how to find an area or an integral that will represent the velocity attained by a body moving under the action of a centripetal force of any kind, a result on which prop. 41 depends.

Motion under the Action of Any Centripetal Forces (Sec. 8); Prop. 41 **6.8**

Sec. 8 takes up the problem of finding the trajectory (or curve of motion) produced by the action of "any centripetal forces." That is, Newton wants to solve the problem of motion under the action of a force which is some unspecified function of the distance. His goal has two parts: to find the radial coordinate of any point P on the curve at any time t, and then to find the angular coordinate θ.

Of the three propositions (props. 40, 41, 42) which constitute sec. 8, prop. 41 is the most celebrated. Here, as in props. 39–42, Newton supposes "the quadratures of [the pertinent] curvilinear figures" and a centripetal force "of any kind," and sets out to find the resultant curved trajectories and "the times of their motions in the trajectories so found."

The actual steps in the demonstration of prop. 41 are expanded below (in §10.12). But a few comments about certain steps are in order here. First of all, the motion under consideration is supposed to begin with some specified velocity at a given point (V) at some angle to the line of action of the force (CV); C is the center of force. The magnitude of the initial velocity (at V) is specified in a manner reminiscent of the way in which Galileo wrote that God (according to Plato) had created the world of the sun and planets (for details, see §6.9 below).

Historians of science have recognized that prop. 41 represents a climax, a high peak, in the development of dynamics in the *Principia*. Here Newton freely uses the direct equivalent of both the differential and the integral calculus and shows himself to be a real master of the subject, dealing with a problem of great generality. There is nothing in the antecedent literature of the science of motion that has this same magnitude or importance. It is fair to say that with prop. 41, the subject of mathematical dynamics achieved its modern maturity for the first time.

Prop. 41 is followed by three corollaries. Corol. 3 explores the inverse-cube force as a particular example of the general force which is the subject of prop. 41. Newton shows that the trajectory in this case is what is known as a Cotesian spiral.[25] Since this is the only example given by Newton to illustrate prop. 41, historians and scientists have often asked why Newton made the odd choice of an inverse-cube force rather than the inverse-square force that is the subject of much

25. For an analysis of corol. 3 and a translation into the language of the calculus, see Whiteside's commentary in *Math. Papers*, vol. 6. A recent study by Herman Ehrlichson is particularly good on corol. 3: "The Visualization of Quadratures in the Mystery of Corollary 3 to Proposition 41 [of book 1] of Newton's *Principia*," *Historia Mathematica* 21 (1994): 148–161.

of book 1 of the *Principia* and also of book 3. Indeed, contemporary critics such as Johann Bernoulli even went so far as to assume that Newton could not solve the problem for the inverse square and produced their own solutions.[26]

There is, however, one very important exception to this general lack of appreciation of Newton's solution of the inverse problem of central forces. As Prof. Niccolò Guicciardini has found, Jacob Hermann admitted in a paper published in the *Giornale dei Letterati d'Italia* for 1711 (7:173–229) that Newton had successfully solved the inverse problem and even showed that his own solution did not differ essentially from Newton's. He thereby plainly distanced himself from the strong criticisms of Newton made by Johann Bernoulli. In his *Phoronomia* (1716, p. 73), Hermann expressed himself strongly on this point, referring to his own paper in the *Giornale*. "This problem," he wrote, "was solved for the first time by the celebrated Newton in prop. 41 of his book, Princ. Phil. Nat. Math. & afterwards by that most perspicacious geometer Joh. Bernoulli by the identical method [*gemino modo*]" and then by Varignon by "different methods [*diversis modis*]."

D. T. Whiteside notes that "Newton's contemporaries were very slow to appreciate the depth and power" of the construction set forth in prop. 41, nor did they "acknowledge the ease with which it might be applied to the particular cases of the inverse-square and inverse-cube orbits."[27] From the very start, there was always a question of why Newton chose an inverse-cube force rather than an inverse-square force to illustrate prop. 41. Clearly, it was within Newton's power to have dealt with the physically more significant case of the inverse square.[28] One suggestion is that perhaps he chose to deal with the spiral orbits produced by an inverse-cube force as a sign that "his refusal to broach the inverse-square case stemmed in no way from his inability to resolve it mathematically."[29] Possibly, furthermore, the choice of the inverse cube rather than the inverse square may be taken as evidence that Newton was more concerned with intellectually interesting examples than in practical illustrations taken from physics and astronomy. It should be pointed out, however, that in book 3, Newton introduces the inverse-cube force as the net gravitational pull of the moon on the oceans to produce tides.[30]

26. The reader will also wish to consult the gloss made by Newton in answer to a request by David Gregory; see *Corresp.* 6:348–354.

27. *Math. Papers* 6:349 n. 209. See also Aiton, "Inverse Problem" and "Contributions" (§5.7, n. 22 above).

28. Requiring only "a couple of easy integrations": Whiteside, "The Mathematical Principles Underlying Newton's *Principia Mathematica*" (§1.2, n. 9 above), p. 18.

29. Ibid.

30. See §8.13 below.

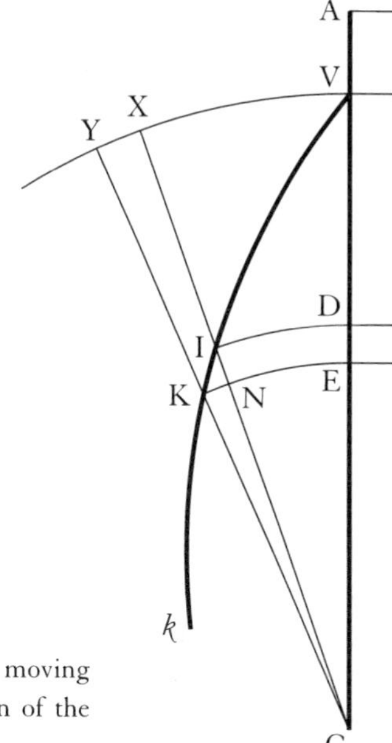

Figure 6.1. Determining the velocity of a body moving under the action of a centripetal force (a portion of the diagram for book 1, prop. 41).

Specifying the Initial Velocity in Prop. 41; The Galileo-Plato Problem **6.9**

In prop. 41, Newton specifies the magnitude of the initial velocity at the point I on the trajectory. He does so in terms of a point A on the line CV (see fig. 6.1), from which he supposes that a body is let fall under the action of the centripetal force directed toward C. In this case, the acceleration is not constant but is determined by the changing force directed toward C, whose magnitude is some function of the distance from C. The point A is specified by the property that when the falling body reaches the point D, the magnitude of its velocity will be the same as the magnitude of the velocity of the body at I moving along the curved path VIK*k*.

Clearly, if the point I is chosen to be the start of the trajectory motion, the point D will coincide with the point V and the magnitude of the velocity of falling from A will be that of the initial velocity of the body whose trajectory is being sought. But if the motion is determined by beginning with a free fall from A to V, then when the body reaches V, it must be made to undergo some change in direction of motion with no diminution or increase in speed.

This mode of attaining the critical velocity resembles a problem apparently devised by Galileo, but attributed by him to Plato and sometimes known today as the Galileo-Plato problem.[31] Here, as in Newton's prop. 41, an object is dropped through a certain distance, falling under the action of an accelerating force; when it reaches a specified point, the falling body (like Newton's falling body in prop. 41) undergoes a change in the direction of its motion without any change in the magnitude of its velocity. There is one very important difference, however, between the Galileo-Plato problem and Newton's prop. 41. Galileo has the object fall with constant acceleration (that is, with a uniformly accelerated motion), whereas Newton has the body fall under the action of a variable force and a consequent variable acceleration.

Galileo's Plato problem is astronomical or cosmological and is concerned with a way in which God could have formed the solar system. Since Galileo is considering orbits of planets around the sun, as in the Copernican system, rather than around the earth, the problem as presented could hardly have been proposed by Plato. According to Galileo, Plato said that there is some point out in the heavens from which God lets each of the planets fall toward the sun. When each planet reaches its proper distance from the sun, it will (according to Galileo) have acquired a velocity equal in magnitude to its regular orbital velocity. At this point, God shifts the direction of motion of the planet, without any change in magnitude, so that it will move out along its orbit. The planet, having gained its orbital speed, then continues in its orbital path.[32]

Newton made a critical analysis of the Galileo-Plato problem shortly after the publication of the *Principia*. First, he had to correct Galileo's assumption that the acceleration of the falling planets would be constant; he knew that the acceleration would be constantly increasing because the solar force varies inversely as the square of the distance. Incidentally, we may see here the enormous gulf between the Galilean and the Newtonian science of motion. Finally, Newton showed that "there is no common place from whence all the Planets being let fall" and then "descending with uniform and equal Gravities (as Galileo supposed)" would gain the velocities with which they revolve in their orbits.[33] For Galileo's method to

31. Despite the efforts of many scholars, no early text, whether by Plato or any other writer, has ever been found in which the Galileo-Plato problem occurs. Accordingly, scholars tend to think that Galileo either invented the problem or mistakenly believed that he had encountered some version of it either in Plato or in some Renaissance Neoplatonic work.

32. The Plato problem occurs in both the *Dialogue* and the *Two New Sciences*. See *Dialogue concerning the Two Chief World Systems, Ptolemaic and Copernican*, trans. Stillman Drake (Berkeley and Los Angeles: University of California Press, 1953), p. 21; [*Discourses and Mathematical Demonstrations concerning*] *Two New Sciences*, trans. Stillman Drake (Madison: University of Wisconsin Press, 1974), pp. 233–234.

33. For details of Galileo's statement of the problem and of Newton's reaction, see I. B. Cohen, "Galileo, Newton, and the Divine Order of the Solar System," in *Galileo, Man of Science*, ed. Ernan McMullin (New York: Basic Books, 1967), pp. 207–231.

succeed, Newton noted, the gravitating force of the sun would have to be halved during the time of fall and then restored to its normal value once the planet reached its orbit. In Newton's words, "the divine power is here required in a double respect, namely, to turn the descending motion of the falling planets into a side motion, and at the same time to double the attractive power of the sun."[34] If the doubling did not occur, Newton added, the planets "would go away from their orbs into the highest Heavens in parabolic Lines."

These results, Newton remarked, "follow from my *Princ. Math., lib.* i, Prop. 33, 34, 36, 37." In props. 32–37, as we have seen, Newton explores the accelerated motion produced by an inverse-square force. In these propositions, Newton shows how to find the speed at any place along the line of descent (props. 36 and 37) and the time of descent.

In book 3 of the *Principia*, there are other examples of falling that are somewhat like one part of Galileo's Plato problem. That is, Newton considers both planets that are allowed to fall in toward the sun and satellites that are allowed to fall in toward a planet, but he does not have these falling bodies change their direction of motion at some point in their descent. For example, in book 3, prop. 4, Newton explores the effect of depriving the moon of its forward component of velocity so that it could fall inward to the earth. In prop. 6, the satellites of Jupiter are let fall toward Jupiter and the circumsolar planets let fall toward the sun.

The foregoing examples of falling differ from the Galileo-Plato problem because in each case the falling motion has a nonuniform acceleration proportional to the gravitating force. But in lem. 11 of book 3 (immediately preceding prop. 41), Newton considers a comet deprived of all its forward motion and let fall toward the sun from some specified height. Here Newton supposes that the comet is "urged toward the sun always by that force, uniformly continued, by which it is urged at the beginning." In part, this is like the Galileo-Plato problem because the force of the sun is "uniformly continued" or constant during the fall.

In the first draft of what we know as book 3,[35] Newton solves a somewhat similar problem. He supposes that any of the planets be deprived of its tangential component of forward motion (or "projectile motion"); then, knowing the distances of the planets from the sun, he computes the time in which the planet would fall into the sun. Using the results of book 1, prop. 36, Newton finds that for each planet, this time of descent would be one-half of the periodic time (or time for a complete revolution) if that planet were in orbit at one-half the normal distance from the sun. That is, if the period of a planet is T, and the time of fall is t, then

34. Newton to Richard Bentley, 17 Jan. 1693, 11 Feb. 1693, *Corresp.* 3:238–240, 244.

35. Concerning this early draft of book 3, the *Essay on the System of the World*, see §1.1, n. 8 above.

$t : T = 1 : 4\sqrt{2}$. Newton computes that Venus would fall to the sun in 40 days, Jupiter in two years and one month, and the earth and moon together in 66 days and 19 hours.[36]

When Newton discussed the Galileo-Plato problem in the early 1690s, he did not at first relate this problem to Galileo, but cited a quite different source. "Blondel," he wrote, "tells us some where in his Book of bombs, that Plato affirms, that . . ." This passage occurs in a letter to Richard Bentley of 17 January 1693. Evidently, even if he had read about this problem in one of Galileo's books, he did not then remember it. We may be quite certain, however, that he had not encountered the Plato problem in either Galileo's *Dialogue* or *Two New Sciences* and so gave Blondel as his source. Such a conclusion would be consistent with the considerable evidence that Newton had never read Galileo's *Two New Sciences* and knew only some selected portions of Thomas Salusbury's English translation of the *Dialogue* on the two systems of the world.[37]

At the time of writing to Bentley, Newton had evidently been reading Blondel's book on bombs. (He does not seem to have had a copy of this book in his personal library.)[38] Blondel does mention this mode of creating the solar system and does ascribe its origin to Plato. In a succeeding letter dated 25 February 1693, Newton did finally refer to Galileo.[39] Very likely he had consulted his copy of Blondel's book and had found that Blondel attributed the problem to Galileo.

At some time after writing the *Principia*, most likely in the late 1690s, Newton did consult the Latin translation of the *Dialogue*, at least with respect to the Plato problem. Our evidence for this is a paragraph written by Newton on the Plato problem which contains an almost verbatim extract on this topic from the Latin translation of the *Dialogue*.[40] This is, so far as is known, the only extract from Galileo to be found among Newton's manuscripts and published writings.[41] A note about Galileo in relation to this problem, based on the *Dialogue*, was entered

36. *Essay on the System of the World*, §27.

37. On Newton's lack of acquaintance with Galileo's *Two New Sciences*, see my "Newton's Attribution of the First Two Laws of Motion to Galileo" (§5.3, n. 8 above). See also my "Galileo and Newton," in *Galileo a Padova, 1592–1610*, vol. 4, *Tribute to Galileo in Padua* (Trieste: Edizioni LINT, 1995), pp. 181–210.

38. John Harrison, *The Library of Isaac Newton* (Cambridge, London, New York: Cambridge University Press, 1978).

39. *Corresp.* 3:255.

40. This extract is quoted in my "Galileo, Newton, and the Divine Order," p. 225.

41. In the *Two New Sciences*, but not in the *Dialogue*, Galileo uses the concept of "sublimity" in presenting the Plato problem. Sublimity is a distance through which bodies fall with constant acceleration in some given time and so is a measure of the acceleration. Since all the evidence indicates that Newton had not read the *Two New Sciences*, it would be historically inaccurate to identify Newton's determination of the initial velocity in prop. 41 with Galileo's sublimity.

by Newton in his personal interleaved copy of the first edition of the *Principia*[42] but never made its way into the second edition.

One final point: Newton used a geometric method of specifying the magnitude of the velocity because this fitted in with his mode of solution to the problem posed in prop. 41. Although the structure and content are analytical, Newton's solution is developed in a geometric framework; that is, as may be seen in §10.12 below, Newton's solution involves differentials and integrals which are expressed in geometric equivalents. Therefore it fitted in with Newton's procedure to specify the magnitude of the velocity geometrically, that is, in terms of the length of a line corresponding to a distance fallen.

Moving Orbits (Sec. 9) and Motions on Smooth Planes and Surfaces (Sec. 10) 6.10

Sec. 9 shifts the discussion from stationary to movable orbits and introduces the problem of the motion of the apsides. This is one of two places in book 1 where Newton produces mathematical results of direct potential use in dealing with the motion of the moon, the other being prop. 66 and its many corollaries. Here, in sec. 9, Newton compares (prop. 44) the forces that, in different bodies, produce the same orbits, one stationary and the other moving. A distinguishing feature of this section is the introduction (in prop. 45, exx. 2 and 3) of the method of infinite series, which Newton calls "an indeterminate series" and which is presented with reference to "our method of converging series." We may recall that in Halley's review of the *Principia*, published in the *Philosophical Transactions* in 1687, he particularly called attention to Newton's "improvements" of "the old and new Geometry" and, specifically, what he calls Newton's "method of infinite *Series*."[43]

Prop. 45 deals with a specific problem in the motion of the moon, the mean motion of the apsides. This motion, known from observations, had not as yet been accounted for. A simplification is made by considering "orbits that differ very little from circles." Newton discusses cases in which the centripetal force is uniform, that is, is as $\frac{A^3}{A^3}$ (ex. 1), and is as $\frac{A^n}{A^3}$ (ex. 2) and as $\frac{bA^m + cA^n}{A^3}$ (ex. 3). Corol. 2 explores the motion of a body in an elliptical orbit, acted on by an inverse-square force directed toward a center and also subject to "any other extraneous force." Much has been made of the fact that Newton, in the third edition, added a sentence at

42. See our Latin edition of the *Principia*, with variant readings (§1.3, n. 45 above).

43. I. B. Cohen, "Halley's Two Essays on Newton's *Principia*," in *Standing on the Shoulders of Giants: A Longer View of Newton and Halley*, ed. Norman Thrower (Berkeley and Los Angeles: University of California Press, 1980), pp. 91–108.

the end of corol. 2, reading: "The [advance of the] apsis of the moon is about twice as swift" as Newton's calculated result of $1°31'14''$ per revolution.[44]

Sec. 10 introduces the motion of bodies on smooth planes under the action of forces whose centers are not on those planes. This leads to motion on curved surfaces and (props. 48 and 49) the "lengths" of the epicycloid and the hypocycloid as a function of time.

Prop. 50 then explores the mathematics of a pendulum swinging under the action of gravity—"in a given cycloid," or hypocycloid. The pendulum in question consists of a flexible (weightless and inextensible) string with a mass point at the extremity; in our translation, we have called this a "simple pendulum." The use of cycloidal cheeks, as Huygens discovered and explained in his *Horologium Oscillatorium* (1673), makes the motion of such a pendulum isochronous, that is, produces a period that is independent of the amplitude of swing. In prop. 52, corol. 1, Newton shows how his study of the oscillations of pendulums may be used to compare "the times of bodies oscillating, falling, and revolving."

George Smith informs me that in sec. 10, "Newton shows that Huygens's isochronism result for constant g ceases to hold for an inverse-square g, but it continues to hold for g proportional to r." Hence, a cycloidal pendulum would be isochronous under the surface of the earth, where g is proportional to r, but not above the surface of the earth, where g is proportional to $\frac{1}{r^2}$. Thus sec. 10 "not only frames the Huygens result in a broader context, but provides an empirical test for distinguishing between Newton's universal gravity and Huygens's constant gravity under the surface of the earth."

6.11 *The Mutual Attractions of Bodies (Sec. 11) and the Newtonian Style*

Sec. 11 introduces a radical change, which at once raises the whole discussion of rational mechanics to a new high plane. The sequence thus far has been as follows: First, Newton explored the motion of a dimensionless mass point of unit mass about a mathematical center of force in a nonresisting medium. He has, to all intents and purposes, been studying a one-body "system,"[45] although the considerations hold as well for a multiple-body system of orbiting bodies so long as the bodies are not interactive and the central body is stationary. Then, Newton passed from stationary to moving orbits, proceeding to an examination of motion

44. See, e.g., Rouse Ball, *Essay*, p. 85, and esp. D. T. Whiteside's comments in *Math. Papers* 6:380–381 n. 260, and 508–537. See §8.16 below.

45. Strictly speaking, there cannot be a one-body "system," although a single body and a mathematical point can be a system.

on certain smooth surfaces (i.e., surfaces which do not offer any resistance.) Now, in sec. 11, Newton will progress from a single body and a center of force to a system of two interacting bodies. He will eventually move on to three (and possibly more) such interacting bodies. Later, he will turn to considerations of the shape and structure or composition of bodies. Finally, he will consider bodies moving in various resisting mediums.

In this sequence, Newton has developed his system of rational mechanics as a mathematician, producing an intellectual or mathematical construct that will mirror some of the principal aspects of the world of nature. This procedure had two very important advantages. First, it allowed Newton to begin by exploring the simplest system, and then to add various degrees of complexity one by one, rather than having to face a set of multiple interrelated problems all at once. Of equal importance was the freedom to explore any kind of mathematical relationship of force without—at this time—having to justify the mode of action and without having to defend his use of a force of attraction of a type that was generally considered to be unacceptable in natural philosophy. That is, Newton was able to consider mathematically the properties of a centrally directed force of attraction even though this was a troublesome concept. He was fully aware that many (perhaps most) natural philosophers of that day would reject any system of rational mechanics and especially of celestial dynamics if it was based on attraction. He nevertheless could consider mathematically a concept banned from scientific discourse by the adherents of the mechanical philosophy. Newton himself, although he believed in the existence of such forces, was never able to accept the idea that forces could act over enormous distances without, as he put it, "the mediation of something else." But, in books 1 and 2 of the *Principia*, he could avoid any such questions—or so he seems to have believed—by restricting the level of discourse to mathematical systems which were analogues of physical systems. It is this procedure which I have called the Newtonian style.[46]

At the start of the *Principia*, Newton presents a mathematical construct,[47] one that is derived of course from nature simplified. Clearly he is considering conditions in a mathematical three-dimensional Euclidean space. In that construct he sets up the artificial (but mathematical) conditions of a dimensionless unit-mass moving about a mathematical point, about a mathematical center of force. Although Newton uses the concept of a centripetal force, and explores its mathematical conditions and consequences, this force is a mathematical concept because—as Newton says in introducing sec. 11—only mathematical forces are directed toward mathematical

46. This concept is developed at length in my *Newt. Revolution*.

47. I shun the use of the word "model," which has philosophical implications that do not apply here.

points; in the world of nature, physical forces arise from and are directed toward physical bodies. That is, despite the language, no considerations of physical aspect or quality are introduced. Nor is any question raised of the mode of action of such a force, whether it is a "pull" toward the center or a "push" (a "vis a tergo") from behind. I have called such a mathematical set of conditions, and the exploration of their mathematical consequences, stage one of the Newtonian style.

There is then a stage two, in which the greatly simplified conditions of stage one are modified as a result of comparing the conditions of the mathematical construct with the external world, by then introducing some factor or factors that can make the original construct in stage one become more nearly an analogue of conditions found in the observed world of nature. This alteration produces a new or revised stage one, a new and more complex mathematical construct, in which Newton once again explores the mathematical consequences of mathematical conditions.

In the introduction to sec. 11, Newton sought to leave no doubt concerning this aspect of his work. He says explicitly that he will now abandon the simple mathematical construct of a single body and a mathematical center of force because, on comparing his construct with the world of nature, he finds that such a situation "hardly exists in the natural world." In that world, "attractions are always directed toward bodies," not toward abstract mathematical points. But if there are two bodies and one of them attracts the other, then by the third law of motion, "the actions of attracting and attracted bodies are always mutual and equal," so that there is a mutual attraction between any two bodies. If the system has more than two bodies, then each body both attracts all the others and is attracted by all the others. Newton says that now he will consider such systems; he will "go on to set forth the motion of bodies that attract one another."

In the Newtonian style, there is a continual counterpoint between the mathematical development of the properties of the systems in his mathematical constructs and an alteration of the systems on the basis of a comparison with the conditions of the external world. In this way we shall see Newton, in the *Principia*, advance from the mathematical construct of a single mass point moving about a mathematical center of force to a system of two interacting such masses and even three and more. Additional alterations, resulting from comparisons with the external world, yield systems with bodies having physical dimensions, then bodies of various composition and shapes. Other comparisons will produce motion in various types of resisting mediums, motions under certain conditions of constraint such as on given surfaces or as in oscillation at the end of strings.

Although I have invented the names "Newtonian style," "stage one," and "stage two," I believe that this characterization accurately describes the way in which Newton addresses his subject in books 1 and 2 of the *Principia*. There is a final "stage three," in which—in book 3—Newton shifts from the mathematical level of analogue by applying his analysis and his results to the phenomena of the world of observation.

We have seen that Newton himself addresses these problems in introducing sec. 11. There he explains that the mathematical construct he has been exploring in the previous ten sections does not correspond to the conditions of the world of nature. Accordingly, he has produced a stage two, a revised stage one, in which the center of attraction will no longer be a mathematical point, but rather a second body. Yet it is to be observed that the two "bodies" are not yet "real" physical bodies but are only mathematical entities in Euclidean space. That is, in secs. 1–10, they do not have physical characteristics of dimensions or size, mass[48] and weight, color, degree of hardness, and so on.

No one who examines Newton's procedure carefully will, I believe, deny that what I have set forth is indeed Newton's method of procedure. But scholarly doubt has been generated by Newton's insistence that in book 1 he is dealing with the subject as a mathematician, that in book 1 his concepts are purely mathematical and not physical. Because he has been creating a mathematical analogue of the world of nature, he has been using the same names for his mathematical concepts that are used commonly for physical entities. As we shall see in a moment, from Newton's day to ours, scientists and philosophers have not been willing to accept Newton's statements about the mathematical level of discourse at their face value. The question must be faced of whether Newton merely adopted a mathematical posture as a subterfuge to deflect criticism—notably for his use of the concept of attraction—or whether he was expressing his own sincere view of his methodology, which I have called the Newtonian style.[49]

In elaborating the properties of "the motion of bodies that attract one another," Newton says that he is "considering centripetal forces as attractions." He explains that he is fully aware that if he were dealing with the realm of physics, rather than with a mathematical construct, he should "perhaps" have to "speak in the language of physics"; in that case, the attractions "might more truly be called impulses." But, he insists, he is here "concerned with mathematics" and not with physics, and so he is "putting aside any debates concerning physics." In his mathematical construct, with two interacting bodies, there is no longer a single center of force. He might still have talked of "mutual centripetal forces acting between bodies"

48. In sec. 11, the bodies are considered to have mass. In prop. 57, Newton writes of distances being inversely "as the bodies" or masses. In prop. 59, he is concerned explicitly with the masses of the revolving bodies.

49. This problem is developed at length in my *Newt. Revolution*.

or some other circumlocution that might appear neutral with respect to physical and philosophical implications. His aim, however, was to put aside "any debates concerning physics" and to have himself be easily understood by mathematicians. And so he has introduced, in a mathematical sense, the simple word "attraction," using—as he said—"familiar language" which he believed would "be more easily understood by mathematical readers." Here he was only repeating what he had said at the end of his discussion of def. 8, that in using "words signifying attraction, impulse, or any sort of propensity toward a center," he was "considering these forces not from a physical but only from a mathematical point of view."

If Newton had in fact limited his presentation to the elaboration of an abstract mathematical theory, and had not used the word "attraction," he would perhaps not have encountered as much hostility as proved to be the case. But it was obvious to all readers, and made clear in the words "natural philosophy" in the title, that Newton's ultimate goal was not mathematics but physics. Even those readers who might accept the procedure I have called the Newtonian style were aware—from the alternations between stage one and stage two—that Newton was fashioning a mathematical construct that was an ever closer analogue of the physical universe. The merest glimpse ahead to book 3 would show that eventually Newton would shift from a mathematically conceived "attraction" to a physical force manifested in the motions observed in the solar system.

In the scholium at the conclusion of sec. 11, Newton stressed the fact that he was using "the word 'attraction'... in a general sense for any endeavor whatever of bodies to approach one another," without any concern for the physical mode of operation of such a force.[50] He insisted once again that here, in book 1 of the *Principia*, he had no concern with "the species of forces and their physical qualities" but only with their "quantities and mathematical proportions." He then explained that on the mathematical level (that is, in his mathematical constructs), the assignment is to investigate "those quantities of forces and their proportions that follow from any conditions that may be supposed." He then contrasted this with "physics, [where] these proportions must be compared with the phenomena, so that it may be found out which conditions [or laws] of forces apply to each kind

50. That is, he was not here concerned with "whether that endeavor occurs as the result of the action of the bodies either drawn toward one another or acting on one another by means of spirits emitted or whether it arises from the action of aether or of air or of any medium whatsoever—whether corporeal or incorporeal—in any way impelling toward one another the bodies floating therein."

In the last proposition (prop. 69) of sec. 11, Newton assumes that law 3 applies to the mutually "attracting" bodies as a necessary condition that the attractions are as the masses of the two attracting bodies. But, as Curtis Wilson has reminded me, law 3 "would not necessarily be applicable if the 'attractions' were due to aether producing a *vis a tergo* to push the bodies toward one another." In this case, since A and P would not be directly interacting, they would not be the consequence of universal gravitation.

of attracting bodies." Here is a statement in Newton's own words about what I have called the three stages of the Newtonian style.

Finally, says Newton, "it will be possible to argue more securely concerning the physical species, physical causes, and physical proportions of these forces." In the final stage three of the Newtonian style, the principles developed in the mathematical construct are assumed to be applicable to the physical world. In book 3, Newton will apply his mathematical principles to the system of the world.

In the beginning of book 3, Newton made much the same point. He said that in the preceding two books of the *Principia*, he had been setting forth "principles of philosophy" (i.e., principles of natural philosophy), which "are not, however, philosophical but strictly mathematical." Of course, the principles ("the laws and conditions of motions and of forces") "especially relate to [natural] philosophy," and so he had illustrated them in scholiums in which philosophical or physical matters were introduced. But he insisted that in books 1 and 2, he had been exploring "principles" only mathematically and not physically.

Newton's critics, however, have never been willing to assume or admit that Newton meant what he said. Even before the *Principia* had been published, Huygens expressed his anxiety concerning what he had heard about the new work. "I have nothing," Huygens wrote, "against his not being a Cartesian, provided he does not give us suppositions like that of attraction."[51] After he had read the *Principia*, Huygens rejected Newton's theory of the tides and "all the other theories he builds upon his Principle of Attraction, which to me seems absurd."[52] He stated as plainly as possible that a Newtonian "attraction is not explainable by any of the principles of Mechanics, or of the rules of motion" and thus cannot be accepted in natural philosophy.[53] Leibniz declared that the Newtonian introduction of "gravitation of matter toward matter" is, "in effect, to return to occult qualities and, even worse, to inexplicable ones."[54] In his official *éloge* of Newton, written for the Royal Academy of Sciences of Paris, Fontenelle (himself a Cartesian) observed that "attraction and vacuum, banished from physics by Descartes, and to all appearances for ever, are now brought back again by Sir Isaac Newton."[55] Fontenelle faulted Newton for arguing that gravity is "really an attraction, an active power of bodies." With specific reference to the discussion in book 1, sec. 11, Fontenelle observed that Newton says that gravity "can . . . act by Impulse." Why, then, he

51. Huygens to Fatio de Duillier, 11 July 1687, in Christiaan Huygens, *Œuvres complètes, publiées par la Société hollandaise des sciences* (The Hague: Martinus Nijhoff, 1888–1950), 9:190.

52. Huygens to Leibniz, in Huygens, *Œuvres* 21:538.

53. "Discours sur la cause de la pesanteur," in Huygens, *Œuvres* 21:471.

54. G. W. Leibniz, *Die philosophischen Schriften* (Berlin: Weidmann, 1875–1890), 5:58 (*Nouveaux essais*).

55. See *Newton's Papers*, ed. Cohen and Schofield (§1.2, n. 41 above), pp. 453ff.

asked, did not Newton give "preference" to "that clearer term"? He did not believe it possible to use both "attraction" and "impulse" indifferently, "since they are so opposite." We must be on our guard, Fontenelle concluded, "lest we imagine that there is any reality" to attraction and thus "expose ourselves to the danger of believing that we comprehend it."[56] The distinguished philosopher and historian of science Alexandre Koyré has refused to accept Newton's position. Commenting on Fontenelle's statement, he remarked that "Fontenelle was right, of course. Words are not neutral. They have, and convey, meanings."[57]

Voltaire gave a classic reply to Fontenelle, in his *Lettres philosophiques sur les Anglais*, addressing the question of why Newton did "not use the word impulsion which we understand so well, instead of attraction which we do not understand." Voltaire phrased a reply for Newton, that the critics "no more understand the word impulsion than attraction."[58] Among other possible replies, Voltaire (quite correctly) would have Newton say, "I use the word attraction only to express an effect that I have discovered in nature, a certain and indisputable effect of an unknown principle, a quality inherent in matter, the cause of which someone other than myself will, perhaps, find." Voltaire concluded that the cause of attraction is "among the *Arcana* of the Almighty"; it "is in the bosom of God." Newton, we now know, would have agreed.

There can, I believe, be little doubt that Newton fully believed in the existence of a universal force of attraction, of which terrestrial gravity is a manifestation. But he was always concerned about how this force could act. As I have mentioned (§3.5) and shall have occasion to mention again, Newton sought until the end of his life for some way to account for the action of this force without having to suppose that it could, of and by itself and without the mediation of anything else, stretch out through enormous reaches of space and still be potent enough to affect the motion of planets or comets. In developing the *Principia*, Newton more or less followed the path indicated by the succession of propositions, in the manner I have described as the Newtonian style, without being in any way troubled or inhibited by having to use the concept of attraction, without being deflected from his main purpose by a need to find a mode of operation for a force of this kind.

I see no reason to believe that Newton was not sincere in his hope that his readers would go along with him and follow his development of mathematical principles, which would prove to be so powerful that their use in natural philosophy could not be denied. The next assignment would then be to find a modus operandi

56. A contemporaneous English version of Fontenelle's *éloge* may be found in *Newton's Papers*, ed. Cohen and Schofield, pp. 453ff.

57. Koyré, *Newtonian Studies* (§1.2, n. 16 above), p. 58.

58. Translated in Koyré, *Newtonian Studies*, pp. 60–61.

for this universal force. In the event, however, some contemporaries (among them Huygens and Leibniz) found the very notion of attraction so abhorrent, even on a mathematical level, that they simply rejected the Newtonian natural philosophy in which they found it embedded. They simply ignored Newton's statements about mathematics and physics and about levels of discourse or refused to accept the Newtonian distinctions. "He could have saved his breath," according to R. S. Westfall. "Fifty such disclaimers—or fifty times fifty—would not have stifled the outrage of mechanical philosophers."[59]

We may gain some measure of the inhibiting force of the abhorrence of attraction by examining a statement made by Huygens after studying the *Principia*. It never had occurred to him, he wrote, to extend "the action of gravity to such great distances as those between the sun and the planets, or between the moon and the earth." The reason was that "the vortices of M. Descartes which formerly seemed to me rather likely, and that I still had in mind, hindered me from doing so." What a pity that he could not relax the strictures of the mechanical philosophy.

Sec. 11: The Theory of Perturbations (Prop. 66 and Its Twenty-Two Corollaries) **6.12**

Sec. 11 begins with a two-body problem (props. 57 and 58), in which two mutually attracting bodies revolve about their common center of gravity, leading to prop. 59, in which Newton compares the periodic time of one body revolving about another with the periodic time of motion about the common center of gravity, the basis for a modification required in Kepler's third law. Props. 60–63 explore further aspects of the system of two bodies. Prop. 64 then examines the motions in a system in which the force of attraction is not the usual inverse-square force but rather is directly proportional to the distance. Newton starts with a system of two bodies, then adds a third body and finally a fourth. In prop. 65, Newton studies two cases in which there is an interaction of three bodies whose attractive forces are as the inverse square of the distance. He shows that, relative to each other, their orbits "will not be very different from ellipses" and that they will, "by radii drawn to the foci," describe areas very nearly "proportional to the times."

Next come prop. 66 and its twenty-two corollaries, in which Newton introduces into rational mechanics one of the most famous problems of that subject: the three-body problem. Prop. 66 differs from prop. 65 in that it deals explicitly with perturbation, including distortion of the elliptical shape. Newton is unable to deal with three mutually attracting bodies in general, although it is apparent to every reader that such a general problem of forces of perturbation would be the logical

59. *Never at Rest*, p. 465.

result of the exercise. In prop. 66, Newton starts out with the perturbing effect of an outer planet on an inner one, but the greater part of prop. 66 by far, including the twenty-two corollaries, deals rather with a more manageable problem, one that closely resembles the motion of the moon around the earth under the perturbing force exerted by the sun.[60]

Here we see a clear example of the Newtonian style in action. A mathematical stage one is enlarged or altered by a stage two, in which some new feature or features of the world of nature are introduced, in this case the force of a third or perturbing body. Of course, Newton's purpose is to produce a complex mathematical construct that will eventually exhibit in analogy the most significant features of the observed system of the world. Sec. 11 shows how Newton's introduction of these features one by one has enabled him to increase the mathematical complexity in relatively small and manageable steps, rather than having to face the system of the world all at once.

In sec. 11, every reader would become fully aware of Newton's ultimate goal: to obtain mathematical results of use in analyzing the system of the world. There could be little doubt in the matter since the central body is denoted by T (for "Terra," the earth), the orbiting body by P (for "Planeta"), and the perturbing body by S (for "Sol," the sun). Newton, we may note, tended to use the noun "planeta" both for what we would call planets and for satellites. Sometimes, but not universally, he refers to planets in our sense as "primary planets," but at other times he uses "planets" without any qualification. In book 3 of the *Principia*, he refers to satellites or moons as "satellites" but also as a kind of "planet" ("planeta"), as in "circumjovial planets" and "circumsaturnian planets" (as opposed to "circumsolar planets").

In the first edition of the *Principia*, Newton did not designate the three interacting bodies by the letters T, P, and S. He rather designated the central by the letter S, rather than T, and then used P for the orbiting body. Evidently, he

60. Rouse Ball (*Essay*, pp. 88–89) has succinctly set forth the principal subjects of the twenty-two corollaries of prop. 66, in which Newton has explored the "disturbing" or perturbing effects of the action of a body like the sun on the motion of a body like the moon with respect to "the motion in longitude, the motion in latitude, and annual equation, the motion of the apse line, the motion of the nodes, the evection, the change of inclination of the lunar orbit, and the precession of the equinoxes; the proposition is also applied to the theory of the tides, and to the determination of the interior constitution of the earth as deduced from the motion of its nodes."

An admirable presentation of this topic is given in D. T. Whiteside, "Newton's Lunar Theory: From High Hope to Disenchantment," *Vistas in Astronomy* 19 (1976): 317–328. Another critical summary of Newton's presentation of the moon's motion may be found in "Newt. Achievement," pp. 262–268. On Newton's theory of the moon, see also *Isaac Newton's Theory of the Moon's Motion (1702)*, introd. Cohen (§1.3, n. 53 above), and the two doctoral dissertations (by Craig Waff and by Philip Chandler) referred to in §8.14, n. 78 below.

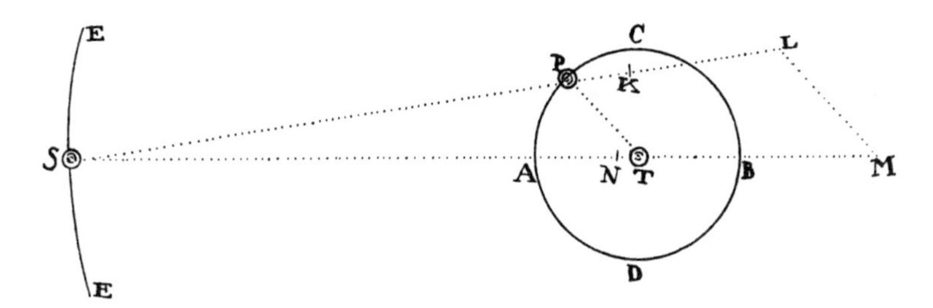

Figure 6.2. The three-body problem (book 1, prop. 66, corollaries). The sun (S = "Sol") exerts a perturbing force on the moon (P = "Planeta," a secondary planet or satellite) as it moves in orbit around the earth (T = "Terra").

simply drew a representation of the Copernican system, unconsciously putting the sun (S for "Sol") at the center and having a planet (P for "Planeta") in orbit around it. Then, continuing the alphabetical sequence, he designated the perturbing body, located beyond P, by Q. In the second edition, he changed the S of the central body to T (for "Terra") but did not call the orbiting body L (for "Luna"), keeping it as P (for secondary planet or planetary satellite). Perhaps the reason was that he had already used L to designate a point in the alphabetical sequence L, M, N, which he did not want to disturb. The third change was in the designation for the perturbing outer body, which now became S (for "Sol").

Prop. 66 is discussed in some detail in §8.15, but the method may be summarized briefly here. Newton first considers the motion of a body P (like the moon) in orbit around a body T (like the earth) under the action of an inverse-square force. (See fig. 6.2.) P will move in an ellipse with center at the center of T, and the radius vector TP will sweep out equal areas in equal times. He then introduces an outer perturbing body S, quite far away, and assumes that all three bodies will attract one another according to the inverse-square law. Next, Newton resolves the force of S on P into two components, one parallel to the direction of a line from P to T, the other directed from S to T. He then examines the effect of each of the three forces. The component parallel to TP (designated by its measure LM) will be added to the centripetal force of T on P, and, not being an inverse-square force, it will cause the net force to be no longer as the inverse square of the distance and so will cause the orbit to deviate from an ellipse. The second component of the force of S on P acts in the direction from T to S, the same direction as the force that S exerts on T, so that when it is combined with the other two forces, it will give a resultant force that no longer is directed from S to P and so will cause a deviation from the elliptical orbit and from motion in orbit according to the law of areas. The net amount of this third force, designated by NM, is the second of the two perturbing forces. Each of these two perturbing forces (LM and NM) varies in relation to the relative positions of S and P, causing a deviation from an ellipse or from the law of areas in accordance with its magnitude. The twenty-two corollaries to prop. 66 explore the consequences of various conditions of S and P. Newton's method, including a variant of the diagram he used, is still part of our current analysis of these two forces of perturbation.[61]

One corollary to prop. 66, corol. 14, is of special interest (and is analyzed in detail in §10.16) because it is used in determining the mass and density of the moon. As we shall see, although Newton refers specifically to corol. 14 in his computation of the density of the moon (in book 3, prop. 37, corol. 3), density is not mentioned explicitly anywhere in book 1, prop. 66, corol. 14.

Another interesting feature of book 1, prop. 66, corol. 14, is that the final result is stated in terms of the "apparent diameter" of the sun. Clearly, the "apparent diameter" must imply a context of observation, which will surely seem anomalous in the apparently purely mathematical level of discourse. Newton needed the result in terms of "apparent diameter" in book 3, prop. 37, corol. 3, however, because he did not want to base the mass and density of the sun on a parameter whose numerical value was uncertain, namely, the solar parallax. The apparent diameter of the sun appears only in the final sentence, suggesting that this sentence was added as an afterthought when Newton later discovered he needed it for book 3, prop. 37, corol. 3.

A final feature of book 1, prop. 66, may be noted. Here Newton makes use of the concept of "absolute force" (or absolute measure of a force), thus bringing in the masses of bodies. In fact, in corol. 14, a specific condition is the proportionality of the "magnitude" or mass of a body and the absolute measure of the force of attraction it exerts. Up until sec. 11, Newton has been exclusively considering unit masses or mass points and so has used only "accelerative" forces (or accelerative measures of forces).[62]

Prop. 67 considers that body S is moving in orbit. The proposition states that a radius vector from S will "more nearly" describe areas proportional to the time with respect to the center of gravity of T and P than with respect to either T or P. Similarly, its orbit will more closely have the shape of an ellipse if a focus is located at that same center of gravity than if the focus were placed in the center of

61. Forest Ray Moulton, *An Introduction to Celestial Mechanics*, 2d ed. (New York: Macmillan Co., 1914; reprint, New York: Dover Publications, 1970), chap. 9, §185, pp. 337ff. Cf. I. B. Cohen, "The *Principia*, the Newtonian Style, and the Newtonian Revolution in Science," in *Action and Reaction: Proceedings of a Symposium to Commemorate the Tercentenary of Newton's "Principia,"* ed. Paul Theerman and Adele Seeff (Newark: University of Delaware Press; London and Toronto: Associated University Presses, 1993), pp. 61–104.

62. On these measures, see defs. 6, 7, and 8 and §4.10 above.

T or P. Prop. 68 and its corollary touch on systems in which "several lesser bodies revolve about a greatest one," and prop. 69 extends the construct to a system of many bodies. The subject of prop. 69 is a system of several bodies attracting one another in proportion to the inverse square of the distance; it is proved that the absolute force exerted by any of the attracting bodies is proportional to the mass.

Sec. 12 (Some Aspects of the Dynamics of Spherical Bodies); Sec. 13 (The Attraction of Nonspherical Bodies); Sec. 14 (The Motion of "Minimally Small" Bodies) **6.13**

In secs. 1–10, Newton has considered bodies to be more or less unit masses or mass points, without dimensions or size, shape, composition, or mass—with the exception that mass was introduced at the end of sec. 11. Now, Newton, in that continued counterpoint between what I have called a stage one and a stage two of the Newtonian style, will add new conditions to his mathematical construct by introducing further aspects of the world of nature. Thus, in sec. 12, Newton introduces physical dimensions and shapes by exploring the properties of spherical bodies. Prop. 70 proves that there is a null force inside a hollow spherical shell whose particles attract one another with an inverse-square force. Under these conditions (prop. 71) a particle or corpuscle outside the shell will be attracted by an inverse-square force directed to the shell's center. The succeeding propositions explore some primary examples of attraction, including the attraction of solid homogeneous spheres and of spheres composed of homogeneous spherical shells, that is, spheres whose density at any internal point depends only on the distance from the center.

The next step is to advance from spherical bodies to "the laws of the attractions of certain other bodies similarly consisting of attracting particles." But, as Newton explains in the concluding scholium to sec. 12, it is not "essential" to his design to "treat these in particular cases." It suffices to set forth "certain more general propositions concerning the forces of bodies of this sort and the motions that arise from such forces." The reason is that such propositions will be "of some use in philosophical questions," that is, in questions of natural philosophy, or physical science. This leads (sec. 13) to the forces of attraction of nonspherical bodies. Among problems studied are the attractive force of bodies composed of particles exerting forces of attraction that vary as the inverse cube (or higher powers) of the distance (prop. 86), the attraction of a uniform circular lamina (prop. 90), and the attraction exerted by a solid of revolution on a corpuscle placed on the axis of revolution, where the "centripetal forces [are] decreasing in any ratio of the distances" (prop. 91). Prop. 93 deals with the attraction of "a solid, plane on one side but infinitely extended on the other sides."

The concluding sec. 14 is concerned with the motion of "minimally small bodies," that is, corpuscles acted on "by centripetal forces tending toward each of the individual parts of some great body." A feature of the propositions in this section (notably props. 94, 95, 96) is a change in direction of motion and in speed that is like the observed refraction of light with respect to initial and terminal paths. In these propositions, the sine law of refraction has the sense that the velocities before and after the "refraction" are in the inverse ratio of the sines of incidence and of refraction, as Newton believed is the case for light. That is, the velocity before incidence is to the velocity after refraction as the sine of the angle of refraction (with respect to the normal) is to the sine of the angle of incidence. This proposition was a feature of the corpuscular theory of light as opposed to the wave theory, which led to the inverse of this proportion. In these propositions, furthermore, the "bending" of the path at the interface is produced gradually rather than at once.

In a scholium Newton explained, in a manner like that adopted in the preceding sections, that there is an "analogy" between the mathematical exploration in sec. 14 of the motion of bodies and the propagation of rays of light. He has, therefore, added a few propositions "for optical uses," while nevertheless insisting that he is not at all entering into any discussion "about the nature of the rays [of light]," that is, "whether they are bodies or not." His sole purpose, he avers, is to determine "trajectories of bodies, which are very similar to the trajectories of rays." This analogy depended on the fairly recent discovery by Ole Rømer that light is transmitted with a finite speed. Newton was concerned with the problems of spherical telescope lenses, which suffer from chromatic aberration. Unless the errors from this latter source are corrected, he concluded, all labors in correcting other errors will be of no avail.

CHAPTER SEVEN

The Structure of Book 2

Some Aspects of Book 2 7.1

Book 2 of the *Principia* differs from books 1 and 3 in a variety of ways. One of the most striking of these is the fact that its major contents, the theoretical and experimental study of the forces of resistance to motion in various types of fluids, is usually not discussed and often not even mentioned by historians of science or even by some historians of mechanics, such as Ernst Mach. Those historians who have studied book 2 have been generally concerned with certain special topics and not with the main theme of book 2: the theoretical and experimental investigation of fluid mechanics. These topics are: (1) lem. 11 (and the accompanying Leibniz scholium), in which Newton set down for the first time in print his methods of the calculus; (2) the concluding scholium, in which Newton offers a proof that Cartesian vortices lead to conclusions which are inconsistent with Kepler's laws; (3) the determination of the speed of sound, in which attention is usually directed solely to Newton's error in supposing an isothermal rather than an adiabatic expansion of air; and (4) Newton's investigations of the solid of least resistance. The mathematical portions of book 2 have, not surprisingly, been carefully studied and analyzed in a most useful way by D. T. Whiteside.[1]

Newton himself was in part responsible for this rather cavalier treatment of book 2 since he suggested to readers that they confine their study to the beginning of book 1 and then skip over to book 3, on the system of the world. For the intellectual historian and philosopher of science, the system of the world has always been a more glamorous field of study than an investigation of the properties of fluids. Book 2 has also suffered from its seemingly hypothetical character, apparently

1. Notably *Math. Papers*, vol. 8 and parts of vol. 6.

based on a sequence of arbitrary suppositions concerning the resistance of fluids as a function of velocity or such imagined fantasies as the consideration of a gushing cataract of water as a solid piece of ice.

Because book 2, unlike book 1, does not lead directly to book 3, it has been treated by historians of science and even by many Newton scholars as an aside, as a needless hurdle placed by the author between the theoretical principles of book 1 and their applications to the system of the world in book 3. As Truesdell has quite correctly observed, book 2 is, to all intents and purposes, "the part of the *Principia* that historians and philosophers, apparently, tear out of their personal copies."[2] In the eighteenth century, the French mathematician Clairaut had a different evaluation, that book 2 "contains a very profound theory of fluids and of the motions of bodies that are immersed in them," but this high estimate is not accepted by historians and scientists today.

Clifford Truesdell, who has been primarily responsible for the extension of scholarly interest from the topics of books 1 and 3 to the history of fluid mechanics, has expressed his opinion of book 2 in his usual forceful manner. Book 2 "is almost entirely original," he writes, "and much of it is false."[3] In his view, "New hypotheses start up at every block; concealed assumptions are employed freely, and the stated assumptions sometimes are not used at all." Truesdell finds that little from book 2 "has found its way into either texts or histories; what has, is often misrepresented." Truesdell sees Euler, rather than Newton, as the founder of modern fluid dynamics or the general subject of the physics of deformable bodies.

Nor is Truesdell wholly alone in his poor opinion of book 2. Even Brougham and Routh, whose nineteenth-century enthusiastic praise for Newton usually knows

2. Clifford Truesdell, "Reactions of Late Baroque Mechanics to Success, Conjectures, Error, and Failure in Newton's *Principia*," in *The Annus Mirabilis of Sir Isaac Newton, 1666–1966*, ed. Robert Palter (Cambridge Mass., and London: MIT Press, 1970), pp. 192–232.

3. Clifford Truesdell, "A Program toward Rediscovering the Rational Mechanics of the Age of Reason," *Archive for History of Exact Sciences* 1 (1960): 3–36, esp. 7.

Truesdell has also written an important book-length study (435 pages) entitled *The Rational Mechanics of Flexible or Elastic Bodies, 1638–1788: Introduction to Leonhardi Euleri Opera Omnia, Vol. X et XI Seriei Secundae*, vol. 11, sec. 2, of *Leonhardi Euleri Opera Omnia*, ser. 2 (Zurich: Orell Füssli, 1960). See also Truesdell's "Rational Fluid Mechanics, 1687–1765 (Editor's Introduction to *Euleri Opera Omnia* II 12)" (125 pages), in *Leonhardi Euleri Opera Omnia*, ser. 2, vol. 12 (Zurich: Orell Füssli, 1954), and his "Editor's Introduction (*Leonhardi Euleri Opera Omnia* II 13)" (118 pages) containing "I. The First Three Sections of Euler's Treatise on Fluid Mechanics (1766); II. The Theory of Aerial Sound, 1687–1788; III. Rational Fluid Mechanics, 1765–1788," in *Leonhardi Euleri Opera Omnia*, ser. 2, vol. 13 (Zurich: Orell Füssli, 1956). In the last of these three works, Truesdell describes book 2, sec. 8, of the *Principia* as "the first attempt at a quantitative theory of sound" (p. xxi). Truesdell recognizes that Newton clearly states the wave formula, that "speed of propagation = (frequency) × (wave length)," and also the principles of "the spreading and diffraction of wave motion."

no bounds, grudgingly admit that in fluid dynamics, the "Newtonian discoveries . . . effected a less considerable change" than "upon Physical Astronomy and the general laws of motion."[4] That is, "the work" which Newton "produced upon that branch of science, did not attain the same perfection under his hands, as the rest of the *Principia*."

There are, however, a number of works with which I am familiar, in which Newton's book 2 is given some serious consideration. The *Analytical View of Sir Isaac Newton's "Principia"* by Brougham and Routh, despite the negative conclusion of its authors, does discuss the actual contents of book 2 in some detail, although in a way that will seem out of date and not entirely illuminating to today's reader. An early-twentieth-century work edited by John R. Freeman contains several essays praising Newton's contributions to "hydraulics" and listing some of his primary original contributions.[5] Among them are the principle of similitude, the principle of reciprocity (or the hydrodynamical equivalence of a body at rest in a moving fluid and that same body moving in a fluid at rest), and the recognition that the oscillatory movement of a fluid in a U-tube is formally equivalent to the vibratory motion of a pendulum. These achievements are recognized by other writers on the technical history of hydraulics or of fluid mechanics. René Dugas does discuss book 2, but not in his rather extensive treatment of Newton in his history of mechanics in the seventeenth century. Nor does he even mention book 2 in the chapter on Newton in his general history of mechanics.[6] In the latter work, however, there is a discussion of some aspects of Newton's investigation of fluid mechanics in the presentation of the ideas of the eighteenth century, attention being given to the ways in which Newton's ideas had to be amended and corrected. The history of hydraulics by Rouse and Ince[7] devotes some considerable space to book 2, but this rather extensive presentation does suffer from the lack of a critical point

4. *Anal. View*, p. 163. Brougham and Routh excuse this lack of effect on the following surprising grounds: "Much more had been accomplished of discovery respecting the dynamics of fluids before the time of Sir Isaac Newton, in proportion to the whole body of the science, than in the other branches of Mechanics." This sentence is the direct opposite of the conclusion reached by studying the significance of the work of Euler and other modern masters of fluid mechanics.

5. John R. Freeman, ed., *Hydraulic Laboratory Practice, Comprising a Translation, Revised to 1929, of Die Wasserbaulaboratorien Europas, Published in 1926 by Verein Deutscher Ingenieure* (New York: American Society of Mechanical Engineers, 1929). See appendix 15, "Dimensional Analysis and the Principle of Similitude . . . ," by Alton C. Chick, esp. pp. 776–777, 796–797.

6. René Dugas, *A History of Mechanics*, trans. J. R. Maddox (Neuchâtel: Editions du Griffon; New York: Central Book Co., 1955); René Dugas, *Mechanics in the Seventeenth Century*, trans. Freda Jacquot (Neuchâtel: Editions du Griffon; New York: Central Book Co., 1958).

7. Hunter Rouse and Simon Ince, *History of Hydraulics* (Ames: Iowa Institute of Hydraulic Research, State University of Iowa, 1957).

of view.[8] A most useful appraisal of Newton's actual achievements and failures is given in the extensive and scholarly "Historical Sketch" written by R. Giacomelli and E. Pistolesi for the six-volume *Aerodynamic Theory* edited by William Frederick Durand.[9] Finally, an important critical analysis of Newton's book 2 forms a conspicuous part of an incomplete survey of the history of hydrodynamics by P. Neméný.[10] Stressing the "main concepts and ideas of fluid mechanics," Neményi tends to concentrate overmuch on inadequacies of Newton's concepts and does not attempt to explore Newton's aims and methods in their own right, nor does he pay much attention to Newton's important experiments. His presentation is nevertheless one of the most thorough and incisive analyses in print of Newton's work on fluids, written by an obvious master of the science. For example, Neményi is the only author I have encountered who has shown the weakness of Newton's "proof," at the end of book 2, that vortices contradict the laws of astronomy. A somewhat different approach to book 2, written from the point of view of an engineer and philosopher of science, is given below in §7.10.

Book 2 of the *Principia* does not on first reading appear to have the homogeneity of book 1 or book 3 and seems rather to be a collection of loosely amalgamated topics. Presented by Newton as *De Motu Corporum Liber Secundus* (On the Motion of Bodies, book 2), the first four of the nine sections deal with motion under several conditions of resistance, and so may be considered to be a complement to book 1, in which the motions studied were in free spaces or in spaces devoid of resistance. This portion of book 2 is an expansion of some propositions on this subject at the end of the original draft of book 1 and go back ultimately to the concluding portion of *De Motu*. In secs. 6 and 7, Newton develops a physical theory of the resistance of fluids. Other sections have as their subjects the motion of pendulums, some aspects of fluid motion, wave motion, theory of sound, and vortices—in short, a variety of aspects of mathematical natural philosophy. The conclusion, that Cartesian vortices produce conditions that contradict Kepler's law of areas and so must be renounced, has continually attracted attention. As Huygens observed, after reading the *Principia*, "Vortices [have been] destroyed by Newton."[11]

8. See the review by C. Truesdell in *Isis* 50 (1959): 69–71.

9. W. F. Durand, ed., *Aerodynamic Theory: A General Review of Progress* (1934–1936; reprint, New York: Dover Publications, 1963), 1:306–394.

10. P. F. Neményi, "The Main Concepts and Ideas of Fluid Dynamics in Their Historical Development," *Archive for History of Exact Sciences* 2 (1962): 52–86. A new analysis of book 2, "The Newtonian Style in Book Two of the *Principia*," by George Smith, may be found in *Newton's Natural Philosophy*, ed. I. B. Cohen and Jed Buchwald (Cambridge: MIT Press, forthcoming).

11. Christiaan Huygens, "Varia Astronomica," in *Œuvres complètes* (§6.11, n. 51 above), 21:437–439; cf. Alexandre Koyré, *Newtonian Studies* (Cambridge: Harvard University Press, 1965), p. 117.

Since most of book 2 is obviously unrelated to the eventual book 3 on the system of the world, and for that reason was not part of the original plan, the question naturally arises as to why Newton wrote this book 2 and included it in the *Principia*. In his commentary on the *Principia*, the French mathematician Clairaut suggested that the purpose of book 2 was "to destroy the system of vortices, even though it is only in the scholium to the final proposition that Newton overtly opposes Descartes" by making it evident "that the celestial motions cannot be produced by vortices." A postil stresses this point: "Newton has written this book in order to destroy the vortices of Descartes."[12]

Newton's goal, in retrospect, was primarily (in secs. 1, 2, and 3) to explore some consequences of conditions of resistance, so that he could deal with the kinds of resistance he believed to be most probable in nature: one a resistance proportional to the velocity, another proportional to the square of the velocity, and another proportional to both of these. In the discussions, Newton takes into account that the resistance which a body encounters in a fluid depends not only on the velocity but also on certain physical properties of the fluid such as density and viscosity. He also makes use of an intuitively obvious principle, but one that is of great importance and that may be taken as a sign of Newton's originality, that the case of a solid moving through a given fluid at a certain velocity is mathematically identical with having the same fluid move with that same velocity with respect to the body kept stationary. One feature of Newton's study of resistance that distinguishes it from the rest of the *Principia* is the extensive use of reports on experiments.

In the long-range history of rational mechanics, looking backward from our own times, book 2 may be more interesting from the point of view of mathematics and engineering than from that of the physics of deformable bodies or fluid mechanics. After all (as we have seen in §5.8), in book 2, sec. 2, lem. 2, Newton's outline of his method of fluxions shows how to obtain the derivative of a polynomial with rational coefficients, and later he actually uses the term "fluxion."[13]

Secs. 1, 2, and 3 of Book 2: Motion under Various Conditions of Resistance 7.2

Using the same kind of progression that we remarked in relation to the Newtonian style in book 1, Newton begins book 2 with the simplest assumption, considering motion in a medium where the resistance (sec. 1) is proportional to the

12. Published in vol. 2 of the marquise du Châtelet's French translation of the *Principia* (Paris, 1756), p. 9, §16.

13. Further to this point, we have seen that in book 1 (e.g., prop. 41) Newton introduces the explicit condition of "granting the quadrature" of specified curves, that is, finding the area under the curve or determining the integral of the function represented by the curve. See, further, §5.8 above.

velocity. Then, having explored the mathematical consequences of this assumption for the motion of bodies, he turns (sec. 2) to the next more complex condition, where the resistance is proportional to the square of the velocity. Then, in sec. 3, he considers motion in a medium where the resistance is partly as the velocity and partly as the square of the velocity.

In a scholium at the end of sec. 1, Newton admits that the condition of resistance being in proportion to the velocity is a hypothesis that "belongs more to mathematics than to nature," and in agreement with our outline of the Newtonian style, he explains how this hypothesis needs to be altered to make it conform more closely to the world of nature. In the third edition, at the conclusion of sec. 3, Newton added a scholium in which he introduces three physical factors in the resistance of a medium to the motion of spherical bodies: the tenacity, the friction, and the density.

In sec. 1, following prop. 1, there is a lemma, basically stating that if

$$a : a - b = b : b - c = c : c - d \dots$$

then

$$a : b = b : c = c : d \dots$$

which he had learned from James Gregory[14] and which had already appeared as lem. 2 in the tract *De Motu*. In the first edition, prop. 4 had only five corollaries; two more (numbered 1 and 2) were inserted in the second edition.

In sec. 2, the calculus makes use of "moments" (proportional to fluxions) of products, quotients, roots, or powers of algebraic quantities (or their geometrical equivalent), which are considered to be "indeterminate and variable," increasing or decreasing "as if by a continual motion or flux" (lem. 2). We have seen (p. 126) that he considers the example of AB, for which the moment (or fluxion) is $a\text{B} + b\text{A}$, and similarly ABC having the moment $a\text{BC} + b\text{AC} + c\text{AB}$. The moments of A^2, A^3, A^4, $\text{A}^{1/2}$, $\text{A}^{3/2}$, $\text{A}^{1/3}$, $\text{A}^{2/3}$, A^{-1}, A^{-2}, and $\text{A}^{-1/2}$ are indicated correctly as $2a\text{A}$, $3a\text{A}^2$, $4a\text{A}^3$, $\frac{1}{2}a\text{A}^{-1/2}$, $\frac{3}{2}a\text{A}^{1/2}$, $\frac{1}{3}a\text{A}^{-2/3}$, $\frac{2}{3}a\text{A}^{-1/3}$, $-a\text{A}^{-2}$, $-2a\text{A}^{-3}$, and $\frac{1}{2}a\text{A}^{-3/2}$, respectively, while the moment of any power $\text{A}^{n/m}$ will be $\frac{n}{m}a\text{A}^{(n-m)/m}$. He also gives the moments of A^2B, $\text{A}^3\text{B}^4\text{C}^2$, and $\frac{\text{A}^3}{\text{B}^2}$. Proofs are given for some specific examples, including AB, A^n, $\frac{1}{\text{A}^n}$, $\text{A}^{m/n}$, and A^mB^n, leading to the next step, to find the moments (and hence the fluxions) of polynomials.

14. See *Math. Papers* 6:33 n. 13.

In the following scholium, Newton (in the first edition, 1687) stated how ten years earlier he had informed Leibniz that he had a method of finding maxima and minima, drawing tangents, and so on, which applied to both rational and irrational quantities. He then printed the key to the cipher in which he had set forth his method in concealed form: "Given an equation involving any number of fluent quantities, to find the fluxions, and vice versa." Newton then asserted that Leibniz replied with a method he had found, which—according to Newton—"hardly differed from mine," save in language and notation. In the second edition, Newton added a phrase to the effect that his method and Leibniz's also differed with respect to "the idea of the generation of quantities." In the third edition, however, Newton introduced a wholly new scholium, quoting from a letter of his to John Collins (10 December 1672), in which the concepts are clearly stated, a letter that figured prominently in Newton's public assessment of the case for his own priority in the invention of the calculus.[15] Now omitting any reference to Leibniz, he referred to "the treatise that I had written on this topic in 1671" and restated that the "foundation of this general method is contained in the preceding lemma."

In the succeeding prop. 10, Newton considers the problem of a projectile with the resistance of a medium varying jointly as the density of the medium and the square of the velocity. Newton revised this proof for the second edition and then almost completely rewrote it after the type had been set and the pages printed off.

Problems with Prop. 10 7.3

Anyone studying the history of the *Principia* will find book 2, prop. 10, to be of special interest. The problem is to find the density of the medium that makes a body move in any given curve under the supposition that gravity is uniform and of constant direction and that the resistance of the medium varies jointly as the density of that medium and the square of the velocity; additionally, the velocity of the body and the resistance of the medium are to be found. This proposition is notable, among other things, for a display of Newtonian fluxions (or "moments"). The text of Newton's manuscript, written out by Humphrey, passed unaltered into type in the first edition with the exception of a few minor changes, substitution of the Latin word "infinite" for Newton's "indefinite"; there is no indication of whether this was Halley's "improvement" or a printer's

15. See Hall, *Philosophers at War* (§1.3 n. 56, above) and the documents and commentaries in *Math. Papers*, vol. 8.

error.[16] When Cotes received the amended and enlarged text of this part of the *Principia* for publication in the second edition, there were only minor alterations in prop. 10. We do not have the actual text sent by Newton to Cotes and then perhaps slightly revised by Cotes and passed on to the printer, but we may reconstruct it—for the most part—on the basis of the revisions entered by Newton in his annotated and interleaved copies of the first edition.[17] These, so far as can be determined, were minor. They include a quite proper change of "Q.E.D." into "Q.E.I." at the end of the demonstration; a substitution of "3XY and 2YG" for "XY and YG" in ex. 3; a substitution of "BN" for "DN" and the insertion of "densitas illa" in ex. 4; a change of $\frac{3nn+3n}{n+2}$ to $\frac{2nn+2n}{n+2}$ in the scholium; and so on. There were thus, apparently, no major modifications of the text. We may gather from the Newton-Cotes correspondence that Cotes found no reason to make any extensive alterations and passed the slightly emended copy to the printer and that it was duly composed and printed off as pages 232–244, comprising the final leaf of signature Gg, the whole of signature Hh, and the first leaf of signature Ii.

In September 1712, Niklaus Bernoulli arrived in London and was introduced by A. de Moivre to Newton, who had him to dinner on two occasions. Bernoulli informed de Moivre that his uncle Johann (I) Bernoulli had found a serious error in prop. 10 with regard to "the motion of a body which describes a circle in a resisting medium."[18] Newton at first thought (or said he had first thought) his error came "simply from having considered a tangent the wrong way" and that "the basis of his calculation and the series of which he made use stand rightly as they are."[19] The error, however, was far more fundamental.[20] Newton then faced a double problem. The first was to recast his mathematical argument and to eliminate any possible error, and the second was to condense his presentation so that it would occupy exactly the same space as the text of prop. 10 that had already been printed off. In fact, when Newton, some time later, sent Cotes "the tenth Proposition of the second book corrected" on 6 January 1712/13, almost all of

16. The printed text of the first edition also differs from the printer's manuscript in some other very minor details, among them (in addition to "infinite" for "indefinite" {255.5}): "dicantur" for "ducantur" {256.28}, the insertion of "ille" {259.2}, "Simili" for "Et simili" {263.10}, and "X" for "V" {264.15}. The numbers in curly braces indicate page and line numbers of the third edition, as in our Latin edition with variant readings.

17. These are all included in our Latin edition with variant readings.

18. De Moivre to Johann Bernoulli, 18 October 1712, quoted in *Math. Papers* 8:52.

19. Compare this account of events with Newton's statement in the unpublished preface (§3.2 above).

20. Whiteside has set forth the details of Newton's error in *Math. Papers* 8:53 n. 180, supplemented by p. 50 n. 175; see his §6, appendix 2.1.

the *Principia* had been printed off,[21] so that the new text had to be fitted into the existing space. Newton reckoned that the substitute material would "require the reprinting of a sheet & a quarter from pag. 230 to pag. 240," comprising the fourth leaf of signature Gg and the whole of signature Hh. As was the custom of those days, the cancel or new leaf (Gg4 = pp. 231/232) was pasted on the stub of the cancellandum, as may plainly be seen on a close examination.[22] This alteration did not escape the notice of the reviewer of the second edition in the *Acta Eruditorum*, who wrote:

> The renowned Bernoulli, in the *Acta* for the year 1713, page 121, noted that Newton, opportunely advised by himself (through his nephew Niklaus Bernoulli), before the completed printing of the new edition, had corrected what he had erroneously stated in the former edition concerning the ratio of the resistance to gravity, and had inserted the [corrected text] into his book on a separate sheet of paper. A comparison of the new edition with the previous one and with Bernoulli's remarks on it shows that this was done; and the sheets cut out and the new ones substituted in their place certainly declare that the errors of the first edition already had crept also into the second.[23]

Newton was readily able to determine that there was an error in a numerical factor in his ex. 1. Our first real glimpse of the extent of Newton's effort to correct his mistake came from a study made by A. Rupert Hall on how Newton corrected the *Principia*.[24] The actual magnitude of Newton's labors with regard to prop. 10 have become fully manifest in the publication, in volume 8 of Newton's *Mathematical Papers*, of the successive manuscript drafts of Newton's revisions and recastings of his argument and their analysis by D. T. Whiteside. Some sense of the magnitude of Newton's labors can be seen in the bare fact that these texts and analyses occupy well over a hundred printed pages in Whiteside's edition. Newton completed the mathematical part of the assignment, followed by tailoring it to fit the pages of the new edition, only some three months after announcing to Cotes the existence of the problem.

21. Actually, by 23 November 1712, Cotes had reported to Newton that the whole treatise had been printed off except for some nineteen (actually twenty) lines of the final sheet Qqq, which was to contain whatever conclusion Newton would add—eventually the General Scholium. See my *Introduction*, pp. 236–237.

22. I have examined many copies of the second edition, in libraries in a number of different countries, and have never encountered one that had the original text of pages 231–240.

23. *Acta Eruditorum*, March 1714:134.

24. A. Rupert Hall, "Correcting the 'Principia,'" *Osiris* 13 (1958): 291–326.

Whiteside's presentation of prop. 10—consisting of a general introduction,[25] texts and translations of successive versions and emendations, and extensive notes—constitutes a book-length monograph.[26] In particular he has recounted and critically analyzed in full the Bernoullis' claims about the inadequacy of Newton's grasp of the calculus, and he has provided explanatory commentaries on Newton's mathematics as exhibited in his presentations. In prop. 10, together with its corollaries and examples, Newton draws heavily on his method of fluxions and their application in infinite series. Brougham and Routh quite appropriately ignored Newton's own style of presentation and rewrote his development of prop. 10 wholly in the direct language of the calculus, using only the Leibnizian and post-Leibnizian algorithms of differentiation and integration.[27]

In the demonstration of prop. 10, Newton introduced momentaneous, or infinitely small, increments of the ordinates and essentially determined the difference between the desired curve and the parabola that would result from a case of no resistance in a uniform force field. In the second paragraph Newton, without making any concession to the reader, writes: "for the abscissas CB, CD, and CE write $-o$, o, and $2o$" and "for MI write any series $Qo+Ro^2+So^3+\ldots$." Here o is the nascent or infinitely small increment of the ordinate. In the demonstration, Newton takes the abscissas CD, CE as positive so that CB (in the opposite direction) is negative. Thus $CD = o$, $CE = (CD + DE) = 2o$, $CB = -o$. As Le Seur and Jacquier and Whiteside find it necessary to explain, Newton here expresses the "differentia fluxionalis" or instantaneous increase MI of the ordinates CH, DN by an infinite series $Qo + Ro^2 + So^3 + \ldots$ in which Q, R, S, ... are (in their more familiar Leibnizian equivalents) $\frac{de}{da}$, $\frac{1}{2}\frac{d^2e}{da^2}$, $\frac{1}{6}\frac{d^3e}{da^3} = \frac{1}{3}\frac{dR}{da}$, ... Newton explains that what he calls "the first term is the term in which the infinitely small quantity o does not exist; the second, the term in which that quantity is of one dimension; the third, the term in which it is of two dimensions; the fourth, the term in which it is of three dimensions; and so on indefinitely." He observes that if the "line-element IN is of a finite magnitude, it will be designated by the third term along with the terms following without limit. But if that line-element is diminished infinitely, the subsequent terms will come out infinitely smaller than the third and thus can be ignored."

One of the features of Whiteside's presentation is his documented analysis of the statements (and misstatements) of the Bernoullis concerning Newton's method

25. *Math. Papers* 8:48–61.

26. Ibid., pp. 312–420.

27. *Anal. View*, pp. 202–205.

of fluxions in general and his analysis in prop. 10 in particular, and of the charges and countercharges between Johann Bernoulli and John Keill. Whiteside takes note that Newton did not, in his preface to the second edition, mention Bernoulli's name or refer to the revised prop. 10 "which the latter had (by way of his nephew) led him to reconstruct with so much trouble." We may agree with Whiteside that this was "at once a meanness—if not spite—and a cowardice which stored up in Bernoulli's mind a hoard of bitter recriminations which, once publicly displayed, both lived on to regret."[28]

Problems with the Diagram for the Scholium to Prop. 10 7.4

The diagram near the end of the scholium to book 2, prop. 10, has had a very curious history. Part of the problem arises from Newton's text which, in all the editions produced by Newton himself, uses the same letter H for two distinct points on a hyperbola, an upper and lower intersection with another curve, a portion of a circle. That is, he writes that "from the two intersections H and H two angles NAH and NAH arise." To eliminate the problem of the two points having the same letter, some later editors and translators have altered the text and the diagram in ways that depart from the strict sense of the original. Had Newton used the notation common in our days, in which one point would be labeled H and the other H′, the problem would not have occurred.

The diagram in question appears in the first edition in the proof of ex. 3 of prop. 10 and again in the following scholium, in the midst of rule 8. This diagram has more elements than are needed at either location. As was common in those days, two diagrams were often combined in order to reduce the number of actual woodblocks that had to be cut. As may be seen in the diagram from the first edition, reproduced here as fig. 7.1, the two intersections of the curves are designated by the same letter H. A dotted line A*h* indicates the tangent to a curve A*k*, a tangent which meets the outer of the two intersecting curves at a point marked *h* that lies between the points H and H. In the second edition, the diagram is unchanged, appearing first in the demonstration of prop. 10 (in the midst of ex. 3) and then in the scholium (in the midst of rule 3 and again in rule 8).

28. *Math. Papers* 8:362 n. 37. On p. 373 n. 1, Whiteside calls attention to the fact that in the second ("corrected") edition of Bernoulli's *Opera Omnia* (Lausanne and Geneva, 1742), the editor, Gabriel Kramer, published (vol. 1, pp. 481–493) in parallel columns the original text of the 1687 edition and the corrected text of the 1713 edition, so that the reader could learn easily and in detail how Newton had profited from Bernoulli's criticism.

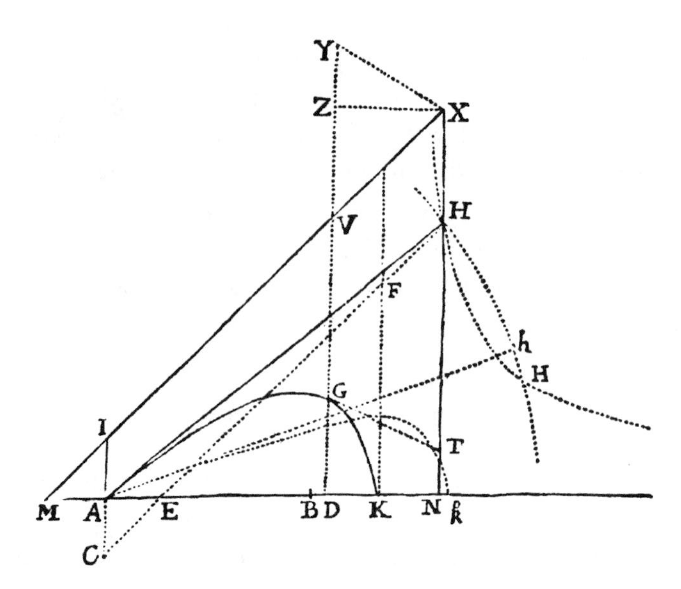

Figure 7.1. The diagram for book 2, prop. 10, and the succeeding scholium, as in the first edition, 1687. (Grace K. Babson Collection, Burndy Library)

In the third edition, however, the portion of the diagram needed for the proof of prop. 10 is a separate diagram from the portion needed for the scholium. The problem with the two H's thus occurs only in the scholium. Here, as may be seen in the diagram for the scholium, reproduced as fig. 7.2, the lower letter H has been removed and there is no letter to designate the bottom intersection of the curves.

The reason why Pemberton eliminated the lower H in the diagram is that when Newton introduces the diagram in the scholium, just before the rules, he mentions only a single point H. The text at this point may seem confusing in any case, because Newton writes that the line "AH produced meets the asymptote NX [to the hyperbola AGK] in H." It would seem that if AH is "produced" or extended, it should go beyond the point H. The first reference to *h* occurs in rule 7. As may be seen in the text, the reference to the two H's occurs only toward the end of rule 8, at the conclusion of the scholium. Accordingly, since the second H does not appear in the main body of the scholium, Pemberton (the editor of the third edition) removed the letter H from the lower intersection so that readers of the main part of the scholium would not be confused by having two different points with the same letter designation. In this altered form the letter H in the diagram unambiguously refers to the upper intersection of the two curves (it lies on the tangent to the curve AK). The lower intersection of the two curves now has no designating letter. Although Newton refers in the text of the third edition, as in the previous ones, to "two intersections H and H" and to "two angles NAH

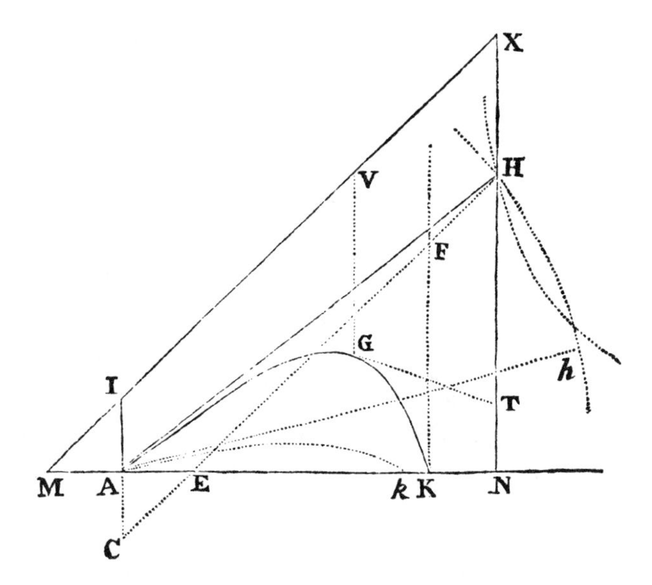

Figure 7.2. The diagram for the scholium following book 2, prop. 10, as in the third edition, 1726. Note the shift in the position of the point *h* and the disappearance of the second H. (Grace K. Babson Collection, Burndy Library)

and NAH," the reader of the third edition would find only one such marked intersection H and one marked angle NAH.

As may be seen in fig. 7.2, Pemberton made a second alteration of the diagram. The position of the dotted line A*h* has been shifted so that the point *h* occurs below the lower intersection of the curves (unlike in the earlier editions, where *h* lies between the two intersections). This alteration has no mathematical significance.

Motte's English translation (1729) gives a faithful rendition of Newton's text, referring (vol. 2, p. 42) to "two intersections H, H," from which "there arise two angles NAH, NAH." In the diagram (vol. 2, plate 2, fig. 6), as in the third Latin edition, only the upper intersection of the two curves is lettered. Again, as in the third Latin edition, the point *h* is very close to the lower intersection of the hyperbola and the circle.

An alteration of both text and diagram appears for the first time in the great Latin edition with commentary, produced in the mid-eighteenth century, by Le Seur and Jacquier. In the first edition (vol. 2, 1740), the text of rule 8 refers to two points "H, H" and two angles "NAH, NAH," accurately reproducing the expressions found in the 1726 edition of the *Principia*, edited by Pemberton. In this 1740 edition, the diagram, which appears several times within the "regulae," has two forms, evidently corresponding to two different woodblocks. In one set of diagrams (p. 102, just before the rules; p. 107, in rule 7; p. 109, in rule 7 once

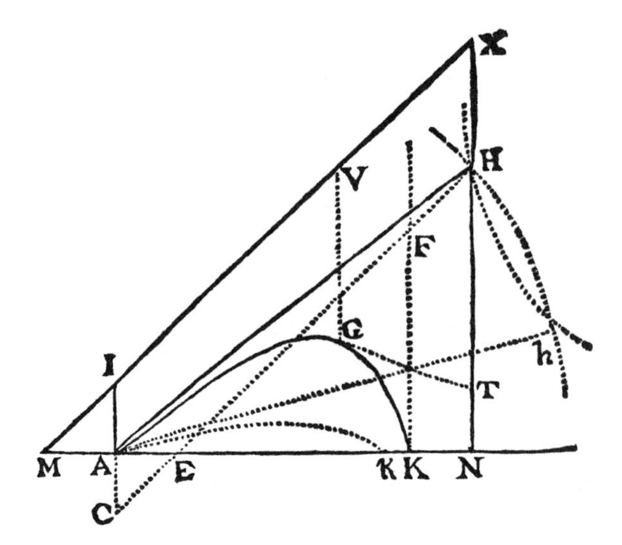

Figure 7.3a. One of the diagrams in the Le Seur and Jacquier edition, vol. 2 (1740). Note that in this version of the diagram, the line A*h* passes just below the lower intersection of the two curves; in the text, however, the letter *h* is used to denote both a point on the line A*h* and the lower intersection of the two curves. (Grace K. Babson Collection, Burndy Library)

again) the line A*h* is drawn, as in the 1726 edition, so as to pass a little below the intersection of the two curves (see fig. 7.3a). But in rule 4 (p. 105) and again in rule 8 (twice: p. 110, p. 112), the last of the "regulae," the diagram appears in the second form (see fig. 7.3b), in which the line A*h* actually passes through the lower intersection of the two curves, which thus becomes a triple-intersection point marked *h*.

In the second edition (vol. 2, 1760), there are two forms of the diagram, just as in the 1740 edition, but they are placed differently. That is, the correct diagram, in which the line A*h* passes below the intersection of the two curves rather than forming a triple intersection, occurs just before the rules (p. 102) and in rule 7 (twice: p. 107, p. 109). The other form of the diagram (with a triple intersection marked *h*) appears in rule 4 (p. 105) and twice in the final rule 8 (p. 110, p. 112).

Thus, in the second edition, the conclusion of rule 8 cannot be presented in the form in which Newton wrote it, since there would then be an anomaly in that the final diagram would not agree with the text. But, as is apparent to anyone who compares this text with the text of the three authorized editions of the *Principia* and of the preceding edition of Le Seur and Jacquier, someone evidently noticed the discrepancy between the final diagram and the text. In the diagram, there is obviously no second intersection "H" of the curves; this intersection is now "*h*." Additionally, there are not two different angles "NAH," but rather an angle NAH

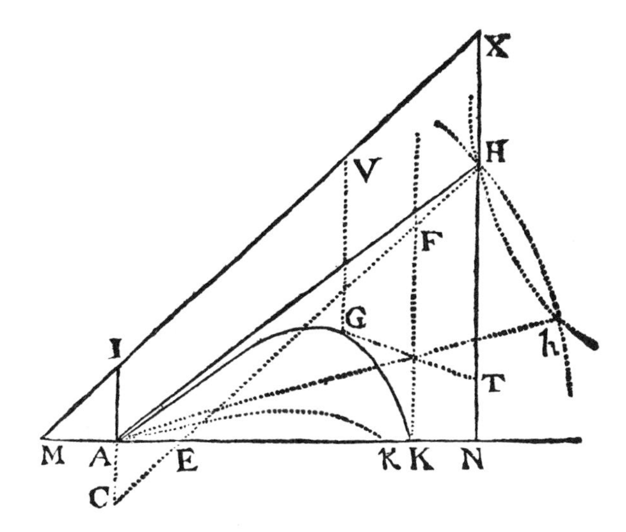

Figure 7.3b. The alternative diagram from the Le Seur and Jacquier edition, vol. 2 (1740). Here the line A*h* passes through the lower intersection of the two curves, so that the letter *h* unquestionably denotes both a point on the line A*h* and the intersection of the two curves. (Grace K. Babson Collection, Burndy Library)

and another NA*h*. Consequently, some repair was needed. Rather than altering the diagram to fit the text, however, someone changed Newton's text so as to have it be in accord with the diagram. Accordingly, in the text of the 1760 edition, Newton's two points "H, H" and two angles "NAH, NAH" have become two points "H, *h*" and two angles "NAH, NA*h*."

The Le Seur and Jacquier edition was twice reprinted in the nineteenth century (Glasgow, 1822, 1833), edited by John M. Wright. A prefatory note alleges that the two early printings (1740, 1760) had "been carefully collated . . . so as to bring to light any errors and discrepancies." Wright, it is declared, "threw out in one fell swoop" the "faults which had lain hidden in the former editions." In fact, in both of these two Glasgow reprints, all of the diagrams are of the form in which the two curves and the line A*h* meet in a single triple intersection marked *h*; the accompanying text is not Newton's but the text of the 1760 edition with its reference to two points "H, *h*" and two angles "NAH, NA*h*."

In the nineteenth-century American editions of Motte's translation, Newton's text is altered as in the second edition of Le Seur and Jacquier. Thus it reads, "from the two intersections H, *h*, there arise two angles NAH, NA*h*." But the diagram is essentially the same as the one that appears in the third Latin edition. Close inspection shows that the point *h* is not one of "the two intersections" of the curves, but rather the intersection of the straight line A*h* and the outer curve. Hence the

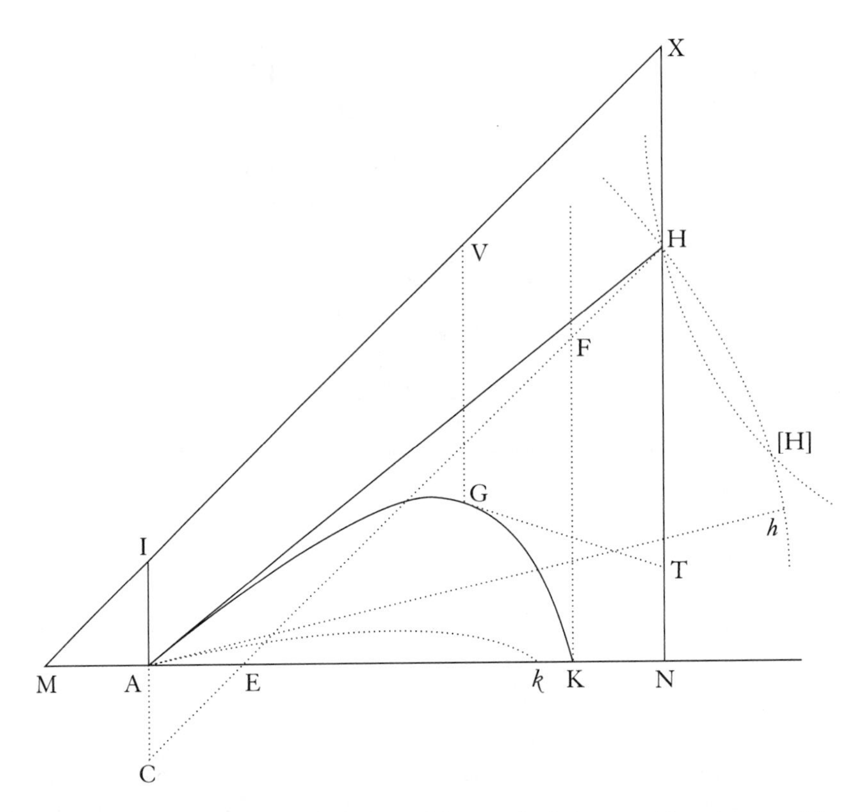

Figure 7.4. The diagram for the scholium following book 2, prop. 10, as in our edition—modified from the version of the third edition. Note the shift in position of the line A*h* and the addition of a second H in square brackets.

text no longer agrees with the diagram. These two intersections, however, occur so close that a first glance may lead to the erroneous conclusion that the line A*h* is drawn to the intersection of the curves (Newton's second H) rather than to a point somewhat lower down.

Finally, in the Motte-Cajori version, the diagram is like the one in the Le Seur and Jacquier editions in which the line A*h* passes through the lower intersection of the two curves. The letter *h* is used to mark a triple intersection which is a confluence of the two curves and the line A*h*. The text is made to agree with the second edition of Le Seur and Jacquier, rather than being a modernization of the Motte translation. It reads, "from two intersections H, *h*, there arise two angles NAH, NA*h*."

In our diagram, as may be seen in fig. 7.4, we have denoted the lower intersection of the two curves by the letter H within square brackets in order to indicate that in the third edition of the *Principia*, on which our translation is based, this letter is missing. We have left the upper intersection marked by H (as in the third edition) and have restored the point *h* to a position below the point [H], where it was placed by Pemberton. It is a matter of some interest that this curious problem of the diagrams and text of prop. 10 and the scholium has not hitherto been noted. The reason would appear to be that, as has been mentioned earlier, scholars have not subjected book 2 to the same minute and scrupulous examination that has been given to books 1 and 3.

7.5 *Secs. 4 and 5: Definition of a Fluid; Newton on Boyle's Law; The Definition of "Simple Pendulum"*

Sec. 4 is devoted to spiral motion in a resisting medium. Newton considers centripetal forces varying as the inverse square of the distance (prop. 15) and as any *n*th power. In sec. 5, Newton defines a fluid as a deformable body, that is, one whose parts "yield to any force applied" and thus are "moved easily with respect to one another."

Props. 21 and 22 explore the forces in fluids whose density is "proportional to the compression." In prop. 23, Newton proves, first, that if a fluid is composed of mutually repelling particles, and the density is as the compression, then the interparticle force will be inversely proportional to the interparticle distance; second, the converse, that, given the law of force, it follows that the density will be as the compression. In the first edition, the order of this proposition and its converse are reversed. In a scholium, Newton shows how different integral values for the law of repelling force produce different relations between compression and density, leading him to a general condition in which the force is as any power n of the distance (D^n).

The law that the density is as the compression had been established experimentally (as Boyle's law), so that the two parts of prop. 23 prove in effect that a law of repulsion $\left(F \propto \frac{1}{D}\right)$, where the repelling force terminates at the neighboring particle, is a necessary and sufficient condition for Boyle's law.[29] As always, however, Newton tries to keep distinct the mathematical and the physical levels of discourse. Thus, in the scholium to prop. 23, he states that whether "elastic fluids" do indeed "consist of particles that repel one another is . . . a question for physics." He has treated the subject only on the level of mathematics, hoping to "provide

29. That is, on the supposition that there *are* forces of repulsion between the particles.

natural philosophers" with the means to examine whether fluids consist of the sort of repelling particles Newton has been studying.

In Newton's discussion of pendulums in book 2, sec. 6, as earlier in book 1, props. 50–52, he specifies the kind of pendulum by introducing the term "corpus funependulum" (spelled "funipendulum" in book 1); thus, the title of book 2, sec. 6, reads "De motu & resistentia corporum funependulorum." The context makes it evident that what Newton is discussing is not what is generally known today as a physical pendulum, but rather a pendulum consisting of a thread or string ("funis") with a weight or particle at the end.[30]

In today's usage, such a pendulum is known as a "simple pendulum." Thus, Bruce Lindsay, in his *Student's Handbook of Elementary Physics* (New York: Dryden Press, 1942, p. 246), defines a "simple pendulum" as a "material particle on the end of a massless string, the other end of which is fastened to a rigid support"; he adds that the "particle is allowed to swing about its equilibrium position under gravity." A simple pendulum is sometimes known as a mathematical or an ideal pendulum. Such a pendulum is contrasted with a "physical pendulum" (ibid., p. 229), said to be a "rigid body supported about some horizontal axis not through the center of mass and allowed to swing about its position of equilibrium under gravity."

In H. C. Plummer, *The Principles of Dynamics* (London: G. Bell and Sons, 1929, p. 110), it is said that a "simple pendulum" consists of "a particle connected to a fixed point by a weightless string" that is "slightly disturbed from the vertical position of equilibrium." According to G. F. Rodwell, *A Dictionary of Science* (London: E. Moxon, Son, and Co., 1871, p. 426), a "simple pendulum" consists of "a small heavy particle" that "is attached by a fine thread to a fixed point."

This designation of "simple pendulum" was used in Newton's day. A document sent by Christiaan Huygens to Thomas Helder in 1686 (*Œuvres complètes* [§6.11, n. 51 above], 9:292), a year before the publication of the *Principia*, was headed "Observatie aengaende de Lenghde van een simpel Pendulum." The accompanying drawing plainly shows a thread fastened to a rigid support, with a tiny ball at the free end. In Huygens's *Horologium Oscillatorium* (Leiden, 1673), a book prized by Newton, def. 3 in part 4 reads: "Pendulum simplex dicatur quod filo vel linea inflexili, gravitatis experte, constare intelligitur, ima sui parte pondus affixum gerente; cujus ponderis gravitas, velut in unum punctum collecta, censenda est." In R. J. Blackwell's English version (Ames: University of Iowa Press, 1986), this is rendered: "A simple pendulum is one which is understood to consist of a string, or an inflexible line devoid of gravity, and of a weight which is attached to

30. The marquise du Châtelet's version uses "corps suspendus à des fils" (prop. 27) and "pendule" (prop. 28); D. T. Whiteside uses "string pendulum."

the lowest part of the string. The gravity of the weight is understood to be located at one point." The reader may wonder why Newton did not follow Huygens in using "pendulum simplex" rather than "funependulum."

In introducing the term "corpus funependulum" (sometimes "corpus funipendulum") in his discussions of pendulums in book 1, props. 50–52, and in book 2, sec. 6, Newton seems to have been alerting the reader to the fact that in these portions of the *Principia* he was not considering common or physical pendulums. In Newton's day and in ours, the primary meaning of the word "pendulum" (without any qualifier) was and remains a physical pendulum, whereas in book 1, props. 51–52, and in sec. 6 of book 2 of the *Principia*, Newton is considering mathematical or idealized pendulums. In John Harris's *Lexicon Technicum* (London, 1704), for example, the lengthy discussion of pendulums deals exclusively with physical pendulums, that is, with material bodies (with physical dimensions) which are made to oscillate.

In the *Principia*, it would have been especially important for Newton to stress that in book 2, sec. 6, he was concerned with mathematical or simple or ideal pendulums and not with physical pendulums, because elsewhere in book 2 he was indeed considering actual pendulums. In the General Scholium, at the end of sec. 7 in the first edition and removed to sec. 6 in the later editions, Newton described in detail various experiments in which he had used actual pendulums in exploring the resistance of fluids.

The word "funipendulum" appears in book 1 in prop. 40, corol. 1, in the first two editions of the *Principia*. The vicissitudes of the text of this corol. 1 indicate that Newton himself was not certain about how to express in Latin what we have called a "simple pendulum." In the third edition, Newton's final word on this subject, he writes of a body that "either oscillates while hanging by a thread or . . . " ("corpus vel oscilletur pendens a filo, vel . . . "), but there were a number of earlier versions of this phrase. The text of the preliminary manuscript version of book 1 (the Lucasian Lectures) shows that Newton first tried "corpus vel pendulum oscilletur, vel . . . ," which he then altered to "corpus filo pendens . . . " and finally changed to "corpus vel funipendulum . . . , vel . . . ," the form that appears in the printer's manuscript and in the first two editions. In book 2, sec. 6, by contrast, the original printer's manuscript and all the printed editions refer simply to "corpora funependula," literally "bodies hanging by a thread [or string]." In the first two editions, the reference to pendulums in book 1, prop. 40, corol. 1, is spelled "funipendulum," whereas in book 2, sec. 6, in all editions, the word is spelled "funependulum." But in the third edition, "funipendulum" does appear twice. In book 3, prop. 20, Newton refers to the research of Richer at Cayenne, in which Richer constructed a "pendulum simplex," or "simple pendulum."

Motte's translation of the *Principia* (1729) refers to "funependulous bodies," but this adjective never made its way into the *Oxford English Dictionary*. The latter, however, does give three occurrences of "funipendulous," of which one (used by the mathematician Augustus De Morgan in 1863) reads as follows: "And so, having shown how the reviewer has hung himself, I leave him funipendulous."

7.6 *Secs. 6 and 7: The Motion of Pendulums and the Resistance of Fluids to the Motions of Pendulums and Projectiles; A General Scholium (Experiments on Resistance to Motion)*

In sec. 6 Newton explores various properties of pendulums. In prop. 24 he shows an awareness of Huygens's work, introducing the center of suspension and the center of oscillation. As has been noted earlier, the opening sentence of the proof states (in an unambiguous manner) the second law for continually acting forces: that the velocity generated by a given force acting on a given mass in a given time is directly as the force and the time and inversely as the mass. Our attention is especially drawn to corol. 7, showing how pendulum experiments can be used to compare the masses of bodies and also to compare the weights of one and the same body in different places so as to determine the variation in gravity. He concludes the corollary by observing, as he has already done in discussing def. 1, "And by making experiments of the greatest possible accuracy, I have always found that the quantity of matter in individual bodies [i.e., the mass of individual bodies] is proportional to the weight." These experiments are presented in book 3, prop. 6.

Following a series of propositions on pendulum motions under a variety of conditions and in relation to certain laws of resistance, there is a General Scholium devoted to experiments on resistance. In the first edition this General Scholium was printed at the end of sec. 7. Newton goes to great length in describing the details of his experiments, giving his data of experiment, and he shows himself to be a superb experimenter. In experiments in air, for example, he makes corrections for the resistance of the thread as well as the bob. He also experimented with pendulums oscillating in water and even with an iron pendulum oscillating in mercury. We may particularly note the care with which Newton prepared the support, designed the form of the ring at the upper end of the pendulum, and worked on the mode of release, so as to make certain that the pendulum oscillated in a plane perpendicular to the edge of the hook.

A major supplementary set of pendulum experiments, described at the conclusion of the General Scholium, was designed to test "the opinion [of some] that there exists a certain aethereal medium, by far the subtlest of all, which quite freely permeates all the pores and passages of all bodies."[31] Unfortunately, as Newton reported, he had lost the paper on which the data had been recorded. He did recall that the initial experiment had problems because the hook was too weak and "yielded to the oscillations of the pendulum." The unquestionable result of these experiments was to show that there exists no such subtle fluid.[32]

Having studied the resistance of fluids in sec. 6 in relation to pendulums, Newton advances in sec. 7 to some general problems of the resistance of fluids to the motion of bodies. This section of the *Principia* seems, in retrospect, to have been the least satisfactory. Rouse Ball, not usually given to direct criticism, admitted that Newton's treatment of the problems of sec. 7 is "incomplete," excusing Newton on the grounds that the "problems treated in this section are far from easy."[33] Rouse Ball nevertheless found in sec. 7 "much that is interesting in studying the way in which Newton attacked questions which seemed to be beyond the analysis at his command." Newton himself was aware that his initial treatment of this subject was unsatisfactory. Accordingly, for the second edition, he discarded props. 34–40 and their corollaries of the first edition and substituted a wholly new text, at which time he transferred the General Scholium from the end of sec. 7 to the end of sec. 6. This General Scholium is relevant to both sec. 6 and the original sec. 7, but in the revised version of the second edition, it was no longer a proper part of sec. 7.[34]

In the new prop. 34, Newton sets up a hypothetical or ideal fluid and compares the resistance encountered by a globe and a cylinder (moving in the direction of its axis), both having the same diameter. He concludes that the globe will encounter a resistance just half that encountered by the cylinder in the case of his ideal rarefied fluids. In a scholium, Newton considers the resistance encountered by solids with other shapes in such fluids.

The Solid of Least Resistance; The Design of Ships; The Efflux of Water 7.7

One topic in book 2 that has attracted considerable scholarly interest is the form of a solid of least resistance, the subject of the scholium to prop. 34. Here Newton first determines the relative proportions or shape of a frustum of a cone of a given base and altitude "which is resisted less than any other frustum constructed

31. In Roger Cotes's index to the *Principia* in the second edition, the holders of this opinion are identified as Descartes and his followers.

32. That is, Newton has shown the limits of detectability of such a medium. In the later Queries of the *Opticks*, he himself advocated an "aethereal medium."

33. *Essay*, p. 99.

34. A translation of props. 39–40, as they appear in the first edition, is given (together with a commentary) in *Newton's Natural Philosophy* (see n. 10 above).

with the same base and height and moving forward along the direction of the axis." Next he considers the solid produced by the revolution of a symmetrical elliptical or oval closed curve, moving along its axis. He then specifies a new solid generated by a particular geometric construction (in which two specified angles are each equal to 135°) and declares—without any proof or hint of the manner of proof—that the solid generated by the revolution of this figure "is less resisted than the former solid." And he adds, "Indeed, I think that this proposition will be of some use for the construction of ships." In a final paragraph, Newton generalizes his result.

On this scholium Rouse Ball remarked that "Newton's determination of the solid of least resistance is deducible from the differential equation of the generating curve, but in the *Principia* he gave no proof." As Rouse Ball explained, "The problem may be solved by the calculus of variations, but it long remained a puzzle to know how Newton had arrived at the result."[35] The key to Newton's reasoning was found in the 1880s, when the earl of Portsmouth gave his family's vast collection of Newton's scientific and mathematical papers to Cambridge University. Among Newton's manuscripts there was found the draft text of a letter, published by J. C. Adams in the *Catalogue*,[36] in which Newton elaborated his mathematical argument. Newton's mathematical reasoning on this topic was never fully understood, however, until the publication of the major manuscript documents by D. T. Whiteside,[37] whose analytical and historical commentary has enabled students of Newton not only to follow fully Newton's path to discovery and proof, but also Newton's later (1694) recomputation of the surface of least resistance.[38]

Newton's remark about the usefulness of his calculation of the solid of least resistance ("for the construction of ships") was originally stated as: "which proposition, indeed, should be of no trifling use in constructing ships." In studying various annotated copies of the *Principia*, I found that the Scots mathematician John Craig entered in his copy, alongside this paragraph, a note to the effect that he had suggested to Newton, while he was visiting Newton in Cambridge, the problem of finding the best shape for ships.[39] Whiteside has found evidence in Newton's

35. Ibid., p. 100.

36. *A Catalogue of the Portsmouth Collection of Books and Papers Written by or Belonging to Sir Isaac Newton* (§1.2, n. 11 above).

37. *Math. Papers* 6:456.

38. Ibid., pp. 470–480.

39. *Introduction*, p. 204; for additional information on this topic, see I. B. Cohen, "Isaac Newton, the Calculus of Variations, and the Design of Ships: An Example of Pure Mathematics in Newton's *Principia*, Allegedly Developed for the Sake of Practical Applications," in *For Dirk Struik: Scientific, Historical and Political Essays in Honor of Dirk J. Struik*, ed. Robert S. Cohen, J. J. Stachel, and M. W. Wartofsky, Boston Studies in the Philosophy of Science, vol. 15 (Dordrecht and Boston: D. Reidel Publishing Co., 1974), pp. 169–187.

manuscripts that Craig was indeed in Cambridge in mid-1685 for "an extended stay" ("diutius commoratus" in Newton's words), and he certainly did see some of Newton's manuscripts. Newton's original draft of this scholium, however, does not show that the sentence in question was a later insertion.

In the first edition, this scholium had a concluding paragraph which Newton omitted from the later editions. Here Newton suggested that his method of studying experimentally the resistance encountered by solids might be used to determine the "aptest shape for ships" at "little cost." That is, models of hulls could be built to scale and placed in tanks of running water. There is no evidence that Newton ever made any such experiments. Had he done so, Whiteside has noted, "he would rapidly have come to appreciate the artificiality of his supposition that resistance to fluid motion varies simply as the sine of the angle of slope," even if he had still been "unable to distinguish such considerable ancillary distorting factors as skin friction and flow disturbance."[40] In the scholium to sec. 7, prop. 34, Newton treated the design of hulls mathematically rather than experimentally. I have mentioned John Craig's claim to have suggested to Newton the problem of the shape of ships.[41] Possibly it was this experiment that he had suggested to Newton and not the application of the mathematics of finding the solid of least resistance.[42]

We do not know why Newton removed the paragraph about ships. One reason may have been the transfer of the General Scholium from sec. 7 to sec. 6. When the General Scholium was at the end of sec. 7, it followed the scholium to prop. 34 in which Newton stated his mathematical results concerning the solid of least resistance and their application to ship design. Thus the reference to experiments quite logically followed the introduction of theory. If the paragraph had remained in the General Scholium after Newton had made the shift from sec. 7 to sec. 6, then the experiments would have been mentioned before, rather than after, the theoretical introduction of this topic.

The subject of prop. 36 is the "motion of water flowing out of a cylindrical vessel through a hole in the bottom." This is so difficult a problem that in 1855 Brougham and Routh introduced it by noting that "it has not yet been completely solved."[43] We should not, they conclude, "be surprised if Newton's solution of the question is not very satisfactory." In the first edition, Newton's presentation of this problem was so totally dissatisfying or erroneous that he completely redid it for the second edition. For example, he had concluded that "the velocity of the efflux

40. *Math. Papers* 6:463 n. 23.

41. See, further, my study of this topic cited in n. 39 above.

42. This suggestion was made by D. T. Whiteside in *Math. Papers* 6:463 n. 23.

43. *Anal. View*, p. 257.

was that due to only *half* the height of the water of the vessel."[44] Even though he succeeded in correcting the error, his "investigation still remains open to very serious objections."[45]

An analysis by Clifford Truesdell leads to a very different conclusion. In the first edition, Truesdell notes, Newton "gave an essentially static argument which in fact sets aside his own laws of motion; after learning that his end result contradicted experiment, he replaced the whole passage in the second edition by an argument far less plausible, based on an *ad hoc* fiction of a cataract of melting ice."[46] Here is evidence that "Newton himself clearly believed then that his first treatment had been at fault, but he did not say how. Instead, he manufactured a maze to lead to a different answer he at that time considered right. In a language where a spade is called a spade, the first treatment of the efflux problem is wrong, while the second is a bluff." Truesdell alleges that Newton's procedure here is much the same as in his "calculation of the velocity of sound." In the latter case, according to Truesdell, Newton in the second edition introduced the "fiction" of the "crassitude of the solid particles of the air" so as to insert what would nowadays be called a "fudge factor" in order to obtain the desired numerical value "from a recalcitrant theory."[47] Truesdell concludes by observing that it "was these two flimsy and unmathematical props, the 'cataract' and the 'solid particles of the air,' no better than the vortices of the Cartesians, that drew the strongest criticism upon the *Principia*."

7.8 *Sec. 8: Wave Motion and the Motion of Sound*

In considering sec. 8, we must take into account that in Newton's day the problem of wave motion was at once physical and mathematical. The fundamental physical difficulty was conceptual: how to understand that a wave motion or disturbance might be transmitted through a medium, even though individual particles of the medium do not move forward but only bob up and down, vibrate back and

44. Ibid., p. 264.

45. Ibid., where some of these objections are set out in detail. For example: "The manner in which Newton deduces the law of resistance from the velocity of efflux is also erroneous. It is very ingenious and wonderful, but at the same time very uncertain. The proposition being false in principle, we cannot expect a corollary founded on that principle to be altogether correct. The reasoning by which the resistances to a sphere and cylinder are shown to be equal can only be correct on the assumption that all the water above the cylinder, sphere, or spheroid, whose fluidity is not necessary to make the passage of water the quickest possible, is congealed. This Newton himself admits. But such an assumption is by no means a legitimate one. It is also certain, by experiment, that the amount of the resistance depends very materially on the form of the surface of the body." See, further, "Note VII," in *Anal. View*, pp. 380–387.

46. Truesdell, "Late Baroque Mechanics" (n. 2 above), p. 201.

47. The velocity of sound is discussed below in §7.8, where the reader will find a somewhat different conclusion (by Laplace) about Newton's erroneous treatment of the velocity of sound.

forth, or move periodically in small circles or ovals. As late as 1736, as Whewell noted, Johann Bernoulli frankly admitted that he could not understand Newton's proposition on sound waves.[48] The conceptual difficulty in understanding wave motion is plainly seen in Huygens's great *Traité de la lumière* (1690). Often considered to be the founding work in the wave theory of light, this treatise postulates that the "waves" of light are not periodic.

Props. 41 and 42, opening book 2, sec. 8, present the now-familiar result that if a wave encounters a barrier with a small opening or aperture, the wave will spread out into the space beyond the barrier. In prop. 43, Newton examines the pulses propagated by a "vibrating body in an elastic medium" so as to produce a wave that moves "straight ahead in every direction." In a nonelastic medium, the motion produced is "circular." Newton then (prop. 44) compares the oscillation of water in a U-shaped pipe and the oscillation of a simple pendulum, leading (corol. 1) to the remarkable conclusion that the time of oscillation of the water is independent of the amplitude of the oscillations. Here is the explicit statement of simple harmonic motion, for which the foundations have been laid in book 1, prop. 52.

Then Newton states (prop. 45), without proof, that the velocity of waves is as the square root of the wavelength, which appears again in corol. 2 of prop. 46, dealing with transverse waves. In prop. 46, Newton defines wavelength ("length of a wave"), much as in today's elementary textbooks of physics, as "the transverse distance either between bottoms of troughs or between tops of crests." Newton notes that he has considered that "the parts of the water go straight up or straight down," but since "this ascent and descent takes place more truly in a circle," he admits that "in this proposition the time [i.e., period] has been determined only approximately."

Props. 48 and 49 deal with the velocity of pulses or waves as a function of the density and elasticity of the medium. Then, in prop. 50, Newton finds the wavelength or "the distances between pulses." He determines the number of vibrations N produced in some time t by a generating oscillator and measures the distance D moved by a generated wave in that same time t. He says that the wavelength (our λ) is the measured distance D divided by the number N. In modern terms, this says that $\lambda = D/N$, which is equivalent to $D = \lambda \times N$, from which it follows that $D/t = \lambda \times N/t$ or $v = \lambda\nu$. So far as I know, this is the first clear statement of the simple wave formula or law, linking wavelength, frequency, and speed.

48. William Whewell, *History of the Inductive Sciences from the Earliest to the Present Time*, 3d ed., 2 vols. (New York: D. Appleton and Co., 1865).

A concluding scholium takes note that the "preceding propositions apply to the motion of light and of sounds." Light is propagated in straight lines, Newton argues; it cannot be a wave phenomenon since (by props. 41 and 42) waves spread out.[49] But sound is a wave phenomenon, consisting of pulses in the air (in accordance with prop. 43), a conclusion justified by experiments or experiences, such as with "loud and deep" drums. In the scholium, Newton concludes that "it is likely that the lengths of the pulses in the sounds of all open pipes are equal to twice the lengths of the pipes."

In this scholium Newton reckons that the velocity of sound is 979 feet/sec.; 968 feet/sec. in the first edition. In the first edition Newton referred to experiments he had performed in the cloister of Trinity College, of which he had been a Fellow and where he lived, which seemed to agree with his calculation. But later, better experiments by others showed the value to be 1,142 English feet/sec. Newton could not satisfactorily explain the discrepancy (about one part in six), the cause of which was found by Laplace and announced in his great *Mécanique céleste*. Laplace showed that Newton had erred in considering that the changes in the elasticity of the air are dependent only on the compression. In the case of sound, the very rapid vibrations produce heat, which causes an additional increase in elasticity. That is, Newton's theoretical deduction of the velocity of sound (props. 48, 49, 50) had the velocity vary as (prop. 48) "the square root of the elastic force directly and the square root of the density [of the medium] inversely." In another formulation (prop. 49, corol. 1), he concluded that the velocity "is that which heavy bodies acquire in falling with a uniformly accelerated motion and describing by their fall half of the height A," where A is the height of a "homogeneous" medium, or the air, taken (in the scholium) to be 29,725 English feet. This yields a velocity of sound of 979 feet/sec. Newton acknowledged the discrepancy between the computed and the observed values and tried to account for his poor value by reference to his not having taken account of the "thickness of the solid particles of air" and other possible factors,[50] but the problem was his neglect of the ways in which the heat of compression and the cold of rarefaction affect the elasticity. Although there were a number of attempts to account for the Newtonian discrepancy, none was truly satisfactory until Laplace found the source of Newton's error and solved the problem. Newton, Laplace quite correctly observed, was the first person to investigate this subject. Newton's "theory, however imperfect, is"—according to Laplace—nevertheless "a monument to his genius."

49. This argument is repeated, with many others, in the *Opticks*, query 28.

50. See §10.17 below.

7.9 *The Physics of Vortices (Sec. 9, Props. 51–53); Vortices Shown to Be Inconsistent with Keplerian Planetary Motion*

Sec. 9 opens with a "Hypothesis" about the resistance of a viscous fluid at any point being directly proportional to the velocity of separation of the particles. This leads to two propositions (51 and 52) about vortices. One such vortex (prop. 51) is created by the rotation of an infinitely long cylinder about its axis in an infinite fluid. The other (prop. 52) invokes the rotation of a sphere about its diameter in such a fluid. Both propositions are actually false, as was pointed out first by G. G. Stokes.

In a scholium, Newton declares that his intention in prop. 52 was "to investigate the properties of vortices in order to test whether the celestial phenomena could be explained in any way by vortices." A principal feature of Newton's demolition of Descartes's system of the world was to reveal that one of its mainstays is physically untenable. That is, Newton shows that—contrary to Descartes's assertion—the vortices which carry the planets and planetary satellites in their respective orbits cannot be self-sustaining.[51] In the conclusion of book 2, he explores the motion of vortices in relation to a specific set of phenomena, the orbital motions of Jupiter's satellites and of the planets, according to Kepler's third law. If the planets are carried around the sun by vortices, and if the satellites of Jupiter are similarly carried around Jupiter by vortices, then—Newton argues—the vortices must themselves revolve in accordance with Kepler's third law.[52] In the motion of vortices, Newton has just found that the periodic times are as the squares of the distances and not as the $3/2$ powers of the distances, as in Kepler's law. Therefore, he concludes, in order to reduce the mathematically determined squared ratio to the observed $3/2$ ratio, either the matter of the vortex must become more fluid with increased distance from the center or the resistance must follow some other rule than the one being considered. In the hypothesis on which this reasoning was based, Newton postulated a resistance proportional to the velocity gradient, but he now points out that the divergence from Kepler's third law is even greater. Newton concludes: "It is therefore up to philosophers to see how that phenomenon of the $3/2$ ratio can be explained by vortices."

A concluding scholium, following prop. 53, is even more directly anti-Cartesian. Here Newton shows that the motion of planetary vortices is inconsistent with the law of areas. He therefore concludes that "the hypothesis of vortices can

51. I owe this important insight to R. S. Westfall.

52. In stating the law, Newton does not explicitly attribute it to Kepler, as he will do in the "phenomena" of book 3.

in no way be reconciled with astronomical phenomena and serves less to clarify the celestial motions than to obscure them."

Neményi has taken note that this "fascinating Scholium which concludes Book II is not convincing" for several reasons. One is that Newton has used "the continuity equation (conservation of fluid mass) in a way that is admissible only in a strictly two-dimensional (plane) flow." Thus, "the imperfections of his analysis of vortices around rotating solids" lead to a serious criticism of his alleged "success" in destroying the system of Descartes.[53] This conclusion was hardly obvious to Newton's contemporaries and immediate successors.

Having disposed of vortices along with the aether, Newton concludes book 2 by informing his readers that he will now proceed to book 3, in which he will make use of the principles elaborated in book 1 to show how the motions of celestial bodies are produced "in free spaces without vortices."

7.10 *Another Way of Considering Book 2: Some Achievements of Book 2* (by George E. Smith)[54]

An appropriate title for book 2 is the one that Newton added before the last two propositions of the augmented version of the tract *De Motu*—"De motu corporum in mediis resistentibus," or "The motion of bodies in resisting mediums." Book 2 has sometimes been considered to be a work in fluid mechanics. But it is more accurately described as an essay on a particular topic within that discipline, fluid resistance. A great part of the unfavorable criticism that has been leveled against book 2 arises from considering its subject to be fluid mechanics (in the sense in which we apply that name to work of Euler, d'Alembert, and their successors) rather than fluid resistance.

Galileo wrote that a science of fluid resistance could never be constructed because too many disparate factors are involved.[55] Three centuries later, we are still unable to calculate the forces of resistance acting on spheres, much less on arbitrary shapes, from theory alone. We have to rely on testing in wind tunnels and other contrivances. Nevertheless, a science of fluid resistance has emerged in the last century, largely in conjunction with the development of aircraft. Book 2

53. Neményi, "Main Concepts" (n. 10 above), pp. 73–74.

54. Unlike books 1 and 3, book 2 has received little systematic study in recent years. D. T. Whiteside has discussed an earlier version of the material in sec. 1, and the treatment of select problems from secs. 4, 6, and 7, in *Math. Papers*, vol. 6, as well as the disputed prop. 10 of sec. 2 in vol. 8 and in "The Mathematical Principles Underlying Newton's *Principia Mathematica*" (§1.2, n. 9 above). Several aspects of the achievements of book 2 are reviewed in various works (see nn. 4–10 above).

55. Galileo, *Two New Sciences*, trans. Drake (§6.9, n. 32 above), pp. 275 ff.

can legitimately be viewed as the first attempt to develop such a science. Although Newton's efforts fell far short of such a science, many elements in book 2 remain fundamental features of our current science.

The last two propositions of *De Motu* provide solutions for the motion of spheres under resistance forces proportional to velocity. Between this tract and the *Principia*, Newton decided, apparently solely on the basis of qualitative physical reasoning, that the component of resistance arising from the inertia of the fluid varies as the density of the medium, the frontal area of the sphere, and the square of the velocity.[56] These three quantities are still used as the dimensional parameters for resistance forces:

$$F_{resist} = ½C_D \rho A_f v^2$$

where C_D is the nondimensional drag coefficient, ρ the density of the fluid, and A_f the frontal area. Newton proposed further that there are two other components of resistance: one arising from the tenacity (or absence of lubricity) of the fluid, which he takes to be independent of velocity, and one from the internal friction of the fluid, which he takes to be proportional to velocity.[57] Newton's approach in book 2, then, is to represent resistance forces within a mathematical framework that superposes these three distinct components:

$$F_{resist} = a + bv + cv^2$$

where c varies as ρA_f, and the factors with which a and b vary remain to be determined.

Newton's mathematical framework thus splits the problem of motion in resisting mediums into two parts. One provides solutions to the equations of motion

56. Huygens had previously developed solutions for motion under resistance proportional to velocity but had not published his results after experiments convinced him that the dominant forces of resistance vary as the square of the velocity. (See pp. 168–176 of the added section of his "Discours de la cause de la pesanteur," in *Œuvres complètes* [§6.11, n. 51 above], vol. 21.) Newton, by contrast, gives no indication of having reached this conclusion from experimental considerations. The only explanation he gives for going beyond the simple case of resistance varying as the velocity is the qualitative physical reasoning given in the scholium at the end of sec. 1.

57. Newton's forces arising from the internal friction of the medium are a close counterpart of modern viscous forces. The modern "no-slip" condition, according to which the fluid particles at the immediate surface of a moving body move at the same speed as that surface, has eliminated from accounts of resistance the Newtonian forces arising from the tenacity of the fluid. Discussions of frictional forces of fluid resistance now refer to the viscous forces.

when the forces acting on the body depend on the unknown velocity (secs. 1 through 4 of book 2). The second must, in effect, provide some basis for determining values of the coefficients a, b, and c. Secs. 5 through 7 are primarily concerned with doing so for c. Secs. 8 and 9 then consider some effects of the equal and opposite reactive forces acting on the fluid medium—sec. 8, the wave motion in fluids produced by bodies moving in them; and sec. 9, the vortex motion in fluids arising from a body rotating in them.

The most celebrated results in the first five sections have been prop. 10 (on projectile motion with resistance varying as v^2) and prop. 23 (which attempts to draw conclusions about the underlying structure of air from Boyle's law). Less celebrated than it deserves to be is Newton's use of similarity principles in the corollaries of props. 5 and 7: he considers two different spheres moving in different mediums which nevertheless display entirely similar motion so long as the value of a certain key parameter remains the same for both. The quest for such similarity principles has been a dominant aspect of research in many areas of fluid mechanics ever since, culminating in the discovery of the Reynolds number, which allows, among other things, for the lift and drag coefficients of aircraft to be determined by testing scale models of them.

Sec. 4 shows that a logarithmic spiral trajectory can result from what would otherwise be uniform circular motion under an inverse-square centripetal force when the motion is occurring in a medium with density varying inversely with radius. This leads to results on the rate of decay of orbital motion under resistance, results that later provide a basis for arguing that there are no resistance forces acting on the planets.[58]

Outwardly, sec. 6 continues the line of thought of secs. 1 through 4 by providing solutions for the motion of cycloidal pendulums under different types of resistance. But the real goal of these solutions is to allow the magnitudes of different types of resistance forces—that is, b and c above—to be determined experimentally from rates of decay of pendulum motion. Newton conducted experiments in which the pendulum bob moved through air, water, and mercury. In the first edition, the results of these experiments (given in a scholium at the end of sec. 7) provide the only basis for determining the magnitudes of the different types of resistance forces. This scholium was moved to the end of sec. 6 in subsequent editions, perhaps because the results were disappointing. In order to achieve a

58. These results, along with the scholium to prop. 10, require information on how the density of our atmosphere and any medium surrounding the sun varies. This leads directly into sec. 5, where the main question is how the density of compressible mediums varies under different types of gravitational forces.

remotely reasonable fit with his data, Newton had to add an odd $v^{3/2}$ resistance term to the v and v^2 terms. He ended up being able to draw only the weak conclusion that the v^2 or inertial term dominates over the range of conditions tested and that once allowances are made for vagaries in the data, this term appears to vary as the density of the medium.[59]

In all three editions, sec. 7 offers hypothetical models in which the inertial component of resistance arises from the impact of fluid particles on moving bodies. Since impact forces depend only on relative motion, the same forces can be obtained from having moving fluid particles impinging on a motionless body. The relative-motion principle Newton puts forward here has been central to research in fluid mechanics ever since, justifying for example wind-tunnel testing of aircraft, in which air flows over a stationary scale model, eliminating the need to make measurements on a moving scale model. Newton uses an impingement model to obtain his much-discussed erroneous results on the solid of least resistance.[60] A model of this sort is also the reason for his interest in the efflux problem—that is, the problem of the flow out of a circular hole in the bottom of a vessel—for he takes the weight of the fluid impinging on a solid circular surface in the middle of the hole to correspond to the inertial resistance.

Most of sec. 7 was rewritten in the second edition. The principal gain was the extension of his impingement models to allow precise values of the inertial resistance force to be determined for spheres, disks, and cylinders (on end), given ρ, A_f, v^2, and the mass of the body. In effect, Newton concludes that c lies between ρA_f and $\frac{\rho A_f}{2}$, multiplied by a factor depending on shape, in the case of an incompressible "rarefied" medium, and by $\frac{\rho A_f}{4}$ in the case of an incompressible "continuous" medium. So long as the viscous forces are negligible, this amounts, in modern terms, to a value between 1.0 and 2.0 for the drag coefficient for spheres in a rarefied medium; and 0.5 for both spheres and disks in a continuous medium. With these coefficients in hand, the vertical-fall solution with resistance varying as v^2 of prop. 9 can be used to predict the time of fall of a sphere from a specified height in a given medium. The last part of sec. 7, added in the second edition

59. Newton originally attributed the disappointing results of these experiments to the unaccounted-for effects of resistance on the pendulum string. While this was a factor, a worse problem was the to-and-fro motion of the fluid induced by the motion of the pendulum. Newton's subsequent vertical-fall experiments showed him that the resistances inferred from the pendulum experiments were much too large to correspond to the simple situation of a sphere moving through a fluid; he then attributed this excess to the induced to-and-fro motion of the fluid.

60. See, e.g., R. Giacomelli and E. Pistolesi, "Historical Sketch," in *Aerodynamic Theory*, ed. Durand (n. 9 above), 1:311 ff., and Neményi, "Main Concepts" (n. 10 above), pp. 70–71.

and extended in the third, compares the times predicted in this way with those observed in a series of vertical-fall experiments conducted by Newton in water and air.[61] From the good agreement between the predicted and observed times Newton drew two conclusions: (1) his impingement model for continuous mediums holds for both water and air; and (2) insofar as the inertial component has now been theoretically accounted for, the small excesses in the observed times of fall over the predicted times offer a means for experimentally investigating the other components of resistance.

In both of these conclusions Newton was mistaken. This became evident in the middle of the eighteenth century when experiments revealed that the resistance on a disk is more than twice as great as that on a sphere of the same cross-sectional area.[62] The loss of interest in Newton's theoretical model of inertial resistance was accompanied by a loss of interest in the vertical-fall experiments described in sec. 7. This was unfortunate, because these experiments produced the first accurate measures of resistance forces. Values for the drag coefficients of spheres can be extracted from these experiments. Taking only the experiments in which Newton expressed confidence, these values range from a low of 0.462 to a high of 0.557. The modern measured values of the drag coefficient for spheres over the range of Reynolds numbers in these experiments vary from a low of 0.38 to a high of 0.51. Moreover, the lowest value Newton obtained occurs at the appropriate value of the Reynolds number. Measuring the drag coefficient for spheres has never been easy, owing to such secondary factors as surface finish and local small irregularities. Additionally, rotation of the sphere produces unwanted effects when it is not fixtured, while fixturing alters the flow about it. Thus, viewed as an attempt to measure drag forces, and not as a test of the hypothetical model, the vertical-fall experiments of sec. 7 represent a real achievement.[63]

61. Newton's experiments in air involved dropping glass balls from the top of the newly completed St. Paul's Cathedral in London. The third edition adds a further such experiment in St. Paul's conducted by Desaguliers. This last experiment was the most carefully performed of all of them and yielded the most accurate results.

62. The key experiments were done by Chevalier de Borda in the 1760s (see *Mémoires de l'Académie des sciences*, 1763 and 1767). He measured resistance by attaching bodies to the circumference of a rotating flywheel. See Dugas, *History of Mechanics* (n. 6 above), pp. 309–313, for a brief summary.

63. One of the few places where Newton's achievement in these experiments has been acknowledged is R. G. Lunnon, "Fluid Resistance to Moving Spheres," *Proceedings of the Royal Society of London* 110 (1926): 302–326, esp. 320–321.

Newton's theoretical model of inertial resistance involves two errors. The first he anticipated. Because the pressure disturbance of a moving body propagates ahead of it, the fluid generally does not impinge on it, but instead forms streamlines around it. The resistance force on the body then results from the net

The one item of sec. 8 that has received the most attention is the 17 percent discrepancy in Newton's theoretical determination of the speed of sound in air. As Laplace discovered more than a century later, when far more was known about the thermal expansion of gases, the discrepancy arose from an implicit assumption of isothermal expansion in pressure waves, rather than adiabatic expansion.[64] Newton's dynamical analysis of wave motion in compressible fluids is otherwise correct. Even more impressive, the same is true of his analysis of wave motion in incompressible fluids, which was based on a friction-free fluid-pendulum model—that is, on oscillating fluid in a U-shaped tube. These results initiated the investigation of oscillatory dynamics in continuous, multidimensional bodies and mediums. They would have been even more historically important if they had not proved so difficult for those immediately following Newton to understand.[65]

Although Newton is correct in his ultimate conclusion that the vortex motion induced by rotating cylinders and spheres is not compatible with Kepler's $3/2$ power rule, the results he obtains in sec. 9 for this vortex motion are not correct. Nevertheless, in the process he did define the problem of Couette flow that has received extensive attention ever since.[66] Sec. 9 also puts forward, in the form of a hypothesis, the notion of what has since come to be known as a Newtonian fluid, namely one in which the resistance force owing to the lack of lubricity of the fluid

---

effect of the fluid pressure at the surface of the body along these streamlines. As a consequence, the shape of the body is always a crucial factor.

He did not anticipate the other error, which is more profound. The drag coefficient on a body of a given shape is not a constant, but varies with the Reynolds number, $\frac{\rho d v}{\mu}$, where d is the diameter in the case of spheres, cylinders, and disks, and μ is the fluid viscosity. The resistance does not arise from two independent components, one associated with the viscosity of the fluid, and the other with its inertia. Instead, there is a single complex mechanism in which the pressure around the surface of a moving body depends on a combination of effects arising from these two, a combination that changes character depending on whether the inertial or the viscous effects dominate. Therefore, any approach that treats resistance as the superposition of two terms, one varying as v and the other as v^2, misrepresents the physics. In any such approach the value of the coefficient of the v^2 term—i.e., c in our presentation of Newton's framework—must depend not only on ρ and A_f, but also on μ!

For more details, see R. P. Feynman, R. B. Leighton, and M. Sands, *The Feynman Lectures on Physics* (Reading, Mass.: Addison-Wesley, 1964), vol. 2, chap. 41; L. D. Landau and E. M. Lifshitz, *Fluid Mechanics*, vol. 6 of *Course of Theoretical Physics*, trans. J. B. Sykes and W. H. Reid (Oxford: Pergamon Press, 1959); or D. J. Tritton, *Physical Fluid Dynamics*, 2d ed. (Oxford: Clarendon Press, 1988).

64. Thus, the correct speed of sound is given by $(\gamma p/\rho)^{1/2}$, where γ is the ratio of the specific heats (1.4 for air), and not by Newtons's $(p/\rho)^{1/2}$. In other words, the relevant modulus is γp, not p.

65. See J. T. Cannon and S. Dostrovsky, *The Evolution of Dynamics: Vibration Theory from 1687 to 1742* (New York: Springer-Verlag, 1981).

66. See, e.g., S. Chandrasekhar, *Hydrodynamic and Hydromagnetic Stability* (Oxford: Clarendon Press, 1961).

is proportional to the velocity gradient.[67] Newton was forced to put this forward as a hypothesis because, in effect, he had not managed to find a way of determining how *b*—that is, the viscous component of the resistance force—varies. This must have been among his principal goals in the second half of book 2. Perhaps his hope of achieving this goal accounts for why he put so much additional effort into sec. 7 in the second edition.

67. In modern terms, the shear stress $= \mu dv/dx$, where x is orthogonal to v.

CHAPTER EIGHT

The Structure of Book 3; The System of the World

The Structure of Book 3 8.1

Book 3 is composed of six distinct parts. The first contains a set of rules ("regulae") for proceeding in natural philosophy; the second the "phenomena" on which the exposition of the system of the world is to be based. Next comes the application of mathematical principles (primarily as developed in book 1) to explain the motion of planets and their satellites by the action of universal gravity. The fourth part sets forth Newton's gravitational theory of the tides. The fifth part of book 3 (props. 22 to 30) contains an analysis of the motion of the moon, at once—as we shall see below—one of the most brilliant and novel parts of the *Principia* and, in some ways, a failure. The sixth and last part (from lem. 4 and prop. 40 to the end) is devoted to the motion of comets. Although these divisions of book 3 become obvious to any reader on the basis of the subject matter of the individual propositions, book 3—unlike books 1 and 2—is not formally divided into sections.

Book 3 is introduced by a general paragraph in which Newton states how this book differs from the preceding ones. Here Newton proposes an abbreviated course of preparation for readers whose primary goal is to learn only enough Newtonian rational mechanics to be able to understand the Newtonian system of the world. They need do no more than "read with care the Definitions, the Laws of Motion, and the first three sections of book 1." Additionally, they may consult "at will" the other propositions of books 1 and 2 "which are referred to" in book 3.

Books 1 and 2, Newton wrote, contain many propositions which would be excessively "time-consuming even for readers who are proficient in mathematics." Then Newton freely admitted that he had purposely made book 3 difficult to read. He had originally written a version of book 3 "in popular form," so that "it might be more widely read" than the eventual printed text.[1] Later, he decided to recast the "substance of the earlier version into propositions in a mathematical style," so as to restrict the readership to "those who have first mastered the principles." He thus hoped to discourage readers who would not be able to "lay aside" their preconceptions; he wanted to "avoid lengthy disputations" on issues not directly related to the straightforward application of his principles to the system of the world.

The introduction also explained that the principles of natural philosophy Newton presented in books 1 and 2 were mathematical and not physical principles.[2] Since these principles relate to physics (or natural philosophy), he had "illustrated them with some philosophical scholiums," that is, scholiums dealing with physical topics. Among the latter he listed "the density and resistance of bodies, spaces void of bodies, and the motion of light and sounds."

Newton's procedure in book 3 is a combination of both the expected and the unexpected. As might be expected, he proceeds from a set of "phenomena" that includes Kepler's law of areas and the harmonic law—but not the law of elliptical orbits—for the planets and for the satellite systems of Jupiter and Saturn (the latter in the second and third editions), assuming that the satellites move in circular orbits. He also states that the moon moves according to the law of areas, but there is no firm evidence for such a statement. Thus he cannot proceed as he has done in book 1, prop. 11, by deducing an inverse-square force from elliptical orbits. Rather, he establishes inverse-square forces for satellite systems, but in a way that is unexpected, and then proves that the gravity which we know on earth extends to the moon. He next argues for an inverse-square solar force acting on planets and then gives reasons why all these inverse-square forces—those acting on satellites and those acting on planets—should be considered instances of the same force, namely, gravity.

Having established the inverse-square force for planets and satellites, he then deduces the elliptical orbits. Thus in prop. 1, he considers the orbits of Jupiter's satellites to be circular and their motion to be uniform. Hence, the law of areas must hold and there must be a centripetal force acting on the satellites, a force that is as the inverse square of the distance. In the second and third editions, the case is the same for the satellites of Saturn.

1. This earlier version was published in an English translation, and then in the Latin original, soon after Newton's death. A new translation, based on the manuscripts, by Anne Whitman and I. Bernard Cohen, is scheduled for publication by the University of California Press under the title *Essay on the System of the World*.

2. See §6.11 above.

With regard to planetary orbital motion, in book 3, prop. 2, Newton invokes phen. 5 (a weak empirical argument for the area law for planetary motion) plus prop. 2 of book 1 to show that there is a centripetal force directed toward the sun. Then, for an inverse-square force, he turns to phen. 4 (Kepler's third law, which is founded directly on empirical evidence) and prop. 4 of book 1 (which establishes the inverse-square law for circular orbits).

Although it might be thought that to consider planetary orbits to be circular is a gross departure from the "true" system of astronomy, the fact is that all the planetary orbits—with the exception of Mercury's—are indeed very nearly circular. Most of us are conditioned to think of planetary orbits as very noticeably flattened ellipses (or as ellipses with a rather high eccentricity), but the reason is that textbooks of astronomy and physical science grossly distort the shape of these orbits so that they can demonstrate in an easily visible fashion the effect of ellipticity on the law of areas. In fact, these textbook ellipses are more like orbits of comets than of planets. What is of primary significance about the planetary orbits is not that they are "grossly" elliptical in shape, but rather that the sun is not directly at or near the center of the orbit and hence that the area law determines a significant variation in planetary velocity from aphelion to perihelion.

In any event, Newton does not rest his case for the inverse-square law for planets on a demonstration using props. 2 and 4 of book 1, but rather turns to a very different kind of argument altogether, one that he says proves the existence of the inverse-square force "with the greatest exactness." This line of proof depends on the observed fact that "the aphelia are at rest." Invoking book 1, prop. 45, corol. 1, he takes note that the "slightest departure" from the ratio of the inverse square would produce a "noticeable motion of the apsides in a single revolution" and an "immense such motion" in many revolutions, quite contrary to observation.

Having established the inverse-square force, and having argued quite convincingly that this force is gravity, Newton observes (in book 3, prop. 13) that his discussion of the orbital motion of planets according to the law of areas has been based on "phenomena." That is, he has based his argument on reasoning that depends on observations to establish the inverse-square law. Now, in prop. 13, he will "deduce the celestial motions . . . a priori" from "the principles of motions" that "have been found." He introduces the elliptical orbits of planets as a deduction based on the inverse-square law and not on phenomena. Toward this end he uses book 1, props. 1 and 11, but especially corol. 1 to book 1, prop. 13. It will seem astonishing, therefore, that the basic theorem for book 3—that an inverse-square force implies an orbit in the shape of a conic—is buried away in a minor position as a corollary following a proposition on parabolic orbits. Even more astonishing may be the fact that this corollary was presented originally without proof and later with only a sketch of a proof in a manner which has been questioned by many critics from Newton's day to ours.

8.2 *From Hypotheses to Rules and Phenomena*

In the first edition of the *Principia*, book 3 opened with a set of "Hypotheses," nine in number. This designation made the *Principia* an easy target for criticism. Since Newton claimed that in the first two books he was concerned primarily with mathematics and since the third book appeared to be founded on "hypotheses," a critic could legitimately complain that the *Principia* did not disclose a system of natural philosophy. This was the harsh line taken by the author of a book review in the *Journal des Sçavans*, probably the Belgian Cartesian Pierre-Sylvain Régis,[3] who concluded that "in order to produce an *opus* as perfect as possible, Newton has only to give us a *Physics* [i.e., a *Natural Philosophy*] as exact as his *Mechanics*." Newton will do so, he added, "when he substitutes true motions for those that he has supposed." Possibly it was in answer to this criticism that Newton altered the opening of book 3 in the second edition. He now divided the "hypotheses" into two major groups. The former hyps. 1 and 2, which are methodological, became the first two rules in a new category, "Regulae Philosophandi," or "Rules for Natural Philosophy," and a new rule 3 was introduced. In the third edition, rule 4 was added.[4]

The second major group of original hypotheses were numbered 5, 6, 7, 8, and 9. In the second edition, these became "phenomena," bearing numbers 1, 3, 4, 5, and 6. A new one, phen. 2, was added at that time. These "phenomena" are indeed phenomenological, dealing with such matters as the observed periodic times of planets and satellites and the third law of Kepler, together with actual tables of data.

The original hyp. 3, that "Every body can be transformed into a body of any other kind and successively take on all the intermediate degrees of qualities," was eliminated in the second edition. Hyp. 4 ("The center of the system of the world is at rest") became "hypothesis 1" in the second edition but was moved from the beginning of book 3 to a position (between props. 10 and 11) where it is needed. In the second edition, Newton also converted lem. 3 of the first edition (following prop. 38) into a "hypothesis 2."

3. Paul Mouy, *Le développement de la physique cartésienne* (Paris: Librairie Philosophique J. Vrin, 1934), p. 256.

4. I. B. Cohen, "Hypotheses in Newton's Philosophy," *Physics* 8 (1966): 63–184.

The "hypotheses" of the first edition constitute a curious grouping. They include procedural rules for natural philosophy, phenomena, and two hypotheses which would be hypotheses in any system of thought or of science. If all these "hypotheses" are to be taken as a group, it is difficult to think of any single neutral name for so motley a collection. In retrospect, this assemblage seems all the more odd in light of the later General Scholium, written in 1713 for the second edition. For there Newton not only will utter the celebrated slogan "Hypotheses non fingo" but will declare that "whatever is not deduced from the phenomena must be called a hypothesis," whereas in the opening part of book 3 in the first edition, a whole set of phenomena are designated as hypotheses.

Newton's "Rules for Natural Philosophy" 8.3

The second edition of the *Principia* contains three "Regulae Philosophandi," or "Rules for Natural Philosophy." We have seen that rules 1 and 2 had been, respectively, hyps. 1 and 2 in the first edition. They declare a traditional principle of parsimony, to admit no more causes of phenomena than are "both true and sufficient" (rule 1) and to assign the same causes, as far as possible, to "natural effects of the same kind" (rule 2).

Rule 3, appearing for the first time in the second edition, is of a different sort. Its message is that there is a certain set of "qualities" that (1) are found in all bodies within our range of direct experience on earth and (2) do not vary, and that these are to be considered qualities of all bodies universally, that is, of bodies everywhere in the universe. This feature of transferring qualities from those on which we can make experiments to those on which we cannot has been called "transdiction" and also "projection."[5] Thus extension (the property of occupying space) and inertia (i.e., force of inertia or mass) are properties of this sort and so may be considered, by rule 3, to be properties of the heavenly bodies: sun, planets, planetary satellites, comets, and stars. It is by this third rule that Newton can assign to the heavenly bodies the property of gravity or of exerting a force "in proportion to the quantity of matter," and it is "by this third rule that all bodies gravitate toward one another."[6]

5. That is, a means of assigning to bodies beyond the immediate range of our sense experience the qualities or properties such as mass and inertia that we observe in bodies close at hand. Nelson Goodman has called this concept "projection," a name that has gained currency among philosophers and has replaced Mandelbaum's "transdiction." In the English-speaking philosophical community, a distinction is commonly made between hypotheses which are taken to be "projectable" and those which are not. See Maurice Mandelbaum, *Philosophy, Science, and Sense Perception* (Baltimore: Johns Hopkins Press, 1954). Philosophers have not generally adopted Mandelbaum's suggestion, preferring the term "projection."

6. Quoted from rule 3.

In stating the third rule, Newton used the words "intendi" and "remitti," the present passive infinitives corresponding to the nouns "intensio" and "remissio," that is, "intension" and "remission," which go back to the late-medieval doctrine of the "latitude of forms." This medieval usage referred to any quality—motion, displacement, and even love and grace—which could be quantified and so undergo an intension or remission by degrees. Motte correctly rendered Newton's text, referring to "qualities of bodies, which admit neither intension nor remission of degrees," but the Cajori version altered "intension" to "intensification" without considering what the expression "intension . . . of degrees" might have meant to a reader in Newton's day. That is, the sense of "intensification" in today's usage is a process of increasing to an extreme degree or becoming more concentrated, whereas "intension" signified a process of becoming greater in magnitude.[7]

John Harris's *Lexicon Technicum* (London, 1704) defines "INTENSION in Natural Philosophy" as "the encrease of the Power or Energy of any Quality, such as *Heat, Cold*, &c. for of all Qualities they say, they are *Intended* and *Remitted*; that is, capable of Increase and Diminution." Harris adds that "the *Intension* of all Qualities, increases reciprocally as the Squares of the Distances from the Centre of the Radiating Quality decreases [*sic*]." A proof "that all Qualities are *Remitted*, or have their Power or Efficacy abated, in a *Duplicate Ratio* of the distance from the Centre of the *Radiation*" is given by Harris under "QUALITY." The proof is taken from John Keill's *Introductio ad Veram Physicam* (Oxford, 1702).

Rule 4 was introduced in the third edition. It is, in many ways, the most important of all. Its purpose is to validate the results of induction against any imagined (and unverified) hypotheses. Inductions, Newton declares, "should be considered either exactly or very nearly true" until new phenomena may make them "either more exact or liable to exceptions."

8.4 *Newton's "Phenomena"*

The "phenomena" which serve as the basis for Newton's system of the world are not phenomena in the sense in which that term is generally understood. Newton does not have in mind a single "occurrence, a circumstance, or a fact that is perceptible to the senses." We may take note that in philosophy, a phenomenon is

7. It may seem odd that this pair of technical terms, used by the medieval "calculatores," is not found in the early edition of the *Principia* (1687) but occurs for the first time in the second edition (1713). A half century or so later, the marquise du Châtelet gave these terms a modern sense, writing of "qualités des corps qui ne sont susceptibles ni d'augmentation ni de diminution."

usually taken to mean that "which appears real to the mind, regardless of whether its underlying existence is proved or its nature understood." In physics, however, a phenomenon is merely "an observable event." But Newton's phenomena are not simply individual observations, the raw data of sense experience or observable events, but are generalizations based upon such data or events and even can be theoretical conclusions that are phenomenologically based. An example is phen. 3, that the orbits of the "five primary planets—Mercury, Venus, Mars, Jupiter, and Saturn—encircle the sun." It is the same for Kepler's third or harmonic law of planetary motion in phen. 4, and the law of areas in phen. 5. While the evidence for phenomena 1–5 is quite convincing, the case for phen. 6, the area law for the moon's orbital motion (reckoned with respect to the center of the earth), is weak, being no more than a correlation of the "apparent motion" with the "apparent diameter."

A notable change was introduced in the content of phen. 1 (formerly hyp. 1) in the second edition. Newton now had better data concerning the satellites of Jupiter. Newton also (phen. 2) introduced the satellites of Saturn, whose existence he had not been willing to acknowledge at the time of the first edition.[8]

In the second edition a correction was made in phen. 4 of an error in the first edition that had escaped Halley's critical eye as well as Newton's. That is, in discussing Kepler's third law in the first edition (in hyp. 7), Newton had said that the periodic times of the planets and the size of their orbits are the same "whether the planets revolve about the earth or about the sun," which is pure nonsense. He obviously meant, as he said in the corrected version in the second edition, that the periodic times and the size of the orbits are the same "whether the sun revolves about the earth, or the earth about the sun"; that is, they are the same in the Copernican as in the Tychonic or Ricciolian systems, but definitely not the same in the Copernican as in the Ptolemaic system.

In stating phen. 4 (formerly hyp. 7), the third law of planetary motion, Newton finally does name Kepler as discoverer. He does not, however, similarly give credit to Kepler in phen. 5 (formerly hyp. 8) for the area law.

Newton's Hyp. 3 8.5

Hyp. 3 of the first edition, eliminated in the second, reads: "Every body can be transformed into a body of any other kind and successively take on

8. In the first edition, Newton took account only of the single satellite of Saturn discovered by Huygens; in the later editions (in phen. 2, and in book 3, prop. 1) he referred to the satellites discovered by Cassini.

all the intermediate degrees of qualities." This hypothesis has been variously interpreted.[9]

In the first edition, hyp. 3 was used explicitly only once, in the proof of book 3, prop. 6, corol. 2. This corollary depends on corol. 1, which (in all editions) reads:

> Hence, the weights of bodies do not depend on their forms and textures. For if the weights could be altered with the forms, they would be, in equal [quantities of] matter, greater or less according to the variety of forms, entirely contrary to experience.

This leads (in the first edition) to corol. 2:

> Therefore all bodies universally that are on or near the earth are heavy [or gravitate] toward the earth, and the weights of all bodies that are equally distant from the center of the earth are as the quantities of matter in them. For if the aether or any other body whatever either were entirely devoid of gravity or gravitated less in proportion to the quantity of its matter, then, since it does not differ from other bodies except in the form of its matter, it could by a change of its form be changed by degrees into a body of the same condition as those that gravitate the most in proportion to the quantity of their matter (by hyp. 3), and, on the other hand, the heaviest bodies, through taking on by degrees the form of the other body, could by degrees lose their gravity. And accordingly the weights would depend on the forms of bodies and could be altered with the forms, contrary to what has been proved in corol. 1.

In the second edition, Newton removed this hyp. 3 and introduced a wholly different corol. 2. He did not, however, alter corol. 1.

The new corollary omits the reference to hyp. 3 and, as Curtis Wilson has remarked, improves the argument "by eliminating the dialectical assumption" of that hypothesis,[10] and by depending on "the rule of inductive generalization" to support the universality of gravity. In the new corollary, Newton says:

> If the aether or any other body whatever either were entirely devoid of gravity or gravitated less in proportion to the quantity of its matter, then, since (according to the opinion of Aristotle, Descartes, and others) it does not differ from other bodies except in the form of its matter, it could by

9. See Dobbs, *Janus Faces* (§3.1, n. 10 above) for detailed information concerning hyp. 3.

10. "Newt. Achievement," p. 258.

> a change of its form be transmuted by degrees into a body of the same condition as those that gravitate the most in proportion to the quantity of their matter; and, on the other hand, the heaviest bodies, through taking on by degrees the form of the other body, could by degrees lose their gravity. And accordingly the weights would depend on the forms of bodies and could be altered with the forms, contrary to what has been proved in corol. 1.

Newton's reference to Aristotle and Descartes constitutes the only direct reference to Aristotle (but not to Descartes) in Newton's *Principia*. We have seen that Newton had drawn directly on Descartes's *Principia* for concepts and even forms of expression in defs. 3 and 4 and in the scholium to the Definitions, and had discussed Descartes's theory of vortices at the end of book 2, but in none of these cases had he either mentioned Descartes by name or referred to Cartesians. Rule 1 (the previous hyp. 1) of book 3, of course, was taken directly from the Aristotelian corpus. On Newton's other references to Descartes, see §3.1 above.

Newton's reference to Aristotle and Descartes ("and others") should be read in the context of a note on this hypothesis which appears in David Gregory's commentary on the *Principia*. Gregory's reports of his conversations with Newton[11] are a major source of our knowledge of the development of Newton's thoughts between the first and second editions of the *Principia* and of Newton's plans during the 1690s for recasting the *Principia*. Newton thought very highly of Gregory's ability, as evidenced by a strong letter of recommendation when Gregory applied for a professorial post in Oxford. Newton must have esteemed Gregory's understanding of the *Principia* since he spoke to him freely of the plans for revising the *Principia*, of the new concepts and explanations he had been developing,[12] and even of his belief in a tradition of ancient learning that encompassed and anticipated some of the most basic principles and laws that Newton himself had discovered.[13]

In Gregory's commentary on the *Principia*, he made the following gloss on hyp. 3: "This the Cartesians will easily concede. But not the Peripatetics, who make a specific difference between celestial and terrestrial matter. Nor the followers of the Epicurean Philosophy, who make atoms and seeds of things immutable."[14] Gregory then takes note that this hypothesis has been revised by the author. He

11. These may be found in *Corresp.*, vol. 3.

12. See my *Introduction*, pp. 181–189.

13. See my "'Quantum in se est'" (§3.1, n. 5 above); also "Hypotheses in Newton's Philosophy" (n. 4 above).

14. On Gregory's commentary, see §6.4, n. 7, above.

enters a lengthy quotation of an early version of the later rule 3, in which Newton begins with a reference to "laws and properties of all bodies" rather than just "qualities."[15]

8.6 *Props. 1–5: The Principles of Motion of Planets and of Their Satellites; The First Moon Test*

In prop. 1 Newton addresses the problem of the centripetal force acting on the satellites of Jupiter. Because (by phen. 1) the orbital motion of the satellites accords with the area law and the harmonic law, he can apply book 1, prop. 2 (or prop. 3) and prop. 4, corol. 6, to show, first, that there is a force directed toward the center of Jupiter and, second, that this force varies inversely as the square of the distance. In the second edition, Newton takes note of Cassini's discovery of satellites of Saturn in addition to the one discovered by Huygens; he extends the results of prop. 1 to these satellites. In prop. 2, he turns to the forces acting on the planets. The proof is the same as for prop. 1, differing only in the use of phen. 5 (the area law for planets) and phen. 4 (the harmonic law for planets). Newton additionally takes note that, by book 1, prop. 45, corol. 1, a small departure from the inverse second power of the distance would introduce a motion of the apsides sufficiently great to be observed in a single revolution and hence very noticeable in many revolutions.

Newton then turns in prop. 3 to the moon. By phen. 6 (the area law for the moon's orbital motion) and book 1, prop. 2 or 3, there must be a centrally directed force. Since there is but one moon encircling the earth, Newton cannot follow the simple path of the harmonic law to establish the inverse-square law. In prop. 3, he applies book 1, prop. 45, corol. 1, to the observed fact that the motion of the lunar apogee is very slow, being about $3°3'$ forward in each revolution or lunar month. According to that corollary, if the ratio of the moon's distance from the center of the earth to the earth's radius is D to 1, then the force producing this motion of the apogee must be as $\frac{1}{\mathrm{D}^n}$, where n has the value of $2\frac{4}{243}$. Newton says in prop. 3 that "this motion of the apogee arises from the action of the sun" and "is to be ignored here." Drawing on book 1, prop. 45, corol. 2, he says that the action of the sun, "insofar as it draws the moon away from the earth," is to the force of the

15. I once suggested that Gregory's statement concerning the Cartesians could indicate that this hyp. 4 may not necessarily have expressed Newton's belief but was being considered in relation to its uses and effects. So long as it was a "hypothesis," Newton's display of it may not necessarily have implied any firm commitment by Newton to its doctrines. I was evidently in error in suggesting any such possible lack of commitment. See Dobbs, *Janus Faces* (§3.1, n. 10 above).

earth on the moon roughly as 2 to 357.45 or as 1 to $178\frac{29}{40}$.[16] Taking away that force leaves the net force of the earth on the moon to be inversely as the square of the distance from the center of the earth. That the law of force is as the inverse square of the distance will be confirmed by prop. 4.

In prop. 4, Newton states the first of the major results which indicate that the system of the world proposed in the *Principia* is radically different from any in existence. This proposition declares that the moon "gravitates" toward the earth and that it is the "force of gravity" which keeps the moon "in its orbit" by always drawing it "back from rectilinear motion." Here Newton is not making a supposition about the nature or mode of action of "gravity," merely assuming that every reader is familiar with "gravity," or weight. The purpose of prop. 4, therefore, is twofold. Newton wants to show, first, that gravity—terrestrial gravity, the gravity with which we on earth are all familiar—extends out to the moon and does so according to the law of the inverse square; and second, that it is this force of gravity which causes the moon to fall inward from its inertial linear path and so keep in its orbit.

In this moon test, Newton first computes how far the moon (or an object placed in the moon's orbit, 60 earth-radii from the earth's center) would actually fall in one minute if deprived of all forward motion. The result is 15 1/12 Paris feet.[17] If gravity diminishes by the inverse-square law and if the earth's gravity extends to the moon, then it follows that a heavy body on the earth's surface should fall freely in one minute through 60×60 or 3,600 of the above 15 1/12 Paris feet in one second (more accurately, 15 feet, 1 inch, 1 4/9 lines). This result agrees so closely with actual terrestrial experiments that Newton can conclude that the force keeping the moon in its orbit is that very force which "we generally call

16. A careful reading of book 1, prop. 45, corol. 2, however, shows that Newton (while considering the action of an "extraneous force . . . added to or taken away from" the centripetal inverse-square force producing an elliptical orbit) writes, "Let us suppose the extraneous force to be 357.45 times less than the other force under the action of which the body revolves in the ellipse." Concerning Newton's claim in book 3, prop. 3, that (on the basis of book 1, prop. 45, corol. 2) the action of the sun in drawing the moon away from the earth is to the centripetal force of the earth on the moon "as roughly 2 to 357.45," see "Newt. Achievement," p. 264.

17. As Curtis Wilson has pointed out to me, "in order to obtain the result of 15 1/12 Paris feet, Newton has to multiply the initial result of his calculation (which I find to equal 15.093 Paris feet/min.) by 178 29/40 divided by 177 29/40. In other words, Newton has, without mentioning the fact, subtracted out what he supposes to be the average subtractive force of the Sun, which counteracts some of the Earth's gravitational effect on the Moon. He has subtracted out too much; see Shinko Aoki, 'The Moon-Test in Newton's *Principia*: Accuracy of Inverse-Square Law of Universal Gravitation,' *Archive for History of Exact Sciences*, 1992, 44: 147–90. The exact agreement between Newton's calculation and the observation of length of fall per second at Paris is fudged. A good enough agreement would have been attained without the fudging." See, further, *Never at Rest*, pp. 732–34.

gravity." An alternative proof, given in the scholium, is based on considerations of a hypothetical terrestrial satellite. Later on, in book 3, prop. 37, corol. 7 (added in the second edition), Newton will present a second such moon test.

Using rule 2, Newton then proves (in prop. 5) that the satellites of Jupiter and of Saturn "gravitate" toward those planets, as do the primary planets toward the sun. Corol. 1 states expressly that "there is gravity toward all planets universally," a force that (prop. 5, corol. 2) is "inversely as the square of the distance." It follows (corol. 3) that "all the planets are heavy toward one another," the proof of which is found in the example of Saturn and Jupiter, near conjunction, where such interactions are said to "sensibly perturb each other's motions." A scholium explains that since the centripetal force producing the orbits of celestial bodies is gravity, he will now use that name.

8.7 *Jupiter's Perturbation of Saturn as Evidence for Universal Gravity*

Props. 1 to 3 of book 3 prove that there must be a centrally directed force that retains planets and planetary satellites in their respective orbits. As we have seen, prop. 4 (the moon test) then shows that this force (in the case of the earth and the moon) can be identified with terrestrial gravity. Prop. 5 then takes a strong step toward the generalization to a force of universal gravity, the assertion that the same force of gravity that keeps planets and satellites in their orbits also acts between every pair of planets. In prop. 5, corol. 3, Newton boldly asserts that "Jupiter and Saturn near conjunction, by attracting each other, sensibly perturb each other's motions."

Newton returns to this topic in prop. 13, where he offers what he alleges is evidence for "the action of Jupiter upon Saturn." Newton hoped to prove the existence of the force of universal gravity by the perturbation of Saturn's motion by Jupiter. Since Jupiter has an enormous mass (more than the combined masses of all the other planets), he could reasonably expect that the force exerted by Jupiter on Saturn would provide a noticeable alteration in Saturn's motion just before and after conjunction. Accordingly, he wrote to the Astronomer Royal, John Flamsteed, seeking evidence of any such alterations of Saturn's orbital motion.[18]

Newton's letter to Flamsteed (30 December 1684), written just when he was beginning the heroic effort that transformed the short tract *De Motu* into the

18. A detailed discussion of the history of the investigation of the interaction of Jupiter and Saturn can be found in Curtis Wilson, "The Great Inequality of Jupiter and Saturn: From Kepler to Laplace," *Archive for History of Exact Sciences* 33 (1985): 15–290. This §8.7 of the Guide draws heavily on notes prepared by George Smith; his own commentary on the problem of planetary perturbation is given below in §8.8.

Principia, asked specifically whether astronomers had observed an irregularity in the motion of Saturn. "The orbit of Saturn as defined by Kepler," he wrote, "is too little for the sesquialterate proportion," that is, for the 3/2 proportion. He continued:

> This Planet so oft as he is in conjunction with Jupiter ought (by reason of Jupiters action upon him) to run beyond his orbit about one or two of the suns semidiameters or a little more & almost all the rest of his motion to run as much or more within it. Perhaps that might be the ground of Keplers defining it too little.

This led him to his basic query:

> I would gladly know if you ever observed Saturn to err considerably from Keplers tables about the time of his conjunction with Jupiter. The greatest error I conceive should be either the year before conjunction when Saturn is 3 or 4 signes from the Sun *in consequentia* or the yeare after when Saturn is as far from the sun *in antecedentia*.[19]

Flamsteed's reply, 5 January 1685, confirmed Newton's conclusion that observations did indeed show a deviation from Keplerian tables for Jupiter and Saturn.[20] Flamsteed, however, could find no evidence to support Newton's hope for a gravitationally caused irregularity in Saturn's motion during the two-year period of near conjunction with Jupiter.

"As for the motion of Saturn," Flamsteed wrote,

> I have found it about 27′ slower in the Antonicall appearances since I came here, then Keplers numbers, & Jupiters about 14 or 15′ swifter as you will find by the account of theire Conjunctions which I published in the last yeares *transactions*.

"The error in Jupiter is not allwayes the same," he continued, "by reason the place of his Aphelion is amisse in Kepler." Furthermore, he wrote, "the fault in Saturn" is not "allwayes the same," but is "lesse in the Quadratures as it ought to be." The "differences in both," however, "are regular & may be easily answered by a small

19. *Corresp.* 2:407.

20. Curtis Wilson has pointed out ("The Great Inequality of Jupiter and Saturn," p. 42) that problems in the motion of Saturn and Jupiter had been noted earlier by Horrocks, and published in 1673 in his *Opera Posthuma*, edited by Flamsteed. Boulliau put Saturn half a degree behind the Rudolphine Tables and a third of a degree behind his own in a published report of his recent observation of an occultation of Saturn by the moon: *Philosophical Transactions* 12, no. 139 (April-May-June 1678): 969–975.

alteration in the Numbers as is found in Saturn by our New Tables which Mr. Halley made at my request & Instigation."

Flamsteed had introduced his own correction factor for Jupiter, with the result that Jupiter "has of late yeares answered my calculus in all places of his orbit." He admitted, however, that he had "not beene strict enough to affirme that there is no such exorbitation as you suggest of Saturn." After "the next terme if not sooner," he promised, he would "inquire diligently" into the matter. But, "to confesse my thoughts freely to you," he admitted, "I can scarce thinke there should be any such influence" as Newton had suggested. The reason was that "the distance of the planets from each other in those positions is neare four Radij of the *Orbis annus*," as a result of which, "in such yeilding matter as our aether, I cannot conceave that any impression made by the one planet upon it can disturbe the motion of the other." In short, Flamsteed was fully cognizant that "Keplers distances of Saturn agree not with the sesquialterate proportion and that Jupiters too ought to be mended." Both would have to "be altered before wee set upon the inquiry whether Jupiters motion had any influence on Saturns."[21]

Plainly, then, when Newton published the first edition of the *Principia*, he knew of no supporting astronomical evidence for his claim of a perturbation of Saturn's motion in its conjunction with Jupiter. Twenty-six years later, in the second edition, Newton made an even stronger case that the action of Jupiter alters the motion of Saturn, as may be seen by placing side by side the same passage from book 3, prop. 13, as it appears (in translation) in the two editions.

First edition

Yet the action of Jupiter upon Saturn is not to be ignored entirely. For the gravity toward Jupiter is to the gravity toward the sun (at equal distances) as 1 to 1,100; and so in the conjunction of Jupiter and Saturn, since the distance of Saturn from Jupiter is to the distance of Saturn from the sun as 4 to 9, the gravity of Saturn toward Jupiter will be to the gravity of Saturn toward the sun as 81 to 16 × 1,100, or roughly 1 to 217.

Second edition

Yet the action of Jupiter upon Saturn is not to be ignored entirely. For the gravity toward Jupiter is to the gravity toward the sun (at equal distances) as 1 to 1,033; and so in the conjunction of Jupiter and Saturn, since the distance of Saturn from Jupiter is to the distance of Saturn from the sun almost as 4 to 9, the gravity of Saturn toward Jupiter will be to the gravity of Saturn toward the sun as 81 to 16

21. *Corresp.* 2:408–409. In his paper in the *Philosophical Transactions* that Flamsteed referred to in the letter, he announced that Saturn's observed longitude was 24′ behind where Kepler's tables put it on 26 January 1683, and Jupiter's 15′41″ ahead.

Nevertheless, all the error in the motion of Saturn around the sun, an error arising from so great a gravitation toward Jupiter, can almost be avoided by placing the focus of the orbit of Saturn at the common center of gravity of Jupiter and the Sun (by book 1, prop. 67); in which case when that error is greatest, it hardly exceeds two minutes.

× 1,033, or roughly 1 to 204. And hence arises a perturbation of the orbit of Saturn in every conjunction of this planet with Jupiter so sensible that astronomers have been at a loss concerning it. According to the different situations of the planet Saturn in these conjunctions, its eccentricity is sometimes increased and at other times diminished, the aphelion sometimes is moved forward and at other times perchance drawn back, and the mean motion is alternately accelerated and retarded. Nevertheless, all the error in its motion around the sun, an error arising from so great a force, can almost be avoided (except in the mean motion) by putting the lower focus of its orbit in the common center of gravity of Jupiter and the sun (by book 1, prop. 67); in which case, when that error is greatest, it hardly exceeds two minutes. And the greatest error in the mean motion hardly exceeds two minutes per year.

In the third edition, the text is the same as in the second, with a substitution of the numbers 1,067 and 211 for 1,033 and 204. In the third edition, Newton then concludes that the difference between the gravity of the sun toward Saturn and the gravity of Jupiter toward Saturn is to the gravity of Jupiter toward the sun as 65 to 156,609 or as 1 to 2,409. In the first edition, this result was given as 65 to 122,342 or 1 to 1,867; in the second edition this became 65 to 124,986 or 1 to 1,923. Since, according to Newton, "the greatest power of Saturn to perturb the motion of Jupiter is proportional to this difference," it follows that "the perturbation of the orbit of Jupiter is far less than that of Saturn's."

We do not know the grounds on which Newton based so assertive a declaration. We do not have documentary evidence that would explain why the second edition contains so much stronger a statement than had been made in the first edition. In the second and third editions of the *Principia*, Newton did not present any specific observations or calculations to justify his claim, nor did he ever identify the "astronomers" to whom he referred. Nor have any data of observations and relevant calculations been identified (if they do in fact exist) among his manuscripts.

Conjunctions of Jupiter and Saturn occur on the average approximately every twenty years. The conjunction following the one of February 1683 took place in August 1702. But eleven years before that event, in August 1691, a few years after the publication of the *Principia*, Newton wrote to Flamsteed of his desire to have the latter's "observations of Jupiter and Saturn for the 4 or 5 next years at least," before he could "think further of their Theory." He would "rather have them for the next 12 or 15 years." He remarked: "If you & I live not long enough, Mr. Gregory and Mr. Halley are young men."[22]

Three years after writing to Flamsteed, however, Newton seems to have become convinced of the reality of the effect of perturbation. In 1694, at a time when he was deep in revising the *Principia* and even planning a reconstruction of a considerable part of that work, he had apparently decided that the earlier conjunction of 1683 had actually provided evidence for the existence of a force of perturbation. In May 1694, David Gregory reported that Newton had unequivocally declared that the "mutual interactions of Saturn and Jupiter were made clear at their very recent conjunction." According to Gregory's memorandum of his conversation with Newton, Newton said that "before their conjunction Jupiter was speeded up and Saturn slowed down, while after their conjunction Jupiter was slowed down and Saturn speeded up." Newton's conclusion was that "corrections of the orbits of Saturn and Jupiter by Halley and Flamsteed" were "afterwards found to be useless and had to be referred to their mutual action."[23]

Newton paid Flamsteed a visit on 1 September 1694, on which occasion (according to Gregory's report) they discussed problems connected with the orbits of the moon and of Jupiter and Saturn.[24] In a subsequent letter to Flamsteed (20 December 1694), Newton wrote: "I intend to determine the Orb of Saturn within a few days & I'le send you the result."[25] In the margin of his copy Flamsteed added that it "was never sent." Flamsteed added his "query" whether Newton had ever "done" it.[26] And so we, like Flamsteed, must end on a note of query concerning

22. *Corresp.* 3:164.

23. Ibid., pp. 314, 318.

24. Ibid. 4:7. Gregory reported Newton as holding that "observations are not sufficient to complete the theory of the Moon. Physical causes must be considered. Flamsteed is about to show him another hundred positions of the Moon. A consideration of physical causes is needed to reconcile the orbits of Jupiter and Saturn with the heavens. Their apses are disturbed by an oscillatory motion." Flamsteed's summary of this meeting does not mention Jupiter and Saturn.

25. Ibid., p. 62.

26. Ibid., p. 63.

the basis for Newton's assertion. As we know today, the effect that Newton sought would have been far too small to have been observed through the instruments available at that time.

Planetary Perturbations: The Interaction of Jupiter and Saturn[27] (BY GEORGE E. SMITH) 8.8

The three-body problem arising from the interaction of two planets orbiting the sun is mathematically very difficult, much more so than the so-called "restricted three-body problem" created by the sun's action on the moon. Forty years of effort were needed from the time Euler first sketched a promising analytical approach to the problem[28] until Laplace obtained the first adequate "solution."[29] Hence, we can easily understand why Newton, decades earlier, was able to offer only a qualitative, and no quantitative, analysis of the effects of Jupiter on the motion of Saturn. Newton's qualitative analysis was announced in proposition 51 of David Gregory's *Astronomiae Physicae & Geometricae Elementa*, published in 1702.[30]

The orbits of Jupiter and Saturn are very nearly circular. But they are not at the same inclination; and their centers are offset from each other, so that their lines of apsides are at an angle with respect to each other; thus they reach their peak velocities at perihelion at different heliocentric longitudes. Consequently their interactions change from one conjunction to the next, just as Newton says in book 3, prop. 13, of the *Principia*. But there was prima facie reason to think that the effects of these second-order vagaries are small compared with the large increase of gravitational interaction when the two planets are near conjunction. Hence, Newton could reasonably have approximated the first-order effects of their interaction by modeling their orbits as concentric circles. In this case the transradial component of the perturbational force of Jupiter on Saturn slows it

27. The major study of this topic is the long article by Curtis Wilson cited in n. 18 above.

28. "Recherches sur le mouvement des corps célestes en général," *Mémoires de l'Académie des sciences de Berlin* [3] (1747): 93–143; reprinted in *Leonhardi Euleri Opera Omnia*, ser. 2, vol. 25 (Zurich, 1960), pp. 1–44. Euler's use of the "F = *ma*" formulation of Newton's second law of motion in this work appears to have been a main influence for its subsequent use. But this was not the first time this formulation of the law had appeared in print. Jacob Hermann had presented the second law in terms of differentials in this form in his *Phoronomia* of 1716 (§5.3, n. 7 above), p. 57.

29. We now know that there is no exact analytical solution to this type of three-body problem. Laplace's perturbational solutions are, strictly speaking, only approximate, insofar as they involve truncation of an infinite series that is not uniformly convergent.

30. An English translation is available as *The Elements of Physical and Geometrical Astronomy*, 2d ed. (London: D. Midwinter, 1726; reprint, New York: Johnson Reprint Corp., 1972); see p. 488.

down before conjunction and speeds it up afterward, while the action of Saturn on Jupiter speeds it up before conjunction and slows it down afterward, just as Newton said and Gregory described.

Newton was aware that irregularities in the motion of Saturn must be defined with respect to some underlying Keplerian orbit. But which Keplerian orbit? In the augmented version of his tract *De Motu*, written at roughly the same time as the letter he wrote to Flamsteed for information about the motion of Saturn, that is, at the end of 1684,[31] Newton himself called attention to the problem of defining appropriate Keplerian orbits in the presence of perturbations. The perturbations must somehow be either identified or filtered out before the appropriate orbit can be defined. So long as significant freedom remained in defining the underlying orbit, different conclusions could have been reached about divagations from this orbit.

From today's vantage point, it is easy to see what was causing confusion. According to the theory of G. W. Hill, set forth in 1890,[32] Saturn is subject to three dominant sinusoidal perturbations, none of which is the least bit obtrusive in a qualitative examination of the problem: one with a period of a little more than 60 years and an amplitude of 7.04′; a second with a period of a little more than 30 years and an amplitude of 11.4′; and a third dwarfing both of these, the "Great Inequality,"[33] with a period of roughly 900 years and an amplitude of 48.49′! Hill's theory of Saturn lists ninety-six other perturbations arising from the action of Jupiter, all with amplitudes below 1′, plus comparably small perturbations arising from the actions of Uranus and Neptune.[34] Newton's 19-year and Flamsteed's conjectured 59-year perturbations are among these, both with amplitude around 7″.

We can obtain a good estimate of the deviation of Saturn from its preferred Keplerian orbit during the 1683 conjunction by summing the three dominant perturbational terms in Hill's theory, plus the principal perturbational terms with periods of less than 30 years:

31. "Leaving aside these fine points, the simple orbit that is the mean between all vagaries will be the ellipse that I have discussed already. If any one shall attempt to determine this ellipse by trigonometrical computations from three observations (as is usual) he will be proceeding without due caution. For those observations will be affected by minutiae of irregular motion that I have neglected here, and so they will cause the ellipse to be somewhat displaced from its proper size and position (which should be the mean between all vagaries); as many ellipses, different from each other, may be obtained as there are sets of three observations. Therefore, as many observations as possible should be combined and brought together in a single procedure so that they even out, and yield an ellipse of mean position and magnitude" (*Unpubl. Sci. Papers*, p. 281).

32. G. W. Hill, "A New Theory of Jupiter and Saturn," in *Astronomical Papers Prepared for the Use of the American Ephemeris and Nautical Almanac* (Washington, D.C.: Government Printing Office, 1890).

33. So named by Laplace.

34. It also includes small perturbational terms associated with the inner planets.

Date	*Discrepancy in heliocentric longitude*	*Annual difference*
Feb. 1680	−44.27′	
		1.57′
Feb. 1681	−43.35	
		1.72
Feb. 1682	−41.63	
		1.88
Feb. 1683	−39.75	
		2.13
Feb. 1684	−37.62	
		2.39
Feb. 1685	−35.23	
		2.64
Feb. 1686	−32.59	

The last column in the table gives the deviation in average annual angular velocity from that in the preferred Keplerian orbit.

At least with respect to its preferred Keplerian orbit, then, the average annual angular velocity of Saturn was increasing throughout the 6-year period bracketing its 1683 conjunction with Jupiter. The perturbational terms with period less than 30 years do show the pattern Newton described, but their net effect yields only a 0.4′/year reduction in velocity from 1680 to conjunction, with a corresponding increase in the 3 years following. This effect, however, is masked over the 6 years in question by the much stronger 30-year perturbational term, which increases the average annual angular velocity by a total of 1.2′/year between 1680 and 1686, and by more than 0.7′/year from 1680 to conjunction.[35]

Needless to say, such small fluctuations in velocity would have been difficult, if not impossible, to detect in Newton's day. But even if they could have been, the numbers above provide no basis for concluding that Saturn perceptibly slowed down before the 1683 conjunction and speeded up after it. This, however, is not to deny that Newton could have performed a calculation in which he found a fluctuation in velocity more nearly symmetric on either side of February 1683, the date of conjunction. For he might have used an underlying orbit and specific observations that gave him a more symmetric pattern. He appears to have been allowing the aphelion of Saturn to oscillate, and we know that variations detected in velocity, over and above those associated with Kepler's area rule, tend to be especially sensitive to where the line of apsides is placed.

About a decade after Newton's death, it became increasingly evident that both Saturn and Jupiter were deviating substantially from Keplerian motion, but it was

35. The effect of the 30-year perturbational term on the angular velocity was smaller in the years bracketing the 1702 conjunction. Consequently it does not mask the symmetric variation arising from the shorter-period perturbational terms nearly so much. But the changes in the average annual angular velocity were nonetheless still well below 1′/year, and therefore scarcely perceptible.

in no way clear what the harmonic content of the deviation was. In response to this, the French Academy of Sciences chose for its prize contest of 1748 the theory of Jupiter and Saturn. Euler's prizewinning essay, "Recherches sur la question des inégalités du mouvement de Saturne et de Jupiter," is the seminal work in the efforts to derive the perturbed motions of planets from gravitation theory via analytical methods. Here Euler introduced trigonometric series into the problem, in addition to employing perturbational approximations in response to the inability to obtain an exact analytical solution to the three-body problem.

The first case Euler presented in the essay is the two-concentric-circle model of Newton announced by Gregory. He concludes:

> This granted, the longitude of Saturn will be
>
> $$\varphi = \text{mean long.} + 4'' \sin \omega - 32'' \sin 2\omega - 7'' \sin 3\omega - 2'' \sin 4\omega - \text{etc.}$$
>
> from which one sees that this disturbance in the motion of Saturn must be all but imperceptible since it very rarely exceeds one-half minute. And as the observed disturbance is often several times greater than $10'$, it is evident that one would not have thought to explain it by the effects of the action of Jupiter. We recognize by the same reason the necessity of the researches that follow where I will introduce into the calculation not only the eccentricity of the orbit of Saturn, but even that of Jupiter.[36]

Yet, when he introduced low-order perturbational corrections involving the eccentricities in the subsequent analysis, he still failed to account for the observed motions of the two planets.

Because of the intractability of the three-body problem, the failure to obtain the motion of Saturn and Jupiter analytically from the theory of gravity was not automatically regarded as a challenge to the theory. Euler was quite outspoken on this point in the introduction to his subsequent essay, "Recherches sur les irrégularités du mouvement de Jupiter et Saturne," which won the prize of the French Academy of Sciences in 1752:

> . . . since Clairaut has made the important discovery that the movement of the apogee of the Moon is perfectly in accord with the Newtonian hypothesis . . . , there no longer remains the least doubt about this proportion . . . . One can now maintain boldly that the two planets Jupiter and Saturn attract each other mutually in the inverse ratio of the squares of their distance, and that all the irregularities that can be discovered in their movement are infallibly caused by this mutual action . . . . And if

36. *Leonhardi Euleri Opera Omnia*, ser. 2, 25:72.

> the calculations that one claims to have drawn from this theory are not found to be in good agreement with the observations, one will always be justified in doubting the correctness of the calculations, rather than the truth of the theory.[37]

Advances were made on the three-body problem over the next three decades, most notably by Lagrange, but also in an initial foray by Laplace. But all attempts to derive the motions of Saturn and Jupiter from gravitation theory remained clearly inadequate until Laplace discovered the Great Inequality in 1785, a century after Newton had made his initial inquiry to Flamsteed.

Specifically, Laplace discovered the crucial ramification of the "near resonance" between 5 times the mean motion of Saturn and 2 times that of Jupiter. Consider terms of the form $C + \sin(5n't - 2nt)$, where n' and n are the mean motions of Saturn and Jupiter respectively. Because $(5n' - 2n)$ is very small, the period of any such term is large—approximately 900 years. But after integration of the equations of motion, the square of this small quantity appears in the denominator of the coefficients of some terms of this form—in particular, terms specifying perturbations in longitude involving powers of the eccentricity above the first. Consequently, even though the eccentricities are small, higher-order correction terms involving them can be huge in comparison with other perturbational terms. Whence the Great Inequality. Laplace followed this discovery, which was announced in November 1785, with his *Théorie de Jupiter et de Saturne* in May 1786, in which the other principal perturbations of Jupiter and Saturn were combined with the Great Inequality to yield the first acceptably adequate account of the complicated motions of these two planets. This account was derived from gravitation theory several decades, if not a century or more, before astronomers had any hope of extracting it from observations.

By the time of Laplace's solution for the motion of Saturn, several major results had been obtained in support of gravitation theory: (1) Clairaut in 1743 had replaced Newton's approximate solution for the shape of the earth with an analytical solution that allowed different "rules" for the variation of gravity, and the findings of the French Academy's expeditions to Lapland and Peru had proved largely consistent with the Newtonian rule;[38] (2) Clairaut had derived the motion

37. *Recueil des pièces qui ont remporté les prix de l'Académie des sciences*, vol. 7 (Paris, 1769), pp. 4–5. The quotation is taken from Curtis Wilson, "Perturbations and Solar Tables from Lacaille to Delambre: The Rapprochement of Observation and Theory, Part I," *Archive for History of Exact Sciences* 22 (1980): 144.

38. See Isaac Todhunter, *A History of the Mathematical Theories of Attraction and the Figure of the Earth from the Time of Newton to That of Laplace* (London, 1873; reprint, New York: Dover Publications, 1962).

of the lunar apsis from gravitation theory in 1749; (3) d'Alembert in 1749, and Euler subsequently, had devised analytical solutions for the rotational motion of the nonspherical earth that accounted not only for the 26,000-year precession of the equinoxes but also for the 18-year nutation induced by the gravity of the moon, which Bradley had discovered in the late 1720s;[39] (4) Halley's comet had returned in 1758, reaching perihelion within the 30-day window that Clairaut had predicted after taking into account the gravitational action of Jupiter and Saturn, as well as that of the sun;[40] and (5) Laplace in the mid-1770s had developed an analytical solution for the tides that included rotation as well as gravitation.[41] So, although Laplace's results on Saturn and Jupiter belong in any such listing of analytical solutions to prominent problems involving gravity that had been treated at best approximately in the *Principia*, they were too late to have been instrumental in establishing gravitation theory.

Nevertheless, these results were of great moment in the history of gravitation theory. For one thing, they showed that this theory offered a more reliable basis for predicting the future motions of the planets than the heretofore standard way of making such predictions, from kinematic models obtained from astronomical observations. In this respect Laplace's results did more than just establish the validity of what Fontenelle had written in his official *éloge* of Newton: "Sometimes these conclusions even foretell events which the astronomers themselves had not remarked." Laplace's results forever altered the relationship between physical theory and observational astronomy.

Moreover, Laplace's success with Saturn and Jupiter surely helped prompt him to write his monumental *Traité de mécanique céleste*, the first volume of which appeared in 1799, and the fifth and final one in 1825. This work did more than just present perturbation-based theories of the motions of all the planets, after the manner of the solution for Jupiter and Saturn. It brought together in one uni-

39. Although Newton's derivation of the precession of the equinoxes successfully identifies the forces at work, it is based on a physically unsound analogy with the motion of the lunar nodes. D'Alembert's and Euler's efforts on the precession are discussed in Curtis Wilson, "D'Alembert versus Euler on the Precession of the Equinoxes and the Mechanics of Rigid Bodies," *Archive for History of Exact Sciences* 37 (1987): 233–273.

40. Specifically, Clairaut determined that the action of Jupiter would delay the return of the comet to perihelion by 518 days, and the action of Saturn, by 100 days. His predicted date of perihelion was 13 April 1759. The subsequently observed date was 13 March 1759. The impact which the return of Halley's comet and Clairaut's prediction had on the reception of gravitation theory is described in Robert Grant, *History of Physical Astronomy from the Earliest Ages to the Middle of the Nineteenth Century* (London, 1852; reprint, New York: Johnson Reprint Corp., 1966), pp. 103–104.

41. For a discussion of the efforts leading up to Laplace's treatment of the tides, see Eric J. Aiton, "The Contributions of Newton, Bernoulli, and Euler to the Theory of the Tides," *Annals of Science* 11 (1955): 206–223.

fied presentation the century of successes in deriving phenomena from gravitation theory.[42] The *Principia* of legend is a book in which a general mechanics is set forth, based on three laws of motion; it, together with Kepler's planetary theory, provides the basis for the theory of universal gravity; and this theory then provides a full detailed account of all celestial motions along with such related matters as the tides, the variation in the acceleration of gravity around the earth, and the precession of the equinoxes. With qualifications, this legend is true. But qualifications cease to be needed when it is taken as describing Laplace's *Celestial Mechanics*.

Props. 6 and 7: Mass and Weight 8.9

Prop. 6 may especially command our interest because here Newton not only states that all bodies gravitate toward each planet, but adds that at any given distance from a planet's center the "weight" of any body "toward that planet" is proportional to the body's "quantity of matter." In simpler language, Newton is saying that at any given location with respect to a planet, the weight of any body is proportional to its mass. We have seen Newton refer to this proposition earlier, in def. 1 at the very opening of the *Principia*. The proof given in prop. 6 is experimental, using a specially devised pair of pendulums with hollow bobs. Newton verified the exactness of the proportion—with an accuracy within one part in a thousand, that is, to within a tenth of a percent—in nine substances: gold, silver, lead, glass, sand, salt, wood, water, and wheat. He observes that this is a more refined version of experiments made with falling bodies.

In prop. 6, Newton says that his pendulum experiments (showing that mass is proportional to weight) provide a confirmation of what had been previously established by experiments with falling bodies. In free fall, the motive force is the weight of the falling body and, according to the second law, this force is jointly proportional to the acceleration and the mass of the falling body. If W is the accelerating force of the body's weight and m its mass, then the second law may be written as $W \propto mA$, where A is the acceleration. The experiments with falling bodies showed that, at any given place, bodies of different weight fall from rest with

42. The opening sentences to the preface of vol. 1 announce Laplace's aim: "Towards the end of the seventeenth century, Newton published his discovery of universal gravitation. Mathematicians have, since that epoch, succeeded in reducing to this great law of nature all the known phenomena of the system of the world, and have thus given to the theories of the heavenly bodies, and to astronomical tables, an unexpected degree of precision. My object is to present a connected view of these theories, which are now scattered in a great number of works" (quoted from Laplace's *Celestial Mechanics*, trans. Nathaniel Bowditch [Bronx, N.Y.: Chelsea Publishing Co., 1966], 1:xxiii).

the same acceleration, save for the slight retardation produced by the resistance of the air. Since A is a constant, it follows that W α m, as Newton says.

For today's reader this experiment is an outstanding example of Newton's deep physico-mathematical insight. He recognized that mass, the concept he had invented, is not only the measure of a body's resistance to a change of state or resistance to being accelerated, but is also the determinant of a body's reaction to a given gravitational field.

In the corollaries to prop. 6, Newton declares that weight does not depend on the form or texture of bodies (corol. 1) and that all bodies on or near the earth gravitate according to the quantity of matter or mass (corol. 2), a property that by rule 3 can be "affirmed of all bodies universally." He also argues against the existence of an aether (corol. 3). In corol. 4, Newton declares that "there must be a vacuum." In corol. 5, he draws a distinction between gravity and the magnetic force. This leads to prop. 7: "Gravity exists in all bodies universally." An important corollary (corol. 1) explains how "the gravity toward the whole planet arises from and is compounded of the gravity toward the individual parts." Referring to book 1, prop. 69, and its corollaries, Newton concludes that "the gravity toward all the planets is proportional to the [quantity of] matter in them."

Prop. 8 deals with the gravitational attraction between two spheres which are composed of homogeneous shells, which is found to be inversely as the square of the distance between their centers. Here Newton addresses the important problem of whether the law of the inverse square would hold exactly or only "nearly so" for a total force compounded of a number of inverse-square forces. This latter problem arose once it had been found (in prop. 7, corol. 1) that the gravity toward a planet taken as a whole "arises from and is compounded of the gravity toward the individual parts" and (corol. 2) that the gravity "toward each of the individual equal particles" is as the inverse square of the distance.

8.10 *Prop. 8 and Its Corollaries (The Masses and Densities of the Sun, Jupiter, Saturn, and the Earth)*

Although prop. 8 is a statement of the utmost importance for the celestial physics of the *Principia*, its corollaries have to some degree assumed a greater significance. Essentially, prop. 8 states that a homogeneous sphere or a sphere made of concentric homogeneous shells will attract gravitationally just as if all its mass were concentrated at its geometric center. Newton, as we have seen, concludes the discussion of prop. 8 with an expression of his concern about whether the inverse-square forces of whole planets would be exactly the sum of the inverse-square forces of the individual particles. He also states a rule, used in the following corollaries, for comparing gravitational forces at different distances.

The corollaries to prop. 8, containing Newton's determination of the masses and densities of the sun and three of the planets, constitute one of the most remarkable portions of the *Principia*. These corollaries contain the supreme expression of the doctrine that one set of physical principles applies to all bodies in the universe, sun and planets as well as terrestrial bodies. By showing how his methods of dynamics produce reasonable quantitative values for these masses and densities, Newton removed any lingering doubt that mass and density were meaningful concepts when applied to celestial objects. Or, to put the matter differently, by showing how to assign numbers to the masses and densities of the sun and planets, along with the earth, Newton established once and for all that these objects do have mass and density (and, therefore, inertia) and accordingly are subject to the laws of physics which he had established.

Newton's tremendous achievement in the corollaries to prop. 8 was recognized almost at once. Four years after the publication of the *Principia*, Christiaan Huygens wrote in his essay "La cause de la pesanteur," published as a supplement to his *Traité de la lumière* (Leiden, 1690), that he was especially pleased to read how Newton, by "supposing the distance from the earth to the sun to be known," had been able to compute "the gravity that the inhabitants of Jupiter and Saturn would feel compared with what we feel here on earth" and "what its measure would be on the surface of the sun"—matters "which hitherto have seemed quite beyond our knowledge."[43] A half-century later, Colin Maclaurin, in his *Account of Sir Isaac Newton's Philosophical Discoveries* (London, 1748), declared that to measure "the matter in the sun and planets was an arduous problem" that, "at first sight, seemed above the reach of human art."[44]

These corollaries are significantly different in the first edition of the *Principia* and the later editions, showing us how Newton's command of this important problem was achieved only by degrees. Newton's procedure may most easily be understood by starting with the third and ultimate edition and then looking at his earlier approach to the problem.

In corol. 1 in the third edition, Newton sets out to find the relative weights toward the sun, Jupiter, Saturn, and the earth of "equal bodies" (that is, bodies of equal mass) at any one distance from the centers of these four central bodies;

43. A translation of this essay has been prepared by Karen Bailey, to be published with a commentary by George Smith.

44. Colin Maclaurin, *An Account of Sir Isaac Newton's Philosophical Discoveries* (London, 1748), p. 288.

he then computes these relative weights at the surfaces of these four bodies. As is usually the case in the *Principia*, Newton's results for these weights, and for the masses and densities of the sun and planets, are relative. Thus the "weights of bodies which are equal [i.e., the weights of bodies of equal mass] and equally distant from the center of the sun, of Jupiter, of Saturn, and of the earth" are given on a scale in which the weight toward the sun is taken arbitrarily as 1. Similarly, the weights on the surfaces of these four bodies are given on a scale in which the weight toward the sun is taken as 10,000.

In the following presentation, an asterisk following the word "weight" indicates that the quantity in question ("weight*") is not the weight-force or the gravitational force as usually understood, but rather the weight-force exerted on bodies of equal mass. In other words, weight*, in the sense of corol. 1, is what Newton calls the "accelerative quantity" or accelerative measure, defined (def. 7) as "the measure of" a "force that is proportional to the velocity which it generates in a given time." Since Newton is concerned with relative masses and relative densities, the test mass can be taken as unity, so that weight* may be considered the weight-force per unit mass, what we would call today the local force field.

In these corollaries, Newton computes the weight-force per unit mass for each of the four members of the solar system then known to have at least one satellite: the sun, Jupiter, Saturn, and the earth. For the sun, the satellite used by Newton is Venus; for Jupiter, it is the outermost satellite (Callisto); for the earth, it is the moon. The satellite used for Saturn is Titan, which Newton calls "Huygens's satellite" or "the Huygenian satellite." At the time of the first edition of the *Principia*, Newton acknowledged the existence of only this one satellite of Saturn, but in the later edition he accepted the additional satellites discovered by Cassini. In his computations, in all editions, Newton uses this first-discovered satellite of Saturn.

In these corollaries, Newton assumes (without saying so explicitly) that the sun, Jupiter, Saturn, and the earth may be regarded as homogeneous bodies or as bodies composed of concentric homogeneous spheres. Furthermore, he takes the orbits of Venus, the satellites of Jupiter and Saturn, and the moon to be circles. Accordingly, he can then apply to them the rule (set forth in book 1, prop. 4, corol. 2) that for a body moving in a circle of radius r with a period T, the centrally directed force is as $\frac{r}{\mathrm{T}^2}$. He then concludes that the same rule holds for the weights of equal bodies "at the surfaces of the planets or at any other distances from the center."

Basically, Newton's method is to compute the radius of the orbit of each of his four satellites (Venus, Callisto, Titan, and the moon) and then divide these radii by the squares of the periods of revolution. He does not give us any details of his computation, but clearly a first assignment is to find the radii of the four orbits and the squares of the periods. To do so requires that he convert all of these periods into some single common unit and then compute the squares. These periods, as given in the first and third editions, are listed in table 8.1.

TABLE 8.1 DATA FOR COMPUTATIONS FOR BOOK 3, PROP. 8

	Sun	*Jupiter*	*Saturn*	*Earth*
Maximum heliocentric elongation of satellite from central body		8′13″	3′20″	
		8′16″	*3′4″*	*(10′33″)*
Period of revolution (T)	$224\frac{2}{3}^{d}$	$16\frac{3}{4}^{d}$	$15^{d}22\frac{2}{3}^{h}$	$17^{d}7^{h}43^{m}$
	$224^{d}16\frac{3}{4}^{h}$	$16^{d}16\frac{8}{15}^{h}$	$15^{d}22\frac{2}{3}^{h}$	$27^{d}7^{h}43^{m}$
Weight* at equal distances from the central body†	1	$\frac{1}{1,100}$	$\frac{1}{2,360}$	$\frac{1}{28,700}$
	1	$\frac{1}{1,067}$	$\frac{1}{3,021}$	$\frac{1}{169,282}$
Semidiameter of planet as seen from sun		$19\frac{3}{4}$″	11″‡	
"True" distance (r) of orbiting body from center of orbit§	72,333	581.96	1,243	946.4
Radius of central body	10,000	1,063	889	208
	10,000	*997*	*791*	*109*
Weight* at surface of central body†	10,000	$804\frac{1}{2}$	536	$805\frac{1}{2}$
	10,000	*943*	*529*	*435*
Quantity of matter (mass) of central body	1	$\frac{1}{1,100}$	$\frac{1}{2,360}$	$\frac{1}{28,700}$
	1	$\frac{1}{1,067}$	$\frac{1}{3,021}$	$\frac{1}{169,282}$ ‖
"True" diameter of central body	10,000	889	1,063	208
	10,000	*997*	*791*	*109*
Density of central body	100	60	76	387
	100	*94½*	*67*	*400*

NOTE: Numbers in roman type are taken from the first edition, those in italic from the third.
†The term "weight*" as used here signifies the weight or gravitational force exerted by the central body (the sun, Jupiter, Saturn, and the earth respectively) on some unit mass.
‡In the first edition, Newton says that he would prefer to make the semidiameter of Saturn as seen from the sun 10″ or 9″.
§The orbiting bodies are, respectively, Venus, Callisto, Titan, and the moon.
‖Newton observes that he has taken the solar parallax to be 10″30‴ and that if the parallax should be greater or less than this value, the mass of the earth will be increased or diminished in "the cubed ratio."

The radii of the orbits are determined in different ways, the distances being given by Newton on a scale in which the earth's mean distance from the sun (the astronomical unit, or AU) is taken as 1. The mean distance of Venus from the sun was known to be 0.724 AU. For Jupiter and Saturn, Newton bases his computation on the maximum heliocentric elongation (that is, the maximum angular elongation as seen from the sun). This was determined astronomically from the maximum elongation as seen from the earth. Then, the radius of each orbit (called by Newton the "true semidiameter") is readily found by multiplying the mean distance of the planet from the sun by the tangent of the angle of maximum heliocentric elongation.

There are special problems associated with finding the radius of the moon's orbit on this same scale. The radius of the moon's orbit cannot be determined by the same method as for the orbits of Callisto and Titan. The reason is that there is no maximum heliocentric elongation that can be computed from a maximum geocentric elongation, as is the case for the satellites of Jupiter and Saturn. Accordingly, the magnitude of the radius of the moon's orbit on the scale used for Venus, Callisto, and Titan requires the distance from the sun to the earth, that is, the value of the astronomical unit or the value of the solar parallax, the angular size of the earth's radius as seen from the sun. Because of the difficulties in determining the solar parallax, there was in Newton's day no agreement on the magnitude of the solar parallax. We shall see how this uncertainty enters Newton's calculations of the mass of the earth.

The $\frac{r}{T^2}$ computation yields the weight* toward the central body of each of these four satellites at its normal orbiting distance from the center. (Newton's results are given in table 8.1.) In a recent study of these corollaries, Robert Garisto has recalculated Newton's results.[45] Using Newton's values for the period of Callisto (16.689 days) and a maximum heliostatic elongation of $8'16''$, plus a distance 5.211 AU of Jupiter from the sun, the orbital radius of Callisto comes out to be

$$5.211 \tan 8'16'' = 0.01253 \text{ AU}.$$

A similar computation yields the size of the radius of the orbit of Titan.

The next step is to convert the weights* toward Jupiter, Saturn, and the earth at orbital distances to the weights* at a common distance from the center, chosen as the radius of the orbit of Venus (VS). For this computation, Newton uses the rule given in prop. 8, according to which the weight or gravitational force W_1 at

45. Robert Garisto, "An Error in Isaac Newton's Determination of Planetary Properties," *American Journal of Physics* 59 (1991): 42–48.

a distance r_1 from a center is to the similar force W_2 at a distance r_2 as the square of r_2 is to the square of r_1.

Now, let JC be the radius of Callisto's orbit and VS the radius of Venus's orbit, and let T_c and T_v be respectively the period of Callisto and of Venus. Then, Newton's calculation proceeds as follows:

$$\frac{\text{weight* toward Jupiter at distance JC}}{\text{weight* toward the sun at distance VS}}$$

$$= \frac{\text{radius of Callisto's orbit}}{(\text{period of Callisto})^2} \Big/ \frac{\text{radius of Venus's orbit}}{(\text{period of Venus})^2}$$

$$= \frac{\text{JC}}{\text{T}_c^2} \times \frac{\text{T}_v^2}{\text{VS}} = \frac{\text{JC}}{\text{VS}} \times \left(\frac{\text{T}_v}{\text{T}_c}\right)^2.$$

Taking the weight* toward the sun at distance VS to be 1 (as Newton does), we next use the rule of prop. 8 to convert the weight* toward Jupiter at distance JC to the similar weight* at distance VS, which involves multiplication by $\left(\frac{\text{JC}}{\text{VS}}\right)^2$. Hence, the ratio of weights* toward Jupiter at distance JC and toward the sun at distance VS will be given by

$$\frac{\text{JC}}{\text{VS}} \times \left(\frac{\text{T}_v}{\text{T}_c}\right)^2 \times \left(\frac{\text{JC}}{\text{VS}}\right)^2 = \left(\frac{\text{JC}}{\text{VS}}\right)^3 \times \left(\frac{\text{T}_v}{\text{T}_c}\right)^2.$$

We have seen that JC = 0.01253 AU; VS = 0.724 AU. The periods are respectively 224.698*d* and 16.698*d*. Therefore the weight* toward Jupiter at distance VS is to the weight* toward the sun at the same distance as

$$\left(\frac{0.01255}{0.724}\right)^3 \times \left(\frac{224.698}{16.689}\right)^2 = 0.00094 = \frac{1}{1{,}064}.$$

On the basis of such calculations, Newton reports that the weights* toward the sun, Jupiter, Saturn, and the earth are as 1, $\frac{1}{1{,}067}$, $\frac{1}{3{,}021}$, and $\frac{1}{169{,}282}$ respectively. Robert Garisto's careful examination of Newton's calculations reveals that the difference between our calculated value of $\frac{1}{1{,}064}$ and Newton's $\frac{1}{1{,}067}$ arises from Newton's premature rounding off.

The next problem is to apply the rule of prop. 8 once again so as to compute the weight* at the surfaces of each of the four central bodies. The distances from the centers to the surfaces (that is, the radii) are given on a scale in which the radius of the sun is taken as 10,000. These distances or radii, as they appear in the third edition, are as 10,000, 997, 791, and 109, yielding the weights* at the surfaces of the sun, Jupiter, Saturn, and the earth as 10,000, 943, 529, and 435. In the first edition, these distances are as 10,000, 1,063, 889, and 208, yielding weights* which are as 10,000, 804½, 536, and 805½. In the second edition, the distances are as 10,000, 1,077, 889, and 104, corresponding to weights* of 10,000, 835, 525, and 410.

In the second and third editions, the computation of the relative masses of the sun and of the three planets known to have satellites is completed in corol. 2. This is a relatively simple matter, and Newton does not even bother to explain his procedure, merely stating that these masses "can . . . be found," and that they are "as their forces at equal distances from their centers." He evidently assumed that readers could understand without further hints that by "their forces" he meant the gravitational forces exerted on bodies of equal masses. Essentially, the calculation consists of multiplying the previously computed weights of equal bodies on the surfaces of the sun, Jupiter, Saturn, and the earth by the squares of the respective radii of the four bodies. This follows at once from the law of gravity. Let m be the mass of these equal bodies, M the mass of any of the central bodies, and r the radius of that body. Then the weight at the surface is given by the law of gravity, $W = G\frac{mM}{r^2}$. In this case, G and m are constants, so that $W = Gm\frac{M}{r^2}$ can be written as $W = k\frac{M}{r^2}$ or $M = \frac{1}{k}Wr^2$, where k is a new constant. The relative masses (M) can be obtained by multiplying the surface weights of equal bodies (W) by the square of the radii (r). (Newton's values for the masses of the sun, Jupiter, Saturn, and the earth are given in table 8.1.)

On comparing Newton's values with those currently accepted (see table 8.2), it will be noted that the masses of Jupiter and of Saturn come out to be reasonably correct. The result for the earth, however, is far from the mark. The primary reason is that Newton has chosen a value for the solar parallax of $10''30'''$, too large by some 10 percent. As Newton pointed out in the conclusion of this corol. 2, "If the parallax of the sun is taken as greater or less than $10''30'''$, the quantity of matter in the earth will have to be increased or decreased in the cubed ratio."

In the succeeding corol. 3, Newton turns to the computing of the densities of the sun, Jupiter, Saturn, and the earth. Here he rather surprisingly does not make direct use of the mass of the earth he has just found in the preceding corollary. Rather, he introduces an argument based on book 1, prop. 72, to the effect that "the weights of equal and homogeneous bodies toward homogeneous spheres are, on the surfaces of the spheres, as the diameters of the spheres." Hence the densities are as "those weights divided by the diameters." Once again, we note that by "weight"

TABLE 8.2 THE MASSES, DIAMETERS, AND DENSITIES

	*According to present values**		*Newton's values (third edition)*		
	Mass	Density	Mass	Diameter	Density
Sun	1.000	100	1	10,000	100
Jupiter	$\frac{1}{1,047}$	94.4	$\frac{1}{1,067}$	997	94½
Saturn	$\frac{1}{3,498}$	50.4	$\frac{1}{3,021}$	791	67
Earth	$\frac{1}{332,946}$	390.1	$\frac{1}{169,282}$	109	400

*As given in Kenneth R. Lang, *Astrophysical Data: Planets and Stars* (New York, Heidelberg: Springer Verlag, 1991); the densities are mean densities.

Newton means the surface gravity or the force field on the surface. Using the same notation as before, this weight W is given by the equation $W = G\frac{M}{r^2}$. Since the density is $\frac{M}{r^3}$, the densities of the planets are as $\frac{W}{r}$, or as the "weights divided by the diameters." In the third edition, the diameters of the sun, Jupiter, Saturn, and the earth are given as 10,000, 997, 791, and 109, respectively, and the corresponding weights are as 10,000, 943, 529, and 435, yielding densities which are as 100, 94½, 67, and 400.

As Newton explains in corol. 3, "The density of the earth that results from this computation does not depend on the parallax of the sun but is determined by the parallax of the moon and therefore is determined correctly." To see why this is so, let M_e be the mass of the earth and M_s the mass of the sun, and let R_e be the radius of the earth and R_s the radius of the sun, and let T_e be the period of revolution of the earth about the sun and T_m the period of the moon. Further, let ES be the distance from the earth to the sun and EM the distance from the earth to the moon. Then, the ratio of the mass of the earth to the mass of the sun (M_e/M_s) will be given by

$$\frac{M_e}{M_s} = \left(\frac{T_e}{T_m}\right)^2 \left(\frac{R_e}{ES}\right)^3 \left(\frac{EM}{R_e}\right)^3.$$

This equation shows how the solar parallax (R_e/ES) does enter into the computation of the ratio of the mass of the earth to the mass of the sun (M_e/M_s). In this computation, the value of the ratio EM/R_e (the inverse of the lunar parallax) also appears. The value of this ratio, approximately 60, has been known quite accurately since antiquity.

The ratio for densities, the ratio of D_e to D_s, is given by a corresponding expression, where D_e is the density of the earth and D_s the density of the sun:

$$\frac{D_e}{D_s} = \left(\frac{T_e}{T_m}\right)^2 \left(\frac{R_s}{ES}\right)^3 \left(\frac{EM}{R_e}\right)^3.$$

This expression does not depend on the solar parallax (R_e/ES) but rather on the lunar parallax, the quantity R_e/EM, and R_s/ES, the radius of the sun as seen from the earth—which is easier to determine than the radius of the earth as seen from the sun (or the solar parallax).

The final corol. 4, as it appears in the second and third editions, begins with a discussion of planetary densities without reference to specific numbers, stating that the planets nearest the sun are smaller than the further ones and also have greater densities. This leads to a discussion of why the planets nearest the sun have the greatest densities. The reason is that they are better able to stand the more intense heat they receive from the sun.

Newton's method of computing densities in corol. 3 exhibits his remarkable insight. Had he determined these densities by using the masses he has just computed in corol. 2, dividing them by the cubes of the respective radii, the result would have depended on the value of the solar parallax in the case of the earth and so have been influenced by the uncertainty concerning the precise value of that quantity. In the case of the earth, this would have introduced an error which, as Newton noted, appears in the cubed power. The method adopted by Newton enabled him to determine the density of the earth, as he notes in corol. 3, in a way that "does not depend on the parallax of the sun . . . and therefore is determined correctly here."

In his manuscript "Notae in Newtoni Principia Mathematica Philosophiae Naturalis," David Gregory recorded his surprise that Newton should have made a statement of such generality. Of course, he explained, in Newton's method "the parallax of the sun is not required for the density of the earth to be compared with the density of other planets." Nevertheless, "this parallax is required in order that the earth's density may be compared with the density of the sun." For this reason, according to Gregory, Newton's statement seemed to be too general.

The soundness of Newton's judgment in this matter may be readily seen by comparing his numerical results with our present values. In the third edition, the mass of the earth is given as $\frac{1}{169{,}282}$ (on a scale in which the mass of the sun is taken as 1), whereas the present value is $\frac{1}{332,946}$, an error of almost 100 percent. The density of the earth, by contrast, is given in the third edition as 400, barely different from our present value of 392. In the first edition, in which the mass of the earth $\left(\frac{1}{28,700}\right)$ is even further from the mark, the density (387) is nevertheless quite close to today's value.

Newton's error in his final determination of the mass of the earth has two causes. One has already been mentioned, the effect of a poor value of the solar parallax. An additional error, which has been found by Garisto, can be traced to a mistake in copying. Newton kept revising his values for the astronomical parameters from edition to edition, constantly making new calculations. Garisto found, on carefully checking all the stages of Newton's work, that a significant part of the error arose from a faulty transcription of the numbers Newton had entered into one of his personal copies of the second edition.

Newton's treatment of the masses and densities of the sun and planets is significantly different in the first edition. Here the methods of calculation are given in more detail, and Newton introduces certain assumptions, which he later abandoned, concerning the value of the solar parallax. He also set forth several conjectural rules which he omitted in the second and third editions, rules which could have enabled him to extend his results for Jupiter, Saturn, and the earth to the three planets not then known to have a satellite: Mercury, Venus, and Mars. The values of the various astronomical parameters are also different.

Additionally, in the first edition the subject is developed in a series of five rather than four corollaries. In the later editions, the original corol. 2 was eliminated, so that the corollaries numbered 3, 4, and 5 in the first edition became corols. 2, 3, and 4 in the later editions.

Corol. 1 in the first edition starts off much as in the later editions. In introducing the greatest heliocentric elongation of the outermost circumjovial satellite, Newton here says that this quantity "at the mean distance of Jupiter from the sun according to the observations of Flamsteed" is 8′13″." Newton refers to Flamsteed a second time, writing, "From the diameter of the shadow of Jupiter as found by eclipses of the satellites, Flamsteed determined that the mean apparent diameter of Jupiter as seen from the sun is to the elongation of the outermost satellite as 1 to 24.9." Hence, Newton concludes that "since that elongation is 8′13″, the semidiameter of Jupiter as seen from the sun will be 19¾″." In the later editions both the reference to the method of determining the radius of Jupiter and the mentions of Flamsteed were eliminated, along with other references to Flamsteed's observations in the "phenomena." Newton's elimination of these mentions of Flamsteed was probably influenced by the bitter quarrel that had arisen between these two titans.

Another difference between the first edition and the later ones is that in the first edition Newton explained that he had computed "the distance of the moon from the earth" on "the hypothesis that the horizontal solar parallax or the semidiameter of the earth as seen from the sun is about 20″." In the later editions, Newton does not mention how he uses the solar parallax in the first corollary. The magnitude of the solar parallax in the first edition is extremely large, about twice the 10″ or 9.5″ determined by Cassini and Flamsteed. We may take note that Newton's adoption of the larger number is but one sign that the astronomical community did not at once accept the results of Cassini and Flamsteed.[46] As we shall see, Newton's choice of this large value for the solar parallax was determined by the needs of an ancillary hypothesis which he later came to reject.

In the first edition, Newton says that the "diameter of Saturn is to the diameter of its ring as 4 to 9, and the diameter of the ring as seen from the sun (by Flamsteed's measurement) is 50″, and thus the semidiameter of Saturn as seen from the sun is 11″." He then makes the astonishing statement, "I would prefer to say 10″ or 9″, because the globe of Saturn is somewhat dilated by a nonuniform refrangibility of light." This is pure nonsense! In his extensive manuscript commentary, "Notae in Newtoni Principia Mathematica Philosophiae Naturalis," David Gregory quite correctly notes, "This also happens with all stars, nay, with all luminous bodies." Accordingly, there is no valid reason why the diameter of Saturn should be "diminished because of this."

As we shall see, Newton nudged this value downward not because of any special optical effect in Saturn, but rather because he wanted to obtain better agreement with a rule he was about to propose concerning the properties of the planets. This rule is set forth in the first edition of the *Principia* in a corol. 2 which was eliminated from the later editions.

The remainder of the original corol. 1 differs from the later versions primarily in the numerical values introduced. Newton concludes that the weight-forces on the surfaces of the sun, Jupiter, Saturn, and the earth are respectively as 10,000, 804½, 536, and 805½. A final comment in this corol. 1 is that "the weights of bodies on the surface of the moon are almost two times less than the weights of bodies on the surface of the earth," as "we shall show below."

Newton's reasons for introducing into the first edition the large value of 20″ for the solar parallax are given in the corol. 2 of that edition. This corollary, no

46. See Albert Van Helden, *Measuring the Universe: Cosmic Dimensions from Aristarchus to Halley* (Chicago and London: University of Chicago Press, 1985), p. 144.

longer present in the later editions, has as its purpose to set up two general rules. Although Newton does not say so explicitly, these rules would make it possible to extend his findings about the masses of planets with satellites to those which have none (Mars, Mercury, and Venus). When, in the second edition, Newton eliminated the original corol. 2, he no longer needed this large value of the solar parallax and so he eliminated it in favor of the smaller number.

At the time of the first edition, Newton was fully aware that 20″ was not a generally accepted value for the solar parallax and that, accordingly, his choice required some discussion. This took two different forms. One was a justification of his adoption of 20″ rather than a smaller number, the other an indication that perhaps the solar parallax is less than 20″.

His defense of the larger solar parallax, as presented in corol. 2 of the first edition, begins with the declaration that "there is as yet no agreement" concerning "the diameter of the earth as seen from the sun." He has taken it to be 40″, corresponding to a solar parallax of 20″. He has chosen this diameter, which was the upper possible limit, he says, according to "the observations of Kepler, Riccioli, and Vendelin," although he was fully aware that "the observations of Horrocks and Flamsteed seem to make it a little smaller." He "preferred," he said, "to err on the side of excess." He was, however, as he reminded his readers, aware of an argument (presented below) that might yield a diameter of the earth as seen from the sun of about 24″ and therefore a solar parallax of about 12″.

The true reason for his having adopted the "slightly larger diameter" was then revealed to the reader. He has done so, he writes, because the large number "agrees better with the rule of this corollary." That is, the larger diameter of the earth as seen from the sun, and hence the larger solar parallax, agrees better with the rule he is about to present concerning the weights of equal bodies on the surface of the earth and planets. These weights, he alleges, are "almost proportional to the square roots of their apparent diameters as seen from the sun."

Newton's second discussion of the value of the solar parallax is also part of the corol. 2 of the first edition. Here Newton suggests the possibility that the diameter of the earth as seen from the sun, and also its mass, might be an average of these two quantities for the whole solar system. He does not, however, actually compute from his formula or rule what the masses of Mars, Venus, and Mercury would be. Nor does he give their apparent diameters as seen from the sun. In his "Notae," Gregory tested Newton's assertion. He did not compute the masses of Mars, Venus, and Mercury, but he did supply a list of the apparent diameters of the planets as seen from the sun: Saturn (18″), Jupiter (39.5″), Mars (8″), Venus (28″), Mercury (20″). With these values it is easy to test part of Newton's statement about averages. The sum of the above five numbers is 113.5 and their average (113.5 divided by

5) is 22.7″. Hence, if the law of averages is true, there would be a corresponding value of a little less than 24″ for the diameter of the earth as seen from the sun, or a solar parallax of about 12″, the value which Newton says is the one with which "Horrocks and Flamsteed pretty nearly concluded."

More than a decade before the second edition, Newton had abandoned the large value of the solar parallax, in part responding to the critique of Huygens in his tract "La cause de la pesanteur," appended to his *Traité de la lumière* (Leiden, 1690), of which Newton had two copies, one of them a gift from the author.[47] Accordingly, he also had to abandon the old corol. 2, since with the new numbers the proposed rule about weights of equal bodies and diameters can no longer be considered valid. At the same time, Newton introduced a new set of values in corol. 1, and he also upgraded the numerical values in the old corols. 3 and 4, now become corols. 2 and 3.

In the third and final edition, the diameters of the sun, Jupiter, Saturn, and the earth are given as 10,000, 997, 791, and 109, and the corresponding weights* are as 10,000, 943, 529, and 435, yielding densities which are as 100, 94½, 67, and 400. These values are quite different in the first edition, where the densities come out to be as 100, 60, 76, and 387. The number 60 is obviously a misprint for 90.

The final corollary (corol. 5 in the first edition, corol. 4 in the others) differs somewhat in the early and later editions. The first portion of the corol. 5 of the first edition was eliminated in the later editions. In the first edition, Newton begins by stating yet another rule, "The densities of the planets, moreover, are to one another nearly in a ratio compounded of the ratio of the distance from the sun and the square roots of the diameters of the planets as seen from the sun." The densities of Saturn, Jupiter, the earth, and the moon (60, 76, 387, and 700 in the first edition) are, he points out, almost as the square roots of the apparent diameters (18″, 39½″, 40″, and 11″) divided by the reciprocals of their distances from the sun $\left(\frac{1}{8{,}538}, \frac{1}{5{,}201}, \frac{1}{1{,}000}, \frac{1}{1{,}000}\right)$. He then applies the rule he has found in corol. 2, to the effect "that the gravities at the surfaces of the planets are approximately as the square roots of their apparent diameters as seen from the sun," and the rule given in the previous corollary (which he mistakenly refers to as "lem. 4" rather than "corol. 4") "that the densities are as the gravities divided by the true diameters." Hence, he concludes, "the densities are almost as the roots of the apparent diameters multiplied by the true diameters," that is, "inversely as the roots of the apparent diameters divided by the distances of the planets from the sun." This leads him to conclude, "Therefore God placed the planets at different distances from the sun

47. See John Harrison, *The Library of Isaac Newton* (Cambridge: Cambridge University Press, 1978).

so that each one might, according to the degree of its density, enjoy a greater or smaller amount of heat from the sun."

In the second and third editions, as we have seen, Newton eliminated the numbers and the reference to rules. He also altered the sentence just quoted so that it would read, "The planets, of course, had to be set at different distances from the sun . . . " All that remained of the numerical introduction to this final corollary was some descriptive sentences (without numbers), stating that the planets nearest the sun are smaller than the further ones and also have greater densities. The remainder of this final corollary is much the same in all editions, a discussion of why the planets nearest the sun have the greatest densities. The reason is that they are better able to stand the more intense heat they receive from the sun.

Some writers, notably Jean-Baptiste Biot, have written that God does not appear in the first edition of the *Principia*, but was introduced only in the second edition in 1713. He had in mind that God appears nowhere in the text proper of the second edition, but only in the concluding General Scholium, published for the first time in that edition. Since the later *Principia* introduced the subject of God and religion, which Biot thought was absent from the first edition of 1687, Biot attributed Newton's religious turn of mind to senility. An examination of the final corollary as it appears in the first edition, however, shows that Newton did mention God in the first edition, in relation to the ordering of the universe, a passage deleted in the second edition and so not known to Biot and others.

Props. 9–20: The Force of Gravity within Planets, the Duration of the Solar System, the Effects of the Inverse-Square Force of Gravity (Further Aspects of Gravity and the System of the World): Newton's Hyp. 1 8.11

Prop. 9 is of special interest in exhibiting the difference between mathematical constructs such as are explored in book 1 and the physical conditions of the external world of nature. In book 1, prop. 73, Newton has shown that in the interior of a uniform sphere, or within a sphere of uniform density, the force of gravity will vary directly as the distance from the center. In book 3, prop. 9, Newton takes note that the planets are not spheres of uniform density and that the force of gravity within a planet is therefore not exactly but only "very nearly" as the distance from the center. Then, in prop. 10, Newton shows that "the motions of the planets can continue in the heavens for a very long time."

In the second and third editions, Newton introduces hyp. 1 (the hyp. 4 of the first edition) just before prop. 11 for which it will be needed. This hypothesis states: "The center of the system of the world is at rest." Who could deny that this is indeed a hypothesis! Concerning it Newton says only: "No one doubts this, although some argue that the earth, others that the sun, is at rest in the center of the system." He at once uses this hypothesis, together with corol. 4 of the laws, to prove that "the common center of gravity of the earth, the sun, and all the planets is at rest." By the relativity of corol. 4 of the laws, the center of gravity must either "be at rest" or "move uniformly straight forward." In the latter case, however, "the center of the universe will also move, contrary to the hypothesis." This is one of the few places in the *Principia* where Newton makes overt reference to absolute space. In prop. 12, Newton discusses the "continual motion" of the sun, which itself "never recedes far from the common center of gravity of all the planets."

Then, in prop. 13, Newton states:

> The planets move in ellipses that have a focus in the center of the sun, and by radii drawn to that center they describe areas proportional to the times.

The discussion of this proposition begins as follows:

> We have already discussed these motions from the phenomena. Now that the principles of motion have been found, we deduce the celestial motions from these principles a priori.

That is, in the opening of book 3, in the section of phenomena, Newton has displayed celestial orbital motion according to the law of areas and the harmonic law. Now he will explore the implications of his prior result (in book 3) that "the weights of the planets toward the sun are inversely as the squares of the distances from the center of the sun." He begins with the conditions that (1) the sun is at rest and (2) "the remaining planets" do not "act upon one another." He then (referring to book 1, props. 1 and 11, and prop. 13, corol. 1) observes that under these conditions, motion according to an inverse-square force will be in "orbits [which] would be elliptical, having the sun in their common focus," and that the planets "would describe areas proportional to the times."[48] He explains that the "actions of the planets upon one another, however, are so very small that they can be ignored, and they perturb the motions of the planets in ellipses about the mobile sun less (by book 1, prop. 66) than if those motions were being performed about the sun at rest." The "action of Jupiter upon Saturn," however, cannot be ignored. He alleges that this perturbing effect is particularly noticeable "in the conjunction of Jupiter and Saturn." He indicates as well that perturbations also noticeably affect the orbit of the moon.

48. Prop. 11 does not, of course, apply here, but only the converse (prop. 13, corol. 1).

The next group of propositions deals with the state of rest of the aphelia and nodes of planetary orbits (prop. 14), the principal diameters of these orbits (prop. 15), and the method for finding the eccentricities and aphelia of the orbits (prop. 16). This leads to prop. 17 on the uniformity of the daily motion of the planets and the libration of the moon.

Props. 18 and 19 then discuss planetary diameters, with consideration given to the oblate-spheroidal shape of the earth. Here Newton introduces a hydraulic analogy in explaining the variation in weight produced by the difference between the spheroidal earth and a perfect sphere, assuming a homogeneity of matter. In this prop. 19 and the succeeding prop. 20, Newton establishes the shape of the earth from theory and the data of the variations observed in weight (in pendulum experiments) as a function of latitude, taking into account the effects of the earth's rotation as well as its shape.

Newton's investigation of the shape of the earth and of Jupiter by a combination of the theory of gravity and terrestrial measurements is one of the many parts of the *Principia* that exhibit both Newton's genius and the enormous intellectual gulf that separates that work from its immediate predecessors. As Laplace wisely remarked, Newton's address to this problem was a "first step" that "must seem immense," even though it "left much to be desired" and had certain "imperfections." The grandeur and immensity of this "first step" will become all the more apparent, he took note, when one takes into account "the importance and the novelty of the propositions which the author establishes concerning the attractions of spheres and spheroids," as well as "the difficulty of the subject matter."

The goal of book 3, prop. 19, is to find the relative dimensions of the axis of rotation of a planet and a diameter perpendicular to it, that is, the ratio of the polar diameter to an equatorial diameter. On the basis of various kinds of measurements, Newton takes the shape of the earth to be an oblate spheroid generated by the rotation of an ellipse about its lesser axis, thus having a flattening at the poles or a bulge at the equator. Newton considers two kinds of forces: (1) the mutual gravitational force of the particles composing the earth, and (2) the effects of the earth's rotation. In his demonstration (see fig. 8.1) he supposes that there is a narrow canal filled with water that runs from one of the poles Q to the center of the earth C and then has a right-angle turn, going on then to a point A on the equator. AC is the equatorial radius, QC the polar radius.[49] Newton seeks to find the dimensions of the two legs of the canal such that the attractions on them

49. The diagram in the printed editions is very confusing because the polar axis is horizontal and the equatorial diameter is vertical.

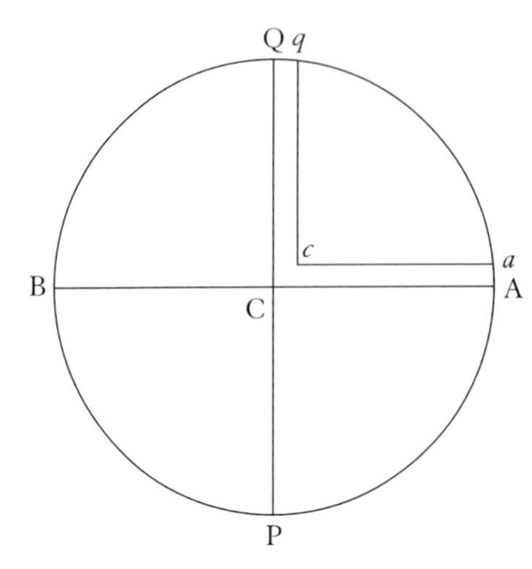

Figure 8.1. Diagram for book 3, prop. 19, on the shape of the earth. The diagram has been rotated through 90 degrees and flipped over, so that the figure of the earth appears in the customary way, with the poles Q and P at the top and bottom and not at the sides.

exerted by the gravitational forces of the earth under conditions of rotation will produce a balance.

Newton concludes that if the centrifugal force of the earth's rotation on every portion of the leg AC (corresponding to an equatorial radius) is to the attraction of the earth on that portion (or the weight of that portion) as 4 to 505, then the "fluid would stay at rest in equilibrium." The ratio of the centrifugal force of every part to the weight of that part, however, is 1 to 289, so that the centrifugal force is $\frac{1}{289}$ parts of the weight, whereas it should be $\frac{4}{505}$ parts of the weight, according to calculations based on book 1, prop. 91, corol. 3. Further considerations lead to the result that "the diameter of the earth at the equator is to its diameter through the poles as 230 to 229." In this case he is not content with an abstract ratio but uses Picard's measurements to conclude that "the earth will be 85,472 feet or 17$\frac{1}{10}$ miles higher at the equator than at the poles." The equatorial radius will thus rise to about 19,658,600 feet, while the polar diameter will be about 19,573,000 feet.[50]

In the first edition these numbers were somewhat different (1 to 290 in place of 1 to 289 and 3 to 689 instead of 1 to 230). Newton took the ellipticity of the earth to be 1 to 230. In obtaining this result he made use of two assumptions: (1) that the earth is homogeneous, and (2) that the shape of the earth would be the same if it were entirely fluid.[51]

In the final part of prop. 19, Newton directs his attention to Jupiter. He concludes that "Jupiter's diameter taken from east to west is to its diameter between

50. Newton says he is "supposing a mile to be 5,000 feet."

51. Newton's argument is rephrased in the language of more modern mathematics in Todhunter, *History*, pp. 17–19, where a detailed presentation is given of the changes in the argument from edition to edition. See esp. p. 23.

the poles very nearly as 10⅓ to 9⅓."[52] At the end of prop. 19, in the first and second editions, there is a paragraph stating that his results follow from "the hypothesis that the matter of planets is uniform." The reason he gives is that "if the matter is denser at the center than at the circumference," then "the diameter which is drawn from east to west will be still greater," that is, if the matter of the earth were not uniform, but increased in density from the circumference to the center, the ellipticity would accordingly be that much greater. As Clairaut showed, however, Newton made a mistake here. If the center is so increased in density as to become a solid, the outer parts of the earth will have a diminished and not an increased ellipticity.[53]

In the third edition, Newton struck out the above two sentences about the matter being denser at the center and substituted in their place the final version: "This argument has been based on the hypothesis that the body of Jupiter is uniformly dense." Then he says, "But if its body is denser toward the plane of the equator than toward the poles, its diameters can be to each other as 12 to 11, or 13 to 12, or even 14 to 13." This is a diminished ellipticity from the above value of 10⅓ to 9⅓. We should take note, however, that Newton has shifted his argument from the condition of the central part of Jupiter being denser than the exterior parts to the condition of the plane of the equator being denser than the polar regions.

In the second edition, where Newton concluded prop. 19 with a discussion of the matter of the planet being denser at the interior than at the circumference, Newton added a sentence to the effect that the observations of Cassini had shown that the polar diameter of Jupiter is less than the other one; he remarked that it would be seen in the following proposition (prop. 20) that the polar diameter of the earth is less than the equatorial diameter. In the third edition, Newton took note that in 1691 Cassini had found that the equatorial diameter of Jupiter exceeded the polar diameter by about a fifteenth part.[54] He then gave a table of

52. Ibid., §29.

53. Ibid.

54. In the second edition, this sentence did not mention either the year of Cassini's observations or the value one-fifteenth. Curtis Wilson notes that it was the astronomer Cassini I (d. 1712) who made the observations on the diameters of Jupiter in the 1690s, but it was certainly Cassini II who finished the geodetic study, ca. 1720. Newton was apparently aware of the idea, emanating from Paris, that the earth was elongated toward the poles. This was supposedly an empirical result, independent both of the shape of Jupiter and of Huygens's as well as Newton's deductions of an oblate shape for the earth. In the 1730s Poleni and Maupertuis echo Newton (second edition) in asserting that the north-south variation in toises/degrees in Cassini's data fell within the range of accuracy of the observational data, and therefore could not be relied on.

recently observed data provided by the astronomer James Pound. These data are mentioned in Newton's preface to the third edition.

In the third edition there is a new pair of paragraphs. The first declares that the astronomical data of Cassini and especially of Pound show how "the theory agrees with the phenomena." Newton then asserts, "Further, the planets are more exposed to the heat of sunlight toward their equators and as a result are somewhat more thoroughly heated there than toward the poles." This sentence has given rise to much confusion. The verb in question is "decoquuntur," which can mean cooked, baked, or even boiled. It has usually been assumed by Newton's translators, beginning with Andrew Motte and the marquise du Châtelet, that by using this term Newton was saying that the intensity of the solar heat at the equator causes the matter to be condensed or increased in density. Andrew Motte, in his translation of 1729, had Newton say that "the Planets are more heated by the Sun's rays towards their equators, and therefore are a little more condensed by that heat, than towards their poles." The marquise du Châtelet added a phrase, so as to make the passage read that because the planets are more exposed at their equators "to the action of the sun, the matter which is there, so to speak, must be more cooked [or baked] and must be denser than toward the poles."

Newton's conclusion must seem odd, since matter tends to expand rather than contract as the result of being heated. Thus, on the basis of considerations of relative heat and cold, one would have expected the polar regions to be more and not less dense than the equatorial regions. The common bewilderment is well expressed by Todhunter, who wrote concerning this passage: "I do not feel certain as to the meaning of this sentence."[55] Since "heat expands bodies," he explained, "it would appear that the equatorial parts ought by the Sun's action to be rendered *less* dense than the polar parts." Indeed, in the following prop. 20, Newton (in the final paragraph) gives evidence of the expansion of an iron rod by the summer heat.[56] Of course, as Newton has shown, if the matter of a planet were more dense at the equator than at the poles, the ellipticity would be diminished so as to accord more closely with the data in Pound's table.

Prop. 20 continues with the subject of the shape of the earth. Essentially Newton takes as his assignment to correlate the various data on weight at different

55. Todhunter, *History*, p. 19.

56. It is possible Newton was implying that the matter at the equator was more "baked" and even more dense as a result of the intense solar heat and so was better adapted to receive that heat. This would be similar to the argument in book 3, prop. 8, corol. 4, to the effect that the denser planets were more suited to receive the intense heat of the sun than the others and so were placed nearer to the sun.

points on the earth's surface with the latitudes.[57] Once again, he begins with a discussion of balanced columns of water. He obtains a theorem, that the "increase of weight in going from the equator to the poles is very nearly as the versed sine of twice the latitude, or (which is the same) as the square of the sine of the latitude."[58]

There are many changes introduced into the presentation of this problem in the second and again in the third edition. In the first edition, Newton checked the accuracy of his theory by calculating the relative weights at Paris, Gorée, Cayenne, and at the equator, using weight determinations from the data on the periods of pendulums. This led to a discussion, revised for the second edition but omitted from the third, in which Newton attempted to account for the factors that needed to be introduced to correct his initial assumption that the earth is uniform. In the second and third editions, many further observations were presented, providing a better agreement between theory and observation. Newton thus eliminated a sentence in the first edition, printed within parentheses, concerning the degree of confidence that could be given to such coarse observations. The difficulties in computing the data tabulated in prop. 20 can be seen in the correspondence of Newton and Cotes, resulting in whole portions having to be corrected as Newton advanced his theory. In particular, Newton was greatly troubled by the problem of determining the ellipticity of the earth, and there is clear evidence of fiddling with numbers in an attempt to make them yield his preferred answers. In the second edition, Newton concludes that correcting the theory by resort to observations concerning pendulums will give the result that the equatorial radius of the earth is 31 7/12 miles greater than the polar radius. In the third edition, he returned to the value given by theory and set this number at approximately 17 miles ("milliarium septendecim circiter").[59] Today's value is a little more than 13 miles.

When Cotes received the revisions for prop. 20, he was aware that the table no longer corresponded to the text, since it had been computed (for the first edition) on the basis of a difference in the terrestrial radii of 17 miles rather than something less than 32 miles. Readers would be better served, he wrote to Newton on 23 February 1712, if a new table were computed on the basis of a difference of 32 miles, and he offered to make the new table. Newton, however, preferred to keep the old table, as he replied on 26 February, even though it "is computed to the excess of 17 miles rather than to that of 32 miles." The reason he gave was that

57. There is a thorough discussion of this portion of the *Principia* in Todhunter, *History*, chap. 1. Todhunter carefully notes the various alterations made in the second and third editions.

58. For a critical analysis of Newton's argument, see Todhunter, *History*, pp. 20–22.

59. On Newton's argument about the shape of the earth, and notably in relation to the later critique of Clairaut, see Todhunter, *History*, pp. 20–22.

"17 is the least that [the excess] can be" and 17 "is certain" on the "supposition that the earth is uniform," while "that of 32 is not yet sufficiently ascertained, & I suspect it is too big."

The concluding portions of prop. 20 differ greatly in the second and the third editions. The final paragraph in the second edition refers to Cassini's terrestrial measurements (and some by Picard) which seemed to indicate that the shape of the earth might be prolate. Newton shows, however, that this hypothesis implies that all bodies would be lighter at the poles than at the equator, that the height of the earth at the poles would be almost 95 miles higher at the poles than at the equator, and that the length of a seconds pendulum would be greater at the equator than at the Observatory in Paris, all contrary to experience.

As can be seen in the notes to the text, prop. 20 originally contained a sentence suggesting that a standard seconds pendulum, that is, a pendulum with a period of two seconds, could serve as a universal standard of length. Such a world standard had already been proposed by Huygens in his *Horologium Oscillatorium* of 1673. There is no evidence that Newton suppressed this notion in the second edition because he had come to reject this new standard. Rather, he seems to have dropped it because he eliminated the lengthy passage in which it was embedded.

8.12 *Prop. 24: Theory of the Tides; The First Enunciation of the Principle of Interference*

Newton's theory of the tides is one of the great achievements of the new natural philosophy based on a force of universal gravity. This theory heralded a wholly new basis for understanding tidal phenomena by introducing physical causes: the combination of gravitational pulls on the waters of the oceans exerted by the sun and by the moon. As is the case for some other gigantic steps taken by Newton, however, Newton was not able to complete the explanation of the tides. In this aspect, Newton's tidal theory resembles his theory of the moon's motion. Both set science on new paths based on physical causes, thus determining new approaches to old problems that we have more or less followed ever since.

Newton's theory of the tides is a logical extension of his theory of gravity. If gravity is, indeed, a universal force, then gravity must affect the waters in the seas as well as planets, moons, apples, and other solid objects. The action of two separate forces, exerted by the sun and by the moon, is in some ways conceptually similar to the analysis of the moon's motion under the action of two forces, exerted by the earth and by the sun. We are not, therefore, surprised that Newton's development of his theory of the tides will draw on book 1, prop. 66, and its many corollaries, in which the joint action of two distinct forces of attraction is explored.

It is at once obvious from Newton's concept that there will be times when the solar and lunar forces will be in phase and reinforce each other, and that there will be other times when the two are out of phase and the resulting force will be their difference. Between these extremes, there will be different combinations depending on the relative orientations of sun and moon.

Newton's theory of the tides is found in three major propositions of book 3: props. 24, 36, and 37. Prop. 24 begins by drawing on book 1, prop. 66, corols. 19–20. In corol. 19, Newton has considered a central body such as the earth, with a satellite such as the moon. Water on the central body (which can be rotating and moving forward) will then be acted on gravitationally by the moon and by the sun, with resulting tidal phenomena. Newton considers, in corol. 20, the case of a central body or globe "which in the regions of the equator is either a little higher than near the poles or consists of matter a little denser" (as in book 3, prop. 19; see §8.10 above). The conclusion is that the waters in the sea ought to rise and fall twice in every day, each such extremity of tide following a transit of the sun or moon across the meridian by less than six hours. This condition, Newton observes, occurs in certain regions of the Atlantic Ocean, in the South Seas, and in the Ethiopic Sea, where the flood tide follows the transit in "the second, third, or fourth hour," except where the movement of the ocean waters passes through shallow channels and may be retarded to "the fifth, sixth, or seventh hour, or later."

Although the sun and the moon each exert a separate force on the waters of the sea, Newton explains, the result is not two separate and independent motions. Rather there is a single motion resulting from a compounding of the two tide-producing forces. These solar and lunar tidal forces, Newton reports, vary according to different periods. In conjunctions and oppositions, the solar and lunar forces act maximally with respect to each other, thus producing the greatest flood and the greatest ebb tide. "In the quadratures," however, "the sun will raise the water while the moon depresses it and will depress the water while the moon raises it." Experience agrees with the conclusion that the effect of the moon is greater than that of the sun, since "the greatest height of the water will occur at about the third lunar hour." Newton discusses other relative situations of the sun and moon in relation to tidal effects, noting in conclusion that this is what happens "in the open sea," whereas the tidal phenomena are different "in the mouths of rivers."

Newton then introduces a quite different factor, the variation in the lunar and solar tidal forces with the distance of these two luminaries from the center of the earth. Another set of considerations involves the effects of the "declination, or distance from the equator," of the sun and moon. Newton also takes account of those "effects of the luminaries" which depend on the "latitude of places" where the tides occur.

Thus far Newton has dealt with general aspects of the tides as they occur in large bodies of open water. He has shown how the theory of gravity may be applied to the waters of the oceans in a general way. The next step is to take account of some possible kinds of deviations from the general rules. Primarily he deals with one such example. Before considering it, however, we should note that Newton's explanation of the role of the gravitational force of the sun and the moon on the waters of the oceans was a tremendous achievement. To understand its significance, we have only to recall that Galileo pilloried Kepler for having suggested that the moon might influence the tides and that Huygens, in Newton's day, could not believe in a force by which the sun or the moon could affect the water so as to produce tides.

And yet, like other breakthroughs in the *Principia*, Newton's theory of the tides, although a tremendous intellectual leap forward, was incomplete and to that extent imperfect. That is, Newton's theory could give satisfactory explanations of many of the principal general features of the tides, but it could not adequately explain the observed departures from two tides per day, nor could it actually predict either the time of occurrence of a high tide at every given location or the height of some designated tide. These failures came from the inadequacy of his gravitational theory to predict and to explain every aspect of tidal phenomena. Newton failed to understand that the analysis of tidal gravitational forces is only part of the problem, the other part being an understanding of the nature of responses to these tidal forces and the effects of the earth's rotation.

One of the examples used by Newton is of special interest because it led him to an explanation embodying the first known use of what was later called the principle of interference. In presenting the concept of interference in the context of the wave theory of light, Thomas Young declared that "there was nothing that could have led" to the law of interference "in any author with whom I am acquainted," except for "some imperfect hints" in "the works of the great Dr. Robert Hooke" and "the Newtonian explanation of the tides in the Port of Batsha."[60] In a later article, Young recognized that Newton's general theory of the tides as well as his explanation of musical "beats" could also be considered examples of interference similar to those Young had proposed in the wave theory of light. "The spring and neap tides, derived from the combination of the simple solar and lunar tides," according to Young, "afford a magnificent example of the interference of two immense waves with each other." That is, the spring tide is "the joint result of the combination when they coincide in time and place, and the neap when they succeed each other at the distance of half an interval."

60. Thomas Young, *Miscellaneous Works* (London: John Murray, 1985), vol. 1.

The tides in the port of Batsha, in what used to be called Indochina but is now Vietnam, were discussed in detail by Newton in book 3, prop. 24. Batsha is situated in the northwest part of the Gulf of Tonkin, at the mouth of the Domea (or principal) branch of the Tonkin River, at about latitude 20°50′ N. An account of the Batsha tides was written by a shipmaster, Francis Davenport, and communicated to the Royal Society of London with a very imperfect explanation by Edmond Halley. At Batsha, there is usually but one flood and one ebb tide every twenty-four hours rather than two of each. When the moon is near the equator, twice in every lunar month, there is about a two-day period in which there appears to be little or no tide. For approximately the first seven days after this tideless period, the tides increase gradually until they attain a maximum; then they fall off gradually for about seven days, after which they reach another stationary period of two days.

It can happen, Newton wrote in explaining this phenomenon in the *Principia*, "that a tide is propagated from the ocean through different channels to the same harbor and passes more quickly through some channels than through others." In this case, according to Newton, "the same tide, divided into two or more tides arriving successively, can compose new motions of different kinds."

> Let us suppose that two equal tides come from different places to the same harbor and that the first precedes the second by a space of six hours and occurs at the third hour after the appulse of the moon to the meridian of the harbor. If the moon is on the equator at the time of this appulse to the meridian, then every six hours there will be equal flood tides coming upon corresponding equal ebb tides and causing those ebb tides to be balanced by the flood tides, and thus during the course of that day they will cause the water to stay quiet and still.

Newton then proceeds to describe how the position of the moon with regard to the equator can cause other combinations of tides in these channels—in such a way as to account for the data of the tides at Batsha. If the tide is considered to be a wave whose crest corresponds to the floods, and whose trough corresponds to the ebb, we see immediately that Newton's explanation consists of all of the varieties of interference from destructive to corroborative.

Newton concludes his presentation by taking into account that there are two inlets to this harbor and two open channels, "the one from the China Sea between the continent and the island of Leuconia, the other from the Indian Ocean between the continent and the island of Borneo." Yet he could not be certain "whether there are tides coming through these channels in twelve hours from the Indian Ocean and in six hours from the China Sea" to "compound motions of this sort," or whether the cause lay in "any other condition of those seas." The validity of his proposed analysis would have "to be determined by observations of the neighboring shores."

Newton's proposed explanation of the tides in the Gulf of Tonkin shows the power of his inventive mind. He did not, however, have the necessary understanding of the several varieties of tidal forces and the different kinds of responses to them of different bodies of water. Today it is recognized that the tide-producing forces of the sun and the moon fall into three general classes: diurnal forces (with a period of approximately 24 hours), semidiurnal forces (with a period of approximately 12 hours), and those having a period of half a month or more. These forces are distributed over the surface of the earth in a rather regular way, varying with the latitude, but the response to them of any particular sea depends on its own physical or geographical features. That is, each body of water has a natural period of oscillation which is a function of its length and depth and accordingly will respond primarily to that tidal force whose period most closely approximates its own natural period. The Atlantic Ocean thus responds primarily to the semidiurnal forces, with the result that there tend to be two ebb and two flood tides daily on the eastern American coast. The Gulf of Tonkin, however, responds almost perfectly to the diurnal forces, with the result that there is usually but one flood and one ebb tide a day, as Newton stated. The variations in the degrees of the tides which Newton explained by a principle of interference are now attributed to the changing orientation of the moon with respect to the equator.[61]

8.13 *Props. 36–38: The Tidal Force of the Sun and the Moon; The Mass and Density (and Dimensions) of the Moon*

Props. 36 and 37 are devoted to a determination of the relative contributions of the sun's and the moon's gravitational force to the net tidal force on the waters in the seas. In prop. 36, Newton first finds the force of the sun on the tides when the sun is at its mean distance from the earth. In other positions, the force is as the versed sine of twice the sun's altitude and inversely as the cube of the distance from the earth. In prop. 37, Newton estimates the moon's contribution to the tidal force by comparing the highest and lowest tides in the Bristol Channel as observed

61. This explanation concerning the causes of the tidal phenomena in the Gulf of Tonkin is based on information supplied to me by J. H. Hawley, former acting director of the U.S. Coast and Geodetic Survey. See, further, H. A. Marmer, *The Tide* (New York: Harper, 1926); also P. C. Whitney, "Some Elementary Facts about the Tide," *Journal of Geography* 34 (1935): 102–108.

and reported by Samuel Sturmy. The highest tide (45 feet) occurs in the vernal and autumnal syzygies, when the two luminaries pull together so that the net force should be the sum of the two separate forces. The lowest tide (25 feet) occurs in the quadratures, when the net force should be the difference between the two. Thus if these two forces are designated by S and L,

$$\frac{L+S}{L-S}=\frac{45}{25}=\frac{9}{5}.$$

This initial ratio requires adjustment, Newton argues, because "the greatest tides do not occur at the syzygies of the luminaries," but at a later time, when the direction of the solar tidal force is no longer the same as (or is no longer parallel to) the lunar tidal force. Another factor causing a diminishing of the lunar force arises from "the declination of the moon from the equator." Thus, Newton finds a corrected ratio of L + 0.7986355S to 0.8570327L − 0.7986355S to be as 9 to 5. A final factor of the moon's distance gives the result that 1.017522L × 0.7986355S is to 0.9830427 × 0.8570327L − 0.7986355S as 9 to 5, which yields the final result that S is to L as 1 to 4.4815. Hence, since (by book 3, prop. 36) the tidal force of the sun is to gravity as 1 to 12,868,200, the tidal force of the moon will be to the force of gravity as 1 to 2,871,400.

In corol. 1 Newton gives examples of open seas in which the tidal phenomena agree with his theory. But he also indicates some exceptions arising from physical features of harbors and straits. In corol. 2, Newton says that because the lunar tidal force is only $\frac{1}{2,871,400}$ of the force of gravity, its effect would be too minute to be appreciable in pendulum or other terrestrial experiments and can be discerned only in the tides.

We should take note that Newton's method does not give reliable results and that today the ratio of the moon's tidal force to the sun's tidal force is found to be about 2.18 to 1, rather than Newton's 4.4815 to 1. This error, of almost 100 percent, had important consequences since it led Newton (in corols. 3 and 4) to an error of equal magnitude in his determination of the mass and the density of the moon. This in turn led to a serious problem (in corol. 7) in his second moon test of the inverse-square law.[62]

Corols. 3, 4, and 5 are devoted to quantitative features of the moon. In corol. 3, in the third edition, Newton begins by noting that the tide-producing forces of the

62. The first edition of the *Principia* contains only corols. 1–5; corols. 6–8 were added in the second edition, corols. 9 and 10 in the third. The second half of corol. 7 was rewritten for the third edition, with new numerical values, and corol. 8 was rather completely recast in the third edition. For details, see our Latin edition with variant readings.

moon and the sun are as 4.4815 to 1. According to book 1, prop. 66, corol. 14, these forces are to each other as the densities and cubes of the apparent diameters. Hence

$$\frac{\text{density of moon}}{\text{density of sun}} = \frac{4.4815}{1} \times \frac{(32'12'')^3}{(32'16\frac{1}{2}'')^3} = \frac{4,891}{1,000}.$$

Since Newton has previously found that the density of the sun is to the density of the earth as 1,000 to 4,000, the density of the moon is to the density of the earth as 4,891 to 4,000, or as 11 to 9. Therefore, Newton finds that "the body of the moon is denser and more earthy than our earth."

Then, in corol. 4, Newton concludes that the mass of the moon is to the mass of the earth as 1 to 39.788. This result is based on a ratio of the "true" diameter of the moon ("from astronomical observations") to the true diameter of the earth, a ratio of 100 to 365. In corol. 5, he uses the foregoing results to compute that the "accelerative gravity" (that is, the accelerative measure of gravity, or the acceleration of free fall) on the surface of the moon is about three times less than on the surface of the earth.

These numerical results are doubly wide of the mark. First of all, as we have seen, they incorporate the error in the relative densities of the sun and the earth (which, however, is very small). This fault is compounded by the error in the determination of the tidal force of the moon, which produces an error in the mass of the moon. Newton's ratio of 1 to 39.788 is in error by about 100 percent, since the currently accepted value is about 1 to 81.

In corol. 6, Newton uses the values he has obtained for the relative masses of the moon and the earth to compute the center of gravity of the earth-moon system, the point which must be considered the center of the moon's orbit and which is known as the barycenter. Because of the faulty value for the moon's mass relative to the earth's, Newton makes a serious error in determining the center of gravity of the earth-moon system. This point is actually within the body of the earth, located at approximately one-quarter of an earth radius (or some 1,000 miles) beneath the surface. In corol. 6, however, Newton finds the distance of this point from the center of the moon to be $\frac{39.788}{40.788}$ of the distance of the center of the moon from the center of the earth. In corol. 7, Newton computes the mean distance from the moon's center to the earth's center to be 1,187,379,440 Paris feet (equivalent to 60.4 earth-radii), and the radius of the moon's orbit, or the distance from the center of the moon to the center of gravity of the earth and moon, to be 1,158,268,534 Paris feet (or 58.9 earth-radii). The center is therefore outside rather than inside the earth.

In corol. 7, Newton makes use of these prior results in a second moon test, a more refined and exact version of the earlier one given in book 3, prop. 4. He computes[63] how far the moon would fall inward toward the center of its orbit in one minute of time if the force on the moon were unaltered but the moon were instantaneously deprived of its forward component of motion. The result is 14.7706353 Paris feet, which is corrected to 14.8538667 Paris feet.

In order to test the accuracy of the inverse-square law, Newton now uses that law to predict how far a heavy body would fall in one second at the earth's surface, considering the differences between the falling of the moon (or of a heavy body in the moon's orbit) and the falling of a heavy body on earth to be as the inverse square of the distances. With certain correction factors, he finds that a heavy body falling at the earth's surface at latitude 45° should fall 15.11175 Paris feet in one second or should fall 15 feet, 1 inch, 4 1/11 lines. He then adjusts this value to the latitude of Paris, to find how far a heavy body does actually fall (with corrections to conditions of a vacuum) at the latitude of Paris. With suitable adjustments, this comes out to be 15 feet, 1 inch, 1 1/2 lines, once the centrifugal effect of the earth's rotation is included.

How did Newton achieve so good a numerical agreement between observed and predicted values if he had made an error of about 100 percent in the mass of the moon which had introduced an error in the location of the center of gravity of the earth-moon system? The answer is that obtaining such close agreement when the data are so poor required a certain number of adjustments, a fiddling with the actual numerical quantities. This example was one of the three case histories used by R. S. Westfall in his celebrated paper "Newton and the Fudge Factor." Westfall gave convincing evidence that Newton had started with his result and worked backward.[64] Later, N. Kollerstrom reanalyzed Newton's procedure and found that Westfall had been "substantially correct," although he claimed that "the accuracy of the computation had been much exaggerated."[65]

In the foregoing presentation, Newton's value of 4,891 to 1,000 for the ratio of the moon's density to the earth's density is given as in the third edition of the

63. For details see N. Kollerstrom's two articles cited in n. 65 below.

64. Richard S. Westfall, "Newton and the Fudge Factor," *Science* 179 (1973): 751–758.

65. N. Kollerstrom, "Newton's Two Moon Tests," *British Journal for the History of Science* 24 (1991): 169–172; "Newton's Lunar Mass Error," *Journal of the British Astronomical Association* 95 (1995): 151–153. The second of these studies is a careful reconstruction of Newton's use of tidal data in computing the mass of the moon and is especially noteworthy for elucidating Newton's calculation of the common center of gravity of the earth-moon system and his use of the data concerning the moon in his attempts to account for precession.

Principia. In the first edition, the ratio of the moon's density to the earth's density is said to be 9 to 5, with a corresponding ratio of the mass of the moon to the mass of the earth of 1 to 26. This ratio is known today to be about 1 to 81, so that Newton's computed value was off by a factor of more than 3. The later value of 1 to 39.788 (1 to 39.37 in the second edition) is still far from the mark, but by a factor of 2, rather than more than 3.

8.14 *Props. 22, 25–35: The Motion of the Moon*

The theory of the moon, or—more exactly—the theory of the moon's motion, occupies a big block of book 3, almost a third of the whole text including prop. 22 and extending from prop. 25 to prop. 35 plus the scholium to prop. 35. In some ways this may be considered one of the most revolutionary parts of the *Principia* since it introduced a wholly new way of analyzing the moon's motion and thereby set the study of the moon in a wholly new direction which astronomers have largely been following ever since. Prior to the *Principia*, all attempts to deal with the moon's motion had consisted of constructing ingenious schemes that would enable astronomers and table-makers to account for and to predict motions and positions with their variations and apparent irregularities. Newton's *Principia* introduced a program to change this part of astronomy from an intricate celestial geometry into a branch of gravitational physics. That is, Newton set up a wholly new way of studying the moon's motion by introducing physical causes, extending the analysis of the two-body problem of earth and moon to a three-body problem by introducing the gravitational perturbations of the sun. Beginning with the foundations established in book 1, prop. 45 and prop. 66 with its twenty-two corollaries, Newton boldly proposed a radical restructuring of the study of the motion of the moon.

In the second edition of the *Principia*, in the new scholium to book 3, prop. 35, Newton boasted that he "wished to show by these computations of the lunar motions that the lunar motions can be computed from their causes by the theory of gravity." This is an echo of his claim, in the first of the propositions introducing the subject of the moon's motion (book 3, prop. 22), that "all the motions of the moon and all the inequalities in its motions" can be deduced from the principles of gravitational dynamics "that have been set forth." Newton's profound alteration of the astronomy of the moon was recognized by the reviewer of the second edition of the *Principia* in the *Acta Eruditorum*. "Indeed," he wrote, "the computation made of the lunar motions from their own causes, by using the theory of gravity, the phenomena being in agreement, proves the divine force of intellect and the outstanding sagacity of the discoverer."[66] Flamsteed recognized the significance of Newton's achievement in a letter to Newton comparing Newton's contributions to Kepler's. "By frequent trials and alterations of his contrivances to answer the appearance [i.e., phenomena] of the heavens," Flamsteed wrote, "Kepler found out the true theory of the planetary motions." Newton, he continued, "need not be ashamed" to admit that he had followed Kepler's "example." But when "the Inequalities are found," Newton "will more easily find the reason of them" than Kepler could, "to whom but little of the Doctrine of gravity was known."[67] Laplace echoed these sentiments when he said that he had no hesitation in considering the application of dynamical principles to the motions of the moon to be "one of the most profound parts of [Newton's] admirable work."

Even though Newton did alter astronomy by introducing a radical new theory of the moon's motion, he did not in fact fully solve the problems he had set himself. He was able to account for some of the known inequalities in the moon's motion and even found some new ones, but he too had to have recourse to pre-dynamical methods, in particular relying on the system of Jeremiah Horrocks, and in some cases even "fudging" the numbers in order to gain agreement of theory with observations. In this sense we may understand why D. T. Whiteside has referred to Newton's lunar theory as "From High Hope to Disenchantment."[68] On the practical side, the application of Newton's rules for calculating the positions of the moon were no improvement over other methods and were not generally adopted by table-makers.[69] In particular, these rules did not help in the proposed use of such tables for accurately determining longitude at sea, which was then the primary goal of applied astronomy and, in particular, of the study of the moon's motion.

The complex subject of Newton's theory of the moon's motion has been studied in recent decades by a number of scholars, among them Philip Chandler, N. Kollerstrom, Craig Waff, R. S. Westfall, D. T. Whiteside, Curtis Wilson, and myself. There are also other important and useful older studies by John Couch Adams, G. B. Airy, Jean-Louis Calandrini, David Gregory, Hugh Godfray, and A. N. Krylov.[70] The textbook from which those of my generation studied celestial mechanics, by F. R. Moulton, presented this subject in a manner very similar

66. *Acta Eruditorum*, March 1714:140.

67. Flamsteed to Newton, 23 July 1695, *Corresp.* 4:153.

68. Whiteside, "Newton's Lunar Theory: From High Hope to Disenchantment" (§6.12, n. 60 above).

69. On the actual use of Newton's methods by table-makers, see the important article by Craig Waff cited in n. 72 below.

70. The works of all of these authors are either cited in the notes to this Guide or listed in Peter Wallis and Ruth Wallis, *Newton and Newtoniana, 1672–1975: A Bibliography* (Folkestone: Dawson, 1977). Calandrini's contribution to the problem of the motion of the moon's apogee was embedded in the

to Newton's (in book 1, prop. 66), consciously drawing on the *Principia*.[71] Curtis Wilson's presentation provides a useful and systematic critical overview of Newton's efforts, while Whiteside's analyses are—as always—incisive and based on an extensive use of manuscript documents.

Newton's initial study of the moon's motion was not wholly satisfactory, and he continued to work on the problem after the first edition of the *Principia* had been published. His first set of new results was rather widely circulated, published in Latin as part of David Gregory's two-volume textbook of astronomy in 1702 and in English as a separate pamphlet in that same year.[72] This new material was included in the second edition of the *Principia* in the revised scholium to book 3, prop. 35. While Newton was planning for a third edition, he received two independent solutions of the problem of the motion of the nodes of the moon's orbit, one by John Machin, the other by Henry Pemberton. He chose to include in the third edition the presentation by Machin rather than the one by Pemberton. Machin's two propositions "show that the mean rate of motion of the sun from the moon's node is a mean proportional between the rates of motion with which the sun separates from the node when in syzygy and when in quadrature respectively."[73]

Newton begins his presentation of the moon's motion with a study (book 3, prop. 25) of the perturbing forces of the sun on the motion of the moon and (prop. 26) of "the hourly increase of the area that the moon, by a radius drawn to the earth, describes in a circular orbit." Then (prop. 27) he turns to "the hourly motion" and the distance of the moon from the earth, leading (prop. 28) to the diameters of a non-eccentric orbit.

Prop. 29 is devoted to Newton's study of the "variation," which was considerably amplified for the second edition. In prop. 30, Newton considers the hourly motion of the nodes of the moon (for a circular orbit), extended in prop. 31 to the orbits obtained in prop. 29. Prop. 32 deals with the mean motion of the nodes, prop. 33 with the true motion. In the third edition, prop. 33 is followed by the alternative method for the motion of the nodes, in the two propositions by John Machin.

Le Seur and Jacquier edition of the *Principia*, as part of the commentary to book 3, prop. 35. On this topic, see the work by Waff, cited in n. 72 below. On Calandrini, see also Robert Palter, "Early Measurements of Magnetic Force," *Isis* 63 (1972): 544–558.

71. Moulton, *Introduction to Celestial Mechanics* (§6.12, n. 61 above), pp. 337–365.

72. Both of these printings are reproduced in facsimile in *Isaac Newton's Theory of the Moon's Motion (1702)*, introd. Cohen (§1.3, n. 53 above). This work also reprints the text as it appeared in the English translation of Gregory's textbook (London, 1715) and in the English version of William Whiston's *Astronomical Lectures* (London, 1728), with a commentary. See the important discussion of this edition by Craig Waff, "Newton and the Motion of the Moon: An Essay Review," *Centaurus* 21 (1977): 64–75.

73. Rouse Ball, *Essay*, p. 108.

Prop. 34 then introduces the hourly variation of the inclination of the moon's orbit to the plane of the ecliptic. In prop. 35, Newton sets out to determine the inclination at any given time. This was followed, in the first edition, by a short scholium which—as has been mentioned—was discarded in the second and replaced by a much longer new scholium on the ways in which Newton claimed that other inequalities in the moon's motion may be derived from the theory of gravity. The material in this scholium had been previously published separately.

This section of book 3, on the general subject of the moon, concludes with a set of four propositions and three lemmas, before advancing to the next topic: the motion and physics of comets. The first of these four propositions (props. 36 and 37) have as their subject, respectively, the solar and lunar tide. We have seen (§8.13 above) that prop. 37 uses the computation of the lunar tidal force as a basis to determine the mass of the moon. This leads to prop. 38 on the spheroidal shape of the moon and Newton's explanation of why the same side of the moon always (except for the effects of libration) faces the earth. The final prop. 39 of the pre-cometary section of book 3 (see §8.17) examines the precession of the equinoxes.

Newton's theory of the moon is actually developed in three separate groups of propositions, corollaries, and scholiums. The first is in book 1, sec. 9, props. 43–45; the second is in book 1, prop. 66, with its twenty-two corollaries; the third is in book 3, props. 22, 25–35, and scholium. In book 1, props. 43–45, he explores the way in which a revolving eccentric orbit may be produced by the action of two forces. One is the centripetal inverse-square force that would produce the orbit at rest; the other is a superposed attractive force which, in the case of the moon's motion, comes from the perturbing action of the sun. In prop. 45, he considers the motion of the apsides when the lunar orbit is very nearly circular.

Newton's presentation of the motion of the moon includes the study of the motion of the apsides. In book 1, prop. 45, he shows how to calculate the forces in moving eccentric orbits, considering some force in addition to the central inverse-square force. In the case of the moon's orbit, this force comes from the perturbing action of the sun. In corol. 1, he concludes that for motion of the upper apsis to be an advance of about 3° in each revolution (which is approximately what is observed on average in the motion of the moon), the combined centripetal force "decreases in a ratio a little greater than that of the square, but 59¾ times closer to that of the square than to that of the cube." In corol. 2, he subtracts from the force on the orbiting body a perturbing force which is a linear function of the distance and is smaller by a factor of $\frac{1}{357.45}$ of it and concludes that under these circumstances the advance of the apsis per revolution will be $1°31'28''$, a quantity which he notes (in an addition appearing in the third edition) is only one-half of the motion of the lunar apsis.[74]

Newton's comment has given rise to much speculation. We should take note, however, that nowhere in the *Principia* does Newton actually calculate the motion of the moon's apsis. He originally planned to include such a calculation and even made a series of attempts to solve what has been called the "most difficult problem of basic lunar theory," the rate of motion of the lunar apogee or apsis of the moon and its mean rate of secular advance. Some hint of Newton's tremendous expenditure of intellectual energy on this problem was given by J. C. Adams in 1881 on the basis of the manuscripts presented to Cambridge University by the earl of Portsmouth.[75] But the full extent of Newton's work was revealed only when the many manuscript fragments were edited for publication and interpreted in volume 6 of D. T. Whiteside's edition of Newton's *Mathematical Papers*.[76]

In Whiteside's edition and commentary we can now follow Newton's attempts to take account of both the radial and the transverse components of the disturbing force of the sun on the moon's orbit. To do so, Newton made use of a "Horrocksian model of the orbit wherein it is assumed that the moon's pristine path is a Keplerian ellipse traversed around the earth at a focus" and that "the effect of solar perturbation is slightly to alter its eccentricity while maintaining the length (but not direction) of its major axis."[77] Whiteside found that Newton did contrive to "arrive ultimately at a close approximation to the observed annual progress of the lunar apogee." In the first edition of the *Principia*, in the scholium to book 3, prop. 35, Newton admitted that his computations seemed "too complicated and encumbered by approximations and not exact enough" to be included in the *Principia*. This is the scholium that was suppressed in the later editions. We should take note that the problem of the lunar apogee engaged the independent efforts of a number of able mathematicians in the post-Newtonian decades, finally being solved by d'Alembert, Clairaut, and Euler.[78]

74. See §6.10 above. On the justification of the number $\frac{1}{357.45}$ as representing the average value of the radial component of the sun's perturbing force, see "Newt. Achievement," p. 264. Wilson notes (ibid., p. 262b), that what is left out of the calculation is the effect of the transradial perturbing force (at right angles to the radius vector).

75. Whiteside (*Math. Papers* 6:537 n. 63) warns us of J. C. Adams's "unsupported conjecture" (*Catalogue* [§1.2, n. 11 above], p. xiii) that Newton derived certain variant values "by a more complete and probably a much more complicated investigation than that contained in the extant MSS."

76. *Math. Papers* 6:508–537.

77. Ibid., p. 509 n. 1; see p. 519 n. 28 on Newton's use of a Horrocksian model.

78. See the intensive study of this problem in two doctoral dissertations: Philip P. Chandler II, "Newton and Clairaut on the Motion of the Lunar Apse" (Ph.D. diss., University of California, San

Newton's work on the motion of the lunar apsis has been neatly summarized by Curtis Wilson, who has provided a clear and concise analysis of the chief features of Newton's theory of the moon.[79] In book 3, props. 26–28, Newton "develops by integration a numerical formula for the rate of description of area as it varies owing to the transradial perturbing force of the Sun"[80] and then proceeds to determine the "variation," the inequality in the motion in longitude which had been discovered by Tycho Brahe.

Newton's analysis of the "variation," basically a fluctuation in the rate at which a radius to the moon sweeps out areas, is one of the most highly praised of Newton's results concerning the motion of the moon. First introduced in book 1, prop. 66, corols. 4–5, this topic is developed numerically in book 3, props. 26–29. In prop. 26, starting with a circular orbit, Newton proceeds by a series of steps, eventually arriving at a value for the variation (prop. 29) of 35.10′. In the first edition, he then noted that an astronomer (unnamed, presumably Tycho Brahe) gave a quite different value and that Halley had recently found the true value to vary from 32′ to 38′. Newton concluded by noting that he had computed the mean value, "ignoring the differences that can arise from the curvature of the earth's orbit and the greater action of the sun upon the sickle-shaped and the new moon than upon the gibbous and the full moon." The text of prop. 29 was considerably amplified in the second edition, in which Newton dropped his reference to Halley having found a maximum value of 38′ for the variation and a minimum value of 32′, the basis on which he sought to excuse his own poor value of 35′ as a "mean value."[81]

Another inequality in the moon's motion which Newton presents in book 3 is the annual inequality (book 3, prop. 22), discovered by Tycho Brahe and by Kepler, and described by Newton in his statement that "the mean motion of the moon is slower in the earth's perihelion than in its aphelion." Newton also studied the evection and an oscillation of the lunar apsis.[82]

As in the case of the mass of the moon, the cause of the tides, and the precession of the equinoxes, Newton boldly charted the future course of investigation, but the achievement of the goal he set had to await the labors of others. Although he failed to produce a complete theory of the moon's motion, he did set the course

---

Diego, 1975); Craig Beale Waff, "Universal Gravitation and the Motion of the Moon's Apogee: The Establishment and Reception of Newton's Inverse-Square Law, 1687–1749" (Ph.D. diss., Johns Hopkins University, 1976).

79. "Newt. Achievement," pp. 262–268.

80. Ibid.

81. For details, see ibid., p. 264b.

82. Ibid., pp. 462b–463b.

for all future studies by grounding this subject firmly on principles of gravitational dynamics and by revealing the principal cause of variations and apparent irregularities. With one bold stroke he set astronomy on a wholly new course in which the motions of the celestial bodies would be analyzed in terms of physical causes. As Newton said on another occasion, "Satis est": it is enough![83] Even though Newton's theory of the motion of the moon was imperfect, yet to have set astronomy on a wholly new course was enough to make his studies of the moon rank as one of the most significant achievements of the *Principia*.

8.15 *Newton and the Problem of the Moon's Motion* (BY GEORGE E. SMITH)

The motion of the moon posed a special problem in mathematical astronomy from Hipparchus on. Since Laplace, we have known that planetary interactions can make the motions of the planets quite complicated. The vagaries in the motions of the planets, however, become evident only over extended periods of time or under refined telescopic observations. The motion of the moon, by contrast, is exceedingly complicated under naked-eye observations over comparatively brief periods of time. As a consequence, the level of accuracy with which the moon's motion was being *described* within mathematical frameworks at the time Newton began the *Principia* was far below the level for the planets. The "problem of the moon" at that time was one of describing its motion to more or less the accuracy Kepler and those after him had achieved for the planets. The first full solution to this problem, the Hill-Brown theory developed from 1877 to 1919, involves some fifteen hundred separate terms.

Tables in textbooks indicate that the moon has a sidereal period of 27.32 days in an elliptical orbit of eccentricity 0.0549, inclined to the ecliptic at an angle of 5°9′. Its line of apsides—that is, the line from apogee to perigee or, equally, from the point of least angular velocity to the point of greatest—precesses in the direction of motion an average of slightly more than 3° per revolution, completing a circuit in roughly 8.85 years. Its line of nodes—the line along which the plane of the moon's orbit intersects that of the earth's—regresses at a little less than half this rate, completing a circuit in roughly 18.60 years. But these facts present only a part of the complexities. The shape of the orbit is not strictly elliptical but varies over time. The eccentricity varies in a complicated manner, becoming as high as 0.0666 and as low as 0.0432. The inclination varies similarly, reaching a high of 5°18′ and a low of 5°. The lines of nodes and apsides both vary from their mean motions in complex ways, so that the location of a node can deviate in either direction from

83. Ibid., p. 262b.

its mean motion by as much as 1°40′, and the location of the apogee can deviate in either direction from its mean motion by as much as 12°20′.[84]

Kinematical models attempting to cover anomalous features of the moon's motion begin with Hipparchus and continue with Ptolemy, Copernicus, Tycho, and Kepler, as well as Boulliau, Wing, Streete, and Horrocks in the years between Kepler and Newton. The goal was always to represent the seemingly irregular motion as arising from the superposition of several underlying regular motions. The problem lay in identifying and characterizing these underlying regularities. Kepler went so far as to contend that the residual irregularities in the moon's motion provided clear evidence that chance physical causes were affecting its motion, making a fully adequate account doubtful.[85]

By Newton's time, four anomalies, or inequalities, in longitude had been singled out. The so-called *first inequality*, identified by Hipparchus, was described in terms of the rotation of the line of apsides in an eccentric orbit. The *second inequality*, as characterized by Ptolemy, involved the combination of a monthly variation in eccentricity and a twice-a-month fluctuation in the motion of the line of apsides.[86] Tycho, in addition to noting inequalities in the motions of the nodes and the inclination, discovered two smaller anomalies in longitude, a twice-a-month fluctuation in speed, *the variation*, in which the moon speeds up in the syzygies and slows down in the quadratures, producing a maximum effect on longitude around 39′ in the octants; and an annual fluctuation in speed, *the annual equation*, with a maximum effect around 11′. Insofar as the diameter of the moon is 30′ in arc, these magnitudes are appreciable.

The modern theory of the lunar orbit includes numerous other inequalities. Some of these, such as the *parallactic inequality* and the *secular acceleration* (discovered by Halley), are significant enough to have acquired names; most, however, remain unnamed.

84. The best recent text on the motion of the moon and the modern history of efforts to deal with it is Alan Cook, *The Motion of the Moon* (Bristol: Adam Hilger, 1988). The description above is adapted from J. M. A. Danby, *Fundamentals of Celestial Mechanics*, 2d ed. (Richmond: Willmann-Bell, 1988), chap. 12. The problem is also discussed in such other standard texts on orbital theory as Moulton, *Introduction to Celestial Mechanics* (§6.12, n. 61 above), chap. 9; Dirk Brouwer and Gerald M. Clemence, *Methods of Celestial Mechanics* (New York: Academic Press, 1961), chap. 12; and A. E. Roy, *Orbital Motion*, 3d ed. (Bristol: Adam Hilger, 1988), chap. 9. The classic text on the motion of the moon is Ernest W. Brown, *An Introductory Treatise on the Lunar Theory* (Cambridge: Cambridge University Press, 1896; reprint, New York: Dover Publications, 1960). The three volumes of *Tables of the Motion of the Moon* by Brown and H. Hedrick (New Haven: Yale University Press, 1919) mark the culmination of the Hill-Brown theory of the moon.

85. Kepler offers remarks to this effect in many places, most prominently at the end of the preface to *Tabulae Rudolphinae*, vol. 10 of *Johannes Keplers Gesammelte Werke* (Munich: C. H. Beck, 1969), p. 44.

86. See O. Neugebauer, *The Exact Sciences in Antiquity* (New York: Harper, 1962), pp. 192–198.

The history of the second inequality, involving both a changing eccentricity and a fluctuation in the motion of the line of apsides, shows the difficulty posed by the descriptive problem.[87] This inequality seems to arise from some systematic displacement of the lunar orbit that does not change the shape of the orbit itself. This is why Boulliau decided to give it the name "evection" (from the Latin "evectio," from the verb "eveho," or "carry forth"), implying that the orbit is carried out of one place to another. The difficulty lay in characterizing this displacement precisely. Ptolemy, Copernicus, Tycho, and Kepler all offered different characterizations. Kepler, in particular, changed the period of the variation in eccentricity from one month to 6.75 months when he discovered that he could obtain a net motion almost equivalent to the one resulting from a monthly period. In modern theory the "evection" refers to an inequality in longitude resulting from a variation in the eccentricity of period 31.8 days, to which is coupled a fluctuation in the motion of the line of apsides. This produces the largest of the anomalies in longitude, of amplitude $1°16'$. Its identification, by means of an expanded harmonic analysis of the disturbing force of the sun on the moon, represents a not insignificant achievement of gravitation theory.

In the last three years of his short life, Jeremiah Horrocks (1618–1641) produced the most successful kinematical model of the moon before Newton.[88] He used a Keplerian ellipse, with motion governed by the area rule. The center of the ellipse, however, revolved on a small circle, producing 6.75-month oscillations both in the eccentricity and in the position of the line of apsides relative to its precessing mean position. In other words, he adopted a version of Kepler's treatment of the second inequality. Using preferred values of its elements from the early eighteenth century, this model reduced the errors in longitude to less than $10'$; while this was still significantly short of the goal, it compared well with the errors in excess of $20'$ from all other models.[89]

Newton took Horrocks's theory to be a basically correct account of the moon's motion. In particular, Newton adopted Horrocks's 6.75-month period for the second inequality, proceeding then to identify the solar disturbing force that could

87. See Curtis Wilson, "On the Origin of Horrocks's Lunar Theory," *Journal for the History of Astronomy* 18 (1987): 77–94, esp. 78–81.

88. Horrocks's account of the moon was published in 1672 in his *Opera Posthuma.* For details, see Curtis Wilson, "Predictive Astronomy in the Century after Kepler," in *Planetary Astronomy from the Renaissance to the Rise of Astrophysics, Part A: Tycho Brahe to Newton*, ed. René Taton and Curtis Wilson, vol. 2A of *The General History of Astronomy* (Cambridge and New York: Cambridge University Press, 1989), pp. 194–201; and "On the Origin of Horrocks's Lunar Theory."

89. By the end of the seventeenth century the goal was $2'$ in astronomical longitude, corresponding to a navigational error of $1°$ in longitude at sea. The errors in astronomical latitude in the Horrocksian model were as high as $4'$.

produce it (book 1, prop. 66, corol. 9). Consequently, even though the evection of modern theory is a product of gravitation theory—and hence an achievement brought about by Newton—Newton's account of the second inequality construes it somewhat differently from modern theory. By contrast, his accounts of the other three major inequalities in longitude agree qualitatively with modern theory. From his account of the perturbing effects of the sun, Newton also identified a few lesser inequalities in longitude that had not been noticed before; though a few of these were spurious, most of them were real. Finally, his accounts of the fluctuation in the inclination and of both the mean and the varying motion of the nodes, which are very much along modern lines, covered the principal inequalities in latitude.

Newton's work on the moon is best viewed as consisting of five separate tracts:[90] (1) prop. 66 of book 1 of the *Principia* and its many corollaries, which remained largely the same in all three editions; (2, 3) the two somewhat different quantitative treatments of lunar inequalities in book 3 in the first and the later editions of the *Principia*;[91] (4) prop. 45 of book 1, along with an unpublished manuscript on the mean motion of the lunar apogee mentioned in the first edition but not in the second and third;[92] and (5) his "Theory of the Moon's Motion," published in 1702 in two versions: in Latin as an appendix to David Gregory's *Astronomiae Physicae & Geometricae Elementa* and in English as a separate pamphlet. This short work was reprinted a number of times, one version having a commentary by William Whiston.[93]

Prop. 66 of book 1 of the *Principia* and its corollaries provide a qualitative account of how perturbational effects of the sun's gravity can cause *every* inequality in the moon's motion depicted by Horrocks's model, and some further, lesser inequalities. Although Newton called his results "imperfect" in the preface to the first edition, they have some claim to being his greatest achievement with the moon. The quantitative results in props. 25 through 35 of book 3 take a concentric circular orbit as the starting point and obtain approximations (ignoring eccentricity) to

90. The best survey of Newton's work on the moon is Whiteside, "Newton's Lunar Theory: From High Hope to Disenchantment" (§6.12, n. 60 above). This article covers unpublished efforts beyond those discussed here.

91. The revisions in the treatment of the moon in book 3 from the second to the third editions of the *Principia* are small compared with those from the first to the second.

92. Newton's unsuccessful efforts on the mean motion of the line of apsides and the subsequent successful treatment of the problem by Clairaut are discussed in §8.16.

93. A facsimile reprint of several of these versions may be found in *Isaac Newton's Theory of the Moon's Motion (1702)*, introd. Cohen (§1.3, n. 53 above). For an account of the use of this work by table-makers in the eighteenth century, see N. Kollerstrom, "A Reintroduction of Epicycles: Newton's 1702 Lunar Theory and Halley's Saros Correction," *The Quarterly Journal of the Royal Astronomical Society* 36 (1995): 357–368. This article contains a flowchart showing the various steps needed to apply Newton's method and also presents an account of the computer program devised by the author as a simulation of Newton's method.

the variation, the variational orbit, the motion of the nodes, and the fluctuation in the inclination. These results agree well with modern results when the effects of eccentricity are ignored.

Where one might expect a comparable approximation for the motion of the apogee in book 3, following prop. 35, the first edition has a brief scholium announcing that similar computations yield decent results for the apogee. Newton said that he had not given the actual computations because they were not exact enough. As noted elsewhere in this Guide, all claims to obtaining reasonably accurate numbers for the mean motion of the apogee disappear from the second and third editions of the *Principia*.

During the 1690s, Newton's efforts on the moon focused on the application of gravitation theory to obtain values of the coefficients for the irregularities in Horrocks's model. His *Theory of the Moon's Motion* (1702) presents the results, which reduce the maximum discrepancies between the model and observation to about 10′. Updated versions of these results are presented in a long scholium following book 3, prop. 35, in the second and third editions. In the second edition this scholium also includes coefficients for four further inequalities that Newton had detected from gravitation theory, some of which, along with two not mentioned in any edition of the *Principia*, had been announced in the *Theory of the Moon's Motion*. One of these four was dropped entirely in the third edition, and two of the remaining ones fail to capture any real irregularity.[94]

As became evident from Euler's and Clairaut's work in the 1740s, if not from Machin's "The Laws of the Moon's Motion according to Gravity" of 1729, published as a supplement to Andrew Motte's translation of the *Principia*, the claims made in Newton's lifetime about his quantitative success with the moon were excessive. Newton himself did not significantly advance the problem of the moon's motion beyond Horrocks. But Newton's contribution to the history of the problem is nonetheless paramount. It is encapsulated in the first sentence of the scholium following book 3, prop. 35, in the second and third editions: "I wished to show by these computations of the lunar motions that the lunar motions can be computed from their causes by the theory of gravity." This was of no small moment to Newton, for he had based his theory of gravity on the Keplerian motion of a handful of celestial bodies, including crucially the moon; and unless its vagaries could be accounted for, the moon threatened to undercut his line of reasoning. After 1750, following Clairaut's success with the motion of the apogee, the "problem of the

94. Curtis Wilson has reviewed Newton's quantitative efforts on the moon in "Newt. Achievement," pp. 262–268, and the results for the variation and the mean motion of the apogee in "Newton on the Moon's Variation and Apsidal Motion: The Need for a Newer 'New Analysis,'" in *Isaac Newton's Natural Philosophy*, ed. Jed Buchwald and I. B. Cohen (Cambridge: MIT Press, forthcoming).

moon" came to designate a special case of the "restricted three-body problem," namely, the motion of the moon under the combined gravitational effects of the sun and the earth. By computing the lunar motions "from their causes by the theory of gravity," the Hill-Brown theory achieved at once an adequate solution of this problem and the original "problem of the moon."

The Motion of the Lunar Apsis (BY GEORGE E. SMITH)[95] 8.16

Few aspects of the *Principia* received more attention during the eighteenth century than the treatment of the sun-moon-earth system. The reason was not only that Newton's account yielded highly celebrated explanations of the tides and the precession of the equinoxes, but also that his account offered the best hope for finally sorting out the complex motion of the moon. The impressive accuracy that Kepler's orbital theory, and variants of it, had achieved in the case of the planets had not been matched in the case of the moon. The problem of devising even so much as a *descriptive* account of the motion of the moon that predicted its observed locations with an accuracy comparable to what had been achieved for other celestial bodies was still unsolved.

The *Principia* gives results on the motion of the moon in two places. Corols. 2 through 11 of prop. 66 of book 1 offer *qualitative* results on how solar gravity causes the angular velocity, the distance from the earth to the moon, the period, the eccentricity of the orbit, and its inclination to vary from their Keplerian ideals, as well as results on the motions of the lines of apsides and nodes. Props. 25 through 35 of book 3 offer *quantitative* results, derived from gravitation theory, on three of the most prominent anomalies in the lunar motion, the "variation" (that is, its speeding up in the syzygies and slowing down in the quadratures), the motion of the nodes, and the varying inclination.

Conspicuously absent from the quantitative analysis in book 3 is the motion of the lunar apsis. On average, the apsis advances a little more than 3 degrees per lunar month, or some 40 degrees per year. This is its mean motion. Its "true" motion varies considerably: in extremes the apsis advances at a rate equivalent to roughly 15 degrees per lunar month, and it regresses at a rate equivalent to roughly 9 degrees per lunar month. Corols. 7, 8, and 9 of prop. 66 provide a qualitative explanation of both the fluctuating and the mean motions of the apsis. Book 3

95. Newton's treatment of the lunar orbit is discussed in "Newt. Achievement," pp. 262–268, and in Whiteside, "Newton's Lunar Theory: From High Hope to Disenchantment." The problem of the motion of the lunar apsis is discussed in detail in Waff, "Universal Gravitation and the Motion of the Moon's Apogee" (n. 78 above); Clairaut's solution is discussed in Wilson, "Perturbations and Solar Tables" (n. 37 above), pp. 133–145.

derives impressively accurate quantitative results from gravitation theory for the analogous problem of the fluctuating and mean motions of the nodes. But it offers no results on the apsis.

The only place where quantitative results are derived for the motion of the line of apsides of orbits is in sec. 9 of book 1. This section addresses a specific, narrow problem: given a body moving in a stationary centripetal-force-governed orbit, what further centripetal force has to be superposed to cause the body to move in an identical revolving orbit. This is a problem about centripetal forces, which is why it belongs in book 1. Props. 43 and 44 give the general answer: a superposed inverse-cube centripetal force. Prop. 45 then gives results pertaining to the rate of motion of the line of apsides for the mathematically simple case of orbits that are nearly circular. The two corollaries of this proposition give specific numerical results that in fact are for the lunar apsis, though this is not announced as such. The first corollary shows that if the line of apsides advances 3 degrees per revolution, then the overall centripetal force varies as $\frac{1}{r^{2\frac{4}{243}}}$ approximately—a number Newton uses in his discussion of the lunar orbit at the beginning of book 3. The second corollary shows that if an extraneous force reduces the centripetal force governing the stationary orbit by a factor of $\frac{1}{357.45}$, then the apsis will advance $1°31'28''$ per revolution. Nothing is said about why 357.45 is an appropriate number to be considering. In the first two editions nothing is even said to indicate that the result is pertinent to the moon. In the third edition, however, Newton added a concluding sentence to the corollary: "The [advance of the] apsis of the moon is about twice as swift."

A close reading of props. 25 and 26 of book 3 reveals where 357.45 comes from. In prop. 25, as in prop. 66 of book 1, Newton prefers to resolve the perturbing force of the sun on the moon into non-orthogonal components, one of which is always directed from the moon toward the earth. Prop. 25 concludes that the mean value of this solar centripetal component is $\frac{1}{178.725}$ of the earth's centripetal force on the moon. In prop. 26 the solar perturbing force is resolved into orthogonal radial and transradial components. The radial component varies from $\frac{1}{178.725}$ of the earth's force, directed *toward* the earth, when the moon is in the quadratures, to $\frac{2}{178.725}$, directed *away* from the earth, when the moon is in the syzygies. This component clearly varies sinusoidally. Therefore its mean value lies midway between these two extremes, that is, $\frac{1}{357.45}$, directed away from the earth. Hence, the $\frac{1}{357.45}$ reduction in the centripetal force used in corol. 2 of prop. 45 represents the mean value of the radial component of the solar perturbing force drawing the moon away from the earth—precisely the appropriate value to use in order to compute a mean motion of the lunar apsis.

Not all readers at the time were able to make this inference from prop. 26 of book 3 to corol. 2 of prop. 45 of book 1. But some did. John Machin announced it in the *Journal Book*, or minutes, of the Royal Society in 1742.[96] Moreover, from what Newton says in prop. 26 of book 3, it was not that difficult to arrive at an analytical statement of the variation of the radial component of the solar perturbing force:

$$\frac{F_R}{F_{\text{moon-earth}}} = \left(\frac{1}{357.45}\right) \times (1 + 3\cos 2\theta)$$

where θ is the angle formed by the sun and moon with the earth as apex.

The mean value of the radial component of the solar perturbing force thus accounts for half of the mean motion of the lunar apsis. The obvious question is, What accounts for the rest? This question was never expressly posed in any of the editions of the *Principia*. It was nevertheless answered in the first edition, in a scholium to prop. 35 of book 3. There Newton indicated that computations had shown that the lunar apogee advances 23′ per day when it is in the syzygies and regresses 16⅓′ per day when it is in the quadratures, resulting in a mean annual motion of about 40 degrees. According to the first edition, Newton did not publish his computations because they were "too complicated and encumbered by approximations" and therefore not exact enough.

The computations Newton was presumably referring to became publicly available for the first time in the 1970s.[97] They use a Keplerian ellipse as a starting point, and they include the transradial, as well as the radial, components of the solar perturbing force. Unfortunately, however, they seem to involve arbitrary factors, the only apparent justification for which is that they yield a mean annual motion of the apsis around 40 degrees. Consequently, the computations do not really show that the missing half of the motion of the lunar apsis is accounted for by the transradial component of the solar gravitational force acting on the moon in an eccentric orbit. At most they show that such an account *may* improve matters, provided that all the steps can be justified by gravitation theory.

All mention of these computations disappeared in the second and third editions of the *Principia*. The original scholium to prop. 35 was replaced by a new, longer scholium, most of which was either taken from or based on Newton's *Theory of*

96. *Journal Book* 17:459–463.

97. *Math. Papers* 6:508–537.

the Moon's Motion, which he had published in 1702.[98] This account of the moon employs a version of the kinematic model that Horrocks had formulated in the late 1630s. As such, it is not derived from gravitation theory, though Newton used gravitation theory in places to determine parameters in the model. The new scholium gives numbers both for the small variation in the motion of the lunar apsis associated with the difference in the solar perturbing force when the earth is at aphelion and perihelion, and for the large variation associated with the difference when the apogee is in the syzygies and the quadratures. But the former of these numbers is based only partly on gravitation theory, and the latter is not based on gravitation theory at all. In Newton's new scholium, no mention is made of deriving the mean motion of the lunar apsis from the solar perturbing force. But, then, no mention has to be made, for the mean motion of the apsis is built into the Horrocksian kinematic model.

Hence, while the new scholium opens with a remark about the goal of computing the lunar motions "from their causes by the theory of gravity," it gives no reason to think that the mean motion of the lunar apsis has ever been computed in this way. The scholium ends with an admonition that the lunar orbit requires further work; in particular, "the mean motions of the moon and of its apogee have not as yet been determined with sufficient exactness." One should therefore not be astonished by Pemberton's statement in his commentary on the *Principia*, "The motion of the apogeon, and the changes in the eccentricity, Sir Isaac Newton has not computed";[99] nor by Machin's 1729 assessment, "Neither is there any method that I have ever met with upon the commonly received principles, which is perfectly sufficient to explain the motion of the Moon's apogee."[100]

Another change Newton made in the second edition added to the confusion. A new corollary to prop. 3 of book 3 indicates the need to augment the centripetal acceleration of the moon toward the earth in the so-called "moon test" of prop. 4—this, in order to remove the extraneous radial acceleration of the moon away from the earth caused by solar gravity. A new portion of the text of prop. 3 gives the requisite value:

> The action of the sun, insofar as it draws the moon away from the earth, is very nearly as the distance of the moon from the earth, and so (from

98. *Isaac Newton's Theory of the Moon's Motion (1702)*, introd. Cohen (§1.3, n. 53 above).

99. Pemberton, *A View of Sir Isaac Newton's Philosophy* (§3.6, n. 65 above), p. 229; cf. Waff, "Universal Gravitation," p. 123.

100. John Machin, "The Laws of the Moon's Motion according to Gravity," appended to Motte's English translation of the *Principia* (1729; reprint, London: Dawsons of Pall Mall, 1968), p. 31. Motte was wise to append this exposition to his translation. Throughout it Machin displays a thorough grasp of Newton's published efforts on the moon.

> what is said in book 1, prop. 45, corol. 2) is to the centripetal force of the moon as roughly 2 to 357.45, or 1 to 178 29/40. And if so small a force of the sun is ignored, the remaining force by which the moon is maintained in its orbit will be inversely as D^2. And this will be even more fully established by comparing this force with the force of gravity as is done in prop. 4 below.

No explanation beyond the reference to corol. 2 of prop. 45 is offered for doubling the $\frac{1}{357.45}$ number used there. One might naturally think that it represents the additional radial force needed to obtain the correct value of $3°3'$ for the mean motion of the lunar apsis from the $1°31'28''$ value calculated there. But in fact the doubling cannot be justified on the basis of the solar perturbing force alone. The mean radial component of this force is $\frac{1}{357.45}$ of the earth's mean force on the moon, not $\frac{2}{357.45}$. So, the reader is left with a phantom doubling of the perturbing radial force. This doubling yields an extraordinary level of agreement in the moon test of prop. 4 in the third edition;[101] and it gives at least the impression that the mean motion of the lunar apsis can be accounted for entirely by the radial component of the solar perturbing force.

Newton's own assessment of the state of the problem of the mean motion of the lunar apsis, as of the second and third editions, must be gleaned from the *Principia*. Nothing in his correspondence helps. But as we have seen, and as Clairaut was quick to point out in 1747, the *Principia* is anything but clear on this score. We do not know why he dropped the idea that the missing half of the mean motion is a consequence of the transradial component of the solar force. He still attributed the large semiannual variation in the motion of the apsis to the transradial component. If he concluded that the mean motion is almost entirely a consequence of a radial perturbing force, as prop. 3 suggests, then the problem was to find a source of a further perturbing force with sufficient strength to produce the motion. Perhaps he was considering sources other than solar gravity.[102]

101. If $\frac{1}{357.45}$ is taken to be the correction in the moon test of prop. 4, then the agreement between the computed value of the fall of the moon in one second at the surface of the earth and Huygens's measured value of this fall is only to the nearest inch: 15 feet 0 inch 7.35 lines computed versus the 15 1/12 feet measured that is quoted in the first and second editions. The extraordinary agreement claimed in the third edition—15 feet 1 inch 1 4/9 lines computed and 15 feet 1 inch 1 7/9 lines measured—takes $\frac{1}{178.725}$ as the correction.

102. Mention of the possibility of the earth's magnetic force affecting the moon appears for the first time in the second edition, in the discussion of the lunar force on the earth in prop. 37. Since Newton thought that magnetic forces decrease with distance cubed (book 3, prop. 6, corol. 5), he would have had reason to think that the earth's magnetic force could cause the lunar apsis to move. Still another possibility is that Newton thought that the sinusoidal variation in the radial force on the moon somehow

An obvious alternative is that Newton had a clear grasp of the problem posed by the mean motion of the lunar apsis but saw no way of solving this problem satisfactorily and so decided to suppress the reference to it in the second and third editions of the *Principia*. In this way he could prevent the opponents of his theory of gravity from dismissing it on grounds that he considered premature. This is not so flagrant as it at first sounds. For even if he did elect to suppress the problem, he left enough clear traces of it to allow the advanced reader to reconstruct it. Machin, among others, did just this.[103]

One point of which we can be certain is that Newton thought it crucial that, in one way or another, the theory of gravity yield an account of the motion of the lunar apsis. In prop. 2 of book 3 he says that the inverse-square character of the centripetal forces on the planets

> is proved with the greatest exactness from the fact that the aphelia are at rest. For the slightest departure from the ratio of the square would (by book 1, prop. 45, corol. 1) necessarily result in a noticeable motion of the apsides in a single revolution and an immense such motion in many revolutions.

In prop. 3 he argues that "the very slow motion of the moon's apogee" similarly shows that the centripetal force on the moon is inverse-square. He acknowledges that the actual mean motion of the lunar apsis entails an exponent of $2\frac{4}{243}$ but attributes this motion, and hence the small fraction, to "the action of the sun (as will be pointed out below) and accordingly is to be ignored here." Therefore, so long as the mean motion of the lunar apsis had not been successfully shown to arise from the perturbing effect of the sun, or from some other extraneous force, the argument for the theory of gravity contained a serious lacuna. In particular, questions could be raised about whether the centripetal force is exactly, and not just approximately, inverse-square.

---

produces an asymmetric effect, causing more forward than backward motion of the apsis. Machin hints at this possibility in his 1729 study (p. 31), published as a supplement to Andrew Motte's translation of the *Principia*, and Clairaut puts it forward explicitly in 1747 in "Remarques sur les articles qui ont rapport au mouvement de l'apogée de la lune, tant dans le livre des Principes mathématiques de la philosophie naturelle de M. Newton que dans le commentaire de cet ouvrage publié par les PP. Jacquier et le Seur," Procès-verbaux de l'Académie royale des sciences 66 (1747): 553–559. The original autograph manuscript of this paper may be found in the "Dossier Clairaut" in the archives of the Académie des Sciences in Paris; a somewhat edited version was copied into the manuscript procès-verbaux of the Académie. For an analysis of this document, see Waff, "Universal Gravitation," chap. 3, and especially p. 153 n. 42 for a discussion of the differences between the two versions.

103. See Waff, "Universal Gravitation," chap. 3, for a review of early-eighteenth-century readings of Newton's passages pertaining to the lunar apsis.

Euler, Clairaut, and d'Alembert all raised such questions in the late 1740s, citing (among other things) calculations that indicated that only half of the mean motion of the lunar apsis could be accounted for by solar gravity.[104] All three discarded Newton's geometrical approach, instead devising analytical approaches in which various small terms were dropped. Clairaut's efforts were most instructive. He formulated analytical expressions for the radial and transradial components of the solar perturbing force, and he took both components into account in calculating a mean motion of the apsis for the case of an eccentric elliptical orbit. Yet he found himself forced to conclude "that the period of the apogee which follows from the attraction reciprocally proportional to the squares of the distance would be about 18 years, instead of a little less than 9 which it really is."[105] He presented this result to the French Academy of Sciences in late 1747, putting it in the context of a thorough review of the evidence for Newton's theory of gravity. He ended by suggesting that gravity diminishes not as $\frac{c}{r^2}$ but as $\frac{c}{r^2} + \frac{d}{r^4}$, where the latter term is too small to be detected in the case of large interplanetary distances.[106]

Clairaut's paper provoked an intense controversy within the French Academy,[107] a controversy that Clairaut himself terminated eighteen months later, after he discovered the correct solution to the problem of the mean motion of the apsis. In his original calculations he had adopted some simplifications, believing the neglected terms too small to make much difference. For instance, the expression he used for the radius vector was for a revolving ellipse. At that time he had noted that if these simplifications were not made, then more exact values could be obtained, but at a price of much longer calculations. More than a year later, he carried through the longer calculations, using such refinements as an expression for the radius vector not for the revolving ellipse, but for the revolving ellipse as already distorted by solar gravity. The result was a calculated mean motion of the lunar apsis in close accord with observation. The mean motion results from the combined effects of the radial and transradial components of the solar perturbing force acting on an eccentric orbit of a perturbed shape. This is tantamount to say-

104. See Waff, "Universal Gravitation," chap. 2.

105. Clairaut, "Du système du monde dans les principes de la gravitation universelle," *Mémoires de l'Académie royale des sciences* (Paris), 1745 (published 1749), p. 336, as quoted in Waff, "Universal Gravitation," p. 71.

106. Clairaut, "Du système du monde," p. 339.

107. Summaries of the controversy can be found in Waff, "Universal Gravitation," chap. 4, and Philip Chandler, "Clairaut's Critique of Newtonian Attraction: Some Insights into His Philosophy of Science," *Vistas in Astronomy* 32 (1975): 369–378, as well as Chandler's doctoral dissertation, "Newton and Clairaut on the Motion of the Lunar Apse" (n. 78 above).

ing that higher-order terms involving eccentricity squared and cubed have to be included in the perturbational corrections.[108] Even in his unpublished calculation Newton had considered only terms linear in eccentricity.

Clairaut's new result was of considerable moment. Euler remarked in a letter of 29 June 1751 to him:

> . . . the more I consider this happy discovery, the more important it seems to me, and in my opinion it is the greatest discovery in the Theory of Astronomy, without which it would be absolutely impossible ever to succeed in knowing the perturbations that the planets cause in each other's motions. For it is certain that it is only since this discovery that one can regard the law of attraction reciprocally proportional to the squares of the distances as solidly established; and on this depends the entire theory of astronomy.[109]

Euler affirmed this view publicly in the introduction to his *Theoria Motus Lunae* of 1753, where, after calling attention to the point about the sensitivity of the line of apsides that Newton had emphasized and the failure of the *Principia* to account for more than half of the motion in question, he remarks:

> Therefore, if by a correctly executed calculation the theory of Newton is shown to imply just that motion of the lunar apogee which is found by the observations actually to occur, namely 40 degrees per year, then surely no stronger argument can be desired by which the truth of this theory might be demonstrated.[110]

The work of Clairaut, the leading French expert on orbital motion, and of Euler, the primary expert outside of France, provided a decisive argument for the law of gravity by showing that an extension of Newton's solution for the motion of the lunar apsis was in precise agreement with observed values.

108. See Wilson, "Perturbations and Solar Tables," pp. 134–139.

109. G. Bigourdan, "Lettres inédites d'Euler à Clairaut," in *Comptes rendus du Congrès des sociétés savantes de Paris et des départements tenu à Lille en 1928, Section des sciences* (Paris: Imprimerie Nationale, 1930), p. 34, as quoted in Wilson, "Perturbations and Solar Tables," p. 143. The commitment Euler expressed here was to the inverse-square law of gravity, not to action-at-a-distance. Curtis Wilson has argued compellingly that Euler never did accept action at a distance; see his "Euler on Action-at-a-Distance and Fundamental Equations in Continuum Mechanics," in *The Investigation of Difficult Things: Essays on Newton and the History of the Exact Sciences*, ed. P. M. Harman and Alan E. Shapiro (Cambridge: Cambridge University Press, 1992), pp. 399–420.

110. Euler's *Theoria Motus Lunae*, in *Leonhardi Euleri Opera Omnia*, ser. 2, vol. 23 (Basel and Zurich, 1969), p. 72, as translated by Wilson, "Perturbations and Solar Tables," pp. 141–142.

Prop. 39: The Shape of the Earth and the Precession of the Equinoxes **8.17**

In retrospect, Newton's explanation of the precession of the equinoxes in relation to the shape of the earth may be reckoned as one of the greatest intellectual achievements of the *Principia*. At one stroke, Newton gave compelling evidence of the power and validity of the law of universal gravity, he provided supporting evidence for the shape of the earth as an oblate spheroid, and he proposed a simple reason founded in the principles of dynamics for a phenomenon known since the second century B.C. but never before reduced to its physical cause. Before the *Principia*, there had never been even a suggestion of a physical cause. Newton not only found out why precession occurs; he even obtained from his theory a value of the mean rate of precession of 50″ per annum, the numerical value then accepted by astronomers. As in the case of the motion of the moon and of the tides, Newton's brilliant insight led him to a position very much like the one we hold today. The case of precession, however, also resembles that of the moon and the tides in that when he shifts from general principles to details, Newton's analysis is deficient in certain major respects. We may thus agree with Curtis Wilson that although Newton's presentation "postulates the true cause," it "arrives at its conclusions by steps that are usually dubious when, upon narrow analysis, they are not plainly wrong."[111]

Basically Newton's explanation of precession depends on his prior conclusions about the shape of the earth. Suppose the earth to be an oblate spheroid, an ellipsoid produced by the revolution of an ellipse about its minor axis, flattened at the poles and bulging at the equator. This shape may be considered to be dynamically equivalent to that of a sphere girded by an equatorial ring of matter. The earth is spinning on its axis at an inclination of 23.5 degrees to the plane of the ecliptic. The moon's orbit is inclined only 5 degrees to the ecliptic. Hence both the sun and the moon will exert a gravitational pull on the earth at nearly the same angle and in a direction at a considerable angle to the plane of the earth's equator. A simple calculation will show that the moon's gravitational force will produce a very different pull on the nearer bulge than on the further one. Since the moon's distance from the earth's center is about 60 earth-radii, these two forces will be as $\frac{1}{(59R)^2}$ to $\frac{1}{(61R)^2}$, so that the difference between the magnitudes of the turning effects of these two forces will be a significant magnitude. There will thus be a net lunar turning force that will tend to shift the direction of the axis of the spinning earth and so cause precession. There is a similar solar force, and hence the net precessional force must be a varying combination of the lunar and solar forces.

111. "Newt. Achievement," p. 269.

The solar precessional force tends to shift the earth's axis of rotation so as to draw the earth's equator into the plane of the ecliptic, whereas the lunar precessional force tends to draw the earth's equator into the plane of the moon's orbit; but as has been mentioned, these two planes are very close.

Newton saw that the net effect of the precessional forces of the sun and the moon would be to produce a conical rotation of the earth's axis. In his computations, he was handicapped by two inadequacies in his knowledge: the mass of the moon and the distance from the earth to the sun. He therefore erred in computing the relative magnitudes of the lunar and solar perturbing forces.

In book 1, prop. 66, corol. 18, Newton supposes not merely a single moon orbiting around the earth but many, in a ring, forming a kind of fluid, all subject to the same perturbing force of the sun which tends to produce a regression of the nodes of the orbit. In corol. 20, the ring of moons becomes hard and the central body expands until there is produced an analogue of an earth with a protruding ring or an equatorial bulge. Newton picks up this argument once again in book 3 with prop. 39, which follows directly on props. 36 and 37 on the tide-producing force of the sun and of the moon and on the mass of the moon, plus prop. 38 on the shape of the moon. Prop. 39 draws on these anterior results, notably on the lunar tidal force as found in prop. 37, and on a series of three lemmas and a hypothesis. In the first edition, there were three lemmas, numbered 1, 2, and 3, following prop. 38 and introducing prop. 39. In the second edition, lem. 1 was considerably rewritten, a new lem. 2 was inserted, followed by the original lem. 2 which became lem. 3. The original lem. 3 was now renamed "hypothesis 2."[112] Prop. 39 makes use of certain corollaries of book 1, prop. 66, mentioned above.

Drawing on book 1, prop. 66, Newton asserts that the effects of the sun's perturbation on a ring around the earth will be the same periodic oscillation of the inclination of the axis and the same regression of the nodes as under the previous conditions. These effects will continue, even if their quantities may not be the same, and even if the ring should become solid. The effects would be the same even if the central body were to be so enlarged that it joined the ring at its equator, in which case the ring would become the equatorial bulge. Suppose, in this case, that "the ring adheres to the globe and communicates to the globe its own motion with which its nodes or equinoctial points regress"; then "the motion that will remain in the ring will be to its former motion" as 4,590 to 489,813, and hence "the motion of the equinoctial points will be diminished in the same ratio."

112. On these changes, along with a recasting of the original "hypotheses" of the first edition, see §8.2 above.

In obtaining this result, Newton has had to make a number of assumptions, among them that a solid ring of moons would exhibit the same features of perturbed motion as a single moon. He has also made use of what is called lem. 3 in the first edition, and hyp. 2 in the second and third editions. In this lemma/hypothesis we are asked to consider a solid ring moving in the earth's orbit around the sun, with all the central matter of the earth supposed to be taken away, having a daily rotation about its axis (which is inclined at 23.5 degrees to the plane of the ecliptic). Under these conditions, Newton asserts, "the motion of the equinoctial points would be the same whether that ring were fluid or consisted of rigid and solid matter." By changing the designation from lem. 3 to hyp. 2, Newton obviously was admitting that he was unable to prove this proposition.[113]

Newton proceeds through further calculations to determine the degree of annual precession of the equinoxes that arises from the gravitational force of the sun. This turns out to be $9''7'''20^{iv}$ per annum.[114] He now computes "the force of the moon to move the equinoxes," which is in the same proportion to the force of the sun to move the equinoxes as "the force of the moon to move the sea [is] to the [similar] force of the sun." In the preceding prop. 37, he has found this proportion to be "as roughly 4.4815 to 1." Hence the annual precession arising from the force of the moon alone will be $40''52'''52^{iv}$. Adding the two together yields $50''00'''12^{iv}$, a "motion of precession [that] agrees with the phenomena."

Wholly apart from the difficulty Newton had in evaluating the tidal force of the moon (in prop. 37), there are serious problems in the three lemmas on which prop. 39 depends. In lem. 1, Newton considers the earth to be a solid sphere surrounded by a "protuberant circumscribed shell"[115] and then compares the turning force (or turning moment) of the sun's force on this outer shell with the force when all the matter of the bulge is packed into a point on the equator most distant from the ecliptic. The first of these is said to be one-half of the second. In lem. 2, a similar comparison is made under the supposition that the matter of the bulge is distributed in a ring along the equator. The proportion is now said to be as 2 to 5. In lem. 3, he compares the turning force of the whole earth and that of a ring with a mass equal to the mass of the whole earth and finds that the two are as 925,275 to 1,000,000.[116] The problem here does not lie in Newton's computation

113. Rouse Ball, *Essay*, p. 110, says of this lemma/hypothesis that "Laplace was the first writer to prove it."

114. On these calculations, see n. 116 below.

115. *Essay*, p. 110.

116. Wilson, "Newt. Achievement," p. 270, has neatly summarized Newton's calculational procedure as follows. Newton starts "from the known regression of the nodes of the lunar orbit caused by the Sun, and on the assumption that the rates of regression for different moons would vary as their periods,

of the comparative value of these forces or turning moments, but rather in the assumption that "the resulting motions will be as the moments, without taking into account the inertial effect of the differential distributions of mass."[117] Because of "this same omission," Curtis Wilson has noted, "the result is wrong" when, in lem. 3, Newton attempts to determine "how the motion of precession in an equatorial ring will be shared with an underlying sphere to which the ring becomes attached." Nowhere here "does Newton take account of the angular momentum of the rotating earth."[118] It should be noted that in lem. 2 Newton uses "the method of fluxions," invoking a quantity whose "fluxion" is given.

8.18 *Comets (The Concluding Portion of Book 3)*

In the early part of book 3, Newton has proved that the gravitational force is universal and that it is an inverse-square force. Accordingly, since comets are made of matter, there must be an inverse-square force of gravity acting on them. According to book 1, prop. 13, corol. 1, therefore, the comets must move in orbits that are conics. In corol. 3 to lem. 4, the first of the lemmas and propositions dealing with comets, Newton concludes that comets "are a kind of planet and revolve in their orbits with a continual motion." In prop. 40, corol. 1, he states the implication of this conclusion: "Hence, if comets revolve in orbits, these orbits will be ellipses." They will move so that their "periodic times will be to the periodic times of the planets as the 3/2 powers of their principal axes," that is, axes of their orbits. And then in prop. 40, corol. 2, Newton points out that cometary orbits, at least in the parts within the solar system, the only parts visible to us, "will be so close to parabolas that parabolas can be substituted for them without sensible errors."

Making use of the accumulated cometary observations of many centuries, Newton was aware that if there are closed cometary orbits, the ellipses in question cannot possibly be nearly circular, as in the case of the planets, but must be extremely elongated so as to carry the comet far out of the visible reaches of the solar system. Thus the consequence of Newton's insight that comets are a sort of planet was not only that some of them may return periodically to the neighborhood of the sun, but that those that do return must have extremely elongated elliptical orbits. In this case, the portions of such elliptical orbits observable by us (that is, the part of the orbit within the visible solar system) are not very distinguishable from

determines the rate of regression for a moon circling once a day at the Earth's surface. This moon is then converted into an equatorial ring and the various ratios derived in the lemmas are applied to it."

117. "Newt. Achievement," p. 270.

118. Ibid.

parabolas. In fact, as Newton states early in the *Principia*, parabolas can be regarded as ellipses stretched out by moving one of the foci very far away. That is, at the end of book 1, sec. 1, in prop. 10, Newton explores motion in an ellipse produced by a force directed toward the ellipse's center; in a scholium he then considers what happens if "the center of the ellipse goes off to infinity," in which case "the ellipse turns into a parabola." Thus Newton will consider the orbits as parabolas, for which the computations are far simpler than those for elliptical orbits. The reason why computations are simpler with parabolas than with ellipses is that all parabolas are similar. Newton also considered the possibility of a hyperbolic orbit, in which case the comet would depart from the solar system. A significant aspect of the elongated shape of cometary ellipses was that gravitational force extends to enormous distances, far beyond the visible range of the solar system.

Newton's interest in comets was greatly stimulated by the appearance of the "great comet" of 1680/81, also known as the comet of 1680. He had only gradually abandoned the old hypothesis, espoused by Kepler among others, that comets move in straight lines toward and away from the sun.[119] His early attempt to deal with cometary orbits in *De Motu* (November 1684), however, was far from satisfactory.[120] By September 1685, as Newton wrote to the Astronomer Royal John Flamsteed, he had not yet succeeded in computing "the orbit of a comet," but he was studying the comet of 1680 and had concluded that probably the comet seen going toward the sun in November and the one seen going away from the sun in December "were the same."[121] Yet so late as June 1686, as he wrote to Halley, he had not yet worked up a satisfactory theory of cometary motion "for want of a good method."[122] He, therefore, had returned to book 1 and had added several "Propositions . . . relating to Comets."[123] By March 1687, when he sent the manuscript of book 3 to Halley for printing, he had developed the needed theory, of which the essential elements are found in book 3, lems. 5–11, and prop. 41.[124] In presenting his results in book 3, prop. 41, Newton remarked that he had tried "many approaches to this exceedingly difficult problem" and had "devised" some propositions in book 1 "intended for its solution." Later on, however, he had "conceived" a "slightly simpler solution." Many commentators, among them Rouse Ball, have pointed out that his earlier

119. There are two doctoral dissertations on Newton and comets: James Alan Ruffner, "The Background and Early Development of Newton's Theory of Comets" (Ph.D. diss., Indiana University, 1966); Sarah Schechner Genuth, "From Monstrous Signs to Natural Causes: The Assimilation of Comet Lore into Natural Philosophy" (Ph.D. diss., Harvard University, 1988).

120. See "Newt. Achievement," pp. 270–273.

121. Newton to Flamsteed, 19 September 1685, *Corresp.*, vol. 2.

122. Newton to Halley, 20 June 1686, *Corresp.*, vol. 2.

123. Ibid.

124. "Newt. Achievement," p. 270.

solution, to be found in the text of the Lucasian Lectures or first draft, "is simpler than the solution printed in the *Principia*."

Newton's treatment of comets is one of the many spectacular parts of the *Principia*. David Gregory reported that Newton told him that his "discussion about comets is the most difficult of the whole book."[125] Newton himself, in introducing his method in book 3, prop. 41, said explicitly that this was an "exceedingly difficult problem." Basically, the problem set by Newton is to determine three special cometary positions, obtained from observations and corrected for the motion of the earth. These positions are to be evenly spaced (and to include the vertex) and to be corrected for the difference in plane between the earth's orbit and the plane of the comet's orbit. From them, Newton determines the parabolic orbit, or the parabola closely fitting the orbit. In the course of this procedure, Newton developed and made use of a method often known today as Newton's approximation formula.[126]

The Russian astronomer Aleksyei Nikolaevich Krylov, author of a Russian translation of the *Principia* with extensive annotations, showed that this result, often known as "Euler's theorem" or "Lambert's theorem" or even "Euler-Lambert's theorem," is clearly stated by Newton in lem. 10 of book 3. It states that in a parabolic orbit of a comet, there is a specific relation between the times in which an arc is traversed, the radii vectores drawn to the extremities of that arc, and the chord. Although most commentators on the *Principia* (e.g., Rouse Ball) agree that Newton's method for determining cometary orbits is "impractical" and never became the basis of standard practice by astronomers, Krylov showed that this method could be easily translated into modern notation and then applied successfully to various comets.[127]

In the *Principia*, Newton not only deals with the mathematical theory of cometary motion but also traces in great detail the actual path of the "great comet" of 1680 (or of 1680/81). This is not, however, as is often mistakenly supposed, Halley's comet, which appeared in 1682. Halley's comet is mentioned in the second edition, but it is not the subject of primary analysis. Newton tells us in the *Principia* that in working out the orbit of the comet of 1680, he proceeded by a graphical method, constructing a huge diagram on a scale in which the earth's radius was 16.33 inches. His successive approximations gave him results which (on the diagram) would agree with some twentieth-century calculations to within

125. *Corresp.* 3:385.

126. See Duncan C. Fraser, "Newton and Interpolation," in *Isaac Newton, 1642–1727: A Memorial Volume Edited for the Mathematical Association*, ed. W. J. Greenstreet (London: G. Bell and Sons, 1927), pp. 45–74.

127. A. N. Kriloff [Krylov], "On Sir Isaac Newton's Method of Determining the Parabolic Orbit of a Comet," *Monthly Notices of the Royal Astronomical Society* 85 (1925): 640–656.

0.0017 inch. Krylov concluded that in order to obtain this result, Newton must have used a second diagram of the central part of the construction, drawn on a much larger scale, unless he "possessed as a draughtsman some *vis prope divina*."

In finding the parabola to fit the cometary orbit, Newton had reference to book 1, prop. 19.[128] Curtis Wilson has remarked that the "complete details of Newton's . . . determinations of the orbital elements of the comet of 1680–81 are not available to us." The methods set forth in prop. 41 "can have furnished no more than a first approximation, which was then refined by *ad hoc* adjustments." Although the "method is not eminently practical, and has seldom been used since," Wilson nevertheless concludes that "Newton's subsumption of cometary motion under the dynamics implied by universal gravitation marks an epoch, setting the stage for all future study of comets."[129]

In the second edition of the *Principia*, Newton reported that, after the first edition had been published, "our fellow countryman Halley determined the orbit [of the comet of 1680/81] more exactly by an arithmetical calculation than could be done graphically." In the third edition, Newton called attention to a remarkable discovery, made by Halley, of a comet with a period of 575 years. This comet had appeared "in September after the murder of Julius Caesar; in A.D. 531 in the consulship of Lampadius and Orestes; in February A.D. 1106; and toward the end of 1680." Newton gave Halley's calculation of the orbit of this comet and published a table showing computations and observations of twenty-five positions of this comet, in which (in the third edition) the errors in longitude do not exceed 2′31″ and those in latitude do not exceed 2′29″.

This led him to the following conclusion:

> The observations of this comet from beginning to end agree no less with the motion of a comet in the orbit just described than the motions of the planets generally agree with planetary theories, and this agreement provides proof that it was one and the same comet which appeared all this time and that its orbit has been correctly determined here.

128. An admirable outline of Newton's method, especially the role of book 3, lem. 8, is given by Wilson, "Newt. Achievement," pp. 270*b*–272*b*. See, further, the article by Krylov (n. 127 above).

129. "Newt. Achievement," p. 273*a*. Some work sheets preliminary to Newton's determination of the comet's orbit in the *Principia* have been published with commentary by D. T. Whiteside in *Math. Papers* 6:498–507. On Newton and the orbits of comets, see two valuable papers in *Standing on the Shoulders of Giants: A Longer View of Newton and Halley*, ed. Norman J. W. Thrower (Berkeley, Los Angeles, London: University of California Press, 1990)—Eric G. Forbes, "The Comet of 1680–1681," pp. 312–324, and David W. Hughes, "Edmund Halley: His Interest in Comets," pp. 324–372—documenting Halley's contribution to Newton's presentation, especially in the new material contained in the second edition. This volume also contains other useful essays, especially Sarah Schechner Genuth's "Newton and the Ongoing Teleological Role of Comets," pp. 299–311.

Later astronomers, however, have invalidated Halley's claims for a supposed comet with a period of 575 years.

Following the discussion of orbits, Newton introduces some questions concerning the physical composition of comets. Newton concluded that "the bodies of comets are solid, compact, fixed, and durable, like the bodies of planets." He also argued that comets receive "an immense heat at [i.e., when near] the sun." He believed that "the tail is nothing other than extremely thin vapor that the head or nucleus of the comet emits by its heat." He did, however, discuss other theories concerning the tails of comets. These are:

> that the tails are the brightness of the sun's light propagated through the translucent heads of comets; that the tails arise from the refraction of light in its progress from the head of the comet to the earth; and finally that these tails are a cloud or vapor continually rising from the head of the comet and going off in a direction away from the sun.

One interesting conclusion reached by Newton is that a study of the tails of comets provides another argument "that the celestial spaces are lacking in any force of resisting, since in them not only the solid bodies of the planets and comets but also the rarest vapors of the tails move very freely and preserve their extremely swift motions for a very long time."[130]

Newton envisioned a purpose for comets, which may seem surprising to readers unacquainted with Newton's thought. In the first edition of the *Principia*, Newton expressed his belief that matter from the tails of comets entering our atmosphere may supply a "vital spirit" needed for sustaining life on earth and also substances to be chemically converted into the matter of animals, vegetables, and minerals. In the second edition, he suggested that comets might refuel the sun and, similarly, the stars. In such an occurrence Newton found a physical means of explaining the otherwise puzzling occurrence of novas. This concept would seem to explain his remark to David Gregory that "Comets are destined for a use other than planets."[131] Gregory also recorded a thought of Newton's, one that was also reported by John Conduitt, that vapors and the matter of light could coalesce so as to produce a moon which then would attract more and more matter until it became a planet, which eventually would turn into a comet and finally fall into

130. Newton pays particular attention to Kepler's theory that the "ascent of the tails of comets from the atmospheres of the heads and the movement of the tails in directions away from the sun" can be ascribed to "the action of rays of light that carry the matter of the tail along with them."

131. Dobbs, *Janus Faces* (§3.1, n. 10 above), pp. 236–237.

the sun in a sort of cyclical cosmos.[132] Newton even imagined that the earth might be destroyed in a kind of cataclysm predicted in Scripture[133] and that God might renew the system by causing a comet to pull a satellite away from Jupiter or Saturn and convert it into a new planet awaiting a new act of creation. Obviously, Newton did not enter such extreme thoughts into the otherwise sober and austere *Principia*.

132. See David C. Kubrin, "Newton and the Cyclical Cosmos: Providence and the Mechanical Philosophy," *Journal of the History of Ideas* 28 (1967): 325–346.

133. See Dobbs, *Janus Faces*; also Sarah Schechner Genuth, "Comets, Teleology, and the Relationship of Chemistry to Cosmology in Newton's Thought," *Annali dell'Istituto e Museo di storia della scienza* 10 (1985): 31–65.

CHAPTER NINE

The Concluding General Scholium

9.1 *The General Scholium: "Hypotheses non fingo"*

The first edition of the *Principia* had no proper conclusion, since Newton suppressed his draft "Conclusio." In the second edition, he planned to have a final discussion about "the attraction of the small particles of bodies," as he wrote to Cotes on 2 March 1712/13,[1] when most of the text had already been printed off. On further reflection, however, he abandoned the temptation to expose his theories of the forces, interactions, structure, and other aspects of particulate matter, and instead composed the concluding General Scholium, in which, he said, he had included a "short Paragraph about that part of Philosophy." One function of the General Scholium was to answer certain critics, notably the Cartesians and other strict adherents of the mechanical philosophy.

The opening paragraph of the General Scholium is a recapitulation of all the arguments in the main text to prove that the celestial phenomena are not compatible with "the hypothesis of vortices." It is followed by a discussion of the emptiness of the "celestial spaces . . . above the atmosphere of the earth," based on the long-term constancy of planetary and cometary motions.

The next topic is the argument from design, the proof of the existence of the creator from the perfection of his creation. Having established the presence of "an intelligent and powerful being," the architect of the universe, Newton proceeds to analyze the names and attributes of this creator. In the second edition, Newton concluded this paragraph by asserting that to discourse of "God" from phenomena is legitimate in "experimental philosophy." In the third edition, this was altered to "natural philosophy." This sentence was not in the original text of the General

1. *Corresp.*, vol. 5.

Scholium that Newton sent to Cotes. Newton wrote to Cotes shortly afterward, directing him to make this addition.

The next paragraph, the penultimate paragraph of the General Scholium, is probably the most discussed portion of all of Newton's writings. Briefly, he here declares that he has explained the phenomena of the heavens and the tides in the oceans "by the force of gravity," but he has "not yet assigned a cause to gravity." Summarizing some of the chief properties he has put forth in book 3, he takes note that this force does not operate "in proportion to the quantity of the *surfaces* of the particles on which it acts," but rather "in proportion to the quantity of *solid* matter." Accordingly, gravity does not act "as mechanical causes are wont to do."[2] Then Newton admits that he has been unable to discover the cause of the properties of gravity from phenomena, and he now declares: "Hypotheses non fingo."

Newton obviously did not mean by this phrase that he never "uses" or "makes" hypotheses, since this statement could easily be belied: for example, by the presence of one "hypothesis" in book 2 and two more in book 3. Alexandre Koyré suggested that by "fingo" Newton probably intended "I feign," in the sense of "inventing a fiction," since the Latin version (1706) of the *Opticks* (1704) used the cognate "confingere" for the English "feign." Thus Newton would be saying that he does not invent or contrive fictions (or "hypotheses") to be offered in place of sound explanations based on phenomena. In this category he included "metaphysical" and "physical" hypotheses, "mechanical" hypotheses, and hypotheses of "occult qualities." In the *Opticks* and in a great number and variety of manuscripts, Newton used the English verb "feign" in the context of showing his disdain for hypotheses.[3]

From manuscript drafts we learn that Newton chose the verb "fingo" with care; it was not the first word to leap into his mind. He tried "fugio," as in "I flee from [or shun] hypotheses"; he tried "sequor," as in "I do not follow [or, perhaps, I am not a follower of] hypotheses."[4] Most English-speaking readers know this phrase in Andrew Motte's translation as "I frame no hypotheses." There is no way of telling exactly what meaning Motte intended the verb "frame" to have. Certainly one use of "frame," in the context of theory, was "to fabricate." In any event, in Newton's day and in Motte's, one of the senses of "to frame" was decidedly pejorative, just as is the case today. In today's usage, however, the pejorative sense of

2. The force of mechanical pressure exerted on a body by a given blast of wind depends only on the area of surface exposed, whereas the effect of a given gravitational field would depend on the quantity of matter or mass.

3. We use the word "feign" today in a fictional sense for "pretend," as to "feign" sleep.

4. For details, see my *Introduction*, pp. 240–245.

"to frame" has a very different signification from what it did in Newton's day, since for us the verb "to frame" means "to concoct false evidence against" (a person). In Samuel Johnson's *Dictionary* (London, 1785), one of the definitions is "To invent, fabricate, in a bad sense: as, to *frame* a story or lie." Johnson gives, as an example, a quotation from Francis Bacon, "Astronomers, to solve the phaenomena, *framed* to their conceit eccentrics and epicycles." Surely, Motte was intelligent enough to recognize that for Newton "fingo" was a derogatory word. Accordingly, Motte would surely have intended that his translation "frame" would equally convey this sense to the reader.[5]

When Cotes received the text of the General Scholium, he was greatly concerned that in book 3, prop. 7, Newton seemed to him to "Hypothesim fingere," and he suggested that Newton write a further discussion in "an *Addendum* to be printed with the Errata."[6] Newton, however, proposed that the end of the penultimate paragraph be altered.

This paragraph, as originally sent to Cotes, concluded as follows:

> Indeed, I have not yet been able to deduce the reason [or cause] of these properties of gravity from phenomena, and I do not feign hypotheses. For whatever is not deduced from phenomena is to be called a hypothesis; and I do not follow *hypotheses*, whether metaphysical or physical, whether of occult qualities or mechanical. It is enough that gravity should really exist and act according to the laws expounded by us, and should suffice for all the motions of the celestial bodies and of our sea.

He now instructed Cotes to alter this and to make it more general. It now would read:

> I have not as yet been able to deduce from phenomena the reason for these properties of gravity, and I do not feign hypotheses. For whatever is not deduced from the phenomena must be called a hypothesis; and hypotheses, whether metaphysical or physical, or based on occult qualities, or mechanical, have no place in experimental philosophy. In this experimental philosophy, propositions are deduced from the phenomena and are made general by induction. The impenetrability, mobility, and impetus of bodies, and the laws of motion and the law of gravity have been found by this method. And it is enough that gravity really exists and acts according to the laws that we have set forth and is sufficient to explain all the motions of the heavenly bodies and of our sea.

5. I. B. Cohen, "The First English Version of Newton's *Hypotheses non fingo*," *Isis* 53 (1962): 379–388.
6. See my *Introduction*, pp. 240–245; *Corresp.*, vol. 3.

This is the text as we know it. In a letter to Cotes, Newton wrote a gloss on the revision of the General Scholium, in which he explained that, "as in Geometry the word Hypothesis is not taken in so large a sense as to include the Axiomes & Postulates, so in Experimental Philosophy it is not to be taken in so large a sense as to include the first Principles or Axiomes which I call the laws of motion." These principles, he explained, "are deduced from Phaenomena & made general by Induction: which is the highest evidence that a Proposition can have in this philosophy." Furthermore, "the word Hypothesis is here used by me to signify only such a Proposition as is not a Phaenomenon nor deduced from any Phaenomena but assumed or supposed without any experimental proof." Newton believed that "the mutual & mutually equal attraction of bodies is a branch of the third Law of motion & . . . this branch is deduced from Phaenomena."[7]

"Satis est": Is It Enough? 9.2

Newton concludes the penultimate paragraph of the General Scholium in a series of assertions. First, in experimental philosophy, "propositions are deduced from the phenomena and are made general by induction." Second, this is the method by which "the impenetrability, mobility, and impetus of bodies" have been found, as also "the laws of motion and the law of gravity." Third, and in conclusion, "And it is enough [*satis est*] that gravity really exists [*revera existat*]" and that gravity (1) "acts according to the laws that we have set forth" and (2) "is sufficient to explain all the motions of the heavenly bodies and of our sea."

On the basis of these expressions, and without further inquiry, Newton was hailed by Ernst Mach as an early positivist, and others have followed Mach in thus attributing to Newton an apparent endorsement of a positivistic position. Anyone who is acquainted with Newton's writings, however, especially his correspondence and his further published and manuscript expressions on this subject, will know that Newton's position was altogether free of any taint of positivistic philosophy. We have seen the many attempts made by Newton, both before and after writing the General Scholium, to find some way of accounting for a force acting at a distance, an endeavor that occupied his attention to varying degrees from at least as early as the time of composing the *Principia* in the 1680s.[8] Not only did he explore possible modes of explanation in various manuscript essays and proposed

7. Newton to Cotes, 28 March 1713, *Corresp.*, vol. 3.

8. The most recent and thorough presentation of Newton's efforts is given in Dobbs, *Janus Faces* (§3.1, n. 10 above).

revisions of the *Principia*; he introduced this topic in the second (revised) English edition of the *Opticks* in 1717/18, where it is mentioned specifically in the preface.[9]

One cannot know with any certainty exactly what Newton intended his readers to understand by "satis est." The evidence leaves no doubt, however, that for Newton, it was never "enough" merely that his explanation was based on a force that "really exists," whose laws he had set forth, and which serves to explain the observed phenomena of heaven and earth. Yet there can be no doubt that Newton firmly believed in the reality of universal gravity and its action in the solar system.

There is no way of telling whether the Newtonian style of initially dealing with the subject on a mathematical rather than a physical plane was only a subterfuge to avoid criticism or a sincere expression of methodological principle. But, from Newton's point of view, this style enabled him to develop the laws of the action of a gravity-like force in a mathematical analogue of the world of nature without having to be concerned with whether or not gravity exists.

The shift from the mathematical level of discourse of books 1 and 2 to the physical level of book 3, with constant reference to deductions or inductions[10] from phenomena, produced an elegant way of ordering or explaining some of the most basic phenomena of nature. In the General Scholium, Newton freely admitted that he had not been able to explain how a force can act over vast reaches of empty space. He had not succeeded in finding out how such a force acts. Nevertheless, he was able to assert that gravity really exists and that gravity does act to produce the principal phenomena of nature.

Mathematical deduction showed that if there are such long-range forces, then the planets move in orbits produced by an inverse-square force. Sound induction led him to the conclusion that the planetary and lunar forces are a variety of terrestrial gravity which, as he wrote, "really exists." It would seem, therefore, that what Newton is saying in the General Scholium is that gravity really exists, that it extends to the moon and beyond, and that it does provide a means of ordering or organizing the observed phenomena of nature. By using this force we gain an explanation of the phenomena of the heavenly bodies, the tides, and much more.

In the General Scholium, then, Newton is not telling his readers to abandon the search for the cause of gravity, nor is he denying the importance of finding out how gravity acts to produce its effects. Indeed, we know how hard Newton himself tried to solve these two problems. What he is saying, rather, is that there are two

9. "In this Second Edition of these Opticks I . . . have added one Question," query 21, concerning the "Cause" of gravity, "chusing to propose it by way of a Question, because I am not yet satisfied about it for want of Experiments."

10. In the General Scholium and elsewhere, Newton refers to both deductions and inductions from phenomena.

jobs ahead. One is to find the cause and mode of operation of the universal force of gravity; the other is to apply the theory of gravity to yet new areas of phenomena, to use our mathematical skills to solve those vexing problems for which he himself had only found imperfect solutions. Among the latter was the complex of problems associated with the moon's motion, including the motion of the lunar apsis.

Newton was concerned by the fact that some scientists, notably those on the Continent, had accepted the inverse-square force for the sun-planet and planet-satellite forces (although not as attractions) but argued on philosophical grounds against accepting a force of universal gravity. To these critics Newton seems to be saying that even though the notion of attraction is not acceptable, that even though we cannot as of now understand the cause and mode of action of universal gravity, it is nevertheless fruitful to move ahead with the science of rational mechanics and celestial dynamics. For this purpose, "it is enough" that gravity does account for so many phenomena; therefore, we may legitimately use this concept to advance the subjects of rational mechanics and celestial dynamics and even to explore new fields of science relating to other forms and varieties of attractions. Such research, especially the discovery of new phenomena, might even produce an explanation of the cause of gravity or of the way in which gravity works. Newton was not a positivist; he did not say "satis est" in the sense that he had no concern for finding causes. But he did not believe that the search for a cause, or for a mode of producing effects, should cause a halt in the application of a useful concept.

The correctness of this reading of the message of the General Scholium seems to be confirmed by the final paragraph. For here, Newton himself is indicating a new path of research that might possibly illuminate the problem of the cause and mode of operation of universal gravity. This research topic that might possibly be fruitful involved the properties and laws of what he conceived to be "a certain very subtle spirit."[11] In the paragraph about this "spirit" he mentioned many types of phenomena concerning it, but he did not include gravity in that list. Perhaps he was not then ready to do more than hint—by implication or by context—that this spirit might somehow be related to gravity. Or, perhaps, he was merely indicating that there are always new areas of research that may illuminate fundamental problems of cause and mode of operation. In various manuscripts, he indicated that perhaps the hoped-for way to account for the action of gravity might be found in some current research in the new field of electricity.[12] Newton's suggestion of a field for fruitful research, as we shall see (in §9.3 below), is similar to his suggestions

11. This "spirit" is discussed in §§9.3, 9.4 below.

12. I believe that my reconstruction has the virtue of explaining why Newton introduced the final paragraph, which, on first inspection, might seem to contravene his warning about hypotheses.

for research at the end of the *Opticks*: both declare a research program for the future.

9.3 *Newton's "Electric and Elastic" Spirit*

The final paragraph of the General Scholium introduces the concept of a "spiritus." Some insight into what Newton meant by his reference to "spirit" in the conclusion to the General Scholium may be gleaned from his explanations in certain drafts of the "Recensio Libri," the review he wrote and published[13] (anonymously) of the Royal Society's report on priority in the invention of the calculus, the *Commercium Epistolicum* (London, 1712). In all these drafts, Newton was especially concerned to explain what this "spirit" is. It is, he wrote, "a subtile spirit or Agent latent in bodies by which Electrical Attraction & many other phaenomena may be performed."[14] His critics ("the Editors of the Acta Eruditorum") had held that "if this Agent be not the subtile matter of the Cartesians it will be looked upon as a trifle." Newton, however, believed that this subject should "be inquired into by experiments," whereas "these Gentlemen" held that we should not seek to inquire "into the properties & effects of the Agent by which electrical attraction is performed, & of that by which light is emitted reflected & refracted, & of that by which sensation is performed & to inform our selves whether they be not one & the same agent; unless we have first explained by an hypothesis what this Agent may be." That is, they believed that "we must not pursue experimental Philosophy by experiments until we founded it upon hypotheses."[15] He added, "And by these indirect practises they [insinuate that *del.*] would have it believed that Mr. Newton was unable to find the infinitesimal method." This led Newton to inveigh against Leibniz by comparing and contrasting their scientific practice and use of the calculus:

> Mr. Leibnitz never found out a new experiment in all his life for proving any thing; . . . Mr. Newton has by a great multitude of new Experiments discovered & proved many things about light & colours & setled a new Theory thereof never to be shaken; . . . Mr. Newton by the infinitesimal Analysis applyed to Geometry & Mechanicks has setled the Theory of the Heavens, & Mr. Leibnitz in his Tentamen de Motuum Coelestium causis has endeavoured to imitate him, but without success for want of skil in this Analysis.

13. See §5.8, n. 25 above.
14. ULC MS Add. 3968, fol. 586v.
15. This passage was marked by Newton for deletion.

In another draft version, Newton rails against the editors of the *Acta Eruditorum* for saying that "Mr. Newton denies that the cause of gravity is mechanical" and for saying that Mr. Newton "laies down a certain new Hypothesis concerning a subtile spirit pervading the pores of bodies (perhaps the same with the Hylarctick principle of D. H. More) & this spirit is represented of less value then Hypotheses are unless it be the Aether or subtile matter of the Cartesians."[16] In reply, Newton would say that he "has no where denyed that the cause of gravity is Mechanical nor affirmed whether that subtile spirit be material or immaterial nor declared any opinion about their causes."

"It appears," Newton wrote, "by Experiments (lately shewed to the R. Society by Mr. Hawksby) that bodies do attract one another at very small distances in such a sense as he uses the word attraction." Hauksbee, furthermore, according to Newton, "suspects that this attraction & electrical attraction may be performed by one & the same agent, the body constantly attracting at very small distances without friction, & the attraction being extended to great distances by friction; & this Agent he calls a subtile spirit." And then Newton adds: "But what is this Agent or spirit & what are the laws by which it acts he leaves to be decided by experiments."

In a passage marked for deletion, Newton added that Hauksbee had "found by some experiments shewed before the R. Society that this Agent or spirit when sufficiently agitated emits light & that light in passing by the edges of bodies at small distances from them is inflected." On the basis of such experiments, Newton concluded, he "has represented in the end of his Principles that light & this spirit may act mutually upon one another for causing heat reflexion refraction inflexion & vision, & that if this spirit may receive impressions from light & convey them into the sensiorium [*sic*] & there act upon that substance which sees & thinks, that substance may mutually act upon this spirit for causing animal motion." These things, however, Newton "only touches upon & leaves them to be further enquired into by experiments."[17]

In yet one more draft version, Newton set forth his procedure in a plain and direct fashion. After he mentioned how the "Editors of the Acta Eruditorum" had accused him "of publishing an Hypothesis in the end of his Principles about the actions of a very subtile spirit," he claimed that he "did not propose it by way of

16. ULC MS Add. 3968, sec. 41, fol. 125.

17. Another canceled statement, ULC MS Add. 3968, sec. 41, fol. 26, contains a somewhat differently worded version of the text quoted at n. 16 above, similarly concluding with a reference to experiments made by Hauksbee "before the Royal Society," which Newton said showed "that electric bodies attract constantly at small distances even without rubbing." He calls the "Agent by which this attraction is performed" a "subtile spirit."

an Hypothesis but in order to [lead to] an inquiry, as by his words may appear to any unprejudiced person." Then he described his procedure as follows:

> After he had shewed the laws, power & effects of Gravity without medling with the cause thereof & from thence deduced all the motions of the great bodies in the system of the Universe [*in an earlier version* Planets, Equinoxes, Comets, & sea], he added in a few words his suspicions about another sort of attraction between the small parts of bodies upon which many Phaenomena might depend, & for want of a sufficient number of experiments left the enquiry to others who might hereafter have time & skill enough to pursue it, & to give them some light into the enquiry mentioned two or three of the principal phaenomena which might arise from the actions of the Spirit or Agent by which this Attraction is performed. He has told his friends that there are sufficient Phaenomena to ground an inquiry upon but not yet sufficient to determin the laws of this attraction.[18]

If we may take Newton at his word, this final paragraph was intended as a presentation of the current state of knowledge and thinking about the consequences of Hauksbee's electrical experiments.[19] Newton was, then, declaring a research program for himself and for other natural philosophers so that there might be discovered the laws by which this "spirit" operates. In this sense, the final paragraph of the General Scholium was to have the same purpose of a research program as the queries with which the *Opticks* concluded, especially in the queries added to the Latin version of 1706. And, indeed, it will be observed that there is a kinship between some of the topics of these final queries and the final paragraph of the General Scholium, especially when taken with the glosses in Newton's manuscript drafts of the "Recensio Libri."

Finally, we may take note that in his interleaved copy of the second edition, containing corrections and emendations for a third edition, Newton qualified the "spirit" by the adjectives "electric and elastic." This emendation did not find its way into the third edition, but it was communicated to Andrew Motte, who inserted these words into his English translation. Newton also indicated later his intention to cancel the whole final paragraph. He no longer considered it to be an accurate presentation of his ideas.

18. ULC MS Add. 3968, fol. 586.

19. On Hauksbee and Newton, see Henry Guerlac, *Essays and Papers on the History of Modern Science* (Baltimore and London: Johns Hopkins University Press, 1977), chap. 8.

9.4 *A Gloss on Newton's "Electric and Elastic" Spirit: An Electrical Conclusion to the* Principia

While Newton was engaged in the final revisions of the *Principia* for the second edition, he composed a conclusion which has never before been published. This essay is of real significance in helping us to understand the final paragraph of the concluding General Scholium because in it Newton elaborates at some length the nature and properties of the "spirit" which is the subject of that final paragraph. There are a number of other manuscripts in which Newton discusses this spirit. Some of these (see §9.3 above) are related to Newton's review of the *Commercium Epistolicum*. Another set is to be found among Newton's optical manuscripts, the collection catalogued as ULC MS Add. 3870, notably those in the region of fols. 427, 599–604. Another is among the *Principia* manuscripts, ULC MS Add. 3965, fols. 356–365, dealing with comets and including drafts of the General Scholium, and fol. 152, among the discussions of comets. Some of the optical manuscripts are related to Newton's final query 22 (second edition), in which electricity is discussed, but others are concerned with a variety of physical phenomena.

In a draft of the General Scholium (ULC MS Add. 3995, fols. 351–352, published by Rupert and Marie Hall),[20] Newton outlined, but did not give details concerning, the attraction among "very small particles," explaining that it is "of the electrical kind." He then lists briefly some of the properties of what he calls "the electric spirit." Since that publication in 1962, there has been no doubt that the "spirit" mentioned in the final paragraph of the General Scholium is electrical.[21]

The document presented here (ULC MS Add. 3965, fols. 351–352; MS 3970, fols. 602–604) is the final portion of a proposed conclusion to the *Principia*. It was written later than 1704, since it refers to the *Opticks* (English edition in 1704, revised Latin version in 1706), and before the General Scholium (1712), most likely in the latter part of this six- to eight-year period. Most of this document is written neatly (in Newton's hand) and is not a rough draft with passages crossed out and rewritten, although the very last part is more tentative. It was apparently written out carefully to be used by the printer. The document is largely a continuation of the discussion of comets with which book 3 concludes in the first edition; it is headed "Pag. 510, post finem adde," or "Add [this] on page 510, after the end." The end or conclusion of book 3 in the first edition occurs on page 510 and is indicated by "FINIS." That this text is indeed a planned new conclusion to the *Principia* is further proved by the near identity, paragraph by paragraph, of this discussion of

20. *Unpubl. Sci. Papers*, pp. 361–362.

21. Further information concerning Newton's "electric spirit" and his speculations about "spirit" in general (and a "vegetative spirit") is available in Dobbs, *Janus Faces* (§3.1, n. 10 above).

comets and the printed conclusion on the subject of comets as it appears in the second and third editions.

The first edition ends with book 3, prop. 42, "To correct a comet's trajectory, once it has been found." The method is set forth in three "operations," followed by a concluding paragraph and "Q.E.I." In the second and third editions, this is followed by several pages of supplement to this problem, containing tables and observations relating to the comet of 1664/65, plus confirming data and a table for the comet that appeared in 1683 and the one in 1682. In the third edition, there is an additional table and data for the comet of 1723. There are further discussions of observations of the comet that was retrograde in 1607, the comet of 1680, and the comet of 1572/73. In both of the later editions, Newton concludes with a discussion of the tails of comets falling into "the atmospheres of the planets" so as to be "condensed and converted into water and humid spirits, and then—by a slow heat—be transformed gradually into salts, sulphurs, tinctures, slime, mud, clay, sand, stones, corals, and other earthy substances."

The manuscript version has the virtually identical text, including the tables neatly set out, but does not conclude—as the printed version does—with a discussion of the transformation of the vaporous tails of comets into "earthy substances." Rather (fols. 601v, 602r, 602v), the paragraph about "earthy substances," which occurs in the middle of the page, is followed without any break in the text by a lengthy discussion of the "spirit," obviously the same "spirit" discussed briefly by Newton in the final paragraph of the General Scholium. It is this text which is printed (in English translation) below.

Since this text was, for the most part, written by Newton in the same neat handwriting as the preceding discussion of comets, it was evidently intended by Newton to be an extension of the discussion of comets and to serve as a conclusion for the new second edition. The very final portion, however, is not written in the same final kind of neat version as the earlier part and appears in short paragraphs that indicate that Newton had not reached the final stage of polishing and completing his revisions. Evidently, Newton later decided to suppress the part of the conclusion dealing with the "electric spirit" and to replace it by the General Scholium and an addendum with a discussion of this "spirit," not identified as electrical. As we have seen, he later planned to add to the third edition an identification of this "spirit" as "electric and elastic" but then decided to cancel the final paragraph altogether.

Readers will note that the draft conclusion for the second edition differs from the General Scholium in a number of features. There is no argument against vortices with which the General Scholium opens, nor is there any discussion of the elegant design of the system of the world "which could not have arisen without the design and dominion of an intelligent and powerful being." In this draft, there is no reference whatever to a divine creator, and hence there is no discussion of the attributes of the deity. In the General Scholium, the single brief paragraph in which this unidentified "spirit" is presented does not contain a single reference to the various kinds of evidence from experiments mentioned in the draft, nor is there even a mention of many of the properties and powers of the "electric spirit" which are presented in the draft. In the paragraph at the end of the General Scholium, Newton refers to the role of this "spirit" in conveying sensation to the brain and in causing the limbs of animals to move "at command of the will," neither of which is mentioned in the draft. Finally, it will be noted that in the draft, Newton begins by comparing and contrasting three forces of attraction—gravity, electricity, and magnetism—a feature that is absent from both the final paragraph of the General Scholium and the preceding paragraph. The draft does not contain any discussion concerning hypotheses in relation to natural philosophy.

In this preliminary version, Newton draws heavily on some experiments performed by Francis Hauksbee (or Hawksbee). The importance of Hauksbee's experiments in the development of Newton's thought has been explored by Henry Guerlac, who was primarily concerned with Hauksbee's influence on the ideas expressed in the later queries of the *Opticks*.[22] Hauksbee's experiments, considered in relation to Newton's ideas about forces of attraction, fall into two main categories. One comprises experiments on electricity, primarily those which seemed to Newton to establish without question the existence of an "electric spirit," which was also "elastic." A principal experiment consisted of rotating an evacuated globe (that is, a globe partially evacuated of air) in a frame while an experimenter held his hand in contact with it. A visible glow was produced. If certain objects were placed near the electrified rotating globe, they would glow or shine with an illumination. These experiments not only confirmed the existence of an electric "spirit" but also revealed a number of its properties.

The second set of experiments had no such direct connection with electricity, although—as the text of the draft shows—Newton interpreted them in terms of the action of such a spirit. These were experiments on capillarity. Hauksbee studied the rise of fluids within capillary tubes, noting that the thickness of the walls of the tube had no influence on the effect. He also performed the experiment with the tubes held at different orientations to the vertical and with a variety of different fluids. Perhaps the most remarkable of his experiments was to show that in an evacuated jar the capillary rise was exactly the same as in full air. In another series

22. Guerlac, *Essays and Papers*, chap. 8, "Francis Hauksbee: Expérimentateur au profit de Newton" (1963); chap. 9, "Newton's Optical Aether: His Draft of a Proposed Addition to His *Opticks*" (1967).

of experiments, he took polished flat plates of glass and other substances, in contact at one edge and set to each other at a very small angle. Here he studied the rise of fluids between the plates. Newton interpreted these experiments as effects of the action of the same "spirit" that produced the phenomena observed in the electrical experiments.[23]

It will be seen that in this essay Newton went on to explore the possibility that the electric "spirit" might be the cause of several other types of phenomena. He was convinced that "glass at small distances always abounds in electric force, even without friction, and therefore abounds in an electric spirit which is diffused through its whole body and always surrounds the body with a small atmosphere," stirred up by friction. This "electric spirit," he noted, "emits light." He even attributed the way that light "is reflected or refracted" to the "action of some spirit which lies hidden in the glass" and which can cause the rays of light to be "inflected in the vicinity of bodies" and "without any contact with the bodies at all." He believed that the rays of light "are agitated by some tremulous spirit in the vicinity of bodies" and that "this spirit moreover is of an electric kind," as is "obvious from what has been said above."[24] He concludes with a brief note about "fermentations and digestions" and the "composition" of "bodies of animals, vegetables, and minerals." In this presentation, as in the General Scholium, Newton does not include the force of gravity among the actions of this spirit. The final impression given a reader is that if more were known about the action of this spirit, then we would understand the nature of attractive forces in general and so be in a better position to understand the action of gravity.

Some four years after the second edition of the *Principia*, Newton published the second edition of the *Opticks* (London, 1717/18), in which he explored the properties and effects of an "aethereal medium," a somewhat different entity from the Cartesian or dense aether which he had rejected earlier. Scholars are divided in their judgment concerning whether this aethereal medium is identical with the electric spirit.[25] An exploration of that question would take us far afield. What is of concern here is not so much to explore all of Newton's speculations about spirit

23. Readers familiar with the later Queries of the *Opticks* will recognize the similarity between Newton's presentation in those Queries and in this draft account of Hauksbee's experiments on electricity and capillarity.

24. These thoughts about "spirit" were later transformed by Newton into the action of an "aethereal medium" and developed in the later Queries of the *Opticks*.

25. See, e.g., R. W. Home, "Force, Electricity, and the Powers of Living Matter in Newton's Mature Philosophy of Nature," in *Religion, Science, and Worldview: Essays in Honor of Richard S. Westfall*, ed. Margaret J. Osler and Paul Lawrence Farber (Cambridge and New York: Cambridge University Press, 1985), pp. 95–107; J. E. McGuire, "Force, Active Principles, and Newton's Invisible Realm," *Ambix* 15 (1968): 154–208, esp. 176; R. W. Home, "Newton on Electricity and the Aether," in *Contemporary Newtonian*

and force, but rather to understand what Newton had in mind when he wrote the paragraph of the General Scholium with which the later editions of the *Principia* conclude. As the draft conclusion printed here shows plainly, and as the documents published by the Halls made clear, Newton's "spirit" is an "electric spirit," whose properties were demonstrated to him by Hauksbee.

The draft leaves no doubt that Newton's thinking about the "electric spirit" was to a high degree bolstered by experiments but was still in a very speculative state. As he admitted in the printed version, there had not been as yet an accurate determination of "the laws governing the actions of this spirit." Accordingly, readers were (and have continued to be) puzzled concerning why Newton introduced this paragraph, all the more so since it almost immediately follows Newton's bold declaration, "Hypotheses non fingo"! We may, I believe, more readily understand why it was that Newton later decided to suppress this paragraph in any future third edition than we can explain why he published it in the first place.

[A Draft Conclusion to the Principia, *Translated from the Latin]*

> I have now set forth the forces, properties, and effects of gravity. It is most certain from phenomena that electric and magnetic attractions also exist. But the laws of these [attractions] are very different from the laws of gravity. Electrical and magnetic attractions are intended and remitted; gravity cannot be intended and remitted. Magnetism and electricity sometimes attract and sometimes repel; gravity always attracts. They act at small distances, gravity at very great distances. Magnetic force is communicated by contact; the other forces are not. All bodies are heavy [i.e., gravitate] in proportion to their quantity of matter; most bodies are electric; only iron bodies are magnetic, but not in proportion to their quantity of matter. Gravity is not at all impeded by the interposition of bodies; electric force is impeded and diminished; magnetic force is not at all impeded by the interposition of cold and nonferrous bodies, but it is impeded by the interposition of ignited bodies and is propagated through interposed iron bodies [*lit.* through the interposition of iron bodies]. Gravity is not changed by friction. Iron, whose parts are strongly

Research, ed. Z. Bechler (Dordrecht: D. Reidel Publishing Co., 1982), pp. 191–214. See also *Never at Rest*, and Dobbs, *Janus Faces*, passim.

A most important addition to our knowledge of Newton's study of the effects of electricity is given in Maurizio Mamiani, "Newton e i fenomeni della vita," *Nuncius: Annali di storia della scienza* 6 (1991): 69–77, with a supplement (pp. 78–87) consisting of the publication of two texts by Newton, "De Motu et Sensatione Animalium" and "De Vita & Morte Vegetabili," edited by Emanuela Trucco, who has provided a commentary (pp. 87–96). A primary feature of this text, as analyzed by Mamiani, is the discussion by Newton of the action of electricity as the agent causing various types of phenomena, some of which are similar to phenomena discussed in the later Queries of the *Opticks*.

agitated by friction or percussion, receives magnetic virtue from a magnet more readily. Electric force in sufficiently electric bodies is greatly excited by friction. For electric spirit is emitted from a rubbed body to a great distance. And this spirit, in proportion to its various agitation, drives about light small bodies such as bits of feathers or of paper or of gold leaf in different ways, sometimes by attracting (and that through curved lines) and at other times by repelling, and at still other times by snatching [a bit of matter] along with itself in a whirling motion like some [sort of] wind.

That spirit is also emitted from some bodies (as from electrum [i.e., amber] and adamant [i.e., the hardest substance, diamond]) by heat alone without friction, and attracts small light bodies.

Furthermore, the same spirit constantly attracts bodies at small distances from the electric body, even without friction and heat, for stagnant liquids ascend in thin glass tubes immersed into them to their lowest parts, and do so in a vacuum just exactly as in open air. And the smaller the internal diameter of the tubes, the higher the liquids ascend. And if two glass planes, well polished and equal, come together at one end perpendicular to the horizon and do so at a very small angle, say of 10 or 20 or 40 minutes, and are immersed to their lowest parts into some stagnant liquid, the liquid will ascend between the glasses; and the narrower the interval is, the higher the liquid will ascend, its upper surface making a curve that is a rectangular hyperbola, of which one asymptote is along the concourse of the glasses and the other on the surface of the stagnant water. And if the concourse of the planes is inclined to the horizon, the hyperbola will come out oblique-angled, the asymptotes always being along the concourse of the glasses and on the surface of the stagnant water. Further, if either plane is set parallel to the horizon, and a drop of oil of oranges or of spirit of turpentine falls into it at one end, and the other glass is so put upon it that at one of its ends it touches the drop [and] at the opposite end comes together with the lower glass, the drop attracted by the glasses—first with a slow motion and then with an accelerated motion—will move toward the concourse of glasses, and meanwhile, if that concourse is elevated a little, the drop will ascend to the same [concourse] with a slower motion; or, if it is elevated more, the drop will rest in equilibrium between the force of its own gravity and the attraction of the glass; or, if the concourse of the glasses is still more elevated, the drop will change its course and will descend, its gravity overcoming the attraction of the glass. And these things are so in a vacuum exactly as in the open air. The first experiment was found by Mr. Taylor, the later one by Mr. Hauksbee; and both were demonstrated before the Royal

Society. When the two glasses were about three inches wide and twenty long, and at one of their ends were a sixteenth of an inch apart and at the other came together, the drop stood in equilibrium at a distance of 18, 16, 14, 12, 10, 8, 6, 5, 4, 3, 2 inches from the concourse of the glasses when the lower glass was inclined to the horizon at an angle of 0°15′, 0°25′, 0°35′, 0°45′, 1°, 1°45′, 2°45′, 4°, 6°, 10°, 22° respectively. Between the distances of three and of two inches from the concourse of the glasses, the drop came out oval and oblong and at length was divided into two drops. One of these, its own weight overcoming the attraction, descended; the other, with a swift and very greatly accelerated motion, ascended by the force of attraction to the concourse of the glasses.

By these experiments it is fully enough clear that glass at small distances always abounds in electric force, even without friction, and therefore abounds in an electric spirit which is diffused through its whole body and always surrounds the body with a small atmosphere, but never goes out far into the air unless it is stirred up by friction. And the case is the same for other electric bodies.

It is clear also that this force is by far strongest at the very surface of the glass. In the latest experiment, the force of attraction came out very nearly inversely in the ratio of the square of the distance [i.e., very nearly as the inverse square of the distance] of the drop of oil of oranges from the concourse of the glasses. And when that distance was only two inches, the force of attraction of the drop directly toward either glass was to the force of its weight (if I have calculated rightly) as about 120 to 1. The perpendicular attraction toward the two glasses was to the attraction tending toward the concourse of the glasses as 20 to 1⁄32, and this attraction equaled the force of weight of the oil for descending next to the plane of the glass, and this force was to the total weight toward the horizon as about 3 to 8. And with the calculations combined, the perpendicular attraction toward the glasses was to the weight toward the horizon as 240 to 1, and the perpendicular attraction toward one of the glasses was to the weight toward the horizon as 120 to 1. In this case let the width of the drop of oil be a third of an inch, and its weight will be about a seventh of a grain and its force of attraction toward the glass to the weight of grain will be as 120 to 7 or about 14 to 1. Further, if the thickness of the oil were one million times smaller, and its width a thousand times greater, the force of the total attraction—now increased (in accordance with the aforesaid experiment) in almost approximately the ratio of the square of the diminished thickness—would equal the weight of 140,000,000,000,000 grains; and the force of attraction of the circular part, whose diameter would be a third of an inch, would equal the weight of 14,000,000 grains or about 30,000 ounces, that is, 2,500 pounds. And this force abundantly suffices for the cohesion of the parts of a body.

And just as the earth takes on a spherical shape by the gravity of its parts toward the center, so the drops of liquids constantly affect spherical shapes by the electrical attraction of their parts toward themselves.

Further, the electric spirit, if strongly excited through the friction of an electric body, emits light. A spherical glass bottle whirled around its axis with a swift motion, and at the same time rubbed by an unmoved hand, shines where it is rubbed; and that is because of the agitation of the electric spirit which it emits through the friction. This spirit, after it has been emitted, quickly ceases to shine; but it will shine again anywhere in the circuit of the whirled-around bottle, and that at a distance of a quarter of an inch and sometimes of half an inch from the glass, provided that—having been whirled about at that distance with the glass—it impinges upon some unmoved body. This spirit, therefore, if sufficiently agitated, emits light, and so emits light in extremely hot bodies and in turn suffers a reaction from the emitted light. For whatever acts on something else suffers a reaction on itself. This is confirmed by the fact that all bodies grow warm in the light of the sun, and when sufficiently warmed emit light in turn. And certain phosphors are aroused in the light of the sun (or even in the light of clouds) to emit light, and that without the heat of a thick body. Between bodies and light there are action and reaction through the mediation of the electric spirit.

But a medium that can emit rays of light, and in turn be agitated by them, can inflect, refract, reflect, and sometimes stop the same rays. For the actions of bending or stopping rays are of the same kind as the actions of emitting them. The same thing that can give motion to rays can change their motion and increase or diminish the motion or take the motion away. When bodies grow warm in the light of the sun, the heat arises from the actions of reflecting, refracting, and stopping the rays. For light does not act upon bodies unless the bodies act at the same time upon the light.

We have shown in the *Opticks* that light is not reflected and refracted in only a single point, by falling upon the thick and solid particles of bodies, but is curved little by little, by a spirit that lies hidden in the bodies. Rays of light passing through glass and falling upon the further surface of the glass are partly reflected and partly refracted, even if the glass is placed in a vacuum. If the glass is immersed into some oil which abounds with the greatest force of reflecting, such as is oil of turpentine or oil of flaxseed or of cinnamon or of sassafras, the light which in a vacuum would be reflected passes out of the glass into the oil, and in nearly straight lines—reflection and refraction ceasing—and therefore is not reflected and refracted by the parts of the glass, but passes directly through the whole glass; and (if the oil is absent) passes into the vacuum before it is reflected or refracted, and therefore is reflected or refracted in a vacuum through the attraction of the glass, that is, through the action of some spirit that lies hidden in the glass and goes out of the glass into the vacuum to some small distance and draws back the rays of light into the glass. Further, that the rays of light are inflected in the vicinity of bodies, and at some distance from the bodies and without any contact with the bodies at all, is most certain from phenomena. They are inflected indeed with a serpentine motion, now approaching the bodies, now receding from them, and this is so whether the bodies are pellucid or opaque. And hence it is concluded also that the rays are agitated by some tremulous spirit in the vicinity of bodies. That this spirit moreover is of an electric kind is obvious from what has been said above.

We showed additionally in the *Opticks*, through most certain experiments, that rays of light in falling upon bodies excite a vibratory motion in that medium by which they are refracted and reflected; and that these vibrations are very short and swifter than light itself and, by successively pursuing and outrunnng the rays, accelerate and retard them successively and dispose them successively toward easy reflection and easy transmission through numberless equal intervals. The intervals of the vibrations of air through which sounds are generated are of one or more feet; but the intervals of the vibrations of this spirit (which it is now established is electric) do not equal a hundred-thousandth of an inch. Whence it is clear that this spirit is by far the most subtle. It penetrates at any rate a body of glass with the greatest of ease, since electrum [amber], if sufficiently excited through friction, attracts straws and similar light bodies through an interposed glass. It is clear also that the same spirit is extremely active and most suitable for warming bodies, since its vibrations are swifter than light itself.

From the things that have been said it can be understood why bodies made warm conserve their heat for a very long time, notwithstanding that the gross and cohering particles of hard bodies are not more suitable for undertaking and continuing motions than stones contiguous to one another and compressed in a great heap. For the electric spirit, which seems to pervade the pores of all bodies, receives vibratory motion very easily and conserves it for a very long time, and does so in the most hard and most dense bodies as well as in the most fluid and most rare, because this spirit must be more abundant in denser bodies, and its vibrations must be propagated through the total uniform spirit as far as the surface of the body, and there not to cease but be reflected and again be propagated through the whole and be reflected and to do this very often.

Hence also the heat of a body is most easily and most quickly propagated into contiguous bodies. For when the electric spirit that lies hidden in two bodies (a hot body and a cold body) becomes continuous through the contact of the bodies, its vibrations in the hot body will not be reflected at the common surface of the bodies but will be propagated into the second body through the continuous spirit.

Hence also bodies, according as their particles flow among themselves more easily or with more difficulty, will through the agitations of this spirit either grow soft and become ductile or flow and be turned into liquids. And some indeed, through the least agitations of this spirit, such as water, oil, spirit of wine, or quicksilver, others through greater agitations, such as wax, tallow [*the Latin is* "cebum"[26]], resin, bismuth, or tin, and still others through very great agitations, such as white clay, stones, copper, silver, gold, will flow and take on the form of liquids. And just as the particles of bodies either cohere more strongly or are separated from one another more easily, some of them through the agitation of this spirit are quickly turned into vapors and fumes, such as water, spirit of wine, spirit of turpentine, or spirit of urine; and others with more difficulty, such as oil, spirit of vitriol, sulphur, or quicksilver; others with the greatest difficulty as lead, copper, iron, stones; yet others with the strongest fire remain fixed, such as gold or adamant.

Through the action of the same spirit, some particles of bodies can attract one another more strongly, others less strongly, and thence can arise the various congregations and separations of particles in fermentations and digestions, especially if the particles are agitated by a slow heat. Heat in any case congregates homogeneous particles on account of greater attraction, and separates heterogeneous particles on account of lesser attraction. And through these operations live bodies attract parts of nutriment similar to themselves and are nourished. Between acid particles and fixed particles deprived of acids by fire there is the greatest attraction. From these are composed the less attractive particles, and from these through slow fermentation and continuous digestion are composed the bodies of animals, vegetables, and minerals.[27]

26. The Latin word for "tallow" is "sebum," but "cebum" is an occasional orthographic variant (see R. E. Latham, *Revised Medieval Latin Word-List* [London: Oxford University Press, 1965], p. 427).

27. A complete Latin transcript of both parts of this conclusion, together with some further information concerning the actual manuscripts, has been deposited in the Burndy Library of the Dibner Institute in Cambridge, Massachusetts.

CHAPTER TEN

How to Read the *Principia*

Some Useful Commentaries and Reference Works; Newton's Directions for Reading the Principia 10.1

The literature concerning Isaac Newton and his *Principia* is vast and ever increasing. Among those many works there are several that can be especially recommended as first guides to anyone who wishes to study the mathematical and technical structure of the *Principia*. These are D. T. Whiteside's "Before the *Principia*: The Maturing of Newton's Thoughts on Dynamical Astronomy" and "The Mathematical Principles Underlying Newton's *Principia*" (§1.2, n. 9 above) and Curtis Wilson's "The Newtonian Achievement" (cited in Abbreviations, pp. 9–10 above). The latter, in a brief compass, gives a splendid overall view of the development of Newton's ideas on dynamics and analyzes the chief astronomical achievements of the *Principia*. For pre-*Principia* texts and for book 1, every serious student of the *Principia* will want to study D. T. Whiteside's monumental edition of Newton's *Mathematical Papers* (Abbreviations, above), primarily volume 6, most of which is devoted to the text and translation of a first draft of book 1 of the *Principia*, together with a running commentary that provides mathematical glosses on individual propositions, and the definitions and laws.[1] Whiteside's notes constitute a major handbook of Newtonian mathematics and physics, with valuable discussions of the origins of Newton's ideas and methods and their reception and later use.

R. S. Westfall's monumental biography, *Never at Rest* (Abbreviations, above), presents in full detail every aspect of Newton's life and thought and is especially

1. Whiteside has also produced a useful facsimile edition of the tracts *De Motu*, the preliminary draft of book 1, and the demonstration sent to Locke: Whiteside, *Preliminary Manuscripts for Isaac Newton's 1687 "Principia"* (§1.1, n. 5 above).

valuable for tracing the development of Newton's scientific concepts and methods of solving problems. In addition to discussing Newton's physics and mathematics, Westfall gives proper weight and consideration to his other intellectual activities, such as prophecy, theology, biblical chronology, alchemy, and much else. For a more rapid overview of Newton's life and thought, there is a new biography by A. Rupert Hall, *Isaac Newton: Adventurer in Thought* (§1.1, n. 2 above). A chronology of Newton's science and mathematics and an overview of his scientific achievements may be found in my own article on Newton in the *Dictionary of Scientific Biography*. A general guide to Newton's thought and to recent scholarship is Derek Gjertsen, *The Newton Handbook* (London and New York: Routledge and Kegan Paul, 1986).

For a history of the *Principia* as a book, that is, the stages of composition, the publication of the three editions, and the reviews and revisions, see my *Introduction to Newton's "Principia"* (Abbreviations, above). The evolution of the text from the final printer's manuscript to the third edition is displayed in our edition of the *Principia* with variant readings (§1.3, n. 45 above).

On many thorny problems, the *Analytical View of Sir Isaac Newton's "Principia"* by Henry Lord Brougham and E. J. Routh (Abbreviations, above), though out of date on many topics, is still valuable. Another useful work, containing extended notes that expand the proofs and help make them intelligible, is the so-called "Jesuits'" edition, actually produced by two Minim Fathers, Thomas Le Seur and François Jacquier (1739–1742, 1760, 1780–1785, 1822, 1833), of which the best edition is the one published in Glasgow in 1822 and in 1833, corrected by J. M. F. Wright, whose two-volume *Commentary on Newton's "Principia"* (1833; §5.4, n. 17 above) is also a useful work. Readers will find a very helpful preparation for reading the *Principia* in Michael Mahoney's presentation of Newton's mathematical methods in proving some principal propositions of book 1.[2]

There are many editions in English of the first three sections of the *Principia*, with valuable commentaries, of which one of the earliest and perhaps still one of the most useful is John Clarke's *A Demonstration of Some of the Principal Sections of Sir Isaac Newton's "Principles of Natural Philosophy"* (1730; see §3.1, n. 14 above). Others, of the nineteenth century, intended for use as textbooks in the universities (e.g., Cambridge and Oxford), were produced by T. Newton (1805, 1825), John Carr (1821, 1825, 1826), J. M. F. Wright (1830), John Harrison Evans (1834–1835, 1837, 1843, 1855, 1871), William Whewell (1838), Harvey Goodwin (1846–1848, 1849, 1853, 1857, 1866), George Leigh Cooke (1850), Percival Frost (1854, 1863, 1878,

2. Michael Mahoney, "Algebraic vs. Geometric Techniques in Newton's Determination of Planetary Orbits," in *Action and Reaction: Proceedings of a Symposium to Commemorate the Tercentenary of Newton's "Principia,"* ed. Paul Theerman and Adele F. Seeff (Newark: University of Delaware Press, 1993), pp. 183–205.

1880, 1883, 1900). Of these, the fifth edition of Evans's work, edited by P. T. Main (1871), and the later editions of Frost's are the most valuable.[3] Frost gave students additional instruction on various aspects of the calculus and its applications to dynamics, with a series of exercises for students.[4]

For those who can read Russian, the translation of the *Principia* (1915–1916, 1936) by A. N. Krylov can be highly recommended for its excellent commentary.[5] Rouse Ball's *An Essay on Newton's "Principia"* (Abbreviations, above) contains a useful outline of the *Principia*, section by section, proposition by proposition (chap. 6).

Special mention needs to be made of Domenico Bertoloni Meli's *Equivalence and Priority: Newton versus Leibniz* (Oxford: Clarendon Press, 1993), which—in addition to new documentary material concerning Newton and Leibniz—contains many important insights concerning the concepts and methods of the *Principia*. Readers who know Spanish are well served by a very accurate new translation, with useful explanatory notes, by Eloy Rada García, *Principios matemáticos de la filosofía natural*, 2 vols. (Madrid: Alianza Editorial, 1987).

Five books, published after our work had been completed and too late to be of use to us, present all or part of the *Principia* either in present-day mathematical dress or in expanded form that will enable the beginner to follow Newton's principal argument. The late S. Chandrasekhar's *Newton's "Principia" for the Common Reader* (Oxford: Clarendon Press, 1995) is an essentially nonhistorical work by one of the world's foremost astrophysicists. Despite its title, it is written for readers who are well grounded in mathematics and celestial mechanics. Chandrasekhar basically devised his own proofs of propositions in books 1 and 3 and then compared his proofs with Newton's. Readers should be warned that Chandrasekhar disdainfully and cavalierly dismisses the whole corpus of historical Newtonian scholarship, relying exclusively on (and quoting extensively from) comments by scientists, many of whose statements on historical issues are long out of date and cannot stand the scrutiny of critical examination. He falls into traps which an examination of the

3. These works generally have two parts; the first an exposition of book 1, secs. 1–3, the second a supplement devoted to sec. 7 (props. 32, 36, 38) and sec. 8 (prop. 40), as in Frost, or sec. 9 (props. 43–45) and sec. 10 (props. 57–60, 64, 66 and corols. 1–17), as in Evans and Main. In the commentaries, some of these works tend to use the calculus. A complete list of these texts and commentaries is given in appendix 8 to our Latin edition of the *Principia* (§1.3, n. 45 above).

4. I own Alfred North Whitehead's copy of Frost, inscribed "A. N. Whitehead/Trin. Coll./March '81"; various exercises are ticked off, assigned for study—among them, p. 29, ex. 1, "Illustrate the terms 'tempore quovis finite' and 'constanter tendunt ad aequalitatem' employed in Lemma I. by taking the case of Lemma III. as an example," and ex. 4, "Shew that the volume of a right cone is one-third of the cylinder on the same base and of the same altitude."

5. I am grateful to two former students, Richard Kotz and Dennis Brezina, who have translated for me many of the valuable notes in the Krylov edition.

historical literature would have helped him to avoid, such as the date of Newton's revisions of book 1, sec. 2, or the form in which Newton expresses the second law. Chandrasekhar incorrectly equates Newton's "change in motion" (or change in quantity of motion or in momentum) with mass × acceleration. It is astonishing that a study of Newton's "mathematical principles" does not draw upon the extensive commentaries in D. T. Whiteside's monumental edition of Newton's *Mathematical Papers*.

An important new book for students of the *Principia* is J. Bruce Brackenridge's *The Key to Newton's Dynamics: The Kepler Problem and the "Principia,"* with English translations by Mary Ann Rossi (Berkeley, Los Angeles, London: University of California Press, 1995). Brackenridge is primarily concerned with the first three sections of book 1 of the *Principia* and the antecedent *De Motu*, his aim being to guide the general reader or beginning student through the sometimes labyrinthine proofs. Starting from the preliminary tract *De Motu*, Brackenridge introduces the reader to Newton's several methods of dealing with orbital motion in general and with Keplerian motion in particular, stressing Newton's method of curvature. Because this presentation draws heavily on the first edition of the *Principia*, a translation is given (prepared by Mary Ann Rossi) of the beginning portions of the first edition.

Of a wholly different sort is Dana Densmore's *Newton's "Principia": The Central Argument—Translations, Notes, and Expanded Proofs*, with translations and illustrations by William H. Donahue (Santa Fe, N.M.: Green Lion Press, 1995), with translated excerpts from books 1 and 3 of the *Principia* (as in the third edition). Newton's proofs and constructions are given in two forms: in Newton's words, with explanations of the several steps, and in a form easier for the nonmathematical reader or student to grasp. Elegant use of typography helps to keep the several kinds of text distinct. Like many earlier introductions to the *Principia* (such as the one by John Clarke in the early eighteenth century), this work presents the definitions, laws, and the first three sections of book 1 and then jumps ahead to book 3. Many readers will be grateful to Densmore for providing the missing steps and unstated assumptions in Newton's constructions and often laconic proofs. They will also be greatly aided by the geometric rules, methods, and procedures from Euclid and Apollonius that Densmore supplies, which Newton supposes readers will know.

The fourth work is François de Gandt's *Force and Geometry in Newton's "Principia,"* translated by Curtis Wilson (Princeton: Princeton University Press, 1995). De Gandt takes the reader through three subjects that are important for understanding the *Principia*: the concepts and methods of *De Motu* (1684), the ways in which forces were conceived by Newton's contemporaries and predecessors before the *Principia*, and Newton's mathematical methods in the *Principia*. In some ways, this work supplements but does not replace R. S. Westfall's classic *Force in Newton's Physics* (New York: American Elsevier, 1971), still a primary resource for anyone wishing to explore this topic.

The fifth work, available only in French, is Michel Blay's short (119 pp.) but elegant *Les "Principia" de Newton* (Paris: Presses universitaires de France, 1995). Blay explains the principles and methods of the first three sections of book 1, working out the demonstrations both in paraphrase of the original and in translation into the algorithm of the calculus. As in other works of this author, the style is concise and clear, with attention paid to both the contemporaneous state of knowledge and the allied philosophical problems.

Readers may look forward to the eventual publication of George Smith's *Companion to Newton's "Principia,"* based on a text prepared for a graduate seminar he has been giving at Tufts University. This work guides the reader through the *Principia* from beginning to end, explaining the concepts and proofs and bringing in information from present-day science. A notable feature is the attention paid to book 2 from a point of view different from that which is usually encountered. I also alert the reader once again to the incisive researches of Michael Nauenberg, which appeared too late to influence this Guide (see §3.9 above).

It has been mentioned earlier (§8.1 above) that at the start of book 3, Newton gives explicit instructions on how to read the *Principia*. His advice was "to read with care the Definitions, the Laws of Motion, and the first three sections of book 1, and then turn to this book 3 on the system of the world." He recommended "consulting at will the other propositions of books 1 and 2 which are referred to here." Newton repeated this advice to Richard Bentley in 1690, when the latter was preparing his Boyle Lectures, in which he advanced arguments from the recently published *Principia* for his goal of the *Confutation of Atheism*.[6]

In his own interleaved copy of the second edition of the *Principia*, Newton entered an emendation of the instructions, evidently intended for a third edition. There would be an additional sentence, reading: "Those who are not mathematically learned can read the Propositions also, and can consult mathematicians concerning the truth of the Demonstrations." It is possible that Newton was thinking of a real person, the philosopher John Locke, when he wrote this sentence. Newton's disciple the physicist J. T. Desaguliers has recorded that the "great Mr. Locke was the first who became a Newtonian Philosopher without the help of Geometry." He cited the example of Locke to prove that although the truth of the *Principia* "is supported by Mathematicks, yet its Physical Discoveries may

6. *Corresp.*, vol. 3; see the essay by Perry Miller in our edition of Newton's *Papers* (§1.2, n. 41 above).

be communicated without." Locke, according to Desaguliers (who reported that he had been told this story "several times by Sir *Isaac Newton* himself"), asked Christiaan Huygens "whether all the mathematical Propositions in Sir Isaac's *Principia* were true." On being told that "he might depend upon their Certainty, he took them for granted, and carefully examined the Reasonings and Corollaries drawn from them." In this way, Desaguliers concluded, Locke "became Master of all the Physics, and was fully convinc'd of the great Discoveries contained in that Book."[7]

10.2 *Some Features of Our Translation; A Note on the Diagrams*

The present translation is based on the third edition of the *Principia*, the final version approved by Isaac Newton. Some of the major differences between the texts of the three editions (1687, 1713, 1726) are indicated in notes, and translations are given of some of the passages where the printed texts of the first or second editions differ in major ways from that of the third edition. No attempt, however, has been made to reproduce all the variants given in our Latin edition with variant readings (1972), nor even to indicate all the places where variants occur. In particular, variant readings from Newton's personal (interleaved and annotated) copies have not been translated or even referenced, except on one or two occasions of particular significance. This is, in short, a rendering into English of the third edition, with some selected references to variant readings in the other two printed editions and with some variant texts given in an English version. This is not, in other words, a complete translation of (or substitute for) our Latin edition with variant readings.

Our goal has been to make the text of Newton's *Principia* available in straightforward English prose, keeping as close to the style and form of the original as possible. We have, however, tended to avoid those English words or phrases that are close or exact translations of Newton's Latin expressions but that have become archaic or are unfamiliar to a modern reader. We have, in other words, striven to produce a version of Newton's *Principia* that is a near English equivalent of the original, one that retains the structure and form of a seventeenth-century work, but that does not slavishly and literally imitate every single feature of the original. In the final stages of revision of our translation (as has been explained in the Preface to this Guide), we made many of our renditions conform to Motte's version to some degree, so as to establish a continuity with what has been a standard text for English readers for almost three centuries.

7. *Introduction*, pp. 147–148.

A problem facing any translator of the *Principia* is Newton's use of such archaic names for ratios as "subsesquiplicate,"[8] or "subsesquialterate," a particular source of bafflement for today's reader. Another is Newton's regular reference to operations with ratios, such as "divisim" or "ex aequo perturbate." A third is Newton's preference for rhetorical rather than symbolic expressions of proportion. A fourth is the use of technical terms that may be unfamiliar, such as "subtense." A fifth is the notation for raising a quantity to a power. A sixth is the compact notation in which ABC means AB × BC.

In dealing with such problems in this translation, the constant goal has been to make Newton's text available to today's readers, to those who cannot work through the Latin text. It has also been our aim to make an English equivalent of the *Principia* much as Newton wrote it, that is, to present Newton's text rather than an overly edited or modernized version of that text. This has two aspects. One is that we have not attempted to impose present-day standards on Newton's prose. For example, rather than impose a strict modern dichotomy between a vector "velocity" and a scalar "speed," we have generally followed Newton and have translated Newton's "velocitas" as "velocity" and "celeritas" as "speed." And it is the same for Newton's "ratio" and "proportio" (for which see §10.4 below).

Wherever possible, we have tried to follow Newton's own choices. For example, on a number of occasions, as in law 1, Newton writes of motion "uniformiter in directum" which we have rendered "uniformly straight forward," where "straight" has the sense of "in a straight or right line." We avoided introducing the words "straight line," which Newton himself had consciously abandoned after having written "in linea recta," in the tract *De Motu* in a preliminary version of the first law.[9] Clearly, motion "uniformly straight forward" must of necessity be rectilinear.

Of course, every translation is a continuous interpretation. In some cases, we have entered an interpretative word or phrase within square brackets to emphasize something that might not seem obvious and was not in Newton's text. In many other cases, however, we have simply expanded the text or added a word without square brackets where the context seemed to require it. For example, in the concluding paragraph of the scholium following the laws of motion, there is a sentence concluding "in omni instrumentorum usu," which we have translated by "in all examples of using devices or machines." The reason is that the context

8. In the third edition, "subsesquiplicate" appears in the phrase "in ratione subsesquiplicata" at the beginning of the proof of book 3, prop. 15. In the first edition, as in the manuscript, this was erroneously given as "sesquialtera," but it was changed in both Newton's annotated and interleaved copies to "subsesquialtera," and then altered to become "subsesquiplicata" in the second and third editions. The meaning is "in the inverse $3/2$ ratio."

9. Whiteside (*Math. Papers* 6:97), using British English, prefers "moving uniformly straight on."

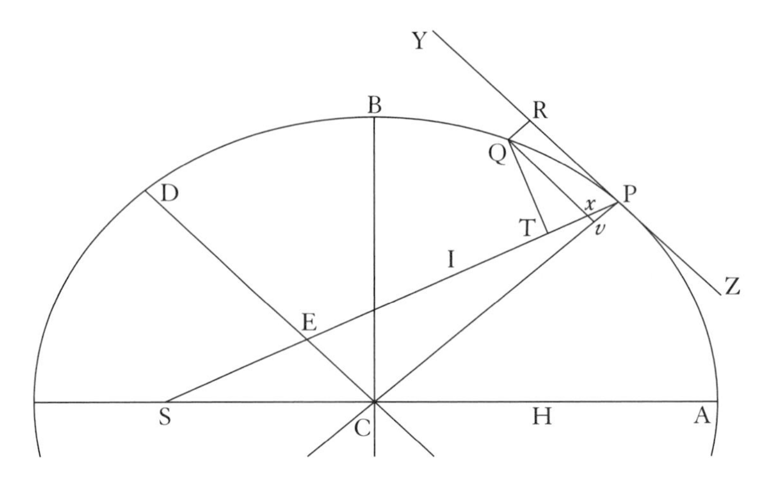

Figure 10.1. A portion of the diagram, as in the first edition of the *Principia*, used for both book 1, prop. 10, and book 1, prop. 11. Note that QR is parallel to CP, as in prop. 10, but in prop. 11 QR should be parallel to SP.

of the passage indicates that Newton is here referring to machines. Indeed, in a preceding sentence he actually does use the word "machinae."

The diagrams used in the present version are taken from the Motte-Cajori edition, published by the University of California Press. In some cases, we have either altered the diagram or introduced a wholly new one. A pair of examples will illustrate the problems we have sought to address: the diagram for book 1, prop. 11, discussed below, and the one for the scholium to book 2, prop. 10, discussed earlier in §7.4.

The diagram for prop. 11 has had a curious history. In the first edition, the same diagram was made to serve both prop. 10 and prop. 11 (and also prop. 16) of book 1. (See fig. 10.1.) The decision to combine the two was evidently made by Halley in order to save money. Many diagrams in the *Principia* do serve more than one proposition and hence have more parts than are needed for any single proposition.

In the case of prop. 10 and prop. 11, however, a single diagram cannot adequately serve both propositions. In prop. 10, the force is directed toward the center of the ellipse, and so the displacement QR should be parallel to the diameter CP; in prop. 11, however, the force is directed toward the focus S, and so the displacement should be parallel to the focal radius PS.

This same misleading use of a single diagram for both propositions occurs also in the second edition. In the third edition, the two diagrams were finally separated. Although a critical reading must take note that in the first two editions the diagrams do not fit all propositions, we should not exaggerate the effect. After

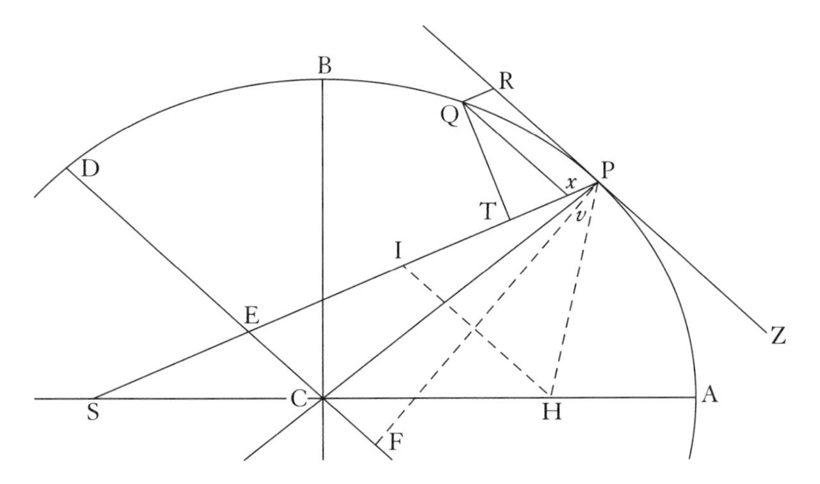

Figure 10.2. A portion of the diagram for book 1, prop. 11, as in the Motte-Cajori version. Among the obvious faults is the failure of line Qxv to extend from x to v.

all, the text makes it plain that in prop. 10, QR is to be taken as parallel to CP and that in prop. 11, QR is parallel to PS. Indeed, it must be assumed that many careful readers (Roger Cotes among them) followed the steps of the proofs as if the diagrams were correct, without even noting the problem of a just fit.

It is much the same for another aspect of prop. 11. In the Motte-Cajori edition, the point v is clearly marked, but a careful examination shows (see fig. 10.2) that the line Qxv does not go all the way to v, even though the text states plainly that this should be the case. Like many other readers, I went through the proof of this proposition many times, assuming that the line Qxv actually goes through to v, not noticing that the line stopped at the point x.[10] Probably most readers have "read" the diagram as if it were correctly drawn and met the requirements of Newton's text. Another fault in the Motte-Cajori diagram is that the angle IPR is not quite equal to the angle HPZ, as stated in the proof of prop. 11. The only solution to these problems was for us to redraw the diagram for book 1, prop. 11.

The portion xv of the line disappeared by stages in the successive nineteenth-century American editions of Motte's translation, finally becoming "enshrined" in the Motte-Cajori version. It is a historical irony that this American error was

10. The foundational study of Newton's diagrams is J. A. Lohne, "The Increasing Corruption of Newton's Diagrams," *History of Science* 6 (1967): 69–89. The diagram for book 1, prop. 11, has been the subject of an extensive scrutiny by J. Bruce Brackenridge, "The Defective Diagram as an Analytical Device in Newton's *Principia*," in *Religion, Science, and Worldview*, ed. Margaret Osler and Paul Lawrence Farber (Cambridge, London, New York: Cambridge University Press, 1985), pp. 61–93.

widely propagated in the first issue of the British pound note that bears Newton's portrait and honors his scientific achievement.[11]

10.3 *Some Technical Terms and Special Translations (including Newton's Use of "Rectangle" and "Solid")*

In dealing with quantities represented geometrically, Newton used the traditional language based upon the properties of geometric figures. Thus, for the product of two equal quantities, represented as lengths, Newton used the word "square," indicating the area of a figure with sides of equal length. Similarly, for the triple product of such a quantity, he used the designation for the volume, "cube." These two terms—"square" and "cube"—are still in common use today, but most people no longer think of the geometric origin and significance of these terms. Newton used two other terms that have a geometric origin. One was "rectangle," originally signifying the area of a rectangle whose sides were not equal and hence denoting the product of two unequal quantities. The other was "solid," similarly signifying the volume of a rectangular parallelepiped of which all three edges were not equal and hence denoting the product of three quantities that were not all equal. Because the *Principia* is a seventeenth-century work, despite its modern contexts and significance, and because our goal was to translate what Newton wrote and not to make a modernized version, we have kept Newton's terms "rectangle" and "solid," which should cause no confusion to a reader. It may be noted that *Webster's New International Dictionary, Second Edition*, states: "As the area of a *rectangle* is the product of its two dimensions, the term *rectangle* has been used for the *product of two factors*; as, the *rectangle* of a and b, that is, ab."[12]

The word "translate" presents a somewhat different problem. For most readers, this word will mean to render something from one language into another. We have, accordingly, tended to render "translate" in a mathematical sense by "transfer."[13]

A translator faces a constant temptation to "improve" Newton's text. For example, Newton sometimes uses "interval" for "line," or "circle" where he obviously intends "circular arc." An example may be found in book 1, prop. 41, of which a gloss has been given below in §10.12. Here Newton refers to a "circulus VR"

11. The problems with the Newton pound note and the difference between the two issues are discussed by Brackenridge in the work cited in the previous note.

12. We have written (e.g., book 1, sec. 1, lem. 11) of "the rectangle of AG and BD" rather than the more pedantically correct and old-fashioned "the rectangle under AG and BD."

13. Of course, these two words come from the same Latin root, one taking the stem of the present indicative "fero," the other the past participle "latus."

as well as "alii quivis circuli ID, KE." In context, it is plain that he is concerned not with whole circles, but only with the circular arcs VR, ID, and KE. We have not, however, believed it necessary to introduce editorial emendations so as to have Newton write of the "circular arc VR" or "any other circular arcs ID and KE."

We have, of course, everywhere expanded Newton's "compact" notation. For example, the proof of book 1, prop. 7, contains the statement that the quantity

$$\frac{\mathrm{QRL} \times \mathrm{PV}\ \textit{quad.}}{\mathrm{AV}\ \textit{quad.}}$$

is equal to "QT *quad.*" Here "QRL" is Newton's "compact" way of writing QR × RL. Accordingly, we have replaced QRL by the expanded form, QR × RL. We have also consistently replaced "*quad.*" or "*q*" by a superscript 2, so that the above expression is rendered by us as

$$\frac{\mathrm{QR} \times \mathrm{RL} \times \mathrm{PV}^2}{\mathrm{AV}^2}.$$

We have similarly introduced a superscript 3 for Newton's "*cub.*" We have also consistently replaced Newton's vinculum by parentheses.

An example of the dangers of using the vinculum as a sign of aggregation may be seen in book 1, prop. 81, ex. 3, where the Latin text of the first edition refers to a quantity

$$\frac{\mathrm{LB}^{1/2} \times \mathrm{SI}^{3/2} - \mathrm{LA}^{1/2} \times \mathrm{SI}^{3/2}}{\sqrt{2}}.$$

In the third edition, this expression was somewhat simplified, so as to become

$$\frac{\mathrm{SI}q}{\sqrt{2\mathrm{SI}}} \text{ in } \overline{\sqrt{\mathrm{LB}} - \sqrt{\mathrm{LA}}},$$

that is,

$$\frac{\mathrm{SI}^2}{\sqrt{(2\mathrm{SI})}} \text{ multiplied by } (\sqrt{\mathrm{LB}} - \sqrt{\mathrm{LA}}).$$

In this edition, as in the eighteenth- and nineteenth-century editions of Motte's translation, the vinculum was used to indicate that $\frac{\mathrm{SI}^2}{\sqrt{(2\mathrm{SI})}}$ was to be multiplied by the quantity $(\sqrt{\mathrm{LB}} - \sqrt{\mathrm{LA}})$ and not just by $\sqrt{\mathrm{LB}}$, following which $\sqrt{\mathrm{LA}}$ was to be subtracted from the product. Because the left end of the vinculum was printed rather close to the top of the square root sign, this expression was misread so that in the Motte-Cajori version (p. 209) $\sqrt{\overline{LB - \surd LA}}$ became

$$\surd(LB - \surd LA)$$

which a simple inspection shows to be incorrect.

Occasionally, Newton uses several different forms of notation in the same passage. For instance, in book 1, prop. 45, ex. 2, we find

$$T^n - nXT^{n-1} + \frac{nn - n}{2} XXT^{n-2} \, \&c.$$

where both index notations and doubling of letters (nn for n^2 and XX for X^2) occur in the same expression for an infinite series. Similarly, in corol. 2 to that same prop. 45, Newton writes, "A^{-2} or $\frac{1}{AA}$" rather than "A^{-2} or $\frac{1}{A^2}$."

As can be seen in this example, Newton himself sometimes did use index notation to indicate raising a quantity to a power. The *Principia* thus contains several different forms of designating a quantity raised to a power, such as A^2, A*q*., *Aquad.*, and AA. In our translation we have reduced all such expressions to a single form, using index notation (G^2, F^2, A^3, . . .) throughout.

A real problem arises in translating Newton's "in infinitum" and its variant, the Latin word "infinite." It would be a disservice to Newton's thought simply to use "infinitely" or "to infinity," although in at least one case Newton does intend that sense. That is, in the scholium to book 1, prop. 10, he writes of an ellipse with its center going off to infinity ("centro in infinitum abeunte"), not so much in the sense of proceeding to a limit but rather the direct transformation of the ellipse into a parabola. In almost all cases, however, Newton merely means (as in the proof of book 1, prop. 1) "indefinitely" or "without limit." That is, he is considering the limit in a mathematical sense. The cognate problem arises with quantities that (as Newton explains at the end of the final scholium to book 1, sec. 1) he speaks of "as minimally small or vanishing or ultimate," by which the reader should always "take care not to understand quantities that are determinate in magnitude" but should "think of quantities that are to be decreased without limit" ("diminuendas sine limite").

In the *Principia*, Newton variously uses "lineola" (e.g., proof of book 1, prop. 41), which we have rendered as "line-element"; "dato tempore quam minimo" (ibid.), which we have rendered as "in a given minimally small time"; "arcus quam minimus" (book 1, prop. 16), which we have rendered as "the minimally small arc"; and "linea minima" (book 1, prop. 14), which we have rendered as "the minimally small line." The sense of infinitesimally small is apparent.

We have tended to keep Newton's "tangunt" as "[they] touch," rather than introduce the circumlocution "are tangent to." Although some writers today use the term "apse," we have preferred "apsis," which has no confusing architectural connotation and also has the sanction of both tradition and modern dictionaries (e.g., *Webster's New International Dictionary, Second Edition; American Heritage Dictionary*).

In some cases, where an older technical term may be unfamiliar, we have added a modern gloss in square brackets, as in the example of planetary motion "in antecedentia" and "in consequentia." John Harris defined these terms in his *Lexicon Technicum* (London, 1704), s.v. "ANTECEDENCE IN, or *in Antecedentia*," as follows: A "Planet is *in Antecedence* when it appears to move, contrary to the usual Course or Order of the Signs of the *Zodiack*, as when it moves from *Taurus* towards *Aries*, &c." Contrariwise, if the planet were to "go from *Aries* to *Taurus* and hence to *Gemini*, &c.," the astronomers "say it goes *in Consequentia* or *in Consequence*."

10.4 *Some Trigonometric Terms ("Sine," "Cosine," "Tangent," "Versed Sine" and "Sagitta," "Subtense," "Subtangent"); "Ratio" versus "Proportion"; "Q.E.D.," "Q.E.F.," "Q.E.I.," and "Q.E.O."*

Today we think of trigonometric functions in terms of angles, but in Newton's day and well into the nineteenth century these functions were conceived and defined in terms of arcs rather than the corresponding angles. The older way of conceiving these terms may be seen in a popular textbook of the nineteenth century, Charles Hutton's *A Course in Mathematics*, originally published in 1798–1801 and reprinted and revised in 1812, 1818, 1841–1843, and then completely revised in 1849 and upgraded once more in 1860 by William Rutherford (London: William Tegg, 1860) for the students of the Royal Military Academy. In the final version, as in the earlier ones, the definition of "sine" (and "right sine") is given in terms of an arc: "The Sine, or Right Sine, of an arc" is "the line drawn from one extremity of the arc, perpendicular to the diameter passing through the other extremity." In Hutton's diagram, reproduced here as fig. 10.3, "BF is the sine of the arc AB, or of the arc BDE. And thus the sine BF is half the chord (BG) of the double arc (BAG)." As the diagram makes plain, these definitions are given for circular arcs.

Similarly, the *cosine* of the arc AB is CF, and AH is its tangent. The length of the *tangent* is defined as follows: the tangent of the arc AB is a line "touching the circle in one extremity of the arc" and "continued from thence to meet a line

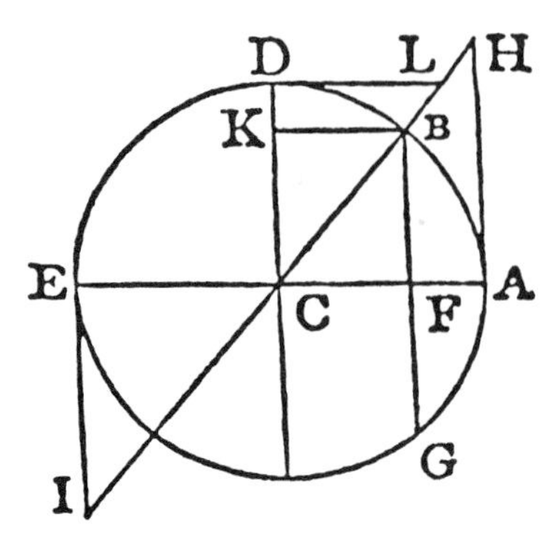

Figure 10.3. The trigonometric functions.

drawn from the centre through the other extremity" of the arc. This latter line (CH) is the *secant* of the arc AB.

A term with which today's readers may not be familiar is "versed sine." This is defined by Hutton as "the part of the diameter intercepted between an arc and its sine." AF is thus the versed sine of the arc AB, while EF is the versed sine of the arc EDB. It is obvious from the diagram that the versed sine plus the cosine adds up to the radius. Hence, since the radius is usually taken to be unity, the versed sine is equal to 1 minus the cosine. And, indeed, in many dictionaries, this latter relation has been taken as the primary definition of the term: the versed sine is simply stated to be 1 minus the cosine.

In more recent usage, these trigonometric functions are defined in terms of angles, rather than arcs. Thus, rather than refer to the sine of the arc AB, a modern definition would invoke the central angle determined by that arc or the angle ACB. Furthermore, a primary definition would more likely be made with respect to a right triangle BCF, where the sine of angle BCF would be the ratio of the sides $\frac{BF}{CB}$, rather than the length BF.

Another term with which today's readers may not be familiar is "sagitta." The traditional definition of that term is a line drawn (at any finite angle) to any curve or arc from the midpoint of the chord of that arc. Hence, in general, and notably in Newton's usage, the sagitta (a Latin word meaning "arrow") would generally be considered in reference to the whole arc, having the appearance of an arrow fitted to a bow. And, indeed, in John Harris's *Lexicon Technicum* (London, 1704), the definition of "sagitta" opens with a statement that it is "so called by some Writers, because 'tis like a Dart or Arrow standing on the Chord of the Ark." From a very strict point of view, there could be two essential differences between the sagitta and the versed sine: (1) the sagitta is defined for curves in general, whereas the versed sine is defined for circular arcs; (2) the versed sine is perpendicular to the chord, whereas the sagitta need not be. In a sense, then, the versed sine can be considered as a special case of a sagitta.

In actual practice, "sagitta" has been used in two quite different senses: (1) a line drawn from the midpoint of an arc to its chord, and (2) a line drawn to an arc from the midpoint of the chord of the arc. The first of these definitions appears in Percival Frost, *Newton's "Principia," with Notes and Illustrations, and a Collection of Problems* (4th ed., London, 1883, p. 100), where the sagitta is said to be the part of a line "intercepted between the chord and the arc" when a line is "drawn from the middle point of an arc of a curve, making a finite angle with the chord." The second occurs in John H. Evans and P. T. Main, *The First Three Sections of Newton's "Principia"* (London, 1871, p. 37), where the "sagitta of an arc" is defined as "a straight line drawn at a finite angle to its chord from the middle point of it to meet the arc."

In book 1 of the *Principia*, Newton declares specifically (prop. 6; see §10.8 below) that "the sagitta of the arc is understood to be drawn so as to bisect the chord." Since there are an unlimited number of such sagittas, Newton has also to specify the angle which the sagitta makes with the chord. He does so in prop. 6 by saying that the sagitta is also drawn so that "when produced," it will be so directed as to "pass through the center of forces."

In Newton's day, and notably in the *Principia*, the concept of the versed sine was extended so that the curve did not have to be the arc of a circle and the angle made with the chord did not have to be a right angle. In short, "versed sine" came to be used interchangeably with "sagitta," and the latter term was more usual. John Harris wrote that the sagitta, "in Mathematicks, is the same as the *Versed Sine* of any Ark." In Andrew Motte's English version and in the derivative edition revised by Florian Cajori, Newton's Latin term "sagitta" is apt to be rendered as "versed sine," as in book 1, prop. 1, corols. 4 and 5, and in book 2, prop. 10 (just before corol. 1).

"Subtense" is a mathematical term that has the general sense of being opposite to, of delimiting, of stretching or extending, and is the noun associated with the verb "to subtend." Examples would be the side of a triangle which "subtends" the opposite angle and a chord which "subtends" an arc and is—in the older terminology—the "subtense" of that arc. In the *Principia*, a subtense appears generally in relation to an arc and an angle. An example is the statement of book 1, lem. 11 (see fig. 10.4), where AB is an arc whose tangent is AD. In this case, the "angle of contact" at the point A is the curvilinear angle between the arc AB and the tangent line AD. Here BD is "the subtense of the angle of contact [angle BAD]," and the line AB is the subtense of the conterminous arc AB. Line AB, the subtense (or chord) of the arc AB, is technically conterminous because its ter-

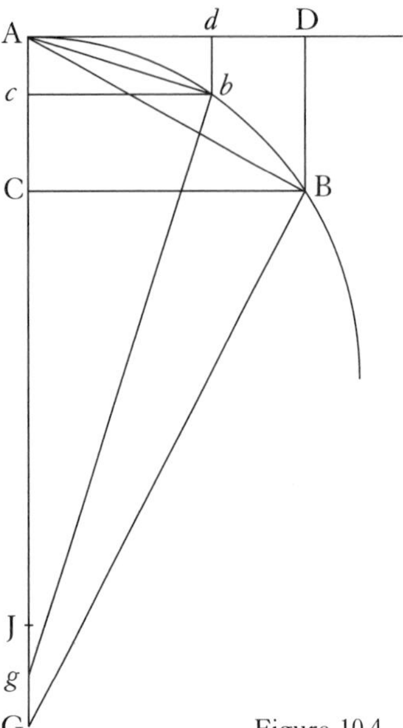

Figure 10.4. The subtense (as in book 1, lem. 11).

minal points A and B are the same as the "point of contact" (A) between the tangent and the curve and the intersection (B) of the curve and the subtense of the angle of contact. In the diagram, the subtense DB is drawn perpendicular to the tangent AD, but it could be drawn according to any fixed rule; the only requirement would be that the rule would have to specify that as the point B approaches the point A in the limit, D would also approach A in some specified manner.

The *subtangent*, not to be confused with the subtense, is that part of the axis of a curve which is contained between the tangent and an ordinate. As John Harris wrote in his *Lexicon Technicum* (vol. 2, London, 1710),[14] the subtangent "is a Line which determines the Intersection of the Tangent in the Axis." Although Harris defines the subtangent in terms of any "curve," his diagram, a redrawn version of which appears as fig. 10.5, shows the curve as a circle. In the diagram, TM is the tangent to the curve at the point M. PM is the ordinate to the axis through the point M. Accordingly, TP is the subtangent of the arc VM. The subtangent determines the point T where the tangent cuts the axis produced beyond V.

In the *Lexicon*, Harris shows that in a parabola (and paraboloids), the subtangent of the curve is positive in sign, so that "the Point of Intersection of the Tangent and the Axis falls on the side of the Ordinate where the Vertex of the

14. Harris did not include the subtangent in the original *Lexicon Technicum* (London, 1704) under "Trigonometry," nor was there a separate entry for it. This was compensated for in the supplementary volume, or vol. 2, which came out six years later.

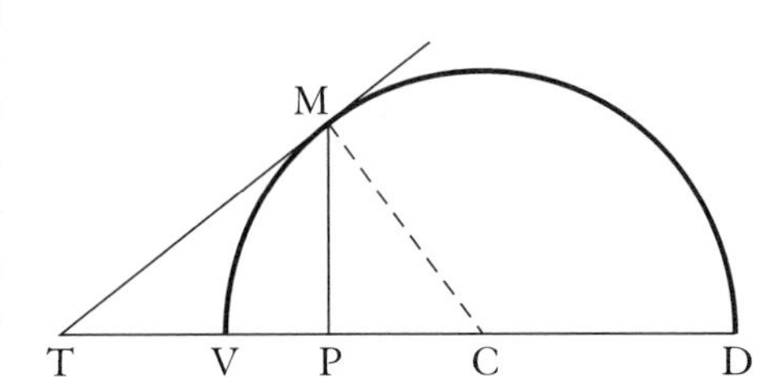

Figure 10.5. The subtangent. According to John Harris's *Lexicon Technicum*, vol. 2 (London, 1710), TM is the tangent to the curve VM at the point M, and PM an ordinate to the axis. Then TP is the subtangent, determining the point T "where the Tangent cuts the Axis produced beyond the Vertex of the Curve V."

Curve lies." But if the subtangent has a negative sign, the point of intersection will be "on the contrary side of the Ordinate, in respect of the Vertex or beginning of the Abscissa," as in the hyperbola and "hyperbiform figures." This pair of examples shows that defining the trigonometric functions in terms of arcs rather than angles enables them to be used readily for arcs of curves other than circles.

Today a distinction is often made between a "ratio" and a "proportion." In his *Lexicon Technicum* (1704), Harris held that a ratio is a comparison between two quantities "in respect of their Greatness or Smallness" and signifies "the *Rate, Reason* or *Proportion* in Quantity, that one hath to the other." Some writers, according to Harris, "confine Ratio or Reason only to 2 Numbers" and "call it *Proportion*, when it is between 3, 4, or more Numbers or Quantities." Nevertheless, he added, "the Word Proportion is often used" by "very good Authors" instead of "Ratio or Reason" to "express the Comparison of one single Quantity to another." Newton would appear to have been one of those "very good Authors" who tended to use "ratio" and "proportion" rather interchangeably. In our translation, we have followed Newton's usage, rendering his *proportio* as "proportion" and his *ratio* as "ratio."

The propositions in the *Principia*, numbered sequentially in each of the three books, are headed by a second category, "theorem" or "problem." The conclusion of the proof of a theorem is generally marked with the traditional "Q.E.D." of Euclidean geometry, standing for "Quod erat demonstrandum," or "What was to be demonstrated." For propositions which are problems, this generally becomes "Q.E.F." or "Quod erat faciendum," that is, "What was to be done." These two abbreviations occur in the editions of Euclid current at the time of the *Principia*, such as the one edited by Isaac Barrow, although there the propositions do not carry the second designation of "theorem" or "problem."

Occasionally, in the *Principia*, in the case of a proposition which is also a "problem," Newton uses the abbreviation "Q.E.I." or "Quod erat inveniendum," "What was to be found," and even "Q.E.O." or "Quod erat ostendendum," "What was to be shown." Some lemmas (e.g., book 1, lem. 16, in "cases" 1 and 3) use

"Q.E.I.," whereas others (e.g., book 1, lem. 11) use "Q.E.D." Some propositions (e.g., book 1, props. 5, 22, and 30; book 2, prop. 26), although designated "problems," use "Q.E.D." rather than "Q.E.F." In some instances (e.g., book 1, prop. 29, prob. 21; book 2, props. 36 and 40), there is neither the usual "Q.E.F." nor any of the other such conclusions. "Q.E.O." appears in book 1, prop. 65 (which is also theor. 25).

Book 3 differs from books 1 and 2 in that most of the propositions are developed by appeals to phenomena, and to the results of propositions in book 1, instead of being presented in terms of mathematical proof. Accordingly, in most of book 3, Newton has abandoned the use of "Q.E.D.," "Q.E.F.," and the others. The reader will note that Newton uses "Q.E.I." in prop. 27 (prob. 8) and uses "Q.E.D." at the end of corols. 1 and 2 to prop. 30 (prob. 11), but not in the proposition itself. "Q.E.I." appears in prop. 34 (prob. 15), prop. 38 (prob. 19), and prop. 41 (prob. 21), as does "Q.E.D." in prop. 35 (prob. 16), lems. 10 and 11, and "Q.E.F." in lem. 7.

10.5 *Newton's Way of Expressing Ratios and Proportions*

In the *Principia*, Newton commonly introduces ratios in terms of "ratio duplicata," "ratio triplicata," and so on, which used to be rendered in English as "duplicate [or doubled] ratio," "triplicate [or tripled] ratio," and so on. These older English forms of stating ratios might not cause too much of a problem for today's reader, who could easily guess that these are the ratios of the square, the cube, and so on. It would not be too difficult to guess that "in the halved ratio" refers to "the square root of the ratio" and means "as the square root of." But the case is quite different for such expressions as "sesquialterate" or "subsesquiplicate" ratios, which would constitute unnecessary chevaux-de-frise in a work that is difficult enough.

We may learn of the awkwardness of the older language for ratios by examining corol. 4 to book 1, sec. 1, lem. 11, where Newton shows us his own awareness of the problems of the language of ratios and actually even gives a preliminary definition that leads him to error. As he himself explains: "Rationem vero sesquiplicatam voco triplicatae subduplicatam, quae nempe ex simplici & subduplicata componitur." That is, "The ratio that I call sesquiplicate is the subduplicate of the triplicate, namely, the one that is compounded of the simple and the subduplicate." The sesquiplicate (or sesquialteral) ratio, this mouthful states, "is the halved of the tripled" or the ratio "compounded of the simple and the halved." In plain English, Newton says that the $3/2$ power (as in $A^{3/2}$) is the square root of the cube, that is, is compounded from the first power and the $1/2$ power (or $A \times A^{1/2} = A^{3/2}$).

This explanation did not appear in the first edition. A preliminary version was written out by Newton in both his annotated and his interleaved copy of the first edition. Here he wrote: "Rationem vero sesquiplicatam voco quae ex triplicata & subduplicata componitur, quamque alias sesquialteram dicunt." That is, "The ratio that I call sesquiplicate is compounded of the triplicate and the subduplicate, another term for which is 'sesquialterate.'" This gloss is confusing because it seems to mean that the $3/2$ power is compounded of the cube and the $1/2$ power ($A^{3/2} = A^3 \times A^{1/2}$), which is not correct and is quite different from the final version, that the $3/2$ power is compounded from the first power and the $1/2$ power.

In producing our translation, we were torn between two conflicting alternatives: to present an English equivalent of Newton's book much as he had written it or to produce a version that would be as readable as possible for today's reader. The main problem was whether to eliminate altogether such expressions as "sesquialterate," "subduplicate," and "doubled" (with respect to ratios) and to substitute their more modern equivalents, "three-halves power," "inverse square," and "square." At one time we contemplated keeping Newton's language and introducing a glossary. This alternative would have required a reader to thumb back and forth through the pages and would have presented yet one more obstacle to reading a book that has difficulties aplenty. In the end, following the advice of colleagues who would be the potential users of our translation, we decided (very reluctantly) to eliminate the older expressions. In justification of this procedure, it may be pointed out that in the *Principia*, Newton himself used both forms of expression. For example, in book 1, prop. 17, the statement of the proposition begins: "Posito quod vis centripeta sit reciproce proportionalis quadrato distantiae locorum a centro . . ." That is, "Supposing that the centripetal force is inversely proportional to the square of the distance of places from the center . . ." Similarly, in book 1, prop. 32, we find it supposed that "vis centripeta sit reciproce proportionalis quadrato distantiae locorum," or that a "force" may have some proportional relation to "the square of the distance of places." Elsewhere we find a quantity being "proportional to the squares of the velocities." Our reluctance to eliminate the older form of stating proportions was based on a desire to prevent the *Principia* from appearing in the guise of a recent rather than an older mathematical treatise, to avoid any semblance of treating his proportions (as one friendly critic put it) as if we were sweeping three hundred years of mathematical history under the rug.

In the end, however, we decided that the primary goal of our work was to make Newton's *Principia* as easily intelligible as possible to a modern reader. Accordingly, we decided to abandon the use of the older language for expressing ratios, eliminating such unintelligible terms as "sesquialterate" and "subduplicate." In short, we have translated this form of Newton's mathematical expressions just as we have translated Newton's Latin text. But we have not attempted to modernize the form of Newton's presentation, for example, by writing his proportions in the form of equations. Nor have we done any editorial rewriting of Newton's text, resisting the temptation to "improve" Newton's style. A particular case of this sort is Newton's reference to a quantity "decreasing" in a given ratio. Thus, book 1, prop. 65, presents a condition of "more than two bodies whose forces decrease as the squares of the distances from their centers." Such a statement, in which a force decreases as the square of the distance, implies that if the distance is doubled (say from 5 to 10), the force will be decreased by a factor of 2^2, or 4. In other words, it would have been somewhat simpler to say that the forces are inversely proportional to the squares of the distances. We, however, have left Newton's text in the form in which he wrote it.

Newton's cumbersome notation could easily lead him into difficulties. In the first edition, the proof of book 1, prop. 6, concludes:

> ... and thus the centripetal force is as QR directly and $SP^2 \times QT^2$ inversely, that is, as $\dfrac{SP^2 \times QT^2}{QR}$ inversely.

The reader will wonder why Newton did not say more simply: "The centripetal force is as $\dfrac{QR}{SP^2 \times QT^2}$." And, indeed, in using this result a little later in prop. 9, Newton himself (in the first edition) forgot the "inversely" (or "reciproce") and wrote: "therefore $\dfrac{QT^2 \times SP^2}{QR}$ is as SP^3 and so (by prop. 6, corol.) the centripetal force is as the cube of the distance SP." Newton corrected this in the later editions to read: "... the centripetal force is inversely as the cube of the distance SP."

In the *Principia*, Newton boldly set forth mixed proportions. Unlike his predecessors, he would state that a quantity is proportional to a quantity of a wholly different kind. That is, he did not restrict himself to proportions of the form: a given velocity is to a second velocity as the first time is to a second time. Newton abandoned the traditional rules, writing such expressions as:

> "the force is as the distance of the body from the center of the ellipse" (book 1, prop. 10, corol. 1);
>
> "the centripetal force is ... inversely as the square of the distance SP" (book 1, prop. 11);
>
> "a centripetal force that is inversely proportional to the square of the distance" (book 1, prop. 13, corol. 1).

In this way Newton freed the mathematical description of nature from the restricting bonds of ancient rules.

The Classic Ratios (as in Euclid's Elements, *Book 5)* **10.6**

Throughout the *Principia*, Newton uses the traditional theory of proportions that is found in book 5 of Euclid's *Elements*. In the following list the reader will find the names used in our version for the various transformations and combinations and other operations given in the Definitions in the *Elements*, together with the names used for them in the Latin and English versions of Euclid of the seventeenth and eighteenth centuries, as well as the name (or names) used by Newton in cases where he did not use the customary name, or where there is more than one name. In addition to a modern symbolic definition, a reference is given to one or more of the places where each appears in the *Principia*. Some examples from the *Principia* are included.

By Alternation

**Alternando*: by alternation (as in book 1, prop. 45, ex. 2). Newton uses *vicissim*; others use *permutando*. Sometimes this appears in English as "by alternate ratio."

If $A : B = C : D$, then

$$A : C = B : D.$$

By Conversion

**Convertendo*: by conversion (as in book 1, prop. 94, case 1; book 2, lem. 1). Newton uses *convertendo*.

If $A : B = C : D$, then

$$A : A - B = C : C - D.$$

By Composition

**Componendo*: by composition (as in book 1, prop. 1; prop. 20, case 2). Newton uses both *componendo* and *composite*.

If $A : B = C : D$, then

$$A + B : B = C + D : D.$$

By Separation

**Dividendo* (or *divisim*): by separation (as in book 1, prop. 20, case 1; book 2, prop. 6). Newton uses both *dividendo* and *divisim*.

If $A : B = C : D$, then

$A - B : B = C - D : D$.

From the Equality of the Ratios

**Ex aequo*: from the equality of the ratios (as in book 1, prop. 39, corol. 3; prop. 71).

If $A : B = C : D$ and

$E : F = G : H$, then

$A \times E : B \times F = C \times G : D \times H$.

In book 1, prop. 71, "PI will be to PF as RI to DF, and *pf* to *pi* as *df* or DF to *ri*, and ex aequo [or, from the equality of the ratios], $PI \times pf$ will be to $PF \times pi$ as RI to *ri*. . . ."

$PI : PF = RI : DF$

$pf : pi = DF : ri$.

Then, ex aequo [or, from the equality of the ratios],

$PI \times pf : PF \times pi = RI \times DF : DF \times ri = RI : ri$.

From the Equality of the Ratios in Inordinate Proportion

**Ex aequo perturbate*: from the equality of the ratios in inordinate proportion (as in book 1, lem. 24; book 2, prop. 30). Here "inordinate" is used in a sense understood by mathematicians in Newton's day and not to be confused with such present meanings as "excessive" or even "without rule or reason." T. L. Heath, in his edition of Euclid, used "perturbed" proportion.

If $A : B = F : G$ and

$B : C = E : F$, then

$A : C = E : G$.

In book 2, prop. 30, "F*g* is to D*d* as DK is to DF. Likewise F*h* is to F*g* as DF to CF, and ex aequo perturbate [from the equality of the ratios in inordinate proportion] F*h* or MN is to D*d* as DK to CF or CM. . . ."

$$\mathrm{F}g : \mathrm{D}d = \mathrm{DK} : \mathrm{DF} \text{ and}$$

$$\mathrm{F}h : \mathrm{F}g = \mathrm{DF} : \mathrm{CF}.$$

Then, ex aequo perturbate [or, from the equality of the ratios in inordinate proportion],

$$\mathrm{F}h : \mathrm{D}d = \mathrm{DK} : \mathrm{CF}.$$

In our translation we have referred to these ratios by both the traditional Latin names and an English equivalent in order to call the reader's attention to the fact that these are technical terms. The English equivalents by themselves, such as "by composition" or "by division," would hardly suggest to today's reader a special operation to be performed on ratios.

The tradition of using Latin names for these operations seems to have been set by Robert Simson's classic English version of Euclid, first published in 1756. Simson's text was "substantially reproduced" in Isaac Todhunter's edition, first published in 1862 and reprinted in the twentieth century in "Everyman's Library."

The designations "ex aequo" and "ex aequo perturbate," used by Newton, follow the Latin versions of Euclid made by Isaac Barrow and in current use in Newton's day. Newton's "ex aequo," however, has been rendered traditionally by "ex aequali," which—according to T. L. Heath—"must apparently mean *ex aequali distantia*." Newton's "ex aequo perturbate" is given in the Simson-Todhunter version as "*ex aequali in proportione perturbata seu inordinata*, from equality in perturbate or disorderly proportion." Heath suggests "in perturbed proportion."

The difference between "ex aequo" and "ex aequo perturbate" may be seen in the following example, in which two sets of ratios both yield the result that $A : D = E : H$.

$$A : B = E : F \qquad A : B = G : H$$

$$B : C = F : G \qquad B : C = F : G$$

$$C : D = G : H \qquad C : D = E : F$$

In the left-hand group (illustrating "ex aequo"), the diagonals from B to B and from C to C and also those from F to F and from G to G all slant from upper right to lower left. In the right-hand group (illustrating "ex aequo perturbate"), the diagonals from F to F and from G to G slope in the opposite direction. The direction from A to D is the same in both, but the direction from E to H in one set is the direct inverse of the direction from E to H in the other.

10.7 *Newton's Proofs; Limits and Quadratures; More on Fluxions in the* Principia

The first stage of reading Newton's proofs and constructions is to follow and try to understand the logic of the individual steps and the sequence of relations, involving often (as in the case of book 1, prop. 11, for which see §10.9 below) a set of cancellations, substitutions, and combinations. This first basic assignment will usually require the reader to transform Newton's rhetorical expressions of ratios and proportions into either equations, introducing a constant of proportionality, or symbolic statements of the form $\mathrm{A} : \mathrm{B} = \mathrm{C} : \mathrm{D}$, $\mathrm{A} : \mathrm{B} :: \mathrm{C} : \mathrm{D}$, or $\frac{\mathrm{A}}{\mathrm{B}} = \frac{\mathrm{C}}{\mathrm{D}}$.

An important part of reading Newton's proofs and constructions is to recognize that many of Newton's statements and diagrams either refer to instantaneous, momentaneous, infinitesimal quantities or hold in the limit as one point (P) and another (Q) come together, or as some quantity is "nascent" or "evanescent." Sometimes Newton gives us the information concerning this infinitesimal or limiting condition by stating explicitly (as in book 1, prop. 11) that at a certain stage P and Q are "coming together."

As has been mentioned (in §5.8 above), in certain propositions, Newton will alert the reader to the mathematical level of discourse by stating (as in book 1, prop. 41) that he is assuming that certain integrations can be performed, that the "quadrature" of certain curves (or finding the area under them) is possible. He also declares the fluxional or infinitesimal character of his argument by the use of such expressions as "minimally small," "indefinitely small," "particle of time," or "infinitely small." In book 1, prop. 41, for example (for which see §10.12 below), Newton refers to "the line-element IK, described in a given minimally small time," which is plainly $\frac{ds}{dt}$ and which Newton says explicitly is "as the velocity." (See §10.12 below.)

In some cases, because Newton did not develop and systematically use an algorithm of the calculus, a first reading may not disclose that certain quantities are fluxions or derivatives. They may appear as varying infinitesimal increments or decrements that can be translated into fluxions of different order, corresponding (as in book 2, prop. 10; see §7.3) to $\frac{de}{da}, \frac{d^2e}{da^2}, \frac{d^3e}{da^3}, \ldots$. This fluxional level of discourse has been noted in many commentaries on the *Principia* (e.g., in the edition of Le Seur and Jacquier); in some of them (e.g., Brougham and Routh) Newton's presentation is wholly recast into the more familiar Leibnizian and post-

Leibnizian algorithms. William Emerson's commentary on the *Principia* explains various propositions in terms of fluxions, both writing about fluxions and also actually introducing dotted letters for fluxions.[15] In D. T. Whiteside's commentaries (in his edition of the *Mathematical Papers*), many of Newton's arguments and results are translated directly into the algorithm of the differential calculus.

The highest level of reading and understanding would have to take account of the state of knowledge in Newton's day and would entail a recognition of the full significance of each proposition together with the ability to detect any flaws and limitations in Newton's arguments and to recognize the hidden assumptions. Additionally, there is required a knowledge of the treatment of the subject at hand in other writings of Newton, both published and in manuscript, plus a critical mathematical ability to assess and to evaluate the hidden as well as the overt assumptions in Newton's arguments.[16] In our own day, only one scholar has achieved anything approaching this kind of mastery of Newton's mathematical thought, D. T. Whiteside. Every serious student of Newton's *Principia* will need to make use of volumes 6 and 8 of his edition of Newton's *Mathematical Papers* and will regret that he has not given us a full commentary, as only he could have done, on the rest of book 2 and on book 3. Let us rejoice that he has made us so rich and unparalleled a scholarly gift.

Example No. 1: Book 1, Prop. 6 (Newton's Dynamical Measure of Force), with Notes on Prop. 1 (A Centrally Directed Force Acting on a Body with Uniform Linear Motion Will Produce Motion according to the Law of Areas) **10.8**

Each of Newton's proofs presents a series of hurdles to the prospective reader. First of all, there is the problem of following the logic of the proof, step by step,

15. Emerson's commentary was published in the third volume of the edition of Motte's translation published in London in 1819 ("carefully revised and corrected by William Davis"). With respect to book 1, prop. 81, ex. 2, he writes: ". . . this is easily calculated by fluxions. LD is a flowing quantity; the fluxion of the area is $\frac{\mathrm{ALB} \times \mathrm{SI}}{2} \times \mathrm{LD}^{-2} \times \mathrm{L\dot{D}}$." Discussing book 1, prop. 90, corols. 1 and 2, he writes: ". . . for the fluxion of the area is as $\mathrm{D}^{-n}\mathrm{D}$," without making any use of dotted letters. But then, in the very next paragraph, he notes that "the fluxion of that area is $= \frac{\mathrm{PF}}{\mathrm{PR}} \times \mathrm{P\dot{F}}$."

In discussing book 1, prop. 31, Emerson introduced $\dot{x}$ and $\dot{z}$ for the first fluxions of x and z. In discussing book 1, prop. 41, corol. 3, Emerson notes that "the force DF being as $\frac{1}{x^3}$, the fluxion of the area ABFD is $\frac{-\dot{x}}{x^3}$, and the fluent as $\frac{1}{2x^2}$; and corrected, the fluent is $= \frac{aa - xx}{2aaxx} =$ area ABFD."

16. On this topic, the comments of Michael Mahoney are especially to the point; see his "Algebraic vs. Geometric Techniques in Newton's Determination of Planetary Orbits" (n. 2 above).

Once again I alert the reader to the current researches of Michael Nauenberg, suggesting revisions of many standard interpretations of Newton's physics and mathematical astronomy.

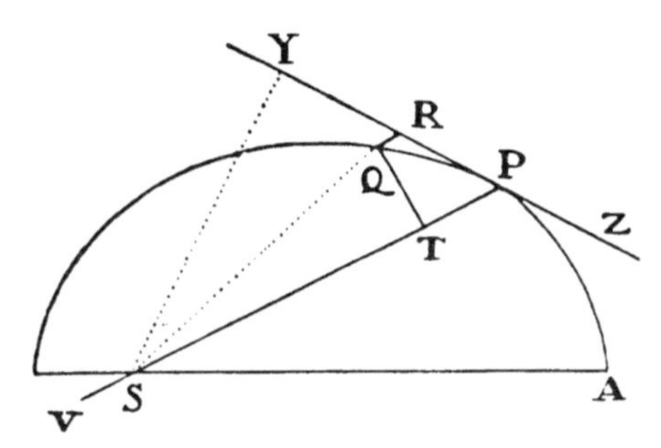

Figure 10.6a. The measure of centripetal force (book 1, prop. 6). This diagram is taken directly from the *Principia*.

which often means expanding each of Newton's statements from his compact or "telegraphic" style, in which he combines several geometric relations or proportions into a single expression. Often, it will be helpful to convert the proof into a modern form as a means of understanding Newton's goals and methods. In what follows, the basic steps of some significant propositions are presented as examples of how to proceed with a first reading of the *Principia*. The first is book 1, prop. 6, on Newton's measure of a force, together with a part of prop. 1 and its corol. 4, used by Newton in the proof of prop. 6. The reader's attention is called to a standard aspect of Newton's procedure: to set up a series of relations and then to determine their fate under some process of considering the limit, of allowing one point to approach another.

As the following examples will show, many of Newton's proofs and constructions fall into two classes. One comprises those that make overt use of the calculus, either introducing fluxions or depending on the quadrature, or integration, of certain specified curves, or that refer to a minimally small increment in a minimally small time. The other class makes use of the method of limits applied to a set of geometric or algebraic relations. Both types have been discussed in previous sections of this Guide. In some cases, a proposition and its proof will draw on previous propositions (or lemmas) that either have made explicit use of limits or include some indication that the conditions are not finite. An example is book 1, prop. 6, presented immediately below.

In prop. 6, Newton considers the motion of a particle along "any just-nascent arc" of a curve during "a minimally small time." The motion is said to occur in a nonresisting medium. The diagram (fig. 10.6a), taken from the *Principia*, shows a curve with the shape like that of an ellipse, with a force directed toward a fixed point S; the text does not specify any particular curve. The particle moves along the curved path because of the action of a centripetal force F directed toward S, causing it to depart from the straight line path PRY, tangent to the curve at P. If there

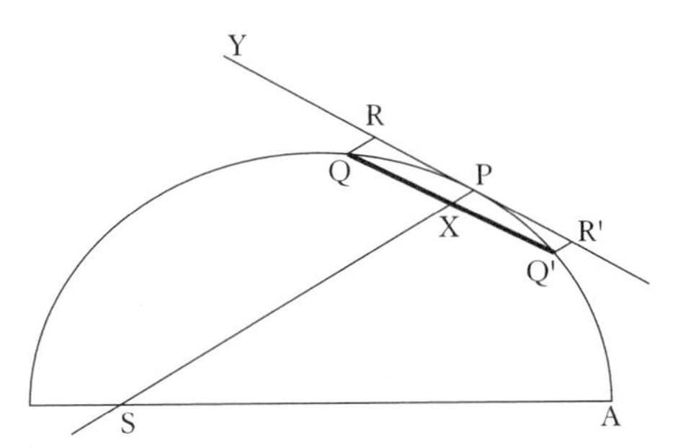

Figure 10.6b. An altered version of the diagram for book 1, prop. 6. Certain lines not needed for prop. 6 (but used in the corollaries) have been eliminated. A point Q′ has been introduced along with the chord QQ′, so that PX is the sagitta used by Newton in the statement of the proposition and in its proof.

were no centrally directed deflecting force, the particle (which is assumed to have a component of inertial motion) would move through some distance PR along the tangent in the minimally small time Δt. Since there is a centrally directed force F, however, the particle does not move in a straight line from P to R, but rather along the curve from P to Q. Basically, the proposition says that the magnitude of the force F may be measured by the amount of deflection from R to Q in a given time Δt. The location of Q must be specified by a line from R parallel to SP which intersects the curve at Q.

In the actual statement of the proposition, however, and in the proof, Newton does not measure the force in terms of the deflection RQ from the tangent; the measure is said to be directly proportional to a certain sagitta and inversely proportional to the square of the time. Then, in corol. 1, this measure is converted into a ratio based on the deflection RQ. The statement of prop. 6 and its proof will seem puzzling to a reader because the diagram (fig. 10.6a) does not contain a sagitta. The sagitta is either a line drawn, at some specified angle, from the midpoint of the chord of an arc to the arc or a line drawn from some point on the chord to the center of the arc. (See the discussion of sagitta and versed sine in §10.4 above.) In Newton's diagram, not only is there an absence of any sagitta, but there is no chord of the arc.

In order to help the reader more easily understand prop. 6 and its proof, the diagram has been modified by the addition of a point Q′, chosen so that the chord QQ′ is divided into two equal parts by the line from the center of forces S to the point P, which intersects the chord in a point X (see fig. 10.6b). The dotted lines SQ and SY and the line QT are not needed for the proof of prop. 6 itself but only for the corollaries; the extension of line PS to V has been discussed in §3.7 above in relation to the method of curvature. As in the case of many other diagrams in the *Principia*, for reasons of economy a single diagram was made to serve more than one proposition or a proposition and its corollaries. Sometimes this was overdone, with unfortunate results, as in having a single diagram serve both prop. 10 and prop. 11 in the first edition (see §10.2, esp. fig. 10.1, also the example of book 2, scholium to prop. 10 in §7.4).

In the modified diagram for prop. 6, the line XP is a sagitta of the arc, a quantity that appears in both the statement of prop. 6 and its proof. Although Newton uses the term "sagitta" both in stating prop. 6 and in proving this proposition, some translators of the *Principia*—from Andrew Motte onward—have changed Newton's "sagitta" to "versed sine." (See §10.4 above.) Newton specifies the sagitta by having it "drawn so as to bisect the chord" and oriented so that "when produced," it will "pass through the center of forces."

In considering the diagram and the proof, the reader should keep in mind that the conditions set by Newton are a "just-nascent arc" and "a minimally small time," the equivalent of considering what happens in the limit as P approaches Q. Under the conditions set by Newton, as we have seen above, he can conclude that the force (or the acceleration which is its measure) is proportional to the distance divided by the square of the time, in this case the deflection RQ (or the sagitta PX, which is equal to it) divided by the square of the time. This is basically what Newton has set out to prove in prop. 6, that "the centripetal force in the middle of the arc" will be "as the sagitta directly" and as "the square of the time" inversely. In terms of the altered diagram, the proposition states that the centripetal force F is proportional to $\frac{\mathrm{PX}}{\Delta t^2}$. This form of analysis has been called the parabolic approximation.

Newton's proof invokes, first of all, prop. 1, corol. 4, which states that the "forces by which any bodies in nonresisting spaces" are deflected from linear paths into curved orbits "are to one another" as the "sagittas of arcs described in equal times." In other words, the sagitta PX "in a given time" is as the force F. Furthermore, according to book 1, lem. 11, corols. 2 and 3, the arc in any small time increases in direct proportion to the time, so that the "sagitta is in the squared ratio of the time in which a body describes the arc."

The sagitta PX, therefore, is jointly proportional to the force F and the square of the time Δt. That is,

$$\mathrm{PX} \propto \mathrm{F} \times \Delta t^2.$$

Hence, as Newton says, "Take away from both sides the squared ratio of the time," whereupon

$$\mathrm{F} \propto \frac{\mathrm{PX}}{\Delta t^2}.$$

The force will be "as the sagitta [PX] directly and as the time twice [or as the square of the time] [Δt^2] inversely."

In a second version of this proof, Newton draws on lem. 11, corols. 2 and 3. By applying corol. 3 to any indefinitely small arc, the displacement will be found to be "as the force and the square of the time jointly." Thus in a time Δt, the sagitta of the arc described will be as the force and the square of the time.

In prop. 6, corol. 1, Newton uses Kepler's area law (which was the subject of the previously proved prop. 1) to find a measure of the time. That is, he takes the time to be proportional to the area of the sector SPQ. The line QT is drawn from Q perpendicular to SP. Under the conditions of the problem, the sector SPR can be considered a triangle, so that its arc may be taken to be proportional to SP × QT. Hence the square of the time is proportional to $SP^2 \times QT^2$, and the force is proportional to $\frac{QR}{SP^2 \times QT^2}$. This is Newton's dynamical measure of a force. We shall see it used below (§10.9) in steps 16 and 17 of the proof of book 1, prop. 11. It is a dynamical measure because it measures the force by its dynamical effect, the rate at which the action of the force causes the moving object to deviate from a linear inertial path. We may take note that in the original parabolic approximation in the 1660s, Newton considered motion along a circle, where the center of the circle is the center of the force. Now, in prop. 6, he is considering any orbital motion. As has been indicated earlier, this measure is not strictly dynamical, because it does not involve the factor of mass.

The completion of the discussion of prop. 6 and its corol. 1 requires a presentation of prop. 1 and its corol. 4. (An outline of the proof of prop. 1 has already been given in §5.4.) In prop. 1, it is assumed (as in prop. 6) that motion occurs in a space in which there is no resistance to motion; also that all bodies in motion have the same mass, or—which is equivalent—that they are all of unit mass. Newton begins by considering a body in uniform motion along a line AB*c*. If there were no force acting, then in equal intervals of time Δt, the body would traverse AB and B*c*, where AB = B*c*. The triangles SAB and SB*c* have equal areas because (fig. 10.7) they have equal bases AB and B*c* and the same altitude (a perpendicular dropped from S to the line AB*c*).

When the body reaches B, however, it does not proceed on to *c* because it is given a thrust or an impulse toward the point S. The subsequent motion is compounded of two parts: one component directed toward S, the other toward *c*. Applying the parallelogram rule, the body's path (fig. 10.8) is found to be along BC. Newton shows that if the body goes from B to C in the time Δt, then the area of triangle SBC will be the same as the area of triangle SB*c* (which is the same as the area of triangle SAB).

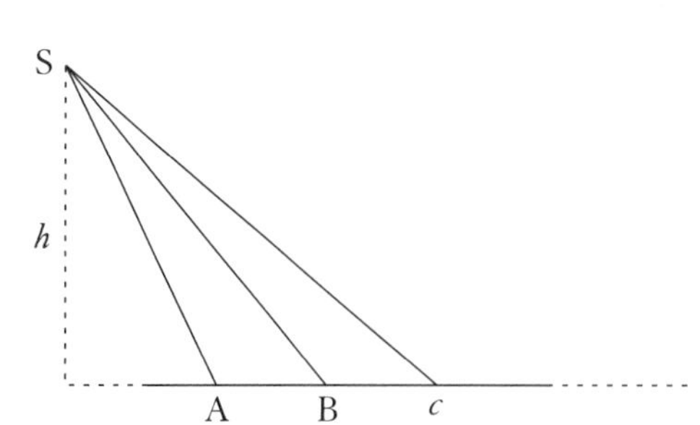

Figure 10.7. The law of inertia and the law of areas (book 1, prop. 1). Suppose a body to be moving along the line . . . AB*c* . . . with uniform motion so that the distances traversed in equal times (AB, B*c*, . . .) are equal. Then, if S is any point not on the line of motion, it follows that triangles ASB, BS*c*, . . . (which have equal bases and a common height *h*) will have the same area.

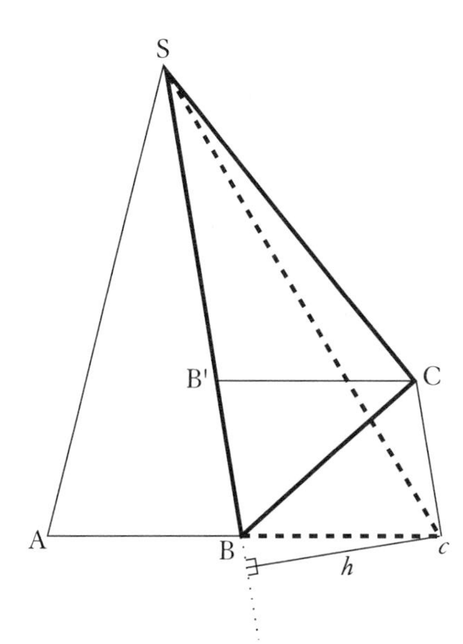

Figure 10.8. An illustration of the law of areas when a change in motion is produced by an impulsive force. In moving from B to C, after the action of an impulsive force at B, the body has two components of uniform motion. One is the original component of inertial motion, which in time *t* would by itself have carried the body from B to *c*; the other is the component of motion toward S, which by itself would in time *t* carry the body from B to B′. The combination of these two components of motion produces a displacement BC along the diagonal of the parallelogram BB′C*c*. It is relatively simple to prove that triangles SB*c* and SBC have equal areas: The two triangles have a common base SB. Since by construction *c*C is parallel to BB′, the two triangles have the same altitude *h*. Therefore, they have the same area.

When the body reaches C, it once again receives an impulse directed toward S and hence in time Δt will move from C to D rather than to *d*. As before, the area (fig. 10.9) of the new triangle SCD is proved to be equal in area to triangle SC*d*. The process continues, and in this way, Newton constructs a polygonal path in which the sides determine triangles of equal area. Newton lets "the number of triangles be increased and their width decreased indefinitely," whereupon the path will be a curve and the force will act "uninterruptedly." This completes Newton's proof that if a centrally directed force acts on a body with uniform linear motion, the result will be a motion according to the law of areas.

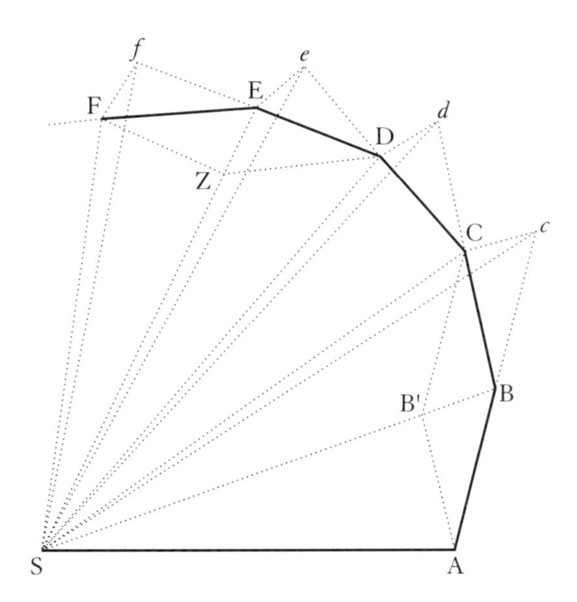

Figure 10.9. The polygonal path produced by a succession of impulsive forces (book 1, prop. 1). The triangles ASB, BSC, CSD, . . . , traced out in equal times, have equal areas.

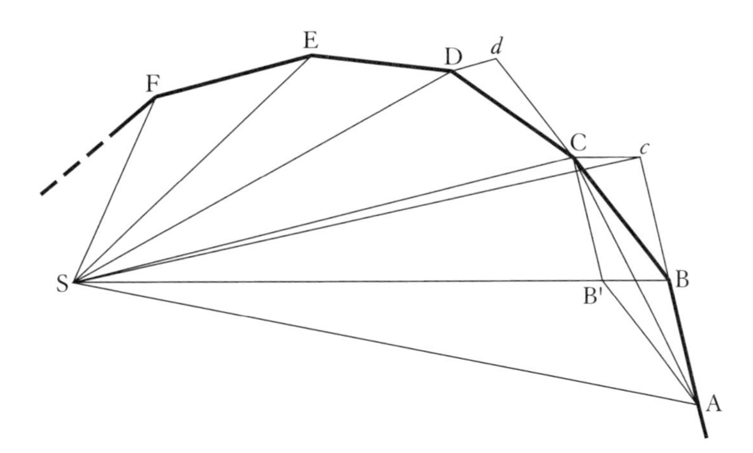

Figure 10.10. Diagram for book 1, prop. 1, corol. 4.

In corol. 4 to prop. 1, the magnitude of the impulsive force is measured by the displacement produced in some time Δt. In the diagram (fig. 10.10) this displacement is *c*C, which by construction is equal to BB′. Draw the line AC. Then, in the limit, as Δt becomes minimally small, AC will become the chord of an arc ABC and ½BB′ will become the sagitta of that arc. This is corol. 4, that in the limit, in an indefinitely small time, the sagittas of the arcs described will be proportional to the forces.

10.9 *Example No. 2: Book 1, Prop. 11 (The Direct Problem: Given an Ellipse, to Find the Force Directed toward a Focus)*

The next example, Newton's proof of prop. 11, may especially command our attention, not only because of its importance in Newton's increasing mastery of celestial dynamics but also because it provides a good illustration of Newton's procedure. In the following presentation of Newton's proof, we have given the justification of each step, so that a beginning student may be able to follow the proof in full detail. The reader will thus be able to see how the proof of book 1, prop. 11, follows a classic pattern, in which certain proportions are established from consideration of the geometry of conics and are then combined and simplified. That is, Newton here draws on principles or results from the geometry of conics as well as from plane geometry in a more or less traditional manner and then jumps ahead almost two millennia by introducing the seventeenth-century method of limits. In a way that is characteristic of the proofs in the *Principia*, this one sets forth certain proportions deriving from the geometry and then introduces the method of limits (no. 12 in the proof) by allowing the point Q to approach the point P and noting that Q*v* and Q*x* are "ultimately" equal.

The method of limits enters the proof not only at the stage of explicitly allowing P and Q to approach each other, but also in making use of the instantaneous measure of centripetal force taken from corols. 1 and 5 of prop. 6. It should be noted (following no. 12 in the proof) that the ease of cross-cancellation shows how powerful the old method of ratios and proportions can be.

The method of limits used in prop. 11 (like the proofs of props. 1 and 4) is intuitively understandable to anyone who can follow the sense of the geometry. Newton does not require that the reader first have mastered the concepts and methods of the differential calculus and have learned how to manipulate a specific form of algorithm $\left(\text{e.g., } \dot{x},\ \dot{y}, \text{ or } \frac{dx}{dt}, \frac{dy}{dt}\right)$. The first three sections of book 1, the part recommended to readers by Newton in the beginning of book 3, prove therefore to be less arcane for the general reader than a presentation using either fluxions or differentials. Of course, for the mathematically literate reader, Newton's method (especially in the later propositions) proved to be unnecessarily cumbersome and even masked the analytic significance of his results.[17]

Prop. 11 requires of the reader only some knowledge of plane geometry and some propositions from the geometry of conics. Additionally, there is needed the

17. An example is book 1, prop. 41; for details see *Math. Papers* 6:349–350 n. 209.

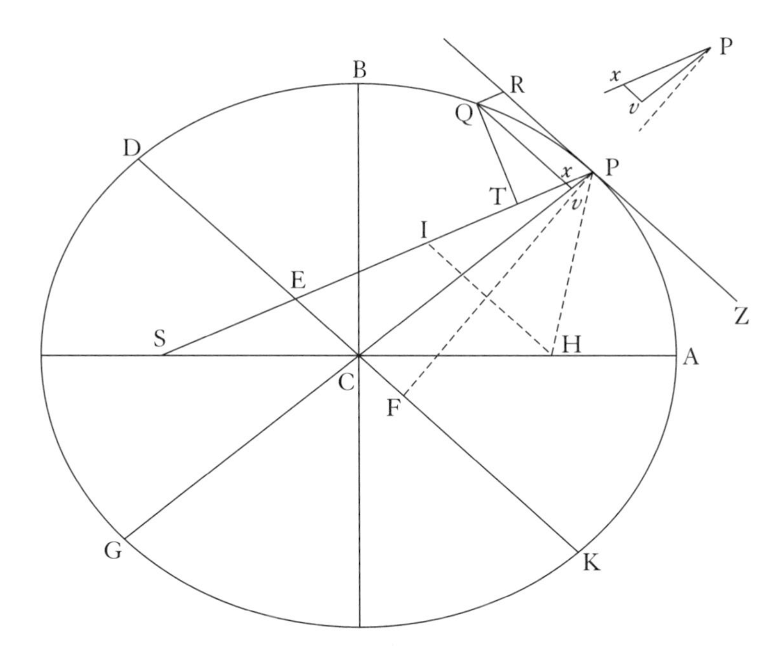

Figure 10.11. Diagram for book 1, prop. 11, the force producing motion in an elliptical orbit (after J. A. Lohne, "The Increasing Corruption of Newton's Diagrams," *History of Science* 6 [1967]: 81). S and H are the two foci of the ellipse; the central body (sun) is at S, the moving body (planet) at P. The diameter PCG is drawn from P through the center C of the ellipse; the other diameter DCK is drawn parallel to the tangent RP. The displacement RQ is drawn parallel to the focal radius PS, and Qxv is parallel to the tangent PR. PF is drawn perpendicular to the diameter DCK.

result of sec. 1 (lem. 7, corol. 2), that the ultimate ratio of an arc, its chord, and its tangent is a ratio of equality. In the first edition of the *Principia*, there was only a single proof of prop. 11. In the second edition, Newton added an alternative proof, as he did in the following prop. 12 (on motion along a hyperbola).

In prop. 11, it is required (see fig. 10.11) to find the force f with which a body moving along an ellipse is drawn toward a focus S. Let the body be at any point P on the ellipse. It is to be shown that

$$f \propto \frac{1}{\mathrm{SP}^2}$$

or that the force f is inversely proportional to SP^2.

At P draw tangent RPZ. Pick any point Q near P and draw the subtense QR $\parallel$ SP and the ordinate Qv $\parallel$ RPZ, meeting SP at x. Draw QT $\perp$ SP, and draw HP, where H is the other focus. Through the center C of the ellipse draw PCG, a diameter, and the conjugate diameter DCK, which (by definition) is parallel to RPZ, meeting SP in E. Finally, from H draw HI $\parallel$ RPZ and PF $\perp$ DK. Then

(1) PE = ½(PS + PI).

Proof: CS = CH, because the foci S and H are equidistant from the center C of the ellipse; therefore SE = EI, since EC $\parallel$ IH; PE = PS − SE = (PI + EI + SE) − SE = PI + EI = ½(2PI + 2EI) = ½[(PI + 2EI) + PI] = ½(PS + PI).

(2) PS + PI = 2AC.

Proof: Since the focal radii SP and HP meet the tangent RPZ at equal angles, $\angle$IPR = $\angle$HPZ, and since IH $\parallel$ RPZ (by construction), it follows that $\angle$HIP = $\angle$IPR = $\angle$HPZ; whence $\angle$HIP = $\angle$IHP, or ΔIPH is isosceles, and PI = PH. Hence PS + PI = PS + PH; but in an ellipse PS + PH = 2AC. Thus, PS + PI = 2AC.

(3) PE = AC (since PS + PI = 2PE).

(4) QR (= Px) : Pv = PE : PC.

Proof: This follows from ΔPxv ~ ΔPEC, a result of Qxv $\parallel$ EC; QR = Px (opposite sides of a parallelogram).

Next, multiply the two left terms of eq. 4 by L, the latus rectum of the ellipse, so that

(5) L × QR : L × Pv = PE : PC.

Whence, by using eq. 3 (PE = AC),

(6) L × QR : L × Pv = AC : PC.

Next, it is obvious that

(7) L × Pv : Gv × Pv = L : Gv

since the two left terms are merely the right-hand pair multiplied by Pv.

(8) $Gv \times Pv : Qv^2 = PC^2 : CD^2$.

Proof: This is an instance of a theorem from the geometry of conics: if the diameter (GP) of an ellipse be cut by an ordinate (Qv), then the product of the two parts of the diameter ($Gv \times Pv$) is to the square of that ordinate (Qv^2) as the square of the whole diameter ($4PC^2$) to the square of its conjugate ($4CD^2$). (See fig. 10.11.)

(9) $Qx^2 : QT^2 = PE^2 : PF^2$.

Proof: Since $Qxv \parallel IH \parallel EF$, $\angle QxT = \angle xEF$; further, $\angle QTx = 90° = \angle PFE$. Hence $\Delta QTx \sim \Delta PFE$, and $Qx : QT = PE : PF$, whence eq. 9 is obtained by squaring.

(10) By substituting $PE = CA$ (eq. 3) into eq. 9, $PE^2 : PF^2 = CA^2 : PF^2$.

(11) $CA^2 : PF^2 = CD^2 : CB^2$.

Proof: Lem. 12 states that all the "parallelograms described about any conjugate diameters of a given ellipse" are equal among themselves. This lemma applies to any ellipse and often takes one of two forms. In current language: the area of the parallelogram formed by tangents at the extremities of any pair of conjugate diameters is constant. Since a tangent drawn at either of the extremities of a diameter is always parallel to the conjugate diameter, in the particular case when the conjugate diameters are the major and minor axis, the parallelogram circumscribing the ellipse is a rectangle. Hence another form of the lemma is: the area of any parallelogram formed by tangents at the extremities of a pair of conjugate diameters is equal to the rectangle (product) of the major and minor axes. In the diagram, the parallelogram with DK as base and PF as altitude has the opposite side along the tangent RPZ; the sides are made up of tangents at D and K drawn from D and from K to the tangent RPZ. The area of this parallelogram is $DK \times PF$. Since this parallelogram is in area just half of the one to which the lemma refers, its area must be equal to half of the product of the major and minor axes (i.e., $2CA \times CB$); hence, $DK \times PF = 2CA \times CB$. But $DK = 2CD$, so that $2CD \times PF = 2CA \times CB$, or $\frac{CA}{PF} = \frac{CD}{CB}$, whence squaring both sides gives eq. 11.

Next, allow the point Q to approach the point P, and observe that by lem. 7, corol. 2, Qv and Qx are *ultimately* equal; hence *ultimately*,

$$(12)\quad Qv^2 : QT^2 = CD^2 : CB^2.$$

This is merely the result of combining eqs. 9, 10, and 11, and taking note that Qv^2 and Qx^2 are *ultimately* equal. Furthermore 2PC and Gv are also *ultimately* equal.

Thus, *ultimately*, the four basic proportions are:

$$*6.\quad L \times QR : L \times Pv = AC : PC$$

$$*7.\quad L \times Pv : Gv \times Pv = L : Gv$$

$$*8.\quad Gv \times Pv : Qv^2 = PC^2 : CD^2$$

$$*12.\quad Qv^2 : QT^2 = CD^2 : CB^2$$

These four proportions may be combined by multiplying together all the first terms, all the second terms, and so on. In the process, quantities that appear on the two sides of a colon (i.e., $L \times Pv$, $Gv \times Pv$; Qv^2; and PC, CD^2) may be canceled out. The result is

$$(13)\quad L \times QR : QT^2 = AC \times L \times PC : Gv \times CB^2$$

The latus rectum L is, by definition, the double ordinate through a focus. It is a well-known theorem (proved in any book on conics) that the semi–latus rectum is a third proportional to AC and BC; that is, $\dfrac{AC}{BC} = \dfrac{BC}{L/2}$, or $AC \times L = 2BC^2$, whence eq. 13 becomes

$$(14)\quad L \times QR : QT^2 = 2BC^2 \times PC : Gv \times CB^2 = 2PC : Gv.$$

Since 2PC and Gv are *ultimately* equal, the quantities proportional to them, $L \times QR$ and QT^2, must also be *ultimately* equal, or $L \times QR = QT^2$, that is,

$$(15)\quad L = \frac{QT^2}{QR}.$$

Now multiply both sides by SP^2, to get

$$(16)\quad L \times SP^2 = \frac{SP^2 \times QT^2}{QR}.$$

But according to corols. 1 and 5 of prop. 6, the centripetal force f will be inversely as $\frac{SP^2 \times QT^2}{QR}$, so that

$$(17) \quad f \propto \frac{1}{L \times SP^2}.$$

Since L is a constant for any given ellipse, the force f in any particular ellipse is inversely proportional to the square of the distance, or

$$(18) \quad f \propto \frac{1}{SP^2}.$$

Another proof

According to prop. 10, corol. 1, the force F directed to the center C is as the distance CP from the center C, or

$$(1) \quad f \propto CP.$$

From this proportionality, it is to be proved that the force f directed to a focus S of the ellipse is

$$(2) \quad f \propto \frac{1}{SP^2}.$$

Proof: Draw CP and SP as before, and the tangent RPZ; also the diameter DK $\parallel$ RPZ intersecting SP at E. Draw CO from C parallel to SP until it intersects tangent RPZ at O. Prop. 7, corol. 3, states that the force F tending to the center C is to the force f tending to any other point S in the following proportion:

$$(3) \quad F : f = CP \times SP^2 : OC^3.$$

Since ECOP is a parallelogram, OC = PE, so that eq. 3 becomes

$$(4) \quad F : f = CP \times SP^2 : PE^3.$$

Rewriting this proportion and introducing eq. 1

$$(5)\quad f = \frac{F \times PE^3}{CP \times SP^2} = \frac{CP \times PE^3}{CP \times SP^2} = \frac{PE^3}{SP^2}.$$

Since in any ellipse PE (= AC, from eq. 3 of the first proof) is a determined quantity, the result is that

$$(6)\quad f \propto \frac{1}{SP^2}.$$

10.10 *Example No. 3: A Theorem on Ellipses from the Theory of Conics, Needed for Book 1, Props. 10 and 11*

In book 1, prop. 10 (and notably in prop. 11), Newton makes use of a property of conics which he presents without proof, merely saying that the result in question comes from "the *Conics*." Here, as elsewhere in the *Principia*, Newton assumes the reader to be familiar with the principles of conics and of Euclid. In the eighteenth and nineteenth centuries, when Newton's treatise was still being read in British universities, authors of books on "conic sections"—for example, W. H. Besant, W. H. Drew, Isaac Milnes—supplied a proof of this particular theorem in order to help readers of the *Principia* who might be baffled by the problem of finding a proof. They even chose letters to designate points on the diagram so that the final result would appear in exactly the same form as in the *Principia*. Such a proof, in adapted form, would be as follows.

Given an ellipse, as in fig. 10.12, let DCK and PCG be conjugate diameters, so that DCK is parallel to the tangent at P and PCG is parallel to the tangent at D. From some point on the ellipse Q (chosen between P and D), draw Qv parallel to Ct and Qu parallel to CP. At Q, draw the tangent TQt meeting CD and CP (produced) in the points T and t. Then, since the diameters are conjugate and according to a property of the ellipse (proved in a previous proposition),

$$(1)\quad \frac{Cv}{CP} = \frac{CP}{CT} \quad \text{or} \quad CP^2 = Cv \times CT$$

$$(2)\quad \frac{Cu}{CD} = \frac{CD}{Ct} \quad \text{or} \quad CD^2 = Cu \times Ct = Qv \times Ct$$

so that

$$(3)\quad \frac{CD^2}{CP^2} = \frac{Qv \times Ct}{Cv \times CT}.$$

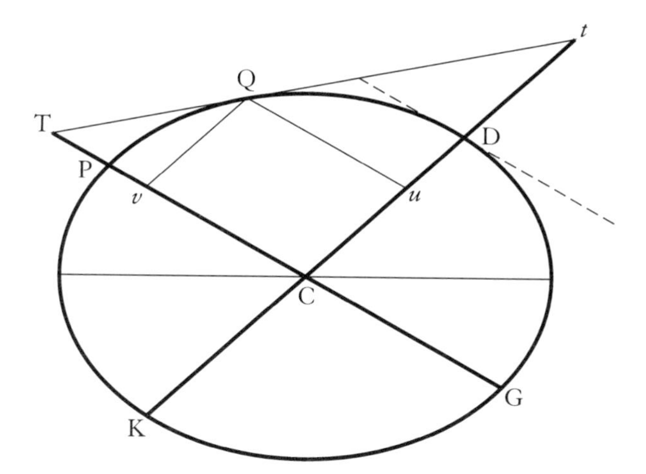

Figure 10.12. Diagram for a theorem on ellipses. The diameter PG is parallel to the tangent at D and Qu; Qv is parallel to the diameter DK. TQt is tangent to the ellipse at Q.

Since $\Delta v\mathrm{TQ} \sim \Delta \mathrm{CT}t$

$$(4) \quad \frac{\mathrm{C}t}{\mathrm{CT}} = \frac{\mathrm{Q}v}{v\mathrm{T}}$$

so that eq. 3 becomes

$$(5) \quad \frac{\mathrm{CD}^2}{\mathrm{CP}^2} = \frac{\mathrm{Q}v \times \mathrm{Q}v}{\mathrm{C}v \times v\mathrm{T}} = \frac{\mathrm{Q}v^2}{\mathrm{C}v \times v\mathrm{T}}.$$

By construction,

$$(6) \quad v\mathrm{T} = \mathrm{CT} - \mathrm{C}v$$

so that

$$(7) \quad \mathrm{C}v \times v\mathrm{T} = \mathrm{C}v \times \mathrm{CT} - \mathrm{C}v^2.$$

But from eq. 1, $\mathrm{C}v \times \mathrm{CT} = \mathrm{CP}^2$, so that eq. 7 becomes

$$(8) \quad \mathrm{C}v \times v\mathrm{T} = \mathrm{CP}^2 - \mathrm{C}v^2.$$

By construction, and since $\mathrm{CP} = \mathrm{CG}$, $\mathrm{CP} + \mathrm{C}v = \mathrm{G}v$, and $\mathrm{CP} - \mathrm{C}v = \mathrm{P}v$, so that

$$(9) \quad \mathrm{CP}^2 - \mathrm{C}v^2 = (\mathrm{CP} + \mathrm{C}v) \times (\mathrm{CP} - \mathrm{C}v) = \mathrm{P}v \times \mathrm{G}v.$$

Substitute the results of eq. 8 and eq. 9 in eq. 5 to get

$$(10)\quad \frac{CD^2}{CP^2} = \frac{Qv^2}{Pv \times Gv},$$

which is Newton's result,

$$(11)\quad Pv \times vG : Qv^2 = PC^2 : CD^2.$$

A slightly different proof is as follows. Again, we start with the same pair of ratios:

$$(1)\quad \frac{Cu}{CD} = \frac{CD}{Ct}$$

$$(2)\quad \frac{Cv}{CP} = \frac{CP}{CT} \quad \text{or} \quad \frac{CP}{Cv} = \frac{CT}{CP}.$$

Applying Euclid 6.20, corol.,

$$(1a)\quad \frac{Cu^2}{CD^2} = \frac{Cu}{Ct}$$

$$(2a)\quad \frac{CP^2}{Cv^2} = \frac{CT}{Cv}.$$

Since $\Delta QTv \sim \Delta tTC$

$$(3)\quad \frac{Tv}{TC} = \frac{Qv}{tC}$$

or, since $Cu = Qv$,

$$(4)\quad \frac{Cu}{Ct} = \frac{Tv}{TC}$$

and eq. 1a becomes

$$(5)\quad \frac{Qv^2}{CD^2} = \frac{Cu}{Ct} = \frac{Tv}{TC}.$$

Eq. 2a may be transformed by "the separation of ratios" (dividendo) to become

$$(6)\quad \frac{CT - Cv}{CT} = \frac{CP^2 - Cv^2}{CP^2}.$$

But, since $CT - Cv = Tv$,

$$(7) \quad \frac{Tv}{CT} = \frac{CP^2 - Cv^2}{CP^2}.$$

As before, we have $CP^2 - Cv^2 = Pv \times Gv$, so that

$$(8) \quad \frac{Tv}{CT} = \frac{Pv \times Gv}{CP^2}.$$

Applying eq. 5 to eq. 8 yields

$$(9) \quad \frac{Qv^2}{CD^2} = \frac{Pv \times Gv}{CP^2},$$

which is Newton's result,

$$(10) \quad Qv^2 : Pv \times Gv = CD^2 : CP^2.$$

Example No. 4: Book 1, Prop. 32 (How Far a Body Will Fall under the Action of an Inverse-Square Force) **10.11**

Prop. 32 sets forth a way to find how far a body will fall vertically downward in a given time (from rest) under the action of a centripetal force varying inversely as the square of the distance. This proposition, already present in *De Motu* as prob. 5, shows how Newton used the methods of geometry combined with the theory of limits to solve a problem in a way that may seem to us to be unnecessarily indirect. Here once again is evidence that the *Principia* is not written in the style or manner of Greek geometry.

Let S be a center of force, as in fig. 10.13. If the body does not fall in a straight line toward S, then (by prop. 13, corol. 1) it will describe a conic with focus at S. Case 1: Let that conic be the ellipse ARPB. On AB (the major axis of the ellipse), draw the semicircle ADB. Let the point C be the position of the falling body and draw CPD perpendicular to AB, cutting the ellipse at P and the semicircle at D; and draw SP, SD, and BD. Because an ellipse is the projection of a circle, the area ASD will be proportional to the area ASP; and since (by the law of areas, props. 1 and 2) ASP is proportional to the time, so will ASD be proportional to the time.

Thus far Newton has used the geometry of conics and the law of areas. But now he introduces the method of limits, the principles set forth in sec. 1 of book 1, on "first and ultimate ratios." Keeping the axis AB of the ellipse constant, let

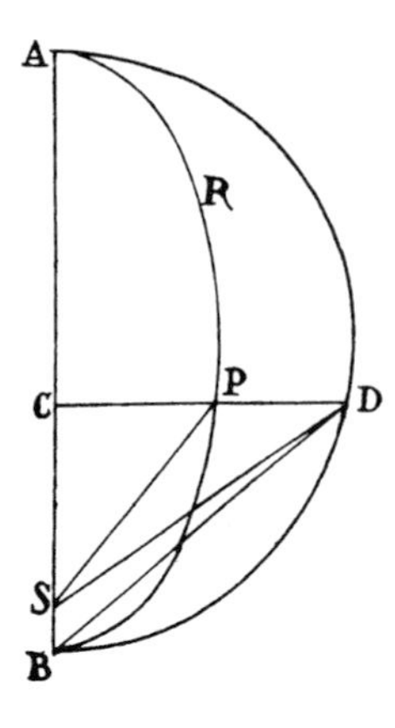

Figure 10.13. Free fall under the action of an inverse-square centripetal force (book 1, prop. 32).

the breadth of the ellipse be continually diminished; the area ASD will always be proportional to the time. In the limit, as the breadth is continually diminished, the orbital path APB will come to coincide with the axis AB, and the endpoint B of the axis will come to coincide with the focus S, so that the body will fall along the straight line AB to the point C and the area ABD (composed of the straight lines AB and BD and the circular arc AD) will be proportional to the time.

Therefore, Newton concludes: to find the distance a body will fall in a given time along the straight line AB under the action of an inverse-square force centered at S, draw a circle with AB as diameter, find a point D such that the area ADB (formed of the diameter AB, the line BD, and the arc AD) will be proportional to the time, and draw CD perpendicular to AB; then the distance AC will be the required distance.

Newton deals separately (case 2 and case 3) with the hyperbola and parabola. The demonstration proceeds along similar lines.

10.12 *Example No. 5: Book 1, Prop. 41 (To Find the Orbit in Which a Body Moves When Acted On by a Given Centripetal Force and Then to Find the Orbital Position at Any Specified Time)*

Prop. 41 is of special interest to readers of the *Principia* for many reasons, some of which have been discussed in §6.8. In some ways one of the most important propositions in the *Principia*, prop. 41 is the high point of Newton's analysis of the motions produced by forces, inasmuch as it makes possible the determination of the trajectory of a particle under a set of very general conditions of force. This proposition has been carefully studied and analyzed by D. T. Whiteside and Eric Aiton, and more recently by Herman Ehrlichson, whose presentations supplant the earlier discussion by Brougham and Routh.[18] As we shall see, Newton uses a somewhat complex geometric construction to solve a problem that ever since the *Principia* has usually been treated analytically. But as we shall see, Newton's seemingly geometric language enables us to translate his presentation rather directly into the more familiar algorithm of the Leibnizian calculus, especially since a careful reading shows how Newton is essentially using geometric expressions for differentials and integrals. Although Newton's proof may seem on a first reading to be like a rebus (to use an expression of René Dugas in relation to another proposition of book 1), nevertheless—as Whiteside observes—"with minimum recasting" Newton's "construction determines the elements of orbit in a known force-field $f(r)$ varying as the distance r, given the speed and direction of motion at some point."[19]

In prop. 41, Newton postulates a centripetal force of any kind (that is, a force that is some unspecified function of the distance) and sets the problem of finding the elements of the trajectory and the time of motion along a part of that trajectory. In order to help the reader to understand and follow Newton's development of prop. 41, I have divided the presentation into several distinct parts or levels. First of all, there is an outline-paraphrase of Newton's procedure in the language of today's calculus, following the reconstructions of D. T. Whiteside and Eric Aiton. Secondly, there is a step-by-step presentation of Newton's own development as presented in the *Principia*. At several stages in the latter there is, thirdly, a translation of certain crucial steps into the familiar language of the Leibnizian algorithm of the calculus.

Newton himself introduces the calculus in this proposition in several ways. First, the statement of the proposition explicitly refers to the ability to make a "quadrature" (or integration) of certain curves. Second, there are infinitesimal quantities and their ratios which are clearly fluxions or instances of the differential calculus. The final result entails specific quadratures or integrations.

In prop. 41, Newton postulates that the "centripetal force" is some unspecified function $f(r)$ of the distance r from the center C of force, and (see fig. 10.14) that the body whose trajectory is being sought has been shot out from some point V

18. Whiteside, "The Mathematical Principles Underlying Newton's *Principia*" (§1.2, n. 9 above); *Math. Papers* 6:344–356 nn. 204–215; Aiton, "The Contributions of Isaac Newton, Johann Bernoulli and Jakob Hermann to the Inverse Problem of Central Forces" (§5.7, n. 22 above), an updated version of an earlier paper, "The Inverse Problem of Central Forces" (§5.7, n. 22 above); see also Aiton's "The Solution of the Inverse Problem of Central Forces in Newton's *Principia*" (§6.4, n. 11 above); *Anal. View*, pp. 79–86. Herman Ehrlichson has addressed this problem in his "The Visualization of Quadratures in the Mystery of Corollary 3 to Proposition 41 [of book 1] of Newton's *Principia*" (§6.8, n. 25 above), which is especially valuable for the discussion of corol. 3 containing Newton's example of an inverse-cube force.

19. Anyone who wishes to make a more detailed study of prop. 41 and its significance should consult Whiteside's notes and commentary, which include a discussion of the reaction of Newton's contemporaries and successors; for the latter, the articles by Aiton (n. 18 above) are especially valuable.

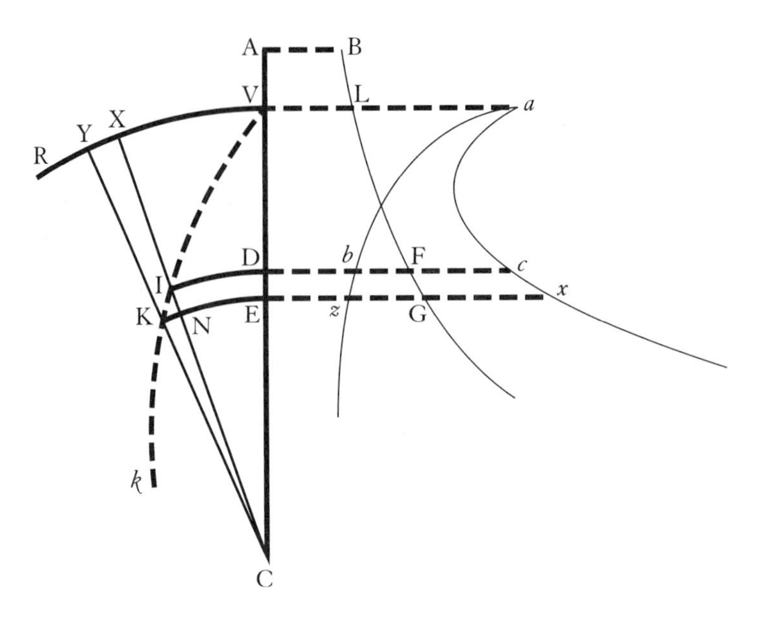

Figure 10.14. The orbit produced by a centripetal force (book 1, prop. 41).

with a specified initial velocity v_0. (Newton's method of determining the initial velocity v_0 has been discussed above in §6.9 and is presented again at step 10 below.) In fact, the actual condition specified by Newton is not the initial velocity per se, but rather the velocity at some point I of the trajectory. Let the distance from V to C be a. In the diagram (see fig. 10.14), VIKk is the resulting trajectory or orbit or curve of motion (Newton calls it the "trajectory"); VR is the arc of a circle with radius CV $= a$ and center at C. Newton picks two points I, K on the curve which are said to be "very close indeed," so that arc IK is an infinitesimal line-element or, in modern terminology, ds, traced out in a "given minimally small time" dt. The instantaneous speed v is $\frac{ds}{dt}$, so that

(i) $\mathrm{IK} = ds = vdt$.

Draw the lines CKY and CIX and let ID and KE be arcs of circles with center at C and with radii CK and CI. We may denote the angle VCX or VCI by the traditional θ, so that angle XCY or ICK will be $d\theta$. Then, if KC $= r$,

(ii) $\mathrm{KN} = rd\theta$

and the triangular area of sector ICK will be

(iii) area ICK $= (1/2)\text{base} \times \text{altitude} = (1/2)r(rd\theta)$.

Since the trajectory is produced by the action of a continually acting centripetal force, then (according to book 1, prop. 1) the time rate of change of area is constant, taken to be $\frac{Q}{2}$, or

$$\text{(iv)} \quad \text{area ICK} = \left(\frac{Q}{2}\right) dt.$$

Combining eq. iii and eq. iv we get

$$\text{(v)} \quad r^2\, d\theta = Q\, dt.$$

From the diagram, $KI^2 = KN^2 + IN^2$ or $ds^2 = (r\,d\theta)^2 + dr^2$. Thus

$$\text{(vi)} \quad v^2 = \left(\frac{ds}{dt}\right)^2 = r^2\left(\frac{d\theta}{dt}\right)^2 + \left(\frac{dr}{dt}\right)^2.$$

Substituting $\frac{d\theta}{dt} = \frac{Q}{r^2}$ from eq. v into eq. vi,

$$\text{(vii)} \quad v^2 = \frac{Q^2}{r^2} + \left(\frac{dr}{dt}\right)^2.$$

Solving for dt, we obtain

$$\text{(viii)} \quad dt = \frac{dr}{\sqrt{\left[v^2 - \left(\frac{Q^2}{r^2}\right)\right]}}.$$

From props. 39 and 40,

$$\text{(ix)} \quad v^2 = v_0^2 + 2\int_r^a f(r)\, dr.$$

Substituting the value for v^2 from eq. ix into eq. viii and then integrating, we get

$$\text{(x)} \quad \int_0^t dt = \int_0^r \frac{-dr}{\sqrt{\left(v_0^2 + 2\int_r^a f(r)dr - \frac{Q^2}{r^2}\right)}}.$$

In order to obtain r as a function of t, it is necessary to perform two integrations, to find two quadratures, the possibility of which has been postulated in the statement of the proposition. The two integrations give us r as a function of t.

The next part of the assignment is to determine θ as a function of t. To do so, we make use of eq. iii, that $\mathrm{KN} = r\,d\theta$, and eq. v, that $\mathrm{KN} = \mathrm{Q}\dfrac{dt}{r}$, to get

$$\text{(xi)}\quad d\theta = \frac{\mathrm{KN}}{r} = \mathrm{Q}\frac{dt}{r^2}$$

$$\text{(xii)}\quad dt = r^2\frac{d\theta}{\mathrm{Q}}.$$

Thus eq. x yields

$$\text{(xiii)}\quad \frac{r^2 d\theta}{\mathrm{Q}} = \frac{-dr}{\sqrt{\left(v_0^2 + 2\int_r^a f(r)dr - \frac{\mathrm{Q}^2}{r^2}\right)}},$$

so that

$$\text{(xiv)}\quad d\theta = \frac{-\mathrm{Q}dr}{r^2\sqrt{\left(v_0^2 + 2\int_r^a f(r)dr - \frac{\mathrm{Q}^2}{r^2}\right)}}.$$

Integrating this final equation gives θ as a function of t. We have thus found an expression for both r and θ as a function of t and have solved the problem of finding the trajectory.

We shall next see how the previous paraphrase is more or less a direct translation of Newton's geometric presentation into the language of the calculus. In doing so, it will be convenient to break up Newton's compact prose into a sequence of individual steps. The reader is reminded that Newton assumes the "quadrature" of certain specified curves and that he makes explicit use of fluxions: for example, in step 11, in which the velocity is quite overtly a fluxion of the displacement s, or $\dfrac{ds}{dt}$. It will be observed how much more compact the construction becomes when translated into the Leibnizian algorithm of the calculus.

The construction is developed in the following steps:

1. Let a centripetal force tend toward C. This force is some unspecified function of the distance r from C.
2. A body is released from a point V with some given initial velocity. Its trajectory is the curve VIK*k*.

3. Let two points, I and K, on the trajectory be "very close indeed to each other" ("sibi invicem vicinissima").
4. Describe the arc VR of a circle with C as center and a radius CV.
5. About C as center draw two additional arcs of circles, ID and KE, cutting the curve in the points I and K.
6. Draw the straight line CV, cutting the above two circles in the points D and E.
7. Draw the straight line CX cutting these two circles in the points I and N and cutting circle VR in the point X.
8. Draw the straight line CKY intersecting circle VR in the point Y.
9. Let the body move from V through I and K to *k*.
10. A point A on the line AVC is now defined as follows. A is the point from which a body is let fall and then will descend under the action of the force centered at C, so that when it reaches the point D, it will have acquired a velocity equal in magnitude to the magnitude of the velocity at I of the body whose trajectory is being sought. Note that I and D are located on the same circle drawn with center at C.

In step 10, Newton has in effect specified the magnitude of the initial velocity of the body whose trajectory is being sought. If we reverse the motions or trace the curve KIV backward, until D and I ultimately come to coincide with V, then the velocity at I will have become the initial velocity and its magnitude will be specified by the magnitude of the "free fall" of the second body from A to V. Of course, as has been noted earlier (in §6.9), the direction of motion of the falling body would have to be changed. The advantage of this geometric method of specifying the magnitude of the initial velocity is that it enables Newton to enter this quantity easily into the geometric frame of the demonstration.

11. Now, making use of prop. 39, "the line-element IK, described in a given minimally small time," will be proportional to the velocity, and so proportional to the square root of the area ABFD. Here "the line-element IK, described in a given minimally small time," is plainly ds. The area of ABDF can be found only by integration, by finding the "quadrature," as mentioned in the conditions of the proposition. We may assume that the value of the constant of proportionality (and all others) may be taken to be unity.[20]

20. On this point, see the article by Ehrlichson (n. 18 above).

Translating this result into the customary symbolism of the calculus, and taking the constant of proportionality to be 1,

$$\frac{ds}{dt} = \sqrt{\text{ABFD}}$$

$$s = \int \sqrt{\text{ABFD}}\,dt.$$

12. Thus the triangle ICK, proportional to the time, will be given.
13. Therefore, KN will be inversely proportional to IC.
14. Let some quantity Q be given and let IC = A.
15. Then $\text{KN} \propto \frac{\text{Q}}{\text{A}}$.
16. For convenience,[21] let $\frac{\text{Q}}{\text{A}}$ be denoted by Z.
17. Suppose Q has such a magnitude that, in some single case,

$$\sqrt{\text{ABFD}} : \text{Z} = \text{IK} : \text{KN}.$$

18. Then, in every case,

$$\sqrt{\text{ABFD}} : \text{Z} = \text{IK} : \text{KN}$$

and

19. $\text{ABFD} : \text{Z}^2 = \text{IK}^2 : \text{KN}^2$.
20. By the principles of ratios[22]

$$\text{ABFD} - \text{Z}^2 : \text{Z}^2 = \text{IK}^2 - \text{KN}^2 : \text{KN}^2.$$

21. But $\text{IK}^2 - \text{KN}^2 = \text{IN}^2$.
22. Therefore,

$$\sqrt{(\text{ABFD} - \text{Z}^2)} : \text{Z} = \text{IN} : \text{KN}.$$

21. As may be seen in our Latin edition of the *Principia*, the substitution of Z for $\frac{\text{Q}}{\text{A}}$ was introduced by Halley into Newton's manuscript before he sent the text of book 1 to the printer. Whiteside has suggested that Halley introduced this substitution of Z for $\frac{\text{Q}}{\text{A}}$ for the convenience of the printer; see *Math. Papers* 6:347 n. 205.

22. See §10.6 above.

23. Since $Z = \frac{Q}{A}$, the relation expressed in 22 becomes

$$\frac{KN}{IN} = \frac{\frac{Q}{A}}{\sqrt{(ABFD - Z^2)}}.$$

24. That is,

$$A \times KN = \frac{Q \times IN}{\sqrt{(ABFD - Z^2)}}.$$

25. Since

$$YX \times XC : A \times KN = CX^2 : A^2,$$

26. $XY \times XC = \dfrac{Q \times IN \times CX^2}{A^2\sqrt{(ABFD - Z^2)}}.$

This rather complex equation contains some familiar elements which can be identified by a process of translation. IN is the infinitesimal change dr in the radial distance r, which Newton denotes by A rather than r. The quantity $\frac{Q}{A}$, or Z, is thus what we would call $\frac{Q}{r}$ and is, in modern terms, the transverse component of the velocity or $r\frac{d\theta}{dt}$, or v_θ, so that

$$\frac{Q}{A^2} = \frac{r\frac{d\theta}{dt}}{r^2} = \frac{1}{r} \times \frac{d\theta}{dt}.$$

The quantity $\sqrt{(ABFD - Z^2)}$ in the denominator is the radial velocity or $\frac{dr}{dt}$, or v_r, since $ABFD = v^2$ and Z^2 is v_θ^2. ABFD has already been found by a process of quadrature or integration (props. 39 and 40).

Thus, in modern terms, as a result of step 26 we have an equation that gives dt in terms of both $d\theta$ and dr. Today we use the standard method of separating variables to get dt as a function of dr and as a function of $d\theta$; we would then solve each of these differential equations by integration. That is, by a separation of variables, we would get a pair of differential equations, one for dr, the other for $d\theta$. One of these would be

$$t = \int \frac{dr}{\sqrt{(ABFD - Z^2)}}.$$

This is essentially what Newton does. He considers the area VIC swept out by the radius vector and which (by book 1, prop. 1) is proportional to the time, and of which the rate is $\frac{Q}{2}$. Therefore,

$$t \times \frac{Q}{2} = \text{area VIC}.$$

Basically, then,

$$\text{area VIC} = \frac{Q}{2} \int \frac{dr}{\sqrt{(\text{ABFD} - Z^2)}},$$

so that the problem of integration is posed as a problem of finding an area VIC. Newton will do so by finding an equivalent area V*Dba*.

The means of getting to the next step is not made transparently clear by Newton. His method, corresponding to our separation of variables, is to define two new variable quantities, of which one is

$$Db = \frac{Q}{2\sqrt{(\text{ABFD} - Z^2)}},$$

the other

$$Dc = \frac{Q \times CX^2}{2A^2\sqrt{(\text{ABFD} - Z^2)}}.$$

In these two equations we may recognize the factors Q and $Q \times \frac{CX^2}{A^2}$ in step 26, with the addition of a 2 in the denominator. The reason for the 2 is that $\frac{Q}{2}$ is the rate at which the area VIC is being swept out. Newton has effectively separated the variables r and θ.

In prop. 41, Newton proceeds to develop the integrals of D*b* and D*c* simultaneously. It will be simpler, however, for us to consider each of them separately. Newton first proves that the integral of D*b* (which is the "generated area" VD*ba*) is equal to the area VIC. Since VIC is proportional to the time, VD*ba* is also proportional to the time. Hence, given any time t, we can find an area VB*ba* proportional to that time, determined by Newton's "Keplerian constant" Q divided by 2. This method of determining the area gives us D*b* as a function of the time t.

Another way of seeing Newton's procedure is to start with the equation $\sqrt{(\mathrm{ABFD} - \mathrm{Z}^2)} = \dfrac{dr}{dt}$ for the radial speed. Hence

$$\frac{dr}{dt} = \sqrt{(\mathrm{ABFD} - \mathrm{Z}^2)}$$

or

$$dt = \frac{dr}{\sqrt{(\mathrm{ABFD} - \mathrm{Z}^2)}}$$

$$t = \int \frac{dr}{\sqrt{(\mathrm{ABFD} - \mathrm{Z}^2)}}.$$

Since

$$\frac{\mathrm{Q}}{2}t = \text{area VIC},$$

$$\text{area VIC} = \frac{\mathrm{Q}}{2}\int \frac{dr}{\sqrt{(\mathrm{ABFD} - \mathrm{Z}^2)}}$$

$$d(\text{area VIC}) = \frac{\mathrm{Q}}{2\sqrt{(\mathrm{ABFD} - \mathrm{Z}^2)}}.$$

The quantity $\dfrac{\mathrm{Q}}{2\sqrt{(\mathrm{ABFD} - \mathrm{Z}^2)}}$ is what Newton has defined as Db.

Newton calls d(area VIC) the "nascent particle" of the area VIC and finds it to be equal to the "nascent particle" of the area VDba. Hence, by integration, VIC is equal to VDba.

Newton adopts a somewhat similar procedure to deal with the circular sector VCX in relation to the variable Bc. In this way he solves the second part of the problem, namely, to find θ as a function of t.

Let us now turn to the actual steps given in the *Principia* for bringing the demonstration to a conclusion after reaching step 26.

27. Erect DF perpendicular to AC.
28. Points b and c have been chosen so that

$$\mathrm{D}b = \frac{\mathrm{Q}}{2\sqrt{(\mathrm{ABFD} - \mathrm{Z}^2)}},$$

$$\mathrm{D}c = \frac{\mathrm{Q} \times \mathrm{CX}^2}{2\mathrm{A}^2\sqrt{(\mathrm{ABFD} - \mathrm{Z}^2)}}.$$

29. These points will trace out the curves[23] ab and ac.
30. At V, draw Va perpendicular to AC, so as to cut off the curvilinear areas VDba and VDca.
31. Draw the ordinates Ez and Ex.
32. Then
    Db × IN = area of DbZE = ½A × KN = area of triangle ICK,
    DC × IN = area of DcxE = ½YX × XC = area of triangle XCY.
33. Next Newton observes that "the nascent particles DbzE and ICK of the areas VDba and VIC are always equal" and "the nascent particles DcxE and XCY of the areas VDca and VCX are always equal."
34. Therefore, the generated area VDba will be equal to the generated area VIC.

That is, since $d(\mathrm{VD}ba) = d(\mathrm{VIC})$, their integrals are equal.

35. But the generated area VIC is proportional to the time, according to book 1, prop. 1, so that the generated area VDba will also be proportional to the time.
36. And therefore the generated area VDca will be equal to the generated sector VCX.
37. Accordingly, "given any time that has elapsed since the body set out from place V, the area VDba proportional to it will be given."
38. And thus "the body's height CD or CI will be given, and the area VDca."

A somewhat similar procedure enables Newton to determine the angle VIC (or θ). That is,

39. The sector VCX (equal to the area VDca), together with its angle VCI, will be given.

23. In the first two editions, the diagram is incorrect, since the curves for Db and Dc meet the extension of the line which—in the third edition—is marked VL in two distinct points, a and d, rather than converging in a single point a. This error was evidently corrected by Pemberton in the third edition. See *Math. Papers* 6:347 n. 297.

In both the first and second editions, the same diagram had been used for both prop. 40 and prop. 41. In the third edition, a wholly new diagram was made for prop. 40, showing only the portion of the whole diagram needed for that proposition.

It is easy to see why the diagram in the early editions is incorrect. The intersections of the curves Db and Dc with VL, which is perpendicular to CV, occur when D coincides with V. Under that condition, the "height" A (equal to CD or CI) will be equal to CX. Accordingly, the factor $\frac{\mathrm{CX}^2}{\mathrm{A}^2}$ will be equal to unity, so that Db = Dc.

40. And thus, "given the angle VCI and the height CI, the place I will be given, in which the body will be found at the end of that time."

Example No. 6: Book 1, Lem. 29 (An Example of the Calculus or of Fluxions in Geometric Form) **10.13**

Props. 79–81 and lem. 29, dealing with various aspects of attractions, may be of special interest in revealing in an overt manner how Newton's evanescent, nascent, or indefinitely small quantities are expressions of fluxions and how Newton's geometric presentations may be read in the more familiar algorithm of the calculus. Lem. 29 and the succeeding props. 79 and 80 of book 1 may serve to indicate both the geometric-fluxional quality of the *Principia* and the actual advantages of the Leibnizian algorithm for problems in mathematical dynamics.

Lem. 29 states (see fig. 10.15) that a circle AEB is to be described with center S; then two additional circles EF and *ef* are to be described with center P, "cutting the first circle in E and *e*, and cutting the line PS in F and *f*." Next, "the perpendiculars ED and *ed* are dropped to PS." Then, "if the distance between the arcs EF and *ef* is supposed to be diminished indefinitely, the ultimate ratio of the evanescent line D*d* to the evanescent line F*f* is the same as that of line PE to line PS."

The infinitesimal level of mathematical discourse is declared in the enunciation of the lemma since the distance F*f* between two circular arcs EF and *ef* is to be indefinitely diminished; then the last ratio of the evanescent line D*d* to the evanescent line F*f* equals the ratio of PE to PS. In the diagram, the evanescent arc E*e* is treated (as in the limit) as the chord. The geometrical conditions yield the result that $\Delta \mathrm{DTE} \sim \Delta d\mathrm{T}e \sim \Delta \mathrm{DES}$. Therefore

$$(1)\quad \frac{\mathrm{D}d}{\mathrm{E}e} = \frac{\mathrm{DT}}{\mathrm{TE}} = \frac{\mathrm{DE}}{\mathrm{ES}}.$$

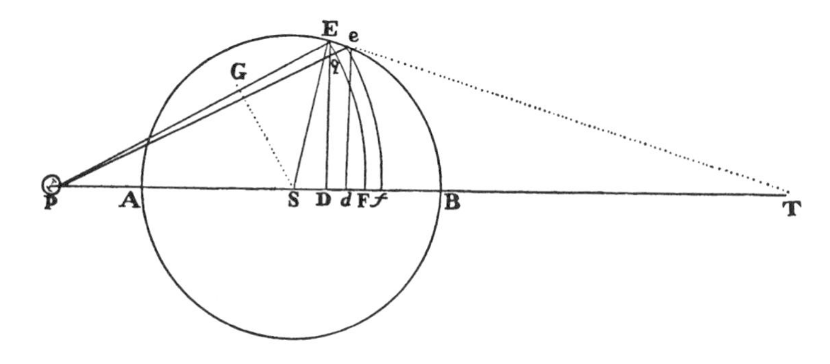

Figure 10.15. Diagram for book 1, lem. 29.

But (by lem. 8 and lem. 7, corol. 3) $\Delta \mathrm{E}eq \sim \Delta \mathrm{ESG}$, so that

$$(2) \quad \frac{\mathrm{E}e}{eq} = \frac{\mathrm{ES}}{\mathrm{SG}} \text{ or } \frac{\mathrm{E}e}{\mathrm{F}f} = \frac{\mathrm{ES}}{\mathrm{SG}}.$$

Multiply the two sets of proportions to get

$$(3) \quad \frac{\mathrm{E}e}{\mathrm{F}f} \times \frac{\mathrm{D}d}{\mathrm{E}e} = \frac{\mathrm{ES}}{\mathrm{SG}} \times \frac{\mathrm{DE}}{\mathrm{ES}},$$

$$(4) \quad \frac{\mathrm{D}d}{\mathrm{F}f} = \frac{\mathrm{DE}}{\mathrm{SG}}.$$

But $\Delta \mathrm{PDE} \sim \Delta \mathrm{PGS}$, so that

$$(5) \quad \frac{\mathrm{PE}}{\mathrm{PS}} = \frac{\mathrm{PD}}{\mathrm{PG}} = \frac{\mathrm{DE}}{\mathrm{SG}}.$$

Hence,

$$(6) \quad \frac{\mathrm{PE}}{\mathrm{PS}} = \frac{\mathrm{D}d}{\mathrm{F}f}.$$

This is what was to be proved.

Newton's result can be readily translated into the more familiar language of the Leibnizian algorithm for the calculus. The conclusion of lem. 29 is:

$$(7) \quad \mathrm{PE} \times \mathrm{F}f = \mathrm{PS} \times \mathrm{D}d.$$

$\mathrm{F}f$ is the infinitesimal or momentaneous increment or decrement dr of the radius ($r = \mathrm{PE}$) of the circle drawn with center P, so that $\mathrm{PE} \times \mathrm{F}f = rdr$. Eq. 7 thus becomes

$$(8) \quad rdr = cdx.$$

Consider the situation in which x, y are the coordinates of any point on a circle with radius r, so that

$$(9) \quad x^2 + y^2 = r^2$$

and

$$(10) \quad xdx + ydy = rdr$$

or

$$(11) \quad xdx - cdx + ydy + cdx = rdr$$

$$(12) \quad rdr = (x - c)dx + ydy + cdx.$$

This is, in fact, a statement for the circle when the center of the circle is placed at a distance c from the origin of coordinates, in which case

$$(13) \quad (x - c)dx + ydy = 0$$

with the result that

$$(14) \quad rdr = cdx.$$

Example No. 7: Book 3, Prop. 19 (The Shape of the Earth) **10.14**

The goal of this proposition is to use theory in conjunction with data (notably measures of the earth and of the acceleration of free fall) to determine the shape of the earth. The final version, presented in the third edition, as noted in §8.11 above, was considerably modified from the earlier version. In this proposition, Newton assumes that the earth has the same shape as if it were a wholly fluid rotating mass. His final result, that the ellipticity of the earth is $\frac{1}{230}$, is obtained by a clever argument. The accepted result today is $\frac{1}{298.275}$.

1. First, Newton assumes that the earth is a perfect sphere. He uses the Picard-Cassini measure of the earth to conclude that the earth's circumference is 123,249,600 Paris feet and that the earth's radius is 19,615,800 Paris feet. In what follows I shall use "feet" to designate Paris feet; a "line" is 1⁄12 of an inch.
2. Adding a small correction factor for the resistance of the air, Newton gives 2,174 lines for the distance a body will fall freely in one second in vacuo at the latitude of Paris. This result had been obtained in Paris by very careful experiments with pendulums.
3. Taking a sidereal day to be $23^{h}56^{m}4^{s}$, Newton computes that a body at the equator of the earth (assumed to revolve uniformly) describes an arc of 1,433.46 feet in one second. The versed sine of this arc is computed to be 7.54064 lines. The versed sine is that portion of the perpendicular bisector of the chord of the arc which extends from the chord to the arc.

4. Therefore, Newton concludes,

$$\frac{\text{force of descent at Paris}}{\text{centrifugal force at the equator}} = \frac{2,174}{7.54064}.$$

Here "force of descent at Paris" refers to the force with which bodies descend at the latitude of Paris; this "force of descent" is measured by the distance a body falls freely in the first second. The "centrifugal force at the equator" is the force at the equator arising from the earth's daily rotation. The proportionality of the centrifugal force to the versed sine of the arc traversed in one second depends ultimately on book 1, prop. 1, corols. 4 and 5.

5. Next, Newton computes the ratio of the centrifugal force at the equator to the centrifugal force at the latitude of Paris. This ratio is as follows:

$$\frac{\text{centrifugal force at equator}}{\text{centrifugal force at Paris}} = \left(\frac{r}{\cos\lambda}\right)^2$$

or

$$\frac{\text{centrifugal force at Paris}}{\text{centrifugal force at equator}} = \left(\frac{\cos\lambda}{r}\right)^2$$

where r is the radius of the earth and λ the latitude of Paris.

The reason is that the radius of rotation at Paris is $r \cos \lambda$ and the centrifugal force is proportional to this radius. We might have expected, therefore, that this ratio would be the square of $\frac{r\cos\lambda}{r}$ rather than of $\frac{\cos\lambda}{r}$. The reason why Newton writes of the square of $\frac{\cos\lambda}{r}$ is that in those days the cosine (and any other trigonometric function) was reckoned with regard to a specific arc on a given sphere (in this case, that is, a sphere of radius r) and not on an angle as at present. Newton thus computes that

$$\frac{\text{centrifugal force at equator}}{\text{centrifugal force at Paris}} = \frac{7.54064}{3.267}.$$

6. At Paris, the force by which bodies are actually observed to descend is not the "total force of gravity" because of the centrifugal effect of the earth's rotation (which is related to the latitude). Therefore, to find the "total force of gravity," it is necessary to add the effect of the centrifugal force to the observed time of free fall in one second. Hence the observed

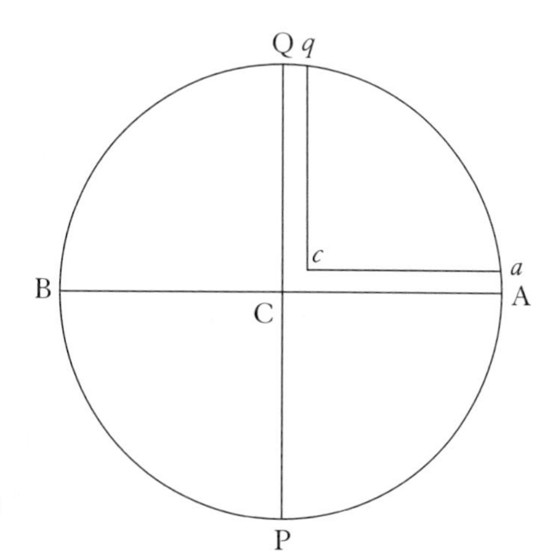

Figure 10.16. Cross section of the earth as a prolate spheroid (book 3, prop. 19).

value of 2,173 7/9 lines, or 15 feet, 1 inch, 1 7/9 lines, must be increased to 2,177.267 lines, or 15 feet, 1 inch, 5.267 lines.

7. Hence

$$\frac{\text{total force of gravity at lat. } \lambda}{\text{centrifugal force at equator}} = \frac{2,177.267}{7.54064} = \frac{289}{1}.$$

8. The undiminished force of gravity at the equator will differ very slightly from the undiminished force of gravity at Paris. Neglecting any small second-order terms, however, Newton has in this manner found the ratio of the undiminished force of gravity at the equator to the centrifugal force at the equator to be 289 to 1, which Newton (in later editions) adapted from Huygens.

Newton next considers the earth to be a prolate spheroid, generated by the rotation of an ellipse about its minor axis PQ. The diagram in the *Principia* is apt to be misleading because the horizontal axis, corresponding to the minor axis, is the axis of rotation whereas we are used to seeing such a diagram turned through 90 degrees and flipped over, that is, with the axis of rotation vertical. For the ease of the reader, therefore, I have here (fig. 10.16) turned the diagram of the *Principia* through 90 degrees.

9. Newton asks us to consider two joined canals QC*cq* and AC*ca*, filled with water. Then, because of the action of the centrifugal force, the weight of the water in the longer canal AC*ca* must (by the result obtained

in step 7 above) be greater than the weight of the water in the smaller canal Q*Ccq* by $\frac{1}{289}$. That is, the weights will be as 289 to 288.

10. Using book 1, prop. 91, corols. 2 and 3, Newton computes that the force of gravity at the pole Q will be to the force of gravity at the equator A as 501 to 500.
11. Next, Newton invokes "the rule of three." If a centrifugal force of $\frac{4}{505}$ will cause a difference in length of the two canals of $\frac{1}{100}$, then a centrifugal force of $\frac{1}{289}$ will produce a difference in lengths of 230 to 229.
12. Hence Newton concludes that the ratio of the equatorial diameter of the earth to the polar diameter is 230 to 229.

With respect to this analysis, Brougham and Routh judge that Newton's analysis here is "manifestly altogether defective" because he "assumes not only that the spheroid is a form of equilibrium, but that the ellipticity is always proportional to the ratio of the centrifugal force to gravity," two assertions which "are indeed true" but "not self-evident." Noting that Colin MacLaurin "first demonstrated their truth," Brougham and Routh conclude: "It is very remarkable in how wonderful a manner Newton often arrives at correct results by means the most inadequate. Of this there are many other instances besides the present one." The examples they cite are that Newton "guessed the mean density of the earth" and "determined by analogy that the velocity of waves varied as the square root of their length."[24]

10.15 *Example No. 8: A Theorem on the Variation of Weight with Latitude (from Book 3, Prop. 20)*

This example differs from the preceding ones, in which the problem was to understand the logic of Newton's proof. In discussing prop. 20, however, Newton states a theorem without any hint of a proof. The burden of prop. 20 is: "To find and compare with one another the weights of bodies in different regions of our earth." Here "weight" ("gravitas") is taken to mean the combination of the force of the earth's gravitation and the centrifugal effect of the earth's rotation. He first concludes that the weights of any objects placed at different locations on the earth's surface are "inversely as the distances of those places from the center" of the earth.

24. *Anal. View*, p. 179.

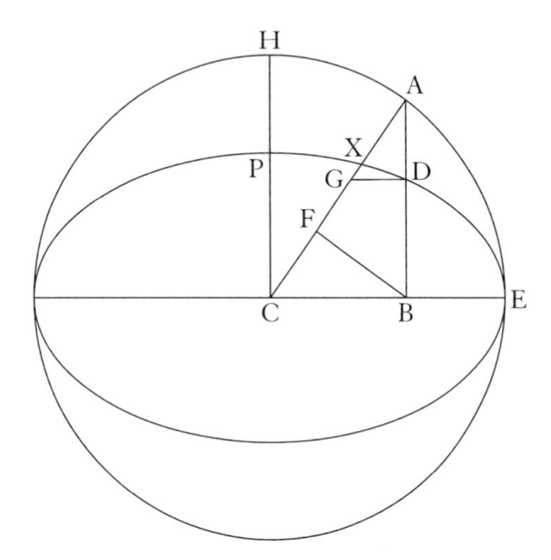

Figure 10.17. A theorem on the increase in weight in going from the equator to the poles (in book 3, prop. 20).

As Todhunter has pointed out, this result should contain the restrictive clause "when resolved along the radius."[25]

The discussion of prop. 20 begins with a summary of the antecedent propositions, including the shape of the earth as an oblate spheroid, squashed at the poles and bulging at the equator. Newton finds that the magnitude of a polar radius compared with the magnitude of an equatorial radius is as 229 to 230. He then states the law that weight depends inversely on the distance from the center of the earth. From this result, he says, a "theorem is deduced." This theorem reads: "The increase of weight [at any place on the earth's surface] in going from the equator to the poles is very nearly as the versed sine of twice the latitude [of that place], or (which is the same) as the square of the sine of the latitude." Newton gives no details concerning the proof, and as is often the case in the *Principia*, the proof is far from obvious to today's readers.

Several writers have supplied the missing proof, among them David Gregory, whose "Notae in Newtoni Principia Mathematica Philosophiae Naturalis" was intended to be a help to readers, proposition by proposition. The following proof is in part adapted from the one given by Gregory.

The diagram (fig. 10.17) shows a cross section of the earth presented as an oblate spheroid (actually an ellipsoid of revolution) with the ellipticity greatly exaggerated. It is inscribed in a concentric sphere whose center is C. P is a pole of the earth and E a point on the equator. X is any point on the earth's surface. The line CA is drawn from C, the center of the earth, through X and extended until it meets the sphere at the point A. AB is drawn from A perpendicular to CE, and

25. Todhunter, *History* (§8.8, n. 38 above), 1:20–21, §33.

BF is drawn from B perpendicular to CXA. The point G is chosen on AC such that AG = PH.

The assignment is to find the increase in weight in going from E to X. Gregory assumes that by "increase" Newton means the increase in weight in going from E to X in relation to the total possible increase in weight in going from the equator all the way to the pole. Thus Newton's theorem states that

$$\frac{\text{weight at X} - \text{weight at E}}{\text{weight at P} - \text{weight at E}} \text{ "is very nearly as" } \sin^2 \lambda.$$

The proof has two parts. The first is to show that the increase in weight in going from the equator (E) to some place (X) is to the increase in weight in going from the equator to a pole (P) "very nearly" as AX to HP, that is,

$$\frac{\text{weight at X} - \text{weight at E}}{\text{weight at P} - \text{weight at E}} \approx \frac{\text{AX}}{\text{HP}}.$$

The second part is to prove that $\frac{\text{AX}}{\text{HP}} = \sin^2 \lambda$.

In the preceding prop. 19, Newton has used an argument based on the balancing of two canals filled with water to show that weight at any place on the surface of the earth is inversely proportional to the distance of that place from the earth's center. Accordingly,

$$\text{weight at E} \propto \frac{1}{\text{CE}}$$

$$\text{weight at X} \propto \frac{1}{\text{CX}}$$

$$\text{weight at P} \propto \frac{1}{\text{CP}}.$$

Therefore,

$$\frac{\text{increase in weight in going from E to X}}{\text{increase in weight in going from E to P}} \propto \frac{\frac{1}{\text{CX}} - \frac{1}{\text{CE}}}{\frac{1}{\text{CP}} - \frac{1}{\text{CE}}}$$

$$= \frac{(\text{CE} - \text{CX})/(\text{CX} \times \text{CE})}{(\text{CE} - \text{CP})/(\text{CP} \times \text{CE})} = \frac{\text{CP}}{\text{CX}} \times \frac{\text{CE} - \text{CX}}{\text{CE} - \text{CP}} = \frac{\text{CP}}{\text{CX}} \times \frac{\text{AX}}{\text{HP}} \approx \frac{\text{AX}}{\text{HP}}.$$

Restated in words, the increase in weight in going from E to X is to the increase in weight in going from E to P "very nearly" (or approximately) as AX to HP. The qualification "very nearly" arises in the last step, where it is assumed that CX may be considered to be "very nearly" (or approximately) equal to CP. According to the preceding prop. 19, the maximum difference between CX and CP (occurring when X is at the pole P) is 1 part in 230, or a little less that one-half a percent. Today's value of about 1 part in 300 would reduce this maximum to about one-third of a percent.

We now proceed to the second part of the proof, to show that

$$\frac{\mathrm{AX}}{\mathrm{HP}} = \sin^2 \lambda.$$

This part of the proof depends on simple geometry.

From the properties of the ellipse,

$$(1)\quad \frac{\mathrm{HP}}{\mathrm{AD}} = \frac{\mathrm{PC}}{\mathrm{DB}}.$$

Since HP = GA by construction and PC = CG, eq. 1 becomes

$$(2)\quad \frac{\mathrm{GA}}{\mathrm{AD}} = \frac{\mathrm{CG}}{\mathrm{DB}}$$

$$(3)\quad \frac{\mathrm{CG}}{\mathrm{GA}} = \frac{\mathrm{DB}}{\mathrm{AD}}.$$

Since CG = AC − AG and DB = AB − AD,

$$(4)\quad \frac{\mathrm{AC} - \mathrm{AG}}{\mathrm{AG}} = \frac{\mathrm{AB} - \mathrm{AD}}{\mathrm{AD}}.$$

Then, by composition of ratios (or componendo),

$$(5)\quad \frac{\mathrm{AC} - \mathrm{AG} + \mathrm{AG}}{\mathrm{AG}} = \frac{\mathrm{AB} - \mathrm{AD} + \mathrm{AD}}{\mathrm{AD}}$$

$$(6)\quad \frac{\mathrm{AC}}{\mathrm{AG}} = \frac{\mathrm{AB}}{\mathrm{AD}}.$$

The four quantities in this equation are corresponding sides of ΔDAG and ΔBCA. Therefore, $\Delta\mathrm{DAG} \sim \Delta\mathrm{BCA}$ and hence

$$(7)\quad \mathrm{GD} \parallel \mathrm{CE}.$$

Because the ellipse is "very nearly" a circle, the arc XD can be taken as a straight line perpendicular to GA. Then, because XD is a perpendicular dropped from a vertex to the hypotenuse GA of ΔDAG and BF is a perpendicular dropped from a vertex to the hypotenuse of ΔBCA, and because ΔDAG ~ ΔBCA,

$$(8) \quad \frac{GA}{XA} = \frac{AC}{AF}.$$

Because $rt\Delta AFB \sim rt\Delta ACB$,

$$(9) \quad \frac{AB}{AF} = \frac{CA}{AB}$$

$$(10) \quad AB^2 = AF \times CA,$$

so that

$$(11) \quad \frac{AB^2}{CA^2} = \frac{AF \times CA}{CA^2}$$

$$(12) \quad \frac{AF}{CA} = \frac{AB^2}{CA^2}.$$

Combining eqs. 8 and 12,

$$(13) \quad \frac{GA}{AX} = \frac{CA}{AF} = \frac{CA^2}{AB^2}.$$

Since AG = PH by construction and CA = HC, eq. 13 becomes

$$(14) \quad \frac{XA}{PH} = \left(\frac{AB}{HC}\right)^2.$$

The quantity $\frac{XA}{PH}$ is the increase in gravity in going from the equator to the point X. The quantity $\frac{AB}{HC}$ $\left(\text{which is equal to } \frac{AB}{CE}\right)$ is the sine (Newton calls it the "right sine") of angle XCE, or the sine of the latitude λ.

Hence, it is proved that the increase in gravity in going from the equator to any point X on the earth's surface is very nearly (or approximately) as the square of the sine of the latitude.

Newton's theorem states the change in weight is "very nearly as the versed sine of twice the latitude," or as the square of the sine of the latitude. This follows from the trigonometric identity

$$\text{vers } 2\lambda = 1 - \cos 2\lambda = \sin^2 \lambda.$$

Example No. 9: Book 1, Prop. 66, Corol. 14, Needed for Book 3, Prop. 37, Corol. 3 **10.16**

Prop. 66, with its twenty-two corollaries, contains Newton's presentation of the three-body problem. There is (see fig. 10.18) a central body T (for "Terra," the earth) with an orbiting satellite or moon P (for "Planeta," or secondary planet or satellite). The perturbing body is S (for "Sol," the sun). As we have seen (§6.12 above), Newton originally designated the central body S, the orbiting body P, and the perturbing body Q. Each of these three bodies attracts the others with a force varying inversely as the square of the distance. Here T is the "greatest body," about which P describes the "inner orbit" around T, while S moves along the "outer orbit." The point K, on the line SP extended, marks the mean distance between bodies S and P. We are to choose some suitable scale so that SK will represent the "accelerative [measure of the] attraction" exerted by S on P at that mean distance. The point L on the line SP is determined by the relation

$$SL : SK = SK^2 : SP^2.$$

SL will thus be the "accelerative [measure of the] attraction" exerted by S on P at the distance SP.

Newton next resolves the force exerted by S on P, and directed along the line SP, into two components. One (LM) is parallel to PT, the force exerted by T on P, that is, has the direction from P to T. The other is represented by SM. Accordingly, the body P is now acted on by three forces. Readers who have studied

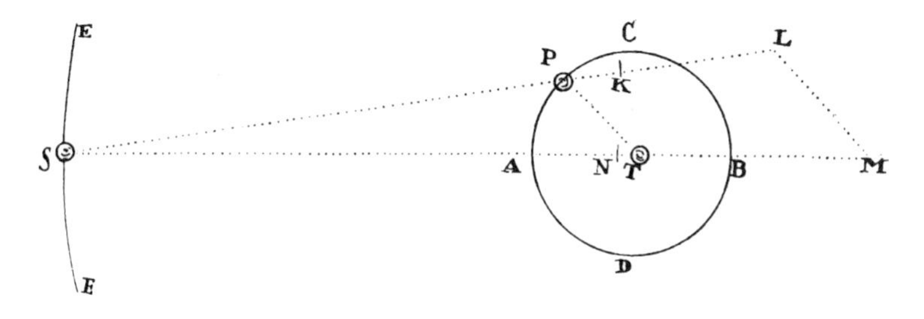

Figure 10.18. Forces of perturbation in the three-body problem (book 1, prop. 66).

some celestial mechanics must be careful not to think of Newton's three forces as the orthogonal forces (the radial component, the component perpendicular to the plane, and the tangential component) with which they are familiar.

The force of T on P, by itself, would cause P to move in an ellipse with the center of body T as focus, sweeping out equal areas in equal times. The second force, LM, coincides in direction with the force of T on P, but it is not inversely proportional to the square of the distance. The result is that the sum of these two forces is not inversely proportional to the square of the distance. Therefore, the combination of these two forces will cause the orbit of the body P to deviate from an ellipse. As before, the magnitude of these two deviations will be greater or less as the perturbing force LM is greater or less. When the third force is added to the other two, it will cause the net force on P to deviate from an attraction toward T in a manner that changes with the relative motions and positions of S and P. The assignment set by Newton is to explore the consequences of these three forces or to explore the effect of the perturbing force of a body like the sun on the motion of a body like the moon in its orbit around a body like the earth.

In corol. 14, Newton explores the forces represented by NM and ML. We have seen that ML is a component of the force of S, actually the component exerted by S on P in the direction PT. SN is defined as the "accelerative [measure of the] attraction" exerted on T by S in the direction TS. In general, SN is not equal to SM, the whole component of force along the line TS. NM is the net component of the force of S exerted in the direction ST; that is, it is the difference between the whole component of S in the direction ST and the portion of that component attracting T. Clearly, the two perturbing forces, ML and NM, will vary as the relative positions of S and P change as a result of their respective orbital motions. What Newton calls their "effect" on the shape of the orbit of P and the motion of P with respect to the area law (their effect in producing deviations from perfect Keplerian motion) will be proportional to their respective magnitudes.

One of the conditions of corol. 14 is that the body S is said to be extremely far away. Newton determines the effects of the mean force SK and the orbital distances PT and ST on the magnitudes of the perturbing forces ("NM and ML"), the two forces that perturb the motion of P relative to T. In what follows, I shall discuss only the magnitude of ML. Newton is interested in the force ML, which the line ML represents and which is proportional to it. He needs the relation

$$\mathrm{ML} \propto \mathrm{SK} \times \frac{\mathrm{PT}}{\mathrm{ST}}.$$

As is the case for many derivations or proofs in the *Principia*, Newton does not say how he establishes this value for ML. Of course, many of the readers in Newton's day and in the succeeding decades were more familiar with geometric techniques than we are and apparently were expected to either fill in the gap or assume that Newton's mathematics was correct. The steps needed in this example are given in full in §10.19 below.

Next Newton supposes that the distance PT is given as is the absolute force of the body S. Then, in the last relation, PT is very nearly constant and is given, the mass of S is fixed or constant, and the mean force of S on P, or SK, is as the inverse square of the distance SP, which—because S is very remote—can be taken as ST. Thus the last relation becomes, as Newton says,

$$\mathrm{ML} \propto \frac{1}{\mathrm{ST}^3}.$$

From this result, Newton proceeds to draw two conclusions. The first declares the result of allowing both the distance ST and the absolute force or mass of body S to vary, the system of bodies T and P remaining unaltered. Under these conditions, NM and ML, and the effects they produce, are "very nearly in a ratio compounded of the direct ratio of the absolute force of body S and the inverse ratio of the cube of the distance ST." The second conclusion is that ML (and NM) and the effects produced by them will be "directly as the cube of the apparent diameter of the distant body S when looked at from body T," and vice versa.

In the first of these conclusions, everything remains the same in the system of P encircling T, but the distance ST and the absolute force of S will become variable quantities. In this case, the final result obtained above must be altered to allow for a variable absolute force of S. Therefore, the last relation becomes

$$\mathrm{ML} \propto (\text{absolute force of S}) \times \left(\frac{1}{\mathrm{ST}}\right)^3.$$

Next, suppose that the system of P encircling T is allowed to move in a circular orbit around S. Then the forces ML and NM "and their effects" will be inversely as the square of the periodic time. This result follows from book 1, prop. 4, corols. 4 and 6. That is, let the period of revolution be t. It has been demonstrated that

$$\mathrm{ML} \propto (\text{absolute force of S})/\mathrm{ST}^3$$

and from prop. 4, corol. 6,

$$t^2 \propto \mathrm{ST}^3,$$

so that

$$\text{ML} \propto (\text{absolute force of S})/t^2.$$

Thus far, the derivation is of the sort that is fully in keeping with the spirit of sec. 11 of the *Principia*. But now there comes a second conclusion, one that invokes the "apparent diameter" of the sun, a purely observational quantity that seems unexpected in the kind of mathematical exercise in which Newton is engaged. As has been mentioned, however, this part of the conclusion is needed for book 3, prop. 37, corol. 3.

In the gloss on this second part of the conclusion of corol. 14, we shall retain the possibility that in the original set of conditions, "the absolute force of body S" may be a variable. That is, the possibility is left open that there may be more than just one single body S, and that the masses and hence the absolute forces of these bodies will differ from one another. In this second part of corol. 14, Newton shows how ML can be expressed in terms of the observed or apparent diameter of S.

Newton assumes that "the magnitude [or mass] of body S is proportional to its absolute force." He begins with the result stated earlier, that ML is jointly proportional to the absolute (measure of the) force exerted by S directly and the inverse cube of the distance ST:

$$\begin{aligned} \text{ML} \quad &\propto \frac{\text{absolute force of S}}{\text{ST}^3} \propto \frac{\text{magnitude (or mass) of S}}{\text{ST}^3} \\ &\propto \frac{\text{density of S} \times (\text{true diameter of S})^3}{\text{ST}^3}. \end{aligned}$$

But

$$\frac{\text{apparent diameter of S}}{\text{true diameter of S}} = \frac{1}{\text{ST}}$$

and so

$$\frac{(\text{true diameter of S})^3}{\text{ST}^3} = (\text{apparent diameter of S})^3$$

and therefore

$$\text{ML} \propto (\text{density of S}) \times (\text{apparent diameter of S})^3.$$

In order to apply corol. 14 to book 3, prop. 37, corol. 3, we may postulate that there are two bodies S, each one acting on the water of T so as to exert a tide-producing force. In this case, the final proportion for ML

$$\text{ML} \propto (\text{density of S}) \times (\text{apparent diameter of S})^3$$

will give the tidal force for each of the bodies. As Newton says in book 3, prop. 37, corol. 3, the tide-producing force of the moon is to the tide-producing force of the sun jointly as their densities and their apparent diameters cubed.

On first examining book 3, prop. 37, corol. 3, the reader is apt to be puzzled because the statement about the tide-producing forces in relation to the densities and cubes of the apparent diameters is justified by a reference to book 1, prop. 66, corol. 14, which does not mention density at all. In fact, this first reading may seem to indicate that there has been an error in Newton's reference, that he had in mind some other corollary. The reason for the apparent lack of correlation is that in book 1, prop. 66, corol. 14, there is only a single body S, and so the density is not a variable but a constant. In this case, Newton quite correctly omits the factor of density in stating his proportions or, what comes to the same thing, absorbs the constant factor of density into the constant of proportionality. Thus, corol. 14 does not mention density, merely concluding that

$$\text{ML} \propto (\text{apparent diameter of S})^3.$$

This particular problem is typical of many of the difficulties that arise in reading the *Principia*. Newton often gives us hardly a clue to the omitted steps, producing a puzzle that may be difficult to unravel until the solution suddenly springs to mind.

The reader's attention is called to the fact that in this problem, Newton uses the "absolute [measure of a] force," whereas in most of book 1 before sec. 11, he uses the accelerative measure. The reason is that in problems such as prop. 66, corol. 14, mass enters the problem, which is not the case for most of the preceding parts of book 1. The reader will also have observed that in corol. 14, the force exerted by S and T, the equivalents of the sun and the earth, is not only an inverse-square force (as per book 1, props. 1, 11, and corol. 2 to props. 11–14) but depends on the mass of the body in which the gravitating force originates. At this stage of the *Principia*, there is no warrant for this assertion, which is proved only later, in book 3. Finally, it should be noted that Newton here invokes a force that varies inversely as the cube of the distance; hence the exploration of the properties of an inverse-cube force in book 1, prop. 41, corol. 3, is not quite so odd as may at first appear.

10.17 *Newton's Numbers: The "Fudge Factor"*

The numerical values given in the *Principia* often vary from edition to edition. In some cases, as in the penultimate sentence of book 1, prop. 45, corol. 2, there was a simple correction of a slip, as from 14″ to 28″. In others, Newton substituted more recent and better observations. For example, in book 3, prop. 19, he introduced in the third edition some recent observations made by the British astronomer James Pound. In the preface to the third edition, he specifically called attention to these "new observations, made by Mr. Pound," on "the proportion of the diameters of Jupiter to each other." The changes in numbers are very extensive in the Phenomena at the beginning of book 3, especially phen. 1 and phen. 2. There were also great changes in the data of observation in the "example" of the comet of 1680, following book 3, prop. 41. Major changes in numbers occur in book 3, prop. 20, where Newton altered the difference between the earth's polar and equatorial diameters from the 17 miles of the first edition, under the idealized assumption of uniform density, to the 31 7/12 miles of the second edition and then returned to the first value ("about 17 miles") in the third edition. In this case, the question was which evidence for the length of a seconds pendulum should be considered the most reliable. Except for a few cases of notable interest, changes in numerical data or numbers used in calculations have not been indicated either in this Guide or in the notes to the translation. They may be found, however, among the variant readings in our Latin edition.

There is an additional class of numbers, however, which were altered from edition to edition and which are not directly related to either errors in calculation or new data of observation. These are numbers which indicate that at times Newton seems to have "fudged" his results. This feature of Newton's *Principia* has been studied in detail and with care by R. S. Westfall, who has reported his findings in the celebrated paper "Newton and the Fudge Factor."[26] Westfall based his study primarily on three case histories: the moon test (book 3, props. 4 and 37), the velocity of sound (book 2, scholium following prop. 50), and the precession of the equinoxes (book 3, lems. 1, 2, 3, and prop. 39). Westfall showed that Newton (at times with the aid of Roger Cotes) sometimes nudged the numbers so that the final computations would yield results consistent with his theory.

Newton's two moon tests are notable because of the close correlation that Newton achieved between two numbers. One was the predicted value of free fall in one second at the earth's surface by applying the inverse-square law to the observed data of the motion of the moon, the other the result of actual terrestrial

26. Westfall, "Newton and the Fudge Factor" (§8.13, n. 64 above).

measurements with pendulums. In making this comparison, Newton had to decide on a mean value for the moon's distance from the earth, and he had to introduce correction factors because the pendulum experiments were made in Paris and not at a mean latitude of 45 degrees; he also had to take account of the centrifugal effect of the earth's rotation. His final two numbers agree to within a fraction of a line, where a line is 1⁄12 inch. The calculations reveal that Newton concluded that the moon would fall in one minute of time through 14.7706353 feet. The mind reels at the thought of precision to within a billionth of a foot. We may easily understand Westfall's concern about an "unsettling combination of decimals carried to seven places and fractions of varying complexity . . . purporting to find a correlation accurate to a fraction of a line, a degree of precision perhaps somewhat better than 1 part in 3000." As has been mentioned, the possibility of close agreement is baffling when one takes into account the consequence of errors arising from Newton's faulty calculations about the moon.

The most blatant of Westfall's examples, however, is the precession of the equinoxes. This example resembles the other two in that here again Newton's theoretical base was not fully up to the requirements of the problem. In the case of precession, Newton was especially handicapped by his inability to deal as successfully with the dynamics of rigid bodies as he had done with the dynamics of a particle. Knowing only the barest essentials of the dynamics of moments, he was also wholly unaware of the concept of moment of inertia, which plays the same role for the dynamics of rigid bodies that the concept of mass, which Newton invented, does for the dynamics of a particle. Whoever reads Westfall's discussion of this example will be struck by his revelation of the degree to which Cotes began to participate in the exercise.

In at least one instance, however, the determination of the velocity of sound, the charge that Newton fudged his numbers seems to be in error. Here (book 2, scholium following prop. 50) the problem of the numbers was simple and straightforward: how to make the theoretical calculations agree with the numbers derived from observation and experiment. We have seen that in the presentation of the velocity of sound, Newton (in the first edition) picked an experimental value most favorable to his theoretical result. He badly needed to obtain a value computed from theory that would agree closely with the data of observation and experiment. In the computation of the velocity of sound, as we have seen, Newton's theory was incomplete because he did not take account of the thermal consequences of the compression and rarefaction of the air, in effect assuming an isothermal rather than an adiabatic process. In this case, Newton did not in my opinion "fudge" his results. That is, he did not try to hide the disagreement between theory and experiment. Rather, he stated the discrepancy and sought for possible numerical factors that would make the theoretical numbers agree with the results of actual measurement. As I have said earlier, he accepted the validity of the data and attempted to find what factors might have been ignored in his theory, what kind of theoretical entities his data might give evidence for. The factors for which he supposed his numbers to provide evidence were "coarseness" and density, for both of which there was no real basis in theory or experiment.[27] We may censure Newton for making a wild guess (or perhaps for making the wrong guess), but we should not on this occasion fault his honesty.[28]

Newton's *Principia* was perhaps the first major work on natural philosophy, as opposed to astronomy, to be based on detailed numbers produced by theory compared with numbers based on experiment and observation. In this feature, among others, the *Principia* was a landmark work. In this sense, the *Principia* was herald to the great transition that marked the birth of modern physics as an exact science, what has been called a shift from "the world of more or less" to a "universe of precision."[29] As in other pathbreaking aspects of science, the *Principia* set a new course for science, one based on quantity or on numbers. In the case of the theory of the moon's motion, and the theory needed for the moon's mass, the velocity of sound, and the precession of the equinoxes, Newton showed the way but did not know how to achieve the final goal he set. So in the use of numbers, when simple procedures failed to yield the needed results, Newton on occasion had resort to improper methods.

10.18 *Newton's Measures*

In the *Principia*, Newton uses both English and French units of measure. English measures include feet ("pedes") and inches ("digiti") and lines ("lineae"), which are twelfths of an inch. Newton also uses Paris feet ("pedes Parisienses"), lines, and toises or hexapeds ("hexapedae"), equal to 6 French feet. A toise was roughly the equivalent of 1.9 meters, or 6⅖ English feet.

Although a "digit" is the width of a finger, or a finger's breadth, and so about three-quarters of an inch, in Newton's day it was used both in English and in Latin as equal to 1 inch. Thus, in 1669, Robert Boyle (in his *New Experiments*) wrote of the height of a column of mercury "of 29 Digits," remarking, "I take Digits for Inches throughout this tract."

27. For details, see Westfall's analysis.

28. For Laplace's view of Newton's attempt to find a cause of the discrepancy between theory and experiment, see §7.8 above.

29. Westfall, "Newton and the Fudge Factor," p. 758. We owe this characterization of the scientific revolution to Alexandre Koyré.

John Harris's *Lexicon Technicum* (London, 1704) gives a comparative table of measures used in many different places. He lists the Paris foot as equal to 1.066 English feet ("according to Dr. Bernard"), but also lists "the Royal Foot" of Paris as 1.068 English feet.

In the General Scholium at the end of book 2, sec. 6, in the second and third editions (but which appeared at the end of sec. 7 in the first edition), Newton describes experiments made with globes of lead. One had a diameter of 2 inches and weighed 26¼ ounces; the other had a diameter of 3⅝ inches and weighed 166¹⁄₁₆ ounces. Newton referred to these weights as being in "Roman ounces" ("globum plumbeum diametro digitorum 2, & pondere unciarum *Romanarum* 26¼"). The question arises whether these are troy weights (12 ounces to the pound) or avoirdupois weights (16 ounces to the pound). In an attempt to answer this question, a calculation can be made using the value of 62.4 pounds per cubic foot as the density of water and 11.34 as the specific gravity of lead. According to John Harris's *Lexicon Technicum* (London, 1704), s.v. "SPECIFICK *Gravity*," the specific gravity of lead is 11⅓; he also gives a second value of 11.42, both of which are close to the present value of 11.34.

Using these values we can compute the weight in pounds of a lead globe with a diameter of 2 inches and of another with a diameter of 3⅝ inches, the two globes mentioned by Newton. It should then be possible to determine whether Newton's ounces correspond to a pound consisting of 12 or 16 ounces. These computations show us an ancillary problem: the values given by Newton for the two globes yield somewhat different densities. This suggests the possibility that one or both were not made of pure lead or possibly the shape was not perfectly spherical.

If 1 cu. ft. of water weighs 62.4 pounds, then 1 cu. ft. of lead should weigh $62.4 \times 11.34 = 707.6$ lb. Accordingly, 1 cu. in. of lead would weigh $\frac{707.6}{12^3} =$ 0.409 lb. Using this result, we can compute the expected weight of spheres with diameters of 2 in. and 3⅝ in.

The volume of a sphere with diameter 2 inches or radius 1 inch is 4.189 cu. in. If 1 cu. in. of lead weighs 0.409 lb., then 4.189 cu. in. of lead ought to weigh $0.409 \times 4.189 = 1.7$ lb. In troy measure, this is $12 \times 1.7 = 20.4$ oz.; in avoirdupois measure, it is $16 \times 1.7 = 27.2$ oz. Newton says that his globe weighed 26.25 ounces, which is within a pound of the computed weight in avoirdupois measure. This calculation would thus seem to indicate that Newton was using avoirdupois rather than troy measure in this General Scholium.

As a check, we may make a similar computation for the second globe. The diameter of this globe is said to have been 3⅝ (or 3.625) inches. Its volume will be $\frac{1}{6}\pi(3.625)^3 = 24.94$ cu. in. Therefore a globe with diameter 3⅝ inches should weigh $0.409 \times 24.94 = 10.2$ lb. In troy measure, this would be $12 \times 10.2 = 122.4$ oz.; in avoirdupois, $16 \times 10.2 = 163.2$ oz. This calculation confirms the conclusion that Newton was using avoirdupois measures in the General Scholium, since he says that the globe weighed $166\frac{1}{16}$ ounces. His value is a little more than our computed value, which would negate the suspicion that his lead was actually an alloy or mixture of lead and some less dense metal and rather suggest either that he had a ball with a not perfectly spherical shape or that he had difficulty in determining the diameter of the balls.

The conclusion seems inescapable that when Newton did the experiments reported in the General Scholium in 1687, he was using avoirdupois ounces and not troy ounces. In the experiments reported in 1713, however, in the new scholium (following prop. 40) for the revised sec. 7, he was evidently using troy ounces, not avoirdupois. The opening paragraph of this new scholium states that a "solid cubic London foot contains 76 Roman pounds of rainwater" while "a solid inch of this foot contains $\frac{19}{36}$ ounce of this pound." As before, Newton does not specify whether the measure is avoirdupois or troy, using the same Latin expression he had used earlier, "76 libras *Romanas*." Since there are $12 \times 12 \times 12$ cu. in. in 1 cu. ft. and since there are 12 ounces to the pound in troy measure, the weight of 1 cu. in. of water will be $76 \times \dfrac{12}{12 \times 12 \times 12}$. Canceling out a 12 from both numerator and denominator yields $\dfrac{76}{12 \times 12}$ or $\dfrac{19}{3 \times 12}$ or the value of $\frac{19}{36}$ reported by Newton.

In both scholiums, Newton used the same adjective, "Roman," even though in one case he was referring to troy measure and in the other to avoirdupois. As a result, in Motte's translation (and therefore in the Motte-Cajori version) both are rendered as troy measure.

Finally, we should note the discrepancy between the values given by Newton in the General Scholium for his two globes of lead. The larger one is off by a little more than 2 percent, while the smaller one is off by about 4 percent. Newton's value for the larger globe is too great, whereas his value for the lesser sphere is too small. Certainly, in those days the weight could be determined with an accuracy greater than 4 percent, and equally it was surely possible to determine the diameter of a sphere to within less than 1.3 percent. In accounting for this discrepancy, we must choose between a number of alternative possibilities. One is that the globes were not perfectly spherical; another, that there was an error in recording the data. Possibly, however, the measurements were made carelessly. It is to be noted that the size and weight of these two globes, as was the case for the wooden and iron globes mentioned in the General Scholium, are not significant numbers in the report of the experiments. That is, they do not figure in any of the calculations. Newton gave these numbers merely to give some idea of the scale of the objects he was using in his experiments. Accordingly, the discrepancies in no way impugn the accuracy of the experiments which Newton reported to have been made with these globes or the care with which the data of experiment were recorded.

A Puzzle in Book 1, Prop. 66, Corol. 14 (BY GEORGE E. SMITH) **10.19**

In book 3 Newton employs specific numerical values for the ratios between his different non-orthogonal components of the solar perturbing force on the moon and the gravitational force of the earth on the moon. He there cites corollaries to prop. 66 of book 1 as his basis for calculating these numerical values, in particular corol. 17. This is the last of four consecutive corollaries to prop. 66 on the subject of the magnitude of the solar perturbing force. The first of these, corol. 14 (see fig. 10.19), opens with the claim that when S is very far away, the perturbing forces ML and NM vary, *quam proxime*, as the force SK (the solar force on the moon when it is at its mean distance from the sun) and PT/ST (the ratio of the distance of the moon and the distance of the sun from the earth); the corollary then goes on to list a series of consequences of this claim. Newton does not indicate how he arrived at this claim about ML and NM. How might he have done so?

An oddity of this claim is that both the forces ML and NM vary with the position of the moon in its orbit, yet the three quantities to which they are said to be related, the force SK and the distances PT and ST, would not at all vary in the case of the circular orbits displayed in Newton's figure. As one can see from book 3, Newton knew that when S is far away the force ML varies little from its mean value, but the force NM varies from +3 times to −3 times the mean value of the force ML as the moon progresses from its nearest to its farthest point from the sun. Even if the circular orbits displayed in the figure were replaced by elliptical orbits, the variation in PT and ST could never account for the variation in the force NM from positive to zero to negative. Hence, when Newton says that

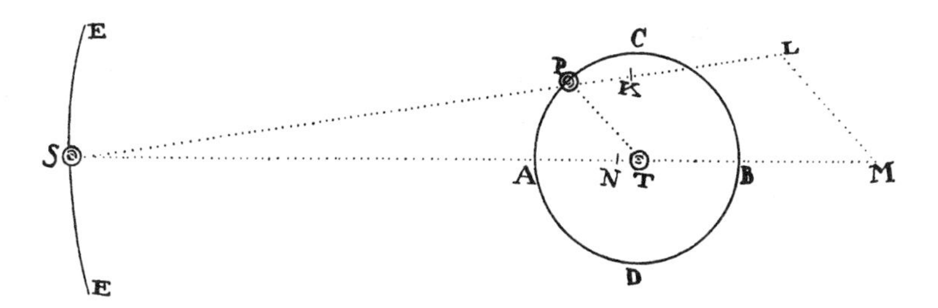

Figure 10.19. The perturbing force of the sun (S) on the Moon (P) in orbit around the earth (T).

the forces ML and NM vary as the force SK compounded with the ratio of PT to ST, he must mean the forces ML and NM at each given position of the moon in its orbit about the earth.

Although Newton indicates in book 3 how to obtain the specific variation in the force NM from positive to zero to negative, he never makes clear how he had confirmed that the force ML deviates negligibly from its mean value. As we shall see, the explanation offered below for how he arrived at his claim about the forces ML and NM also shows how he likely confirmed that the force ML varies negligibly.

While SK is the mean distance from the sun to the moon, it also serves to represent the solar force on the moon at this mean distance, and in this guise it is used to fix the scale of the solar forces SL on the moon and SN on the earth. Specifically,

$$\frac{\text{force SL}}{\text{force SK}} = \frac{\text{SK}^2}{\text{SP}^2}$$

and

$$\frac{\text{force SN}}{\text{force SK}} = \frac{\text{SK}^2}{\text{ST}^2}.$$

At the same time, the triangle of forces SLM is similar to the triangle SPT, so that

$$\frac{\text{force ML}}{\text{force SL}} = \frac{\text{PT}}{\text{SP}}$$

and

$$\frac{\text{force SM}}{\text{force SL}} = \frac{\text{ST}}{\text{SP}}.$$

Newton defines the force NM to be the force SM minus the force SN. Combining these equations then yields

$$\frac{\text{force ML}}{\text{force SK}} = \frac{\text{PT} \times \text{SK}^2}{\text{SP}^3}$$

and

$$\frac{\text{force NM}}{\text{force SK}} = \frac{\text{SK}^2}{\text{SP}^3}\left(\frac{\text{ST}^3 - \text{SP}^3}{\text{ST}^2}\right).$$

The problem now is to replace SP^3 in these equations with an expression in terms of PT and ST. Using modern notation,

$$ST = SP\cos(\angle PST) + PT\cos(\angle PTS),$$

so that

$$SP = \frac{ST - PT\cos(\angle PTS)}{\cos(\angle PST)}.$$

Newton himself would not have used the cosine function here. Instead, he would have drawn a line PX perpendicular to ST intersecting the latter at X, and then would have used the ratio of PX to PT in place of $\cos(\angle PTS)$ and the ratio SX to SP in place of $\cos(\angle PST)$. An example of his representing the cosine in this way can be found in his treatment of the force NM—there designated MT—in prop. 26 of book 3. Thus he would have had

$$SP = \frac{ST - PT(PX/PT)}{(SX/SP)},$$

with appropriate changes in sign when the moon P is in the two quadrants away from the sun. For simplicity, however, we will retain the cosine function below.

If S is very far away, $\cos(\angle PST) \approx 1.0$, so that

$$SP^3 \approx ST^3 - 3ST^2PT\cos(\angle PTS) + 3PT^2ST\cos^2(\angle PTS) - PT^3\cos^3(\angle PTS)$$

and hence

$$\frac{SP^3}{ST^2} \approx ST\left\{1 - \frac{3PT\cos(\angle PTS)}{ST} + \frac{3PT^2\cos^2(\angle PTS)}{ST^2} - \frac{PT^3\cos^3(\angle PTS)}{ST^3}\right\}.$$

But if S is very far away, $PT/ST \ll 1$, so that

$$\frac{SP^3}{ST^2} \approx ST.$$

Let us return now to the forces ML and NM. If S is very far away, $SK \approx ST$, so that

$$\frac{\text{force ML}}{\text{force SK}} = \frac{PT \times SK^2}{SP^3} \approx \frac{PT \times ST^2}{SP^3} \approx \frac{PT}{ST}.$$

It follows immediately from this that ML deviates little from its mean value if S is very far away, for the cos(∠PTS) has dropped out of the equation.

Similarly, in the case of the force NM we have

$$\frac{\text{force NM}}{\text{force SK}} = \frac{\text{SK}^2}{\text{SP}^3}\left(\frac{\text{ST}^3 - \text{SP}^3}{\text{ST}^2}\right),$$

$$\frac{\text{force NM}}{\text{force SK}} \approx \frac{\text{ST}^2}{\text{SP}^3}\left(3\text{PT}\cos(\angle\text{PTS}) - \frac{3\text{PT}^2\cos^2(\angle\text{PTS})}{\text{ST}} + \frac{\text{PT}^3\cos^3(\angle\text{PTS})}{\text{ST}^2}\right),$$

$$\frac{\text{force NM}}{\text{force SK}} \approx \frac{\text{PT}\cos(\angle\text{PTS})}{\text{ST}} \times\left(3 - \frac{3\text{PT}\cos(\angle\text{PTS})}{\text{ST}} + \frac{\text{PT}^2\cos^2(\angle\text{PTS})}{\text{ST}^2}\right),$$

$$\frac{\text{force NM}}{\text{force SK}} \approx 3\frac{\text{PT}\cos(\angle\text{PTS})}{\text{ST}}.$$

From this we also obtain as a corollary

$$\frac{\text{force NM}}{\text{force SK}} \approx 3(\text{force ML})\cos(\angle\text{PTS}),$$

a result Newton deploys in book 3.

There are three important things to notice here. First, this entire derivation is based on the similarity of the force triangle SLM and the triangle SPT, a similarity that is emphatically evident in Newton's figure. Second, no advanced mathematical methods were needed to obtain the results for forces ML and NM—nothing beyond geometry plus the approximation obtained when S is very far away, that is, PT/ST is small. Finally, notice that the framework can also be used to obtain exact rather than approximate results if the assumption that S is very far away is not adopted.

CHAPTER ELEVEN

Conclusion

Newton's *Principia* is a book of mathematical principles applied to nature insofar as nature is revealed by experiment and observation. As such, it is a treatise based on evidence. Never before had a treatise on natural philosophy so depended on an examination of numerical predictions and numerical evidence. The significance of Newton's numbers, in the context of the mission of the *Principia*, was not only to provide convincing evidence of the correctness of the mathematical principles being applied to natural philosophy, but to explore the theoretical significance of possible conflicts between simple—perhaps overly simple—theory and the evidential universe. An example would be the difference between a system in which Kepler's laws would be exactly true and the "real" evidential or experiential world, in which there are two or more interacting bodies. In his zeal to have the numerical evidence conform exactly to theory, as in the case of the velocity of sound, Newton at times (especially in the second edition of the *Principia*) introduced factors for which he had no real evidential basis. His program needed confirming evidence that terrestrial gravity, varying as the inverse square of the distance, extends to the moon. What may seem most astonishing to today's reader about the moon tests may not be the numbers introduced by Newton so much as the very fact that he had no instinctive feeling for significant figures, that he could carry out calculations to so many places. Here, in fact, is a signal alerting us to the absolute novelty of a general natural philosophy that was not only mathematical but numerical. There were no guides to follow, no standard procedures.[1] In the case of precession, so great was Newton's zeal that he ended up by manufacturing

1. I refer here to works devoted to a mathematical natural philosophy. Kepler, to be sure, had made extensive use of numerical data in relation to theory in his *Astronomia Nova* (1609), but anyone who examines that work will hardly recommend it as a model; in any event, its subject was orbital astronomy, not natural philosophy. Huygens's *Horologium Oscillatorium* (1673) was the nearest thing to such a model, but it dealt with a restricted topic and not with natural philosophy at large.

or manipulating numbers in order to provide the evidence to support his brilliant inspiration concerning the cause, an insight—we must not forget—that was essentially correct in its basic concept of the cause, even if Newton did not have any understanding of the dynamics of rigid bodies that would have enabled him to follow through the implications of his own suggestion.

The age of Newton was the age of reason, but for some of the titans of science it plainly was not in every respect the age of absolute candor. Newton invented chronologies of discovery, he wrote the report on the priority of invention of the calculus which appeared under the name of a committee of the Royal Society, and he then composed and published an anonymous review of the report that appeared in the Royal Society's official journal and was then prefixed to a reprint of the report. But he was not alone in conduct beyond the canons of acceptable behavior of later times. Leibniz's standards of conduct were on a par with Newton's. For example, he published an epitome of his own celestial physics, rushing into print—so he said—because he had read the lengthy review of the *Principia* in the *Acta Eruditorum* but had not yet seen the book itself. But some recently discovered documents provide evidence that Leibniz had carefully read and made notes on the *Principia* before composing that piece.[2] Leibniz's standards of conduct in matters of science, like Newton's, should not be taken as norms for the age of the Scientific Revolution. Rather, their actions should be interpreted as signs of the mightiness of two great intellects, on occasion attempting to make history conform to their wishes just as on occasion there was an attempt to coerce nature into yielding confirming evidence. Our judgments of their conduct, some three centuries later, can in no way minimize the majesty of their achievements. Newton was the greatest scientist of his age and Leibniz the most intelligent thinker of the century.

When the queen of Prussia in 1701 asked Leibniz what he thought of Newton, it is reported that "Leibniz said that taking Mathematicks from the beginning of the world to the time of Sir I[saac], What he had done was much the better half."[3] We ourselves can say much the same of the *Principia* in relation to the history of physics and astronomy. Whoever studies the *Principia* in awareness of the works of Newton's predecessors will share the high value assigned this work ever since its first publication in 1687 and will rejoice that the human mind has been able to produce so magnificent a creation.

2. Domenico Bertoloni Meli, *Equivalence and Priority: Newton versus Leibniz* (Oxford: Oxford University Press, 1993).

3. Report by Sir A. Fontaine of a dinner conversation with Leibniz at the royal palace in Berlin, Keynes MS. 130.7, sheet 2, King's College, Cambridge, transcribed by R. S. Westfall in *Never at Rest*, p. 721. This recollection must have been written some time after the event since the reported conversation occurred in 1701 and Newton was not knighted until 1705. This note about Leibniz and Newton may have been composed for John Conduitt after Newton's death, when Conduitt began to collect information and recollections to be used in his planned biography of his famous uncle by marriage.

THE *PRINCIPIA*

Translated by I. Bernard Cohen and Anne Whitman
with the assistance of Julia Budenz

NEWTONI

PRINCIPIA

PHILOSOPHIÆ.

The third edition of Newton's *Principia* begins with this half title followed by two inserted leaves. One is a portrait of Newton engraved by George Vertue from a painting by John Vanderbank, the other the "privilege" or license for publication, dated 25 March 1726. (Grace K. Babson Collection, Burndy Library)

NEWTON'S

PRINCIPLES

OF PHILOSOPHY

PHILOSOPHIÆ

NATURALIS

PRINCIPIA

MATHEMATICA.

AUCTORE

ISAACO NEWTONO, Eq. Aur.

Editio tertia aucta & emendata.

LONDINI:

Apud Guil. & Joh. Innys, Regiæ Societatis typographos.

MDCCXXVI.

Title page of the third edition of Newton's *Principia*. In the original, the words PHILOSOPHIAE and PRINCIPIA are printed in red, as are ISAACO NEWTONO and LONDINI. (Grace K. Babson Collection, Burndy Library)

MATHEMATICAL
PRINCIPLES
OF NATURAL
PHILOSOPHY.

WRITTEN BY
Sir ISAAC NEWTON.

Third edition, enlarged & revised.

LONDON:
WILL. & JNO. INNYS, printers to the Royal Society.
M DCC XXVI.

ILLUSTRISSIMÆ

SOCIETATI REGALI

A

SERENISSIMO REGE

CAROLO II

AD PHILOSOPHIAM PROMOVENDAM

FUNDATÆ,

ET

AUSPICIIS

SERENISSIMI REGIS

GEORGII

FLORENTI

TRACTATUM HUNC D.D.D.

IS. NEWTON.

TO THE MOST ILLUSTRIOUS

ROYAL SOCIETY,

FOUNDED

FOR THE PROMOTION OF PHILOSOPHY

BY

HIS MOST SERENE MAJESTY

CHARLES II,

AND

FLOURISHING

UNDER THE PATRONAGE OF

HIS MOST SERENE MAJESTY

GEORGE,

THIS TREATISE IS DEDICATED.

IS. NEWTON.

The Latin dedication to the third edition (*opposite*; Grace K. Babson Collection, Burndy Library) describes the Royal Society as "ad philosophiam promovendam," in the sense of the promotion of natural philosophy or science. In this expression, Newton was producing a variant of the official name, "The Royal Society of London for Promoting Natural Knowledge." The Latin original of this English version, however, is "Regalis Societas Londini pro scientia naturali promovenda," as stated in the third charter. In the first edition of the *Principia*, the latter part of the dedication reads: "and flourishing under the patronage of the Most Powerful Monarch James II"; additionally, it is said that this treatise is "most humbly" ("humillime") dedicated. In the second edition, the latter part of the dedication reads: "and flourishing under the patronage of the Most August Queen Anne."

Ode on This Splendid Ornament of Our Time and Our Nation, the Mathematico-Physical Treatise by the Eminent Isaac Newton

Behold the pattern of the heavens, and the balances of the divine structure;
Behold Jove's calculation and the laws
That the creator of all things, while he was setting the beginnings of the world,
would not violate;
Behold the foundations he gave to his works.
Heaven has been conquered and its innermost secrets are revealed;
The force that turns the outermost orbs around is no longer hidden.
The Sun sitting on his throne commands all things
To tend downward toward himself, and does not allow the chariots of the
heavenly bodies to move
Through the immense void in a straight path, but hastens them all along
In unmoving circles around himself as center.
Now we know what curved path the frightful comets have;
No longer do we marvel at the appearances of a bearded star.
From this treatise we learn at last why silvery Phoebe moves at an unequal pace,
Why, till now, she has refused to be bridled by the numbers of any astronomer,
Why the nodes regress, and why the upper apsides move forward.
We learn also the magnitude of the forces with which wandering Cynthia
Impels the ebbing sea, while its weary waves leave the seaweed far behind
And the sea bares the sands that sailors fear, and alternately beat high up on
the shores.
The things that so often vexed the minds of the ancient philosophers
And fruitlessly disturb the schools with noisy debate
We see right before our eyes, since mathematics drives away the cloud.

Error and doubt no longer encumber us with mist;
For the keenness of a sublime intelligence has made it possible for us to enter
The dwellings of the gods above and to climb the heights of heaven.

Mortals arise, put aside earthly cares,
And from this treatise discern the power of a mind sprung from heaven,
Far removed from the life of beasts.
He who commanded us by written tablets to abstain from murder,
Thefts, adultery, and the crime of bearing false witness,
Or he who taught nomadic peoples to build walled cities, or he who enriched the nations with the gift of Ceres,
Or he who pressed from the grape a solace for cares,
Or he who with a reed from the Nile showed how to join together
Pictured sounds and to set spoken words before the eyes,
Exalted the human lot less, inasmuch as he was concerned with only a few comforts of a wretched life,
And thus did less than our author for the condition of mankind.
But we are now admitted to the banquets of the gods;
We may deal with the laws of heaven above; and we now have
The secret keys to unlock the obscure earth; and we know the immovable order of the world
And the things that were concealed from the generations of the past.

O you who rejoice in feeding on the nectar of the gods in heaven,
Join me in singing the praises of NEWTON, who reveals all this,
Who opens the treasure chest of hidden truth,
NEWTON, dear to the Muses,
The one in whose pure heart Phoebus Apollo dwells and whose mind he has filled with all his divine power;
No closer to the gods can any mortal rise.

Edm. Halley

Author's Preface to the Reader

SINCE THE ANCIENTS (according to Pappus) considered *mechanics* to be of the greatest importance in the investigation of nature and science and since the moderns—rejecting substantial forms and occult qualities—have undertaken to reduce the phenomena of nature to mathematical laws, it has seemed best in this treatise to concentrate on *mathematics* as it relates to natural philosophy. The ancients divided *mechanics* into two parts: the *rational*, which proceeds rigorously through demonstrations, and the *practical*.[a] *Practical mechanics* is the subject that comprises all the manual arts, from which the subject of *mechanics* as a whole has adopted its name. But since those who practice an art do not generally work with a high degree of exactness, the whole subject of *mechanics* is distinguished from *geometry* by the attribution of exactness to *geometry* and of anything less than exactness to *mechanics*. Yet the errors do not come from the art but from those who practice the art. Anyone who works with less exactness is a more imperfect mechanic, and if anyone could work with the greatest exactness, he would be the most perfect mechanic of all. For the description of straight lines and circles, which is the foundation of *geometry*, appertains to *mechanics*. *Geometry*

All notes to the translation are keyed to the text by superscript letters. When a note is introduced by two letters, such as "aa," it refers to that part of the text enclosed between an opening superscript "a" and a final or closing "a."

These notes are, for the most part, extracts from variant passages or expressions as found in the first two editions. The glosses and explanations of the text are to be found in the Guide, the text of which follows the order of Newton's presentation in the *Principia*.

a. Newton's comparison and contrast between the subject of rational or theoretical mechanics and practical mechanics was a common one at the time of the *Principia*. Thus John Harris in his Newtonian *Lexicon Technicum* (London, 1704), citing the authority of John Wallis, made a distinction between the two as follows. One was a "Geometry of Motion," a "Mathematical Science which shews the Effects of *Powers*, or moving Forces," and "demonstrates the Laws of Motion." The other is "commonly taken for those *Handy-crafts*, which require as well the Labour of the Hands, as the Study of the Brain." The subject of the *Principia* became generally known as "rational mechanics" following Newton's use of that name in his Preface.

does not teach how to describe these straight lines and circles, but postulates such a description. For *geometry* postulates that a beginner has learned to describe lines and circles exactly before he approaches the threshold of *geometry*, and then it teaches how problems are solved by these operations. To describe straight lines and to describe circles are problems, but not problems in *geometry. Geometry* postulates the solution of these problems from *mechanics* and teaches the use of the problems thus solved. And *geometry* can boast that with so few principles obtained from other fields, it can do so much. Therefore *geometry* is founded on mechanical practice and is nothing other than that part of *universal mechanics* which reduces the art of measuring to exact propositions and demonstrations. But since the manual arts are applied especially to making bodies move, *geometry* is commonly used in reference to magnitude, and *mechanics* in reference to motion. In this sense *rational mechanics* will be the science, expressed in exact propositions and demonstrations, of the motions that result from any forces whatever and of the forces that are required for any motions whatever. The ancients studied this part of *mechanics* in terms of the *five powers* that relate to the manual arts [i.e., the five mechanical powers] and paid hardly any attention to gravity (since it is not a manual power) except in the moving of weights by these powers. But since we are concerned with natural philosophy rather than manual arts, and are writing about natural rather than manual powers, we concentrate on aspects of gravity, levity, elastic forces, resistance of fluids, and forces of this sort, whether attractive or impulsive. And therefore our present work sets forth mathematical principles of natural philosophy. For the basic problem [*lit.* whole difficulty[b]] of philosophy seems to be to discover the forces of nature from the phenomena of motions and then to demonstrate the other phenomena from these forces. It is to these ends that the general propositions in books 1 and 2 are directed, while in book 3 our explanation of the system of the world illustrates these propositions. For in book 3, by means of propositions demonstrated mathematically in books 1 and 2, we derive from celestial phenomena the gravitational forces by which bodies tend toward the sun and toward the individual planets. Then the motions of the planets, the comets, the moon, and the sea are deduced from these forces by propositions that are also mathematical. If only we could derive the other phenomena of nature from mechanical principles by the same kind of reasoning! For many things lead me to have a suspicion that all phenomena may depend on certain forces by which the particles of bodies, by causes not yet known, either are impelled toward one another and cohere in regular figures, or are repelled

b. Newton would seem to be expressing in Latin more or less the same concept that later appears in English (in query 28 of the *Opticks*) as "the main Business of natural Philosophy."

from one another and recede. Since these forces are unknown, philosophers have hitherto made trial of nature in vain. But I hope that the principles set down here will shed some light on either this mode of philosophizing or some truer one.

In the publication of this work, Edmond Halley, a man of the greatest intelligence and of universal learning, was of tremendous assistance; not only did he correct the typographical errors and see to the making of the woodcuts, but it was he who started me off on the road to this publication. For when he had obtained my demonstration of the shape of the celestial orbits, he never stopped asking me to communicate it to the Royal Society, whose subsequent encouragement and kind patronage made me begin to think about publishing it. But after I began to work on the inequalities of the motions of the moon, and then also began to explore other aspects of the laws and measures of gravity and of other forces, the curves that must be described by bodies attracted according to any given laws, the motions of several bodies with respect to one another, the motions of bodies in resisting mediums, the forces and densities and motions of mediums, the orbits of comets, and so forth, I thought that publication should be put off to another time, so that I might investigate these other things and publish all my results together. I have grouped them together in the corollaries of prop. 66 the inquiries (which are imperfect) into lunar motions, so that I might not have to deal with these things one by one in propositions and demonstrations, using a method more prolix than the subject warrants, which would have interrupted the sequence of the remaining propositions. There are a number of things that I found afterward which I preferred to insert in less suitable places rather than to change the numbering of the propositions and the cross-references. I earnestly ask that everything be read with an open mind and that the defects in a subject so difficult may be not so much reprehended as investigated, and kindly supplemented, by new endeavors of my readers.

Trinity College, Cambridge Is. Newton
8 May 1686

Author's Preface to the Second Edition

IN THIS SECOND EDITION of the *Principles*, many emendations have been made here and there, and some new things have been added. In sec. 2 of book 1, the finding of forces by which bodies could revolve in given orbits has been made easier and has been enlarged. In sec. 7 of book 2, the theory of the resistance of fluids is investigated more accurately and confirmed by new experiments. In book 3 the theory of the moon and the precession of the equinoxes are deduced more fully from their principles; and the theory of comets is confirmed by more examples of their orbits, calculated with greater accuracy.

London
28 March 1713

Is. Newton

Editor's Preface to the Second Edition

THE LONG-AWAITED NEW EDITION of Newton's *Principles of Natural Philosophy* is presented to you, kind reader, with many corrections and additions. The main topics of this celebrated work are listed in the table of contents and the index prepared for this edition. The major additions or changes are indicated in the author's preface. Now something must be said about the method of this philosophy.

Those who have undertaken the study of natural science can be divided into roughly three classes. There have been those who have endowed the individual species of things with specific occult qualities, on which—they have then alleged—the operations of individual bodies depend in some unknown way. The whole of Scholastic doctrine derived from Aristotle and the Peripatetics is based on this. Although they affirm that individual effects arise from the specific natures of bodies, they do not tell us the causes of those natures, and therefore they tell us nothing. And since they are wholly concerned with the names of things rather than with the things themselves, they must be regarded as inventors of what might be called philosophical jargon, rather than as teachers of philosophy.

Therefore, others have hoped to gain praise for greater carefulness by rejecting this useless hodgepodge of words. And so they have held that all matter is homogeneous, and that the variety of forms that is discerned in bodies all arises from certain very simple and easily comprehensible attributes of the component particles. And indeed they are right to set up a progression from simpler things to more compounded ones, so long as they do not give those primary attributes of the particles any characteristics other than those given by nature itself. But when they take the liberty of imagining that the unknown shapes and sizes of the particles are whatever they please, and of assuming their uncertain positions and motions, and even further of feigning certain occult fluids that permeate the pores of bodies very freely, since they are endowed with an omnipotent subtlety and are acted on by occult motions: when they do this, they are drifting off into dreams, ignoring the true constitution of things, which is obviously to be sought in vain from false conjectures, when it can scarcely be found out even by the most certain observations. Those who take the foundation of their speculations from hypotheses, even if they then proceed most rigorously according to mechanical laws, are merely putting together a romance, elegant perhaps and charming, but nevertheless a romance.

There remains then the third type, namely, those whose natural philosophy is based on experiment. Although they too hold that the causes of all things are to be derived from the simplest possible principles, they assume nothing as a principle that has not yet been thoroughly proved from phenomena. They do not contrive hypotheses, nor do they admit them into natural science otherwise than as questions whose truth may be discussed. Therefore they proceed by a twofold method, analytic and synthetic. From certain selected phenomena they deduce by analysis the forces of nature and the simpler laws of those forces, from which they then give the constitution of the rest of the phenomena by synthesis. This is that incomparably best way of philosophizing which our most celebrated author thought should be justly embraced in preference to all others. This alone he judged worthy of being cultivated and enriched by the expenditure of his labor. Of this therefore he has given a most illustrious example, namely, the explication of the system of the world most successfully deduced from the theory of gravity. That the force of gravity is in all bodies universally, others have suspected or imagined; Newton was the first and only one who was able to demonstrate it [universal gravity] from phenomena and to make it a solid foundation for his brilliant theories.

I know indeed that some men, even of great reputation, unduly influenced by certain prejudices, have found it difficult to accept this new principle [of gravity] and have repeatedly preferred uncertainties to certainties. It is not my intention to carp at their reputation; rather, I wish to give you in brief, kind reader, the basis for making a fair judgment of the issue for yourself.

Therefore, to begin our discussion with what is simplest and nearest to us, let us briefly consider what the nature of gravity is in terrestrial bodies, so that when we come to consider celestial bodies, so very far removed from us, we may proceed more securely. It is now agreed among all philosophers that all bodies on or near the earth universally gravitate toward the earth. Manifold experience has long confirmed that there are no truly light bodies. What is called relative levity is not true levity, but only apparent, and arises from the more powerful gravity of contiguous bodies.

Furthermore, just as all bodies universally gravitate toward the earth, so the earth in turn gravitates equally toward the bodies; for the action of gravity is mutual and is equal in both directions. This is shown as follows. Let the whole body of the earth be divided into any two parts, whether equal or in any way unequal; now, if the weights of the parts toward each other were not equal, the lesser weight would yield to the greater, and the parts, joined together, would proceed to move straight on without limit in the direction toward which the greater weight tends, entirely contrary to experience. Therefore the necessary conclusion is that the weights of the parts are in equilibrium—that is, that the action of gravity is mutual and equal in both directions.

The weights of bodies equally distant from the center of the earth are as the quantities of matter in the bodies. This is gathered from the equal acceleration of all bodies falling from rest by the force of their weights; for the forces by which unequal bodies are equally accelerated must be proportional to the quantities of matter to be moved. Now, that all falling bodies universally are equally accelerated is evident from this, that in the vacuum produced by Boyle's air pump (that is, with the resistance of the air removed), they describe, in falling, equal spaces in equal times, and this is proved more exactly by experiments with pendulums.

The attractive forces of bodies, at equal distances, are as the quantities of matter in the bodies. For, since bodies gravitate toward the earth, and the earth in turn gravitates toward the bodies, with equal moments [i.e., strengths or powers], the weight of the earth toward each body, or the force by which the body attracts the earth, will be equal to the weight of the body toward the earth. But, as mentioned above, this weight is as the quantity of matter in the body, and so the force by which each body attracts the earth, or the absolute force of the body, will be as its quantity of matter.

Therefore the attractive force of entire bodies arises and is compounded from the attractive force of the parts, since (as has been shown), when the amount of matter is increased or diminished, its force is proportionally increased or diminished. Therefore the action of the earth must result from the combined actions of its parts; hence all terrestrial bodies must attract one another by absolute forces that are proportional to the attracting matter. This is the nature of gravity on earth; let us now see what it is in the heavens.

Every body perseveres in its state either of being at rest or of moving uniformly straight forward, except insofar as it is compelled by impressed forces to change that state: this is a law of nature accepted by all philosophers. It follows that bodies that move in curves, and so continually deviate from straight lines tangent to their orbits, are kept in a curvilinear path by some continually acting force. Therefore, for the planets to revolve in curved orbits, there will necessarily be some force by whose repeated actions they are unceasingly deflected from the tangents.

Now, it is reasonable to accept something that can be found by mathematics and proved with the greatest certainty: namely, that all bodies moving in some curved line described in a plane, which by a radius drawn to a point (either at rest or moving in any way) describe areas about that point proportional to the times, are urged by forces that tend toward that same point. Therefore, since it is agreed among astronomers that the primary planets describe areas around the sun proportional to the times, as do the secondary planets around their own primary planets, it follows that the force by which they are continually pulled away from rectilinear tangents and are compelled to revolve in curvilinear orbits is directed toward the bodies that are situated in the centers of the orbits. Therefore this force can, appropriately, be called centripetal with respect to the revolving body, and attractive with respect to the central body, from whatever cause it may in the end be imagined to arise.

The following rules must also be accepted and are mathematically demonstrated. If several bodies revolve with uniform motion in concentric circles, and if the squares of the periodic times are as the cubes of the distances from the common center, then the centripetal forces of the revolving bodies will be inversely as the squares of the distances. Again, if the bodies revolve in orbits that are very nearly circles, and if the apsides of the orbits are at rest, then the centripetal forces of the revolving bodies will be inversely as the squares of the distances. Astronomers agree that one or the other case holds for all the planets, [both primary and secondary]. Therefore the centripetal forces of all the planets are inversely as the squares of the distances from the centers of the orbits. If anyone objects that the apsides of the planets, especially the apsides of the moon, are not completely at rest but are carried progressively forward [or in consequentia] with a slow motion, it can be answered that even if we grant that this very slow motion arises from a slight deviation of the centripetal force from the proportion of the inverse square, this difference can be found by mathematical computation and is quite insensible. For the ratio of the moon's centripetal force itself, which should deviate most of all from the square, will indeed exceed the square by a very little, but it will be about sixty times closer to it than to the cube. But our answer to the objection will be truer if we say that this progression of the apsides does not arise from a deviation from the proportion of the [inverse] square but from another and entirely different cause, as is admirably shown in Newton's philosophy. As a result, the centripetal forces by which the primary planets tend toward the sun, and the secondary planets toward their primaries, must be exactly as the squares of the distances inversely.

From what has been said up to this point, it is clear that the planets are kept in their orbits by some force continually acting upon them, that this force is always directed toward the centers of the orbits, and that its efficacy is increased in approaching the center and decreased in receding from the center—actually increased in the same proportion in which the square of the distance is decreased, and decreased in the same proportion in which the square of the distance is increased. Let us now, by comparing the centripetal forces of the planets and the force of gravity, see whether or not they might be of the same kind. They will be of the same kind if the same laws and the same attributes are found in both. Let us first, therefore, consider the centripetal force of the moon, which is closest to us.

When bodies are let fall from rest, and are acted on by any forces whatever, the rectilinear spaces described in a given time at the very beginning of the motion are proportional to the forces themselves; this of course follows from mathematical reasoning. Therefore the centripetal force of the moon revolving in its orbit will be to the force of gravity on the earth's surface as the space that the moon would describe in a minimally small time in descending toward the earth by its centripetal force—supposing it to be deprived of all circular motion—is to the space that a heavy body describes in the same minimally small time in the vicinity of the earth, in falling by the force of its own gravity. The first of these spaces is equal to the versed sine of the arc described by the moon during the same time, inasmuch as this versed sine measures the departure of the moon from the tangent caused by centripetal force and thus can be calculated if the moon's periodic time and its distance from the center of the earth are both given. The second space is found by experiments with pendulums, as Huygens has shown. Therefore, the result of the calculation will be that the first space is to the second space, or the centripetal force of the moon revolving in its orbit is to the force of gravity on the surface of the earth, as the square of the semidiameter of the earth is to the square of the semidiameter of the orbit. By what is shown above, the same ratio holds for the centripetal force of the moon revolving in its orbit and the centripetal force of the moon if it were near the earth's surface. Therefore this centripetal force near the earth's surface is equal to the force of gravity. They are not, therefore, different forces, but one and the same; for if they were different, bodies acted on by both forces together would fall to the earth twice as fast as from the force of gravity alone. And therefore it is clear that this centripetal force by which the moon is continually either drawn or impelled from the tangent and is kept in its orbit is the very force of terrestrial gravity extending as far as the moon. And indeed it is reasonable for this force to extend itself to enormous distances, since one can observe no sensible diminution of it even on the highest peaks of mountains. Therefore the moon gravitates toward the earth. Further, by mutual action, the earth in turn gravitates equally toward the moon, a fact which is abundantly confirmed in this philosophy, when we deal with the tide of the sea and the precession of the equinoxes, both of which arise from the action of both the moon and the sun upon the earth. Hence finally we learn also by what law the force of gravity decreases at greater distances from the earth. For since gravity is not different from the moon's centripetal force, which is inversely proportional to the square of the distance, gravity will also be diminished in the same ratio.

Let us now proceed to the other planets. The revolutions of the primary planets about the sun and of the secondary planets about Jupiter and Saturn are phenomena of the same kind as the revolution of the moon about the earth; furthermore, it has been demonstrated that the centripetal forces of the primary planets are directed toward the center of the sun, and those of the secondary planets toward the centers of Jupiter and of Saturn, just as the moon's centripetal force is directed toward the center of the earth; and, additionally, all these forces are inversely as the squares of the distances from the centers, just as the force of the moon is inversely as the square of the distance from the earth. Therefore it must be concluded that all of these primary and secondary planets have the same nature. Hence, as the moon gravitates toward the earth, and the earth in turn gravitates toward the moon, so also all the secondary planets will gravitate toward their primaries, and the primaries in turn toward the secondaries, and also all the primary planets will gravitate toward the sun, and the sun in turn toward the primary planets.

Therefore the sun gravitates toward all the primary and secondary planets, and all these toward the sun. For the secondary planets, while accompanying their primaries, revolve with them around the sun. By the same argument, therefore, both kinds of planets gravitate toward the sun, and the sun toward them. Additionally, that the secondary planets gravitate toward the sun is also abundantly clear from the inequalities of the moon, concerning which a most exact theory is presented with marvelous sagacity in the third book of this work.

The motion of the comets shows very clearly that the attractive force of the sun is propagated in every direction to enormous distances and is diffused to every part of the surrounding space, since the comets, starting out from immense distances, come into the vicinity of the sun and sometimes approach so very close to it that in their perihelia they all seemingly touch its globe. Astronomers until now have tried in vain to find the theory of these comets; now at last, in our time, our most illustrious author has succeeded in finding the theory and has demonstrated it with the greatest certainty from observations. It is therefore evident that the comets move in conic sections having their foci in the center of the sun and by radii drawn to the sun describe areas proportional to the times. From these phenomena it is manifest and it is mathematically proved that the forces by which the comets are kept in their orbits are directed toward the sun and are inversely as the squares of their distances from its center. Thus the comets gravitate toward the sun; and so the attractive force of the sun reaches not only to the bodies of the planets, which are at fixed distances and in nearly the same plane, but also to the comets, which are in the most diverse regions of the heavens and at the most diverse distances. It is the nature of gravitating bodies, therefore, that they propagate their forces at all distances to all other gravitating bodies. From this it follows that all planets and comets universally attract one another and are heavy toward one another—which is also confirmed by the perturbation of Jupiter and Saturn, known to astronomers and arising from the actions of these planets upon each other; it is also confirmed by the very slow motion of the apsides that was mentioned above and that arises from an entirely similar cause.

We have at last reached the point where it must be acknowledged that the earth and the sun and all the celestial bodies that accompany the sun attract one another. Therefore every least particle of each of them will have its own attractive force in proportion to the quantity of matter, as was shown above for terrestrial bodies. And at different distances their forces will also be in the squared ratio of the distances inversely; for it is mathematically demonstrated that particles attracting by this law must constitute globes attracting by the same law.

The preceding conclusions are based upon an axiom which is accepted by every philosopher, namely, that effects of the same kind—that is, effects whose known properties are the same—have the same causes, and their properties which are not yet known are also the same. For if gravity is the cause of the fall of a stone in Europe, who can doubt that in America the cause of the fall is the same? If gravity is mutual between a stone and the earth in Europe, who will deny that it is mutual in America? If in Europe the attractive force of the stone and the earth is compounded of the attractive forces of the parts, who will deny that in America the force is similarly compounded? If in Europe the attraction of the earth is propagated to all kinds of bodies and to all distances, why should we not say that in America it is propagated in the same way? All philosophy is based on this rule, inasmuch as, if it is taken away, there is then nothing we can affirm about things universally. The constitution of individual things can be found by observations and experiments; and proceeding from there, it is only by this rule that we make judgments about the nature of things universally.

Now, since all terrestrial and celestial bodies on which we can make experiments or observations are heavy, it must be acknowledged without exception that gravity belongs to all bodies universally. And just as we must not conceive of bodies that are not extended, mobile, and impenetrable, so we should not conceive of any that are not heavy. The extension, mobility, and impenetrability of bodies are known only through experiments; it is in exactly the same way that the gravity of bodies is known. All bodies for which we have observations are extended and mobile and impenetrable; and from this we conclude that all bodies universally are extended and mobile and impenetrable, even those for which we do not have observations. Thus all bodies for which we have observations are heavy; and from this we conclude that all bodies universally are heavy, even those for which we do not have observations. If anyone were to say that the bodies of the fixed stars are not heavy, since their gravity has not yet been observed, then by the same argument one would be able to say that they are neither extended nor mobile nor impenetrable, since these properties of the fixed stars have not yet been observed. Need I go on? Among the primary qualities of all bodies universally, either gravity will have a place, or extension, mobility, and impenetrability will not. And the nature of things either will be correctly explained by the gravity of bodies or will not be correctly explained by the extension, mobility, and impenetrability of bodies.

I can hear some people disagreeing with this conclusion and muttering something or other about occult qualities. They are always prattling on and on to the effect that gravity is something occult, and that occult causes are to be banished completely from philosophy. But it is easy to answer them: occult causes are not those causes whose existence is very clearly demonstrated by observations, but only those whose existence is occult, imagined, and not yet proved. Therefore gravity is not an occult cause of celestial motions, since it has been shown from phenomena that this force really exists. Rather, occult causes are the refuge of those who assign the governing of these motions to some sort of vortices of a certain matter utterly fictitious and completely imperceptible to the senses.

But will gravity be called an occult cause and be cast out of natural philosophy on the grounds that the cause of gravity itself is occult and not yet found? Let those who so believe take care lest they believe in an absurdity that, in the end, may overthrow the foundations of all philosophy. For causes generally proceed in a continuous chain from compound to more simple; when you reach the simplest cause, you will not be able to proceed any further. Therefore no mechanical explanation can be given for the simplest cause; for if it could, the cause would not yet be the simplest. Will you accordingly call these simplest causes occult, and banish them? But at the same time the causes most immediately depending on them, and the causes that in turn depend on these causes, will also be banished, until philosophy is emptied and thoroughly purged of all causes.

Some say that gravity is preternatural and call it a perpetual miracle. Therefore they hold that it should be rejected, since preternatural causes have no place in physics. It is hardly worth spending time on demolishing this utterly absurd objection, which of itself undermines all of philosophy. For either they will say that gravity is not a property of all bodies—which cannot be maintained—or they will assert that gravity is preternatural on the grounds that it does not arise from other affections of bodies and thus not from mechanical causes. Certainly there are primary affections of bodies, and since they are primary, they do not depend on others. Therefore let them consider whether or not all these are equally preternatural, and so equally to be rejected, and let them consider what philosophy will then be like.

There are some who do not like all this celestial physics just because it seems to be in conflict with the doctrines of Descartes and seems scarcely capable of being reconciled with these doctrines. They are free to enjoy their own opinion, but they ought to act fairly and not deny to others the same liberty that they demand for themselves. Therefore, we should be allowed to adhere to the Newtonian philosophy, which we consider truer, and to prefer causes proved by phenomena to causes imagined and not yet proved. It is the province of true philosophy to derive the natures of things from causes that truly exist, and to seek those laws by which the supreme artificer willed to establish this most beautiful order of the world, not those laws by which he could have, had it so pleased him. For it is in accord with reason that the same effect can arise from several causes somewhat different from one another; but the true cause will be the one from which the effect truly and actually does arise, while the rest have no place in true philosophy. In mechanical clocks one and the same motion of the hour hand can arise from the action of a suspended weight or an internal spring. But if the clock under discussion is really activated by a weight, then anyone will be laughed at if he imagines a spring and on such a premature hypothesis undertakes to explain the motion of the hour hand; for he ought to have examined the internal workings of the machine more thoroughly, in order to ascertain the true principle of the motion in question. The same judgment or something like it should be passed on those philosophers who have held that the heavens are filled with a certain most subtle matter, which is endlessly moved in vortices. For even if these philosophers could account for the phenomena with the greatest exactness on the basis of their hypotheses, still they cannot be said to have given us a true philosophy and to have found the true causes of the celestial motions until they have demonstrated either that these causes really do exist or at least that others do not exist. Therefore if it can be shown that the attraction of all bodies universally has a true place in the nature of things, and if it further can be shown how all the celestial motions are solved by that attraction, then it would be an empty and ridiculous objection if anyone said that those motions should be explained by vortices, even if we gave our fullest assent to the possibility of such an explanation. But we do not give our assent; for the phenomena can by no means be explained by vortices, as our author fully proves with the clearest arguments. It follows that those who devote their fruitless labor to patching up a most absurd figment of their imagination and embroidering it further with new fabrications must be overly indulging their fantasies.

If the bodies of the planets and the comets are carried around the sun by vortices, the bodies carried around must move with the same velocity and in the same direction as the immediately surrounding parts of the vortices, and must have the same density or the same force of inertia in proportion to the bulk of the matter.

But it is certain that planets and comets, while they are in the same regions of the heavens, move with a variety of velocities and directions. Therefore it necessarily follows that those parts of the celestial fluid that are at the same distances from the sun revolve in the same time in different directions with different velocities; for there will be need of one direction and velocity to permit the planets to move through the heavens, and another for the comets. Since this cannot be accounted for, either it will have to be confessed that all the celestial bodies are not carried by the matter of a vortex, or it will have to be said that their motions are to be derived not from one and the same vortex, but from more than one, differing from one another and going through the same space surrounding the sun.

If it is supposed that several vortices are contained in the same space and penetrate one another and revolve with different motions, then—since these motions must conform to the motions of the bodies being carried around, motions highly regular in conic sections that are sometimes extremely eccentric and sometimes very nearly circular—it will be right to ask how it can happen that these same vortices keep their integrity without being in the least perturbed through so many centuries by the interactions of their matter. Surely, if these imaginary motions are more complex and more difficult to explain than the true motions of the planets and comets, I think it pointless to admit them into natural philosophy; for every cause must be simpler than its effect. Granted the freedom to invent any fiction, let someone assert that all the planets and comets are surrounded by atmospheres, as our earth is, a hypothesis that will certainly seem more reasonable than the hypothesis of vortices. Let him then assert that these atmospheres, of their own nature, move around the sun and describe conic sections, a motion that can surely be much more easily conceived than the similar motion of vortices penetrating one another. Finally, let him maintain that it must be believed that the planets themselves and the comets are carried around the sun by their atmospheres, and let him celebrate his triumph for having found the causes of the celestial motions. Anyone who thinks that this fiction should be rejected will also reject the other one; for the hypothesis of atmospheres and the hypothesis of vortices are as alike as two peas in a pod.

Galileo showed that when a stone is projected and moves in a parabola, its deflection from a rectilinear path arises from the gravity of the stone toward the earth, that is, from an occult quality. Nevertheless it can happen that some other philosopher, even more clever, may contrive another cause. He will accordingly imagine that a certain subtle matter, which is not perceived by sight or by touch or by any of the senses, is found in the regions that are most immediately contiguous to the surface of the earth. He will argue, moreover, that this matter is carried in different directions by various and—for the most part—contrary motions and that it describes parabolic curves. Finally he will beautifully show how the stone is deflected and will earn the applause of the crowd. The stone, says he, floats in that subtle fluid and, by following the course of that fluid, cannot but describe the same path. But the fluid moves in parabolic curves; therefore the stone must move in a parabola. Who will not now marvel at the most acute genius of this philosopher, brilliantly deducing the phenomena of nature from mechanical causes [i.e., matter and motion]—at a level comprehensible even to ordinary people! Who indeed will not jeer at that poor Galileo, who undertook by a great mathematical effort once more to bring back occult qualities, happily excluded from philosophy! But I am ashamed to waste any more time on such trifles.

It all finally comes down to this: the number of comets is huge; their motions are highly regular and observe the same laws as the motions of the planets. They move in conic orbits; these orbits are very, very eccentric. Comets go everywhere into all parts of the heavens and pass very freely through the regions of the planets, often contrary to the order of the signs. These phenomena are confirmed with the greatest certainty by astronomical observations and cannot be explained by vortices. Further, these phenomena are even inconsistent with planetary vortices. There will be no room at all for the motions of the comets unless that imaginary matter is completely removed from the heavens.

For if the planets are carried around the sun by vortices, those parts of the vortices that most immediately surround each planet will be of the same density as the planet, as has been said above. Therefore all the matter that is contiguous to the perimeter of the earth's orbit will have the same density as the earth, while all the matter that lies between the earth's orbit and the orbit of Saturn will have either an equal or a greater density. For, in order that the constitution of a vortex may be able to last, the less dense parts must occupy the center, and the more dense parts must be further away from the center. For since the periodic times of the planets are as the $\frac{3}{2}$ powers of the distances from the sun, the periods of the parts of the vortex should keep the same ratio. It follows that the centrifugal forces of these parts will be inversely as the squares of the distances. Therefore those parts that are at a greater distance from the center strive to recede from it by a smaller force; accordingly, if they should be less dense, it would be necessary for them to yield to the greater force by which the parts nearer to the center endeavor to ascend. Therefore the denser parts will ascend, the less dense will descend, and a mutual exchange of places will occur, until the fluid matter of the whole vortex has been arranged in such order that it can now rest in equilibrium [i.e., its parts are completely at rest with respect to one another or no longer have any motion of ascent or descent]. If two fluids of different density are contained in the same vessel, certainly it will happen that the fluid whose density is greater will go to the lowest place under the action of its greater force of gravity, and by similar reasoning it must be concluded that the denser parts of the vortex will go to the highest place under the action of their greater centrifugal force. Therefore the whole part of the vortex that lies outside the earth's orbit (much the greatest part) will have a density and so a force of inertia (proportional to the quantity of matter) that will not be smaller than the density and force of inertia of the earth. From this will arise a huge and very noticeable resistance to the comets as they pass through, not to say a resistance that rightly seems to be able to put a complete stop to their motion and absorb it entirely. It is however clear from the altogether regular motion of comets that they encounter no resistance that can be in the least perceived, and thus that they do not come upon any matter that has any force of resistance, or accordingly that has any density or force of inertia. For the resistance of mediums arises either from the inertia of fluid matter or from its friction.[a] That which arises from friction is extremely slight and indeed can scarcely be observed in commonly known fluids, unless they are very tenacious like oil and honey. The resistance that is encountered in air, water, quicksilver, and nontenacious fluids of this sort is almost wholly of the first kind and cannot be decreased in subtlety by any further degree, if the fluid's density or force of inertia—to which this resistance is always proportional—remains the same. This is most clearly demonstrated by our author in his brilliant theory of the resistance of fluids, which in this second edition is presented in a somewhat more accurate manner and is more fully confirmed by experiments with falling bodies.

As bodies move forward, they gradually communicate their motion to a surrounding fluid, and by communicating their motion lose it, and by losing it are retarded. Therefore the retardation is proportional to the motion so communicated, and the motion communicated (where the velocity of the moving body is given) is as the density of the fluid; therefore the retardation or resistance will also be as the density of the fluid and cannot be removed by any means unless the fluid, returning to the back of the body, restores the lost motion. But this cannot be the case unless the force of the fluid on the rear of the body is equal to the force the body exerts on the fluid in front, that is, unless the relative velocity with which the fluid pushes the body from behind is equal to the velocity with which the body pushes the fluid, that is, unless the absolute velocity of the returning fluid is twice as great as the absolute velocity of the fluid pushed forward, which cannot happen. Therefore there is no way in which the resistance of fluids that arises from their density and force of inertia can be taken away. And so it must be concluded that the celestial fluid has no force of inertia, since it has no force of resistance; it has

a. Literally, lack of lubricity or slipperiness.

no force by which motion may be communicated, since it has no force of inertia; it has no force by which any change may be introduced into one or more bodies, since it has no force by which motion may be communicated; it has no efficacy at all, since it has no faculty to introduce any change. Surely, therefore, this hypothesis, plainly lacking in any foundation and not even marginally useful to explain the nature of things, may well be called utterly absurd and wholly unworthy of a philosopher. Those who hold that the heavens are filled with fluid matter, but suppose this matter to have no inertia, are saying there is no vacuum but in fact are assuming there is one. For, since there is no way to distinguish a fluid matter of this sort from empty space, the whole argument comes down to the names of things and not their natures. But if anyone is so devoted to matter that he will in no way admit a space void of bodies, let us see where this will ultimately lead him.

For such people will say that this constitution of the universe as everywhere full, which is how they imagine it, has arisen from the will of God, so that a very subtle aether pervading and filling all things would be there to facilitate the operations of nature; this cannot be maintained, however, since it has already been shown from the phenomena of comets that this aether has no efficacy. Or they will say that this constitution has arisen from the will of God for some unknown purpose, which ought not to be said either, since a different constitution of the universe could equally well be established by the same argument. Or finally they will say that it has not arisen from the will of God but from some necessity of nature. And so at last they must sink to the lowest depths of degradation, where they have the fantasy that all things are governed by fate and not by providence, that matter has existed always and everywhere of its own necessity and is infinite and eternal. On this supposition, matter will also be uniform everywhere, for variety of forms is entirely inconsistent with necessity. Matter will also be without motion; for if by necessity matter moves in some definite direction with some definite velocity, by a like necessity it will move in a different direction with a different velocity; but it cannot move in different directions with different velocities; therefore it must be without motion. Surely, this world—so beautifully diversified in its forms and motions—could not have arisen except from the perfectly free will of God, who provides and governs all things.

From this source, then, have all the laws that are called laws of nature come, in which many traces of the highest wisdom and counsel certainly appear, but no traces of necessity. Accordingly we should not seek these laws by using untrustworthy conjectures, but learn them by observing and experimenting. He who is confident that he can truly find the principles of physics, and the laws of things, by relying only on the force of his mind and the internal light of his reason should maintain either that the world has existed from necessity and follows the said laws from the same necessity, or that although the order of nature was constituted by the will of God, nevertheless a creature as small and insignificant as he is has a clear understanding of the way things should be. All sound and true philosophy is based on phenomena, which may lead us—however unwilling and reluctant—to principles in which the best counsel and highest dominion of an all-wise and all-powerful being are most clearly discerned; these principles will not be rejected because certain men may perhaps not like them. These men may call the things that they dislike either miracles or occult qualities, but names maliciously given are not to be blamed on the things themselves, unless these men are willing to confess at last that philosophy should be based on atheism. Philosophy must not be overthrown for their sake, since the order of things refuses to be changed.

Therefore honest and fair judges will approve the best method of natural philosophy, which is based on experiments and observations. It need scarcely be said that this way of philosophizing has been illumined and dignified by our illustrious author's well-known book; his tremendous genius, enodating each of the most difficult problems and reaching out beyond the accepted limits of the human, is justly admired and esteemed by all who are more than superficially versed in these matters. Having unlocked the gates, therefore, he has opened our way to the most beautiful mysteries of nature. He has finally so clearly revealed a most elegant structure of the system of the world for our further scrutiny that even were King Alfonso himself to come to life again, he would not find it wanting either in simplicity or in grace of harmony. And hence it is now possible to have a closer view of the majesty of nature, to enjoy the sweetest contemplation, and to worship and venerate more zealously the maker and lord of all; and this is by far the greatest fruit of philosophy. He must be blind who does not at once see, from the best and wisest structures of things, the infinite wisdom and goodness of their almighty creator; and he must be mad who refuses to acknowledge them.

Therefore Newton's excellent treatise will stand as a mighty fortress against the attacks of atheists; nowhere else will you find more effective ammunition against that impious crowd. This was understood long ago, and was first splendidly demonstrated in learned discourses in English and in Latin, by a man of universal learning and at the same time an outstanding patron of the arts, Richard Bentley, a great ornament of his time and of our academy, the worthy and upright master of our Trinity College. I must confess that I am indebted to him on many grounds; you as well, kind reader, will not deny him due thanks. For, as a long-time intimate friend of our renowned author (he considers being celebrated by posterity for this friendship to be of no less value than becoming famous for his own writings, which are the delight of the learned world), he worked simultaneously for the public recognition of his friend and for the advancement of the sciences. Therefore, since the available copies of the first edition were extremely rare and very expensive, he tried with persistent demands to persuade Newton (who is distinguished as much by modesty as by the highest learning) and finally—almost scolding him—prevailed upon Newton to allow him to get out this new edition, under his auspices and at his own expense, perfected throughout and also enriched with significant additions. He authorized me to undertake the not unpleasant duty of seeing to it that all this was done as correctly as possible.

Cambridge, 12 May 1713

Roger Cotes,
Fellow of Trinity College,
Plumian Professor of Astronomy
and Experimental Philosophy

Author's Preface to the Third Edition

IN THIS THIRD EDITION, supervised by Henry Pemberton, M.D., a man greatly skilled in these matters, some things in the second book concerning the resistance of mediums are explained a little more fully than previously, and new experiments are added concerning the resistance of heavy bodies falling in air. In the third book, the argument proving that the moon is kept in its orbit by gravity is presented a little more fully; and new observations, made by Mr. Pound, on the proportion of the diameters of Jupiter to each other have been added. There are also added some observations of the comet that appeared in 1680, which were made in Germany during the month of November by Mr. Kirk, and which recently came into our hands; these observations make it clear how closely parabolic orbits correspond to the motions of comets. The orbit of that comet, by Halley's computations, is determined a little more accurately than heretofore, and in an ellipse. And it is shown that the comet traversed its course through nine signs of the heavens in this elliptical orbit just as exactly as the planets move in the elliptical orbits given by astronomy. There is also added the orbit of the comet that appeared in the year 1713, which was calculated by Mr. Bradley, professor of astronomy at Oxford.

London,
12 Jan. 1725/6.

Is. Newton

[In the third edition, the final Author's Preface was followed by a two-page table of contents and a list of corrigenda.]

MATHEMATICAL PRINCIPLES OF NATURAL PHILOSOPHY

(PHILOSOPHIAE NATURALIS PRINCIPIA MATHEMATICA)

DEFINITIONS

MASS

Definition 1

[a]*Quantity of matter is a measure of matter that arises from its density and volume jointly.*[a]

[b]If the density of air is doubled in a space that is also doubled, there is four times as much air, and there is six times as much if the space is tripled.[b] The case is the same for snow and powders condensed by compression or liquefaction, and also for all bodies that are condensed in various ways by any causes whatsoever. For the present, I am not taking into account any medium, if there should be any, freely pervading the interstices between the parts of

aa. In translating def. 1, we have rendered Newton's "Quantitas materiae est mensura ejusdem . . ." as "Quantity of matter is a measure of matter . . ." rather than the customary ". . . is the measure . . ." The indefinite article is more in keeping with the Latin usage, with its absence of articles, and accords better with the sense in which we have interpreted this definition. See the Guide, §4.2. It should be noted that the indefinite article permits the possibility of the sense of either a definite or an indefinite article, whereas a definite article precludes the possibility of the sense of an indefinite article.

bb. Ed. 3 reads literally: "Air, if the density is doubled, in a space also doubled, becomes quadruple; in [a space] tripled, sextuple." The printer's manuscript for ed. 1 and the printed text of ed. 1 have: "Air twice as dense in twice the space is quadruple." Newton's interleaved copy of ed. 1 has: "Air twice as dense in twice the space is quadruple; in three times [the space], sextuple." Newton's annotated copy of ed. 1 has first: "Air twice as dense in twice the space becomes quadruple; in three times [the space], sextuple." This is then deleted and replaced with: "Air, by doubling the density, in the same container becomes double; in a container twice as large, quadruple; in one three times as large, sextuple; and by tripling the density, it becomes triple in the same container and sextuple in a container twice as large," but the last clause, "and by tripling . . . large," is then deleted.

The manuscript errata at the end of the annotated copy have: "For this quantity, if the density is given [or fixed], is as the volume and, if the volume is given, is as the density and therefore, if neither is given, is as the product of both. Thus indeed Air, if the density is doubled, in a space also doubled, becomes quadruple; in [a space] tripled, sextuple." The first sentence, "For this . . . product of both," and the following two words, "Thus indeed," are inserted over a caret preceding "Air."

An interleaf of the interleaved copy of ed. 1 and then the printed text of ed. 2 have exactly the same phrasing as ed. 3.

bodies. Furthermore, I mean this quantity whenever I use the term "body" or "mass" in the following pages. It can always be known from a body's weight, for—by making very accurate experiments with pendulums—I have found it to be proportional to the weight, as will be shown below.

Definition 2 *Quantity of motion is a measure of motion that arises from the velocity and the quantity of matter jointly.*

The motion of a whole is the sum of the motions of the individual parts, and thus if a body is twice as large as another and has equal velocity there is twice as much motion, and if it has twice the velocity there is four times as much motion.

Definition 3 *Inherent force of matter is the power of resisting by which every body,* [a]*so far as it is able,*[a] *perseveres in its state either of resting or of moving* [b]*uniformly straight forward.*[b]

This force is always proportional to the body and does not differ in any way from the inertia of the mass except in the manner in which it is conceived. Because of the inertia of matter, every body is only with difficulty put out of its state either of resting or of moving. Consequently, inherent force may also be called by the very significant name of force of inertia.[c] Moreover, a body exerts this force only during a change of its state, caused by another force impressed upon it, and this exercise of force is, depending on the viewpoint, both resistance and impetus: resistance insofar as the body, in order to maintain its state, strives against the impressed force, and impetus insofar as the same body, yielding only with difficulty to the force of a resisting obstacle, endeavors to change the state of that obstacle. Resistance is commonly attributed to resting bodies and impetus to moving bodies; but

aa. Newton's Latin clause is "quantum in se est," which here means "to the degree that it can of and by itself." See I. Bernard Cohen, " 'Quantum in se est': Newton's Concept of Inertia in Relation to Descartes and Lucretius," *Notes and Records of the Royal Society of London* 19 (1964): 131–155.

bb. Newton's "in directum" (used together with "uniformiter" ["uniformly"]) has the sense of moving straight on, of going continuously straight forward, and therefore in a straight line. In an earlier version, Newton had used the phrase "in linea recta" ("in a right line" or "in a straight line"), but by the time of the *Principia* he had rejected this expression in favor of "in directum." For details, see the Guide, §10.2. On Newton's "vis insita" and our rendition, see the Guide, §4.7.

c. Newton's interleaved copy of ed. 2 adds the following, which was never printed: "I do not mean Kepler's force of inertia, by which bodies tend toward rest, but a force of remaining in the same state either of resting or of moving."

motion and rest, in the popular sense of the terms, are distinguished from each other only by point of view, and bodies commonly regarded as being at rest are not always truly at rest.

Definition 4

Impressed force is the action exerted on a body to change its state either of resting or of moving uniformly straight forward.

This force consists solely in the action and does not remain in a body after the action has ceased. For a body perseveres in any new state solely by the force of inertia. Moreover, there are various sources of impressed force, such as percussion, pressure, or centripetal force.

Definition 5

Centripetal force is the force by which bodies are drawn from all sides, are impelled, or in any way tend, toward some point as to a center.

One force of this kind is gravity, by which bodies tend toward the center of the earth; another is magnetic force, by which iron seeks a lodestone; and yet another is that force, whatever it may be, by which the planets are continually drawn back from rectilinear motions and compelled to revolve in curved lines.

[a]A stone whirled in a sling endeavors to leave the hand that is whirling it, and by its endeavor it stretches the sling, doing so the more strongly the more swiftly it revolves; and as soon as it is released, it flies away. The force opposed to that endeavor, that is, the force by which the sling continually draws the stone back toward the hand and keeps it in an orbit, I call centripetal, since it is directed toward the hand as toward the center of an orbit. And the same applies to all bodies [b]that are made to move in orbits.[b] They all endeavor to recede from the centers of their orbits, and unless some force opposed to that endeavor is present, restraining them and keeping them in orbits and hence called by me centripetal, they will go off in straight lines with uniform motion. If a projectile were deprived of the force of gravity, it would not be deflected toward the earth but would go off in a straight line into the heavens and do so with uniform motion, provided that the resistance of the air were removed. The projectile, by its gravity, is drawn back from a rectilinear course and continually deflected toward the earth, and this is so

aa. Ed. 1 lacks this.

bb. See the Guide, §2.4.

to a greater or lesser degree in proportion to its gravity and its velocity of motion. The less its gravity in proportion to its quantity of matter, or the greater the velocity with which it is projected, the less it will deviate from a rectilinear course and the farther it will go. If a lead ball were projected with a given velocity along a horizontal line from the top of some mountain by the force of gunpowder and went in a curved line for a distance of two miles before falling to the earth, then the same ball projected with twice the velocity would go about twice as far and with ten times the velocity about ten times as far, provided that the resistance of the air were removed. And by increasing the velocity, the distance to which it would be projected could be increased at will and the curvature of the line that it would describe could be decreased, in such a way that it would finally fall at a distance of 10 or 30 or 90 degrees or even go around the whole earth or, lastly, go off into the heavens and continue indefinitely in this motion. And in the same way that a projectile could, by the force of gravity, be deflected into an orbit and go around the whole earth, so too the moon, whether by the force of gravity—if it has gravity—or by any other force by which it may be urged toward the earth, can always be drawn back toward the earth from a rectilinear course and deflected into its orbit; and without such a force the moon cannot be kept in its orbit. If this force were too small, it would not deflect the moon sufficiently from a rectilinear course; if it were too great, it would deflect the moon excessively and draw it down from its orbit toward the earth. In fact, it must be of just the right magnitude, and mathematicians have the task of finding the force by which a body can be kept exactly in any given orbit with a given velocity and, alternatively, to find the curvilinear path into which a body leaving any given place with a given velocity is deflected by a given force.[a]

The quantity of centripetal force is of three kinds: absolute, accelerative, and motive.

Definition 6 *The absolute quantity of centripetal force is the measure of this force that is greater or less in proportion to the efficacy of the cause propagating it from a center through the surrounding regions.*

An example is magnetic force, which is greater in one lodestone and less in another, in proportion to the bulk or potency of the lodestone.

Definition 7

The accelerative quantity of centripetal force is the measure of this force that is proportional to the velocity which it generates in a given time.

One example is the potency of a lodestone, which, for a given lodestone, is greater at a smaller distance and less at a greater distance. Another example is the force that produces gravity, which is greater in valleys and less on the peaks of high mountains and still less (as will be made clear below) at greater distances from the body of the earth, but which is everywhere the same at equal distances, because it equally accelerates all falling bodies (heavy or light, great or small), provided that the resistance of the air is removed.

Definition 8

The motive quantity of centripetal force is the measure of this force that is proportional to the motion which it generates in a given time.

An example is weight, which is greater in a larger body and less in a smaller body; and in one and the same body is greater near the earth and less out in the heavens. This quantity is the centripetency, or propensity toward a center, of the whole body, and (so to speak) its weight, and it may always be known from the force opposite and equal to it, which can prevent the body from falling.

These quantities of forces, for the sake of brevity, may be called motive, accelerative, and absolute forces, and, for the sake of differentiation, may be referred to bodies seeking a center, to the places of the bodies, and to the center of the forces: that is, motive force may be referred to a body as an endeavor of the whole directed toward a center and compounded of the endeavors of all the parts; accelerative force, to the place of the body as a certain efficacy diffused from the center through each of the surrounding places in order to move the bodies that are in those places; and absolute force, to the center as having some cause without which the motive forces are not propagated through the surrounding regions, whether this cause is some central body (such as a lodestone in the center of a magnetic force or the earth in the center of a force that produces gravity) or whether it is some other cause which is not apparent. This concept is purely mathematical, for I am not now considering the physical causes and sites of forces.

Therefore, accelerative force is to motive force as velocity to motion. For quantity of motion arises from velocity and quantity of matter jointly, and motive force from accelerative force and quantity of matter jointly. For the sum of the actions of the accelerative force on the individual particles of a body is the motive force of the whole body. As a consequence, near the surface of the earth, where the accelerative gravity, or the force that produces gravity, is the same in all bodies universally, the motive gravity, or weight, is as the body, but in an ascent to regions where the accelerative gravity becomes less, the weight will decrease proportionately and will always be as the body and the accelerative gravity jointly. Thus, in regions where the accelerative gravity is half as great, a body one-half or one-third as great will have a weight four or six times less.

Further, it is in this same sense that I call attractions and impulses accelerative and motive. Moreover, I use interchangeably and indiscriminately words signifying attraction, impulse, or any sort of propensity toward a center, considering these forces not from a physical but only from a mathematical point of view. Therefore, let the reader beware of thinking that by words of this kind I am anywhere defining a species or mode of action or a physical cause or reason, or that I am attributing forces in a true and physical sense to centers (which are mathematical points) if I happen to say that centers attract or that centers have forces.

Scholium Thus far it has seemed best to explain the senses in which less familiar words are to be taken in this treatise. Although time, space, place, and motion are very familiar to everyone, it must be noted that these quantities are popularly conceived solely with reference to the objects of sense perception. And this is the source of certain preconceptions; to eliminate them it is useful to distinguish these quantities into absolute and relative, true and apparent, mathematical and common.

1. Absolute, true, and mathematical time, in and of itself and of its own nature, without reference to anything external, flows uniformly and by another name is called duration. Relative, apparent, and common time is any sensible and external measure [a](precise or imprecise)[a] of duration by means of motion; such a measure—for example, an hour, a day, a month, a year—is commonly used instead of true time.

2. Absolute space, of its own nature without reference to anything external, always remains homogeneous and immovable. Relative space is any

aa. Newton uses the phrase "seu accurata seu inaequabilis"—literally, "exact or nonuniform."

movable measure or dimension of this absolute space; such a measure or dimension is determined by our senses from the situation of the space with respect to bodies and is popularly used for immovable space, as in the case of space under the earth or in the air or in the heavens, where the dimension is determined from the situation of the space with respect to the earth. Absolute and relative space are the same in species and in magnitude, but they do not always remain the same numerically. For example, if the earth moves, the space of our air, which in a relative sense and with respect to the earth always remains the same, will now be one part of the absolute space into which the air passes, now another part of it, and thus will be changing continually in an absolute sense.

3. Place is the part of space that a body occupies, and it is, depending on the space, either absolute or relative. I say the part of space, not the position of the body or its outer surface. For the places of equal solids are always equal, while their surfaces are for the most part unequal because of the dissimilarity of shapes; and positions, properly speaking, do not have quantity and are not so much places as attributes of places. The motion of a whole is the same as the sum of the motions of the parts; that is, the change in position of a whole from its place is the same as the sum of the changes in position of its parts from their places, and thus the place of a whole is the same as the sum of the places of the parts and therefore is internal and in the whole body.

4. Absolute motion is the change of position of a body from one absolute place to another; relative motion is change of position from one relative place to another. Thus, in a ship under sail, the relative place of a body is that region of the ship in which the body happens to be or that part of the whole interior of the ship which the body fills and which accordingly moves along with the ship, and relative rest is the continuance of the body in that same region of the ship or same part of its interior. But true rest is the continuance of a body in the same part of that unmoving space in which the ship itself, along with its interior and all its contents, is moving. Therefore, if the earth is truly at rest, a body that is relatively at rest on a ship will move truly and absolutely with the velocity with which the ship is moving on the earth. But if the earth is also moving, the true and absolute motion of the body will arise partly from the true motion of the earth in unmoving space and partly from the relative motion of the ship on the earth. Further, if the body is also moving relatively on the ship, its true motion will arise partly from

the true motion of the earth in unmoving space and partly from the relative motions both of the ship on the earth and of the body on the ship, and from these relative motions the relative motion of the body on the earth will arise. For example, if that part of the earth where the ship happens to be is truly moving eastward with a velocity of 10,010 units, and the ship is being borne westward by sails and wind with a velocity of 10 units, and a sailor is walking on the ship toward the east with a velocity of 1 unit, then the sailor will be moving truly and absolutely in unmoving space toward the east with a velocity of 10,001 units and relatively on the earth toward the west with a velocity of 9 units.

In astronomy, absolute time is distinguished from relative time by the equation of common time. For natural days, which are commonly considered equal for the purpose of measuring time, are actually unequal. Astronomers correct this inequality in order to measure celestial motions on the basis of a truer time. It is possible that there is no uniform motion by which time may have an exact measure. All motions can be accelerated and retarded, but the flow of absolute time cannot be changed. The duration or perseverance of the existence of things is the same, whether their motions are rapid or slow or null; accordingly, duration is rightly distinguished from its sensible measures and is gathered from them by means of an astronomical equation. Moreover, the need for using this equation in determining when phenomena occur is proved by experience with a pendulum clock and also by eclipses of the satellites of Jupiter.

Just as the order of the parts of time is unchangeable, so, too, is the order of the parts of space. Let the parts of space move from their places, and they will move (so to speak) from themselves. For times and spaces are, as it were, the places of themselves and of all things. All things are placed in time with reference to order of succession and in space with reference to order of position. It is of the essence of spaces to be places, and for primary places to move is absurd. They are therefore absolute places, and it is only changes of position from these places that are absolute motions.

But since these parts of space cannot be seen and cannot be distinguished from one another by our senses, we use sensible measures in their stead. For we define all places on the basis of the positions and distances of things from some body that we regard as immovable, and then we reckon all motions with respect to these places, insofar as we conceive of bodies as being changed in position with respect to them. Thus, instead of absolute places and motions we use relative ones, which is not inappropriate in ordinary human affairs, although in philosophy abstraction from the senses is required. For it is possible that there is no body truly at rest to which places and motions may be referred.

Moreover, absolute and relative rest and motion are distinguished from each other by their properties, causes, and effects. It is a property of rest that bodies truly at rest are at rest in relation to one another. And therefore, since it is possible that some body in the regions of the fixed stars or far beyond is absolutely at rest, and yet it cannot be known from the position of bodies in relation to one another in our regions whether or not any of these maintains a given position with relation to that distant body, true rest cannot be defined on the basis of the position of bodies in relation to one another.

It is a property of motion that parts which keep given positions in relation to wholes participate in the motions of such wholes. For all the parts of bodies revolving in orbit endeavor to recede from the axis of motion, and the impetus of bodies moving forward arises from the joint impetus of the individual parts. Therefore, when bodies containing others move, whatever is relatively at rest within them also moves. And thus true and absolute motion cannot be determined by means of change of position from the vicinity of bodies that are regarded as being at rest. For the exterior bodies ought to be regarded not only as being at rest but also as being truly at rest. Otherwise all contained bodies, besides being subject to change of position from the vicinity of the containing bodies, will participate in the true motions of the containing bodies and, if there is no such change of position, will not be truly at rest but only be regarded as being at rest. For containing bodies are to those inside them as the outer part of the whole to the inner part or as the shell to the kernel. And when the shell moves, the kernel also, without being changed in position from the vicinity of the shell, moves as a part of the whole.

A property akin to the preceding one is that when a place moves, whatever is placed in it moves along with it, and therefore a body moving away from a place that moves participates also in the motion of its place. Therefore, all motions away from places that move are only parts of whole and absolute motions, and every whole motion is compounded of the motion of a body away from its initial place, and the motion of this place away from its place, and so on, until an unmoving place is reached, as in the above-mentioned example of the sailor. Thus, whole and absolute motions can be determined only by means of unmoving places, and therefore in what has preceded I have referred such motions to unmoving places and relative motions to movable places. Moreover, the only places that are unmoving are those that all keep given positions in relation to one another from infinity to infinity and therefore always remain immovable and constitute the space that I call immovable.

The causes which distinguish true motions from relative motions are the forces impressed upon bodies to generate motion. True motion is neither generated nor changed except by forces impressed upon the moving body itself, but relative motion can be generated and changed without the impression of forces upon this body. For the impression of forces solely on other bodies with which a given body has a relation is enough, when the other bodies yield, to produce a change in that relation which constitutes the relative rest or motion of this body. Again, true motion is always changed by forces impressed upon a moving body, but relative motion is not necessarily changed by such forces. For if the same forces are impressed upon a moving body and also upon other bodies with which it has a relation, in such a way that the relative position is maintained, the relation that constitutes the relative motion will also be maintained. Therefore, every relative motion can be changed while the true motion is preserved, and can be preserved while the true one is changed, and thus true motion certainly does not consist in relations of this sort.

The effects distinguishing absolute motion from relative motion are the forces of receding from the axis of circular motion. For in purely relative circular motion these forces are null, while in true and absolute circular motion they are larger or smaller in proportion to the quantity of motion. If a bucket is hanging from a very long cord and is continually turned around until the cord becomes twisted tight, and if the bucket is thereupon filled with water and is at rest along with the water and then, by some sudden force, is made to turn around in the opposite direction and, as the cord unwinds, perseveres for a while in this motion; then the surface of the water will at first be level, just as it was before the vessel began to move. But after the vessel, by the force gradually impressed upon the water, has caused the water also to begin revolving perceptibly, the water will gradually recede from the middle and rise up the sides of the vessel, assuming a concave shape (as experience has shown me), and, with an ever faster motion, will rise further and further until, when it completes its revolutions in the same times as the vessel, it is relatively at rest in the vessel. The rise of the water reveals its endeavor to recede from the axis of motion, and from such an endeavor one can find out and measure the true and absolute circular motion of the water, which here is the direct opposite of its relative motion. In the beginning, when the relative motion of the water in the vessel was greatest, that motion was not giving rise to any endeavor to recede from the axis; the water did not seek the circumference by rising up the sides of the vessel but remained level, and therefore its true circular motion had not yet begun. But afterward, when the relative motion of the water decreased, its rise up the sides of the vessel revealed its endeavor to recede from the axis, and this endeavor showed the true circular motion of the water to be continually increasing and finally becoming greatest when the water was relatively at rest in the vessel. Therefore, that endeavor does not depend on the change of position of the water with respect to surrounding bodies, and thus true circular motion cannot be determined by means of such changes of position. The truly circular motion of each revolving body is unique, corresponding to a unique endeavor as its proper and sufficient effect, while relative motions are innumerable in accordance with their varied relations to external bodies and, like relations, are completely lacking in true effects except insofar as they participate in that true and unique motion. Thus, even in the system of those who hold that our heavens revolve below the heavens of the fixed stars and carry the planets around with them, the individual parts of the heavens, and the planets that are relatively at rest in the heavens to which they belong, are truly in motion. For they change their positions relative to one another (which is not the case with things that are truly at rest), and as they are carried around together with the heavens, they participate in the motions of the heavens and, being parts of revolving wholes, endeavor to recede from the axes of those wholes.

Relative quantities, therefore, are not the actual quantities whose names they bear but are those sensible measures of them (whether true or erroneous) that are commonly used instead of the quantities being measured. But if the meanings of words are to be defined by usage, then it is these sensible measures which should properly be understood by the terms "time," "space," "place," and "motion," and the manner of expression will be out of the ordinary and purely mathematical if the quantities being measured are understood here. Accordingly those who there interpret these words as referring to the quantities being measured do violence to the Scriptures. And they no less corrupt mathematics and philosophy who confuse true quantities with their relations and common measures.

It is certainly very difficult to find out the true motions of individual bodies and actually to differentiate them from apparent motions, because the parts of that immovable space in which the bodies truly move make no impression on the senses. Nevertheless, the case is not utterly hopeless. For it is possible to draw evidence partly from apparent motions, which are the differences between the true motions, and partly from the forces that are the causes and effects of the true motions. For example, if two balls, at a given distance from each other with a cord connecting them, were revolving about a common center of gravity, the endeavor of the balls to recede from the axis of motion could be known from the tension of the cord, and thus the quantity of circular motion could be computed. Then, if any equal forces were simultaneously impressed upon the alternate faces of the balls to increase or decrease their circular motion, the increase or decrease of the motion could be known from the increased or decreased tension of the cord, and thus, finally, it could be discovered which faces of the balls the forces would have to be impressed upon for a maximum increase in the motion, that is, which were the posterior faces, or the ones that are in the rear in a circular motion. Further, once the faces that follow and the opposite faces that precede were known, the direction of the motion would be known. In this way both the quantity and the direction of this circular motion could be found in any immense vacuum, where nothing external and sensible existed with which the balls could be compared. Now if some distant bodies were set in that space and maintained given positions with respect to one another, as the fixed stars do in the regions of the heavens, it could not, of course, be known from the relative change of position of the balls among the bodies whether the motion was to be attributed to the bodies or to the balls. But if the cord was examined and its tension was discovered to be the very one which the motion of the balls required, it would be valid to conclude that the motion belonged to the balls and that the bodies were at rest, and then, finally, from the change of position of the balls among the bodies, to determine the direction of this motion. But in what follows, a fuller explanation will be given of how to determine true motions from their causes, effects, and apparent differences, and, conversely, of how to determine from motions, whether true or apparent, their causes and effects. For this was the purpose for which I composed the following treatise.

AXIOMS, OR THE LAWS OF MOTION

Law 1 *Every body perseveres in its state of being at rest or of moving* [a]*uniformly straight forward,*[a] *except insofar as* [b]*it*[b] *is compelled to change* [c]*its*[c] *state by forces impressed.*

Projectiles persevere in their motions, except insofar as they are retarded by the resistance of the air and are impelled downward by the force of gravity. A spinning hoop,[d] which has parts that by their cohesion continually draw one another back from rectilinear motions, does not cease to rotate, except insofar as it is retarded by the air. And larger bodies—planets and comets—preserve for a longer time both their progressive and their circular motions, which take place in spaces having less resistance.

Law 2 *A change in motion is proportional to the motive force impressed and takes place along the straight line in which that force is impressed.*

If some force generates any motion, twice the force will generate twice the motion, and three times the force will generate three times the motion, whether the force is impressed all at once or successively by degrees. And if the body was previously moving, the new motion (since motion is always in the same direction as the generative force) is added to the original motion if that motion was in the same direction or is subtracted from the original motion if it was in the opposite direction or, if it was in an oblique direction,

aa. See note bb to def. 3.

bb. Ed. 1 and ed. 2 lack the pronoun "illud," which, by expressing the subject, renders it somewhat more emphatic than it is when conveyed only by the form of the verb ("is compelled") and which makes more explicit the reference to an antecedent noun ("body").

cc. Ed. 1 and ed. 2 have "that."

d. The Latin word is "trochus," i.e., a top or some kind of spinner.

is combined obliquely and compounded with it according to the directions of both motions.

Law 3

To any action there is always an opposite and equal reaction; in other words, the actions of two bodies upon each other are always equal and always opposite in direction.

Whatever presses or draws something else is pressed or drawn just as much by it. If anyone presses a stone with a finger, the finger is also pressed by the stone. If a horse draws a stone tied to a rope, the horse will (so to speak) also be drawn back equally toward the stone, for the rope, stretched out at both ends, will urge the horse toward the stone and the stone toward the horse by one and the same endeavor to go slack and will impede the forward motion of the one as much as it promotes the forward motion of the other. If some body impinging upon another body changes the motion of that body in any way by its own force, then, by the force of the other body (because of the equality of their mutual pressure), it also will in turn undergo the same change in its own motion in the opposite direction. By means of these actions, equal changes occur in the motions, not in the velocities—that is, of course, if the bodies are not impeded by anything else.[a] For the changes in velocities that likewise occur in opposite directions are inversely proportional to the bodies because the motions are changed equally. This law is valid also for attractions, as will be proved in the next scholium.

Corollary 1

A body acted on by [two] forces acting jointly describes the diagonal of a parallelogram in the same time in which it would describe the sides if the forces were acting separately.

Let a body in a given time, by force M alone impressed in A, be carried with uniform motion from A to B, and, by force N alone impressed in the same place, be carried from A to C; then complete the parallelogram ABDC, and by both forces the body will be carried in the same time along the diagonal from A to D. For, since force N acts along the line AC parallel to

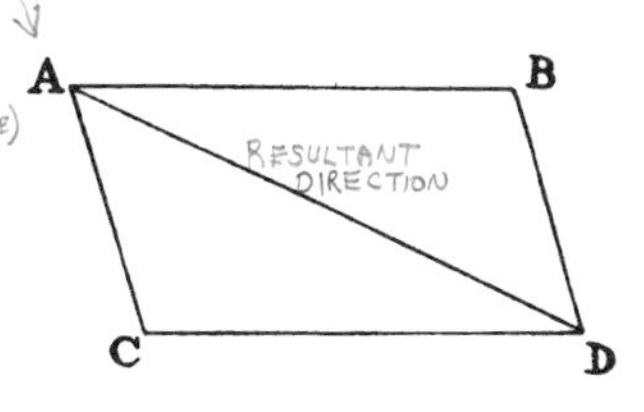

a. By "body" Newton means quantity of matter or mass (def. 1) and by "motion" he means quantity of motion (def. 2) or momentum.

BD, this force, by law 2, will make no change at all in the velocity toward the line BD which is generated by the other force. Therefore, the body will reach the line BD in the same time whether force N is impressed or not, and so at the end of that time will be found somewhere on the line BD. By the same argument, at the end of the same time it will be found somewhere on the line CD, and accordingly it is necessarily found at the intersection D of both lines. And, by law 1, it will go with [uniform] rectilinear motion from A to D.

Corollary 2 *And hence the composition of a direct force* AD *out of any oblique forces* AB *and* BD *is evident, and conversely the resolution of any direct force* AD *into any oblique forces* AB *and* BD. *And this kind of composition and resolution is indeed abundantly confirmed from mechanics.*

For example, let OM and ON be unequal spokes going out from the center O of any wheel, and let the spokes support the weights A and P by means of the cords MA and NP; it is required to find the forces of the weights to move the wheel. Draw the straight line KOL through the center O, so as to meet the cords perpendicularly at K and L; and with center O and radius OL, which is the greater of OK and OL, describe a circle meeting the cord MA at D; and draw the straight line OD, and let AC be drawn parallel to it and DC perpendicular to it. Since it makes no difference

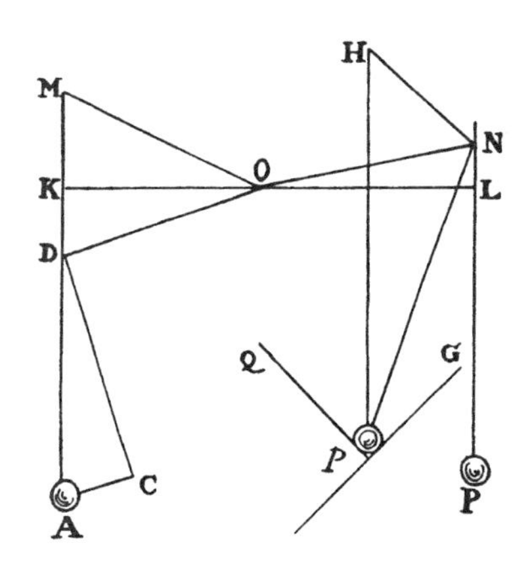

whether points K, L, and D of the cords are attached or not attached to the plane of the wheel, the weights will have the same effect whether they are suspended from the points K and L or from D and L. And if now the total force of the weight A is represented by line AD, it will be resolved into forces [i.e., components] AC and CD, of which AC, drawing spoke OD directly from the center, has no effect in moving the wheel, while the other force DC, drawing spoke DO perpendicularly, has the same effect as if it were drawing spoke OL (equal to OD) perpendicularly; that is, it has the same effect as the weight P, provided that the weight P is to the weight A as the force DC is to the force DA; that is (because triangles ADC and

DOK are similar), as OK to OD or OL. Therefore, the weights A and P, which are inversely as the spokes OK and OL (which are in a straight line), will be equipollent and thus will stand in equilibrium, which is a very well known property of the balance, the lever, and the wheel and axle. But if either weight is greater than in this ratio, its force to move the wheel will be so much the greater.

But if the weight *p*, equal to the weight P, is partly suspended by the cord N*p* and partly lies on the oblique plane *p*G, draw *p*H perpendicular to the plane of the horizon and NH perpendicular to the plane *p*G; then if the force of the weight *p* tending downward is represented by the line *p*H, it can be resolved into the forces [i.e., components] *p*N and HN. If there were some plane *p*Q perpendicular to the cord *p*N and cutting the other plane *p*G in a line parallel to the horizon, and the weight *p* were only lying on these planes *p*Q and *p*G, the weight *p* would press these planes perpendicularly with the forces *p*N and HN—plane *p*Q, that is, with force *p*N and plane *p*G with force HN. Therefore, if the plane *p*Q is removed, so that the weight stretches the cord, then—since the cord, in sustaining the weight, now takes the place of the plane which has been removed—the cord will be stretched by the same force *p*N with which the plane was formerly pressed. Thus the tension of this oblique cord will be to the tension of the other, and perpendicular, cord PN as *p*N to *p*H. Therefore, if the weight *p* is to the weight A in a ratio that is compounded of the inverse ratio of the least distances of their respective cords *p*N and AM from the center of the wheel and the direct ratio of *p*H to *p*N, the weights will have the same power to move the wheel and so will sustain each other, as anyone can test.

Now, the weight *p*, lying on those two oblique planes, has the role of a wedge between the inner surfaces of a body that has been split open; and hence the forces of a wedge and hammer can be determined, because the force with which the weight *p* presses the plane *p*Q is to the force with which weight *p* is impelled along the line *p*H toward the planes, whether by its own gravity or by the blow of a hammer, as *p*N is to *p*H, and because the force with which *p* presses plane *p*Q is to the force by which it presses the other plane *p*G as *p*N to NH. Furthermore, the force of a screw can also be determined by a similar resolution of forces, inasmuch as it is a wedge impelled by a lever. Therefore, this corollary can be used very extensively, and the variety of its applications clearly shows its truth, since the whole of mechanics—demonstrated in different ways by those who have written on this subject—depends on what has just now been said. For from this are easily derived the forces of machines, which are generally composed of wheels, drums, pulleys, levers, stretched strings, and weights, ascending directly or obliquely, and the other mechanical powers, as well as the forces of tendons to move the bones of animals.

Corollary 3 *The quantity of motion, which is determined by adding the motions made in one direction and subtracting the motions made in the opposite direction, is not changed by the action of bodies on one another.*

For an action and the reaction opposite to it are equal by law 3, and thus by law 2 the changes which they produce in motions are equal and in opposite directions. Therefore, if motions are in the same direction, whatever is added to the motion of the first body [*lit.* the fleeing body] will be subtracted from the motion of the second body [*lit.* the pursuing body] in such a way that the sum remains the same as before. But if the bodies meet head-on, the quantity subtracted from each of the motions will be the same, and thus the difference of the motions made in opposite directions will remain the same.

For example, suppose a spherical body A is three times as large as a spherical body B and has two parts of velocity, and let B follow A in the same straight line with ten parts of velocity; then the motion of A is to the motion of B as six to ten. Suppose that their motions are of six parts and ten parts respectively; the sum will be sixteen parts. When the bodies collide, therefore, if body A gains three or four or five parts of motion, body B will lose just as many parts of motion and thus after reflection body A will continue with nine or ten or eleven parts of motion and B with seven or six or five parts of motion, the sum being always, as originally, sixteen parts of motion. Suppose body A gains nine or ten or eleven or twelve parts of motion and so moves forward with fifteen or sixteen or seventeen or eighteen parts of motion after meeting body B; then body B, by losing as many parts of motion as A gains, will either move forward with one part, having lost nine parts of motion, or will be at rest, having lost its forward motion of ten parts, or will move backward with one part of motion, having lost its motion and (if I may say so) one part more, or will move backward with two parts of motion because a forward motion of twelve parts has been subtracted. And thus the sums, $15+1$ or $16+0$, of the motions in the same direction and the differences, 17 − 1 and 18 − 2, of the motions in opposite directions will always be sixteen parts of motion, just as before the bodies met and were reflected. And since the motions with which the bodies will continue to move after reflection are known, the velocity of each will be found, on the supposition that it is to the velocity before reflection as the motion after reflection is to the motion before reflection. For example, in the last case, where the motion of body A was six parts before reflection and eighteen parts afterward, and its velocity was two parts before reflection, its velocity will be found to be six parts after reflection on the basis of the following statement: as six parts of motion before reflection is to eighteen parts of motion afterward, so two parts of velocity before reflection is to six parts of velocity afterward.

But if bodies that either are not spherical or are moving in different straight lines strike against each other obliquely and it is required to find their motions after reflection, the position of the plane by which the colliding bodies are touched at the point of collision must be determined; then (by corol. 2) the motion of each body must be resolved into two motions, one perpendicular to this plane and the other parallel to it. Because the bodies act upon each other along a line perpendicular to this plane, the parallel motions [i.e., components] must be kept the same after reflection; and equal changes in opposite directions must be attributed to the perpendicular motions in such a way that the sum of the motions in the same direction and the difference of the motions in opposite directions remain the same as before the bodies came together. The circular motions of bodies about their own centers also generally arise from reflections of this sort. But I do not consider such cases in what follows, and it would be too tedious to demonstrate everything relating to this subject.

Corollary 4

The common center of gravity of two or more bodies does not change its state whether of motion or of rest as a result of the actions of the bodies upon one another; and therefore the common center of gravity of all bodies acting upon one another (excluding external actions and impediments) either is at rest or moves uniformly straight forward.

For if two points move forward with uniform motion in straight lines, and the distance between them is divided in a given ratio, the dividing point either is at rest or moves forward uniformly in a straight line. This is demonstrated below in lem. 23 and its corollary for the case in which the motions of the points take place in the same plane, and it can be demonstrated by the same reasoning for the case in which those motions do not take place in the same plane. Therefore, if any number of bodies move uniformly in straight lines, the common center of gravity of any two either is at rest or moves forward uniformly in a straight line, because any line joining these bodies through their centers—which move forward uniformly in straight lines—is divided by this common center in a given ratio. Similarly, the common center of gravity of these two bodies and any third body either is at rest or moves forward uniformly in a straight line, because the distance between the common center of the two bodies and the center of the third body is divided in a given ratio by the common center of the three. In the same way, the common center of these three and of any fourth body either is at rest or moves forward uniformly in a straight line, because that common center divides in a given ratio the distance between the common center of the three and the center of the fourth body, and so on without end. Therefore, in a system of bodies in which the bodies are entirely free of actions upon one another and of all other actions impressed upon them externally, and in which each body accordingly moves uniformly in its individual straight line, the common center of gravity of them all either is at rest or moves uniformly straight forward.

Further, in a system of two bodies acting on each other, since the distances of their centers from the common center of gravity are inversely as the bodies, the relative motions of these bodies, whether of approaching that center or of receding from it, will be equal. Accordingly, as a result of equal changes in opposite directions in the motions of these bodies, and consequently as a result of the actions of the bodies on each other, that center is neither accelerated nor retarded nor does it undergo any change in its state of motion or of rest. In a system of several bodies, the common center of gravity of any two acting upon each other does not in any way change its state as a result of that action, and the common center of gravity of the rest of the bodies (with which that action has nothing to do) is not affected by that action; the distance between these two centers is divided by the common center of gravity of all the bodies into parts inversely proportional to the total sums of the bodies whose centers they are, and (since those two centers maintain their state of moving or of being at rest) the common center of all maintains its state also—for all these reasons it is obvious that this common center of all never changes its state with respect to motion and rest as a result of the actions of two bodies upon each other. Moreover, in such a system all the actions of bodies upon one another either occur between two bodies or are compounded of such actions between two bodies and therefore never introduce any change in the state of motion or of rest of the common center of all. Thus, since that center either is at rest or moves forward uniformly in some straight line, when the bodies do not act upon one another, that center will, notwithstanding the actions of the bodies upon one another, continue either to be always at rest or to move always uniformly straight forward, unless it is driven from this state by forces impressed on the system from outside. Therefore, the law is the same for a system of several bodies as for a single body with respect to perseverance in a state of motion or of rest. For the progressive motion, whether of a single body or of a system of bodies, should always be reckoned by the motion of the center of gravity.

Corollary 5

When bodies are enclosed in a given space, their motions in relation to one another are the same whether the space is at rest or whether it is moving uniformly straight forward without circular motion.

For in either case the differences of the motions tending in the same direction and the sums of those tending in opposite directions are the same at the beginning (by hypothesis), and from these sums or differences there arise the collisions and impulses [*lit.* impetuses] with which the bodies strike one another. Therefore, by law 2, the effects of the collisions will be equal in both cases, and thus the motions with respect to one another in the one case will remain equal to the motions with respect to one another in the other case. This is proved clearly by experience: on a ship, all the motions are the same with respect to one another whether the ship is at rest or is moving uniformly straight forward.

Corollary 6

If bodies are moving in any way whatsoever with respect to one another and are urged by equal accelerative forces along parallel lines, they will all continue to move with respect to one another in the same way as they would if they were not acted on by those forces.

For those forces, by acting equally (in proportion to the quantities of the bodies to be moved) and along parallel lines, will (by law 2) move all the bodies equally (with respect to velocity), and so will never change their positions and motions with respect to one another.

Scholium The principles I have set forth are accepted by mathematicians and confirmed by experiments of many kinds. By means of the first two laws and the first two corollaries Galileo found that the descent of heavy bodies is in the squared ratio of the time and that the motion of projectiles occurs in a parabola, as experiment confirms, except insofar as these motions are somewhat retarded by the resistance of the air. [a]When a body falls, uniform gravity, by acting equally in individual equal particles of time, impresses equal forces upon that body and generates equal velocities; and in the total time it impresses a total force and generates a total velocity proportional to the time. And the spaces described in proportional times are as the velocities and the times jointly, that is, in the squared ratio of the times. And when a body is projected upward, uniform gravity impresses forces and takes away velocities proportional to the times; and the times of ascending to the greatest heights are as the velocities to be taken away, and these heights are as the velocities and the times jointly, or as the squares of the velocities. And when a body is projected along any straight line, its motion arising from the projection is compounded with the motion arising from gravity.

For example, let body A by the motion of projection alone describe the straight line AB in a given time, and by the motion of falling alone describe the vertical distance AC in the same time; then complete the parallelogram ABDC, and by the compounded motion the body will be found in place D at the end of the time; and the curved line AED which the body will describe will be a parabola which the straight line AB touches at A and whose ordinate BD is as AB^2.[a]

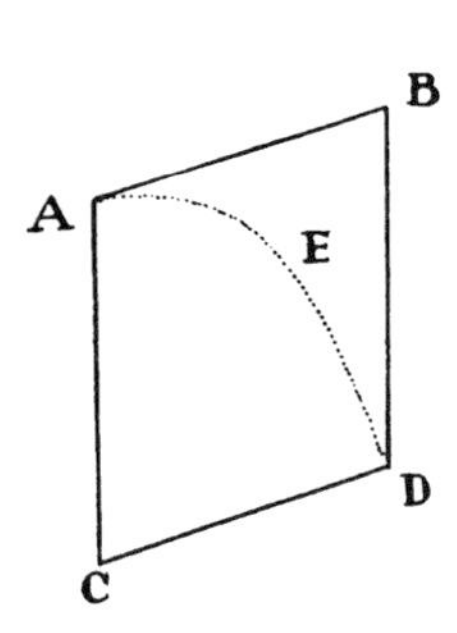

What has been demonstrated concerning the times of oscillating pendulums depends on the same first two laws and first two corollaries, and this is supported by daily experience with clocks. From the same laws and corollaries and law 3, Sir Christopher Wren, Dr. John Wallis, and Mr. Christiaan Huygens, easily the foremost geometers of the previous generation, independently found the rules of the collisions and reflections of hard bodies, and communicated them to the Royal Society at nearly the same time, entirely agreeing with one another (as to these rules); and Wallis was indeed the

aa. Ed. 1 and ed. 2 lack this.

first to publish what had been found, followed by Wren and Huygens. But Wren additionally proved the truth of these rules before the Royal Society by means of an experiment with pendulums, which the eminent Mariotte soon after thought worthy to be made the subject of a whole book.

However, if this experiment is to agree precisely with the theories, account must be taken of both the resistance of the air and the elastic force of the colliding bodies. Let the spherical bodies A and B be suspended from centers C and D by parallel and equal cords AC and BD. With these centers and with those distances as radii describe semicircles EAF and GBH bisected by radii CA and DB. Take away body B, and let body A be brought to any point R of the arc

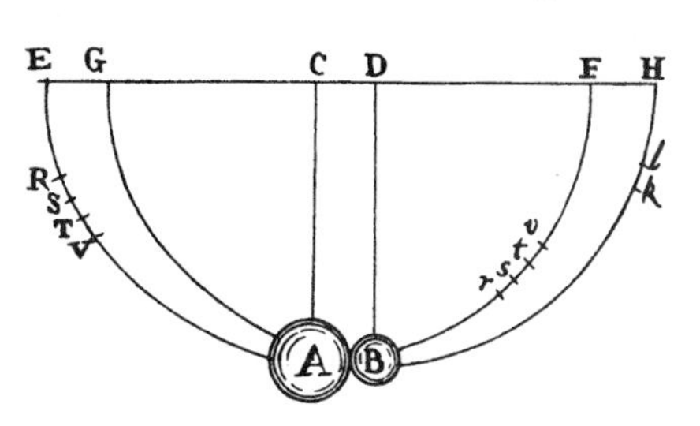

EAF and be let go from there, and let it return after one oscillation to point V. RV is the retardation arising from the resistance of the air. Let ST be a fourth of RV and be located in the middle so that RS and TV are equal and RS is to ST as 3 to 2. Then ST will closely approximate the retardation in the descent from S to A. Restore body B to its original place. Let body A fall from point S, and its velocity at the place of reflection A, without sensible error, will be as great as if it had fallen in a vacuum from place T. Therefore let this velocity be represented by the chord of the arc TA. For it is a proposition very well known to geometers that the velocity of a pendulum in its lowest point is as the chord of the arc that it has described in falling. After reflection let body A arrive at place *s*, and body B at place *k*. Take away body B and find place *v* such that if body A is let go from this place and after one oscillation returns to place *r*, *st* will be a fourth of *rv* and be located in the middle, so that *rs* and *tv* are equal; and let the chord of the arc *t*A represent the velocity that body A had in place A immediately after reflection. For *t* will be that true and correct place to which body A must have ascended if there had been no resistance of the air. By a similar method the place *k*, to which body B ascends, will have to be corrected, and the place *l*, to which that body must have ascended in a vacuum, will have to be found. In this manner it is possible to make all our experiments, just as if we were in a vacuum. Finally body A will have to be multiplied (so to speak) by the chord of the arc TA, which represents its velocity, in order to get its motion in place A immediately before reflection, and then by the chord of the arc *t*A in order to get its motion in place A immediately after reflection. And thus body B will have to be multiplied by the chord of the arc B*l* in order to get its motion immediately after reflection. And by a similar method, when two bodies are let go simultaneously from different places, the motions of both will have to be found before as well as after reflection, and then finally the motions will have to be compared with each other in order to determine the effects of the reflection.

On making a test in this way with ten-foot pendulums, using unequal as well as equal bodies, and making the bodies come together from very large distances apart, say of eight or twelve or sixteen feet, I always found—within an error of less than three inches in the measurements—that when the bodies met each other directly, the changes of motions made in the bodies in opposite directions were equal, and consequently that the action and reaction were always equal. For example, if body A collided with body B, which was at rest, with nine parts of motion and, losing seven parts, proceeded after reflection with two, body B rebounded with those seven parts. If the bodies met head-on, A with twelve parts of motion and B with six, and A rebounded with two, B rebounded with eight, fourteen parts being subtracted from each. Subtract twelve parts from the motion of A and nothing will remain; subtract another two parts, and a motion of two parts in the opposite direction will be produced; and so, subtracting fourteen parts from the six parts of the motion of body B, eight parts will be produced in the opposite direction. But if the bodies moved in the same direction, A more quickly with fourteen parts and B more slowly with five parts, and after reflection A moved with five parts, then B moved with fourteen, nine parts having been transferred from A to B. And so in all other cases. As a result of the meeting and collision of bodies, the quantity of motion—determined by adding the motions in the same direction and subtracting the motions in opposite directions—was never changed. I would attribute the error of an inch or two in the measurements to the difficulty of doing everything with sufficient accuracy. It was difficult both to release the pendulums simultaneously in such a way that the bodies would impinge upon each other in the lowest place AB, and to note the places *s* and *k* to which the bodies ascended after colliding. But also, with respect to the pendulous bodies themselves, errors were introduced by the unequal density of the parts and by irregularities of texture arising from other causes.

Further, lest anyone object that the rule which this experiment was designed to prove presupposes that bodies are either absolutely hard or at least perfectly elastic and thus of a kind which do not occur [b]naturally,[b] I add that the experiments just described work equally well with soft bodies and with hard ones, since surely they do not in any way depend on the condition of hardness. For if this rule is to be tested in bodies that are not perfectly hard, it will only be necessary to decrease the reflection in a fixed proportion to the quantity of elastic force. In the theory of Wren and Huygens, absolutely hard bodies rebound from each other with the velocity with which they have collided. This will be affirmed with more certainty of perfectly elastic bodies. In imperfectly elastic bodies the velocity of rebounding must be decreased together with the elastic force, because that force (except when the parts of the bodies are damaged as a result of collision, or experience some sort of extension such as would be caused by a hammer blow) is fixed and determinate (as far as I can tell) and makes the bodies rebound from each other with a relative velocity that is in a given ratio to the relative velocity with which they collide. I have tested this as follows with tightly wound balls of wool strongly compressed. First, releasing the pendulums and measuring their reflection, I found the quantity of their elastic force; then from this force I determined what the reflections would be in other cases of their collision, and the experiments which were made agreed with the computations. The balls always rebounded from each other with a relative velocity that was to the relative velocity of their colliding as 5 to 9, more or less. Steel balls rebounded with nearly the same velocity and cork balls with a slightly smaller velocity, while with glass balls the proportion was roughly 15 to 16. And in this manner the third law of motion—insofar as it relates to impacts and reflections—is proved by this theory, which plainly agrees with experiments.

I demonstrate the third law of motion for attractions briefly as follows. Suppose that between any two bodies A and B that attract each other any obstacle is interposed so as to impede their coming together. If one body A is more attracted toward the other body B than that other body B is attracted toward the first body A, then the obstacle will be more strongly pressed by body A than by body B and accordingly will not remain in equilibrium. The stronger pressure will prevail and will make the system of the two bodies and

bb. Evidently "in natural compositions" or "in natural bodies."

the obstacle move straight forward in the direction from A toward B and, in empty space, go on indefinitely with a motion that is always accelerated, which is absurd and contrary to the first law of motion. For according to the first law, the system will have to persevere in its state of resting or of moving uniformly straight forward, and accordingly the bodies will urge the obstacle equally and on that account will be equally attracted to each other. I have tested this with a lodestone and iron. If these are placed in separate vessels that touch each other and float side by side in still water, neither one will drive the other forward, but because of the equality of the attraction in both directions they will sustain their mutual endeavors toward each other, and at last, having attained equilibrium, they will be at rest.

[c]In the same way gravity is mutual between the earth and its parts. Let the earth FI be cut by any plane EG into two parts EGF and EGI; then their weights toward each other will be equal. For if the greater part EGI is cut into two parts EGKH and HKI by another plane HK parallel to the first plane EG, in such a way that HKI is equal to the part EFG that has been cut off earlier, it is manifest that the middle part EGKH will not preponderate toward either of the outer parts but will, so to speak, be suspended in equilibrium between both and will be at rest. Moreover, the outer part HKI will press upon the middle part with all its weight and will urge it toward the other outer part EGF, and therefore the force by which EGI, the sum of the parts HKI and EGKH, tends toward the third part EGF is equal to the weight of the part HKI, that is, equal to the weight of the third part EGF. And therefore the weights of the two parts EGI and EGF toward each other are equal, as I set out to demonstrate. And if these weights were not equal, the whole earth, floating in an aether free of resistance, would yield to the greater weight and in receding from it would go off indefinitely.[c]

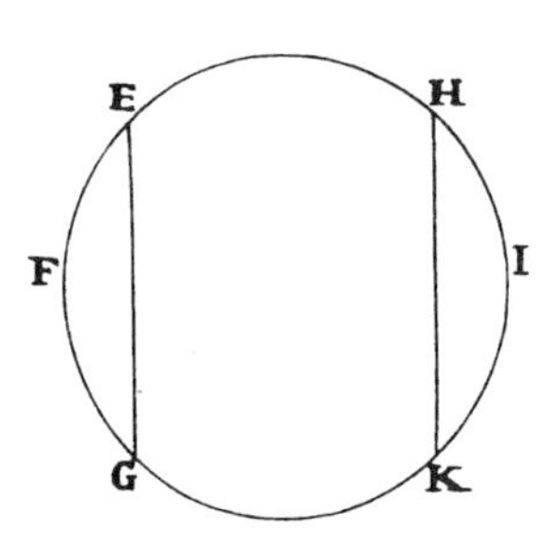

As bodies are equipollent in collisions and reflections if their velocities are inversely as their inherent forces [i.e., forces of inertia], so in the motions of machines those agents [i.e., acting bodies] whose velocities (reckoned in the direction of their forces) are inversely as their inherent forces are equipol-

cc. Ed. 1 lacks this.

lent and sustain one another by their contrary endeavors. Thus weights are equipollent in moving the arms of a balance if during oscillation of the balance they are inversely as their velocities upward and downward; that is, weights which move straight up and down are equipollent if they are inversely as the distances between the axis of the balance and the points from which they are suspended; but if such weights are interfered with by oblique planes or other obstacles that are introduced and thus ascend or descend obliquely, they are equipollent if they are inversely as the ascents and descents insofar as these are reckoned with respect to a perpendicular, and this is so because the direction of gravity is downward. Similarly, in a pulley or combination of pulleys, the weight will be sustained by the force of the hand pulling the rope vertically, which is to the weight (ascending either straight up or obliquely) as the velocity of the perpendicular ascent to the velocity of the hand pulling the rope. In clocks and similar devices, which are constructed out of engaged gears, the contrary forces that promote and hinder the motion of the gears will sustain each other if they are inversely as the velocities of the parts of the gears upon which they are impressed. The force of a screw to press a body is to the force of a hand turning the handle as the circular velocity of the handle, in the part where it is urged by the hand, is to the progressive velocity of the screw toward the pressed body. The forces by which a wedge presses the two parts of the wood that it splits are to the force of the hammer upon the wedge as the progress of the wedge (in the direction of the force impressed upon it by the hammer) is to the velocity with which the parts of the wood yield to the wedge along lines perpendicular to the faces of the wedge. And the case is the same for all machines.

The effectiveness and usefulness of all machines or devices consist wholly in our being able to increase the force by decreasing the velocity, and vice versa; in this way the problem is solved in the case of any working machine or device: "To move a given weight by a given force" or to overcome any other given resistance by a given force. For if machines are constructed in such a way that the velocities of the agent [or acting body] and the resistant [or resisting body] are inversely as the forces, the agent will sustain the resistance and, if there is a greater disparity of velocities, will overcome that resistance. Of course the disparity of the velocities may be so great that it can also overcome all the resistance which generally arises from the friction of contiguous bodies sliding over one another, from the cohesion of continuous bodies that are to be separated from one another, or from the weights of bodies to be raised; and if all this resistance is overcome, the remaining force will produce an acceleration of motion proportional to itself, partly in the parts of the machine, partly in the resisting body.[d]

But my purpose here is not to write a treatise on mechanics. By these examples I wished only to show the wide range and the certainty of the third law of motion. For if the action of an agent is reckoned by its force and velocity jointly, and if, similarly, the reaction of a resistant is reckoned jointly by the velocities of its individual parts and the forces of resistance arising from their friction, cohesion, weight, and acceleration, the action and reaction will always be equal to each other in all examples of using devices or machines. And to the extent to which the action is propagated through the machine and ultimately impressed upon each resisting body, its ultimate direction will always be opposite to the direction of the reaction.

d. Newton writes of "instrumentorum" (literally, "equipment") and of "instrumentis mechanicis" (literally, "mechanical instruments"), as well as "machinae." See §5.7 of the Guide.

BOOK 1

THE MOTION OF BODIES

SECTION 1

The method of first and ultimate ratios, for use in demonstrating what follows

Lemma 1

Quantities, and also ratios of quantities, which in [a]*any finite time*[a] *constantly tend to equality, and which before the end of that time approach so close to one another that their difference is less than any given quantity, become ultimately equal.*

If you deny this, [b]let them become ultimately unequal, and[b] let their ultimate difference be D. Then they cannot approach so close to equality that their difference is less than the given difference D, contrary to the hypothesis.

Lemma 2

If in any figure A*ac*E, *comprehended by the straight lines* A*a and* AE *and the curve ac*E, *any number of parallelograms* A*b*, B*c*, C*d*, . . . *are inscribed upon equal bases* AB, BC, CD, . . . *and have sides* B*b*, C*c*, D*d*, . . . *parallel to the side* A*a of the figure; and if the parallelograms a*K*bl*, *b*L*cm*, *c*M*dn*, . . . *are completed; if then the width of these parallelograms is diminished and their number increased indefinitely, I say that the ultimate ratios which the inscribed figure* AK*b*L*c*M*d*D, *the circumscribed figure* A*albmcndo*E, *and the curvilinear figure* A*abcd*E *have to one another are ratios of equality.*

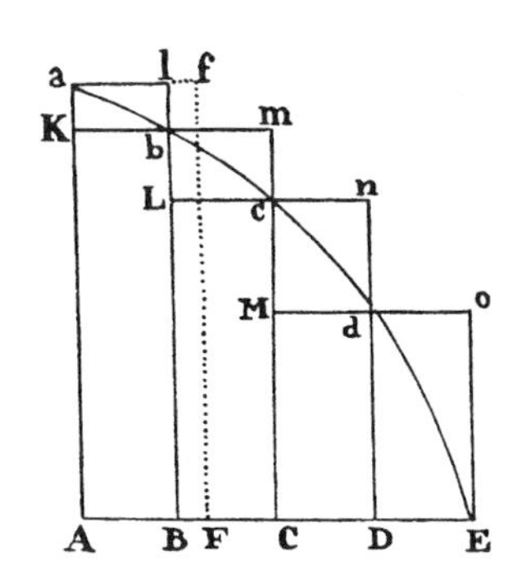

For the difference of the inscribed and circumscribed figures is the sum of the parallelograms K*l*, L*m*, M*n*, and D*o*, that is (because they all have equal bases), the rectangle having as base K*b* (the base of one of them) and as altitude A*a* (the sum of the altitudes), that is, the rectangle AB*la*. But this rectangle, because its width AB is diminished indefinitely, becomes less than any given rectangle. Therefore (by lem. 1) the inscribed figure and the circumscribed figure and, all the more, the intermediate curvilinear figure become ultimately equal. Q.E.D.

Lemma 3

The same ultimate ratios are also ratios of equality when the widths AB, BC, CD, . . . *of the parallelograms are unequal and are all diminished indefinitely.*

aa. Ed. 1 has "a given time."
bb. Ed. 1 lacks this.

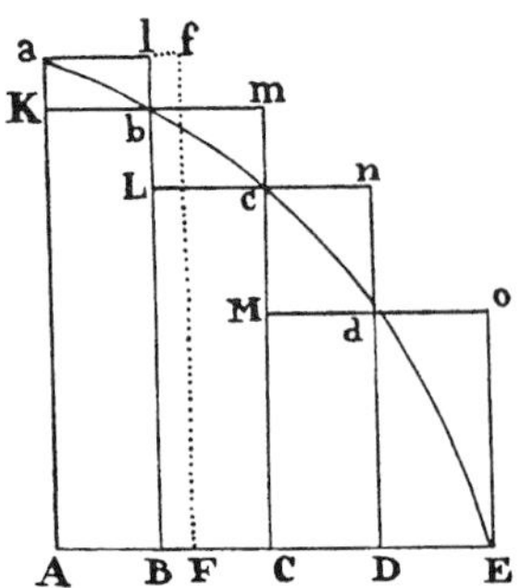

For let AF be equal to the greatest width, and let the parallelogram FA*af* be completed. This parallelogram will be greater than the difference of the inscribed and the circumscribed figures; but if its width AF is diminished indefinitely, it will become less than any given rectangle. Q.E.D.

COROLLARY 1. Hence the ultimate sum of the vanishing parallelograms coincides with the curvilinear figure in its every part.

COROLLARY 2. And, all the more, the rectilinear figure that is comprehended by the chords of the vanishing arcs *ab*, *bc*, *cd*, ... coincides ultimately with the curvilinear figure.

COROLLARY 3. And it is the same for the circumscribed rectilinear figure that is comprehended by the tangents of those same arcs.

COROLLARY 4. And therefore these ultimate figures (with respect to their perimeters *ac*E) are not rectilinear, but curvilinear limits of rectilinear figures.

Lemma 4 *If in two figures* A*ac*E *and* P*pr*T *two series of parallelograms are inscribed (as above) and the number of parallelograms in both series is the same; and if, when their widths are diminished indefinitely, the ultimate ratios of the parallelograms in one figure to the corresponding parallelograms in the other are the same; then I say that the two figures* A*ac*E *and* P*pr*T *are to each other in that same ratio.*

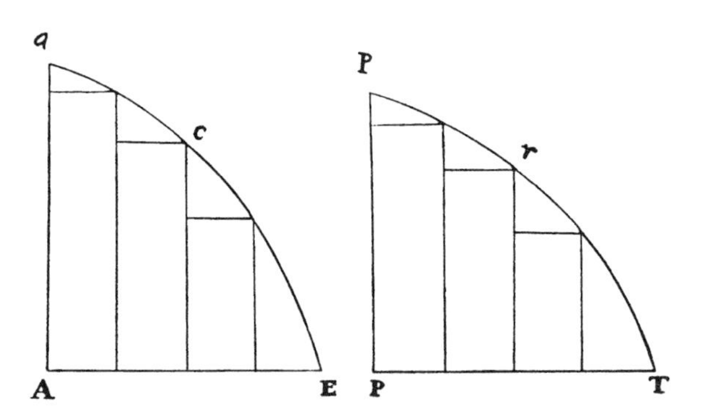

For as the individual parallelograms in the one figure are to the corresponding individual parallelograms in the other, so (by composition [or componendo]) will the sum of all the parallelograms in the one become to the sum of all the parallelograms in the other, and so also the one figure to the other—the first figure, of course, being (by lem. 3) to the first sum, and the second figure to the second sum, in a ratio of equality. Q.E.D.

COROLLARY. Hence, if two quantities of any kind are divided in any way into the same number of parts, and these parts—when their number is increased and their size is diminished indefinitely—maintain a given ratio to one another, the first to the first, the second to the second, and so on in sequence, then the wholes will be to each other in the same given ratio. For if the parallelograms in the figures of this lemma are taken in the same proportion to one another as those parts, the sums of the parts will always be as the sums of the parallelograms; and therefore, when the number of parts and parallelograms is increased and their size diminished indefinitely, the sums of the parts will be in the ultimate ratio of a parallelogram in one figure to a corresponding parallelogram in the other, that is (by hypothesis), in the ultimate ratio of part to part.

Lemma 5

All the mutually corresponding sides—curvilinear as well as rectilinear—of similar figures are proportional, and the areas of such figures are as the squares of their sides.

Lemma 6

If any arc ACB, *given in position, is subtended by the chord* AB *and at some point* A, *in the middle of the continuous curvature, is touched by the straight line* AD, *produced in both directions, and if then points* A *and* B *approach each other and come together, I say that the angle* BAD *contained by the chord and the tangent will be indefinitely diminished and will ultimately vanish.*

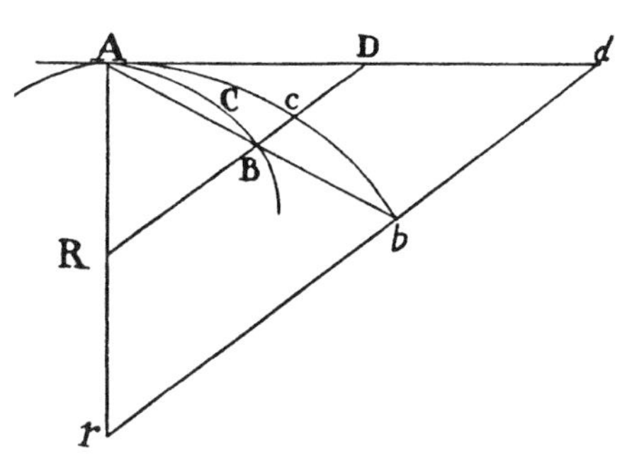

For [a]if that angle does not vanish, the angle contained by the arc ACB and the tangent AD will be equal to a rectilinear angle, and therefore the curvature at point A will not be continuous, contrary to the hypothesis.[a]

aa. Ed. 1 has "produce AB to *b* and AD to *d*; then, since points A and B come together and thus no part AB of A*b* still lies within the curve, it is obvious that this straight line A*b* will either coincide with the tangent A*d* or be drawn between the tangent and the curve. But the latter case is contrary to the nature of curvature; therefore, the former obtains. Q.E.D."

Lemma 7 *With the same suppositions, I say that the ultimate ratios of the arc, the chord, and the tangent to one another are ratios of equality.*

For while point B approaches point A, let AB and AD be understood always to be produced to the distant points *b* and *d*; and let *bd* be drawn parallel to secant BD. And let arc A*cb* be always similar to arc ACB. Then as points A and B come together, the angle *d*A*b* will vanish (by lem. 6), and thus the straight lines A*b* and A*d* (which are always finite) and the intermediate arc A*cb* will coincide and therefore will be equal. Hence, the straight lines AB and AD and the intermediate arc ACB (which are always proportional to the lines A*b* and A*d* and the arc A*cb* respectively) will also vanish and will have to one another an ultimate ratio of equality. Q.E.D.

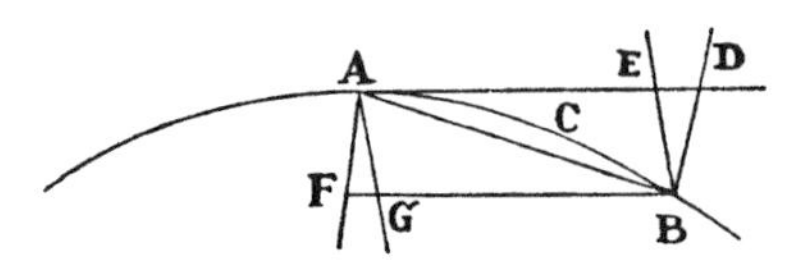

Corollary 1. Hence, if BF is drawn through B parallel to the tangent and always cutting at F any straight line AF passing through A, then BF will ultimately have a ratio of equality to the vanishing arc ACB, because, if parallelogram AFBD is completed, BF always has a ratio of equality to AD.

Corollary 2. And if through B and A additional straight lines BE, BD, AF, and AG are drawn cutting the tangent AD and its parallel BF, the ultimate ratios of all the abscissas AD, AE, BF, and BG and of the chord and arc AB to one another will be ratios of equality.

Corollary 3. And therefore all these lines can be used for one another interchangeably in any argumentation concerning ultimate ratios.

Lemma 8 *If the given straight lines* AR *and* BR, *together with the arc* ACB, *its chord* AB, *and the tangent* AD, *constitute three triangles* RAB, RACB, *and* RAD, *and if then points* A *and* B *approach each other, I say that the triangles as they vanish are similar in their ultimate form, and that their ultimate ratio is one of equality.*

For while point B approaches point A, let AB, AD, and AR be understood always to be produced to the distant points *b*, *d*, and *r*, and *rbd* to be drawn parallel to RD; and let arc A*cb* be always similar to arc ACB. Then as points A and B come together, the angle *b*A*d* will

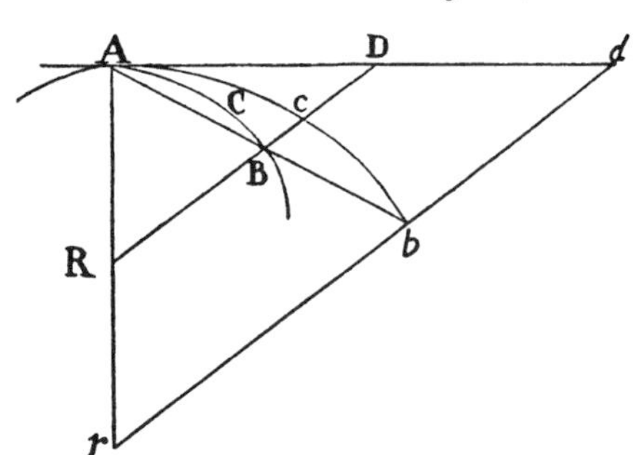

vanish, and therefore the three triangles *r*A*b*, *r*A*cb*, and *r*A*d*, which are always finite, will coincide and on that account are similar and equal. Hence also RAB, RACB, and RAD, which will always be similar and proportional to these, will ultimately become similar and equal to one another. Q.E.D.

COROLLARY. And hence those triangles can be used for one another interchangeably in any argumentation concerning ultimate ratios.

Lemma 9

If the straight line AE *and the curve* ABC, *both given in position, intersect each other at a given angle* A, *and if* BD *and* CE *are drawn as ordinates to the straight line* AE *at another given angle and meet the curve in* B *and* C, *and if then points* B *and* C *simultaneously approach point* A, *I say that the areas of the triangles* ABD *and* ACE *will ultimately be to each other as the squares of the sides.*

For while points B and C approach point A, let AD be understood always to be produced to the distant points *d* and *e*, so that A*d* and A*e* are proportional to AD and AE; and erect ordinates *db* and *ec* parallel to ordinates DB and EC and meeting AB and AC, produced, at *b* and *c*. Understand the curve A*bc* to be drawn similar to ABC, and the straight line A*g* to be drawn touching both curves at A and cutting the ordinates DB, EC, *db*, and *ec* at F, G, *f*, and *g*. Then, with the length A*e* remaining the same, let points B and C come together with point A; and as the angle *c*A*g* vanishes, the curvilinear areas A*bd* and A*ce* will coincide with the rectilinear areas A*fd* and A*ge*, and thus (by lem. 5) will be in the squared ratio of the sides A*d* and A*e*. But areas ABD and ACE are always proportional to these areas, and sides AD and AE to these sides. Therefore areas ABD and ACE also are ultimately in the squared ratio of the sides AD and AE. Q.E.D.

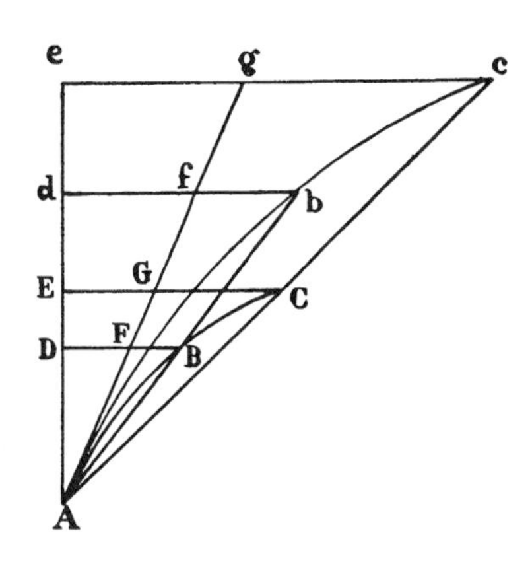

Lemma 10

The spaces which a body describes when urged by any [a]*finite*[a] *force,* [b]*whether that force is determinate and immutable or is continually increased or continually*

aa. Ed. 1 has "regular."
bb. Ed. 1 lacks this.

decreased,[b] *are at the very beginning of the motion in the squared ratio of the times.*

Let the times be represented by lines AD and AE, and the generated velocities by ordinates DB and EC; then the spaces described by these velocities will be as the areas ABD and ACE described by these ordinates, that is, at the very beginning of the motion these spaces will be (by lem. 9) in the squared ratio of the times AD and AE. Q.E.D.

COROLLARY 1. And hence it is easily concluded that when bodies describe similar parts of similar figures in proportional times, the errors that are generated by any equal forces similarly applied to the bodies, and that are measured by the distances of the bodies from those points on the similar figures at which the same bodies would arrive in the same proportional times without these forces, are very nearly as the squares of the times in which they are generated.

COROLLARY 2. But the errors that are generated by proportional forces similarly applied to similar parts of similar figures are as the forces and the squares of the times jointly.

[c]COROLLARY 3. The same is to be understood of any spaces which bodies describe when different forces urge them. These spaces are, at the very beginning of the motion, as the forces and the squares of the times jointly.

COROLLARY 4. And thus the forces are as the spaces described at the very beginning of the motion directly and as the squares of the times inversely.

COROLLARY 5. And the squares of the times are directly as the spaces described and inversely as the forces.

Scholium If indeterminate quantities of different kinds are compared with one another and any one of them is said to be directly or inversely as any other, the meaning is that the first one is increased or decreased in the same ratio as the second or as its reciprocal. And if any one of them is said to be as two or more others, directly or inversely, the meaning is that the first is increased or decreased in a ratio that is compounded of the ratios in which the others, or the reciprocals of the others, are increased or decreased. For example, if A is said to be as B directly and C directly and D inversely, the meaning is

cc. Ed. 1 lacks corols. 3–5 and scholium.

that A is increased or decreased in the same ratio as $B \times C \times \frac{1}{D}$, that is, that A and $\frac{BC}{D}$ are to each other in a given ratio.[c]

Lemma 11

In all curves having a finite curvature at the point of contact, the vanishing subtense of the angle of contact is ultimately in the squared ratio of the subtense of the conterminous arc.

CASE 1. Let AB be the arc, AD its tangent, BD the subtense of the angle of contact perpendicular to the tangent [angle BAD], and [the line] AB the subtense [i.e., the conterminous chord] of the arc [AB]. Erect BG and AG perpendicular to this subtense AB and tangent AD and meeting in G; then let points D, B, and G approach points *d*, *b*, and *g*, and let J be the intersection of lines BG and AG, which ultimately occurs when points D and B reach A. It is evident that the distance GJ can be less than any assigned distance. And (from the nature of the circles passing through points A, B, G and *a*, *b*, *g*) AB^2 is equal to $AG \times BD$, and Ab^2 is equal to $Ag \times bd$, and thus the ratio of AB^2 to Ab^2 is compounded of the ratios of AG to A*g* and BD to *bd*. But since GJ can be taken as less than any assigned length, it can happen that the ratio of AG to A*g* differs from the ratio of equality by less than any assigned difference, and thus that the ratio of AB^2 to Ab^2 differs from the ratio of BD to *bd* by less than any assigned difference. Therefore, by lem. 1, the ultimate ratio of AB^2 to Ab^2 is the same as the ultimate ratio of BD to *bd*. Q.E.D.

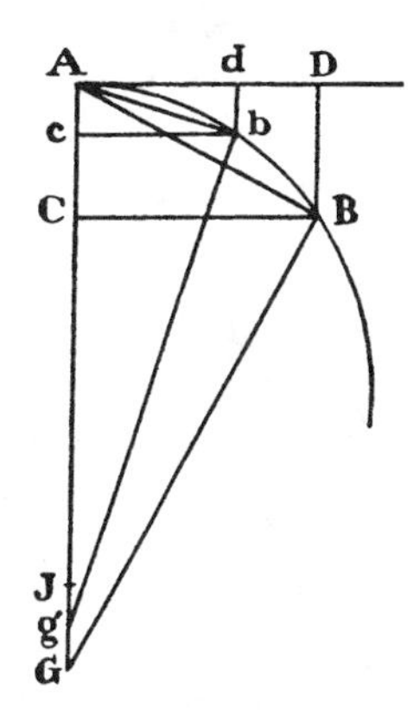

CASE 2. Now let BD be inclined to AD at any given angle, and the ultimate ratio of BD to *bd* will always be the same as before and thus the same as AB^2 to Ab^2. Q.E.D.

CASE 3. And even when angle D is not given, if the straight line BD converges to a given point or is drawn according to any other specification, still the angles D and *d* (constructed according to the specification common to both) will always tend to equality and will approach each other so closely that their difference will be less than any assigned quantity, and thus will ultimately be equal, by lem. 1; and therefore lines BD and *bd* are in the same ratio to each other as before. Q.E.D.

Corollary 1. Hence, since tangents AD and A*d*, arcs AB and A*b*, and their sines BC and *bc* become ultimately equal to chords AB and A*b*, their squares will also be ultimately as the subtenses BD and *bd*.

[a]Corollary 2. The squares of these tangents, arcs, and sines are also ultimately as the sagittas of the arcs, which bisect the chords and converge to a given point. For these sagittas are as the subtenses BD and *bd*.

Corollary 3. And thus the sagitta is in the squared ratio of the time in which a body describes the arc with a given velocity.[a]

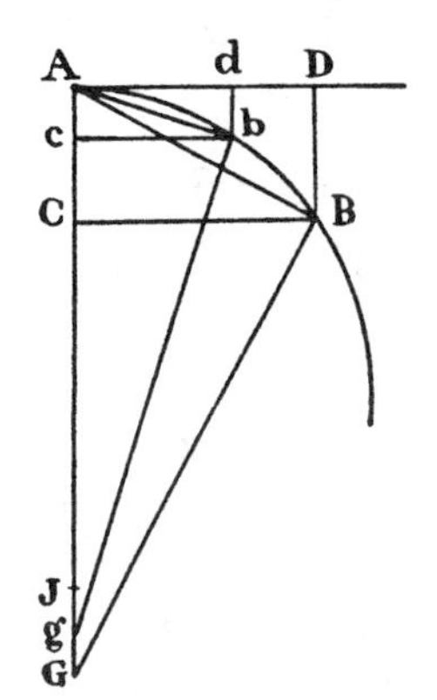

Corollary 4. The rectilinear triangles ADB and A*db* are ultimately in the cubed ratio of the sides AD and A*d*, and in the sesquialteral ratio [i.e., as the 3/2 power] of the sides DB and *db*, inasmuch as these triangles are in a ratio compounded of the ratios of AD and DB to A*d* and *db*. So also the triangles ABC and A*bc* are ultimately in the cubed ratio of the sides BC and *bc*. [b]The ratio that I call sesquialteral is the halved of the tripled, namely, the one that is compounded of the simple and the halved.[b]

Corollary 5. And since DB and *db* are ultimately parallel and in the squared ratio of AD and A*d*, the ultimate curvilinear areas ADB and A*db* will be (from the nature of the parabola) two-thirds of the rectilinear triangles ADB and A*db*; and the segments AB and A*b* will be thirds of these same triangles. And hence these areas and segments will be in the cubed ratio of both of the tangents AD and A*d* and of the chords AB and A*b* and their arcs.

Scholium But we suppose throughout that the angle of contact is neither infinitely greater nor infinitely less than the angles of contact that circles contain with their tangents, that is, that the curvature at point A is neither infinitely small nor infinitely great—in other words, that the distance AJ is of a finite magnitude. For DB can be taken proportional to AD^3, in which case no circle can be drawn through point A between tangent AD and curve AB, and accordingly the angle of contact will be infinitely less than those of

aa. Ed. 1 lacks corols. 2 and 3.

bb. Ed. 1 lacks this sentence.

circles. And, similarly, if DB is made successively proportional to AD^4, AD^5, AD^6, AD^7, . . . , there will be a sequence of angles of contact going on to infinity, any succeeding one of which is infinitely less than the preceding one. And if DB is made successively proportional to AD^2, $AD^{3/2}$, $AD^{4/3}$, $AD^{5/4}$, $AD^{6/5}$, $AD^{7/6}$, . . . , there will be another infinite sequence of angles of contact, the first of which is of the same kind as those of circles, the second infinitely greater, and any succeeding one infinitely greater than the preceding one. Moreover, between any two of these angles a sequence of intermediate angles, going on to infinity in both directions, can be inserted, any succeeding one of which will be infinitely greater or smaller than the preceding one—as, for example, if between the terms AD^2 and AD^3 there were inserted the sequence $AD^{13/6}$, $AD^{11/5}$, $AD^{9/4}$, $AD^{7/3}$, $AD^{5/2}$, $AD^{8/3}$, $AD^{11/4}$, $AD^{14/5}$, $AD^{17/6}$, ... And again, between any two angles of this sequence a new sequence of intermediate angles can be inserted, differing from one another by infinite intervals. And nature knows no limit.

What has been demonstrated concerning curved lines and the [plane] surfaces comprehended by them is easily applied to curved surfaces and their solid contents. In any case, I have presented these lemmas before the propositions in order to avoid the tedium of working out [c]lengthy[c] proofs by *reductio ad absurdum* in the manner of the ancient geometers. Indeed, proofs are rendered more concise by the method of indivisibles. But since the hypothesis of indivisibles is [d]problematical[d] and this method is therefore accounted less geometrical, I have preferred to make the proofs of what follows depend on the ultimate sums and ratios of vanishing quantities and the first sums and ratios of nascent quantities, that is, on the limits of such sums and ratios, and therefore to present proofs of those limits beforehand as briefly as I could. For the same result is obtained by these as by the method of indivisibles, and we shall be on safer ground using principles that have been proved. Accordingly, whenever in what follows I consider quantities as consisting of particles or whenever I use curved line-elements [or minute curved lines] in

cc. For "lengthy" (Lat. "longas") ed. 1 and ed. 2 have "complicated" (Lat. "perplexas"), which Newton inserted with his own hand into the manuscript of ed. 1. Motte gives "perplexed," thus obviously using ed. 2, and Cajori has "involved," revealing that the Latin text was not consulted at this point. But in *A History of the Conceptions of Limits and Fluxions in Great Britain from Newton to Woodhouse* (Chicago and London: Open Court Publishing Co., 1919), Cajori notes on p. 5 that "in the third edition 'longas' takes the place of 'perplexas,' " and on p. 8 he uses Thorp's translation ("long").

dd. Newton uses the adjective "durior," which is traditionally translated by "rather harsh."

place of straight lines, I wish it always to be understood that I have in mind not indivisibles but evanescent divisibles, and not sums and ratios of definite parts but the limits of such sums and ratios, and that the force of such proofs always rests on the method of the preceding lemmas.

It may be objected that there is no such thing as an ultimate proportion of vanishing quantities, inasmuch as before vanishing the proportion is not ultimate, and after vanishing it does not exist at all. But by the same argument it could equally be contended that there is no ultimate velocity of a body reaching a certain place at which the motion ceases; for before the body arrives at this place, the velocity is not the ultimate velocity, and when it arrives there, there is no velocity at all. But the answer is easy: to understand the ultimate velocity as that with which a body is moving, neither before it arrives at its ultimate place and the motion ceases, nor after it has arrived there, but at the very instant when it arrives, that is, the very velocity with which the body arrives at its ultimate place and with which the motion ceases. And similarly the ultimate ratio of vanishing quantities is to be understood not as the ratio of quantities before they vanish or after they have vanished, but the ratio with which they vanish. Likewise, also, the first ratio of nascent quantities is the ratio with which they begin to exist [or come into being]. And the first and the ultimate sum is the sum with which they begin and cease to exist (or to be increased or decreased). There exists a limit which their velocity can attain at the end of the motion, but cannot exceed. This is their ultimate velocity. And it is the same for the limit of all quantities and proportions that come into being and cease existing. And since this limit is certain and definite, the determining of it is properly a geometrical problem. But everything that is geometrical is legitimately used in determining and demonstrating whatever else may be geometrical.

It can also be contended that if the ultimate ratios of vanishing quantities are given, their ultimate magnitudes will also be given; and thus every quantity will consist of indivisibles, contrary to what Euclid had proved concerning incommensurables in the tenth book of his *Elements*. But this objection is based on a false hypothesis. Those ultimate ratios with which quantities vanish are not actually ratios of ultimate quantities, but limits which the ratios of quantities decreasing without limit are continually approaching, and which they can approach so closely that their difference is less than any given quantity, but which they can never exceed and can never reach before the quantities are decreased indefinitely. This matter will be understood more clearly in the case of quantities that are indefinitely great. If two quantities whose difference is given are increased indefinitely, their ultimate ratio will be given, namely the ratio of equality, and yet the ultimate or maximal quantities of which this is the ratio will not on this account be given. Therefore, whenever, to make things easier to comprehend, I speak in what follows of quantities as minimally small or vanishing or ultimate, take care not to understand quantities that are determinate in magnitude, but always think of quantities that are to be decreased without limit.

SECTION 2

To find centripetal forces

Proposition 1[a] *The areas which bodies* [b]*made to move in orbits*[b] *describe by radii drawn to an*
Theorem 1 *unmoving center of forces lie in unmoving planes and are proportional to the times.*

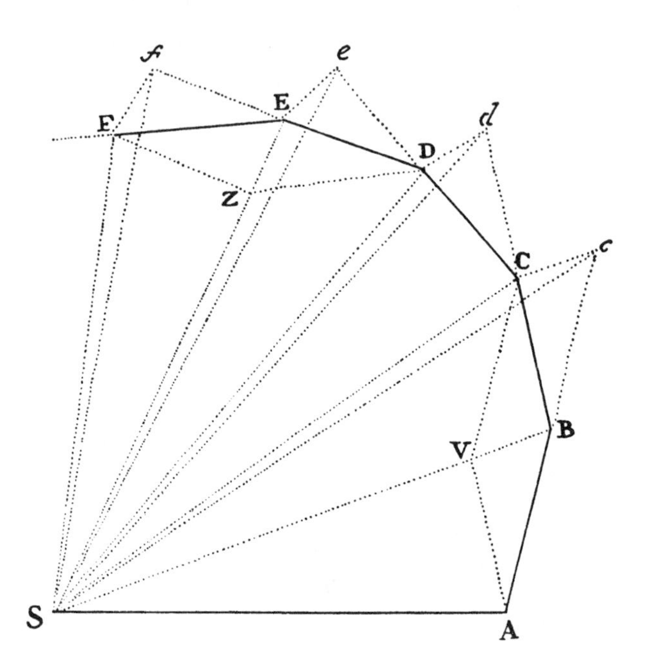

Let the time be divided into equal parts, and in the first part of the time let a body by its inherent force describe the straight line AB. In the second part of the time, if nothing hindered it, this body would (by law 1) go straight on to *c*, describing line B*c* equal to AB, so that—when radii AS, BS, and *c*S were drawn to the center—the equal areas ASB and BS*c* would be described. But when the body comes to B, let a centripetal force act with a single but great impulse and make the body deviate from the straight line B*c* and proceed in the straight line BC. Let *c*C be drawn parallel to BS and meet BC at C; then, when the second part of the time has been completed, the body (by corol. 1 of the laws) will be found at C in the same plane as

a. For a gloss on this proposition see the Guide, §10.8.

bb. In the statement of prop. 1, Newton uses the phrase "in gyros acta"; see the Guide, §2.4.

triangle ASB. Join SC; and because SB and C*c* are parallel, triangle SBC will be equal to triangle SB*c* and thus also to triangle SAB. By a similar argument, if the centripetal force acts successively at C, D, E, . . . , making the body in each of the individual particles of time describe the individual straight lines CD, DE, EF, . . . , all these lines will lie in the same plane; and triangle SCD will be equal to triangle SBC, SDE to SCD, and SEF to SDE. Therefore, in equal times equal areas are described in an unmoving plane; and by composition [or componendo], any sums SADS and SAFS of the areas are to each other as the times of description. Now let the number of triangles be increased and their width decreased indefinitely, and their ultimate perimeter ADF will (by lem. 3, corol. 4) be a curved line; and thus the centripetal force by which the body is continually drawn back from the tangent of this curve will act uninterruptedly, while any areas described, SADS and SAFS, which are always proportional to the times of description, will be proportional to those times in this case. Q.E.D.

[c]COROLLARY 1. In nonresisting spaces, the velocity of a body attracted to an immobile center is inversely as the perpendicular dropped from that center to the straight line which is tangent to the orbit. For the velocities in those places A, B, C, D, and E are respectively as the bases of the equal triangles AB, BC, CD, DE, and EF, and these bases are inversely as the perpendiculars dropped to them.

COROLLARY 2. If chords AB and BC of two arcs successively described by the same body in equal times in nonresisting spaces are completed into the parallelogram ABCV, and diagonal BV (in the position that it ultimately has when those arcs are decreased indefinitely) is produced in both directions, it will pass through the center of forces.[c]

[d]COROLLARY 3. If chords AB, BC and DE, EF of arcs described in equal times in nonresisting spaces are completed into parallelograms ABCV and DEFZ, then the forces at B and E are to each other in the ultimate ratio of the diagonals BV and EZ when the arcs are decreased indefinitely. For the motions BC and EF of the body are (by corol. 1 of the laws) compounded of the motions B*c*, BV and E*f*, EZ; but in the proof of this

cc. In ed. 1, corols. 1 and 2 are earlier versions of prop. 2, corols. 1 and 2, and the corols. 1 and 2 of ed. 2 and ed. 3 are lacking.

dd. Ed. 1 lacks corols. 3–6.

proposition BV and EZ, equal to C*c* and F*f*, were generated by the impulses of the centripetal force at B and E, and thus are proportional to these impulses.

COROLLARY 4. The forces by which any bodies in nonresisting spaces are drawn back from rectilinear motions and are deflected into curved orbits are to one another as those sagittas of arcs described in equal times which converge to the center of forces and bisect the chords when the arcs are decreased indefinitely. For these sagittas are halves of the diagonals with which we dealt in corol. 3.

COROLLARY 5. And therefore these forces are to the force of gravity as these sagittas are to the sagittas, perpendicular to the horizon, of the parabolic arcs that projectiles describe in the same time.

COROLLARY 6. All the same things hold, by corol. 5 of the laws, when the planes in which the bodies are moving, together with the centers of forces that are situated in those planes, are not at rest but move uniformly straight forward.[d]

Proposition 2
Theorem 2

Every body that moves in some curved line described in a plane and, by a radius drawn to a point, either unmoving or moving uniformly forward with a rectilinear motion, describes areas around that point proportional to the times, is urged by a centripetal force tending toward that same point.

CASE 1. For every body that moves in a curved line is deflected from a rectilinear course by some force acting upon it (by law 1). And that force by which the body is deflected from a rectilinear course and in equal times is made to describe, about an immobile point S, the equal minimally small triangles SAB, SBC, SCD, . . . , acts in place B along a line parallel to *c*C (by book 1, prop. 40, of the *Elements*, and law 2), that is, along the line BS; and in place C, the force acts along a line parallel to *d*D, that is, along the line SC, ... Therefore it always acts along lines tending toward that unmoving point S. Q.E.D.

CASE 2. And, by corol. 5 of the laws, it makes no difference whether the surface on which the body describes a curvilinear figure is at rest or whether it moves uniformly straight forward, together with the body, the figure described, and the point S.

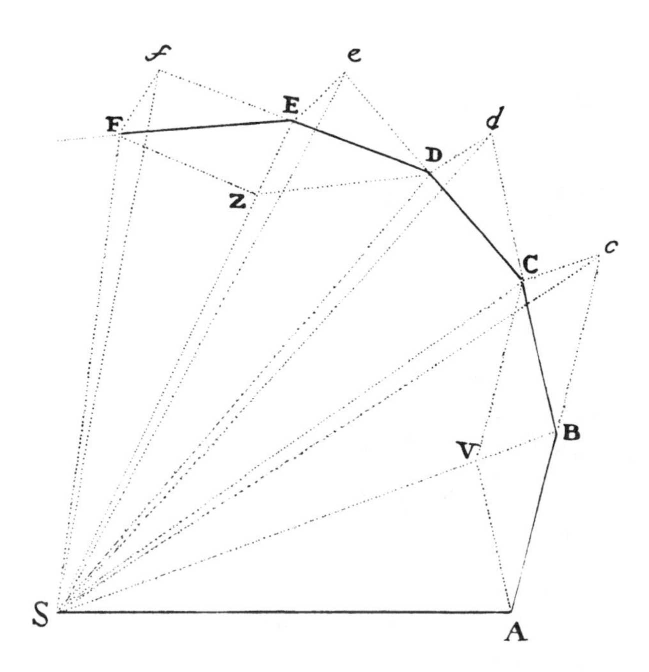

[a]Corollary 1. [b]In nonresisting spaces or mediums, if the areas are not proportional to the times, the forces do not tend toward the point where the radii meet but deviate forward [or in consequentia] from it, that is, in the direction toward which the motion takes place, provided that the description of the areas is accelerated; but if it is retarded, they deviate backward [or in antecedentia, i.e., in a direction contrary to that in which the motion takes place].[b]

Corollary 2. [c]In resisting mediums also, if the description of areas is accelerated, the directions of the forces deviate from the point where the radii meet in the direction toward which the motion takes place.[a c]

Scholium

A body can be urged by a centripetal force compounded of several forces. In this case the meaning of the proposition is that the force which is compounded of all the forces tends toward point S. Further, if some force acts

aa. In ed. 1, prop. 2 has no corollaries. Corols. 1 and 2 of ed. 2 and ed. 3 are basically revised versions of corols. 1 and 2 to prop. 1 of ed. 1.

bb. Ed. 1 has (as prop. 1, corol. 1): "In nonresisting mediums, if the areas are not proportional to the times, the forces do not tend toward the point where the radii meet."

cc. Ed. 1 has (as prop. 1, corol. 2): "In all mediums, if the description of areas is accelerated, the forces do not tend toward the point where the radii meet but deviate forward [or in consequentia] from it."

continually along a line perpendicular to the surface described, it will cause the body to deviate from the plane of its motion, but it will neither increase nor decrease the quantity of the surface-area described and is therefore to be ignored in the compounding of forces.

Proposition 3
Theorem 3

[a]*Every body that, by a radius drawn to the center of a second body moving in any way whatever, describes about that center areas that are proportional to the times is urged by a force compounded of the centripetal force tending toward that second body and of the whole accelerative force by which that second body is urged.*

Let the first body be L, and the second body T; and (by corol. 6 of the laws) if each of the two bodies is urged along parallel lines by a new force that is equal and opposite to the force by which body T is urged, body L will continue to describe about body T the same areas as before; but the force by which body T was urged will now be annulled by an equal and opposite force, and therefore (by law 1) body T, now left to itself, either will be at rest or will move uniformly straight forward; and body L, since the difference of the forces [i.e., the remaining force] is urging it, will continue to describe areas proportional to the times about body T. Therefore, the difference of the forces tends (by theor. 2) toward the second body T as center. Q.E.D.

Corollary 1. Hence, if a body L, by a radius drawn to another body T, describes areas proportional to the times, and from the total force (whether simple or compounded of several forces according to corol. 2 of the laws) by which body L is urged there is subtracted (according to the same corol. 2 of the laws) the total accelerative force by which body T is urged, the whole remaining force by which body L is urged will tend toward body T as center.

Corollary 2. And if the areas are very nearly proportional to the times, the remaining force will tend toward body T very nearly.

aa. In both the statement and the demonstration of the proposition and also in the corollaries, ed. 1 lacks letters to designate the two bodies. In Newton's annotated copy of ed. 1, the letters L and T are added in all of these parts of the proposition. In Newton's interleaved copy of ed. 1, letters are added in all of these sections but are then deleted from the statement of the proposition, where the letters written in might have first been A and B and then been changed to L and T before being crossed out. In the first sentence of the demonstration in this interleaved copy, the first two letters added at the beginning of the sentence were originally A and B, which were then altered to L and T. It is these letters, L and T, that were added elsewhere and were kept in the demonstration and corollaries.

COROLLARY 3. And conversely, if the remaining force tends very nearly toward body T, the areas will be very nearly proportional to the times.

COROLLARY 4. If body L, by a radius drawn to another body T, describes areas which, compared with the times, are extremely unequal, and body T either is at rest or moves uniformly straight forward, either there is no centripetal force tending toward body T or the action of the centripetal force is mixed and compounded with the very powerful actions of other forces; and the total force compounded of all the forces, if there are several, is directed toward another center (whether fixed or moving). The same thing holds when the second body moves with any motion whatever, if the centripetal force is what remains after subtraction of the total force acting upon body T.[a]

Scholium

Since the uniform description of areas indicates the center toward which that force is directed by which a body is most affected and by which it is drawn away from rectilinear motion and kept in orbit, why should we not in what follows use uniform description of areas as a criterion for a center about which all orbital motion takes place in free spaces?

Proposition 4
Theorem 4

The centripetal forces of bodies that describe different circles with uniform motion tend toward the centers of those circles and are to one another as the squares of the arcs described in the same time divided by the radii of the circles.

[a]These forces tend toward the centers of the circles by prop. 2 and prop. 1, corol. 2, and are to another as the versed sines of the arcs described in

aa. Ed. 1 has: "Let bodies B and *b*, revolving in the circumferences of circles BD and *bd*, describe arcs BD and *bd* in the same time. Since by their inherent force alone they would describe tangents BC and *bc* equal to these arcs, it is obvious that centripetal forces are the ones which continually draw the bodies back from the tangents to the circumferences of the circles, and thus these forces are to each other in the first ratio of the nascent spaces CD and *cd*, and they tend toward the centers of the circles, by theor. 2, because the areas described by the radii are supposed proportional to the times. [Newton is using "first ratio" here in the special sense developed in sec. 1 above, where he introduces the concept of "first" and "ultimate" ratio.] Let figure *tkb* be similar to DCB and, by lem. 5, line-element CD will be to line-element *kt* as arc BD to arc *bt*, and also, by lem. 11, the nascent line-element *tk* will be to the nascent line-element *dc* as bt^2 to bd^2 and, from the equality of the ratios [or ex aequo], the nascent line-element DC will be to the nascent line-element *dc* as BD × *bt* to bd^2 or,

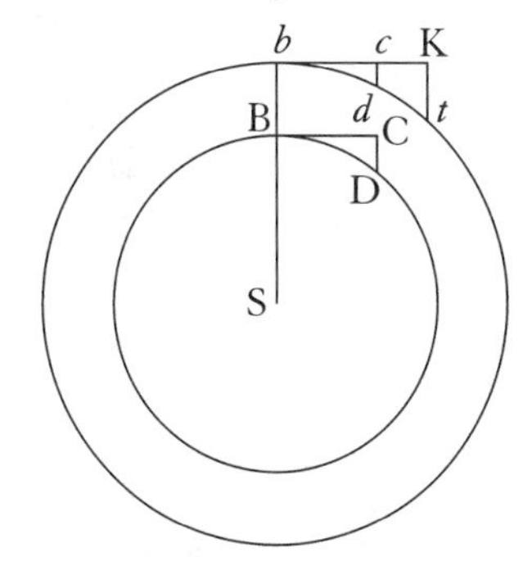

minimally small equal times, by prop. 1, corol. 4, that is, as the squares of those arcs divided by the diameters of the circles, by lem. 7; and therefore, since these arcs are as the arcs described in any equal times and the diameters are as their radii, the forces will be as the squares of any arcs described in the same time divided by the radii of the circles. Q.E.D.[a]

[b]Corollary 1. [c]Since those arcs are as the velocities of the bodies, the centripetal forces will be in a ratio compounded of the squared ratio of the velocities directly and the simple ratio of the radii inversely.[c]

Corollary 2. [d]And since the periodic times are in a ratio compounded of the ratio of the radii directly and the ratio of the velocities inversely, the centripetal forces are in a ratio compounded of the ratio of the radii directly and the squared ratio of the periodic times inversely.[d]

Corollary 3. Hence, if the periodic times are equal and therefore the velocities are as the radii, the centripetal forces also will be as the radii; and conversely.

---

what comes to the same thing, as $\mathrm{BD} \times \frac{bt}{\mathrm{S}b}$ to $\frac{bd^2}{\mathrm{S}b}$ and thus (because the ratios $\frac{bt}{\mathrm{S}b}$ and $\frac{\mathrm{BD}}{\mathrm{SB}}$ are equal) as $\frac{\mathrm{BD}^2}{\mathrm{SB}}$ to $\frac{bd^2}{\mathrm{S}b}$. Q.E.D."

Here, as in the very similar earlier formulation of *De Motu* and in a later handwritten revision of ed. 1, the sentence specifying centrifugal forces has some ambiguity because the grammatical structure can indicate that Newton is redefining these forces whereas the context shows that he is giving one of their properties.

bb. Different versions of corols. 1, 2, 4, 5, and 6 exemplify interesting variations in basic mathematical terminology, as is indicated in the following notes.

cc. In ed. 1 this corollary reads: "Hence the centripetal forces are as the squares of the velocities divided by the radii of the circles." In manuscript revisions of ed. 1 "Hence" is deleted and the sentence begins with an additional clause: "Whence, since the arcs described in the same time are directly as the velocities and inversely as the periodic times." Ed. 2 reads: "Therefore, since those arcs are as the velocities of the bodies, the centripetal forces are as the squares of the velocities divided by the radii of the circles; that is, to express it as the geometers do, the forces are in a ratio compounded of the squared ratio of the velocities directly and the simple ratio of the radii inversely." And then, in ed. 3, Newton decides to eliminate the first formulation and express his result only "as the geometers do."

dd. In ed. 1 this corollary reads: "And inversely as the squares of the periodic times divided by the radii so are these forces to one another. That is (to express it as the geometers do), these forces are in a ratio compounded of the squared ratio of the velocities directly and the simple ratio of the radii inversely, and also in a ratio compounded of the simple ratio of the radii directly and the squared ratio of the periodic times inversely." The inversion in the first sentence suggests that originally it was not a full sentence but a continuation from corol. 1, as comparison with the earlier *De Motu* shows to be true. Ed. 2 reads: "And since the periodic times are in a ratio compounded of the ratio of the radii directly and the ratio of the velocities inversely, the centripetal forces are inversely as the squares of the periodic times divided by the radii of the circles: that is, in a ratio compounded of the ratio of the radii directly and the squared ratio of the periodic times inversely."

COROLLARY 4. [e]If both the periodic times and the velocities are as the square roots of the radii, the centripetal forces will be equal to one another; and conversely.[e]

COROLLARY 5. [f]If the periodic times are as the radii, and therefore the velocities are equal, the centripetal forces will be inversely as the radii; and conversely.[f]

COROLLARY 6. [g]If the periodic times are as the 3/2 powers of the radii, and therefore the velocities are inversely as the square roots of the radii, the centripetal forces will be inversely as the squares of the radii; and conversely.[b] [g]

[h]COROLLARY 7. And universally, if the periodic time is as any power R^n of the radius R, and therefore the velocity is inversely as the power R^{n-1} of the radius, the centripetal force will be inversely as the power R^{2n-1} of the radius; and conversely.

COROLLARY 8. In cases in which bodies describe similar parts of any figures that are similar and have centers similarly placed in those figures, all the same proportions with respect to the times, velocities, and forces follow from applying the foregoing demonstrations to these cases. And the application is made by substituting the uniform description of areas for uniform motion, and by using the distances of bodies from the centers for the radii.

COROLLARY 9. From the same demonstration it follows also that the arc which a body, in revolving uniformly in a circle with a given centripetal force, describes in any time is a mean proportional between the diameter of the circle and the distance through which the body would fall under the action of the same given force and in the same time.[h]

ee. In ed. 1 this corollary reads: "If the squares of the periodic times are as the radii, the centripetal forces are equal, and the velocities are in the halved ratio of the radii, and vice versa."

ff. In ed. 1 this corollary reads: "If the squares of the periodic times are as the squares of the radii, the centripetal forces are inversely as the radii, and the velocities are equal, and vice versa." After "of the radii," handwritten revisions of ed. 1 add "that is, the times [are] as the radii."

gg. In ed. 1 this corollary reads: "If the squares of the periodic times are as the cubes of the radii, the centripetal forces are inversely as the squares of the radii, but the velocities are in the halved ratio of the radii, and vice versa."

hh. Ed. 1 lacks corols. 7 and 9, and in corol. 8, which is numbered 7, it lacks "in those figures" in the first sentence and all of the second sentence.

Scholium [i]The case of corol. 6 holds for the heavenly bodies (as our compatriots Wren, Hooke, and Halley have also found out independently). Accordingly, I have decided that in what follows I shall deal more fully with questions relating to the centripetal forces that decrease as the squares of the distances from centers [i.e., centripetal forces that vary inversely as the squares of the distances].

Further, with the help of the preceding proposition and its corollaries the proportion of a centripetal force to any known force, such as that of gravity, may also be determined. [j]For if a body revolves by the force of its gravity in a circle concentric with the earth, this gravity is its centripetal force. Moreover, by prop. 4, corol. 9, both the time of one revolution and the arc described in any given time are given from the descent of heavy bodies.[j] And by propositions of this sort Huygens in his excellent treatise *On the Pendulum Clock* compared the force of gravity with the centrifugal forces of revolving bodies.

This proposition can also be demonstrated in the following manner. In any circle, suppose that a polygon of any number of sides is described. And if a body moving with a given velocity along the sides of the polygon is reflected from the circle at each of the angles of the polygon, the force with which it impinges upon the circle at each reflection will be as its velocity; and therefore the sum of the forces in a given time will be as that velocity and the number of reflections jointly; that is (if the sides and angles of the polygon are specified), as the length described in that given time and increased or decreased in the ratio of the length to the radius of the above-mentioned circle, that is, as the square of that length divided by the radius. And, therefore,

ii. In the printer's manuscript of ed. 1 the scholium originally consisted of a single sentence, corresponding to the first sentence of ed. 3 but without the parenthesis containing the three proper names. A separate sheet in this manuscript and the printed text of ed. 1 contain the entire scholium, but in the addition to the manuscript the names are listed as Wren, Halley, and Hooke, whereas in ed. 1 they appear in the order retained in ed. 3. We cannot tell by whose authority Hooke's name was moved to a position before Halley's, but we can infer that the alteration was made in proof (and so presumably by Halley), since the handwritten addition to the manuscript as sent by Newton to Halley and by Halley to the printer is unaltered. It is very probable that Halley put Hooke's name ahead of his own because he did not want Hooke to be offended.

jj. Ed. 1 has: "For since the former force, in the time in which a body traverses arc BC, impels the body through space CD, which at the very beginning of the motion is equal to the square of that arc BD divided by the diameter of the circle, and since every body, by the same force continued always in the same direction, describes spaces that are in the squared ratio of the times, that force, in the time in which the revolving body describes any given arc, will cause the body as it advances directly forward to describe a space equal to the square of that arc divided by the diameter of the circle and thus is to the force of gravity as that space is to the space which a heavy body in falling describes in the same time."

if the sides are diminished indefinitely, the polygon will coincide with the circle, and the sum of the forces in a given time will be as the square of the arc described in the given time divided by the radius. This is the centrifugal force with which the body urges the circle; and the opposite force, with which the circle continually repels the body toward the center, is equal to this centrifugal force.[i]

Proposition 5
Problem 1

Given, in any places, the velocity with which a body describes a given curve when acted on by forces tending toward some common center, to find that center.

Let the curve so described be touched in three points P, Q, and R by three straight lines PT, TQV, and VR, meeting in T and V. Erect PA, QB, and RC perpendicular to the tangents and inversely proportional to the velocities of the body at the points P, Q, and R from which the perpendiculars are erected—that is, so that PA is to QB as the velocity at Q to the velocity at P, and QB to RC as the velocity at R to the velocity at Q. Through the ends A, B, and C of the perpendiculars draw AD, DBE, and EC at right angles to those perpendiculars, and let them meet in D and E; then TD and VE, when drawn and produced, will meet in the required center S.

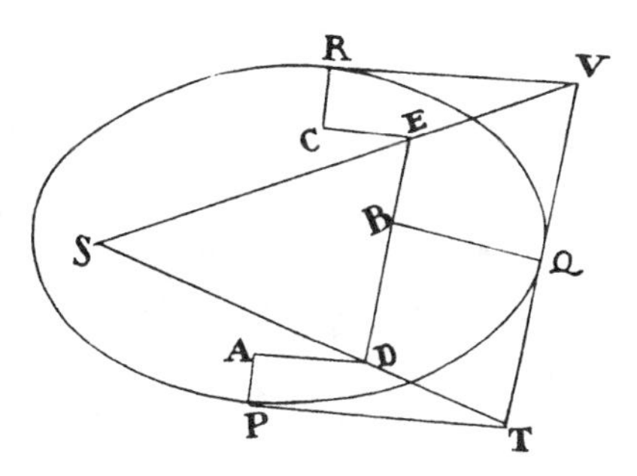

For the perpendiculars dropped from center S to tangents PT and QT are (by prop. 1, corol. 1) inversely as the velocities of the body at points P and Q, and therefore, by the construction, as the perpendiculars AP and BQ directly, that is, as the perpendiculars dropped from point D to the tangents. Hence it is easily gathered that points S, D, and T are in one straight line. And, by a similar argument, the points S, E, and V are also in one straight line; and therefore the center S is at the point where the straight lines TD and VE meet. Q.E.D.

Proposition 6[a]
Theorem 5

[b]*If in a nonresisting space a body revolves in any orbit about an immobile center and describes any just-nascent arc in a minimally small time, and if the sagitta of*

a. For a gloss on this proposition see the Guide, §10.8.

bb. In ed. 1 there is a different prop. 6, with its proof and single unnumbered corollary. In ed. 2 and ed. 3 the statement of this proposition becomes corol. 1 to the new prop. 6 and the single corollary

the arc is understood to be drawn so as to bisect the chord and, when produced, to pass through the center of forces, the centripetal force in the middle of the arc will be as the sagitta directly and as the time twice [i.e., as the square of the time] inversely.

For the sagitta in a given time is as the force (by prop. 1, corol. 4), and if the time is increased in any ratio, then—because the arc is increased in the same ratio—the sagitta is increased in that ratio squared (by lem. 11, corols. 2 and 3) and therefore is as the force once and the time twice [i.e., as the force and the square of the time jointly]. Take away from both sides the squared ratio of the time, and the force will become as the sagitta directly and as the time twice [or as the square of the time] inversely. Q.E.D.

This proposition is also easily proved by lem. 10, corol. 4.

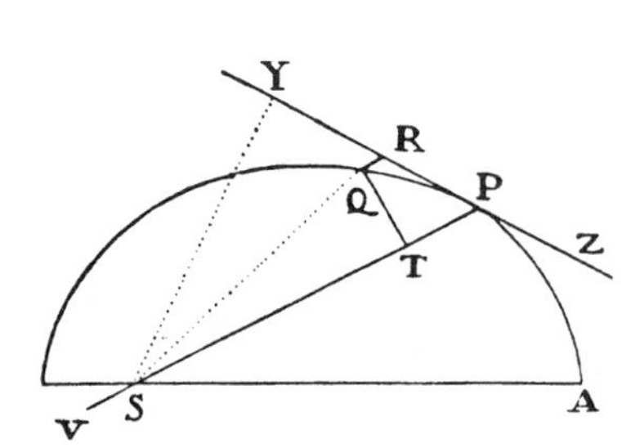

COROLLARY 1. If a body P, revolving about a center S, describes the curved line APQ, while the straight line ZPR touches the curve at any point P; and QR, parallel to distance SP, is drawn to the tangent from any other point Q of the curve, and QT is drawn perpendicular to that distance SP; then the centripetal force will be inversely as the solid $\frac{SP^2 \times QT^2}{QR}$, provided that the magnitude of that solid is always taken as that which it has ultimately when the points P and Q come together. For QR is equal to the sagitta of an arc that is twice the length of arc QP, with P being in the middle; and twice the triangle SQP (or SP × QT) is proportional to the time in which twice that arc is described and therefore can stand for the time.

COROLLARY 2. By the same argument the centripetal force is inversely as the solid $\frac{SY^2 \times QP^2}{QR}$, provided that SY is a perpendicular dropped from the

becomes corol. 5. The proof in ed. 1 reads as follows: "For in the indefinitely small figure QRPT the nascent line-element QR, if the time is given, is as the centripetal force (by law 2) and, if the force is given, is as the square of the time (by lem. 10) and thus, if neither is given, is as the centripetal force and the square of the time jointly, and thus the centripetal force is as the line-element QR directly and the square of the time inversely. But the time is as the area SPQ, or its double SP × QT, that is, as SP and QT jointly, and thus the centripetal force is as QR directly and SP^2 times QT^2 inversely, that is, as $\frac{SP^2 \times QT^2}{QR}$ inversely. Q.E.D." (The figure for prop. 6 in ed. 1 is the same as in eds. 2 and 3, except that the line PS does not extend below the line SA, so that there is no point V.)

center of forces to the tangent PR of the orbit. For the rectangles $SY \times QP$ and $SP \times QT$ are equal.

COROLLARY 3. If the orbit APQ either is a circle or touches a circle concentrically or cuts it concentrically—that is, if it makes with the circle an angle of contact or of section which is the least possible—and has the same curvature and the same radius of curvature at point P, and if the circle has a chord drawn from the body through the center of forces, then the centripetal force will be inversely as the solid $SY^2 \times PV$. For PV is equal to $\frac{QP^2}{QR}$.

COROLLARY 4. Under the same conditions [as corol. 3], the centripetal force is directly as the square of the velocity and inversely as the chord. For, by prop. 1, corol. 1, the velocity is inversely as the perpendicular SY.

COROLLARY 5. Hence, if there is given any curvilinear figure APQ and on it there is given also point S, to which the centripetal force is continually directed, the law of the centripetal force can be found by which any body P, continually drawn away from a rectilinear course, will be kept in the perimeter of that figure and will describe it as an orbit. That is, the solid $\frac{SP^2 \times QT^2}{QR}$ or the solid $SY^2 \times PV$, inversely proportional to this force, is to be found by computation. We will give examples of this in the following problems.[b]

Proposition 7
Problem 2

Let a body revolve in the circumference of a circle; it is required to find the law of the centripetal force tending toward any given point.

Let VQPA be the circumference of the circle, S the given point toward which the force tends as to its center, P the body revolving in the circumference, Q the place to which it will move next, and PRZ the tangent of the circle at the previous place. Through point S draw chord PV; and when the diameter VA of the circle has been drawn, join AP; and to SP drop perpendicular

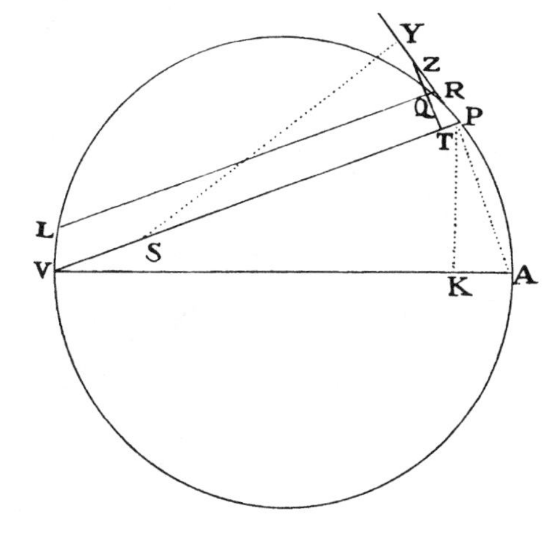

QT, which when produced meets the tangent PR at Z; and finally through point Q draw LR parallel to SP and meeting both the circle at L and the tangent PZ at R. Then because the triangles ZQR, ZTP, and VPA are similar, RP^2 (that is, $QR \times RL$) will be to QT^2 as AV^2 to PV^2. And therefore $\frac{QR \times RL \times PV^2}{AV^2}$ is equal to QT^2. Multiply these equals by $\frac{SP^2}{QR}$ and, the points P and Q coming together, write PV for RL. Thus $\frac{SP^2 \times PV^3}{AV^2}$ will become equal to $\frac{SP^2 \times QT^2}{QR}$. Therefore (by prop. 6, corols. 1 and 5), the centripetal force is inversely as $\frac{SP^2 \times PV^3}{AV^2}$, that is (because AV^2 is given), inversely as the square of the distance or altitude SP and the cube of the chord PV jointly. Q.E.I.

Another solution

Draw SY perpendicular to the tangent PR produced; then, because triangles SYP and VPA are similar, AV will be to PV as SP to SY, and thus $\frac{SP \times PV}{AV}$ will be equal to SY, and $\frac{SP^2 \times PV^3}{AV^2}$ will be equal to $SY^2 \times PV$. And therefore (by prop. 6, corols. 3 and 5), the centripetal force is inversely as $\frac{SP^2 \times PV^3}{AV^2}$, that is, because AV is given, inversely as $SP^2 \times PV^3$. Q.E.I.

COROLLARY 1. Hence, if the given point S to which the centripetal force always tends is located in the circumference of this circle, say at V, the centripetal force will be inversely as the fifth power of the altitude SP.

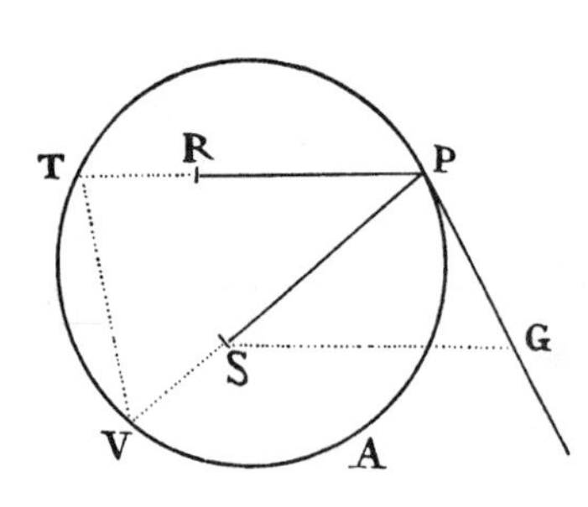

COROLLARY 2. The force by which body P revolves in the circle APTV around the center of forces S is to the force by which the same body P can revolve in the same circle and in the same periodic time around any other center of forces R as $RP^2 \times SP$ to the cube of the straight line SG, which is drawn from the first center of forces S to the tangent of the orbit PG and is parallel to the distance of the body from the second center of forces. For by the construction of this proposition the first force is

to the second force as $RP^2 \times PT^3$ to $SP^2 \times PV^3$, that is, as $SP \times RP^2$ to $\frac{SP^3 \times PV^3}{PT^3}$, or (because the triangles PSG and TPV are similar) to SG^3.

COROLLARY 3. The force by which body P revolves in any orbit around the center of forces S is to the force by which the same body P can revolve in the same orbit and in the same periodic time around any other center of forces R as the solid $SP \times RP^2$—contained under the distance of the body from the first center of forces S and the square of its distance from the second center of forces R—to the cube of the straight line SG, which is drawn from the first center of forces S to the tangent of the orbit PG and is parallel to the distance RP of the body from the second center of forces. For the forces in this orbit at any point of it P are the same as in a circle of the same curvature.

Proposition 8
Problem 3

Let a body move in the semicircle PQA; *it is required to find the law of the centripetal force for this effect, when the centripetal force tends toward a point* S *so distant that all the lines* PS *and* RS *drawn to it can be considered parallel.*

From the center C of the semicircle draw the semidiameter CA, intersecting those parallels perpendicularly at M and N, and join CP. Because triangles CPM, PZT, and RZQ are similar, CP^2 is to PM^2 as PR^2 to QT^2, and from the nature of a circle PR^2 is equal to the rectangle $QR \times (RN + QN)$, or, the points P and Q coming together, to the rectangle $QR \times 2PM$. Therefore, CP^2 is to PM^2 as $QR \times 2PM$ to QT^2, and thus $\frac{QT^2}{QR}$ is equal to $\frac{2PM^3}{CP^2}$, and $\frac{QT^2 \times SP^2}{QR}$ is equal to $\frac{2PM^3 \times SP^2}{CP^2}$. Therefore (by prop. 6, corols. 1 and 5), the centripetal force is inversely as $\frac{2PM^3 \times SP^2}{CP^2}$, that is $\left(\text{neglecting the determinate}^{a} \text{ ratio } \frac{2SP^2}{CP^2}\right)$, inversely as PM^2. Q.E.I.

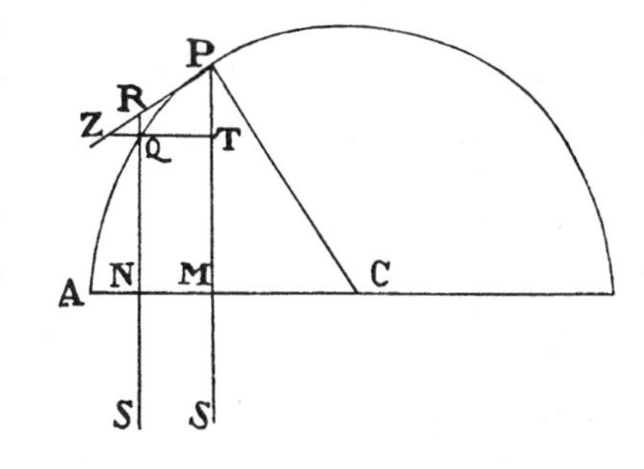

The same is easily gathered also from the preceding proposition.

a. CP is the radius of the semicircle, and SP may be considered constant.

Scholium And by a not very different argument, a body will be found to move in an ellipse, or even in a hyperbola or a parabola, under the action of a centripetal force that is inversely as the cube of the ordinate tending toward an extremely distant center of forces.

Proposition 9
Problem 4 *Let a body revolve in a spiral* PQS *intersecting all its radii* SP, SQ, . . . , *at a given angle; it is required to find the law of the centripetal force tending toward the center of the spiral.*

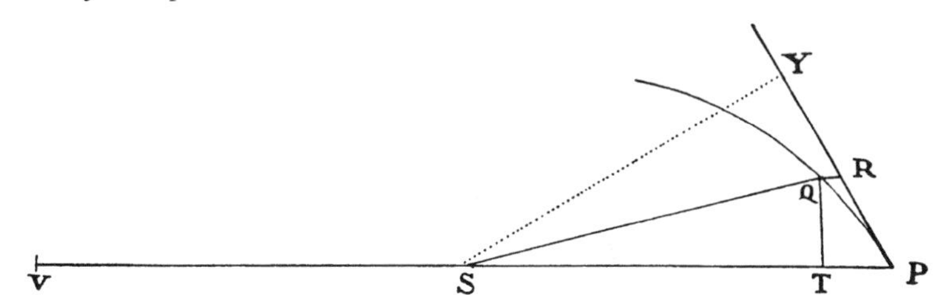

Let the indefinitely small angle PSQ be given, and because all the angles are given, the species [i.e., the ratio of all the parts] of the figure SPRQT will be given. Therefore, the ratio $\frac{QT}{QR}$ is given, and $\frac{QT^2}{QR}$ is as QT, that is (because the species of the figure is given), as SP. Now change the angle PSQ in any way, and the straight line QR subtending the angle of contact QPR will be changed (by lem. 11) as the square of PR or QT. Therefore, $\frac{QT^2}{QR}$ will remain the same as before, that is, as SP. And therefore $\frac{QT^2 \times SP^2}{QR}$ is as SP^3, and thus (by prop. 6, corols. 1 and 5) the centripetal force is inversely as the cube of the distance SP. Q.E.I.

Another solution

The perpendicular SY dropped to the tangent, and the chord PV of the circle cutting the spiral concentrically, are to the distance SP in given ratios; and thus SP^3 is as $SY^2 \times PV$, that is (by prop. 6, corols. 3 and 5), inversely as the centripetal force.

Lemma 12 *All the parallelograms described about any conjugate diameters of a given ellipse or hyperbola are equal to one another.*

This is evident from the *Conics*.

Proposition 10
Problem 5

Let a body revolve in an ellipse; it is required to find the law of the centripetal force tending toward the center of the ellipse.

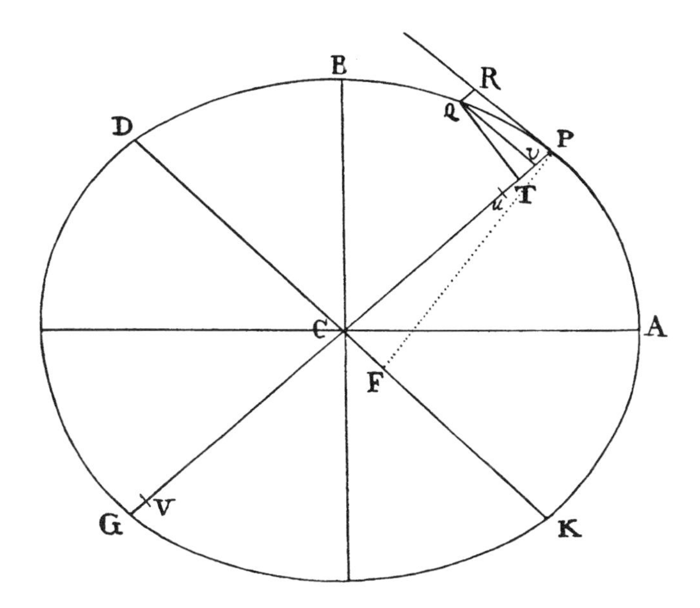

Let CA and CB be the semiaxes of the ellipse, GP and DK other conjugate diameters, PF and QT perpendiculars to those diameters, Qv an ordinate to diameter GP; then, if parallelogram QvPR is completed, the rectangle Pv × vG will (from the *Conics*[a]) be to $\mathrm{Q}v^2$ as PC^2 to CD^2, and (because triangles QvT and PCF are similar) $\mathrm{Q}v^2$ is to QT^2 as PC^2 to PF^2, and, when these ratios are combined, the rectangle Pv × vG is to QT^2 as PC^2 to CD^2 and PC^2 to PF^2; that is, vG is to $\frac{\mathrm{QT}^2}{\mathrm{P}v}$ as PC^2 to $\frac{\mathrm{CD}^2 \times \mathrm{PF}^2}{\mathrm{PC}^2}$. Write QR for P$v$ and (by lem. 12) BC × CA for CD × PF, and also (points P and Q coming together) 2PC for vG, and, multiplying the extremes and means together, $\frac{\mathrm{QT}^2 \times \mathrm{PC}^2}{\mathrm{QR}}$ will become equal to $\frac{2\mathrm{BC}^2 \times \mathrm{CA}^2}{\mathrm{PC}}$. Therefore (by prop. 6, corol. 5), the centripetal force is as $\frac{2\mathrm{BC}^2 \times \mathrm{CA}^2}{\mathrm{PC}}$ inversely, that is (because $2\mathrm{BC}^2 \times \mathrm{CA}^2$ is given), as $\frac{1}{\mathrm{PC}}$ inversely, that is, as the distance PC directly. Q.E.I.

a. Concerning this reference to "the *Conics*," see the Guide, §10.10.

Another solution

On the straight line PG take a point u on the other side of point T, so that Tu is equal to Tv; then take uV such that it is to vG as DC^2 is to PC^2. And since (from the *Conics*) Qv^2 is to $Pv \times vG$ as DC^2 to PC^2, Qv^2 will be equal to $Pv \times uV$. Add the rectangle $uP \times Pv$ to both sides, and the square of the chord of arc PQ will come out equal to the rectangle $VP \times Pv$; and therefore a circle that touches the conic section at P and passes through point Q will also pass through point V. Let points P and Q come together, and the ratio of uV to vG, which is the same as the ratio of DC^2 to PC^2, will become the ratio of PV to PG or PV to 2PC; and therefore PV will be equal to $\frac{2DC^2}{PC}$. Accordingly, the force under the action of which body P revolves in the ellipse will (by prop. 6, corol. 3) be as $\frac{2DC^2}{PC} \times PF^2$ inversely, that is (because $2DC^2 \times PF^2$ is given), as PC directly. Q.E.I.

Corollary 1. Therefore, the force is as the distance of the body from the center of the ellipse; and, conversely, if the force is as the distance, the body will move in an ellipse having its center in the center of forces, or perhaps it will move in a circle, into which an ellipse can be changed.

Corollary 2. And the periodic times of the revolutions made in all ellipses universally around the same center will be equal. For in similar ellipses those times are equal (by prop. 4, corols. 3 and 8), while in ellipses having a common major axis they are to one another as the total areas of the ellipses directly and the particles of the areas described in the same time inversely; that is, as the minor axes directly and the velocities of bodies in their principal vertices inversely; that is, as those minor axes directly and the ordinates to the same point of the common axis inversely; and therefore (because of the equality of the direct and inverse ratios) in the ratio of equality.

Scholium If the center of the ellipse goes off to infinity, so that the ellipse turns into a parabola, the body will move in this parabola, and the force, now tending toward an infinitely distant center, will prove to be uniform. This is Galileo's theorem. And if (by changing the inclination of the cutting plane to the cone) the parabolic section of the cone turns into a hyperbola, the body will move in the perimeter of the hyperbola, with the centripetal force turned into a centrifugal force. And just as in a circle or an ellipse, if the forces tend toward a figure's center located in the abscissa, and if the ordinates are increased or decreased in any given ratio or even if the angle of the inclination of the ordinates to the abscissa is changed, these forces are always increased or decreased in the ratio of the distances from the center, provided that the periodic times remain equal; so also in all figures universally, if the ordinates are increased or decreased in any given ratio or the angle of inclination of the ordinates is changed in any way while the periodic time remains the same, the forces tending toward any center located in the abscissa are, for each individual ordinate, increased or decreased in the ratio of the distances from the center.

SECTION 3

The motion of bodies in eccentric conic sections

Proposition 11[a] **Problem 6** *Let a body revolve in an ellipse; it is required to find the law of the centripetal force tending toward a focus of the ellipse.*

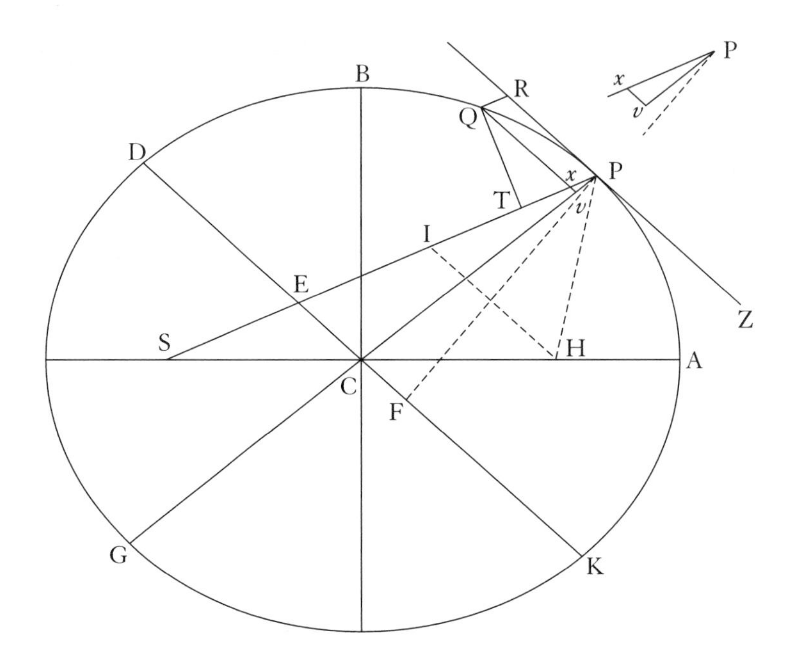

Let S be a focus of the ellipse. Draw SP cutting both the diameter DK of the ellipse in E and the ordinate Qv in x, and complete the parallelogram QxPR. It is evident that EP is equal to the semiaxis major AC because when line HI is drawn parallel to EC from the other focus H of the ellipse, ES and EI are equal because CS and CH are equal; so that EP is the half-sum of PS and PI, that is (because HI and PR are parallel and angles IPR and HPZ are equal), the half-sum of PS and PH (which taken together equal the whole axis 2AC). Drop QT perpendicular to SP, and if L denotes the principal latus rectum of the ellipse $\left(\text{or } \frac{2\text{BC}^2}{\text{AC}}\right)$, L × QR will be to L × Pv as QR to Pv, that is, as PE or AC to PC; and L × Pv will be to Gv × vP as

a. For a gloss on this proposition see the Guide, §10.9.

L to Gv; and[b] G$v \times v$P will be to Qv^2 as PC^2 to CD^2; and (by lem. 7, corol. 2) the ratio of Qv^2 to Qx^2, with the points Q and P coming together, is the ratio of equality; and Qx^2 or Qv^2 is to QT^2 as EP^2 to PF^2, that is, as CA^2 to PF^2 or (by lem. 12) as CD^2 to CB^2. And when all these ratios are combined, L $\times$ QR will be to QT^2 as AC $\times$ L $\times$ PC^2 $\times$ CD^2, or as 2CB^2 $\times$ PC^2 $\times$ CD^2 to PC $\times$ Gv $\times$ CD^2 $\times$ CB^2, or as 2PC to Gv. But with the points Q and P coming together, 2PC and Gv are equal. Therefore, L$\times$QR and QT^2, which are proportional to these, are also equal. Multiply these equals by $\frac{\mathrm{SP}^2}{\mathrm{QR}}$, and L $\times$ SP^2 will become equal to $\frac{\mathrm{SP}^2 \times \mathrm{QT}^2}{\mathrm{QR}}$. Therefore (by prop. 6, corols. 1 and 5) the centripetal force is inversely as L $\times$ SP^2, that is, inversely as the square of the distance SP. Q.E.I.

Another solution

The force which tends toward the center of the ellipse, and by which body P can revolve in that ellipse, is (by prop. 10, corol. 1) as the distance CP of the body from the center C of the ellipse; hence, if CE is drawn parallel to the tangent PR of the ellipse and if CE and PS meet at E, then the force by which the same body P can revolve around any other point S of the ellipse will (by prop. 7, corol. 3) be as $\frac{\mathrm{PE}^3}{\mathrm{SP}^2}$; that is, if point S is a focus of the ellipse, and therefore PE is given, this force will be inversely as SP^2. Q.E.I.

This solution could be extended to the parabola and the hyperbola as concisely as in prop. 10, but because of the importance of this problem and its use in what follows, it will not be too troublesome to confirm each of these other cases by a separate demonstration.

Proposition 12
Problem 7

Let a body move in a hyperbola; it is required to find the law of the centripetal force tending toward the focus of the figure.

Let CA and CB be the semiaxes of the hyperbola, PG and KD other conjugate diameters, PF a perpendicular to the diameter KD, and Qv an ordinate to the diameter GP. Draw SP cutting diameter DK in E and ordinate Qv in x, and complete the parallelogram QRPx. It is evident that EP is

b. This result is given in prop. 10 with reference to "the *Conics*"; see the Guide, §10.10.

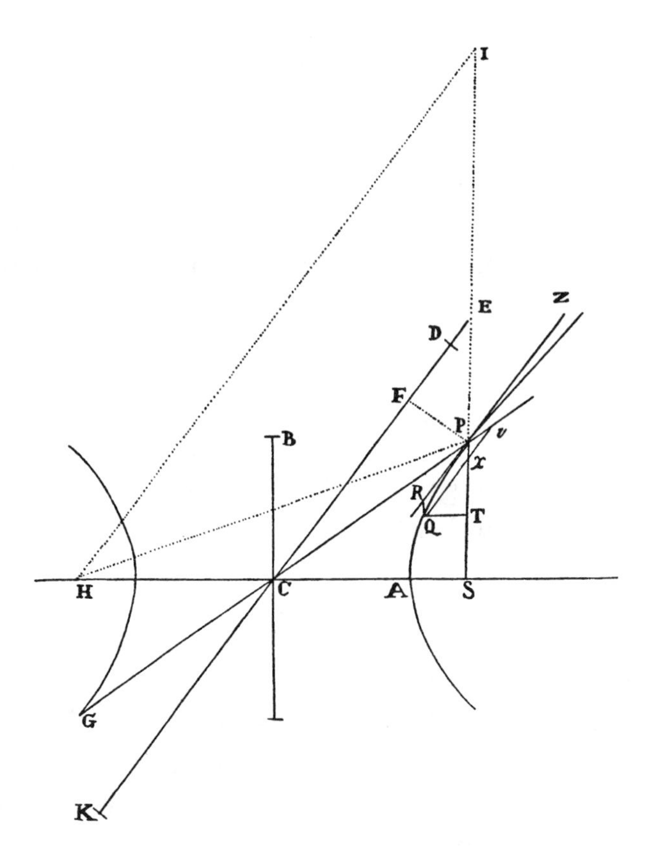

equal to the transverse semiaxis AC, because when line HI is drawn parallel to EC from the other focus H of the hyperbola, ES and EI are equal because CS and CH are equal; so that EP is the half-difference of PS and PI, that is (because IH and PR are parallel and the angles IPR and HPZ are equal), of PS and PH, the difference of which equals the whole axis 2AC. Drop QT perpendicular to SP. Then, if L denotes the principal latus rectum of the hyperbola $\left(\text{or } \dfrac{2\text{BC}^2}{\text{AC}}\right)$, L × QR will be to L × Pv as QR to Pv, or Px to Pv, that is (because the triangles Pxv and PEC are similar), as PE to PC, or AC to PC. L × Pv will also be to Gv × Pv as L to Gv; and (from the nature of conics) the rectangle Gv × vP will be to Qv^2 as PC^2 to CD^2; and (by lem. 7, corol. 2) the ratio of Qv^2 to Qx^2, the points Q and P coming together, comes to be the ratio of equality; and Qx^2 or Qv^2 is to AT^2 as EP^2 to PF^2, that is, as CA^2 to PF^2, or (by lem. 12) as CD^2 to CB^2; and if all these ratios are combined, L × QR will be to QT^2 as AC × L × PC^2 × CD^2 or 2CB^2 × PC^2 × CD^2 to PC × Gv × CD^2 × CB^2, or as 2PC to Gv. But, the points

P and Q coming together, 2PC and Gv are equal. Therefore, $L \times QR$ and QT^2, which are proportional to these, are also equal. Multiply these equals by $\frac{SP^2}{QR}$, and $L \times SP^2$ will become equal to $\frac{SP^2 \times QT^2}{QR}$. Therefore (by prop. 6, corols. 1 and 5), the centripetal force is inversely as $L \times SP^2$, that is, inversely as the square of the distance SP. Q.E.I.

Another solution

Find the force that tends from the center C of the hyperbola. This will come out proportional to the distance CP. And hence (by prop. 7, corol. 3) the force tending toward the focus S will be as $\frac{PE^3}{SP^2}$, that is, because PE is given, inversely as SP^2. Q.E.I.

It is shown in the same way that if this centripetal force is turned into a centrifugal force, a body will move in the opposite branch of the hyperbola.

Lemma 13

In a parabola the latus rectum belonging to any vertex is four times the distance of that vertex from the focus of the figure.

This is evident from the *Conics*.

Lemma 14

A perpendicular dropped from the focus of a parabola to its tangent is a mean proportional between the distance of the focus from the point of contact and its distance from the principal vertex of the figure.

For let AP be the parabola, S its focus, A the principal vertex, P the point of contact, PO an ordinate to the principal diameter, PM a tangent meeting the principal diameter in M, and SN a perpendicular line from the focus to the tangent. Join AN, and because MS and SP, MN and NP, and MA and AO are equal, the straight lines AN and OP will be parallel; and hence triangle SAN will be right-angled at A and similar to the equal triangles SNM and SNP; therefore, PS is to SN as SN to SA. Q.E.D.

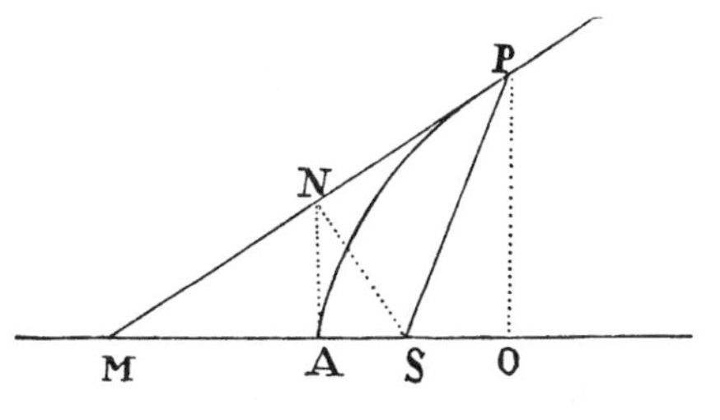

Corollary 1. PS^2 is to SN^2 as PS to SA.

Corollary 2. And because SA is given, SN^2 is as PS.

COROLLARY 3. And the point where any tangent PM meets the straight line SN, which is drawn from the focus perpendicular to that tangent, occurs in the straight line AN, which touches the parabola in the principal vertex.

Proposition 13
Problem 8

Let a body move in the perimeter of a parabola; it is required to find the law of the centripetal force tending toward a focus of the figure.

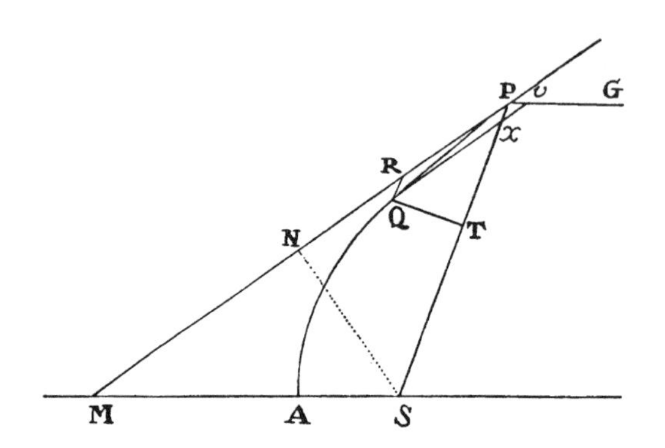

Let the construction be the same as in lem. 14, and let P be the body in the perimeter of the parabola; from the place Q into which the body moves next, draw QR parallel and QT perpendicular to SP and draw Qv parallel to the tangent and meeting both the diameter PG in v and the distance SP in x. Now, because triangles Pxv and SPM are similar and the sides SM and SP of the one are equal, the sides Px or QR and Pv of the other are equal. But from the *Conics* the square of the ordinate Qv is equal to the rectangle contained by the latus rectum and the segment Pv of the diameter, that is (by lem. 13), equal to the rectangle 4PS × Pv, or 4PS × QR, and, the points P and Q coming together, the ratio of Qv to Qx (by lem. 7, corol. 2) becomes the ratio of equality. Therefore, in this case Qx^2 is equal to the rectangle 4PS × QR. Moreover (because triangles QxT and SPN are similar), Qx^2 is to QT^2 as PS^2 to SN^2, that is (by lem. 14, corol. 1), as PS to SA, that is, as 4PS × QR to 4SA × QR, and hence (by Euclid's *Elements*, book 5, prop. 9) QT^2 and 4SA × QR are equal. Multiply these equals by $\frac{SP^2}{QR}$, and $\frac{SP^2 \times QT^2}{QR}$ will become equal to SP^2 × 4SA; and therefore (by prop. 6, corols. 1 and 5) the centripetal force is inversely as SP^2 × 4SA, that is, because 4SA is given, inversely as the square of the distance SP. Q.E.I.

COROLLARY 1. From the last three propositions it follows that if any body P departs from the place P along any straight line PR with any velocity whatever and is at the same time acted upon by a centripetal force that is inversely proportional to the square of the distance of places from the center, this body will move in some one of the conics having a focus in the center of forces; and conversely. For if the focus and the point of contact and the position of the tangent are given, a conic can be described that will have a given curvature at that point. But the curvature is given from the given centripetal force and velocity of the body; and two different orbits touching each other cannot be described with the same centripetal force and the same velocity.

COROLLARY 2. If the velocity with which the body departs from its place P is such that the line-element PR can be described by it in some minimally small particle of time, and if the centripetal force is able to move the same body through space QR in that same time, this body will move in some conic whose principal latus rectum is the quantity $\frac{QT^2}{QR}$ which ultimately results when the line-elements PR and QR are diminished indefinitely. In these corollaries I include the circle along with the ellipse, but not for the case where the body descends straight down to a center.

Proposition 14
Theorem 6

If several bodies revolve about a common center and the centripetal force is inversely as the square of the distance of places from the center, I say that the principal latera recta of the orbits are as the squares of the areas which the bodies describe in the same time by radii drawn to the center.

For (by prop. 13, corol. 2) the latus rectum L is equal to the quantity $\frac{QT^2}{QR}$ that results ultimately when points P and Q come together. But the minimally small line QR is in a given time as the generating centripetal force, that is (by hypothesis), inversely as SP^2. Therefore, $\frac{QT^2}{QR}$ is as $QT^2 \times SP^2$, that is, the latus rectum L is as the square of the area $QT \times SP$. Q.E.D.

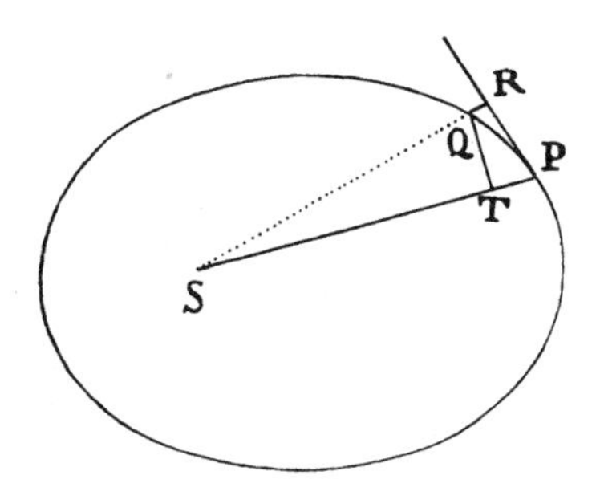

COROLLARY. Hence the total area of the ellipse and, proportional to it, the rectangle contained by the axes is as the square root of the latus rectum and as the periodic time. For the total area is as the area QT × SP, which is described in a given time, multiplied by the periodic time.

Proposition 15
Theorem 7

Under the same suppositions as in prop. 14, I say that the squares of the periodic times in ellipses are as the cubes of the major axes.

For the minor axis is a mean proportional between the major axis and the latus rectum, and thus the rectangle contained by the axes is as the square root of the latus rectum and as the $3/2$ power of the major axis. But this rectangle (by prop. 14, corol.) is as the square root of the latus rectum and as the periodic time. Take away from both sides [i.e., divide through by] the square root of the latus rectum, and the result will be that the squares of the periodic times are as the cubes of the major axes. Q.E.D.

COROLLARY. Therefore the periodic times in ellipses are the same as in circles whose diameters are equal to the major axes of the ellipses.

Proposition 16
Theorem 8

Under the same suppositions as in prop. 15, if straight lines are drawn to the bodies in such a way as to touch the orbits in the places where the bodies are located, and if perpendiculars are dropped from the common focus to these tangents, I say that the velocities of the bodies are inversely as the perpendiculars and directly as the square roots of the principal latera recta.

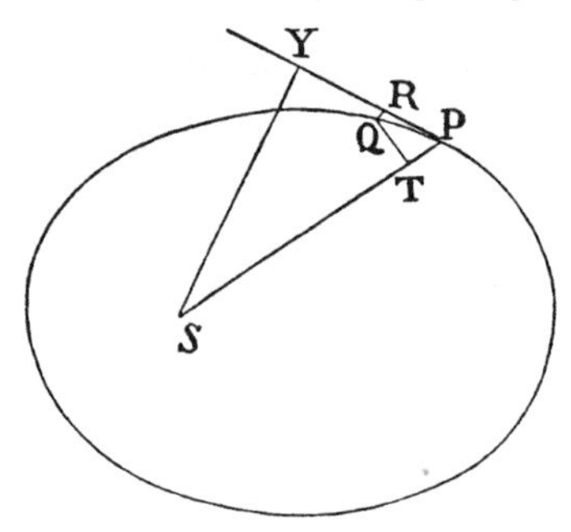

From focus S to tangent PR drop perpendicular SY, and the velocity of body P will be inversely as the square root of $\frac{SY^2}{L}$. For this velocity is as the minimally small arc PQ described in a given particle of time, that is (by lem. 7), as the tangent PR, that is—because the proportion of PR to QT is as SP to SY—as $\frac{SP \times QT}{SY}$, or as SY inversely and SP × QT directly; and SP × QT is as the area described in the given time, that is (by prop. 14), as the square root of the latus rectum. Q.E.D.

COROLLARY 1. The principal latera recta are as the squares of the perpendiculars and as the squares of the velocities.

COROLLARY 2. The velocities of bodies at their greatest and least distances from the common focus are inversely as the distances and directly as the square roots of the principal latera recta. For the perpendiculars are now the distances themselves.

COROLLARY 3. And thus the velocity in a conic, at the greatest or least distance from the focus, is to the velocity with which the body would move in a circle, at the same distance from the center, as the square root of the principal latus rectum is to the square root of twice that distance.

COROLLARY 4. The velocities of bodies revolving in ellipses are, at their mean distances from the common focus, the same as those of bodies revolving in circles at the same distances, that is (by prop. 4, corol. 6), inversely as the square roots of the distances. For the perpendiculars now coincide with the semiaxes minor, and these are as mean proportionals between the distances and the latera recta. Compound this ratio [of the semiaxes] inversely with the square root of the ratio of the latera recta directly, and it will become the square root of the ratio of the distances inversely.

COROLLARY 5. In the same figure, or even in different figures whose principal latera recta are equal, the velocity of a body is inversely as the perpendicular dropped from the focus to the tangent.

COROLLARY 6. In a parabola the velocity is inversely as the square root of the distance of the body from the focus of the figure; in an ellipse the velocity varies in a ratio that is greater than this, and in a hyperbola in a ratio that is less. For (by lem. 14, corol. 2) the perpendicular dropped from the focus to the tangent of a parabola is as the square root of that distance. In a hyperbola the perpendicular is smaller, and in an ellipse greater, than in this ratio.

COROLLARY 7. In a parabola the velocity of a body at any distance from the focus is to the velocity of a body revolving in a circle at the same distance from the center as the square root of the ratio of 2 to 1; in an ellipse it is smaller and in a hyperbola greater than in this ratio. For by corol. 2 of this proposition the velocity in the vertex of a parabola is in this ratio, and—by corol. 6 of this proposition and by prop. 4, corol. 6—the same proportion is kept at all distances. Hence, also, in a parabola the velocity everywhere is equal to the velocity of a body revolving in a circle at half the distance; in an ellipse it is smaller and in a hyperbola greater.

COROLLARY 8. The velocity of a body revolving in any conic is to the velocity of a body revolving in a circle at a distance of half the principal latus rectum of the conic as that distance is to the perpendicular dropped from the focus to the tangent of the conic. This is evident by corol. 5.

Corollary 9. Hence, since (by prop. 4, corol. 6) the velocity of a body revolving in this circle is to the velocity of a body revolving in any other circle inversely in the ratio of the square roots of the distances, it follows from the equality of the ratios [or ex aequo] that the velocity of a body revolving in a conic will have the same ratio to the velocity of a body revolving in a circle at the same distance that a mean proportional between that common distance and half of the principal latus rectum of the conic has to the perpendicular dropped from the common focus to the tangent of the conic.

Proposition 17
Problem 9

Supposing that the centripetal force is inversely proportional to the square of the distance of places from the center and that the absolute quantity of this force is known, it is required to find the line which a body describes when going forth from a given place with a given velocity along a given straight line.

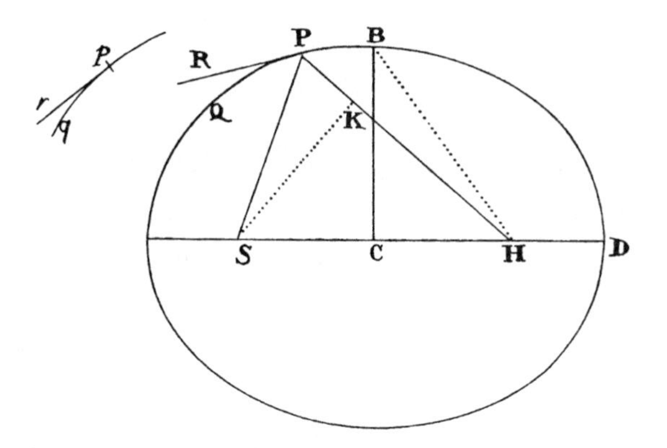

Let the centripetal force tending toward a point S be such that a body p revolves by its action in any given orbit pq, and let its velocity in the place p be found out. Let body P go forth from place P along line PR with a given velocity and thereupon be deflected from that line into a conic PQ under the compulsion of the centripetal force. Therefore the straight line PR will touch this conic at P. Let some straight line pr likewise touch the orbit pq at p, and if perpendiculars are understood to be dropped from S to these tangents, the principal latus rectum of the conic will (by prop. 16, corol. 1) be to the principal latus rectum of the orbit in a ratio compounded of the squares of the perpendiculars and the squares of the velocities and thus is given. Let L be the latus rectum of the conic. The focus S of the conic is also given. Let angle RPH be the complement of angle RPS to two right angles [i.e., the supplement of angle RPS]; then the line PH, on which the other focus H is located, will be given in position. Drop the perpendicular SK to PH and understand the conjugate semiaxis BC to be erected; then $SP^2 - 2KP \times PH + PH^2 = SH^2 = 4CH^2 = 4BH^2 - 4BC^2 = (SP + PH)^2 - L \times (SP + PH) = SP^2 + 2SP \times PH + PH^2 - L \times (SP + PH)$. Add to each side $2(KP \times PH) - SP^2 - PH^2 + L \times (SP + PH)$, and $L \times (SP + PH)$ will become $= 2(SP \times PH) + 2(KP \times PH)$, or $SP + PH$ will be to PH as $2SP + 2KP$ to L. Hence PH is given in length as well as in position. Specifically, if the velocity of the body at P is such that the latus rectum L is less than $2SP + 2KP$, PH will lie on the same side of the tangent PR as the line PS; and thus the figure will be an ellipse and will be given from the given foci S and H and the given principal axis $SP + PH$. But if the velocity of the body is so great that the latus rectum L is equal to $2SP + 2KP$, the length PH will be infinite; and accordingly the figure will be a parabola having its axis SH parallel to the line PK, and hence will be given. But if the body goes forth from its place P with a still greater velocity, the length PH will have to be taken on the other side of the tangent; and thus, since the tangent goes between the foci, the figure will be a hyperbola having its principal axis equal to the difference of the lines SP and PH, and hence will be given. For if the body in these cases revolves in a conic thus found, it has been demonstrated in props. 11, 12, and 13 that the centripetal force will be inversely as the square of the distance of the body from the center of forces S; and thus the line PQ is correctly determined, which a body will describe under the action of such a force, when it goes forth from a given place P with a given velocity along a straight line PR given in position. Q.E.F.

COROLLARY 1. Hence in every conic, given the principal vertex D, the latus rectum L, and a focus S, the other focus H is given when DH is taken to DS as the latus rectum is to the difference between the latus rectum and 4DS. For the proportion $SP + PH$ to PH as $2SP + 2KP$ to L in the case of this corollary becomes $DS + DH$ to DH as 4DS to L and, by separation [or dividendo], becomes DS to DH as $4DS - L$ to L.

COROLLARY 2. Hence, given the velocity of a body in the principal vertex D, the orbit will be found expeditiously, namely, by taking its latus rectum to twice the distance DS as the square of the ratio of this given velocity to the velocity of a body revolving in a circle at a distance DS (by prop. 16, corol. 3), and then taking DH to DS as the latus rectum to the difference between the latus rectum and 4DS.

COROLLARY 3. Hence also, if a body moves in any conic whatever and is forced out of its orbit by any impulse, the orbit in which it will afterward pursue its course can be found. For by compounding the body's own motion with that motion which the impulse alone would generate, there will be found the motion with which the body will go forth from the given place of impulse along a straight line given in position.

COROLLARY 4. And if the body is continually perturbed by some force impressed from outside, its trajectory can be determined very nearly, by noting the changes which the force introduces at certain points and estimating from the order of the sequence the continual changes at intermediate places.[a]

Scholium If a body P, under the action of a centripetal force tending toward any given point R, moves in the perimeter of any given conic whatever, whose center is C, and the law of the centripetal force is required, let CG be drawn parallel to the radius RP and meeting the tangent PG of the orbit at G; then the force (by prop. 10, corol. 1 and schol.; and prop. 7, corol. 3) will be as $\frac{CG^3}{RP^2}$.

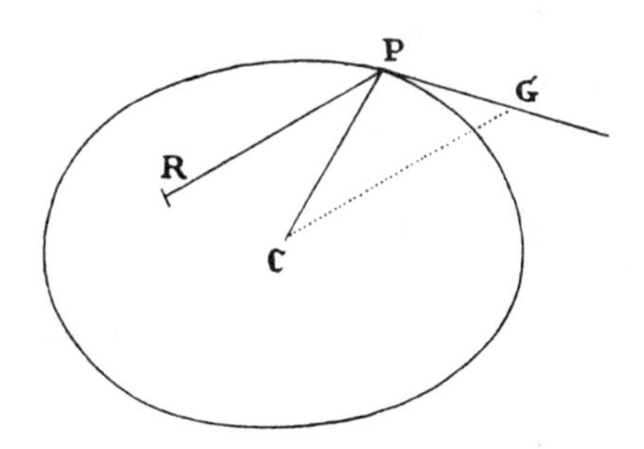

a. The sense of corol. 4 is that Newton can determine "the changes which the [impressed] force will make at certain points" and, by interpolation, estimate the changes continually made at intermediary points.

SECTION 4[a]

To find elliptical, parabolic, and hyperbolic orbits, given a focus

Lemma 15

If from the two foci S *and* H *of any ellipse or hyperbola two straight lines* SV *and* HV *are inclined to any third point* V, *one of the lines* HV *being equal to the principal axis of the figure, that is, to the axis on which the foci lie, and the other line* SV *being bisected in* T *by* TR *perpendicular to it, then the perpendicular* TR *will touch the conic at some point; and conversely, if* TR *touches the conic,* HV *will be equal to the principal axis of the figure.*

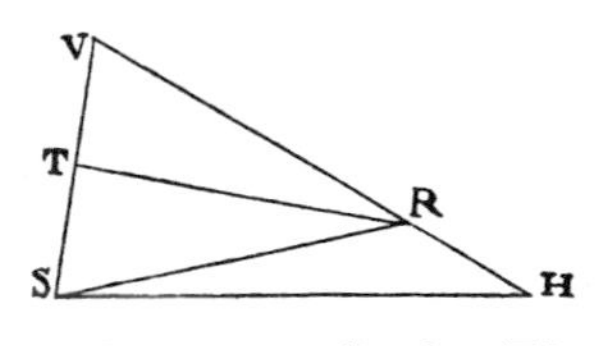

For let the perpendicular TR cut the straight line HV (produced, if need be) in R; and join SR. Because TS and TV are equal, the straight lines SR and VR and the angles TRS and TRV will be equal. Hence the point R will be on the conic, and the perpendicular TR will touch that conic, and conversely. Q.E.D.

Proposition 18
Problem 10

Given a focus and the principal axes, to describe elliptical and hyperbolic trajectories that will pass through given points and will touch straight lines given in position.

Let S be the common focus of the figures, AB the length of the principal axis of any trajectory, P a point through which the trajectory ought to pass, and TR a straight line which it ought to touch. Describe the circle HG with P as center and AB − SP as radius if the orbit is an ellipse, or AB + SP if it is a hyperbola. Drop the perpendicular ST to the tangent TR and produce ST to V so that TV is equal to ST, and with center V and radius AB describe the circle FH. By this method, whether two points P and *p* are given, or two tangents TR and *tr*, or a point P and a tangent TR, two circles are to be described. Let H be their common intersection, and with foci S and H and the given axis, describe the trajectory. I say that the problem has been solved. For the trajectory described (because PH + SP in an ellipse, or PH − SP in a hyperbola,

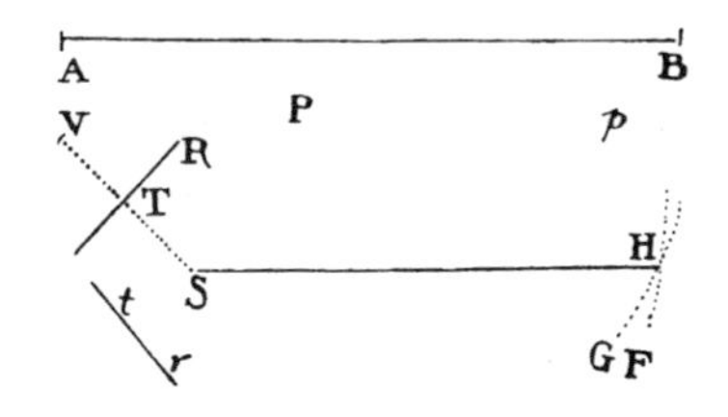

a. For Newton's statement of the reason for including secs. 4 and 5 in book 1, see book 3, prop. 41 (p. 901).

is equal to the axis) will pass through point P and (by lem. 15) will touch the straight line TR. And by the same argument, this trajectory will pass through the two points P and *p* or will touch the two straight lines TR and *tr*. Q.E.F.

Proposition 19
Problem 11

To describe about a given focus a parabolic trajectory that will pass through given points and will touch straight lines given in position.

Let S be the focus, P a given point, and TR a tangent of the trajectory to be described. With center P and radius PS describe the circle FG. Drop the perpendicular ST from the focus to the tangent and produce ST to V, so that TV is equal to ST. In the same manner, if a second point *p* is given, a second circle *fg* is to be described; or if a second tangent *tr* is given, or a second point *v* is to be found, then the straight line IF is to be drawn touching the two circles FG and *fg* if the two points P and *p* are given, or passing through the two points V and *v* if the two tangents TR and *tr* are given, or touching the circle FG and passing through the point V if the point P and tangent TR are given. To FI drop the perpendicular SI, and bisect it in K; and with axis SK and principal vertex K describe a parabola. I say that the problem has been solved. For, because SK and IK are equal, and SP and FP are equal, the parabola will pass through point P; and (by lem. 14, corol. 3) because ST and TV are equal and the angle STR is a right angle, the parabola will touch the straight line TR. Q.E.F.

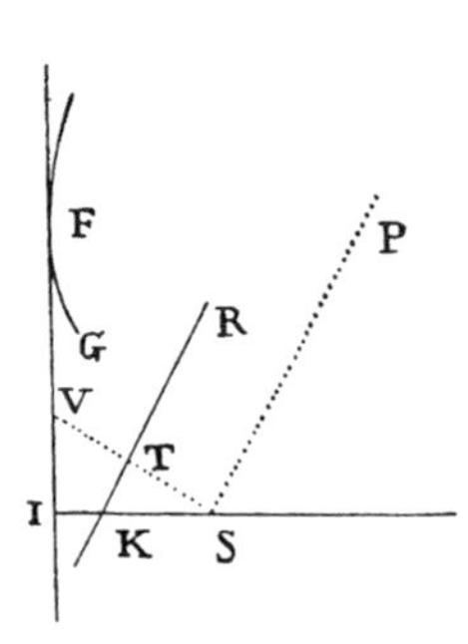

Proposition 20
Problem 12

To describe about a given focus any trajectory, given in species [i.e., of given eccentricity], that will pass through given points and will touch straight lines given in position.

Case 1. Given a focus S, let it be required to describe a trajectory ABC through two points B and C. Since the trajectory is given in species, the ratio of the principal axis to the distance between the foci will be given. Take KB to BS in this ratio and also LC to CS. With centers B and C and radii BK and CL, describe two circles, and drop the

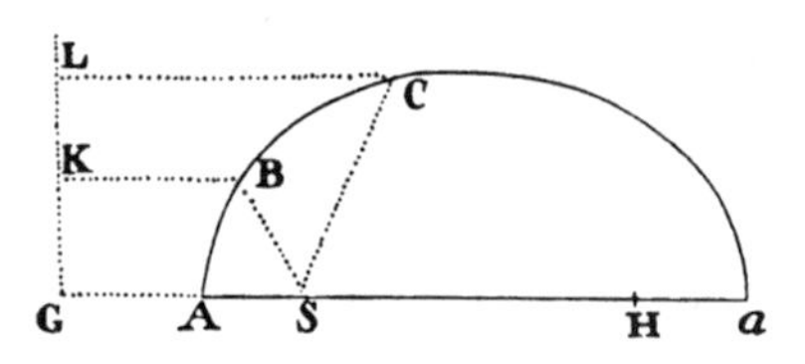

perpendicular SG to the straight line KL, which touches those circles in K and L, and cut SG in A and *a* so that GA is to AS, and G*a* to *a*S, as KB is to BS; and describe a trajectory with axis A*a* and vertices A and *a*. I say that the problem has been solved. For let H be the other focus of the figure described, and since GA is to AS as G*a* to *a*S, then by separation [or dividendo] G*a* − GA or A*a* to *a*S − AS or SH will be in the same ratio and thus in the ratio which the principal axis of the figure that was to be described has to the distance between its foci; and therefore the figure described is of the same species as the one that was to be described. And since KB to BS and LC to CS are in the same ratio, this figure will pass through the points B and C, as is manifest from the *Conics*.

CASE 2. Given a focus S, let it be required to describe a trajectory which somewhere touches the two straight lines TR and *tr*. Drop the perpendiculars ST and S*t* from the focus to the tangents and produce ST and S*t* to V and *v*, so that TV and *tv* are equal to TS and *t*S. Bisect V*v* in O, and erect the indefinite perpendicular OH, and cut the straight line VS, indefinitely produced, in K and *k*, so that VK is to KS and V*k* to *k*S as the principal axis of the trajectory to be described is to the distance between the foci. On the diameter K*k* describe a circle cutting OH in H; and with foci S and H and a principal axis equal to VH, describe a trajectory. I say that the problem has been solved. For bisect K*k* in X, and draw HX, HS, HV, and H*v*. Since VK is to KS as V*k* to *k*S and, by composition [or componendo], as VK + V*k* to KS + *k*S and, by separation [or dividendo], as V*k* − VK to *k*S − KS, that is, as 2VX to 2KX and 2KX to 2SX and thus as VX to HX and HX to SX, the triangles VXH and HXS will be similar, and therefore VH will be to SH as VX to XH and thus as VK to KS. Therefore the principal axis VH of the trajectory which has been described has the same ratio to the distance SH between its foci as the principal axis of the trajectory to be described has to the distance between its foci and is therefore of the same species. Besides, since VH and *v*H are equal to the principal axis and since VS and *v*S are perpendicularly bisected by the straight lines TR and *tr*, it is clear (from lem. 15) that these straight lines touch the trajectory described. Q.E.F.

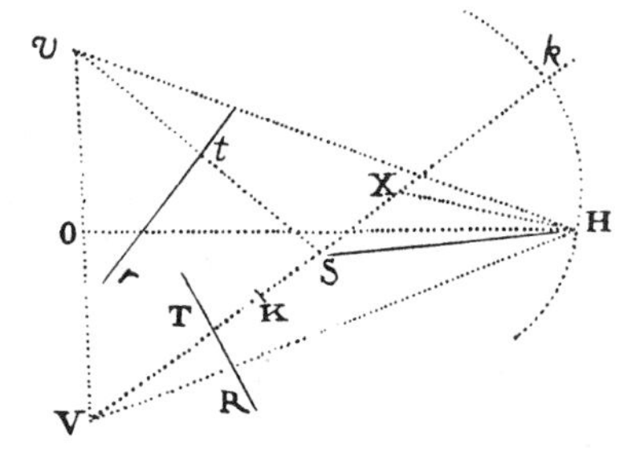

CASE 3. Given a focus S, let it be required to describe a trajectory which will touch the straight line TR in a given point R. Drop the perpendicular ST to the straight line TR and produce ST to V so that TV is equal to ST. Join VR and cut the straight line VS, indefinitely produced, in K and k so that VK is to SK and Vk to Sk as the principal axis of the ellipse to be described is to the distance between the foci; and after describing a circle on the diameter Kk, cut the straight line VR, produced, in H, and with foci S and H and a principal axis equal to the straight line VH, describe a trajectory. I say that the problem has been solved. For, from what has been demonstrated in case 2, it is evident that VH is to SH as VK to SK and thus as the principal axis of the trajectory which was to be described to the distance between its foci, and therefore the trajectory which was described is of the same species as the one which was to be described, while it is evident from the *Conics* that the straight line TR by which the angle VRS is bisected touches the trajectory at point R. Q.E.F.

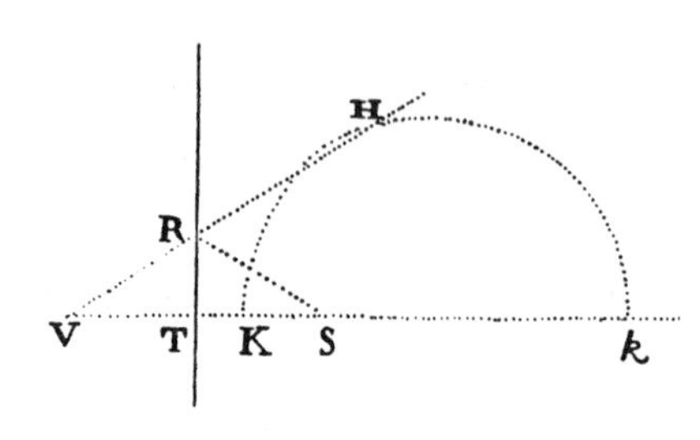

CASE 4. About a focus S let it be now required to describe a trajectory APB which touches the straight line TR and passes through any point P outside the given tangent and which is similar to the figure *apb* described with principal axis *ab* and foci *s* and *h*. Drop the perpendicular ST to the tangent TR and produce ST to V so that TV is equal to ST. Next make

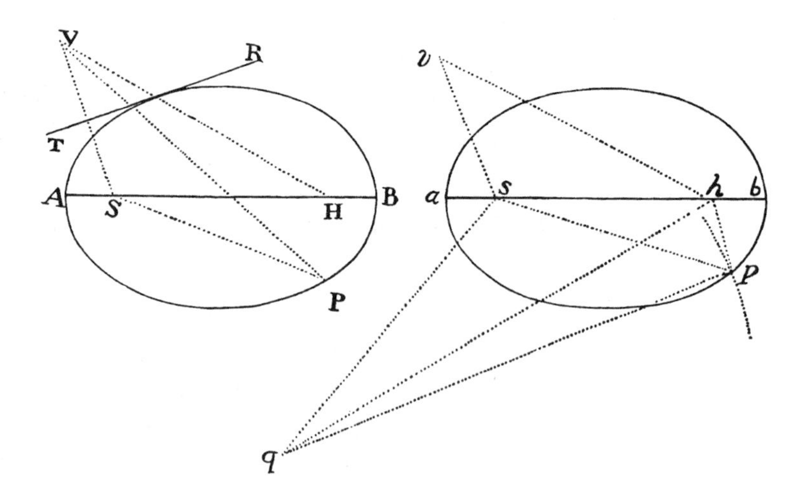

the angles *hsq* and *shq* equal to the angles VSP and SVP; and with center *q* and a radius that is to *ab* as SP to VS, describe a circle cutting the figure *apb* in *p*. Join *sp* and draw SH such that it is to *sh* as SP is to *sp* and makes the angle PSH equal to the angle *psh* and the angle VSH equal to the angle *psq*. Finally with foci S and H and with principal axis AB equaling the distance VH, describe a conic. I say that the problem has been solved. For if SV is drawn such that it is to *sp* as *sh* is to *sq*, and makes the angle *vsp* equal to the angle *hsq* and the angle *vsh* equal to the angle *psq*, the triangles *svh* and *spq* will be similar, and therefore *vh* will be to *pq* as *sh* is to *sq*, that is (because the triangles VSP and *hsq* are similar), as VS is to SP or *ab* to *pq*. Therefore *vh* and *ab* are equal. Furthermore, because the triangles VSH and *vsh* are similar, VH is to SH as *vh* to *sh*; that is, the axis of the conic just described is to the distance between its foci as the axis *ab* to the distance *sh* between the foci; and therefore the figure just described is similar to the figure *apb*. But because the triangle PSH is similar to the triangle *psh*, this figure passes through point P; and since VH is equal to the axis of this figure and VS is bisected perpendicularly by the straight line TR, the figure touches the straight line TR. Q.E.F.

Lemma 16

From three given points to draw three slanted straight lines to a fourth point, which is not given, when the differences between the lines either are given or are nil.

CASE 1. Let the given points be A, B, and C, and let the fourth point be Z, which it is required to find; because of the given difference of the lines AZ and BZ, point Z will be located in a hyperbola whose foci are A and B and whose principal axis is the given difference. Let the axis be MN. Take PM to MA as MN is to AB, and let PR be erected perpendicular to AB and let ZR be dropped perpendicular to PR; then, from the nature of this hyperbola, ZR will be to AZ as MN is to AB. By a similar process, point Z will be located in another hyperbola, whose foci are A and C and whose principal axis is the difference between AZ and CZ; and QS can be drawn perpendicular to AC,

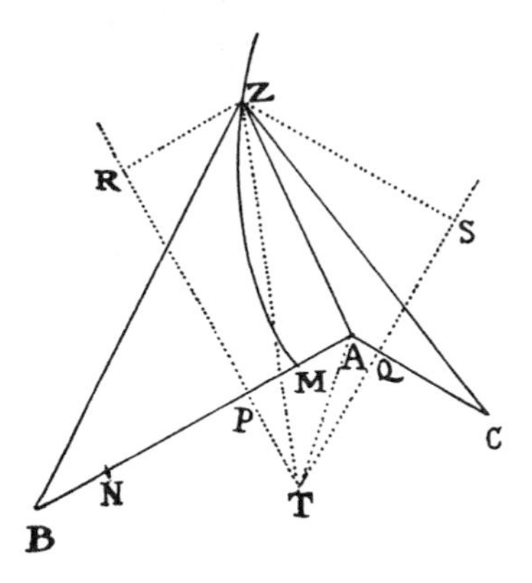

whereupon, if the normal ZS is dropped to QS from any point Z of this hyperbola, ZS will be to AZ as the difference between AZ and CZ is to AC. Therefore the ratios of ZR and ZS to AZ are given, and consequently the ratio of ZR and ZS to each other is given; and thus if the straight lines RP and SQ meet in T, and TZ and TA are drawn, the figure TRZS will be given in species, and the straight line TZ, in which point Z is somewhere located, will be given in position. The straight line TA will also be given, as will also the angle ATZ; and because the ratios of AZ and TZ to ZS are given, their ratio to each other will be given; and hence the triangle ATZ, whose vertex is the point Z, will be given. Q.E.I.

CASE 2. If two of the three lines, say AZ and BZ, are equal, draw the straight line TZ in such a way that it bisects the straight line AB; then find the triangle ATZ as above.

CASE 3. If all three lines are equal, point Z will be located in the center of a circle passing through points A, B, and C. Q.E.I.

The problem dealt with in this lemma is also solved by means of Apollonius's book *On Tangencies*, restored by Viète.

Proposition 21
Problem 13

To describe about a given focus a trajectory that will pass through given points and will touch straight lines given in position.

Let a focus S, a point P, and a tangent TR be given; the second focus H is to be found. Drop the perpendicular ST to the tangent and produce ST to Y so that TY is equal to ST, and YH will be equal to the principal axis. Join SP and also HP, and SP will be the difference between HP and the principal axis. In this way, if more tangents TR or more points P are given, there will always be the same number of lines YH or PH, which can be drawn from the said points Y or P to the focus H, and which either are equal to the axes or differ from them by given lengths SP and so either are equal to one another or have given differences; and hence, by lem. 16, that second focus H is given. And once the foci are found, together with the length of the axis (which length is either YH, or PH + SP if the trajectory is an ellipse, but PH − SP if the trajectory is a hyperbola), the trajectory is found. Q.E.I.

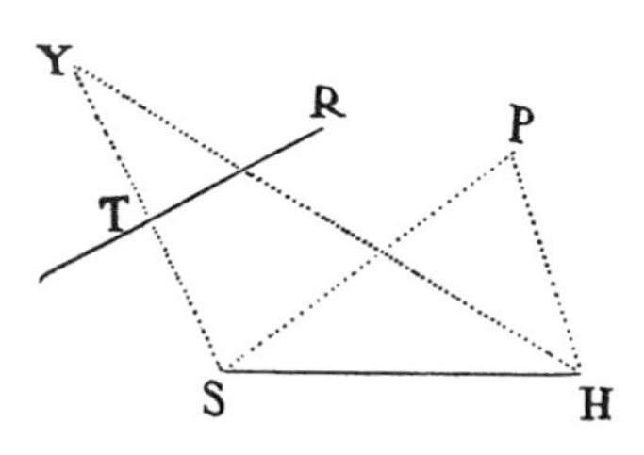

When the trajectory is a hyperbola, I do not include the opposite branch of the hyperbola as part of the trajectory. For a body going on with an uninterrupted motion cannot pass from one branch of a hyperbola into the other.

Scholium

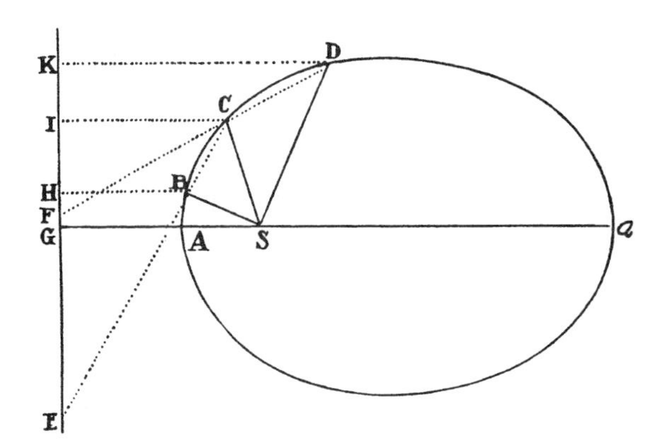

The case in which three points are given is solved more speedily as follows: Let the points B, C, and D be given. Join BC and also CD and produce them to E and F so that EB is to EC as SB to SC and FC is to FD as SC to SD. Draw EF, and drop the normals SG and BH to EF produced, and on GS indefinitely produced take GA to AS and G*a* to *a*S as HB is to BS; then A will be the vertex and A*a* the principal axis of the trajectory. According as GA is greater than, equal to, or less than AS, this trajectory will be an ellipse, a parabola, or a hyperbola, with point *a* in the first case falling on the same side of the line GF as point A, in the second case going off to infinity, in the third falling on the other side of the line GF. For if the perpendiculars CI and DK are dropped to GF, IC will be to HB as EC to EB, that is, as SC to SB; and by alternation [or alternando], IC will be to SC as HB to SB or as GA to SA. And by a similar argument it will be proved that KD is to SD in the same ratio. Therefore points B, C, and D lie in a conic described about the focus S in such a way that all the straight lines drawn from the focus S to the individual points of the conic are to the perpendiculars dropped from the same points to the straight line GF in that given ratio.

By a method that is not very different, the eminent geometer La Hire presents a solution of this problem in his *Conics*, book 8, prop. 25.

SECTION 5

To find orbits when neither focus is given

Lemma 17 *If four straight lines* PQ, PR, PS, *and* PT *are drawn at given angles from any point* P *of a given conic to the four indefinitely produced sides* AB, CD, AC, *and* DB *of some quadrilateral* ABDC *inscribed in the conic, one line being drawn to each side, the rectangle* PQ × PR *of the lines drawn to two opposite sides will be in a given ratio to the rectangle* PS × PT *of the lines drawn to the other two opposite sides.*

CASE 1. Let us suppose first that the lines drawn to opposite sides are parallel to either one of the other sides, say PQ and PR parallel to side AC, and PS and PT parallel to side AB. In addition, let two of the opposite sides, say AC and BD, be parallel to each other. Then the straight line which bisects those parallel sides will be one of the diameters of the conic and will bisect RQ also. Let O be the point in which RQ is bisected, and PO will be an ordinate to that diameter. Produce PO to K so that OK is equal to PO, and OK will be the ordinate on the opposite side of the diameter. Therefore, since points A, B, P, and K are on the conic and PK cuts AB at a given angle, the rectangle PQ × QK will be to the rectangle AQ × QB in a given ratio (by book 3, props. 17, 19, 21, and 23, of the *Conics* of Apollonius). But QK and PR are equal, inasmuch as they are differences of the equal lines OK and OP, and OQ and OR, and hence also the rectangles PQ × QK and PQ × PR are equal, and therefore the rectangle PQ × PR is to the rectangle AQ × QB, that is, to the rectangle PS × PT, in a given ratio. Q.E.D.

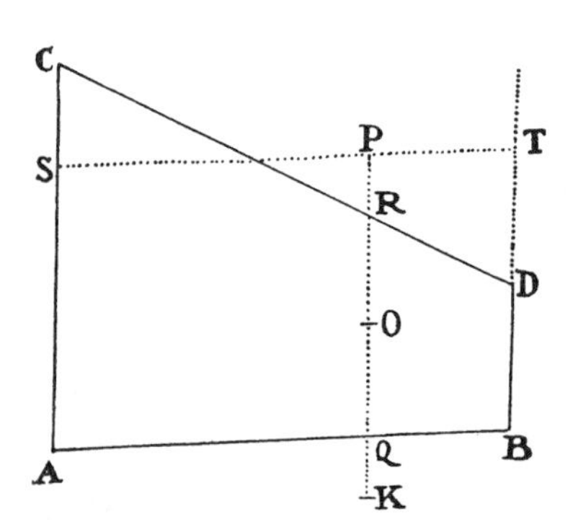

CASE 2. Let us suppose now that the opposite sides AC and BD of the quadrilateral are not parallel. Draw Bd parallel to AC, meeting the straight line ST in t and the conic in d. Join Cd cutting PQ in r; and draw DM parallel to PQ, cutting Cd in M and AB in N. Now, because triangles BTt and DBN are similar, Bt or PQ is to Tt as DN to NB. So also Rr is to AQ or PS as DM to AN. Therefore, multiplying the antecedents by the

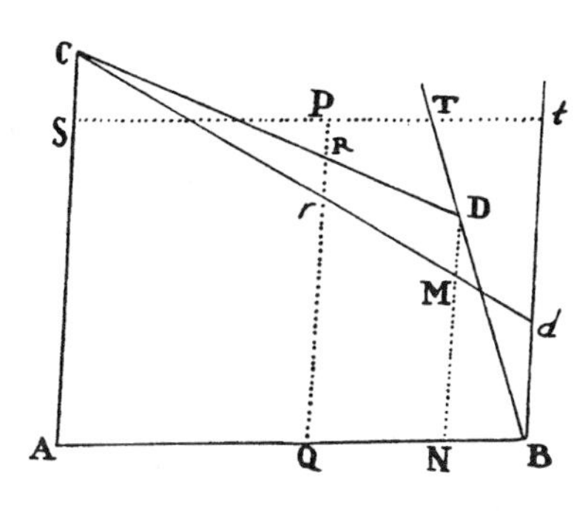

antecedents and the consequents by the consequents, the rectangle PQ × R*r* is to the rectangle PS × T*t* as the rectangle ND × DM is to the rectangle AN × NB, and (by case 1) as the rectangle PQ × P*r* is to the rectangle PS × P*t*, and by separation [or dividendo] as the rectangle PQ × PR is to the rectangle PS × PT. Q.E.D.

CASE 3. Let us suppose finally that the four lines PQ, PR, PS, and PT are not parallel to the sides AC and AB, but are inclined to them in any way whatever. In place of these lines draw P*q* and P*r* parallel to AC, and P*s* and P*t* parallel to AB; then because the angles of the triangles PQ*q*, PR*r*, PS*s*, and PT*t* are given, the ratios of PQ to P*q*, PR to P*r*, PS to P*s*, and PT to P*t* will be given, and thus the compound ratios of PQ × PR to P*q* × P*r*, and PS × PT to P*s* × P*t*. But, by what has been demonstrated above, the ratio of P*q* × P*r* to P*s* × P*t* is given, and therefore also the ratio of PQ × PR to PS × PT. Q.E.D.

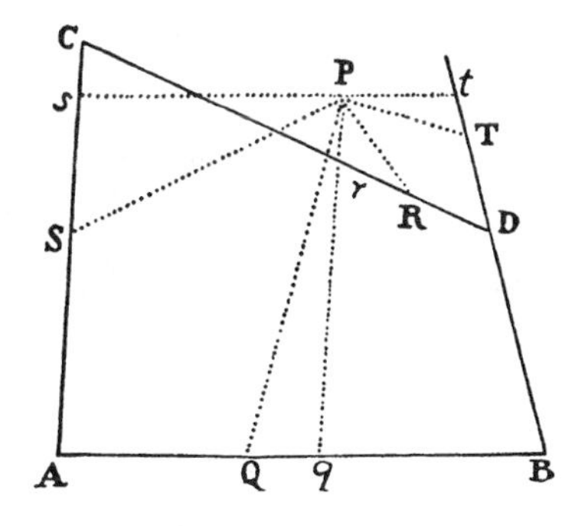

Lemma 18

With the same suppositions as in lem. 17, if the rectangle PQ × PR *of the lines drawn to two opposite sides of the quadrilateral is in a given ratio to the rectangle* PS × PT *of the lines drawn to the other two sides, the point* P *from which the lines are drawn will lie on a conic circumscribed about the quadrilateral.*

Suppose that a conic is described through points A, B, C, D, and some one of the infinite number of points P, say *p*; I say that point P always lies on this conic. If you deny it, join AP cutting this conic in some point other than P, if possible, say in *b*. Therefore, if from these points *p* and *b* the straight lines *pq*, *pr*, *ps*, *pt* and *bk*, *bn*, *bf*, *bd* are drawn at given angles to the sides of the quadrilateral, then *bk* × *bn* will be to *bf* × *bd* as (by lem. 17) *pq* × *pr* is to *ps* × *pt*, and as (by hypothesis) PQ × PR is to PS × PT. Also, because the quadrilaterals *bk*A*f* and PQAS are similar, *bk* is to *bf* as PQ to PS. And therefore, if the terms of the previous

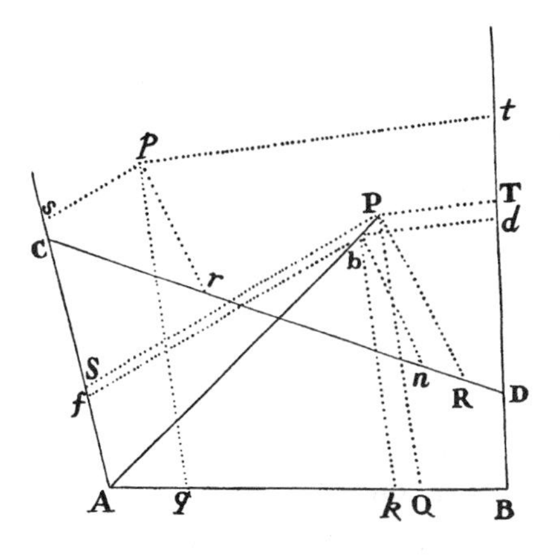

proportion are divided by the corresponding terms of this one, *bn* will be to *bd* as PR to PT. Therefore the angles of the quadrilateral D*nbd* are respectively equal to the angles of quadrilateral DRPT and the quadrilaterals are similar, and consequently their diagonals D*b* and DP coincide. And thus *b* falls upon the intersection of the straight lines AP and DP and accordingly coincides with point P. And therefore point P, wherever it is taken, falls on the assigned conic. Q.E.D.

COROLLARY. Hence if three straight lines PQ, PR, and PS are drawn at given angles from a common point P to three other straight lines given in position, AB, CD, and AC, one line being drawn to each of the other lines, and if the rectangle PQ × PR of two of the lines drawn is in a given ratio to the square of the third line PS, then the point P, from which the straight lines are drawn, will be located in a conic which touches lines AB and CD at A and C, and conversely. For let line BD coincide with line AC, while the position of the three lines AB, CD, and AC remains the same, and let line PT also coincide with line PS; then the rectangle PS × PT will come to be PS^2, and the straight lines AB and CD, which formerly cut the curve in points A and B, C and D, can no longer cut the curve in those points which now coincide, but will only touch it.

Scholium The term "conic" [or "conic section"] is used in this lemma in an extended sense, so as to include both a rectilinear section passing through the vertex of a cone and a circular section parallel to the base. For if point *p* falls on a straight line which joins points A and D or C and B, the conic section will turn into twin straight lines, one of which is the straight line on which point *p* falls and the other the straight line which joins the other two of the four points.

If two opposite angles of the quadrilateral, taken together, are equal to two right angles, and the four lines PQ, PR, PS, and PT are drawn to its sides either perpendicularly or at any equal angles, and the rectangle PQ × PR of two of the lines drawn is equal to the rectangle PS × PT of the other two, the conic will turn out to be a circle. The same will happen if the four lines are drawn at any angles and the rectangle PQ × PR of two of the lines drawn is to the rectangle PS × PT of the other two as the rectangle of the sines of the angles S and T, at which the last two lines PS and PT are drawn, is to the rectangle of the sines of the angles Q and R, at which the first two lines PQ and PR are drawn.

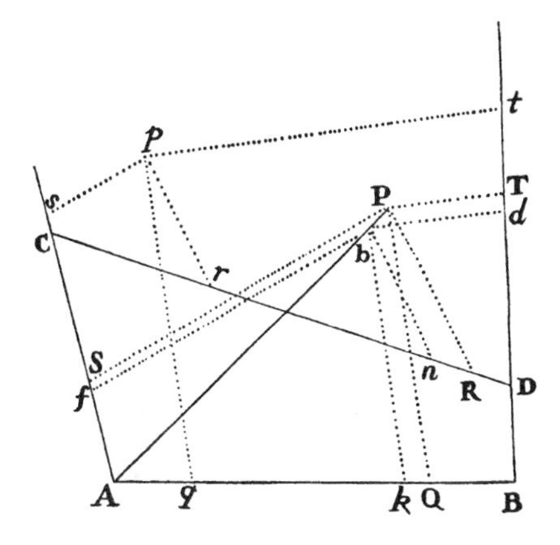

In the other cases the locus of point P will be some one of the three figures that are commonly called conic sections [or conics]. In place of the quadrilateral ABCD, however, there can be substituted a quadrilateral whose two opposite sides decussate each other as diagonals do. But also, one or two of the four points A, B, C, and D can go off to infinity, and in this way the sides of the figure which converge to these points can turn out to be parallel, in which case the conic will pass through the other points and will go off to infinity in the direction of the parallels.

Lemma 19

To find a point P *such that if four straight lines* PQ, PR, PS, *and* PT *are drawn from it at given angles to four other straight lines* AB, CD, AC, *and* BD *given in position, one line being drawn from the point* P *to each of the four other straight lines, the rectangle* PQ × PR *of two of the lines drawn will be in a given ratio to the rectangle* PS × PT *of the other two.*

Let lines AB and CD, to which the two straight lines PQ and PR containing one of the rectangles are drawn, meet the other two lines given in position in the points A, B, C, and D. From some one of them A draw any straight line AH, in which you wish point P to be found. Let this line AH cut the opposite lines BD and CD—that is, BD in H and CD in I—and because all the angles of the figure are given, the ratios of PQ to PA and PA to PS, and consequently the ratio

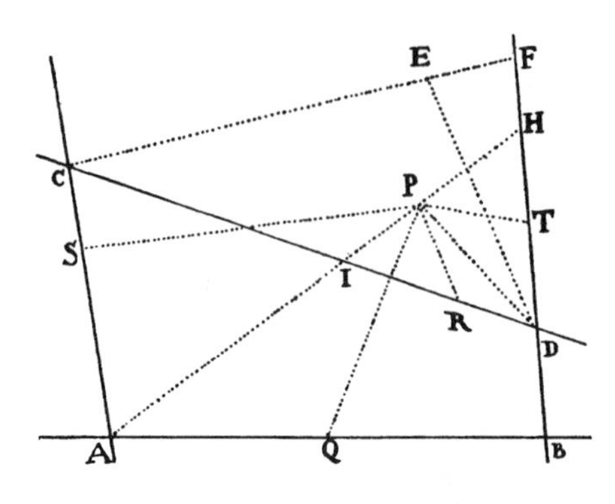

of PQ to PS, will be given. On eliminating this ratio of PQ to PS from the given ratio of PQ × PR to PS × PT, the ratio of PR to PT will be given; and when the given ratios of PI to PR and PT to PH are combined, the ratio of PI to PH, and thus the point P, will be given. Q.E.I.

Corollary 1. Hence also a tangent can be drawn to any point D of the locus of the infinite number of points P. For when points P and D come together—that is, when AH is drawn through the point D—the chord PD becomes a tangent. In this case the ultimate ratio of the vanishing lines IP and PH will be found as above. Therefore, draw CF parallel to AD and meeting BD in F and being cut in E in that ultimate ratio; then DE will be a tangent, because CF and the vanishing line IH are parallel and are similarly cut in E and P.

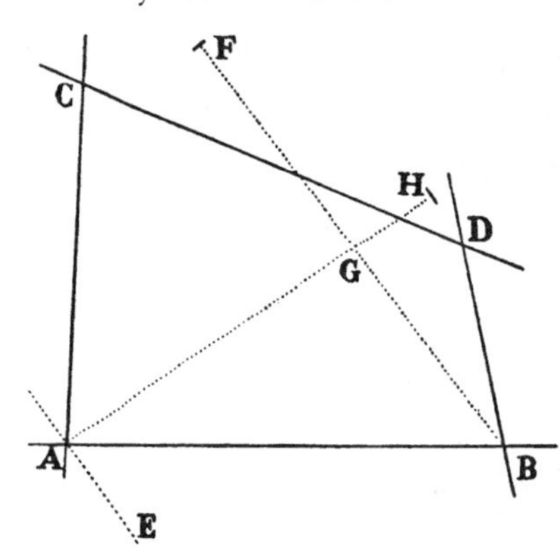

Corollary 2.[a] Hence also, the locus of all the points P can be determined. Through any one of the points A, B, C, D—say A—draw the tangent AE of the locus, and through any other point B draw BF parallel to the tangent and meeting the locus in F. The point F will be found by means of lem. 19. Bisect BF in G, and the indefinite line AG, when drawn, will be the position of the diameter to which BG and FG are ordinates. Let this line AG meet the locus in H, and AH will be the diameter or la-

a. In the index prepared by Cotes for ed. 2 and retained in ed. 3, this corollary is keyed under "Problematis" ("of the problem") and characterized as follows: "Geometrical synthesis of the classical problem of four lines made famous by Pappus and attempted by Descartes through algebraic computation." As this description makes explicit, Newton's rejection of "an [analytical] computation" in favor of "a geometrical synthesis" is directed at Descartes, who reduced the four-line locus to a curve defined algebraically by an equation of the second degree. See *The Mathematical Papers of Isaac Newton*, ed. D. T. Whiteside (Cambridge: Cambridge University Press, 1967–1981), 6:252–254 n. 35, 4:291 n. 17, 4:274–282, esp. 274–276.

tus transversum [i.e., transverse diameter] to which the latus rectum will be as BG^2 to $AG \times GH$. If AG nowhere meets the locus, the line AH being indefinitely produced, the locus will be a parabola, and its latus rectum corresponding to the diameter AG will be $\frac{BG^2}{AG}$. But if AG does meet the locus somewhere, the locus will be a hyperbola when points A and H are situated on the same side of G, and an ellipse when G is between points A and H, unless angle AGB happens to be a right angle and additionally BG^2 is equal to the rectangle $AG \times GH$, in which case the locus will be a circle.

And thus there is exhibited in this corollary not an [analytical] computation but a geometrical synthesis, such as the ancients required, of the classical problem of four lines, which was begun by Euclid and carried on by Apollonius.

Lemma 20

If any parallelogram ASPQ *touches a conic in points* A *and* P *with two of its opposite angles* A *and* P, *and if the sides* AQ *and* AS, *indefinitely produced, of one of these angles meet the said conic in* B *and* C, *and if from the meeting points* B *and* C *two straight lines* BD *and* CD *are drawn to any fifth point* D *of the conic, meeting the other two indefinitely produced sides* PS *and* PQ *of the parallelogram in* T *and* R; *then* PR *and* PT, *the parts cut off from the sides, will always be to each other in a given ratio. And conversely, if the parts which are cut off are to each other in a given ratio, the point* D *will touch a conic passing through the four points* A, B, C, *and* P.

CASE 1. Join BP and also CP, and from point D draw two straight lines DG and DE, the first of which (DG) is parallel to AB and meets PB, PQ, and CA at H, I, and G, while the second (DE) is parallel to AC and meets PC, PS, and AB at F, K, and E; then (by lem. 17) the rectangle $DE \times DF$ will be to the rectangle $DG \times DH$ in a given ratio. But PQ is to DE (or IQ) as PB to

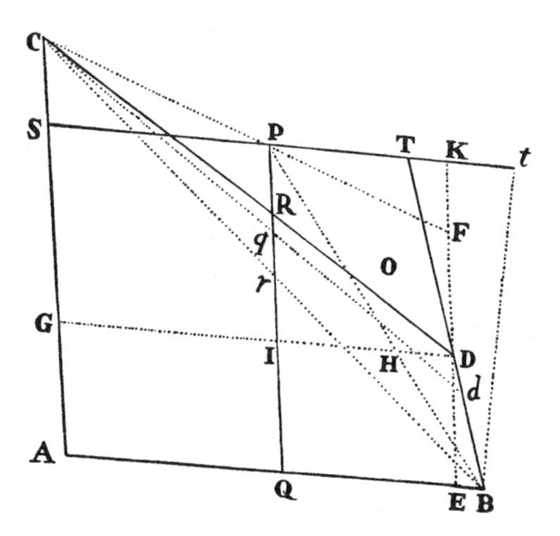

HB and thus as PT to DH; and by alternation [or alternando] PQ is to PT as DE to DH. Additionally, PR is to DF as RC to DC and hence as (IG or) PS to DG; and by alternation [or alternando] PR is to PS as DF to DG; and when the ratios are combined, the rectangle PQ × PR comes to be to the rectangle PS × PT as the rectangle DE × DF to the rectangle DG × DH, and hence in a given ratio. But PQ and PS are given, and therefore the ratio of PR to PT is given. Q.E.D.

CASE 2. But if PR and PT are supposed in a given ratio to each other, then on working backward with a similar argument, it will follow that the rectangle DE × DF is to the rectangle DG × DH in a given ratio and consequently that point D (by lem. 18) lies in a conic passing through points A, B, C, and P. Q.E.D.

COROLLARY 1. Hence, if BC is drawn cutting PQ in *r*, and if P*t* is taken on PT in the ratio to P*r* which PT has to PR, B*t* will be a tangent of the conic at point B. For conceive of point D as coming together with point B in such a way that, as chord BD vanishes, BT becomes a tangent; then CD and BT will coincide with CB and B*t*.

COROLLARY 2. And vice versa, if B*t* is a tangent and BD and CD meet in any point D of the conic, PR will be to PT as P*r* to P*t*. And conversely, if PR is to PT as P*r* to P*t*, BD and CD will meet in some point D of the conic.

COROLLARY 3. One conic does not intersect another conic in more than four points. For, if it can be done, let two conics pass through five points A, B, C, P, and O, and let the straight line BD cut these conics in points D and *d*, and let the straight line C*d* cut PQ in *q*. Then PR is to PT as P*q* to PT; hence PR and P*q* are equal to each other, contrary to the hypothesis.

Lemma 21 *If two movable and infinite straight lines* BM *and* CM, *drawn through given points* B *and* C *as poles, describe by their meeting-point* M *a third straight line* MN *given in position, and if two other infinite straight lines* BD *and* CD *are drawn, making given angles* MBD *and* MCD *with the first two lines at those given points* B *and* C; *then I say that the point* D, *where these two lines* BD *and* CD *meet, will describe a conic passing through points* B *and* C. *And conversely, if the point* D, *where the straight lines* BD *and* CD *meet, describes a conic passing through the given points* B, C, *and* A, *and the angle* DBM *is always equal to the*

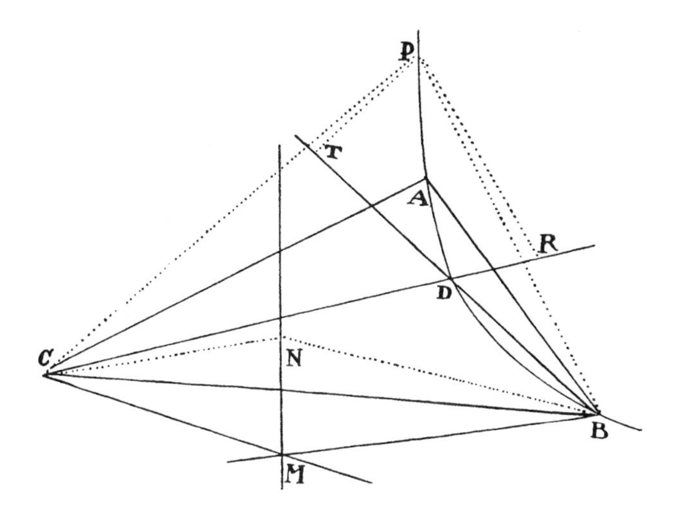

given angle ABC, *and the angle* DCM *is always equal to the given angle* ACB; *then point* M *will lie in a straight line given in position.*

For let point N be given in the straight line MN; and when the movable point M falls on the stationary point N, let the movable point D fall on the stationary point P. Draw CN, BN, CP, and BP, and from point P draw the straight lines PT and PR meeting BD and CD in T and R and forming an angle BPT equal to the given angle BNM, and an angle CPR equal to the given angle CNM. Since therefore (by hypothesis) angles MBD and NBP are equal, as are also angles MCD and NCP, take away the angles NBD and NCD that are common, and there will remain the equal angles NBM and PBT, NCM and PCR; and therefore triangles NBM and PBT are similar, as are also triangles NCM and PCR. And therefore PT is to NM as PB to NB, and PR is to NM as PC to NC. But the points B, C, N, and P are stationary. Therefore, PT and PR have a given ratio to NM and accordingly a given ratio to each other; and thus (by lem. 20) the point D, the perpetual meeting-point of the movable straight lines BT and CR, lies in a conic passing through points B, C, and P. Q.E.D.

And conversely, if the movable point D lies in a conic passing through the given points B, C, and A; and if angle DBM is always equal to the given angle ABC, and the angle DCM is always equal to the given angle ACB; and if, when point D falls successively on any two stationary points *p* and P of the conic, the movable point M falls successively on the two stationary points *n* and N; then through these same points *n* and N draw the straight

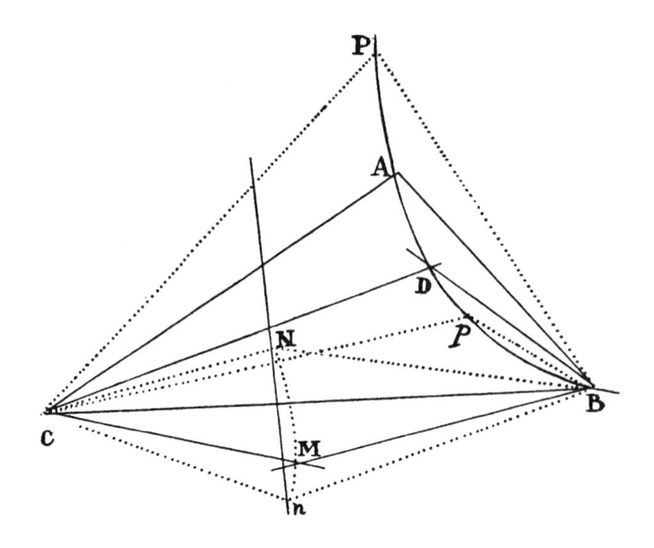

line nN, and this will be the perpetual locus of the movable point M. For, if it can be done, let point M move in some curved line. Then the point D will lie in a conic passing through the five points B, C, A, p, and P when the point M perpetually lies in a curved line. But from what has already been demonstrated, point D will also lie in a conic passing through the same five points B, C, A, p, and P when point M perpetually lies in a straight line. Therefore, two conics will pass through the same five points, contrary to lem. 20, corol. 3. Therefore, it is absurd to suppose the point M to be moving in a curved line. Q.E.D.

Proposition 22
Problem 14

To describe a trajectory through five given points.

Let five points A, B, C, P, and D be given. From one of them A to any other two B and C (let B and C be called poles), draw the straight lines AB and AC, and parallel to these draw TPS and PRQ through the fourth point P. Then from the two poles B and C draw two indefinite lines BDT and CRD through the fifth point D, BDT meeting the line TPS (just drawn) in T, and CRD meeting PRQ in R. Finally, draw the straight line tr parallel to TR, and cut off from the straight lines PT and PR any straight lines Pt and Pr proportional to PT and PR; then, if through their ends t and r and poles B and C the lines Bt and Cr are drawn meeting in d, that point d will be located in the required trajectory. For that point d (by lem. 20) lies in a conic passing through the four points A, B, C, and P; and, the lines Rr and

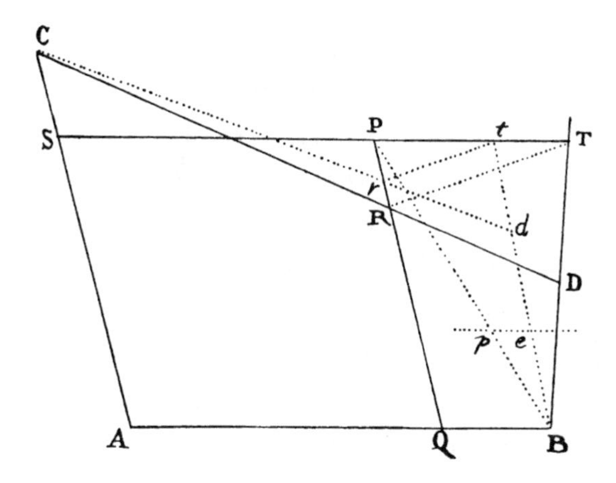

T*t* vanishing, point *d* coincides with point D. Therefore, the conic section passes through the five points A, B, C, P, and D. Q.E.D.

Another solution

Join any three of the given points, A, B, and C; and, rotating the angles ABC and ACB, given in magnitude, around two of these points B and C as poles, apply the legs BA and CA first to point D and then to point P, and note the points M and N in which the other legs BL and CL cross in each

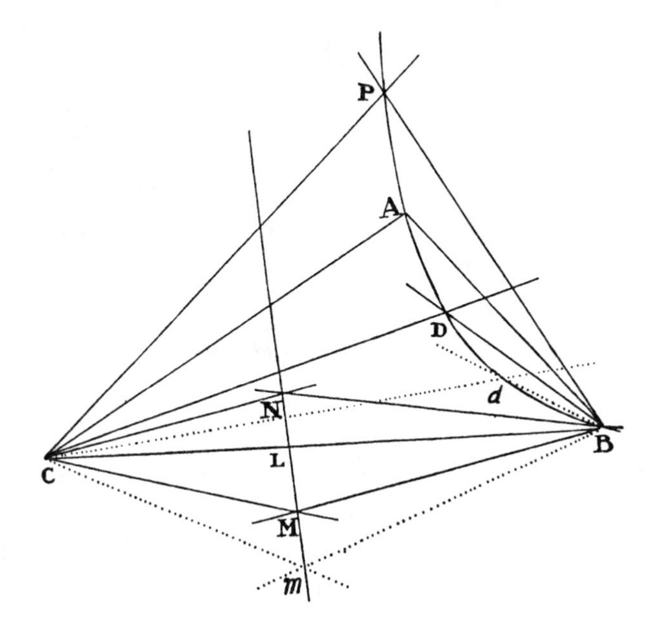

case. Draw the indefinite straight line MN, and rotate these movable angles around their poles B and C in such a way that the intersection of the legs BL and CL or BM and CM (which now let be *m*) always falls on that indefinite straight line MN; and the intersection of the legs BA and CA or BD and CD (which now let be *d*) will trace out the required trajectory PAD*d*B. For point *d* (by lem. 21) will lie in a conic passing through points B and C: and when point *m* approaches points L, M, and N, point *d* (by construction) will approach points A, D, and P. Therefore a conic will be described passing through the five points A, B, C, P, and D. Q.E.F.

COROLLARY 1. Hence a straight line can readily be drawn that will touch the required trajectory in any given point B. Let point *d* approach point B, and the straight line B*d* will come to be the required tangent.

COROLLARY 2. Hence also the centers, diameters, and latera recta of the trajectories can be found, as in lem. 19, corol. 2.

Scholium The first of the constructions of prop. 22 will become a little simpler by joining BP, producing it if necessary, and in it taking B*p* to BP as PR is to PT, and then drawing through *p* the indefinite straight line *pe* parallel to SPT and in it always taking *pe* equal to P*r*, and then drawing the straight lines B*e* and C*r* meeting in *d*. For since the ratios P*r* to P*t*, PR to PT, *p*B to PB, and *pe* to P*t* are equal, *pe* and P*r* will always be equal. By this method the points of the trajectory are found most readily, unless you prefer to describe the curve mechanically, as in the second construction.

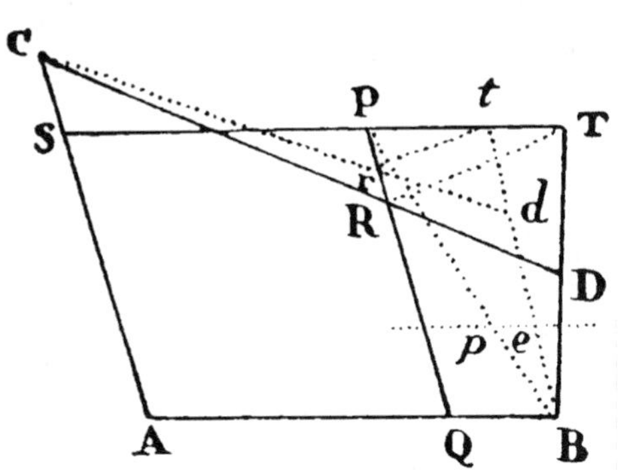

Proposition 23
Problem 15 *To describe a trajectory that will pass through four given points and touch a straight line given in position.*

CASE 1. Let the tangent HB, the point of contact B, and three other points C, D, and P be given. Join BC, and by drawing PS parallel to the straight line BH, and PQ parallel to the straight line BC, complete the parallelogram BSPQ. Draw BD cutting SP in T, and CD cutting PQ in R. Finally, by drawing any line *tr* parallel to TR, cut off P*r* and P*t* from PQ

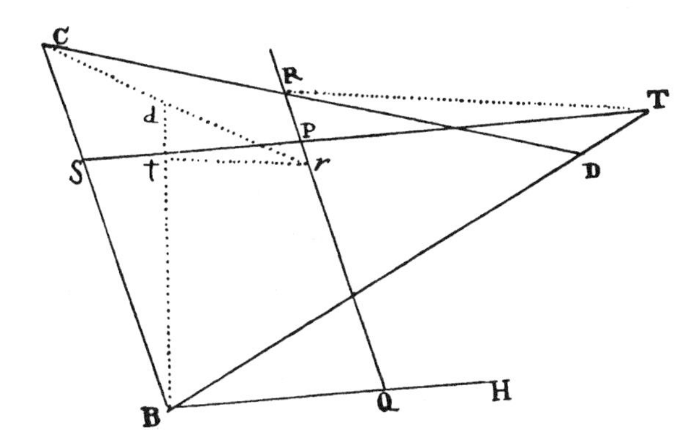

and PS in such a way that Pr and Pt are proportional respectively to PR and PT; then draw Cr and Bt, and their meeting-point d (by lem. 20) will always fall on the trajectory to be described.

Another solution

Revolve the angle CBH, given in magnitude, about the pole B, and revolve about the pole C any rectilinear radius DC, produced at both ends. Note the points M and N at which the leg BC of the angle cuts that radius when the other leg BH meets the same radius in points P and D. Then draw the indefinite line MN, and let that radius CP or CD and the leg BC of the angle meet perpetually in the line MN; and the meeting-point of the other leg BH with the radius will trace out the required trajectory.

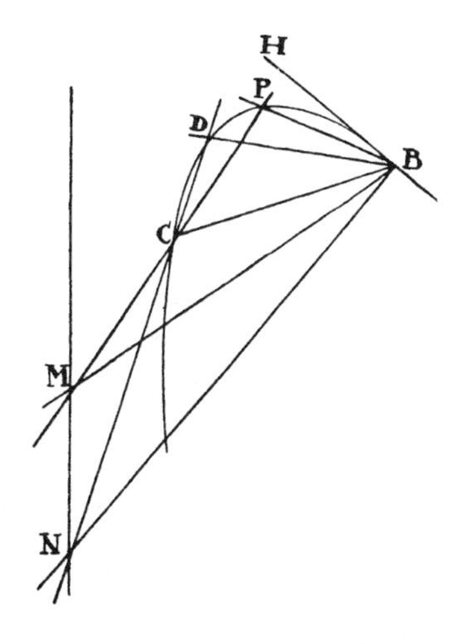

For if, in the constructions of prop. 22, point A approaches point B, lines CA and CB will coincide, and line AB in its ultimate position will come to be the tangent BH; and therefore the constructions set forth in prop. 22 will come to be the same as the constructions described in this proposition. Therefore, the meeting-point of the leg BH with the radius will trace out a conic passing through points C, D, and P and touching the straight line BH in point B. Q.E.F.

Case 2. Let four points B, C, D, and P be given, situated outside the tangent HI. Join them in pairs by the lines BD and CP coming together in

G and meeting the tangent in H and I. Cut the tangent in A in such a way that HA is to IA as the rectangle of the mean proportional between CG and GP and the mean proportional between BH and HD is to the rectangle of the mean proportional between DG and GB and the mean proportional between PI and IC, and A will be the point of contact. For if HX, parallel to the straight line PI, cuts the trajectory in any points X and Y, then (from the *Conics*) point A will have to be so placed that HA^2 is to AI^2 in a ratio compounded of the ratio of the rectangle $XH \times HY$ to the product $BH \times HD$, or of the rectangle $CG \times GP$ to the rectangle $DG \times GD$, and of the ratio of the rectangle $BH \times HD$ to the rectangle $PI \times IC$. And once the point of contact A has been found, the trajectory will be described as in the first case. Q.E.F.

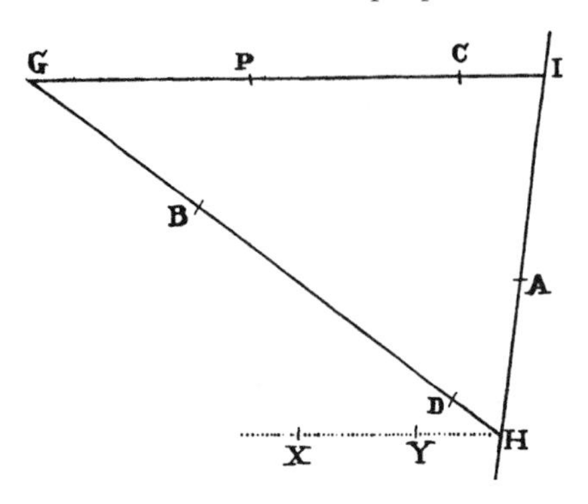

But point A can be taken either between points H and I or outside them, and accordingly two trajectories can be described as solutions to the problem.

Proposition 24
Problem 16

To describe a trajectory that will pass through three given points and touch two straight lines given in position.

Let tangents HI and KL and points B, C, and D be given. Through any two of the points, B and D, draw an indefinite straight line BD meeting the tangents in points H and K. Then, likewise, through any two other points, C and D, draw the indefinite straight line CD meeting the tangents in points I and L. Cut BD in R and CD in S in such a way that HR will be to KR as the mean proportional between BH and HD is to the mean proportional between BK and KD and that IS will be to LS as the mean proportional between CI and ID is to the mean proportional between CL and LD. And cut these lines at will either between points K and H, and between I and L, or outside them; then draw RS cutting the tangents in A and P, and A and P will be

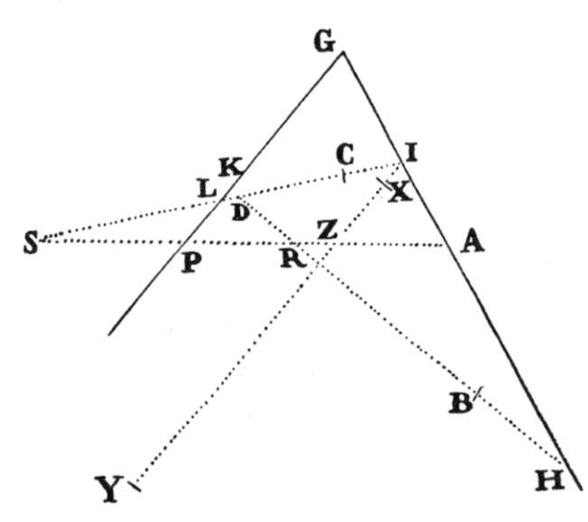

the points of contact. For if A and P are supposed to be the points of contact situated anywhere on the tangents, and if through any one of the points H, I, K, and L, say I, situated in either tangent HI, the straight line IY is drawn parallel to the other tangent KL and meeting the curve in X and Y; and if in this line, IZ is taken so as to be the mean proportional between IX and IY; then, from the *Conics*, the rectangle XI × IY or IZ^2 will be to LP^2 as the rectangle CI × ID to the rectangle CL × LD, that is (by construction), as SI^2 to SL^2, and thus IZ will be to LP as SI to SL. Therefore the points S, P, and Z lie in one straight line. Furthermore, since the tangents meet in G, the rectangle XI × IY or IZ^2 will be to IA^2 (from the *Conics*) as GP^2 to GA^2, and hence IZ will be to IA as GP to GA. Therefore, the points P, Z, and A lie in one straight line, and thus the points S, P, and A are in one straight line. And by the same argument it will be proved that the points R, P, and A are in one straight line. Therefore the points of contact A and P lie in the straight line RS. And once these points have been found, the trajectory will be described as in prop. 23, case 1. Q.E.F.

In this proposition and in prop. 23, case 2, the constructions are the same whether or not the straight line XY cuts the trajectory in X and Y, and they do not depend on this cut. But once the constructions have been demonstrated for the case in which the straight line does cut the trajectory, the constructions for the case in which it does not cut the trajectory also can be found; and for the sake of brevity I do not take the time to demonstrate them further.

Lemma 22

To change figures into other figures of the same class.

Let it be required to transmute any figure HGI. Draw at will two parallel straight lines AO and BL cutting in A and B any third line AB, given in position; and from any point G of the figure draw to the straight line AB any other straight line GD parallel to OA. Then from some point O, given in line OA, draw to the point D the straight line OD meeting BL at *d*, and from the meeting-point erect the straight line *dg* containing any given angle with the straight line BL and having the same ratio to O*d* that DG has to OD; and *g* will be the point in the new figure *hgi* corresponding to point G. By the same method, each of the points in the first figure will yield a corresponding point in the new figure. Therefore, suppose point G to be running through all the points in the first figure with a continual motion;

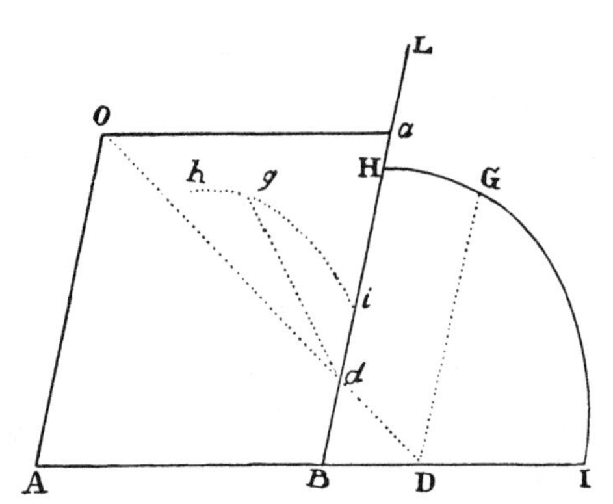

then point g—also with a continual motion—will run through all the points in the new figure and will describe that figure. For the sake of distinction let us call DG the first ordinate, dg the new ordinate, AD the first abscissa, ad the new abscissa, O the pole, OD the abscinding radius, OA the first ordinate radius, and Oa (which completes the parallelogram OAba) the new ordinate radius.

I say now that if point G traces a straight line given in position, point g will also trace a straight line given in position. If point G traces a conic, point g will also trace a conic. I here count a circle among the conic sections. Further, if point G traces a curved line of the third analytic order, point g will likewise trace a curved line of the third order; and so on with curves of higher orders, the two curved lines which points G and g trace will always be of the same analytic order. For as ad is to OA, so are Od to OD, dg to DG, and AB to AD; and hence AD is equal to $\frac{\mathrm{OA} \times \mathrm{AB}}{ad}$, and DG is equal to $\frac{\mathrm{OA} \times dg}{ad}$. Now, if point G traces a straight line and consequently, in any equation which gives the relation between the abscissa AD and the ordinate DG, the indeterminate lines AD and DG rise to only one dimension, and if in this equation $\frac{\mathrm{OA} \times \mathrm{AB}}{ad}$ is written for AD and $\frac{\mathrm{OA} \times dg}{ad}$ for DG, then the result will be a new equation in which the new abscissa ad and the new ordinate dg will rise to only one dimension and which therefore designates a straight line. But if AD and DG or either one of them rose to two dimensions in the first equation, then ad and dg will also rise to two dimensions in the second equation. And so on for three or more dimensions. The indeterminates ad and dg in the second equation, and AD and DG in the first, will always rise to the same number of dimensions, and therefore the lines which points G and g trace are of the same analytic order.

I say further that if some straight line touches a curved line in the first figure, this straight line—after being transferred into the new figure in the same manner as the curve—will touch that curved line in the new figure; and conversely. For if any two points of the curve approach each other and come together in the first figure, the same points—after being transferred—will approach each other and come together in the new figure; and thus the straight lines by which these points are joined will simultaneously come to be tangents of the curves in both figures.

The demonstrations of these assertions could have been composed in a more geometrical style. But I choose to be brief.

Therefore, if one rectilinear figure is to be transmuted into another, it is only necessary to transfer the intersections of the straight lines of which it is made up and to draw straight lines through them in the new figure. But if it is required to transmute a curvilinear figure, then it is necessary to transfer the points, tangents, and other straight lines which determine the curved line. Moreover, this lemma is useful for solving more difficult problems by transmuting the proposed figures into simpler ones. For any converging straight lines are transmuted into parallels by using for the first ordinate radius any straight line that passes through the meeting-point of the converging lines; and this is so because the meeting-point goes off this way to infinity, and lines that nowhere meet are parallel. Moreover, after the problem is solved in the new figure, if this figure is transmuted into the first figure by the reverse procedure, the required solution will be obtained.

This lemma is useful also for solving solid problems. For whenever two conics occur by whose intersection a problem can be solved, either one of them, if it is a hyperbola or parabola, can be transmuted into an ellipse; then the ellipse is easily changed into a circle. Likewise, in constructing plane problems, a straight line and a conic are turned into a straight line and a circle.

Proposition 25
Problem 17

To describe a trajectory that will pass through two given points and touch three straight lines given in position.

Through the meeting-point of any two tangents with each other and the meeting-point of a third tangent with the straight line that passes through two given points, draw an indefinite straight line; and using it as the first ordinate radius, transmute the figure, by lem. 22, into a new figure. In this figure the two tangents will come to be parallel to each other, and the third tangent will become parallel to the straight line passing through the two given points. Let hi and kl be the two parallel tangents,

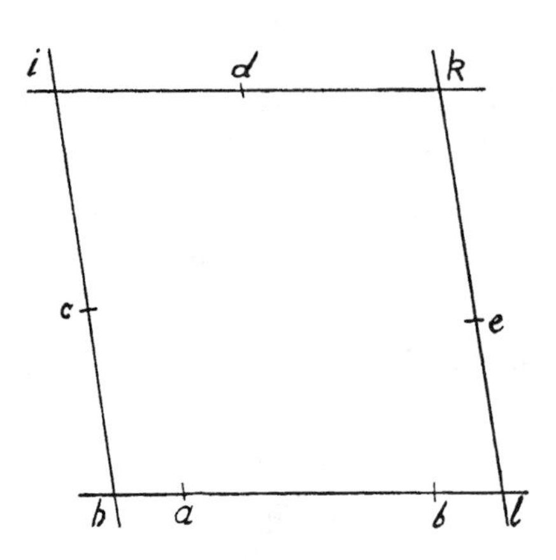

ik the third tangent, and *hl* the straight line parallel to it, passing through the points *a* and *b* through which the conic ought to pass in this new figure and completing the parallelogram *hikl*. Cut the straight lines *hi*, *ik*, and *kl* in *c*, *d*, and *e*, so that *hc* is to the square root of the rectangle $ah \times hb$, and *ic* is to *id*, and *ke* is to *kd*, as the sum of the straight lines *hi* and *kl* is to the sum of three lines, of which the first is the straight line *ik* and the other two are the square roots of the rectangles $ah \times hb$ and $al \times lb$; then *c*, *d*, and *e* will be the points of contact. For, from the *Conics*, hc^2 is to the rectangle $ah \times hb$ in the same ratio as ic^2 to id^2, and ke^2 to kd^2, and el^2 to the rectangle $al \times lb$; and therefore *hc* is to the square root of $ah \times hb$, and *ic* is to *id*, and *ke* is to *kd*, and *el* is to the square root of $al \times lb$, as the square root of that ratio and hence, by composition [or componendo], in the given ratio of all the antecedents *hi* and *kl* to all the consequents, which are the square root of the rectangle $ah \times hb$, the straight line *ik*, and the square root of the rectangle $al \times lb$ [i.e., in the given ratio of $hi + kl$ to $\sqrt{(ah \times hb)} + ik + \sqrt{(al \times lb)}$]. Therefore, the points of contact *c*, *d*, and *e* in the new figure are obtained from that given ratio. By the reverse procedure of lem. 22, transfer these points to the first figure, and there (by prop. 22) the trajectory will be described. Q.E.F.

But according as points *a* and *b* lie between points *h* and *l* or lie outside them, points *c*, *d*, and *e* must be taken either between points *h*, *i*, *k*, and *l*, or outside them. If either one of the points *a* and *b* falls between points *h* and *l*, and the other outside, the problem is impossible.

Proposition 26
Problem 18

To describe a trajectory that will pass through a given point and touch four straight lines given in position.

From the common intersection of any two of the tangents to the common intersection of the other two, draw an indefinite straight line; then, using this as the first ordinate radius, transmute the figure (by lem. 22) into

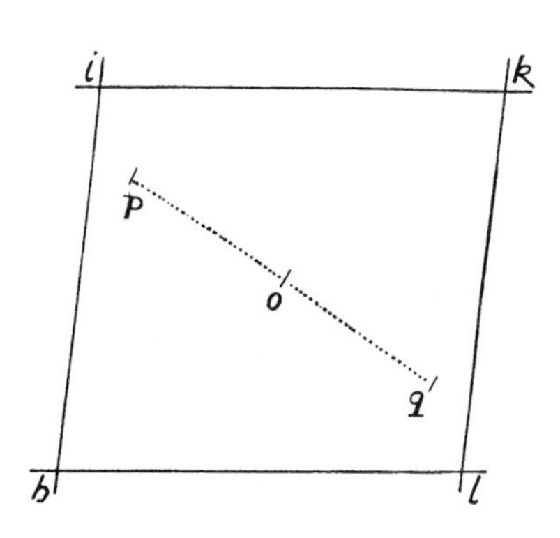

a new figure; then the tangents, which formerly met in the first ordinate radius, will now come to be parallel in pairs. Let those tangents be *hi* and *kl*, *ik* and *hl*, forming the parallelogram *hikl*. And let *p* be the point in this new figure corresponding to the given point in the first figure. Through the center O of the figure draw *pq*, and, on O*q* being equal to O*p*, *q* will be another point through which the conic must pass in this new figure. By the reverse procedure of lem. 22 transfer this point to the first figure, and in that figure two points will be obtained through which the trajectory is to be described. And that trajectory can be described through these same points by prop. 25.

Lemma 23

If two straight lines AC *and* BD, *given in position, terminate at the given points* A *and* B *and have a given ratio to each other; and if the straight line* CD, *by which the indeterminate points* C *and* D *are joined, is cut in* K *in a given ratio; I say that point* K *will be located in a straight line given in position.*

For let the straight lines AC and BD meet in E, and in BE take BG to AE as BD is to AC, and let FD always be equal to the given line EG; then, by construction, EC will be to GD, that is, to EF, as AC to BD, and thus in a given ratio, and therefore the species of the triangle EFC will be given. Cut CF in L so that CL is to CF in the ratio of CK to CD; then, because that ratio is given, the species of the triangle EFL will also be given, and accordingly point L will be located in the straight line EL given in position. Join LK, and the triangles CLK and CFD will be similar; and because FD and the ratio of LK to FD are given, LK will be given. Take EH equal to LK, and ELKH will always be a parallelogram. Therefore, point K is located in the side HK, given in position, of the parallelogram. Q.E.D.

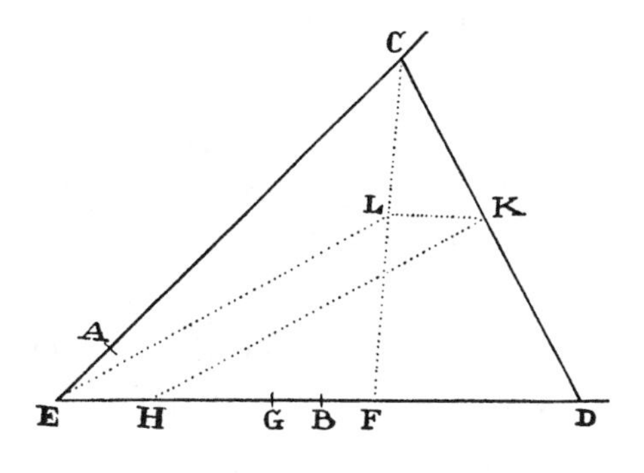

COROLLARY. Because the species of the figure EFLC is given, the three straight lines EF, EL, and EC (that is, GD, HK, and EC) have given ratios to one another.

Lemma 24 *If three straight lines, two of which are parallel and given in position, touch any conic section, I say that the semidiameter of the section which is parallel to the two given parallel lines is a mean proportional between their segments that are intercepted between the points of contact and the third tangent.*

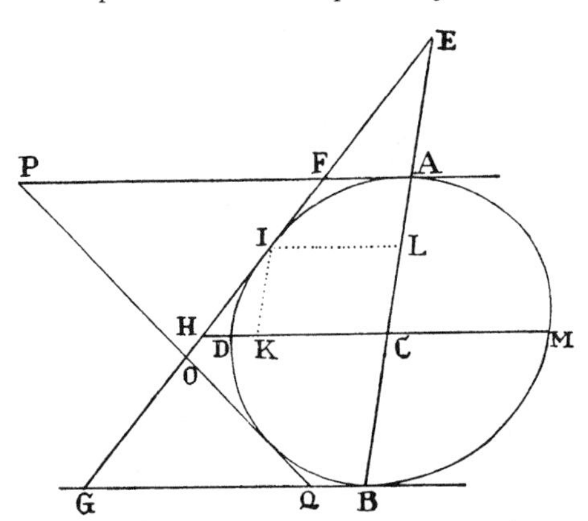

Let AF and GB be two parallel lines touching the conic ADB in A and B; and let EF be a third straight line touching the conic in I and meeting the first tangents in F and G; and let CD be the semidiameter of the figure parallel to the tangents; then I say that AF, CD, and BG are continually proportional.

For if the conjugate diameters AB and DM meet the tangent FG in E and H and cut each other in C, and the parallelogram IKCL is completed, then, from the nature of conics, EC will be to CA as CA to CL, and by separation [or dividendo] as EC − CA to CA − CL, or EA to AL; and by composition [or componendo], EA will be to EA + AL or EL as EC to EC + CA or EB; and therefore, because the triangles EAF, ELI, ECH, and EBG are similar, AF will be to LI as CH to BG. And likewise, from the nature of conics, LI or CK is to CD as CD to CH and therefore from the equality of the ratios in inordinate proportion [or ex aequo perturbate] AF will be to CD as CD to BG. Q.E.D.

COROLLARY 1. Hence if two tangents FG and PQ meet the parallel tangents AF and BG in F and G, P and Q, and cut each other in O; then, from the equality of the ratios in inordinate proportion [or ex aequo perturbate] AF will be to BQ as AP to BG, and by separation [or dividendo] as FP to GQ, and thus as FO to OG.

COROLLARY 2. Hence also, two straight lines PG and FQ drawn through points P and G, F and Q, will meet in the straight line ACB that passes through the center of the figure and the points of contact A and B.

Lemma 25

If the indefinitely produced four sides of a parallelogram touch any conic and are intercepted at any fifth tangent, and if the intercepts of any two conterminous sides are taken so as to be terminated at opposite corners of the parallelogram; I say that either intercept is to the side from which it is intercepted as the part of the other conterminous side between the point of contact and the third side is to the other intercept.

Let the four sides ML, IK, KL, and MI of the parallelogram MLIK touch the conic section in A, B, C, and D, and let a fifth tangent FQ cut those sides in F, Q, H, and E; and take the intercepts ME and KQ of the sides MI and KI or the intercepts KH and MF of the sides KL and ML; I

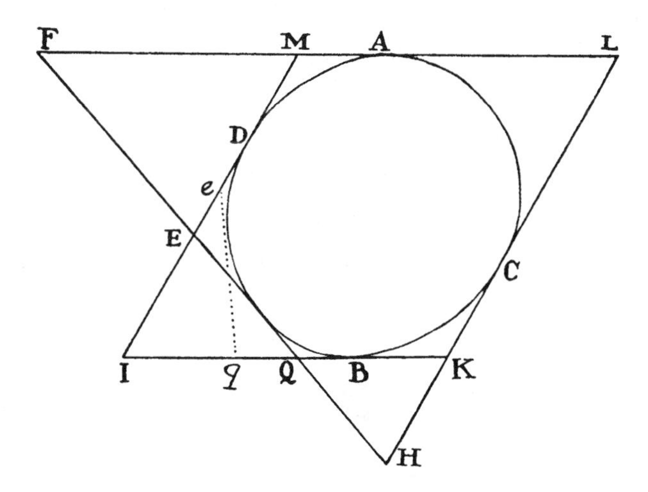

say that ME is to MI as BK to KQ, and KH is to KL as AM to MF. For by lem. 24, corol. 1, ME is to EI as AM or BK to BQ, and by composition [or componendo] ME is to MI as BK to KQ. Q.E.D. Likewise, KH is to HL as BK or AM to AF, and by separation [or dividendo] KH is to KL as AM to MF. Q.E.D.

COROLLARY 1. Hence if the parallelogram IKLM is given, described about a given conic, the rectangle KQ × ME will be given, as will also the rectangle KH × MF equal to it. For those rectangles are equal because the triangles KQH and MFE are similar.

COROLLARY 2. And if a sixth tangent eq is drawn meeting the tangents KI and MI at q and e, the rectangle KQ × ME will be equal to the rectangle Kq × Me, and KQ will be to Me as Kq to ME, and by separation [or dividendo] as Qq to Ee.

Corollary 3. Hence also, if Eq and eQ are drawn and bisected and a straight line is drawn through the points of bisection, this line will pass through the center of the conic. For since Qq is to Ee as KQ to Me, the same straight line will (by lem. 23) pass through the middle of all the lines Eq, eQ, and MK, and the middle of the straight line MK is the center of the section.

Proposition 27
Problem 19

To describe a trajectory that will touch five straight lines given in position.

Let the tangents ABG, BCF, GCD, FDE, and EA be given in position. Bisect in M and N the diagonals AF and BE of the quadrilateral figure ABFE formed by any four of those tangents, and (by lem. 25, corol. 3) the straight line MN drawn through the points of bisection will pass through the center of the trajectory. Again, bisect in P and Q the diagonals (as I call them) BD and GF of the quadrilateral figure BGFD formed by any other four tangents; then the straight line PQ drawn through the points of

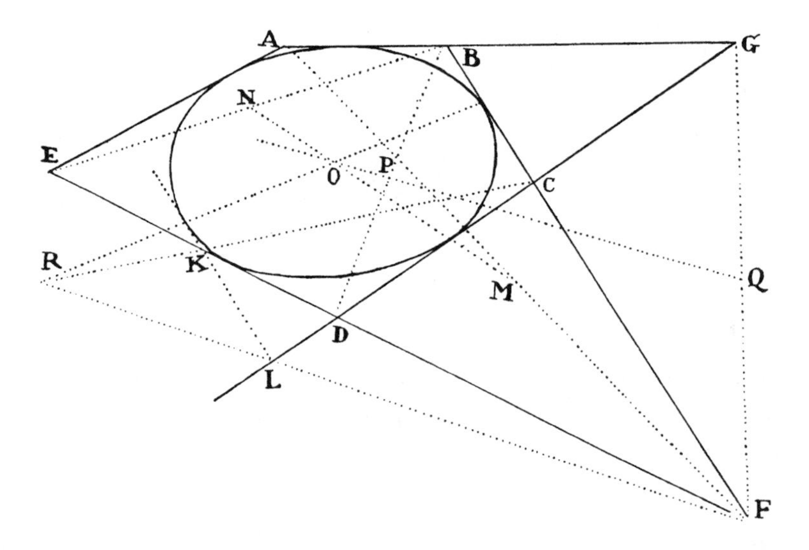

bisection will pass through the center of the trajectory. Therefore, the center will be given at the meeting-point of the bisecting lines. Let that center be O. Parallel to any tangent BC draw KL at such a distance that the center O is located midway between the parallels, and KL so drawn will touch the trajectory to be described. Let this line KL cut any other two tangents GCD and FDE in L and K. Through the meeting-points C and K, F and L, of these nonparallel tangents CL and FK with the parallels CF and KL, draw

CK and FL meeting in R, and the straight line OR, drawn and produced, will cut the parallel tangents CF and KL in the points of contact. This is evident by lem. 24, corol. 2. By the same method other points of contact may be found, and then finally the trajectory may be described by the construction of prop. 22. Q.E.F.

Scholium

What has gone before includes problems in which either the centers or the asymptotes of trajectories are given. For when points and tangents are given together with the center, the same number of other points and tangents are given equally distant from the center on its other side. Moreover, an asymptote is to be regarded as a tangent, and its infinitely distant end-point (if it is permissible to speak of it in this way) as a point of contact. Imagine the point of contact of any tangent to go off to infinity, and the tangent will be turned into an asymptote, and the constructions of the preceding problems will be turned into constructions in which the asymptote is given.

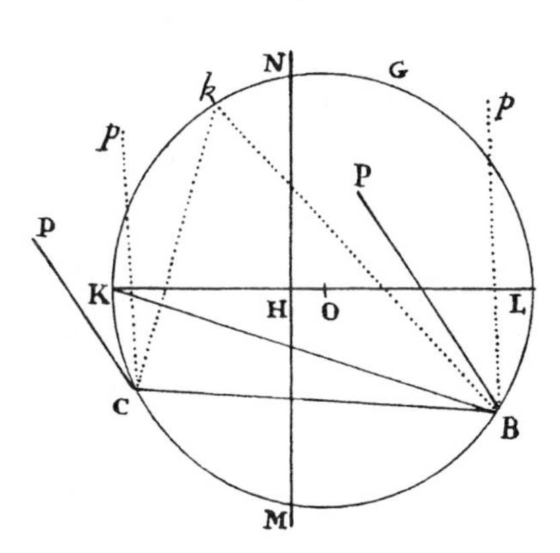

After the trajectory has been described, its axes and foci may be found by the following method. In the construction and figure of lem. 21 make the legs BP and CP (by the meeting of which the trajectory was there described) of the mobile angles PBN and PCN be parallel to each other, and let them—while maintaining that position—revolve about their poles B and C in that figure. Meanwhile, let the circle BGKC be described by the point K or *k* in which the other legs CN and BN of those angles meet. Let the center of this circle be O. From this center to the ruler MN, at which those other legs CN and BN met while the trajectory was being described, drop the normal OH meeting the circle in K and L. And when those other legs CK and BK meet in K, the point that is nearer to the ruler, the first legs CP and BP will be parallel to the major axis and perpendicular to the minor axis; and the converse will occur if the same legs meet in the farther point L. Hence, if the center of a trajectory is given, the axes will be given. And when these are given, the foci are apparent.

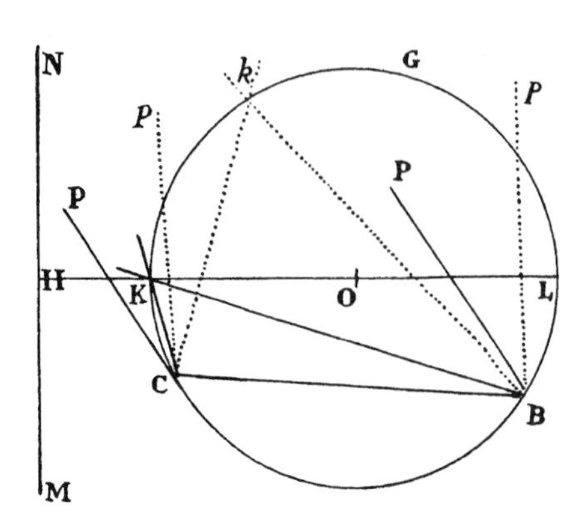

But the squares of the axes are to each other as KH to LH, and hence it is easy to describe a trajectory, given in species, through four given points. For if two of the given points constitute the poles C and B, a third will give the mobile angles PCK and PBK; and once these are given, the circle BGKC can be described. Then, because the species of the trajectory is given, the ratio of OH to OK, and thus OH itself, will be given. With center O and radius OH describe another circle, and the straight line that touches this circle and passes through the meeting-point of the legs CK and BK when the first legs CP and BP meet in the fourth given point will be that ruler MN by means of which the trajectory will be described. Hence, in turn, a quadrilateral given in species can (except in certain impossible cases) also be inscribed in any given conic.

There are also other lemmas by means of which trajectories given in species can be described if points and tangents are given. An example: if a straight line, drawn through any point given in position, intersects a given conic in two points, and the distance between the intersections is bisected, the point of bisection will lie on another conic that is of the same species as the first one and that has its axes parallel to the axes of the first. But I pass quickly to what is more useful.

Lemma 26 *To place the three corners of a triangle given in species and magnitude on three straight lines given in position and not all parallel, with one corner on each line.*

[a]Three indefinite straight lines, AB, AC, and BC, are given in position, and it is required to place triangle DEF in such a way that its corner D touches line AB, corner E line AC, and corner F line BC.[a] On DE, DF, and

aa. In all three editions, and in the preliminary manuscripts (see *The Mathematical Papers of Isaac Newton*, ed. D. T. Whiteside [Cambridge: Cambridge University Press, 1967–1981], 6:287), there is a minor discrepancy between the text and the accompanying diagram. The text refers (in the opening sentence) to "triangle DEF," but the corresponding diagram would indicate that this should rather be "triangle *def*," and similarly "corner [*lit.* vertex] D" and "corner F" should be respectively "corner *d*" and "corner *f*." At the end of the paragraph, however, and in the succeeding paragraph, Newton introduces lowercase letters *a*, *b*, *c* for the triangle *abc*.

EF describe three segments DRE, DGF, and EMF of circles, containing angles equal respectively to angles BAC, ABC, and ACB. And let these segments be described on those sides of the lines DE, DF, and EF that will make the letters DRED go round in the same order as the letters BACB, the letters DGFD in the same order as ABCA, and the letters EMFE in the same order as ACBA; then complete these segments into full circles. Let the first two circles cut each other in G, and let their centers be P and Q. Joining GP and also PQ, take G*a* to AB as GP is to PQ; and with center G and radius G*a* describe a circle that cuts the first circle DGE in *a*. Join *a*D cutting the second circle DFG in *b*, and *a*E cutting the third circle EMF in *c*. And now the figure ABC*def* may be constructed similar and equal to the figure *abc*DEF. This being done, the problem is solved.

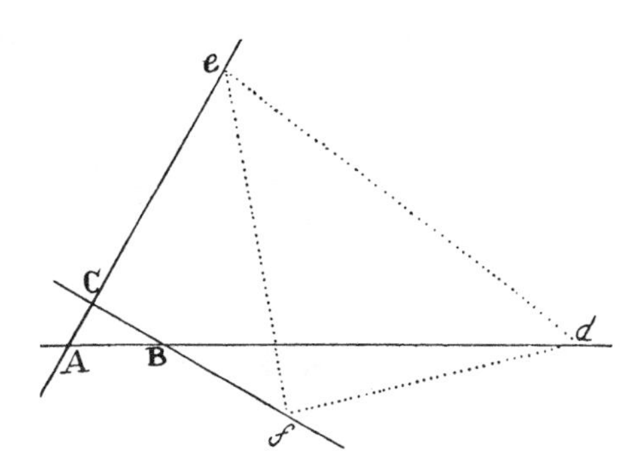

For draw F*c* meeting *a*D in *n*, and join *a*G, *b*G, QG, QD, and PD. By construction, angle E*a*D is equal to angle CAB, and angle *ac*F is equal to angle ACB, and thus the angles of triangle *anc* are respectively equal to the angles of triangle ABC. Therefore angle *anc* or F*n*D is equal to angle

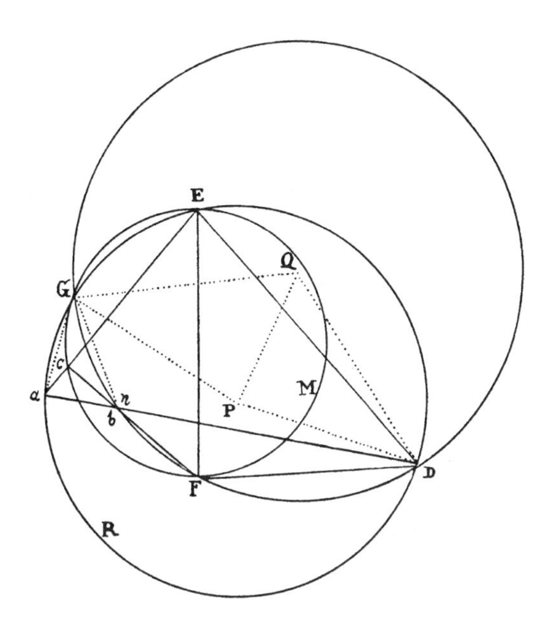

ABC, and hence equal to angle F*b*D; and therefore point *n* coincides with point *b*. Further, angle GPQ, which is half of angle GPD at the center, is equal to angle G*a*D at the circumference; and angle GQP, which is half of angle GQD at the center, is equal to the supplement of angle G*b*D at the circumference, and hence equal to angle G*ba*; and therefore triangles GPQ and G*ab* are similar, and G*a* is to *ab* as GP to PQ, that is (by construction), as G*a* to AB. And thus *ab* and AB are equal; and therefore triangles *abc* and ABC, which we have just proved to be similar, are also equal. Hence, since in addition the corners D, E, and F of the triangle DEF touch the sides *ab*, *ac*, and *bc* respectively of the triangle *abc*, the figure ABC*def* can be completed similar and equal to the figure *abc*DEF; and by its completion the problem will be solved. Q.E.F.

COROLLARY. Hence a straight line can be drawn whose parts given in length will lie between three straight lines given in position. Imagine that triangle DEF, with point D approaching side EF and sides DE and DF placed in a straight line, is changed into a straight line whose given part DE is to be placed between the straight lines AB and AC given in position and whose given part DF is to be placed between the straight lines AB and BC given in position; then, by applying the preceding construction to this case, the problem will be solved.

Proposition 28
Problem 20

To describe a trajectory given in species and magnitude, whose given parts will lie between three straight lines given in position.

Let it be required to describe a trajectory that is similar and equal to the curved line DEF and that will be cut by three straight lines AB, AC, and BC, given in position, into parts similar and equal to the given parts DE and EF of this curved line.

Draw the straight lines DE, EF, and DF, place one of the corners D, E, and F of this triangle DEF on each of those straight lines given in position

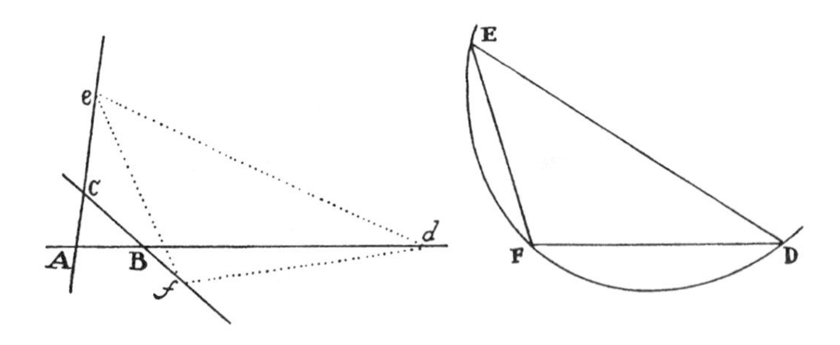

(by lem. 26); then about the triangle describe a trajectory similar and equal to the curve DEF. Q.E.F.

Lemma 27

To describe a quadrilateral given in species, whose corners will lie on four straight lines, given in position, which are not all parallel and do not all converge to a common point—each corner lying on a separate line.

Let four straight lines ABC, AD, BD, and CE be given in position, the first of which cuts the second in A, cuts the third in B, and cuts the fourth in C; let it be required to describe a quadrilateral *fghi* which is similar to the quadrilateral FGHI and whose corner *f*, equal to the given corner F, touches the straight line ABC, and whose other corners *g*, *h*, and *i*, equal to the other given corners G, H, and I, touch the other lines AD, BD, and CE respectively. Join FH, and on FG, FH, and FI describe three segments of circles, FSG, FTH, and FVI, of which the first (FSG) contains an angle equal to angle BAD, the second (FTH) contains an angle equal to angle CBD, and the third (FVI) contains an angle equal to angle ACE. The segments ought, moreover, to be described on those sides of the lines FG, FH, and FI that will make the circular order of the letters FSGF the same as that of the letters BADB, and will make the letters FTHF go round in the same order as CBDC, and the letters FVIF in the same order as ACEA.

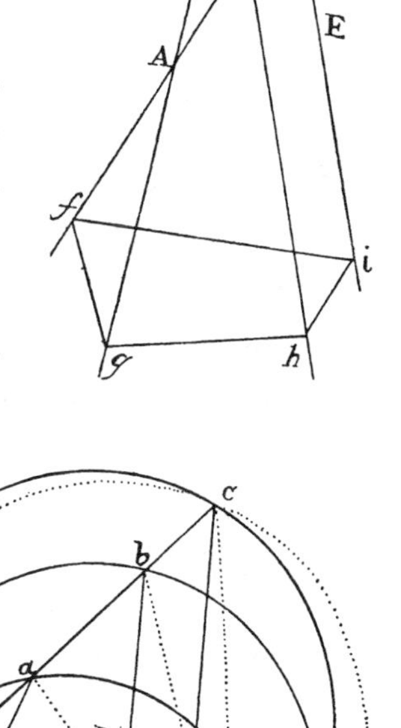

Complete the segments into whole circles, and let P be the center of the first circle FSG, and Q the center of the second circle FTH. Join PQ and produce it in both directions; and in it take QR in the ratio to PQ that BC has to AB. And take QR on the side of the point Q which makes the order of the letters P, Q, and R the same as that of the letters A, B, and C; and then with center R and radius RF describe a fourth circle FN*c* cutting the third circle FVI in *c*. Join F*c* cutting the first circle in *a* and the second in *b*. Draw *a*G, *b*H, and *c*I, and the figure ABC*fghi* can be constructed similar to the figure *abc*FGHI. When this is done, the quadrilateral *fghi* will be the very one which it was required to construct.

For let the first two circles FSG and FTH intersect each other in K. Join PK, QK, RK, *a*K, *b*K, and *c*K, and produce QP to L. The angles F*a*K, F*b*K, and F*c*K at the circumferences are halves of the angles FPK, FQK, and FRK at the centers, and hence are equal to the halves LPK, LQK, and LRK of these angles. Therefore, the angles of figure PQRK are respectively equal to the angles of figure *abc*K, and the figures are similar; and hence *ab* is to *bc* as PQ to QR, that is, as AB to BC. Besides, the angles *f*A*g*, *f*B*h*, and *f*C*i* are (by construction) equal to the angles F*a*G, F*b*H, and F*c*I. Therefore, ABC*fghi*, a figure similar to the figure *abc*FGHI, can be completed. When this is done, the quadrilateral *fghi* will be constructed similar to the quadrilateral FGHI with its corners *f*, *g*, *h*, and *i* touching the straight lines ABC, AD, BD, and CE. Q.E.F.

COROLLARY. Hence a straight line can be drawn whose parts, intercepted in a given order between four straight lines given in position, will have a given proportion to one another. Increase the angles FGH and GHI until the straight lines FG, GH, and HI lie in a single straight line; and by constructing the problem in this case, a straight line *fghi* will be drawn, whose parts *fg*, *gh*, and *hi*, intercepted between four straight lines given in position, AB and AD, AD and BD, BD and CE, will be to one another as the lines FG, GH, and HI, and will keep the same order with respect to one another. But the same thing is done more expeditiously as follows.

Produce AB to K and BD to L so that BK is to AB as HI to GH, and DL to BD as GI to FG; and join KL meeting the straight line CE in *i*. Produce *i*L to M, so that LM is to *i*L as GH to HI; and draw MQ parallel to LB and meeting the straight line AD in *g*, and draw *gi* cutting AB and BD in *f* and *h*. I declare it done.

For let M*g* cut the straight line AB in Q, and let AD cut the straight line KL in S, and draw AP parallel to BD and meeting *i*L in P; then *g*M will be to L*h* (*gi* to *hi*, M*i* to L*i*, GI to HI, AK to BK) and AP to BL in the same ratio. Cut DL in R so that DL is to RL in that same ratio; then, because *g*S to *g*M, AS to AP, and DS to DL are proportional, from the equality of

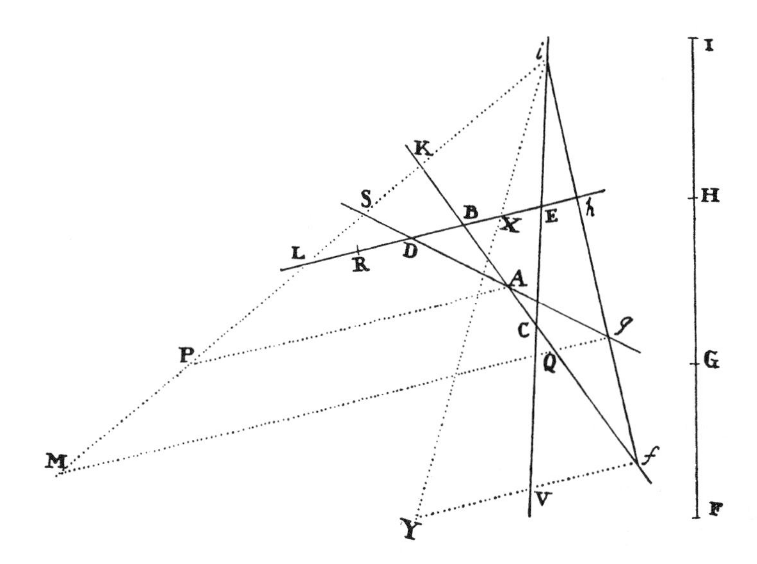

the ratios [or ex aequo] AS*a* will be to BL, and DS to RL, as *g*S to L*h*, and by a mixture of operations BL − RL will be to L*h* − BL as AS − DS to *g*S − AS. That is, BR will be to B*h* as AD to A*g* and thus as BD to *g*Q. And by alternation [or alternando] BR is to BD as B*h* to *g*Q or as *fh* to *fg*. But by construction the line BL was cut in D and R in the same ratio as the line FI in G and H; and therefore BR is to BD as FH to FG. As a result, *fh* is to *fg* as FH to FG. Therefore, since *gi* is also to *hi* as M*i* to L*i*, that is, as GI to HI, it is evident that the lines FI and *fi* are similarly cut in *g* and *h*, G and H. Q.E.F.

In the construction of this corollary, after LK is drawn cutting CE in *i*, it is possible to produce *i*E to V, so that EV is to E*i* as FH to HI, and then to draw V*f* parallel to BD. It comes to the same thing if with center *i* and radius IH a circle is described cutting BD in X, and if *i*X is produced to Y, so that *i*Y is equal to IF, and if Y*f* is drawn parallel to BD.

Other solutions of this problem were devised some time ago by Wren and Wallis.

Proposition 29
Problem 21

To describe a trajectory, given in species, which four straight lines given in position will cut into parts given in order, species, and proportion.

Let it be required to describe a trajectory that is similar to the curved line FGHI and whose parts, similar and proportional to the parts FG, GH, and HI of the curve, are intercepted between the straight lines AB and AD,

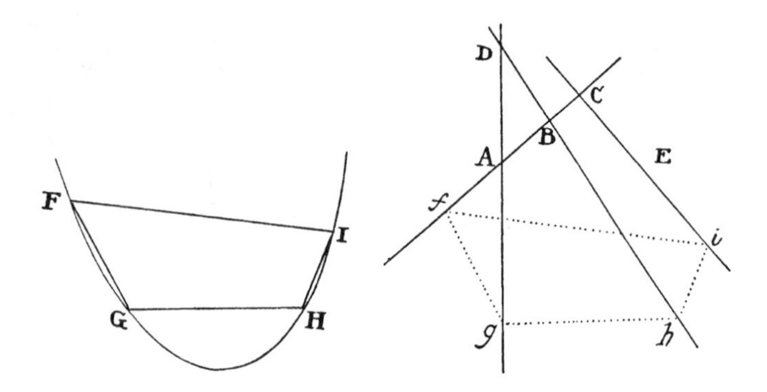

AD and BD, BD and CE given in position, the first part between the first two lines, the second between the second two lines, and the third between the third two lines. After drawing the straight lines FG, GH, HI, and FI, describe (by lem. 27) a quadrilateral *fghi* that is similar to the quadrilateral FGHI and whose corners *f*, *g*, *h*, and *i* touch the straight lines AB, AD, BD, and CE, given in position, each corner touching a separate line in the order stated. Then about this quadrilateral describe a trajectory exactly similar to the curved line FGHI.

Scholium This problem can also be constructed as follows. After joining FG, GH, HI, and FI, produce GF to V, join FH and IG, and make angles CAK and DAL equal to angles FGH and VFH. Let AK and AL meet the straight line BD in K and L, and from these points draw KM and LN, of which KM makes an angle AKM equal to angle GHI and is to AK as HI is to GH, and LN makes an angle ALN equal to angle FHI and is to AL as HI to FH. And draw AK, KM, AL, and LN on those sides of the lines AD, AK, and AL that will make the letters CAKMC, ALKA, and DALND go round in the same order as the letters FGHIF; and draw MN meeting the straight line CE in *i*. Make angle *i*EP equal to the angle IGF, and let PE

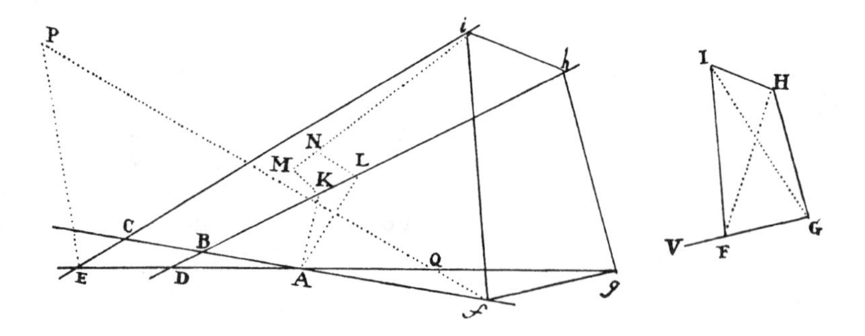

be to E*i* as FG to GI; and through P draw PQ*f*, which with the straight line ADE contains the angle PQE equal to the angle FIG and meets the straight line AB in *f*; and join *fi*. Now draw PE and PQ on those sides of the lines CE and PE that will make the circular order of the letters PE*i*P and PEQP the same as that of the letters FGHIF; and then, if on line *fi* a quadrilateral *fghi* similar to the quadrilateral FGHI is constructed (with the same order of the letters), and a trajectory given in species is circumscribed about the quadrilateral, the problem will be solved.

So much for the finding of orbits. It remains to determine the motions of bodies in the orbits that have been found.

SECTION 6

To find motions in given orbits

Proposition 30
Problem 22

If a body moves in a given parabolic trajectory, to find its position at an assigned time.

Let S be the focus and A the principal vertex of the parabola, and let $4AS \times M$ be equal to the parabolic area APS to be cut off, which either was described by the radius SP after the body's departure from the vertex or is to be described by that radius before the body's arrival at the vertex. The quantity of that area to be cut off can be found from the time, which is proportional to it. Bisect AS in G, and erect the perpendicular GH equal to 3M, and a circle described with center H and radius HS will cut the parabola in the required place P. For, when the perpendicular PO has been dropped to the axis and PH has been drawn, then $AG^2 + GH^2$ $(= HP^2 = (AO - AG)^2 + (PO - GH)^2) =$ $AO^2 + PO^2 - 2GA \times AO - 2GH \times PO + AG^2 + GH^2$. Hence $2GH \times PO$ $(= AO^2 + PO^2 - 2GA \times AO) = AO^2 + \frac{3}{4}PO^2$. For AO^2 write $\frac{AO \times PO^2}{4AS}$, and if all the terms are divided by 3PO and multiplied by 2AS, it will result that $\frac{4}{3}GH \times AS \left(= \frac{1}{6}AO \times PO + \frac{1}{2}AS \times PO = \frac{AO + 3AS}{6} \times PO = \frac{4AO - 3SO}{6} \times PO = \text{area } (APO - SPO)\right) = \text{area } APS$. But GH was 3M, and hence $\frac{4}{3}GH \times AS$ is $4AS \times M$. Therefore, the area APS that was cut off is equal to the area $4AS \times M$ that was to be cut off. Q.E.D.

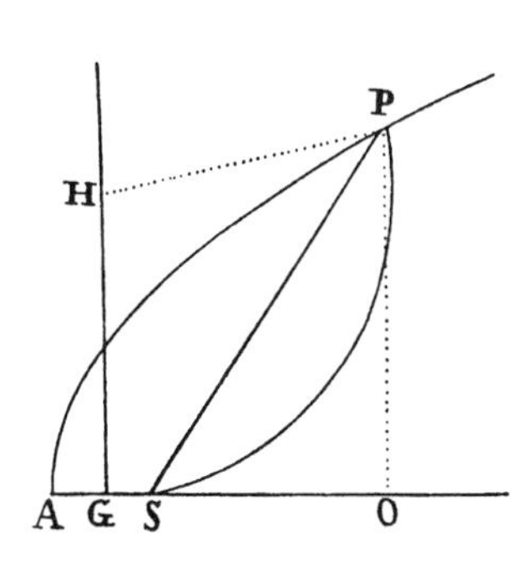

Corollary 1. Hence GH is to AS as the time in which the body described the arc AP is to the time in which it described the arc between the vertex A and a perpendicular erected from the focus S to the axis.

Corollary 2. And if a circle ASP continually passes through the moving body P, the velocity of point H is to the velocity which the body had at the vertex A as 3 to 8, and thus the line GH is also in this ratio to the straight line which the body could describe in the time of its motion from A to P with the velocity which it had at the vertex A.

COROLLARY 3. Hence also, conversely, the time can be found in which the body described any assigned arc AP. Join AP and at its midpoint erect a perpendicular meeting the straight line GH in H.

Lemma 28

No oval figure exists whose area, cut off by straight lines at will, can in general be found by means of equations finite in the number of their terms and dimensions.

Within an oval let any point be given about which, as a pole, a straight line revolves continually with uniform motion, and meanwhile in that straight line let a mobile point go out from the pole and proceed always with the velocity that is as the square of that straight line within the oval. By this motion that point will describe a spiral with an infinite number of gyrations. Now, if the portion of the area of the oval cut off by that straight line can be found by means of a finite equation, there will also be found by the same equation the distance of the point from the pole, a distance that is proportional to this area, and thus all the points of the spiral can be found by means of a finite equation; and therefore the intersection of any straight line, given in position, with the spiral can also be found by means of a finite equation. But every infinitely produced straight line cuts a spiral in an infinite number of points; and the equation by which some intersection of two lines [i.e., curved lines] is found gives all their intersections by as many roots [as there are intersections] and therefore rises to as many dimensions as there are intersections. Since two circles cut each other in two points, one intersection will not be found except by an equation of two dimensions, by which the other intersection may also be found. Since two conics can have four intersections, one of these intersections cannot generally be found except by an equation of four dimensions, by means of which all four of the intersections may be found simultaneously. For if those intersections are sought separately, since they all have the same law and condition, the computation will be the same in each case, and therefore the conclusion will always be the same, which accordingly must comprehend all the intersections together and give them indiscriminately. Hence also the intersections of conics and of curves of the third power, because there can be six such intersections, are found simultaneously by equations of six dimensions; and intersections of two curves of the third power, since there can be nine of them, are found simultaneously by equations of nine dimensions. If this did not happen necessarily, all solid problems might be reduced to plane problems, and higher than solid to solid problems. I am speaking here of curves with a power that cannot be reduced. For if the equation by which the curve is defined can be reduced to a lower power, the curve will not be simple, but will be compounded of two or more curves whose intersections can be found separately by different computations. In the same way, the pairs of intersections of straight lines and conics are always found by equations of two dimensions; the trios of intersections of straight lines and of irreducible curves of the third power, by equations of three dimensions; the quartets of intersections of straight lines and of irreducible curves of the fourth power, by equations of four dimensions; and so on indefinitely. Therefore, the intersections of a straight line and of a spiral, which are infinite in number (since this curve is simple and cannot be reduced to more curves), require equations infinite in the number of their dimensions and roots, by which all the intersections can be given simultaneously. For they all have the same law and computation. For if a perpendicular is dropped from the pole to the intersecting straight line, and the perpendicular, together with the intersecting straight line, revolves about the pole, the intersections of the spiral will pass into one another, and the one that was the first or the nearest to the pole will be the second after one revolution, and after two revolutions will be third, and so on; nor will the equation change in the meantime except insofar as there is a change in the magnitude of the quantities by which the position of the intersecting line is determined. Hence, since the quantities return to their initial magnitudes after each revolution, the equation will return to its original form, and thus one and the same equation will give all the intersections and therefore will have an infinite number of roots by which all of the intersections can be given. Therefore, it is not possible for the intersection of a straight line and a spiral to be found universally by means of a finite equation, and on that account no oval exists whose area, cut off by prescribed straight lines, can universally be found by such an equation.

By the same argument, if the distance between the pole and the point by which the spiral is described is taken proportional to the intercepted part of the perimeter of the oval, it can be proved that the length of the perimeter cannot universally be found by a finite equation. [a]But here I am speaking of ovals that are not touched by conjugate figures extending out to infinity.[a]

aa. This concluding sentence appeared for the first time in ed. 2.

Corollary. Hence the area of an ellipse that is described by a radius drawn from a focus to a moving body cannot be found, from a time that has been given, by means of a finite equation, and therefore cannot be determined by describing geometrically rational curves. I call curves "geometrically rational" when all of their points can be determined by lengths defined by equations, that is, by involved ratios of lengths, and I call the other curves (such as spirals, quadratrices, and cycloids) "geometrically irrational." For lengths that are or are not as integer to integer (as in book 10 of the *Elements*) are arithmetically rational or irrational. Therefore I cut off an area of an ellipse proportional to the time by a geometrically irrational curve as follows.

Proposition 31[a]
Problem 23

If a body moves in a given elliptical trajectory, to find its position at an assigned time.

Let A be the principal vertex of the ellipse APB, S a focus, and O the center, and let P be the position of the body. Produce OA to G so that OG is to OA as OA to OS. Erect the perpendicular GH, and with center O and radius OG describe the circle GEF; then, along the rule GH as a base let the wheel GEF move progressively forward, revolving about its own axis,

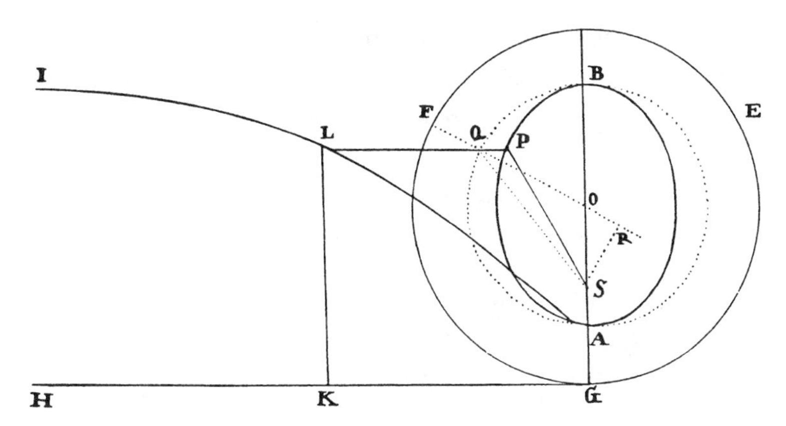

while the point A on the wheel describes the cycloid ALI. When this has been done, take GK so that it will have the same ratio to the perimeter GEFG of the wheel as the time in which the body, in moving forward from A, described the arc AP has to the time of one revolution in the ellipse.

a. In the index prepared by Cotes for ed. 2 and retained in ed. 3, this proposition is keyed under "Problematis" ("of the problem") and characterized as follows: "Solution of Kepler's problem by the cycloid and by approximations."

Erect the perpendicular KL meeting the cycloid in L; and when LP has been drawn parallel to KG, it will meet the ellipse in the required position P of the body.

For with center O and radius OA describe the semicircle AQB, and let LP, produced if necessary, meet the arc AQ in Q, and join SQ and also OQ. Let OQ meet the arc EFG in F, and drop the perpendicular SR to OQ. Area APS is as area AQS, that is, as the difference between sector OQA and triangle OQS, or as the difference of the rectangles ½OQ × AQ and ½OQ × SR, that is, because ½OQ is given, as the difference between the arc AQ and the straight line SR, and hence (because of the equality of the given ratios of SR to the sine of the arc AQ, OS to OA, OA to OG, AQ to GF, and so by separation [or dividendo] AQ − SR to GF − the sine of the arc AQ) as GK, the difference between the arc GF and the sine of the arc AQ. Q.E.D.

Scholium But the description of this curve is difficult; hence it is preferable to use a solution that is approximately true. Find a certain angle B that is to the angle of 57.29578° (which an arc equal to the radius subtends) as the distance SH between the foci is to the diameter AB of the ellipse; and also find a certain length L that is to the radius in the inverse of that ratio. Once these have been found, the problem can thereupon be solved by the following analysis.

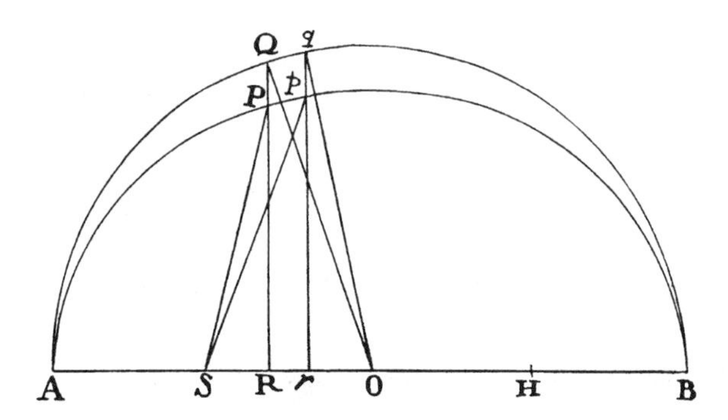

By any construction, or by making any kind of guess, find the body's position P very close to its true position *p*. Then, when the ordinate PR has been dropped to the axis of the ellipse, the ordinate RQ of the circumscribed circle AQB will be given from the proportion of the diameters of the ellipse, where the ordinate RQ is the sine of the angle AOQ (AO being the radius)

and cuts the ellipse in P. It is sufficient to find this angle AOQ approximately by a rough numerical calculation. Also find the angle proportional to the time, that is, the angle that is to four right angles as the time in which the body described the arc A*p* is to the time of one revolution in the ellipse. Let that angle be N. Then take an angle D that will be to angle B as the sine of angle AOQ is to the radius, and also take an angle E that will be to angle N − AOQ + D as the length L is to this same length L minus the cosine of angle AOQ when that angle is less than a right angle, but plus that cosine when it is greater. Next take an angle F that will be to angle B as the sine of angle AOQ + E is to the radius, and take an angle G that will be to angle N − AOQ − E + F as the length L is to this same length minus the cosine of angle AOQ + E when that angle is less than a right angle, and plus that cosine when it is greater. Thirdly, take an angle H that will be to angle B as the sine of angle AOQ + E + G is to the radius, and take an angle I that will be to angle N − AOQ − E − G + H as the length L is to this same length L minus the cosine of angle AOQ + E + G when that angle is less than a right angle, but plus that cosine when it is greater. And so on indefinitely. Finally take angle AO*q* equal to angle AOQ + E + G + I + · · · . And from its cosine O*r* and ordinate *pr*, which is to its sine *qr* as the minor axis of the ellipse to the major axis, the body's corrected place *p* will be found. If the angle N − AOQ + D is negative, the + sign of E must everywhere be changed to −, and the − sign to +. The same is to be understood of the signs of G and I when the angles N − AOQ − E + F and N − AOQ − E − G + H come out negative. But the infinite series AOQ + E + G + I + · · · converges so very rapidly that it is scarcely ever necessary to proceed further than the second term E. And the computation is based on this theorem: that the area APS is as the difference between the arc AQ and the straight line dropped perpendicularly from the focus S to the radius OQ.

In the case of a hyperbola the problem is solved by a similar computation. Let O be its center, A a vertex, S a focus, and OK an asymptote. Find the quantity of the area to be cut off, which is proportional to the time. Let this quantity be A, and guess the position of the straight line SP that cuts off an approximately true area APS. Join OP, and from A and P to the asymptote OK draw AI and PK parallel to the second asymptote; then a table of logarithms will give the area AIKP and the equal area OPA,

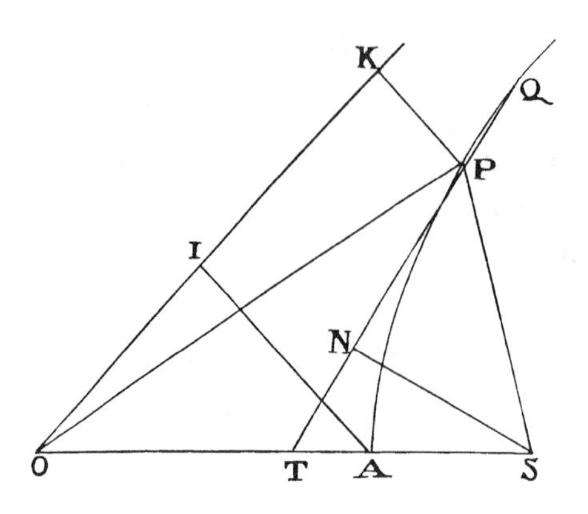

which, on being subtracted from the triangle OPS, will leave the cut-off area APS. Divide 2APS − 2A or 2A − 2APS (twice the difference of the area A to be cut off and the cut-off area APS) by the line SN, which is perpendicular to the tangent TP from the focus S, so as to obtain the length of the chord PQ. Now, draw the chord PQ between A and P if the cut-off area APS is greater than the area A to be cut off, but otherwise draw PQ on the opposite side of point P, and then the point Q will be a more accurate position of the body. And by continually repeating the computation, a more and more accurate position will be obtained.

And by these computations a general analytical solution of the problem is achieved. But the particular computation that follows is more suitable for astronomical purposes. Let AO, OB, and OD be the semiaxes of the ellipse, and L its latus rectum, and D the difference between the semiaxis minor OD and half of the latus rectum ½L; find an angle Y, whose sine is to the radius as the rectangle of that difference D and the half-sum of the axes AO+OD is to the square of the major axis AB; and find also an angle Z, whose sine is to the radius as twice the rectangle of the distance SH between the foci and the difference D is to three times the square of the semiaxis major AO. Once these angles have been found, the position of the body will thereupon be determined as follows: Take an angle T proportional to the time in which arc BP was described, or equal to the mean motion (as it is called); and an angle V (the first equation of the mean motion) that shall be to angle Y (the greatest first equation) as the sine of twice angle T is to the radius; and an angle X (the second equation) that shall be to angle Z (the greatest second equation) as the cube of the sine of angle T is to the cube of the radius. Then take the angle BHP (the equated mean motion) equal either to the sum T + X + V of angles T, X, and V if angle T is less than a right angle, or to the difference T + X − V if angle

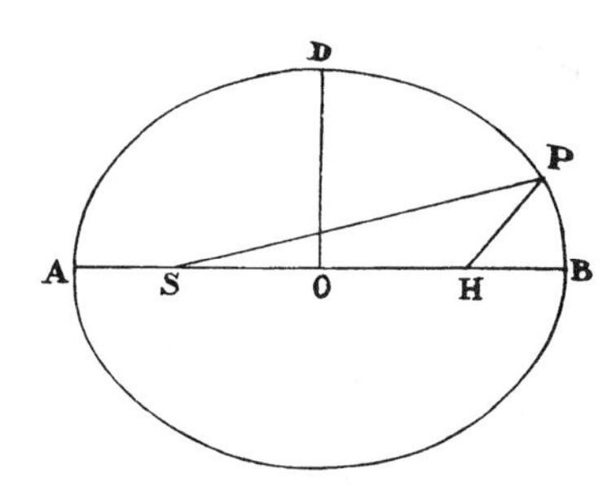

T is greater than a right angle and less than two right angles; and if HP meets the ellipse in P, SP (when drawn) will cut off the area BSP very nearly proportional to the time.

This technique seems expeditious enough because it is sufficient to find the first two or three figures of the extremely small angles V and X (reckoned in seconds, if it is agreeable). This technique is also accurate enough for the theory of the planets. For even in the orbit of Mars itself, whose greatest equation of the center is ten degrees, the error will hardly exceed one second. But when the angle BHP of equated mean motion has been found, the angle BSP of true motion and the distance SP are readily found by the very well known method.

So much for the motion of bodies in curved lines. It can happen, however, that a moving body descends straight down or rises straight up; and I now go on to expound what relates to motions of this sort.

SECTION 7

The rectilinear ascent and descent of bodies

Proposition 32[a] **Problem 24** *Given a centripetal force inversely proportional to the square of the distance of places from its center, to determine the spaces which a body in falling straight down describes in given times.*

CASE 1. If the body does not fall perpendicularly, it will (by prop. 13, corol. 1) describe some conic having a focus coinciding with the center of forces. Let the conic be ARPB, and its focus S. And first, if the figure is an ellipse, on its major axis AB describe the semicircle ADB, and let the straight line DPC pass through the falling body and be perpendicular to the axis; and when DS and PS have been drawn, area ASD will be proportional to area ASP and thus also to the time. Keeping the axis AB fixed, continually diminish the width of the ellipse, and area ASD will always remain proportional to the time. Diminish that width indefinitely; and, the orbit APB now coming to coincide with the axis AB, and the focus S with the terminus B of the axis, the body will descend in the straight line AC, and the area ABD will become proportional to the time. Therefore the space AC will be given, which the body in falling perpendicularly from place A describes in a given time, provided that area ABD is taken proportional to that time and the perpendicular DC is dropped from point D to the straight line AB. Q.E.I.

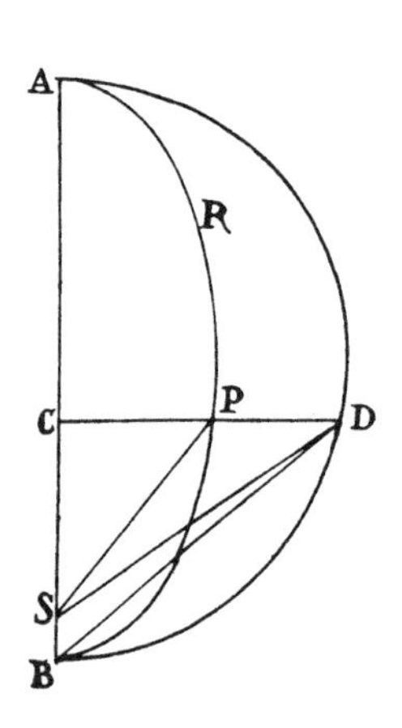

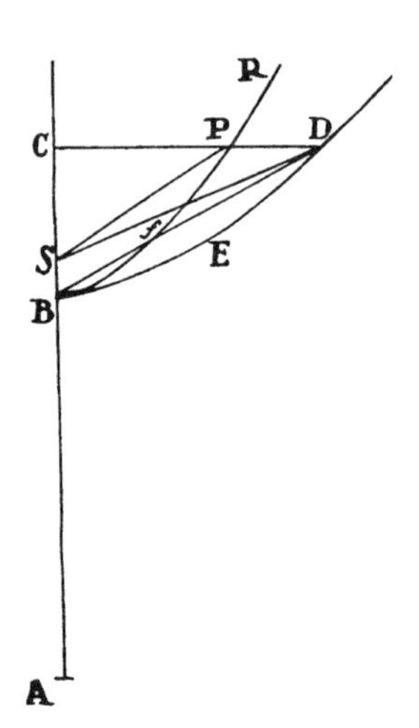

CASE 2. If the figure RPB is a hyperbola, describe the rectangular hyperbola BED on the same principal diameter AB; and since the areas CSP, CBfP, and SPfB are respectively to the areas CSD, CBED, and SDEB in the given ratio of the distances CP and CD, and the area SPfB is

a. For a gloss on this proposition see the Guide, §10.11.

proportional to the time in which body P will move through the arc P*f*B, the area SDEB will also be proportional to that same time. Diminish the latus rectum of the hyperbola RPB indefinitely, keeping the principal diameter fixed, and the arc PB will coincide with the straight line CB, and the focus S with the vertex B, and the straight line SD with the straight line BD. Accordingly, the area BDEB will be proportional to the time in which body C, falling straight down, describes the line CB. Q.E.I.

CASE 3. And by a similar argument, let the figure RPB be a parabola and let another parabola BED with the same principal vertex B be described and always remain given, while the latus rectum of the first parabola (in whose perimeter the body P moves) is diminished and reduced to nothing, so that this parabola comes to coincide with the line CB; then the parabolic segment BDEB will become proportional to the time in which the body P or C will descend to the center S or B. Q.E.I.

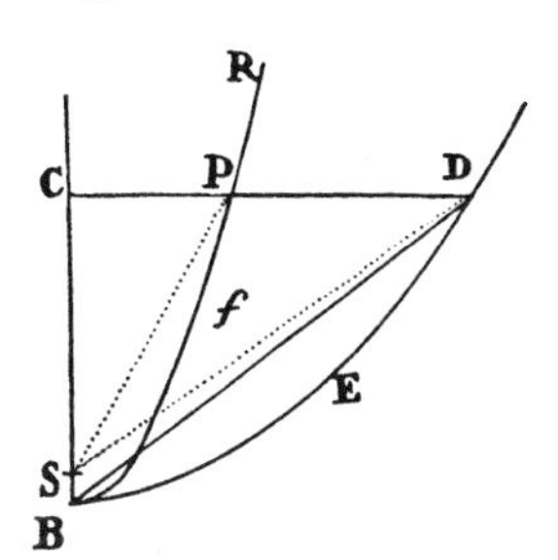

Proposition 33
Theorem 9

Supposing what has already been found, I say that the velocity of a falling body at any place C *is to the velocity of a body describing a circle with center* B *and radius* BC *as the square root of the ratio of* AC *(the distance of the body from the further vertex* A *of the circle or rectangular hyperbola) to* ½AB *(the principal semidiameter of the figure).*

Bisect AB, the common diameter of both figures RPB and DEB, in O; and draw the straight line PT touching the figure RPB in P and also cutting the common diameter AB (produced if necessary) in T, and let SY be perpendicular to this straight line and BQ be perpendicular to this diameter, and take the latus rectum of the figure RPB to be L. It is established by prop. 16, corol. 9, that at any place P the velocity of a body moving about the center S in the [curved] line RPB is to the velocity of a body describing a circle about the same center with the radius SP as the square root of the ratio of the rectangle $\frac{1}{2}L \times SP$ to SY^2. But from the *Conics*, $AC \times CB$ is to CP^2 as 2AO to L, and thus $\frac{2CP^2 \times AO}{AC \times CB}$ is equal to L. Therefore, the velocities are to each other as the square root of the ratio of $\frac{CP^2 \times AO \times SP}{AC \times CB}$ to SY^2. Further, from the *Conics*, CO is to BO as BO to TO, and by composition [or componendo] or by separation [or dividendo], as CB to BT. Hence, either

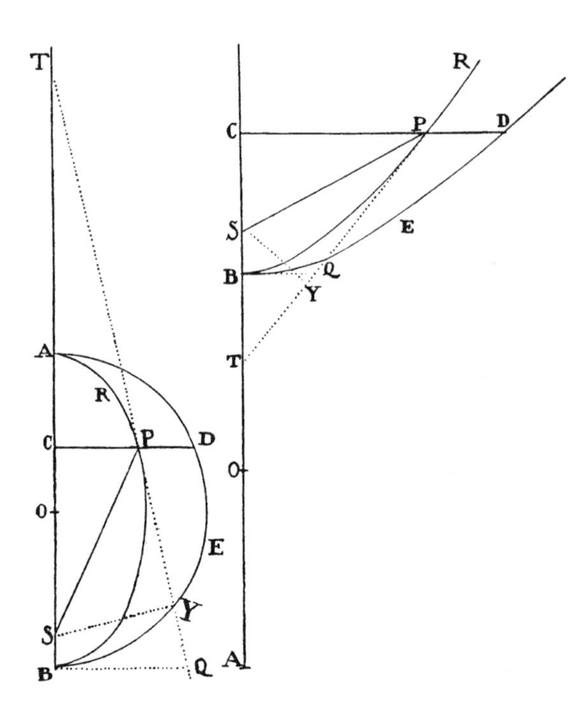

by separation or by composition, $BO \mp CO$ becomes to BO as CT to BT, that is, AC to AO as CP to BQ; and hence $\frac{CP^2 \times AO \times SP}{AC \times CB}$ is equal to $\frac{BQ^2 \times AC \times SP}{AO \times BC}$. Now let the width CP of the figure RPB be diminished indefinitely, in such a way that point P comes to coincide with point C and point S with point B and the line SP with the line BC and the line SY with the line BQ; then the velocity of the body now descending straight down in the line CB will become to the velocity of a body describing a circle with center B and radius BC as the square root of the ratio of $\frac{BQ^2 \times AC \times SP}{AO \times BC}$ to SY^2, that is (neglecting the ratios of equality SP to BC and BQ^2 to SY^2), as the square root of the ratio of AC to AO or ½AB. Q.E.D.

COROLLARY 1. When the points B and S come to coincide, TC becomes to TS as AC to AO.

COROLLARY 2. A body revolving in any circle at a given distance from the center will, when its motion is converted to an upward motion, ascend to twice that distance from the center.

Proposition 34
Theorem 10

If the figure BED *is a parabola, I say that the velocity of a falling body at any place* C *is equal to the velocity with which a body can uniformly describe a circle with center* B *and a radius equal to one-half of* BC.

For at any place P the velocity of a body describing the parabola RPB about the center S is (by prop. 16, corol. 7) equal to the velocity of a body uniformly describing a circle about the same center S with a radius equal to half of the interval SP. Let the width CP of the parabola be diminished indefinitely, so that the parabolic arc P*f*B will come to coincide with the straight line CB, the center S with the vertex B, and the interval SP with the interval BC, and the proposition will be established. Q.E.D.

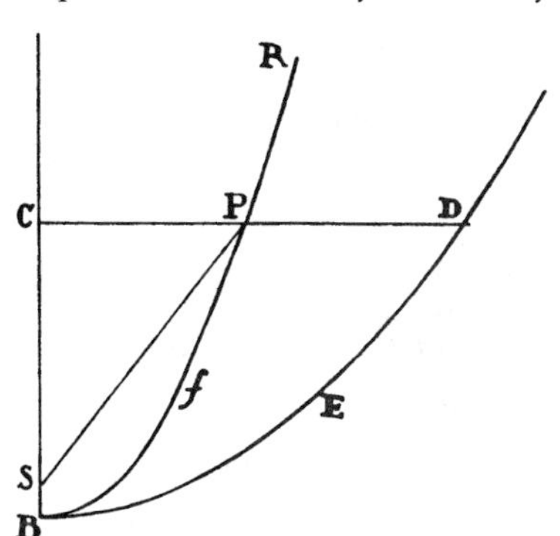

Proposition 35
Theorem 11

Making the same suppositions, I say that the area of the figure DES *described by the indefinite radius* SD *is equal to the area that a body revolving uniformly in orbit about the center* S *can describe in the same time by a radius equal to half of the latus rectum of the figure* DES.

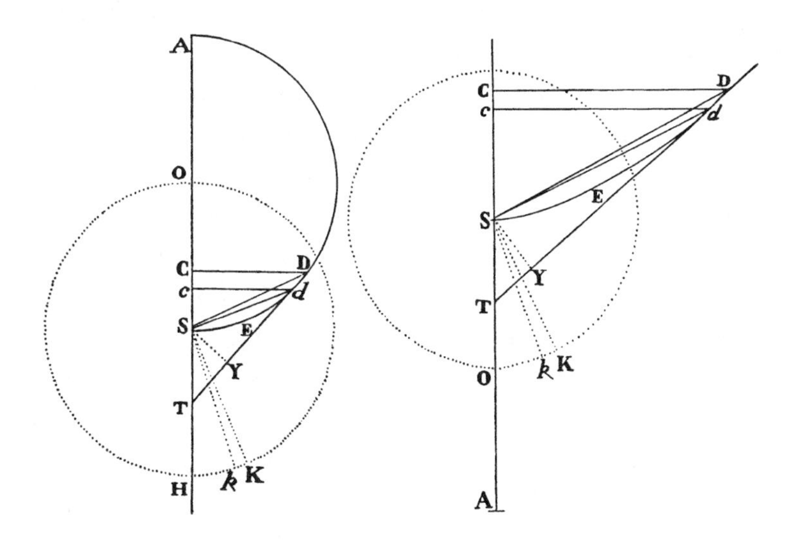

For suppose that body C falling in a minimally small particle of time describes the line-element C*c* while another body K, revolving uniformly in the circular orbit OK*k* about the center S, describes the arc K*k*. Erect the perpendiculars CD and *cd* meeting the figure DES in D and *d*. Join SD, S*d*, SK, and S*k*, and draw D*d* meeting the axis AS in T, and drop the perpendicular SY to D*d*.

CASE 1. Now, if the figure DES is a circle or a rectangular hyperbola, bisect its transverse diameter AS in O, and SO will be half of the latus rectum. And since TC is to TD as C*c* to D*d*, and TD to TS as CD to SY, from the equality of the ratios [or ex aequo] TC will be to TS as CD × C*c* to SY × D*d*. But (by prop. 33, corol. 1) TC is to TS as AC to AO, if, say, in the coming together of points D and *d* the ultimate ratios of the lines are taken. Therefore, AC is to AO or SK as CD × C*c* is to SY × D*d*. Further, the velocity of a descending body at C is to the velocity of a body describing a circle about the center S with radius SC as the square root of the ratio of AC to AO or SK (by prop. 33). And this velocity is to the velocity of a body describing the circle OK*k* as the square root of the ratio of SK to SC (by prop. 4, corol. 6), and from the equality of the ratios [or ex aequo] the first velocity is to the ultimate velocity, that is, the line-element C*c* is to the arc K*k*, as the square root of the ratio of AC to SC, that is, in the ratio of AC to CD. Therefore, CD × C*c* is equal to AC × K*k*, and thus AC is to SK as AC × K*k* to SY × D*d*, and hence SK × K*k* is equal to SY × D*d*, and ½K × K*k* is equal to ½SY × D*d*, that is, the area KS*k* is equal to the area SD*d*. Therefore, in each particle of time, particles KS*k* and SD*d* of the two areas are generated such that, if their magnitude is diminished and their number increased indefinitely, they obtain the ratio of equality; and therefore (by lem. 4, corol.), the total areas generated in the same times are always equal. Q.E.D.

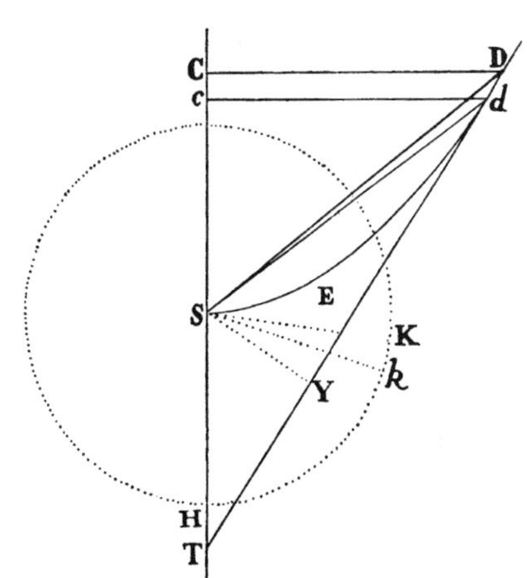

CASE 2. But if the figure DES is a parabola, then it will be found that, as above, CD × C*c* is to SY × D*d* as TC to TS, that is, as 2 to 1, and thus ¼CD × C*c* will be equal to ½SY × D*d*. But the velocity of the falling body at C is equal to the velocity with which a circle can be described uniformly with the radius ½SC (by prop. 34). And this velocity is to the velocity with which a circle

can be described with the radius SK, that is, the line-element C*c* is to the arc K*k* (by prop. 4, corol. 6), as the square root of the ratio of SK to ½SC, that is, in the ratio of SK to ½CD. And therefore ½SK × K*k* is equal to ¼CD × C*c* and thus equal to ½SY × D*d*; that is, the area KS*k* is equal to the area SD*d*, as above. Q.E.D.

Proposition 36
Problem 25

To determine the times of descent of a body falling from a given place A.

Describe a semicircle ADS with diameter AS (the distance of the body from the center at the beginning of the descent), and about the center S describe a semicircle OKH equal to ADS. From any place C of the body erect the ordinate CD. Join SD, and construct the sector OSK equal to the area ASD. It is evident by prop. 35 that the body in falling will describe the space AC in the same time in which another body, revolving uniformly in orbit about the center S, can describe the arc OK. Q.E.F.

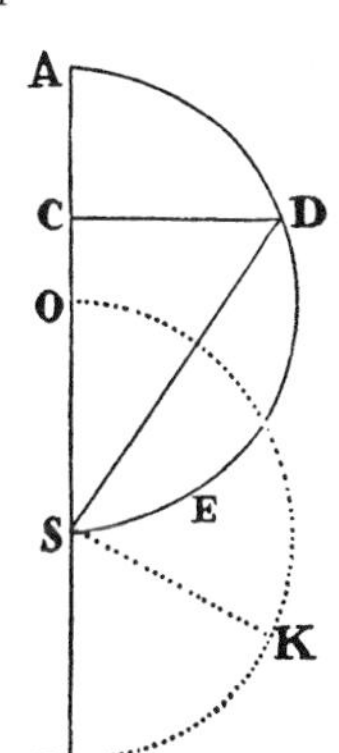

Proposition 37
Problem 26

To define the times of the ascent or descent of a body projected upward or downward from a given place.

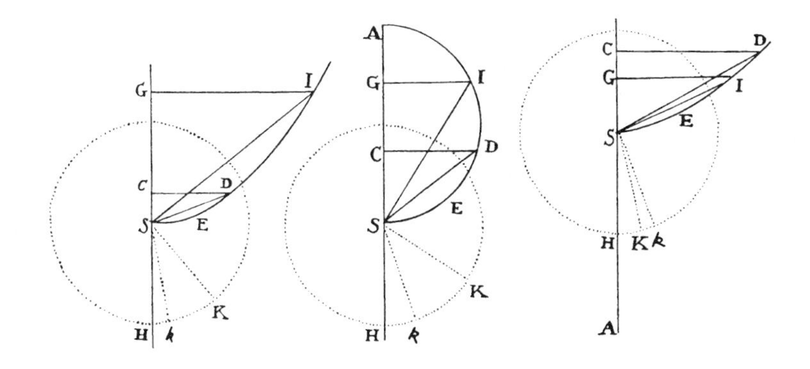

Let the body depart from the given place G along the line GS with any velocity whatever. Take GA to ½AS as the square of the ratio of this velocity to the uniform velocity in a circle with which the body could revolve about the center S at the given interval (or distance) SG. If that ratio is as 2 to 1, point A is infinitely distant, in which case a parabola is to be described with vertex S, axis SG, and any latus rectum, as is evident by prop. 34. But if that ratio is smaller or greater than the ratio of 2 to 1, then in the former case a circle, and in the latter case a rectangular hyperbola, must be described on the diameter SA, as is evident by prop. 33. Then, with center S and a radius equaling half of the latus rectum, describe the circle H*k*K, and to the place G of the descending or ascending body and to any other place C, erect the perpendiculars GI and CD meeting the conic or the circle in I and D. Then joining SI and SD, let the sectors HSK and HS*k* be made equal to the segments SEIS and SEDS, and by prop. 35 the body G will describe the space GC in the same time as the body K can describe the arc K*k*. Q.E.F.

Proposition 38
Theorem 12

Supposing that the centripetal force is proportional to the height or distance of places from the center, I say that the times of falling bodies, their velocities, and the spaces described are proportional respectively to the arcs, the right sines, and the versed sines.

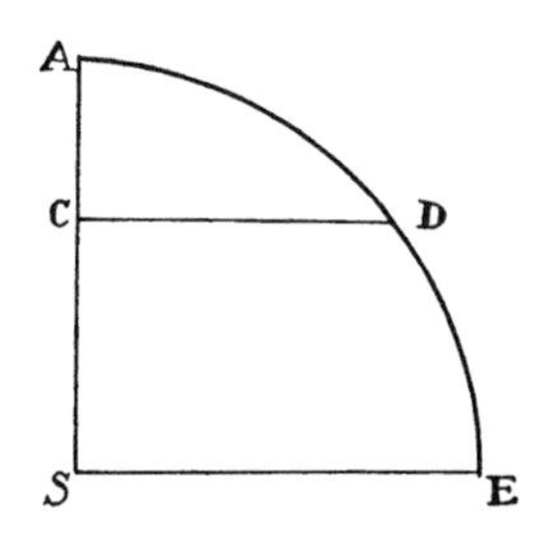

Let a body fall from any place A along the straight line AS; and with center of forces S and radius AS describe the quadrant AE of a circle, and let CD be the right sine of any arc AD; then the body A, in the time AD, will in falling describe the space AC and at place C will acquire the velocity CD.

This is demonstrated from prop. 10 in the same way that prop. 32 was demonstrated from prop. 11.

Corollary 1. Hence the time in which one body, falling from place A, arrives at the center S is equal to the time in which another body, revolving, describes the quadrantal arc ADE.

Corollary 2. Accordingly, all the times are equal in which bodies fall from any places whatever as far as to the center. For all the periodic times of revolving bodies are (by prop. 4, corol. 3) equal.

Proposition 39
Problem 27

Suppose a centripetal force of any kind, and grant the quadratures of curvilinear figures; it is required to find, for a body ascending straight up or descending straight down, the velocity in any of its positions and the time in which the body will reach any place; and conversely.

Let a body E fall from any place A whatever in the straight line ADEC, and let there be always erected from the body's place E the perpendicular

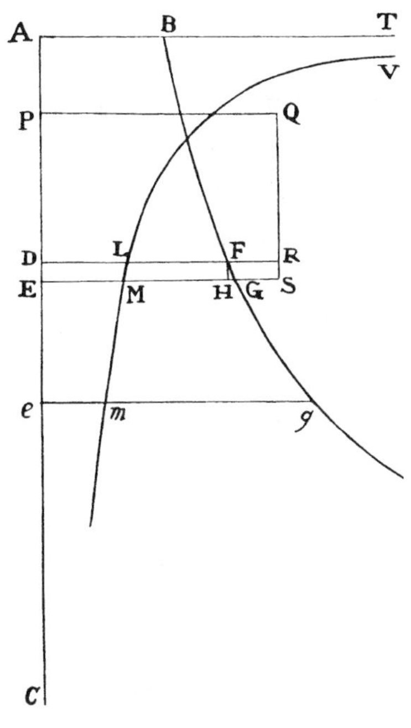

EG, proportional to the centripetal force in that place tending toward the center C; and let BFG be the curved line which the point G continually traces out. At the very beginning of the motion let EG coincide with the perpendicular AB; then the velocity of the body in any place E will be as the straight line whose square is equal to the curvilinear area ABGE. Q.E.I.

In EG take EM inversely proportional to the straight line whose square is equal to the area ABGE, and let VLM be a curved line which the point M continually traces out and whose asymptote is the straight line AB produced; then the time in which the body in falling describes the line AE will be as the curvilinear area ABTVME. Q.E.I.

For in the straight line AE take a minimally small line DE of a given length, and let DLF be the location of the line EMG when the body was at D; then, if the centripetal force is such that the straight line whose square is equal to the area ABGE is as the velocity of the descending body, the area itself will be as the square of the velocity, that is, if V and V + I are written for the velocities at D and E, the area ABFD will be as V^2, and the area ABGE as $V^2 + 2VI + I^2$, and by separation [or dividendo] the area DFGE will be as $2VI + I^2$, and thus $\frac{DFGE}{DE}$ will be as $\frac{2VI + I^2}{DE}$, that is, if the first ratios of nascent quantities are taken, the length DF will be as the quantity $\frac{2VI}{DE}$, and thus also as half of that quantity, or $\frac{I \times V}{DE}$. But the time in which the body in falling describes the line-element DE is as that line-element directly and the velocity V inversely, and the force is as the increment I of the velocity directly and the time inversely, and thus—if the first ratios of nascent quantities are taken—as $\frac{I \times V}{DE}$, that is, as the length DF. Therefore a force proportional to DF or EG makes the body descend with the velocity that is as the straight line whose square is equal to the area ABGE. Q.E.D.

Moreover, since the time in which any line-element DE of a given length is described is as the velocity inversely, and hence inversely as the straight line whose square is equal to the area ABFD, and since DL (and hence the nascent area DLME) is as the same straight line inversely, the time will be as the area DLME, and the sum of all the times will be as the sum of all the areas, that is (by lem. 4, corol.), the total time in which the line AE is described will be as the total area ATVME. Q.E.D.

COROLLARY 1. Let P be the place from which a body must fall so that, under the action of some known uniform centripetal force (such as gravity is commonly supposed to be), it will acquire at place D a velocity equal to the velocity that another body, falling under the action of any force whatever, acquired at the same place D. In the perpendicular DF take DR such that it is to DF as that uniform force is to the other force at the place D. Complete the rectangle PDRQ and cut off the area ABFD equal to it. Then A will be the place from which the other body fell.

For, when the rectangle DRSE has been completed, the area ABFD is to the area DFGE as V^2 to 2VI and hence as ½V to I, that is, as half of the total velocity to the increment of the velocity of the body falling under the action of the nonuniform force; and similarly, the area PQRD is to the area DRSE as half of the total velocity is to the increment of the velocity of the body falling under the action of the uniform force, and those increments (because the nascent times are equal) are as the generative forces, that is, as the ordinates DF and DR, and thus as the nascent areas DFGE and DRSE. Therefore, the total areas ABFD and PQRD will then from the equality of the ratios [or ex aequo] be to each other as halves of the total velocities and therefore are equal because the velocities are equal.

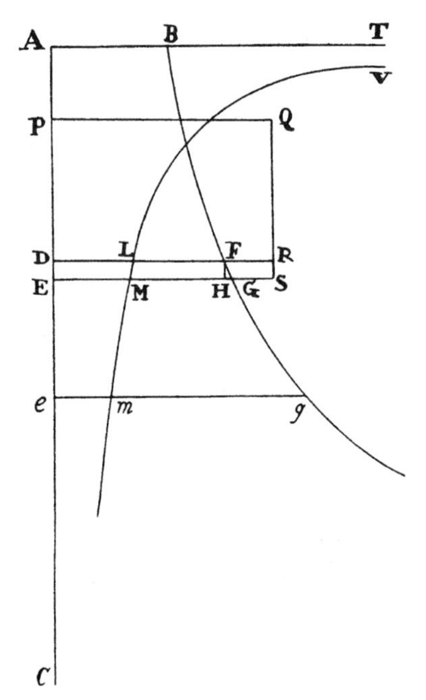

COROLLARY 2. Hence if any body is projected with a given velocity either upward or downward from any place D and the law of centripetal force is given, the velocity of the body at any other

place *e* will be found by erecting the ordinate *eg* and taking that velocity at place *e* to the velocity at place D as the straight line whose square is equal to the rectangle PQRD, either increased by the curvilinear area DF*ge* (if place *e* is lower than place D) or diminished by DF*ge* (if place *e* is higher), is to the straight line whose square is equal to the rectangle PQRD alone.

COROLLARY 3. The time, also, will be determined by erecting the ordinate *em* inversely proportional to the square root of PQRD ± DF*ge*, and by taking the time in which the body described the line D*e* to the time in which the other body fell under the action of a uniform force from P and (by so falling) reached D as the curvilinear area DL*me* to the rectangle 2PD × DL. For the time in which the body descending under the action of a uniform force described the line PD is to the time in which the same body described the line PE as the square root of the ratio of PD to PE, that is (the line-element DE being just now nascent), in the ratio of PD to PD + ½DE or 2PD to 2PD + DE and by separation [or dividendo] to the time in which the same body described the line-element DE as 2PD to DE, and thus as the rectangle 2PD × DL to the area DLME; and the time in which each of the two bodies described the line-element DE is to the time in which the second body with nonuniform motion described the line D*e* as the area DLME to the area DL*me*, and from the equality of the ratios [or ex aequo] the first time is to the ultimate time as the rectangle 2PD × DL to the area DL*me*.

SECTION 8

To find the orbits in which bodies revolve when acted upon by any centripetal forces

Proposition 40
Theorem 13

If a body, under the action of any centripetal force, moves in any way whatever, and another body ascends straight up or descends straight down, and if their velocities are equal in some one instance in which their distances from the center are equal, their velocities will be equal at all equal distances from the center.

Let some body descend from A through D and E to the center C, and let another body move from V in the curved line VIK*k*. With center C and any radii describe the concentric circles DI and EK meeting the straight line AC in D and E and the curve VIK in I and K. Join IC meeting KE in N, and to IK drop the perpendicular NT, and let the interval DE or IN between the circumferences of the circles be minimally small, and let the bodies have equal velocities at D and I. Since the distances CD and CI are equal, the centripetal forces at D and I will be equal. Represent these forces by the equal line-elements DE and IN; then, if one of these forces IN is (by corol. 2 of the laws) resolved into two, NT and IT, the force NT, acting along the line NT perpendicular to the path ITK of the body, will in no way change the velocity of the body in that path but will only draw the body back from a rectilinear path and make it turn aside continually from the tangent of the orbit and move forward in the curvilinear path ITK*k*. That whole force will be spent in producing this effect, while the whole of the other force IT, acting along the body's path, will accelerate the body and in a given minimally small time will generate an acceleration proportional to itself. Accordingly, the accelerations of the bodies at D and I that are made in equal times (if the first ratios of the nascent lines DE, IN, IK, IT, and NT are taken) are as the lines DE and IT, but in unequal times they are as those lines and the times jointly. Now, the times in which DE and IK are described are as the described paths DE and IK (because the velocities are equal), and hence the accelerations in the path of the bodies along the lines DE and IK are jointly as DE and IT, DE and IK, that is, as DE^2 and the rectangle $IT \times IK$.

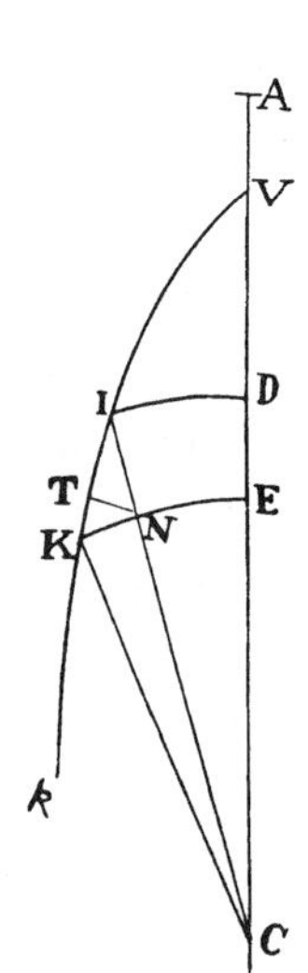

But the rectangle $IT \times IK$ is equal to IN^2, that is, equal to DE^2, and therefore the accelerations generated in the passing of the bodies from D and I to E and K are equal. Therefore the velocities of the bodies at E and K are equal, and by the same argument they will always be found equal at subsequent equal distances. Q.E.D.

But also by the same argument bodies that have equal velocities and are equally distant from the center will be equally retarded in ascending to equal distances. Q.E.D.

COROLLARY 1. Hence if a body either oscillates while hanging by a thread or is compelled by any very smooth and perfectly slippery impediment to move in a curved line, and another body ascends straight up or descends straight down, and their velocities are equal at any identical height, their velocities at any other equal heights will be equal. For the thread of the pendent body or the impediment of an absolutely slippery vessel produces the same effect as the transverse force NT. The body is neither retarded nor accelerated by these, but only compelled to depart from a rectilinear course.

COROLLARY 2. Now let the quantity P be the greatest distance from the center to which a body, either oscillating or revolving in any trajectory whatever, can ascend when projected upward from any point of the trajectory with the velocity that it has at that point. Further, let the quantity A be the distance of the body from the center at any other point of the orbit. And let the centripetal force be always as any power A^{n-1} of A, the index $n-1$ being any number n diminished by unity. Then the velocity of the body at every height A [i.e., distance A] will be as $\sqrt{(P^n - A^n)}$ and therefore is given. For the velocity of a body ascending straight up and descending straight down is (by prop. 39) in this very ratio.

Proposition 41[a]
Problem 28

Supposing a centripetal force of any kind and granting the quadratures of curvilinear figures, it is required to find the trajectories in which bodies will move and also the times of their motions in the trajectories so found.

Let any force tend toward a center C; and let it be required to find the trajectory VIK*k*. Let the circle VR be given, described about the center C with any radius CV; and about the same center let there be described any other circles ID and KE cutting the trajectory in I and K and cutting the

a. For a gloss on this proposition see the Guide, §10.12.

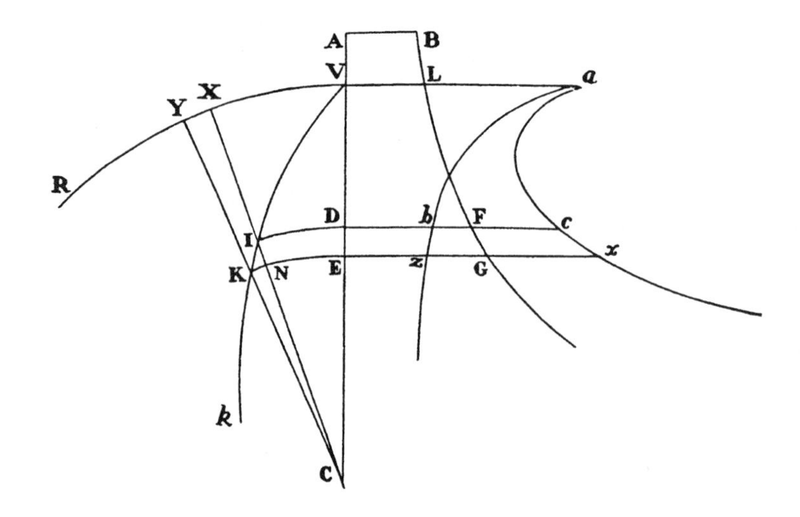

straight line CV in D and E. Then draw the straight line CNIX cutting the circles KE and VR in N and X, and also draw the straight line CKY meeting the circle VR in Y. Let the points I and K be very close indeed to each other, and let the body proceed from V through I and K to *k*; and let point A be the place from which another body must fall so as to acquire at place D a velocity equal to the velocity of the first body at I. And with everything remaining as it was in prop. 39, the line-element IK, described in a given minimally small time, will be as the velocity and hence as the straight line whose square equals the area ABFD, and the triangle ICK proportional to the time will be given; and therefore KN will be inversely as the height IC, that is, if some quantity Q is given and the height IC is called A, as $\frac{Q}{A}$. Let us denote this quantity $\frac{Q}{A}$ by Z, and let us suppose the magnitude of Q to be such that in some one case $\sqrt{ABFD}$ is to Z as IK is to KN, and in every case $\sqrt{ABFD}$ will be to Z as IK to KN and ABFD to Z^2 as IK^2 to KN^2, and by separation [or dividendo] $ABFD - Z^2$ will be to Z^2 as IN^2 to KN^2, and therefore $\sqrt{(ABFD - Z^2)}$ will be to Z, or $\frac{Q}{A}$, as IN to KN, and therefore $A \times KN$ will be equal to $\frac{Q \times IN}{\sqrt{(ABFD - Z^2)}}$. Hence, since $YX \times XC$ is to $A \times KN$ as CX^2 to A^2, the rectangle $XY \times XC$ will be equal to $\frac{Q \times IN \times CX^2}{A^2\sqrt{(ABFD - Z^2)}}$. Therefore, in the perpendicular DF take D*b* and D*c* always equal respectively to $\frac{Q}{2\sqrt{(ABFD - Z^2)}}$ and $\frac{Q \times CX^2}{2A^2\sqrt{(ABFD - Z^2)}}$, and describe the curved lines *ab* and *ac* which the points *b* and *c* continually trace out, and from point V erect V*a* perpendicular to the line AC so as to cut off the curvilinear areas VD*ba* and VD*ca*, and also erect the ordinates E*z* and E*x*. Then, since the rectangle D*b* × IN or D*bz*E is equal to half of the rectangle A × KN or is equal to the triangle ICK, and the rectangle D*c* × IN or D*cx*E is equal to half of the rectangle YX × XC or is equal to the triangle XCY—that is, since the nascent particles D*bz*E and ICK of the areas VD*ba* and VIC are always equal, and the nascent particles D*cx*E and XCY of the areas VD*ca* and VCX are always equal—the generated area VD*ba* will be equal to the generated area VIC and hence will be proportional to the time, and the generated area VD*ca* will be equal to the generated sector VCX. Therefore, given any time that has elapsed since the body set out from place V, the area VD*ba* proportional to it will be given and hence the body's height CD or CI will be given, and the area VD*ca* and, equal to that area, the sector VCX along with its angle VCI. And given the angle VCI and the height CI, the place I will be given, in which the body will be found at the end of that time. Q.E.I.

COROLLARY 1. Hence the greatest and least heights of bodies (that is, the apsides of their trajectories) can be found expeditiously. For the apsides are those points in which the straight line IC drawn through the center falls perpendicularly upon the trajectory VIK, which happens when the straight lines IK and NK are equal, and thus when the area ABFD is equal to Z^2.

COROLLARY 2. The angle KIN, in which the trajectory anywhere cuts the line IC, is also expeditiously found from the given height IC of the body, namely, by taking its sine to the radius as KN to IK, that is, as Z to the square root of the area ABFD.

COROLLARY 3. If with center C and principal vertex V any conic VRS is described, and from any point R of it the tangent RT is drawn so as to meet the axis CV, indefinitely produced, at point T; and, joining CR, there is drawn the straight line CP, which is equal to the abscissa CT and makes an angle VCP proportional to the sector VCR; then, if a centripetal

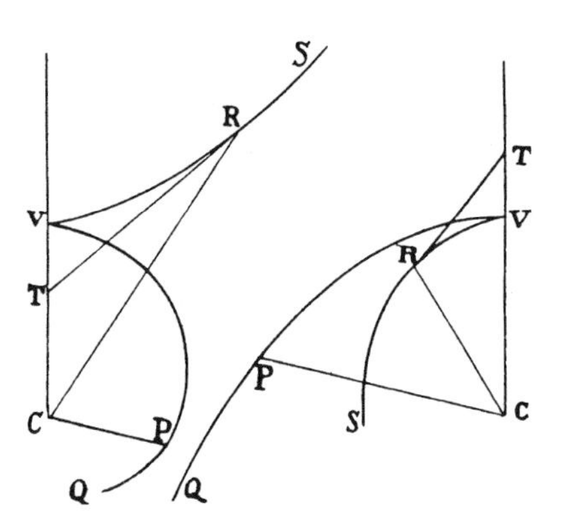

force inversely proportional to the cube of the distance of places from the center tends toward that center C, and the body leaves the place V with the proper velocity along a line perpendicular to the straight line CV, the body will move forward in the trajectory VPQ which point P continually traces out; and therefore, if the conic VRS is a hyperbola, the body will descend to the center. But if the conic is an ellipse, the body will ascend continually and will go off to infinity.

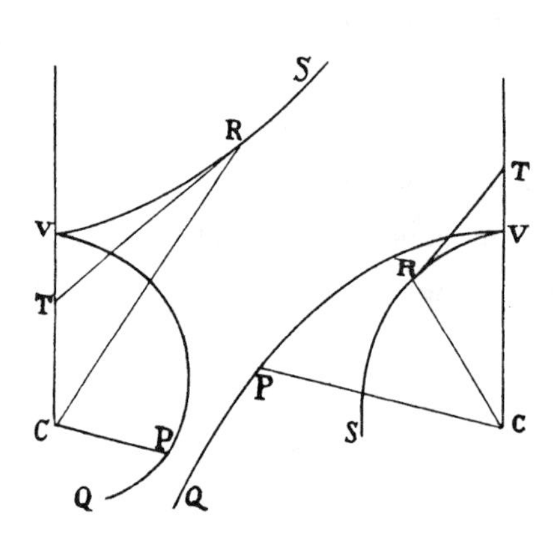

And conversely, if the body leaves the place V with any velocity and, depending on whether the body has begun either to descend obliquely to the center or to ascend obliquely from it, the figure VRS is either a hyperbola or an ellipse, the trajectory can be found by increasing or diminishing the angle VCP in some given ratio. But also, if the centripetal force is changed into a centrifugal force, the body will ascend obliquely in the trajectory VPQ, which is found by taking the angle VCP proportional to the elliptic sector VRC and by taking the length CP equal to the length CT, as above. All this follows from the foregoing (prop. 41), by means of the quadrature of a certain curve, the finding of which, as being easy enough, I omit for the sake of brevity.

Proposition 42
Problem 29

Let the law of centripetal force be given; it is required to find the motion of a body setting out from a given place with a given velocity along a given straight line.

With everything remaining as it was in the three preceding propositions, let the body go forth from the place I along the line-element IK, with the velocity which another body, falling from the place P under the action of some uniform centripetal force, could acquire at D; and let this uniform force be to the force with which the first body is urged at I as DR to DF. Let the body go on toward *k*; and with center C and radius C*k* describe the circle *ke* meeting the straight line PD at *e*, and erect the ordinates *eg*, *ev*, and *ew* of the curves BF*g*, *abv*, and *acw*. From the given rectangle PDRQ and the given law of the centripetal force acting on the first body, the curved line BF*g* is given by the construction of prop. 39 and its corol. 1. Then, from

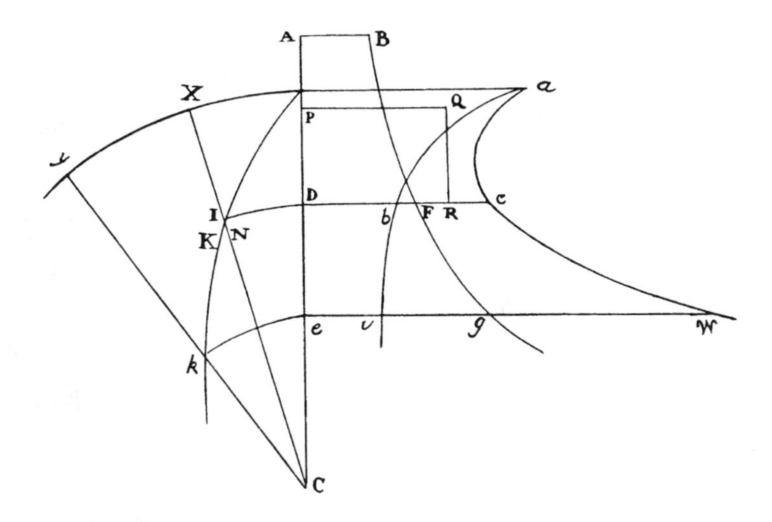

the given angle CIK, the proportion of the nascent lines IK and KN is given, and hence, by the construction of prop. 41, the quantity Q is given, along with the curved lines *abv* and *acw*; and therefore, when any time D*bve* is completed, the body's height C*e* or C*k* is given and the area D*cwe* and the sector XC*y* equal to it and the angle IC*k* and the place *k* in which the body will then be. Q.E.I.

In these propositions we suppose that the centripetal force in receding from the center varies according to any law which can be imagined, but that at equal distances from the center it is everywhere the same. And so far we have considered the motion of bodies in nonmoving orbits. It remains for us to add a few things about the motion of bodies in orbits that revolve about a center of forces.

SECTION 9

The motion of bodies in mobile orbits, and the motion of the apsides

Proposition 43
Problem 30

[a]*It is required to find the force that makes a body capable of moving in any trajectory that is revolving about the center of forces in the same way as another body in that same trajectory at rest.*[a]

Let a body P revolve in the orbit VPK given in position, moving forward from V toward K. From center C continually draw Cp equal to CP and making the angle VCp which is proportional to the angle VCP; and the area that the line Cp describes will be to the area VCP that the line CP simultaneously describes as the velocity of the describing line Cp to the velocity of the describing line CP, that is, as the angle VCp to the angle VCP and thus in a given ratio and therefore proportional to the time. Since the area that line Cp describes in the immobile plane is proportional to the time, it is manifest that the body, under the action of a centripetal force of just the right quantity, can revolve along with point p in the curved line that the same point p, in the manner just explained, describes in an immobile plane. Let the angle VCu be made equal to the angle PCp, and the line Cu equal to the line CV, and the figure uCp equal to the figure VCP; then the body, being always at the point p, will move in the perimeter of the revolving figure uCp, and will describe its arc up in the same time in which another body P can describe the arc VP, similar and equal to up, in the figure VPK at rest. Determine, therefore, by prop. 6, corol. 5, the centripetal force by which a body can revolve in the curved line that point p describes in an immobile plane, and the problem will be solved. Q.E.F.

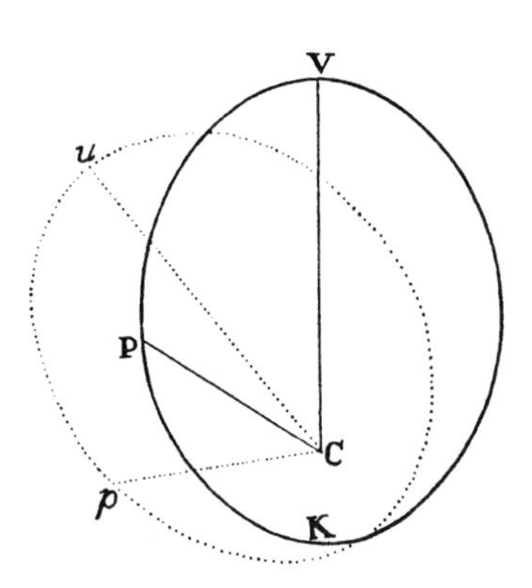

aa. Newton does not use the word "force" in the statement of prop. 43, but he does so in the conclusion of the demonstration. A literal translation of prop. 43 would read: "It is required to make it happen [*or*, It is to be effected] that a body may be able to move in any trajectory that is revolving about the center of forces exactly as another body moves in that same trajectory at rest." The force in question must be centripetal.

Proposition 44
Theorem 14

The difference between the forces under the action of which two bodies are able to move equally—one in an orbit that is at rest and the other in an identical orbit that is revolving—is inversely as the cube of their common height.

Let the parts *up* and *pk* of the revolving orbit be similar and equal to the parts VP and PK of the orbit at rest; and let it be understood that the distance between points P and K is minimally small. From point *k* drop the perpendicular *kr* to the straight line *p*C, and produce *kr* to *m* so that *mr* is to *kr* as the angle VC*p* to the angle VCP. Since the heights PC and *p*C, KC and *k*C, of the bodies are always equal, it is manifest that the increments and decrements of the lines PC and *p*C are always equal, and hence, if the motions of each of these bodies, when they are at places P and *p*, are resolved (by corol. 2 of the laws) into two components, of which one is directed toward the center, or along the line PC or *p*C, and the second is transverse to the first and has a direction along a line perpendicular to PC or *p*C, the components of motion toward the center will be equal, and the transverse component of motion of body *p* will be to the transverse component of motion of body P as the angular motion of line *p*C to the angular motion of line PC, that is, as the angle VC*p* to the angle VCP. Therefore, in the same time in which body P by the two components of its motion reaches point K, body *p* by its equal component of motion toward the center will move equally from *p*

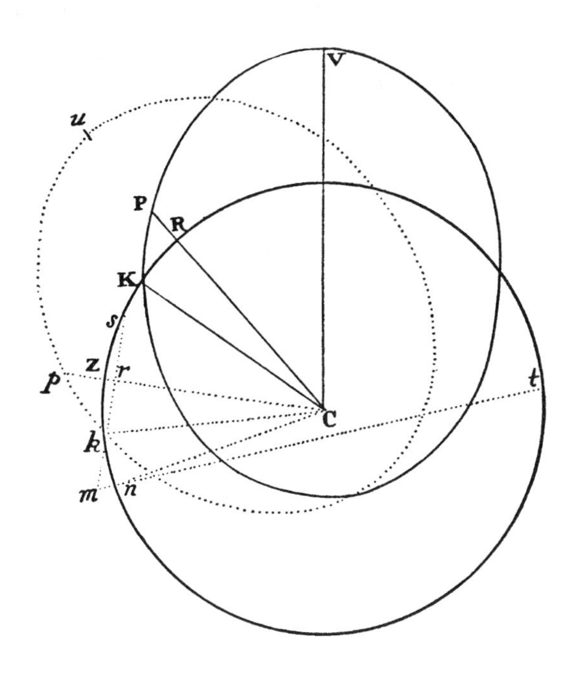

toward C and thus, when that time is completed, will be found somewhere on the line *mkr* (which is perpendicular to the line *p*C through point *k*) and by its transverse motion will reach a distance from the line *p*C that is to the distance from the line PC (which the other body P reaches) as the transverse motion of body *p* is to the transverse motion of the other body P. Therefore, since *kr* is equal to the distance from the line PC which body P reaches, and since *mr* is to *kr* as the angle VC*p* to the angle VCP, that is, as the transverse motion of body *p* to the transverse motion of body P, it is manifest that body *p*, at the completion of the time, will be found at the place *m*.

This will be the case when bodies *p* and P move equally along lines *p*C and PC and thus are urged along those lines by equal forces. But now, take the angle *p*C*n* to the angle *p*C*k* as the angle VC*p* is to the angle VCP, and let *n*C be equal to *k*C, and then body *p*—at the completion of the time—will actually be found at the place *n*; and thus body *p* is urged by a greater force than that by which body P is urged, provided that the angle *n*C*p* is greater than the angle *k*C*p*, that is, if the orbit *upk* either moves forward [or in consequentia] or moves backward [or in antecedentia] with a speed greater than twice that with which the line CP is carried forward [or in consequentia] and it is urged by a smaller force if the orbit moves backward [or in antecedentia] more slowly. And the difference between the forces is as the intervening distance *mn* through which the body *p* ought to be carried by the action of that difference in the given space of time.

Understand that a circle is described, with center C and radius C*n* or C*k*, cutting in *s* and *t* the lines *mr* and *mn* produced; then the rectangle $mn \times mt$ will be equal to the rectangle $mk \times ms$, and thus *mn* will be equal to $\frac{mk \times ms}{mt}$. But since the triangles *p*C*k* and *p*C*n* are, in a given time, given in magnitude, *kr* and *mr* and their difference *mk* and sum *ms* are inversely as the height *p*C, and thus the rectangle $mk \times ms$ is inversely as the square of the height *p*C. Also, *mt* is directly as $\frac{1}{2}mt$, that is, as the height *p*C. These are the first ratios of the nascent lines; and hence $\frac{mk \times ms}{mt}$ (that is, the nascent line-element *mn* and, proportional to it, the difference between the forces) becomes inversely as the cube of the height *p*C. Q.E.D.

COROLLARY 1. Hence the difference of the forces in the places P and *p* or K and *k* is to the force by which a body would be able to revolve with circular motion from R to K in the same time in which body P in an immobile orbit describes the arc PK as the nascent line-element mn is to the versed sine of the nascent arc RK, that is, as $\frac{mk \times ms}{mt}$ to $\frac{rk^2}{2kC}$ or as $mk \times ms$ to rk^2, that is, if the given quantities F and G are taken in the ratio to each other that the angle VCP has to the angle VCp, as $G^2 - F^2$ to F^2. And therefore, if with center C and any radius CP or Cp a circular sector is described equal to the total area VPC which the body P revolving in an immobile orbit has described in any time by a radius drawn to the center, the difference between the forces by which body P in an immobile orbit and body p in a mobile orbit revolve will be to the centripetal force by which some body, by a radius drawn to the center, would have been able to describe that sector uniformly in the same time in which the area VPC was described, as $G^2 - F^2$ to F^2. For that sector and the area pCk are to each other as the times in which they are described.

Corollary 2. If the orbit VPK is an ellipse having a focus C and an upper apsis V, and the mobile ellipse upk is supposed similar and equal to it, so that pC is always equal to PC and the angle VCp is to the angle VCP in the given ratio of G to F; and if A is written for the height PC or pC, and 2R is put for the latus rectum of the ellipse; then the force by which

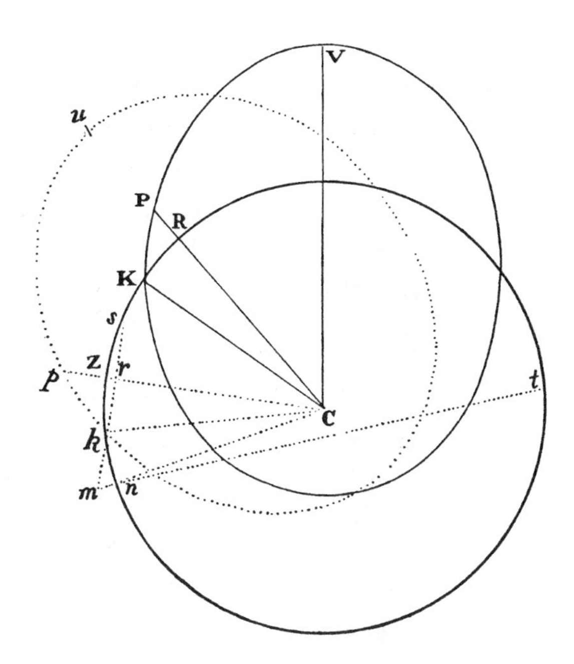

a body can revolve in the mobile ellipse will be as $\frac{F^2}{A^2} + \frac{R(G^2 - F^2)}{A^3}$, and conversely. For let the force by which a body revolves in the unmoving ellipse be represented by the quantity $\frac{F^2}{A^2}$, and then the force at V will be $\frac{F^2}{CV^2}$. But the force by which a body could revolve in a circle at the distance CV with the velocity that a body revolving in an ellipse has at V is to the force by which a body revolving in an ellipse is urged at the apsis V as half of the latus rectum of the ellipse to the semidiameter CV of the circle, and thus has the value $\frac{R \times F^2}{CV^3}$; and the force that is to this as $G^2 - F^2$ to F^2 has the value $\frac{R(G^2 - F^2)}{CV^3}$; and this force (by corol. 1 of this prop.) is the difference between the forces at V by which body P revolves in the unmoving ellipse VPK and body *p* revolves in the mobile ellipse *upk*. Hence, since (by this proposition) that difference at any other height A is to itself at the height CV as $\frac{1}{A^3}$ to $\frac{1}{CV^3}$, the same difference at every height A will have the value $\frac{R(G^2 - F^2)}{A^3}$. Therefore, add the excess $\frac{R(G^2 - F^2)}{A^3}$ to the force $\frac{F^2}{A^2}$ by which a body can revolve in the immobile ellipse VPK, and the result will be the total force $\frac{F^2}{A^2} + \frac{R(G^2 - F^2)}{A^3}$ by which a body may be able to revolve in the same times in the mobile ellipse *upk*.

COROLLARY 3. In the same way it will be gathered that if the immobile orbit VPK is an ellipse having its center at the center C of forces, and a mobile ellipse *upk* is supposed similar, equal, and concentric with it; and if 2R is the principal latus rectum of this ellipse, and 2T the principal diameter or major axis, and the angle VC*p* is always to the angle VCP as G to F; then the forces by which bodies can revolve in equal times in the immobile ellipse and the mobile ellipse will be as $\frac{F^2A}{T^3}$ and $\frac{F^2A}{T^3} + \frac{R(G^2 - F^2)}{A^3}$ respectively.

COROLLARY 4. And universally, if the greatest height CV of a body is called T; and the radius of the curvature which the orbit VPK has at V (that is, the radius of a circle of equal curvature) is called R; and the centripetal force by which a body can revolve in any immobile trajectory VPK at place V is called $\frac{VF^2}{T^2}$ and at other places P is indefinitely styled X, while the height CP is called A; and if G is taken to F in the given ratio of the angle

VCp to the angle VCP; then the centripetal force by which the same body can complete the same motions in the same times in the same trajectory *upk* which is moving circularly will be as the sum of the forces $X + \frac{VR(G^2 - F^2)}{A^3}$.

COROLLARY 5. Therefore, given the motion of a body in any immobile orbit, its angular motion about the center of forces can be increased or diminished in a given ratio, and hence new immobile orbits can be found in which bodies may revolve by new centripetal forces.

COROLLARY 6. Therefore, if on the straight line CV, given in position, there is erected the perpendicular VP of indeterminate length, and CP is joined, and Cp is drawn equal to it making the angle VCp that is to the angle VCP in a given ratio; then the force by which a body can revolve in the curve Vpk which the point p continually traces out will be inversely as the cube of the height Cp. For body P, by its own force of inertia, and with no other force urging it, can move forward uniformly in the straight line VP.

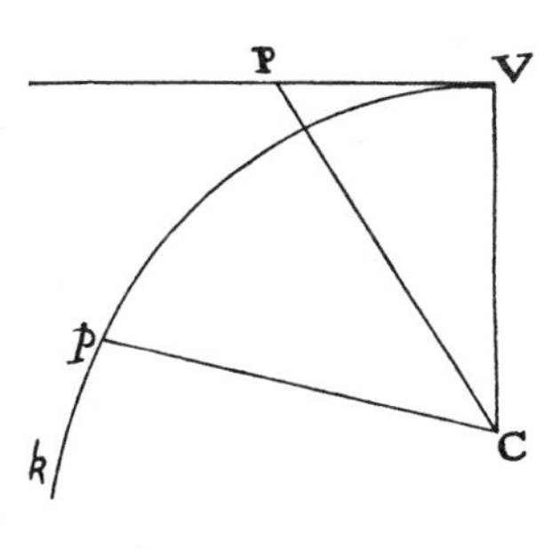

Add the force toward the center C, inversely proportional to the cube of the height CP or Cp, and (by what has just been demonstrated) the rectilinear motion will be bent into the curved line Vpk. But this curve Vpk is the same as the curve VPQ found in prop. 41, corol. 3, and (as we said there) bodies attracted by forces of this kind ascend obliquely in this curve.

Proposition 45
Problem 31

It is required to find the motions of the apsides of orbits that differ very little from circles.

This problem is solved arithmetically by taking the orbit that is described in an immobile plane by a body revolving in a mobile ellipse (as in prop. 44, corol. 2 or 3) and making it approach the form of the orbit whose apsides are required, and by seeking the apsides of the orbit which that body describes in an immobile plane. Orbits will acquire the same shape if the centripetal forces with which those orbits are described, when compared with each other, are made proportional at equal heights. Let point V be the upper apsis, and write T for the greatest height CV, A for any other height CP or Cp, and X for the difference CV − CP of the heights; then the force by which a body moves in an ellipse revolving about its own focus C (as in corol. 2)—and which in corol. 2 was as $\frac{F^2}{A^2} + \frac{RG^2 - RF^2}{A^3}$, that is, as $\frac{F^2A + RG^2 - RF^2}{A^3}$—will, when T − X is substituted for A, be as $\frac{RG^2 - RF^2 + TF^2 - F^2X}{A^3}$. Any other centripetal force is similarly to be reduced to a fraction whose denominator is A^3; and the numerators are to be made analogous [i.e., made proportional in the same degree] by bringing together homologous terms [i.e., corresponding terms, or terms of the same degree]. All of this will be clarified by the following examples.

Example 1. Let us suppose the centripetal force to be uniform and thus as $\frac{A^3}{A^3}$, or (writing T − X for A in the numerator) as

$$\frac{T^3 - 3T^2X + 3TX^2 - X^3}{A^3};$$

and by bringing together the corresponding [or homologous] terms of the numerators (namely, given ones with given ones, and ones not given with ones not given), $RG^2 - RF^2 + TF^2$ to T^3 will come to be as $-F^2X$ to $-3T^2X + 3TX^2 - X^3$ or as $-F^2$ to $-3T^2 + 3TX - X^2$. Now, since the orbit is supposed to differ very little from a circle, let the orbit come to coincide with a circle; and because R and T become equal and X is diminished indefinitely, the ultimate ratios will be RG^2 to T^3 as $-F^2$ to $-3T^2$, or G^2 to T^2 as F^2 to $3T^2$, and by alternation [or alternando] G^2 to F^2 as T^2 to $3T^2$, that is, as 1 to 3; and therefore G is to F, that is, the angle VCp is to the angle VCP, as 1 to $\sqrt{3}$. Therefore, since a body in an immobile ellipse, in descending from the upper apsis to the lower apsis, completes the angle VCP (so to speak) of 180 degrees, another body in the mobile ellipse (and hence in the immobile orbit with which we are dealing) will, in descending from the upper apsis to the lower apsis, complete the angle VCp of $\frac{180}{\sqrt{3}}$ degrees; this is so because of the similarity of this orbit, which the body describes under the action of a uniform centripetal force, to the orbit which a body completing its revolutions in a revolving ellipse describes in a plane at rest. By the above collation of terms, these orbits are made similar, not universally but at the time when they very nearly approach a circular form. Therefore a body revolving with uniform centripetal force in a very nearly circular orbit will always complete an angle of $\frac{180}{\sqrt{3}}$ degrees between the upper apsis and the lower apsis, or $103°55'23''$ at the center, arriving at the lower apsis from the upper apsis when it has completed this angle once, and returning from the lower to the upper apsis when it has completed the same angle again, and so on without end.

EXAMPLE 2. Let us suppose the centripetal force to be as the height A raised to any power, as A^{n-3} (that is, $\frac{A^n}{A^3}$), where $n - 3$ and n signify any indices of powers whatsoever—integral or fractional, rational or irrational, positive or negative. On reducing the numerator $A^n = (T - X)^n$ to an indeterminate series by our method of converging series, the result is $T^n - nXT^{n-1} + \frac{n^2 - n}{2}X^2T^{n-2} \ldots$. And by collating the terms of this with the terms of the other numerator $RG^2 - RF^2 + TF^2 - F^2X$, the result is that $RG^2 - RF^2 + TF^2$ is to T^n as $-F^2$ to $-nT^{n-1} + \frac{n^2 - n}{2}XT^{n-2} \ldots$. And after taking the ultimate ratios that result when the orbits approach circular form, RG^2 will be to T^n as $-F^2$ to $-nT^{n-1}$, or G^2 to T^{n-1} as F^2 to nT^{n-1}, and by alternation [or alternando] G^2 is to F^2 as T^{n-1} to nT^{n-1}, that is, as 1 to n; and therefore G is to F, that is, the angle VCp is to the angle VCP as 1 to $\sqrt{n}$. Therefore, since the angle VCP, completed in the descent of a body from the upper apsis to the lower apsis in an ellipse, is 180 degrees, the angle VCp, completed in the descent of a body from the upper apsis to the lower apsis in the very nearly circular orbit which any body describes under the action of a centripetal force proportional to A^{n-3}, will be equal to an angle of $\frac{180}{\sqrt{n}}$ degrees; and when this angle is repeated, the body will return from the lower apsis to the upper apsis, and so on without end.

For example, if the centripetal force is as the distance of the body from the center, that is, as A or $\frac{A^4}{A^3}$, n will be equal to 4 and $\sqrt{n}$ will be equal to 2; and therefore the angle between the upper apsis and the lower apsis will be equal to $\frac{180°}{2}$ or $90°$. Therefore, at the completion of a quarter of a revolution the body will arrive at the lower apsis, and at the completion of another quarter the body will arrive at the upper apsis, and so on by turns without end. This is also manifest from prop. 10. For a body urged by this centripetal force will revolve in an immobile ellipse whose center is in the center of forces. But if the centripetal force is inversely as the distance, that is, directly as $\frac{1}{A}$ or $\frac{A^2}{A^3}$, n will be equal to 2, and thus the angle between the upper and the lower apsis will be $\frac{180}{\sqrt{2}}$ degrees, or 127°16′45″, and therefore a body revolving under the action of such a force will—by the continual repetition of this angle—go alternately from the upper apsis to the lower and from the lower to the upper forever. Further, if the centripetal force is inversely as the fourth root of the eleventh power of the height, that is, inversely as $A^{11/4}$ and thus directly as $\frac{1}{A^{11/4}}$ or as $\frac{A^{1/4}}{A^3}$, n will be equal to ¼, and $\frac{180°}{\sqrt{n}}$ will be equal to 360°; and therefore a body, setting out from the upper apsis and continually descending from then on, will arrive at the lower apsis when it has completed an entire revolution, and then, completing another entire revolution by continually ascending, will return to the upper apsis; and so on by turns forever.

EXAMPLE 3. Let m and n be any indices of powers of the height, and let b and c be any given numbers, and let us suppose that the centripetal force is as $\frac{bA^m + cA^n}{A^3}$, that is, as $\frac{b(T - X)^m + c(T - X)^n}{A^3}$ or (again by our method of converging series) as

$$\frac{bT^m + cT^n - mbXT^{m-1} - ncXT^{n-1} + \frac{m^2 - m}{2}bX^2T^{m-2} + \frac{n^2 - n}{2}cX^2T^{n-2} \ldots}{A^3},$$

then, if the terms of the numerators are collated, the result will be $RG^2 - RF^2 + TF^2$ to $bT^m + cT^n$ as $-F^2$ to $-mbT^{m-1} - ncT^{n-1} + \frac{m^2 - m}{2}bXT^{m-2} + \frac{n^2 - n}{2}cXT^{n-2} \cdots$. And after taking the ultimate ratios that result when the orbits approach circular form, G^2 will be to $bT^{m-1} + cT^{n-1}$ as F^2 to $mbT^{m-1} + ncT^{n-1}$, and by alternation [or alternando] G^2 will be to F^2 as $bT^{m-1} + cT^{n-1}$ to $mbT^{n-1} + ncT^{n-1}$. This proportion, if the greatest height CV or T is expressed arithmetically by unity, becomes G^2 to F^2 as $b + c$ to $mb + nc$ and thus as 1 to $\frac{mb + nc}{b + c}$. Hence G is to F, that is, the angle VCp is to the angle VCP, as 1 to $\sqrt{\frac{mb + nc}{b + c}}$. And therefore, since the angle VCP between the upper apsis and the lower apsis in the immobile ellipse is 180 degrees, the angle VCp between the same apsides, in the orbit described by a body under the action of a centripetal force proportional to the quantity $\frac{bA^m + cA^n}{A^3}$, will be equal to an angle of $180\sqrt{\frac{b+c}{mb+nc}}$ degrees. And by the same argument, if the centripetal force is as $\frac{bA^m - cA^n}{A^3}$, the angle between the apsides will be found to be $180\sqrt{\frac{b-c}{mb-nc}}$ degrees. And the problem will be resolved in just the same way in more difficult cases. The quantity to which the centripetal force is proportional must always be resolved into converging series having the denominator A^3. Then the ratio of the given part of the numerator (resulting from that operation) to its other part, which is not given, is to be made the same as the ratio of the given part of this numerator $RG^2 - RF^2 + TF^2 - F^2X$ to its other part, which is not given; and when the superfluous quantities are taken away and unity is written for T, the proportion of G to F will be obtained.

COROLLARY 1. Hence, if the centripetal force is as some power of the height, that power can be found from the motion of the apsides, and conversely. That is, if the total angular motion with which the body returns to the same apsis is to the angular motion of one revolution, or 360 degrees, as some number m to another number n, and the height is called A, the force will be as the power of the height $A^{\frac{n^2}{m^2}-3}$ whose index is $\frac{n^2}{m^2} - 3$. This is manifest by the instances in ex. 2. Hence it is clear that the force, in receding from the center, cannot decrease in a ratio greater than that of the cube of the height; if a body revolving under the action of such a force and setting out from an apsis begins to descend, it will never reach the lower apsis or minimum height but will descend all the way to the center, describing that curved line which we treated in prop. 41, corol. 3. But if the body, on setting out from an apsis, begins to ascend even the least bit, it will ascend indefinitely and will never reach the upper apsis. For it will describe the curved line treated in the above-mentioned corol. 3 and in prop. 44, corol. 6. So also, when the force, in receding from the center, decreases in a ratio greater than that of the cube of the height, a body setting out from an apsis (depending on whether it begins to descend or to ascend) either will descend all the way to the center or will ascend indefinitely. But if the force, in receding from the center, either decreases in a ratio less than that of the cube of the height or increases in any ratio of the height whatever, the body will never descend all the way to the center, but will at some time reach a lower apsis; and conversely, if a body descending and ascending alternately from apsis to apsis never gets to the center, either the force in receding from the center will be increased or it will decrease in a ratio less than that of the cube of the height; and the more swiftly the body returns from apsis to apsis, the farther the ratio of the forces will recede from that of the cube.

For example, if by alternate descent and ascent a body returns from upper apsis to upper apsis in 8 or 4 or 2 or 1½ revolutions, that is, if m is to n as 8 or 4 or 2 or 1½ to 1, and therefore $\frac{n^2}{m^2} - 3$ has the value $\frac{1}{64} - 3$ or $\frac{1}{16} - 3$ or $\frac{1}{4} - 3$ or $\frac{4}{9} - 3$, the force will be as $A^{\frac{1}{64}-3}$ or $A^{\frac{1}{16}-3}$ or $A^{\frac{1}{4}-3}$ or $A^{\frac{4}{9}-3}$, that is, inversely as $A^{3-\frac{1}{64}}$ or $A^{3-\frac{1}{16}}$ or $A^{3-\frac{1}{4}}$ or $A^{3-\frac{4}{9}}$. If the body returns in each revolution to the same unmoving apsis, m will be to n as 1 to 1, and thus $A^{\frac{n^2}{m^2}-3}$ will be equal to A^{-2} or $\frac{1}{A^2}$; and therefore the decrease in force will be as the square of the height, as has been demonstrated in the preceding propositions. If the body returns to the same apsis in three-quarters or two-thirds or one-third or one-quarter of a single revolution, m will be to n as ¾ or ⅔ or ⅓ or ¼ to 1, and so $A^{\frac{n^2}{m^2}-3}$ will be equal to $A^{\frac{16}{9}-3}$ or $A^{\frac{9}{4}-3}$ or A^{9-3} or A^{16-3}; and therefore the force will be either inversely as $A^{\frac{11}{9}}$ or $A^{\frac{3}{4}}$, or directly as A^6 or A^{13}. Finally, if the body in proceeding from upper apsis to upper apsis completes an entire revolution and an additional three degrees (and therefore, during each revolution of the body, that apsis moves three degrees forward [or in consequentia]), m will be to n as 363° to 360° or as 121 to 120, and thus $A^{\frac{n^2}{m^2}-3}$ will be equal to $A^{-\frac{29,523}{14,641}}$, and therefore the centripetal force will be inversely as $A^{\frac{29,523}{14,641}}$ or inversely as $A^{2\frac{4}{243}}$ approximately. Therefore the centripetal force decreases in a ratio a little greater than that of the square, but 59¾ times closer to that of the square than to that of the cube.

Corollary 2. Hence also if a body, under the action of a centripetal force that is inversely as the square of the height, revolves in an ellipse having a focus in the center of forces, and any other extraneous force is added to or taken away from this centripetal force, the motion of the apsides that will arise from that extraneous force can be found out (by instances in ex. 3), and conversely. For example, if the force under the action of which the body revolves in the ellipse is as $\frac{1}{A^2}$ and the extraneous force which has been taken away is as cA, and hence the remaining force is as $\frac{A - cA^4}{A^3}$, then (as in ex. 3) b will be equal to 1, m will be equal to 1, and n will be equal to 4, and therefore the angle of the revolution between apsides will be equal to an angle of $180\sqrt{\frac{1-c}{1-4c}}$ degrees. Let us suppose the extraneous force to be 357.45 times less than the other force under the action of which the body revolves in the ellipse, that is, let us suppose c to be $\frac{100}{35{,}745}$, A or T being equal to 1, and then $180\sqrt{\frac{1-c}{1-4c}}$ will come to be $180\sqrt{\frac{35{,}645}{35{,}345}}$, or 180.7623, that is, 180°45′44″. Therefore a body, setting out from the upper apsis, will reach the lower apsis by an angular motion of 180°45′44″ and will return to the upper apsis if this angular motion is doubled; and thus in each revolution the upper apsis will move forward through 1°31′28″. [a]The [advance of the] apsis of the moon is about twice as swift.[a]

So much concerning the motion of bodies in orbits whose planes pass through the center of forces. It remains for us to determine additionally those motions which occur in planes that do not pass through the center of forces. For writers who deal with the motion of heavy bodies are wont to consider the oblique ascents and descents of weights in any given planes as well as perpendicular ascents and descents, and there is equal justification for considering here the motion of bodies that tend to centers under the action of any forces whatever and are supported by eccentric planes. We suppose, however, that these planes are highly polished and absolutely slippery, so as not to retard the bodies. Further, in these demonstrations, in place of the planes on which bodies rest and which they touch by resting on them, we even make use of planes parallel to them, in which the centers of the bodies move and by so moving describe orbits. And by the same principle we then determine the motions of bodies performed in curved surfaces.

aa. Ed. 1 and ed. 2 lack this, but it appears both in the interleaved copy and in the annotated copy of ed. 2. The interleaved copy also has: "Query: Can this motion arise from twice the external force?" See further the Guide to the present translation, §6.10.

SECTION 10

The motion of bodies on given surfaces and the oscillating motion of [a]*simple pendulums*[a]

Proposition 46
Problem 32

Suppose a centripetal force of any kind, and let there be given both the center of force and any plane in which a body revolves, and grant the quadratures of curvilinear figures; it is required to find the motion of a body setting out from a given place with a given velocity along a given straight line in that plane.

Let S be the center of force, SC the least distance of this center from the given plane, P a body setting out from place P along the straight line PZ, Q the same body revolving in its trajectory, and PQR the required trajectory described in the given plane. Join CQ and also QS, and if SV is taken in

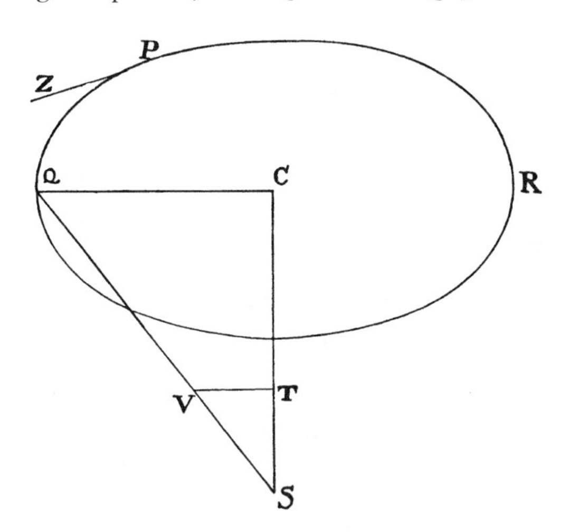

QS and is proportional to the centripetal force by which the body is drawn toward the center S, and VT is drawn parallel to CQ and meeting SC in T, then the force SV will be resolved (by corol. 2 of the laws) into the forces ST and TV, of which ST, by drawing the body along a line perpendicular to the plane, does not at all change the body's motion in this plane. But the other

aa. We use the term "simple pendulum" in its classical and technical sense. For example, according to Brougham and Routh, "A *simple pendulum* consists of a material particle suspended from a fixed point by an inflexible inextensible string without weight" (Henry Lord Brougham and E. J. Routh, *Analytical View of Sir Isaac Newton's "Principia"* [1855; reprint, with an introd. by I. Bernard Cohen, New York and London: Johnson Reprint Corp., 1972], pp. 240–241). See, further, §7.5 of the Guide.

force TV, by acting along the position of the plane, draws the body directly toward [i.e., along a line directed toward] the given point C in the plane and thus causes the body to move in this plane just as if the force ST were removed and as if the body revolved in free space about the center C under the action of the force TV alone. But, given the centripetal force TV under the action of which the body Q revolves in free space about the given center C, there are also given (by prop. 42) not only the trajectory PQR described by the body, but also the place Q in which the body will be at any given time, and finally the velocity of the body in that place Q; and conversely. Q.E.I.

Proposition 47
Theorem 15

Suppose that a centripetal force is proportional to the distance of a body from a center; then all bodies revolving in any planes whatever will describe ellipses and will make their revolutions in equal times; and bodies that move in straight lines, by oscillating to and fro, will complete in equal times their respective periods of going and returning.

For, under the same conditions as in prop. 46, the force SV, by which the body Q revolving in any plane PQR is drawn toward the center S, is as the distance SQ; and thus—because SV and SQ, TV and CQ are proportional—the force TV, by which the body is drawn toward the given point C in the plane of the orbit, is as the distance CQ. Therefore, the forces by which bodies that are in the plane PQR are drawn toward point C are, in proportion to the distances, equal to the forces by which bodies are drawn from all directions toward the center S; and thus in the same times the bodies will move in the same figures in any plane PQR about the point C as they would move in free spaces about the center S; and hence (by prop. 10, corol. 2, and prop. 38, corol. 2) in times which are always equal, they will either describe ellipses [i.e., complete a whole revolution in such ellipses] in that plane about the center C or will complete periods of oscillating to and fro in straight lines drawn through the center C in that plane. Q.E.D.

Scholium

The ascents and descents of bodies in curved surfaces are very closely related to the motions just discussed. Imagine that curved lines are described in a plane, that they then revolve around any given axes passing through the center of force and describe curved surfaces by this revolution, and then that bodies move in such a way that their centers are always found in these surfaces. If those bodies, in ascending and descending obliquely, oscillate to and fro, their motions will be made in planes passing through the axis and hence in curved lines by whose revolution those curved surfaces were generated. In these cases, therefore, it is sufficient to consider the motion in those curved lines.

Proposition 48
Theorem 16

If a wheel stands upon the outer surface of a globe at right angles to that surface and, rolling as wheels do, moves forward in a great circle [in the globe's surface], the length of the curvilinear path traced out by any given point in the perimeter [or rim] of the wheel from the time when that point touched the globe (a curve which may be called a cycloid or epicycloid) will be to twice the versed sine of half the arc [of the rim of the wheel] which during the time of rolling has been in contact with the globe's surface as the sum of the diameters of the globe and wheel is to the semidiameter of the globe.

Proposition 49
Theorem 17

If a wheel stands upon the inner surface of a hollow globe at right angles to that surface and, rolling as wheels do, moves forward in a great circle [in the globe's surface], the length of the curvilinear path traced out by any given point in the perimeter [or rim] of the wheel from the time when that point touched the globe will be to twice the versed sine of half the arc [of the rim of the wheel] which during the time of rolling has been in contact with the globe's surface as the difference of the diameters of the globe and wheel is to the semidiameter of the globe.

Let ABL be the globe, C its center, BPV the wheel standing upon it, E the center of the wheel, B the point of contact, and P the given point in the perimeter of the wheel. Imagine that this wheel proceeds in the great circle ABL from A through B toward L and, while rolling, rotates in such a way that the arcs AB and PB are always equal to each other and that the given point P in the perimeter of the wheel is meanwhile describing the curvilinear path AP. Now, let AP be the whole curvilinear path described since the wheel was in contact with the globe at A, and the length AP of this path will be to twice the versed sine of the arc ½PB as 2CE to CB. For let the straight line CE (produced if need be) meet the wheel in V, and join CP, BP, EP, VP, and drop the normal VF to CP produced. Let PH and

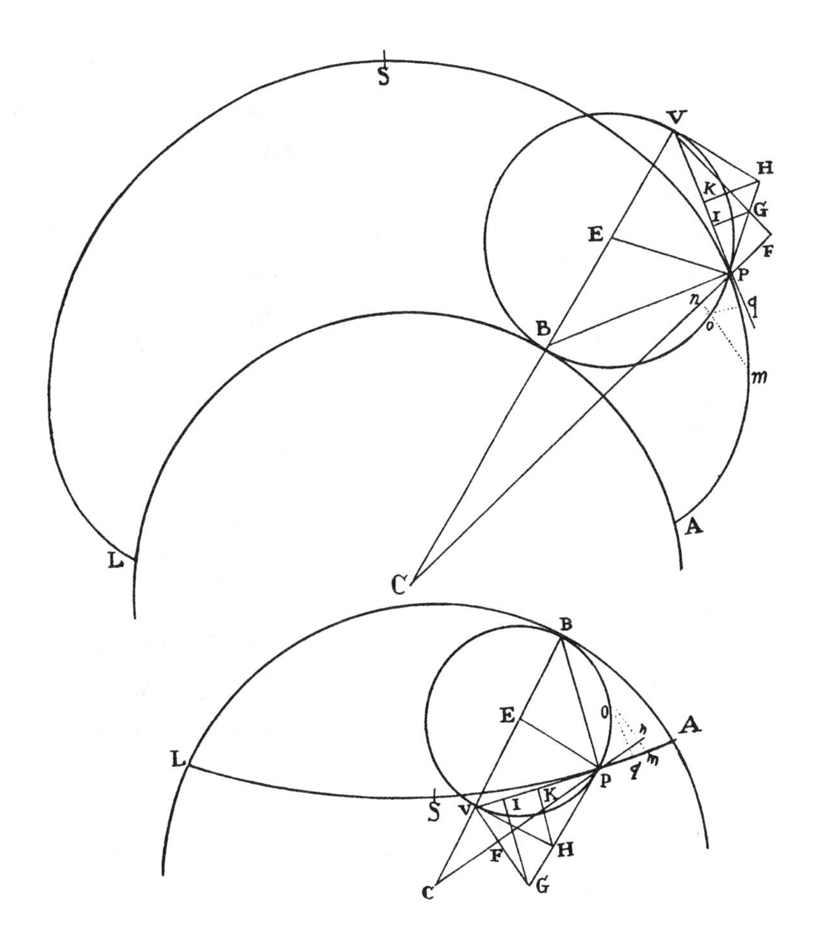

VH, meeting in H, touch the circle in P and V, and let PH cut VF in G, and drop the normals GI and HK to VP. With the same center C and with any radius whatever describe the circle *nom* cutting the straight line CP in *n*, the wheel's perimeter BP in *o*, and the curvilinear path AP in *m*; and with center V and radius V*o* describe a circle cutting VP produced in *q*.

Since the wheel, in rolling, always revolves about the point of contact B, it is manifest that the straight line BP is perpendicular to the curved line AP described by the wheel's point P, and therefore that the straight line VP will touch this curve in point P. Let the radius of the circle *nom* be gradually increased or decreased, and so at last become equal to the distance CP; then, because the evanescent figure P*nomq* and the figure PFGVI are similar, the ultimate ratio of the evanescent line-elements P*m*, P*n*, P*o*, and P*q*, that is, the ratio of the instantaneous changes of the curve AP, the straight line CP, the circular arc BP, and the straight line VP, will be the same as that of the lines PV, PF, PG, and PI respectively. But since VF is perpendicular to CF, and VH is perpendicular to CV, and the angles HVG and VCF are therefore equal, and the angle VHG is equal to the angle CEP (because the angles of the quadrilateral HVEP are right angles at V and P), the triangles VHG and CEP will be similar; and hence it will come about that EP is to CE as HG to HV or HP and as KI to KP, and by composition [or componendo] or by separation [or dividendo] CB is to CE as PI to PK, and—by doubling of the consequents—CB is to 2CE as PI to PV and as Pq to Pm. Therefore the decrement of the line VP, that is, the increment of the line BV − VP, is to the increment of the curved line AP in the given ratio of CB to 2CE, and therefore (by lem. 4, corol.) the lengths BV − VP and AP, generated by those increments, are in the same ratio. But since BV is the radius, VP is the cosine of the angle BVP or ½BEP, and therefore BV − VP is the versed sine of the same angle; and therefore in this wheel, whose radius is ½BV, BV − VP will be twice the versed sine of the arc ½BP. And thus AP is to twice the versed sine of the arc ½BP as 2CE to CB. Q.E.D.

For the sake of distinction, we shall call the curved line AP in prop. 48 a *cycloid outside the globe*, and the curved line AP in prop. 49 a *cycloid inside the globe*.

COROLLARY 1. Hence, if an entire cycloid ASL is described and is bisected in S, the length of the part PS will be to the length VP (which is twice the sine of the angle VBP, where EB is the radius) as 2CE to CB, and thus in a given ratio.

COROLLARY 2. And the length of the semiperimeter AS of the cycloid will be equal to a straight line that is to the diameter BV of the wheel as 2CE to CB.

Proposition 50
Problem 33

To make a pendulum bob oscillate in a given cycloid.

Within a globe QVS described with center C, let the cycloid QRS be given, bisected in R and with its end-points Q and S meeting the surface of the globe on the two sides. Draw CR bisecting the arc QS in O, and produce CR to A, so that CA is to CO as CO to CR. Describe an outer globe DAF with center C and radius CA; and inside this globe let two half-cycloids AQ and AS be described by means of a wheel whose diameter is AO, and let these two half-cycloids touch the inner globe at Q and S and meet the outer globe in A. Let a body T hang from the point A by a thread APT

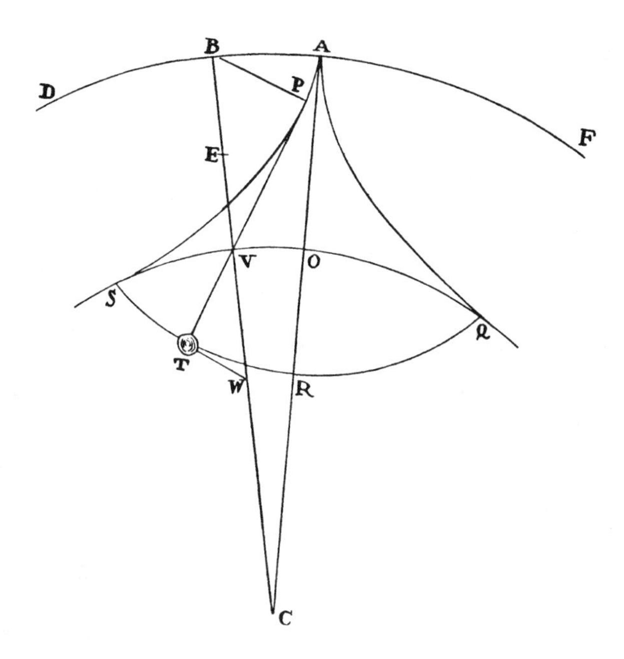

equal to the length AR, and let this body T oscillate between the two half-cycloids AQ and AS in such a way that each time the pendulum departs from the perpendicular AR, the upper part AP of the thread comes into contact with that half-cycloid APS toward which the motion is directed, and is bent around it as an obstacle, while the other part PT of the thread, to which the half-cycloid is not yet exposed, stretches out in a straight line; then the weight T will oscillate in the given cycloid QRS. Q.E.F.

For let the thread PT meet the cycloid QRS in T and the circle QOS in V, and draw CV; and from the end-points P and T of the straight part PT of the thread, erect BP and TW perpendicular to PT, meeting the straight line CV in B and W. It is evident, from the construction and the generation of the similar figures AS and SR, that the perpendiculars PB and TW cut off from CV the lengths VB and VW equal respectively to OA and OR, the diameters of the wheels. Therefore, TP is to VP (which is twice the sine of the angle VBP, where ½BV is the radius) as BW to BV, or AO + OR to AO, that is (since CA is proportional to CO, CO to CR, and by separation [or dividendo] AO to OR), as CA + CO to CA, or, if BV is bisected in E, as 2CE to CB. Accordingly (by prop. 49, corol. 1), the length of the straight part PT of the thread is always equal to the arc PS of the cycloid, and the whole thread APT is always equal to the half-arc APS of the cycloid, that is (by prop. 49, corol. 2), to the length AR. And therefore, conversely, if the thread always remains equal to the length AR, point T will move in the given cycloid QRS. Q.E.D.

Corollary. The thread AR is equal to the half-cycloid AS and thus has the same ratio to the semidiameter AC of the outer globe that the half-cycloid SR, similar to it, has to the semidiameter CO of the inner globe.

Proposition 51
Theorem 18

If a centripetal force tending from all directions to the center C *of a globe is in each individual place as the distance of that place from the center; and if, under the action of this force alone, the body* T *oscillates (in the way just described) in the perimeter of the cycloid* QRS; *then I say that the times of the oscillations, however unequal the oscillations may be, will themselves be equal.*

For let the perpendicular CX fall to the indefinitely produced tangent TW of the cycloid and join CT. Now the centripetal force by which the body T is impelled toward C is as the distance CT, and CT may be resolved (by corol. 2 of the laws) into the components CX and TX, of which CX (by impelling the body directly from P) stretches the thread PT and is wholly nullified by the resistance of the thread and produces no other effect, while the other component TX (by urging the body transversely or toward X) directly accelerates the motion of the body in the cycloid; hence it is manifest that the body's acceleration, which is proportional to this accelerative force, is at each individual moment as the length TX, that is (because CV and WV—and TX and TW, proportional to them—are given), as the length TW, that is (by prop. 49, corol. 1), as the length of the arc of the cycloid TR. Therefore, if the two pendulums APT and A*pt* are drawn back unequally from the perpendicular [or vertical] AR and are let go simultaneously, their accelerations will always be as the respective arcs to be described TR and *t*R. But

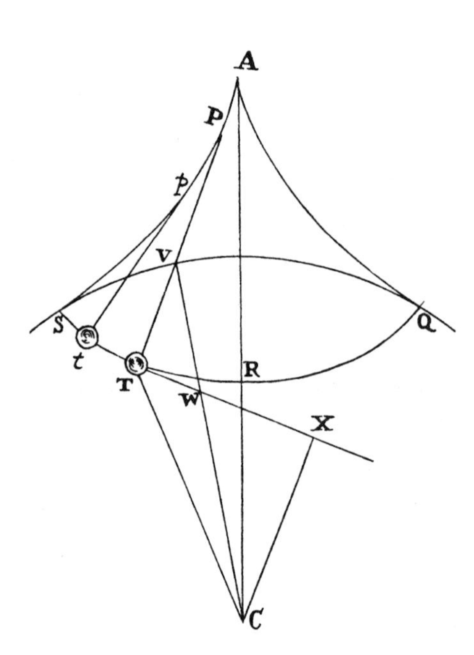

the parts of these arcs described at the beginning of the motion are as the accelerations, that is, as the whole arcs to be described at the beginning, and therefore the parts that remain to be described and the subsequent accelerations proportional to these parts are also as the whole arcs, and so on. Therefore the accelerations—and hence the velocities generated, the parts of the arcs described with these velocities, and the parts to be so described—are always as the whole arcs; and therefore the parts to be described, preserving a given ratio to one another, will vanish simultaneously, that is, the two oscillating bodies will arrive at the same time at the perpendicular [or vertical] AR. And since, conversely, the ascents of the pendulums, made from the lowest place R through the same cycloidal arcs with a reverse motion, are retarded in individual places by the same forces by which their descents were accelerated, it is evident that the velocities of the ascents and descents made through the same arcs are equal and hence occur in equal times; and therefore, since the two parts RS and RQ of the cycloid, each lying on a different side of the perpendicular [or vertical], are similar and equal, the two pendulums will always make their whole oscillations as well as their half-oscillations in the same times. Q.E.D.

COROLLARY. The force by which body T is accelerated or retarded in any place T of the cycloid is to the total weight of body T in the highest place S or Q as the arc TR of the cycloid to its arc SR or QR.

Proposition 52
Problem 34

To determine both the velocities of pendulums in individual places and the times in which complete oscillations, as well as the separate parts of oscillations, are completed.

With any center G and with a radius GH equal to the arc RS of the cycloid, describe the semicircle HKM bisected by the semidiameter GK. And if a centripetal force proportional to the distances of places from the center tends toward that center G, and if in the perimeter HIK that force is equal to the centripetal force in the perimeter of the globe QOS tending toward its center, and if, at the same time that the pendulum T is let go from its highest place S, some other body L falls from H to G; then, since the forces by which the bodies are urged are equal at the beginning of the motion, and are always proportional to the spaces TR and LG which are to be described, and are therefore equal in the places T and L if TR and LG are equal, it is evident that the two bodies describe the equal spaces ST and HL at the

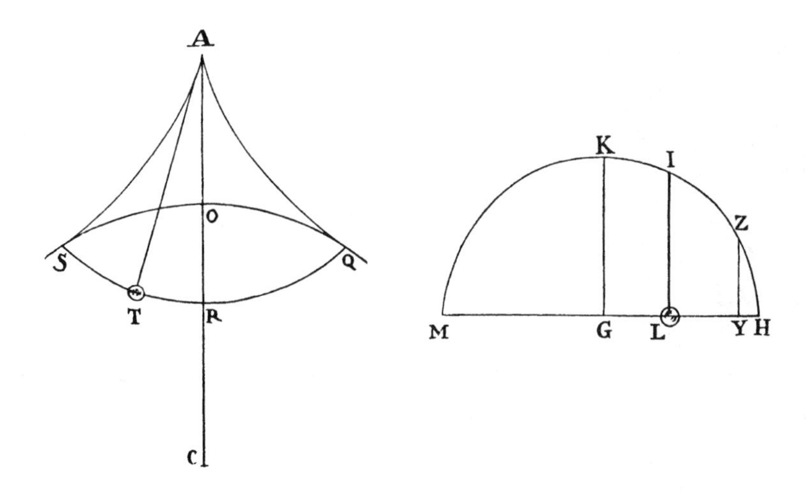

beginning of the motion and thus will proceed thereafter to be equally urged and to describe equal spaces. Therefore (by prop. 38), the time in which the body describes the arc ST is to the time of one oscillation as the arc HI (the time in which the body H will reach L) to the semiperiphery HKM (the time in which the body H will reach M). And the velocity of the pendulum bob at the place T is to its velocity at the lowest place R (that is, the velocity of body H in the place L to its velocity in the place G, or the instantaneous increment of the line HL to the instantaneous increment of the line HG, where the arcs HI and HK increase with a uniform flow) as the ordinate LI to the radius GK, or as $\sqrt{(SR^2 - TR^2)}$ to SR. Hence, since in unequal oscillations arcs proportional to the total arcs of the oscillations are described in equal times, both the velocities and the arcs described in all oscillations universally can be found from the given times. As was first to be found.

Now let simple pendulums oscillate in different cycloids described within different globes, whose absolute forces are also different; and if the absolute force of any globe QOS is called V, the accelerative force by which the pendulum is urged in the circumference of this globe, when it begins to move directly toward its center, will be jointly as the distance of the pendulum bob from that center and the absolute force of the globe, that is, as $CO \times V$. Therefore the line-element HY (which is as this accelerative force $CO \times V$) will be described in a given time; and if the normal YZ is erected so as to meet the circumference in Z, the nascent arc HZ will denote that given time. But this nascent arc HZ is as the square root of the rectangle $GH \times HY$, and thus as $\sqrt{(GH \times CO \times V)}$. Hence the time of a complete oscillation in the cycloid QRS (since it is directly as the semiperiphery HKM, which denotes that complete oscillation, and inversely as the arc HZ, which similarly denotes the given time) will turn out to be as GH directly and $\sqrt{(GH \times CO \times V)}$ inversely, that is, because GH and SR are equal, as $\sqrt{\frac{SR}{CO \times V}}$, or (by prop. 50, corol.) as $\sqrt{\frac{AR}{AC \times V}}$. Therefore the oscillations in all globes and cycloids, made with any absolute forces whatever, are as the square root of the length of the thread directly and as the square root of the distance between the point of suspension and the center of the globe inversely and also as the square root of the absolute force of the globe inversely. Q.E.I.

Corollary 1. Hence also the times of bodies oscillating, falling, and revolving can be compared one with another. For if the diameter of the wheel by which a cycloid is described within a globe is made equal to the semidiameter of the globe, the cycloid will turn out to be a straight line passing through the center of the globe, and the oscillation will now be a descent and subsequently an ascent in this straight line. Hence the time of the descent from any place to the center is given, as well as the time (equal to that time of descent) in which a body, by revolving uniformly about the center of the globe at any distance, describes a quadrantal arc. For this time (by the second case [that is, according to the second paragraph above]) is to the time of a half-oscillation in any cycloid QRS as 1 to $\sqrt{\frac{AR}{AC}}$.

Corollary 2. Hence also there follows what Wren and Huygens discovered about the common cycloid. For if the diameter of the globe is increased indefinitely, its spherical surface will be changed into a plane, and the centripetal force will act uniformly along lines perpendicular to this plane, and our cycloid will turn into a common cycloid. But in that case the length of the arc of the cycloid between that plane and the describing point will come out equal to four times the versed sine of half of the arc of the wheel between that same plane and the describing point, as Wren discovered; and a pendulum between two cycloids of this sort will oscillate in a similar and equal cycloid in equal times, as Huygens demonstrated. But also the descent of heavy bodies during the time of one oscillation will be the descent which Huygens indicated.

Moreover, the propositions that we have demonstrated fit the true constitution of the earth, insofar as wheels, moving in the earth's great circles, describe cycloids outside this globe by the motion of nails fastened in their perimeters; and pendulums suspended lower down in mines and caverns of the earth must oscillate in cycloids within globes in order that all their oscillations may be isochronous. For gravity (as will be shown in book 3) decreases in going upward from the surface of the earth as the square of the distance from the earth's center, and in going downward from the surface is as the distance from that center.

Proposition 53
Problem 35

Granting the quadratures of curvilinear figures, it is required to find the forces by whose action bodies moving in given curved lines will make oscillations that are always isochronous.

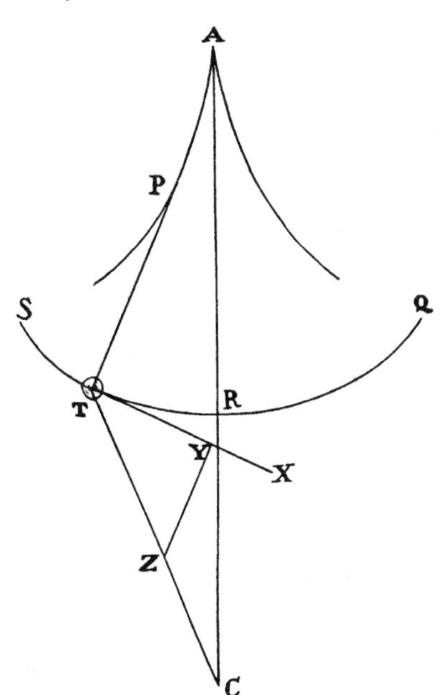

Let a body T oscillate in any curved line STRQ whose axis is AR passing through the center of forces C. Draw TX touching that curve in any place T of the body, and on this tangent TX take TY equal to the arc TR. [This may be done] since the length of that arc can be known from the quadratures of figures by commonly used methods. From point Y draw the straight line YZ perpendicular to the tangent. Draw CT meeting the perpendicular in Z, and the centripetal force will be proportional to the straight line TZ. Q.E.I.

For if the force by which the body is drawn from T toward C is represented by the straight line TZ taken proportional to it, this will be resolved into the forces TY and YZ, of which YZ, by drawing the body along the length of the thread PT, does not change its motion at all, while the other force TY directly accelerates or directly retards its motion in the curve STRQ. Accordingly, since this force is as the projection TR to be described, the body's accelerations or retardations in describing proportional parts of two oscillations (a greater and a lesser oscillation) will always be as those parts, and will therefore cause those parts to be described simultaneously. And bodies that in the same time describe parts always proportional to the wholes will describe the wholes simultaneously. Q.E.D.

COROLLARY 1. Hence, if body T, hanging by a rectilinear thread AT from the center A, describes the circular arc STRQ and meanwhile is urged downward along parallel lines by some force that is to the uniform force of gravity as the arc TR to its sine TN, the times of any single oscillations will be equal. For, because TZ and AR are parallel, the triangles ATN and ZTY will be similar; and therefore TZ will be to AT as TY to TN; that is, if the uniform force of gravity is represented by the given length AT, the force TZ, by the action of which the oscillations will turn out to be isochronous, will be to the force of gravity AT as the arc TR (equal to TY) to the sine TN of that arc.

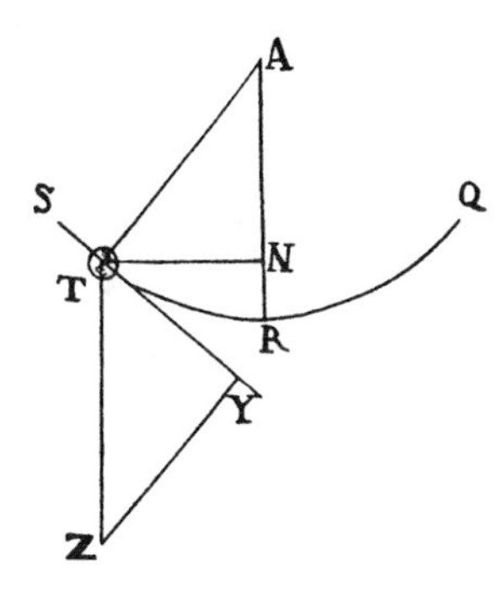

COROLLARY 2. And therefore in [pendulum] clocks, if the forces impressed by the mechanism upon the pendulum to maintain the motion can be compounded with the force of gravity in such a way that the total force downward is always as the line that arises from dividing the rectangle of the arc TR and the radius AR by the sine TN, all the oscillations will be isochronous.

Proposition 54
Problem 36

Granting the quadratures of curvilinear figures, to find the times in which bodies under the action of any centripetal force will descend and ascend in any curved lines described in a plane passing through the center of forces.

Let a body descend from any place S through any curved line ST*t*R given in a plane passing through the center of forces C. Join CS and divide it into innumerable equal parts, and let D*d* be some one of those parts. With center C and radii CD and C*d*, describe the circles DT and *dt*, meeting the curved line ST*t*R in T and *t*. Then, since both the law of centripetal force and the height CS from which the body has fallen are given, the velocity of the body at any other height CT will be given (by prop. 39). Moreover, the time in which the body describes the line-element T*t* is as the length of this line-element (that is, as the secant of the angle *t*TC) directly and as the velocity inversely. Let the ordinate DN be proportional to this time and perpendicular to the straight line CS through point D; then, because D*d* is

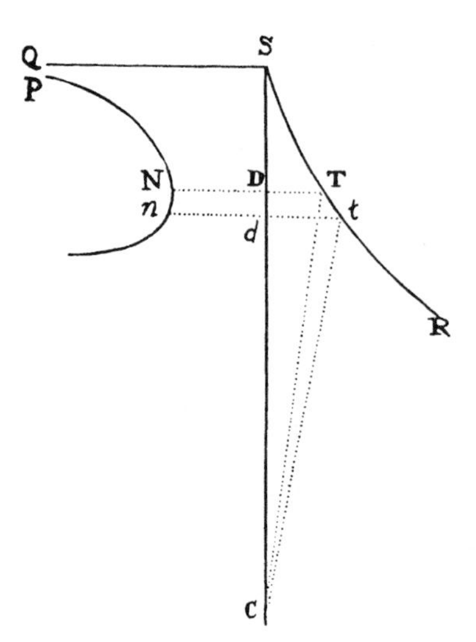

given, the rectangle $Dd \times DN$, that is, the area DN*nd*, will be proportional to that same time. Therefore if PN*n* is the curved line that point N continually traces out, [a]and its asymptote is the straight line SQ standing perpendicularly upon the straight line CS,[a] the area SQPND will be proportional to the time in which the body, by descending, has described the line ST; and accordingly, when that area has been found, the time will be given. Q.E.I.

Proposition 55
Theorem 19

If a body moves in any curved surface whose axis passes through a center of forces, and a perpendicular is dropped from the body to the axis, and a straight line parallel and equal to the perpendicular is drawn from any given point of the axis; I say that the parallel will describe an area proportional to the time.

Let BKL be the curved surface, T the body revolving in it, STR the trajectory which the body describes in it, S the beginning of the trajectory, OMK the axis of the curved surface, TN the perpendicular straight line dropped from the body to the axis; and let OP be the straight line parallel and equal to TN and drawn from a point O that is given in the axis, AP the path described by point P in the plane AOP of the revolving line OP, A the beginning of the projection (corresponding to point S); and let TC be a straight line drawn

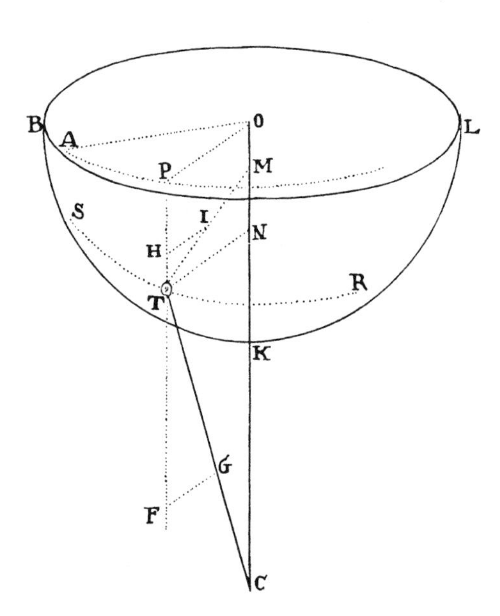

aa. A clarification by Pemberton after he had called Newton's attention to the incorrect diagrams in eds. 1 and 2 (cf. *The Mathematical Papers of Isaac Newton*, ed. D. T. Whiteside [Cambridge: Cambridge University Press, 1967–1981], 6:409, nn. 308–309).

from the body to the center, TG the part of TC that is proportional to the centripetal force by which the body is urged toward the center C, TM a straight line perpendicular to the curved surface, TI the part of TM proportional to the force of pressure by which the body urges the surface and is in turn urged by the surface toward M; and let PTF be a straight line parallel to the axis and passing through the body, and GF and IH straight lines dropped perpendicularly from the points G and I to the parallel PHTF. I say now that the area AOP, described by the radius OP from the beginning of the motion, is proportional to the time. For the force TG (by corol. 2 of the laws) is resolved into the forces TF and FG, and the force TI into the forces TH and HI. But the forces TF and TH, by acting along the line PF perpendicular to the plane AOP, change the body's motion only insofar as it is perpendicular to this plane. And therefore the body's motion, insofar as it takes place in the position of the plane—that is, the motion of point P, by which the projection AP of the trajectory is described in this plane—is the same as if the forces TF and TH were taken away and the body were acted on by the forces FG and HI alone; that is, it is the same as if the body were to describe the curve AP in the plane AOP under the action of a centripetal force tending toward the center O and equal to the sum of the forces FG and HI. But by the action of such a force the area AOP is (by prop. 1) described proportional to the time.[a] Q.E.D.

COROLLARY. By the same argument, if a body, acted on by forces tending toward two or more centers in any one given straight line CO, described any curved line ST in free space, the area AOP would always be proportional to the time.

Proposition 56
Problem 37

Granting the quadratures of curvilinear figures, and given both the law of centripetal force tending toward a given center and a curved surface whose axis passes through that center, it is required to find the trajectory that a body will describe in that same surface when it has set out from a given place with a given velocity, in a given direction in that surface.

Assuming the same constructions as in prop. 55, let body T go forth from the given place S, along a straight line given in position, in the required

a. In this proposition, Newton's "vestigium," literally, "a trace," has been translated as "projection," following D. T. Whiteside.

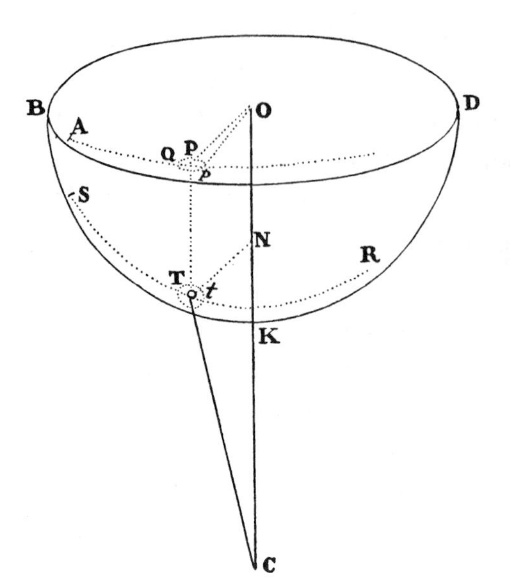

trajectory STR, and let the projection of this trajectory in the plane BLO be AP. And since the velocity of the body is given at the height SC, its velocity at any other height TC will be given. With this velocity, let the body in a given minimally small time describe the particle T*t* of its trajectory, and let P*p* be its projection described in the plane AOP. Join O*p*, and let the projection (in the plane AOP) of the little circle described with center T and radius T*t* in the curved surface be the ellipse *p*Q. Then, because the little circle T*t* is given in magnitude, and its distance TN or PO from the axis CO is given, the ellipse *p*Q will be given in species and in magnitude, as well as in its position with respect to the straight line PO. And since the area PO*p* is proportional to the time and therefore given because the time is given, the angle PO*p* will be given. And hence the common intersection *p* of the ellipse and the straight line O*p* will be given, along with the angle OP*p* in which the projection AP*p* of the trajectory cuts the line OP. And accordingly (by consulting prop. 41 with its corol. 2) the way of determining the curve AP*p* is readily apparent. Then, erecting perpendiculars to the plane AOP one by one, from the points P of the projection, so as to meet the curved surface in T, the points T of the trajectory will be given one by one. Q.E.I.

SECTION 11

The motion of bodies drawn to one another by centripetal forces

Up to this point, I have been setting forth the motions of bodies attracted toward an immovable center, such as, however, hardly exists in the natural world. For attractions are always directed toward bodies, and—by the third law—the actions of attracting and attracted bodies are always mutual and equal; so that if there are two bodies, neither the attracting nor the attracted body can be at rest, but both (by corol. 4 of the laws) revolve about a common center of gravity as if by a mutual attraction; and if there are more than two bodies that either are all attracted by and attract a single body or all attract one another, these bodies must move with respect to one another in such a way that the common center of gravity either is at rest or moves uniformly straight forward. For this reason I now go on to set forth the motion of bodies that attract one another, considering centripetal forces as attractions, although perhaps—if we speak in the language of physics—they might more truly be called impulses. For here we are concerned with mathematics; and therefore, putting aside any debates concerning physics, we are using familiar language so as to be more easily understood by mathematical readers.

Proposition 57
Theorem 20

Two bodies that attract each other describe similar figures about their common center of gravity and also about each other.

For the distances of these bodies from their common center of gravity are inversely proportional to the masses of the bodies and therefore in a given ratio to each other and, by composition [or componendo], in a given ratio to the total distance between the bodies. These distances, moreover, rotate about their common end-point with an equal angular motion because, since they always lie in the same straight line, they do not change their inclination toward each other. And straight lines that are in a given ratio to each other and that rotate about their end-points with an equal angular motion describe entirely similar figures about the end-points in planes that, along with these end-points, either are at rest or move with any motion that is not angular. Accordingly, the figures described by the rotation of these distances are similar. Q.E.D.

Proposition 58
Theorem 21

If two bodies attract each other with any forces whatever and at the same time revolve about their common center of gravity, I say that by the action of the same forces there can be described around either body if unmoved a figure similar and equal to the figures that the bodies so moving describe around each other.

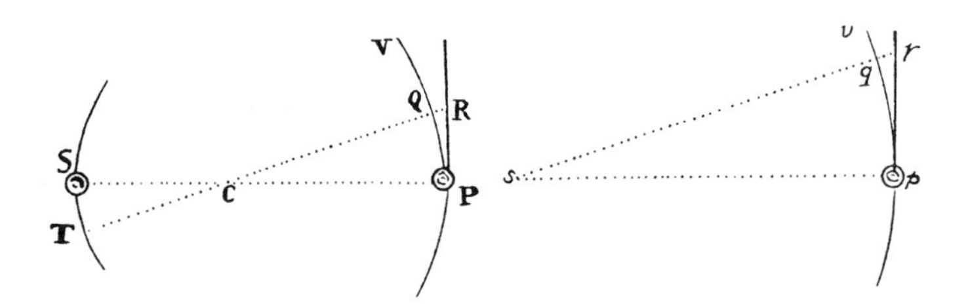

Let bodies S and P revolve about their common center of gravity C, going from S to T and from P to Q. From a given point *s* let *sp* and *sq* be drawn always equal and parallel to SP and TQ; then the curve *pqv*, which the point *p* describes by revolving around the motionless point *s*, will be similar and equal to the curves that bodies S and P describe around each other; and accordingly (by prop. 57) this curve *pqv* will be similar to the curves ST and PQV, which the same bodies describe around their common center of gravity C; and this is so because the proportions of the lines SC, CP, and SP or *sp* to one another are given.

Case 1. The common center of gravity C (by corol. 4 of the laws) either is at rest or moves uniformly straight forward. Let us suppose first that it is at rest, and at *s* and *p* let two bodies be placed, a motionless one at *s* and a moving one at *p*, similar and equal to bodies S and P. Then let the straight lines PR and *pr* touch the curves PQ and *pq* in P and *p*, and let CQ and *sq* be produced to R and *r*. Then, because the figures CPRQ and *sprq* are similar, RQ will be to *rq* as CP to *sp* and thus in a given ratio. Accordingly, if the force with which body P is attracted toward body S, and therefore toward the intermediate center C, were in that same given ratio to the force with which body *p* is attracted toward center *s*, then in equal times these forces would always attract the bodies from the tangents PR and *pr* to the arcs PQ and *pq* through the distances RQ and *rq* proportional to them; and therefore the latter force would cause body *p* to revolve in orbit in the curve *pqv*, which would be similar to the curve PQV, in which the former force causes body P to revolve in orbit, and the revolutions would be completed in the same times. But those forces are not to each other in the ratio CP to *sp* but are equal to each other (because bodies S and *s*, P and *p* are similar and equal, and distances SP and *sp* are equal); therefore, the bodies will in equal times be equally drawn away from the tangents; and therefore, for the second body *p* to be attracted through the greater distance *rq*, a greater time is required, which is as the square root of the distances, because (by lem. 10) the spaces described at the very beginning of the motion are as the squares of the times. Therefore, let the velocity of body *p* be supposed to be to the velocity of body P as the square root of the ratio of the distance *sp* to the distance CP, so that the arcs *pq* and PQ, which are in a simple ratio, are described in times which are as the square roots of the distances. Then bodies P and *p*, being always attracted by equal forces, will describe around the centers C and *s* at rest the similar figures PQV and *pqv*, of which *pqv* is similar and equal to the figure that body P describes around the moving body S. Q.E.D.

CASE 2. Let us suppose now that the common center of gravity, along with the space in which the bodies are moving with respect to each other, is moving uniformly straight forward; then (by corol. 6 of the laws) all motions in this space will occur as in case 1. Hence the bodies will describe around each other figures which are the same as before and which therefore will be similar and equal to the figure *pqv*. Q.E.D.

COROLLARY 1. Hence (by prop. 10) two bodies, attracting each other with forces proportional to their distance, describe concentric ellipses, both around their common center of gravity and also around each other; and, conversely, if such figures are described, the forces are proportional to the distance.

COROLLARY 2. And (by props. 11, 12, and 13) two bodies, under the action of forces inversely proportional to the square of the distance, describe—around their common center of gravity and also around each other—conics having their focus in that center about which the figures are described. And, conversely, if such figures are described, the centripetal forces are inversely proportional to the square of the distance.

COROLLARY 3. Any two bodies revolving in orbit around a common center of gravity describe areas proportional to the times, by radii drawn to that center and also to each other.

Proposition 59
Theorem 22

The periodic time of two bodies S *and* P *revolving about their common center of gravity* C *is to the periodic time of one of the two bodies* P, *revolving in orbit about the other body* S *which is without motion, and describing a figure similar and equal to the figures that the bodies describe around each other, as the square root of the ratio of the mass of the second body* S *to the sum of the masses of the bodies* S + P.

For, from the proof of prop. 58, the times in which any similar arcs PQ and *pq* are described are as the square roots of the distances CP and SP or *sp*, that is, as the square root of the ratio of body S to the sum of the bodies S + P [or, as $\sqrt{S}$ to $\sqrt{(S + P)}$]. And by composition [or componendo] the sums of the times in which all the similar arcs PQ and *pq* are described, that is, the whole times in which the whole similar figures are described, are in that same ratio. Q.E.D.

Proposition 60
Theorem 23

If two bodies S *and* P, *attracting each other with forces inversely proportional to the square of the distance, revolve about a common center of gravity, I say that the principal axis of the ellipse which one of the bodies* P *describes by this motion about the other body* S *will be to the principal axis of the ellipse which the same body* P *would be able to describe in the same periodic time about the other body* S *at rest as the sum of the masses of the two bodies* S + P *is to the first of two mean proportionals between this sum and the mass of the other body* S.[a]

For if the ellipses so described were equal to each other, the periodic times would (by prop. 59) be as the square root of the mass of body S is to the square root of the sum of the masses of the bodies S + P. Let the periodic time in the second ellipse be decreased in this same ratio, and then the periodic times will become equal; but the principal axis of the second ellipse (by prop. 15) will be decreased as the 3/2 power of the former ratio, that is, in the ratio of which the ratio S to S + P is the cube; and therefore the principal axis of the second ellipse will be to the principal axis of the first ellipse as the first of two mean proportionals between S + P and S to S + P. And inversely, the principal axis of the ellipse described about the body in motion will be to the principal axis of the ellipse described about the body not in motion as S + P to the first of two mean proportionals between S + P and S. Q.E.D.

Proposition 61
Theorem 24

If two bodies, attracting each other with any kind of forces and not otherwise acted on or impeded, move in any way whatever, their motions will be the same as if

a. That is, as (S + P) to the cube root of $S \times (S + P)^2$.

they were not attracting each other but were each being attracted with the same forces by a third body set in their common center of gravity. And the law of the attracting forces will be the same with respect to the distance of the bodies from that common center and with respect to the total distance between the bodies.

For the forces with which the bodies attract each other, in tending toward the bodies, tend toward a common center of gravity between them and therefore are the same as if they were emanating from a body between them. Q.E.D.

And since there is given the ratio of the distance of either of the two bodies from that common center to the distance between the bodies, there will also be given the ratio of any power of one such distance to the same power of the other distance, as well as the ratio that any quantity derived in any manner from one such distance together with given quantities has to another quantity derived in the same manner from the other distance together with the same number of given quantities having that given ratio of distances to the former ones. Accordingly, if the force with which one body is attracted by the other is directly or inversely as the distance of the bodies from each other or as any power of this distance or finally as any quantity derived in any manner from this distance and given quantities, the same force with which the same body is attracted to the common center of gravity will be likewise directly or inversely as the distance of the attracted body from that common center or as the same power of this distance or finally as a quantity derived in the same manner from this distance and analogous given quantities. That is, the law of the attracting force will be the same with respect to either of the distances. Q.E.D.

Proposition 62
Problem 38

To determine the motions of two bodies that attract each other with forces inversely proportional to the square of the distance and are let go from given places.

These bodies will (by prop. 61) move just as if they were being attracted by a third body set in their common center of gravity; and by hypothesis, that center will be at rest at the very beginning of the motion and therefore (by corol. 4 of the laws) will always be at rest. Accordingly, the motions of the bodies are (by prop. 36) to be determined just as if they were being urged by forces tending toward that center, and the motions of the bodies attracting each other will then be known. Q.E.I.

Proposition 63
Problem 39

To determine the motions of two bodies that attract each other with forces inversely proportional to the square of the distance and that set out from given places with given velocities along given straight lines.

From the given motions of the bodies at the beginning the uniform motion of the common center of gravity is given, as well as the motion of the space that moves along with this center uniformly straight forward, and also the initial motions of the bodies with respect to this space. Now (by corol. 5 of the laws and prop. 61), the subsequent motions take place in this space just as if the space itself, along with that common center of gravity, were at rest, and as if the bodies were not attracting each other but were being attracted by a third body situated in that center. Therefore the motion of either body in this moving space, setting out from a given place with a given velocity along a given straight line and pulled by a centripetal force tending toward that center, is to be determined by props. 17 and 37), and at the same time the motion of the other body about the same center will be known. This motion is to be compounded with that uniform progressive motion (found above) of the system of the space and bodies revolving in it, and the absolute motion of the bodies in an unmoving space will be known. Q.E.I.

Proposition 64
Problem 40

If the forces with which bodies attract one another increase in the simple ratio of the distances from the centers, it is required to find the motions of more than two bodies in relation to one another.

Suppose first that two bodies T and L have a common center of gravity D. These bodies will (by prop. 58, corol. 1) describe ellipses that have their centers at D and that have magnitudes which become known by prop. 10.

Now let a third body S attract the first two bodies T and L with accelerative forces ST and SL, and let it be attracted by those bodies in turn. The force ST (by corol. 2 of the laws) is resolved into forces SD and DT, and the force SL into forces SD and DL. Moreover, the forces DT and DL, which are as their sum TL and therefore as the accelerative forces with which

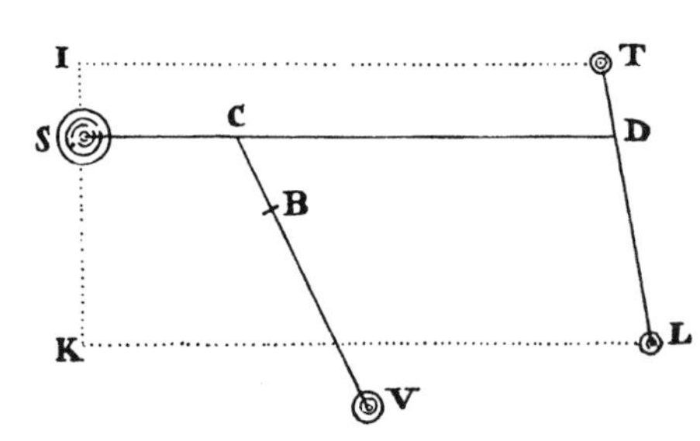

bodies T and L attract each other, when added respectively to those forces of bodies T and L, compose forces proportional to the distances DT and

DL, as before, but greater than those former forces, and therefore (by prop. 10, corol. 1, and prop. 4, corols. 1 and 8) they cause those bodies to describe ellipses as before, but with a swifter motion. The remaining accelerative forces, each of which is SD, by attracting those bodies T and L equally and along lines TI and LK (which are parallel to DS) with motive actions SD × T and SD × L (which are as the bodies), do not at all change the situations of those bodies in relation to one another, but make them equally approach line IK, which is to be conceived as drawn through the middle of body S, perpendicular to the line DS. That approach to line IK, however, will be impeded by causing the system of bodies T and L on one side and body S on the other to revolve in orbit with just the right velocities about a common center of gravity C. Body S describes an ellipse about that same point C with such a motion, because the sum of the motive forces SD × T and SD × L, which are proportional to the distance CS, tends toward the center C; and because CS and CD are proportional, point D will describe a similar ellipse directly opposite. But bodies T and L, being attracted respectively by motive forces SD × T and SD × L equally and along the parallel lines TI and LK (as has been said), will (by corols. 5 and 6 of the laws) proceed to describe their own ellipses about the moving center D, as before. Q.E.I.

Now let a fourth body V be added, and by a similar argument it will be concluded that this point and point C describe ellipses about B, the common center of gravity of all the bodies, while the motions of the former bodies T, L, and S about centers D and C remain the same as before, but accelerated. And by the same method it will be possible to add more bodies. Q.E.I.

These things are so, even if bodies T and L attract each other with accelerative forces that are greater or less than those by which they attract the rest of the bodies in proportion to the distance. Let the mutual accelerative attractions of all the bodies to one another be as the distances multiplied by the attracting bodies; then, from what has gone before, it will be easily deduced that all the bodies describe different ellipses in equal periodic times about B, the common center of gravity of them all, in a motionless plane. Q.E.I.

Proposition 65
Theorem 25

More than two bodies whose forces decrease as the squares of the distances from their centers are able to move with respect to one another in ellipses and, by radii drawn to the foci, are able to describe areas proportional to the times very nearly.

In prop. 64 the case was demonstrated in which the several motions occur exactly in ellipses. The more the law of force departs from the law there supposed, the more the bodies will perturb their mutual motions; nor can it happen that bodies will move exactly in ellipses while attracting one another according to the law here supposed, except by maintaining a fixed proportion of distances one from another. In the following cases, however, the orbits will not be very different from ellipses.

Case 1. Suppose that several lesser bodies revolve about some very much greater one at various distances from it, and that absolute forces proportional to these bodies [i.e., their masses] tend toward each and every one of them. Then, since the common center of gravity of them all (by corol. 4 of the laws) either is at rest or moves uniformly straight forward, let us imagine that the lesser bodies are so small that the greater body never is sensibly distant from this center. In this case, the greater body will—without any sensible error—either be at rest or move uniformly straight forward, while the lesser ones will revolve about this greater one in ellipses and by radii drawn to it will describe areas proportional to the times, except insofar as there are errors introduced either by a departure of the greater body from that common center of gravity or by the mutual actions of the lesser bodies on one another. The lesser bodies, however, can be diminished until that departure and the mutual actions are less than any assigned values, and therefore until the orbits square with ellipses and the areas correspond to the times without any error that is not less than any assigned value. Q.E.O.

Case 2. Let us now imagine a system of lesser bodies revolving in the way just described around a much greater one, or any other system of two bodies revolving around each other, to be moving uniformly straight forward and at the same time to be urged sideways by the force of another very much greater body, situated at a great distance. Then, since the equal accelerative forces by which the bodies are urged along parallel lines do not change the situations of the bodies in relation to one another, but cause the whole system to be transferred simultaneously, while the motions of the parts with respect to one another are maintained; it is manifest that no change whatsoever of the motion of the bodies attracted among themselves will result from their attractions toward the greater body, unless such a change comes either from the inequality of the accelerative attractions or from the inclination to one another of the lines along which the attractions take place. Suppose, therefore, that all the accelerative attractions toward the greater body are with respect to one another inversely as the squares of the distances; then by increasing the distance of the greater body until the differences (with respect to their length) among the straight lines drawn from this body to the other bodies and their inclinations with respect to one another are less than any assigned values, the motions of the parts of the system with respect to one another will persevere without any errors that are not less than any assigned values. And since, because of the slight distance of those parts from one another, the whole system is attracted as if it were one body, that system will be moved by this attraction as if it were one body; that is, by its center of gravity it will describe about the greater body some conic (namely, a hyperbola or parabola if the attraction is weak, an ellipse if the attraction is stronger) and by a radius drawn to the greater body will describe areas proportional to the times without any errors except the ones that may be produced by the distances between the parts, and these are admittedly slight and may be diminished at will. Q.E.O.

By a similar argument one can go on to more complex cases indefinitely.

COROLLARY 1. In case 2, the closer the greater body approaches to the system of two or more bodies, the more the motions of the parts of the system with respect to one another will be perturbed, because the inclinations to one another of the lines drawn from this great body to those parts are now greater, and the inequality of the proportion is likewise greater.

COROLLARY 2. But these perturbations will be greatest if the accelerative attractions of the parts of the system toward the greater body are not to one another inversely as the squares of the distances from that greater body, especially if the inequality of this proportion is greater than the inequality of the proportion of the distances from the greater body. For if the accelerative force, acting equally and along parallel lines, in no way perturbs the motions of the parts of the system with respect to one another, it will necessarily cause a perturbation to arise when there is an inequality in its action, and such perturbation will be greater or less according as this inequality is greater or less. The excess of the greater impulses acting on some bodies, but not acting on others, will necessarily change the situation of the bodies with respect to one another. And this perturbation, added to the perturbation that arises from the inclination and inequality of the lines, will make the total perturbation greater.

Corollary 3. Hence, if the parts of this system—without any significant perturbation—move in ellipses or circles, it is manifest that these parts either are not urged at all (except to a very slight degree indeed) by accelerative forces tending toward other bodies, or are all urged equally and very nearly along parallel lines.

Proposition 66[a]
Theorem 26

Let three bodies—whose forces decrease as the squares of the distances—attract one another, and let the accelerative attractions of any two toward the third be to each other inversely as the squares of the distances, and let the two lesser ones revolve about the greatest. Then I say that if that greatest body is moved by these attractions, the inner body [of the two revolving bodies] will describe about the innermost and greatest body, by radii drawn to it, areas more nearly proportional to the times and a figure more closely approaching the shape of an ellipse (having its focus in the meeting point of the radii) than would be the case if that greatest body were not attracted by the smaller ones and were at rest, or if it were much less or much more attracted and were acted on either much less or much more.

This is sufficiently clear from the demonstration of the second corollary of prop. 65, but it is proved as follows by a more lucid and more generally convincing argument.

Case 1. Let the lesser bodies P and S revolve in the same plane about a greatest body T, and let P describe the inner orbit PAB, and S the outer orbit ESE. Let SK be the mean distance between bodies P and S, and let the accelerative attraction of body P toward S at that mean distance be represented by that same line SK. Let SL be taken to SK as SK^2 to SP^2, and SL will be the accelerative attraction of body P toward S at any distance SP. Join PT, and parallel to it draw LM meeting ST in M; then the attraction SL will be resolved (by corol. 2 of the laws) into attractions SM and LM. And

a. In ed. 1, Newton used a different system of letters. In imitation of the usual form of Copernican diagram, the central body was labeled S (for "Sol," the sun) and the encircling body was P (for "Planeta," or planet). The next or outer body continued the sequence from P to Q. In ed. 2, as in ed. 3, the central body is T (suggesting "Terra" for the earth), the encircling body is still P (but now secondary planet or planetary satellite), while the outermost or perturbing body is S (suggesting "Sol"). In this way, in ed. 2 and ed. 3, Newton quite properly alerts the reader to the fact that he is basically analyzing mathematically a form of the three-body problem, exemplified by the moon moving in orbit around the earth while being perturbed by the gravitational force of the distant sun. The corollaries will not only serve for the discussion of the moon's motion in book 3 but also be used in determining the mass of the moon in book 3, prop. 37, corol. 3.

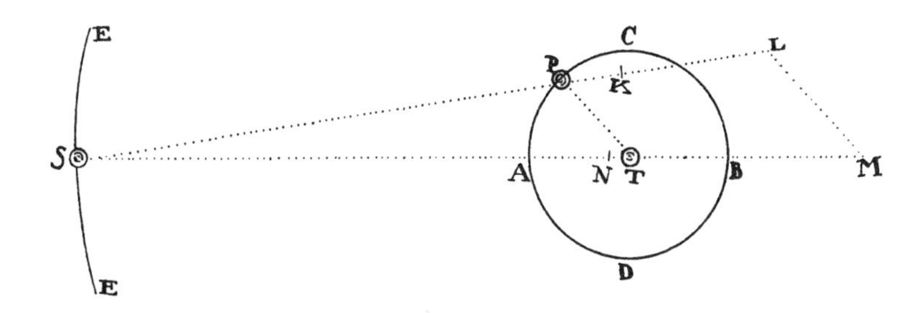

thus body P will be urged by a threefold accelerative force. One such force tends toward T and arises from the mutual attraction of bodies T and P. By this force alone (whether T is motionless or is moved by this attraction), body P must, by a radius PT, describe around body T areas proportional to the times and must also describe an ellipse whose focus is in the center of body T. This is clear from prop. 11 and prop. 58, corols. 2 and 3.

The second force is that of the attraction LM, which (since it tends from P to T) will, when added to the first of these forces, coincide with it and will thus cause areas to be described that are still proportional to the times, by prop. 58, corol. 3. But since this force is not inversely proportional to the square of the distance PT, it will, together with the first force, compose a force differing from this proportion—and the more so, the greater the proportion of this force is to that first force, other things being equal. Accordingly, since (by prop. 11 and by prop. 58, corol. 2) the force by which an ellipse is described about the focus T must tend toward that focus and be inversely proportional to the square of the distance PT, that composite force, by differing from this proportion, will cause the orbit PAB to deviate from the shape of an ellipse having its focus in T, and the more so the greater the difference from this proportion; and the difference from this proportion will be greater according as the proportion of the second force LM to the first force is greater, other things being equal.

But now the third force SM, by attracting body P along a line parallel to ST, will, together with the former forces, compose a force which is no longer directed from P to T and which deviates from this direction the more, the greater the proportion of this third force is to the former forces, other things being equal; and this compound force therefore will make body P describe, by a radius TP, areas no longer proportional to the times and will make the divergence from this proportionality be the greater, the greater the proportion of this third force is to the other forces. This third force will increase the deviation of the orbit PAB from the aforesaid elliptical shape for two reasons: not only is this force not directed from P to T, but also it is not inversely proportional to the square of the distance PT. Once these things have been understood, it is manifest that the areas will be most nearly proportional to the times when this third force is least, the other forces remaining the same as they were; and that the orbit PAB approaches closest to the aforesaid elliptical shape when both the second force and the third (but especially the third force) are least, the first force remaining the same as it was.

Let the accelerative attraction of body T toward S be represented by line SN; and if the accelerative attractions SM and SN were equal, they would, by attracting bodies T and P equally and along parallel lines, not at all change the situation of those two bodies with respect to each other. In this case, their motions with respect to each other would (by corol. 6 of the laws) be the same as it would be without these attractions. And for the same reason, if the attraction SN were smaller than the attraction SM, it would take away the part SN of the attraction SM, and only the part MN would remain, by which the proportionality of the times and areas and the elliptical shape of the orbit would be perturbed. And similarly, if the attraction SN were greater than the attraction SM, the perturbation of the proportionality and of the orbit would arise from the difference MN alone. Thus SM, the third attraction above, is always reduced by the attraction SN to the attraction MN, the first and second attractions remaining completely unchanged; and therefore the areas and times approach closest to proportionality, and the orbit PAB approaches closest to the aforesaid elliptical shape, when the attraction MN is either null or the least possible—that is, when the accelerative attractions of bodies P and T toward body S approach as nearly as possible to equality, in other words, when the attraction SN is neither null nor less than the least of all the attractions SM, but is a kind of mean between the maximum and minimum of all those attractions SM, that is, not much greater and not much smaller than the attraction SK. Q.E.D.

CASE 2. Now let the lesser bodies P and S revolve about the greatest body T in different planes; then the force LM, acting along a line PT situated in the plane of orbit PAB, will have the same effect as before, and will not draw body P away from the plane of its orbit. But the second force NM, acting along a line that is parallel to ST (and therefore, when body S is outside the line of the nodes, is inclined to the plane of orbit PAB), besides

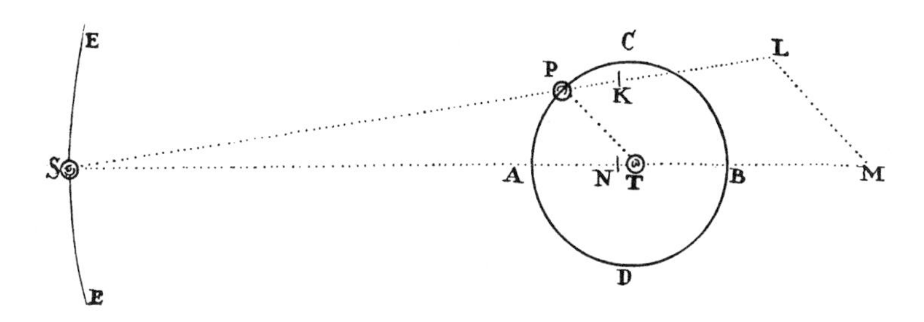

the perturbation of its motion in longitude, already set forth above, will introduce a perturbation of the motion in latitude, by attracting body P out of the plane of its orbit. And this perturbation, in any given situation of bodies P and T with respect to each other, will be as the generating force MN, and therefore becomes least when MN is least, that is (as I have already explained), when the attraction SN is not much greater and not much smaller than the attraction SK. Q.E.D.

COROLLARY 1. Hence it is easily gathered that if several lesser bodies P, S, R, . . . revolve about a greatest body T, the motion of the innermost body P will be least perturbed by the attractions of the outer bodies when the greatest body T is attracted and acted on as much by the other bodies (according to the ratio of the accelerative forces) as the other bodies are by one another.

COROLLARY 2. In a system of three bodies T, P, and S, if the accelerative attractions of any two toward the third are to each other inversely as the squares of the distances, body P will describe, by a radius PT, an area about body T more swiftly near their conjunction A and their opposition B than near the quadratures C and D. For every force by which body P is urged and body T is not, and which does not act along line PT, accelerates or retards the description of areas, according as its direction is forward and direct [or in consequentia] or retrograde [or in antecedentia]. Such is the force NM. In the passage of body P from C to A, this force is directed forward [or in consequentia] and accelerates the motion; afterward, as far as D, it is retrograde [or in antecedentia] and retards the motion; then forward up to B, and finally retrograde in passing from B to C.

COROLLARY 3. And by the same argument it is evident that body P, other things being the same, moves more swiftly in conjunction and opposition than in the quadratures.

Corollary 4. The orbit of body P, other things being the same, is more curved in the quadratures than in conjunction and opposition. For swifter bodies are deflected less from a straight path. And besides, in conjunction and opposition the force KL, or NM, is opposite to the force with which body T attracts body P and therefore diminishes that force, while body P will be deflected less from a straight path when it is less urged toward body T.

Corollary 5. Accordingly, body P, other things being the same, will recede further from body T in the quadratures than in conjunction and opposition. These things are so if the motion of [i.e., change in] eccentricity is neglected. For if the orbit of body P is eccentric, its eccentricity (as will shortly be shown in corol. 9 of this proposition) will come out greatest when the apsides are in the syzygies; and thus it can happen that body P, arriving at the upper apsis, may be further away from body T in the syzygies than in the quadratures.

Corollary 6. Since the centripetal force of the central body T, which keeps body P in its orbit, is increased in the quadratures by the addition of the force LM and is diminished in the syzygies by the subtraction of the force KL and, because of the magnitude of the force KL [which is greater than LM], is more diminished than increased; and since that centripetal force (by prop. 4, corol. 2) is in a ratio compounded of the simple ratio of the radius TP directly and the squared ratio of the periodic time inversely [i.e., the force is directly as the radius and inversely as the square of the periodic time], it is evident that this compound ratio is diminished by the action of the force KL, and therefore that the periodic time (assuming the radius TP of the orbit to remain unchanged) is increased as the square root of the ratio in which that centripetal force is diminished. It is therefore further evident that, assuming this radius to be increased or diminished, the periodic time is increased more or diminished less than as the $3/2$ power of this radius, by prop. 4, corol. 6. If the force of the central body were gradually to weaken, body P, attracted always less and less, would continually recede further and further from the center T; and on the contrary, if the force were increased, body P would approach nearer and nearer. Therefore, if the action of the distant body S, whereby the force is diminished, is alternately increased and diminished, radius TP will at the same time also be alternately increased and diminished, and the periodic time will be increased and diminished in a ratio compounded of the $3/2$ power of the ratio of the radius and the square root of the ratio in which the centripetal force of the central body T is diminished or increased by the increase or decrease of the action of the distant body S.

COROLLARY 7. From what has gone before, it follows also that with respect to angular motion the axis of the ellipse described by body P, or the line of the apsides, advances and regresses alternately, but nevertheless advances more than it regresses and is carried forward [or in consequentia] by the excess of its direct forward motion. For the force whereby body P is

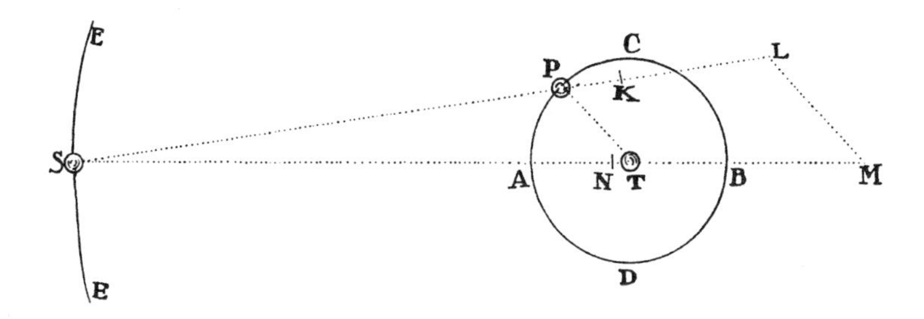

urged toward body T in the quadratures, when the force MN vanishes, is compounded of the force LM and the centripetal force with which body T attracts body P. If the distance PT is increased, the first force LM is increased in about the same ratio as this distance, and the latter force is decreased as the square of that ratio, and so the sum of these forces is decreased in a less than squared ratio of the distance PT, and therefore (by prop. 45, corol. 1) causes the auge, or upper apsis, to regress. But in conjunction and opposition the force whereby body P is urged toward body T is the difference between the force by which body T attracts body P and the force KL; and that difference, because the force KL is increased very nearly in the ratio of the distance PT, decreases in a ratio of the distance PT that is greater than the square of the distance PT, and so (by prop. 45, corol. 1) causes the upper apsis to advance. In places between the syzygies and quadratures the motion of the upper apsis depends on both of these causes jointly, so that according to the excess of the one or the other it advances or regresses. Accordingly, since the force KL in the syzygies is roughly twice as large as the force LM in the quadratures, the excess will have the same sense as the force KL and will carry the upper apsis forward [or in consequentia]. The truth of this corollary and its predecessor will be easily understood by supposing that a system of two bodies T and P is surrounded on all sides by more bodies S, S, S, . . . that are in an orbit ESE. For by the actions of these bodies, the action of T will be diminished on all sides and will decrease in a ratio greater than the square of the distance.

COROLLARY 8. Since, however, the advance or retrogression of the apsides depends on the decrease of the centripetal force, a decrease occurring in a ratio of the distance TP that is either greater or less than the square of the ratio of the distance TP, in the passage of the body from the lower to the upper apsis, and also depends on a similar increase in its return to the lower apsis, and therefore is greatest when the proportion of the force in the upper apsis to the force in the lower apsis differs most from the ratio of the inverse squares of the distances, it is manifest that KL or NM − LM, the force that subtracts, will cause the apsides to advance more swiftly in their syzygies and that LM, the force that adds, will cause them to recede more slowly in their quadratures. And because of the length of time in which the swiftness of the advance or slowness of the retrogression is continued, this inequality becomes by far the greatest.

COROLLARY 9. If a body, by the action of a force inversely proportional to the square of its distance from a center, were to revolve about this center in an ellipse, and if then, in its descent from the upper apsis or auge to the lower apsis, that force—because of the continual addition of a new force—were increased in a ratio that is greater than the square of the diminished distance, it is manifest that that body, being always impelled toward the center by the continual addition of that new force, would incline toward this center more than if it were urged only by a force increasing as the square of the diminished distance, and therefore would describe an orbit inside the elliptical orbit and in its lower apsis would approach nearer to the center than before. Therefore by the addition of this new force, the eccentricity of the orbit will be increased. Now if, during the receding of the body from the lower to the upper apsis, the force were to decrease by the same degrees by which it had previously increased, the body would return to its former distance; and so, if the force decreases in a greater ratio, the body, now attracted less, will ascend to a greater distance, and thus the eccentricity of its orbit will be increased still more. And therefore, if the ratio of the increase and decrease of the centripetal force is increased in each revolution, the eccentricity will always be increased; and contrariwise, the eccentricity will be diminished if that ratio decreases.

Now, in the system of bodies T, P, and S, when the apsides of the orbit PAB are in the quadratures, this ratio of the increase and decrease is least, and it becomes greatest when the apsides are in the syzygies. If the apsides are in the quadratures, the ratio near the apsides is smaller and near the syzygies is greater than the squared ratio of the distances, and from that greater ratio arises the forward or direct motion of the upper apsis, as has already been stated. But if one considers the ratio of the total increase or decrease in the forward motion between the apsides, this ratio is smaller than the squared ratio of the distances. The force in the lower apsis is to the force in the upper apsis in a ratio that is less than the squared ratio of the distance of the upper apsis from the focus of the ellipse to the distance of the lower apsis from that same focus; and conversely, when the apsides are in the syzygies, the force in the lower apsis is to the force in the upper apsis in a ratio greater than that of the squares of the distances.

For the forces LM in the quadratures, added to the forces of body T, compose forces in a smaller ratio, and the forces KL in the syzygies, subtracted from the forces of body T, leave forces in a greater ratio. Therefore, the ratio of the total decrease and increase during the passage between apsides is least in the quadratures and greatest in the syzygies; and therefore, during the passage of the apsides from quadratures to syzygies, this ratio is continually increased and it increases the eccentricity of the ellipse; and in the passage from syzygies to quadratures, this ratio is continually diminished and it diminishes the eccentricity.

COROLLARY 10. To give an account of the errors in latitude, let us imagine that the plane of the orbit EST remains motionless; then from the cause of errors just expounded, it is manifest that of the forces NM and ML (which are the entire cause of these errors) the force ML, always acting in the plane of the orbit PAB, never perturbs the motions in latitude. It is likewise manifest that when the nodes are in the syzygies, the force NM, also acting in

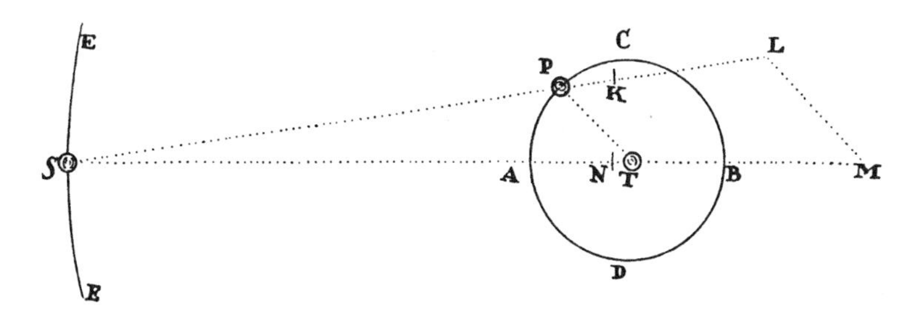

the same plane of the orbit, does not perturb these motions; but when the nodes are in the quadratures, this force perturbs those motions to the greatest extent, and—by continually attracting body P away from the plane of its orbit—diminishes the inclination of the plane during the passage of the body from quadratures to syzygies and increases that inclination in turn during the passage from syzygies to quadratures. Hence it happens that when the body is in the syzygies the inclination turns out to be least of all, and it returns approximately to its former magnitude when the body comes to the next node. But if the nodes are situated in the octants after the quadratures, that is, between C and A, or D and B, it will be understood from what has just been explained that in the passage of body P from either node to a position 90 degrees from there, the inclination of the plane is continually diminished; then, in its passage through the next 45 degrees to the next quadrature, the inclination is increased; and afterward, in its next passage through another 45 degrees to the next node, it is diminished. Therefore, the inclination is diminished more than it is increased, and hence it is always less in each successive node than in the immediately preceding one. And by a similar reasoning, it follows that the inclination is increased more than it is diminished when the nodes are in the other octants between A and B, or B and C. Thus, when the nodes are in the syzygies, the inclination is greatest of all. In the passage of the nodes from syzygies to quadratures, the inclination is diminished in each appulse of the body to the nodes, and it becomes least of all when the nodes are in the quadratures and the body is in the syzygies; then it increases by the same degrees by which it had previously decreased, and at the appulse of the nodes to the nearest syzygies it returns to its original magnitude.

Corollary 11. When the nodes are in the quadratures, the body P is continually attracted away from the plane of its orbit in the direction toward S, during its passage from the node C through the conjunction A to the node D, and in the opposite direction in its passage from node D through

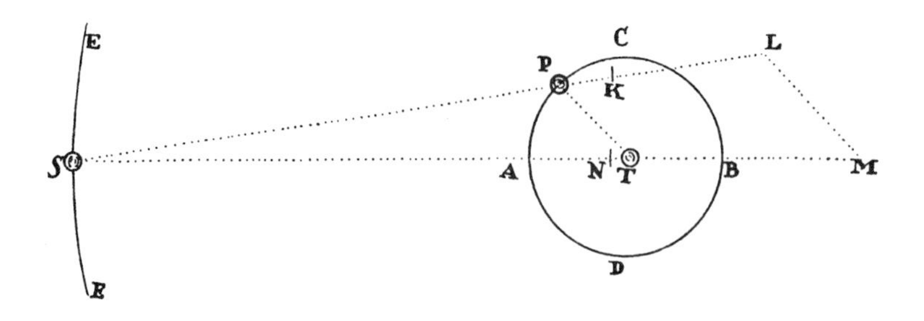

opposition B to node C; hence it is manifest that the body, in its motion from node C, continually recedes from the first plane CD of its orbit until it has reached the next node; and therefore at this node, being at the greatest distance from that first plane CD, it passes through EST, the plane of the orbit, not in the other node D of that plane but in a point that is closer to body S and which accordingly is a new place of the node, behind its former place. By a similar argument the nodes will continue to recede in the passage of the body from this node to the next node. Hence the nodes, when situated in the quadratures, continually recede; in the syzygies, when the motion in latitude is not at all perturbed, the nodes are at rest; in the intermediate places, since they share in both conditions, they recede more slowly; and therefore, since the nodes always either have a retrograde motion or are stationary, they are carried backward [or in antecedentia] in each revolution.

COROLLARY 12. All the errors described in these corollaries are slightly greater in the conjunction of bodies P and S than in their opposition; and this occurs because then the generating forces NM and ML are greater.

COROLLARY 13. And since the proportions in these corollaries do not depend on the magnitude of the body S, all the preceding statements are valid when the magnitude of body S is assumed to be so great that the system of two bodies T and P will revolve about it. And from this increase of body S, and consequently the increase of its centripetal force (from which the errors of body P arise), all those errors will—at equal distances—come out greater in this case than in the other, in which body S revolves around the system of bodies P and T.

COROLLARY 14.[b] When body S is extremely far away, the forces NM and ML are very nearly as the force SK and the ratio of PT to ST jointly (that is, if both the distance PT and the absolute force of body S are given, as ST^3 inversely), and those forces NM and ML are the causes of all the errors and effects that have been dealt with in the preceding corollaries; hence it is manifest that all these effects—if the system of bodies T and P stays the same and only the distance ST and the absolute force of body S are changed—are very nearly in a ratio compounded of the direct ratio of the absolute force of body S and the inverse ratio of the cube of the distance ST. Accordingly, if the system of bodies T and P revolves about the distant body S, those forces NM and ML and their effects will (by prop. 4, corols. 2 and 6) be inversely

b. For a gloss on this corollary see the Guide, §10.16.

as the square of the periodic time. And hence also, if the magnitude of body S is proportional to its absolute force, those forces NM and ML and their effects will be directly as the cube of the apparent diameter of the distant body S when looked at from body T, and conversely. For these ratios are the same as the above-mentioned compounded ratio.

Corollary 15. If the magnitudes of the orbits ESE and PAB are changed, while their forms and their proportions and inclinations to each other remain the same, and if the forces of bodies S and T either remain the same or are changed in any given ratio, then these forces (that is, the force of body T, by whose action body P is compelled to deflect from a straight path into an orbit PAB; and the force of body S, by whose action that same body P is compelled to deviate from that orbit) will always act in the same way and in the same proportion; thus it will necessarily be the case that all the effects will be similar and proportional and that the times for these effects will be proportional as well—that is, all the linear errors will be as the diameters of the orbits, the angular errors will be the same as before, and the times of similar linear errors or of equal angular errors will be as the periodic times of the orbits.

Corollary 16. And hence, if the forms of the orbits and their inclination to each other are given, and the magnitudes, forces, and distances of the bodies are changed in any way, then from the given errors and given times of errors in one case there can be found the errors and times of errors in any other case very nearly. This may be done more briefly, however, by the following method. The forces NM and ML, other things remaining the same, are as the radius TP, and their periodic effects are (by lem. 10, corol. 2) jointly as the forces and the square of the periodic time of body P. These are the linear errors of body P, and hence the angular errors as seen from the center T (that is, the motions of the upper apsis and of the nodes, as well as all the apparent errors in longitude and latitude) are in any revolution of body P very nearly as the square of the time of revolution. Let these ratios

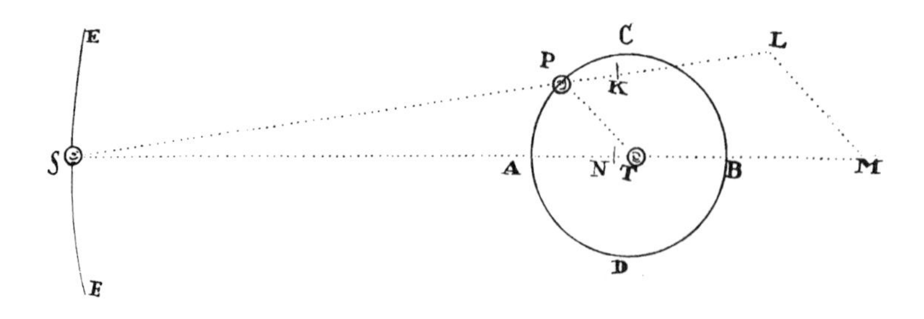

be compounded with the ratios of corol. 14; then in any system of bodies T, P, and S, in which P revolves around T which is near to it and T revolves around a distant S, the angular errors of body P, as seen from the center T, will—in each revolution of that body P—be as the square of the periodic time of body P directly and the square of the periodic time of body T inversely. And thus the mean motion of the upper apsis will be in a given ratio to the mean motion of the nodes, and each of the two motions will be as the periodic time of body P directly and the square of the periodic time of body T inversely. By increasing or decreasing the eccentricity and inclination of the orbit PAB, the motions of the upper apsis and of the nodes are not changed sensibly, except when the eccentricity and inclination are too great.

COROLLARY 17. Since, however, the line LM is sometimes greater and sometimes less than the radius PT, let the mean force LM be represented by that radius PT; then this force will be to the mean force SK or SN (which can be represented by ST) as the length PT to the length ST. But the mean force SN or ST by which body T is kept in its orbit around S is to the force by which body P is kept in its orbit around T in a ratio compounded of the ratio of the radius ST to the radius PT and the square of the ratio of the periodic time of body P around T to the periodic time of body T around S. And from the equality of the ratios [or ex aequo] the mean force LN is to the force by which a body P is kept in its orbit around T (or by which the same body P could revolve in the same periodic time around any immobile point T at a distance PT) in the same squared ratio of the periodic times. Therefore, if the periodic times are given, along with the distance PT, the mean force LM is also given; and if the force LM is given, the force MN is also given very nearly by the proportion of lines PT and MN.

COROLLARY 18. Let us imagine many fluid bodies to move around body T at equal distances from it according to the same laws by which body P revolves around the same body T; then let a ring—fluid, round, and concentric to body T—be produced by making these individual fluid bodies come into contact with one another; these individual parts of the ring, carrying out all their motions according to the law of body P, will approach closer to body T and will move more swiftly in the conjunction and opposition of themselves and body S than in the quadratures. The nodes of this ring, or its intersections with the plane of the orbit of body S or T, will be at rest in the syzygies, but outside the syzygies they will move backward [or in antecedentia], and do so most swiftly in the quadratures and more slowly in other places. The inclination of the ring will also vary, and its axis will oscillate in each revolution; and when a revolution has been completed, it will return to its original position except insofar as it is carried around by the precession of the nodes.

Corollary 19. Now imagine the globe T, which consists of nonfluid matter, to be so enlarged as to extend out to this ring, and to have a channel to contain water dug out around its whole circumference; and imagine this new globe to revolve uniformly about its axis with the same periodic motion. This water, being alternately accelerated and retarded (as in the previous corollary), will be swifter in the syzygies and slower in the quadratures than the surface of the globe itself, and thus will ebb and flow in the channel just as the sea does. If the attraction of body S is taken away, the water—now revolving about the quiescent center of the globe—will acquire no motion of ebb and flow. This is likewise the case for a globe advancing uniformly straight forward and meanwhile revolving about its own center (by corol. 5 of the laws), and also for a globe uniformly attracted away from a rectilinear path (by corol. 6 of the laws). But let body S now draw near, and by its nonuniform attraction of the water, the water will soon be disturbed. For its attraction of the nearer water will be greater and that of the more distant water will be smaller. Moreover, the force LM will attract the water downward in the quadratures and will make it descend as far as the syzygies, and the force KL will attract this same water upward in the syzygies and will prevent its further descent and will make it ascend as far as the quadratures, except insofar as the motion of ebb and flow is directed by the channel of water and is somewhat retarded by friction.

Corollary 20. If the ring now becomes hard and the globe is diminished, the motion of ebb and flow will cease; but the oscillatory motion of the inclination and the precession of the nodes will remain. Let the globe have the same axis as the ring and complete its revolutions in the same times, and let its surface touch the inside of the ring and adhere to it; then, with the

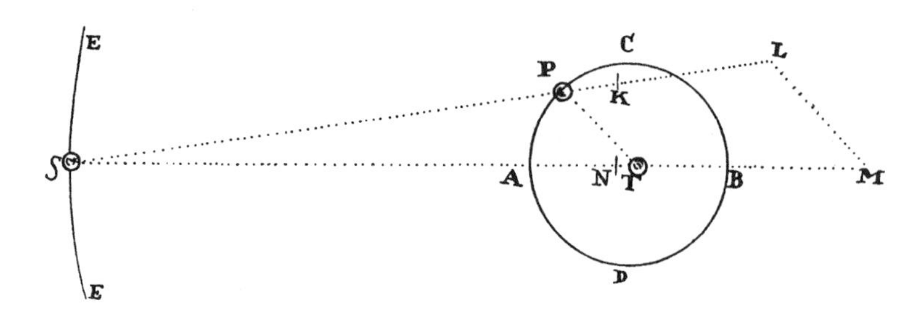

globe participating in the motion of the ring, the structure of the two will oscillate and the nodes will regress. For the globe, as will be shown presently, is susceptible to all impressions equally. The greatest angle of inclination of the ring alone, with the globe removed, occurs when the nodes are in the syzygies. From there in the forward motion of the nodes to the quadratures it endeavors to diminish its inclination and by that endeavor impresses a motion upon the whole globe. The globe keeps this impressed motion until the ring removes this motion by an opposite endeavor and impresses a new motion in the opposite direction; and in this way the greatest motion of the decreasing inclination occurs when the nodes are in the quadratures, and the least angle of inclination occurs in the octants after the quadratures; and the greatest motion of reclination occurs in the syzygies, and the greatest angle in the next octants. And this is likewise the case for a globe which has no such ring and which in the regions of the equator is either a little higher than near the poles or consists of matter a little denser. For that excess of matter in the regions of the equator takes the place of a ring. And although, by increasing the centripetal force of this globe in any way whatever, all its parts are supposed to tend downward, as the gravitating parts of the earth do, nevertheless the phenomena of this corollary and of corol. 19 will scarcely be changed on that account, except that the places of the greatest and least height of the water will be different. For the water is now sustained and remains in its orbit not by its own centrifugal force but by the channel in which it is flowing. And besides, the force LM attracts the water downward to the greatest degree in the quadratures, and the force KL or NM − LM attracts the same water upward to the greatest degree in the syzygies. And these forces conjoined cease to attract the water downward and begin to attract the water upward in the octants before the syzygies, and they cease to attract the water upward and begin to attract the water downward in the octants after the syzygies. As a result, the greatest height of the water can occur very nearly in the octants after the syzygies, and the least height can occur very nearly in the octants after the quadratures, except insofar as the motion of ascent or descent impressed on the water by these forces either perseveres a little longer because of the inherent force of the water or is stopped a little more swiftly because of the impediments of the channel.

COROLLARY 21. In the same way that the excess matter of a globe near its equator makes the nodes regress (and thus the retrogression is increased by increase of equatorial matter and is diminished by its diminution and is removed by its removal), it follows that if more than the excess matter is removed, that is, if the globe near the equator is made either more depressed or more rare than near the poles, there will arise a motion of the nodes forward [or in consequentia].

COROLLARY 22. And thus, in turn, from the motion of the nodes the constitution of a globe can be found. That is to say, if a globe constantly preserves the same poles and there occurs a motion backward [or in antecedentia], there is an excess of matter near the equator; if there occurs a motion forward [or in consequentia], there is a deficiency. Suppose that a uniform and perfectly spherical globe is at first at rest in free space; then is propelled by any impetus whatever delivered obliquely upon its surface, from which it takes on a motion that is partly circular [i.e., rotational] and partly straight forward. Because the globe is indifferent to all axes passing through its center and does not have a greater tendency to turn around any one axis or an axis at any particular inclination, it is clear that the globe, by its own force alone, will never change its axis and the inclination of the axis. Now let the globe be impelled obliquely by any new impulse whatever, delivered to that same part of the surface as before; then, since the effect of an impulse is in no way changed by its being delivered sooner or later, it is manifest that the same motion will be produced by these two impulses being successively impressed as if they had been impressed simultaneously, that is, the resultant motion will be the same as if the globe had been impelled by a simple force compounded of these two (by corol. 2 of the laws), and hence will be a simple motion about an axis of a given inclination. This is likewise the case for a second impulse impressed in any other place on the equator of the first motion; and also for a first impulse impressed in any place on the equator of the motion which the second impulse would generate without the first, and hence for both impulses impressed in any places whatever. These two impulses will generate the same circular motion as if they had been impressed together and all at once in the place of intersection of the equators of the motions which each of them would generate separately. Therefore a homogeneous and perfect globe does not retain several distinct motions but compounds all the motions impressed on it and reduces them to one; and insofar as it can in and of itself, it always rotates with a simple and uniform motion about a single axis of a given and always invariable inclination. A centripetal force cannot change either this inclination of the axis or the velocity of rotation.

If a globe is thought of as divided into two hemispheres by any plane passing through the center of the globe and the center toward which a force is directed, that force will always urge both hemispheres equally and therefore will not cause the globe—as regards its motion of rotation—to incline in any direction. Let some new matter, heaped up in the shape of a mountain, be added to the globe anywhere between the pole and the equator; then this matter, by its continual endeavor to recede from the center of its motion, will disturb the motion of the globe and will make its poles wander over its surface and continually describe circles about themselves and the point opposite to them. And this tremendous wandering of the poles will not be corrected, save by placing the mountain either in one of the two poles, in which case (by corol. 21) the nodes of the equator will advance, or on the equator, in which case (by corol. 20) the nodes will regress, or finally by placing on the other side of the axis some additional matter by which the mountain is balanced in its motion, and in this way the nodes will either advance or regress, according as the mountain and this new matter are closer to a pole or to the equator.

Proposition 67
Theorem 27

With the same laws of attraction being supposed, I say that with respect to the common center of gravity O *of the inner bodies* P *and* T, *the outer body* S*—by radii drawn to that center—describes areas more nearly proportional to the times, and an orbit more closely approaching the shape of an ellipse having its focus in that same center, than it can describe about the innermost and greatest body* T *by radii drawn to that body.*

For the attractions of body S toward T and P compose its absolute attraction, which is directed more toward the common center of gravity O of bodies T and P than toward the greatest body T, and which is more nearly inversely proportional to the square of the distance SO than to the square of the distance ST, as will easily be seen by anyone carefully considering the matter.

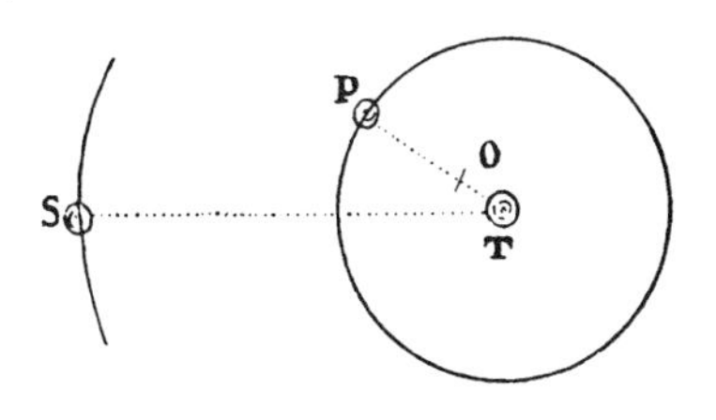

Proposition 68
Theorem 28

With the same laws of attraction being supposed, I say that with respect to the common center of gravity O *of the inner bodies* P *and* T, *the outer body* S*—by* *radii drawn to that center—describes areas more nearly proportional to the times, and an orbit more closely approaching the shape of an ellipse having its focus in the same center, if the innermost and greatest body is acted on by these attractions just as the others are, than would be the case if it is either not attracted and is at rest or is much more or much less attracted or much more or much less moved.*

This is demonstrated in almost the same way as prop. 66, but the proof is more prolix and I therefore omit it. The following considerations should suffice.

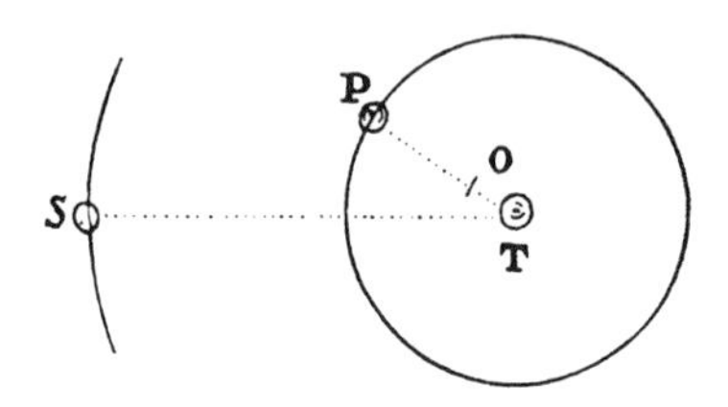

From the demonstration of the last proposition it is apparent that the center toward which body S is urged by both forces combined is very near to the common center of gravity of the other bodies P and T. If this center were to coincide with the common center of those two bodies, and the common center of gravity of all three bodies were to be at rest, body S on the one hand and the common center of the other two bodies on the other would describe exact ellipses about the common center of them all which is at rest. This is clear from the second corollary of prop. 58 compared with what is demonstrated in props. 64 and 65. Such an exact elliptical motion is perturbed somewhat by the distance of the center of the two bodies from the center toward which the third body S is attracted. Let a motion be given, in addition, to the common center of the three, and the perturbation will be increased. Accordingly, the perturbation is least when the common center of the three is at rest, that is, when the innermost and greatest body T is attracted by the very same law as the others; and it always becomes greater when the common center of the three bodies, by a diminution of the motion of body T, begins to be moved and thereupon acted on more and more.

Corollary. And hence, if several lesser bodies revolve about a greatest one, it can be found that the orbits described will approach closer to elliptical orbits, and the descriptions of areas will become more uniform, if all the bodies attract and act on one another by accelerative forces that are directly as their absolute forces and inversely as the squares of the distances, and if the focus of each orbit is located in the common center of gravity of all the inner bodies (that is to say, with the focus of the first and innermost orbit in the center of gravity of the greatest and innermost body; the focus of the second orbit in the common center of gravity of the two innermost bodies; the focus of the third in the common center of gravity of the three inner bodies; and so on), than if the innermost body is at rest and is set at the common focus of all the orbits.

Proposition 69
Theorem 29

If, in a system of several bodies A, B, C, D, . . . , *some body* A *attracts all the others,* B, C, D, . . . , *by accelerative forces that are inversely as the squares of the distances from the attracting body; and if another body* B *also attracts the rest of the bodies* A, C, D, . . . , *by forces that are inversely as the squares of the distances from the attracting body; then the absolute forces of the attracting bodies* A *and* B *will be to each other in the same ratio as the bodies [i.e., the masses]* A *and* B *themselves to which those forces belong.*

For, at equal distances, the accelerative attractions of all the bodies B, C, D, . . . toward A are equal to one another by hypothesis; and similarly, at equal distances, the accelerative attractions of all the bodies toward B are equal to one another. Moreover, at equal distances, the absolute attractive force of body A is to the absolute attractive force of body B as the accelerative attraction of all the bodies toward A is to the accelerative attraction of all the bodies toward B at equal distances; and the accelerative attraction of body B toward A is also in the same proportion to the accelerative attraction of body A toward B. But the accelerative attraction of body B toward A is to the accelerative attraction of body A toward B as the mass of body A is to the mass of body B, because the motive forces—which (by defs. 2, 7, and 8) are as the accelerative forces and the attracted bodies jointly—are in this case (by the third law of motion) equal to each other. Therefore the absolute attractive force of body A is to the absolute attractive force of body B as the mass of body A is to the mass of body B. Q.E.D.

COROLLARY 1. Hence if each of the individual bodies of the system A, B, C, D, . . . , considered separately, attracts all the others by accelerative forces that are inversely as the squares of the distances from the attracting body, the absolute forces of all those bodies will be to one another in the ratios of the bodies [i.e., the masses] themselves.

COROLLARY 2. By the same argument, if each of the individual bodies of the system A, B, C, D, . . . , considered separately, attracts all the others by accelerative forces that are either inversely or directly as any powers whatever of the distances from the attracting body, or that are defined in terms of the distances from each one of the attracting bodies according to any law common to all these bodies; then it is evident that the absolute forces of those bodies are as the bodies [i.e., the masses].

COROLLARY 3. If, in a system of bodies whose forces decrease in the squared ratio of the distances [i.e., vary inversely as the squares of the distances], the lesser bodies revolve about the greatest one in ellipses as exact as they can be, having their common focus in the center of that greatest body, and—by radii drawn to the greatest body—describe areas as nearly as possible proportional to the times, then the absolute forces of those bodies will be to one another, either exactly or very nearly, as the bodies, and conversely. This is clear from the corollary of prop. 68 compared with corol. 1 of this proposition.

Scholium By these propositions we are directed to the analogy between centripetal forces and the central bodies toward which those forces tend. For it is reasonable that forces directed toward bodies depend on the nature and the quantity of matter of such bodies, as happens in the case of magnetic bodies. And whenever cases of this sort occur, the attractions of the bodies must be reckoned by assigning proper forces to their individual particles and then taking the sums of these forces.

I use the word "attraction" here in a general sense for any endeavor whatever of bodies to approach one another, whether that endeavor occurs as a result of the action of the bodies either drawn toward one another or acting on one another by means of spirits emitted or whether it arises from the action of aether or of air or of any medium whatsoever—whether corporeal or incorporeal—in any way impelling toward one another the bodies floating therein. I use the word "impulse" in the same general sense, considering in this treatise not the species of forces and their physical qualities but their quantities and mathematical proportions, as I have explained in the definitions.

Mathematics requires an investigation of those quantities of forces and their proportions that follow from any conditions that may be supposed. Then, coming down to physics, these proportions must be compared with the phenomena, so that it may be found out which conditions [or laws] of forces apply to each kind of attracting bodies. And then, finally, it will be possible to argue more securely concerning the physical species, physical causes, and physical proportions of these forces. Let us see, therefore, what the forces are by which spherical bodies, consisting of particles that attract in the way already set forth, must act upon one another, and what sorts of motions result from such forces.

SECTION 12

The attractive forces of spherical bodies

Proposition 70
Theorem 30

If toward each of the separate points of a spherical surface there tend equal centripetal forces decreasing as the squares of the distances from the point, I say that a corpuscle placed inside the surface will not be attracted by these forces in any direction.

Let HIKL be the spherical surface, and P the corpuscle placed inside. Through P draw to this surface the two lines HK and IL intercepting minimally small arcs HI and KL; and because triangles HPI and LPK are similar (by lem. 7, corol. 3), those arcs will be proportional to the distances HP and LP; and any particles of the spherical surface at HI and KL, terminated everywhere by straight lines passing through point P, will be in that proportion squared. Therefore the forces exerted on body P by these particles of surface are equal to one another. For they are as the particles directly and the squares of the distances inversely. And these two ratios, when compounded, give the ratio of equality. The attractions, therefore, being made equally in opposite directions, annul each other. And by a similar argument, all the attractions throughout the whole spherical surface are annulled by opposite attractions. Accordingly, body P is not impelled by these attractions in any direction. Q.E.D.

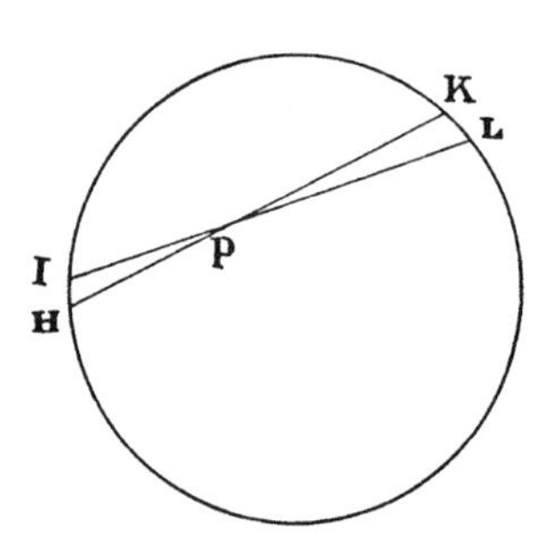

Proposition 71
Theorem 31

With the same conditions being supposed as in prop. 70, I say that a corpuscle placed outside the spherical surface is attracted to the center of the sphere by a force inversely proportional to the square of its distance from that same center.

Let AHKB and *ahkb* be two equal spherical surfaces, described about centers S and *s* with diameters AB and *ab*, and let P and *p* be corpuscles located outside those spheres in those diameters produced. From the corpuscles draw lines PHK, PIL, *phk*, and *pil*, so as to cut off from the great circles AHB and *ahb* the equal arcs HK and *hk*, and IL and *il*. And onto these lines drop perpendiculars SD and *sd*, SE and *se*, IR and *ir*, of which SD and *sd* cut PL and *pl* at F and *f*. Also drop perpendiculars IQ and

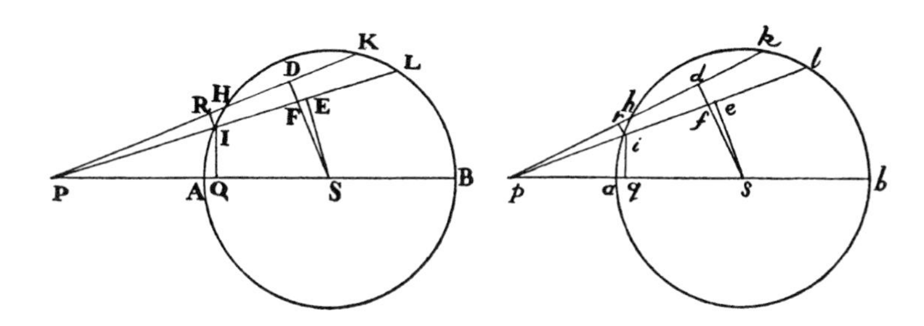

iq onto the diameters. Let angles DPE and *dpe* vanish; then, because DS and *ds*, ES and *es* are equal, lines PE, PF and *pe*, *pf* and the line-elements DF and *df* may be considered to be equal, inasmuch as their ultimate ratio, when angles DPE and *dpe* vanish simultaneously, is the ratio of equality.

On the basis of these things, therefore, PI will be to PF as RI to DF, and *pf* to *pi* as *df* or DF to *ri*, and from the equality of the ratios [or ex aequo] $\text{PI} \times pf$ will be to $\text{PF} \times pi$ as RI to *ri*, that is (by lem. 7, corol. 3), as the arc IH to the arc *ih*. Again, PI will be to PS as IQ to SE, and *ps* will be to *pi* as *se* or SE to *iq*; and from the equality of the ratios [or ex aequo] $\text{PI} \times ps$ will be to $\text{PS} \times pi$ as IQ to *iq*. And by compounding these ratios, $\text{PI}^2 \times pf \times ps$ will be to $pi^2 \times \text{PF} \times \text{PS}$ as $\text{IH} \times \text{IQ}$ to $ih \times iq$; that is, as the circular surface that the arc IH will describe by the revolution of the semicircle AKB about the diameter AB to the circular surface that the arc *ih* will describe by the revolution of the semicircle *akb* about the diameter *ab*. And the forces by which these surfaces attract the corpuscles P and *p* (along lines tending to these same surfaces) are (by hypothesis) as these surfaces themselves directly and the squares of the distances of these surfaces from the bodies inversely, that is, as $pf \times ps$ to $\text{PF} \times \text{PS}$.

Now (once the resolution of the forces has been made according to corol. 2 of the laws), these forces are to their oblique parts, which tend along the lines PS and *ps* toward the centers, as PI to PQ and *pi* to *pq*; that is (because the triangles PIQ and PSF, *piq* and *psf* are similar), the forces are to their oblique parts as PS to PF and *ps* to *pf*. Hence, from the equality of the ratios [or ex aequo] the attraction of this corpuscle P toward S becomes to the attraction of the corpuscle *p* toward *s* as $\dfrac{\text{PF} \times pf \times ps}{\text{PS}}$ to $\dfrac{pf \times \text{PF} \times \text{PS}}{ps}$, that is, as ps^2 to PS^2. And by a similar argument, the forces by which the surface described by the revolution of the arcs KL and *kl* attract the corpuscles will be as ps^2 to PS^2. And the same ratio will hold for the forces of all the spherical surfaces into which each of the two spherical surfaces can be divided by taking *sd* always equal to SD and *se* equal to SE. And by composition [or componendo] the forces of the total spherical surfaces exercised upon the corpuscles will be in the same ratio. Q.E.D.

Proposition 72
Theorem 32

If toward each of the separate points of any sphere there tend equal centripetal forces, decreasing in the squared ratio of the distances from those points, and there are given both the density of the sphere and the ratio of the diameter of the sphere to the distance of the corpuscle from the center of the sphere, I say that the force by which the corpuscle is attracted will be proportional to the semidiameter of the sphere.

For imagine that two corpuscles are attracted separately by two spheres, one corpuscle by one sphere, and the other corpuscle by the other sphere, and that their distances from the centers of the spheres are respectively proportional to the diameters of the spheres, and that the two spheres are resolved into particles that are similar and similarly placed with respect to the corpuscles. Then the attractions of the first corpuscle, made toward each of the separate particles of the first sphere, will be to the attractions of the second toward as many analogous particles of the second sphere in a ratio compounded of the direct ratio of the particles and the inverse squared ratio of the distances [i.e., the attractions will be to one another as the particles directly and the squares of the distances inversely]. But the particles are as the spheres, that is, they are in the cubed ratio of the diameters, and the distances are as the diameters; and thus the first of these ratios directly combined with the second ratio taken twice inversely becomes the ratio of diameter to diameter. Q.E.D.

Corollary 1. Hence, if corpuscles revolve in circles about spheres consisting of equally attractive matter, and their distances from the centers of the spheres are proportional to the diameters of the spheres, the periodic times will be equal.

Corollary 2. And conversely, if the periodic times are equal, the distances will be proportional to the diameters. These two corollaries are evident from prop. 4, corol. 3.

Corollary 3. If toward each of the separate points of any two similar and equally dense solids there tend equal centripetal forces decreasing in the squared ratio of the distances from those points, the forces by which corpuscles will be attracted by those two solids, if they are similarly situated with regard to them, will be to each other as the diameters of the solids.

Proposition 73
Theorem 33

If toward each of the separate points of any given sphere there tend equal centripetal forces decreasing in the squared ratio of the distances from those points, I say that a corpuscle placed inside the sphere is attracted by a force proportional to the distance of the corpuscle from the center of the sphere.

Let a corpuscle P be placed inside the sphere ABCD, described about center S; and about the same center S with radius SP, suppose that an inner sphere PEQF is described. It is manifest (by prop. 70) that the concentric spherical surfaces of which the difference AEBF of the spheres is composed do not act at all upon body P, their attractions having been annulled by opposite attractions. There remains only the attraction of the inner sphere PEQF. And (by prop. 72) this is as the distance PS. Q.E.D.

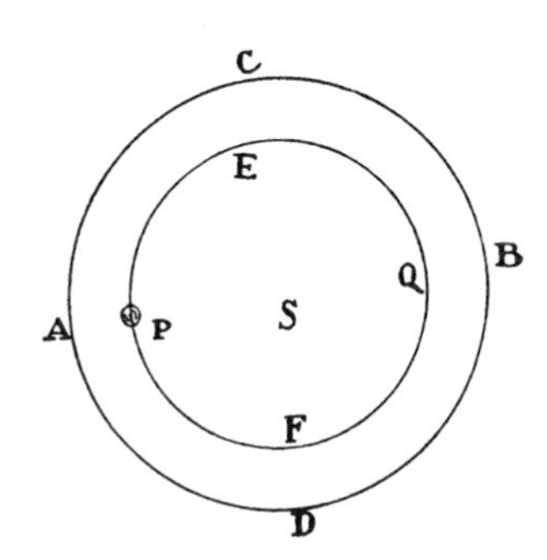

Scholium

The surfaces of which the solids are composed are here not purely mathematical, but orbs [or spherical shells] so extremely thin that their thickness is as null: namely, evanescent orbs of which the sphere ultimately consists when the number of those orbs is increased and their thickness diminished indefinitely. Similarly, when lines, surfaces, and solids are said to be composed of points, such points are to be understood as equal particles of a magnitude so small that it can be ignored.

Proposition 74
Theorem 34

With the same things being supposed as in prop. 73, I say that a corpuscle placed outside a sphere is attracted by a force inversely proportional to the square of the distance of the corpuscle from the center of the sphere.

For let the sphere be divided into innumerable concentric spherical surfaces; then the attractions of the corpuscle that arise from each of the surfaces will be inversely proportional to the square of the distance of the corpuscle from the center (by prop. 71). And by composition [or componendo] the sum of the attractions (that is, the attraction of the corpuscle toward the total sphere) will come out in the same ratio. Q.E.D.

COROLLARY 1. Hence at equal distances from the centers of homogeneous spheres the attractions are as the spheres themselves. For (by prop. 72) if the distances are proportional to the diameters of the spheres, the forces will be as the diameters. Let the greater distance be diminished in that ratio; and, the distances having now become equal, the attraction will be increased in that ratio squared, and thus will be to the other attraction in that ratio cubed, that is, in the ratio of the spheres.

COROLLARY 2. At any distances the attractions are as the spheres divided by the squares of the distances.

COROLLARY 3. If a corpuscle placed outside a homogeneous sphere is attracted by a force inversely proportional to the square of the distance of the corpuscle from the center of the sphere, and the sphere consists of attracting particles, the force of each particle will decrease in the squared ratio of the distance from the particle.

Proposition 75
Theorem 35

If toward each of the points of a given sphere there tend equal centripetal forces decreasing in the squared ratio of the distances from the points, I say that this sphere will attract any other homogeneous sphere with a force inversely proportional to the square of the distance between the centers.[a]

For the attraction of any particle is inversely as the square of its distance from the center of the attracting sphere (by prop. 74), and therefore is the same as if the total attracting force emanated from one single corpuscle situated in the center of this sphere. Moreover, this attraction is as great as the attraction of the same corpuscle would be if, in turn, it were attracted by each of the individual particles of the attracted sphere with the same force by which it attracts them. And that attraction of the corpuscle (by prop. 74) would be inversely proportional to the square of its distance from the center of the sphere; and therefore the sphere's attraction, which is equal to the attraction of the corpuscle, is in the same ratio. Q.E.D.

COROLLARY 1. The attractions of spheres toward other homogeneous spheres are as the attracting spheres [i.e., as the masses of the attracting spheres] divided by the squares of the distances of their own centers from the centers of those that they attract.

a. Newton writes of a "sphaera quaevis alia similaris," literally, "any other like [or similar] sphere," but the context (see prop. 74, corols. 1 and 3) is that of a homogeneous sphere.

Corollary 2. The same is true when the attracted sphere also attracts. For its individual points will attract the individual points of the other with the same force by which they are in turn attracted by them; and thus, since in every attraction the attracting point is as much urged (by law 3) as the attracted point, the force of the mutual attraction will be duplicated, the proportions remaining the same.

Corollary 3. Everything that has been demonstrated above concerning the motion of bodies about the focus of conics is valid when an attracting sphere is placed in the focus and the bodies move outside the sphere.

Corollary 4. And whatever concerns the motion of bodies around the center of conics applies when the motions are performed inside the sphere.

Proposition 76
Theorem 36

If spheres are in any way nonhomogeneous (as to the density of their matter and their attractive force) going from the center to the circumference, but are uniform throughout in every spherical shell at any given distance from the center, and the attractive force of each point decreases in the squared ratio of the distance of the attracted body, I say that the total force by which one sphere of this sort attracts another is inversely proportional to the square of the distance between their centers.

[a]Let there be any number of concentric homogeneous spheres [i.e., hollow spheres, or spherical shells or surfaces] AB, CD, EF, . . . ; and suppose that the addition of one or more inner ones to the outer one or ones forms a sphere composed of matter more dense, or the taking away leaves it less

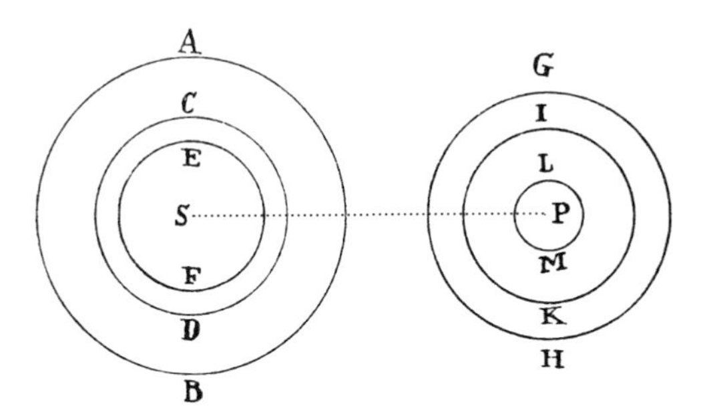

dense, toward the center than at the circumference. Then these spheres will together (by prop. 75) attract any number of other concentric homogeneous spheres GH, IK, LM, . . . , each sphere of one set attracting every one of

aa. The text of this proof has been translated somewhat freely, and in part expanded, for greater ease in comprehension.

the other set with forces inversely proportional to the square of the distance SP. And by adding up these forces (or by the reverse process when spheres are taken away) the sum of all those forces (or the excess of any one—or of some—of them above the others); that is, the force with which the whole sphere AB, composed of any concentric spheres (or the difference between some concentric spheres and others which have been taken away), attracts the whole sphere GH, composed of any concentric spheres (or the differences between some such concentric spheres and others)—will be in the same inverse ratio of the square of the distance SP. Let the number of concentric spheres be increased indefinitely, in such a way that the density of the matter, together with the force of attraction, may—on going from the circumference to the center—increase or decrease according to any law whatever; and by the addition of non-attracting matter, let the deficiencies in density be supplied wherever needed so that the spheres may acquire any desired form; then the force with which one of these spheres attracts the other will still be, by the former argument, in the same inverse ratio of the square of the distance.[a] Q.E.D.

COROLLARY 1. Hence, if many spheres of this sort, similar to one another in all respects, attract one another, the accelerative attraction of any one to any other of them, at any equal distances between the centers, will be as the attracting spheres.

COROLLARY 2. And at any unequal distances, as the attracting sphere divided by the square of the distances between the centers.

COROLLARY 3. And the motive attractions, or the weights of spheres toward other spheres, will—at equal distances from the centers—be as the attracting and the attracted spheres jointly, that is, as the products produced by multiplying the spheres by each other.

COROLLARY 4. And at unequal distances, as those products directly and the squares of the distances between the centers inversely.

COROLLARY 5. These results are valid when the attraction arises from each sphere's force of attraction being mutually exerted upon the other sphere. For the attraction is duplicated by both forces acting, the proportion remaining the same.

COROLLARY 6. If some spheres of this sort revolve about others at rest, one sphere revolving about each sphere at rest, and the distances between the centers of the revolving spheres and those at rest are proportional to the diameters of those at rest, the periodic times will be equal.

COROLLARY 7. And conversely, if the periodic times are equal, the distances will be proportional to those diameters.

COROLLARY 8. Everything that has been demonstrated above about the motion of bodies around the foci of conics holds when the attracting sphere, of any form and condition that has already been described, is placed in the focus.

COROLLARY 9. As also when the bodies revolving in orbit are also attracting spheres of any condition that has already been described.

Proposition 77
Theorem 37

If toward each of the individual points of spheres there tend centripetal forces proportional to the distances of the points from attracted bodies, I say that the composite force by which two spheres will attract each other is as the distance between the centers of the spheres.

CASE 1. Let AEBF be a sphere, S its center, P an attracted exterior corpuscle, PASB that axis of the sphere which passes through the center of the corpuscle, EF and *ef* two planes by which the sphere is cut and which are perpendicular to this axis and equally distant on both sides from the center of the sphere, G and *g* the intersections of the planes and the axis, and H any point in the plane EF. The centripetal force of point H upon the corpuscle P, exerted along the line PH, is as the distance PH; and (by corol. 2 of the laws) along the line PG, or toward the center S, as the length PG. Therefore the force of all the points in the plane EF (that is, of the total plane) by which the corpuscle P is attracted toward the center S is as the distance PG multiplied by the number of such points, that is, as the solid contained by that plane EF itself and the distance PG [i.e., as the product of the plane EF and the distance PG]. And similarly the force of the plane *ef*, by which the corpuscle P is attracted toward the center S, is as that plane multiplied by its distance P*g*, or as the plane EF equal thereto multiplied by that distance P*g*; and the sum of the forces of both planes is as the plane EF multiplied by the sum of the distances PG + P*g*; that is, as that plane multiplied by twice the distance

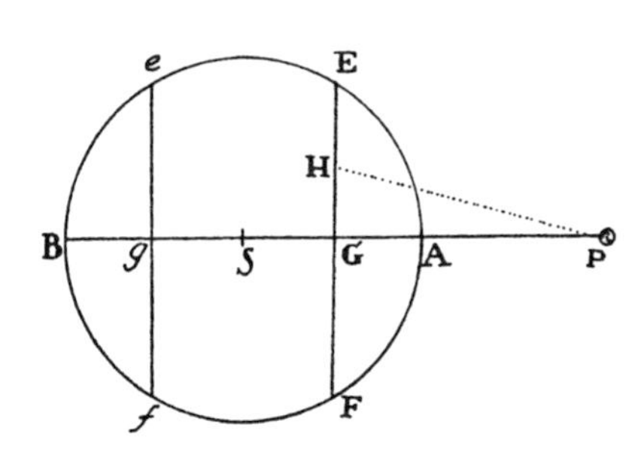

PS between the center S and the corpuscle P; that is, as twice the plane EF multiplied by the distance PS, or as the sum of the equal planes EF + *ef* multiplied by that same distance. And by a similar argument, the forces of all the planes in the whole sphere, equally distant on both sides from the center of the sphere, are as the sum of those planes multiplied by the distance PS, that is, as the whole sphere and the distance PS jointly. Q.E.D.

CASE 2. Now let the corpuscle P attract the sphere AEBF. Then by the same argument it can be proved that the force by which that sphere is attracted will be as the distance PS. Q.E.D.

CASE 3. Now let a second sphere be composed of innumerable corpuscles P; then, since the force by which any one corpuscle is attracted is as the distance of the corpuscle from the center of the first sphere and as that same sphere jointly, and thus is the same as if all the force came from one single corpuscle in the center of the sphere, the total force by which all the corpuscles in the second sphere are attracted (that is, by which that whole sphere is attracted) will be the same as if that sphere were attracted by a force coming from one single corpuscle in the center of the first sphere, and therefore is proportional to the distance between the centers of the spheres. Q.E.D.

CASE 4. Let the spheres attract each other mutually; then the force, now duplicated, will keep the former proportion. Q.E.D.

CASE 5. Now let a corpuscle *p* be placed inside the sphere AEBF. Then, since the force of the plane *ef* upon the corpuscle is as the solid contained by

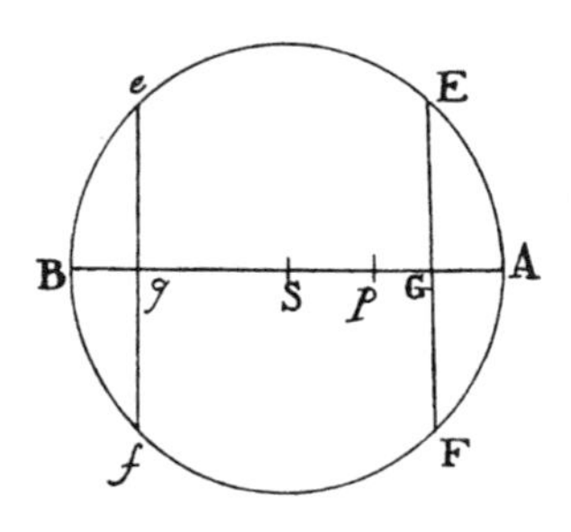

[or the product of] that plane and the distance *pg*; and the opposite force of the plane EF is as the solid contained by [or the product of] that plane and the distance *p*G; the force compounded of the two will be as the difference of the solids [or the products], that is, as the sum of the equal planes multiplied by half of the difference of the distances, that is, as that sum multiplied by *p*S, the distance of the corpuscle from the center of the sphere. And by a similar argument, the attraction of all the planes EF and *ef* in the whole sphere (that is, the attraction of the whole sphere) is jointly as the sum of all the planes (or the whole sphere) and as *p*S, the distance of the corpuscle from the center of the sphere. Q.E.D.

Case 6. And if from innumerable corpuscles *p* a new sphere is composed, placed inside the former sphere AEBF, then it can be proved as above that the attraction, whether the simple attraction of one sphere toward the other, or a mutual attraction of both toward each other, will be as the distance *p*S between the centers. Q.E.D.

Proposition 78
Theorem 38

If spheres, on going from the center to the circumference, are in any way nonhomogeneous and nonuniform, but in every concentric spherical shell at any given distance from the center are homogeneous throughout; and the attracting force of each point is as the distance of the body attracted; then I say that the total force by which two spheres of this sort attract each other is proportional to the distance between the centers of the spheres.

This is demonstrated from prop. 77 in the same way that prop. 76 was demonstrated from prop. 75.

Corollary. Whatever was demonstrated above in props. 10 and 64 on the motion of bodies about the centers of conics is valid when all the attractions take place by the force of spherical bodies of the condition already described, and when the attracted bodies are spheres of the same condition.

Scholium

I have now given explanations of the two major cases of attractions, namely, when the centripetal forces decrease in the squared ratio of the distances or increase in the simple ratio of the distances, causing bodies in both cases to revolve in conics, and composing centripetal forces of spherical bodies that decrease or increase in proportion to the distance from the center according to the same law—which is worthy of note. It would be tedious to go one by one through the other cases which lead to less elegant conclusions. I prefer to comprehend and determine all the cases simultaneously under a general method as follows.

Lemma 29[a]

If any circle AEB *is described with center* S*; and then two circles* EF *and* ef *are described with center* P*, cutting the first circle in* E *and* e*, and cutting the line* PS *in* F *and* f*; and if the perpendiculars* ED *and* ed *are dropped to* PS*; then I say that if the distance between the arcs* EF *and* ef *is supposed to be diminished*

a. For a gloss on this lemma see the Guide, §10.13.

indefinitely, the ultimate ratio of the evanescent line D*d* *to the evanescent line* F*f* *is the same as that of line* PE *to line* PS.

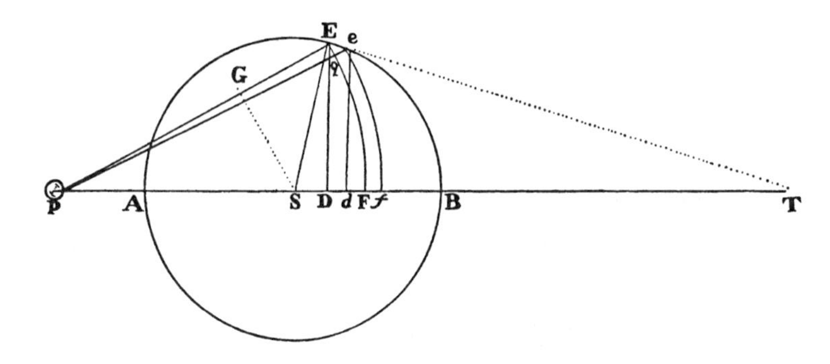

For if line P*e* cuts arc EF in *q*, and the straight line E*e*, which coincides with the evanescent arc E*e*, when produced meets the straight line PS in T, and the normal SG is dropped from S to PE; then because the triangles DTE, *d*T*e*, and DES are similar, D*d* will be to E*e* as DT to TE, or DE to ES; and because the triangles E*eq* and ESG (by sec. 1, lem. 8 and lem. 7, corol. 3) are similar, E*e* will be to *eq* or F*f* as ES to SG; and from the equality of the ratios [or ex aequo] D*d* will be to F*f* as DE to SG—that is (because the triangles PDE and PGS are similar), as PE to PS. Q.E.D.

Proposition 79
Theorem 39

If the surface EF*fe*, *just now vanishing because its width has been indefinitely diminished, describes by its revolution about the axis* PS *a concavo-convex spherical solid, toward each of whose individual equal particles there tend equal centripetal forces; then I say that the force by which that solid attracts an exterior corpuscle located in* P *is in a ratio compounded of the ratio of the solid [or product]* $DE^2 \times Ff$ *and the ratio of the force by which a given particle at the place* F*f* *would attract the same corpuscle.*

For if we first consider the force of the spherical surface FE, which is generated by the revolution of the arc FE and is cut anywhere by the line *de* in *r*, the annular part of this surface generated by the revolution of the arc *r*E will be as the line-element D*d*, the radius PE of the sphere remaining the same (as Archimedes demonstrated in his book on the Sphere and Cylinder). And the force of that surface, exerted along lines PE or P*r*, placed everywhere in the surface of a cone, will be as this annular part of the surface—that is, as the line-element D*d* or, what comes to the same thing, as the rectangle of the given radius PE of the sphere and that line-element D*d*;

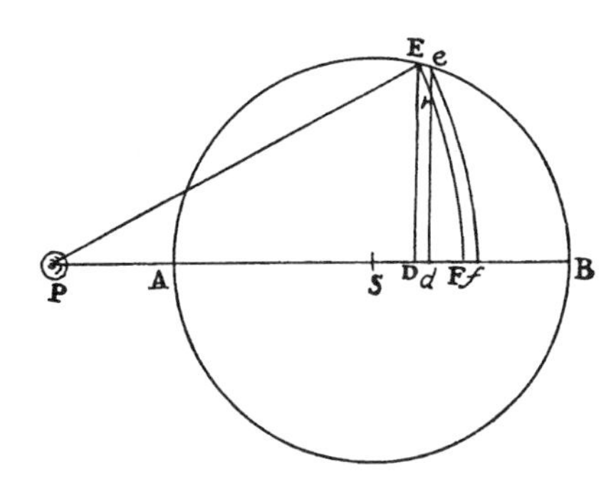

but along the line PS tending toward the center S, this force will be smaller in the ratio of PD to PE, and hence this force will be as PD × D*d*. Now suppose the line DF to be divided into innumerable equal particles, and let each of them be called D*d*; then the surface FE will be divided into the same number of equal rings, whose total forces will be as the sum of all the products PD × D*d*, that is, as $\frac{1}{2}PF^2 - \frac{1}{2}PD^2$, and thus as DE^2. Now multiply the surface FE by the altitude F*f*, and the force of the solid EF*fe* exerted upon the corpuscle P will become as $DE^2 \times Ff$, if there is given the force that some given particle F*f* exerts on the corpuscle P at the distance PF. But if that force is not given, the force of the solid EF*fe* will become as the solid $DE^2 \times Ff$ and that non-given force jointly. Q.E.D.

Proposition 80
Theorem 40

If equal centripetal forces tend toward each of the individual equal particles of some sphere ABE, *described about a center* S; *and if from each of the individual points* D *to the axis* AB *of the sphere, in which some corpuscle* P *is located, there are erected the perpendiculars* DE, *meeting the sphere in the points* E; *and if, on these perpendiculars, the lengths* DN *are taken, which are jointly as the quantity* $\frac{DE^2 \times PS}{PE}$ *and as the force that a particle of the sphere, located on the axis, exerts at the distance* PE *upon the corpuscle* P; *then I say that the total force with which the corpuscle* P *is attracted toward the sphere is as the area* ANB *comprehended by the axis* AB *of the sphere and the curved line* ANB, *which the point* N *traces out.*

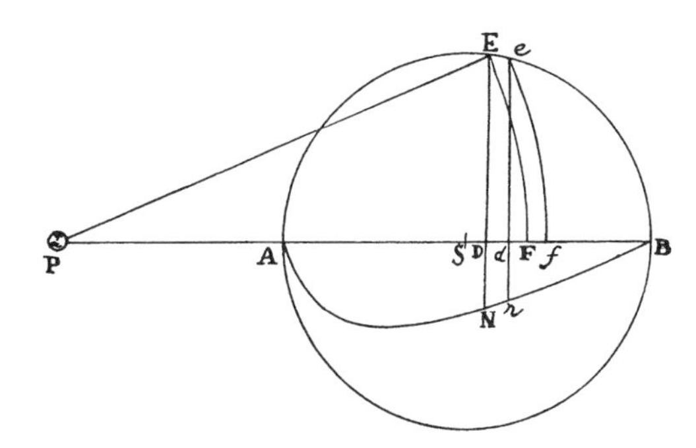

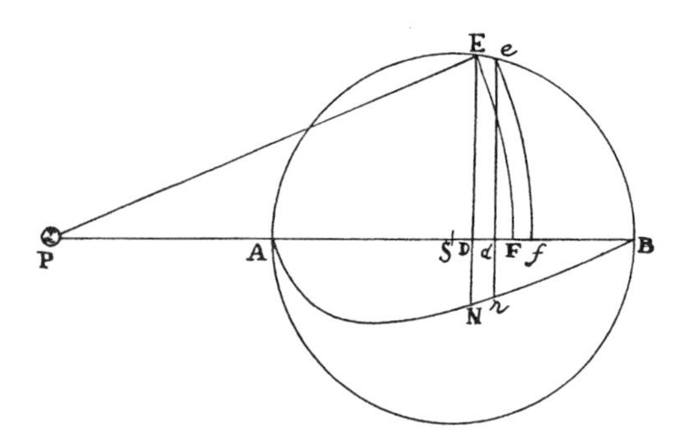

For, keeping the same constructions as in lem. 29 and prop. 79, suppose the axis AB of the sphere to be divided into innumerable equal particles D*d*, and the whole sphere to be divided into as many spherical concavo-convex laminae EF*fe*; and erect the perpendicular *dn*. By prop. 79, the force with which the lamina EF*fe* attracts the corpuscle P is jointly as $DE^2 \times Ff$ and the force of one particle exerted at the distance PE or PF. But D*d* is to F*f* (by lem. 29) as PE to PS, and hence F*f* is equal to $\frac{PS \times Dd}{PE}$, and $DE^2 \times Ff$ is equal to $Dd\frac{DE^2 \times PS}{PE}$; and therefore the force of the lamina EF*fe* is jointly as $Dd\frac{DE^2 \times PS}{PE}$ and the force of a particle exerted at the distance PF; that is (by hypothesis) as $DN \times Dd$, or as the evanescent area DN*nd*. Therefore the forces upon body P exerted by all the laminae are as all the areas DN*nd*, that is, the total force of the sphere is as the total area ANB. Q.E.D.

Corollary 1. Hence, if the centripetal force tending toward each of the individual particles always remains the same at all distances, and DN is taken proportional to $\frac{DE^2 \times PS}{PE}$, the total force by which the corpuscle P is attracted by the sphere will be as the area ANB.

Corollary 2. If the centripetal force of the particles is inversely as the distance of the attracted corpuscle, and DN is taken proportional to $\frac{DE^2 \times PS}{PE^2}$, the force by which the corpuscle P is attracted by the whole sphere will be as the area ANB.

Corollary 3. If the centripetal force of the particles is inversely as the cube of the distance of the attracted corpuscle, and DN is taken proportional to $\frac{DE^2 \times PS}{PE^4}$, the force by which the corpuscle is attracted by the whole sphere will be as the area ANB.

COROLLARY 4. And universally, if the centripetal force tending toward each of the individual particles of a sphere is supposed to be inversely as the quantity V, and DN is taken proportional to $\frac{DE^2 \times PS}{PE \times V}$, the force by which a corpuscle is attracted by the whole sphere will be as the area ANB.

Proposition 81
Problem 41

Under the same conditions as before, it is required to measure the area ANB.

From point P draw the straight line PH touching the sphere in H; and, having dropped the normal HI to the axis PAB, bisect PI in L; then (by book 2, prop. 12, of Euclid's *Elements*) PE^2 will be equal to $PS^2+SE^2+2(PS \times SD)$.

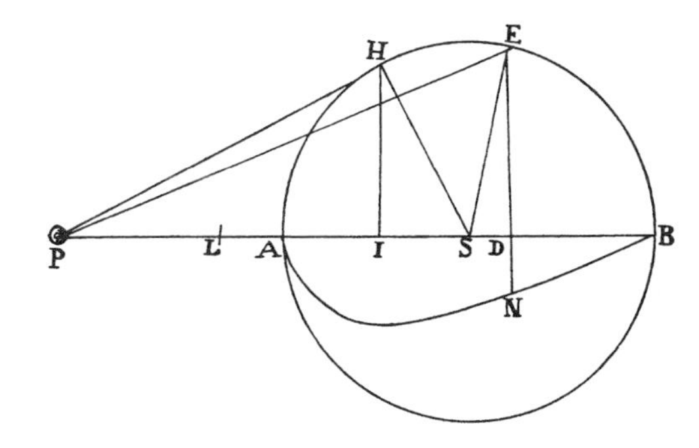

Moreover, SE^2 or SH^2 (because the triangles SPH and SHI are similar) is equal to the rectangle $PS \times SI$. Therefore PE^2 is equal to the rectangle of PS and $PS + SI + 2SD$, that is, of PS and $2LS + 2SD$, that is, of PS and 2LD. Further, DE^2 is equal to $SE^2 - SD^2$, or $SE^2 - LS^2 + 2(SL \times LD) - LD^2$, that is, $2(SL \times LD) - LD^2 - AL \times LB$. For $LS^2 - SE^2$ or $LS^2 - SA^2$ (by book 2, prop. 6, of the *Elements*) is equal to the rectangle $AL \times LB$. Write, therefore, $2(SL \times LD) - LD^2 - AL \times LB$ for DE^2; and the quantity $\frac{DE^2 \times PS}{PE \times V}$, which (according to corol. 4 of the preceding prop. 80) is as the length of the ordinate DN, will resolve itself into the three parts $\frac{2(SL \times LD \times PS)}{PE \times V} - \frac{LD^2 \times PS}{PE \times V} - \frac{AL \times LB \times PS}{PE \times V}$: where, if for V we write the inverse ratio of the centripetal force, and for PE the mean proportional between PS and 2LD, those three parts will become ordinates of as many curved lines, whose areas can be found by ordinary methods. Q.E.F.

EXAMPLE 1. If the centripetal force tending toward each of the individual particles of the sphere is inversely as the distance, write the distance PE in place of V, and then $2PS \times LD$ in place of PE^2, and DN will become as $SL - \frac{1}{2}LD - \frac{AL \times LB}{2LD}$. Suppose DN equal to its double $2SL - LD - \frac{AL \times LB}{LD}$; and the given part 2SL of that ordinate multiplied by the length AB will describe a rectangular area $2SL \times AB$, and the indefinite part LD multiplied perpendicularly by the same length AB in a continual motion (according to the rule that, while moving, either by increasing or decreasing, it is always equal to the length LD) will describe an area $\frac{LB^2 - LA^2}{2}$, that is, the area $SL \times AB$, which, subtracted from the first area $2SL \times AB$, leaves the area $SL \times AB$. Now the third part $\frac{AL \times LB}{LD}$, likewise multiplied perpendicularly by the same length AB in a local [i.e., continual] motion, will describe a hyperbolic area, which subtracted from the area $SL \times AB$ will leave the required area ANB. Hence, there arises the following construction of the problem.

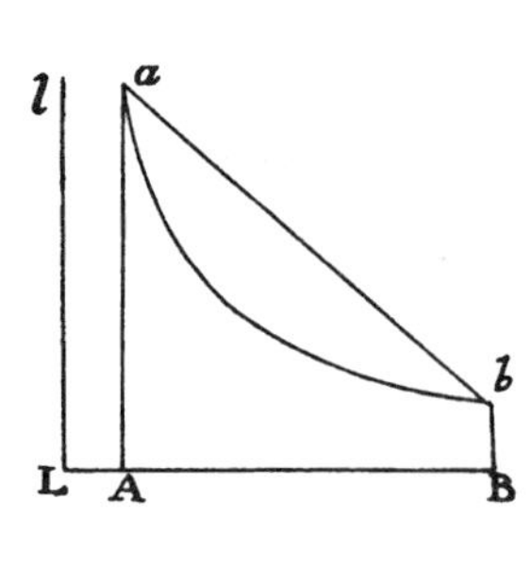

At points L, A, and B erect perpendiculars L*l*, A*a*, and B*b*, of which A*a* is equal to LB, and B*b* to LA. With asymptotes L*l* and LB, through points *a* and *b* describe the hyperbola *ab*. Then the chord *ba*, when drawn, will enclose the area *aba* equal to the required area ANB.

EXAMPLE 2. If the centripetal force tending toward each of the individual particles of the sphere is inversely as the cube of the distance, or (which comes to the same thing) as that cube divided by any given plane, write $\frac{PE^3}{2AS^2}$ for V, and then $2PS \times LD$ for PE^2, and DN will become as $\frac{SL \times AS^2}{PS \times LD} - \frac{AS^2}{2PS} - \frac{AL \times LB \times AS^2}{2PS \times LD^2}$, that is (because PS, AS, and SI are continually proportional [or PS is to AS as AS to SI]), as $\frac{LS \times SI}{LD} - \frac{1}{2}SI - \frac{AL \times LB \times SI}{2LD^2}$. If the three parts of this quantity are multiplied by the length AB, the first, $\frac{LS \times SI}{LD}$, will generate a hyperbolic area; the second, $\frac{1}{2}SI$, will generate the area $\frac{1}{2}AB \times SI$; the third, $\frac{AL \times LB \times SI}{2LD^2}$, will generate the area $\frac{AL \times LB \times SI}{2LA} - \frac{AL \times LB \times SI}{2LB}$, that is, $\frac{1}{2}AB \times SI$.

From the first subtract the sum of the second and third, and the required area ANB will remain.

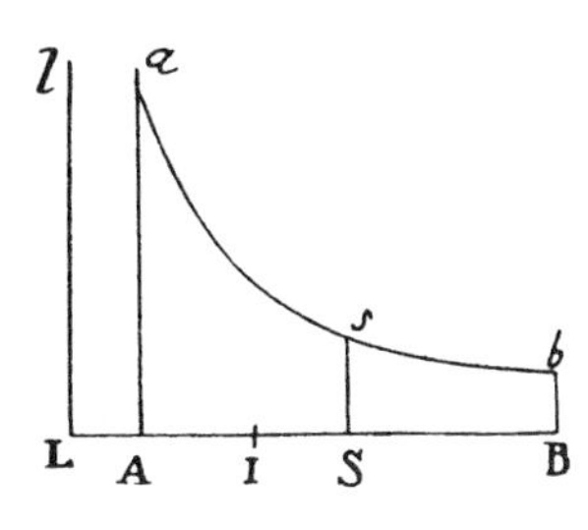

Hence there arises the following construction of the problem. At the points L, A, S, and B erect the perpendiculars L*l*, A*a*, S*s*, and B*b*, of which S*s* is equal to SI; and through the point *s*, with asymptotes L*l* and LB, describe the hyperbola *asb* meeting the perpendiculars A*a* and B*b* in *a* and *b*; then the rectangle 2AS × SI subtracted from the hyperbolic area A*asb*B will leave the required area ANB.

EXAMPLE 3. If the centripetal force tending toward each of the individual particles of the sphere decreases as the fourth power of the distance from those particles, write $\frac{PE^4}{2AS^3}$ for V, and then $\sqrt{(2PS \times LD)}$ for PE, and DN will become as $\frac{SI^2 \times SL}{\sqrt{(2SI)}} \times \frac{1}{\sqrt{LD^3}} - \frac{SI^2}{2\sqrt{(2SI)}} \times \frac{1}{\sqrt{LD}} - \frac{SI^2 \times AL \times LB}{2\sqrt{(2SI)}} \times \frac{1}{\sqrt{LD^5}}$. Those three parts, multiplied by the length AB, produce three areas, namely $\frac{2SI^2 \times SL}{\sqrt{(2SI)}}$ multiplied by $\left(\frac{1}{\sqrt{LA}} - \frac{1}{\sqrt{LB}}\right)$; $\frac{SI^2}{\sqrt{(2SI)}}$ multiplied by $(\sqrt{LB} - \sqrt{LA})$; and $\frac{SI^2 \times AL \times LB}{3\sqrt{(2SI)}}$ multiplied by $\left(\frac{1}{\sqrt{LA^3}} - \frac{1}{\sqrt{LB^3}}\right)$. And these, after the due reduction, become $\frac{2SI^2 \times SL}{LI}$, SI^2, and $\frac{SI^2 + 2SI^3}{3LI}$. And when the latter two quantities are subtracted from the first one, the result comes out to be $\frac{4SI^3}{3LI}$. Accordingly, the total force by which the cor-

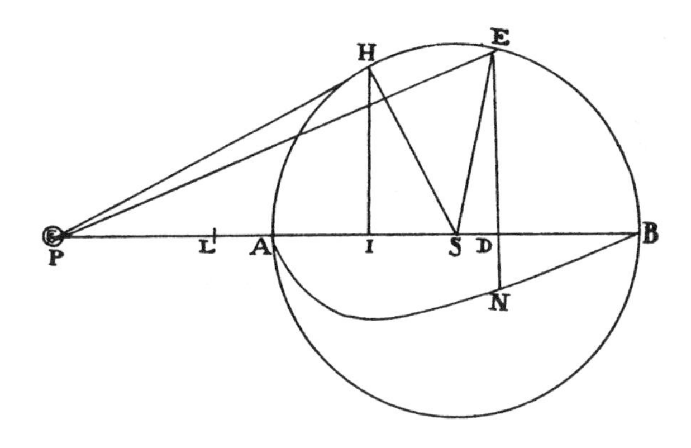

puscle P is attracted to the center of the sphere is as $\frac{SI^3}{PI}$, that is, inversely as $PS^3 \times PI$. Q.E.I.

The attraction of a corpuscle located inside a sphere can be determined by the same method, but more expeditiously by means of the following proposition.

Proposition 82
Theorem 41

If—in a sphere described about center S *with radius* SA—SI, SA, *and* SP *are taken continually proportional [i.e.,* SI *to* SA *as* SA *to* SP*], I say that the attraction of a corpuscle inside the sphere at any place* I *is to its attraction outside the sphere at place* P *in a ratio compounded of the square root of the ratio of the distances* IS *and* PS *from the center, and the square root of the ratio of the centripetal forces, tending at those places* P *and* I *toward the center.*

If, for example, the centripetal forces of the particles of the sphere are inversely as the distances of the corpuscle attracted by them, the force by which the corpuscle situated at I is attracted by the total sphere will be to the force by which it is attracted at P in a ratio compounded of the square

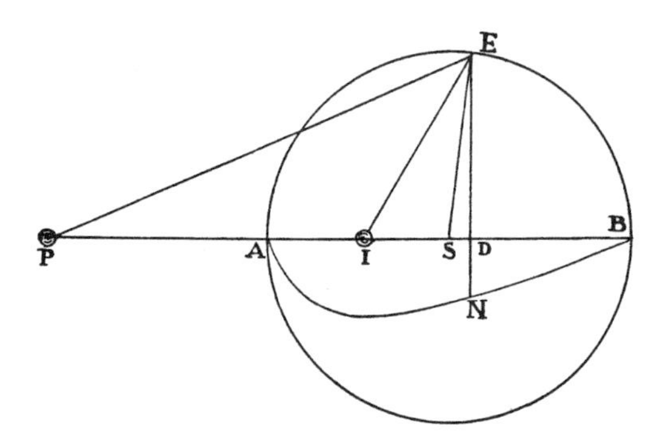

root of the ratio of the distance SI to the distance SP and the square root of the ratio of the centripetal force at place I arising from some particle in the center to the centripetal force at place P arising from the same particle in the center, which is the square root of the ratio of the distances SI and SP to each other inversely. Compounding these two square roots of ratios gives the ratio of equality, and therefore the attractions produced at I and P by the whole sphere are equal. By a similar computation, if the forces of the particles of the sphere are inversely in the squared ratio of the distances, it will be seen that the attraction at I is to the attraction at P as the distance

SP is to the semidiameter SA of the sphere. If those forces are inversely in the cubed ratio of the distances, the attractions at I and P will be to each other as SP^2 to SA^2; if as the inverse fourth power, as SP^3 to SA^3. Hence, since—in this last case [of the inverse fourth power, as in the final ex. 3 of prop. 81]—the attraction at P was found to be inversely as $PS^3 \times PI$, the attraction at I will be inversely as $SA^3 \times PI$, that is (because SA^3 is given), inversely as PI. And the progression goes on in the same way indefinitely. Moreover, the theorem is demonstrated as follows.

With the same construction and with the corpuscle being in any place P, the ordinate DN was found to be as $\frac{DE^2 \times PS}{PE \times V}$. Therefore, if IE is drawn, that ordinate for any other place I of the corpuscle will—*mutatis mutandis* [i.e., by substituting I for P in the considerations and arguments that have previously been applied to P]—come out as $\frac{DE^2 \times IS}{IE \times V}$. Suppose the centripetal forces emanating from any point E of the sphere to be to each other at the distances IE and PE as PE^n to IE^n (where let the number n designate the index of the powers of PE and IE); then those ordinates will become as $\frac{DE^2 \times PS}{PE \times PE^n}$ and $\frac{DE^2 \times IS}{IE \times IE^n}$, whose ratio to each other is as $PS \times IE \times IE^n$ to $IS \times PE \times PE^n$. Because SI, SE, and SP are continually proportional, the triangles SPE and SEI are similar, and hence IE becomes to PE as IS to SE or SA; for the ratio of IE to PE, write the ratio of IS to SA, and the ratio of the ordinates will come out $PS \times IE^n$ to $SA \times PE^n$. But PS to SA is the square root of the ratio of the distances PS and SI, and IE^n to PE^n (because IE is to PE as IS to SA) is the square root of the ratio of the forces at the distances PS and IS. Therefore the ordinates, and consequently the areas that the ordinates describe and the attractions proportional to them, are in a ratio compounded of the foregoing square-root ratios. Q.E.D.

Proposition 83
Problem 42

To find the force by which a corpuscle located in the center of a sphere is attracted toward any segment of it whatever.

Let P be the corpuscle in the center of the sphere, and RBSD a segment of the sphere contained by the plane RDS and the spherical surface KBS.

Let DB be cut at F by the spherical surface EFG described about the center P, and divide that segment into the parts BREFGS and FEDG. But

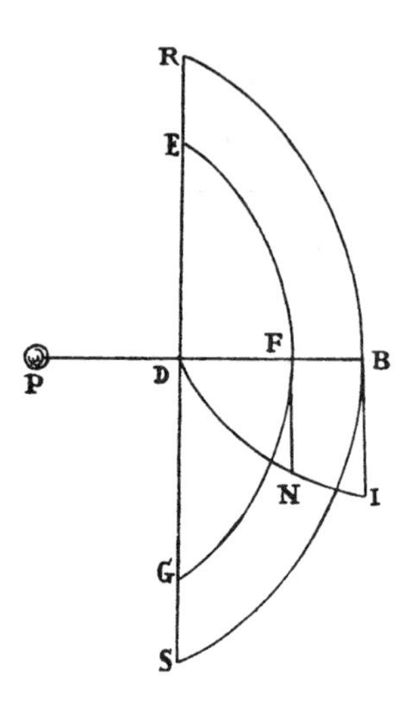

let that surface be taken to be not purely mathematical, but physical, having a minimally small thickness. Call that thickness O, and this surface (by what Archimedes has demonstrated), will be as $PF \times DF \times O$. Let us suppose, additionally, the attractive forces of the particles of the sphere to be inversely as that power of the distances whose index is n; then the force by which the surface EFG attracts the body P will be (by prop. 79) as $\frac{DE^2 \times O}{PF^n}$, that is, as $\frac{2DF \times O}{PF^{n-1}} - \frac{DF^2 \times O}{PF^n}$. Let the perpendicular FN drawn in [the thickness] O be proportional to this quantity; then the curvilinear area BDI, as described by the ordinate FN, drawn in a continual motion, applied to the length DB, will be as the whole force by which the whole segment RBSD attracts the corpuscle P. Q.E.I.

Proposition 84
Problem 43 *To find the force with which a corpuscle is attracted by a segment of a sphere when it is located on the axis of the segment beyond the center of the sphere.*

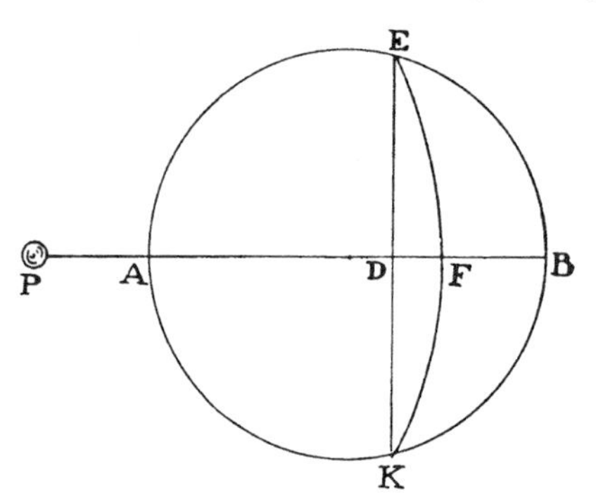

Let corpuscle P, located on the axis ADB of the segment EBK, be attracted by that segment. About center P and with radius PE describe the spherical surface EFK, which divides the segment into two parts EBKFE and EFKDE. Find the force of the first part by prop. 81 and the force of the second part by prop. 83, and the sum of these two forces will be the force of the whole segment EBKDE. Q.E.I.

Scholium Now that the attractions of spherical bodies have been explained, it would be possible to go on to the laws of the attractions of certain other bodies similarly consisting of attracting particles, but to treat these in particular cases is not essential to my design. It will be enough to subjoin certain more general propositions concerning the forces of bodies of this sort and the motions that arise from such forces, because these propositions are of some use in philosophical questions [i.e., questions of natural philosophy, or physical science].

SECTION 13

The attractive forces of nonspherical bodies

Proposition 85
Theorem 42

If the attraction of an attracted body is far stronger when it is contiguous to the attracting body than when the bodies are separated from each other by even a very small distance, then the forces of the particles of the attracting body decrease, as the attracted body recedes, in a more than squared ratio of the distances from the particles.

For if the forces decrease in the squared ratio of the distances from the particles, the attraction toward a spherical body will not be sensibly increased by contact, because (by prop. 74) it is inversely as the square of the distance of the attracted body from the center of the sphere; and still less will it be increased by contact, if the attraction decreases in a smaller ratio as the attracted body recedes. Therefore, this proposition is evident in the case of attracting spheres. It is the same for concave spherical orbs[a] attracting external bodies. And it is much more established in the case of orbs attracting bodies placed inside of them, since the attractions spreading out through the concavities of the orbs are annulled by opposite attractions (by prop. 70), and therefore the attracting forces are null, even in contact. But if any parts remote from the place of contact are taken away from these spheres and spherical orbs, and new parts are added anywhere away from the place of contact, the shapes of these attracting bodies can be changed at will; and yet the parts added or subtracted will not notably increase the excess of attraction that arises from contact, since they are remote from the place of contact. Therefore the proposition is established concerning bodies of all shapes. Q.E.D.

Proposition 86
Theorem 43

If the forces of the particles composing an attracting body decrease, as an attracted body recedes, in the cubed or more than cubed ratio of the distances from the particles, the attraction will be far stronger in contact than when the attracting body and attracted body are separated from each other by even a very small distance.

a. Here, as elsewhere in the *Principia*, Newton uses the word "orb" for what we would more precisely call a spherical shell.

For by the solution of prop. 81 given in exx. 2 and 3, it is established that the attraction is increased indefinitely in the approach of an attracted corpuscle to an attracting sphere of this sort. By the combination of those examples and prop. 82, the same result is easily inferred concerning the attractions of bodies toward concavo-convex orbs whether the attracted bodies are placed outside those orbs or in the cavities inside the orbs. But the proposition will also be established concerning all bodies universally by adding some attractive matter to these spheres and orbs, or taking some away from them, anywhere away from the place of contact, so that the attracting bodies take on any desired shape. Q.E.D.

Proposition 87
Theorem 44

If two bodies, similar to each other and consisting of equally attracting matter, separately attract corpuscles proportional to those bodies and similarly placed with respect to them, then the accelerative attractions of the corpuscles toward the whole bodies will be as the accelerative attractions of those corpuscles toward particles of those bodies proportional to the wholes and similarly situated in those whole bodies.

For if the bodies are divided into particles that are proportional to the whole bodies and similarly placed in those whole bodies, then the attraction toward an individual particle of the first body will be to the attraction toward the corresponding individual particle of the second body as the attractions toward any given particles of the first body are to the attractions toward the corresponding particles of the second body, and by compounding, the attraction toward the whole first body will be to the attraction toward the whole second body in that same ratio. Q.E.D.

COROLLARY 1. Therefore, if the attracting forces of the particles, on increasing the distances of the attracted corpuscles, decrease in the ratio of any power of those distances, the accelerative attractions toward the whole bodies will be as the bodies directly and those powers of the distances inversely. For example, if the forces of the particles decrease in the squared ratio of the distances from the attracted corpuscles, and the bodies are as A^3 and B^3, and thus both the cube roots of the bodies and the distances of the attracted corpuscles from the bodies are as A and B, the accelerative attractions toward the bodies will be as $\frac{A^3}{A^2}$ and $\frac{B^3}{B^2}$, that is, as those cube roots A and B of the bodies. If the forces of the particles decrease in the cubed ratio of the distances from the attracted corpuscles, the accelerative attractions toward the whole bodies will be as $\frac{A^3}{A^3}$ and $\frac{B^3}{B^3}$, that is, will be equal. If the forces decrease in the fourth power of the distance, the attractions toward the bodies will be as $\frac{A^3}{A^4}$ and $\frac{B^3}{B^4}$, that is, inversely as the cube roots A and B. And so on.

COROLLARY 2. Hence, on the other hand, from the forces with which similar bodies attract corpuscles similarly placed with respect to such bodies, there can be gathered the ratio of the decrease of the forces of the attracting particles, as the attracted corpuscle recedes, so long as that decrease is directly or inversely in some ratio of the distances.

Proposition 88
Theorem 45

If the attracting forces of equal particles of any body are as the distances of places from the particles, the force of the whole body will tend toward its center of gravity, and will be the same as the force of a globe consisting of entirely similar and equal matter and having its center in that center of gravity.

Let the particles A and B of the body RSTV attract some corpuscle Z by forces which, if the particles are equal to each other, are as the distances AZ and BZ: but if the particles are supposed unequal, are as these particles and their distances AZ and BZ jointly, or (so to speak) as these particles multiplied respectively by their distances AZ and BZ. And let the forces be represented by those solids [or products] A × AZ and B × BZ. Join AB, and let it be cut in G so that AG is to BG as the particle B to the particle A; then G will be the common center of gravity of the particles A and B. The force A × AZ (by corol. 2 of the laws) is resolved into the forces A × GZ and A × AG, and the force B × BZ into the forces B × GZ and B × BG. But the forces A × AG and B × BG are equal (because A is to B as BG to AG); and therefore, since they tend in opposite directions, they nullify each other. There remain the forces A × GZ and B × GZ. These tend from Z toward the center G and compose the force (A + B) × GZ—that is, the same force as if the attracting particles A and B were situated in their common center of gravity G and there composed a globe.

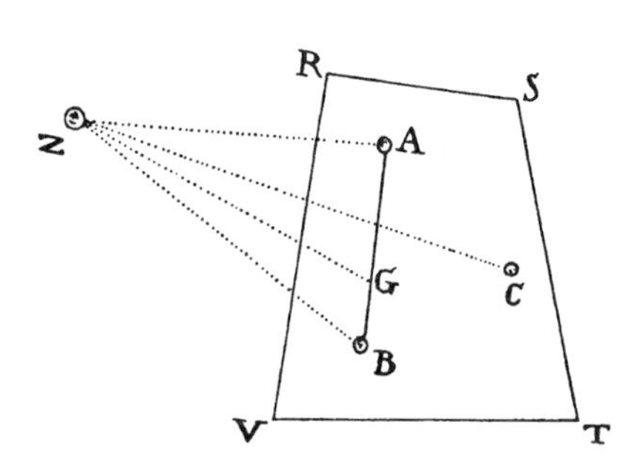

By the same argument, if a third particle C is added, and its force is compounded with the force $(A + B) \times GZ$ tending toward the center G, the force thence arising will tend toward the common center of gravity of the globe (at G) and the particle C (that is, toward the common center of gravity of the three particles A, B, and C), and will be the same as if the globe and the particle C were situated in their common center, there composing a greater globe. And so on indefinitely. Therefore the whole force of all the particles of any body RSTV is the same as if that body, while maintaining the same center of gravity, were to assume the shape of a globe. Q.E.D.

COROLLARY. Hence the motion of the attracted body Z will be the same as if the attracting body RSTV were spherical; and therefore, if that attracting body either is at rest or progresses uniformly straight forward, the attracted body will move in an ellipse having its center in the center of gravity of the attracting body.

Proposition 89
Theorem 46

If there are several bodies consisting of equal particles whose forces are as the distances of places from each individual particle, the force—compounded of the forces of all these particles—by which any corpuscle is attracted will tend toward the common center of gravity of the attracting bodies and will be the same as if those attracting bodies, while maintaining their common center of gravity, were united together and were formed into a globe.

This is demonstrated in the same way as the preceding proposition.

COROLLARY. Therefore the motion of an attracted body will be the same as if the attracting bodies, while maintaining their common center of gravity, came together and were formed into a globe. And hence, if the common center of gravity of the attracting body either is at rest or progresses uniformly in a straight line, the attracted body will move in an ellipse having its center in the common center of gravity of the attracting bodies.

Proposition 90
Problem 44

If equal centripetal forces, increasing or decreasing in any ratio of the distances, tend toward each of the individual points of any circle, it is required to find the force by which a corpuscle is attracted when placed anywhere on the straight line that stands perpendicularly upon the plane of the circle at its center.

Suppose a circle to be described with center A and any radius AD in a plane to which the straight line AP is perpendicular; then it is required to find the force by which any corpuscle P is attracted toward the circle. From

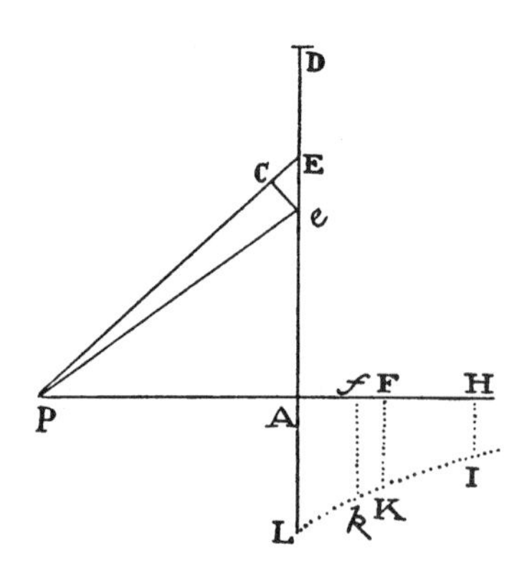

any point E of the circle, draw the straight line PE to the attracted corpuscle P. In the straight line PA take PF equal to PE, and erect the normal FK so that it will be as the force by which the point E attracts the corpuscle P. And let IKL be the curved line that the point K traces out. Let that line meet the plane of the circle in L. In PA take PH equal to PD, and erect the perpendicular HI meeting the aforesaid curve at I, and the attraction of the corpuscle P toward the circle will be as the area AHIL multiplied by the altitude AP. Q.E.I.

For on AE take the minimally small line E*e*. Join P*e*, and in PE and PA take PC and P*f* equal to P*e*. And since the force by which any point E of the ring described with center A and radius AE in the aforesaid plane attracts body [i.e., corpuscle] P toward itself has been supposed to be as FK, and hence the force by which that point attracts body P toward A is as $\frac{\mathrm{AP} \times \mathrm{FK}}{\mathrm{PE}}$; and the force by which the whole ring attracts body P toward A is as the ring and $\frac{\mathrm{AP} \times \mathrm{FK}}{\mathrm{PE}}$ jointly; and that ring is as the rectangle of the radius AE and the width E*e*, and this rectangle (because PE is to AE as E*e* to CE) is equal to the rectangle PE × CE or PE × F*f*; it follows that the force by which that ring attracts body P toward A will be as PE × F*f* and $\frac{\mathrm{AP} \times \mathrm{FK}}{\mathrm{PE}}$ jointly, that is, as the solid [or product] F*f* × FK × AP, or as the area FK*kf* multiplied by AP. And therefore the sum of the forces by which all the rings in the circle that is described with center A and radius AD attract body P toward A is as the whole area AHIKL multiplied by AP. Q.E.D.

COROLLARY 1. Hence, if the forces of the points decrease in the squared ratio of the distances, that is, if FK is as $\frac{1}{\mathrm{PF}^2}$ $\Big($and thus the area AHIKL is as $\frac{1}{\mathrm{PA}} - \frac{1}{\mathrm{PH}}\Big)$, the attraction of the corpuscle P toward the circle will be as $1 - \frac{\mathrm{PA}}{\mathrm{PH}}$, that is, as $\frac{\mathrm{AH}}{\mathrm{PH}}$.

COROLLARY 2. And universally, if the forces of the points at the distances D are inversely as any power D^n of the distances $\left(\text{that is, if FK is as } \frac{1}{D^n}, \text{ and hence the area AHIKL is as } \frac{1}{PA^{n-1}} - \frac{1}{PH^{n-1}}\right)$, the attraction of the corpuscle P toward the circle will be as $\frac{1}{PA^{n-2}} - \frac{PA}{PH^{n-1}}$.

COROLLARY 3. And if the diameter of the circle is increased indefinitely and the number n is greater than unity, the attraction of the corpuscle P toward the whole indefinitely extended plane will be inversely as PA^{n-2}, because the other term, $\frac{PA}{PH^{n-1}}$, will vanish.

Proposition 91
Problem 45

To find the attraction of a corpuscle placed in the axis of a round solid, to each of whose individual points there tend equal centripetal forces decreasing in any ratio of the distances.

Let corpuscle P, placed in the axis AB of the solid DECG, be attracted toward that same solid. Let this solid be cut by any circle RFS perpendicular to this axis, and in its semidiameter FS, in a plane PALKB passing through the axis, take (according to prop. 90) the length FK proportional to the force by which the corpuscle P is attracted toward that circle. Let point K touch the curved line LKI meeting the planes of the outermost circles AL and BI at L and I, and the attraction of the corpuscle P toward the solid will be as the area LABI. Q.E.I.

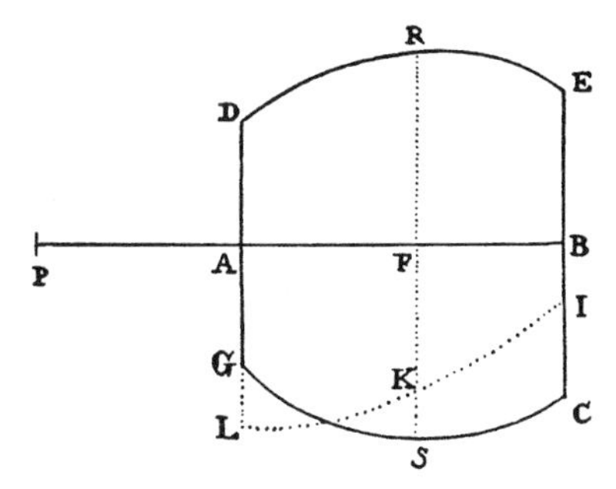

COROLLARY 1. Hence, if the solid is a cylinder described by parallelogram ADEB revolving about the axis AB, and the centripetal forces tending toward each of its individual points are inversely as the squares of the distances from the points, the attraction of the corpuscle P toward this cylinder will be as AB − PE + PD. For the ordinate FK (by prop. 90, corol. 1) will be as

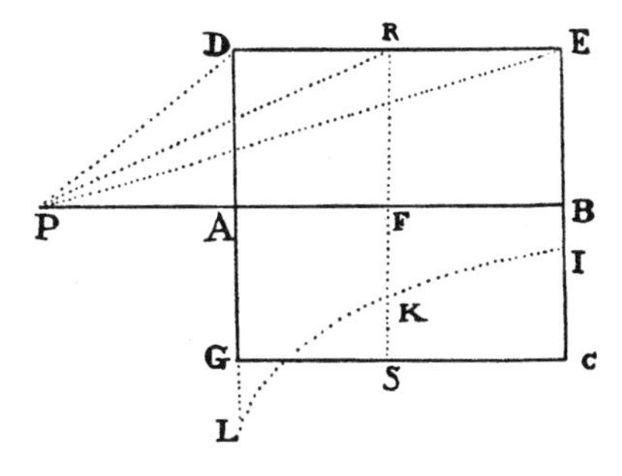

$1 - \frac{PF}{PR}$. The unit part of this $\left[\text{or the quantity 1 in } 1 - \frac{PF}{PR}\right]$ multiplied by the length AB describes the area $1 \times AB$, and the other part $\frac{PF}{PR}$ multiplied by the length PB describes the area $1 \times (PE - AD)$, which can easily be shown from the quadrature of the curve LKI; and similarly the same part $\frac{PF}{PR}$ multiplied by the length PA describes the area $1 \times (PD - AD)$, and multiplied by the difference AB of PB and PA describes the difference of the areas $1 \times (PE - PD)$. From the first product $1 \times AB$ take away the last product $1 \times (PE - PD)$, and there will remain the area LABI equal to $1 \times (AB - PE + PD)$. Therefore the force proportional to this area is as $AB - PE + PD$.

COROLLARY 2. Hence also the force becomes known by which a spheroid AGBC attracts any body P, situated outside the spheroid in its axis AB. Let NKRM be a conic whose ordinate ER, perpendicular to PE, is always

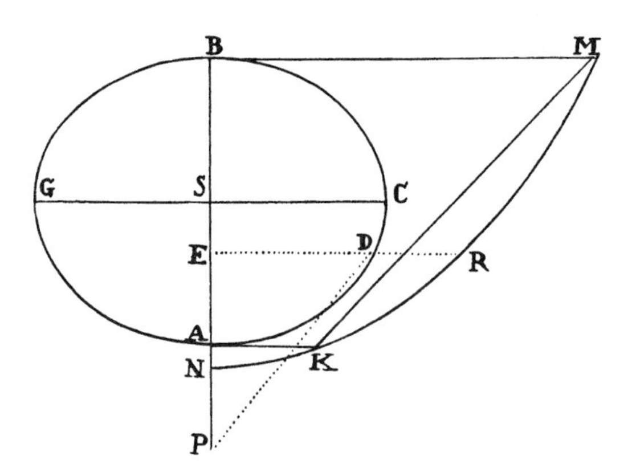

equal to the length of the line PD, which is drawn to the point D in which the ordinate cuts the spheroid. From the vertices A and B of the spheroid, erect AK and BM perpendicular to the axis AB of the spheroid and equal respectively to AP and BP, and therefore meeting the conic in K and M; and join KM cutting off the segment KMRK from the conic. Let the center of the spheroid be S, and its greatest semidiameter SC. Then the force by which the spheroid attracts the body P will be to the force by which a sphere described with diameter AB attracts the same body as $\frac{AS \times CS^2 - PS \times KMRK}{PS^2 + CS^2 - AS^2}$ to

$\frac{AS^3}{3PS^2}$. And by the same mode of computation it is possible to find the forces of the segments of the spheroid.

COROLLARY 3. But if the corpuscle is located inside the spheroid and in its axis, the attraction will be as its distance from the center. This is seen more easily by the following argument, whether the particle is in the axis or in any other given diameter. Let AGOF be the attracting spheroid, S its center, and P the attracted body. Through that body P draw both the semidiameter SPA and any two straight lines DE and FG meeting the spheroid in D and F on one side and in E and G on the other; and let PCM and HLN be the surfaces of two inner spheroids, similar to and concentric with the outer spheroid; and let the first of these pass through the body P and cut the straight lines DE and FG in B and C, and let the latter cut the same straight lines in H, I and K, L. Let all the spheroids have a common axis, and the parts of the straight lines intercepted on the two sides, DP and BE, FP and CG, DH and IE, FK and LG will be equal to one another, because the straight lines DE, PB, and HI are bisected in the same point, as are also the straight lines FG, PC, and KL. Now suppose that DPF and EPG designate opposite cones described with the infinitely small vertical angles DPF and EPG, and that the lines DH and EI also are infinitely small; then the particles of the cones—that is, the particles DHKF and GLIE—cut off by the surfaces of the spheroids will (because of the equality of the lines DH and EI) be to each other as the squares of their distances from the corpuscle P, and therefore will attract the corpuscle equally. And by a like reasoning, if the spaces DPF and EGCB are divided into particles by the surfaces of innumerable similar concentric spheroids, having a common axis, then all of these particles will attract the body P in opposite directions equally on both sides. Therefore the forces of the cone DPF and of the conical segment [or truncated cone] EGCB are equal, and—being opposite—annul each other. And it is the same with regard to the forces of all the matter outside the innermost spheroid PCBM. Therefore the body P is attracted only by the innermost spheroid PCBM, and accordingly (by prop. 72, corol. 3) its attraction is to the force by which

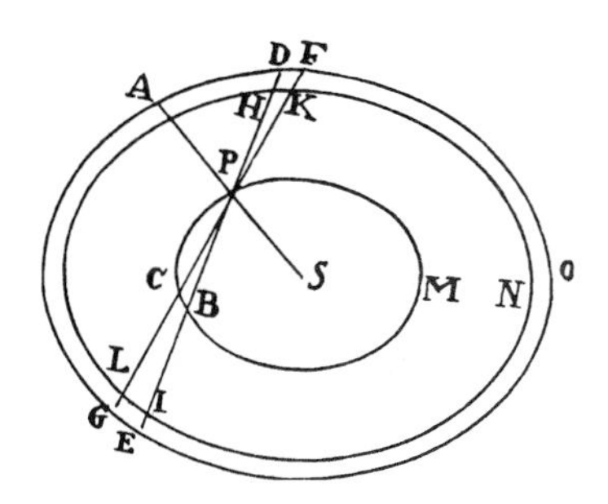

the body A is attracted by the whole spheroid AGOD as the distance PS to the distance AS. Q.E.D.

Proposition 92
Problem 46

Given an attracting body, it is required to find the ratio by which the centripetal forces tending toward each of its individual points decrease [i.e., decrease as a function of distance].

From the given body a sphere or cylinder or other regular figure is to be formed, whose law of attraction—corresponding to any ratio of decrease [in relation to distance]—can be found by props. 80, 81, and 91. Then, by making experiments, the force of attraction at different distances is to be found; and the law of attraction toward the whole that is thus revealed will give the ratio of the decrease of the forces of the individual parts, which was required to be found.

Proposition 93
Theorem 47

*If a solid, plane on one side but infinitely extended on the other sides, consists of equal and equally attracting particles, whose forces—in receding from the solid—decrease in the ratio of any power of the distances that is more than the square; and if a corpuscle set on either side of the plane is attracted by the force of the whole solid; then I say that that force of attraction of the solid in receding from its plane surface will decrease in the ratio of the distance of the corpuscle from the plane raised to a power whose index is less by 3 units than that of the power of the distances in the law of attractive force [*lit. *will decrease in the ratio of the power whose base is the distance of the corpuscle from the plane and whose index is less by 3 than the index of the power of the distances].*

Case 1. Let LG*l* be the plane by which the solid is terminated. Let the solid lie on the side of this plane toward I, and let it be resolved into innumerable planes *m*HM, *n*IN, *o*KO, . . . parallel to GL. And first let the attracted body C be placed outside the solid. Draw CGHI perpendicular to those innumerable planes, and let the forces of attraction of the points of the solid decrease in the ratio of a power of the distances whose index is the number *n* not

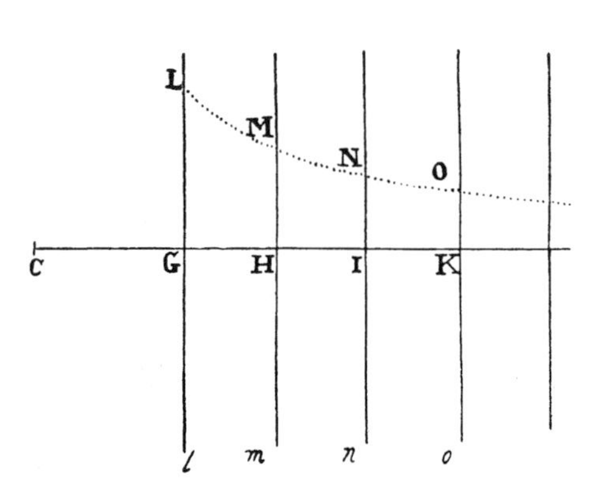

smaller than 3. Therefore (by prop. 90, corol. 3) the force by which any plane *m*HM attracts the point C is inversely as CH^{n-2}. In the plane *m*HM take the length HM inversely proportional to CH^{n-2}, and that force will be as HM. Similarly, on each of the individual planes *l*GL, *n*IN, *o*KO, . . . , take the lengths GL, IN, KO, . . . inversely proportional to CG^{n-2}, CI^{n-2}, CK^{n-2}, . . . ; then the forces of these same planes will be as the lengths taken, and thus the sum of the forces will be as the sum of the lengths; that is, the force of the whole solid will be as the area GLOK produced infinitely in the direction OK. But that area (by the well-known methods of quadratures) is inversely as CG^{n-3}, and therefore the force of the whole solid is inversely as CG^{n-3}. Q.E.D.

CASE 2. Now let the corpuscle C be placed on the side of the plane *l*GL inside the solid, and take the distance CK equal to the distance CG. Then the part LG*lo*KO of this solid, terminated by the parallel planes *l*GL and *o*KO, will not attract the corpuscle C (situated in the middle) in any direction, the opposite actions of opposite points annulling each other because of their equality. Accordingly, corpuscle C is attracted only by the force of the solid situated beyond the plane OK. But this force (by case 1) is inversely as CK^{n-3}, that is (because CG and CK are equal), inversely as CG^{n-3}. Q.E.D.

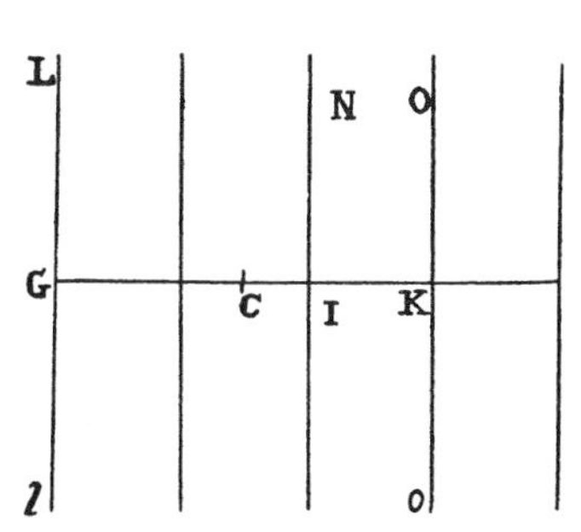

COROLLARY 1. Hence, if the solid LGIN is terminated on both sides by two infinitely extended and parallel planes LG and IN, its force of attraction becomes known by subtracting from the force of attraction of the whole infinitely extended solid LGKO the force of attraction of the further part NIKO produced infinitely in the direction KO.

COROLLARY 2. If the more distant part of this infinitely extended solid is ignored, since its attraction compared with the attraction of the nearer part is of almost no moment, then the attraction of that nearer part, with an increase of the distance, will decrease very nearly in the ratio of the power CG^{n-3}.

COROLLARY 3. And hence, if any body that is finite and plane on one side attracts a corpuscle directly opposite the middle of that plane, and the distance between the corpuscle and the plane is exceedingly small compared with the dimensions of the attracting body, and the attracting body consists of homogeneous particles whose forces of attraction decrease in the ratio of any power of the distances that is more than the fourth; the force of attraction of the whole body will decrease very nearly in the ratio of a power of that exceedingly small distance, whose index is less by 3 than the index of the stated power. This assertion is not valid for a body consisting of particles whose forces of attraction decrease in the ratio of the third power of the distances, because in this case the attraction of the more distant part of the infinitely extended body in corol. 2 is always infinitely greater than the attraction of the nearer part.

Scholium If a body is attracted perpendicularly toward a given plane, and the motion of the body is required to be found from the given law of attraction, the problem will be solved by seeking (by prop. 39) the motion of the body descending directly to this plane and by compounding this motion (according to corol. 2 of the laws) with a uniform motion performed along lines parallel to the same plane. And conversely, if it is required to find the law of an attraction made toward the plane along perpendicular lines, under the condition that the attracted body moves in any given curved line whatever, the problem will be solved by the operations used in the third problem [i.e., prop. 8].

The procedure can be shortened by resolving the ordinates into converging series. For example, if B is the ordinate to the base A at any given angle, and is as any power $A^{\frac{m}{n}}$ of that base, and the force is required by which a body that is either attracted toward the base or repelled away from the base (according to the position of the ordinate) can move in a curved line that the upper end of the ordinate traces out; I suppose the base to be increased by a minimally small part O, and I resolve the ordinate $(A+O)^{\frac{m}{n}}$ into the infinite series

$$A^{\frac{m}{n}} + \frac{m}{n}OA^{\frac{m-n}{n}} + \frac{m^2 - mn}{2n^2}O^2A^{\frac{m-2n}{n}} \ldots,$$

and I suppose the force to be proportional to the term of this series in which O is of two dimensions, that is, to the term $\frac{m^2 - mn}{2n^2}O^2A^{\frac{m-2n}{n}}$. Therefore the required force is as $\frac{m^2 - mn}{n^2}A^{\frac{m-2n}{n}}$, or, which is the same, as $\frac{m^2 - mn}{n^2}B^{\frac{m-2n}{m}}$. For example, if the ordinate traces out a parabola, where $m = 2$ and $n = 1$, the force will become as the given quantity $2B^o$, and thus will be given. Therefore with a given [i.e., constant] force the body will move in a parabola, as Galileo demonstrated. But if the ordinate traces out a hyperbola, where $m = 0 - 1$ and $n = 1$, the force will become as $2A^{-3}$ or $2B^3$; and therefore with a force that is as the cube of the ordinate, the body will move in a hyperbola. But putting aside propositions of this sort, I go on to certain others on motion which I have not as yet considered.

SECTION 14

The motion of minimally small bodies that are acted on by centripetal forces tending toward each of the individual parts of some great body

Proposition 94
Theorem 48

If two homogeneous mediums are separated from each other by a space terminated on the two sides by parallel planes, and a body passing through this space is attracted or impelled perpendicularly toward either medium and is not acted on or impeded by any other force, and the attraction at equal distances from each plane (taken on the same side of that plane) is the same everywhere; then I say that the sine of the angle of incidence onto either plane will be to the sine of the angle of emergence from the other plane in a given ratio.

CASE 1. Let A*a* and B*b* be the two parallel planes. Let the body be incident upon the first plane A*a* along line GH, and in all its passage through the intermediate space let it be attracted or impelled toward the medium of incidence, and by this action let it describe the curved line HI and emerge along the line IK. To the plane of emergence B*b* erect the perpendicular IM meeting the line of incidence GH produced in M and the plane of incidence A*a* in R; and let the line of emergence KI produced meet HM in L. With

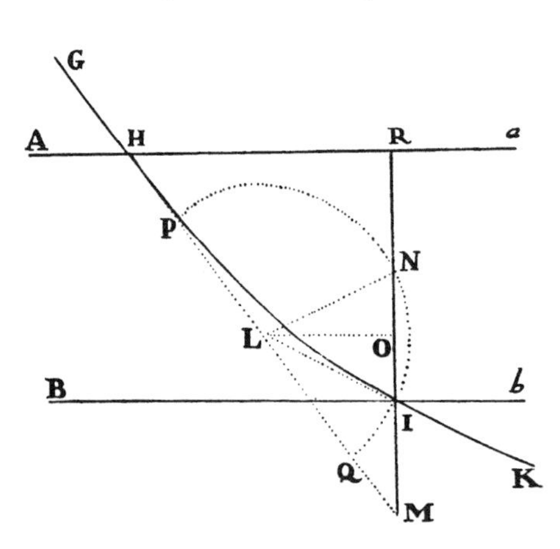

center L and radius LI describe a circle cutting HM in P and Q, as well as MI produced in N. Then first, if the attraction or impulse is supposed uniform, the curve HI (from what Galileo demonstrated) will be a parabola, of which this is a property: that the rectangle of its given latus rectum and the line IM is equal to HM squared; but also the line HM will be bisected in L. Hence, if the perpendicular LO is dropped to MI, MO and OR will be equal; and when the equals ON and OI have been added to these quantities, the totals MN and IR will become equal. Accordingly, since IR is given, MN is also given; and the rectangle NM × MI is to the rectangle of the latus rectum and IM (that is, to HM^2) in a given ratio. But the rectangle NM × MI is equal to the rectangle PM × MQ, that is, to the difference of the squares ML^2 and PL^2 or LI^2; and HM^2 has a given ratio to its fourth part ML^2: therefore the ratio of $ML^2 - LI^2$ to ML^2 is given, and by conversion [or convertendo] the ratio LI^2 to ML^2 is given, and also the square root of that ratio, LI to ML. But in every triangle LMI, the sines of the angles are proportional to the opposite sides. Therefore the ratio of the sine of the angle of incidence LMR to the sine of the angle of emergence LIR is given. Q.E.D.

CASE 2. Now let the body pass successively through several spaces terminated by parallel planes, A*ab*B, B*bc*C, . . . , and be acted on by a force that is uniform in each of the individual spaces considered separately but is different in each of the different spaces. Then by what has just been demonstrated, the sine of the angle of incidence upon the first plane A*a* will be to the sine of the angle of emergence from the second plane B*b* in a given ratio; and this sine, which is the sine of the angle of incidence upon the second plane B*b*, will be to the sine of the angle of emergence from the third plane C*c* in a given ratio; and this sine will be in a given ratio to the sine of the angle of emergence from the fourth plane D*d*; and so on indefinitely. And from the equality of the ratios [or ex aequo] the sine of the angle of incidence upon the first plane will be in a given ratio to the sine of the angle of emergence from the last plane. Now let the distances between the planes be diminished and their number increased indefinitely, so that the action of attraction or of impulse, according to any assigned law whatever, becomes continuous; then the ratio of the sine of the angle of incidence upon the first plane to the sine of the angle of emergence from the last plane, being always given, will still be given now. Q.E.D.

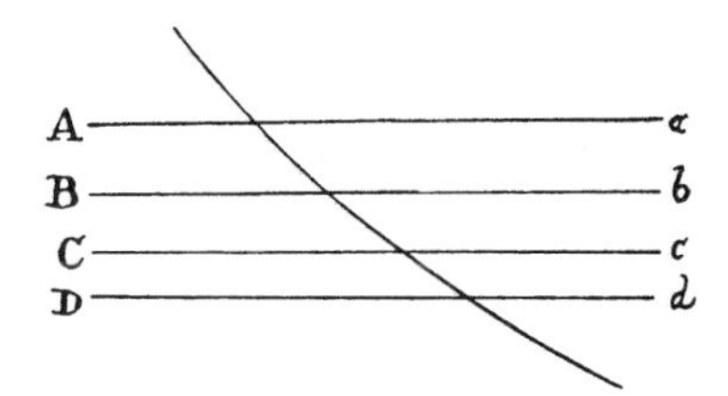

Proposition 95
Theorem 49

With the same suppositions as in prop. 94, I say that the velocity of the body before incidence is to its velocity after emergence as the sine of the angle of emergence to the sine of the angle of incidence.

Let AH be taken equal to I*d*, and erect the perpendiculars AG and *d*K meeting the lines of incidence and emergence GH and IK in G and K. In GH take TH equal to IK, and drop T*v* perpendicular to the plane A*a*. And (by corol. 2 of the laws) resolve the motion of the body into two motions, one

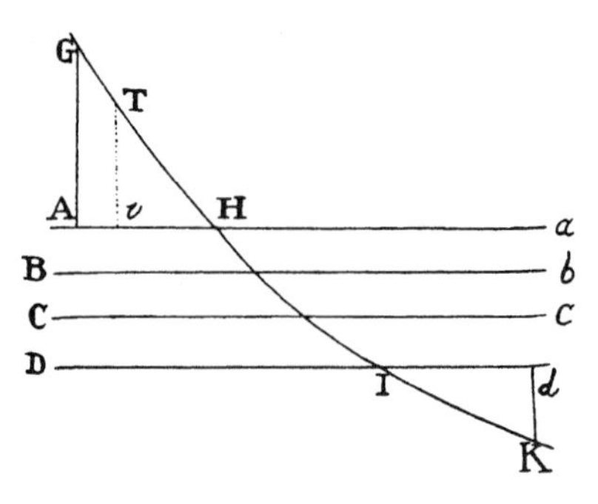

perpendicular, the other parallel, to the planes A*a*, B*b*, C*c*, ... The [component of the] force of attraction or of impulse acting along perpendicular lines does not at all change the motion in the direction of the parallels; and therefore the body, by this latter motion, will in equal times pass through equal distances along parallels between the line AG and the point H, and between the point I and the line *d*K, that is, it will describe the lines GH and IK in equal times. Accordingly, the velocity before incidence is to the velocity after emergence as GH to IK or TH; that is, as AH or I*d* to *v*H, that is (with respect to the radius TH or IK), as the sine of the angle of emergence to the sine of the angle of incidence. Q.E.D.

Proposition 96
Theorem 50

With the same suppositions, and supposing also that the motion before incidence is faster than afterward, I say that as a result of [a]*changing the inclination*[a] *of the line of incidence, the body will at last be reflected, and the angle of reflection will become equal to the angle of incidence.*

For suppose the body to describe parabolic arcs between the parallel planes A*a*, B*b*, C*c*, . . . , as before; and let those arcs be HP, PQ, QR, ...

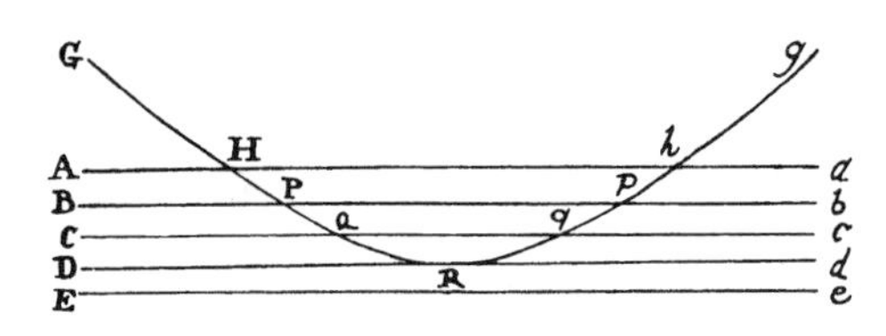

And let the obliquity of the line of incidence GH to the first plane A*a* be such that the sine of the angle of incidence is to the radius of the circle whose sine it is in the ratio which that same sine of the angle of incidence has to the sine of the angle of emergence from the plane D*d* into the space D*de*E; then, because the sine of the angle of emergence will now have become equal to the radius, the angle of emergence will be a right angle, and hence the line of emergence will coincide with the plane D*d*. Let the body arrive at this plane at the point R; and since the line of emergence coincides with

aa. The sense of Newton's "changing the inclination" is that of increasing the angle of incidence.

that same plane, it is obvious that the body cannot go any further toward the plane E*e*. But neither can it go on in the line of emergence R*d*, because it is continually attracted or impelled toward the medium of incidence. Therefore, this body will be turned back between the planes C*c* and D*d*, describing an arc of the parabola QR*q*, whose principal vertex (according to what Galileo demonstrated) is at R, and will cut the plane C*c* in the same angle at *q* as formerly at Q; and then, proceeding in the parabolic arcs *qp*, *ph*, . . . , similar and equal to the former arcs QP and PH, this body will cut the remaining planes in the same angles at *p*, *h*, . . . , as formerly at P, H, . . . , and will finally emerge at *h* with the same obliquity with which it was incident upon the plane at H. Now suppose the distances between the planes A*a*, B*b*, C*c*, D*d*, E*e*, . . . to be diminished and their number increased indefinitely, so that the action of attraction or impulse, according to any assigned law whatever, is made to be continuous; then the angle of emergence, being always equal to the angle of incidence, will still remain equal to it now. Q.E.D.

Scholium

These attractions are very similar to the reflections and refractions of light made according to a given ratio of the secants, as Snel discovered, and consequently according to a given ratio of the sines, as Descartes set forth. For the fact that light is propagated successively [i.e., in time and not instantaneously] and comes from the sun to the earth in about seven or eight minutes is now established by means of the phenomena of the satellites of Jupiter, confirmed by the observations of various astronomers. Moreover, the rays of light that are in the air (as Grimaldi recently discovered, on admitting light into a dark room through a small hole—something I myself have also tried) in their passing near the edges of bodies, whether opaque or transparent (such as are the circular-rectangular edges of coins minted from gold, silver, and bronze, and the sharp edges of knives, stones, or broken glass), are inflected around the bodies, as if attracted toward them; and those of the rays that in such passing approach closer to the bodies are inflected the more, as if more attracted, as I myself have also diligently observed. And those that pass at greater distances are less inflected, and at still greater distances are inflected somewhat in the opposite direction and form three bands of colors.

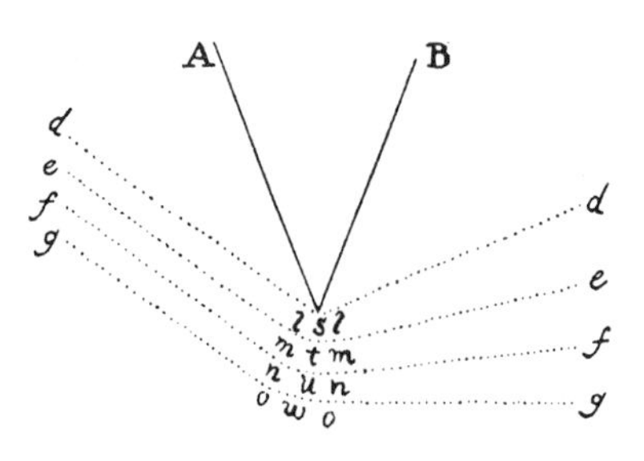

In the figure, *s* designates the sharp edge of a knife or of any wedge A*s*B, and *gowog*, *fnunf*, *emtme*, and *dlsld* are rays, inflected in the arcs *owo*, *nun*, *mtm*, and *lsl* toward the knife, more so or less so according to their distance from the knife. Moreover, since such an inflection of the rays takes place in the air outside the knife, the rays which are incident upon the knife must also be inflected in the air before they reach it. And the case is the same for those rays incident upon glass. Therefore refraction takes place not at the point of incidence, but gradually by a continual inflection of the rays, made partly in the air before the rays touch the glass, and partly (if I am not mistaken) within the glass after they have entered it, as has been delineated in the rays *ckzc*, *biyb*, and *ahxa* incident upon the glass at *r*, *q*, and *p*, and inflected between *k* and *z*, *i* and *y*, *h* and *x*. Therefore because of the analogy that exists between the propagation of rays of light and the motion of bodies, I have decided to subjoin the following propositions for optical uses, meanwhile not arguing at all about the nature of the rays (that is, whether they are bodies or not), but only determining the trajectories of bodies, which are very similar to the trajectories of rays.

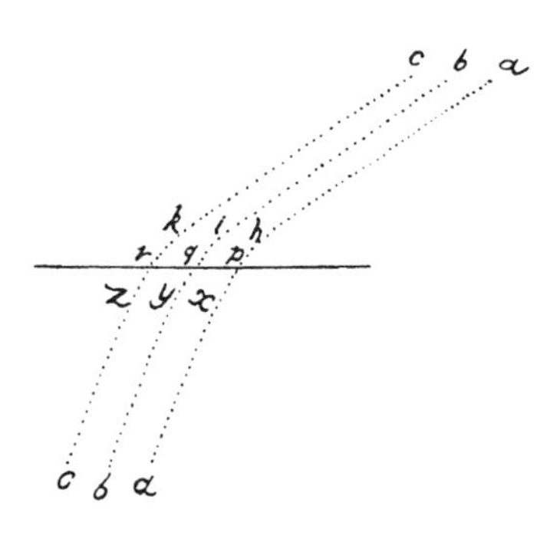

Proposition 97
Problem 47

Supposing that the sine of the angle of incidence upon some surface is to the sine of the angle of emergence in a given ratio, and that the inflection of the paths of bodies in close proximity to that surface takes place in a very short space, which can be considered to be a point; it is required to determine the surface that may make all the corpuscles emanating successively from a given place converge to another given place.

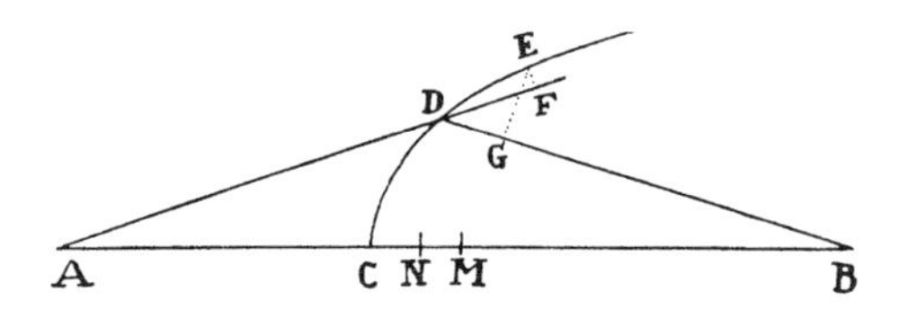

Let A be the place from which the corpuscles diverge, B the place to which they should converge, CDE the curved line that—by revolving about the axis

AB—describes the required surface, D and E any two points of that curve, and EF and EG perpendiculars dropped to the paths AD and DB of the body. Let point D approach point E; then the ultimate ratio of the line DF (by which AD is increased) to the line DG (by which DB is decreased) will be the same as that of the sine of the angle of incidence to the sine of the angle of emergence. Therefore the ratio of the increase of the line AD to the decrease of the line DB is given; and as a result, if a point C is taken anywhere on the axis AB, this being a point through which the curve CDE should pass, and the increase CM of AC is taken in that given ratio to the decrease CN of BC, and if two circles are described with centers A and B and radii AM and BN and cut each other at D, that point D will touch the required curve CDE, and by touching it anywhere whatever will determine that curve. Q.E.I.

COROLLARY 1. But by making point A or B in one case go off indefinitely, in another case move to the other side of point C, all the curves which Descartes exhibited with respect to refractions in his treatises on optics and geometry will be traced out. Since Descartes concealed the methods of finding these, I have decided to reveal them by this proposition.

COROLLARY 2. If a body, incident upon any surface CD along the straight line AD drawn according to any law, emerges along any other straight line DK; and if from point C the curved lines CP and CQ, always perpendicular to AD and DK, are understood to be drawn; then the increments of the lines PD and QD, and hence the lines themselves PD and QD generated by those increments, will be as the sines of the angles of incidence and emergence to each other, and conversely.

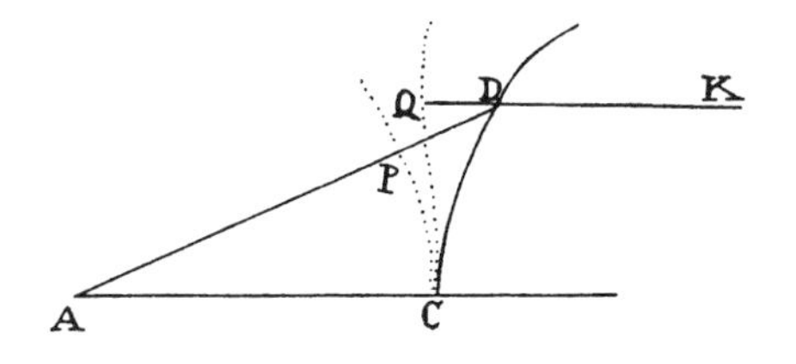

Proposition 98
Problem 48

The same conditions being supposed as in prop. 97, and supposing that there is described about the axis AB *any attracting surface* CD, *regular or irregular, through which the bodies coming out from a given place* A *must pass; it is required to find a second attracting surface* EF *that will make the bodies converge to a given place* B.

Join AB and let it cut the first surface in C and the second in E, point D being taken in any way whatever. And supposing that the sine of the angle

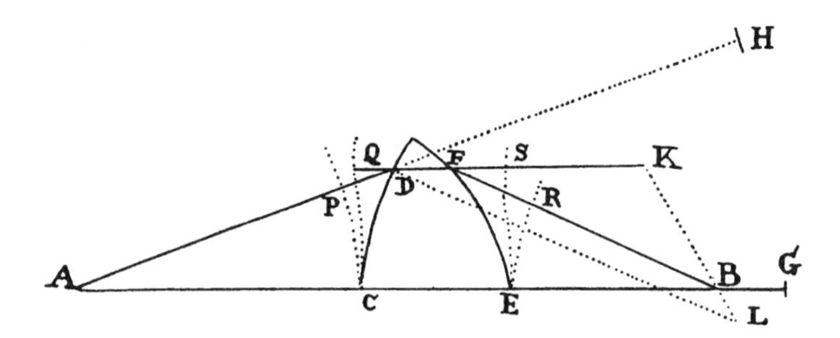

of incidence upon the first surface is to the sine of the angle of emergence from that first surface, and that the sine of the angle of emergence from the second surface is to the sine of the angle of incidence upon the second surface, as some given quantity M is to another given quantity N; produce AB to G so that BG is to CE as M − N to N, and produce AD to H so that AH is equal to AG, and also produce DF to K so that DK is to DH as N to M. Join KB, and with center D and radius DH describe a circle meeting KB produced in L, and draw BF parallel to DL; then the point F will touch the line EF, which—on being revolved about the axis AB—will describe the required surface. Q.E.F.

Now suppose the lines CP and CQ to be everywhere perpendicular to AD and DF respectively, and the lines ER and ES to be similarly perpendicular to FB and FD, with the result that QS is always equal to CE; then (by prop. 97, corol. 2) PD will be to QD as M to N, and therefore as DL to DK or FB to FK; and by separation [or dividendo] as DL − FB or PH − PD − FB to FD or FQ − QD, and by composition [or componendo] as PH − FB to FQ, that is (because PH and CG, QS and CE are equal), as CE + BG − FR to CE − FS. But (because BG is proportional to CE and M − N is proportional to N) CE + BG is also to CE as M to N, and thus by separation [or dividendo] FR is to FS as M to N; and therefore (by prop. 97, corol. 2) the surface EF compels a body incident upon it along the line DF to go on in the line FR to the place B. Q.E.D.

Scholium It would be possible to use the same method for three surfaces or more. But for optical uses spherical shapes are most suitable. If the objective lenses of telescopes are made of two lenses that are spherically shaped and water is enclosed between them, it can happen that errors of the refractions that take place in the extreme surfaces of the lenses are accurately enough corrected by the refractions of the water. Such objective lenses are to be preferred to elliptical and hyperbolical lenses, not only because they can be formed more easily and more accurately but also because they more accurately refract the pencils of rays situated outside the axis of the glass. Nevertheless, the differing refrangibility of different rays [i.e., of rays of different colors] prevents optics from being perfected by spherical or any other shapes. Unless the errors arising from this source can be corrected, all labor spent in correcting the other errors will be of no avail.

BOOK 2

THE MOTION OF BODIES

SECTION 1

The motion of bodies that are resisted in proportion to their velocity

Proposition 1
Theorem 1

If a body is resisted in proportion to its velocity, the motion lost as a result of the resistance is as the space described in moving.

For since the motion lost in each of the equal particles of time is as the velocity, that is, as a particle of the path described, then, by composition [or componendo], the motion lost in the whole time will be as the whole path. Q.E.D.

Corollary. Therefore, if a body, devoid of all gravity, moves in free spaces by its inherent force alone and if there are given both the whole motion at the beginning and also the remaining motion after some space has been described, the whole space that the body can describe in an infinite time will be given. For that space will be to the space already described as the whole motion at the beginning is to the lost part of that motion.

Lemma 1

Quantities proportional to their differences are continually proportional.

Let A be to A − B as B to B − C and C to C − D, . . . ; then, by conversion [or convertendo], A will be to B as B to C and C to D, ... Q.E.D.

Proposition 2
Theorem 2

If a body is resisted in proportion to its velocity and moves through a homogeneous medium by its inherent force alone and if the times are taken as equal, the velocities at the beginnings of the individual times are in a geometric progression, and the spaces described in the individual times are as the velocities.

Case 1. Divide the time into equal particles; and if, at the very beginnings of the particles, a force of resistance which is as the velocity acts with a single impulse, the decrease of the velocity in the individual particles of time will be as that velocity. The velocities are therefore proportional to their differences and thus (by book 2, lem. 1) are continually proportional. Accordingly, if any equal times are compounded of an equal number of particles, the velocities at the very beginnings of the times will be as the terms in a continual progression in which some have been skipped, omitting an equal number of intermediate terms in each interval. The ratios of these terms are indeed compounded of equally repeated equal ratios of the intermediate terms, and therefore these compound ratios are also equal to one another.

Therefore, since the velocities are proportional to these terms, they are in a geometric progression. Now let those equal particles of times be diminished, and their number increased indefinitely, so that the impulse of the resistance becomes continual; then the velocities at the beginnings of equal times, which are always continually proportional, will also be continually proportional in this case. Q.E.D.

Case 2. And by separation [or dividendo] the differences of the velocities (that is, the parts of them which are lost in the individual times) are as the wholes, while the spaces described in the individual times are as the lost parts of the velocities (by book 2, prop. 1) and are therefore also as the wholes. Q.E.D.

Corollary. Hence, if a hyperbola BG is described with respect to the rectangular asymptotes AC and CH and if AB and DG are perpendicular to asymptote AC and if both the velocity of the body and the resistance of the medium are represented, at the very beginning of the motion, by any given line AC, but after some time has elapsed, by the indefinite line DC, then the time can be represented by area ABGD, and the space described in that time can be represented by line AD.

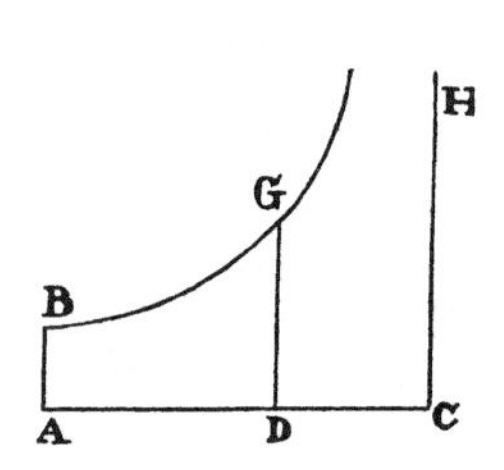

For if the area is increased uniformly by the motion of point D, in the same manner as the time, the straight line DC will decrease in a geometric ratio in the same way as the velocity, and the parts of the straight line AC described in equal times will decrease in the same ratio.

Proposition 3
Problem 1

To determine the motion of a body which, while moving straight up or down in a homogeneous medium, is resisted in proportion to the velocity, and which is acted on by uniform gravity.

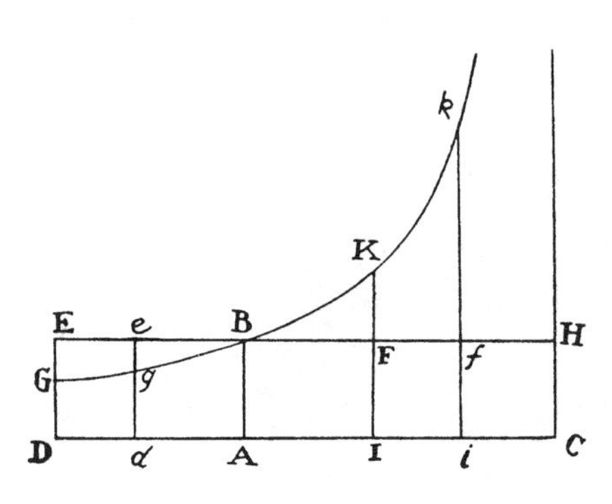

When the body is moving up, represent the gravity by any given rectangle BACH, and the resistance of the medium at the beginning of the ascent by the rectangle BADE taken on the other side of the straight line AB. With respect to the rectangular asymptotes AC and CH, describe a hyperbola through point B, cutting perpendiculars DE and *de* in G and *g*; then the body, by ascending in the time DG*gd*, will describe the space EG*ge*; and in the time DGBA will describe the space of the total ascent EGB; and in the time ABKI will describe the space of descent BFK; and in the time IK*ki* will describe the space of descent KF*fk*; and the body's velocities (proportional to the resistance of the medium) in these periods of time will be ABED, AB*ed*, null, ABFI, and AB*fi* respectively; and the greatest velocity that the body can attain in descending will be BACH.

For resolve the rectangle BACH into innumerable rectangles A*k*, K*l*, L*m*, M*n*, . . . , which are as the increases of the velocities, occurring in the same number of equal times; then nil, A*k*, A*l*, A*m*, A*n*, . . . will be as the total velocities, and thus (by hypothesis) as the resistances of the medium at the beginning of each of the equal times. Make AC to AK, or ABHC to AB*k*K, as the force of gravity to the resistance at the beginning of the second time, and subtract the resistances from the force of gravity; then the remainders ABHC, K*k*HC, L*l*HC, M*m*HC, . . . will be as the absolute forces by which the body is urged at the beginning of each of the times, and thus (by the second law of motion) as the increments of the velocities, that is, as the rectangles A*k*, K*l*, L*m*, M*n*, . . . , and therefore (by book 2, lem. 1) in a geometric progression. Therefore, if the straight lines K*k*, L*l*, M*m*, N*n*, . . . , produced, meet the hyperbola in *q*, *r*, *s*, *t*, . . . , areas AB*q*K, K*qr*L, L*rs*M, M*st*N, . . . will be equal, and thus proportional both to the times and to the forces of gravity, which are always equal. But area AB*q*K (by book 1, lem. 7, corol. 3, and lem. 8) is to area B*kq* as K*q* to ½*kq* or AC to ½AK, that is, as the force of gravity to the resistance in the middle of the first time. And by a similar argument, areas *q*KL*r*, *r*LM*s*, *s*MN*t*, . . . are to areas *qklr*, *rlms*, *smnt*, . . . as the force of gravity to the resistance in the middle of the second time, of the third, of the fourth, ... Accordingly, since the equal areas BAK*q*, *q*KL*r*, *r*LM*s*, *s*MN*t*, . . . are proportional to the forces of gravity, areas B*kq*, *qklr*, *rlms*, *smnt*, . . . will be proportional to the resistance in the middle of each of the times, that is (by hypothesis), to the velocities, and thus to the spaces described. Take the sums of the pro-

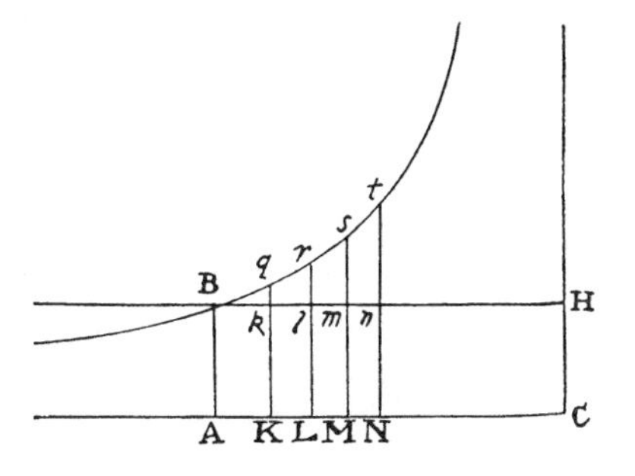

portional quantities; then areas B*kq*, B*lr*, B*ms*, B*nt*, ... will be proportional to the total spaces described, and areas AB*q*K, AB*r*L, AB*s*M, AB*t*N, ... will be proportional to the times. Therefore the body, while descending in any time AB*r*L, describes the space B*lr*, and in the time L*rt*N describes the space *rlnt*. Q.E.D. And the proof is similar for an ascending motion. Q.E.D.

COROLLARY 1. Therefore the greatest velocity that a body can acquire in falling is to the velocity acquired in any given time as the given force of gravity by which the body is continually urged to [a]the force of the resistance by which it is impeded at the end of that time.[a]

COROLLARY 2. If the time is increased in an arithmetic progression, the sum of that greatest velocity and of the velocity in the ascent, and also their difference in the descent, decreases in a geometric progression.

COROLLARY 3. The differences of the spaces which are described in equal differences of the times decrease in the same geometric progression.

COROLLARY 4. The space described by a body is the difference of two spaces, of which one is as the time reckoned from the beginning of the descent, and the other is as the velocity; and these spaces are equal to each other at the very beginning of the descent.

Proposition 4
Problem 2

Supposing that the force of gravity in some homogeneous medium is uniform and tends perpendicularly toward the plane of the horizon, it is required to determine the motion of a projectile in that medium, while it is resisted in proportion to the velocity.

From any place D let a projectile go forth along any straight line DP, and represent its velocity at the beginning of the motion by the length DP. Drop the perpendicular PC from point P to the horizontal line DC, and cut DC in A so that DA is to AC as the resistance of the medium arising from the upward motion at the beginning is to the force of gravity; or (which comes to the same thing) so that the rectangle of DA and DP is to the rectangle of AC and CP as the whole resistance at the beginning of the motion is to the force of gravity. Describe any hyperbola GTBS with asymptotes DC and CP which cuts the perpendiculars DG and AB in G and B; and complete the parallelogram DGKC, whose side GK cuts AB in Q. Take the line N

aa. Ed. 2 has "the excess of this force over the force by which it is resisted at the end of that time."

in the same ratio to QB as DC to CP, and at any point R of the straight line DC erect the perpendicular RT which meets the hyperbola in T and the straight lines EH, GK, and DP in I, *t*, and V, and then on RT take V*r* equal to $\frac{t\mathrm{GT}}{\mathrm{N}}$, or (which comes to the same thing) take R*r* equal to $\frac{\mathrm{GTIE}}{\mathrm{N}}$. Then in the time DRTG the projectile will arrive at point *r*, describing the curved line D*ra*F which point *r* traces out, reaching its greatest height *a* in the perpendicular AB, and afterward always approaching the asymptote PC. And its velocity at any point *r* is as the tangent *r*L of the curve. Q.E.I.

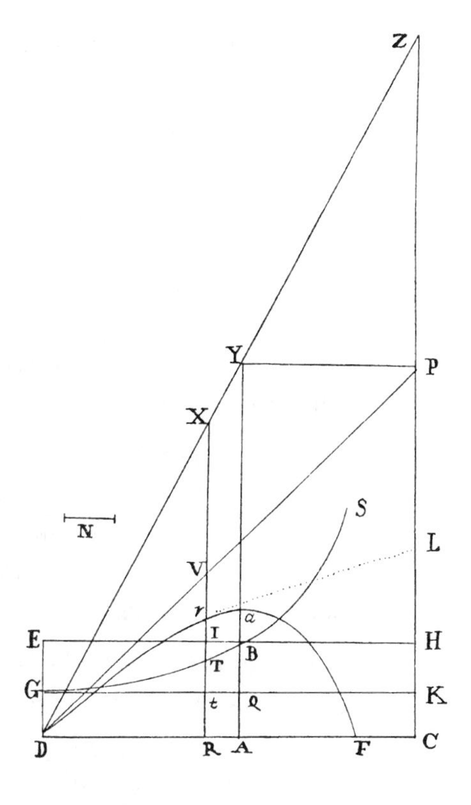

For N is to QB as DC to CP or DR to RV, and thus RV is equal to $\frac{\mathrm{DR} \times \mathrm{QB}}{\mathrm{N}}$, and R*r* $\left(\text{that is, RV} - \text{V}r\text{, or } \frac{\mathrm{DR} \times \mathrm{QB} - t\mathrm{GT}}{\mathrm{N}}\right)$ is equal to $\frac{\mathrm{DR} \times \mathrm{AB} - \mathrm{RDGT}}{\mathrm{N}}$. Now represent the time by area RDGT, and (by corol. 2 of the laws) divide the motion of the body into two parts, one upward and the other lateral. Since the resistance is as the motion, it also will be divided into two parts proportional to and opposite to the parts of the motion; and thus the distance described by the lateral motion will be (by book 2, prop. 2) as line DR, and the distance described by the upward motion will be (by book 2, prop. 3) as the area DR × AB − RDGT, that is, as line R*r*. But at the very beginning of the motion the area RDGT is equal to the rectangle DR × AQ, and thus that line R*r* $\left(\text{or } \frac{\mathrm{DR} \times \mathrm{AB} - \mathrm{DR} \times \mathrm{AQ}}{\mathrm{N}}\right)$ is then to DR as AB − AQ or QB to N, that is, as CP to DC, and hence as the upward motion to the lateral motion at the beginning. Since, therefore, R*r* is always as the distance upward, and DR is always as the distance sideward, and R*r*

is to DR at the beginning as the distance upward to the distance sideward, R*r* must always be DR as the distance upward to the distance sideward, and therefore the body must move in the line D*ra*F, which the point *r* traces out. Q.E.D.

Corollary 1. R*r* is therefore equal to $\frac{\text{DR} \times \text{AB}}{\text{N}} - \frac{\text{RDGT}}{\text{N}}$; and thus, if RT is produced to X so that RX is equal to $\frac{\text{DR} \times \text{AB}}{\text{N}}$, that is, if the parallelogram ACPY is completed, and DY is joined cutting CP in Z, and RT is produced until it meets DY in X, then X*r* will be equal to $\frac{\text{RDGT}}{\text{N}}$, and therefore will be proportional to the time.

Corollary 2. Hence, if innumerable lines CR are taken (or, which comes to the same thing, innumerable lines ZX) in a geometric progression, then as many lines X*r* will be in an arithmetic progression. And hence it is easy to draw curve D*ra*F with the help of a table of logarithms.

Corollary 3. If a parabola is constructed with vertex D and diameter DG (produced downward) and a latus rectum that is to 2DP as the whole resistance at the very beginning of the motion is to the force of gravity, then the velocity with which a body must go forth from place D along the straight line DP in order to describe curve D*ra*F in a uniform resisting medium will be the very one with which it must go forth from the same place D along the same straight line DP in order to describe a parabola in a nonresisting space.

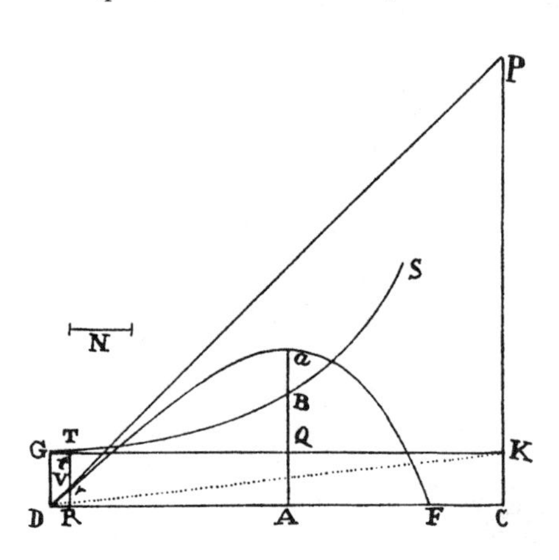

For the latus rectum of this parabola, at the very beginning of the motion, is $\frac{\text{DV}^2}{\text{V}r}$; and V*r* is $\frac{t\text{GT}}{\text{N}}$ or $\frac{\text{DR} \times \text{T}t}{2\text{N}}$. But the straight line that, if it were drawn, would touch the hyperbola GTS in G is parallel to DK, and thus T*t* is $\frac{\text{CK} \times \text{DR}}{\text{DC}}$, and N has been taken as $\frac{\text{QB} \times \text{DC}}{\text{CP}}$. Therefore V*r* is $\frac{\text{DR}^2 \times \text{CK} \times \text{CP}}{2\text{DC}^2 \times \text{QB}}$, that is (because DR is to DC as DV is to DP),

$\frac{DV^2 \times CK \times CP}{2DP^2 \times QB}$; and the latus rectum $\frac{DV^2}{Vr}$ comes out $\frac{2DP^2 \times QB}{CK \times CP}$, that is (because QB is to CK as DA is to AC), $\frac{2DP^2 \times DA}{AC \times CP}$, and thus is to 2DP as DP × DA to CP × AC—that is, as the resistance to the gravity. Q.E.D.

COROLLARY 4. Hence, if a body is projected from any place D with a given velocity along any straight line DP given in position, and the resistance of the medium at the very beginning of the motion is given, the curve D*ra*F which the body will describe can be found. For from the given velocity the latus rectum of the parabola is given, as is well known. And if 2DP is taken to that latus rectum as the force of gravity to the force of resistance, DP is given. Then, if DC is cut in A so that CP × AC is to DP × DA in that same ratio of gravity to resistance, point A will be given. And hence curve D*ra*F is given.

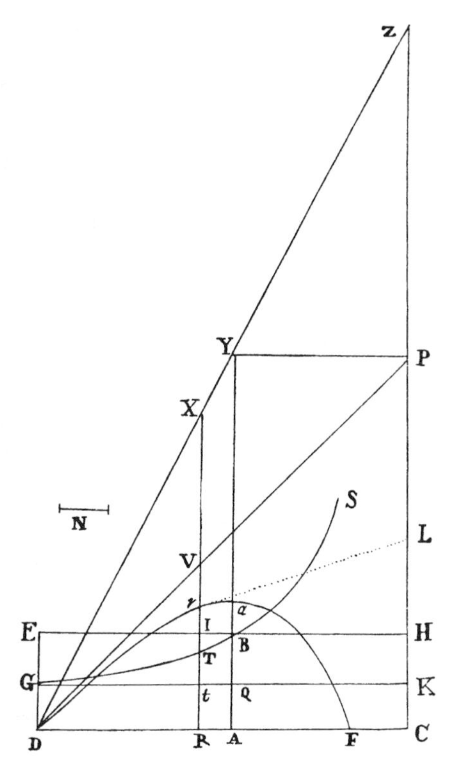

COROLLARY 5. And conversely, if curve D*ra*F is given, both the velocity of the body and the resistance of the medium in each of the places *r* will be given. For since the ratio of CP × AC to DP × DA is given, both the resistance of the medium at the beginning of the motion and the latus rectum of the parabola are also given; and hence the velocity at the beginning of the motion is also given. Then from the length of the tangent *r*L, both the velocity (which is proportional to it) and the resistance (which is proportional to the velocity) are given in any place *r*.

COROLLARY 6. The length 2DP is to the latus rectum of the parabola as the gravity to the resistance at D; and when the velocity is increased the resistance is increased in the same ratio, but the latus rectum of the parabola is increased in the square of that ratio; hence it is evident that the length

2DP is increased in the simple ratio and thus is always proportional to the velocity and is not increased or decreased when the angle CDP is changed unless the velocity is also changed.

COROLLARY 7. Hence the method is apparent for determining the curve D*ra*F from phenomena approximately and for obtaining thereby the resistance and the velocity with which the body is projected. Project two similar and equal bodies with the same velocity from place D along the different angles CDP and CD*p*, and let the places F and *f* where they fall upon the horizontal plane DC be known. Then, taking any length for DP or D*p*, suppose that the resistance at D is to the gravity in any ratio, and represent that ratio by any length SM. Then, by computation, find the lengths DF and D*f* from that assumed length DP, and from the ratio $\frac{\mathrm{F}f}{\mathrm{DF}}$ (found by computation) take away the same ratio (found by experiment), and represent the difference by the perpendicular MN. Do the same thing a second and a third time, always taking a new ratio SM of resistance to gravity, and obtain a new difference MN. But draw the positive differences on one side of the straight line SM and the negative differences on the other, and through points N, N, N draw the regular curve NNN cutting the straight line SMMM in X, and then SX will be the true ratio of the resistance to the gravity, which it was required to find. From this ratio the length DF is to be obtained by calculation; then the length that is to the assumed length DP as the length DF (found out by experiment) to the length DF (just found by computation) will be the true length DP. When this is found, there will be known both the curved line D*ra*F that the body describes and the body's velocity and resistance in every place.

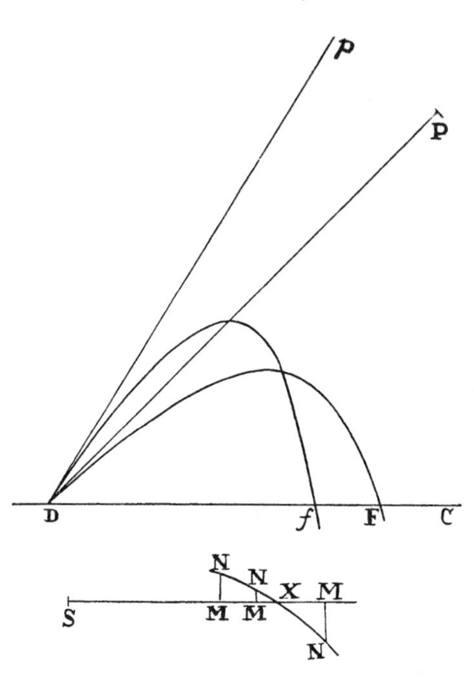

Scholium

However, the hypothesis that the resistance encountered by bodies is in the ratio of the velocity belongs more to mathematics than to nature.[a] In mediums wholly lacking in rigidity, the resistances encountered by bodies are as the squares of the velocities. For by the action of a swifter body, a motion that is greater in proportion to that greater velocity is communicated to a given quantity of the medium in a smaller time; and thus in an equal time, because a greater quantity of the medium is disturbed, a greater motion is communicated in proportion to the square of the velocity, and (by the second and third laws of motion) the resistance is as the motion communicated. Let us see, therefore, what kinds of motions arise from this law of resistance.

a. Ed. 1 and ed. 2 have an additional sentence: "This ratio obtains very nearly when bodies are moving very slowly in mediums having some rigidity." In Newton's annotated copy of ed. 2, "very nearly" is changed to "more closely."

SECTION 2

The motion of bodies that are resisted as the squares of the velocities

Proposition 5
Theorem 3

If the resistance of a body is proportional to the square of the velocity and if the body moves through a homogeneous medium by its inherent force alone and if the times are taken in a geometric progression going from the smaller to the greater terms, I say that the velocities at the beginning of each of the times are inversely in that same geometric progression and that the spaces described in each of the times are equal.

For since the resistance of the medium is proportional to the square of the velocity, and the decrement of the velocity is proportional to the resistance, if the time is divided into innumerable equal particles, the squares of the velocities at each of the beginnings of the times will be proportional to the differences of those same velocities. Let the particles of time be AK, KL, LM, . . . , taken in the straight line CD, and erect perpendiculars AB, K*k*, L*l*, M*m*, . . . , meeting the hyperbola B*klm*G (described with center C and rectangular asymptotes CD and CH) in B, *k*, *l*, *m*, . . . ; then AB will be to K*k* as CK to CA, and by separation [or dividendo] AB − K*k* to K*k* as AK to CA, and by alternation [or alternando] AB − K*k* to AK as K*k* to CA, and thus as AB × K*k* to AB × CA. Hence, since AK and AB × CA are given, AB − K*k* will be as AB × K*k*; and ultimately, when AB and K*k* come together, as AB^2. And by a similar argument K*k* − L*l*, L*l* − M*m*, . . . will be as Kk^2, Ll^2, ... The squares of lines AB, K*k*, L*l*, M*m*, therefore, are as their differences; and on that account, since the squares of the velocities were also as their differences, the progression of both will be similar. It follows from what has been proved that the areas described by these lines are also in a progression entirely similar to that of the spaces described by the velocities. Therefore, if the velocity at the beginning of the first time AK is represented by line AB, and the velocity at the beginning of the second time KL by line K*k*, and the length described in the first time is represented by area AK*k*B, then all the subsequent velocities will be represented by the subsequent lines L*l*, M*m*, . . . ,

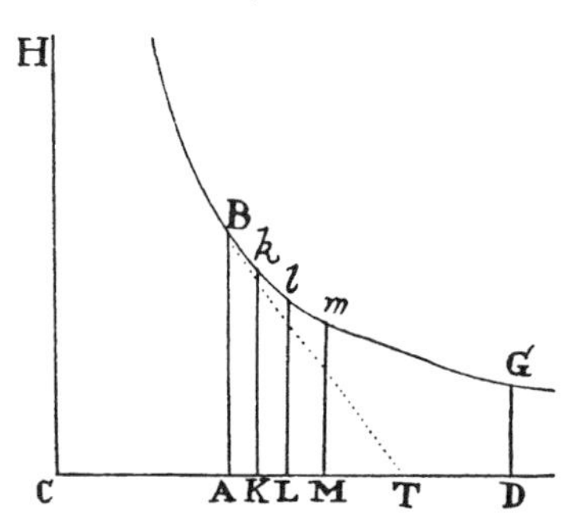

and the lengths described will be represented by areas Kl, Lm, ... And by composition [or componendo], if the whole time is represented by the sum of its parts AM, the whole length described will be represented by the sum of its parts AMmB. Now imagine time AM to be divided into parts AK, KL, LM, . . . in such a way that CA, CK, CL, CM, . . . are in a geometric progression; then those parts will be in the same progression, and the velocities AB, Kk, Ll, Mm, . . . will be in the same progression inverted, and the spaces described Ak, Kl, Lm, . . . will be equal. Q.E.D.

Corollary 1. Therefore it is evident that if the time is represented by any part AD of the asymptote, and the velocity at the beginning of the time by ordinate AB, then the velocity at the end of the time will be represented by ordinate DG, and the whole space described will be represented by the adjacent hyperbolic area ABGD; and furthermore, the space that a body in a nonresisting medium could describe in the same time AD, with the first velocity AB, will be represented by the rectangle AB × AD.

Corollary 2. Hence the space described in a resisting medium is given by taking that space to be in the same proportion to the space which could be described simultaneously with a uniform velocity AB in a nonresisting medium as the hyperbolic area ABGD is to the rectangle AB × AD.

Corollary 3. The resistance of the medium is also given by setting it to be, at the very beginning of the motion, equal to the uniform centripetal force that in a nonresisting medium could generate the velocity AB in a falling body in the time AC. For if BT is drawn, touching the hyperbola in B and meeting the asymptote in T, the straight line AT will be equal to AC and will represent the time in which the first resistance uniformly continued could annul the whole velocity AB.

Corollary 4. And hence the proportion of this resistance to the force of gravity or to any other given centripetal force is also given.

Corollary 5. And conversely, if the proportion of the resistance to any given centripetal force is given, the time AC is given in which a centripetal force equal to the resistance could generate any velocity AB; and hence point B is given, through which the hyperbola with asymptotes CH and CD must be described, as is also the space ABGD which the body, beginning its motion with that velocity AB, can describe in any time AD in a homogeneous resisting medium.

Proposition 6
Theorem 4

Equal homogeneous spherical bodies that are resisted in proportion to the square of the velocity, and are carried forward by their inherent forces alone, will, in times that are inversely as the initial velocities, always describe equal spaces, and lose parts of their velocities proportional to the wholes.

Describe any hyperbola B*b*E*e*, with rectangular asymptotes CD and CH, which cuts perpendiculars AB, *ab*, DE, and *de* in B, *b*, E, and *e*; and represent the initial velocities by perpendiculars AB and DE and the times by lines A*a* and D*d*. Therefore A*a* is to D*d* as (by hypothesis) DE is to AB, and as (from the nature of the hyperbola) CA is to CD, and by composition [or componendo] as C*a* is to C*d*. Hence areas AB*ba* and DE*ed*, that is, the spaces described, are equal to each other, and the first velocities AB and DE are proportional to the ultimate velocities *ab* and *de*, and therefore, by separation [or dividendo], also to the lost parts of those velocities AB − *ab* and DE − *de*. Q.E.D.

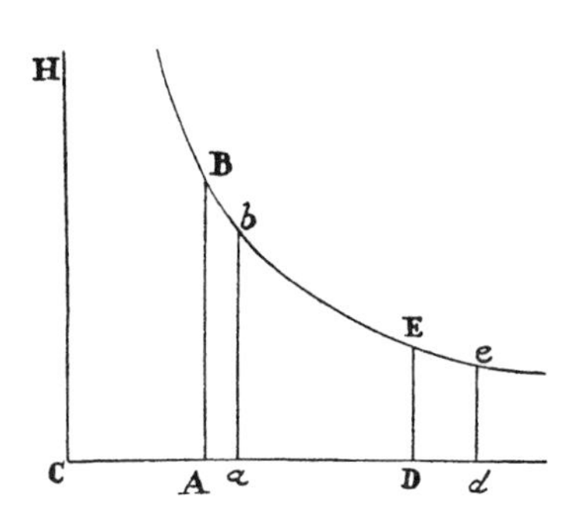

Proposition 7
Theorem 5

Spherical bodies that are resisted in proportion to the squares of the velocities will, in times that are as the first motions directly and the first resistances inversely, lose parts of the motions proportional to the wholes and will describe spaces proportional to those times and the first velocities jointly.

For the lost parts of the motions are as the resistances and the times jointly. Therefore, for those parts to be proportional to the wholes, the resistance and time jointly must be as the motion. Accordingly, the time will be as the motion directly and the resistance inversely. Therefore, if the particles of times are taken in this ratio, the bodies will always lose particles of their motions proportional to the wholes and thus will retain velocities always proportional to their first velocities. And because the ratio of the velocities is given, they will always describe spaces that are as the first velocities and the times jointly. Q.E.D.

COROLLARY 1. Therefore, if equally swift bodies are resisted in proportion to the squares of their diameters, then homogeneous globes moving with any velocities will, in describing spaces proportional to their diameters, lose parts of their motions proportional to the wholes. For the motion of each globe will be as its velocity and mass jointly, that is, as its velocity and the cube of its diameter; the resistance (by hypothesis) will be as the square of the diameter and the square of the velocity jointly; and the time (by this proposition) is in the former ratio directly and the latter ratio inversely, that is, as the diameter directly and the velocity inversely; and thus the space, being proportional to the time and the velocity, is as the diameter.

Corollary 2. If equally swift bodies are resisted in proportion to the $\frac{3}{2}$ powers of the diameters, then homogeneous globes moving with any velocities will, in describing spaces that are as the $\frac{3}{2}$ powers of the diameters, lose parts of motions proportional to the wholes.

Corollary 3. And universally, if equally swift bodies are resisted in the ratio of any power of the diameters, the spaces in which homogeneous globes moving with any velocities will lose parts of their motions proportional to the wholes will be as the cubes of the diameters divided by that power. Let the diameters be D and E; and if the resistances, when the velocities are supposed equal, are as D^n and E^n, then the spaces in which the globes, moving with any velocities, will lose parts of their motions proportional to the wholes will be as D^{3-n} and E^{3-n}. And therefore homogeneous globes, in describing spaces proportional to D^{3-n} and E^{3-n}, will retain velocities in the same ratio to each other that they had at the beginning.

Corollary 4. But if the globes are not homogeneous, the space described by the denser globe must be augmented in proportion to the density. For the motion, with an equal velocity, is greater in proportion to the density, and the time (by this proposition) is increased in proportion to the motion directly, and the space described is increased in proportion to the time.

Corollary 5. And if the globes move in different mediums, the space in the medium that, other things being equal, resists more will have to be decreased in proportion to the greater resistance. For the time (by this proposition) will be decreased in proportion to the increase of the resistance, and the space will be decreased in proportion to the time.

Lemma 2[a] *The moment of a generated quantity is equal to the moments of each of the generating roots multiplied continually by the exponents of the powers of those roots and by their coefficients.*

I call "generated" every quantity that is, without addition or subtraction, generated from any roots or terms: in arithmetic by multiplication, division, or extraction of roots; in geometry by the finding either of products and roots or of extreme and mean proportionals. Quantities of this sort are products, quotients, roots, rectangles, squares, cubes, square roots, cube roots, and the like.[b] I here consider these quantities as indeterminate and variable, and increasing or decreasing as if by a continual motion or flux; and it is their

a. Newton's use of "terminus" and "latus" for "root" is of particular interest in lem. 2 and its cases, corollaries, and scholium. "Radix" appears only twice and is unchanged from edition to edition, but Newton tends to replace the "terminus" ("term," "root") of ed. 1 with the "latus" ("side," "root") of ed. 2 and ed. 3. In the statement of the lemma, for example, ed. 1 has "momentis Terminorum singulorum generantium" ("the moments of the individual generating terms," i.e., "the moments of each of the generating roots") and "eorundem laterum indices dignitatum" ("the exponents of the powers of the same sides," i.e., "the exponents of the powers of those roots"). Thus "terminus" and "latus" are obviously synonyms. In ed. 2 and ed. 3, however, "laterum" ("sides," "roots") is substituted for "Terminorum" ("terms," "roots"). In the first sentence of the explanation, where ed. 1 has "ex Terminis quibuscunque" ("from any terms," i.e., "from any roots"), ed. 2 and ed. 3 have "ex lateribus vel terminis quibuscunque" ("from any sides or terms," i.e., "from any roots or terms"). As the explanation proceeds, ed. 1 has, like ed. 2 and ed. 3, "extractionem radicum" ("extraction of roots"), "contentorum & laterum" ("of products and roots"), "Radices" ("roots"), and "latera quadrata, latera cubica" ("square roots, cube roots"), but ed. 1 has "Termini" and "Terminum" where ed. 2 and ed. 3 have "Lateris" and "latus" in the last sentence of the first paragraph: "And the coefficient of each generating root is the quantity that results from dividing the generated quantity by this root." In corol. 1, on the other hand, all the editions have "terminus" (with the ordinary sense of "term," not with the sense of "root"), while all have "latus" (with the sense of "root") in cases 1 and 2 and corols. 2 and 3. "Terminus" also occurs, in the phrase "in terminis surdis" ("in surd terms"), in the scholium of ed. 1 and ed. 2, which is, as will be seen below, very different from that of ed. 3, where, however, "quantitatibus surdis" ("surd quantities") is at least comparable, especially since "surdis" ("surd") appears nowhere else in all the editions of the *Principia*.

b. In the Latin text of this lemma, Newton referred to roots in two senses. The first occurs in the opening sentence, where he writes of "extraction of roots," using the Latin term "radix," or "root." The second occurs in the next sentence, where he writes of "products, quotients, roots, rectangles, squares, cubes, square roots, cube roots, and the like." Here both senses of "roots" appear in a single sentence, the first as "radices" (or "roots"), the second as "latera quadrata, latera cubica" (*lit.* "square sides" and "cubic sides"). In the geometric language of algebra, in which a "rectangle" of A and B indicates the product of two unequal quantities A and B as the area of a rectangle whose sides are A and B, the square root and cube root have similar geometric expression. Thus the square root of A is the side of a square whose area is A, while the cube root of A is the "side" (actually the edge) of a cube whose volume is A.

In his *Lexicon Technicum* (London, 1704), John Harris explained these two different mathematical senses of the word "root." An "Unknown Quantity in an Algebraick Equation," he wrote, "is often called the Root." This is the sense of the word as it appears in the first sentence of the lemma. But, as Harris explained, a root is also "whatever Quantity being multiplied by it self produces a Square" and when

instantaneous increments or decrements that I mean by the word "moments," in such a way that increments are considered as added or positive moments, and decrements as subtracted or negative moments. But take care: do not understand them to be finite particles! [c]Finite particles are not moments, but the very quantities generated from the moments.[c] They must be understood to be the just-now nascent beginnings of finite magnitudes. For in this lemma the magnitude of moments is not regarded, but only their first proportion when nascent. It comes to the same thing if in place of moments there are used either the velocities of increments and decrements (which it is also possible to call motions, mutations, and fluxions of quantities) or any finite quantities proportional to these velocities. And the coefficient of each generating root is the quantity that results from dividing the generated quantity by this root.

Therefore, the meaning of this lemma is that if the moments of any quantities A, B, C, ... increasing or decreasing by a continual motion, or the velocities of mutation which are proportional to these moments are called a, b, c, ..., then the moment or mutation of the generated rectangle AB would be $a\mathrm{B} + b\mathrm{A}$, and the moment of the generated solid ABC would be $a\mathrm{BC} + b\mathrm{AC} + c\mathrm{AB}$, and the moments of the generated powers A^2, A^3, A^4, $\mathrm{A}^{1/2}$, $\mathrm{A}^{3/2}$, $\mathrm{A}^{1/3}$, $\mathrm{A}^{2/3}$, A^{-1}, A^{-2}, and $\mathrm{A}^{-1/2}$ would be $2a\mathrm{A}$, $3a\mathrm{A}^2$, $4a\mathrm{A}^3$, $\frac{1}{2}a\mathrm{A}^{-1/2}$, $\frac{3}{2}a\mathrm{A}^{1/2}$, $\frac{1}{3}a\mathrm{A}^{-2/3}$, $\frac{2}{3}a\mathrm{A}^{-1/3}$, $-a\mathrm{A}^{-2}$, $-2a\mathrm{A}^{-3}$, and $-\frac{1}{2}a\mathrm{A}^{-3/2}$ respectively. And generally, the moment of any power $\mathrm{A}^{\frac{n}{m}}$ would be $\frac{n}{m}a\mathrm{A}^{\frac{n-m}{m}}$. Likewise, the moment of the generated quantity $\mathrm{A}^2\mathrm{B}$ would be $2a\mathrm{AB} + b\mathrm{A}^2$, and the moment of the generated quantity $\mathrm{A}^3\mathrm{B}^4\mathrm{C}^2$ would be $3a\mathrm{A}^2\mathrm{B}^4\mathrm{C}^2 +$

---

once again "multiplied by that first Quantity produces a Cube, *&c.*" These, he said, are called "Square, Cube . . . *Root*."

Even without any knowledge of the geometric sense of algebra, one might easily guess that Newton is referring to square and cube roots in the phrase "products, quotients, . . . squares, cubes, square sides, cube sides, and the like." Yet Andrew Motte, in his English translation (London, 1729), rendered these terms literally as "products, quotients, roots, rectangles, squares, cubes, square and cubic sides, and the like," which was carried over into the Motte-Cajori version. The marquise du Châtelet knew better and in her French translation (Paris, 1756) wrote, just as we would today, of "les produits, les quotiens, les racines, les rectangles, les quarrés, les cubes, les racines quarrées, & les racines cubes."

cc. Ed. 1 has: "Moments, as soon as they are of finite magnitude, cease to be moments. For being finite is somewhat incompatible with their continual increment or decrement." When one reads the "somewhat" ("aliquatenus": "to a certain extent," "in some respects") in the second of these sentences, one can understand why Newton decided to revise this portion of his explanation.

$4bA^3B^3C^2 + 2cA^3B^4C$, and the moment of the generated quantity $\frac{A^3}{B^2}$ or A^3B^{-2} would be $3aA^2B^{-2} - 2bA^3B^{-3}$, and so on. The lemma is proved as follows.

CASE 1. Any rectangle AB increased by continual motion, when the halves of the moments, $\frac{1}{2}a$ and $\frac{1}{2}b$, were lacking from the sides A and B, was $A - \frac{1}{2}a$ multiplied by $B - \frac{1}{2}b$, or $AB - \frac{1}{2}aB - \frac{1}{2}bA + \frac{1}{4}ab$, and as soon as the sides A and B have been increased by the other halves of the moments, it comes out $A + \frac{1}{2}a$ multiplied by $B + \frac{1}{2}b$, or $AB + \frac{1}{2}aB + \frac{1}{2}bA + \frac{1}{4}ab$. Subtract the former rectangle from this rectangle, and there will remain the excess $aB + bA$. Therefore by the total increments a and b of the sides there is generated the increment $aB + bA$ of the rectangle. Q.E.D.

CASE 2. Suppose that AB is always equal to G; then the moment of the solid ABC or GC (by case 1) will be $gC + cG$, that is (if AB and $aB + bA$ are written for G and g), $aBC + bAC + cAB$. And the same is true of the solid contained under any number of sides [or the product of any number of terms]. Q.E.D.

CASE 3. Suppose that the sides A, B, and C are always equal to one another; then the moment $aB + bA$ of A^2, that is, of the rectangle AB, will be $2aA$, while the moment $aBC + bAC + cAB$ of A^3, that is, of the solid ABC, will be $3aA^2$. And by the same argument, the moment of any power A^n is naA^{n-1}. Q.E.D.

CASE 4. Hence, since $\frac{1}{A}$ multiplied by A is 1, the moment of $\frac{1}{A}$ multiplied by A together with $\frac{1}{A}$ multiplied by a will be the moment of 1, that is, nil. Accordingly, the moment of $\frac{1}{A}$ or of A^{-1} is $-\frac{a}{A^2}$. And in general, since $\frac{1}{A^n}$ multiplied by A^n is 1, the moment of $\frac{1}{A^n}$ multiplied by A^n together with $\frac{1}{A^n}$ multiplied by naA^{n-1} will be nil. And therefore the moment of $\frac{1}{A^n}$ or A^{-n} will be $-\frac{na}{A^{n+1}}$. Q.E.D.

CASE 5. And since $A^{\frac{1}{2}}$ multiplied by $A^{\frac{1}{2}}$ is A, the moment of $A^{\frac{1}{2}}$ multiplied by $2A^{\frac{1}{2}}$ will be a, by case 3; and thus the moment of $A^{\frac{1}{2}}$ will be $\frac{a}{2A^{\frac{1}{2}}}$ or $\frac{1}{2}aA^{-\frac{1}{2}}$. And in general, if $A^{\frac{m}{n}}$ is supposed equal to B, A^m will be equal to B^n, and hence maA^{m-1} will be equal to nbB^{n-1}, and maA^{-1} will be equal to nbB^{-1} or $nbA^{\frac{-m}{n}}$, and thus $\frac{m}{n}aA^{\frac{m-n}{n}}$ equal to b, that is, equal to the moment of $A^{\frac{m}{n}}$. Q.E.D.

CASE 6. Therefore the moment of any generated quantity A^mB^n is the moment of A^m multiplied by B^n, together with the moment of B^n multiplied by A^m, that is, $maA^{m-1}B^n+nbB^{n-1}A^m$; and this is so whether the exponents m and n of the powers are whole numbers or fractions, whether positive or negative. And it is the same for a solid contained by more than two terms raised to powers. Q.E.D.

COROLLARY 1. Hence in continually proportional quantities, if one term is given, the moments of the remaining terms will be as those terms multiplied by the number of intervals between them and the given term. Let A, B, C, D, E, and F be continually proportional; then, if the term C is given, the moments of the remaining terms will be to one another as −2A, −B, D, 2E, and 3F.

COROLLARY 2. And if in four proportionals the two means are given, the moments of the extremes will be as those same extremes. The same is to be understood of the sides of any given rectangle.

COROLLARY 3. And if the sum or difference of two squares is given, the moments of the sides will be inversely as the sides.

Scholium

[d]In a certain letter written to our fellow Englishman Mr. J. Collins on 10 December 1672, when I had described a method of tangents that I suspected to be the same as Sluse's method, which at that time had not yet been made public, I added: "This is one particular, or rather a corollary, of a general

dd. In ed. 1 this scholium reads: "In correspondence which I carried on ten years ago with the very able geometer G. W. Leibniz, I indicated that I was in possession of a method of determining maxima and minima, drawing tangents, and performing similar operations, and that the method worked for surd as well as rational terms. I concealed this method under an anagram comprising this sentence: 'Given an equation involving any number of fluent quantities, to find the fluxions, and vice versa.' The distinguished gentleman wrote back that he too had come upon a method of this kind, and he communicated his method, which hardly differed from mine except in the forms of words and notations. The foundation of both methods is contained in this lemma." In ed. 2 the scholium is exactly the same except that "and the concept of the generation of quantities" is added at the end of the penultimate sentence.

For the principal texts with interpretative comments on the Newton-Leibniz controversy over priority in the invention of the calculus, see *The Mathematical Papers of Isaac Newton*, ed. D. T. Whiteside (Cambridge: Cambridge University Press, 1967–1981), vol. 8, esp. pp. 469–697; also A. Rupert Hall, *Philosophers at War: The Quarrel between Newton and Leibniz* (Cambridge: Cambridge University Press, 1980).

method, which extends, without any troublesome calculation, not only to the drawing of tangents to all curve lines, whether geometric or mechanical or having respect in any way to straight lines or other curves, but also to resolving other more abstruse kinds of problems concerning curvatures, areas, lengths, centers of gravity of curves, . . . , and is not restricted (as Hudde's method of maxima and minima is) only to those equations which are free from surd quantities. I have interwoven this method with that other by which I find the roots of equations by reducing them to infinite series." So much for the letter. And these last words refer to the treatise that I had written on this topic in 1671. The foundation of this general method is contained in the preceding lemma.[d]

Proposition 8
Theorem 6

If a body, acted on by gravity uniformly, goes straight up or down in a uniform medium, and the total space described is divided into equal parts, and the absolute forces at the beginnings of each of the parts are found (adding the resistance of the medium to the force of gravity when the body is ascending, or subtracting it when the body is descending), I say that those absolute forces are in a geometric progression.

Represent the force of gravity by the given line AC; the resistance, by the indefinite line AK; the absolute force in the descent of the body, by the difference KC; the velocity of the body, by the line AP, which is the mean proportional between AK and AC, and thus is as the square root of the resistance; the increment of the resistance occurring in a given particle of time, by the line-element KL; and the simultaneous increment of the velocity, by the line-element PQ; then with center C and rectangular asymptotes CA and CH, describe any hyperbola BNS, meeting the erected perpendiculars AB, KN, and LO in B, N, and O. Since AK is as AP^2, the moment KL of AK will be as the moment $2AP \times PQ$ of AP^2, that is, as AP multiplied by KC, since the increment PQ of the velocity (by the second law of motion) is proportional to the generating force KC. Compound the ratio of KL with the ratio of KN, and the rectangle $KL \times KN$ will become as $AP \times KC \times KN$—that is, because the rectangle $KC \times KN$ is given, as

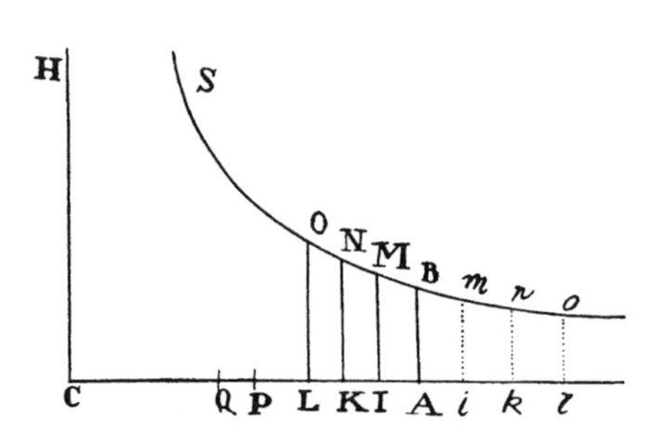

AP. But the ultimate ratio of the hyperbolic area KNOL to the rectangle KL × KN, when points K and L come together, is the ratio of equality. Therefore that evanescent hyperbolic area is as AP. Hence the total hyperbolic area ABOL is composed of the particles KNOL, which are always proportional to the velocity AP, and therefore this area is proportional to the space described with this velocity. Now divide that area into equal parts ABMI, IMNK, KNOL, . . . , and the absolute forces AC, IC, KC, LC, . . . will be in a geometric progression. Q.E.D.

And by a similar argument, if—in the ascent of the body—equal areas AB*mi*, *imnk*, *knol*, . . . are taken on the opposite side of point A, it will be manifest that the absolute forces AC, *i*C, *k*C, *l*C, . . . are continually proportional. And thus, if all the spaces in the ascent and descent are taken equal, all the absolute forces *l*C, *k*C, *i*C, AC, IC, KC, LC, . . . will be continually proportional. Q.E.D.

COROLLARY 1. Hence, if the space described is represented by the hyperbolic area ABNK, the force of gravity, the velocity of the body, and the resistance of the medium can be represented by lines AC, AP, and AK respectively, and vice versa.

COROLLARY 2. And line AC represents the greatest velocity that the body can ever acquire by descending infinitely.

COROLLARY 3. Therefore, if for a given velocity the resistance of the medium is known, the greatest velocity will be found by taking its ratio to the given velocity as the square root of the ratio of the force of gravity to that known resistance of the medium.[a]

Proposition 9
Theorem 7

Given what has already been proved, I say that if the tangents of the angles of a sector of a circle and of a hyperbola are taken proportional to the velocities, the radius being of the proper magnitude, the whole time [a]of ascending to the highest

a. Ed. 1 has two additional corollaries as follows: "Corol. 4. But also the particle of time wherein the minimally small particle of space NKLO is described in descent is as the rectangle KN × PQ. For since the space NKLO is as the velocity multiplied by the particle of time, the particle of time will be as that space divided by the velocity, that is, as the minimally small rectangle KN × KL divided by AP. For KL, above, was as AP × PQ. Therefore the particle of time is as KN × PQ, or what comes to the same, as $\frac{PQ}{CK}$. Q.E.D."

"Corollary 5. By the same argument the particle of time wherein the particle of space *nklo* is described in ascent is as $\frac{pq}{Ck}$."

aa. Ed. 1 and ed. 2 have "of the future ascent."

place[a] *will be as the sector of the circle, and the whole time* [b]*of descending from the highest place*[b] *will be as the sector of the hyperbola.*

Draw AD perpendicular and equal to the straight line AC, which represents the force of gravity. With center D and semidiameter AD describe the quadrant AtE of a circle and the rectangular hyperbola AVZ having axis AX, principal vertex A, and asymptote DC. Draw Dp and DP, and the sector AtD of the circle will be as [c]the whole time of ascending to the highest place,[c] and the sector ATD of the hyperbola will be as [d]the whole time of descending from the highest place,[d] provided that the tangents Ap and AP of the sectors are as the velocities.

CASE 1. Draw Dvq cutting off the moments or the minimally small particles tDv and qDp, described simultaneously, of the sector ADt and of the triangle ADp. Since those particles, because of the common angle D, are

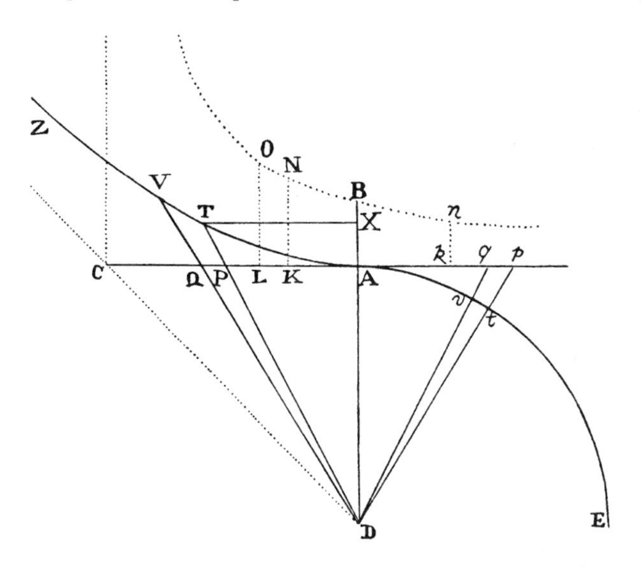

as the squares of the sides, particle tDv will be as $\frac{q\mathrm{D}p \times t\mathrm{D}^2}{p\mathrm{D}^2}$, that is, because tD is given, as $\frac{q\mathrm{D}p}{p\mathrm{D}^2}$. But $p\mathrm{D}^2$ is $\mathrm{AD}^2 + \mathrm{A}p^2$, that is, $\mathrm{AD}^2 + \mathrm{AD} \times \mathrm{A}k$, or $\mathrm{AD} \times \mathrm{C}k$; and $q\mathrm{D}p$ is $\frac{1}{2}\mathrm{AD} \times pq$. Therefore particle tDv of the sector is as $\frac{pq}{\mathrm{C}k}$, that is, directly as the minimally small decrement pq of the velocity and

bb. Ed. 1 and ed. 2 have "of the past descent."

cc. Ed. 1 and ed. 2 have "the time of the whole future ascent."

dd. Ed. 1 and ed. 2 have "the time of the whole past descent."

inversely as the force Ck that decreases the velocity, and thus as the particle of time corresponding to the decrement of the velocity. And by composition [or componendo] the sum of all the particles tDv in the sector ADt will be as the sum of the particles of time corresponding to each of the lost particles pq of the decreasing velocity Ap, until that velocity, decreased to nil, has vanished; that is, the whole sector ADt is as [e]the whole time of ascending to the highest place.[e] Q.E.D.

CASE 2. Draw DQV cutting off the minimally small particles TDV and PDQ of the sector DAV and of the triangle DAQ; and these particles will be to each other as DT^2 to DP^2, that is (if TX and AP are parallel), as DX^2 to DA^2 or TX^2 to AP^2, and by separation [or dividendo] as $DX^2 - TX^2$ to $DA^2 - AP^2$. But from the nature of the hyperbola, $DX^2 - TX^2$ is AD^2, and by hypothesis AP^2 is $AD \times AK$. Therefore the particles are to each other as AD^2 to $AD^2 - AD \times AK$, that is, as AD to AD − AK or AC to CK; and thus the particle TDV of the sector is $\frac{PDQ \times AC}{CK}$, and hence, because AC and AD are given, as $\frac{PQ}{CK}$, that is, directly as the increment of the velocity and inversely as the force generating the increment, and thus as the particle of time corresponding to the increment. And by composition [or componendo] the sum of the particles of time in which all the particles PQ of the velocity AP are generated will be as the sum of the particles of the sector ATD, that is, the whole time will be as the whole sector. Q.E.D.

COROLLARY 1. Hence, if AB is equal to a fourth of AC, the space that a body describes by falling in any time will be in the same ratio to the space that the body can describe by progressing uniformly in that same time with its greatest velocity AC as the ratio of area ABNK (which represents the space described in falling) to area ATD (which represents the time). For, since AC is to AP as AP to AK, it follows (by book 2, lem. 2, corol. 1) that LK will be to PQ as 2AK to AP, that is, as 2AP to AC, and hence LK will be to ½PQ as AP to ¼AC or AB; KN is also to AC or AD as AB to CK; and thus, from the equality of the ratios [or ex aequo], LKNO will be to DPQ as AP to CK. But DPQ was to DTV as CK to AC. Therefore, once again by the equality of the ratios [or ex aequo], LKNO is to DTV as AP to AC, that is, as the velocity of the falling body to the greatest velocity that the

ee. Ed. 1 and ed. 2 have "the time of the whole future ascent."

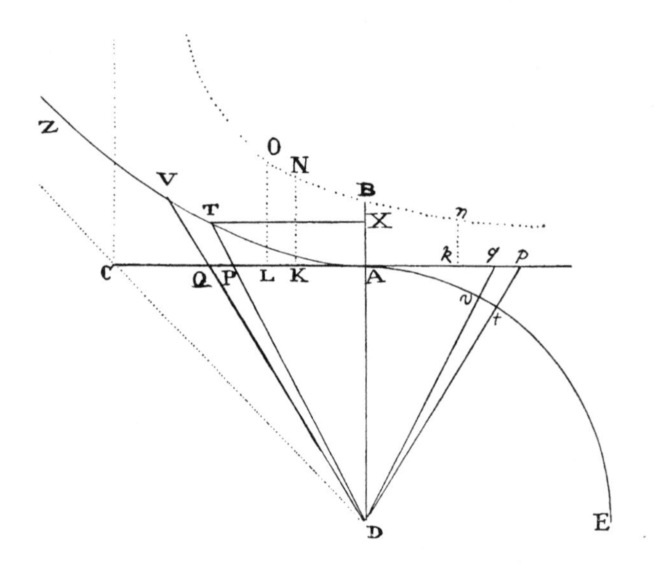

body can acquire in falling. Since, therefore, the moments LKNO and DTV of areas ABNK and ATD are as the velocities, all the parts of those areas generated simultaneously will be as the spaces described simultaneously, and thus the whole areas ABNK and ATD generated from the beginning will be as the whole spaces described from the beginning of the descent. Q.E.D.

COROLLARY 2. The same result follows for the space described in ascent: namely, the whole space is to the space described in the same time with a uniform velocity AC as area AB*nk* is to sector AD*t*.

COROLLARY 3. The velocity of a body falling in time ATD is to the velocity that it would acquire in the same time in a nonresisting space as the triangle APD to the hyperbolic sector ATD. For the velocity in a nonresisting medium would be as time ATD, and in a resisting medium is as AP, that is, as triangle APD. And the velocities at the beginning of the descent are equal to each other, as are those areas ATD and APD.

COROLLARY 4. By the same argument, the velocity in the ascent is to the velocity with which the body in the same time in a nonresisting space could lose its whole ascending motion as the triangle A*p*D is to the sector A*t*D of the circle, or as the straight line A*p* is to the arc A*t*.

COROLLARY 5. Therefore the time in which a body, by falling in a resisting medium, acquires the velocity AP is to the time in which it could acquire its greatest velocity AC, by falling in a nonresisting space, as sector ADT to triangle ADC; and the time in which it could lose the velocity A*p* by ascending in a resisting medium is to the time in which it could lose the same velocity by ascending in a nonresisting space as arc A*t* is to its tangent A*p*.

COROLLARY 6. Hence, from the given time, the space described by ascent or descent is given. For the greatest velocity of a body descending infinitely is given (by book 2, prop. 8, corols. 2 and 3), and hence the time is given in which a body could acquire that velocity by falling in a nonresisting space. And if sector ADT or AD*t* is taken to be to triangle ADC in the ratio of the given time to the time just found, there will be given both the velocity AP or A*p* and the area ABNK or AB*nk*, which is to the sector ADT or AD*t* as the required space is to the space that can be described uniformly in the given time with that greatest velocity which has already been found.

COROLLARY 7. And working backward, the time AD*t* or ADT will be given from the given space AB*nk* or ABNK of ascent or descent.

Proposition 10
Problem 3

Let a uniform force of gravity tend straight toward the plane of the horizon, and let the resistance be as the density of the medium and the square of the velocity jointly; it is required to find, in each individual place, the density of the medium that makes the body move in any given curved line and also the velocity of the body and resistance of the medium.

[a]Let PQ be the plane of the horizon, perpendicular to the plane of the figure; PFHQ a curved line meeting this plane in points P and Q; G, H, I,

aa. Ed. 1 has: "Let AK be the plane of the horizon, perpendicular to the plane of the figure; ACK a curved line; C a body moving along the line; and FC*f* a straight line touching it in C. And suppose that body C now goes forward from A to K along the line ACK and now goes back along the same line and that in going forward it is impeded by the medium and in going back is equally assisted, so that in the same places the velocity of the body as it goes forward and back is always the same.

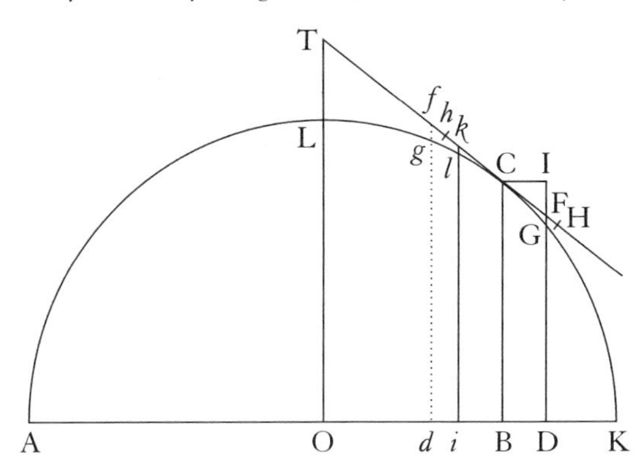

"And in equal times let the body as it goes forward describe the minimally small arc CG, and let the body as it goes back describe arc C*g*, and let CH and C*h* be equal rectilinear lengths which bodies

and K four places of the body as it goes in the curve from F to Q; and GB, HC, ID, and KE four parallel ordinates dropped from these points to the horizon and standing upon the horizontal line PQ at points B, C, D, and E; and let BC, CD, and DE be distances between the ordinates equal to one another. From points G and H draw the straight lines GL and HN touching the curve in G and H, and meeting in L and N the ordinates CH and DI produced upward; and complete the parallelogram HCDM. Then the times in which the body describes arcs GH and HI will be as the square roots of the distances LH and NI which the body could describe in those times by falling from the tangents; and the velocities will be directly as GH and HI (the lengths described) and inversely as the times. Represent the times by T and *t*,

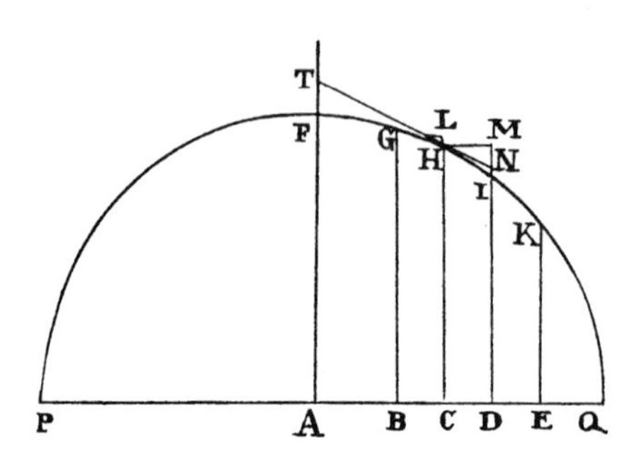

---

moving away from place C would describe in these times without the actions of the medium and of gravity, and from points C, G, and *g* to the horizontal plane AK drop perpendiculars CB, GD, and *gd*, letting GD and *gd* meet the tangent in F and *f*. Through the resistance of the medium it comes about that the body as it goes forward describes, instead of length CH, only length CF, and through the force of gravity the body is transferred from F to G, and thus line-element HF and line-element FG are generated simultaneously, the first by the force of resistance and the second by the force of gravity. Accordingly (by book 1, lem. 10), line-element FG is as the force of gravity and the square of the time jointly and thus (since the gravity is given) as the square of the time, and line-element HF is as the resistance and the square of the time, that is, as the resistance and line-element FG. And hence the resistance comes to be as HF directly and FG inversely, or as $\frac{\mathrm{HF}}{\mathrm{FG}}$. This is so in the case of nascent line-elements. For in the case of line-elements of finite magnitude these ratios are not accurate.

"And by a similar argument *fg* is as the square of the time and thus, since the times are equal, is equal to FG, and the impulse by which the body going back is urged is as $\frac{hf}{fg}$. But the impulse upon the body as it goes back and the resistance to it as it goes forward are equal at the very beginning of the motion, and thus also $\frac{hf}{fg}$ and $\frac{\mathrm{HF}}{\mathrm{FG}}$, proportional to them, are equal, and therefore, because *fg* and FG are equal, *hf* and HF are also equal, and thus CF, CH (or C*h*), and C*f* are in arithmetic progression, and hence HF is half the difference between C*f* and CF, and the resistance, which above was as $\frac{\mathrm{HF}}{\mathrm{FG}}$, is as $\frac{\mathrm{C}f - \mathrm{CF}}{\mathrm{FG}}$.

"But the resistance is as the density of the medium and the square of the velocity. And the velocity is as the described length CF directly and the time $\sqrt{\mathrm{FG}}$ inversely, that is, as $\frac{\mathrm{CF}}{\sqrt{\mathrm{FG}}}$, and thus the square of the velocity is as $\frac{\mathrm{CF}^2}{\mathrm{FG}}$. Therefore the resistance, proportional to $\frac{\mathrm{C}f - \mathrm{CF}}{\mathrm{FG}}$, is as the density of the medium and the velocities by $\frac{GH}{T}$ and $\frac{HI}{t}$; and the decrement of the velocity occurring in time t will be represented by $\frac{GH}{T} - \frac{HI}{t}$. This decrement arises from the resistance retarding the body and from the gravity accelerating the body. In a body falling and describing in its fall the space NI, gravity generates a velocity by which twice that space could have been described in the same time, as Galileo proved, that is, the velocity $\frac{2NI}{t}$; but in a body describing arc HI, gravity increases the arc by only the length HI − HN or $\frac{MI \times NI}{HI}$, and thus generates only the velocity $\frac{2MI \times NI}{t \times HI}$. Add this velocity to the above decrement, and the result is the decrement of the velocity arising from the resistance alone, namely $\frac{GH}{T} - \frac{HI}{t} + \frac{2MI \times NI}{t \times HI}$. And accordingly, since gravity generates the velocity $\frac{2NI}{t}$ in the same time in a falling body, the resistance will be to the gravity as $\frac{GH}{T} - \frac{HI}{t} + \frac{2MI \times NI}{t \times HI}$ to $\frac{2NI}{t}$, or as $\frac{t \times GH}{T} - HI + \frac{2MI \times NI}{HI}$ to 2NI.

---

and $\frac{CF^2}{FG}$ jointly, and hence the density of the medium is as $\frac{Cf - CF}{FG}$ directly and $\frac{CF^2}{FG}$ inversely, that is, as $\frac{Cf - CF}{CF^2}$. Q.E.I.

"Corollary 1. And hence it is gathered that if Ck on Cf is taken as equal to CF and the perpendicular ki is dropped to the horizontal plane AK, cutting the curve ACK in l, the density of the medium will come to be as $\frac{FG - kl}{CF \times (FG + kl)}$. For fC will be to kC as $\sqrt{fg}$ or $\sqrt{FG}$ to $\sqrt{kl}$, and by separation [or dividendo] fk will be to kC, that is, $Cf - CF$ to CF, as $\sqrt{FG} + \sqrt{kl}$ to $\sqrt{kl}$, that is, if both terms are multiplied by $\sqrt{FG} + \sqrt{kl}$, as $FG - kl$ to $kl + \sqrt{(FG \times kl)}$, or to $FG + kl$. For the first ratio of the nascent quantities $kl + \sqrt{(FG \times kl)}$ and $FG + kl$ is that of equality. And so let $\frac{FG - kl}{FG + kl}$ be written for $\frac{Cf - CF}{CF}$, and the density of the medium, which was as $\frac{Cf - CF}{CF^2}$, will turn out to be as $\frac{FG - kl}{CF \times (FG + kl)}$.

"Corollary 2. Hence, since 2HF and Cf − CF are equal and FG and kl (because of the ratio of equality) compose 2FG, 2HF will be to CF as $FG - kl$ to 2FG, and hence HF will be to FG, that is, the resistance will be to the gravity, as the rectangle $CF \times (FG - kl)$ to $4FG^2$."

The demonstration in ed. 1 is incorrect, and the error was brought to Newton's attention only after the corresponding pages in ed. 2 had been printed off. For details see the Guide to the present translation, §7.3; also *The Mathematical Papers of Isaac Newton*, ed. D. T. Whiteside (Cambridge: Cambridge University Press, 1967–1981), 8:312–424; *The Correspondence of Isaac Newton*, vol. 5, ed. A. Rupert Hall and Laura Tilling (Cambridge: published for the Royal Society by Cambridge University Press, 1975); A. Rupert Hall, "Correcting the *Principia*," *Osiris* 13 (1958): 291–326; I. Bernard Cohen, *Introduction to Newton's "Principia"* (Cambridge, Mass.: Harvard University Press; Cambridge: Cambridge University Press, 1971), pp. 236–238.

Now for the abscissas CB, CD, and CE write $-o$, o, and $2o$. For the ordinate CH write P, and for MI write any series $Qo + Ro^2 + So^3 + \cdots$. And all the terms of the series after the first, namely $Ro^2 + So^3 + \cdots$, will be NI, and the ordinates DI, EK, and BG will be $P - Qo - Ro^2 - So^3 - \cdots$, $P - 2Qo - 4Ro^2 - 8So^3 - \cdots$, and $P + Qo - Ro^2 + So^3 - \cdots$ respectively. And by squaring the differences of the ordinates BG − CH and CH − DI and by adding to the resulting squares the squares of BC and CD, there will result the squares of the arcs GH and HI: $o^2 + Q^2o^2 - 2QRo^3 + \cdots$ and $o^2 + Q^2o^2 + 2QRo^3 + \cdots$. The roots of these, $o\sqrt{(1+Q^2)} - \frac{QRo^2}{\sqrt{(1+Q^2)}}$ and $o\sqrt{(1+Q^2)} + \frac{QRo^2}{\sqrt{(1+Q^2)}}$, are the arcs GH and HI. Furthermore, if from ordinate CH half the sum of ordinates BG and DI is subtracted, and from ordinate DI half the sum of ordinates CH and EK is subtracted, the remainders will be the sagittas Ro^2 and $Ro^2 + 3So^3$ of arcs GI and HK. And these are proportional to the line-elements LH and NI, and thus as the squares of the infinitely small times T and t; and hence the ratio $\frac{t}{T}$ is $\sqrt{\frac{R + 3So}{R}}$ or $\frac{R + \frac{3}{2}So}{R}$; and if the values just found of $\frac{t}{T}$, GH, HI, MI, and NI are substituted in $\frac{t \times GH}{T} - HI + \frac{2MI \times NI}{HI}$, the result will be $\frac{3So^2}{2R}\sqrt{(1+Q^2)}$. And since 2NI is $2Ro^2$, the resistance will now be to the gravity as $\frac{3So^2}{2R}\sqrt{(1+Q^2)}$ to $2Ro^2$, that is, as $3S\sqrt{(1+Q^2)}$ to $4R^2$.

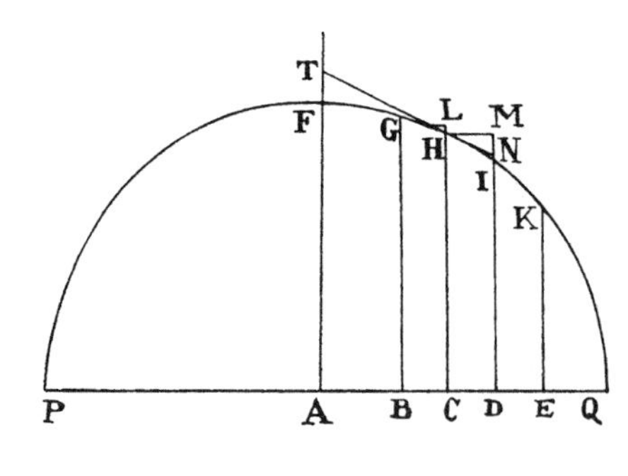

And the velocity is that with which a body going forth from any place H along tangent HN can then move in a vacuum in a parabola having a diameter HC and a latus rectum $\frac{HN^2}{NI}$ or $\frac{1+Q^2}{R}$.

And the resistance is as the density of the medium and the square of the velocity jointly, and therefore the density of the medium is as the resistance directly and the square of the velocity inversely, that is, as $\frac{3S\sqrt{(1+Q^2)}}{4R^2}$ directly and $\frac{1+Q^2}{R}$ inversely, that is, as $\frac{S}{R\sqrt{(1+Q^2)}}$. Q.E.I.

COROLLARY 1. If the tangent HN is produced in both directions until it meets any ordinate AF in T, $\frac{\text{HT}}{\text{AC}}$ will be equal to $\surd(1 + \text{Q}^2)$ and thus can be written for $\surd(1 + \text{Q}^2)$ above. And so the resistance will be to the gravity as $3\text{S} \times \text{HT}$ to $4\text{R}^2 \times \text{AC}$, the velocity will be as $\frac{\text{HT}}{\text{AC}\surd\text{R}}$, and the density of the medium will be as $\frac{\text{S} \times \text{AC}}{\text{R} \times \text{HT}}$.[a]

[b]COROLLARY 2. And hence, if the curved line PFHQ is defined by the relation between the base or abscissa AC and the ordinate CH, as is customary, and the value of the ordinate is resolved into a converging series, then the problem will be solved readily by means of the first terms of the series, as in the following examples.[b]

EXAMPLE 1. Let line PFHQ be a semicircle described on the diameter PQ, and let it be required to find the density of the medium that would make a projectile move in this semicircle.

Bisect diameter PQ in A; call AQ, n; AC, a; CH, e; and CD, o. Then DI^2 or $\text{AQ}^2 - \text{AD}^2$ will be $= n^2 - a^2 - 2ao - o^2$, or $e^2 - 2ao - o^2$, and when the root has been extracted by our method, DI will become $= e - \frac{ao}{e} - \frac{o^2}{2e} - \frac{a^2o^2}{2e^3} - \frac{ao^3}{2e^3} - \frac{a^3o^3}{2e^5} - \cdots$. Here write n^2 for $e^2 + a^2$, and DI will come out $= e - \frac{ao}{e} - \frac{n^2o^2}{2e^3} - \frac{an^2o^3}{2e^5} - \cdots$.

I divide series of this sort into successive terms in the following manner. What I call the first term is the term in which the infinitely small quantity o does not exist; the second, the term in which that quantity is of one dimension; the third, the term in which it is of two dimensions; the fourth, the term in which it is of three dimensions; and so on indefinitely. And the first term, which here is e, will always denote the length of the ordinate CH, standing at the beginning of the indefinite quantity o. The second term, which here is $\frac{ao}{e}$, will denote the difference between CH and DN, that is, the line-element MN, which is cut off by completing the parallelogram HCDM and thus always determines the position of the tangent HN; as, for example, in this case by taking MN to HM as $\frac{ao}{e}$ is to o, or a to e. The third term,

bb. In ed. 1 this is, with some variants, corol. 3.

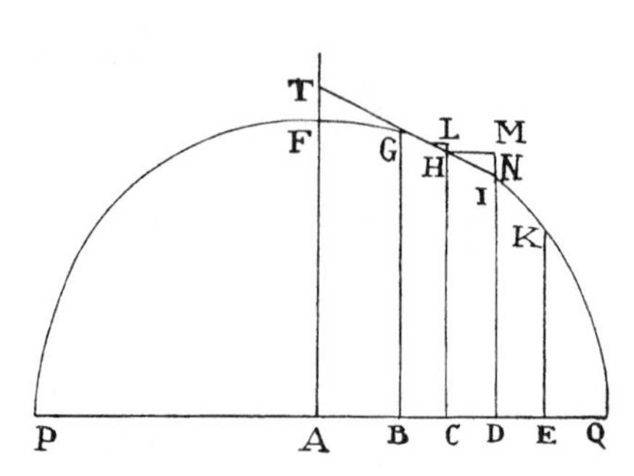

which here is $\frac{n^2o^2}{2e^3}$, will designate the line-element IN, which lies between the tangent and the curve and thus determines the angle of contact IHN or the curvature that the curved line has in H. If that line-element IN is of a finite magnitude, it will be designated by the third term along with the terms following without limit. But if that line-element is diminished infinitely, the subsequent terms will come out infinitely smaller than the third and thus can be ignored. The fourth term determines the variation of the curvature, the fifth the variation of the variation, and so on. Hence, by the way, one can see clearly the not inconsiderable usefulness of these series in the solution of problems that depend on tangents and the curvature of curves.

[c]Now compare the series $e - \frac{ao}{e} - \frac{n^2o^2}{2e^3} - \frac{an^2o^3}{2e^5} - \cdots$ with the series $P - Qo - Ro^2 - So^3 - \cdots$, and in the same manner for P, Q, R, and S write e, $\frac{a}{e}$, $\frac{n^2}{2e^3}$, and $\frac{an^2}{2e^5}$, and for $\sqrt{(1+Q^2)}$ write $\sqrt{\left(1+\frac{a^2}{e^2}\right)}$ or $\frac{n}{e}$; then the density of the medium will come out[c] as $\frac{a}{ne}$, that is (because n is given), as $\frac{a}{e}$, or $\frac{AC}{CH}$, that is, as the tangent's length HT terminated at the semidiameter

cc. Ed. 1 has: "Besides, CF is the square root of CI^2 and IF^2, that is, of BD^2 and the square of the second term. And $FG + kl$ is equal to twice the third term, and $FG - kl$ is equal to twice the fourth. For the value of DG is converted into the value of il, and the value of FG into the value of kl, by writing Bi for BD, or $-o$ for $+o$. Accordingly, since FG is $-\frac{n^2o^2}{2e^3} - \frac{an^2o^3}{2e^5} \ldots$, kl will be $= -\frac{n^2o^2}{2e^3} + \frac{an^2o^3}{2e^5} \ldots$. And the sum of these is $-\frac{n^2o^2}{e^3}$; the difference, $-\frac{an^2o^3}{e^5}$. The fifth and following terms I ignore here as infinitely less than such as come under consideration in this problem. And so if the series is universally designated by the terms $\mp Qo - Ro^2 - So^3 \ldots$, CF will be equal to $\sqrt{(o^2 + Q^2o^2)}$, $FG + kl$ will be equal to $2Ro^2$, and $FG - kl$ will be equal to $2So^3$. For CF, $FG + kl$, and $FG - kl$, write these values of theirs, and the density of the medium, which was as $\frac{FG - kl}{CF \times (FG + kl)}$, will now be as $\frac{S}{R\sqrt{(1+Q^2)}}$. Therefore by reducing each problem to a converging series and here writing for Q, R, and S the terms of the series corresponding to these and then supposing the resistance of the medium in any place G to be to the gravity as $S\sqrt{(1+Q^2)}$ to $2R^2$, and the velocity to be the same as that with which a body, departing from place C along straight line CF, could subsequently move in a parabola having diameter CB and latus rectum $\frac{1+Q^2}{R}$, the problem will be solved.

"Thus, in now solving the problem, if $\sqrt{\left(1+\frac{a^2}{e^2}\right)}$ or $\frac{n}{e}$ is written for $\sqrt{(1+Q^2)}$, $\frac{n^2}{2e^3}$ for R, and $\frac{an^2}{2e^5}$ for S, the density of the medium will come out."

AF, which stands perpendicularly upon PQ; and the resistance will be to the gravity as $3a$ to $2n$, that is, as 3AC to the diameter PQ of the circle, while the velocity will be as $\sqrt{CH}$. Therefore, if the body goes forth from place F with the proper velocity along a line parallel to PQ, and the density of the medium in each place H is as the length of the tangent HT, and the resistance, also in some place H, is to the force of gravity as 3AC to PQ, then that body will describe the quadrant FHQ of a circle. Q.E.I.

But if the same body were to go forth from place P along a line perpendicular to PQ and were to begin to move in an arc of the semicircle PFQ, AC or a would have to be taken on the opposite side of center A, and therefore its sign would have to be changed, and $-a$ would have to be written for $+a$. Thus the density of the medium would come out as $-\frac{a}{e}$. But nature does not admit of a negative density, that is, a density that accelerates the motions of bodies; and therefore it cannot naturally happen that a body by ascending from P should describe the quadrant PF of a circle. For this effect the body would have to be accelerated by an impelling medium, not impeded by a resisting medium.

EXAMPLE 2. Let the line PFQ be a parabola having its axis AF perpendicular to the horizon PQ, and let it be required to find the density of the medium that would make a projectile move in that parabola.

From the nature of the parabola, the rectangle PD × DQ is equal to the rectangle of the ordinate DI and some given straight line. Let that straight line be called b; PC, a; PQ, c; CH, e; and CD, o. Then the rectangle $(a+o)\times(c-a-o)$, or $ac-a^2-2ao+co-o^2$, is equal to the rectangle $b\times$DI, and thus DI is equal to $\frac{ac-a^2}{b}+\frac{c-2a}{b}o-\frac{o^2}{b}$. Now the second term $\frac{c-2a}{b}o$ of this series should be written for Qo, the third term $\frac{o^2}{b}$ likewise for Ro^2. But since there are not more terms, the coefficient S of the fourth will have to vanish, and therefore the quantity $\frac{S}{R\sqrt{(1+Q^2)}}$, to which the density of the medium is proportional, will be nil. Therefore, if the density of the medium is null, a projectile will move in a parabola, as Galileo once proved. Q.E.I.

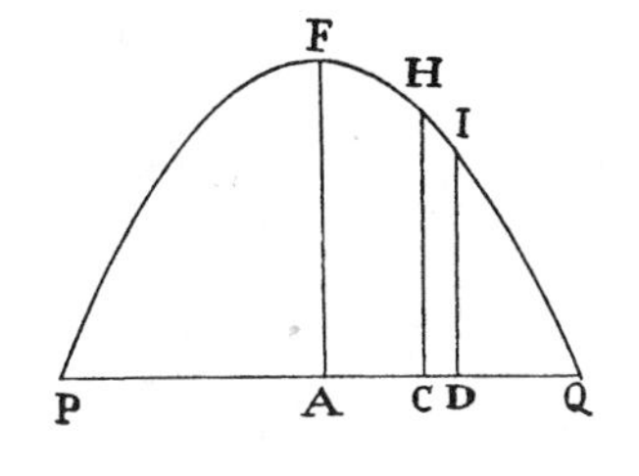

EXAMPLE 3. Let line AGK be a hyperbola having an asymptote NX perpendicular to the horizontal plane AK; and let it be required to find the density of the medium that would make a projectile move in this hyperbola.

Let MX be the other asymptote, meeting in V the ordinate DG produced; and from the nature of the hyperbola, the rectangle XV × VG will be given. Moreover, the ratio of DN to VX is given, and therefore the rectangle DN × VG is given also. Let this rectangle be b^2. And after completing the parallelogram DNXZ, call BN a; BD, o; NX, c; and suppose the given ratio of VZ to ZX or DN to be $\frac{m}{n}$. Then DN will be equal to $a - o$,

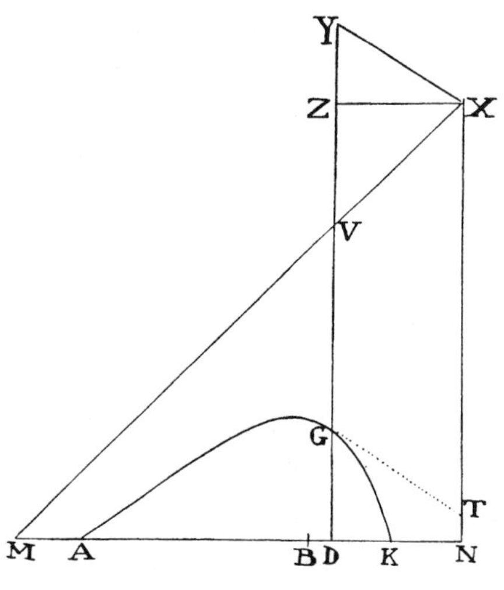

VG will be equal to $\frac{b^2}{a-o}$, VZ will be equal to $\frac{m}{n}(a-o)$, and GD or NX − VZ − VG will be equal to $c - \frac{m}{n}a + \frac{m}{n}o - \frac{b^2}{a-o}$. Resolve the term $\frac{b^2}{a-o}$ into the converging series $\frac{b^2}{a} + \frac{b^2}{a^2}o + \frac{b^2}{a^3}o^2 + \frac{b^2}{a^4}o^3 \ldots$, and GD will become equal to $c - \frac{m}{n}a - \frac{b^2}{a} + \frac{m}{n}o - \frac{b^2}{a^2}o - \frac{b^2}{a^3}o^2 - \frac{b^2}{a^4}o^3 \ldots$. The second term $\frac{m}{n}o - \frac{b^2}{a^2}o$ of this series is to be used for Qo, the third (with the sign changed) $\frac{b^2}{a^3}o^2$ for Ro^2, and the fourth (with the sign also changed) $\frac{b^2}{a^4}o^3$ for So^3, and their coefficients $\frac{m}{n} - \frac{b^2}{a^2}$, $\frac{b^2}{a^3}$, and $\frac{b^2}{a^4}$ are to be written in the above rule for Q, R, and S. When this is done, the density of the medium comes out as $\dfrac{\frac{b^2}{a^4}}{\frac{b^2}{a^3}\sqrt{\left(1 + \frac{m^2}{n^2} - \frac{2mb^2}{na^2} + \frac{b^4}{a^4}\right)}}$ or $\dfrac{1}{\sqrt{\left(a^2 + \frac{m^2}{n^2}a^2 - \frac{2mb^2}{n} + \frac{b^4}{a^2}\right)}}$, that is (if in VZ, VY is taken equal to VG), as $\frac{1}{XY}$. For a^2 and $\frac{m^2}{n^2}a^2 - \frac{2mb^2}{n} + \frac{b^4}{a^2}$ are the squares of XZ and ZY. And the resistance is found to have the same ratio to gravity that 3XY has to 2YG; and the velocity is that with which the body would go in a parabola having vertex G, diameter DG, and latus rectum $\frac{XY^2}{VG}$. Therefore suppose that the densities of the medium in each of the individual places G are inversely as the distances XY and that the resistance in some place G is to gravity as 3XY to 2YG; then a body sent forth from place A with the proper velocity will describe that hyperbola AGK. Q.E.I.

EXAMPLE 4. Suppose generally that line AGK is a hyperbola described with center X and asymptotes MX and NX with the condition that when the rectangle XZDN is described, whose side ZD cuts the hyperbola in G and its asymptote in V, VG would be inversely as some power DN^n (whose index is the number n) of ZX or DN; and let it be required to find the density of the medium in which a projectile would progress in this curve.

For BN, BD, and NX write A, O, and C respectively, and let VZ be to XZ or DN as d to e, and let VG be equal to $\frac{b^2}{DN^n}$; then DN will be equal to A − O, $VG = \frac{b^2}{(A-O)^n}$, $VZ = \frac{d}{e}(A-O)$, and GD or NX − VZ − VG will be equal to $C - \frac{d}{e}A + \frac{d}{e}O - \frac{b^2}{(A-O)^n}$. Resolve the term $\frac{b^2}{(A-O)^n}$ into the infinite series $\frac{b^2}{A^n} + \frac{nb^2}{A^{n+1}}O + \frac{n^2+n}{2A^{n+2}}b^2O^2 + \frac{n^3+3n^2+2n}{6A^{n+3}}b^2O^3 \ldots$, and GD will become equal to $C - \frac{d}{e}A - \frac{b^2}{A^n} + \frac{d}{e}O - \frac{nb^2}{A^{n+1}}O - \frac{+n^2+n}{2A^{n+2}}b^2O^2 - \frac{+n^3+3n^2+2n}{6A^{n+3}}b^2O^3 \ldots$. The second term of this series $\frac{d}{e}O - \frac{nb^2}{A^{n+1}}O$ is to be used for Qo, the third term $\frac{n^2+n}{2A^{n+2}}b^2O^2$ for Ro^2, the fourth term $\frac{n^3+3n^2+2n}{6A^{n+3}}b^2O^3$ for So^3. And hence the density of the medium, $\frac{S}{R\sqrt{(1+Q^2)}}$, in any place G, becomes

$$\frac{n+2}{3\sqrt{\left(A^2+\dfrac{d^2}{e^2}A^2-\dfrac{2dnb^2}{eA^n}A+\dfrac{n^2b^4}{A^{2n}}\right)}},$$

and thus if in VZ, VY is taken equal to $n \times$ VG, the density is inversely as XY. For A^2 and $\frac{d^2}{e^2}A^2 - \frac{2dnb^2}{eA^n}A + \frac{n^2b^4}{A^{2n}}$ are the squares of XZ and ZY. Moreover, the resistance in the same place G becomes to the gravity as $3S \times \frac{XY}{A}$ is to $4R^2$, that is, as XY to $\frac{2n^2+2n}{n+2}$VG. And the velocity in the same place is the very velocity with which a projected body would go in a parabola having vertex G, diameter GD, and latus rectum $\frac{1+Q^2}{R}$ or $\frac{2XY^2}{(n^2+n) \times VG}$. Q.E.I.

Scholium [d]In the same way in which the density of the medium turned out to be as $\frac{S \times AC}{R \times HT}$ in corol. 1, if the resistance is supposed to be as any power V^n of the velocity V, the density of the medium will turn out to be as $\frac{S}{R^{\frac{4-n}{2}}} \times \left(\frac{AC}{HT}\right)^{n-1}$. And therefore if a curve can be found under the condition that there would be given the ratio of $\frac{S}{R^{\frac{4-n}{2}}}$ to $\left(\frac{HT}{AC}\right)^{n-1}$, or $\frac{S^2}{R^{4-n}}$ to $(1+Q^2)^{n-1}$, a body will move in this curve in a uniform medium with a resistance that is as the power V^n of the velocity. But let us return to simpler curves.[d]

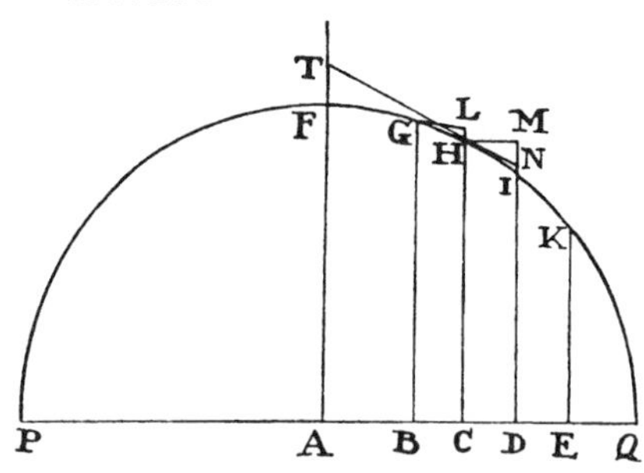

Since motion does not take place in a parabola except in a nonresisting medium, but does take place in the hyperbola here described if there is a continual resistance, it is obvious that the line which a projectile describes in a uniformly resisting medium approaches closer to these hyperbolas than to a parabola. At any rate, that line is of a hyperbolic kind, but about its vertex it is more distant from the asymptotes, and in

dd. Ed. 1 lacks this.

those parts that are further from the vertex it approaches the asymptotes more closely, than the hyperbolas which I have described here. But the difference between them is not so great that one cannot be conveniently used in place of the other in practice. And the hyperbolas which I have been describing will perhaps prove to be more useful than a hyperbola that is more exact and at the same time more compounded. And they will be brought into use as follows.

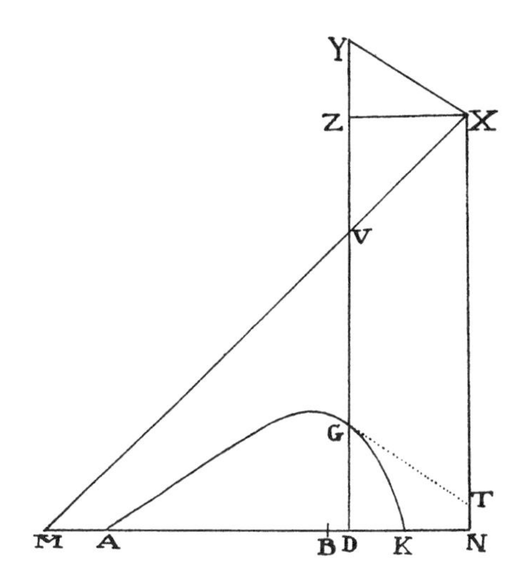

Complete the parallelogram XYGT, and the straight line GT will touch the hyperbola in G, and thus the density of the medium in G is inversely as the tangent GT, and the velocity in the same place is as $\sqrt{\frac{\text{GT}^2}{\text{GV}}}$, while the resistance is to the force of gravity as GT to $\frac{2n^2 + 2n}{n + 2} \times \text{GV}$.

Accordingly, if a body projected from place A along the straight line AH describes the hyperbola AGK and if AH produced meets the asymptote NX in H and if AI drawn parallel to NX meets the other asymptote MX in I, then the density of the medium in A will be inversely as AH, and the velocity of the body will be as $\sqrt{\frac{\text{AH}^2}{\text{AI}}}$, and the resistance in the same place will be to the gravity as AH to $\frac{2n^2 + 2n}{n + 2} \times \text{AI}$. Hence the following rules.

Rule 1. If both the density of the medium at A and the velocity with which the body is projected remain the same, and angle NAH is changed, lengths AH, AI, and HX will remain the same. And thus, if those lengths are found in some one case, the hyperbola can then be determined readily from any given angle NAH.

Rule 2. If both angle NAH and the density of the medium at A remain the same, and the velocity with which the body is projected is changed, the length AH will remain the same, and AI will be changed in the ratio of the inverse square of the velocity.

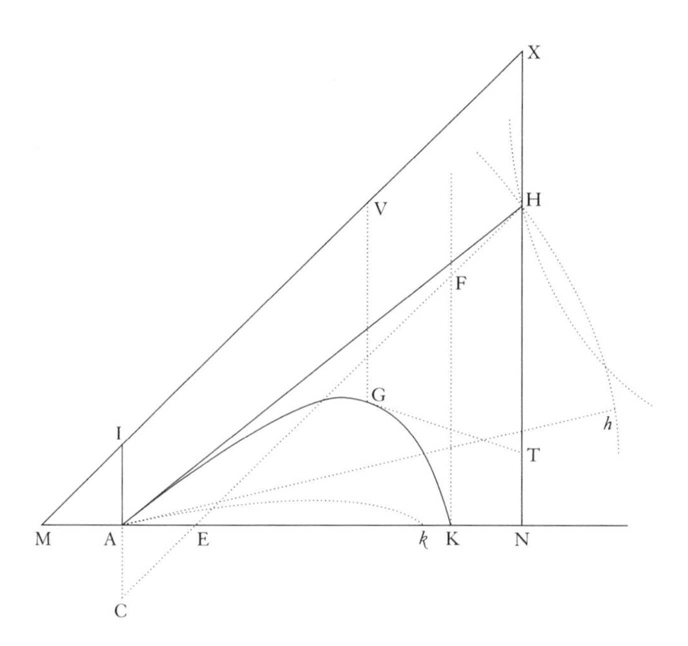

RULE 3. If angle NAH, the velocity of the body at A, and the accelerative gravity remain the same, and the proportion of the resistance at A to the motive gravity is increased in any ratio, the proportion of AH to AI will be increased in the same ratio, and the latus rectum of the above parabola as well as the length $\frac{\mathrm{AH}^2}{\mathrm{AI}}$ (proportional to it) will remain the same; and therefore AH will be decreased in the same ratio, and AI will be decreased as the square of that ratio. But the proportion of the resistance to the weight is increased when the specific gravity (the volume remaining constant) becomes smaller, or the density of the medium becomes greater, or the resistance (as a result of the decreased volume) is decreased in a smaller ratio than the weight.

RULE 4. The density of the medium near the vertex of the hyperbola is greater than at place A; hence, in order to have the mean density, the ratio of the least of the tangents GT to tangent AH must be found, and the density at A must be increased in a slightly greater ratio than that of half the sum of these tangents to the least of the tangents GT.

RULE 5. If lengths AH and AI are given, and it is required to describe the figure AGK, produce HN to X so that HX is to AI as $n + 1$ to 1, and with center X and asymptotes MX and NX, describe a hyperbola through point A in such a way that AI is to any VG as XV^n to XI^n.

RULE 6. The greater the number n, the more exact are these "hyperbolas" in the ascent of the body from A, and the less exact in its descent to K, and conversely. A conic hyperbola holds a mean ratio between them and is simpler than the others. Therefore, if the hyperbola is of this kind, and if it is required to find point K, where the projected body will fall upon any straight line AN passing through point A, let AN produced meet asymptotes MX and NX in M and N, and take NK equal to AM.

RULE 7. And hence a ready method of determining this kind of hyperbola from the phenomena is clear. Project two similar and equal bodies with the same velocity in different angles HAK and hAk, and let them fall upon the plane of the horizon in K and k, and note the proportion of AK to Ak (let this be d to e). Then, having erected a perpendicular AI of any length, assume length AH or Ah in any way and from this determine graphically lengths AK and Ak by rule 6. If the ratio of AK to Ak is the same as the ratio of d to e, length AH was correctly assumed. But if not, then on the indefinite straight line SM take a length SM equal to the assumed AH, and erect perpendicular MN equal to the difference of the ratios, $\frac{AK}{Ak} - \frac{d}{e}$, multiplied by any given straight line. From several assumed lengths AH find several points N by a similar method [e]and through them all draw a regular curved line NNXN cutting the straight line SMMM in X. Finally, assume AH equal to abscissa SX, and from this find length AK again; then the lengths that are to the assumed length AI and this last length AH as that length AK (found by experiment) is to the length AK (last found) will be those true lengths AI and AH which it was required to find. And these being given, the resistance of the medium in place A will also be given, inasmuch as it is to the force of gravity as AH to 2AI. The density of the medium, moreover, must be increased (by rule 4), and the resistance just found, if it is increased in the same ratio, will become more exact.[e]

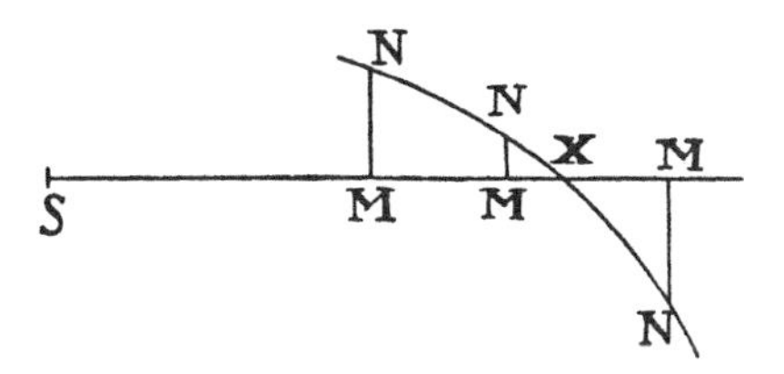

ee. Ed. 1 has: "and then finally, if a regular curved line NN × N is drawn through them all, this will cut off SX equal to the required length AH. For mechanical purposes it suffices to keep the same lengths AH, AI in all angles HAK. But if the figure must be determined more exactly in order to find the resistance of the medium, these lengths must always be corrected (by rule 4)."

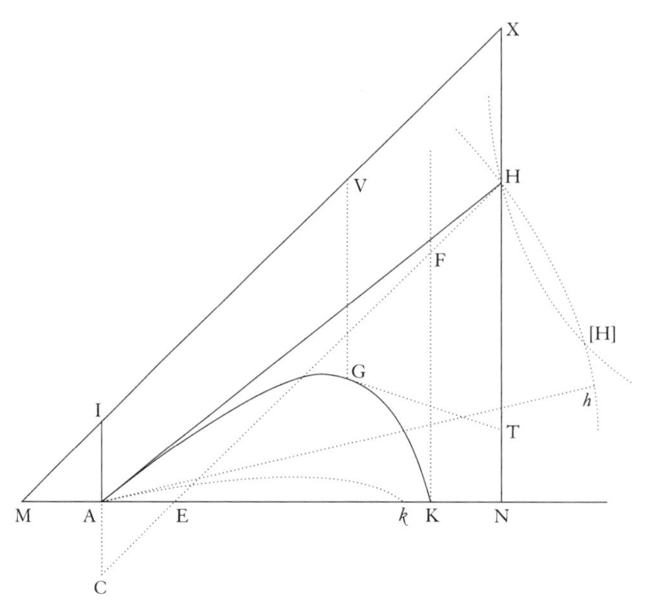

[In ed. 1, as well as in ed. 2, the same letter H is used for both the upper and the lower intersection of the two curves on the right side of the diagram, but in ed. 3 only the upper intersection is lettered. For the sake of clarity, we have introduced an [H] to designate the lower intersection and we have decreased the inclination of the tangent A*h* so that *h* is quite distinct from [H]. For further details, see the Guide, §7.4.]

RULE 8. If the lengths AH and HX have been found, and the position of the straight line AH is now desired along which a projectile sent forth with that given velocity falls upon any point K, erect at points A and K the straight lines AC and KF perpendicular to the horizon, of which AC tends downward and is equal to AI or ½HX. With asymptotes AK and KF describe a hyperbola whose conjugate passes through point C, and with center A and radius AH describe a circle cutting that hyperbola in point H; then a projectile sent forth along the straight line AH will fall upon point K. Q.E.I.

For point H, because length AH is given, is located somewhere in the circle described. Draw CH meeting AK and KF, the former in E, the latter in F; then, because CH and MX are parallel and AC and AI are equal, AE will be equal to AM, and therefore also equal to KN. But CE is to AE as FH to KN, and therefore CE and FH are equal. Point H therefore falls upon the hyperbola described with asymptotes AK and KF whose conjugate passes through point C, and thus H is found in the common intersection of this hyperbola and the circle described. Q.E.D.

It is to be noted, moreover, that this operation is the same whether the straight line AKN is parallel to the horizon or is inclined to the horizon at any angle, and that from the two intersections H and H two angles NAH and NAH arise, and that in a mechanical operation it is sufficient to describe a circle once, then to apply the indeterminate rule CH to point C in such a way that its part FH, placed between the circle and the straight line FK, is equal to its part CE situated between point C and the straight line AK.

What has been said about hyperbolas is easily applied to parabolas. For if XAGK designates a parabola that the straight line XV touches in vertex X and if ordinates IA and VG are as any powers XI^n and XV^n of abscissas XI and XV, draw XT, GT, and AH, of which XT is parallel to VG, and GT and AH touch the parabola in G and A; then a body projected with the proper velocity from any place A along the straight line AH (produced) will describe this parabola, provided that the density of the medium in each individual place G is inversely as tangent GT. The velocity in G, however, will be that with which a projectile would go, in a nonresisting space, in a conic parabola having vertex G, diameter VG produced downward, and latus rectum $\frac{2GT^2}{(n^2 - n) \times VG}$. And the resistance in G will be to the force of gravity as GT to $\frac{2n^2 - 2n}{n - 2}VG$. Hence, if NAK designates a horizontal line and if, while both the density of the medium in A and the velocity with which the body is projected remain the same, the angle NAH is changed in any way, then lengths AH, AI, and HX will remain the same; and hence vertex X of the parabola and the position of the straight line XI are given, and, by taking VG to IA as XV^n to XI^n, all the points G of the parabola, through which the projectile will pass, are given.

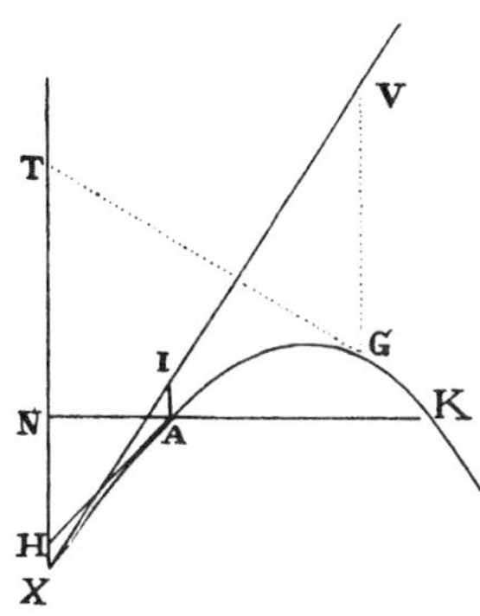

SECTION 3

The motion of bodies that are resisted partly in the ratio of the velocity and partly in the squared ratio of the velocity

Proposition 11
Theorem 8

If a body is resisted partly in the ratio of the velocity and partly in the squared ratio of the velocity and moves in a homogeneous medium by its inherent force alone, and if the times are taken in an arithmetic progression, then quantities inversely proportional to the velocities and increased by a certain given quantity will be in a geometric progression.

With center C and rectangular asymptotes CAD*d* and CH, describe a hyperbola BE*e*, and let AB, DE, and *de* be parallel to asymptote CH. Let points A and G be given in asymptote CD. Then if the time is represented by the hyperbolic area ABED increasing uniformly, I say that the velocity can be represented by the length DF, whose reciprocal GD together with the given quantity CG composes the length CD increasing in a geometric progression.

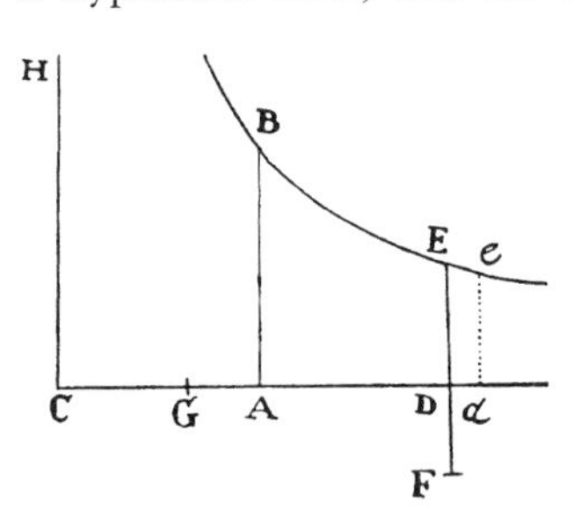

For let the area-element DE*ed* be a minimally small given increment of time; then D*d* will be inversely as DE and thus directly as CD. And the decrement of $\frac{1}{\mathrm{GD}}$, which (by book 2, lem. 2) is $\frac{\mathrm{D}d}{\mathrm{GD}^2}$, will be as $\frac{\mathrm{CD}}{\mathrm{GD}^2}$ or $\frac{\mathrm{CG}+\mathrm{GD}}{\mathrm{GD}^2}$, that is, as $\frac{1}{\mathrm{GD}}+\frac{\mathrm{CG}}{\mathrm{GD}^2}$. Therefore, when the time ABED increases uniformly by the addition of the given particles ED*de*, $\frac{1}{\mathrm{GD}}$ decreases in the same ratio as the velocity. For the decrement of the velocity is as the resistance, that is (by hypothesis), as the sum of two quantities, of which one is as the velocity and the other is as the square of the velocity; and the decrement of $\frac{1}{\mathrm{GD}}$ is as the sum of the quantities $\frac{1}{\mathrm{GD}}$ and $\frac{\mathrm{CG}}{\mathrm{GD}^2}$, of which the former is $\frac{1}{\mathrm{GD}}$ itself and the latter $\frac{\mathrm{CG}}{\mathrm{GD}^2}$ is as $\frac{1}{\mathrm{GD}^2}$. Accordingly, because the decrements are analogous, $\frac{1}{\mathrm{GD}}$ is as the velocity. And if the quantity GD, which is inversely proportional to $\frac{1}{\mathrm{GD}}$, is increased by the given quantity

CG, then as the time ABED increases uniformly, the sum CD will increase in a geometric progression. Q.E.D.

Corollary 1. Therefore if, given the points A and G, the time is represented by the hyperbolic area ABED, the velocity can be represented by $\frac{1}{GD}$, the reciprocal of GD.

Corollary 2. And by taking GA to GD as the reciprocal of the velocity at the beginning to the reciprocal of the velocity at the end of any time ABED, point G will be found. And when G has been found, then if any other time is given, the velocity can be found.

Proposition 12
Theorem 9

With the same suppositions, I say that if the spaces described are taken in an arithmetic progression, the velocities increased by a certain given quantity will be in a geometric progression.

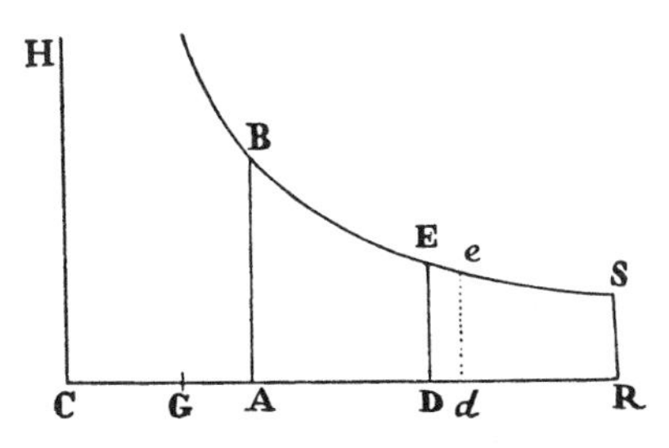

Let point R be given in the asymptote CD, and after erecting perpendicular RS meeting the hyperbola in S, represent the described space by the hyperbolic area RSED; then the velocity will be as the length GD, which with the given quantity CG composes the length CD decreasing in a geometric progression while space RSED is increased in an arithmetic progression.

For, because the increment ED*de* of the space is given, the line-element D*d*, which is the decrement of GD, will be inversely as ED and thus directly as CD, that is, as the sum of GD and the given length CG. But the decrement of the velocity, in the time inversely proportional to it in which the given particle D*de*E of space is described, is as the resistance and the time jointly, that is, directly as the sum of two quantities (of which one is as the velocity and the other is as the square of the velocity) and inversely as the velocity; and thus is directly as the sum of two quantities, of which one is given and the other is as the velocity. Therefore the decrement of the velocity as well as of line GD is as a given quantity and a decreasing quantity jointly; and because the decrements are analogous, the decreasing quantities will always be analogous, namely, the velocity and the line GD. Q.E.D.

COROLLARY 1. If the velocity is represented by the length GD, the space described will be as the hyperbolic area DESR.

COROLLARY 2. And if point R is taken at will, point G will be found by taking GR to GD as the velocity at the beginning is to the velocity after any space RSED has been described. And when point G has been found, the space is given from the given velocity, and conversely.

COROLLARY 3. Hence, since (by prop. 11) the velocity is given from the given time, and by this prop. 12 the space is given from the given velocity, the space will be given from the given time, and conversely.

Proposition 13
Theorem 10

Supposing that a body attracted downward by uniform gravity ascends straight up or descends straight down and is resisted partly in the ratio of the velocity and partly in the squared ratio of the velocity, I say that if straight lines parallel to the diameters of a circle and a hyperbola are drawn through the ends of the conjugate diameters and if the velocities are as certain segments of the parallels, drawn from a given point, then the times will be as the sectors of areas cut off by straight lines drawn from the center to the ends of the segments, and conversely.

CASE 1. Let us suppose first that the body is ascending. With center D and any semidiameter DB describe the quadrant BETF of a circle, and through the end B of semidiameter DB draw the indefinite line BAP parallel to semidiameter DF. Let point A be given in that line, and take segment AP proportional to the velocity. Since one part of the resistance is as the velocity and the other part is as the square of the velocity, let the whole resistance be as $AP^2 + 2BA \times AP$. Draw DA and DP cutting the circle in E and T, and represent the gravity by DA^2 in such a way that the gravity is to the resistance as DA^2 to $AP^2 + 2BA \times AP$; and the time of the whole ascent will be as sector EDT of the circle.

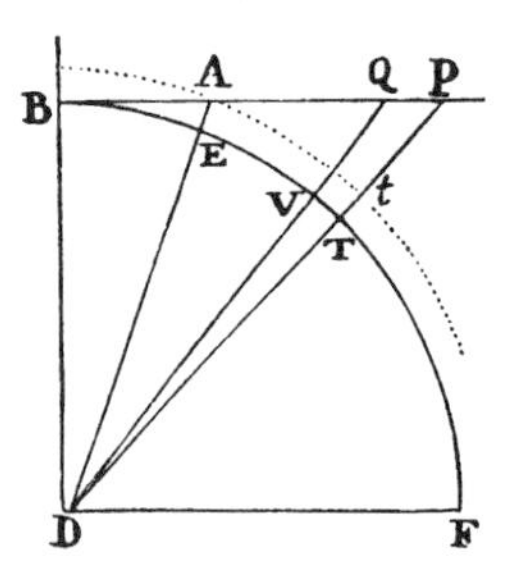

For draw DVQ cutting off both the moment PQ of velocity AP and the moment DTV (corresponding to a given moment of time) of sector DET; then that decrement PQ of the velocity will be as the sum of the forces of the gravity DA^2 and the resistance $AP^2 + 2BA \times AP$, that is (by book 2, prop. 12 of the *Elements*), as DP^2. Accordingly, the area DPQ, which is proportional to PQ, is as DP^2, and the area DTV, which is to the area DPQ as DT^2 to DP^2, is as the given quantity DT^2. The area EDT therefore decreases uniformly as the remaining time, by the subtraction of the given particles DTV, and therefore is proportional to the time of the whole ascent. Q.E.D.

CASE 2. If the velocity in the ascent of the body is represented by the length AP as in case 1, and the resistance is supposed to be as $AP^2 + 2BA \times AP$, and if the force of gravity is less than what could be represented by DA^2, take BD of such a length that $AB^2 - BD^2$ is proportional to the gravity, and let DF be perpendicular and equal to DB, and through the vertex F describe the hyperbola FTVE, whose conjugate semidiameters are DB and DF and which cuts DA in E and cuts DP and DQ in T and V; then the time of the whole ascent will be as the sector TDE of the hyperbola.

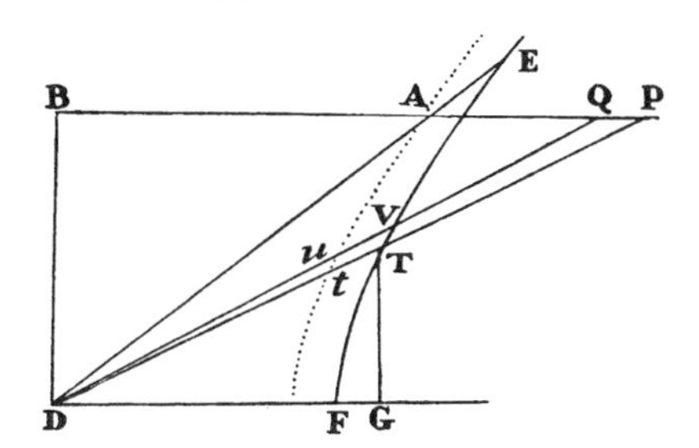

For the decrement PQ of the velocity occurring in a given particle of time is as the sum of the resistance $AP^2 + 2BA \times AP$ and the gravity $AB^2 - BD^2$, that is, as $BP^2 - BD^2$. But area DTV is to area DPQ as DT^2 to DP^2 and thus, if a perpendicular GT is dropped to DF, is as GT^2 or $GD^2 - DF^2$ to BD^2, and as GD^2 to BP^2, and by separation [or dividendo] as DF^2 to $BP^2 - BD^2$. Therefore, since area DPQ is as PQ, that is, as $BP^2 - BD^2$, area DTV will be as DF^2, which is given. Area EDT therefore decreases uniformly in each equal particle of time, by the subtraction of the same number of given particles DTV, and therefore is proportional to the time. Q.E.D.

CASE 3. Let AP be the velocity in the descent of the body, and $AP^2 + 2BA \times AP$ the resistance, and $BD^2 - AB^2$ the force of gravity, angle DBA being a right angle. And if with center D and principal vertex B the rectangular hyperbola BETV is described, cutting the produced lines DA, DP, and DQ in E, T, and V, then sector DET of this hyperbola will be as the whole time of descent.

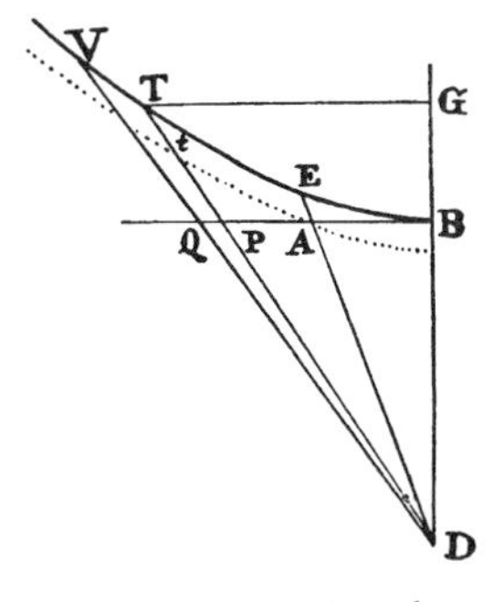

For the increment PQ of the velocity, and the area DPQ proportional to it, is as the excess of the gravity over the resistance, that is, as $BD^2 - AB^2 - 2BA \times AP - AP^2$ or $BD^2 - BP^2$. And area DTV is to area DPQ as DT^2 to DP^2 and thus as GT^2 or $GD^2 - BD^2$ to BP^2, and as GD^2 to BD^2, and by separation [or dividendo] as BD^2 to $BD^2 - BP^2$. Therefore, since area DPQ is as $BD^2 - BP^2$, area DTV will be as BD^2, which is given. Therefore area EDT increases uniformly in each equal particle of time, by the addition of the same number of given particles DTV, and therefore is proportional to the time of descent. Q.E.D.

COROLLARY. If with center D and semidiameter DA, the arc A*t* similar to arc ET and similarly subtending angle ADT is drawn through vertex A, then the velocity AP will be to the velocity that the body in time EDT in a nonresisting space could lose by ascending, or acquire by descending, as the area of triangle DAP to the area of sector DA*t* and thus is given from the given time. For in a nonresisting medium the velocity is proportional to the time and thus proportional to this sector; in a resisting medium the velocity is as the triangle; and in either medium, when the velocity is minimally small, it approaches the ratio of equality just as the sector and the triangle do.

Scholium[a] The case could also be proved in the ascent of the body, where the force of gravity is less than what can be represented by DA^2 or $AB^2 + BD^2$ and greater than what can be represented by $AB^2 - BD^2$, and must be represented by AB^2. But I hasten to other topics.

Proposition 14
Theorem 11

With the same suppositions, I say that the space described in the ascent or descent is as the difference between the area which represents the time and a certain other area that increases or decreases in an arithmetic progression, if the forces compounded of the resistance and the gravity are taken in a geometric progression.

Take AC (in the three figures) proportional to the gravity, and AK proportional to the resistance. And take them on the same side of point A if the body is descending, otherwise on opposite sides. Erect A*b*, which is to DB as DB^2 to $4BA \times AC$; and when the hyperbola *b*N has been described with respect to the rectangular asymptotes CK and CH, and KN has been erected perpendicular to CK, area A*b*NK will be increased or decreased in an arithmetic progression while the forces CK are taken in a geometric progression. I say therefore that the distance of the body from its greatest height is as the excess of area A*b*NK over area DET.

a. Ed. 1 and ed. 2 lack the scholium.

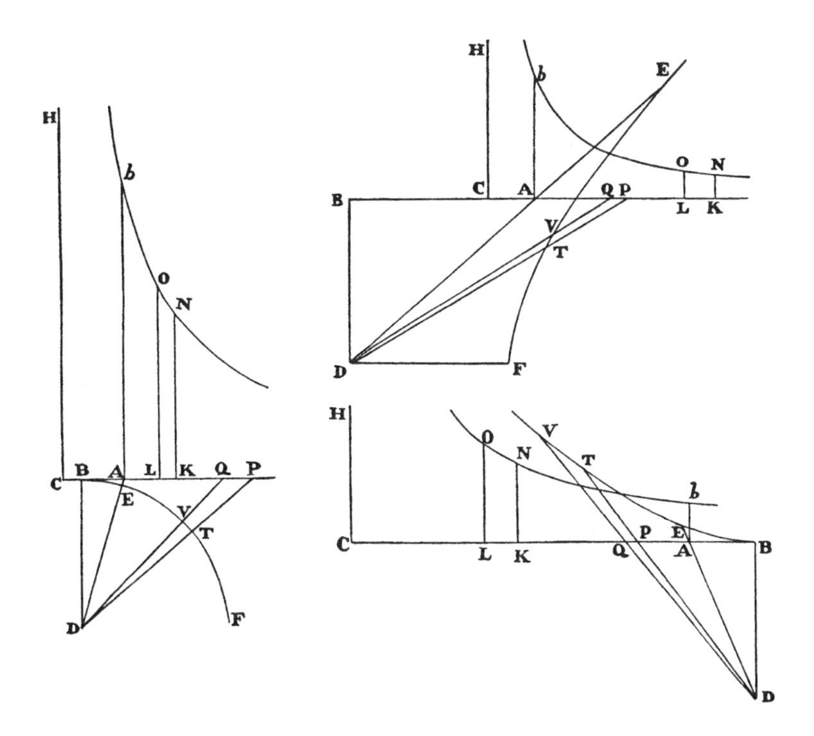

For since AK is as the resistance, that is, as $AP^2 + 2BA \times AP$, assume any given quantity Z, and suppose AK equal to $\dfrac{AP^2 + 2BA \times AP}{Z}$, and (by book 2, lem. 2) the moment KL of AK will be equal to $\dfrac{2AP \times PQ + 2BA \times PQ}{Z}$ or $\dfrac{2BP \times PQ}{Z}$, and the moment KLON of area A*b*NK will be equal to $\dfrac{2BP \times PQ \times LO}{Z}$ or $\dfrac{BP \times PQ \times BD^3}{2Z \times CK \times AB}$.

Case 1. Now, if the body is ascending and the gravity is as $AB^2 + BD^2$, BET being a circle (in the first figure), then line AC, which is proportional to the gravity, will be $\dfrac{AB^2 + BD^2}{Z}$, and DP^2 or $AP^2 + 2BA \times AP + AB^2 + BD^2$ will be $AK \times Z + AC \times Z$ or $CK \times Z$; and thus area DTV will be to area DPQ as DT^2 or DB^2 or $CK \times Z$.

Case 2. But if the body is ascending and the gravity is as $AB^2 - BD^2$, then line AC (in the second figure) will be $\dfrac{AB^2 - BD^2}{Z}$, and DT^2 will be to DP^2 as DF^2 or DB^2 to $BP^2 - BD^2$ or $AP^2 + 2BA \times AP + AB^2 - BD^2$, that is, to $AK \times Z + AC \times Z$ or $CK \times Z$. And thus area DTV will be to area DPQ as DB^2 to $CK \times Z$.

CASE 3. And by the same argument, if the body is descending and therefore the gravity is as $BD^2 - AB^2$, and line AC (in the third figure) is equal to $\frac{BD^2 - AB^2}{Z}$, then area DTV will be to area DPQ as DB^2 to $CK \times Z$, as above.

Since, therefore, those areas are always in this ratio, if for area DTV, which represents the moment of time always equal to it, any determinate rectangle is written, say $BD \times m$, then area DPQ, that is, $\frac{1}{2}BD \times PQ$, will be to $BD \times m$ as $CK \times Z$ to BD^2. And hence $PQ \times BD^3$ becomes equal to $2BD \times m \times CK \times Z$, and the moment KLON (found above) of area A*b*NK becomes $\frac{BP \times BD \times m}{AB}$. Take away the moment DTV or $BD \times m$ of area DET, and there will remain $\frac{AP \times BD \times m}{AB}$. Therefore the difference of the moments, that is, the moment of the difference of the areas, is equal to $\frac{AP \times BD \times m}{AB}$, and therefore (because $\frac{BD \times m}{AB}$ is given) is as the velocity AP, that is, as the moment of the space that the body describes in ascending or descending. And thus that space and the difference of the areas, increasing or decreasing by proportional moments and beginning simultaneously or vanishing simultaneously, are proportional. Q.E.D.

COROLLARY. If the length that results from dividing area DET by the line BD is called M, and another length V is taken in the ratio to length M that line DA has to line DE, then the space that a body describes in its whole ascent or descent in a resisting medium will be to the space that the body can describe in the same time in a nonresisting medium, by falling from a state of rest, as the difference of the above areas to $\frac{BD \times V^2}{AB}$, and thus is given from the given time. For the space in a nonresisting medium is in the squared ratio of the time, or as V^2, and, because BD and AB are given, as $\frac{BD \times V^2}{AB}$. [a]This area is equal to area $\frac{DA^2 \times BD \times M^2}{DE^2 \times AB}$, and the moment of

aa. Ed. 1 has: "But the time is as DET or $\frac{1}{2}BD \times ET$, and the moments of these areas are as $\frac{BD \times V}{2AB}$ multiplied by the moment of V and $\frac{1}{2}BD$ multiplied by the moment of ET, that is, as $\frac{BD \times V}{2AB} \times \frac{DA^2 \times 2m}{DE^2}$ and $\frac{1}{2}BD \times 2m$, or as $\frac{BD \times V \times DA^2 \times m}{AB \times DE^2}$ and $BD \times m$. And therefore the moment of area V^2 is to the moment of the difference of areas DET and AKN*b* as $\frac{BD \times V \times DA \times m}{AB \times DE}$

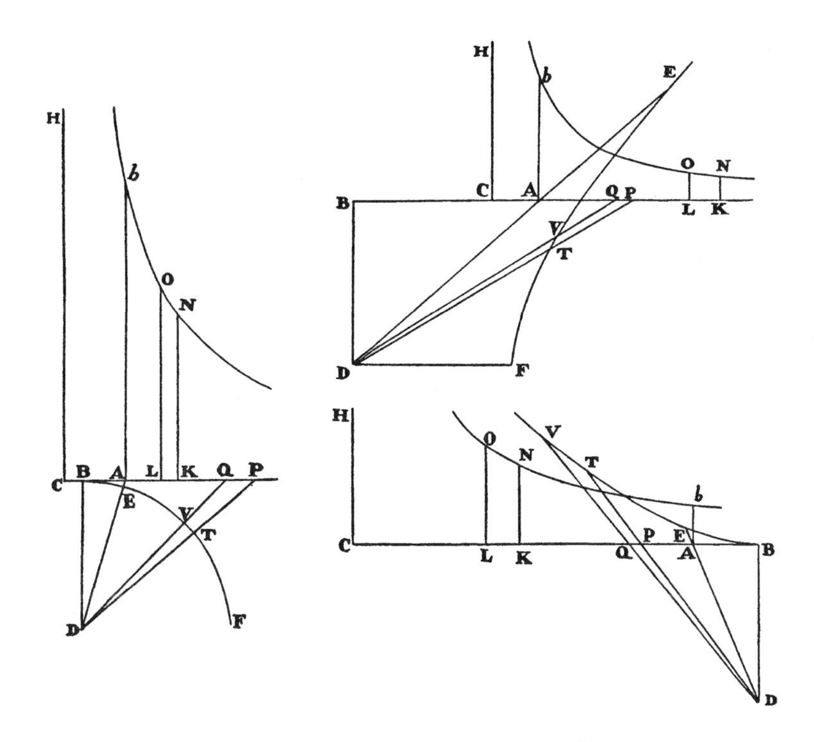

M is m; and therefore the moment of this area is $\frac{DA^2 \times BD \times 2M \times m}{DE^2 \times AB}$. But this moment is to the moment of the difference of the above areas DET and A*b*NK $\left(\text{that is, to } \frac{AP \times BD \times m}{AB}\right)$ as $\frac{DA^2 \times BD \times M}{DE^2}$ is to ½BD × AP, or as $\frac{DA^2}{DE^2} \times DET$ is to DAP; and thus, when areas DET and DAP are minimally small, in the ratio of equality. Therefore area $\frac{BD \times V^2}{AB}$ and the difference of areas DET and A*b*NK, when all these areas are minimally

---

to $\frac{AP \times BD \times m}{AB}$ or as $\frac{V \times DA}{DE}$ to AP and thus, when V and AP are minimally small, in the ratio of equality. Therefore the minimally small area $\frac{BD \times V^2}{4AB}$ is equal to the minimally small difference of areas DET and AKN*b*. Hence, since the spaces described simultaneously in both mediums at the beginning of the descent or at the end of the ascent approach equality and thus are then to one another as area $\frac{BD \times V^2}{4AB}$ and the difference of areas DET and AKN*b*, it follows that, because of their analogous increments, in any equal times they must be to one another as the area $\frac{BD \times V^2}{4AB}$ and the difference of areas DET and AKN*b*. Q.E.D." In ed. 2 the passage is the same as in ed. 1 except that AKN*b* is A*b*NK and the first two sentences read: "The moment of this area or of its equivalent, $\frac{DA^2 \times BD \times M^2}{DE^2 \times AB}$, is to the moment of the difference of areas DET and A*b*NK as $\frac{DA^2 \times BD \times 2M \times m}{DE^2 \times AB}$ to $\frac{AP \times BD \times m}{AB}$,

small, have equal moments and thus are equal. Hence, since the velocities, and therefore also the spaces described simultaneously in both mediums at the beginning of the descent or the end of the ascent, approach equality and thus are then to one another as area $\frac{BD \times V^2}{AB}$ and the difference of areas DET and A*b*NK; and furthermore since the space in a nonresisting medium is always as $\frac{BD \times V^2}{AB}$, and the space in a resisting medium is always as the difference of areas DET and A*b*NK; it follows that the spaces described in both mediums in any equal times must be to one another as the area $\frac{BD \times V^2}{AB}$ and the difference of areas DET and A*b*NK. Q.E.D.[a]

Scholium[b] The resistance encountered by spherical bodies in fluids arises partly from the tenacity, partly from the friction, and partly from the density of the medium. And we have said that the part of the resistance that arises from the density of the fluid is in the squared ratio of the velocity; the other part, which arises from the tenacity of the fluid, is uniform, or as the moment of the time; and thus it would now be possible to proceed to the motion of bodies that are resisted partly by a uniform force or in the ratio of the moments of the time and partly in the squared ratio of the velocity. But it is sufficient to have opened the way to the examination of this subject in the preceding props. 8 and 9 and their corollaries. In these propositions and corollaries, in place of the uniform resistance of the ascending body, which arises from its gravity, there can be substituted the uniform resistance that arises from the tenacity of the medium, when the body is moved by its inherent force alone; and when the body is ascending straight up, it is possible to add this uniform resistance to the force of gravity, and to subtract it when the body is descending straight down. It would also be possible to proceed to the motion of bodies that are resisted partly uniformly, partly in the ratio of the

---

that is, as $\frac{DA^2 \times BD \times M}{DE^2}$ to $\frac{1}{2}BD \times AP$, or as $\frac{DA^2}{DE^2} \times DET$ to DAP, and thus, when the areas DET and DAP are minimally small, in the ratio of equality." In both eds. 1 and 2 the fraction $\frac{BD \times V^2}{AB}$, which occurs just before this passage, is $\frac{BD \times V^2}{4AB}$.

b. Ed. 1 and ed. 2 lack the scholium.

velocity, and partly in the squared ratio of the velocity. And I have opened the way in the preceding props. 13 and 14, in which the uniform resistance that arises from the tenacity of the medium can also be substituted for the force of gravity, or can be compounded with it as before. But I hasten to other topics.

SECTION 4

The revolving motion of bodies in resisting mediums

Lemma 3 *Let* PQR *be a spiral that cuts all the radii* SP, SQ, SR, . . . *in equal angles. Draw the straight line* PT *touching the spiral in any point* P *and cutting the radius* SQ *in* T; *erect* PO *and* QO *perpendicular to the spiral and meeting in* O, *and join* SO. *I say that if points* P *and* Q *approach each other and coincide, angle* PSO *will come out a right angle, and the ultimate ratio of rectangle* $TQ \times 2PS$ *to* PQ^2 *will be the ratio of equality.*

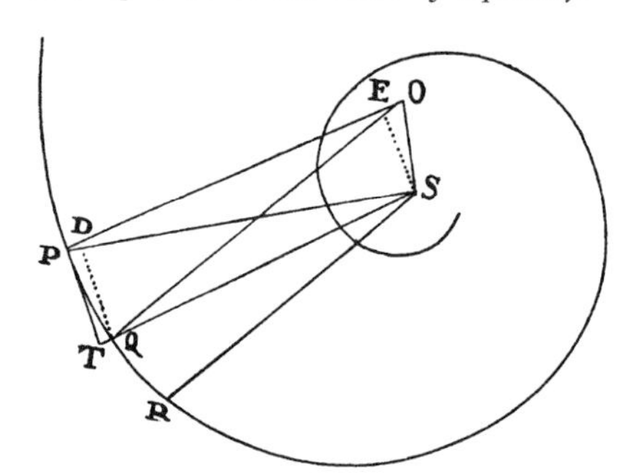

For, from the right angles OPQ and OQR subtract the equal angles SPQ and SQR, and the equal angles OPS and OQS will remain. Therefore a circle that passes through points O, S, and P will also pass through point Q. Let points P and Q come together, and this circle will touch the spiral in the place PQ where they coincide, and thus will cut the straight line OP perpendicularly. OP will therefore become a diameter of this circle, and OSP, an angle in a semicircle, will become a right angle. Q.E.D.

Drop perpendiculars QD and SE to OP, and the ultimate ratios of the lines will be as follows: TQ will be to PD as TS (or PS) to PE, or 2PO to 2PS; likewise, PD will be to PQ as PQ to 2PO; and from the equality of the ratios in inordinate proportion [or ex aequo perturbate] TQ will be to PQ as PQ to 2PS. Hence PQ^2 becomes equal to $TQ \times 2PS$. Q.E.D.

Proposition 15
Theorem 12 *If the density of a medium in every place is inversely as the distance of places from a motionless center and if the centripetal force is in the squared ratio of the density, I say that a body can revolve in a spiral that intersects in a given angle all the radii drawn from that center.*

Let the same things be supposed as in lemma 3, and produce SQ to V, so that SV is equal to SP. In any time, in a resisting medium, let a body describe the minimally small arc PQ, and in twice the time, the minimally small arc PR; then the decrements of these arcs arising from the resistance, that is,

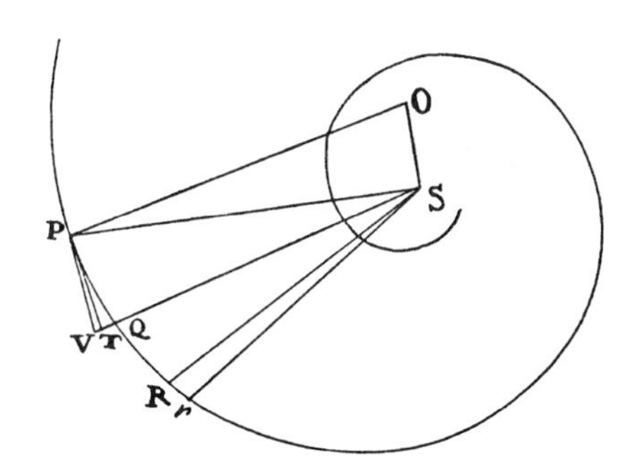

the differences between these arcs and the arcs that would be described in the same times in a nonresisting medium, will be to each other as the squares of the times in which they are generated. The decrement of arc PQ is therefore a fourth of the decrement of arc PR. Hence also, if area QS*r* is taken equal to area PSQ, the decrement of arc PQ will be equal to half of the line-element R*r*; and thus the force of resistance and the centripetal force are to each other as the line-elements ½R*r* and TQ that they simultaneously generate. Since the centripetal force by which the body is urged in P is inversely as SP^2; and since (by book 1, lem. 10) the line-element TQ, which is generated by that force, is in a ratio compounded of the ratio of this force and the squared ratio of the time in which arc PQ is described (for I ignore the resistance in this case, as being infinitely smaller than the centripetal force); then it follows that $TQ \times SP^2$, that is (by lem. 3), $\frac{1}{2}PQ^2 \times SP$, will be in the squared ratio of the time, and thus the time is as $PQ \times \sqrt{SP}$; and the body's velocity with which arc PQ is described in that time will be as $\frac{PQ}{PQ \times \sqrt{SP}}$ or $\frac{1}{\sqrt{SP}}$, that is, as the square root of SP inversely. And by a similar argument, the velocity with which arc QR is described is as the square root of SQ inversely. But these arcs PQ and QR are as the velocities of description to each other, that is, as $\sqrt{SQ}$ to $\sqrt{SP}$, or as SQ to $\sqrt{(SP \times SQ)}$; and because angles SPQ and SQ*r* are equal and areas PSQ and QS*r* are equal, arc PQ is to arc Q*r* as SQ to SP. Take the differences of the proportional consequents, and arc PQ will become to arc R*r* as SQ to $SP - \sqrt{(SP \times SQ)}$, or ½VQ. For, points P and Q coming together, the ultimate ratio of $SP - \sqrt{(SP \times SQ)}$ to ½VQ is the ratio of equality. [a]Since the decrement of arc PQ arising from the resistance, or its double R*r*, is as the resistance and the square of the time jointly, the resistance will be as $\frac{Rr}{PQ^2 \times SP}$.[a] But PQ was to R*r* as SQ to ½VQ, and

aa. Ed. 1 has: "In a nonresisting medium, equal areas PSQ, QS*r* would (by book 1, theor. 1) have to be described in equal times. From the resistance arises the difference RS*r* of the areas, and therefore the resistance is as decrement R*r* of line-element Q*r* compared with the square of the time in which it is generated. For line-element R*r* (by book 1, lem. 10) is as the square of the time. Therefore the resistance is as $\frac{Rr}{PQ^2 \times SP}$."

hence $\frac{Rr}{PQ^2 \times SP}$ becomes as $\frac{\frac{1}{2}VQ}{PQ \times SP \times SQ}$, or as $\frac{\frac{1}{2}OS}{OP \times SP^2}$. For, points P and Q coming together, SP and SQ coincide, and angle PVQ becomes a right angle; and because triangles PVQ and PSO are similar, PQ becomes to ½VQ as OP to ½OS. Therefore $\frac{OS}{OP \times SP^2}$ is as the resistance, that is, in the ratio of the density of the medium at P and the squared ratio of the velocity jointly. Take away the squared ratio of the velocity, namely the ratio $\frac{1}{SP}$, and the result will be that the density of the medium at P is as $\frac{OS}{OP \times SP}$. Let the spiral be given, and because the ratio of OS to OP is given, the density of the medium at P will be as $\frac{1}{SP}$. Therefore in a medium whose density is inversely as the distance SP from the center, a body can revolve in this spiral. Q.E.D.

COROLLARY 1. The velocity in any place P is always the velocity with which a body in a nonresisting medium, under the action of the same centripetal force, can revolve in a circle at the same distance SP from the center.

COROLLARY 2. The density of the medium, if the distance SP is given, is as $\frac{OS}{OP}$; but if that distance is not given, it is as $\frac{OS}{OP \times SP}$. And hence a spiral can be made to conform to any density of the medium.

COROLLARY 3. The force of resistance in any place P is to the centripetal force in the same place as ½OS to OP. For those forces are to each other as ½Rr and TQ or as $\frac{\frac{1}{4}VQ \times PQ}{SQ}$ and $\frac{\frac{1}{2}PQ^2}{SP}$, that is, as ½VQ and PQ, or ½OS and OP. Given the spiral, therefore, the proportion of the resistance to the centripetal force is given; and conversely, from that given proportion the spiral is given.

COROLLARY 4. The body, therefore, cannot revolve in this spiral except when the force of resistance is less than half of the centripetal force. Let the resistance become equal to half of the centripetal force; then the spiral will coincide with the straight line PS, and the body will descend to the center in this straight line with a velocity that is (as we proved in book 1, prop. 34) to the velocity with which the body descends in a nonresisting medium in the case of a parabola in the ratio of 1 to $\sqrt{2}$. [b]And the times of descent will here be inversely as the velocities, and thus are given.[b]

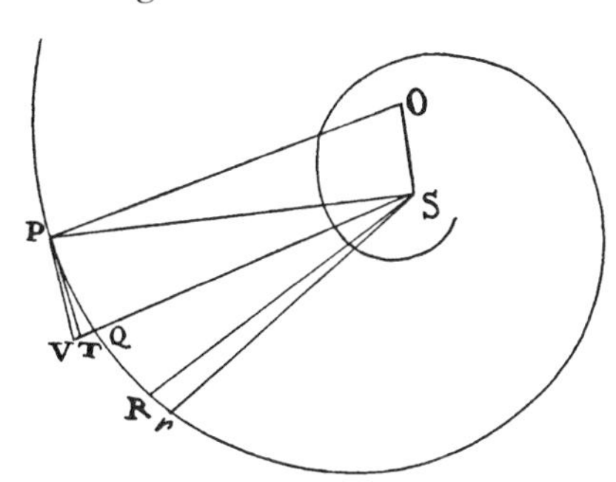

Corollary 5. And since at equal distances from the center the velocity is the same in the spiral PQR as in the straight line SP, and since the length of the spiral is in a given ratio to the length of the straight line PS, namely the ratio of OP to OS, the time of descent in the spiral will be to the time of descent in the straight line SP in that same given ratio, and accordingly is given.

Corollary 6. If, with center S and any two given radii, two circles are described, and if—these circles remaining the same—the angle that the spiral contains with radius PS is changed in any way, then the number of revolutions that the body can complete between the circumferences of the circles, by revolving in the spiral from one circumference to the other, is as $\frac{PS}{OS}$, or as the tangent of the angle that the spiral contains with radius PS. And the time of those revolutions is as $\frac{OP}{OS}$, that is, as the secant of that angle, or inversely as the density of the medium.

Corollary 7. If a body, in a medium whose density is inversely as the distance of places from the center, has made a revolution about that center in any curve AEB and has cut the first radius AS in the same angle in B as it did previously in A, with a velocity that was to its prior velocity in A inversely as the square roots of distances from the center—that is, as AS to a mean proportional between AS and BS—then that body will make innumerable entirely similar revolutions BFC, CGD, . . . , and by the intersections will divide the radius AS into the continually proportional parts AS, BS, CS, DS, ... And the times of revolution will be as the perimeters of the orbits AEB, BFC, CGD, . . . , directly, and the velocities in the beginnings A, B, C, inversely—that is, as $AS^{3/2}$, $BS^{3/2}$, $CS^{3/2}$. And the whole time in which the body will reach the center will be to the time of the first revolution as the

bb. Ed. 1 has: "Hence the times of descent will here be twice as great as those times and so are given."

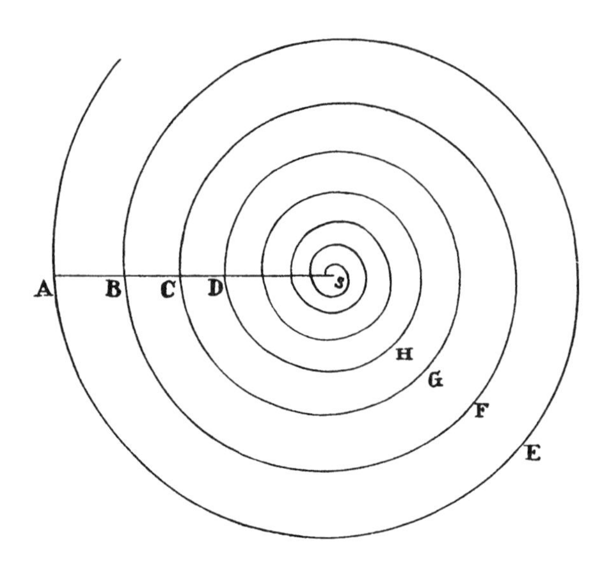

sum of all the continually proportional quantities $AS^{3/2}$, $BS^{3/2}$, $CS^{3/2}$, going on indefinitely, is to the first term $AS^{3/2}$—that is, as that first term $AS^{3/2}$ is to the difference of the first two terms $AS^{3/2} - BS^{3/2}$, or very nearly as $\frac{2}{3}$AS to AB. In this way the whole time is readily found.

COROLLARY 8. From what has been presented, it is also possible to determine approximately the motions of bodies in mediums whose density either is uniform or accords with any other assigned law. With center S and radii SA, SB, SC, . . . which are continually proportional, describe any number of circles. And suppose that the time of the revolutions between the perimeters of any two of these circles in the medium treated in corol. 7 is to the time of revolutions between those perimeters in the proposed medium very nearly as the mean density of the proposed medium between those circles is to the mean density of the medium in corol. 7 between those same circles; and suppose additionally that the secant of the angle by which the spiral in corol. 7, in the medium treated in that corollary, cuts the radius AS is in the same ratio to the secant of the angle by which the new spiral cuts that same radius in the proposed medium; and also that the numbers of all the revolutions between those same two circles are very nearly as the tangents of those same angles. If this is done throughout between every pair of circles, the motion will be continued through all the circles. And thus we can imagine without difficulty in what ways and in what times bodies would have to revolve in any regular medium.

COROLLARY 9. And even if the motions are eccentric, being performed in spirals approaching an oval shape, nevertheless by conceiving that the single revolutions of those spirals are the same distance apart from one another and approach the center by the same degrees as the spiral described above, we shall also understand how the motions of bodies are performed in spirals of this sort.

Proposition 16
Theorem 13

If the density of the medium in every place is inversely as the distance of places from a motionless center and if the centripetal force is inversely as any power of that distance, I say that a body can revolve in a spiral that intersects in a given angle all the radii drawn from that center.

This is proved by the same method as prop. 15. For if the centripetal force in P is inversely as any power SP^{n+1} (whose index is $n+1$) of the distance SP, then it will be gathered, as above, that the time in which the body describes any arc PQ will be as $PQ \times PS^{\frac{1}{2}n}$, and the resistance in P will be as $\frac{Rr}{PQ^2 \times SP^n}$, or as $\frac{(1 - \frac{1}{2}n) \times VQ}{PQ \times SP^n \times SQ}$, and thus as $\frac{(1 - \frac{1}{2}n) \times OS}{OP \times SP^{n+1}}$, that is, because $\frac{(1 - \frac{1}{2}n) \times OS}{OP}$ is given, inversely as SP^{n+1}. And therefore, since the velocity is inversely as $SP^{\frac{1}{2}n}$, the density in P will be inversely as SP.

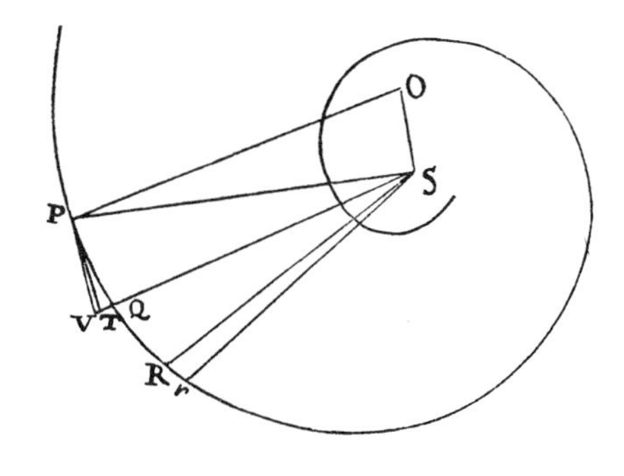

COROLLARY 1. The resistance is to the centripetal force as $(1 - \frac{1}{2}n) \times OS$ to OP.

COROLLARY 2. If the centripetal force is inversely as SP^3, $1 - \frac{1}{2}n$ will be $= 0$, and thus the resistance and density of the medium will be null, as in book 1, prop. 9.

COROLLARY 3. If the centripetal force is inversely as some power of the radius SP whose index is greater than the number 3, positive resistance will be changed to negative.

Scholium

But this proposition and the previous ones, which relate to unequally dense mediums, are to be understood of the motion of bodies so small that no consideration need be taken of a greater density of the medium on one side of the body than on the other. I also suppose the resistance, other things being equal, to be proportional to the density. Hence, in mediums whose force of resisting is not as the density, the density ought to be increased or decreased to such an extent that either the excess of the resistance may be taken away or its defect supplied.

Proposition 17
Problem 4

To find both the centripetal force and the resistance of the medium by means of which a body can revolve in a given spiral, if the law of the velocity is given.

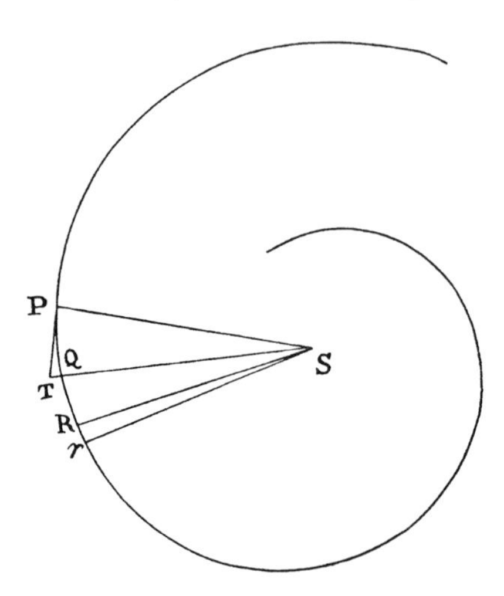

Let the spiral be PQR. The time will be given from the velocity with which the body traverses the minimally small arc PQ, and the force will be given from the height TQ, which is as the centripetal force and the square of the time. Then the retardation of the body will be given from the difference RS*r* of the areas PSQ and QSR traversed in equal particles of time, and the resistance and density of the medium will be found from the retardation.

Proposition 18
Problem 5

Given the law of the centripetal force, it is required to find in every place the density of the medium with which a body will describe a given spiral.

The velocity in every place is to be found from the centripetal force; then the density of the medium is to be sought from the retardation of the velocity, as in prop. 17.

I have presented the method of dealing with these problems in book 2, prop. 10 and lem. 2, and I do not wish to detain the reader any longer in complex inquiries of this sort. Some things must now be added on the forces of bodies in their forward motion, and on the density and resistance of the mediums in which the motions hitherto explained and motions related to these are performed.

SECTION 5

The density and compression of fluids, and hydrostatics

Definition of a Fluid

A fluid is any body whose parts yield to any force applied to it and yielding are moved easily with respect to one another.

Proposition 19
Theorem 14

All the parts of a homogeneous and motionless fluid that is enclosed in any motionless vessel and is compressed on all sides (apart from considerations of condensation, gravity, and all centripetal forces) are equally pressed on all sides and remain in their places without any motion arising from that pressure.

CASE 1. Let a fluid be enclosed in the spherical vessel ABC and be uniformly compressed on all sides; I say that no part of this fluid will move as a result of that pressure. For if some one part D moves, all the parts of this sort, standing on all sides at the same distance from the center, must move simultaneously with a similar motion; and this is so because the pressure on them all is similar and equal, and every motion is supposed excluded except that which arises from the pressure. But they cannot all approach closer to the center unless the fluid is condensed at the center, contrary to the hypothesis. They cannot recede farther from it unless the fluid is condensed at the circumference, also contrary to the hypothesis. They cannot move in any direction and keep their distance from the center, since by a like reasoning they will move in the opposite direction, and the same part cannot move in opposite directions at the same time. Therefore no part of the fluid will move from its place. Q.E.D.

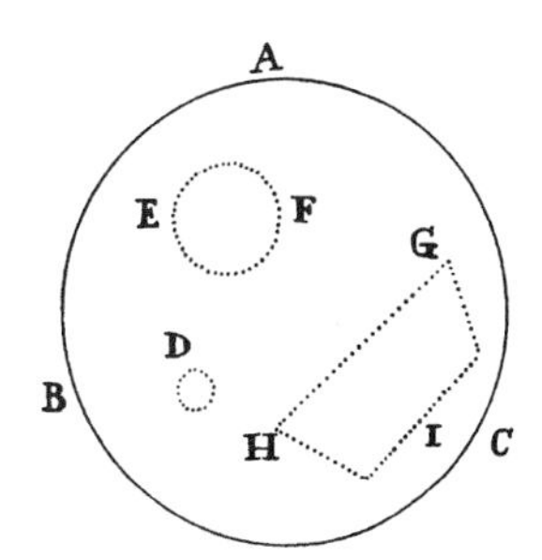

CASE 2. I say additionally that all the spherical parts of this fluid are equally pressed on all sides. For let EF be a spherical part of the fluid; if this part is not pressed equally on all sides, let the lesser pressure be increased until this part is pressed equally on all sides; then its parts, by case 1, will remain in their places. But before the increase of the pressure they will remain in their places, also by case 1, and by the addition of new pressure they will be moved out of their places, by the definition of a fluid. These two results

are contradictory. Therefore it was false to say that the sphere EF was not pressed equally on all sides. Q.E.D.

Case 3. I say furthermore that there is equal pressure on different spherical parts. For contiguous spherical parts press one another equally in the point of contact, by the third law of motion. But by case 2, they are also pressed on all sides by the same force. Therefore any two noncontiguous spherical parts will be pressed by the same force, since an intermediate spherical part can touch both. Q.E.D.

Case 4. I say also that all the parts of the fluid are equally pressed on every side. For any two parts can be touched by spherical parts in any points, and there they press those spherical parts equally, by case 3, and in turn are equally pressed by them, by the third law of motion. Q.E.D.

Case 5. Since, therefore, any part GHI of the fluid is enclosed in the remaining fluid as if in a vessel and is pressed equally on all sides, while its parts press one another equally and are at rest with respect to one another, it is manifest that all the parts of any fluid GHI which is pressed equally on all sides press one another equally and are at rest with respect to one another. Q.E.D.

Case 6. Therefore, if that fluid is enclosed in a vessel that is not rigid and is not pressed equally on all sides, it will yield to a greater pressure, by the definition of a fluid.

Case 7. And thus in a rigid vessel a fluid will not sustain a pressure that is greater on one side than on another, but will yield to it, and will do so in an instant of time, since the rigid side of the vessel does not follow the yielding liquid. And by yielding, it will press the opposite side, and thus the pressure will tend on all sides to equality. And since, as soon as the fluid endeavors to recede from the part that is pressed more, it is hindered by the resistance of the vessel on the opposite side, the pressure will be reduced on all sides to equality in an instant of time without local motion; and thereupon the parts of the fluid, by case 5, will press one another equally and will be at rest with respect to one another. Q.E.D.

Corollary. Hence the motions of the parts of the fluid with respect to one another cannot be changed by pressure applied to the fluid anywhere on the external surface, except insofar as either the shape of the surface is changed somewhere or all the parts of the fluid, by pressing one another more intensely or more remissly [i.e., by pressing one another more strongly or less strongly], flow among themselves with more or less difficulty.

Proposition 20
Theorem 15

If every part of a fluid that is spherical and homogeneous at equal distances from the center and rests upon a concentric spherical bottom gravitates toward the center of the whole, then the bottom will sustain the weight of a cylinder whose base is equal to the surface of the bottom and whose height is the same as that of the fluid resting upon it.

Let DHM be the surface of the bottom, and AEI the upper surface of the fluid. Divide the fluid into equally thick concentric spherical shells[a] by innumerable spherical surfaces BFK, CGL; and suppose the force of gravity to act only upon the upper surface of each spherical shell, and the actions upon equal parts of all the surfaces to be equal. The highest surface AE is pressed, therefore, by the simple force of its own gravity, by which also all the parts of the highest spherical shell, and the second surface BFK (by prop. 19), are equally pressed in accordance with their measure. The second surface BFK is pressed additionally by the force of its own gravity, which, added to the previous force, makes the pressure double. The third surface CGL is acted on by this pressure, in accordance with its measure, and additionally by the force of its gravity, that is, by a triple pressure. And similarly the fourth surface is urged by a quadruple pressure, the fifth by a quintuple, and so on. The pressure by which any one surface is urged is therefore not as the solid quantity of the fluid lying upon it, but as the number of spherical shells up to the top of the fluid, and is equal to the gravity of the lowest spherical shell multiplied by the number of shells; that is, it is equal to the gravity of a solid whose ultimate ratio to the cylinder specified above will become that of equality—provided that the number of shells is increased and their thickness decreased indefinitely, in such a way that the action of gravity is made continuous from the lowest surface to the highest. The lowest surface therefore sustains the weight of the cylinder specified above. Q.E.D. And

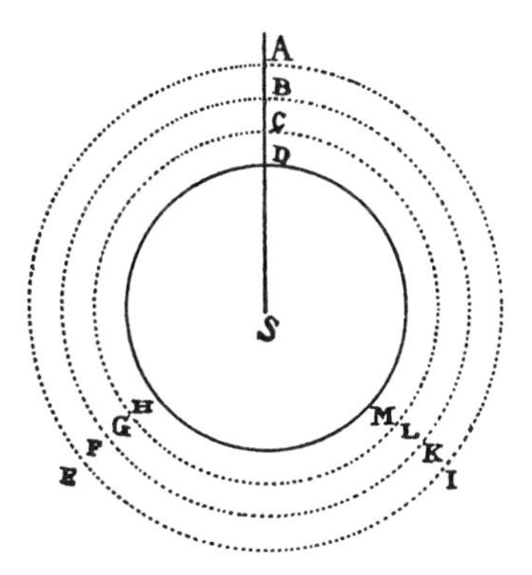

a. Here, as elsewhere in the *Principia*, Newton uses the noun "orbis" (orb) for a spherical shell.

by a similar argument this proposition is evident when the gravity decreases in any assigned ratio of the distance from the center, and also when the fluid is rarer upward and denser downward. Q.E.D.

COROLLARY 1. Therefore the bottom is not pressed by the whole weight of the incumbent fluid, but sustains only that part of the weight which is described in this proposition, the rest of the weight being sustained by the vaulted shape of the fluid.

COROLLARY 2. At equal distances from the center, moreover, the quantity of pressure is always the same, whether the pressed surface is parallel to the horizon or perpendicular or oblique, or whether the fluid—continued upward from the pressed surface—rises perpendicularly along a straight line or snakes obliquely through twisted cavities and channels, regular or extremely irregular, wide or very narrow. That the pressure is not at all changed by these circumstances is gathered by applying the proof of this theorem to the various cases of fluids.

COROLLARY 3. By the same proof it is also gathered (by prop. 19) that the parts of a heavy fluid acquire no motion with respect to one another as a result of the pressure of the incumbent weight, provided that the motion arising from condensation is excluded.

COROLLARY 4. And therefore, if another body, in which there is no condensation, of the same specific gravity is submerged in this fluid, it will acquire no motion as a result of the pressure of the incumbent weight; it will not descend, it will not ascend, and it will not be compelled to change its shape. If it is spherical, it will remain spherical despite the pressure; if it is square, it will remain square; and it will do so whether it is soft or very fluid, whether it floats freely in the fluid or lies on the bottom. For any internal part of a fluid is in the same situation as a submerged body, and the case is the same for all submerged bodies of the same size, shape, and specific gravity. If a submerged body, while keeping its weight, were to liquefy and assume the form of a fluid, then, if it were formerly ascending or descending or assuming a new shape as a result of pressure, it would also now ascend or descend or be compelled to assume a new shape, and would do so because its gravity and the other causes of motions remain fixed. But (by prop. 19, case 5) this body would now be at rest and would maintain its shape. Hence, this would also be the case under the earlier conditions.

Corollary 5. Accordingly, a body that is of a greater specific gravity than a fluid contiguous to it will sink, and a body that is of a lesser specific gravity will ascend, and will acquire as much motion and change of shape as that excess or deficiency of gravity can bring about. For that excess or deficiency acts like an impulse by which the body, otherwise in equilibrium with the parts of the fluid, is urged; and it can be compared with the excess or deficiency of weight in either of the scales of a balance.

Corollary 6. The gravity of bodies in fluids is therefore twofold: the one, true and absolute; the other, apparent, common, and relative. Absolute gravity is the whole force with which a body tends downward; relative or common gravity is the excess of gravity with which the body tends downward more than the surrounding fluid. By absolute gravity the parts of all fluids and bodies gravitate in their places, and thus the sum of the individual weights is the weight of the whole. For every whole is heavy, as can be tested in vessels full of liquids, and the weight of the whole is equal to the sum of the weights of all the parts, and thus is composed of them. By relative gravity bodies do not gravitate in their places; that is, compared with one another, one is not heavier than another, but each one opposes the endeavors of the others to descend, and they remain in their places just as if they had no gravity. Whatever is in the air and does not gravitate more than the air is not commonly considered to be heavy. Things that do gravitate more are commonly considered to be heavy, inasmuch as they are not sustained by the weight of the air. Weight as commonly conceived is nothing other than the excess of the true weight over the weight of the air. Bodies are commonly called light which are less heavy than the surrounding air and, by yielding to that air, which gravitates more, move upward. They are, however, only comparatively light and not truly so, since they descend in a vacuum. Similarly, bodies in water that descend or ascend because of their greater or smaller gravity are comparatively and apparently heavy or light, and their comparative and apparent heaviness or lightness is the excess or deficiency by which their true gravity either exceeds the gravity of the water or is exceeded by it. And bodies that neither descend by gravitating more nor ascend by yielding to water which gravitates more—even though they increase the weight of the whole by their own true weights—nevertheless, comparatively and as commonly understood, do not gravitate in water. For the demonstration of all these cases is similar.

Corollary 7. What has been demonstrated concerning gravity is valid for any other centripetal forces.

Corollary 8. Accordingly, if the medium in which some body moves is urged either by its own gravity or by any other centripetal force, and the body is urged more strongly by the same force, then the difference between the forces is that motive force which we have considered to be the centripetal force in the preceding propositions. But if the body is urged more lightly by that force, the difference between the forces should be considered a centrifugal force.

Corollary 9. Since fluids, moreover, do not change the external shapes of enclosed bodies that they press upon, it is evident in addition (by prop. 19, corol.) that fluids will not change the situation of the internal parts with respect to one another; and accordingly, if animals are immersed, and if all sensation arises from the motion of the parts, fluids will neither harm these immersed bodies nor excite any sensation, except insofar as these bodies can be condensed by compression. And the case is the same for any system of bodies that is surrounded by a compressing fluid. All the parts of the system will be moved with the same motions as if they were in a vacuum and retained only their relative gravity, except insofar as the fluid either resists their motions somewhat or is needed to make them cohere by compression.

Proposition 21
Theorem 16

Let the density of a certain fluid be proportional to the compression, and let its parts be drawn downward by a centripetal force inversely proportional to their distances from the center; I say that if the distances are taken continually proportional, the densities of the fluid at these distances will also be continually proportional.

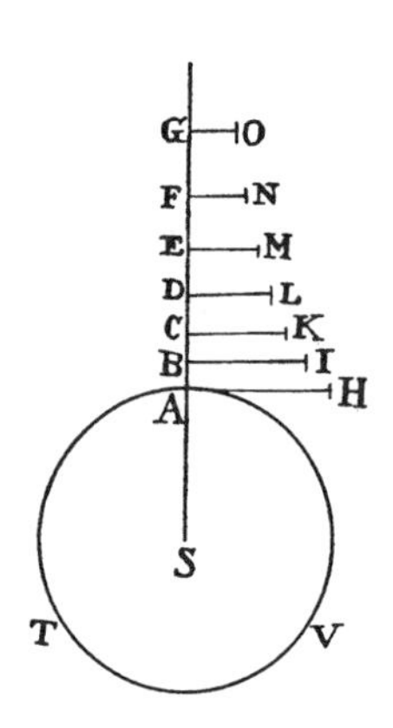

Let ATV designate the spherical bottom on which the fluid lies, S the center, and SA, SB, SC, SD, SE, SF, ... the continually proportional distances. Erect perpendiculars AH, BI, CK, DL, EM, FN, ..., which are as the densities of the medium in places A, B, C, D, E, F; then the specific gravities in those places will be as $\frac{AH}{AS}, \frac{BI}{BS}, \frac{CK}{CS}, \ldots$, or—which is the same—as $\frac{AH}{AB}, \frac{BI}{BC}, \frac{CK}{CD}, \ldots$. Suppose first that these gravities continue uniformly, the first from A to B, the second from B to C, the third from C to D, . . . , the decrements thus occurring by degrees at points B, C, D, ... Then these specific gravities multiplied by the heights AB, BC, CD, . . . will give the pressures AH, BI, CK, . . . , by which the bottom ATV (according to prop. 20) is pressed. The particle A therefore sustains all the pressures AH, BI, CK, DL, going on indefinitely; and the particle B, all the pressures except the first, AH; and the particle C, all except the first two, AH and BI; and so on. And thus the density AH of the first particle A is to the density BI of the second particle B as the sum of all the AH + BI + CK + DL indefinitely, to the sum of all the BI + CK + DL . . . . And the density BI of the second particle B is to the density CK of the third particle C as the sum of all the BI + CK + DL . . . to the sum of all the CK + DL . . . . Those sums are therefore proportional to their differences AH, BI, CK, . . . , and thus are continually proportional (by book 2, lem. 1); and accordingly the differences AH, BI, CK, . . . , proportional to those sums, are also continually proportional. Therefore, since the densities in places A, B, C, . . . are as AH, BI, CK, . . . , these also will be continually proportional. Proceed now by jumps, and from the equality of the ratios [or ex aequo], at the continually proportional distances SA, SC, SE, the densities AH, CK, EM will be continually proportional. And by the same argument, at any continually proportional distances SA, SD, SG, the densities AH, DL, GO will be continually proportional. Now let points A, B, C, D, E, . . . come together so that the progression of the specific gravities is made continual from the bottom A to the top of the fluid; and at any continually proportional distances SA, SD, SG, the densities AH, DL, GO, being always continually proportional, will still remain continually proportional now. Q.E.D.

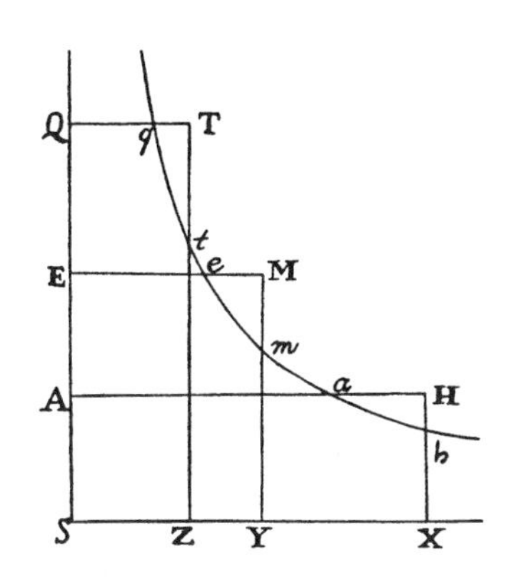

Corollary. Hence, if the density of a fluid is given in two places, say A and E, its density in any other place Q can be determined. With center S and rectangular asymptotes SQ and SX describe a hyperbola cutting perpendiculars AH, EM, and QT in *a*, *e*, and *q*, and also perpendiculars HX, MY, and TZ, dropped to asymptote SX, in *h*, *m*, and *t*. Make the area Y*mt*Z be to the given area Y*mh*X as the given area E*eq*Q is to the given area $EeaA$; and the line Zt produced will cut off the line QT proportional to the density. For if lines SA, SE, and SQ are continually proportional, areas $EeqQ$ and $EeaA$ will be equal, and hence the areas proportional to these, $YmtZ$ an $XhmY$, will also be equal, and lines SX, SY, and SZ—that is, AH, EM, and QT—will be continually proportional, as they ought to be. And if lines SA, SE, and SQ obtain any other order in the series of continually proportional quantities, lines AH, EM, and QT, because the hyperbolic areas are proportional, will obtain the same order in another series of continually proportional quantities.

Proposition 22
Theorem 17

Let the density of a certain fluid be proportional to the compression, and let its parts be drawn downward by a gravity inversely proportional to the squares of their distances from the center; I say that if the distances are taken in a harmonic progression, the densities of the fluid at these distances will be in a geometric progression.

Let S designate the center, and SA, SB, SC, SD, and SE the distances in a geometric progression. Erect perpendiculars AH, BI, CK, . . . , which are as the densities of the fluid in places A, B, C, D, E, . . . ; then the specific gravities in those places will be $\frac{AH}{SA^2}, \frac{BI}{SB^2}, \frac{CK}{SC^2}, \ldots$. Imagine these specific gravities to be uniformly continued, the first from A to B, the second from B to C, the third from C to D, ... Then these, multiplied by the heights AB, BC,

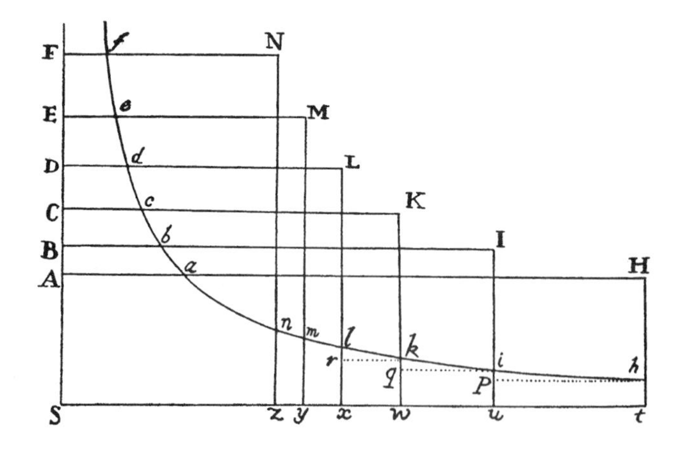

CD, DE, . . .—or, which is the same, by the distances SA, SB, SC, . . . , proportional to those heights—will yield $\frac{\mathrm{AH}}{\mathrm{SA}}, \frac{\mathrm{BI}}{\mathrm{SB}}, \frac{\mathrm{CK}}{\mathrm{SC}}, \ldots$, which represent the pressures. Therefore, since the densities are as the sums of these pressures, the differences $(\mathrm{AH}-\mathrm{BI}, \mathrm{BI}-\mathrm{CK}, \ldots)$ of the densities will be as the differences $\left(\frac{\mathrm{AH}}{\mathrm{SA}}, \frac{\mathrm{BI}}{\mathrm{SB}}, \frac{\mathrm{CK}}{\mathrm{SC}}, \ldots\right)$ of the sums. With center S and asymptotes SA and S*x* describe any hyperbola that cuts the perpendiculars AH, BI, CK, . . . in *a*, *b*, *c*, . . . and also cuts in *h*, *i*, and *k* the perpendiculars H*t*, I*u*, and K*w*, dropped to asymptote S*x*; then the differences *tu*, *uw*, . . . between the densities will be as $\frac{\mathrm{AH}}{\mathrm{SA}}, \frac{\mathrm{BI}}{\mathrm{SB}}, \ldots$. And the rectangles $tu \times th$, $uw \times ui$, . . . , or *tp*, *uq*, . . . , will be as $\frac{\mathrm{AH} \times th}{\mathrm{SA}}, \frac{\mathrm{BI} \times ui}{\mathrm{SB}}, \ldots$, that is, as A*a*, B*b*, ... For, from the nature of the hyperbola, SA is to AH or S*t* as *th* to A*a*, and thus $\frac{\mathrm{AH} \times th}{\mathrm{SA}}$ is equal to A*a*. And by a similar argument, $\frac{\mathrm{BI} \times ui}{\mathrm{SB}}$ is equal to B*b*, ... Moreover, A*a*, B*b*, C*c*, . . . are continually proportional, and therefore proportional to their differences A*a* − B*b*, B*b* − C*c*, . . . ; and thus the rectangles *tp*, *uq*, . . . are proportional to these differences, and also the sums of the rectangles $tp + uq$ or $tp + uq + wr$ are proportional to the sums of the differences A*a* − C*c* or A*a* − D*d*. Let there be as many terms of this sort as you wish; then the sum of all the differences, say A*a* − F*f*, will be proportional to the sum of all the rectangles, say *zthn*. Increase the number of terms and decrease the distances of points A, B, C, . . . , indefinitely; then these rectangles will come out equal to the hyperbolic area *zthn*, and thus the difference A*a* − F*f* is proportional to this area. Now take any distances, say SA, SD, SF, in a harmonic progression, and the differences A*a* − D*d* and D*d* − F*f* will be equal; and therefore the areas *thlx* and *xlnz* which are proportional to these differences will be equal to each other, and the densities S*t*, S*x*, and S*z* (that is, AH, DL, and FN) will be continually proportional. Q.E.D.

COROLLARY. Hence, if any two densities of a fluid are given, say AH and BI, the area *thiu* corresponding to their difference *tu* will be given; and accordingly the density FN at any height SF will be found by taking the area *thnz* to be to that given area *thiu* as the difference Aa − Ff is to the difference Aa − Bb.

Scholium Similarly, it can be proved that if the gravity of the particles of a fluid is decreased as the cubes of the distances from the center, and if the reciprocals of the squares of the distances SA, SB, SC, . . . $\left(\text{namely, } \frac{SA^3}{SA^2}, \frac{SA^3}{SB^2}, \frac{SA^3}{SC^2}\right)$ are taken in an arithmetic progression, then the densities AH, BI, CK, . . . will be in a geometric progression. And if the gravity is decreased as the fourth power of the distances, and if the reciprocals of the cubes of the distances $\left(\text{say, } \frac{SA^4}{SA^3}, \frac{SA^4}{SB^3}, \frac{SA^4}{SC^3}, \ldots\right)$ are taken in an arithmetic progression, the densities AH, BI, CK, . . . will be in a geometric progression. And so on indefinitely. Again, if the gravity of the particles of a fluid is the same at all distances, and if the distances are in an arithmetic progression, the densities will be in a geometric progression, as the distinguished gentleman Edmond Halley has found. If the gravity is as the distance, and if the squares of the distances are in an arithmetic progression, the densities will be in a geometric progression. And so on indefinitely.

These things are so when the density of a fluid condensed by compression is as the force of the compression or, which is the same, when the space occupied by the fluid is inversely as this force. Other laws of condensation can be imagined, as, for example, that the cube of the compressing force is as the fourth power of the density, or that the force ratio cubed is the same as the density ratio to the fourth power. In this case, if the gravity is inversely as the square of the distance from the center, the density will be inversely as the cube of the distance. Imagine that the cube of the compressing force is as the fifth power of the density; then, if the gravity is inversely as the square of the distance, the density will be inversely as the $\frac{3}{2}$ power of the distance. Imagine that the compressing force is as the square of the density, and that the gravity is inversely as the square of the distance; then the density will be inversely as the distance. It would be tedious to cover all cases. But it is established by experiments that the density of air is either exactly or at least very nearly as the compressing force; and therefore the density of the air in the earth's atmosphere is as the weight of the whole incumbent air, that is, as the height of the mercury in a barometer.

Proposition 23
Theorem 18

[a]*If the density of a fluid composed of particles that are repelled from one another is as the compression, the centrifugal forces [or forces of repulsion] of the particles are inversely proportional to the distances between their centers. And conversely, particles that are repelled from one another by forces that are inversely proportional to the distances between their centers constitute an elastic fluid whose density is proportional to the compression.*[a]

Suppose a fluid to be enclosed in the cubic space ACE, and then by compression to be reduced into the smaller cubic space *ace*; then the distances between particles maintaining similar positions with respect to one another in the two spaces will be as the edges AB and *ab* of the cubes; and the densities of the mediums will be inversely as the containing spaces AB^3 and ab^3. On the plane side ABCD of the larger cube take the square DP equal to the plane side of the smaller cube *db*; then (by hypothesis) the pressure by which the square DP urges the enclosed fluid will be to the pressure by which the square *db* urges the enclosed fluid as the densities of the medium to each other, that is, as ab^3 to AB^3. But the pressure by which the square DB urges the enclosed fluid is to the pressure by which the square DP urges that same fluid as the square DB to the square DP, that is, as AB^2 to ab^2. Therefore, from the equality of the ratios [or ex aequo] the pressure by which the square DB urges the fluid is to the pressure by which the square *db* urges the fluid as *ab* to AB. Divide the fluid into two parts by planes FGH and *fgh* drawn through the middles of the cubes; then these parts will press each other with the same forces with which they are pressed by planes AC and *ac*, that is, in the proportion of *ab* to AB; and thus the centrifugal forces [or forces of repulsion] by which these pressures are sustained are in the same ratio. Because in both cubes the number of particles is the same and their situation similar, the forces that all the particles along planes FGH and *fgh* exert upon all the others are as the forces that each individual particle exerts upon every other particle. Therefore the forces that each particle exerts upon every other particle along the plane FGH in the larger cube are to the forces that individual particles exert on the particle next to them along the plane *fgh* in

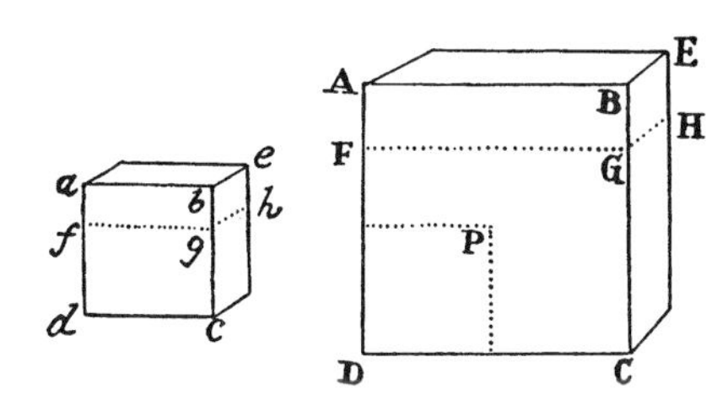

aa. In ed. 1 the order of the two sentences is reversed.

the smaller cube as *ab* to AB, that is, inversely as the distances between the particles are to one another. Q.E.D.

And conversely, if the forces of the individual particles are inversely as the distances, that is, inversely as the edges AB and *ab* of the cubes, the sums of the forces will be in the same ratio, and the pressures of the sides DB and *db* will be as the sums of the forces; and the pressure of the square DP will be to the pressure of the side DB as ab^2 to AB^2. And from the equality of the ratios [or ex aequo] the pressure of the square DP will be to the pressure of the side *db* as ab^3 to AB^3; that is, the one force of compression will be to the other force of compression as the one density to the other density. Q.E.D.

Scholium By a similar argument, if the centrifugal forces [or forces of repulsion] of the particles are inversely as the squares of the distances between the centers, the cubes of the compressing forces will be as the fourth powers of the densities. If the centrifugal forces are inversely as the third or fourth powers of the distances, the cubes of the compressing forces will be as the fifth or sixth powers of the densities. And universally, if D is the distance, and E the density of the compressed fluid, and if the centrifugal forces are inversely as any power of the distance D^n, whose index is the number n, then the compressing forces will be as the cube roots of the powers E^{n+2}, whose index is the number $n + 2$; and conversely. In all of this, it is supposed that the centrifugal forces of particles are terminated in the particles which are next to them or do not extend far beyond them. We have an example of this in magnetic bodies. Their attractive virtue [or power] is almost terminated in bodies of their own kind which are next to them. The virtue of a magnet is lessened by an interposed plate of iron and is almost terminated in the plate. For bodies farther away are drawn not so much by the magnet as by the plate. In the same way, if particles repel other particles of their own kind that are next to them but do not exert any virtue upon more remote particles,[b] particles of this sort are the ones of which the fluids treated in this proposition will be composed. But if the virtue of each particle is propagated indefinitely, a greater force will be necessary for the equal condensation of a

b. Ed. 1 has in addition: "except perhaps through the increase of the intermediate particles by that virtue."

greater quantity of the fluid.[c] Whether elastic fluids consist of particles that repel one another is, however, a question for physics. We have mathematically demonstrated a property of fluids consisting of particles of this sort so as to provide natural philosophers with the means with which to treat that question.

c. Ed. 1 has in addition: "For example, if each particle by its own force, which is inversely as the distance of places from its center, repels all other particles indefinitely, the forces by which the fluid can be equally compressed and condensed in similar vessels will be as the squares of the diameters of the vessels, and thus the force by which the fluid is compressed in the same vessel will be inversely as the cube root of the fifth power of the density."

SECTION 6

Concerning the motion of [a]*simple pendulums*[a] *and the resistance to them*

Proposition 24
Theorem 19

In simple pendulums whose centers of oscillation are equally distant from the center of suspension, the quantities of matter are in a ratio compounded of the ratio of the weights and the squared ratio of the times of oscillation in a vacuum.

For the velocity that a given force can generate in a given time in a given quantity of matter is as the force and the time directly and the matter inversely. The greater the force, or the greater the time, or the less the matter, the greater the velocity that will be generated. This is manifest from the second law of motion. Now if the pendulums are of the same length, the motive forces in places equally distant from the perpendicular are as the weights; and thus if two oscillating bodies describe equal arcs and if the arcs are divided into equal parts, then, since the times in which the bodies describe single corresponding parts of the arcs are as the times of the whole oscillations, the velocities in corresponding parts of the oscillations will be to one another as the motive forces and the whole times of the oscillations directly and the quantities of matter inversely; and thus the quantities of matter will be as the forces and the times of the oscillations directly and the velocities inversely. But the velocities are inversely as the times, and thus the times are directly, and the velocities are inversely, as the squares of the times, and therefore the quantities of matter are as the motive forces and the squares of the times, that is, as the weights and the squares of the times. Q.E.D.

COROLLARY 1. And thus if the times are equal, the quantities of matter in the bodies will be as their weights.

COROLLARY 2. If the weights are equal, the quantities of matter will be as the squares of the times.

COROLLARY 3. If the quantities of matter are equal, the weights will be inversely as the squares of the times.

COROLLARY 4. Hence, since the squares of the times, other things being equal, are as the lengths of the pendulums, the weights will be as the lengths of the pendulums if both the times and the quantities of matter are equal.

aa. Newton uses the term "corpora funependula," literally "bodies hanging by a thread [or string]," which we have translated as "simple pendulums"; see the Guide, §7.5.

Corollary 5. And universally, the quantity of matter in a bob of a simple pendulum is as the weight and the square of the time directly and the length of the pendulum inversely.

Corollary 6. But in a nonresisting medium also, the quantity of matter in the bob of a simple pendulum is as the relative weight and the square of the time directly and the length of the pendulum inversely. For the relative weight is the motive force of a body in any heavy medium, as I have explained above, and thus fulfills the same function in such a nonresisting medium as absolute weight does in a vacuum.

Corollary 7. And hence a method is apparent both for comparing bodies with one another with respect to the quantity of matter in each, and for comparing the weights of one and the same body in different places in order to find out the variation in its gravity. And by making experiments of the greatest possible accuracy, I have always found that the quantity of matter in individual bodies is proportional to the weight.

Proposition 25
Theorem 20

The bobs of simple pendulums that are resisted in any medium in the ratio of the moments of time, and those that move in a nonresisting medium of the same specific gravity, perform oscillations in a cycloid in the same time and describe proportional parts of arcs in the same time.

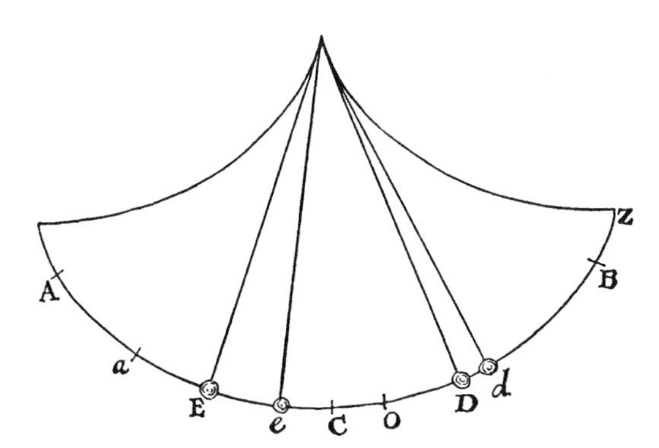

Let AB be the arc of a cycloid, which body D describes by oscillating in a nonresisting medium in any time. Bisect the arc AB in C so that C is its lowest point; then the accelerative force by which the body is urged in any place D or *d* or E will be as the length of arc CD or C*d* or CE. Represent that force by the appropriate arc [CD or C*d* or CE], and since

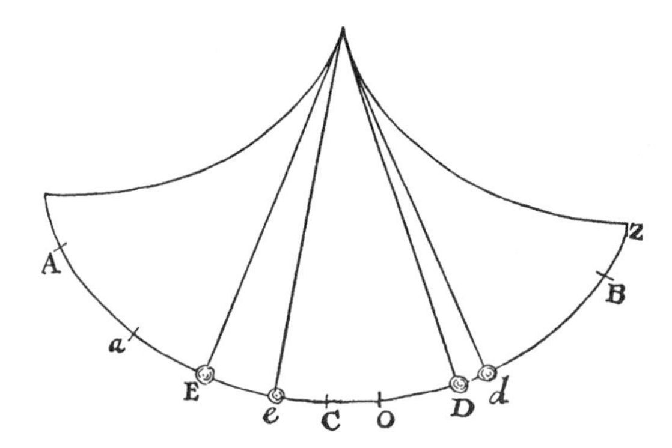

the resistance is as the moment of time, and thus is given, represent it by a given part CO of the arc of the cycloid, taking arc O*d* in the ratio to arc CD that arc OB has to arc CB; then the force by which the body at *d* is urged in the resisting medium (since it is the excess of the force C*d* over the resistance CO) will be represented by arc O*d*, and thus will be to the force by which body D is urged in a nonresisting medium in place D as arc O*d* to arc CD, and therefore also in place B as arc OB to arc CB. Accordingly, if two bodies D and *d* leave place B and are urged by these forces, then, since the forces at the beginning are as arcs CB and OB, the first velocities and the arcs first described will be in the same ratio. Let those arcs be DO and B*d*; then the remaining arcs CD and O*d* will be in the same ratio. Accordingly the forces, being proportional to CD and O*d*, will remain in the same ratio as at the beginning, and therefore the bodies will proceed simultaneously to describe arcs in the same ratio. Therefore the forces and the velocities and the remaining arcs CD and O*d* will always be as the whole arcs CB and OB, and therefore those remaining arcs will be described simultaneously. Therefore the two bodies D and *d* will arrive simultaneously at places C and O, the one in the nonresisting medium at place C, and the one in the resisting medium at place O. And since the velocities in C and O are as arcs CB and OB, the arcs that the bodies describe in the same time by going on further will be in the same ratio. Let those arcs be CE and O*e*. The force by which body D in the nonresisting medium is retarded in E is as CE, and the force by which body *d* in the resisting medium is retarded in *e* is as the sum of the force C*e* and the resistance CO, that is, as O*e*; and thus the forces by which the bodies are retarded are as arcs CB and OB, which are proportional to arcs CE and O*e*; and accordingly the velocities, which are retarded in that given ratio, remain in that same given ratio. The velocities, therefore, and the arcs described with those velocities are always to one another in the given ratio of arcs CB and OB; and therefore, if the whole arcs AB and *a*B are taken in the same ratio, bodies D and *d* will describe these arcs together and will simultaneously lose all motion in places A and *a*. The whole oscillations are therefore isochronal, and any parts of the arcs, BD and B*d* or BE and B*e*, that are described in the same time are proportional to the whole arcs BA and B*a*. Q.E.D.

COROLLARY. Therefore the swiftest motion in the resisting medium does not occur at the lowest point C, but is found in that point O by which *a*B, the whole arc described, is bisected. And the body, proceeding from that point to *a*, is retarded at the same rate by which it was previously accelerated in its descent from B to O.

Proposition 26
Theorem 21

If simple pendulums are resisted in the ratio of the velocities, their oscillations in a cycloid are isochronal.

For if two oscillating bodies equally distant from the centers of suspension describe unequal arcs and if the velocities in corresponding parts of the arcs are to one another as the whole arcs, then the resistances, being proportional to the velocities, will also be to one another as the same arcs. Accordingly, if these resistances are taken away from (or added to) the motive forces arising from gravity, which are as the same arcs, the differences (or sums) will be to one another in the same ratio of the arcs; and since the increments or decrements of the velocities are as these differences or sums, the velocities will always be as the whole arcs. Therefore, if in some one case the velocities are as the whole arcs, they will always remain in that ratio. But in the beginning of the motion, when the bodies begin to descend and to describe those arcs, the forces—since they are proportional to the arcs—will generate velocities proportional to the arcs. Therefore the velocities will always be as the whole arcs to be described, and therefore those arcs will be described in the same time. Q.E.D.

Proposition 27
Theorem 22

If simple pendulums are resisted as the squares of the velocities, the differences between the times of the oscillations in a resisting medium and the times of the oscillations in a nonresisting medium of the same specific gravity will be very nearly proportional to the arcs described during the oscillations.

For let the unequal arcs A and B be described by equal pendulums in a resisting medium; then the resistance to the body in arc A will be to the resistance to the body in the corresponding part of arc B very nearly in the squared ratio of the velocities, that is, as A^2 to B^2. If the resistance in arc B were to the resistance in arc A as AB to A^2, the times in arcs A and B would be equal, by the previous proposition. And thus the resistance A^2 in arc A, or AB in arc B, produces an excess of time in arc A over the time in a nonresisting medium; and the resistance B^2 produces an excess of time in arc B over the time in a nonresisting medium. And those excesses are very nearly as the forces AB and B^2 that produce them, that is, as arcs A and B. Q.E.D.

COROLLARY 1. Hence from the times of the oscillations made in a resisting medium in unequal arcs, the times of the oscillations in a nonresisting medium of the same specific gravity can be found. For the difference between these times will be to the excess of the time in the smaller arc over the time in the nonresisting medium as the difference between the arcs is to the smaller arc.

COROLLARY 2. Shorter oscillations are more isochronal, and the shortest are performed in very nearly the same times as in a nonresisting medium. In fact, the times of those that are made in greater arcs are a little greater, because the resistance in the descent of the body (by which the time is prolonged) is greater in proportion to the length described in the descent than the resistance in the subsequent ascent (by which the time is shortened). But also the time of short as well as long oscillations seems to be somewhat prolonged by the motion of the medium. For retarded bodies are resisted a little less in proportion to the velocity, and accelerated bodies a little more, than those that progress uniformly; and this is so because the medium, going in the same direction as the bodies with the motion that it has received from them, is in the first case more agitated, in the second less, and accordingly concurs to a greater or to a less degree with the moving bodies. The medium therefore resists the pendulums more in the descent, and less in the ascent, than in proportion to the velocity, and the time is prolonged as a result of both causes.

Proposition 28
Theorem 23

If a simple pendulum oscillating in a cycloid is resisted in the ratio of the moments of time, its resistance will be to the force of gravity as the excess of the arc described in the whole descent over the arc described in the subsequent ascent is to twice the length of the pendulum.

Let BC designate the arc described in the descent, C*a* the arc described in the ascent, and A*a* the difference between the arcs; then, with the same constructions and proofs as in prop. 25, the force by which the oscillating body is urged in any place D will be to the force of resistance as arc CD to arc CO, which is half of that difference A*a*. And thus the force by which the

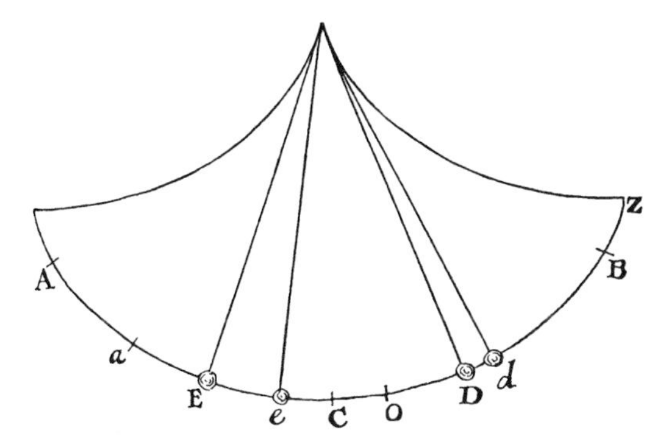

oscillating body is urged in the beginning (or highest point) of the cycloid—that is, the force of gravity—will be to the resistance as the arc of the cycloid between that highest point and the lowest point C is to arc CO, that is (if the arcs are doubled), as the arc of the whole cycloid, or twice the length of the pendulum, is to arc A*a*. Q.E.D.

Proposition 29
Problem 6

Supposing that a body oscillating in a cycloid is resisted as the square of the velocity, it is required to find the resistance in each of the individual places.

Let B*a* be the arc described in an entire oscillation, and let C be the lowest point of the cycloid, and let CZ be half of the arc of the whole cycloid and be equal to the length of the pendulum; and let it be required to find the resistance to the body in any place D. Cut the indefinite straight line OQ in points O, S, P, and Q, with the conditions that—if perpendiculars OK, ST, PI, and QE are erected; and if, with center O and asymptotes OK and OQ, hyperbola TIGE is described so as to cut perpendiculars ST, PI, and QE in T, I, and E; and if, through point I, KF is drawn parallel to asymptote OQ and meeting asymptote OK in K and perpendiculars ST and QE in L and F—the hyperbolic area PIEQ is to the hyperbolic area PITS as the arc BC described during the body's descent is to the arc C*a* described during the

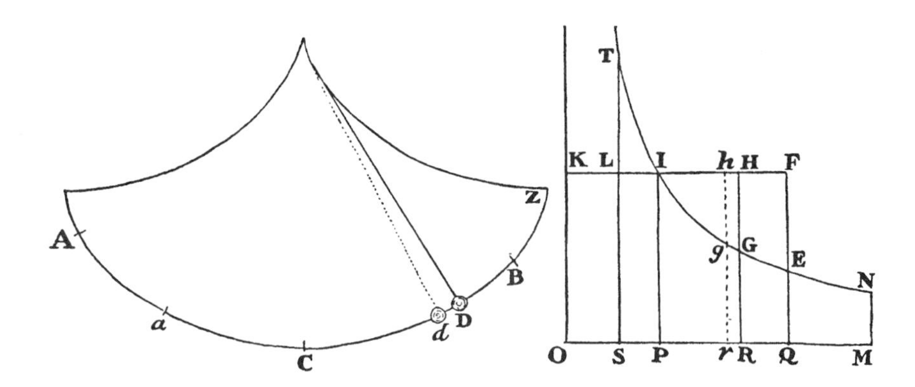

ascent, and area IEF is to area ILT as OQ to OS. Then let perpendicular MN cut off the hyperbolic area PINM, which is to the hyperbolic area PIEQ as arc CZ is to the arc BC described in the descent. And if perpendicular RG cuts off the hyperbolic area PIGR, which is to area PIEQ as any arc CD is to the arc BC described in the whole descent, then the resistance in place D will be to the force of gravity as the area $\frac{\text{OR}}{\text{OQ}}\text{IEF} - \text{IGH}$ to the area PINM.

For, since the forces which arise from gravity and by which the body is urged in places Z, B, D, and *a* are as arcs CZ, CB, CD, and C*a*, and those arcs are as areas PINM, PIEQ, PIGR, and PITS, let the arcs and the forces be represented by these areas respectively. In addition, let D*d* be the minimally small space described by the body while descending, and represent it by the minimally small area RG*gr* comprehended between the parallels RG and *rg*; and produce *rg* to *h*, so that GH*hg* and RG*gr* are decrements of areas IGH and PIGR made in the same time. And the increment $\text{GH}hg - \frac{\text{R}r}{\text{OQ}}\text{IEF}$, or $\text{R}r \times \text{HG} - \frac{\text{R}r}{\text{OQ}}\text{IEF}$, of area $\frac{\text{OR}}{\text{OQ}}\text{IEF} - \text{IGH}$ will be to the decrement RG*gr*, or $\text{R}r \times \text{RG}$, of area PIGR as $\text{HG} - \frac{\text{IEF}}{\text{OQ}}$ is to RG, and thus as $\text{OR} \times \text{HG} - \frac{\text{OR}}{\text{OQ}}\text{IEF}$ is to $\text{OR} \times \text{GR}$ or $\text{OP} \times \text{PI}$, that is (because $\text{OR} \times \text{HG}$, or $\text{OR} \times \text{HR} - \text{OR} \times \text{GR}$, $\text{ORHK} - \text{OPIK}$, PIHR, and $\text{PIGR} + \text{IGH}$ are equal), as $\text{PIGR} + \text{IGH} - \frac{\text{OR}}{\text{OQ}}\text{IEF}$ is to OPIK. Therefore, if area $\frac{\text{OR}}{\text{OQ}}\text{IEF} - \text{IGH}$ is called Y, and if the decrement RG*gr* of area PIGR is given, then the increment of area Y will be as $\text{PIGR} - \text{Y}$.

But if V designates the force arising from gravity, by which the body is urged in D and which is proportional to the arc CD to be described, and if R represents the resistance, then V − R will be the whole force by which the body is urged in D. The increment of the velocity is therefore jointly as V−R and that particle of time in which the increment is made. But furthermore the velocity itself is directly as the increment of the space described in the same time and inversely as that same particle of time. Hence, since the resistance, by hypothesis, is as the square of the velocity, the increment of the resistance (by lem. 2) will be as the velocity and the increment of the velocity jointly, that is, as the moment of the space and V−R jointly; and thus, if the moment of the space is given, as V − R; that is, as PIGR − Z, if for the force V there is written PIGR (which represents it), and if the resistance R is represented by some other area Z.

Therefore, as area PIGR decreases uniformly by the subtraction of the given moments, area Y increases in the ratio of PIGR − Y, and area Z increases in the ratio of PIGR − Z. And therefore, if areas Y and Z begin simultaneously and are equal at the beginning, they will continue to be equal by the addition of equal moments and, thereafter decreasing by moments that are likewise equal, will vanish simultaneously. And conversely, if they begin simultaneously and vanish simultaneously, they will have equal moments and will always be equal; and this is so because, if the resistance Z is increased, the velocity will be decreased along with that arc C*a* which is described in the body's ascent, and as the point in which there is a cessation of all motion and resistance approaches closer to point C, the resistance will vanish more quickly than area Y. And the contrary will happen when the resistance is decreased.

Now area Z begins and ends where the resistance is nil, that is, in the beginning of the motion where arc CD is equal to arc CB and the straight line RG falls upon the straight line QE, and in the end of the motion where arc CD is equal to arc C*a* and RG falls upon the straight line ST. And area Y or $\frac{\text{OR}}{\text{OQ}}\text{IEF} - \text{IGH}$ begins and ends where the resistance is nil, and thus where $\frac{\text{OR}}{\text{OQ}}\text{IEF}$ and IGH are equal; that is (by construction), where the straight line RG falls successively upon the straight lines QE and ST. And accordingly those areas begin simultaneously and vanish simultaneously and

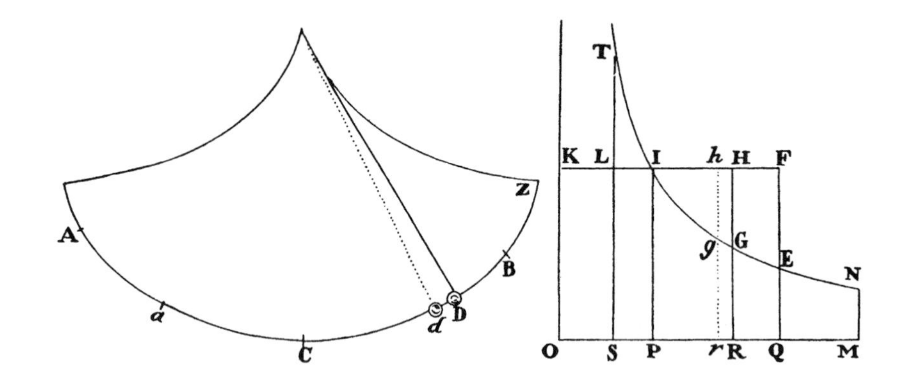

therefore are always equal. Therefore area $\frac{OR}{OQ}IEF - IGH$ is equal to area Z (which represents the resistance) and therefore is to area PINM (which represents the gravity) as the resistance is to the gravity. Q.E.D.

COROLLARY 1. The resistance in the lowest place C is, therefore, to the force of gravity as area $\frac{OP}{OQ}IEF$ is to area PINM.

COROLLARY 2. And this resistance becomes greatest when area PIHR is to area IEF as OR is to OQ. For in that case its moment (namely, PIGR − Y) comes out nil.

COROLLARY 3. Hence also the velocity in each of the individual places can be known, inasmuch as it is as the square root of the resistance, and at the very beginning of the motion is equal to the velocity of the body oscillating without any resistance in the same cycloid.

But because the computation by which the resistance and velocity are to be found by this proposition is difficult, it seemed appropriate to add the following proposition.[a]

Proposition 30
Theorem 24

If the straight line aB *is equal to a cycloidal arc that is described by an oscillating body, and if at each of its individual points* D *perpendiculars* DK *are erected that are to the length of the pendulum as the resistance encountered by the body in corresponding points of the arc is to the force of gravity, then I say that the difference between the arc described in the whole descent and the arc described in the whole subsequent ascent multiplied by half the sum of those same arcs will be equal to the area* BK*a occupied by all the perpendiculars* DK.

a. Ed. 1 and ed. 2 have in addition: "which is both more general and more than exact enough for use in natural philosophy."

Represent the cycloidal arc described in an entire oscillation by the straight line *a*B equal to it, and represent the arc that would be described in a vacuum by the length AB. Bisect AB in C, and point C will represent

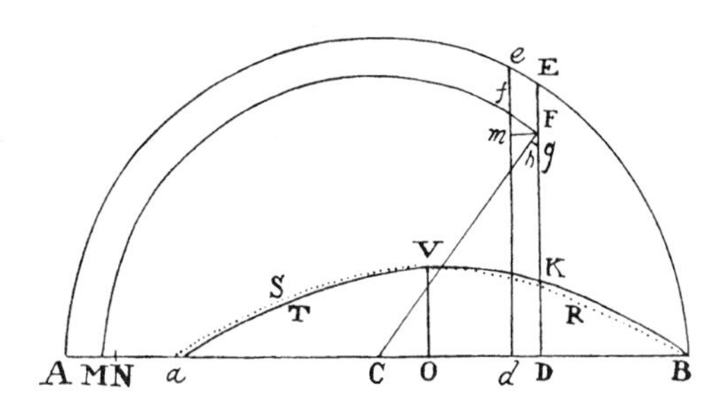

the lowest point of the cycloid, and CD will be as the force arising from gravity (by which the body at D is urged along the tangent of the cycloid) and will have the ratio to the length of the pendulum that the force at D has to the force of gravity. Therefore represent that force by the length CD, and the force of gravity by the length of the pendulum; then, if DK is taken in DE in the ratio to the length of the pendulum that the resistance has to the gravity, DK will represent the resistance. With center C and radius CA or CB construct semicircle BE*e*A. And let the body describe space D*d* in a minimally small time; then, when perpendiculars DE and *de* have been erected, meeting the circumference in E and *e*, they will be as the velocities that the body in a vacuum would acquire in places D and *d* by descending from point B. This is evident by book 1, prop. 52. Therefore represent these velocities by perpendiculars DE and *de*, and let DF be the velocity that the body acquires in D by falling from B in the resisting medium. And if with center C and radius CF circle F*f*M is described, meeting the straight lines *de* and AB in *f* and M, then M will be the place to which the body would then ascend if there were no further resistance, and *df* will be the velocity that it would acquire in *d*. Hence also, if F*g* designates the moment of velocity that body D, in describing the minimally small space D*d*, loses as a result of the resistance of the medium, and if CN is taken equal to C*g*, then N will be the place to which the body would then ascend if there were no further resistance, and MN will be the decrement of the ascent arising from the loss of that velocity. Drop perpendicular F*m* to *df*, and the decrement F*g* (generated by the resistance DK) of the velocity DF will be to the increment *fm* (generated by the force CD) of that same velocity as the generating force DK is to the generating force CD. Furthermore, because triangles F*mf*, F*hg*, and FDC are similar, *fm* is to F*m* or D*d* as CD is to DF and from the equality of the ratios [or ex aequo] F*g* is to D*d* as DK is to DF. [a]Likewise F*h* is to F*g* as DF to CF, and from the equality of the ratios in inordinate proportion [or ex aequo perturbate] F*h* or MN is to D*d* as DK to CF or CM; and thus the sum of all the MN × CM will be equal to the sum of all the D*d* × DK. Suppose that a rectangular ordinate is always erected at the moving point M, equal to the indeterminate CM, which in its continual motion is multiplied by the whole length A*a*; then the quadrilateral described as a result of that motion—or the rectangle equal to it, A*a* × ½*a*B—will become equal to the sum of all the MN × CM, and thus equal to the sum of all the D*d* × DK, that is, equal to area BKVT*a*. Q.E.D.[a]

COROLLARY. Hence from the law of the resistance and the difference A*a* of arcs C*a* and CB, the proportion of the resistance to the gravity can be determined very nearly.

For if the resistance DK is uniform, the figure BKT*a* will be equal to the rectangle of B*a* and DK; and hence the rectangle of ½B*a* and A*a* will be equal to the rectangle of B*a* and DK, and DK will be equal to ½A*a*. Therefore, since DK represents the resistance, and the length of the pendulum represents the gravity, the resistance will be to the gravity as ½A*a* is to the length of the pendulum, exactly as was proved in prop. 28.

aa. Ed. 1 has: "Likewise F*g* is to F*h* as CF to DF, and from the equality of the ratios in inordinate proportion [or ex aequo perturbate] F*h* or MN is to D*d* as DK to CF. Take DR to ½*a*B as DK to CF, and MN will be to D*d* as DR to ½*a*B, and thus the sum of all the MN × ½*a*B, that is, A*a* × ½*a*B, will be equal to the sum of all the D*d* × DR, that is, to area BR*r*S*a*, which all the rectangles D*d* × DR or DR*rd* compose. Bisect A*a* and *a*B in P and O, and ½*a*B or OB will be equal to CP, and thus DR is to DK as CP to CF or CM, and by separation [or dividendo] KR will be to DR as PM to CP. And thus, since point M, when the body is in the midpoint O of the oscillation, falls approximately on point P and in the earlier part of the oscillation is between A and P and in the later part is between P and *a*, in both cases deviating equally from point P in opposite directions, it follows that point K, at about the midpoint of the oscillation, that is, over against point O, say in point V, will fall on point R and in the earlier part of the oscillation will lie between R and E and in the later part between R and D, in both cases deviating equally from point R in opposite directions. Accordingly, the area which line KR describes will in the earlier part of the oscillation lie outside area BRS*a* and in the later part within it and will do so within ranges nearly equal to each other on each of the two sides and therefore, when added to area BRS*a* in the first case and subtracted from it in the second, will result in area BKT*a* very nearly equal to area BRS*a*. Therefore the rectangle A*a* × ½*a*B, or A*a*O, will, since it is equal to area BRS*a*, also be very nearly equal to area BKT*a*. Q.E.D."

If the resistance is as the velocity, the figure BKT*a* will be very nearly an ellipse. For if a body in a nonresisting medium were to describe the length BA in a whole oscillation, the velocity in any place D would be as the ordinate DE of a circle described with diameter AB. Accordingly, since B*a* in the resisting medium, and BA in a nonresisting medium, are described in roughly equal times, and the velocities in the individual points of B*a* are thus very nearly to the velocities in the corresponding points of the length BA as B*a* is to BA, the velocity in point D in the resisting medium will be as the ordinate of a circle or ellipse described upon diameter B*a*; and thus the figure BKVT*a* will be very nearly an ellipse. Since the resistance is supposed proportional to the velocity, let OV represent the resistance in the midpoint O; then ellipse BRVS*a*, described with center O and semiaxes OB and OV, will be very nearly equal to the figure BKVT*a* and the rectangle equal to it, A*a* × BO. A*a* × BO is therefore to OV × BO as the area of this ellipse is to OV × BO; that is, A*a* is to OV as the area of the semicircle is to the square of the radius, or as 11 to 7, roughly; and therefore $^7/_{11}$A*a* is to the length of the pendulum as the resistance of the oscillating body in O is to its gravity.

But if the resistance DK is as the square of the velocity, the figure BKVT*a* will be almost a parabola having vertex V and axis OV, and thus will be very nearly equal to the rectangle contained by $^2/_3$B*a* and OV. The rectangle contained by $^1/_2$B*a* and A*a* is therefore equal to the rectangle contained by $^2/_3$B*a* and OV, and thus OV is equal to $^3/_4$A*a*; and therefore the resistance on the oscillating body in O is to its gravity as $^3/_4$A*a* is to the length of the pendulum.

And I judge that these conclusions are more than accurate enough for practical purposes. For, since the ellipse or parabola BRVS*a* and the figure BKVT*a* have the same midpoint V, if it is greater than that figure on either side BRV or VS*a*, it will be smaller than it on the other side, and thus will be very nearly equal to it.

Proposition 31
Theorem 25

If the resistance encountered by an oscillating body in each of the proportional parts of the arcs described is increased or decreased in a given ratio, the difference between the arc described in the descent and the arc described in the subsequent ascent will be increased or decreased in the same ratio.

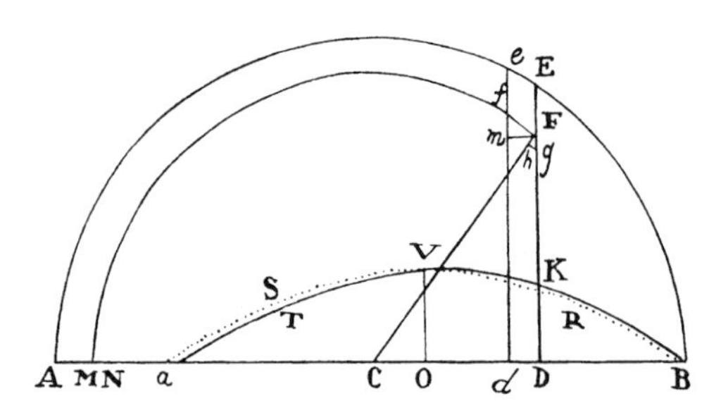

For that difference arises from the retardation of the pendulum by the resistance of the medium, and thus is as the whole retardation and the retarding resistance, which is proportional to it. In the previous proposition the rectangle contained under the straight line ½*a*B and the difference A*a* of arcs CB and C*a* was equal to area BKT*a*. And that area, if the length *a*B remains the same, is increased or decreased in the ratio of the ordinates DK, that is, in the ratio of the resistance, and thus is as the length *a*B and the resistance jointly. And accordingly the rectangle contained by A*a* and ½*a*B is as *a*B and the resistance jointly, and therefore A*a* is as the resistance. Q.E.D.

COROLLARY 1. Hence, if the resistance is as the velocity, the difference of the arcs in the same medium will be as the whole arc described; and conversely.

COROLLARY 2. If the resistance is in the squared ratio of the velocity, that difference will be in the squared ratio of the whole arc; and conversely.

COROLLARY 3. And universally, if the resistance is in the cubed or any other ratio of the velocity, the difference will be in the same ratio of the whole arc; and conversely.

COROLLARY 4. And if the resistance is partly in the simple ratio of the velocity and partly in the squared ratio of the velocity, the difference will be partly in the simple ratio of the whole arc and partly in the squared ratio of it; and conversely. The law and ratio of the resistance in relation to the velocity will be the same as the law and ratio of that difference of the arcs in relation to the length of the arc itself.

COROLLARY 5. And thus if a pendulum successively describes unequal arcs and there can be found the ratio of the increment and decrement of this difference [i.e., the difference of the arcs] in relation to the length of the arc described, then there will also be had the ratio of the increment and decrement of the resistance in relation to a greater or smaller velocity.

General Scholium[a]

From these propositions it is possible to find the resistance of any mediums by means of pendulums oscillating in those mediums. In fact, I have investigated the resistance of air by the following experiments. I suspended a wooden ball by a fine thread from a sufficiently firm hook in such a way that the distance between the hook and the center of oscillation of the ball was 10½ feet; the ball weighed 57⁷⁄₂₂ ounces avoirdupois and had a diameter of 6⅞ London inches. I marked a point on the thread 10 feet and 1 inch distant from the center of suspension; and at a right angle at that point I placed a ruler divided into inches, by means of which I might note the lengths of the arcs described by the pendulum. Then I counted the oscillations during which the ball would lose an eighth of its motion. When the pendulum was drawn back from the perpendicular to a distance of 2 inches and was let go from there, so as to describe an arc of 2 inches in its whole descent and to describe an arc of about 4 inches in the first whole oscillation (composed of the descent and subsequent ascent), it then lost an eighth of its motion in 164 oscillations, so as to describe an arc of 1¾ inches in its final ascent. When it described an arc of 4 inches in its first descent, it lost an eighth of its motion in 121 oscillations, so as to describe an arc of 3½ inches in its final ascent. When it described an arc of 8, 16, 32, or 64 inches in its first descent, it lost an eighth of its motion in 69, 35½, 18½, and 9⅔ oscillations respectively. Therefore the difference between the arcs described in the first descent and the final ascent, in the first, second, third, fourth, fifth, and sixth cases, was ¼, ½, 1, 2, 4, and 8 inches respectively. Divide these differences by the number of oscillations in each case, and in one mean oscillation—in which an arc of 3¾, 7½, 15, 30, 60, and 120 inches was described—the difference between the arcs described in the descent and subsequent ascent will be ¹⁄₆₅₆, ¹⁄₂₄₂, ¹⁄₆₉, ⁴⁄₇₁, ⁸⁄₃₇, and ²⁴⁄₂₉ parts of an inch respectively. In the greater oscillations, moreover, these are very nearly in the squared ratio of the arcs described, while in the smaller oscillations they are a little greater than in that ratio; and therefore (by book 2, prop. 31, corol. 2) the resistance of the ball when it moves more swiftly is very nearly in the squared ratio of the velocity; when more slowly, a little greater than in that ratio.

a. In ed. 1 the general scholium appears at the end of book 2, sec. 7, with some variations, primarily in the numerical values.

Now let V designate the greatest velocity in any oscillation, and let A, B, and C be given quantities, and let us imagine the difference between the arcs to be $AV+BV^{3/2}+CV^2$. In a cycloid the greatest velocities are as halves of the arcs described in oscillating, but in a circle they are as the chords of halves of these arcs, and thus with equal arcs are greater in a cycloid than in a circle in the ratio of halves of the arcs to their chords, while the times in a circle are greater than in a cycloid in the inverse ratio of the velocity. Hence it is evident that the differences between the arcs (differences which are as the resistance and the square of the time jointly) would be very nearly the same in both curves. For those differences in the cycloid would have to be increased along with the resistance in roughly the squared ratio of the arc to the chord (because the velocity is increased in the simple ratio of the arc to the chord) and would have to be decreased along with the square of the time in that same squared ratio. Therefore, in order to reduce all of this to the cycloid, take the same differences between the arcs that were observed in the circle, while supposing the greatest velocities to correspond to the arcs, whether halved or entire, that is, to the numbers ½, 1, 2, 4, 8, and 16. In the second, fourth, and sixth cases, therefore, let us write the numbers 1, 4, and 16 for V; and the difference between the arcs will come out $\frac{1/2}{121}=A+B+C$ in the second case; $\frac{2}{35\frac{1}{2}}=4A+8B+16C$ in the fourth case; and $\frac{8}{9\frac{2}{3}}=16A+64B+256C$ in the sixth case. And by the proper analytic reduction of these equations taken together, A becomes = 0.0000916, B = 0.0010847, and C = 0.0029558. The difference between the arcs is therefore as $0.0000916V+0.0010847V^{3/2}+0.0029558V^2$; and therefore—since (by prop. 30, corol., applied to this case) the resistance of the ball in the middle of the arc described by oscillating, where the velocity is V, is to its weight as $\frac{7}{11}AV+\frac{7}{10}BV^{3/2}+\frac{3}{4}CV^2$ is to the length of the pendulum—if the numbers found are written for A, B, and C, the resistance of the ball will become to its weight as $0.0000583V+0.0007593V^{3/2}+0.0022169V^2$ is to the length of the pendulum between the center of suspension and the ruler, that is, to 121 inches. Hence, since V in the second case has the value 1, in the fourth 4, and in the sixth 16, the resistance will be to the weight of the ball in the second case as 0.0030345 to 121, in the fourth as 0.041748 to 121, and in the sixth as 0.61705 to 121.

The arc which in the sixth case was described by the point marked on the thread was $120 - \frac{8}{9\frac{2}{3}}$ or $119\frac{5}{29}$ inches. And therefore, since the radius was 121 inches, and the length of the pendulum between the point of suspension and the center of the ball was 126 inches, the arc that the center of the ball described was $124\frac{3}{31}$ inches. Since, because of the resistance of the air, the greatest velocity of an oscillating body does not occur at the lowest point of the arc described but is located near the midpoint of the whole arc, that velocity will be roughly the same as if the ball in its whole descent in a nonresisting medium described half that arc ($62\frac{3}{62}$ inches) and did so in a cycloid, to which we have above reduced the motion of the pendulum; and therefore that velocity will be equal to the velocity which the ball could acquire by falling perpendicularly and describing in its fall a space equal to the versed sine of that arc. But that versed sine in the cycloid is to that arc ($62\frac{3}{62}$) as that same arc is to twice the length of the pendulum (252) and thus is equal to 15.278 inches. Therefore the velocity is the very velocity that the body could acquire by falling and describing in its fall a space of 15.278 inches. With such a velocity, then, the ball encounters a resistance that is to its weight as 0.61705 to 121, or (if only that part of the resistance is considered which is in the squared ratio of the velocity) as 0.56752 to 121.

By a hydrostatic experiment, I found that the weight of this wooden ball was to the weight of a globe of water of the same size as 55 to 97; and therefore, since 121 is to 213.4 in the same ratio as 55 to 97, the resistance of a globe of water moving forward with the above velocity will be to its weight as 0.56752 to 213.4, that is, as 1 to $376\frac{1}{50}$. The weight of the globe of water, in the time during which the globe describes a length of 30.556 inches with a uniformly continued velocity, could generate all that velocity in the globe if it were falling; hence it is manifest that in the same time the force of resistance uniformly continued could take away a velocity smaller in the ratio of 1 to $376\frac{1}{50}$, that is, $\frac{1}{376\frac{1}{50}}$ of the whole velocity. And therefore in the same time in which the globe, with that velocity uniformly continued, could describe the length of its own semidiameter, or $3\frac{7}{16}$ inches, it would lose $\frac{1}{3,342}$ of its motion.

I also counted the oscillations in which the pendulum lost a fourth of its motion. In the following table the top numbers denote the length of the arc described in the first descent, expressed in inches and parts of an inch; the middle numbers signify the length of the arc described in the final ascent; and at the bottom stand the numbers of oscillations. I have described this experiment because it is more accurate than when only an eighth of the motion was lost. Let anyone who wishes test the computation.

First descent	2	4	8	16	32	64
Final ascent	1½	3	6	12	24	48
Number of oscillations	374	272	162½	83⅓	41⅔	22⅔

Later, using the same thread, I suspended a lead ball with a diameter of 2 inches and a weight of 26¼ ounces avoirdupois, in such a way that the distance between the center of the ball and the point of suspension was 10½ feet, and I counted the oscillations in which a given part of the motion was lost. The first of the following tables shows the number of oscillations in which an eighth of the whole motion was lost; the second shows the number of oscillations in which a fourth of it was lost.

First descent	1	2	4	8	16	32	64
Final ascent	⅞	7/4	3½	7	14	28	56
Number of oscillations	226	228	193	140	90½	53	30
First descent	1	2	4	8	16	32	64
Final ascent	¾	1½	3	6	12	24	48
Number of oscillations	510	518	420	318	204	121	70

Select the third, fifth, and seventh observations from the first table and represent the greatest velocities in these particular observations by the numbers 1, 4, and 16 respectively, and generally by the quantity V as above; then it will be the case that in the third observation $\frac{\frac{1}{2}}{193} = A+B+C$, in the fifth $\frac{2}{90\frac{1}{2}} = 4A+8B+16C$, in the seventh $\frac{8}{30} = 16A+64B+256C$. The reduction of these equations gives $A = 0.001414$, $B = 0.000297$, $C = 0.000879$. Hence the resistance of the ball moving with velocity V comes out to have the ratio to its own weight (26¼ ounces) that $0.0009V + 0.000208V^{3/2} + 0.000659V^2$ has to the pendulum's length (121 inches). And if we consider only that part of the resistance which is in the squared ratio of the velocity, it will be to the weight of the ball as $0.000659V^2$ is to 121 inches. But in the first experiment this part of the resistance was to the weight of the wooden ball (57 7/22 ounces) as $0.002217V^2$ to 121; and hence the resistance of the wooden ball becomes to the resistance of the lead ball (their velocities being equal) as $57\frac{7}{22} \times 0.002217$ to $26\frac{1}{4} \times 0.000659$, that is, as $7\frac{1}{3}$ to 1. The diameters of the two balls were $6\frac{7}{8}$ and 2 inches, and the squares of these are to each other as $47\frac{1}{4}$ and 4, or $11\frac{13}{16}$ and 1, very nearly. Therefore the resistances of equally swift balls were in a smaller ratio than the squared ratio of the diameters. But we have not yet considered the resistance of the thread, which certainly was very great and ought to be subtracted from the resistance of the pendulum that has been found. I could not determine this resistance of the thread accurately, but nevertheless I found it to be greater than a third of the whole resistance of the smaller pendulum; and I learned from this that the resistances of the balls, taking away the resistance of the thread, are very nearly in the squared ratio of the diameters. For the ratio of $7\frac{1}{3} - \frac{1}{3}$ to $1 - \frac{1}{3}$, or $10\frac{1}{2}$ to 1, is very close to the squared ratio of the diameters $11\frac{13}{16}$ to 1.

Since the resistance of the thread is of less significance in larger balls, I also tried the experiment in a ball whose diameter was $18\frac{3}{4}$ inches. The length of the pendulum between the point of suspension and the center of oscillation was $122\frac{1}{2}$ inches; between the point of suspension and a knot in the thread, $109\frac{1}{2}$ inches. The arc described by the knot in the first descent of the pendulum was 32 inches. The arc described by that same knot in the final ascent after five oscillations was 28 inches. The sum of the arcs, or the whole arc described in a mean oscillation, was 60 inches. The difference between the arcs was 4 inches. A tenth of it, or the difference between the descent and the ascent in a mean oscillation, was $\frac{2}{5}$ inch. The ratio of the radius $109\frac{1}{2}$ to the radius $122\frac{1}{2}$ is the same as the ratio of the whole arc of 60 inches described by the knot in a mean oscillation to the whole arc of $67\frac{1}{8}$ inches described by the center of the ball in a mean oscillation, and is equal to the ratio of the difference $\frac{2}{5}$ to the new difference 0.4475. If the length of the pendulum were to be increased in the ratio of 126 to $122\frac{1}{2}$ while the length of the arc described remained the same, the time of oscillation would be increased and the velocity of the pendulum would be decreased as the square root of that ratio, while the difference 0.4475 between the arcs described in a descent and subsequent ascent would remain the same. Then, if the arc described were to be increased in the ratio of $124\frac{3}{31}$ to $67\frac{1}{8}$, that difference 0.4475 would be increased as the square of that ratio, and thus would come out 1.5295. These things would be so on the hypothesis that the resistance of the pendulum was in the squared ratio of the velocity. Therefore, if the pendulum were to describe a whole arc of $124\frac{3}{31}$ inches, and its length between the point of suspension and the center of oscillation were 126 inches, the difference between the arcs described in a descent and subsequent ascent would be 1.5295 inches. And this difference multiplied by the weight of the ball of the pendulum, which was 208 ounces, yields the product 318.136. Again, when the above-mentioned pendulum (made with a wooden ball) described a whole arc of $124\frac{3}{31}$ inches by its center of oscillation (which was 126 inches distant from the point of suspension), the difference between the arcs described in the descent and ascent was $\frac{126}{121} \times \frac{8}{9\frac{2}{3}}$, which multiplied by the weight of the ball (which was $57\frac{7}{22}$ ounces) yields the product 49.396. And I multiplied these differences by the weights of the balls in order to find their resistances. For the differences arise from the resistances and are as the resistances directly and the weights inversely. The resistances are therefore as the numbers 318.136 and 49.396. But the part of the resistance of the smaller ball that is in the squared ratio of the velocity was to the whole resistance as 0.56752 to 0.61675, that is, as 45.453 to 49.396; and the similar part of the resistance of the larger ball is almost equal to its whole resistance; and thus those parts are very nearly as 318.136 and 45.453, that is, as 7 and 1. But the diameters of the balls are $18\frac{3}{4}$ and $6\frac{7}{8}$, and the squares of these diameters, $351\frac{9}{16}$ and $47\frac{17}{64}$, are as 7.438 and 1, that is, very nearly as the resistances 7 and 1 of the balls. The difference between the ratios is no greater than what could have arisen from the resistance of the thread. Therefore, those parts of the resistances that are (the balls being equal) as the squares of the velocities are also (the velocities being equal) as the squares of the diameters of the balls.

The largest ball that I used in these experiments, however, was not perfectly spherical, and therefore for the sake of brevity I have ignored certain minutiae in the above computation, being not at all worried about a computation being exact when the experiment itself was not sufficiently exact. Therefore, since the demonstration of a vacuum depends on such experiments, I wish that they could be tried with more, larger, and more exactly spherical balls. If the balls are taken in geometric proportion, say with diameters of 4, 8, 16, and 32 inches, it will be discovered from the progression of the experiments what ought to happen in the case of still larger balls.

To compare the resistances of different fluids with one another, I made the following experiments. I got a wooden box four feet long, one foot wide, and one foot deep. I took off its lid and filled it with fresh water, and I immersed pendulums in the water and made them oscillate. A lead ball weighing 166⅙ ounces, with a diameter of 3⅝ inches, moved as in the following table, that is, with the length of the pendulum from the point of suspension to a certain point marked on the thread being 126 inches, and to the center of oscillation being 134⅜ inches.

Arc described by the point marked on the thread in the first descent	64″	32″	16″	8″	4″	2″	1″	½″	¼″
Arc described in the final ascent	48″	24″	12″	6″	3″	1½″	¾″	⅜″	3/16″
Difference between the arcs, proportional to the motion lost	16″	8″	4″	2″	1″	½″	¼″	⅛″	1/16″
Number of oscillations in water			29/60	1⅕	3	7	11¼	12⅔	13⅓
Number of oscillations in air	85½	287	535						

In the experiment recorded in the fourth column, equal motions were lost in 535 oscillations in air, and 1⅕ in water. The oscillations were indeed a little quicker in air than in water. But if the oscillations in water were accelerated in such a ratio that the motions of the pendulums in both mediums would become equally swift, the number 1⅕ oscillations in water during which the same motion would be lost as before would remain the same because the resistance is increased and the square of the time simultaneously decreased in that same ratio squared. With equal velocities of the pendulums, therefore, equal motions were lost, in air in 535 oscillations and in water in 1⅕ oscillations; and thus the resistance of the pendulum in water is to its resistance in air as 535 to 1⅕. This is the proportion of the whole resistances in the case of the fourth column.

Now let $AV + CV^2$ designate the difference between the arcs described (in a descent and subsequent ascent) by the ball moving in air with the greatest velocity V; and since the greatest velocity in the case of the fourth column is to the greatest velocity in the case of the first column as 1 to 8, and since that difference between the arcs in the case of the fourth column is to the difference in the case of the first column as $\frac{2}{535}$ to $\frac{16}{85\frac{1}{2}}$, or as 85½ to 4,280, let us write 1 and 8 for the velocities in these cases and 85½ and 4,280 for the differences between the arcs; then $A + C$ will become $= 85\frac{1}{2}$

and $8A + 64C = 4,280$ or $A + 8C = 535$; and hence, by reduction of the equations, 7C will become $= 449\frac{1}{2}$ and $C = 64\frac{3}{14}$ and $A = 21\frac{2}{7}$; and thus the resistance, since it is as $\frac{7}{11}AV + \frac{3}{4}CV^2$, will be as $13\frac{6}{11}V + 48\frac{9}{56}V^2$. Therefore, in the case of the fourth column, where the velocity was 1, the whole resistance is to its part proportional to the square of the velocity as $13\frac{6}{11} + 48\frac{9}{56}$ or $61\frac{12}{17}$ to $48\frac{9}{56}$; and on that account the resistance of the pendulum in water is to that part of the resistance in air which is proportional to the square of the velocity (and which alone comes into consideration in swifter motions) as $61\frac{12}{17}$ to $48\frac{9}{56}$ and 535 to $1\frac{1}{5}$ jointly, that is, as 571 to 1. If the whole thread of the pendulum oscillating in water had been immersed, its resistance would have been still greater, to such an extent that the part of the resistance of the pendulum oscillating in water which is proportional to the square of the velocity (and which alone comes into consideration in swifter bodies) is to the resistance of that same whole pendulum oscillating in air, with the same velocity, as about 850 to 1, that is, very nearly as the density of water to the density of air.

In this computation also, that part of the resistance of the pendulum in water which would be as the square of the velocity ought to be taken into consideration, but (which may perhaps seem strange) the resistance in water was increased in more than the squared ratio of the velocity. In searching for the reason, I hit upon this: that the box was too narrow in proportion to the size of the ball of the pendulum, and because of its narrowness overly impeded the motion of the water as it yielded to the oscillation of the ball. For if a ball of a pendulum whose diameter was one inch was immersed, the resistance was increased in very nearly the squared ratio of the velocity. I tested this by constructing a pendulum out of two balls, so that the lower and smaller of them oscillated in the water, and the higher and larger one was fastened to the thread just above the water and, by oscillating in the air, aided the pendulum's motion and made it last longer. And the experiments made with this pendulum came out as in the following table.

Arc described in the first descent	16″	8″	4″	2″	1″	½″	¼″
Arc described in the final ascent	12″	6″	3″	1½″	¾″	⅜″	$\frac{3}{16}$″
Difference between the arcs, proportional to the motion lost	4″	2″	1″	½″	¼″	⅛″	$\frac{1}{16}$″
Number of oscillations	$3\frac{3}{8}$	$6\frac{1}{2}$	$12\frac{1}{12}$	$21\frac{1}{5}$	34	53	$62\frac{1}{5}$

In comparing the resistances of the mediums with one another I also caused iron pendulums to oscillate in quicksilver. The length of the iron wire was about three feet, and the diameter of the ball of the pendulum was about ⅓ inch. And to the wire just above the mercury there was fastened another lead ball large enough to continue the motion of the pendulum for a longer time. Then I filled a small vessel (which held about three pounds of quicksilver) with quicksilver and common water successively, so that as the pendulum oscillated first in one and then in the other of the two fluids I might find the proportion of the resistances; and the resistance of the quicksilver came out to the resistance of the water as about 13 or 14 to 1, that is, as the density of quicksilver to the density of water. When I used [b]a slightly larger pendulum ball, say one whose diameter would be about ⅓ or ⅔ inch,[b] the resistance of the quicksilver came out in the ratio to the resistance of the water that the number 12 or 10 has to 1, roughly. But the former experiment is more trustworthy because in the latter the vessel was too narrow in proportion to the size of the immersed ball. With the ball enlarged, the vessel also would have to be enlarged. Indeed, I had determined to repeat experiments of this sort in larger vessels and in molten metals and certain other liquids, hot as well as cold; but there is not time to try them all, and from what has already been described it is clear enough that the resistance of bodies moving swiftly is very nearly proportional to the density of the fluids in which they move. I do not say exactly proportional. For the more viscous fluids, of an equal density, doubtless resist more than the more liquid fluids—as, for example, cold oil more than hot, hot oil more than rainwater, water more than spirit of wine. But in the liquids that are sufficiently fluid to the senses—as in air, in water (whether fresh or salt), in spirits of wine, of turpentine, and of salts, in oil freed of its dregs by distillation and then heated, and in oil of vitriol and in mercury, and in liquefied metals, and any others there may be which are so fluid that when agitated in vessels they conserve for some time a motion impressed upon them and when poured out are quite freely broken

bb. Here Newton makes a puzzling statement, namely, that the diameter of this ball, "about ⅓ or ⅔ inch," was larger than the one mentioned earlier, which was "about ⅓ inch." The source of this puzzling "about ⅓ or ⅔ inch" may be seen by comparing the various editions, as is done in our Latin edition. In the printer's manuscript and in ed. 1, the larger ball is said to have a diameter of "about ½ or ⅔ inch," which in ed. 2 was wrongly printed as "about ⅓ or ⅔ inch." In Newton's annotated copy of the *Principia*, it was noted that this should be corrected to "about ½ or ⅔ inch," but this was not done in ed. 3.

up into falling drops—in all these I have no doubt that the above rule holds exactly enough, especially if the experiments are made with pendulums that are larger and move more swiftly.

Finally, since [c]some people are of the opinion[c] that there exists a certain aethereal medium, by far the subtlest of all, which quite freely permeates all the pores and passages of all bodies, and that a resistance ought to arise from such a medium flowing through the pores of bodies, I devised the following experiment so that I might test whether the resistance that we experience in moving bodies is wholly on their external surface or whether the internal parts also encounter a perceptible resistance on their own surfaces. I suspended a round firwood box by a cord eleven feet long from a sufficiently strong steel hook, by means of a steel ring. The upper arc of the ring rested on the very sharp concave edge of the hook so that it might move very freely. And the cord was attached to the lower arc of the ring. I drew this pendulum away from the perpendicular to a distance of about six feet, and did so along the plane perpendicular to the edge of the hook, so that the ring, as the pendulum oscillated, would not slide to and fro on the edge of the hook. For the point of suspension, in which the ring touches the hook, ought to remain motionless. I marked the exact place to which I had drawn back the pendulum and then, letting the pendulum fall, marked another three places: those to which it returned at the end of the first, second, and third oscillations. I repeated this quite often, so that I might find those places as exactly as possible. Then I filled the box with lead and some of the other heavier metals that were at hand. But first I weighed the empty box along with the part of the cord that was wound around the box and half of the remaining part that was stretched between the hook and the suspended box. For a stretched cord always urges with half of its weight a pendulum drawn aside from the perpendicular. To this weight I added the weight of the air that the box contained. And the whole weight was about 1/78 of the box full of metals. Then, since the box full of metals increased the length of the pendulum as a result of stretching the cord by its weight, I shortened the

cc. This reads literally: "the opinion of some is." Ed. 1 and ed. 2 have: "the most widely accepted opinion of the philosophers of this age is." The index prepared by Cotes for ed. 2 and retained for ed. 3 keys this opinion under "Materia" ("Matter") and specifies the "philosophers" (and hence the later "some") by thus describing the paragraph: "The subtle matter of the Cartesians is subjected to a certain examination."

cord so that the length of the pendulum now oscillating would be the same as before. Then, drawing the pendulum back to the first marked place and letting it fall, I counted about 77 oscillations until the box returned to the second marked place, and as many thereafter until the box returned to the third marked place, and again as many until the box on its return reached the fourth place. Hence I conclude that the whole resistance of the full box did not have a greater proportion to the resistance of the empty box than 78 to 77. For if the resistances of both were equal, the full box, because its inherent force was 78 times greater than the inherent force of the empty box, ought to conserve its oscillatory motion that much longer, and thus always return to those marked places at the completion of 78 oscillations. But it returned to them at the completion of 77 oscillations.

Let A therefore designate the resistance of the box on its external surface, and B the resistance of the empty box on its internal parts; then, if the resistances of equally swift bodies on their internal parts are as the matter, or the number of particles that are resisted, 78B will be the resistance of the full box on its internal parts; and thus the whole resistance A + B of the empty box will be to the whole resistance A + 78B of the full box as 77 to 78, and by separation [or dividendo] A + B will be to 77B as 77 to 1, and hence A + B will be to B as 77 × 77 to 1, and by separation [or dividendo] A will be to B as 5,928 to 1. The resistance encountered by the empty box on its internal parts is therefore more than 5,000 times smaller than the similar resistance on the external surface. This argument depends on the hypothesis that the greater resistance encountered by the full box does not arise from some other hidden cause but only from the action of some subtle fluid upon the enclosed metal.

I have reported this experiment from memory. For the paper on which I had once described it is lost. Hence I have been forced to omit certain fractions of numbers which have escaped my memory.

There is no time to try everything again. The first time, since I had used a weak hook, the full box was retarded more quickly. In seeking the cause, I found that the hook was so weak as to give way to the weight of the box and to be bent in this direction and that as it yielded to the oscillations of the pendulum. I got a strong hook, therefore, so that the point of suspension would remain motionless, and then everything came out as we have described it above.

SECTION 7[a]

The motion of fluids and the resistance encountered by projectiles

Proposition 32
Theorem 26

Let two similar systems of bodies consist of an equal number of particles, and let each of the particles in one system be similar and proportional to the corresponding particle in the other system, and let the particles be similarly situated with respect to one another in the two systems and have a given ratio of density to one another. And let them begin to move similarly with respect to one another in proportional times (the particles that are in the one system with respect to the particles in that system, and the particles in the other with respect to those in the other). Then, if the particles that are in the same system do not touch one another except in instants of reflection and do not attract or repel one another except by accelerative forces that are inversely as the diameters of corresponding particles and directly as the squares of the velocities, I say that the particles of the systems will continue to move similarly with respect to one another in proportional times.

I say that bodies which are similar and similarly situated move similarly with respect to one another in proportional times when their situations in relation to one another are always similar at the end of the times—for instance, if the particles of one system are compared with the corresponding particles of another. Hence the times in which similar and proportional parts of similar figures are described by corresponding particles will be proportional. Therefore, if there are two systems of this sort, the corresponding particles, because of the similarity of their motions at the beginning, will continue to move similarly until they meet one another. For if they are acted upon by no forces, they will, by the first law of motion, move forward uniformly in straight lines. If they act upon one another by some forces and if those forces are as the diameters of the corresponding particles inversely and the squares of the velocities directly, then, since the situations of the particles are similar and the forces proportional, the whole forces by which the corresponding particles are acted upon, compounded of the separate acting forces (by corol. 2 of the laws), will have similar directions, just as if they tended to centers similarly

a. In ed. 1, sec. 7 is very different. Props. 32–34 (32–35 in ed. 1) underwent partial alteration, including the suppression of the original prop. 34. The remainder of sec. 7 was completely rewritten for ed. 2 and essentially retained, with only minor changes, in ed. 3. For details, see the Guide to the present translation, §7.6.

placed among the particles, and those whole forces will be to one another as the separate component forces, that is, as the diameters of the corresponding particles inversely and the squares of the velocities directly, and therefore they will cause corresponding particles to continue describing similar figures. This will be so (by book 1, prop. 4, corols. 1 and 8) provided that the centers are at rest. But if they move, since their situations with respect to the particles of the systems remain similar (because the transferences are similar), similar changes will be introduced in the figures which the particles describe. The motions of corresponding similar particles will therefore be similar until they first meet, and therefore the collisions will be similar and the reflections similar, and then (by what has already been shown) the motions of the particles with respect to one another will be similar until they encounter one another again, and so on indefinitely. Q.E.D.

COROLLARY 1. Hence, if any two bodies that are similar and similarly situated (in relation to the corresponding particles of the systems) begin to move similarly with respect to the particles in proportional times, and if their volumes and densities are to each other as the volumes and densities of the corresponding particles, the bodies will continue to move similarly in proportional times. For the case is the same for the larger parts of both systems as for the particles.

COROLLARY 2. And if all the similar and similarly situated parts of the systems are at rest with respect to one another, and if two of them, which are larger than the others and correspond to each other in the two systems, begin to move in any way with a similar motion along lines similarly situated, they will cause similar motions in the remaining parts of the systems and will continue to move similarly with respect to them in proportional times and thus will continue to describe spaces proportional to their own diameters.

Proposition 33
Theorem 27

If the same suppositions are made, I say that the larger parts of the systems are resisted in a ratio compounded of the squared ratio of their velocities and the squared ratio of the diameters and the simple ratio of the density of the parts of the systems.

For the resistance arises partly from the centripetal or centrifugal forces with which the particles of the systems act upon one another and partly from the collisions and reflections of the particles and the larger parts. Resistances of the first kind, moreover, are to one another as the whole motive forces from which they arise, that is, as the whole accelerative forces and the quantities of matter in corresponding parts, that is (by hypothesis), as the squares of the velocities directly and the distances of the corresponding particles inversely and the quantities of matter in the corresponding parts directly. Thus, since the distances of the particles of the one system are to the corresponding distances of the particles of the other as the diameter of a particle or part in the first system to the diameter of the corresponding particle or part in the other, and since the quantities of matter are as the densities of the parts and the cubes of the diameters, the resistances are to one another as the squares of the velocities, the squares of the diameters, and the densities of the parts of the systems. Q.E.D.

Resistances of the second kind are as the numbers and forces of corresponding reflections jointly. The number of reflections in any one case, moreover, is to the number in any other as the velocities of the corresponding parts directly and the spaces between their reflections inversely. And the forces of the reflections are as the velocities and volumes and densities of the corresponding parts jointly, that is, as the velocities, the cubes of the diameters, and the densities of the parts. And if all these ratios are compounded, the resistances of the corresponding parts are to one another as the squares of the velocities, the squares of the diameters, and the densities of the parts, jointly. Q.E.D.

COROLLARY 1. Therefore, if the systems are two elastic fluids such as air and if their parts are at rest with respect to one another, and if two bodies which are similar and are proportional (with regard to volume and density) to the parts of the fluids and are similarly situated with respect to those parts are projected in any way along lines similarly situated, and if the accelerative forces with which the particles of the fluids act upon one another are as the diameters of the projected bodies inversely and the squares of the velocities directly, then the bodies will cause similar motions in the fluids in proportional times and will describe spaces that are similar and are proportional to their diameters.

COROLLARY 2. Accordingly, in the same fluid a swift projectile encounters a resistance that is very nearly in the squared ratio of the velocity. For if the forces with which distant particles act upon one another were increased in the squared ratio of the velocity, the resistance would be exactly in the squared ratio of the velocity; and thus, in a medium whose parts act upon one another with no forces because they are far apart, the resistance is exactly in the squared ratio of the velocity. Let A, B, and C, therefore, be three mediums consisting of parts that are similar and equal and regularly distributed at equal distances. Let the parts of mediums A and B recede from one another with forces that are to one another as T and V, and let the parts of medium C be entirely without forces of this sort. Then, let four equal bodies D, E, F, and G move in these mediums, the first two bodies D and E in the first two mediums A and B respectively, and the other two bodies F and G in the third medium C; and let the velocity of body D be to the velocity of body E, and let the velocity of body F be to the velocity of body G, as the square root of the ratio of the forces T to the forces V [i.e., as $\sqrt{T}$ to $\sqrt{V}$]; then the resistance of body D will be to the resistance of body E, and the resistance of body F to the resistance of body G, in the squared ratio of the velocities; and therefore the resistance of body D will be to the resistance of body F as the resistance of body E to the resistance of body G. Let bodies D and F have equal velocities, and also bodies E and G; then, if the velocities of bodies D and F are increased in any ratio and the forces of the particles of medium B are decreased in the same ratio squared, medium B will approach the form and condition of medium C as closely as is desired, and on that account the resistances of the equal and equally swift bodies E and G in these mediums will continually approach equality, in such a way that their difference finally comes out less than any given difference. Accordingly, since the resistances of bodies D and F are to each other as the resistances of bodies E and G, these also will similarly approach the ratio of equality. Therefore, the resistances of bodies D and F, when they move very swiftly, are very nearly equal, and therefore, since the resistance of body F is in the squared ratio of the velocity, the resistance of body D will be very nearly in the same ratio.

COROLLARY 3. The resistance of a body moving very swiftly in any elastic fluid is about the same as if the parts of the fluid lacked their centrifugal forces and did not recede from one another, provided that the elastic force of the fluid arises from the centrifugal forces of the particles and that the velocity is so great that the forces do not have enough time to act.

COROLLARY 4. Accordingly, since the resistances of similar and equally swift bodies, in a medium whose parts (being far apart) do not recede from one another, are as the squares of the diameters, the resistances of equally swift and very quickly moving bodies in an elastic fluid are also very nearly as the squares of the diameters.

Corollary 5. And since similar, equal, and equally swift bodies, in mediums which have the same density and whose particles do not recede from one another, impinge upon an equal quantity of matter in equal times (whether the particles are more and smaller or fewer and larger) and impress upon it an equal quantity of motion and in turn (by the third law of motion) undergo an equal reaction from it (that is, are equally resisted), it is manifest also that in elastic fluids of the same density, when the bodies move very swiftly, the resistances they encounter are very nearly equal, whether those fluids consist of coarser particles or are made of the most subtle particles of all. The resistance to projectiles moving very quickly is not much diminished as a result of the subtlety of the medium.

Corollary 6. These statements all hold for fluids whose elastic force originates in the centrifugal forces [i.e., forces of repulsion] of the particles. But if that force arises from some other source, such as the expansion of the particles in the manner of wool or the branches of trees, or from any other cause which makes the particles move less freely with respect to one another, then the resistance will be greater than in the preceding corollaries because the medium is less fluid.

Proposition 34
Theorem 28

In a rare medium consisting of particles that are equal and arranged freely at equal distances from one another, let a sphere and a cylinder—described with equal diameters—move with equal velocity along the direction of the axis of the cylinder; then the resistance of the sphere will be half the resistance of the cylinder.

For since the action of a medium on a body is (by corol. 5 of the laws) the same whether the body moves in a medium at rest or the particles of the medium impinge with the same velocity on the body at rest, let us consider the body to be at rest and see with what force it will be urged by the moving medium. Let ABKI, therefore, designate a spherical body described with center C and semidiameter CA, and let the particles of the

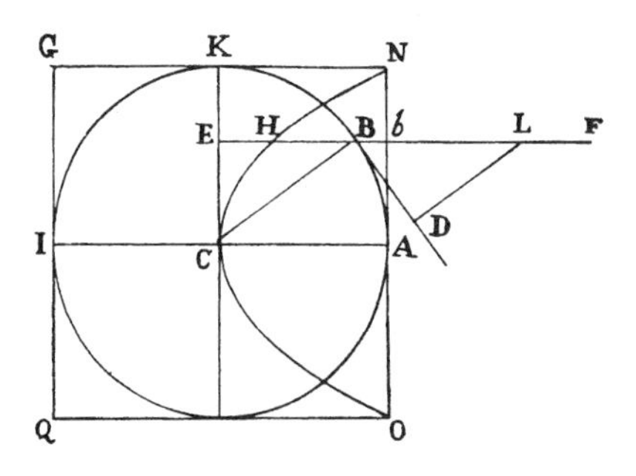

medium strike the spherical body with a given velocity along straight lines parallel to AC; and let FB be such a straight line. On FB take LB equal to the semidiameter CB, and draw BD touching the sphere in B. To KC and BD drop the perpendiculars BE and LD; then the force with which a particle of the medium, obliquely incident along the straight line FB, strikes the sphere at B will be to the force with which the same particle would strike the cylinder ONGQ (described with axis ACI about the sphere) perpendicularly at *b* as LD to LB or BE to BC. Again, the efficacy of this force to move the sphere along the direction FB (or AC) of its incidence is to its efficacy to move the sphere along the direction of its determination—that is, along the direction of the straight line BC in which it urges the sphere directly [a direction through the center of the sphere]—as BE to BC. And, compounding the ratios, if a particle strikes the sphere obliquely along the straight line FB, its efficacy to move the sphere along the direction of its incidence is to the efficacy of the same particle to move the cylinder in the same direction, when striking the cylinder perpendicularly along the same straight line, as BE^2 to BC^2. Therefore, if in *b*E, which is perpendicular to the circular base NAO of the cylinder and equal to the radius AC, *b*H is taken equal to $\frac{BE^2}{CB}$, then *b*H will be to *b*E as the effect of a particle upon the sphere to the effect of the particle upon the cylinder. And therefore the solid that is composed of all the straight lines *b*H will be to the solid that is composed of all the straight lines *b*E as the effect of all the particles upon the sphere to the effect of all the particles upon the cylinder. But the first solid is a paraboloid described with vertex C, axis CA, and latus rectum CA, and the second solid is a cylinder circumscribed around the paraboloid; and it is known that a paraboloid is half of the circumscribed cylinder. Therefore the whole force of the medium upon the sphere is half of its whole force upon the cylinder. And therefore, if the particles of the medium were at rest and the cylinder and the sphere were moving with equal velocity, the resistance of the sphere would be half the resistance of the cylinder. Q.E.D.

Scholium

By the same method other figures can be compared with one another with respect to resistance, and those that are more suitable for continuing their motions in resisting mediums can be found. For example, let it be required

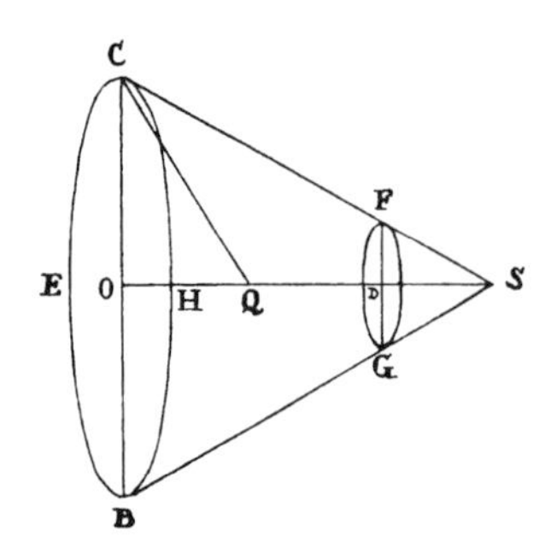

to construct a frustum CBGF of a cone with the circular base CEBH (which is described with center O and radius OC) and with the height OD, which is resisted less than any other frustum constructed with the same base and height and moving forward along the direction of the axis toward D; bisect the height OD in Q, and produce OQ to S so that QS is equal to QC, and S will be the vertex of the cone whose frustum is required.

Note in passing that since the angle CSB is always acute, it follows that if the solid ADBE is generated by a revolution of the elliptical or oval figure ADBE about the axis AB, and if the generating figure is touched by the three straight lines FG, GH, and HI in points F, B, and I, in such a way that GH is perpendicular to the axis in the point of contact B, and FG and HI meet the said line GH at the angles FGB and BHI of 135 degrees, then the solid that is generated by the revolution of the figure ADFGHIE about the same axis AB is less resisted than the former solid, provided that each of the two moves forward along the direction of its axis AB, and the end B of each one is in front. Indeed, I think that this proposition will be of some use for the construction of ships.

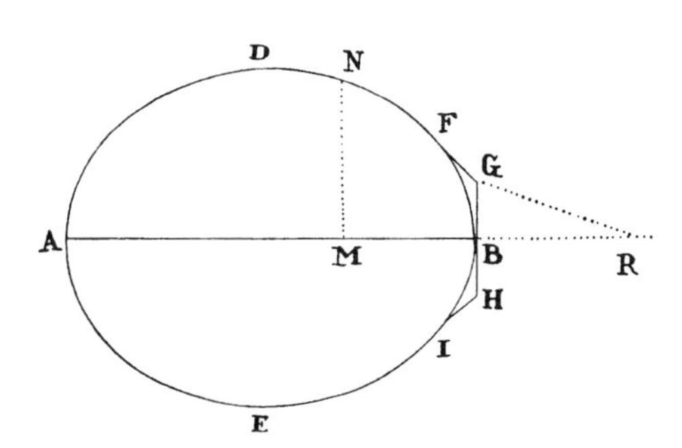

But suppose the figure DNFG to be a curve of such a sort that if the perpendicular NM is dropped from any point N of that curve to the axis AB, and if from the given point G the straight line GR is drawn, which is parallel to a straight line touching the figure in N and cuts the axis (produced) in R, then MN would be to GR as GR^3 to $4BR \times GB^2$. Then, in this case, the solid that is described by a revolution of this figure about the axis AB will, in moving in the aforesaid rare medium from A toward B, be resisted less than any other solid of revolution described with the same length and width.

Proposition 35[a]
Problem 7

If a rare medium consists of minimally small equal particles that are at rest and arranged freely at equal distances from one another, it is required to find the resistance encountered by a sphere moving forward uniformly in this medium.

CASE 1. Let a cylinder described with the same diameter and height as before move forward with the same velocity along the length of its own axis in the same medium. And let us suppose that the particles of the medium upon which the sphere or cylinder impinges rebound with the greatest possible force of reflection. Then the resistance of the sphere (by prop. 34) is half the resistance of the cylinder, and the sphere is to the cylinder as 2 to 3, and the cylinder in impinging perpendicularly upon the particles and reflecting them as greatly as possible communicates twice its own velocity to them. Therefore, the cylinder, in the time in which it describes half the length of its axis by moving uniformly forward, will communicate to the particles a motion which is to the whole motion of the cylinder as the density of the medium is to the density of the cylinder; and the sphere, in the time in which it describes the whole length of its diameter by moving uniformly forward, will communicate the same motion to the particles, and in the time in which it describes ⅔ of its diameter it will communicate to the particles a motion which is to the whole motion of the sphere as the density of the medium to the density of the sphere. And therefore the sphere encounters a resistance that is to the force by which its whole motion could be either destroyed or generated, in the time in which it describes ⅔ of its diameter by moving uniformly forward, as the density of the medium is to the density of the sphere.

CASE 2. Let us suppose that the particles of the medium impinging upon the sphere or cylinder are not reflected; then the cylinder, in impinging perpendicularly upon the particles, will communicate its whole velocity to them and thus encounters half the resistance which it met in the former case, and the resistance encountered by the sphere will also be half of what it was before.

CASE 3. Let us suppose that the particles of the medium rebound from the sphere with a force of reflection that is neither the greatest nor nil but

a. A translation of the versions of book 2, props. 35–40, that appear in the first edition has been made by I. Bernard Cohen and Anne Whitman and will be published, together with a commentary by George Smith, in *Newton's Natural Philosophy*, ed. Jed Buchwald and I. Bernard Cohen (Cambridge: MIT Press, forthcoming).

some intermediate force; then the resistance encountered by the sphere will also be intermediate between the resistance in case 1 and the resistance in case 2. Q.E.I.

COROLLARY 1. Hence, if the sphere and the particles are infinitely hard without any elastic force and therefore also without any force of reflection, the resistance encountered by the sphere will be to the force by which its whole motion could be either destroyed or generated, in the time in which the sphere describes ⅓ of its diameter, as the density of the medium is to the density of the sphere.

COROLLARY 2. The resistance encountered by the sphere, other things being equal, is in the squared ratio of the velocity.

COROLLARY 3. The resistance encountered by the sphere, other things being equal, is in the squared ratio of the diameter.

COROLLARY 4. The resistance encountered by the sphere, other things being equal, is as the density of the medium.

COROLLARY 5. The resistance encountered by the sphere is in a ratio that is compounded of the squared ratio of the velocity and the squared ratio of the diameter, and the simple ratio of the density of the medium.

COROLLARY 6. And the motion of the sphere with the resistance it encounters can be represented as follows. Let AB be the time in which the sphere can lose its whole motion when the resistance is continued uniformly. Erect AD and BC perpendicular to AB. And let BC be the whole motion, and through point C with asymptotes AD and AB describe the hyperbola CF. Produce AB to any point E. Erect the perpendicular EF meeting the hyperbola in F. Complete the parallelogram CBEG, and draw AF meeting BC in H. Then, if the sphere, in any time BE, when its first motion BC is continued uniformly, in a nonresisting medium, describes the space CBEG represented by the area of the parallelogram, it will in a resisting medium describe the space CBEF represented by the area of the hyperbola, and its motion at the end of that time will be represented by the ordinate EF of the hyperbola, with loss of part FG of its motion. And the resistance at the end of the same time will be represented by the length BH, with loss of part CH of the resistance. All of this is evident by book 2, prop. 5, corols. 1 and 3.

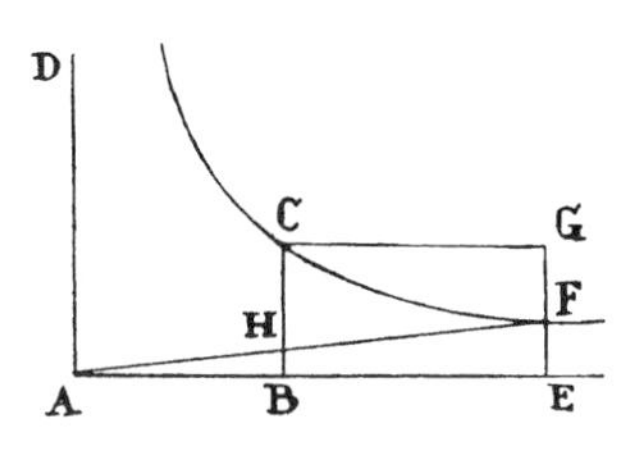

COROLLARY 7. Hence, if in time T, when the resistance R is continued uniformly, the sphere loses its whole motion M, then in time t in a resisting medium, when the resistance R decreases in the squared ratio of the velocity, the sphere will lose part $\frac{tM}{T+t}$ of its motion M without loss of part $\frac{TM}{T+t}$; and the sphere will describe a space that is to the space described by the uniform motion M, in the same time t, as the logarithm of the number $\frac{T+t}{T}$ multiplied by the number 2.302585092994 is to the number $\frac{t}{T}$, because the hyperbolic area BCFE is in this proportion to the rectangle BCGE.

Scholium

In this proportion I have set forth the resistance and retardation encountered by spherical projectiles in noncontinuous mediums, and I have shown that this resistance is to the force by which the whole motion of a sphere could be either destroyed or generated in the time in which the sphere describes ⅔ of its diameter, with a velocity continued uniformly, as the density of the medium is to the density of the sphere, provided that the sphere and the particles of the medium are highly elastic and possess the greatest force of reflecting, and I have shown that this force is half as great when the sphere and the particles of the medium are infinitely hard and devoid of all force of reflecting. Moreover, in continuous mediums such as water, hot oil, and quicksilver, in which the sphere does not impinge directly upon all the particles of the fluid which generate resistance but presses only the nearest particles, and these press others and these still others, the resistance is half as great as in the second case. In extremely fluid mediums of this sort the sphere encounters a resistance that is to the force by which its whole motion could be either destroyed or generated, in the time in which it describes 8/3 of its diameter with the motion continued uniformly, as the density of the medium is to the density of the sphere. We will try to show this in what follows.

Proposition 36
Problem 8

To determine the motion of water flowing out of a cylindrical vessel through a hole in the bottom.

Let ACDB be the cylindrical vessel, AB its upper opening, CD its bottom parallel to the horizon, EF a circular hole in the middle of the bottom, G the center of the hole, and GH the cylinder's axis perpendicular to the horizon.

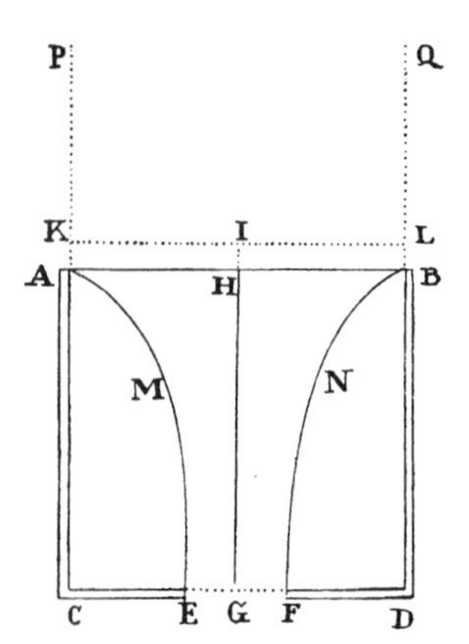

And imagine that a cylinder of ice APQB is of the same width as the interior of the vessel, has the same axis, and descends continually with a uniform motion. Imagine also that its parts liquefy as soon as they touch the surface AB, that when they have turned into water they flow down into the vessel as a result of their gravity, and that in falling these parts form a cataract or column of water ABNFEM and pass through the hole EF and fill it exactly. And let the uniform velocity of the descending ice, as well as that of the contiguous water in the circle AB, be the velocity which the water can acquire in falling and describing by its fall the space IH, and let IH and HG lie in a straight line, and through point I draw the straight line KL parallel to the horizon and meeting the sides of the ice in K and L. Then the velocity of the water flowing out through the hole EF will be that which the water can acquire in falling from I and describing by its fall the space IG. And thus, by Galileo's theorems, IG will be to IH as the square of the ratio of the velocity of the water flowing out through the hole to the velocity of the water in the circle AB, that is, as the square of the ratio of the circle AB to the circle EF, for these circles are inversely as the velocities of the water passing through them in the same time and with an equal quantity, filling them both exactly. Here it is the velocity of the water toward the horizon that is of concern. And the motion parallel to the horizon by which the parts of the falling water approach one another is not considered here, since it does not arise from gravity or change the motion perpendicular to the horizon that does arise from gravity. Indeed, we are supposing that the parts of the water cohere somewhat and that by their cohesion they approach one another with motions parallel to the horizon as they fall, so that they form only one single cataract and are not divided into several cataracts, but here we are not considering the motion parallel to the horizon arising from that cohesion.

CASE 1. Now suppose that the interior of the vessel around the falling water ABNFEM is filled with ice, so that the water passes through the ice as if through a funnel. Then, if the water does not quite touch the ice, or (what comes to the same thing) if it touches it and, because of the great smoothness of the ice, slides through it with the greatest possible freedom and without any resistance, the water will flow down through the hole EF with the same velocity as before, and the whole weight of the column of water ABNFEM will be used in generating its downflow as before, and the bottom of the vessel will sustain the weight of the ice surrounding the column.

Now let the ice liquefy in the vessel; then the flow of the water will remain the same as before with respect to velocity. It will not be less, since the melted ice will endeavor to descend; and not greater, since the melted ice cannot descend without impeding an equal descent of the original water. The same force ought to generate the same velocity in the flowing water [i.e., since the force is the same, the velocity that it generates will also be the same].

But the hole in the bottom of the vessel, because of the oblique motions of the particles of the flowing water, ought to be a little larger than before. For now the particles of water do not all pass through the hole perpendicularly but, flowing together from all the sides of the vessel and converging into the hole, pass through with oblique motions and, turning their course downward, unite into a stream of water gushing out which is narrower a little below the hole than in the hole itself, its diameter being to the diameter of the hole as 5 to 6, or 5½ to 6½ very nearly, provided that I measured the diameters correctly. At any rate, I obtained a very thin flat plate perforated in the middle, the diameter of the circular hole being ⅝ inch. And so that the stream of water gushing out might not be accelerated in falling and made narrower by the acceleration, I fastened this plate not to the bottom but to the side of the vessel in such a way that the stream went out along a line parallel to the horizon. Then, when the vessel was full of water, I opened the hole so that the water might flow out, and the diameter of the stream, measured as accurately as possible at a distance of about ½ inch from the hole, came out 21/40 inch. The diameter of this circular hole, therefore, was to the diameter of the stream very nearly as 25 to 21. Therefore the water in passing through the hole converges from all directions, and after flowing out of the vessel the stream is made narrower by converging and is accelerated by narrowing until it has reached a distance of ½ inch from the hole and at that distance becomes narrower and swifter than it is in the hole itself in the ratio of 25 × 25 to 21 × 21 or very nearly 17 to 12, that is, roughly as the square root of the ratio of 2 to 1. And experiments prove that the quantity of water that flows out in a given time through a circular hole in the bottom of a vessel is the quantity that ought to flow out in the same time, with the velocity mentioned above, not through that hole but through a circular hole whose diameter is to the diameter of that hole as 21 to 25. And thus the flowing water has the downward velocity in the hole itself that a heavy body can acquire very nearly in falling and describing by its fall a space equal to half the height of the water standing in the vessel. But after the water has gone out of the vessel, it is accelerated by converging until it has reached a distance from the hole almost equal to the diameter of the hole and has acquired a velocity that is greater approximately as the square root of the ratio of 2 to 1, which is, as a matter of fact, very nearly the velocity that a heavy body can acquire in falling and describing by its fall a space equal to the whole height of the water standing in the vessel.

In what follows, therefore, let the diameter of the stream be designated by that smaller hole which we have called EF. And suppose that another higher plane VW is drawn parallel to the plane of the hole EF at a distance about equal to the diameter of the hole and pierced by a larger hole ST, and through this let a stream fall that exactly fills the lower hole EF and thus has a diameter which is to the diameter of this lower hole as about 25 to 21. For thus the stream will pass perpendicularly through the lower hole,

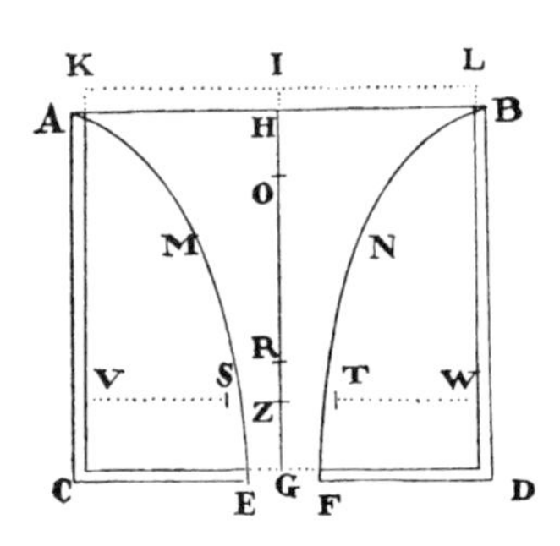

and the quantity of the water flowing out, depending on the size of this hole, will be very nearly that which the solution of the problem demands. Now, the space which is enclosed by the two planes and the falling stream can be considered to be the bottom of the vessel. But so that the solution of the problem may be simpler and more mathematical, it is preferable to use only the lower plane for the bottom of the vessel and to imagine that the water which flowed down through the ice as if through a funnel and came out of the vessel through the hole EF in the lower plane keeps its motion continually and that the ice keeps its state of rest. In what follows, therefore, let ST be the diameter of a circular hole described with center Z, through which a cataract flows out of the vessel when all the water in the vessel is fluid. And let EF be the diameter of the hole which the cataract fills exactly when falling through it, whether the water comes out of the vessel through the upper hole ST or falls through the middle of the ice in the vessel as if through a funnel. And let the diameter of the upper hole ST be to the diameter of the lower hole EF as about 25 to 21, and let the perpendicular distance between the planes of the holes be equal to the diameter of the smaller hole EF. Then the downward velocity of the water coming out of the vessel through the hole ST will in the hole itself be that which a body can acquire in falling from half of the height IZ; and the velocity of both falling cataracts will, in the hole EF, be that which a body will acquire in falling from the whole height IG.

CASE 2. If the hole EF is not in the middle of the bottom of the vessel, but the bottom is perforated elsewhere, the water will flow out with the same velocity as before, provided that the size of the hole is the same. For a heavy body does descend to the same depth in a greater time along an oblique line than along a perpendicular line, but in descending it acquires the same velocity in either case, as Galileo proved.

CASE 3. The velocity of the water flowing out through a hole in the side of the vessel is the same. For if the hole is small, so that the distance between the surfaces AB and KL vanishes, so far as the senses can tell, and the stream of water gushing out horizontally forms a parabolic figure, it will be found from the latus rectum of this parabola that the velocity of the water flowing out is that which a body could have acquired by falling from the height HG or IG of the water standing in the vessel. Indeed, by making an experiment I found that when the height of the standing water above the hole was 20 inches and the height of the hole above a plane parallel to the horizon was also 20 inches, the stream of water gushing forth would fall upon the plane at a distance of about 37 inches, taken from a perpendicular that was dropped to the plane from the hole. For in the absence of resistance the stream would have had to fall upon the plane at a distance of 40 inches, the latus rectum of the parabolic stream being 80 inches.

CASE 4. Further, if the water flowing out has an upward motion, it comes out with the same velocity. For a small stream of water gushing out ascends with a perpendicular motion to the height GH or GI of the water standing in the vessel, except insofar as its ascent is somewhat impeded by the resistance of the air; and accordingly it flows out with the velocity that it could have acquired in falling from that height. Any one particle of the standing water (by book 2, prop. 19) is pressed equally from all sides and, yielding to the pressure, goes with equal force in every direction, whether it descends through a hole in the bottom of the vessel or flows out horizontally through a hole in its side or comes out into a channel and ascends from there through a small hole in the upper part of the channel. And that the velocity with which the water flows out is that which we have designated in this proposition is not only found by reason but is also manifest from the well-known experiments already described.

CASE 5. The velocity of the water flowing out is the same whether the hole is circular or square or triangular or of any other shape equal in area to the circular one. For the velocity of the water flowing out does not depend on the shape of the hole but on the height of the water in relation to the plane KL.

CASE 6. If the lower part of the vessel ABDC is immersed in standing water, and the height of the standing water above the bottom of the vessel is GR, the velocity with which the water in the vessel will flow out through the hole EF into the standing water will be that which the water can acquire in falling and describing by its fall the space IR. For the weight of all the water in the vessel that is lower than the surface of the standing water will be sustained in equilibrium by the weight of the standing water and thus will not at all accelerate the motion of the descending water in the vessel. This case can also be shown by experiments, by measuring the times in which the water flows out.

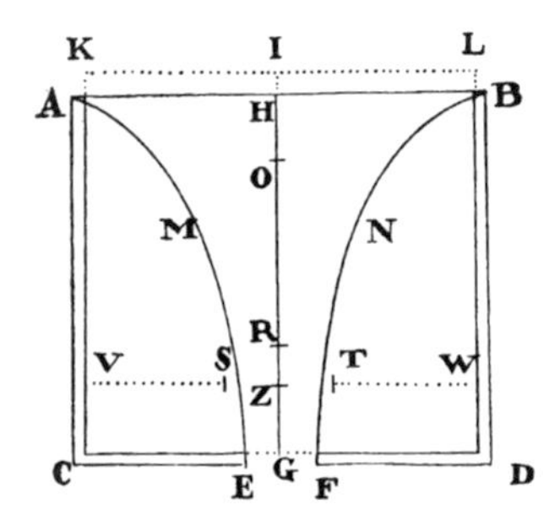

COROLLARY 1. Hence, if the height CA of the water is produced to K, so that AK is to CK in the squared ratio of the area of a hole made in any part of the bottom to the area of the circle AB, the velocity of the water flowing out will be equal to the velocity that the water can acquire in falling and describing by its fall the space KC.

COROLLARY 2. And the force by which the whole motion of the water gushing out can be generated is equal to the weight of a cylindrical column of water whose base is the hole EF and whose height is 2GI or 2CK. For the gushing water, in the time in which it equals this column, can acquire in falling (by its weight) from the height GI the very velocity with which it gushes out.

Corollary 3. The weight of all the water in the vessel ABDC is to the part of the weight that is used in making the water flow down as the sum of the circles AB and EF to twice the circle EF. For let IO be a mean proportional between IH and IG; then the water coming out through the hole EF, in the time in which a drop could describe a space equal to the height IG in falling from I, will be equal to a cylinder whose base is the circle EF and whose height is 2IG, that is, to a cylinder whose base is the circle AB and whose height is 2IO, for the circle EF is to the circle AB as the square root of the ratio of the height IH to the height IG, that is, in the simple ratio of the mean proportional IO to the height IG, and in the time in which a drop can describe a space equal to the height IH in falling from I, the water coming out will be equal to a cylinder whose base is the circle AB and whose height is 2IH, and in the time in which a drop describes a space equal to the difference HG between the heights in falling from I through H to G, the water coming out—that is, all the water in the solid ABNFEM—will be equal to the difference between the cylinders, that is, equal to a cylinder whose base is AB and whose height is 2HO. And therefore all the water in the vessel ABDC is to all the water falling in the solid ABNFEM as HG to 2HO, that is, as HO + OG to 2HO, or IH + IO to 2IH. But the weight of all the water in the solid ABNFEM is used in making the water flow down, and accordingly the weight of all the water in the vessel is to the part of the weight that is used in making the water flow down as IH + IO to 2IH and thus as the sum of the circles EF and AB to twice the circle EF.

Corollary 4. And hence the weight of all the water in the vessel ABDC is to the part of the weight sustained by the bottom of the vessel as the sum of the circles AB and EF is to the difference between these circles.

Corollary 5. And the part of the weight sustained by the bottom of the vessel is to the part of the weight used in making the water flow down as the difference between the circles AB and EF is to twice the smaller circle EF, or as the area of the bottom to twice the hole.

Corollary 6. And the part of the weight which alone presses upon the bottom is to the weight of all the water resting perpendicularly on the bottom as the circle AB is to the sum of the circles AB and EF, or as the circle AB is to the amount by which twice the circle AB exceeds the bottom. For the part of the weight which alone presses upon the bottom is to the weight of all the water in the vessel as the difference between the circles AB and EF is to the sum of these circles, by corol. 4; and the weight of all the water in the vessel is to the weight of all the water resting perpendicularly on the bottom as the circle AB is to the difference between the circles AB and EF. Therefore, from the equality of the ratios in inordinate proportion [or ex aequo perturbate], the part of the weight which alone presses upon the bottom is to the weight of all the water resting perpendicularly on the bottom as the circle AB is to the sum of the circles AB and EF, or to the amount by which twice the circle AB exceeds the bottom.

Corollary 7. If in the middle of the hole EF there is placed a little circle PQ described with center G and parallel to the horizon, the weight of the water which that little circle sustains is greater than the weight of ⅓ of a cylinder of water whose base is that little circle and whose height is GH. For let ABNFEM be a cataract or column of falling water, with axis GH as above, and suppose that there has been a freezing of all the water in the vessel (around the cataract as well as above the little circle) whose fluidity is not required for the very ready and very swift descent of the water. And let PHQ be the frozen column of water above the little circle, having vertex H and height GH. And imagine that this cataract falls with its whole weight and does not rest or press on PHQ but slides past freely and without friction, except perhaps at the very vertex of the ice, where at the very beginning of falling the cataract begins to be concave. And just as the frozen water (AMEC and BNFD) which is around the cataract is convex on the inner surface (AME and BNF) toward the falling cataract, so also this column PHQ will be convex toward the cataract, and therefore will be greater than a cone whose base is the little circle PQ and whose height is GH, that is, greater than ⅓ of a cylinder described with the same base and height. And the little circle sustains the weight of this column, that is, a weight that is greater than the weight of the cone or of ⅓ of the cylinder.

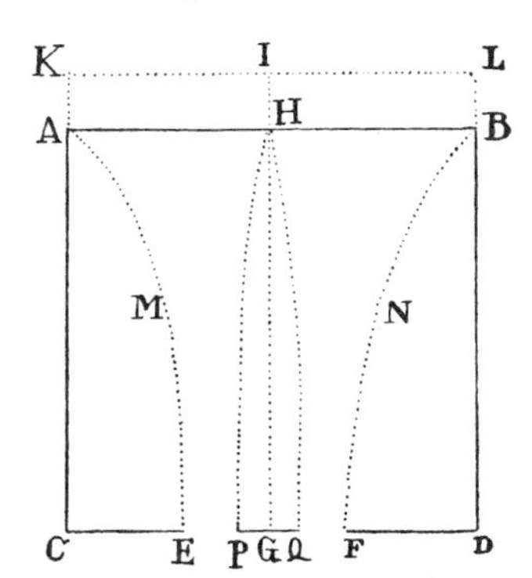

Corollary 8. The weight of the water sustained by the little circle PQ, when it is extremely small, appears to be less than the weight of ⅔ of a cylinder of water whose base is that little circle and whose height is HG. Keeping the same suppositions, imagine that half a spheroid is described,

whose base is the little circle and whose semiaxis or height is HG. Then this figure will be equal to ⅔ of that cylinder and will comprehend the column of frozen water PHQ whose weight that little circle sustains. In order that the motion of the water may be straight down, the outer surface of this column must meet the base PQ in a somewhat acute angle, because the water in falling is continually accelerated and the acceleration makes the column become narrower; and since that angle is less than a right angle, the lower parts of this column will lie within the half-spheroid. But higher up, the column will be acute or pointed, for otherwise the horizontal motion of the water at the vertex of the spheroid would be infinitely swifter than its motion toward the horizon. And the smaller the little circle PQ, the more acute the vertex of the column; and if the little circle is diminished indefinitely, the angle PHQ will be diminished indefinitely, and therefore the column will lie within the half-spheroid. That column is therefore less than the half-spheroid, or less than ⅔ of a cylinder whose base is that little circle and whose height is GH. Moreover, the little circle sustains the water's force equal to the weight of this column, since the weight of the surrounding water is used in making it flow down.

COROLLARY 9. The weight of the water sustained by the little circle PQ, when it is extremely small, is very nearly equal to the weight of a cylinder of water whose base is that little circle and whose height is ½GH. For this weight is an arithmetical mean between the weights of the cone and the said half-spheroid. If, however, the little circle is not extremely small but is increased until it equals the hole EF, it will sustain the weight of all the water resting perpendicularly on it, that is, the weight of a cylinder of water whose base is that little circle and whose height is GH.

COROLLARY 10. And (as far as I can tell) the weight that the little circle sustains always has the proportion to the weight of a cylinder of water whose base is that little circle and whose height is ½GH that EF^2 has to $EF^2 - \frac{1}{2}PQ^2$, or that the circle EF has to the excess of this circle over half of the little circle PQ, very nearly.

Lemma 4

The resistance of a cylinder moving uniformly forward in the direction of its length is not changed by an increase or decrease in length and thus is the same as the resistance of a circle described with the same diameter and moving forward with the same velocity along a straight line perpendicular to its plane.

For the sides of a cylinder offer no opposition to its motion, and a cylinder is turned into a circle if its length is decreased indefinitely.

Proposition 37
Theorem 29

If a cylinder moves uniformly forward in a compressed, infinite, and nonelastic fluid in the direction of its own length, its resistance arising from the magnitude of its transverse section is to the force by which its whole motion can be either destroyed or generated, while it is describing four times its length, very nearly as the density of the medium is to the density of the cylinder.

For if the bottom CD of the vessel ABDC touches the surface of stagnant water, and if water flows out of this vessel into the stagnant water through the cylindrical channel EFTS perpendicular to the horizon, and if the little circle PQ is placed parallel to the horizon anywhere in the middle of the channel, and if CA is produced to K so that CK is to AK in the squared ratio of the circle AB to the amount by which the opening of the channel EF exceeds the little circle PQ, then it is obvious (by prop. 36, case 5, case 6, and corol. 1) that the velocity of the water passing through the annular space between the little circle and the sides of the vessel will be that which the water can acquire in falling and describing by its fall a space equal to the height KC or IG.

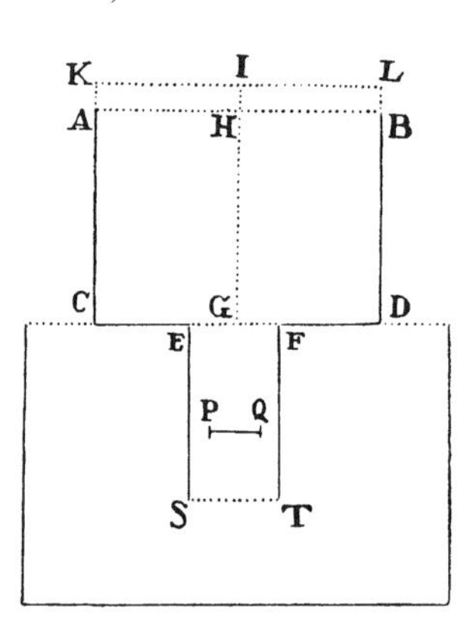

And (by prop. 36, corol. 10) if the width of the vessel is infinite, so that the line-element HI vanishes and the heights IG and HG are equal, then the force of the water flowing down into the little circle will be to the weight of a cylinder whose base is that little circle, and whose height is ½IG, very nearly as EF^2 to $EF^2 - \frac{1}{2}PQ^2$. For the force of the water flowing down through the whole channel with uniform motion will be the same upon the little circle PQ in whatever part of the channel it is placed.

Now let the openings EF and ST of the channel be closed, and let the little circle ascend in the fluid compressed on all sides, and by its ascent let it make the upper water descend through the annular space between the little circle and the sides of the channel; then the velocity of the ascending little circle will be to the velocity of the descending water as the difference between the circles EF and PQ is to the circle PQ, and the velocity of the ascending little circle will be to the sum of the velocities (that is, to the relative velocity of the descending water, with which it flows past the ascending little circle) as the difference between the circles EF and PQ is to the circle EF, or as $EF^2 - PQ^2$ to EF^2. Let that relative velocity be equal to the velocity with which (as shown above) the water passes through the same annular space while the little circle remains unmoved, that is, to the velocity that the water can acquire in falling and describing by its fall a space equal to the height IG; then the force of the water upon the ascending little circle will be the same as before (by corol. 5 of the laws), that is, the resistance of the ascending little circle will be to the weight of a cylinder of water whose base is that little circle, and whose height is ½IG, very nearly as EF^2 to $EF^2 - \frac{1}{2}PQ^2$. And the velocity of the little circle will be to the velocity that the water acquires in falling, and describing by its fall a space equal to the height IG, as $EF^2 - PQ^2$ to EF^2.

Let the breadth of the channel be increased indefinitely; then those ratios between $EF^2 - PQ^2$ and EF^2 and between EF^2 and $EF^2 - \frac{1}{2}PQ^2$ will ultimately approach ratios of equality. And therefore the velocity of the little circle will now be that which the water can acquire in falling and describing by its fall a space equal to the height IG, and its resistance will come out equal to the weight of a cylinder whose base is that little circle and whose height is half of the height IG from which the cylinder must fall in order to acquire the velocity of the ascending little circle, and with this velocity the cylinder will, in the time of falling, describe four times its own length. And the resistance of the cylinder, moving forward with this velocity in the direction of its length, is the same as the resistance of the little circle (by lem. 4) and thus is very nearly equal to the force by which its motion can be generated while it is describing four times its length.

If the length of the cylinder is increased or decreased, its motion, and also the time in which it describes four times its length, will be increased or decreased in the same ratio; and thus that force by which the increased or decreased motion, in a time equally increased or decreased, could be generated or destroyed will not be changed and accordingly is under these circumstances still equal to the resistance of the cylinder; for this also remains unchanged, by lem. 4.

If the density of the cylinder is increased or decreased, its motion, and also the force by which the motion can be generated or destroyed in the same time, will be increased or decreased in the same ratio. The resistance, therefore, of any cylinder to the force by which its whole motion could be either generated or destroyed, while it is describing four times its length, will be very nearly as the density of the medium to the density of the cylinder. Q.E.D.

A fluid must be compressed in order to be continuous, and it must be continuous and nonelastic in order that every pressure arising from its compression may be propagated instantaneously and, by acting equally upon all parts of a moving body, not change the resistance. The pressure arising from the body's motion is of course used in generating the motion of the parts of the fluid and creates resistance. But the pressure arising from the compression of the fluid, however strong it may be, if it is propagated instantaneously, generates no motion in the parts of a continuous fluid, introduces no change of motion at all, and thus neither increases nor decreases the resistance. Certainly the action of a fluid that arises from its compression cannot be stronger upon the back of a moving body than upon the front and thus cannot decrease the resistance described in this proposition; and the action will not be stronger upon the front than upon the back provided that its propagation is infinitely swifter than the motion of the body pressed. And the action will be infinitely swifter and will be propagated instantaneously provided that the fluid is continuous and nonelastic.

COROLLARY 1. The resistances to cylinders that move uniformly forward in the direction of their lengths in infinite and continuous mediums are in a ratio compounded of the squared ratio of the velocities and the squared ratio of the diameters and the ratio of the density of the mediums.

COROLLARY 2. If the breadth of the channel is not increased indefinitely, but the cylinder moves forward in the direction of its own length in an enclosed medium at rest, and meanwhile its axis coincides with the axis of the channel, then the resistance to the cylinder will be to the force by which its whole motion could be either generated or destroyed, in the time in which it describes four times its length, in a ratio compounded of the simple ratio of EF^2 to $EF^2 - \frac{1}{2}PQ^2$ and the squared ratio of EF^2 to $EF^2 - PQ^2$ and the ratio of the density of the medium to the density of the cylinder.

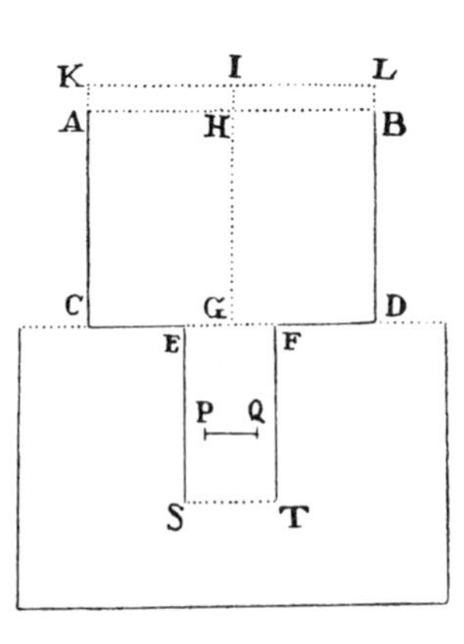

Corollary 3. With the same suppositions, let the length L be to four times the length of the cylinder in a ratio compounded of the simple ratio of $EF^2 - \frac{1}{2}PQ^2$ to EF^2 and the squared ratio of $EF^2 - PQ^2$ to EF^2; then the resistance of the cylinder will be to the force by which its whole motion could be either destroyed or generated, while it is describing the length L, as the density of the medium to the density of the cylinder.

Scholium

In this proposition we have investigated the resistance arising solely from the magnitude of the transverse section of a cylinder, without considering the part of the resistance that can arise from the obliquity of the motions. In prop. 36, case 1, the flow of the water through the hole EF was impeded by the obliquity of the motions with which the parts of the water in the vessel converged from all sides into the hole. Similarly, in this proposition, the obliquity of the motions with which the parts of the water pressed by the front end of the cylinder yield to the pressure and diverge on all sides has these effects: it retards the passage of those motions through the places around that front end toward the back of the cylinder, it makes the fluid move to a greater distance, and it increases the resistance in nearly the ratio with which it decreases the flow of the water from the vessel, that is, in the squared ratio of 25 to 21, roughly.

In case 1 of prop. 36 we made the parts of the water pass through the hole EF perpendicularly and in the greatest abundance by supposing that all the water in the vessel that had been frozen around the cataract, and whose motion was oblique and useless, remained without motion. Similarly, in this proposition, in order that the obliquity of the motions may be annulled, and the parts of the water, by yielding with the most direct and rapid motion, may provide the easiest passage to the cylinder, and in order that only the resistance may remain that arises from the magnitude of the transverse section and that cannot be decreased except by decreasing the diameter of the cylinder, it must be understood that the parts of the fluid whose motions are oblique and useless and create resistance are at rest with respect to one another at both ends of the cylinder and cohere and are joined to the cylinder. Let ABDC be a rectangle, and let AE and BE be two parabolic arcs described with axis AB

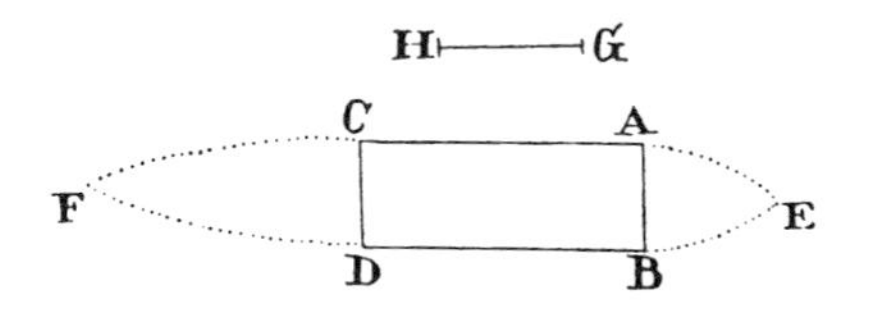

and with a latus rectum that is to the space HG, which is to be described by the falling cylinder while it is acquiring its velocity, as HG to ½AB. Additionally, let CF and DF be two other parabolic arcs, described with axis CD and a latus rectum that is four times the former latus rectum; and by the revolution of the figure about its axis EF, let a solid be generated whose middle ABDC is the cylinder with which we are dealing, and whose extremities ABE and CDF contain the parts of the fluid which are at rest with respect to one another and solidified into two rigid bodies that adhere to the cylinder at the ends as head and tail. Then the resistance to the solid EACFDB moving forward in the direction of its axis FE from F toward E will be very nearly that which we have described in this proposition. That is, the density of the fluid is to the density of the cylinder very nearly in the ratio of this resistance to the force by which the whole motion of the cylinder could be either destroyed or generated, while the length 4AC is being described with that motion continued uniformly. And with this force the resistance cannot be less than in the ratio of 2 to 3, by prop. 36, corol. 7.

Lemma 5 *If a cylinder, a sphere, and a spheroid, whose widths are equal, are placed successively in the middle of a cylindrical channel in such a way that their axes coincide with the axis of the channel, these bodies will equally impede the flow of water through the channel.*

For the spaces through which the water passes between the channel and the cylinder, sphere, and spheroid are equal; and water passes equally through equal spaces.

This is so on the hypothesis that all the water is frozen which is above the cylinder, sphere, or spheroid, and whose fluidity is not required for the very swift passage of the water, as I have explained in prop. 36, corol. 7.

Lemma 6 *With the same suppositions, these bodies are equally urged by the water flowing through the channel.*

This is evident by lem. 5 and the third law of motion. Of course, the water and the bodies act equally upon one another.

Lemma 7 *If the water is at rest in the channel, and these bodies go through the channel with equal velocity in opposite directions, the resistances will be equal to one another.*

This is clear from lem. 6; for the relative motions remain the same with respect to one another.

Scholium

It is the same for all convex round bodies whose axes coincide with the axis of the channel. Some difference can arise from a greater or lesser friction; but in these lemmas we are supposing that the bodies are very smooth, that the tenacity and friction of the medium are nil, and that the parts of the fluid which by their oblique and superfluous motions can perturb, impede, and retard the flow of the water through the channel are at rest with respect to one another as if icebound and adhere to the front and back of the bodies, as I have explained in the scholium to prop. 37. For what follows deals with the least possible resistance of round bodies described with the greatest given transverse sections.

Bodies moving straight ahead in fluids make the fluid ascend in front of them and subside in back of them, especially if they are blunt in shape; and hence they encounter a slightly greater resistance than if they had pointed heads and tails. And bodies moving in elastic fluids, if they are blunt in front and in back, condense the fluid a little more at the front and make it a little less dense at the back; and hence they encounter a slightly greater resistance than if they had pointed heads and tails. But in these lemmas and propositions we are not dealing with elastic fluids but with nonelastic fluids, not with bodies floating on the surface of the fluid but with bodies deeply immersed. And once the resistance of bodies in nonelastic fluids is known, this resistance will have to be increased somewhat for elastic fluids such as air as well as for the surfaces of stagnant fluids such as seas and swamps.

Proposition 38
Theorem 30

The resistance to a sphere moving uniformly forward in an infinite and nonelastic compressed fluid is to the force by which its whole motion could either be destroyed or generated, in the time in which it describes 8/3 of its diameter, very nearly as the density of the fluid to the density of the sphere.

For a sphere is to the circumscribed cylinder as 2 to 3, and therefore the force that could take away all the motion of a cylinder, while the cylinder is describing a length of four diameters, will take away all the motion of the sphere while the sphere describes 2/3 of this length, that is, 8/3 of its own diameter. And the resistance of the cylinder is to this force very nearly as the density of the fluid to the density of the cylinder or sphere, by prop. 37, and the resistance of the sphere is equal to the resistance of the cylinder, by lems. 5, 6, and 7. Q.E.D.

COROLLARY 1. The resistances of spheres in infinite compressed mediums are in a ratio compounded of the squared ratio of the velocity and the squared ratio of the diameter and the ratio of the density of the mediums.

COROLLARY 2. The greatest velocity with which a sphere, by the force of its own relative weight, can descend in a resisting fluid is that which the same sphere with the same weight can acquire in falling without resistance and describing by its fall a space that is to 4/3 of its diameter as the density of the sphere to the density of the fluid. For the sphere in the time of its fall, with the velocity acquired in falling, will describe a space that will be to 8/3 of its diameter as the density of the sphere to the density of the fluid; and the force of its weight generating this motion will be to the force that could generate the same motion, in the time in which the sphere describes 8/3 of its diameter with the same velocity, as the density of the fluid to the density of the sphere; and thus, by this proposition, the force of its weight will be equal to the force of resistance and therefore cannot accelerate the sphere.

COROLLARY 3. Given both the density of the sphere and its velocity at the beginning of the motion, and also the density of the compressed fluid at rest in which the sphere moves, then by prop. 35, corol. 7, the velocity of the sphere, its resistance, and the space described by it are given for any time.

COROLLARY 4. A sphere moving in a compressed fluid at rest, having the same density as itself, will, by the same corol. 7, lose half of its motion before it has described the length of two of its diameters.

Proposition 39
Theorem 31

The resistance to a sphere moving uniformly forward through a fluid enclosed and compressed in a cylindrical channel is to the force by which its whole motion could be either generated or destroyed, while it describes 8/3 of its diameter, in a ratio compounded of three ratios, very nearly: the ratio of the opening of the channel to the excess of this opening over half of a great circle of the sphere, the squared ratio of the opening of the channel to the excess of this opening over a great circle of the sphere, and the ratio of the density of the fluid to the density of the sphere.

This is evident by prop. 37, corol. 2, and the proof proceeds as in prop. 38.

Scholium

In the last two propositions (as in lem. 5) I assume that all the water which is in front of the sphere, and whose fluidity increases the resistance to the sphere, is frozen. If all that water liquefies, the resistance will be somewhat increased. But in these propositions the increase will be small and can be ignored because the convex surface of the sphere has almost the same effect as ice.

Proposition 40
Problem 9

To find from phenomena the resistance of a sphere moving forward in a compressed, very fluid medium.

Let A be the weight of the sphere in a vacuum, B its weight in a resisting medium, D the diameter of the sphere, F a space that is to $\frac{4}{3}$D as the density of the sphere to the density of the medium (that is, as A to A−B), G the time in which the sphere in falling by its weight B without resistance describes the space F, and H the velocity that the sphere acquires by this fall. Then H will be the greatest velocity with which the sphere can descend by its weight B in the resisting medium, by prop. 38, corol. 2, and the resistance that the sphere encounters while descending with this velocity will be equal to its weight B; and the resistance that it encounters with any other velocity will be to the weight B as the square of the ratio of this velocity to the greatest velocity H, by prop. 38, corol. 1.

This is the resistance that arises from the inertia of matter of the fluid. And that which arises from the elasticity, tenacity, and friction of its parts can be investigated as follows.

Drop the sphere so that it descends in the fluid by its own weight B; and let P be the time of falling, in seconds if the time G is in seconds. Find the absolute number N that corresponds to the logarithm $0.4342944819\frac{2P}{G}$, and let L be the logarithm of the number $\frac{N+1}{N}$, then the velocity acquired in falling will be $\frac{N-1}{N+1}H$, and the space described will be $\frac{2PF}{G} - 1.3862943611F + 4.605170186LF$.

If the fluid is sufficiently deep, the term 4.605170186LF can be ignored, and $\frac{2PF}{G} - 1.3862943611F$ will be the space described, very nearly. These things are evident by book 2, prop. 9 and its corollaries, on the hypothesis that the sphere encounters no other resistance than that which arises from the inertia of matter. But if it encounters another resistance in addition, the descent will be slower, and the quantity of this resistance can be found from the retardation.

Times P	*Velocities of body falling in fluid*	*Spaces described by falling in fluid*	*Spaces described by greatest motion*	*Spaces described by falling in vacuum*
0.001G	$99999\frac{29}{30}$	0.000001F	0.002F	0.000001F
0.01G	999967	0.0001F	0.02F	0.0001F
0.1G	9966799	0.0099834F	0.2F	0.01F
0.2G	19737532	0.0397361F	0.4F	0.04F
0.3G	29131261	0.0886815F	0.6F	0.09F
0.4G	37994896	0.1559070F	0.8F	0.16F
0.5G	46211716	0.2402290F	1.0F	0.25F
0.6G	53704957	0.3402706F	1.2F	0.36F
0.7G	60436778	0.4545405F	1.4F	0.49F
0.8G	66403677	0.5815071F	1.6F	0.64F
0.9G	71629787	0.7196609F	1.8F	0.81F
1G	76159416	0.8675617F	2F	1F
2G	96402758	2.6500055F	4F	4F
3G	99505475	4.6186570F	6F	9F
4G	99932930	6.6143765F	8F	16F
5G	99990920	8.6137964F	10F	25F
6G	99998771	10.6137179F	12F	36F
7G	99999834	12.6137073F	14F	49F
8G	99999980	14.6137059F	16F	64F
9G	99999997	16.6137057F	18F	81F
10G	$99999999\frac{3}{5}$	18.6137056F	20F	100F

So that the velocity and descent of a body falling in a fluid may be found more easily, I have put together the accompanying table, in which the first column denotes the times of descent, the second shows the velocities acquired in falling (the greatest velocity being 100,000,000), the third shows the spaces described in falling in those times (2F being the space that the body describes in the time G with the greatest velocity), and the fourth shows the spaces described in the same times with the greatest velocity. The numbers in the fourth column are $\frac{2P}{G}$, and by subtracting the number 1.3862944 − 4.6051702L, the numbers in the third column are found, and these numbers must be multiplied by the space F in order to get the spaces described in falling. There has been added to these a fifth column, which contains the spaces described in the same times by a body falling in a vacuum by the force of its relative weight B.

Scholium In order to investigate the resistances of fluids by experiments, I got a square wooden vessel, with an internal length and width of 9 inches (of a London foot), and a depth of 9½ feet, and I filled it with rainwater; and making balls of wax with lead inside, I noted the times of descent of the balls, the space of the descent being 112 inches. A solid cubic London foot contains 76 Roman pounds [troy] of rainwater, and a solid inch of this foot contains 19/36 ounce of this pound or 253⅓ grains; and a sphere of water described with a diameter of 1 inch contains 132.645 grains in air, or 132.8 grains in a vacuum; and any other ball is as the excess of its weight in a vacuum over its weight in water.

EXPERIMENT 1. A ball which weighed 156¼ grains in air and 77 grains in water described the whole space of 112 inches [when dropped in water] in a time of 4 seconds. And when the experiment was repeated, the ball again fell in the same time of 4 seconds.

The weight of the ball in a vacuum is 156 13/38 grains, and the excess of this weight over the weight of the ball in water is 79 13/38 grains. And hence the diameter of the ball comes out 0.84224 inch. That excess is to the weight of the ball in a vacuum as the density of water to the density of the ball, and as 8/3 of the diameter of the ball (that is, 2.24597 inches) to the space 2F, which accordingly will be 4.4256 inches. In a time of 1 second the ball will fall in a vacuum by its whole weight of 156 13/38 grains through 193⅓ inches; and by a weight of 77 grains falling in water without resistance, it will in the same time describe 95.219 inches; and in the time G, which is to 1 second as the square root of the ratio of the space F or 2.2128 inches to 95.219 inches, it will describe 2.2128 inches and will attain the greatest velocity H with which it can descend in water. Therefore the time G is 0.15244 seconds. And in this time G, with that greatest velocity H, the ball will describe a space 2F of 4.4256 inches; and thus in the time of 4 seconds it will describe a space of 116.1245 inches. Subtract the space 1.3862944F or 3.0676 inches and there will remain a space of 113.0569 inches which the ball will describe in falling in water in a very wide vessel in the time of 4 seconds. This space, because of the narrowness of the wooden vessel, must be decreased in a ratio which is compounded of the square root of the ratio of the opening of the vessel to the excess of this opening over a great semicircle of the ball, and of the simple ratio of that same opening to its excess over a great circle of the ball, that is, in the ratio of 1 to 0.9914. When this has been done, the result will be a space of 112.08 inches which the ball should, according to the theory, have very nearly described in falling in water in this wooden vessel in the time of 4 seconds. And it described 112 inches in the experiment.

EXPERIMENT 2. Three equal balls, each of which weighed $76\frac{1}{3}$ grains in air and $5\frac{1}{16}$ grains in water, were dropped successively in water, and in a time of 15 seconds each one fell through 112 inches.

By computation the weight of a ball in a vacuum is $76\frac{5}{12}$ grains; the excess of this weight over the weight in water is $71\frac{17}{48}$ grains; the diameter of the ball is 0.81296 inch; $\frac{8}{3}$ of this diameter is 2.16789 inches; the space 2F is 2.3217 inches; the space that a ball describes in falling by a weight of $5\frac{1}{16}$ grains in the time of 1 second without resistance is 12.808 inches; and the time G is 0.301056 second. The ball, therefore, with the greatest velocity with which it can descend in water by the force of the weight of $5\frac{1}{16}$ grains, will describe in a time of 0.301056 second a space of 2.3217 inches, and in the time of 15 seconds a space of 115.678 inches. Subtract the space 1.3862944F or 1.609 inches, and there will remain a space of 114.069 inches which accordingly the ball ought to describe in falling in the same time in a very wide vessel. Because of the narrowness of our vessel a space of roughly 0.895 inch must be taken away. And thus there will remain a space of 113.174 inches which the ball, according to the theory, should have very nearly described in falling in this vessel in the time of 15 seconds. And it described 112 inches in the experiment. The difference is imperceptible.

EXPERIMENT 3. Three equal balls, each of which weighed 121 grains in air and 1 grain in water, were dropped successively in water, and in times of 46 seconds, 47 seconds, and 50 seconds, fell 112 inches.

According to the theory, these balls should have fallen in a time of roughly 40 seconds. I am uncertain whether their falling more slowly is to be attributed to the smaller proportion of the resistance that arises from the force of inertia in slow motions to the resistance that arises from other causes, or rather to some little bubbles adhering to the ball, or to the rarefaction of the wax from the heat either of the weather or of the hand dropping the ball, or even to imperceptible errors in weighing the balls in water. And thus the weight of the ball in water ought to be more than 1 grain, so that the experiment may be made certain and trustworthy.

EXPERIMENT 4. I began the experiments thus far described in order to investigate the resistances of fluids before formulating the theory set forth in the immediately preceding propositions. Afterward, in order to examine that theory, I obtained a wooden vessel with an internal width of 8⅔ inches and a depth of 15⅓ feet. Then I made four balls out of wax with lead inside, each one weighing 139¼ grains in air and 7⅛ grains in water. And I let them fall in water in order to measure the times of falling, using a pendulum oscillating in half-seconds. When the balls were being weighed, and afterward when they were falling, they were cold and had remained cold for some time, because heat rarefies the wax and by the rarefaction diminishes the weight of the ball in water, and the rarefied wax is not immediately brought back to its original density by chilling. Before they fell, they were entirely immersed in water, so that their descent might not be accelerated at the beginning by the weight of some part projecting out of the water. And when totally immersed and at rest, they were let fall as carefully as possible, so as not to receive some impulse from the hand letting them fall. And they fell successively in the times of 47½, 48½, 50, and 51 oscillations, describing a space of 15 feet 2 inches. But the weather was now a little colder than when the balls were weighed, and so I repeated the experiment on another day, and the balls fell in the times of 49, 49½, 50, and 53 oscillations, and on a third day in the times of 49½, 50, 51, and 53 oscillations. The experiment was made quite often, and the balls for the most part fell in the times of 49½ and 50 oscillations. When they fell more slowly, I suspect that they were retarded by hitting against the sides of the vessel.

Now by computation according to the theory, the weight of a ball in a vacuum is 139⅖ grains; the excess of this weight over the weight of the ball in water is 132 11/40 grains; the diameter of the ball is 0.99868 inch; 8/3 of the diameter is 2.66315 inches; the space 2F is 2.8066 inches; the space that a ball describes in falling with a weight of 7⅛ grains in the time of 1 second without resistance is 9.88164 inches; and the time G is 0.376843 second. The ball, therefore, with the greatest velocity with which it can descend in water by a force of weight of 7⅛ grains, describes in the time of 0.376843 second a space of 2.8066 inches; in the time of 1 second a space of 7.44766 inches; and in the time of 25 seconds, or 50 oscillations, a space of 186.1915 inches. Subtract the space 1.386294F, or 1.9454 inches, and there will remain the space of 184.2461 inches which the ball will describe in the same time in a very wide vessel. Because of the narrowness of our vessel, decrease this space in a ratio that is compounded of the square root of the ratio of the opening of the vessel to the excess of this opening over a great semicircle of the ball, and the simple ratio of that same opening to its excess over a great circle of the ball, and the result will be the space of 181.86 inches which the ball, according to the theory, should very nearly have described in this vessel in the time of 50 oscillations. And in the experiment it described a space of 182 inches in the time of 49½ or 50 oscillations.

Experiment 5. Four balls weighing 154⅜ grains in air and 21½ grains in water were dropped often and fell in the times of 28½, 29, 29½, and 30 oscillations, and sometimes 31, 32, and 33, describing a space of 15 feet 2 inches.

By the theory they ought to have fallen in the time of very nearly 29 oscillations.

Experiment 6. Five balls weighing 212⅜ grains in air and 79½ in water were dropped often and fell in the times of 15, 15½, 16, 17, and 18 oscillations, describing a space of 15 feet 2 inches.

By the theory they ought to have fallen in the time of very nearly 15 oscillations.

Experiment 7. Four balls weighing 293⅜ grains in air and 35⅞ grains in water were dropped often and fell in the times of 29½, 30, 30½, 31, 32, and 33 oscillations, describing a space of 15 feet 1½ inches.

By the theory they ought to have fallen in the time of very nearly 28 oscillations.

In investigating the reason why some of the balls which were of the same weight and size fell more quickly and others more slowly, I hit upon this: that when the balls were first dropped and were beginning to fall, the side which happened to be heavier descended first and generated an oscillatory motion, so that they oscillated around their centers. For by its oscillations a ball communicates a greater motion to the water than if it were descending without oscillations, and in the process loses part of its own motion with which it should descend; and it is retarded more or retarded less in proportion to the greatness or smallness of the oscillation. Further, the ball always recedes from that side which is descending in the oscillation and, by receding, approaches the sides of the vessel and sometimes strikes against the sides. In the case of heavier balls, this oscillation is stronger, and with larger balls, it agitates the water more. Therefore, in order to reduce the oscillation of the balls, I constructed new balls of wax and lead, fixing the lead into one side of the ball near its surface; and I dropped the ball in such a way that the heavier side, as far as possible, was lowest at the beginning of the descent. Thus the oscillations became much smaller than before, and the balls fell in less unequal times, as in the following experiments.

Experiment 8. Four balls, weighing 139 grains in air and 6½ in water, were dropped often and fell in the times of not more than 52 oscillations, and not fewer than 50, and for the most part in the time of roughly 51 oscillations, describing a space of 182 inches.

By the theory they ought to have fallen in the time of roughly 52 oscillations.

Experiment 9. Four balls, weighing 273¼ grains in air and 140¾ in water, were dropped often and fell in the times of not fewer than 12 oscillations and not more than 13, describing a space of 182 inches.

And by the theory these balls ought to have fallen in the time of very nearly 11⅓ oscillations.

Experiment 10. Four balls, weighing 384 grains in air and 119½ in water, were dropped often and fell in the times of 17¾, 18, 18½, and 19 oscillations, describing a space of 181½ inches. And when they fell in the time of 19 oscillations, I sometimes heard them strike the sides of the vessel before they reached the bottom.

And by the theory they ought to have fallen in the time of very nearly 15⁵⁄₉ oscillations.

Experiment 11. Three equal balls, weighing 48 grains in air and 3²⁹⁄₃₂ in water, were dropped often and fell in the times of 43½, 44, 44½, 45, and 46 oscillations, and for the most part 44 and 45, describing a space of very nearly 182½ inches.

By the theory they ought to have fallen in the time of roughly 46⁵⁄₉ oscillations.

Experiment 12. Three equal balls, weighing 141 grains in air and 4⅜ in water, were dropped several times and fell in the times of 61, 62, 63, 64, and 65 oscillations, describing a space of 182 inches.

And by the theory they ought to have fallen in the time of very nearly 64½ oscillations.

From these experiments it is obvious that when the balls fell slowly (as in the second, fourth, fifth, eighth, eleventh, and twelfth experiments), the times of falling are shown correctly by the theory, but that when the balls fell more quickly (as in the sixth, ninth, and tenth experiments), the resistance was a little greater than in the squared ratio of the velocity. For the balls oscillate somewhat while falling, and this oscillation—in balls that are lighter and fall more slowly—ceases swiftly because the motion is weak, while in heavier and larger balls, because the motion is strong, the oscillation lasts longer and can be checked by the surrounding water only after more oscillations. Additionally, the swifter the balls, the less they are pressed by the fluid in back of them; and if the velocity is continually increased, they will at length leave an empty space behind, unless the compression of the fluid is simultaneously increased. The compression of the fluid, moreover, ought (by props. 32 and 33) to be increased in the squared ratio of the velocity in order for the resistance also to be in a squared ratio. Since this does not happen, the swifter balls are pressed a little less from behind, and because of this diminished pressure their resistance becomes a little greater than in the squared ratio of the velocity.

The theory therefore agrees with the phenomena of bodies falling in water; it remains for us to examine the phenomena of bodies falling in air.

Experiment 13. [a]From the top of St. Paul's Cathedral in London[a] in June 1710, glass balls were dropped simultaneously in pairs, one full of quicksilver, the other full of air; and in falling they described a space of 220 London feet. A wooden platform was suspended at one end by iron pivots, and at the other was supported by a wooden peg. The two balls were placed upon this platform and were let fall simultaneously by pulling out the peg by means of an iron wire extending to the ground, so that the platform, resting on the iron pivots alone, might swing downward upon the pivots and at the same moment a seconds pendulum, pulled by that iron wire, might be released and begin to oscillate. The diameters and weights of the balls and the times of falling are shown in the following table.

However, the observed times need to be corrected. For balls filled with mercury will (by Galileo's theory) describe 257 London feet in 4 seconds, and 220 feet in only 3 seconds 42 thirds. The wooden platform, when the

aa. In expt. 13, Newton writes of weights being dropped "a culmine ecclesiae Sancti Pauli, in urbe Londini." Newton is not referring to St. Paul's Church in Covent Garden, as is obvious from the fact that the distance through which the weights are let fall is 220 London feet. The only house of worship that tall (about twenty stories) was St. Paul's Cathedral. That these experiments were conducted in St. Paul's Cathedral is evident from the fact that in the cathedral there is a balcony, just below the cupola, at a height corresponding to Newton's 220 London feet. See, below, the note to expt. 14.

Balls full of mercury			*Balls full of air*		
Weights	Diameters	Times of falling	Weights	Diameters	Times of falling
grains	*inches*	*seconds*	*grains*	*inches*	*seconds*
908	0.8	4	510	5.1	8½
983	0.8	4−	642	5.2	8
866	0.8	4	599	5.1	8
747	0.75	4+	515	5.0	8¼
808	0.75	4	483	5.0	8½
784	0.75	4+	641	5.2	8

peg was withdrawn, swung downward more slowly than it should have [i.e., more slowly than in free fall] and as a result impeded the descent of the balls at the start. For the balls were lying upon the platform near its center, and were in fact a little closer to the pivots than to the peg. And hence the times of falling were prolonged by roughly 18 thirds and so need to be corrected by taking away those thirds, especially in the larger balls, which because of the magnitude of their diameters remained a little longer upon the platform as it swung downward. When this has been done, the times in which the six larger balls fell will come out 8 sec. 12 thirds, 7 sec. 42 thirds, 7 sec. 42 thirds, 7 sec. 57 thirds, 8 sec. 12 thirds, and 7 sec. 42 thirds.

Therefore the fifth of those balls filled with air, with a diameter of 5 inches and a weight of 483 grains, fell in the time of 8 sec. 12 thirds, describing the space of 220 feet. The weight of water equal to this ball is 16,600 grains; and the weight of air equal to it is $\frac{16,600}{860}$ grains, or 19³⁄10 grains, and thus the weight of the ball in a vacuum is 502³⁄10 grains, and this weight is to the weight of air equal to the ball as 502³⁄10 to 19³⁄10, as is the ratio of 2F to ⁸⁄₃ of the diameter of the ball (that is, 2F to 13⅓ inches). And hence 2F comes out 28 feet 11 inches. The ball in falling in a vacuum, with its whole weight of 502³⁄10 grains, in the time of one second describes 193⅓ inches as above, and with a weight of 483 grains describes 185.905 inches, and with the same weight of 483 grains also in a vacuum describes the space F, or 14 feet 5½ inches, in the time of 57 thirds 58 fourths, and attains the greatest velocity with which it could descend in air. With this velocity the ball, in the time of 8 sec. 12 thirds, will describe a space of 245 feet 5⅓ inches. Take away 1.3863F, or 20 feet ½ inch, and there will remain 225 feet 5 inches. It is this space, therefore, that the ball should, by the theory, have described in falling in the time of 8 sec. 12 thirds. And it described a space of 220 feet in the experiment. The difference is negligible.

Applying similar computations also to the remaining balls filled with air, I constructed the following table.

Weights of the balls	*Diameters*	*Times of falling from a height of 220 feet*		*Spaces to be described by the theory*		*Excesses*	
grains	*inches*	*seconds*	*thirds*	*feet*	*inches*	*feet*	*inches*
510	5.1	8	12	226	11	6	11
642	5.2	7	42	230	9	10	9
599	5.1	7	42	227	10	7	10
515	5	7	57	224	5	4	5
483	5	8	12	225	5	5	5
641	5.2	7	42	230	7	10	7

EXPERIMENT 14. In July 1719, Dr. Desaguliers made experiments of this sort again, making hogs' bladders into a round shape by means of a concave wooden sphere, which the moist bladders, inflated with air, were forced to fill; after they were dried and taken out, they were dropped [b]from the lantern at the top of the cupola of the same cathedral, that is, from a height of 272 feet,[b] and at the same moment a lead ball was also dropped, whose weight was roughly two pounds troy. And meanwhile some people standing in the highest part of St. Paul's where the balls were released noted the whole times of falling, and others standing on the ground noted the difference between the times of fall of the lead ball and of the bladder. And the times were measured by half-second pendulums. And one of those who were standing on the ground had a clock with an oscillating spring, vibrating four times per second; someone else had another machine ingeniously constructed with a pendulum also vibrating four times per second. And one of those who were standing in the gallery of the cupola had a similar device. And these instruments were so constructed that their motions might begin or be stopped at will. The lead ball fell in a time of roughly 4¼ seconds. And by adding

bb. Newton here writes of weights dropped "ab altiore loco in templi ejusdem turri rotunda fornicata, nempe ab altitudine pedum 272," that is, "from a higher place in the round arched tower [i.e., from the lantern at the top of the cupola] of the same cathedral." This position corresponds to the height given by Newton, 272 feet.

this time to the aforesaid difference between the times, the whole time in which the bladder fell was determined. The times in which the five bladders continued to fall after the lead ball had completed its fall were 14¾ sec., 12¾ sec., 14⅝ sec., 17¾ sec., and 16⅞ sec. the first time, and 14½ sec., 14¼ sec., 14 sec., 19 sec., and 16¾ sec. the second time. Add 4¼ sec., the time in which the lead ball fell, and the whole times in which the five bladders fell were 19 sec., 17 sec., 18⅞ sec., 22 sec., and 21⅛ sec. the first time, and 18¾ sec., 18½ sec., 18¼ sec., 23¼ sec., and 21 sec. the second time. And the times noted from the cupola were 19⅜ sec., 17¼ sec., 18¾ sec., 22⅛ sec., and 21⅝ sec. the first time, and 19 sec., 18⅝ sec., 18⅜ sec., 24 sec., and 21¼ sec. the second time. But the bladders did not always fall straight down, but sometimes flew about and oscillated to and fro while falling. And the times of falling were prolonged and increased by these motions, sometimes by one-half of one second, sometimes by a whole second. The second and fourth bladders, moreover, fell straighter down the first time, as did the first and third the second time. The fifth bladder was wrinkled and was somewhat retarded by its wrinkles. I calculated the diameters of the bladders from their circumferences, measured by a very thin thread wound round them twice. And I compared the theory with the experiments in the following table, assuming the density of air to be to the density of rainwater as 1 to 860, and calculating the spaces that the balls should, by the theory, have described in falling.

Weights of bladders	*Diameters*	*Times of falling from a height of 272 feet*	*Spaces to be described in those same times, according to the theory*		*Difference between theory and experiments*	
grains	*inches*	*seconds*	*feet*	*inches*	*feet*	*inches*
128	5.28	19	271	11	− 0	1
156	5.19	17	272	0½	+ 0	0½
137½	5.3	18½	272	7	+ 0	7
97½	5.26	22	277	4	+ 5	4
99⅛	5	21⅛	282	0	+10	0

Therefore almost all the resistance encountered by balls moving in air as well as in water is correctly shown by our theory, and is proportional to the density of the fluids—the velocities and sizes of the balls being equal.

In the scholium at the end of sec. 6, we showed by experiments with pendulums that the resistances encountered by equal and equally swift balls moving in air, water, and quicksilver are as the densities of the fluids. We have shown the same thing here more accurately by experiments with bodies falling in air and water. For pendulums in each oscillation arouse in the fluid a motion always opposite to the motion of the pendulum when it returns; and the resistance arising from this motion, and also the resistance to the cord by which the pendulum was suspended, made the whole resistance to the pendulum greater than the resistance found by the experiments with falling bodies. For by the experiments with pendulums set forth in that scholium, a ball of the same density as water ought, in describing the length of its own semidiameter in air, to lose $\frac{1}{3,342}$ of its motion. But by the theory set forth in this seventh section and confirmed by experiments with falling bodies, that same ball ought, in describing that same length, to lose only $\frac{1}{4,586}$ of its motion, supposing that the density of water is to the density of air as 860 to 1. The resistances therefore were found to be greater by the experiments with pendulums (for the reasons already described) than by the experiments with falling balls, and in a ratio of roughly 4 to 3. But since the resistances to pendulums oscillating in air, water, and quicksilver are increased similarly by similar causes, the proportion of the resistances in these mediums will be shown correctly enough by the experiments with pendulums as well as by the experiments with falling bodies. And hence it can be concluded that the resistances encountered by bodies moving in any fluids that are very fluid, other things being equal, are as the densities of the fluids.

On the basis of what has been established, it is now possible to predict very nearly what part of the motion of any ball projected in any fluid will be lost in a given time. Let D be the diameter of the ball, and V its velocity at the beginning of the motion, and T the time in which the ball will—with velocity V in a vacuum—describe a space that is to the space $\frac{8}{3}$D as the density of the ball to the density of the fluid; then the ball projected in that fluid will, in any other time t, lose the part $\frac{tV}{T+t}$ of its velocity $\left(\text{the part } \frac{TV}{T+t} \text{ remaining}\right)$ and will describe a space that is to the space described in a vacuum in the same time with the uniform velocity V as the logarithm of the number $\frac{T+t}{T}$ multiplied by the number 2.302585093 is to the number t/T, by prop. 35, corol. 7. In slow motions the resistance can be a little less, because the shape of a ball is a little more suitable for motion than the shape of a cylinder described with the same diameter. In swift motions the resistance can be a little greater, because the elasticity and the compression of the fluid are not increased in the squared ratio of the velocity. But here I am not considering petty details of this sort.

And even if air, water, quicksilver, and similar fluids, by some infinite division of their parts, could be subtilized and become infinitely fluid mediums, they would not resist projected balls any the less. For the resistance which is the subject of the preceding propositions arises from the inertia of matter; and the inertia of matter is essential to bodies and is always proportional to the quantity of matter. By the division of the parts of a fluid, the resistance that arises from the tenacity and friction of the parts can indeed be diminished, but the quantity of matter is not diminished by the division of its parts; and since the quantity of matter remains the same, its force of inertia—to which the resistance discussed here is always proportional—remains the same. For this resistance to be diminished, the quantity of matter in the spaces through which bodies move must be diminished. And therefore the celestial spaces, through which the globes of the planets and comets move continually in all directions very freely and without any sensible diminution of motion, are devoid of any corporeal fluid, except perhaps the very rarest vapors and rays of light transmitted through those spaces.

Projectiles, of course, arouse motion in fluids by going through them, and this motion arises from the excess of the pressure of the fluid on the front of the projectile over the pressure on the back, and cannot be less in infinitely fluid mediums than in air, water, and quicksilver in proportion to the density of matter in each. And this excess of pressure, in proportion to its quantity, not only arouses motion in the fluid but also acts upon the projectile to retard its motion; and therefore the resistance in every fluid is as the motion excited in the fluid by the projectile, and it cannot be less in the most subtle aether, in proportion to the density of the aether, than in air, water, and quicksilver, in proportion to the densities of these fluids.

SECTION 8

Motion propagated through fluids

Proposition 41
Theorem 32

Pressure is not propagated through a fluid along straight lines, unless the particles of the fluid lie in a straight line.

If the particles *a*, *b*, *c*, *d*, and *e* lie in a straight line, a pressure can indeed be propagated directly from *a* to *e*; but the particle *e* will urge the obliquely placed particles *f* and *g* obliquely, and those particles *f* and *g* will not sustain the pressure brought upon them unless they are supported by the further particles *h* and *k*; but to the extent that they are supported, they press the supporting particles, and these will not sustain the pressure unless they are supported by the further particles *l* and *m* and press them, and so on indefinitely. Therefore, as soon as a pressure is propagated to particles which do not lie in a straight line, it will begin to spread out and will be obliquely propagated indefinitely; and after the pressure begins to be propagated obliquely, if it should impinge upon further particles which do not lie in a straight line, it will spread out again, and will do so as often as it impinges upon particles not lying exactly in a straight line. Q.E.D.

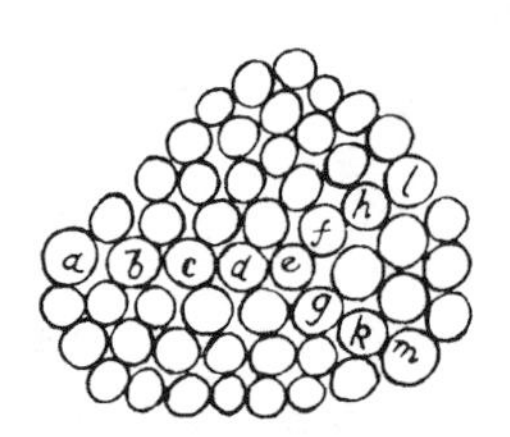

Corollary. If some part of a pressure propagated through a fluid from a given point is intercepted by an obstacle, the remaining part (which is not intercepted) will spread out into the spaces behind the obstacle. This can be proved as follows. From point A let a pressure be propagated in any direction and, if possible, along straight lines; and by the obstacle NBCK, perforated in BC, let all the pressure be intercepted except the cone-shaped part APQ, which passes through the circular hole BC. By transverse planes *de*, *fg*, and *hi*, divide the cone APQ into frusta; then, while the cone ABC, by propagating the pressure, is urging the further conic frustum *degf* on the surface *de*, and this frustum is urging the next frustum *fgih* on the surface *fg*, and that frustum is urging a third frustum, and so on indefinitely, obviously (by the third law of motion) the first frustum *defg* will be as much urged and pressed on the surface *fg* by the reaction of the second frustum *fghi* as it urges and presses the second frustum. Therefore the frustum *degf* between the cone A*de* and the frustum *fhig* is compressed on both sides,

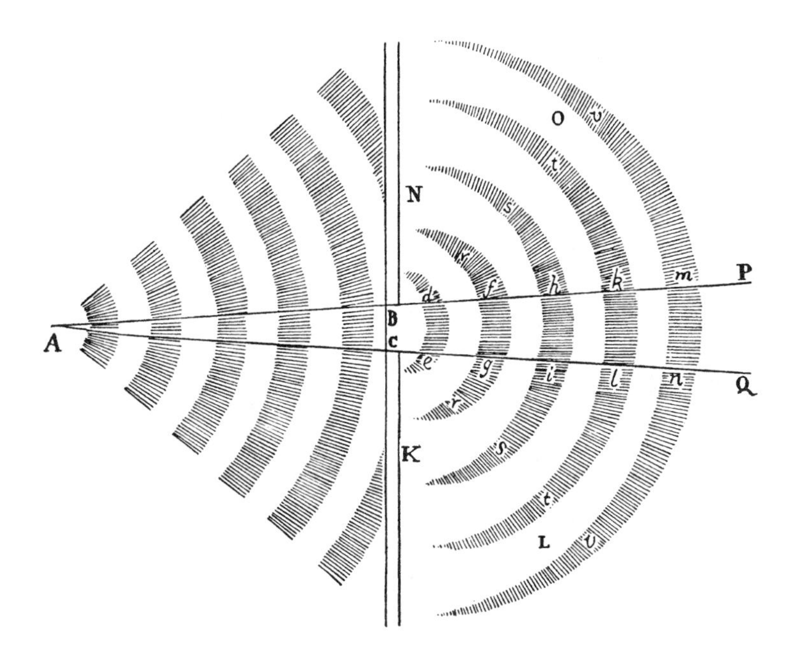

and therefore (by book 2, prop. 19, case 6) it cannot keep its figure unless it is compressed by the same force on all sides. With the same force, therefore, with which it is pressed on the surfaces *de* and *fg*, it will endeavor to yield at the sides *df* and *eg*; and there (since it is not rigid, but altogether fluid) it will run out and expand, unless a surrounding fluid is present to restrain that endeavor. Accordingly, by the endeavor to run out, it will press the surrounding fluid at the sides *df* and *eg*, as well as the frustum *fghi*, with the same force; and therefore the pressure will be no less propagated from the sides *df* and *eg* into the spaces NO on one side and KL on the other, than it is propagated from the surface *fg* toward PQ. Q.E.D.

Proposition 42
Theorem 33

All motion propagated through a fluid diverges from a straight path into the motionless spaces.

CASE 1. Let a motion be propagated from point A through a hole BC, and let it proceed, if possible, in the conic space BCQP along straight lines diverging from point A. And let us suppose first that this motion is that of waves on the surface of stagnant water. And let *de*, *fg*, *hi*, *kl*, . . . be the highest parts of the individual waves, separated from one another by the same number of intermediate troughs. Therefore, since the water is higher in the crests of the waves than in the motionless parts LK and NO of the

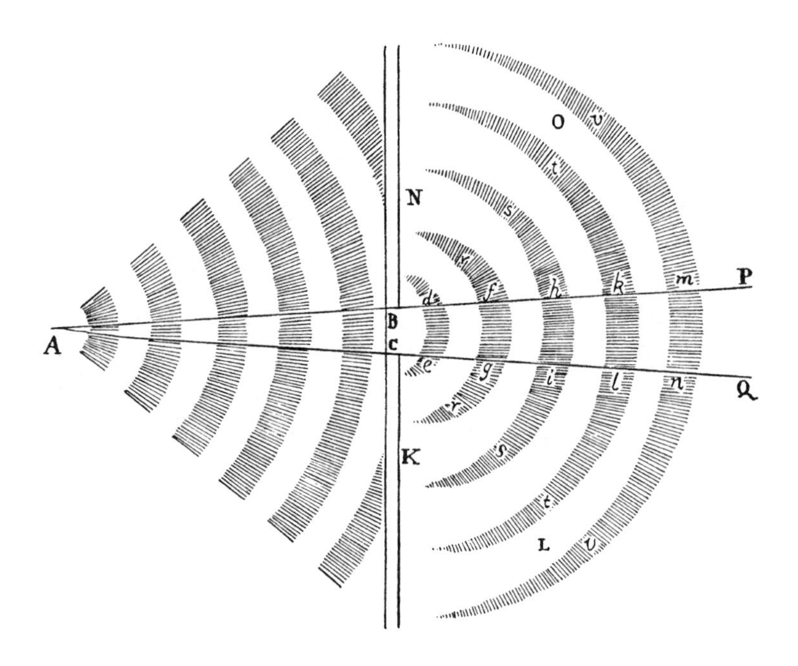

fluid, it will flow down from *e*, *g*, *i*, *l*, . . . , and *d*, *f*, *h*, *k*, . . . , the ends of the crests, toward KL on one side and NO on the other; and since it is lower in the troughs of the waves than in the motionless parts KL and NO of the fluid, it will flow down from those motionless parts into the troughs of the waves. In one case the crests of the waves, and in the other their troughs, are expanded and propagated toward KL on one side and NO on the other. And since the motion of the waves from A toward PQ takes place by the continual flowing down of the crests into the nearest troughs, and thus is not quicker than in proportion to the quickness of the descent, and since the descent of the water toward KL on one side and NO on the other ought to occur with the same velocity, the expansion of the waves will be propagated toward KL on one side and NO on the other with the same velocity with which the waves themselves progress directly from A toward PQ. And accordingly the whole space toward KL on one side and NO on the other will be occupied by the expanded waves *rfgr*, *shis*, *tklt*, *vmnv*, ... Q.E.D. Anyone can test this in stagnant water.

Case 2. Now let us suppose that *de*, *fg*, *hi*, *kl*, and *mn* designate pulses successively propagated from point A through an elastic medium. Think of the pulses as propagated by successive condensations and rarefactions of the medium, in such a way that the densest part of each pulse occupies a spherical surface described about the center A, and that the spaces which come between successive pulses are equal. And let *de*, *fg*, *hi*, *kl*, . . . designate the densest parts of the pulses, parts which are propagated through the hole BC. And since the medium is denser there than in the spaces toward KL on one side and NO on the other, it will expand toward those spaces KL and NO situated on both sides as well as toward the rarer intervals between the pulses; and thus, always becoming rarer next to the intervals and denser next to the pulses, the medium will participate in their motion. And since the progressive motion of the pulses arises from the continual slackening of the denser parts toward the rarer intervals in front of them, and since the pulses ought to slacken with nearly the same speed into the medium's parts KL on one side and NO on the other, which are at rest, those pulses will expand on all sides into the motionless spaces KL and NO with nearly the same speed with which they are propagated straight forward from the center A, and thus will occupy the whole space KLON. Q.E.D. We find this by experience in the case of sounds, which are heard when there is a mountain in the way or which expand into all parts of a room when let in through a window and are heard in all corners, being not so much reflected from the opposite walls as propagated directly from the window, as far as the senses can tell.

CASE 3. Finally, let us suppose that a motion of any kind is propagated from A through the hole BC. That propagation does not occur except insofar as the parts of the medium that are nearer to the center A urge and move the further parts; and the parts that are urged are fluid and thus recede in every direction into regions where they are less pressed, and so will recede toward all the parts of the medium that are at rest, the parts KL and NO on the sides as well as the parts PQ in front. And therefore all the motion, as soon as it has passed through the hole BC, will begin to spread out and to be propagated directly from there into all parts as if from an origin and center. Q.E.D.

Proposition 43
Theorem 34

Every vibrating body in an elastic medium will propagate the motion of the pulses straight ahead in every direction, but in a nonelastic medium will produce a circular motion.

CASE 1. For the parts of a vibrating body, by going forward and returning alternately, will in their going urge and propel the parts of the medium that are nearest to them and by that urging will compress and condense them; then in their return they will allow the compressed parts to recede [i.e., to move apart from one another] and expand. Thus the parts of the medium that are nearest to the vibrating body will go and return alternately, like the parts of the vibrating body; and just as the parts of this body acted upon the parts of the medium, so the latter, acted upon by similar vibrations, will act upon the parts nearest to them, and these, similarly acted upon, will act upon further parts, and so on indefinitely. And just as the first parts of the medium condense in going and rarefy in returning, so the remaining parts will condense whenever they go and will expand [i.e., rarefy] whenever they return. And therefore they will not all go and return at the same time (for thus, by keeping determined distances from one another, they would not rarefy and condense alternately), but by approaching one another when they condense and moving apart when they rarefy, some of them will go while others return, and these conditions will alternate indefinitely. And the parts that are going and that condense in going (because of their forward motion with which they strike obstacles) are pulses; and therefore successive pulses will be propagated straight ahead from every vibrating body, and they will be so propagated at roughly equal distances from one another, because of the equal intervals of time in which the body produces each pulse by each of its vibrations. And even if the parts of the vibrating body go and return in some fixed and determined direction, nevertheless the pulses propagated from there through the medium will (by prop. 42) expand sideways and will be propagated in all directions from the vibrating body as if from a common center, in surfaces almost spherical and concentric. We have an example of this in waves, which, if they are produced by a wagging finger, not only will go to and fro according to the finger's motion but will immediately surround the finger like concentric circles and will be propagated in all directions. For the gravity of the waves takes the place of the elastic force.

CASE 2. But if the medium is not elastic, then, since its parts, pressed by the oscillating parts of the vibrating body, cannot be condensed, the motion will be propagated instantly to the parts where the medium yields most easily, that is, to the parts that the vibrating body would otherwise leave empty behind it. The case is the same as the case of a body projected in any medium. A medium, in yielding to projectiles, does not recede indefinitely, but goes with a circular motion to the spaces that the body leaves behind it. Therefore, whenever a vibrating body goes toward any place [or in any direction], the medium, in yielding, will go with a circular motion to the spaces that the body leaves; and whenever the body returns to its former place, the medium will be forced out and will return to its former place. And even though the vibrating body is not rigid but completely pliant, if it nevertheless remains of a fixed size, then, since it cannot urge the medium by its vibrations in any one place without simultaneously yielding to it in another, that body will make the medium, by receding from the parts where it is pressed, go always with a circular motion to the parts that yield to it. Q.E.D.

COROLLARY. Therefore it is a delusion to believe that the agitation of the parts of flame conduces to the propagation of a pressure along straight lines through a surrounding medium. A pressure of this sort must be derived not only from the agitation of the parts of the flame but from the dilation of the whole.

Proposition 44
Theorem 35

If water ascends and descends alternately in the vertical arms KL *and* MN *of a tube, and if a pendulum is constructed whose length between the point of suspension and the center of oscillation is equal to half of the length of the water in the tube, then I say that the water will ascend and descend in the same times in which the pendulum oscillates.*

I measure the length of the water along the axes of the tube and the arms and make it equal to the sum of these axes, and I do not here consider the resistance of the water that arises from the friction of the tube. Let AB and CD therefore designate the mean height of the water in the two arms, and when the water in the arm KL ascends to the height EF, the water in the arm MN will have descended to the height GH. Moreover, let P be a pendulum bob, VP the cord, V the point of suspension, RPQS the

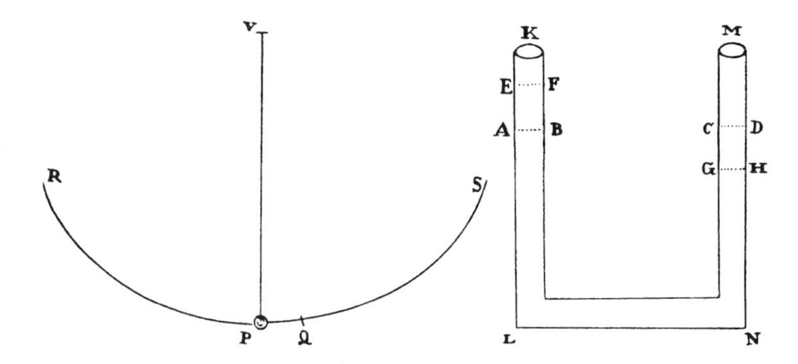

cycloid described by the pendulum, P its lowest point, and PQ an arc equal to the height AE. The force by which the motion of the water is alternately accelerated and retarded is the amount by which the weight of the water in one of the two arms exceeds the weight in the other. And thus, when the water in the arm KL ascends to EF, and in the other arm descends to GH, that force is twice the weight of the water EABF and therefore is to the weight of all the water as AE or PQ to VP or PR. Furthermore, the force by which the weight P in any place Q is accelerated and retarded in the cycloid is (by book 1, prop. 51, corol.) to its whole weight as its distance PQ from the lowest place P to the length PR of the cycloid. Therefore the motive forces of the water and the pendulum, describing the equal spaces AE and PQ, are as the weights that are to be moved; and thus, if the water and the pendulum are at rest in the beginning, those forces will move them equally in equal times and will cause them to go and return synchronously with an alternating motion. Q.E.D.

Corollary 1. Therefore all the alternations of the ascending and descending water are isochronous, whether the motion is of greater intension or greater remission.[a]

Corollary 2. If the length of all the water in the tube is $6\frac{1}{9}$ Paris feet, the water will descend in the time of one second and will ascend in another second and will continue to alternate in this way indefinitely. For a pendulum $3\frac{1}{18}$ feet long oscillates in the time of one second.

Corollary 3. When the length of the water is increased or decreased, moreover, the time of alternation is increased or decreased as the square root of the length.

Proposition 45
Theorem 36

The velocity of waves is as the square roots of the lengths.

This follows from the construction of the following proposition.

Proposition 46
Problem 10

To find the velocity of waves.

Set up a pendulum whose length between the point of suspension and the center of oscillation is equal to the length of the waves; and in the same time in which the pendulum performs each of its oscillations, the waves as they move forward will traverse nearly their own lengths.

a. Newton evidently is referring to amplitude.

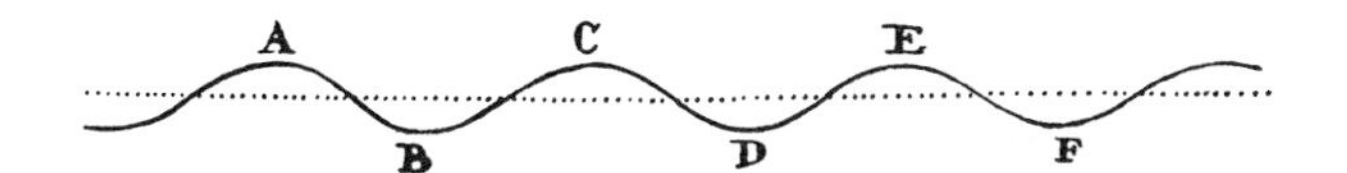

By length of a wave I mean the transverse distance either between bottoms of troughs or between tops of crests. Let ABCDEF designate the surface of stagnant water ascending and descending in successive waves; and let A, C, E, . . . be the crests of the waves, and B, D, F, . . . the troughs in between. Since the motion of the waves is caused by the successive ascent and descent of the water, in such a way that its parts, A, C, E, . . . , which now are highest, soon become lowest, and since the motive force by which the highest parts descend and the lowest ascend is the weight of the elevated water, the alternate ascent and descent will be analogous to the alternating motion of the water in the tube and will observe the same laws with respect to times; and therefore (by prop. 44), if the distances between the highest places A, C, and E of the waves and the lowest, B, D, and F, are equal to twice the length of a pendulum, the highest parts A, C, and E will in the time of one oscillation come to be the lowest, and in the time of a second oscillation will ascend once again. Therefore there will be a time of two oscillations between successive waves; that is, a wave will describe its own length in the time in which the pendulum oscillates twice; but in the same time a pendulum whose length is four times as great, and thus equals the length of the waves, will oscillate once. Q.E.I.

COROLLARY 1. Therefore waves with a length of 3⅟18 Paris feet will move forward through their own length in the time of one second and thus in the time of one minute will traverse 183⅓ feet, and in the space of an hour very nearly 11,000 feet.

COROLLARY 2. And the velocity of waves of greater or smaller length will be increased or decreased as the square root of the length.

What has been said is premised on the hypothesis that the parts of the water go straight up or straight down; but this ascent and descent takes place more truly in a circle, and thus I admit that in this proposition the time has been determined only approximately.

Proposition 47
Theorem 37

If pulses are propagated through a fluid, the individual particles of the fluid, going and returning with a very short alternating motion, are always accelerated and retarded in accordance with the law of an oscillating pendulum.

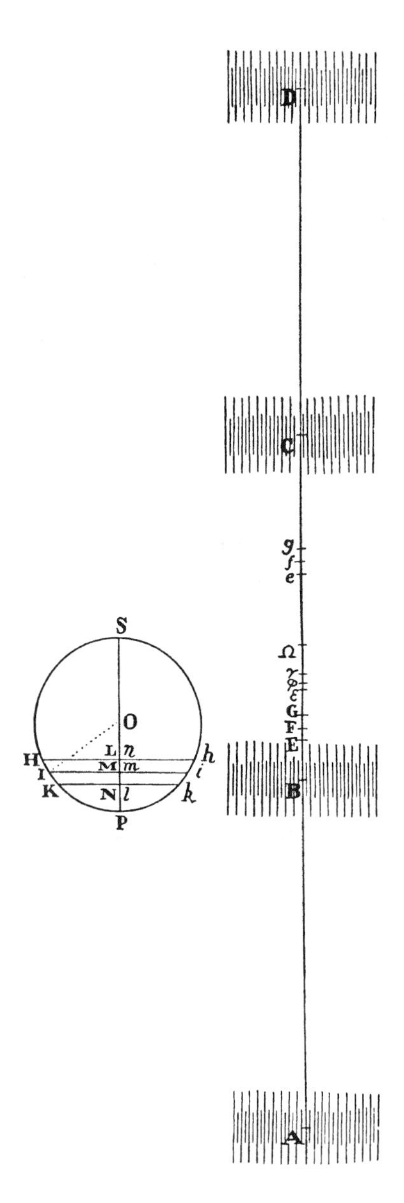

Let AB, BC, CD, . . . designate the equal distances between successive pulses; ABC the line of motion of the pulses, propagated from A toward B; E, F, and G three physical points in the medium at rest, situated at equal intervals along the straight line AC; E*e*, F*f*, and G*g* very short equal spaces through which those points go and return in each vibration with an alternating motion; ε, φ, γ any intermediate positions of those same points; and EF and FG physical line-elements or linear parts of the medium, put between those points and successively transferred into the places $\varepsilon\varphi$, $\varphi\gamma$ and *ef*, *fg*. Draw the straight line PS equal to the straight line E*e*. Bisect PS in O, and with center O and radius OP describe the circle SIP*i*.

Let the whole circumference of this circle with its parts represent the whole time of one vibration with its proportional parts, in such a way that when any time PH or PHS*h* is completed, if the perpendicular HL or *hl* is dropped to PS, and if Eε is taken equal to PL or P*l*, then the physical point E is found in ε. By this law any point E, in going from E through ε to *e* and returning from there through ε to E, will perform each vibration with the same degrees of acceleration and retardation as the oscillating pendulum. It is to be proved that each of the physical points of the medium must move in such a way. Let us imagine, therefore, that there is such a motion in the medium, arising from any cause, and see what follows.

In the circumference PHS*h* take the equal arcs HI and IK or *hi* and *ik*, having the ratio to the whole circumference that the equal straight lines

EF and FG have to the whole interval BC between pulses. Drop the perpendiculars IM and KN and also *im* and *kn*. Then the points E, F, and G are successively agitated with similar motions and carry out their complete vibrations (consisting of a going and returning) while a pulse is transferred from B to C; accordingly, if PH or PHS*h* is the time from the beginning of the motion of point E, PI or PHS*i* will be the time from the beginning of the motion of point F, and PK or PHS*k* will be the time from the beginning of the motion of point G; and therefore Eε, Fφ, and Gγ will be equal respectively to PL, PM, and PN in the going of the points, or to P*l*, P*m*, and P*n* in the returning of the points. Hence $\varepsilon\gamma$ or EG + Gγ − Eε will be equal to EG − LN in the going of the points, and will be equal to EG + *ln* in their returning. But $\varepsilon\gamma$ is the width or expansion of the part of the medium EG in the place $\varepsilon\gamma$; and therefore the expansion of that part in the going is to its mean expansion as EG − LN to EG, and in the returning is as EG + *ln* or EG + LN to EG. Therefore, since LN is to KH as IM to the radius OP, and KH is to EG as the circumference PHS*h*P to BC, that is (if V is put for the radius of a circle having a circumference equal to the interval between the pulses BC), as OP to V, and since, from the equality of the ratios [or ex aequo], LN is to EG as IM to V, the expansion of the part EG or of the physical point F in the place $\varepsilon\gamma$ will be to the mean expansion which that part has in its own first place EG as V − IM to V in the going, and as V + *im* to V in the returning. Hence the elastic force of point F in the place $\varepsilon\gamma$ is to its mean elastic force in the place EG as $\frac{1}{\mathrm{V}-\mathrm{IM}}$ to $\frac{1}{\mathrm{V}}$ in the going, and as $\frac{1}{\mathrm{V}+im}$ to $\frac{1}{\mathrm{V}}$ in the returning. And by the same argument the elastic forces of the physical points E and G in the going are as $\frac{1}{\mathrm{V}-\mathrm{HL}}$ and $\frac{1}{\mathrm{V}-\mathrm{KN}}$ to $\frac{1}{\mathrm{V}}$; and the difference between the forces is to the mean elastic force of the medium as $\frac{\mathrm{HL}-\mathrm{KN}}{\mathrm{V}^2-\mathrm{V}\times\mathrm{HL}-\mathrm{V}\times\mathrm{KN}+\mathrm{HL}\times\mathrm{KN}}$ to $\frac{1}{\mathrm{V}}$, that is, as $\frac{\mathrm{HL}-\mathrm{KN}}{\mathrm{V}^2}$ to $\frac{1}{\mathrm{V}}$, or as HL − KN to V, provided that (because of the narrow limits of the vibrations) we suppose HL and KN to be indefinitely smaller than the quantity V. Therefore, since the quantity V is given, the difference between the forces is as HL − KN, that is, as OM (because HL − KN is proportional to HK and OM to OI or OP; and HK and OP are given)—that is, if F*f* is bisected in Ω, as $\Omega\varphi$. And by the same argument the difference between the elastic forces of the physical points ε and γ, in the returning of the physical line-element $\varepsilon\gamma$, is as $\Omega\varphi$. But that difference (that is, the amount by which the elastic force of point ε exceeds the elastic force of point γ) is the force by which the intervening physical line-element $\varepsilon\gamma$ of the medium is accelerated in the going and retarded in the returning; and therefore the accelerative force of the physical line-element $\varepsilon\gamma$ is as its distance from the midpoint Ω of the vibration. Accordingly, the time (by book 1, prop. 38) is correctly represented by the arc PI, and the linear part $\varepsilon\gamma$ of the medium moves by the law previously mentioned, that is, by the law of an oscillating pendulum; and the same is true of all the linear parts of which the whole medium is composed. Q.E.D.

COROLLARY. Hence it is evident that the number of pulses propagated is the same as the number of vibrations of the vibrating body and does not increase as the pulses move forward. For as soon as the physical line-element $\varepsilon\gamma$ has returned to its first place, it will be at rest and will not move afterward unless it receives a new motion either by the impact of the vibrating body or by the impact of pulses that are propagated from the vibrating body. It will be at rest, therefore, as soon as the pulses cease to be propagated from the vibrating body.

Proposition 48
Theorem 38

The velocities of pulses propagated in an elastic fluid are as the square root of the elastic force directly and the square root of the density inversely, provided that the elastic force of the fluid is proportional to its condensation.

CASE 1. If the mediums are homogeneous and the distances between pulses in these mediums are equal to one another, but the motion in one medium is more intense, then the contractions and expansions of corresponding parts will be as the motions. In fact, this proportion is not exact. Even so, unless the contractions and expansions are extremely intense, the error will not be perceptible, and thus the proportion can be considered physically exact. But the motive elastic forces are as the contractions and expansions; and the velocities—generated in the same time—of equal parts are as the forces. And thus equal and corresponding parts of corresponding pulses will go and return together through spaces proportional to the contractions and expansions, with velocities that are as the spaces; and therefore the pulses, which advance through their own length in the time of one going and returning and which always succeed into the places of the immediately preceding pulses, will progress in both mediums with an equal velocity, because of the equality of the distances.

CASE 2. But if the distances between pulses, or their lengths, are greater in one medium than in the other, let us suppose that the corresponding parts by going and returning in each alternation describe spaces proportional to the lengths of the pulses; then their contractions and expansions will be equal. And thus if the mediums are homogeneous, those motive elastic forces by which they are agitated with an alternating motion will also be equal. But the matter to be moved by these forces is as the length of the pulses; and the space through which they must move by going and returning in each alternation is in the same ratio. And the time of going and returning is jointly proportional to the square root of the matter and the square root of the space and thus is as the space. But the pulses advance through their own lengths in the times of one going and returning, that is, traverse spaces proportional to the times, and therefore have equal velocities.

CASE 3. In mediums of the same density and elastic force, therefore, all pulses have equal velocities. But if either the density or the elastic force of the medium is intended [i.e., increased], then, since the motive force is increased in the ratio of the elastic force, and the matter to be moved is increased in the ratio of the density, the time in which the same motions as before can be performed will be increased as the square root of the density and will be decreased as the square root of the elastic force. And therefore the velocity of the pulses will be jointly proportional to the square root of the density of the medium inversely and the square root of the elastic force directly. Q.E.D.

This proposition will be clearer from the construction of the following proposition.

Proposition 49
Problem 11

Given the density and elastic force of a medium, it is required to find the velocity of the pulses.

Let us imagine the medium to be compressed, as our air is, by an incumbent weight and let A be the height of a homogeneous medium whose weight is equal to the incumbent weight and whose density is the same as the density of the compressed medium in which the pulses are propagated.

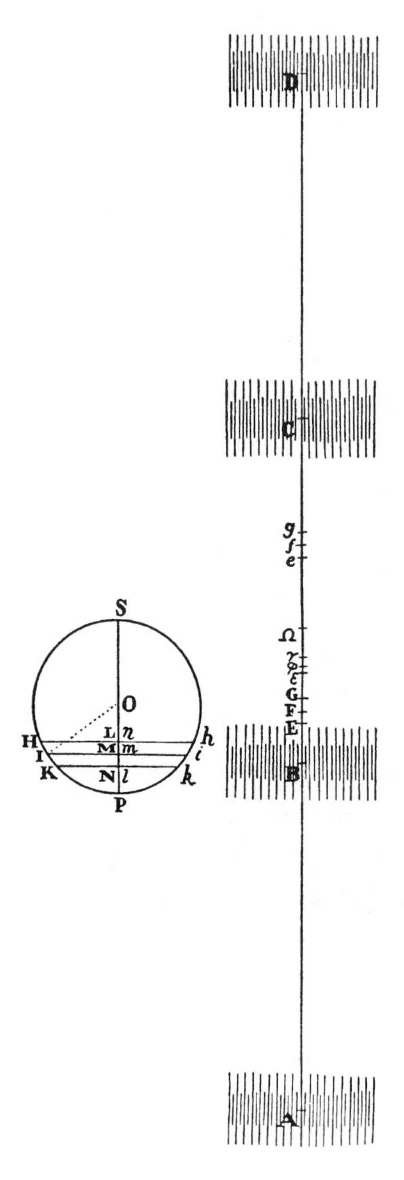

And suppose that a pendulum is set up, whose length between the point of suspension and the center of oscillation is A; then, in the same time in which that pendulum performs an entire oscillation composed of a going and a returning, a pulse will advance through a space equal to the circumference of a circle described with radius A.

For with the same constructions as in prop. 47, if any physical line EF, describing the space PS in each single vibration, is urged in the extremities P and S of each going and returning by an elastic force that is equal to its weight, it will perform each single vibration in the time in which it could oscillate in a cycloid whose whole perimeter is equal to the length PS; and this is so because equal forces will simultaneously impel equal corpuscles through equal spaces. Therefore, since the times of the oscillations are as the square root of the length of the pendulums, and since the length of the pendulum is equal to half the arc of the whole cycloid, the time of one vibration would be to the time of oscillation of a pendulum whose length is A as the square root of the length ½PS or PO to the length A. But the elastic force by which the physical line-element EG is urged in its extremities P and S was (in the proof of prop. 47) to its whole elastic force as HL − KN to V, that is (since point K now falls upon P), as HK to V; and that whole force, that is, the incumbent weight by which the line-element EG is compressed, is to the weight of the line-element as the height A of the incumbent weight to the length EG of the line-element; and thus from the equality of the ratios [or ex aequo] the force by which the line-element EG is urged in its places P and S is to the weight of that line-element as HK × A to V × EG, or as PO × A to V^2 (for HK was to EG as PO to V). Therefore, since the times in which equal bodies are impelled through equal spaces are inversely as the square root of the forces, the time of one vibration under the action of that elastic force will be to the time of the vibration, under the action of the force of weight, as the square root of V^2 to PO × A, and thus will be to the time of oscillation of a pendulum having a length A as $\sqrt{\frac{V^2}{PO \times A}}$ and $\sqrt{\frac{PO}{A}}$ jointly, that is, as V to A. But in the time of one vibration, composed of a going and returning, a pulse advances through its own length BC. Therefore the time in which the pulse traverses the space BC is to the time of one oscillation (composed of a going and returning) as V to A, that is, as BC to the circumference of a circle whose radius is A. But the time in which the pulse will traverse the space BC is in the same ratio to the time in which it will traverse a length equal to this circumference; and thus in the time of such an oscillation the pulse will traverse a length equal to this circumference. Q.E.D.

COROLLARY 1. The velocity of the pulses is that which heavy bodies acquire in falling with a uniformly accelerated motion and describing by their fall half of the height A. For in the time of this fall, with the velocity acquired in falling, the pulse will traverse a space equal to the whole height A; and thus in the time of one oscillation (composed of a going and returning) it will traverse a space equal to the circumference of a circle described with radius A; for the time of fall is to the time of oscillation as the radius of the circle to its circumference.

COROLLARY 2. Hence, since that height A is as the elastic force of the fluid directly and its density inversely, the velocity of the pulses will be as the square root of the density inversely and the square root of the elastic force directly.

Proposition 50
Problem 12

To find the distances between pulses.

In a given time, find the number of vibrations of the body by whose vibration the pulses are excited. Divide by that number the space that a pulse could traverse in the same time, and the part found will be the length of one pulse. Q.E.I.

Scholium The preceding propositions apply to the motion of light and of sounds. For since light is propagated along straight lines, it cannot consist in action alone (by props. 41 and 42). And because sounds arise from vibrating bodies, they are nothing other than propagated pulses of air (by prop. 43). This is confirmed from the vibrations that they excite in bodies exposed to them, provided that they are loud and deep, such as the sounds of drums. For swifter and shorter vibrations are excited with more difficulty. But it is also well known that any sounds impinging upon strings in unison with the sonorous bodies excite vibrations in them. It is confirmed also from the velocity of sounds. For since the specific weights of rainwater and quicksilver are to each other as roughly 1 to 13⅔, and since, when the mercury in a barometer reaches a height of 30 English inches, the specific weight of the air and that of rainwater are to each other as roughly 1 to 870, the specific weights of air and quicksilver will be as 1 to 11,890. Accordingly, since the height of the quicksilver is 30 inches, the height of uniform air whose weight could compress our air lying beneath it will be 356,700 inches, or 29,725 English feet. And this height is the very one that we called A in the construction of prop. 49. The circumference of a circle described with a radius of 29,725 feet is 186,768 feet. And since a pendulum 39⅕ inches long completes an oscillation composed of a going and returning in the time of 2 seconds, as is known, a pendulum 29,725 feet or 356,700 inches long must complete an entirely similar oscillation in the time of 190¾ seconds. In that time, therefore, sound will advance 186,768 feet, and thus in the time of one second, 979 feet.

[a]But in this computation no account is taken of the thickness of the solid particles of air, a thickness through which sound is of course propagated

aa. Ed. 1 has: "Mersenne writes in prop. 35 of his *Ballistics* that he found by making experiments that sound travels 1,150 French toises (that is, 6,900 French feet) in 5 seconds. Hence, since a French foot is to an English foot as 1,068 to 1,000, sound will have to travel 1,474 English feet in the time of 1 second. Mersenne also writes that the eminent geometer Roberval observed during the siege of Thionville that the noise of cannons was heard 13 or 14 seconds after the fire was seen, although he was scarcely half a league away from the cannons. A French league contains 2,500 toises, and thus, according to Roberval's observation, in the time of 13 or 14 seconds sound traveled 7,500 Paris feet, and in the time of 1 second 560 Paris feet, or about 600 English feet. These observations are very different from one another, and our computation falls in the middle. In the cloister of our college, which is 208 feet long, a sound excited at either end makes a fourfold echo in four returnings. And by making experiments I found that at each returning of the sound a pendulum of about 6 or 7 inches completed an oscillation, starting at the first returning of the sound and completing its oscillation at the second one. I was not able to determine the length of the pendulum exactly enough, but I judged that with a length of 4 inches the oscillations were

instantaneously. Since the weight of air is to the weight of water as 1 to 870, and since salts are nearly twice as dense as water, if the particles of air are supposed to be of roughly the same density as the particles of either water or salts, and if the rarity of air arises from the distances between the particles, the diameter of a particle of air will be to the distance between the centers of the particles roughly as 1 to 9 or 10, and to the distance between the particles as 1 to 8 or 9. Accordingly, to the 979 feet which sound will travel in the time of 1 second according to the above calculation, $\frac{979}{9}$ feet or roughly 109 feet may be added, because of the density of the particles of air; and thus sound will travel roughly 1,088 feet in the time of 1 second.

Additionally, the vapors lying hidden in the air, since they are of another elasticity and another tone, participate scarcely or not at all in the motion of the true air by which sounds are propagated. And when these vapors are at rest, that motion will be propagated more swiftly through the true air alone

---

too fast and that with a length of 9 inches they were too slow. Hence in going and returning the sound traveled 416 feet in a smaller time than that in which a pendulum of 9 inches oscillates and in a greater time than a pendulum of 4 inches, that is, in a smaller time than 28¾ thirds and a greater than 19⅙. And therefore in the time of 1 second the sound travels more than 866 English feet and fewer than 1,272 and thus is faster than according to Roberval's observation and slower than according to Mersenne's. Further, by more accurate observations made afterward, I determined that the length of the pendulum ought to be greater than 5½ inches and less than 8 inches and thus that sound in the time of 1 second traveled more than 920 English feet and fewer than 1,085. Therefore the motion of sounds, being between these limits according to the geometrical calculation given above, squares with the phenomena insofar as it has been possible to test it up to now. Accordingly, since this motion depends on the density of the whole air, it follows that sounds consist not in the motion of aether or of some more subtle air but in the agitation of the whole air.

"Certain experiments concerning sound propagated in vessels empty of air seem to contradict this, but vessels can scarcely be emptied of all air; and when they are sufficiently emptied, sounds are noticeably diminished. For example, if only a hundredth of the whole air remains in the vessel, a sound will have to be a hundred times weaker and thus should not be less audible than if someone, hearing the same sound excited in free air, immediately withdrew to ten times the distance from the sonorous body. Two equally sonorous bodies therefore must be compared, of which one is in an emptied vessel and the other is in free air and whose distances from the hearer are as the square roots of the densities of the air, and if the sound of the former body does not exceed the sound of the latter, the objection will cease.

"Once the velocity of sounds has been found, the intervals between the pulses can also be found. Mersenne writes (*Harmonics*, book 1, prop. 4) that (by making certain experiments which he describes in the same place) he found that a stretched musical string vibrates 104 times in the space of 1 second when it makes a unison with an open four-foot organ pipe or a stopped two-foot pipe, which organists call C fa ut. Accordingly, there are 104 pulses in a space of 968 feet, the distance which sound travels in the time of 1 second, and thus one pulse occupies a space of roughly 9¼ feet, that is, roughly twice the length of the pipe. Hence it is likely that the lengths of the pulses in the sounds of all open pipes are equal to twice the lengths of the pipes."

as the square root of the ratio of the total atmosphere of air and vapor to the matter of the particles of air alone. For example, if the atmosphere consists of 10 parts of true air and 1 part of vapors, the motion of sounds will be swifter as the square root of the ratio of 11 to 10, or in roughly the ratio of 21 to 20, than if it were propagated through 11 parts of true air; and thus the motion of sounds that was found above will have to be increased in this ratio. Thus in the time of 1 second, sound will travel 1,142 feet.

These things ought to be so in the springtime and autumn, when the air is rarefied by the temperate heat and its elastic force is somewhat intended [i.e., increased]. But in winter, when the air is condensed by the cold, and its elastic force is remitted [i.e., decreased], the motion of sounds should be slower as the square root of the density; and alternately, in summer it should be swifter.

It is established by experiments, moreover, that in the time of 1 second sounds advance through more or less 1,142 London feet, or 1,070 Paris feet.

Once the velocity of sounds has been found, the intervals between the pulses can also be found. Sauveur found by making experiments that an open pipe, whose length is more or less 5 Paris feet, produces a sound with the same pitch as the sound of a string that vibrates a hundred times in 1 second. Accordingly, there are more or less 100 pulses in the space of 1,070 Paris feet, the distance which sound travels in the time of 1 second, and thus 1 pulse occupies a space of about $10\frac{7}{10}$ Paris feet, that is, roughly twice the length of the pipe. Hence it is likely that the lengths of the pulses in the sounds of all open pipes are equal to twice the lengths of the pipes.[a]

Furthermore, it is evident from book 2, prop. 47, corol., why sounds immediately cease when the motion of the sonorous body ceases, and why they are not heard for a longer time when we are very far distant from the sonorous bodies than when we are very close. Why sounds are very much increased in megaphones is also manifest from the principles set forth. For every reciprocal motion is increased at each reflection by the generating cause. And the motion is lost more slowly and is reflected more strongly in tubes that impede the expansion of sounds, and therefore is more increased by the new motion impressed at each reflection. And these are the major phenomena of sounds.

SECTION 9

The circular motion of fluids

Hypothesis

*The resistance that arises from the friction [*lit. *lack of lubricity or slipperiness] of the parts of a fluid is, other things being equal, proportional to the velocity with which the parts of the fluid are separated from one another.*

Proposition 51
Theorem 39

If an infinitely long solid cylinder revolves with a uniform motion in a uniform and infinite fluid about an axis given in position, and if the fluid is made to revolve by only the impulse of the cylinder, and if each part of the fluid perseveres uniformly in its motion, then I say that the periodic times of the parts of the fluid are as their distances from the axis of the cylinder.

Let AFL be the cylinder made to revolve uniformly about the axis S, and divide the fluid into innumerable concentric solid cylindrical orbs[a] of the same thickness by the concentric circles BGM, CHN, DIO, EKP, ... Then, since the fluid is homogeneous, the impressions that contiguous orbs make upon one another will (by hypothesis) be as their relative displacements and the contiguous surfaces on which the impressions are made. If the impression upon some orb is greater or less on its concave side than on its convex side, the stronger impression will prevail and will either accelerate or retard the motion of the orb, according as it is directed the same way as its motion or the opposite way. Consequently, so that each orb may persevere uniformly in its motion, the impressions on each of the two sides should be equal and be made in opposite directions. Hence,

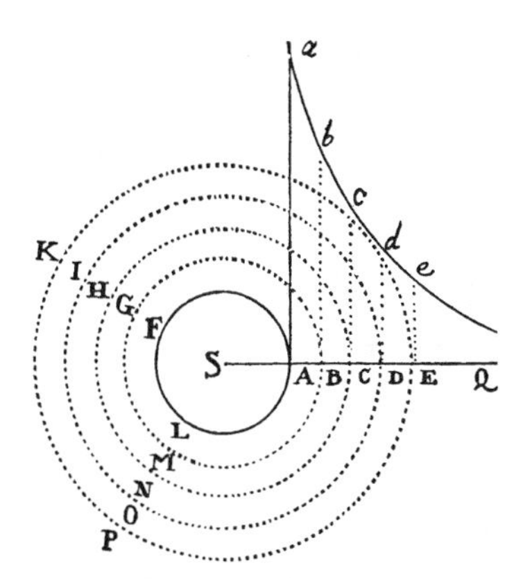

a. In props. 51 and 52, Newton is using the word "orb" in two closely related senses. One is that of a series of nested hollow spheres or orbs, much as in the older Aristotelian universe, where the orbits of the planets were considered to be embedded in a set of nesting or concentric hollow spherical shells or orbs. In prop. 52, Newton writes of a set of "innumerable concentric orbs of the same thickness." In prop. 51, a similar concept is introduced for a cylinder, which Newton says is to be divided into "innumerable concentric solid cylindrical orbs of the same thickness." Today it would not be usual to call such cylindrical shells "orbs" as Newton did; nevertheless, we have kept Newton's "orbs" in prop. 51 so as to keep it in harmony with the language of prop. 52.

since the impressions are as the contiguous surfaces and their relative velocities, the relative velocities will be inversely as the surfaces, that is, inversely as the distances of the surfaces from the axis. And the differences between the angular motions about the axis are as these relative velocities divided by the distances, or as the relative velocities directly and the distances inversely—that is, if the ratios are compounded, as the squares of the distances inversely. Therefore, if the perpendiculars A*a*, B*b*, C*c*, D*d*, E*e*, . . . , inversely proportional to the squares of SA, SB, SC, SD, SE, . . . , are erected to each of the parts of the infinite straight line SABCDEQ and if a hyperbolic curve is understood to be drawn through the ends of the perpendiculars, then the sums of the differences, that is, the whole angular motions, will be as the corresponding sums of the lines A*a*, B*b*, C*c*, D*d*, E*e*; that is, if, in order to make the medium uniformly fluid, the number of orbs is increased and their width decreased indefinitely, as the hyperbolic areas A*a*Q, B*b*Q, C*c*Q, D*d*Q, E*e*Q, . . . , corresponding to these sums. And the times, which are inversely proportional to the angular motions, will also be inversely proportional to these areas. The periodic time of any particle D, therefore, is inversely as the area D*d*Q, that is (by the known quadratures of curves), directly as the distance SD. Q.E.D.

COROLLARY 1. Hence the angular motions of the particles of the fluid are inversely as the distances of the particles from the axis of the cylinder, and the absolute velocities are equal.

COROLLARY 2. If the fluid is contained in a cylindrical vessel of an infinite length and contains another inner cylinder, and if both cylinders revolve about a common axis, and the times of the revolutions are as the semidiameters of the cylinders, and each part of the fluid perseveres in its motion, then the periodic times of the individual parts will be as their distances from the axis of the cylinders.

COROLLARY 3. If any common angular motion is added to, or taken away from, the cylinder and the fluid moving in this way, then, since the mutual friction of the parts of the fluid is not changed by this new motion, the motions of the parts with respect to one another will not be changed. For the relative velocities of the parts depend upon the friction. Any part will persevere in that motion which is not more accelerated than retarded by the friction on opposite sides in opposite directions.

Corollary 4. Hence, if all the angular motion of the outer cylinder is taken away from the whole system of the cylinders and fluid, the result will be the motion of the fluid in the cylinder at rest.

Corollary 5. Therefore, if, while the fluid and outer cylinder are at rest, the inner cylinder revolves uniformly, a circular motion will be communicated to the fluid and will be propagated little by little through the whole fluid, and it will not cease to be increased until the individual parts of the fluid acquire the motion defined in corol. 4.

Corollary 6. And since the fluid endeavors to propagate its own motion even further, its force will make the outer cylinder also revolve, unless that cylinder is forcibly held in place, and the motion of that cylinder will be accelerated until the periodic times of both cylinders become equal. But if the outer cylinder is forcibly held in place, it will endeavor to retard the motion of the fluid, and unless the inner cylinder conserves that motion by some force impressed from outside, the outer cylinder will cause the motion to cease little by little.

All of this can be tested in deep stagnant water.

Proposition 52
Theorem 40

If a solid sphere revolves with a uniform motion in a uniform and infinite fluid about an axis given in position, and if the fluid is made to revolve by only the impulse of this sphere, and if each part of the fluid perseveres uniformly in its motion, then I say that the periodic times of the parts of the fluid will be as the squares of the distances from the center of the sphere.

Case 1. Let AFL be a sphere made to revolve uniformly about the axis S, and divide the fluid into innumerable concentric orbs[a] of the same thickness by means of the concentric circles BGM, CHN, DIO, EKP, ... And imagine the orbs to be solid; then, since the fluid is homogeneous, the impressions that the contiguous orbs make upon one another will (by the hypothesis) be as their relative velocities and

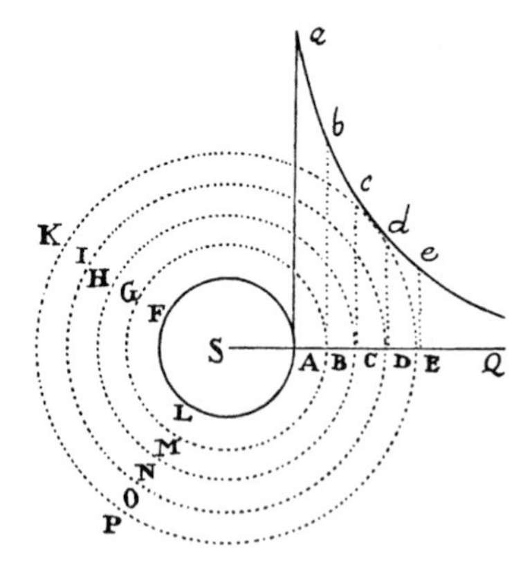

a. On the use of "orbs" in prop. 52, and in the antecedent prop. 51, see the note to prop. 51.

the contiguous surfaces on which the impressions are made. If the impression upon some orb is greater or less on the concave side than on the convex side, the stronger impression will prevail and will either accelerate or retard the velocity of the orb, according as it is directed the same way as the motion of the orb or the opposite way. Consequently, so that each orb may persevere uniformly in its motion, the impressions on each of the two sides will have to be equal and to be made in opposite directions. Hence, since the impressions are as the contiguous surfaces and their relative velocities, the relative velocities will be inversely as the surfaces, that is, inversely as the squares of the distances of the surfaces from the center. But the differences in the angular motions about the axis are as these relative velocities divided by the distances, or as the relative velocities directly and the distances inversely—that is, if the ratios are compounded, as the cubes of the distances inversely. Therefore, if to each of the parts of the infinite straight line SABCDEQ there are erected the perpendiculars A*a*, B*b*, C*c*, D*d*, E*e*, . . . , inversely proportional to the cubes of SA, SB, SC, SD, SE, . . . , then the sums of the differences, that is, the whole angular motions, will be as the corresponding sums of the lines A*a*, B*b*, C*c*, D*d*, E*e*—that is (if, to make the medium uniformly fluid, the number of orbs is increased and their width decreased indefinitely), as the hyperbolic areas A*a*Q, B*b*Q, C*c*Q, D*d*Q, E*e*Q, . . . , corresponding to these sums. And the periodic times, inversely proportional to the angular motions, will also be inversely proportional to these areas. Therefore the periodic time of any orb DIO is inversely as the area D*d*Q, that is (by the known methods of quadratures of curves), directly as the square of the distance SD. And this is what I wanted to prove in the first place.

CASE 2. From the center of the sphere draw as many infinite straight lines as possible which with the axis contain given angles exceeding one another by equal differences, and imagine the orbs to be cut into innumerable rings by the revolution of these straight lines about the axis; then each ring will have four rings contiguous to it, one inside, another outside, and two at the sides. Each ring cannot be urged equally and in opposite directions by the friction of the inner ring and of the outer ring, except in a motion made according to the law of case 1. This is evident from the proof of case 1. And therefore any series of rings proceeding straight from the sphere indefinitely will be moved in accordance with the law of case 1, except insofar as it is impeded by the friction of the rings at the sides. But in motion made

according to this law the friction of the rings at the sides is nil, and thus it will not impede the motion from being made according to this law. If rings equally distant from the center revolved either more quickly or more slowly near the poles than near the ecliptic, the slower rings would be accelerated and the swifter would be retarded by mutual friction, and thus the periodic times would always tend toward equality, in accordance with the law of case 1. This friction, therefore, does not prevent the motion from being made according to the law of case 1, and therefore that law will hold good; that is, the periodic time of each of the rings will be as the square of its distance from the center of the sphere. This is what I wanted to prove in the second place.

CASE 3. Now let each ring be divided by transverse sections into innumerable particles constituting an absolutely and uniformly fluid substance; then, since these sections have no relation to the law of circular motion but contribute only to the constitution of the fluid, the circular motion will continue as before. As a result of this sectioning, all the minimally small rings either will not change the unevenness and the force of their mutual friction or will change them equally. Furthermore, since the proportion of the causes remains the same, the proportion of the effects—that is, the proportion of the motions and periodic times—will also remain the same. Q.E.D.

But since the circular motion, along with the centrifugal force arising from it, is greater at the ecliptic than at the poles, there must be some cause by which each of the particles is kept in its circle; otherwise the matter at the ecliptic would always recede from the center and move on the outside of the vortex to the poles, and return from there along the axis to the ecliptic with a continual circulation.

COROLLARY 1. Hence the angular motions of the parts of the fluid about the axis of the sphere are inversely as the squares of the distances from the center of the sphere, and the absolute velocities are inversely as those same squares divided by the distances from the axis.

COROLLARY 2. If a sphere, in a homogeneous and infinite fluid at rest, revolves with a uniform motion about an axis given in position, it will communicate a motion to the fluid like that of a vortex, and this motion will be propagated little by little without limit, and this motion will not cease to be accelerated in each part of the fluid until the periodic time of each of the parts is as the squares of the distances from the center of the sphere.

COROLLARY 3. Since the inner parts of a vortex, because of their greater velocity, rub and push the outer parts and continually communicate motion to them by this action, and since those outer parts simultaneously transfer the same quantity of motion to others still further out and by this action conserve the quantity of their motion completely unaltered, it is evident that the motion is continually transferred from the center to the circumference of the vortex and is absorbed in that limitless circumference. The matter between any two spherical surfaces concentric with the vortex will never be accelerated, because all of the motion it receives from the inner matter is continually transferred to the outer matter.

COROLLARY 4. Accordingly, for a vortex to conserve the same state of motion constantly, some active principle is required from which the sphere may always receive the same quantity of motion that it impresses on the matter of the vortex. Without such a principle, it is necessary for the sphere and the inner parts of the vortex, always propagating their motion to outer parts and not receiving any new motion, to slow down little by little and cease to be carried around.

COROLLARY 5. If a second sphere were to be immersed in this vortex at a certain distance from the center, and meanwhile by some force were to revolve constantly about an axis given in inclination, then the fluid would be drawn into a vortex by the motion of this sphere; and first this new and tiny vortex would revolve along with the sphere about the center of the first vortex, and meanwhile its motion would spread more widely and little by little would be propagated without limit, in the same way as the first vortex. And for the same reason that the sphere of the new vortex was drawn into the motion of the first vortex, the sphere of the first vortex would also be drawn into the motion of this new vortex, in such a way that the two spheres would revolve about some intermediate point and because of that circular motion would recede from each other unless constrained by some force. Afterward, if the continually impressed forces by which the spheres persevere in their motions were to cease, and everything were left to the laws of mechanics, the motion of the spheres would weaken little by little (for the reason assigned in corols. 3 and 4), and the vortices would at last be completely at rest.

COROLLARY 6. If several spheres in given places revolved continually with certain velocities around axes given in position, the same number of vortices, going on without limit, would be made. For all of the spheres, for the same reason that any one of them propagates its motion without limit, will also propagate their motions without limit, in such a way that each part of the infinite fluid is agitated by that motion which results from the actions of all the spheres. Hence the vortices will not be limited by fixed bounds but will little by little run into one another, and the spheres will be continually moved from their places by the actions of the vortices upon one another, as was explained in corol. 5; nor will they keep any fixed position with respect to one another, unless constrained by some force. And when those forces, which conserve the motions by being continually impressed upon the spheres, cease, the matter—for the reason assigned in corols. 3 and 4—will little by little come to rest and will no longer be made to move in vortices.

COROLLARY 7. If a homogeneous fluid is enclosed in a spherical vessel and is made to revolve in a vortex by the uniform rotation of a sphere placed in the center, and if the sphere and the vessel revolve in the same direction about the same axis, and if their periodic times are as the squares of the semidiameters, then the parts of the fluid will not persevere in their motions without acceleration and retardation until their periodic times are as the squares of the distances from the center of the vortex. No other constitution of a vortex can be stable.

COROLLARY 8. If the vessel, the enclosed fluid, and the sphere conserve this motion and additionally revolve with a common angular motion about any given axis, then, since the friction of the parts of the fluid upon one another is not changed by this new motion, the motions of the parts with respect to one another will not be changed. For the relative velocities of the parts with respect to one another depend upon friction. Any part will persevere in that motion by which the friction on one side does not retard it more than the friction on the other accelerates it.

COROLLARY 9. Hence, if the vessel is at rest, and if the motion of the sphere is given, the motion of the fluid will be given. For imagine that a plane passes through the axis of the sphere and revolves with an opposite motion, and suppose that the sum of the time of the revolution of the plane and the revolution of the sphere is to the time of the revolution of the sphere as the square of the semidiameter of the vessel to the square of the semidiameter of the sphere; then the periodic times of the parts of the fluid with respect to the plane will be as the squares of their distances from the center of the sphere.

Corollary 10. Accordingly, if the vessel moves with any velocity either about the same axis as the sphere or about some different axis, the motion of the fluid will be given. For if the angular motion of the vessel is taken away from the whole system, all the motions with respect to one another will remain the same as before, by corol. 8. And these motions will be given by corol. 9.

Corollary 11. If the vessel and the fluid are at rest, and if the sphere revolves with a uniform motion, then the motion will be propagated little by little through the whole fluid to the vessel, and the vessel will be driven around unless forcibly constrained, and the fluid and vessel will not cease to be accelerated until their periodic times are equal to the periodic times of the sphere. But if the vessel is constrained by some force or revolves with any continual and uniform motion, the medium will little by little come to the state of the motion defined in corols. 8, 9, and 10, nor will it ever persevere in any other state. But then if, when those forces cease by which the vessel and the sphere were revolving with fixed motions, the whole system is left to the laws of mechanics, the vessel and the sphere will act upon each other by means of the intervening fluid and will not cease to propagate their motions to each other through the fluid until their periodic times are equal and the whole system revolves together like one solid body.

Scholium In the preceding propositions, I have been supposing the fluid to consist of matter which is uniform in density and fluidity. The fluid is such that a given sphere, set anywhere in it, would with a given motion in a given interval of time be able to propagate similar and equal motions, at distances always equal from itself. Indeed, matter endeavors by its circular motion to recede from the axis of a vortex and therefore presses all the further matter. From this pressure the friction of the parts becomes stronger and their separation from one another more difficult, and consequently the fluidity of the matter is decreased. Again, if there is any place where the parts of the fluid are thicker or larger, the fluidity will be less there, because the surfaces separating the parts from one another are fewer. In cases of this sort, I suppose the deficiency in fluidity to be supplied either by the slipperiness of the parts or by their pliancy or by some other condition. If this does not happen, the matter will cohere more and will be more sluggish where it is less fluid, and thus will receive motion more slowly and will propagate it further than according to the ratio assigned above. If the shape of the vessel is not spherical, the particles will move in paths which are not circular but correspond to the shape of the vessel, and the periodic times will be very nearly as the squares of the mean distances from the center. In the parts between the center and the circumference where the spaces are wider, the motions will be slower, and where the spaces are narrower the motions will be swifter, and yet the swifter particles will not seek the circumference. For they will describe less-curved arcs, and the endeavor to recede from the center will not be less decreased by the decrement of this curvature than it will be increased by the increment of the velocity. In going from the narrower spaces into the wider, they will recede a little further from the center, but they will be retarded by this receding, and afterward in approaching the narrower spaces from the wider ones they will be accelerated, and thus each of the particles will forever alternately be retarded and accelerated. All of this will be so in a rigid vessel. For the constitution of vortices in an infinite fluid can be found by corol. 6 of this proposition.

Moreover, in this proposition I have tried to investigate the properties of vortices in order to test whether the celestial phenomena could be explained in any way by vortices. For it is a phenomenon that the periodic times of the secondary planets that revolve about Jupiter are as the $3/2$ powers of the distances from the center of Jupiter; and the same rule applies to the planets that revolve about the sun. Moreover, these rules apply to both the primary and the secondary planets very exactly, as far as astronomical observations have shown up to now. And thus if those planets are carried along by vortices revolving about Jupiter and the sun, the vortices will also have to revolve according to the same law. But the periodic times of the parts of a vortex turned out to be in the squared ratio of the distances from the center of motion, and that ratio cannot be decreased and reduced to the $3/2$ power, unless either the matter of the vortex is the more fluid the further it is from the center, or the resistance arising from a deficiency in the slipperiness of the parts of the fluid (as a result of the increased velocity by which the parts of the fluid are separated from one another) is increased in a greater ratio than the ratio in which the velocity is increased. Yet neither of these seems reasonable. The thicker and less-fluid parts, if they are not heavy toward the center, will seek the circumference; and although—for the sake of the proofs—I proposed at the beginning of this section a hypothesis in which the resistance would be proportional to the velocity, it is nevertheless likely that the resistance is in a lesser ratio than that of the velocity. If this is conceded, then the periodic times of the parts of a vortex will be in a ratio greater than the squared ratio of the distances from its center. But if vortices (as is the opinion of some) move more quickly near the center, then more slowly up to a certain limit, then again more quickly near the circumference, certainly neither the $3/2$ power nor any other fixed and determinate ratio can hold. It is therefore up to philosophers to see how that phenomenon of the $3/2$ power can be explained by vortices.

Proposition 53
Theorem 41

Bodies that are carried along in a vortex and return in the same orbit have the same density as the vortex and move according to the same law as the parts of the vortex with respect to velocity and direction.

For if some tiny part of the vortex is composed of particles or physical points which preserve a given situation with respect to one another and is supposed to be frozen, then this part will move according to the same law as before, since it is not changed with respect to its density, or its inherent force or figure. And conversely, if a frozen and solid part of the vortex has the same density as the rest of the vortex and is resolved into a fluid, this part will move according to the same law as before, except insofar as its particles, which have now become fluid, move with respect to one another. Therefore, the motion of the particles with respect to one another may be ignored as having no relevance to the progressive motion of the whole, and the motion of the whole will be the same as before. But this motion will be the same as the motion of other parts of the vortex that are equally distant from the center, because the solid resolved into a fluid becomes a part of the vortex similar in every way to the other parts. Therefore, if a solid is of the same density as the matter of the vortex, it will move with the same motion as the parts of the vortex and will be relatively at rest in the immediately surrounding matter. But if the solid is denser, it will now endeavor to recede from the center of the vortex more than before; and thus, overcoming that force of the vortex by which it was formerly kept in its orbit as if set in equilibrium, it will recede from the center and in revolving will describe a spiral and will no longer return into the same orbit. And by the same argument, if the solid is rarer, it will approach the center. Therefore, the solid will not return into the same orbit unless it is of the same density as the fluid. And it has been shown that in this case the solid would revolve according to the same law as the parts of the fluid that are equally distant from the center of the vortex. Q.E.D.

COROLLARY 1. Therefore a solid that revolves in a vortex and always returns into the same orbit is relatively at rest in the fluid in which it is immersed.

COROLLARY 2. And if the vortex is of a uniform density, the same body can revolve at any distance from the center of the vortex.

Scholium

Hence it is clear that the planets are not carried along by corporeal vortices. For the planets, which—according to the Copernican hypothesis—move about the sun, revolve in ellipses having a focus in the sun, and by radii drawn to the sun describe areas proportional to the times. But the parts of a vortex cannot revolve with such a motion. Let AD, BE, and CF designate three orbits described about the sun S, of which let the outermost CF be a circle concentric with the sun, and let A and B be the aphelia of the two inner ones, and D and E their perihelia. Therefore, a body that revolves in the orbit CF, describing areas proportional to the times by a radius drawn to the sun, will move with a uniform motion. And a body that revolves in the orbit BE will, according to the laws of astronomy, move more slowly in the aphelion B and more swiftly in the perihelion E, although according to the laws of mechanics the matter of the vortex ought to move more swiftly in the narrower space between A and C than in the wider space between D and F, that is, more swiftly in the aphelion than in the perihelion. These two statements are contradictory. Thus in the beginning of the sign of Virgo, where the aphelion of Mars now is, the distance between the orbits of Mars and Venus is to the distance between these orbits in the beginning of the sign of Pisces as roughly 3 to 2, and therefore the matter of the vortex between these orbits in the beginning of Pisces must move more swiftly than in the beginning of Virgo in the ratio of 3 to 2. For the narrower the space through which a given quantity of matter passes in the given time of one revolution,

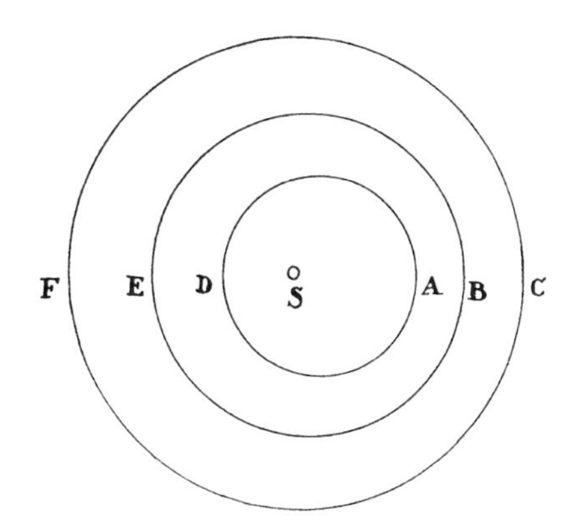

the greater the velocity with which it must pass. Therefore, if the earth, relatively at rest in this celestial matter, were carried by it and revolved along with it about the sun, its velocity in the beginning of Pisces would be to its velocity in the beginning of Virgo as 3 to 2. Hence the apparent daily motion of the sun in the beginning of Virgo would be greater than 70 minutes, and in the beginning of Pisces less than 48 minutes, although (as experience bears witness) the apparent motion of the sun is greater in the beginning of Pisces than in the beginning of Virgo, and thus the earth is swifter in the beginning of Virgo than in the beginning of Pisces. Therefore the hypothesis of vortices can in no way be reconciled with astronomical phenomena and serves less to clarify the celestial motions than to obscure them. But how those motions are performed in free spaces without vortices can be understood from book 1 and will now be shown more fully in book 3 on the system of the world.

BOOK 3

THE SYSTEM OF THE WORLD

In the preceding books I have presented principles of philosophy[a] that are not, however, philosophical but strictly mathematical—that is, those on which the study of philosophy can be based. These principles are the laws and conditions of motions and of forces, which especially relate to philosophy. But in order to prevent these principles from seeming sterile, I have illustrated them with some philosophical scholiums [i.e., scholiums dealing with natural philosophy], treating topics that are general and that seem to be the most fundamental for philosophy, such as the density and resistance of bodies, spaces void of bodies, and the motion of light and sounds. It still remains for us to exhibit the system of the world from these same principles. On this subject I composed an earlier version of book 3 in popular form, so that it might be more widely read. But those who have not sufficiently grasped the principles set down here will certainly not perceive the force of the conclusions, nor will they lay aside the preconceptions to which they have become accustomed over many years; and therefore, to avoid lengthy disputations, I have translated the substance of the earlier version into propositions in a mathematical style, so that they may be read only by those who have first mastered the principles. But since in books 1 and 2 a great number of propositions occur which might be too time-consuming even for readers who are proficient in mathematics, I am unwilling to advise anyone to study every one of these propositions. It will be sufficient to read with care the Definitions, the Laws of Motion, and the first three sections of book 1, and then turn to this book 3 on the system of the world, consulting at will the other propositions of books 1 and 2 which are referred to here.

a. In this introduction to book 3, Newton uses "philosophy" and its adjective "philosophical" to refer to "natural philosophy." According to John Harris's *Lexicon Technicum* (London, 1704), natural philosophy is that "Science which contemplates the Powers of Nature, the Properties of Natural Bodies, and their mutual Action one upon another." The half title of the third edition of the *Principia* reads "Newtoni Principia Philosophiae" ("Newton's Principles of Philosophy"). The dedication page of the *Principia*, in all editions, refers to the Royal Society as founded "ad philosophiam promovendam" ("for the promotion of philosophy").

[a]RULES FOR THE STUDY OF NATURAL PHILOSOPHY

Rule 1 *No more causes of natural things should be admitted than are both true and sufficient to explain their phenomena.*

As the philosophers say: Nature does nothing in vain, and more causes are in vain when fewer suffice. For nature is simple and does not indulge in the luxury of superfluous causes.

aa. Ed. 1 has nine numbered "Hypotheses," most of which ed. 2 converts into two categories, now called "Rules for Natural Philosophy" and "Phenomena." Hyps. 1 and 2 become rules 1 and 2; hyp. 3 is discarded, to be replaced by rule 3; hyp. 4 becomes hyp. 1 and is transferred to a location between prop. 10 and prop. 11; hyps. 5–9 become phen. 1, 3–6, while phen. 2 is new in ed. 2. Ed. 3 further introduces rule 4. These changes may be tabulated as follows:

Ed. 1	*Ed. 2*	*Ed. 3*
hypothesis 1	rule 1	rule 1
hypothesis 2	rule 2	rule 2
hypothesis 3	—	—
—	rule 3	rule 3
—	—	rule 4
hypothesis 4	hypothesis 1*	hypothesis 1*
hypothesis 5	phenomenon 1	phenomenon 1
—	phenomenon 2	phenomenon 2
hypothesis 6	phenomenon 3	phenomenon 3
hypothesis 7	phenomenon 4	phenomenon 4
hypothesis 8	phenomenon 5	phenomenon 5
hypothesis 9	phenomenon 6	phenomenon 6

*between prop. 10 and prop. 11

Ed. 2 also has additions of explanatory phrases and sentences, alterations in wording, and, for the phenomena, revisions of numerical data and references to observers. Ed. 3 further expands or adds some explanatory sentences. For details see the Guide to the present translation, §8.2. Cf. also Alexandre Koyré, "Newton's 'Regulae Philosophandi,'" in his *Newtonian Studies* (Cambridge, Mass.: Harvard University Press, 1965), pp. 261–272; I. Bernard Cohen, "Hypotheses in Newton's Philosophy," *Physis: Rivista inter-*

Rule 2

Therefore, the causes assigned to natural effects of the same kind must be, so far as possible, the same.

Examples are the cause of respiration in man and beast, or of the falling of stones in Europe and America, or of the light of a kitchen fire and the sun, or of the reflection of light on our earth and the planets.

[b]Rule 3

Those qualities of bodies that cannot be intended and remitted [i.e., qualities that cannot be increased and diminished] and that belong to all bodies on which experiments can be made should be taken as qualities of all bodies universally.

For the qualities of bodies can be known only through experiments; and therefore qualities that square with experiments universally are to be regarded as universal qualities; and qualities that cannot be diminished cannot be taken away from bodies. Certainly idle fancies ought not to be fabricated recklessly against the evidence of experiments, nor should we depart from the analogy of nature, since nature is always simple and ever consonant with itself. The extension of bodies is known to us only through our senses, and yet there are bodies beyond the range of these senses; but because extension is found in all sensible bodies, it is ascribed to all bodies universally. We know by experience that some bodies are hard. Moreover, because the hardness of the whole arises from the hardness of its parts, we justly infer from this not only the hardness of the undivided particles of bodies that are accessible to our senses, but also of all other bodies. That all bodies are impenetrable we gather not by reason but by our senses. We find those bodies that we handle to be impenetrable, and hence we conclude that impenetrability is a property of all bodies universally. That all bodies are movable and persevere in motion or in rest by means of certain forces (which we call forces of inertia) we infer from finding these properties in the bodies that we have seen. The extension, hardness, impenetrability, mobility, and force of inertia of the whole arise from the extension, hardness, impenetrability, mobility, and force of inertia of each of the parts; and thus we conclude that every one of the least parts

---

nazionale di storia della scienza 8 (1966): 163–184, reprinted in *Proceedings of the Boston Colloquium for the Philosophy of Science 1966/1968*, ed. Robert S. Cohen and Marx W. Wartofsky, Boston Studies in the Philosophy of Science, vol. 5 (Dordrecht: D. Reidel Publishing Co., 1969), pp. 304–326; I. Bernard Cohen, *Introduction to Newton's "Principia"* (Cambridge, Mass.: Harvard University Press; Cambridge: Cambridge University Press, 1971), pp. 23–26, 240–245.

bb. Ed. 1 has: "Hypothesis 3. Every body can be transformed into a body of any other kind and successively take on all the intermediate degrees of qualities." Cf. prop. 6, corol. 2, below.

of all bodies is extended, hard, impenetrable, movable, and endowed with a force of inertia. And this is the foundation of all natural philosophy. Further, from phenomena we know that the divided, contiguous parts of bodies can be separated from one another, and from mathematics it is certain that the undivided parts can be distinguished into smaller parts by our reason. But it is uncertain whether those parts which have been distinguished in this way and not yet divided can actually be divided and separated from one another by the forces of nature. But if it were established by even a single experiment that in the breaking of a hard and solid body, any undivided particle underwent division, we should conclude by the force of this third rule not only that divided parts are separable but also that undivided parts can be divided indefinitely.

Finally, if it is universally established by experiments and astronomical observations that all bodies on or near the earth gravitate [*lit.* are heavy] toward the earth, and do so in proportion to the quantity of matter in each body, and that the moon gravitates [is heavy] toward the earth in proportion to the quantity of its matter, and that our sea in turn gravitates [is heavy] toward the moon, and that all planets gravitate [are heavy] toward one another, and that there is a similar gravity [heaviness] of comets toward the sun, it will have to be concluded by this third rule that all bodies gravitate toward one another. Indeed, the argument from phenomena will be even stronger for universal gravity than for the impenetrability of bodies, for which, of course, we have not a single experiment, and not even an observation, in the case of the heavenly bodies. Yet I am by no means affirming that gravity is essential to bodies. By inherent force I mean only the force of inertia. This is immutable. Gravity is diminished as bodies recede from the earth.[b]

Rule 4 *In experimental philosophy, propositions gathered from phenomena by induction should be considered either exactly or very nearly true notwithstanding any contrary hypotheses, until yet other phenomena make such propositions either more exact or liable to exceptions.*

This rule should be followed so that arguments based on induction may not be nullified by hypotheses.

PHENOMENA

Phenomenon 1

The circumjovial planets [or satellites of Jupiter], by radii drawn to the center of Jupiter, describe areas proportional to the times, and their periodic times—the fixed stars being at rest—are as the ³⁄₂ powers of their distances from that center.

This is established from astronomical observations. The orbits of these planets do not differ sensibly from circles concentric with Jupiter, and their motions in these circles are found to be uniform. Astronomers agree that their periodic times are as the ³⁄₂ power of the semidiameters of their orbits, and this is manifest from the following table.

Periodic times of the satellites of Jupiter			
$1^d 18^h 27^m 34^s$	$3^d 13^h 13^m 42^s$	$7^d 3^h 42^m 36^s$	$16^d 16^h 32^m 9^s$

Distances of the satellites from the center of Jupiter, in semidiameters of Jupiter				
	1	2	3	4
From the observations of				
Borelli	5⅔	8⅔	14	24⅔
Townly, by a micrometer	5.52	8.78	13.47	24.72
Cassini, by a telescope	5	8	13	23
Cassini, by eclipses of the satellites	5⅔	9	14²³⁄₆₀	25³⁄₁₀
From the periodic times	5.667	9.017	14.384	25.299

Using the best micrometers, Mr. Pound has determined the elongations of the satellites of Jupiter and the diameter of Jupiter in the following way.

The greatest heliocentric elongation of the fourth satellite from the center of Jupiter was obtained with a micrometer in a telescope 15 feet long and came out roughly 8′16″ at the mean distance of Jupiter from the earth. That of the third satellite was obtained with a micrometer in a telescope 123 feet long and came out 4′42″ at the same distance of Jupiter from the earth. The greatest elongations of the other satellites, at the same distance of Jupiter from the earth, come out 2′56″47‴ and 1′51″6‴, on the basis of the periodic times.

The diameter of Jupiter was obtained a number of times with a micrometer in a telescope 123 feet long and, when reduced to the mean distance of Jupiter from the sun or the earth, always came out smaller than 40″, never smaller than 38″, and quite often 39″. In shorter telescopes this diameter is 40″ or 41″. For the light of Jupiter is somewhat dilated by its nonuniform refrangibility, and this dilation has a smaller ratio to the diameter of Jupiter in longer and more perfect telescopes than in shorter and less perfect ones. The times in which two satellites, the first and the third, crossed the disk of Jupiter, from the beginning of their entrance [i.e., from the moment of their beginning to cross the disk] to the beginning of their exit and from the completion of their entrance to the completion of their exit, were observed with the aid of the same longer telescope. And from the transit of the first satellite, the diameter of Jupiter at its mean distance from the earth came out 37⅛″ and, from the transit of the third satellite, 37⅜″. The time in which the shadow of the first satellite passed across the body of Jupiter was also observed, and from this observation the diameter of Jupiter at its mean distance from the earth came out roughly 37″. Let us assume that this diameter is very nearly 37¼″; then the greatest elongations of the first, second, third, and fourth satellites will be equal respectively to 5.965, 9.494, 15.141, and 26.63 semidiameters of Jupiter.

Phenomenon 2 *The circumsaturnian planets [or satellites of Saturn], by radii drawn to the center of Saturn, describe areas proportional to the times, and their periodic times—the fixed stars being at rest—are as the $^{3}/_{2}$ powers of their distances from that center.*

Cassini, in fact, from his own observations has established their distances from the center of Saturn and their periodic times as follows.

Periodic times of the satellites of Saturn				
$1^d21^h18^m27^s$	$2^d17^h41^m22^s$	$4^d12^h25^m12^s$	$15^d22^h41^m14^s$	$79^d7^h48^m00^s$

Distances of the satellites from the center of Saturn, in semidiameters of the ring					
From the observations	$1\frac{19}{20}$	$2\frac{1}{2}$	$3\frac{1}{2}$	8	24
From the periodic times	1.93	2.47	3.45	8	23.35

Observations yield a value of the greatest elongation of the fourth satellite from the center of Saturn that is very near eight semidiameters. But the greatest elongation of this satellite from the center of Saturn, as determined by an excellent micrometer in Huygens's 123-foot telescope, came out $8\frac{7}{10}$ semidiameters. And from this observation and the periodic times, the distances of the satellites from the center of Saturn are, in semidiameters of the ring, 2.1, 2.69, 3.75, 8.7, and 25.35. The diameter of Saturn in the same telescope was to the diameter of the ring as 3 to 7, and the diameter of the ring on the 28th and 29th day of May 1719 came out 43″. And from this the diameter of the ring at the mean distance of Saturn from the earth is 42″, and the diameter of Saturn is 18″. These are the results obtained with the longest and best telescopes, because the apparent magnitudes of heavenly bodies, as seen in longer telescopes, have a greater proportion to the dilation of light at the edges of these bodies than when seen in shorter telescopes. If all erratic light [i.e., dilated light] is disregarded, the diameter of Saturn will not be greater than 16″.

Phenomenon 3

The orbits of the five primary planets—Mercury, Venus, Mars, Jupiter, and Saturn—encircle the sun.

That Mercury and Venus revolve about the sun is proved by their exhibiting phases like the moon's. When these planets are shining with a full face, they are situated beyond the sun; when half full, to one side of the sun; when horned, on this side of the sun; and they sometimes pass across the sun's disk like spots. Because Mars also shows a full face when near conjunction with the sun, and appears gibbous in the quadratures, it is certain that Mars goes around the sun. The same thing is proved also with respect to Jupiter and Saturn from their phases being always full; and in these two planets, it is manifest from the shadows that their satellites project upon them that they shine with light borrowed from the sun.

Phenomenon 4 *The periodic times of the five primary planets and of either the sun about the earth or the earth about the sun—the fixed stars being at rest—are as the 3/2 powers of their mean distances from the sun.*

This proportion, which was found by Kepler, is accepted by everyone. In fact, the periodic times are the same, and the dimensions of the orbits are the same, whether the sun revolves about the earth, or the earth about the sun. There is universal agreement among astronomers concerning the measure of the periodic times. But of all astronomers, Kepler and Boulliau have determined the magnitudes of the orbits from observations with the most diligence; and the mean distances that correspond to the periodic times as computed from the above proportion do not differ sensibly from the distances that these two astronomers found [from observations], and for the most part lie between their respective values, as may be seen in the following table.

Periodic times of the planets and of earth about the sun with respect to the fixed stars, in days and decimal parts of a day

♄	♃	♂	♁	♀	☿
10759.275	4332.514	686.9785	365.2565	224.6176	87.9692

Mean distances of the planets and of the earth from the sun

	♄	♃	♂	♁	♀	☿
According to Kepler	951000	519650	152350	100000	72400	38806
According to Boulliau	954198	522520	152350	100000	72398	38585
According to the periodic times	954006	520096	152369	100000	72333	38710

There is no ground for dispute about the distances of Mercury and Venus from the sun, since these distances are determined by the elongations of the planets from the sun. Furthermore, with respect to the distances of the superior planets from the sun, any ground for dispute is eliminated by the eclipses of the satellites of Jupiter. For by these eclipses the position of the shadow that Jupiter projects is determined, and this gives the heliocentric longitude of Jupiter. And from a comparison of the heliocentric and geocentric longitudes, the distance of Jupiter is determined.

Phenomenon 5

The primary planets, by radii drawn to the earth, describe areas in no way proportional to the times but, by radii drawn to the sun, traverse areas proportional to the times.

For with respect to the earth they sometimes have a progressive [direct or forward] motion, they sometimes are stationary, and sometimes they even have a retrograde motion; but with respect to the sun they move always forward, and they do so with a motion that is almost uniform—but, nevertheless, a little more swiftly in their perihelia and more slowly in their aphelia, in such a way that the description of areas is uniform. This is a proposition very well known to astronomers and is especially provable in the case of Jupiter by the eclipses of its satellites; by means of these eclipses we have said that the heliocentric longitudes of this planet and its distances from the sun are determined.

Phenomenon 6

The moon, by a radius drawn to the center of the earth, describes areas proportional to the times.

This is evident from a comparison of the apparent motion of the moon with its apparent diameter. Actually, the motion of the moon is somewhat perturbed by the force of the sun, but in these phenomena I pay no attention to minute errors that are negligible.[a]

PROPOSITIONS

Proposition 1
Theorem 1

The forces by which the circumjovial planets [or satellites of Jupiter] are continually drawn away from rectilinear motions and are maintained in their respective orbits are directed to the center of Jupiter and are inversely as the squares of the distances of their places from that center.

The first part of the proposition is evident from phen. 1 and from prop. 2 or prop. 3 of book 1, and the second part from phen. 1 and from corol. 6 to prop. 4 of book 1.

The same is to be understood for the planets that are Saturn's companions [or satellites] by phen. 2.

Proposition 2
Theorem 2

The forces by which the primary planets are continually drawn away from rectilinear motions and are maintained in their respective orbits are directed to the sun and are inversely as the squares of their distances from its center.

The first part of the proposition is evident from phen. 5 and from prop. 2 of book 1, and the latter part from phen. 4 and from prop. 4 of the same book. But this second part of the proposition is proved with the greatest exactness from the fact that the aphelia are at rest. For the slightest departure from the ratio of the square would (by book 1, prop. 45, corol. 1) necessarily result in a noticeable motion of the apsides in a single revolution and an immense such motion in many revolutions.

Proposition 3
Theorem 3

The force by which the moon is maintained in its orbit is directed toward the earth and is inversely as the square of the distance of its places from the center of the earth.

The first part of this statement is evident from phen. 6 and from prop. 2 or prop. 3 of book 1, and the second part from the very slow motion of the moon's apogee. For that motion, which in each revolution is only three degrees and three minutes forward [or in consequentia, i.e., in an easterly direction] can be ignored. For it is evident (by book 1, prop. 45, corol. 1) that if the distance of the moon from the center of the earth is to the semidiameter of the earth as D to 1, then the force from which such a motion may arise is inversely as $D^{2\frac{4}{243}}$, that is, inversely as that power of D of which the index is $2\frac{4}{243}$; that is, the proportion of the force to the distance is inversely as a little greater than the second power of the distance, but is $59\frac{3}{4}$ times closer to the square than to the cube. Now this motion of the apogee arises from the action of the sun (as will be pointed out below) and accordingly is to be ignored here. The action of the sun, insofar as it draws the moon away from the earth, is very nearly as the distance of the moon from the earth, and so (from what is said in book 1, prop. 45, corol. 2) is to the centripetal force of the moon as roughly 2 to 357.45, or 1 to $178\frac{29}{40}$. And if so small a force of the sun is ignored, the remaining force by which the moon is maintained in its orbit will be inversely as D^2. And this will be even more fully established by comparing this force with the force of gravity as is done in prop. 4 below.

Corollary. If the mean centripetal force by which the moon is maintained in its orbit is increased first in the ratio of $177\frac{29}{40}$ to $178\frac{29}{40}$, then also in the squared ratio of the semidiameter of the earth to the mean distance of the center of the moon from the center of the earth, the result will be the lunar centripetal force at the surface of the earth, supposing that that force, in descending to the surface of the earth, is continually increased in the ratio of the inverse square of the height.

Proposition 4
Theorem 4

The moon gravitates toward the earth and by the force of gravity is always drawn back from rectilinear motion and kept in its orbit.

The mean distance of the moon from the earth in the syzygies is, according to Ptolemy and most astronomers, 59 terrestrial semidiameters, 60 according to Vendelin and Huygens, $60\frac{1}{3}$ according to Copernicus, $60\frac{2}{5}$ according to Street, and $56\frac{1}{2}$ according to Tycho. But Tycho and all those who follow his tables of refractions, by making the refractions of the sun and moon (entirely contrary to the nature of light) be greater than those of the fixed stars—in fact greater by about four or five minutes—have increased the parallax of the moon by that many minutes, that is, by about a twelfth or fifteenth of the whole parallax. Let that error be corrected, and the distance will come to be roughly $60\frac{1}{2}$ terrestrial semidiameters, close to the value that has been assigned by others. Let us assume a mean distance of 60 semidiameters in the syzygies; and also let us assume that a revolution of the moon with respect to the fixed stars is completed in 27 days, 7 hours, 43 minutes, as has been established by astronomers; and that the circumference of the earth is 123,249,600 Paris feet, according to the measurements made by the French. If now the moon is imagined to be deprived of all its motion and to be let fall so that it will descend to the earth with all that force urging it by which (by prop. 3, corol.) it is [normally] kept in its orbit, then in the space of one minute, it will by falling describe 15¹⁄₁₂ Paris feet. This is determined by a calculation carried out either by using prop. 36 of book 1 or (which comes to the same thing) by using corol. 9 to prop. 4 of book 1. For the versed sine of the arc which the moon would describe in one minute of time by its mean motion at a distance of 60 semidiameters of the earth is roughly 15¹⁄₁₂ Paris feet, or more exactly 15 feet, 1 inch, and 1⁴⁄₉ lines [or twelfths of an inch]. Accordingly, since in approaching the earth that force is increased as the inverse square of the distance, and so at the surface of the earth is 60 × 60 times greater than at the moon, it follows that a body falling with that force, in our regions, ought in the space of one minute to describe 60 × 60 × 15¹⁄₁₂ Paris feet, and in the space of one second 15¹⁄₁₂ feet, or more exactly 15 feet, 1 inch, and 1⁴⁄₉ lines. And heavy bodies do actually descend to the earth with this very force. For a pendulum beating seconds in the latitude of Paris is 3 Paris feet and 8½ lines in length, as Huygens observed. And the height that a heavy body describes by falling in the time of one second is to half the length of this pendulum as the square of the ratio of the circumference of a circle to its diameter (as Huygens also showed), and so is 15 Paris feet, 1 inch, 1⁷⁄₉ lines. And therefore that force by which the moon is kept in its orbit, in descending from the moon's orbit to the surface of the earth, comes out equal to the force of gravity here on earth, and so (by rules 1 and 2) is that very force which we generally call gravity. For if gravity were different from this force, then bodies making for the earth by both forces acting together would descend twice as fast, and in the space of one second would by falling describe 30⅙ Paris feet, entirely contrary to experience.

This calculation is founded on the hypothesis that the earth is at rest. For if the earth and the moon move around the sun and in the meanwhile also revolve around their common center of gravity, then, the law of gravity remaining the same, the distance of the centers of the moon and earth from each other will be roughly 60½ terrestrial semidiameters, as will be evident to anyone who computes it. And the computation can be undertaken by book 1, prop. 60.

Scholium

The proof of the proposition can be treated more fully as follows. If several moons were to revolve around the earth, as happens in the system of Saturn or of Jupiter, their periodic times (by the argument of induction) would observe the law which Kepler discovered for the planets, and therefore their centripetal forces would be inversely as the squares of the distances from the center of the earth, by prop. 1 of this book 3. And if the lowest of them were small and nearly touched the tops of the highest mountains, its centripetal force, by which it would be kept in its orbit, would (by the preceding computation) be very nearly equal to the gravities of bodies on the tops of those mountains. And this centripetal force would cause this little moon, if it were deprived of all the motion with which it proceeds in its orbit, to descend to the earth—as a result of the absence of the centrifugal force with which it had remained in its orbit—and to do so with the same velocity with which heavy bodies fall on the tops of those mountains, because the forces with which they descend are equal. And if the force by which the lowest little moon descends were different from gravity and that little moon also were heavy toward the earth in the manner of bodies on the tops of mountains, this little moon would descend twice as fast by both forces acting together. Therefore, since both forces—namely, those of heavy bodies and those of the moons—are directed toward the center of the earth and are similar to each other and equal, they will (by rules 1 and 2) have the same cause. And therefore that force by which the moon is kept in its orbit is the very one that we generally call gravity. For if this were not so, the little moon at the top of a mountain must either be lacking in gravity or else fall twice as fast as heavy bodies generally do.

Proposition 5
Theorem 5

The circumjovial planets [or satellites of Jupiter] gravitate toward Jupiter, the circumsaturnian planets [or satellites of Saturn] gravitate toward Saturn, and the circumsolar [or primary] planets gravitate toward the sun, and by the force of their gravity they are always drawn back from rectilinear motions and kept in curvilinear orbits.

For the revolutions of the circumjovial planets about Jupiter, of the circumsaturnian planets about Saturn, and of Mercury and Venus and the other circumsolar planets about the sun are phenomena of the same kind as the revolution of the moon about the earth, and therefore (by rule 2) depend on causes of the same kind, especially since it has been proved that the forces on which those revolutions depend are directed toward the centers of Jupiter, Saturn, and the sun, and decrease according to the same ratio and law (in receding from Jupiter, Saturn, and the sun) as the force of gravity (in receding from the earth).

COROLLARY 1. Therefore, there is gravity toward all planets universally. For no one doubts that Venus, Mercury, and the rest [of the planets, primary and secondary,] are bodies of the same kind as Jupiter and Saturn. And since, by the third law of motion, every attraction is mutual, Jupiter will gravitate toward all its satellites, Saturn toward its satellites, and the earth will gravitate toward the moon, and the sun toward all the primary planets.

COROLLARY 2. The gravity that is directed toward every planet is inversely as the square of the distance of places from the center of the planet.

COROLLARY 3. All the planets are heavy toward one another by corols. 1 and 2. And hence Jupiter and Saturn near conjunction, by attracting each other, sensibly perturb each other's motions, the sun perturbs the lunar motions, and the sun and moon perturb our sea, as will be explained in what follows.

Scholium Hitherto we have called "centripetal" that force by which celestial bodies are kept in their orbits. It is now established that this force is gravity, and therefore we shall call it gravity from now on. For the cause of the centripetal force by which the moon is kept in its orbit ought to be extended to all the planets, by rules 1, 2, and 4.

Proposition 6
Theorem 6

All bodies gravitate toward each of the planets, and at any given distance from the center of any one planet the weight of any body whatever toward that planet is proportional to the quantity of matter which the body contains.

Others have long since observed that the falling of all heavy bodies toward the earth (at least on making an adjustment for the inequality of the retardation that arises from the very slight resistance of the air) takes place in equal times, and it is possible to discern that equality of the times, to a very high degree of accuracy, by using pendulums. I have tested this with gold, silver, lead, glass, sand, common salt, wood, water, and wheat. I got two wooden boxes, round and equal. I filled one of them with wood, and I suspended the same weight of gold (as exactly as I could) in the center of oscillation of the other. The boxes, hanging by equal eleven-foot cords, made pendulums exactly like each other with respect to their weight, shape, and air resistance. Then, when placed close to each other [and set into vibration], they kept swinging back and forth together with equal oscillations for a very long time. Accordingly, the amount of matter in the gold (by book 2, prop. 24, corols. 1 and 6) was to the amount of matter in the wood as the action of the motive force upon all the gold to the action of the motive force upon all the [added] wood—that is, as the weight of one to the weight of the other. And it was so for the rest of the materials. In these experiments, in bodies of the same weight, a difference of matter that would be even less than a thousandth part of the whole could have been clearly noticed. Now, there is no doubt that the nature of gravity toward the planets is the same as toward the earth. For imagine our terrestrial bodies to be raised as far as the orbit of the moon and, together with the moon, deprived of all motion, to be released so as to fall to the earth simultaneously; and by what has already been shown, it is certain that in equal times these falling terrestrial bodies will describe the same spaces as the moon, and therefore that they are to the quantity of matter in the moon as their own weights are to its weight. Further, since the satellites of Jupiter revolve in times that are as the $\frac{3}{2}$ power of their distances from the center of Jupiter, their accelerative gravities toward Jupiter will be inversely as the squares of the distances from the center of Jupiter, and, therefore, at equal distances from Jupiter their accelerative gravities would come out equal. Accordingly, in equal times in falling from equal heights [toward Jupiter] they would describe equal spaces, just as happens with heavy bodies on this earth of ours. And by the same argument the circumsolar [or primary] planets, let fall from equal distances from the sun, would describe equal spaces in equal times in their descent to the sun. Moreover, the forces by which unequal bodies are equally accelerated are as the bodies; that is, the weights [of the primary planets toward the sun] are as the quantities of matter in the planets. Further, that the weights of Jupiter and its satellites toward the sun are proportional to the quantities of their matter is evident from the extremely regular motion of the satellites, according to book 1, prop. 65, corol. 3. For if some of these were more strongly attracted toward the sun in proportion to the quantity of their matter than the rest, the motions of the satellites (by book 1, prop. 65, corol. 2) would be perturbed by that inequality of attraction. If, at equal distances from the sun, some satellite were heavier [or gravitated more] toward the sun in proportion to the quantity of its matter than Jupiter in proportion to the quantity of its own matter, in any given ratio, say d to e, then the distance between the center of the sun and the center of the orbit of the satellite would always be greater than the distance between the center of the sun and the center of Jupiter and these distances would be to each other very nearly as the square root of d to the square root of e, as I found by making a certain calculation. And if the satellite were less heavy [or gravitated less] toward the sun in that ratio of d to e, the distance of the center of the orbit of the satellite from the sun would be less than the distance of the center of Jupiter from the sun in that same ratio of the square root of d to the square root of e. And so if, at equal distances from the sun, the accelerative gravity of any satellite toward the sun were greater or smaller than the accelerative gravity of Jupiter toward the sun, by only a thousandth of the whole gravity, the distance of the center of the orbit of the satellite from the sun would be greater or smaller than the distance of Jupiter from the sun by $\frac{1}{2,000}$ of the total distance, that is, by a fifth of the distance of the outermost satellite from the center of Jupiter; and this eccentricity of the orbit would be very sensible indeed. But the orbits of the satellites are concentric with Jupiter, and therefore the accelerative gravities of Jupiter and of the satellites toward the sun are equal to one another. And by the same argument the weights [or gravities] of Saturn and its companions toward the sun, at equal distances from the sun, are as the quantities of matter in them; and the weights of the moon and earth toward the sun are either nil or exactly proportional to their masses. But they do have some weight, according to prop. 5, corols. 1 and 3.

But further, the weights [or gravities] of the individual parts of each planet toward any other planet are to one another as the matter in the individual parts. For if some parts gravitated more, and others less, than in proportion to their quantity of matter, the whole planet, according to the kind of parts in which it most abounded, would gravitate more or gravitate less than in proportion to the quantity of matter of the whole. But it does not matter whether those parts are external or internal. For if, for example, it is imagined that bodies on our earth are raised to the orbit of the moon and compared with the body of the moon, then, if their weights were to the weights of the external parts of the moon as the quantities of matter in them, but were to the weights of the internal parts in a greater or lesser ratio, they would be to the weight of the whole moon in a greater or lesser ratio, contrary to what has been shown above.

COROLLARY 1. Hence, the weights of bodies do not depend on their forms and textures. For if the weights could be altered with the forms, they would be, in equal matter, greater or less according to the variety of forms, entirely contrary to experience.

COROLLARY 2. [a]All bodies universally that are on or near the earth are heavy [or gravitate] toward the earth, and the weights of all bodies that are equally distant from the center of the earth are as the quantities of matter in them. This is a quality of all bodies on which experiments can be performed and therefore by rule 3 is to be affirmed of all bodies universally. If the aether or any other body whatever either were entirely devoid of gravity or gravitated less in proportion to the quantity of its matter, then, since (according to the opinion of Aristotle, Descartes, and others) it does not differ from other bodies except in the form of its matter, it could by a change of its form be transmuted by degrees into a body of the same condition as those that gravitate the most in proportion to the quantity of their matter; and, on the other hand, the heaviest bodies, through taking on by degrees the form of the other body, could by degrees lose their gravity. And accordingly the weights would depend on the forms of bodies and could be altered with the forms, contrary to what has been proved in corol. 1.[a]

aa. Ed. 1 has: "Therefore all bodies universally that are on or near the earth are heavy [or gravitate] toward the earth, and the weights of all bodies that are equally distant from the center of the earth are as the quantities of matter in them. For if the aether or any other body whatever either were entirely devoid of gravity or gravitated less in proportion to the quantity of its matter, then, since it does not differ from other bodies except in the form of its matter, it could by a change of its form be changed by degrees into a body of the same condition as those that gravitate the most in proportion to the quantity of their matter (by hyp. 3), and, on the other hand, the heaviest bodies, through taking on by degrees the form of the other body, could by degrees lose their gravity. And accordingly the weights would depend on the forms of bodies and could be altered with the forms, contrary to what has been proved in corol. 1."

Some of the handwritten notes to Newton's copies of ed. 1 show various other alterations that never appeared in printed editions at this point. In one, for example, everything after the first sentence is replaced by "This is evident by hyp. 3, provided that this hypothesis holds here," while another has the substitution "This follows from the preceding proposition by hyp. 3, provided that this hypothesis holds here." See further the notes to the Rules and Phenomena above.

[b]Corollary 3. All spaces are not equally full. For if all spaces were equally full, the specific gravity of the fluid with which the region of the air would be filled, because of the extreme density of its matter, would not be less than the specific gravity of quicksilver or of gold or of any other body with the greatest density, and therefore neither gold nor any other body could descend in air. For bodies do not ever descend in fluids unless they have a greater specific gravity. But if the quantity of matter in a given space could be diminished by any rarefaction, why should it not be capable of being diminished indefinitely?

Corollary 4. If all the solid particles of all bodies have the same density and cannot be rarefied without pores, there must be a vacuum. I say particles have the same density when their respective forces of inertia [or masses] are as their sizes.[b]

Corollary 5. The force of gravity is of a different kind from the magnetic force. For magnetic attraction is not proportional to the [quantity of] matter attracted. Some bodies are attracted [by a magnet] more [than in proportion to their quantity of matter], and others less, while most bodies are not attracted [by a magnet at all]. And the magnetic force in one and the same body can be intended and remitted [i.e., increased and decreased] and is sometimes far greater in proportion to the quantity of matter than the force of gravity; and this force, in receding from the magnet, decreases not as the square but almost as the cube of the distance, as far as I have been able to tell from certain rough observations.

Proposition 7
Theorem 7

Gravity exists in all bodies universally and is proportional to the quantity of matter in each.

We have already proved that all planets are heavy [or gravitate] toward one another and also that the gravity toward any one planet, taken by itself, is inversely as the square of the distance of places from the center of the planet. And it follows (by book 1, prop. 69 and its corollaries) that the gravity toward all the planets is proportional to the matter in them.

Further, since all the parts of any planet A are heavy [or gravitate] toward any planet B, and since the gravity of each part is to the gravity of the whole

bb. In place of corols. 3 and 4, ed. 1 has a single corol. 3: "And thus a vacuum is necessary. For if all spaces were full, the specific gravity of the fluid with which the region of the air would be filled, because of the extreme density of its matter, would not be less than the specific gravity of quicksilver or of gold or of any other body with the greatest density, and therefore neither gold nor any other body could descend in air. For bodies do not ever descend in fluids unless they have a greater specific gravity."

as the matter of that part to the matter of the whole, and since to every action (by the third law of motion) there is an equal reaction, it follows that planet B will gravitate in turn toward all the parts of planet A, and its gravity toward any one part will be to its gravity toward the whole of the planet as the matter of that part to the matter of the whole. Q.E.D.

COROLLARY 1. Therefore the gravity toward the whole planet arises from and is compounded of the gravity toward the individual parts. We have examples of this in magnetic and electric attractions. For every attraction toward a whole arises from the attractions toward the individual parts. This will be understood in the case of gravity by thinking of several smaller planets coming together into one globe and composing a larger planet. For the force of the whole will have to arise from the forces of the component parts. If anyone objects that by this law all bodies on our earth would have to gravitate toward one another, even though gravity of this kind is by no means detected by our senses, my answer is that gravity toward these bodies is far smaller than what our senses could detect, since such gravity is to the gravity toward the whole earth as [the quantity of matter in each of] these bodies to the [quantity of matter in the] whole earth.

COROLLARY 2. The gravitation toward each of the individual equal particles of a body is inversely as the square of the distance of places from those particles. This is evident by book 1, prop. 74, corol. 3.

Proposition 8
Theorem 8

If two globes gravitate toward each other, and their matter is homogeneous on all sides in regions that are equally distant from their centers, then the weight of either globe toward the other will be inversely as the square of the distance between the centers.

After I had found that the gravity toward a whole planet arises from and is compounded of the gravities toward the parts and that toward each of the individual parts it is inversely proportional to the squares of the distances from the parts, I was still not certain whether that proportion of the inverse square obtained exactly in a total force compounded of a number of forces, or only nearly so. For it could happen that a proportion which holds exactly enough at very great distances might be markedly in error near the surface of the planet, because there the distances of the particles may be unequal and their situations dissimilar. But at length, by means of book 1, props. 75 and 76 and their corollaries, I discerned the truth of the proposition dealt with here.

[a]Corollary 1. Hence the weights of bodies toward different planets can be found and compared one with another. For the weights of equal bodies revolving in circles around planets are (by book 1, prop. 4, corol. 2) as the diameters of the circles directly and the squares of the periodic times inversely, and weights at the surfaces of the planets or at any other distances from the center are greater or smaller (by the same proposition) as the inverse squares of the distances. I compared the periodic times of Venus around the sun (224

aa. The text of the first part of corol. 1 as it appears in the later editions—that is, the first two sentences and part of the third sentence up to "of the moon around the earth (27 days, 7 hours, 43 minutes)"—is almost the same as in the version in ed. 1, except that the later editions have a more complete reference to prop. 4 (the addition of "corol. 2") and have more exact values for the periods of Venus (224 days and 16¾ hours) and of the outermost satellite of Jupiter (16 days and 16⁸⁄₁₅ hours). In the remainder of the text, however, the later versions are notably different from the earlier one. (For a gloss on this corollary, see the Guide, §8.10.) In ed. 1, corol. 1 reads as follows:

"Corollary 1. Hence the weights of bodies toward different planets can be found and compared one with another. For the weights of equal bodies [i.e., bodies with equal masses] revolving in circles around planets are (by book 1, prop. 4) as the diameters of the circles directly and the squares of the periodic times inversely, and weights at the surfaces of the planets or at any other distances from the center are greater or smaller (by the same proposition) inversely as the squared ratio of the distances. I compared the periodic times of Venus around the sun (224⅔ days), of the outermost circumjovial satellite around Jupiter (16¾ days), of Huygens's satellite around Saturn (15 days and 22⅔ hours), and of the moon around the earth (27 days, 7 hours, 43 minutes) respectively with the mean distance of Venus from the sun, with the greatest heliocentric elongation of the outermost circumjovial satellite, which (at the mean distance of Jupiter from the sun according to the observations of Flamsteed) is 8′13″, with the greatest heliocentric elongation of the satellite of Saturn (3′20″), and with the distance of the moon from the earth, on the hypothesis that the horizontal solar parallax or the semidiameter of the earth as seen from the sun is about 20″.

In this way I found by calculation that the weights of bodies which are equal and equally distant from the sun, Jupiter, Saturn, and the earth as directed toward the sun, Jupiter, Saturn, and the earth, were to one another as 1, $\frac{1}{1,100}$, $\frac{1}{2,360}$, and $\frac{1}{28,700}$. But the mean apparent semidiameter of the sun is about 16′6″. From the diameter of the shadow of Jupiter as found by eclipses of the satellites, Flamsteed determined that the mean apparent diameter of Jupiter as seen from the sun is to the elongation of the outermost satellite as 1 to 24.9, and since that elongation is 8′13″, the semidiameter of Jupiter as seen from the sun will be 19¾″. The diameter of Saturn is to the diameter of its ring as 4 to 9, and the diameter of the ring as seen from the sun (by Flamsteed's measurement) is 50″, and thus the semidiameter of Saturn as seen from the sun is 11″. I would prefer to say 10″ or 9″, because the globe of Saturn is somewhat dilated by a nonuniform refrangibility of light.

Thus, when the calculation is made, the true semidiameters of the sun, Jupiter, Saturn, and the earth to one another come out as 10,000, 1,063, 889, and 208. Whence, because the weights of bodies which are equal and equally distant from the centers of the sun, Jupiter, Saturn, and the earth are, respectively, toward the sun, Jupiter, Saturn, and the earth as 1, $\frac{1}{1,000}$, $\frac{1}{2,360}$, $\frac{1}{28,700}$, and because, when the distances are increased or decreased, the weights are decreased or increased in the squared ratio, [it follows that] the weights of the same equal bodies toward the sun, Jupiter, Saturn, and the earth at distances of 10,000, 1,063, 889, and 208 from their centers, and hence their weights at the surfaces, will be as 10,000, 804½, 536, and 805½ respectively. We shall show below that the weights of bodies on the surface of the moon are almost two times less than the weights of bodies on the surface of the earth."

days and 16¾ hours), of the outermost circumjovial satellite around Jupiter (16 days and 16⁸⁄₁₅ hours), of Huygens's satellite around Saturn (15 days and 22⅔ hours), and of the moon around the earth (27 days, 7 hours, 43 minutes) respectively with the mean distance of Venus from the sun, and with the greatest heliocentric elongations of the outermost circumjovial satellite from the center of Jupiter (8′16″), of Huygens's satellite from the center of Saturn (3′4″), and of the moon from the center of the earth (10′33″). In this way I found by computation that the weights of bodies which are equal and equally distant from the center of the sun, of Jupiter, of Saturn, and of the earth are respectively toward the sun, Jupiter, Saturn, and the earth as 1, $\frac{1}{1,067}$, $\frac{1}{3,021}$, and $\frac{1}{169,282}$. And when the distances are increased or decreased, the weights are decreased or increased as the squares of the distances. The weights of equal bodies toward the sun, Jupiter, Saturn, and the earth at distances of 10,000, 997, 791, and 109 respectively from their centers (and hence their weights on the surfaces) will be as 10,000, 943, 529, and 435. What the weights of bodies are on the surface of the moon will be shown below.[a]

[b]COROLLARY 2. The quantity of matter in the individual planets can also be found. For the quantities of matter in the planets are as their forces at equal distances from their centers; that is, in the sun, Jupiter, Saturn, and the earth, they are as 1, $\frac{1}{1,067}$, $\frac{1}{3,021}$, and $\frac{1}{169,282}$ respectively. If the parallax of the sun is taken as greater or less than 10″30‴, the quantity of matter in the earth will have to be increased or decreased in the cubed ratio.[b]

bb. In ed. 1, there was an additional corollary numbered 2, so that the corollaries numbered 2, 3, and 4 in the later editions were originally numbered 3, 4, and 5. (For a gloss on this corollary see the Guide, §8.10.) The corol. 2 of the first edition reads as follows:

"Corollary 2. Therefore the weights of equal bodies [i.e., bodies with equal masses], on the surfaces of the earth and of the planets, are almost proportional to the square roots of their apparent diameters as seen from the sun. With respect to the diameter of the earth as seen from the sun there is as yet no agreement. I have taken it to be 40″, because the observations of Kepler, Riccioli, and Vendelin do not permit it to be much greater; the observations of Horrocks and Flamsteed seem to make it a little smaller. And I have preferred to err on the side of excess. But if perhaps that diameter and the gravity on the surface of the earth are a mean among the diameters of the planets and the gravities on their surfaces, then, since the diameters of Saturn, Jupiter, Mars, Venus, and Mercury are about 18″, 39½″, 8″, 28″, 20″, the diameter of the earth will be about 24″ and therefore the parallax of the sun about 12″, as Horrocks and Flamsteed pretty nearly concluded. But a slightly larger diameter agrees better with the rule of this corollary." That is, the larger diameter of the earth as seen from the sun, and hence the larger solar parallax, agrees better with the rule about the weights of equal bodies on the surface of the earth and planets being "almost proportional to the square roots of their apparent diameters as seen from the sun."

COROLLARY 3. The densities of the planets can also be found. For the weights of equal and homogeneous bodies toward homogeneous spheres are, on the surfaces of the spheres, as the diameters of the spheres, by book 1, prop. 72; and therefore the densities of heterogeneous spheres are as those weights divided by the diameters of the spheres. Now, the true diameters of the sun, Jupiter, Saturn, and the earth were found to be to one another as 10,000, 997, 791, and 109, and the weights toward them are as 10,000, 943, 529, and 435 respectively, and therefore the densities are as 100, 94½, 67, and 400. The density of the earth that results from this computation does not depend on the parallax of the sun but is determined by the parallax of the moon and therefore is determined correctly here. Therefore the sun is a little denser than Jupiter, and Jupiter denser than Saturn, and the earth four times denser than the sun. For the sun is rarefied by its great heat. And the moon is denser than the earth, as will be evident from what follows [i.e., prop. 37, corol. 3].

[c]COROLLARY 4. Therefore, other things being equal, the planets that are smaller are denser. For thus the force of gravity on their surfaces approaches closer to equality. But, other things being equal, the planets that are nearer to the sun are also denser; for example, Jupiter is denser than Saturn, and the earth is denser than Jupiter. The planets, of course, had to be set at different distances from the sun so that each one might, according to the degree of its density, enjoy a greater or smaller amount of heat from the sun.[c] If the earth were located in the orbit of Saturn, our water would freeze; in the orbit of Mercury, it would immediately go off in a vapor. For the light of the sun, to which its heat is proportional, is seven times denser in the orbit of Mercury than on earth, and I have found with a thermometer that water boils at seven

cc. In place of this portion of corol. 4, ed. 1 has:

"Corollary 5. The densities of the planets, moreover, are to one another nearly in a ratio compounded of the ratio of the distance from the sun and the square roots of the diameters of the planets as seen from the sun. For the densities of Saturn, Jupiter, the earth, and the moon (60, 76, 387, and 700) are almost as the roots of the apparent diameters (18″, 39½″, 40″, and 11″) divided by the reciprocals of their distances from the sun $\left(\frac{1}{8{,}538}, \frac{1}{5{,}201}, \frac{1}{1{,}000}, \frac{1}{1{,}000}\right)$. We said, moreover [utique], in corol. 2, that the gravities at the surfaces of the planets are approximately as the square roots of their apparent diameters as seen from the sun; and in lem. 4 [i.e., corol. 4] that the densities are as the gravities divided by the true diameters; and so the densities are almost as the roots of the apparent diameters multiplied by the true diameters—that is, inversely as the roots of the apparent diameters divided by the distances of the planets from the sun. Therefore God placed the planets at different distances from the sun so that each one might, according to the degree of its density, enjoy a greater or smaller amount of heat from the sun."

times the heat of the summer sun. And there is no doubt that the matter of the planet Mercury is adapted to its heat and therefore is denser than this matter of our earth, since all denser matter requires a greater heat for the performance of the operations of nature.

Proposition 9
Theorem 9

In going inward from the surfaces of the planets, gravity decreases very nearly in the ratio of the distances from the center.

If the matter of the planets were of uniform density, this proposition would hold true exactly, by book 1, prop. 73. Therefore the error is as great as can arise from the nonuniformity of the density.

Proposition 10
Theorem 10

The motions of the planets can continue in the heavens for a very long time.

In the scholium to prop. 40, book 2, it was shown that a globe of frozen water moving freely in our air would, as a result of the resistance of the air, lose $\frac{1}{4,586}$ of its motion in describing the length of its own semidiameter. And the same proportion obtains very nearly in any globes, however large they may be and however swift their motions. Now, I gather in the following way that the globe of our earth is denser than if it consisted totally of water. If this globe were wholly made of water, whatever things were rarer than water would, because of their smaller specific gravity, emerge from the water and float on the surface. And for this reason a globe made of earth that was covered completely by water would emerge somewhere, if it were rarer than water; and all the water flowing away from there would be gathered on the opposite side. And this is the case for our earth, which is in great part surrounded by seas. If the earth were not denser than the seas, it would emerge from those seas and, according to the degree of its lightness, a part of the earth would stand out from the water, while all those seas flowed to the opposite side. By the same argument the spots on the sun are lighter than the solar shining matter on top of which they float. And in whatever way the planets were formed, at the time when the mass was fluid, all heavier matter made for the center, away from the water. Accordingly, since the ordinary matter of our earth at its surface is about twice as heavy as water, and a little lower down, in mines, is found to be about three or four or even five times heavier than water, it is likely that the total amount of matter in the earth is about five or six times greater than it would be if the whole earth consisted of water, especially since it has already been shown above that the earth is about four times denser than Jupiter. Therefore, if Jupiter is a little denser than water, then in the space of thirty days (during which this planet describes a length of 459 of its semidiameters) it would, in a medium of the same density as our air, lose almost a tenth of its motion. But since the resistance of mediums decreases in the ratio of their weight and density (so that water, which is 13⅗ times lighter than quicksilver, resists 13⅗ times less; and air, which is 860 times lighter than water, resists 860 times less), it follows that up in the heavens, where the weight of the medium in which the planets move is diminished beyond measure, the resistance will nearly cease. We showed in the scholium to prop. 22, book 2, that at a height of two hundred miles above the earth, the air would be rarer than on the surface of the earth in a ratio of 30 to 0.0000000000003998, or 75,000,000,000,000 to 1, roughly. And hence the planet Jupiter, revolving in a medium with the same density as that upper air, would not, in the time of a million years, lose a millionth of its motion as a result of the resistance of the medium. In the spaces nearest to the earth, of course, nothing is found that creates resistance except air, exhalations, and vapors. If these are exhausted with very great care from a hollow cylindrical glass vessel, heavy bodies fall within the glass vessel very freely and without any sensible resistance; gold itself and the lightest feather, dropped simultaneously, fall with equal velocity and, in falling through a distance of four or six or eight feet, reach the bottom at the same time, as has been found by experiment. And therefore in the heavens, which are void of air and exhalations, the planets and comets, encountering no sensible resistance, will move through those spaces for a very long time.

Hypothesis 1 *The center of the system of the world is at rest.*

No one doubts this, although some argue that the earth, others that the sun, is at rest in the center of the system. Let us see what follows from this hypothesis.

Proposition 11
Theorem 11

The common center of gravity of the earth, the sun, and all the planets is at rest.

For that center (by corol. 4 of the Laws) either will be at rest or will move uniformly straight forward. But if that center always moves forward, the center of the universe will also move, contrary to the hypothesis.

Proposition 12
Theorem 12

The sun is engaged in continual motion but never recedes far from the common center of gravity of all the planets.

For since (by prop. 8, corol. 2) the matter in the sun is to the matter in Jupiter as 1,067 to 1, and the distance of Jupiter from the sun is to the semidiameter of the sun in a slightly greater ratio, the common center of gravity of Jupiter and the sun will fall upon a point a little outside the surface of the sun. By the same argument, since the matter in the sun is to the matter in Saturn as 3,021 to 1, and the distance of Saturn from the sun is to the semidiameter of the sun in a slightly smaller ratio, the common center of gravity of Saturn and the sun will fall upon a point a little within the surface of the sun. And continuing the same kind of calculation, if the earth and all the planets were to lie on one side of the sun, the distance of the common center of gravity of them all from the center of the sun would scarcely be a whole diameter of the sun. In other cases the distance between those two centers is always less. And therefore, since that center of gravity is continually at rest, the sun will move in one direction or another, according to the various configurations of the planets, but will never recede far from that center.

COROLLARY. Hence the common center of gravity of the earth, the sun, and all the planets is to be considered the center of the universe. For since the earth, sun, and all the planets gravitate toward one another and therefore, in proportion to the force of the gravity of each of them, are constantly put in motion according to the laws of motion, it is clear that their mobile centers cannot be considered the center of the universe, which is at rest. If that body toward which all bodies gravitate most had to be placed in the center (as is the commonly held opinion), that privilege would have to be conceded to the sun. But since the sun itself moves, an immobile point will have to be chosen for that center from which the center of the sun moves away as little as possible and from which it would move away still less, supposing that the sun were denser and larger, in which case it would move less.

Proposition 13
Theorem 13

The planets move in ellipses that have a focus in the center of the sun, and by radii drawn to that center they describe areas proportional to the times.

We have already discussed these motions from the phenomena. Now that the principles of motions have been found, we deduce the celestial motions from these principles a priori. Since the weights of the planets toward the sun are inversely as the squares of the distances from the center of the sun, it follows (from book 1, props. 1 and 11, and prop. 13, corol. 1) that if the sun were at rest and the remaining planets did not act upon one another, their orbits would be elliptical, having the sun in their common focus, and they would describe areas proportional to the times. The actions of the planets upon one another, however, are so very small that they can be ignored, and they perturb the motions of the planets in ellipses about the mobile sun less (by book 1, prop. 66) than if those motions were being performed about the sun at rest.

Yet the action of Jupiter upon Saturn is not to be ignored entirely. For the gravity toward Jupiter is to the gravity toward the sun (at equal distances) as 1 to 1,067; and so in the conjunction of Jupiter and Saturn, since the distance of Saturn from Jupiter is to the distance of Saturn from the sun almost as 4 to 9, the gravity of Saturn toward Jupiter will be to the gravity of Saturn toward the sun as 81 to $16 \times 1{,}067$, or roughly as 1 to 211. And hence arises a perturbation of the orbit of Saturn in every conjunction of this planet with Jupiter so sensible that astronomers have been at a loss concerning it. According to the different situations of the planet Saturn in these conjunctions, its eccentricity is sometimes increased and at other times diminished, the aphelion sometimes is moved forward and at other times perchance drawn back, and the mean motion is alternately accelerated and retarded. Nevertheless, all the error in its motion around the sun, an error arising from so great a force, can almost be avoided (except in the mean motion) by putting the lower focus of its orbit in the common center of gravity of Jupiter and the sun (by book 1, prop. 67); in which case, when that error is greatest, it hardly exceeds two minutes. And the greatest error in the mean motion hardly exceeds two minutes per year. But in the conjunction of Jupiter and Saturn the accelerative gravities of the sun toward Saturn, of Jupiter toward Saturn, and of Jupiter toward the sun are almost as 16, 81, and $\frac{16 \times 81 \times 3{,}021}{25}$, or 156,609, and so the difference of the gravities of the sun toward Saturn and of Jupiter toward Saturn is to the gravity of Jupiter toward the sun as 65 to 156,609, or 1 to 2,409. But the greatest power of Saturn to perturb the motion of Jupiter is proportional to this difference, and therefore the perturbation of the orbit of Jupiter is far less than that of Saturn's. The perturbations of the remaining orbits are still less by far, except that the orbit of the earth is sensibly perturbed by the moon. The common center of gravity of the earth and the moon traverses an ellipse about the sun, an ellipse in which the sun is located at a focus, and this center of gravity, by a radius drawn to the sun, describes areas (in that ellipse) proportional to the times; the earth, during this time, revolves around this common center with a monthly motion.

Proposition 14
Theorem 14

The aphelia and nodes of the [planetary] orbits are at rest.

The aphelia are at rest, by book 1, prop. 11, as are also the planes of the orbits, by prop. 1 of the same book; and if these planes are at rest, the nodes are also at rest. But yet from the actions of the revolving planets and comets upon one another some inequalities will arise, which, however, are so small that they can be ignored here.

COROLLARY 1. The fixed stars also are at rest, because they maintain given positions with respect to the aphelia and nodes.

COROLLARY 2. And so, since the fixed stars have no sensible parallax arising from the annual motion of the earth, their forces will produce no sensible effects in the region of our system, because of the immense distance of these bodies from us. Indeed, the fixed stars, being equally dispersed in all parts of the heavens, by their contrary attractions annul their mutual forces, by book 1, prop. 70.

Scholium

Since the planets nearer to the sun (namely, Mercury, Venus, the earth, and Mars) act but slightly upon one another because of the smallness of their bodies [i.e., because their masses are small], their aphelia and nodes will be at rest, except insofar as they are disturbed by the forces of Jupiter, Saturn, and any bodies further away. And by the theory of gravity it follows that their aphelia move slightly forward [or in consequentia] with respect to the fixed stars, and do this as the $3/2$ powers of the distances of these planets from the sun. For example, if in a hundred years the aphelion of Mars is carried forward [or in consequentia] $33'20''$ with respect to the fixed stars, then in a hundred years the aphelia of the earth, Venus, and Mercury will be carried forward $17'40''$, $10'53''$, and $4'16''$ respectively. And these motions are ignored in this proposition because they are so small.

Proposition 15
Problem 1

To find the principal diameters of the [planetary] orbits.

These diameters are to be taken as the $2/3$ powers of the periodic times by book 1, prop. 15; and then each one is to be increased in the ratio of the sum of the masses of the sun and each revolving planet to the first of two mean proportionals between that sum and the sun, by book 1, prop. 60.

Proposition 16
Problem 2

To find the eccentricities and aphelia of the [planetary] orbits.

The problem is solved by book 1, prop. 18.

Proposition 17
Theorem 15

The daily motions of the planets are uniform, and the libration of the moon arises from its daily motion.

This is clear from the first law of motion and book 1, prop. 66, corol. 22. With respect to the fixed stars Jupiter revolves in 9^h56^m, Mars in 24^h39^m, Venus in about 23 hours, the earth in 23^h56^m, the sun in 25½ days, and the moon in $27^d7^h43^m$. That these things are so is clear from phenomena. With respect to the earth, the spots on the body of the sun return to the same place on the sun's disc in about 27½ days; and therefore with respect to the fixed stars the sun revolves in about 25½ days. Now, since a lunar day (the moon revolving uniformly about its own axis) is a month long [i.e., is equal to a lunar month, the periodic time of the moon's revolution in its orbit], the same face of the moon will always very nearly look in the direction of the further focus of its orbit, and therefore will deviate from the earth on one side or the other according to the situation of that focus. This is the moon's libration in longitude; for the libration in latitude arises from the latitude of the moon and the inclination of its axis to the plane of the ecliptic. Mr. N. Mercator, in his book on astronomy, published in the beginning of the year 1676, set forth this theory of the moon's libration more fully on the basis of a letter from me.

The outermost satellite of Saturn seems to revolve about its own axis with a motion similar to our moon's, constantly presenting the same aspect toward Saturn. For in revolving about Saturn, whenever it approaches the eastern part of its own orbit, it is just barely seen and for the most part disappears from sight; and possibly this occurs because of certain spots in that part of its body which is then turned toward the earth, as Cassini noted. The outermost satellite of Jupiter also seems to revolve about its own axis with a similar motion, because in the part of its body turned away from Jupiter it has a spot which, whenever the satellite passes between Jupiter and our eyes, appears as if it were on the body of Jupiter.

Proposition 18
Theorem 16

The axes of the planets are smaller than the diameters that are drawn perpendicularly to those axes.

If it were not for the daily circular motion of the planets, then, because the gravity of their parts is equal on all sides, they would have to assume a spherical figure. Because of that circular motion it comes about that those parts, by receding from the axis, endeavor to ascend in the region of the equator. And therefore if the matter is fluid, it will increase the diameters at the equator by ascending, and will decrease the axis at the poles by descending. Thus the diameter of Jupiter is found by astronomical observations to be shorter between the poles than from east to west. By the same argument, if our earth were not a little higher around the equator than at the poles, the seas would subside at the poles and, by ascending in the region of the equator, would flood everything there.

Proposition 19[a]
Problem 3

To find the proportion of a planet's axis to the diameters perpendicular to that axis.

[bc]Our fellow countryman Norwood, in about the year 1635, measured a distance of 905,751 London feet between London and York and observed the

a. For a gloss on this proposition see the Guide, §10.14.

bb. Ed. 1 has: "The solution of this problem requires a complex computation, which is shown more easily by example than by precept. Through making the calculation, therefore, I find, by book 1, prop. 4, that the centrifugal force of the parts of the earth at the equator, arising from the daily motion, is to the force of gravity as 1 to 290⅘."

cc. Ed. 2 has: "Picard measured an arc of 1°22′55″ between Amiens and Malvoisine and found an arc of one degree to be 57,060 Paris toises. Hence the circumference of the earth is 123,249,600 Paris feet, as above. But since an error of four hundredths of an inch, either in the construction of the instruments or in their application to making observations, is imperceptible and, in the ten-foot sector by which the French observed the latitudes of places, corresponds to four seconds and, in single observations, can fall upon the center of the sector as well as on its circumference, and since errors in smaller arcs are of greater significance, therefore Cassini by the king's order undertook the measurement of the earth by means of greater intervals between the places (see the History of the Royal Academy of Sciences for the year 1700) and in the process, by using the distance between the Royal Paris Observatory and the village of Collioure in Roussillon and the difference of 6°18′ between the latitudes and supposing that the earth's shape is spherical, found one degree to be 57,292 toises, nearly as our fellow countryman Norwood had found earlier. For Norwood, in about the year 1635, measured a distance of 905,751 London feet between London and York and observed the difference of latitudes between those places to be 2°28′ and thereby found the measure of one degree to be 367,196 London feet, that is, 57,300 Paris toises. Because of the magnitude of the interval measured by Cassini, I shall use 57,292 toises for the measure of one degree in the middle of that interval, that is, between the latitudes of 45° and 46°. Hence, if the earth is spherical, its semidiameter will be 19,695,539 Paris feet.

"The length of a seconds pendulum oscillating in the latitude of Paris is three Paris feet and 8⁵⁄₉ lines. And the length which a heavy body describes by falling in the time of one second is to half the length of

difference of latitudes between those places to be 2°28′ and thereby found the measure of one degree to be 367,196 London feet, that is, 57,300 Paris toises. Picard measured an arc of 1°22′55″ along the meridian between Amiens and Malvoisine and found an arc of one degree to be 57,060 Paris toises. The elder Cassini [Gian Domenico or Jean-Dominique] measured the distance along the meridian from the town of Collioure in Roussillon to the Paris observatory; and his son [Jacques] added the distance from the observatory to the tower of the city of Dunkerque. The total distance was 486,156½ toises, and the difference in latitudes between the town of Collioure and the city of Dunkerque was 8°31′11⅚″. Thus an arc of one degree comes out to be 57,061 Paris toises. And from these measures the circumference of the earth is found to be 123,249,600 Paris feet, and its semidiameter 19,615,800 feet, on the hypothesis that the earth is spherical.

At the latitude of Paris, a heavy body falling in the time of one second describes 15 Paris feet 1 inch 1 7/9 lines as has been mentioned above, that is, 2,173 7/9 lines. The weight of a body is diminished by the weight of the surrounding air. Let us suppose that the weight lost in this way is an eleven-thousandth part of the total weight; then such a heavy body falling in a vacuum will describe a space of 2,174 lines in the time of one second.[c]

A body revolving uniformly in a circle at a distance of 19,615,800 feet from the center, making a revolution in a single sidereal day of $23^{h}56^{m}4^{s}$, will describe an arc of 1,433.46 feet in the time of one second, an arc whose versed sine is 0.0523656 feet, or 7.54064 lines. And therefore the force by which heavy bodies descend at the latitude of Paris is to the [d]centrifugal[d] force of bodies on the equator (which arises from the daily motion of the earth) as 2,174 to 7.54064.

The centrifugal force of bodies on the earth's equator is to the centrifugal force by which bodies recede rectilinearly from the earth at the latitude of Paris (48°50′10″) as the square of the radius to the square of the cosine of that latitude, that is, as 7.54064 to 3.267. Let this force be added to the force by which heavy bodies descend at the latitude of Paris; then a body falling at that latitude with the total force of gravity will, in the time of one second, describe 2,177.267 lines, or 15 Paris feet 1 inch and 5.267 lines. And the total

---

this pendulum or as the square of the ratio of the circumference of the circle to its diameter (as Huygens indicated) and thus is 15 Paris feet, 1 inch, 2 1/18 lines, or 2,174 1/18 lines."

dd. Ed. 2 has "centripetal."

force of gravity at that latitude will be to the [e]centrifugal[e] force of bodies on the earth's equator as 2,177.267 to 7.54064 or 289 to 1.[b]

Therefore, if APBQ represents the figure of the earth, which is now no longer spherical but generated by the rotation of an ellipse about its minor axis PQ; and if ACQ*qca* is a channel full of water, going from the pole Q*q* to the center C*c* and from that center out to the equator A*a*; then the weight of the water in the leg AC*ca* will have to be to the weight of the water in the other leg QC*cq* as 289 to 288, because the centrifugal force arising from the circular motion will sustain and take away one of the 289 parts of weight of the water in the leg AC*ca*, and consequently the 288 parts of weight of the water in the leg QC*cq* will sustain the 288 parts remaining in the leg AC*ca*. Further, on making the computation (according to book 1, prop. 91, corol. 2), I find that if the earth were composed of uniform matter and were deprived of all its motion, and its axis PQ were to its diameter AB as 100 to 101, then the gravity in place Q toward the earth would be to the gravity in the same place Q toward a sphere described about the center C with a radius PC or QC as 126 to 125. And by the same argument, the gravity in place A toward a spheroid generated by the rotation of the ellipse APBQ about the axis AB is to the gravity in the same place A toward a sphere described about a center C with a radius AC as 125 to 126. Moreover, the gravity in place A toward the earth is a mean proportional between the gravity toward the spheroid and the gravity toward the sphere, because the sphere, when its diameter PQ is diminished in the ratio of 101 to 100, is transformed into the figure of the earth; and this figure, when a third diameter (perpendicular to the two given diameters AB and PQ) is diminished in the same ratio, is transformed into the said spheroid; and the gravity in A, in either case, is diminished in very nearly the same ratio. Therefore the gravity in A toward a sphere described about the center C with a radius AC is to the gravity in A toward the earth as 126 to 125½; and the gravity in place Q, toward a sphere described about the center C with a radius QC, is to the gravity in place A, toward a sphere described about the center C with a radius AC, in the ratio of the diameters (by book 1, prop. 72),

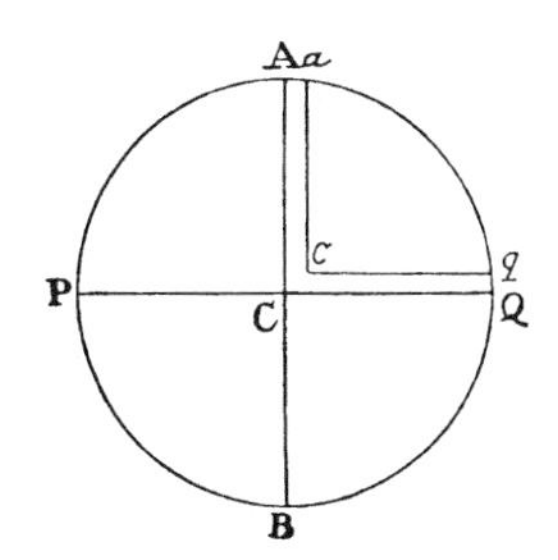

ee. Ed. 2 has "centripetal."

that is, as 100 to 101. Now let these three ratios (126 to 125, 126 to 125½, and 100 to 101) be combined, and the gravity in place Q toward the earth will become to the gravity in place A toward the earth as 126 × 126 × 100 to 125 × 125½ × 101, or as 501 to 500.

Now, since (by book 1, prop. 91, corol. 3) the gravity in either leg AC*ca* or QC*cq* of the channel is as the distance of places from the earth's center, if those legs are separated by transverse, equidistant surfaces into parts proportional to the wholes, the weights of any number of these individual parts in the leg AC*ca* will be to the weights of the same number of individual parts in the other leg as their magnitudes and accelerative gravities jointly, that is, as 101 to 100 and 500 to 501, which is as 505 to 501. And accordingly, if the centrifugal force of each part of the leg AC*ca* (which force arises from the daily motion) had been to the weight of the same part as 4 to 505, so that it would take away four parts from the weight of each part (supposing it to be divided into 505 parts), the weights would remain equal in each leg, and therefore the fluid would stay at rest in equilibrium. But the centrifugal force of each part is to the weight of the same part as 1 to 289; that is, the [f]centrifugal[f] force, which ought to have been $\frac{4}{505}$ of the weight, is only $\frac{1}{289}$ of it. And therefore I say, according to the golden rule [or rule of three], that if a centrifugal force of $\frac{4}{505}$ of the weight makes the height of the water in the leg AC*ca* exceed the height of the water in the leg QC*cq* by a hundredth of its total height, the centrifugal force of $\frac{1}{289}$ of the weight will make the excess of the height in the leg AC*ca* be only $\frac{1}{229}$ of the height of the water in the other leg QC*cq*. Therefore the diameter of the earth at the equator is to its diameter through the poles as 230 to 229. And thus, since the mean semidiameter of the earth, according to Picard's measurement, is 19,615,800 Paris feet, or 3,923.16 miles (supposing a mile to be 5,000 feet), the earth will be 85,472 feet or 17⅒ miles higher at the equator than at the poles. And its height at the equator will be about 19,658,600 feet, and at the poles will be about 19,573,000 feet.

If a planet is larger or smaller than the earth, while its density and periodic time of daily revolution remain the same, the ratio of centrifugal force

ff. Ed. 1 has "centripetal."

to gravity will remain the same, and therefore the ratio of the diameter between the poles to the diameter at the equator will also remain the same. But if the daily motion is accelerated or retarded in any ratio, the centrifugal force will be increased or decreased in that same ratio squared, and therefore the difference between the diameters will be increased or decreased very nearly in the same squared ratio. And if the density of a planet is increased or decreased in any ratio, the gravity tending toward the planet will also be increased or decreased in the same ratio, and the difference between the diameters in turn will be decreased in the ratio of the increase in the gravity or will be increased in the ratio of the decrease in the gravity. Accordingly, since the earth revolves [i.e., rotates] with respect to the fixed stars in 23^h56^m, and Jupiter in 9^h56^m, and the squares of their periodic times are as 29 to 5, and the densities of these revolving bodies are as 400 to 94½, the difference between the diameters of Jupiter will be to its smaller diameter as $\frac{29}{5} \times \frac{440}{94\frac{1}{2}} \times \frac{1}{229}$ to 1, or very nearly as 1 to 9⅓. Therefore Jupiter's diameter taken from east to west is to its diameter between the poles very nearly as 10⅓ to 9⅓. [g]Thus, since its larger diameter is $37''$, its smaller diameter (which lies between the poles) will be $33''25'''$. Because of the erratic light let about $3''$ be added, and the apparent diameters of this planet will come out to be $40''$ and $36''25'''$, which are to each other nearly as 11⅙ to 10⅙. This argument has been based on the hypothesis that the body of Jupiter is uniformly dense. But if its body is denser toward the plane of the equator than toward the poles, its diameters can be to each other as 12 to 11, or 13 to 12, or even 14 to 13. As a matter of fact, Cassini observed in the year 1691 that the diameter of Jupiter extending from east to west would exceed its other diameter by about a fifteenth part of itself. Moreover, our fellow countryman Pound, with a 123-foot-long telescope and the best micrometer, measured the diameters of Jupiter in the year 1719 with the following results.

gg. Ed. 1 and ed. 2 have: "These things are so on the hypothesis that the matter of the planets is uniform. For if the matter is denser at the center than at the circumference, the diameter which is drawn from east to west will be still greater." In ed. 1 the proposition ends here, but in ed. 2 it continues: "Indeed, Cassini observed long ago that the diameter of Jupiter which lies between its poles is smaller than the other diameter, and it will be apparent from what is said in prop. 20 below that the diameter of the earth between the poles is smaller than the other diameter."

Times			*Largest diameter*	*Smallest diameter*	*Diameters to each other*
	days	*hours*	*parts*	*parts*	
January	28	6	13.40	12.28	as 12 to 11
March	6	7	13.12	12.20	13¾ to 12¾
March	9	7	13.12	12.08	12⅔ to 11⅔
April	9	9	12.32	11.48	14½ to 13½

Therefore the theory agrees with the phenomena. Further, the planets are more exposed to the heat of sunlight toward their equators and as a result [h]are somewhat more thoroughly heated there[h] than toward the poles.

Even further, it will be apparent—from the experiments with pendulums reported in prop. 20 below—that gravity is decreased at the equator by the daily rotation of our earth, and therefore that the earth (supposing its matter to be uniformly dense) rises higher there than at the poles.[g]

Proposition 20
Problem 4

To find and compare with one another the weights of bodies in different regions of our earth.

Since the weights of the unequal legs of the water-channel ACQ*qca* are equal, and the weights of any parts that are proportional to the whole legs and similarly situated in those legs are to one another as the weights of the wholes, and thus are also equal to one another, the weights of parts that are equal and similarly situated in the legs will be inversely as the legs, that is, inversely as 230 to 229. This is likewise the case for any homogeneous equal bodies that are similarly situated in the legs of the channel. The weights of these bodies are inversely as the legs, that is, inversely as the distances of the bodies from the earth's center. Accordingly, if the bodies are in the topmost parts of the channels, or on the surface of the earth, their weights will be to one another inversely as their distances from the center. And by the same argument, weights that are in any other regions whatever, anywhere on the whole surface of the earth, are inversely as

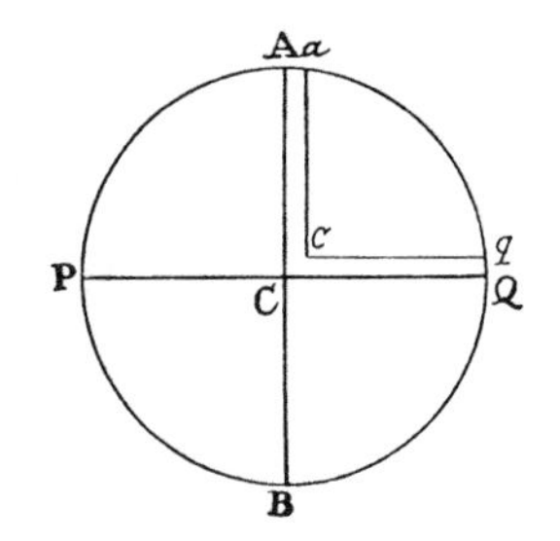

hh. The Latin text reads "paulo magis ibi decoquuntur," employing the verb "decoquo," which could mean, literally, "boil down" or "cook" or "bake." The problems of interpretation are discussed in the Guide to the present translation, §8.11.

the distances of those places from the center; and therefore, on the hypothesis that the earth is a spheroid, the proportion of those weights is given.

From this the following theorem is deduced:[a] The increase of weight in going from the equator to the poles is very nearly as the versed sine of twice the latitude, or (which is the same) as the square of the sine of the latitude. [b]And the arcs of degrees of latitude on a meridian are increased in about the same ratio. Now, the latitude of Paris is 48°50′, the latitude of places on the equator 00°00′, and that of places at the poles 90°; the versed sines of twice those arcs of latitude are 11,334 and 00,000 and 20,000 (the radius being taken to be 10,000); the gravity at the pole is to the gravity at the equator as 230 to 229; and the excess of the gravity at the pole to the gravity at the equator is

a. See the Guide, §10.15.

bb. In place of the remaining part of this proposition, ed. 1 has: "For example, the latitude of Paris is 48°45′, that of the island of Gorée near Cape Verde 14°15′, that of Cayenne off the coast of Guiana about 50°, that of places at the pole 90°. If the arcs of latitude are doubled, they are 97.5°, 28.5°, 10°, and 180°. The versed sines are 11,305, 1,211, 152, and 2,000. Furthermore, since the gravity at the pole is to the gravity at the equator as 692 to 689 and the excess of gravity at the pole is to the gravity at the equator as 3 to 689, the excess of gravity at Paris, on the island of Gorée, and at Cayenne will be to the gravity at the equator as $\frac{3 \times 11,305}{2,000}$, $\frac{3 \times 1,211}{2,000}$, and $\frac{3 \times 152}{2,000}$ to 689, or as 33,915, 3,633, and 456 to 13,780,000, and therefore the total gravities in these places will be to one another as 13,813,915, 13,783,633, 13,780,456, and 13,780,000. Therefore, since the lengths of pendulums oscillating with equal periods are as the gravities, and the length of a seconds pendulum at Paris is 3 Paris feet and 17⁄24 inches, the lengths of seconds pendulums on the island of Gorée, at Cayenne, and at the equator will be surpassed by the length of a Paris pendulum by excesses of $\frac{81}{1,000}$, $\frac{89}{1,000}$, and $\frac{90}{1,000}$ inches. All of these things will be so on the hypothesis that the earth consists of uniform matter. For if the matter at the center is a little denser than at the surface, those excesses will be a little greater. The reason is that if the superabundant matter at the center, by which the density there is rendered greater, is taken away and considered separately, the gravity toward the rest of the earth, which is uniformly dense, will be inversely as the distance of a weight from the center, but toward the superabundant matter inversely as the square of the distance from that matter very nearly. Therefore, gravity at the equator will be less toward that superabundant matter than as in the above computation, and therefore the earth there, on account of the deficiency of gravity, will rise a little higher than has been determined above. Indeed, the French by making experiments have already found that the length of seconds pendulums at Paris is greater than on the island of Gorée by 1⁄10 of an inch and greater than at Cayenne by 1⁄8. These differences are a little greater than the differences $\frac{81}{1,000}$ and $\frac{89}{1,000}$ which resulted from the above computation, and therefore (if one can have enough confidence in these rough observations) the earth will be somewhat higher at the equator than according to the above calculation and denser at the center than in mines near the surface. If the excess of gravity in these northern places over the gravity at the equator is finally determined exactly by experiments conducted with greater diligence, and then its excess is everywhere taken in the ratio of the versed sine of twice the latitude, then there will be determined a universal measure, an equalizing of the time of equal pendulums in the different places indicated, and also the proportion of the diameters of the earth and its density at the center, on the hypothesis that that density, as one goes to the circumference, decreases uniformly. And indeed this hypothesis, even though it is not exact, can be assumed for undertaking such a calculation."

as 1 to 229. Hence the excess of the gravity at the latitude of Paris will be to the gravity at the equator as $1 \times \frac{11,334}{20,000}$ to 229, or 5,667 to 2,290,000. And therefore the total gravities in these places will be to each other as 2,295,667 to 2,290,000. And thus, since the lengths of pendulums oscillating with equal periods are as the gravities, and at the latitude of Paris the length of a seconds pendulum is 3 Paris feet and 8½ lines (or rather, because of the weight of the air, 8⁵⁄₉ lines), the length of a pendulum at the equator will be shorter than the length of a pendulum with the same period at Paris in the amount of 1.087 lines. And a similar computation yields the following table.

Latitude of the place	*Length of the pendulum*		*Measure of one degree on the meridian*	*Latitude of the place*	*Length of the pendulum*		*Measure of one degree on the meridian*
degrees	*feet*	*lines*	*toises*	*degrees*	*feet*	*lines*	*toises*
0	3	7.468	56637	45	3	8.428	57010
5	3	7.482	56642	6	3	8.461	57022
10	3	7.526	56659	7	3	8.494	57035
15	3	7.596	56687	8	3	8.528	57048
20	3	7.692	56724	9	3	8.561	57061
25	3	7.812	56769	50	3	8.594	57074
30	3	7.948	56823	55	3	8.756	57137
35	3	8.099	56882	60	3	8.907	57196
40	3	8.261	56945	65	3	9.044	57250
1	3	8.294	56958	70	3	9.162	57295
2	3	8.327	56971	75	3	9.258	57332
3	3	8.361	56984	80	3	9.329	57360
4	3	8.394	56997	85	3	9.372	57377
				90	3	9.387	57382

[c] Moreover, it is established by this table that the inequality [in the length] of degrees [at different latitudes] is so small that in geographical matters the

cc. Ed. 2 has: "and that the inequality of the diameters of the earth can be ascertained more easily and more surely by experiments with pendulums or even by eclipses of the moon than by arcs measured geographically on the meridian.

"These things are so on the hypothesis that the earth consists of uniform matter. For if the matter at the center is a little denser than at the surface, the differences of pendulums and degrees on a meridian will be a little greater than according to the preceding table. The reason is that if superabundant matter at the center, by which the density there is rendered greater, is taken away and regarded separately, the gravity toward the rest of the earth, which is uniformly dense, will be inversely as the distance of a weight from the center; but toward the superabundant matter, it will be inversely as the square of the distance from that matter very nearly. Therefore gravity at the equator is less toward that superabundant matter than according to the above computation, and therefore the earth there, on account of the deficiency of

shape of the earth can be considered to be spherical, especially if the earth is a little denser toward the plane of the equator than toward the poles.[c]

Now some astronomers, sent to distant regions to make astronomical observations, have observed that their pendulum clocks went more slowly near the equator than in our regions. And indeed M. Richer first observed this in the year 1672 on the island of Cayenne. For while he was observing the transit of the fixed stars across the meridian in the month of August, he found that his clock was going more slowly than in its proper proportion to the mean motion of the sun, the difference being $2^{m}28^{s}$ every day. Then by constructing a simple pendulum that would oscillate in seconds as measured by the best clock, he noted the length of the simple pendulum, and he did this frequently, every week for ten months. Then, when he had returned to France, he compared the length of this pendulum with the length of a seconds pendulum at Paris (which was 3 Paris feet and 8⅗ lines long) and found that it was shorter than the Paris pendulum, the difference being 1¼ lines.[d]

Afterward, our fellow countryman Halley, sailing in about the year 1677 to the island of St. Helena, found that his pendulum clock went more slowly there than in London, but he did not record the difference. He made the pendulum of his clock shorter by more than ⅛ of an inch, or 1½ lines. And to effect this, since the length of the threaded part at the lower end of the pendulum rod was not sufficient, he put a wooden ring between the nut (on the threaded part) and the weight at the end of the pendulum.

Then in the year 1682 M. Varin and M. Des Hayes found that the length of a seconds pendulum in the Royal Observatory at Paris was 3 feet 8⁵⁄₉ lines. And on the island of Gorée they found by the same method that the length of a pendulum with the same period was 3 feet 6⁵⁄₉ lines, the difference in lengths being 2 lines. And sailing in the same year to the islands of Guadeloupe and Martinique, they found that on these islands the length of a pendulum with the same period was 3 feet 6½ lines.

Afterward, in July 1697, M. Couplet the younger adjusted his pendulum clock to the mean motion of the sun in the Royal Observatory at Paris in such a way that for quite a long time the clock agreed with the motion of

gravity, will rise a little higher, and the excesses of the lengths of pendulums and of the degrees at the poles will be a little greater than has been determined above."

d. Here ed. 2 has an additional sentence: "But from the slowness of the pendulum clock in Cayenne, the difference of the pendulums is gathered to be 1½ lines."

the sun. Then sailing to Lisbon, he found that by the next November the clock went more slowly than before, the difference being $2^{m}13^{s}$ in 24 hours. And sailing to Paraíba in the following March, he found that his clock went more slowly there than in Paris, the difference being $4^{m}12^{s}$ in 24 hours. And he declares that a seconds pendulum was 2½ lines shorter at Lisbon and 3⅔ lines shorter at Paraíba than at Paris. He might more correctly have put these differences as 1⅓ and 2⁵⁄₉; these are the differences that correspond to the differences in times of $2^{m}13^{s}$ and $4^{m}12^{s}$. He is less trustworthy because of the crudity of his observations.

In the next years (1699 and 1700) M. Des Hayes, again sailing to America, determined that on the islands of Cayenne and Grenada the length of a seconds pendulum was a little less than 3 feet 6½ lines, that on the island of St. Kitts that length was 3 feet 6¾ lines, and that on the island of Santo Domingo it was 3 feet 7 lines.

And in the year 1704 Father Feuillée found that in Portobello in America the length of a seconds pendulum was 3 Paris feet and only 5⁷⁄₁₂ lines, that is, about 3 lines shorter than at Paris, but he made an error in his observation. For, sailing afterward to the island of Martinique, he found that the length of a pendulum with the same period was 3 Paris feet and 5¹⁰⁄₁₂ lines.

Moreover, the latitude of Paraíba is 6°38′ S, and that of Portobello is 9°33′ N; and the latitudes of the islands of Cayenne, Gorée, Guadeloupe, Martinique, Grenada, St. Kitts, and Santo Domingo are respectively 4°55′, 14°40′, 14°00′, 14°44′, 12°6′, 17°19′, and 19°48′ N. And the excesses of the length of the pendulum at Paris over the observed lengths of pendulums with the same period in these latitudes are a little greater than they would be according to the table of pendulum lengths computed above. And therefore the earth is somewhat higher at the equator than according to the above computation, and is denser toward the center than in mines near the surface, unless perhaps the heat in the torrid zone somewhat increased the length of the pendulums.

M. Picard, at any rate, observed that an iron rod, which in wintertime when the weather was freezing was 1 foot long, came to be 1 foot and ¼ of a line long when heated by a fire. Later M. La Hire observed that an iron rod, which in an exactly similar winter was 6 feet long, came to be 6 feet and ⅔ of a line long when it was exposed to the summer sun. The heat [i.e., temperature] was greater in the first example than in the second, and in the second it was greater than that of the external parts of the human body. For metals grow extremely hot in the summer sun. But the pendulum rod in a pendulum clock is ordinarily never exposed to the heat of the summer sun, and never acquires a heat equal to that of the external surface of the human body. And, therefore, although a 3-foot-long pendulum rod in a clock will indeed be a little longer in summertime than in wintertime, this increase will scarcely surpass 1/4 of 1 line. Accordingly, all of the difference in the length of pendulums with the same period in different regions cannot be attributed to differences in heat. Nor can this difference be attributed to errors made by the astronomers sent from France. For although their observations do not agree perfectly with one another, the errors are so small that they can be ignored. And in this they all agree: that at the equator, pendulums are shorter than pendulums with the same period at the Royal Observatory in Paris, [e]the difference being neither less than 1 1/4 lines nor more than 2 2/3 lines. By the observations of M. Richer made in Cayenne the difference was 1 1/4 lines. By those of M. Des Hayes that difference when corrected became 1 1/2 or 1 3/4 lines. By the less accurate observations made by others, this difference came out as more or less 2 lines. And this discrepancy could have arisen partly from errors in observations, partly from the dissimilitude of the internal parts of the earth and from the height of mountains, and partly from the differences in heat [i.e., temperatures] of the air.

As far as I can tell, in England an iron rod 3 feet long is 1/6 of 1 line shorter in the wintertime than in the summertime. Let this quantity be sub-

ee. Ed. 2 has: "... the difference being about 2 lines or 1/6 of an inch. By the observations of M. Richer made in Cayenne, the difference was 1 1/2 lines. An error of half a line is easily committed. And M. Des Hayes afterward, by his observations made on the same island, corrected the error, finding a difference of 2 1/18 lines. But also by observations made on the islands of Gorée, Guadeloupe, Martinique, Grenada, St. Kitts, and Santo Domingo and reduced to the equator, that difference came out to be scarcely smaller than 1 19/20 of a line and scarcely greater than 2 1/2 lines. And the mean quantity between these limits is 2 9/40 lines. Because of the heat of places in the torrid zone, let us ignore 9/40 of a line, and a difference of 2 lines will remain.

"Therefore, since that difference, by the preceding table, on the hypothesis that the earth consists of uniformly dense matter, is only $1\frac{87}{1{,}000}$ of a line, the excess of the height of the earth at the equator over its height at the poles, which was 17 1/6 miles, being now increased in the ratio of the differences, will become 31 7/12 miles. For the slowness of a pendulum at the equator proves the deficiency of the gravity; and the lighter the matter is, the greater its height must be in order that by its weight it may hold in equilibrium the matter at the poles.

"Hence the shape of the earth's shadow, which is to be determined by eclipses of the moon, will not be entirely circular, but its diameter drawn from east to west will be greater than its diameter drawn

tracted (because of the heat at the equator) from the difference of 1¼ lines observed by Richer, and there will remain 1 1/12 lines, in excellent agreement with the $1\frac{87}{1,000}$ lines already found from the theory. Moreover, Richer repeated his observations in Cayenne every week during a ten-month period, and compared the lengths he found there for a pendulum consisting of an iron rod with its lengths similarly found in France [i.e., with its lengths adjusted in Paris so as to have the same period]. This diligence and caution seem to have been lacking in other observers. If his observations are to be trusted, the earth will be higher at the equator than at the poles by an excess of about seventeen miles, as came out above by the theory.[b e]

Proposition 21
Theorem 17

The equinoctial points regress, and the earth's axis, by a nutation in every annual revolution, inclines twice toward the ecliptic and twice returns to its former position.

This is clear by book 1, prop. 66, corol. 20. This motion of nutation, however, must be very small—either scarcely or not at all perceptible.

Proposition 22
Theorem 18

All the motions of the moon and all the inequalities in its motions follow from the principles that have been set forth.

---

from south to north by an excess of about 55″. And the greatest longitudinal parallax of the moon will be a little greater than its greatest latitudinal parallax. And the greatest semidiameter of the earth will be 19,767,630 Paris feet, the least, 19,609,820 feet, and the mean, 19,688,725 feet, very nearly.

"Since one degree by Picard's measurement is 57,060 toises but by Cassini's measurement is 57,292 toises, some suspect that each degree, as one goes southward through France, is greater than the preceding degree by 72 toises more or less, or $\frac{1}{800}$ of one degree, the earth being an oblong spheroid whose parts are highest at the poles. Under this supposition, all bodies at the earth's poles would be lighter than at the equator, and the height of the earth at the poles would exceed its height at the equator by nearly 95 miles, and isochronous pendulums would be longer at the equator than in the Royal Observatory at Paris by an excess of about half an inch, as will be easily seen by anyone comparing the proportions set forth here with the proportions set forth in the preceding table. But also the diameter of the earth's shadow drawn from south to north would be greater than its diameter drawn from east to west by an excess of 2′46″, or 1/12 of the moon's diameter. Experience contradicts all this. Certainly Cassini, in determining that one degree is 57,292 toises, took a mean between all his measurements, on the hypothesis of the equality of degrees. And although Picard on the northern border of France found a degree to be a little smaller, yet our compatriot Norwood in more northern reigons, by measuring a greater interval, found a degree to be a little greater than Cassini had found. And Cassini himself, when he attempted to determine the measurement of one degree by using a far greater interval, judged Picard's measurement to be insufficiently certain and exact because of the smallness of the interval measured. But the differences among the measurements of Cassini, Picard, and Norwood are nearly imperceptible and could easily have arisen from imperceptible errors in observations, not to mention the nutation of the earth's axis."

That the major planets, while they are being carried about the sun, can carry other or minor planets [or satellites], revolving around them, and that those minor planets must revolve in ellipses having their foci in the centers of the major planets, is evident from book 1, prop. 65. Moreover, their motions will be perturbed in many ways by the sun's action, and they will be influenced by those inequalities that are observed in our moon. Our moon, in any case (by book 1, prop. 66, corols. 2, 3, 4, and 5), moves more swiftly, and by a radius drawn to the earth describes an area greater for the time, and has a less curved orbit, and therefore approaches closer to the earth, in the syzygies than in the quadratures, except insofar as these effects are hindered by the motion of eccentricity. For the eccentricity is greatest (by book 1, prop. 66, corol. 9) when the moon's apogee is in the syzygies, and least when it stands in the quadratures; and thus the moon in its perigee is swifter and closer to us, while in its apogee it is slower and more remote, in the syzygies than in the quadratures. Additionally, the apogee advances and the nodes regress, but with a nonuniform motion. And indeed the apogee (by prop. 66, corols. 7 and 8) advances more swiftly in its syzygies, regresses more slowly in the quadratures, and by the excess of the advance over the regression is annually carried forward [or in consequentia, i.e., from east to west in the direction of the signs]. But the nodes (by prop. 66, corol. 2) are at rest in their syzygies and regress most swiftly in the quadratures. The moon's greatest latitude is also greater in its quadratures (by prop. 66, corol. 10) than in its syzygies, and (by prop. 66, corol. 6) the mean motion of the moon is slower in the earth's perihelion than in its aphelion. And these are the more significant inequalities [of the moon's motion] taken note of by astronomers.

There are also certain other inequalities not observed by previous astronomers, by which the lunar motions are so perturbed that until now these motions have not been reducible, by any law, to any definite rule. For the velocities or hourly motions of the moon's apogee and nodes, and their equations, and also the difference between the greatest eccentricity in the syzygies and the least in the quadratures, and that inequality which is called the variation, are increased and decreased annually (by prop. 66, corol. 14) as the cube of the sun's apparent diameter. And, additionally, the variation is increased or decreased very nearly as the square of the time between the quadratures (by book 1, lem. 10, corols. 1 and 2, and prop. 66, corol. 16), but in astronomical calculations this inequality is generally included under the moon's prosthaphaeresis [or equation of the center] and confounded with it.

Proposition 23
Problem 5

To derive the unequal motions [i.e., the inequalities in the motions] of the satellites of Jupiter and of Saturn from the motions of our moon.

From the motions of our moon the analogous motions of the moons or satellites of Jupiter are derived as follows. The mean motion of the nodes of Jupiter's outermost satellite is (by book 1, prop. 66, corol. 16) to the mean motion of the nodes of our moon in a ratio compounded of the square of the ratio of the earth's periodic time about the sun to Jupiter's periodic time about the sun, and of the simple ratio of the satellite's periodic time about Jupiter to the moon's periodic time about the earth, and so in one hundred years that node completes 8°24′ backward [or in antecedentia, i.e., counter to the order of the signs]. The mean motions of the nodes of the inner satellites are (by the same corollary) to the motion of this outermost satellite as the periodic times of those inner satellites are to the periodic time of the outermost satellite and hence are given. Moreover (by the same corollary), the forward [or direct] motion of the upper apsis of each satellite [or its motion in consequentia] is to the backward [or retrograde] motion of its nodes [or the motion in antecedentia] as the motion of the apogee of our moon to the motion of its nodes, and hence is also given. However, the motion of the upper apsis found in this way must be decreased in the ratio of 5 to 9, or about 1 to 2, for a reason which would take too much time to explain here. The greatest equations of the nodes and upper apsis of each satellite are approximately to the greatest equations of the nodes and upper apsis of our moon respectively as the motions of the nodes and upper apsis of the satellites in the time of one revolution of the former equations are to the motions of the nodes and apogee of our moon in the time of one revolution of the latter equations. By the same corollary, the variation of a satellite as it would be observed from Jupiter is to the variation of our moon in the same proportion as the total motions of their nodes during the times in which respectively the satellite and our moon revolve as reckoned in relation to the sun; and therefore in the outermost satellite the variation does not exceed 5″12‴.

Proposition 24
Theorem 19

The ebb and flow of the sea arise from the actions of the sun and moon.

It is clear from book 1, prop. 66, corols. 19 and 20, that the sea should twice rise and twice fall in every day, lunar as well as solar, and also that the greatest height of the water, in deep and open seas, should occur less than six hours after the appulse of the luminaries to the meridian of a place, as happens in the whole eastern section of the Atlantic Ocean and the Ethiopic [or South Atlantic] Sea between France and the Cape of Good Hope, and also on the Chilean and Peruvian shore of the Pacific Ocean; on all these shores the tide comes in at about the second, third, or fourth hour, except in cases when the motion has been propagated from the deep ocean through shallow places and is delayed until the fifth, sixth, or seventh hour, or later. I number the hours from the appulse of either luminary to the meridian of a place, below the horizon as well as above, and by hours of a lunar day I mean twenty-fourths of that time in which the moon, by its apparent daily motion, returns to the meridian of the place. The force of the sun or moon to raise the sea is greatest in the very appulse of the luminary to the meridian of the place. But the force impressed upon the sea at that time remains for a while and is increased by a new force subsequently impressed, until the sea has ascended to its greatest height, which will happen in one or two hours, but more frequently at the shores in about three hours or even more if the sea is shallow.

Moreover, the two motions which the two luminaries excite will not be discerned separately but will cause what might be called a mixed motion. In the conjunction or the opposition of the luminaries their effects will be combined, and the result will be the greatest ebb and flow. In the quadratures the sun will raise the water while the moon depresses it and will depress the water while the moon raises it; and the lowest tide of all will arise from the difference between these two effects. And since, as experience shows, the effect of the moon is greater than that of the sun, the greatest height of the water will occur at about the third lunar hour. Outside of the syzygies and quadratures, the highest tide, which by the lunar force alone would always have to occur at the third lunar hour, and by the solar force alone at the third solar hour, will occur, as a result of the combining of the lunar and solar forces, at some intermediate time which is closer to the third lunar hour [than to the third solar hour]; and thus in the transit of the moon from the syzygies to the quadratures, when the third solar hour precedes the third lunar hour, the greatest height of the water will also precede the third lunar hour, and will do so by the greatest interval a little after the octants of the moon; and the highest tide will follow the third lunar hour with the same intervals in the transit of the moon from the quadratures to the syzygies. This is what happens in the open sea. For in the mouths of rivers the higher tides, other things being equal, will come to their peaks later.

Additionally, the effects of the luminaries depend on their distances from the earth. For at smaller distances their effects are greater, and at greater distances smaller, and this varies as the cubes of their apparent diameters. Therefore the sun in wintertime, when it is in its perigee, produces greater effects and makes the tides a little higher in the syzygies and a little lower (other things being equal) in the quadratures than in summertime; and the moon in its perigee every month produces higher tides than fifteen days before or after, when it is in its apogee. Accordingly, it happens that the two very highest tides do not follow each other in successive syzygies.

The effect of each luminary depends also on its declination, or distance from the equator. For if the luminary should be at one of the poles, it would constantly draw the individual parts of water, without intension and remission of action, and thus would produce no reciprocation of motion. Therefore the luminaries, in receding from the equator toward a pole, will lose their effects by degrees, and for this reason will produce lower tides in the solstitial syzygies than in the equinoctial syzygies. In the solstitial quadratures, however, they will produce higher tides than in the equinoctial quadratures, because the effect of the moon, which is now at the equator, most exceeds the effect of the sun. Therefore the highest tides occur at the syzygies of the luminaries, and the lowest at their quadratures, at about the times of either of the two equinoxes. And the highest tide in the syzygies is always acompanied by the lowest tide in the quadratures, as has been learned by experience. Moreover, as a result of the smaller distance of the sun from the earth in winter than in summer, it comes about that the highest and lowest tides more often precede the vernal equinox than follow it, and more often follow the autumnal equinox than precede it.

The effects of the luminaries depend also on the latitude of places. Let A*p*EP represent the earth covered everywhere with deep waters, C its center, P and *p* the poles, AE the equator, F any place not on the equator, F*f* the parallel of that place, D*d* the parallel corresponding to it on the other side of the equator, L the place that the moon was occupying three hours earlier, H the place on the earth situated perpendicularly beneath L, *h* the place opposite H, K and *k* places 90 degrees distant from H and *h*, CH and C*h* the greatest heights of the sea (measured from the center of the earth), and CK and C*k* the least heights. If an ellipse is described with axes H*h*, K*k*, and then if by the revolution of this ellipse about the major axis H*h* a spheroid HPK*hpk* is described, this spheroid will represent the figure of the sea very nearly, and CF, C*f*, CD, C*d* will be the heights of the sea at places F, *f*, D, *d*. Further, if in the aforesaid revolution of the ellipse any point N describes a circle NM which cuts parallels F*f*, D*d* in any places R, T, and cuts the equator AE in S, CN will be the height of the sea in all places R, S, T located on this circle. Hence, in the daily revolution of any place F, the greatest flood tide will be in F at the third hour after the appulse of the moon to the meridian above the horizon; afterward, the greatest ebb tide will be in Q at the third hour after the setting of the moon; then the greatest flood tide will be in *f* at the third hour after the appulse of the moon to the meridian below the horizon; finally, the greatest ebb tide will be in Q at the third hour after the rising of the moon; and the latter flood tide in *f* will be smaller than the former flood tide in F.

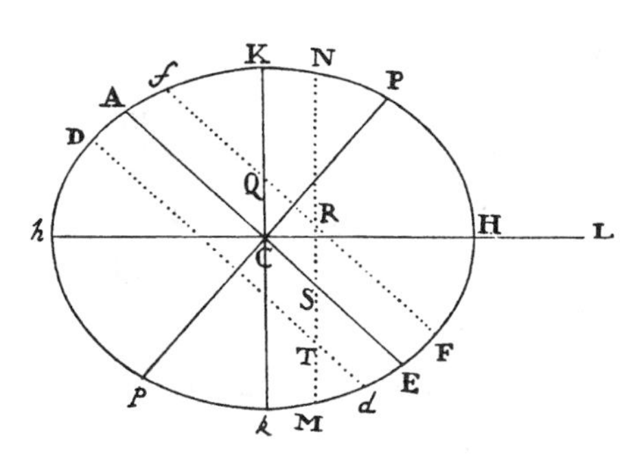

For the whole sea is divided into just two hemispherical flows [or flowing bodies of water], one in the hemisphere KH*k* verging to the north, the other in the opposite hemisphere K*hk*; and these may therefore be called the northern flow and the southern flow. These flowing bodies of waters, which are always opposite to each other, come by turns to the meridian of every single place, with an interval of twelve lunar hours between them. And since the northern regions partake more of the northern flow, and the southern regions more of the southern flow, higher and lower tides arise from them alternately, in every single place not on the equator in which the luminaries rise and set. Moreover, the higher tide, when the moon declines toward the vertex of the place, will occur at about the third hour after the appulse of the moon to the meridian above the horizon, and when the moon changes its declination[a], this higher tide will be turned into a lower one. And the greatest difference between these tides will occur at the times of the solstices, especially if the ascending node of the moon is in the first of Aries. Thus it has been found by experience that in winter, morning tides exceed evening tides and that in summer, evening tides exceed morning tides, at Plymouth by a height of about one foot, and at Bristol by a height of fifteen inches, according to the observations of Colepress and Sturmy.

Moreover, the motions hitherto described are changed somewhat by the force of reciprocation of the waters, by which a tide of the sea, even if the actions of the luminaries were to cease, would be able to persevere for a while. This conservation of impressed motion lessens the difference between alternate tides; and it makes the tides immediately after the syzygies higher and makes those immediately after the quadratures lower. Hence it happens that alternate tides at Plymouth and Bristol do not differ from each other by much more than a height of one foot or fifteen inches, and that the very highest tides in those same harbors are not the first tides after the syzygies but the third. All the motions are made slower also in their passing through shallows, to such an extent that the very highest tides, in certain straits and the mouths of rivers, are the fourth or even the fifth after the syzygies.

Further, it can happen that a tide is propagated from the ocean through different channels to the same harbor and passes more quickly through some channels than through others; in this case the same tide, divided into two or more tides arriving successively, can compose new motions of different kinds. Let us suppose that two equal tides come from different places to the same harbor and that the first precedes the second by a space of six hours and occurs at the third hour after the appulse of the moon to the meridian of the harbor. If the moon is on the equator at the time of this appulse to the meridian, then every six hours there will be equal flood tides coming upon corresponding equal ebb tides and causing those ebb tides to be balanced by the flood tides, and thus during the course of that day they will cause the water to stay quiet and still. If at that time the moon is declining from the equator, there will be alternately higher and lower tides in the ocean, as has been said; and from the ocean, two higher and two lower tides will each be alternately propagated toward this harbor. Moreover, the two greater flood

a. Motte adds, "to the other side of the equator."

tides will produce the highest water in the middle time between them; the greater and lesser flood tides will make the water ascend to its mean height in the middle time between them; and between the two lesser flood tides the water will ascend to its least height. Thus in the space of twenty-four hours, the water will only once reach its greatest height, not twice as usually happens, and will only once reach its least height; and the greatest height, if the moon is declining toward the pole above the horizon of the place, will occur at either the sixth or the thirtieth hour after the appulse of the moon to the meridian; and when the moon changes its declination, this flood tide will be changed into an ebb tide. An example of all these things has been given by Halley, on the basis of sailors' observations, in Batsha harbor in the kingdom of Tonkin at a latitude of 20°50′ N. There the water stays still on the day following the transit of the moon over the equator; then, when the moon declines toward the north, the water begins to ebb and flow—not twice, as in other harbors, but only once every day; and the flood tide occurs at the setting of the moon, and the greatest ebb tide at its rising. This flood tide increases with the declination of the moon until the seventh or eighth day; then during the next seven days it decreases at the same rate at which it had previously increased. And when the moon changes its declination, the flood ceases and is then turned into an ebb. For thereafter the ebb tide occurs at the setting of the moon and the flood tide at its rising, until the moon again changes its declination. There are two different approaches from the ocean into this harbor and the neighboring channels, the one from the China Sea between the continent and the island of Leuconia, the other from the Indian Ocean between the continent and the island of Borneo. But whether there are tides coming through these channels in twelve hours from the Indian Ocean and in six hours from the China Sea, which thus occurring at the third and ninth lunar hours compound motions of this sort, or whether there is any other condition of those seas, I leave to be determined by observations of the neighboring shores.

Hitherto I have given the causes of the motions of the moon and seas. It is now proper to subjoin some things about the quantity of those motions.

Proposition 25
Problem 6

To find the forces of the sun that perturb the motions of the moon.

Let S designate the sun, T the earth, P the moon, CADB the orbit of the moon. On SP take SK equal to ST; and let SL be to SK as SK^2 to SP^2, and

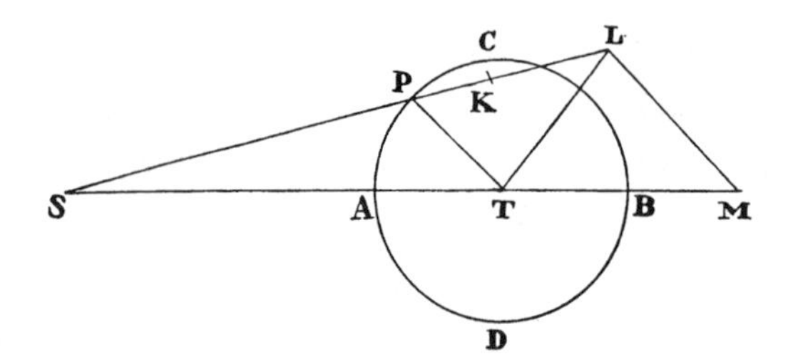

draw LM parallel to PT; and if the accelerative gravity of the earth toward the sun is represented by the distance ST or SK, SL will be the accelerative gravity of the moon toward the sun. This is compounded of the parts SM and LM, of which LM and the part TM of SM perturb the motion of the moon, as has been set forth in book 1, prop. 66 and its corollaries. Insofar as the earth and moon revolve around their common center of gravity, the motion of the earth about that center will also be perturbed by entirely similar forces; but it is possible to refer the sums of the forces and the sums of the motions to the moon, and to represent the sums of the forces by the lines TM and ML that correspond to them. The force ML, in its mean quantity, is to the centripetal force by which the moon could revolve in its orbit, about an earth at rest at a distance PT, as the square of the ratio of the periodic time of the moon about the earth to that of the earth about the sun (by book 1, prop. 66, corol. 17), that is, as the square of the ratio of $27^d7^h43^m$ to $365^d6^h9^m$, that is, as 1,000 to 178,725, or 1 to 178 29/40. But we found in prop. 4 of this book 3 that if the earth and moon revolve about their common center of gravity, their mean distance from each other will be very nearly 60½ mean semidiameters of the earth. And the force by which the moon could revolve in orbit about the earth at rest at a distance PT of 60½ terrestrial semidiameters is to the force by which it could revolve in the same time at a distance of 60 semidiameters as 60½ to 60; and this force is to the force of gravity on the earth as 1 to 60×60 very nearly. And so the mean force ML is to the force of gravity on the surface of the earth as $1 \times 60\frac{1}{2}$ to $60 \times 60 \times 60 \times 178\frac{29}{40}$, or as 1 to 638,092.6. From this and from the proportion of the lines TM and ML, the force TM is also given; and these are the forces of the sun by which the motions of the moon are perturbed. Q.E.I.

Proposition 26
Problem 7
To find the hourly increase of the area that the moon, by a radius drawn to the earth, describes in a circular orbit.

We have said that the area which the moon describes by a radius drawn to the earth is proportional to the time, except insofar as the motion of the moon is disturbed by the action of the sun. We propose to investigate here the inequality of the moment, or of the hourly increase [under the foregoing condition of disturbance]. To make the computation easier, let us imagine

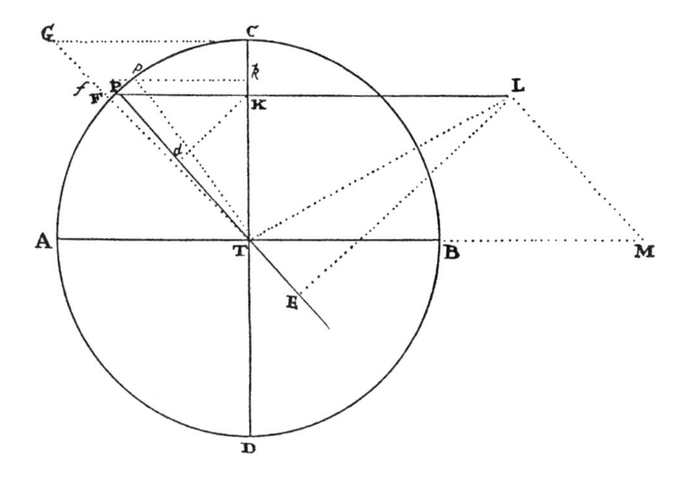

that the orbit of the moon is circular, and let us ignore all inequalities with the sole exception of the one under discussion here. Because of the enormous distance of the sun, let us suppose also that the lines SP and ST are parallel to each other. By this means the force LM will always be reduced to its mean quantity TP, and so will the force TM be reduced to its mean quantity 3PK. These forces (by corol. 2 of the laws of motion) compose the force TL; and if a perpendicular LE is dropped to the radius TP, this force is resolved into the forces TE and EL, of which TE, always acting along the radius TP, neither accelerates nor retards the description of the area TPC made by that radius TP; and EL, acting along the perpendicular to the radius, accelerates or retards the description of the area, as much as it accelerates or retards the moon. That acceleration of the moon, made in each individual moment of time, in the transit of the moon from the quadrature C to the conjunction A, is as the accelerating force itself EL, that is, as

$$\frac{3PK \times TK}{TP}.$$

Let the time be represented by the mean motion of the moon or (which comes to about the same thing) by the angle CTP or by the arc CP. On CT erect a normal CG (equal to CT). And when the quadrantal arc AC has been divided into innumerable equal particles $Pp, \ldots,$ by which the same innumerable quantity of equal particles of time can be represented, and when a perpendicular *pk* has been drawn to CT, draw TG meeting KP and *kp* (produced) in F and *f*; and FK will be equal to TK, and K*k* will be to PK as P*p* to T*p*, that is, in a given ratio; and therefore FK × K*k*, or the area FK*kf*, will be as $\frac{3PK \times TK}{TP}$, that is, as EL; and, by compounding, the total area GCKF will be as the sum of all the forces EL impressed on the moon in the total time CP, and so also as the velocity generated by this sum, that is, as the acceleration of the description of the area CTP, or the increase of its moment. The force by which the moon could revolve in its periodic time CADB of $27^d7^h43^m$ about the earth at rest, at the distance TP, would make a body, by falling in the time CT, describe the space ½CT, and at the same time acquire a velocity equal to the velocity with which the moon moves in its orbit. This is evident from book 1, prop. 4, corol. 9. However, since the perpendicular K*d* dropped to TP is a third of EL, and is equal to a half of TP or ML in the octants, the force EL in the octants (where it is greatest) will exceed the force ML in the ratio of 3 to 2, and so will be to that force by which the moon could revolve in its periodic time about the earth at rest as 100 to ⅔ × 17,872½, or 11,915, and should in the time CT generate a velocity which would be $\frac{100}{11,915}$ of the moon's velocity; but in the time CPA this force would generate a greater velocity in the ratio of CA to CT or TP. Let the greatest force EL in the octants be represented by the area FK × K*k* equal to the rectangle ½TP × P*p*. And the velocity which that greatest force could generate in any time CP will be to the velocity which any other lesser force EL generates, in the same time, as the rectangle ½TP × CP to the area KCGF; but the velocities generated in the whole time CPA will be to each other as the rectangle ½TP × CA to the triangle TCG, or as the quadrantal arc CA to the radius TP. And so (by book 5, prop. 9 of the *Elements*) the latter velocity generated in the whole time will be $\frac{100}{11,915}$ of the velocity of the moon. Change this velocity of the moon, which corresponds to the mean moment of the area, by adding and subtracting half of the other velocity; and if the mean moment is represented by the number 11,915, the sum 11,915 + 50 (or 11,965) will represent the greatest moment of the area in the syzygy A, and the difference 11,915 − 50 (or 11,865) the least moment of the same area in the quadratures. Therefore the areas which are described in equal times in the syzygies and quadratures are to each other as 11,965 to 11,865. To the least moment 11,865 add the moment that is to the difference (100) of the two above-mentioned moments as the quadrilateral FKCG is to the triangle TCG or, which comes to the same thing, as the square of the sine PK to the square of the radius TP (that is, as P*d* to TP); then the sum will represent the moment of the area when the moon is in any intermediate place P.

All these things are so on the hypothesis that the sun and earth are at rest, and that the moon has a synodic period of revolution of $27^{d}7^{h}43^{m}$. But since the moon's synodic period is actually $29^{d}12^{h}44^{m}$, the increments of the moments should be increased in the ratio of the time, that is, in the ratio of 1,080,853 to 1,000,000. In this way the total increment, which was $\frac{100}{11,915}$ of the mean moment, will now become $\frac{100}{11,023}$ of it. And so the moment of the area in the quadrature of the moon will be to its moment in the syzygy as $11,023 - 50$ to $11,023 + 50$, or as 10,973 to 11,073; and to its moment, when the moon is in any other intermediate place P, as 10,973 to $10,973 + \text{P}d$, taking TP to be equal to 100.

Therefore the area that the moon, by a radius drawn to the earth, describes in every equal particle of time is very nearly as the sum of the number 219.46 and of the versed sine of twice the distance of the moon from the nearest quadrature, with respect to a circle whose radius is unity. These things are so when the variation in the octants is at its mean magnitude. But if the variation there is greater or less, that versed sine should be increased or decreased in the same ratio.

Proposition 27
Problem 8

From the hourly motion of the moon, to find its distance from the earth.

The area that the moon, by a radius drawn to the earth, describes in every moment of time is as the hourly motion of the moon and the square of the distance of the moon from the earth jointly. And therefore the distance of the moon from the earth is directly proportional to the square root of the area and inversely proportional to the square root of the hourly motion. Q.E.I.

COROLLARY 1. Hence the apparent diameter of the moon is given, since it is inversely as the distance of the moon from the earth. Let astronomers test how accurately this rule agrees with the phenomena.

COROLLARY 2. Hence also the lunar orbit can be defined more exactly from the phenomena than could have been done before now.

Proposition 28
Problem 9

To find the diameters of the orbit in which the moon would have to move, if there were no eccentricity.

The curvature of the trajectory that a moving body describes, if it is attracted in a direction which is everywhere perpendicular to that trajectory, is as the attraction directly and the square of the velocity inversely. I reckon the curvatures of lines as being among themselves in the ultimate ratio of the sines or of the tangents of the angles of contact, with respect to equal radii, when those radii are diminished indefinitely. Now, the attraction of the moon toward the earth in the syzygies is the excess of its gravity toward the earth over the solar force 2PK (as in the figure to prop. 25), by which force the accelerative gravity of the moon toward the sun exceeds the accelerative gravity of the earth toward the sun or is exceeded by it. In the quadratures that attraction is the sum of the gravity of the moon toward the earth and the solar force KT (which draws the moon toward the earth). And these attractions, if $\frac{AT + CT}{2}$ is called N, are very nearly as $\frac{178{,}725}{AT^2} - \frac{2{,}000}{CT \times N}$ and $\frac{178{,}725}{CT^2} + \frac{1{,}000}{AT \times N}$, or as $178{,}725N \times CT^2 - 2{,}000AT^2 \times CT$ and $178{,}725N \times AT^2 + 1{,}000CT^2 \times AT$. For if the accelerative gravity of the moon toward the earth is represented by the number 178,725, then the mean force ML, which in the quadratures is PT or TK and draws the moon toward the earth, will be 1,000, and the mean force TM in the syzygies will be 3,000; if the mean force ML is subtracted from that, there will remain the force 2,000 by which the moon in the syzygies is drawn apart from the earth and which I have called 2PK above. Now, the velocity of the moon in the syzygies (A and B) is to its velocity in the quadratures (C and D) jointly as CT is to AT and as the moment of the area that the moon (by a radius drawn to the earth) describes in the syzygies is to the moment of that same area as described in the quadratures, that is, as 11,073CT to 10,973AT. Take this ratio squared inversely and the above ratio directly, and the curvature of the moon's orbit in the syzygies will become to its curvature in the quadratures as $120{,}406{,}729 \times 178{,}725AT^2 \times CT^2 \times N - 120{,}406{,}729 \times 2{,}000AT^4 \times CT$ to $122{,}611{,}329 \times 178{,}725AT^2 \times CT^2 \times N + 122{,}611{,}329 \times 1{,}000CT^4 \times AT$, that is, as $2{,}151{,}969\text{AT} \times \text{CT} \times \text{N} - 24{,}081\text{AT}^3$ to $2{,}191{,}371\text{AT} \times \text{CT} \times \text{N} + 12{,}261\text{CT}^3$.

Since the figure of the lunar orbit is unknown, in its place let us assume an ellipse DBCA, in whose center T the earth is placed, and let its major axis DC lie between the quadratures and its minor axis AB between the syzygies. And since the plane of this ellipse revolves about the earth with an angular motion, and since the trajectory whose curvature we are considering ought to be described in a plane that is entirely devoid of any angular motion, we must consider the figure that the moon, while revolving in that ellipse, describes in this place, that is, the figure C*pa*, whose individual points *p* are found by taking any point P on the ellipse to represent the place of the moon, and by drawing T*p* equal to TP in such a way that the angle PT*p* is equal to the apparent motion of the sun since the time of quadrature C, or (which comes to almost the same thing) in such a way that the angle CT*p* is to the angle CTP as the time of a synodic revolution of the moon is to the time of a periodic revolution, or as $29^{\text{d}}12^{\text{h}}44^{\text{m}}$ to $27^{\text{d}}7^{\text{h}}43^{\text{m}}$. Therefore, take the angle CT*a* in this same ratio to the right angle CTA, and let the length T*a* be equal to the length TA, then *a* will be the lower apsis and *c* the upper apsis of this orbit C*pa*. And by making calculations I find that the difference between the curvature of the orbit C*pa* at the vertex *a* and the curvature of the circle described with center T and radius TA has a ratio to the difference between the curvature of the ellipse at the vertex A and the curvature of that circle which is equal to the ratio of the square of the angle CTP to the square of the angle CT*p* and that the curvature of the ellipse at A is to the curvature of that circle in the ratio of TA^2 to TC^2; and the curvature of that circle is to the curvature of a circle described with center T and radius TC as TC to TA; but this curvature is to the curvature of the ellipse at C in the ratio of TA^2 to TC^2; and the difference between the curvature of the ellipse at the vertex C and the curvature of this last circle is to the difference between the curvature of the figure T*pa* at the vertex C and the curvature of the

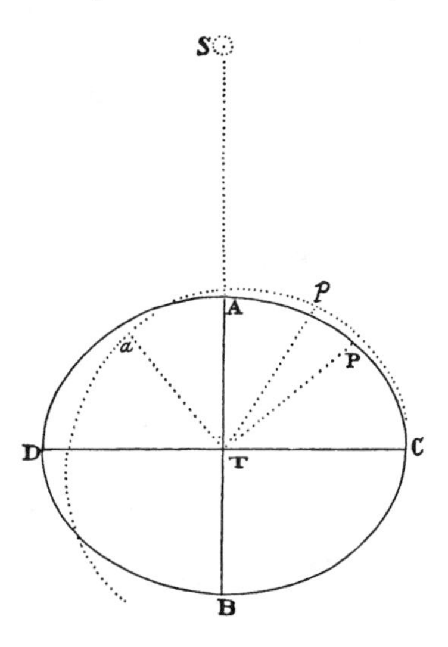

same circle in the ratio of the square of the angle CTp to the square of the angle CTP. And these ratios are easily gathered from the sines of the angles of contact and of the differences of the angles. Moreover, by comparing these, the curvature of the figure Cpa at a comes out to its curvature at C as $AT^3 + \frac{16,824}{100,000}CT^2 \times AT$ to $CT^3 + \frac{16,824}{100,000}AT^2 \times CT$; where the factor $\frac{16,824}{100,000}$ represents the difference of the squares of the angles CTP and CTp divided by the square of the smaller angle CTP, or (which is the same) the difference of the squares of the times $27^d7^h43^m$ and $29^d12^h44^m$ divided by the square of the time $27^d7^h43^m$.

Therefore, since a designates the syzygy of the moon and C its quadrature, the proportion just found must be the same as the proportion of the curvature of the orbit of the moon in the syzygies to its curvature in the quadratures, which we found above. Accordingly, to find the proportion of CT to AT, I multiply the extremes by the means. And the resulting terms divided by $AT \times CT$ become $2,062.79CT^4 - 2,151,969N \times CT^3 + 368,676N \times AT \times CT^2 + 36,342AT^2 \times CT^2 - 362,047N \times AT^2 \times CT + 2,191,371N \times AT^3 + 4,051.4AT^4 = 0$. When I take the half-sum N of the terms AT and CT to be 1, and their half-difference to be x, there results $CT = 1 + x$ and $AT = 1 - x$; and when these values are put into the equation and the resulting equation is resolved, x is found equal to 0.00719, and hence the semidiameter CT comes out 1.00719 and the semidiameter AT 0.99281. These numbers are very nearly as $70\frac{1}{24}$ and $69\frac{1}{24}$. Therefore the distance of the moon from the earth in the syzygies is to its distance in the quadratures (setting aside, that is, any consideration of eccentricity) as $69\frac{1}{24}$ to $70\frac{1}{24}$, or in round numbers as 69 to 70.

Proposition 29
Problem 10

To find the variation of the moon.

This inequality arises partly from the elliptical form of the orbit of the moon and partly from the inequality of the moments of the area that the moon describes by a radius drawn to the earth. If the moon P moved in the ellipse DBCA about the earth at rest in the center of the ellipse and, by a radius TP drawn to the earth, described the area CTP proportional to the time, and if furthermore the greatest semidiameter CT of the ellipse were to the least semidiameter TA as 70 to 69, then the tangent of the angle CTP would be to the tangent of the angle of the mean motion (reckoned from the quadrature C) as the semidiameter TA of the ellipse to its semidiameter TC, or as 69 to 70. Moreover, the description of the area CTP ought, in the progress of the moon from quadrature to syzygy, to be accelerated in such a way that the moment of this area in the syzygy of the moon will be to its moment in its quadrature as 11,073 to 10,973, and in such a way that the excess of the moment in any intermediate place P over the moment in the quadrature will be as the square of the sine of the angle CTP. And this will occur exactly enough if the tangent of angle CTP is diminished in the ratio of $\sqrt{10,973}$ to $\sqrt{11,073}$, or in the ratio of 68.6877 to 69. In this way the tangent of angle CTP will now be to the tangent of the mean motion as 68.6877 to 70; and the angle CTP in the octants, where the mean motion is 45°, will be found to be 44°27′28″, which, when subtracted from the angle of the mean motion of 45°, leaves the greatest variation 32′32″. These things would be so if the moon, in going from quadrature to syzygy, described an angle CTA of only 90°. But because of the motion of the earth, by which the sun is transferred forward [or in consequentia] in its apparent motion, the moon, before it reaches the sun, describes an angle CT*a* greater than a right angle, in the ratio of the time of a synodic revolution of the moon to the time of its periodic revolution, that is, in the ratio of $29^{d}12^{h}44^{m}$ to $27^{d}7^{h}43^{m}$. And in this way all the angles about the center T are enlarged in the same ratio; and the greatest variation, which would otherwise be 32′32″, now increased in the same ratio, becomes 35′10″.

This is the magnitude of the greatest variation at the mean distance of the sun from the earth, ignoring the differences that can arise from the curvature of the earth's orbit and the greater action of the sun upon the sickle-shaped and the new moon than upon the gibbous and the full moon. At other distances of the sun from the earth, the greatest variation is directly as the square of the time of synodic revolution and inversely as the cube of the distance of the sun from the earth. And therefore in the apogee of the sun the greatest variation is 33′14″, and in its perigee 37′11″, provided that the eccentricity of the sun is to the transverse semidiameter of the great orbit [i.e., the earth's orbit] as $16\frac{15}{16}$ to 1,000.

Hitherto we have investigated the variation in a noneccentric orbit, in which the moon in its octants is always at its mean distance from the earth. If the moon, because of its eccentricity, is more distant or less distant from the earth than if it were placed in this orbit, the variation can be a little greater or a little less than according to the rule asserted here; but I leave the excess or deficiency for astronomers to determine from phenomena.

Proposition 30
Problem 11

To find the hourly motion of the nodes of the moon in a circular orbit.

Let S designate the sun, T the earth, P the moon, N*P*n the orbit of the moon, Npn the projection of the orbit in the plane of the ecliptic; N and n the nodes, nTNm the line of the nodes, indefinitely produced; PI and PK perpendiculars dropped to the lines ST and Qq; Pp a perpendicular

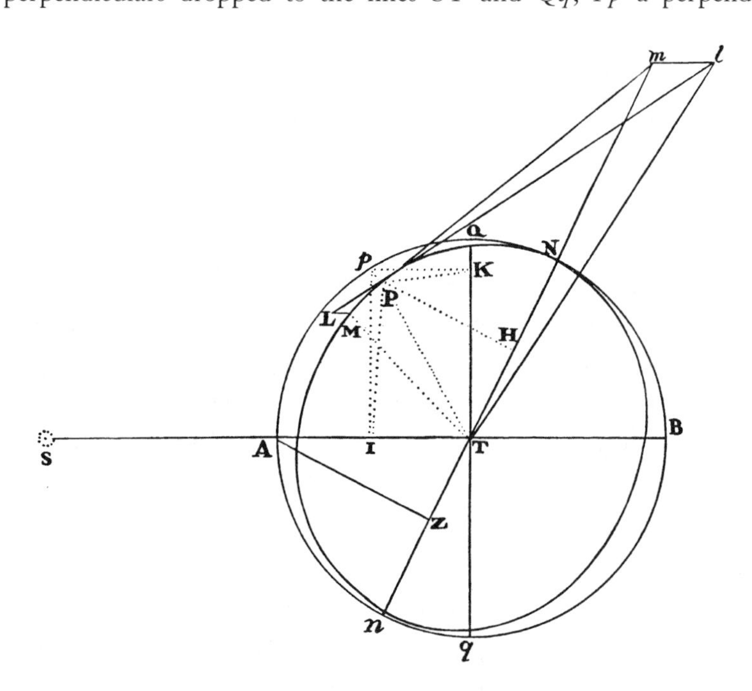

dropped to the plane of the ecliptic; A and B the syzygies of the moon in the plane of the ecliptic; AZ a perpendicular to the line of the nodes Nn; Q and q the quadratures of the moon in the plane of the ecliptic; and pK a perpendicular to the line Qq, which lies between the quadratures. The force of the sun to perturb the motion of the moon has (by prop. 25) two components, one proportional to the line LM in the figure of that proposition, the other proportional to the line MT in that same figure. And the moon is attracted toward the earth by the first of these forces, and by the second it is attracted toward the sun along a line parallel to the straight line ST drawn from the earth to the sun. The first force LM acts in the plane of the moon's orbit and therefore can make no change in the position of that plane. Therefore this force is to be ignored. The second force MT, by which the plane of the lunar orbit is perturbed, is the same as the force 3PK or 3IT. And this force (by prop. 25) is to the force by which the moon could revolve uniformly in a circle in its periodic time about the earth at rest as 3IT to the radius of the circle multiplied by the number 178.725, or as IT to the radius multiplied by 59.575. But in this calculation and in what follows, I consider all lines drawn from the moon to the sun to be parallel to the line drawn from the earth to the sun, because the inclination diminishes all effects in some cases nearly as much as it increases them in others; and we are here seeking the mean motions of the nodes, ignoring those niceties of detail which would make the calculation too cumbersome.

Now let PM represent the arc that the moon describes in a minimally small given time, and ML the line-element one-half of which the moon could describe in the same time by the impulse of the above-mentioned force 3IT. Draw PL and MP, and produce them to *m* and *l*, and let them cut the plane of the ecliptic there, and upon T*m* drop the perpendicular PH. Since the straight line ML is parallel to the plane of the ecliptic and so cannot meet with the straight line *ml* (which lies in that plane) and yet these straight lines lie in a common plane LMP*ml*, these straight lines will be parallel, and therefore the triangles LMP and *lm*P will be similar. Now, since MP*m* is in the plane of the orbit in which the moon was moving while in place P, the point *m* will fall upon the line N*n* drawn through the nodes N and *n* of that orbit. The force by which half of the line-element LM is generated—if all of it were impressed all at once in place P—would generate that whole line and would cause the moon to move in an arc whose chord would be LP, and so would transfer the moon from the plane MP*m*T into the plane LP*l*T; therefore the angular motion of the nodes that is generated by that force will be equal to the angle *m*T*l*. Moreover, *ml* is to *m*P as ML is to MP, and so, since MP is given (because the time is given), *ml* is as the rectangle ML × *m*P, that is, as the rectangle IT × *m*P. And, provided that the angle T*ml* is a right angle, the angle *m*T*l* is as $\frac{ml}{\mathrm{T}m}$, and therefore as $\frac{\mathrm{IT} \times \mathrm{P}m}{\mathrm{T}m}$, that is (because T*m* is to *m*P as TP is to PH), as $\frac{\mathrm{IT} \times \mathrm{PH}}{\mathrm{TP}}$; and so, because TP is given, as IT × PH. But if the angle T*ml* or STN is oblique, the angle *m*T*l* will be still smaller, in the ratio of the sine of the angle STN to the radius, or of AZ to AT. Therefore the velocity of the nodes is as IT × PH × AZ, or as the solid contained by [or the product of] the sines of the three angles TPI, PTN, and STN.

If those angles are right angles, as happens when the nodes are in the quadratures and the moon is in the syzygy, the line-element *ml* will go off indefinitely and the angle *m*T*l* will become equal to the angle *m*P*l*. But in this case the angle *m*P*l* is to the angle PTM, which the moon describes about the earth in the same time by its apparent motion, as 1 to 59.575. For the angle *m*P*l* is equal to the angle LPM, that is, to the angle of the deflection of the moon from the straight-line path that the aforesaid solar force 3IT could generate by itself in that given time, if the gravity of the moon were then to cease; and the angle PTM is equal to the angle of the deflection of the moon from the straight-line path that the force by which the moon is kept in its orbit would generate in the same time, if the solar force 3IT were then to cease. And these forces, as we have said above, are to each other as 1 to 59.575. Therefore, since the mean hourly motion of the moon with respect to the fixed stars is $32'56''27'''12^{iv}\frac{1}{2}$, the hourly motion of the node in this case will be $33''10'''33^{iv}12^{v}$. But in other cases this hourly motion will be to $33''10'''33^{iv}12^{v}$ as the solid contained by [or the product of] the sines of the three angles TPI, PTN, and STN (or the distance of the moon from the quadrature, of the moon from the node, and of the node from the sun) to the cube of the radius. And whenever the sign of any of the angles is changed from positive to negative and from negative to positive, retrograde motion will have to be changed into progressive motion, and progressive into retrograde. Hence it happens that the nodes advance whenever the moon is between either of the quadratures and the node nearest to that quadrature. In other cases, the nodes are retrograde, and they are carried backward [or in antecedentia] each month by the excess of the retrograde motion over the progressive.

Corollary 1. Hence, if from the ends P and M of the minimally small given arc PM, the perpendiculars PK and M*k* are dropped to the line Q*q* that joins the quadratures, and these perpendiculars are produced until they cut the line of the nodes N*n* in D and *d*, then the hourly motion of the nodes will be as the area MPD*d* and the square of the line AZ jointly. For let PK, PH, and AZ be the above-mentioned three sines—namely, PK the sine of the distance of the moon from the quadrature, PH the sine of the distance of the moon from the node, and AZ the sine of the distance of the node from the sun—then the velocity of the node will be as the solid [or product] PK × PH × AZ. But PT is to PK as PM to K*k*, and so, because PT and PM are given, K*k* is proportional to PK. Also, AT is to PD as AZ to PH, and therefore PH is proportional to the rectangle PD × AZ; and, combining these ratios, PK × PH is as the solid K*k* × PD × AZ, and PK × PH × AZ is as K*k* × PD × AZ^2, that is, as the area PD*d*M and AZ^2 jointly. Q.E.D.

COROLLARY 2. In any given position of the nodes, the mean hourly motion is half of their hourly motion in the moon's syzygies, and thus is to $16''55'''16^{iv}36^{v}$ as the square of the sine of the distance of the nodes from the syzygies is to the square of the radius, or as AZ^2 to AT^2. For if the moon

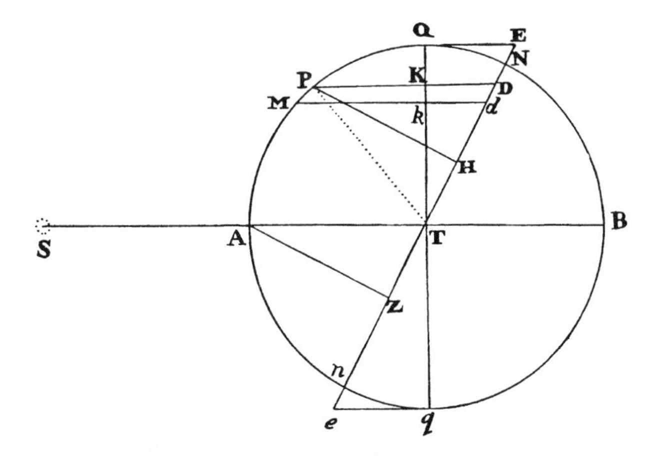

traverses the semicircle QA*q* with uniform motion, the sum of all the areas PD*d*M during the time in which the moon goes from Q to M will be the area QM*d*E, which is terminated at the tangent QE of the circle; and in the time in which the moon reaches point *n*, that sum will be the total area EQA*n*, which the line PD describes; then as the moon goes from *n* to *q*, the line PD will fall outside the circle and will describe the area *nqe* (which is terminated at the tangent *qe* of the circle)—which, since the nodes were previously retrograde but now are progressive, must be subtracted from the former area, and (since it is equal to the area QEN) will leave the semicircle NQA*n*. Therefore, during the time in which the moon describes a semicircle, the sum of all the areas PD*d*M is the area of that semicircle; and in the time in which the moon describes a circle, the sum of all those areas is the area of the whole circle. But the area PD*d*M, when the moon is in the syzygies, is the rectangle of the arc PM and the radius PT; and in the time in which the moon describes a circle, the sum of all the areas that are equal to this one is the rectangle of the whole circumference and the radius of the circle; and this rectangle, since it is equal to two circles, is twice as large as the former rectangle. Accordingly, if the nodes moved with the same velocity uniformly continued that they have in the lunar syzygies, they would describe a space twice as large as the space which they really describe; and therefore the mean motion—with which, if it were continued uniformly, they would describe the space that they really cover with their nonuniform motion—is one-half of the motion which they have in the moon's syzygies. Hence, since the greatest hourly motion of the nodes, if the nodes are in the quadratures, is $33''10'''33^{iv}12^{v}$, their mean hourly motion in this case will be $16''35'''16^{iv}36^{v}$. And since the hourly motion of the nodes is always as AZ^2 and the area PD*d*M jointly, and therefore the hourly motion of the nodes in the moon's syzygies is as AZ^2 and the area PD*d*M jointly, that is (because the area PD*d*M described in the syzygies is given), as AZ^2, the mean motion will also be as AZ^2; and hence this motion, when the nodes are outside the quadratures, will be to $16''35'''16^{iv}36^{v}$ as AZ^2 to AT^2. Q.E.D.

Proposition 31
Problem 12

To find the hourly motion of the nodes of the moon in an elliptical orbit.

Let Q*pmaq* represent an ellipse described with a major axis Q*q* and a minor axis *ab*, QA*q*B a circle circumscribed about this ellipse, T the earth in the common center of both, S the sun, *p* the moon moving in the ellipse, and *pm* the arc that the moon describes in a minimally small given particle of time, N and *n* the nodes joined by the line N*m*, *p*K and *mk* perpendiculars dropped to the axis Q*q* and produced on both sides until they meet the circle at P and M and the line of the nodes at D and *d*. And if the moon, by a radius drawn to the earth, describes an area proportional to the time, the hourly motion of the node in the ellipse will be as the area *p*D*dm* and AZ^2 jointly.

To demonstrate this, let PF touch the circle at P and, produced, meet TN at F; let *pf* touch the ellipse at *p* and, produced, meet the same TN at *f*; and let these tangents come together on the axis TQ at Y. And let ML designate the space that the moon, revolving in a circle, would describe by a transverse motion under the action and impulse of the aforesaid force 3IT or

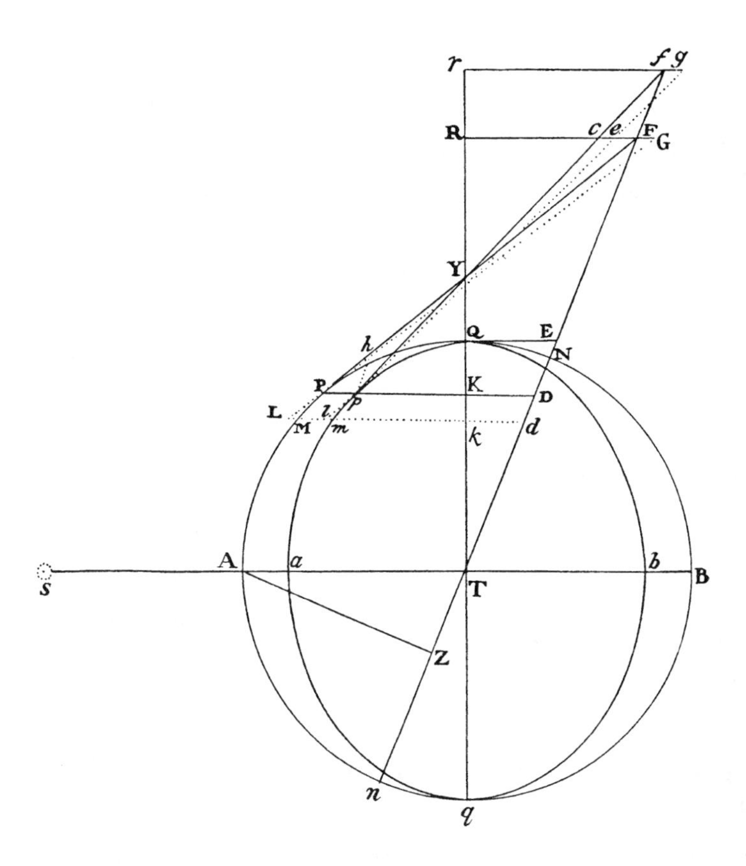

3PK, while it describes the arc PM; and let *ml* designate the space that the moon, revolving in an ellipse, could describe in the same time, also under the action of the force 3IT or 3PK. Further, let L*p* and *lp* be produced until they meet the plane of the ecliptic at G and *g*; and let FG and *fg* be drawn, of which let FG produced cut *pf*, *pg*, and TQ at *c*, *e*, and R respectively; and let *fg* produced cut TQ at *r*. Then, since the force 3IT or 3PK in the circle is to the force 3IT or 3*p*K in the ellipse as PK is to *p*K, or as AT to *a*T, the space ML generated by the first force will be to the space *ml* generated by the second force as PK to *p*K, that is (because the figures PYK*p* and FYR*c* are similar), as FR to *c*R. Moreover, ML is to FG (because the triangles PLM and PGF are similar) as PL to PG, that is (because L*k*, PK, and GR are parallel), as *pl* to *pe*, that is (because the triangles *plm* and *cpe* are similar), as *lm* to *ce*; and thus LM is to *lm*, or FR is to *c*R, as FG is to *ce*. And therefore if *fg* were to *ce* as *f*Y to *c*Y, that is, as *fr* to *c*R (that is, as *fr* to FR and FR to *c*R jointly, that is, as *f*T to FT and FG to ce jointly), then, since the ratio FG to ce taken away from both sides leaves the ratios fg to FG and fT to FT, the ratio fg to FG would be as fT to FT, and so the angles that FG and fg would subtend at the earth T would be equal to each other. But these angles (by what we have set forth in the preceding prop. 30) are the motions of the nodes in the time in which the moon traverses the arc PM in the circle, and the arc pm in the ellipse; and therefore the motions of the nodes in the circle and in the ellipse would be equal to each other. These things would be so, if only fg were to ce as fY to cY, that is, if fg were equal to $\frac{ce \times f\mathrm{Y}}{c\mathrm{Y}}$. But because the triangles fgp and cep are similar, fg is to ce as fp to cp, and so fg is equal to $\frac{ce \times fp}{cp}$; and therefore the angle that fg really subtends is to the former angle that FC subtends (that is, the motion of the nodes in the ellipse is to the motion of the nodes in the circle) as this fg or $\frac{ce \times fp}{cp}$ to the former fg or $\frac{ce \times f\mathrm{Y}}{c\mathrm{Y}}$, that is, as $fp \times c\mathrm{Y}$ to $f\mathrm{Y} \times cp$, or as fp to fY and cY to cp; that is (if ph, parallel to TN, meets FP at h), as Fh to FY and FY to FP; that is, as Fh to FP or Dp to DP, and so as the area Dpmd to the area DPMd. And therefore, since (by prop. 30, corol. 1) the latter area and AZ^2 jointly are proportional to the hourly motion of the nodes in the circle, the former area and AZ^2 jointly will be proportional to the hourly motion of the nodes in the ellipse. Q.E.D.

Corollary. Therefore, since in any given position of the nodes, the sum of all the areas pDdm, in the time in which the moon goes from the quadrature to any place m, is the area mpQEd, which is terminated at the tangent QE of the ellipse, and the sum of all those areas in a complete revolution is the area of the whole ellipse, the mean motion of the nodes in the ellipse will be to the mean motion of the nodes in the circle as the ellipse to the circle, that is, as Ta to TA, or as 69 to 70. And therefore, since (by prop. 30, corol. 2) the mean hourly motion of the nodes in the circle is to $16''35'''16^{\mathrm{iv}}36^{\mathrm{v}}$ as AZ^2 to AT^2 if the angle $16''21'''3^{\mathrm{iv}}30^{\mathrm{v}}$ is taken to the angle $16''35'''16^{\mathrm{iv}}36^{\mathrm{v}}$ as 69 to 70, the mean hourly motion of the nodes in the ellipse will be to $16''21'''3^{\mathrm{iv}}30^{\mathrm{v}}$ as AZ^2 to AT^2, that is, as the square of the sine of the distance of the node from the sun to the square of the radius.

But the moon, by a radius drawn to the earth, describes an area more swiftly in the syzygies than in the quadratures, and on that account the time is shortened in the syzygies and lengthened in the quadratures, and along with the time the motion of the nodes is increased and decreased. Now, the moment of an area in the quadratures of the moon was to its moment in the syzygies as 10,973 to 11,073; and therefore the mean motion in the octants is to the excess in the syzygies and to the deficiency in the quadratures as the half-sum 11,023 of the numbers is to their half-difference 50. Accordingly, since the time of the moon in each equal particle of its orbit is inversely as its velocity, the mean time in the octants will be to the excess of time in the quadratures and its deficiency in the syzygies, arising from this cause, as 11,023 to 50 very nearly. With regard to positions of the moon between the quadratures and the syzygies, I find that the excess of the moments of the area in any one of these positions over the least moment in the quadratures is very nearly as the square of the sine of the distance of the moon from the quadratures; and therefore the difference between the moment in any position and the mean moment in the octants is as the difference between the square of the sine of the distance of the moon from the quadratures and the square of the sine of 45°, or half of the square of the radius; and the increase of the time in any one of the positions between the octants and the quadratures, and its decrease between the octants and the syzygies, is in the same ratio. But the motion of the nodes, in the time in which the moon traverses each equal particle of its orbit, is accelerated or retarded as the square of the time.

For that motion, while the moon traverses PM, is (other things being equal) as ML, and ML is in the squared ratio of the time. Therefore, the motion of the nodes in the syzygies, a motion completed in the time in which the moon traverses given particles of its orbit, is diminished as the square of the ratio of the number 11,073 to the number 11,023; and the decrement is to the remaining motion as 100 to 10,973 and to the total motion as 100 to 11,073 very nearly. But the decrement in positions between the octants and syzygies and the increment in positions between the octants and quadratures are to this decrement very nearly as [*i*] the total motion in those positions to the total motion in the syzygies and as [*ii*] the difference between the square of the sine of the distance of the moon from the quadrature and half of the square of the radius to half of the square of the radius, jointly. Hence, if the nodes are in the quadratures and two positions are taken equally distant from the octant, one on one side and one on the other, and another two are taken at the same distance from the syzygy and from the quadrature, and if from the decrements of the motions in the two positions between the syzygy and octant are subtracted the increments of the motions in the remaining two positions that are between the octant and quadrature, the remaining decrement will be equal to the decrement in the syzygy, as will be easily apparent upon examination. And accordingly the mean decrement, which must be subtracted from the mean motion of the nodes, is a fourth of the decrement in the syzygy. The total hourly motion of the nodes in the syzygies (when it was supposed the moon described, by a radius drawn to the earth, an area proportional to the time) was $32''42'''7^{iv}$. And according to what we have just said, the decrement of the motion of the nodes, in the time when the moon—now moving more swiftly—describes the same space, is to this motion as 100 to 11,073; and so the decrement is $17'''43^{iv}11^{v}$, of which a fourth ($4'''25^{iv}48^{v}$) subtracted from the mean hourly motion found above ($16''21'''3^{iv}30^{v}$) leaves $16''16'''37^{iv}42^{v}$, the corrected mean hourly motion.

If the nodes are beyond the quadratures and two places equally distant from the syzygies are considered, one on one side and one on the other, the sum of the motions of the nodes when the moon is in these positions will be to the sum of their motions when the moon is in the same positions and the nodes are in the quadratures as AZ^2 to AT^2. And the decrements of the motions, arising from the causes just now set forth, will be to each other as the motions themselves, and therefore the remaining motions will be to each other as AZ^2 to AT^2, and the mean motions will be as the remaining motions. Therefore the corrected mean hourly motion, in any given situation of the nodes, is to $16''16'''37^{iv}42^{v}$ as AZ^2 to AT^2, that is, as the square of the sine of the distance of the nodes from the syzygies to the square of the radius.

Proposition 32
Problem 13

To find the mean motion of the nodes of the moon.

The mean annual motion is the sum of all the mean hourly motions in a year. Suppose that the node is in N and that as each hour is completed, it is drawn back into its former place so that, notwithstanding its own proper motion, it always maintains some given position with respect to the fixed stars.

And suppose that during this same time the sun S, as a result of the motion of the earth, advances from the node and completes its apparent annual course with a uniform apparent motion. Moreover, let A*a* be the minimally small given arc that the straight line TS, always drawn to the sun, describes in a minimally small given time by its intersection with the circle NA*n*; then (by what has already been shown) the mean hourly motion will be as AZ^2, that is (because AZ and ZY are proportional), as the rectangle of AZ and ZY, that is, as the area AZY*a*. And the sum of all the mean hourly motions from the beginning will be as the sum of all the areas *a*YZA, that is, as the area NAZ. Moreover, the greatest area AZY*a* is equal to the rectangle of the arc A*a* and the radius of the circle; and therefore the sum of all such rectangles in the whole circle will be to the sum of the same number of greatest rectangles as the area of the whole circle to the rectangle of the whole circumference and the radius, that is, as 1 to 2. Now, the hourly motion corresponding to the greatest rectangle was $16''16'''37^{iv}42^{v}$, and this motion, in a whole sidereal year of $365^{d}6^{h}9^{m}$, adds up to $39°38'7''50'''$. And so half of this, $19°49'3''55'''$, is the mean motion of the nodes that corresponds to the whole circle. And the motion of the nodes in the time during which the sun goes from N to A is to $19°49'3''55'''$ as the area NAZ is to the whole circle.

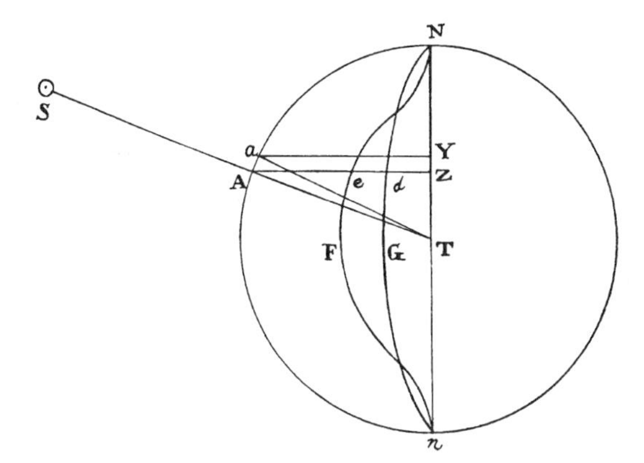

These things are so on the hypothesis that each hour the node is drawn back to its former place, in such a way that when a whole year is completed, the sun returns to the same node from which it had initially departed. But as a result of the motion of that node, it comes about that the sun returns to the node more quickly; and now this shortening of the time must be computed. Since in a total year the sun travels through 360°, and in the same time the node with its greatest motion would travel through 39°38′7″50‴, or 39.6355°, and the mean motion of the node in any place N is to its mean motion in its quadratures as AZ^2 to AT^2, the motion of the sun will be to the motion of the node in N as $360AT^2$ to $39.6355AZ^2$, that is, as $9.0827646AT^2$ to AZ^2. Hence, if the circumference NA*n* of the whole circle is divided into equal particles A*a*, then the time in which the sun traverses the particle A*a* (the circle being at rest) will be to the time in which it traverses the same particle (if the circle revolves along with the nodes about the center T) inversely as $9.0827646AT^2$ to $9.0827646AT^2 + AZ^2$. For the time is inversely as the velocity with which the particle [of arc] is traversed, and this velocity is the sum of the velocities of the sun and of the node. Let the sector NTA represent the time in which the sun, without the motion of the node, would traverse the arc NA, and let the particle AT*a* of the sector represent the particle of time in which it would traverse the minimally small arc A*a*; furthermore, drop a perpendicular *a*Y to N*n* and on AZ take *d*Z of a length such that the rectangle of *d*Z and ZY is to the particle AT*a* of the sector as AZ^2 is to $9.0827646AT^2 + AZ^2$ (that is, such that *d*Z is to ½AZ as AT^2 is to $9.0827646AT^2 + AZ^2$); then the rectangle of *d*Z and ZY will designate the decrement of time arising from the motion of the node during the total time in which the arc A*a* is traversed. And if the point *d* touches the curve N*d*G*n*,[a] the curvilinear area N*d*Z will be the total decrement in the time in which the whole arc NA is traversed; and therefore the excess of the sector NAT over the area NAZ will be that total time. And since the motion of the node in a smaller time is smaller in proportion to the time, the area A*a*YZ also will have to be diminished in the same proportion. And this will happen if on AZ the length *e*Z is taken, which is to the length AZ as AZ^2 is to $9.0827646AT^2 + AZ^2$. For thus the rectangle of *e*Z and ZY will be to the area AZY*a* as the decrement of the time in which the arc A*a* is traversed is to the total time in which it would be traversed if the node were at rest; and therefore that rectangle will correspond to the decrement of the motion of the node. And if the point *e* touches the curve N*e*F*n*,[b] the total area N*e*Z, which is the sum of all the decrements of that motion, will correspond to the total decrement in the time during which the arc AN

a. That is, if the point *d* traces out the curve N*d*G*n*, or if N*d*G*n* is the curve which is the locus of the point *d*.

b. That is, if the curve N*e*F*n* is the locus of point *e*.

is traversed, and the remaining area NAe will correspond to the remaining motion, which is the true motion of the node in the time in which the total arc NA is traversed by the joint motions of the sun and the node. Now, the area of the semicircle is to the area of the figure NeFn, found by the method of infinite series, nearly as 793 to 60. And the motion that corresponds to the whole circle was 19°49′3″55‴, and therefore the motion that corresponds to double the figure NeFn is 1°29′58″2‴. Subtracting this from the former motion leaves 18°19′5″53‴, the total motion of the node with respect to the fixed stars between its successive conjunctions with the sun; and this motion, subtracted from the annual motion of the sun of 360°, leaves 341°40′54″7‴, the motion of the sun between the same conjunctions. And this motion is to the annual motion of 360° as the motion of the node just found (18°19′5″53‴) to its annual motion, which will therefore be 19°18′1″23‴. This is the mean motion of the nodes in a sidereal year. From the astronomical tables this is 19°21′21″50‴. The difference is less than $\frac{1}{300}$ of the total motion and seems to arise from the eccentricity of the moon's orbit and its inclination to the plane of the ecliptic. By the eccentricity of the orbit, the motion of the nodes is too much accelerated; and on the other hand, by its inclination it is retarded somewhat, and reduced to its correct velocity.

Proposition 33
Problem 14

To find the true motion of the nodes of the moon.

In the time which is as the area NTA − NdZ (in the preceding figure), that motion is as the area NAe, and hence is given. But because the calculation is too difficult, it is preferable to use the following construction of the problem. With center C and any interval CD as radius, describe a circle BEFD. Produce DC to A so that AB is to AC as the mean motion is to half

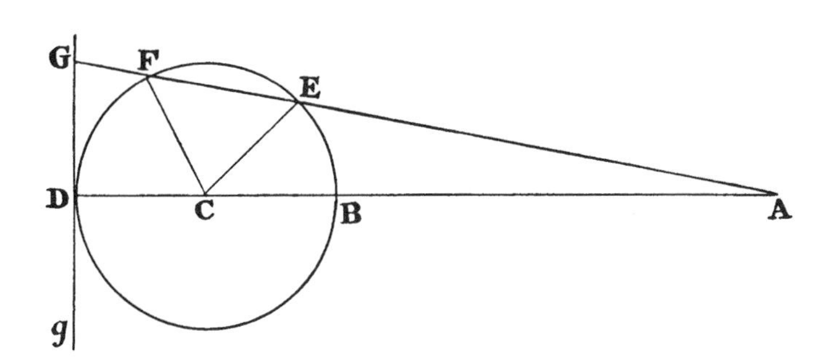

of the true mean motion when the nodes are in the quadratures (that is, as $19°18'1''23'''$ to $19°49'3''55'''$); and thus BC will be to AC as the difference of the motions ($0°31'2''32'''$) to the latter motion ($19°49'3''55'''$), that is, as 1 to $38\frac{3}{10}$. Next, through point D draw the indefinite line G*g*, touching the circle in D; and let the angle BCE or BCF be taken equal to twice the distance of the sun from the place of the node, as found from the mean motion, and let AE or AF be drawn cutting the perpendicular DG in G. The true motion of the nodes will be found if now an angle is taken that is to the total motion of the node between its syzygies (that is, to $9°11'3''$) as the tangent DG is to the total circumference of the circle BED, and if that angle (for which the angle DAG can be used) is added to the mean motion of the nodes when the nodes are passing from quadratures to syzygies and is subtracted from the same mean motion when they are passing from syzygies to quadratures. For the true motion thus found will agree very nearly with the true motion which results from representing the time by the area NTA − N*d*Z and the motion of the node by the area NA*e*, as will be evident to anyone considering the matter and performing the computations. This is the semimonthly equation of the motion of the nodes. There is also a monthly equation, but it is not at all needed in order to find the latitude of the moon. For, since the variation of the inclination of the moon's orbit to the plane of the ecliptic is subject to a double inequality, one semimonthly and the other monthly, the monthly inequality of the variation and the monthly equation of the nodes so moderate and correct each other that both can be ignored in determining the latitude of the moon.

COROLLARY. From this and the preceding proposition it is clear that the nodes are stationary in their syzygies; in the quadratures, however, they regress by an hourly motion of $16''19'''26^{iv}$. It is also clear that the equation of the motion of nodes in the octants is $1°30'$. This all squares exactly with celestial phenomena.

Scholium J. Machin, Gresham Professor of Astronomy, and Henry Pemberton, M.D., have independently found the motion of the nodes by yet another method. Some mention of the latter's method has been made elsewhere. And the papers (which I have seen) of both men contained two propositions, which agreed with each other. Here I shall present Mr. Machin's paper, since it was the first to come into my hands.

On the Motion of the Nodes of the Moon

Proposition 1

The mean motion of the sun from the node is defined by a mean geometrical proportional between the mean motion of the sun and that mean motion with which the sun recedes most swiftly from the node in the quadratures.

Let T be the place where the earth is, N*n* the line of the nodes of the moon at any given time, KTM a line drawn at right angles to this line, and TA a straight line revolving around the center with the angular velocity with which the sun and the node recede from each other, in such a way that the angle between the straight line N*m* (which is at rest) and TA (which is revolving) is always equal to the distance between the places of the sun and of the node. Now, if any straight line TK is divided into parts TS and SK, which are to each other as the hourly mean motion of the sun is to the hourly mean motion of the nodes in the quadratures, and if the straight line TH is taken so as to be a mean proportional between the part TS and the whole TK, this straight line among the rest will be proportional to the mean motion of the sun from the node.

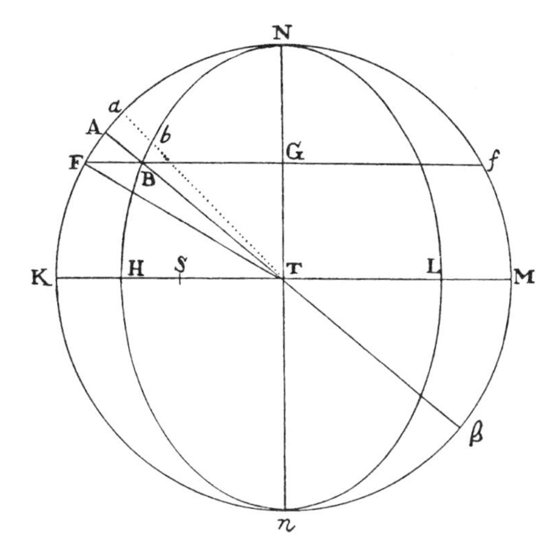

For describe a circle NK*n*M with center T and radius TK, and with the same center and the semiaxes TH and TN describe an ellipse NH*n*L, and in the time in which the sun recedes from the node through the arc N*a*, if the straight line T*ba* is drawn, the area of the sector NT*a* will represent the sum of the motions of the node and sun in that same time. Therefore let A*a* be the minimally small arc that the straight line T*ba*—revolving according to the aforesaid law—uniformly describes in a given particle of time, and the minimally small sector TA*a* will be as the sum of the velocities whereby the sun and the node are carried separately in that time. The velocity of the sun, however, is nearly uniform, since its small inequality introduces scarcely any variation in the mean motion of the nodes. The other part of this sum, namely the velocity of the node in its mean quantity, increases in receding from the syzygies as the square of the sine of its distance from the sun (by *Principia*, book 3, prop. 31, corol.) and, when it is greatest in the quadratures to the sun at K, has the same ratio to the velocity of the sun that SK has to TS; that is, it is as (the difference of the squares of TK and TH or) the rectangle KH × HM is to TH^2. But the ellipse NBH divides the sector AT*a*, which represents the sum of these two velocities, into two parts AB*ba* and BT*b*, which are proportional to the velocities. For, produce BT to the circle in β, and, from point B to the major axis, drop a perpendicular BG which, produced in both directions, meets the circle in points F and *f*. Then, since the space AB*ba* is to the sector TB*b* as the rectangle AB × Bβ to BT^2 (for that rectangle is equal to the difference of the squares of TA and TB because the straight line Aβ is cut equally at T and unequally at B), this ratio—when the space AB*ba* is greatest at K—will be the same as the ratio of the rectangle KH × HM to HT^2; but the greatest mean velocity of the node was [previously shown to be] in this ratio to the velocity of the sun. Therefore, in the quadratures, the sector AT*a* is divided into parts proportional to the velocities. And since the rectangle KH × HM is to HT^2 as FB × B*f* to BG^2 and since the rectangle AB × Bβ is equal to the rectangle FB × B*f*, it follows that the area-element AB*ba* when it is greatest will be to the remaining sector TB*b* as the rectangle AB × Bβ to BG^2. But the ratio of the area-elements was always as the rectangle AB × Bβ to BT^2; and therefore the area-element AB*ba* in the place A is smaller than the corresponding area-element in the quadratures, in the ratio of BG^2 to BT^2, that is, as the square of the sine of the distance of the sun from the node. And accordingly the sum of all the area-elements AB*ba* (namely, the space ABN) will be as the motion of the node in the time in which the sun departs from the node and passes through the arc NA. And the remaining space (namely, the elliptical sector NTB) will be as the mean motion of the sun in the same time. And, therefore, since the mean annual motion of the node is the motion that it makes in the time during which the sun has completed its period, the mean motion of the node from the sun will be to the mean motion of the sun itself as the area of the circle to the area of the ellipse, that is, as the straight line TK to the straight line TH (which is the mean proportional between TK and TS); or, which comes to the same thing, as the mean proportional TH to the straight line TS.

Proposition 2

Given the mean motion of the nodes of the moon, to find the true motion.

Let the angle A be the distance of the sun from the mean place of the node, or the mean motion of the sun from the node. Then if angle B is taken so that its tangent is to the tangent of angle A as TH to TK—that is, as the square root of the ratio of the mean

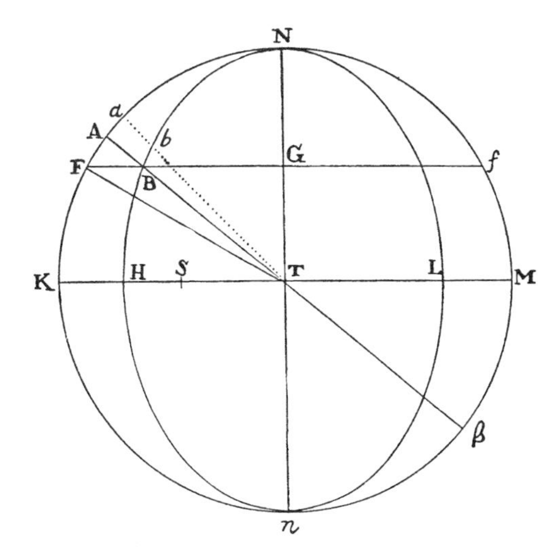

hourly motion of the sun to the mean hourly motion of the sun from the node when the node is in the quadratures—that same angle B will be the distance of the sun from the true place of the node. For draw FT, and (by the proof of the previous proposition) the angle FTN will be the distance of the sun from the mean place of the node, while the angle ATN will be the distance from the true place, and the tangents of these angles are to each other as TK to TH.

Corollary. Hence the angle FTA is the equation of the moon's nodes, and the sine of this angle, when it is greatest in the octants, is to the radius as KH to TK + TH. And the sine of this equation in any other place A is to the greatest sine as the sine of the sum of the angles FTN + ATN is to the radius—that is, nearly as the sine of 2FTN (that is, twice the distance of the sun from the mean place of the node) is to the radius.

Scholium

If the mean hourly motion of the nodes in the quadratures is $16''16'''37^{iv}42^{v}$ (that is, $39°38'7''50'''$ in a whole sidereal year), then TH will be to TK as the square root of the ratio of the number 9.0827646 to the number 10.0827646, that is, as 18.6524761 to 19.6524761. And therefore TH is to HK as 18.6524761 to 1, that is, as the motion of the sun in a sidereal year to the mean motion of the node, which is $19°18'1''23\frac{2}{3}'''$.

But if the mean motion of the nodes of the moon in twenty Julian years is $386°50'15''$, as is deduced from observations used in the theory of the moon, the mean motion of the nodes in a sidereal year will be $19°20'31''58'''$. And TH will be to HK as 360° to $19°20'31''58'''$, that is, as 18.61214 to 1. Hence the mean hourly motion of the nodes in the quadratures will come out $16''18'''48^{iv}$. And the greatest equation of the nodes in the octants will be $1°29'57''$.

Proposition 34
Problem 15

To find the hourly variation of the inclination of the lunar orbit to the plane of the ecliptic.

Let A and *a* represent the syzygies, Q and *q* the quadratures, N and *n* the nodes, P the place of the moon in its orbit, *p* the projection of that place on the plane of the ecliptic, and *m*T*l* the momentaneous motion of the nodes as above. Drop the perpendicular PG to the line T*m*, join *p*G and produce it until it meets T*l* in *g*, and also join P*g*; then the angle PG*p* will be the inclination of the moon's orbit to the plane of the ecliptic when the moon is in P, and the angle P*gp* will be the inclination of the same orbit after a moment of time has been completed; and thus the angle GP*g* will be the momentaneous variation of the inclination. But this angle GP*g* is to the angle GT*g* as TG to PG and P*p* to PG jointly. And therefore, if an hour is substituted for the moment of time, then—since the angle GT*g* (by prop. 30) is to the angle $33''10'''33^{iv}$ as IT × PG × AZ to AT^3—the angle GP*g* (or

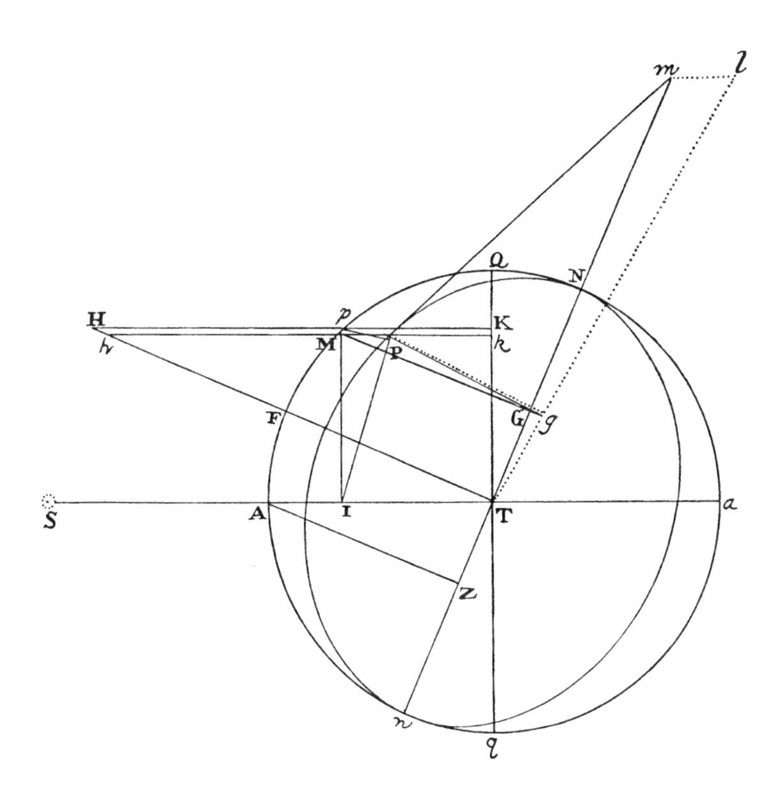

the hourly variation of the inclination) will be to the angle $33''10'''33^{iv}$ as $\mathrm{IT} \times \mathrm{AZ} \times \mathrm{TG} \times \dfrac{\mathrm{P}p}{\mathrm{PG}}$ to AT^3. Q.E.I.

These things are so on the hypothesis that the moon revolves uniformly in a circular orbit. But if that orbit is elliptical, the mean motion of the nodes will be diminished in the ratio of the minor axis to the major axis, as has been set forth above. And the variation of the inclination will also be diminished in the same ratio.

COROLLARY 1. If the perpendicular TF is erected on N*n*, and *p*M is the hourly motion of the moon in the plane of the ecliptic, and if the perpendiculars *p*K and M*k* are dropped to QT and produced in both directions to meet TF at H and *h*, then IT will be to AT as K*k* to M*p*, and TG to H*p* as TZ to AT, and so IT × TG will be equal to $\dfrac{\mathrm{K}k \times \mathrm{H}p \times \mathrm{TZ}}{\mathrm{M}p}$, that is, equal to the area H*p*M*h* multiplied by the ratio $\dfrac{\mathrm{TZ}}{\mathrm{M}p}$; and therefore the hourly variation of the inclination will be to $33''10'''33^{iv}$ as H*p*M*h* multiplied by $\mathrm{AZ} \times \dfrac{\mathrm{TZ}}{\mathrm{M}p} \times \dfrac{\mathrm{P}p}{\mathrm{PG}}$ is to AT^3.

COROLLARY 2. And so, if the earth and the nodes, as each hour is completed, were drawn back from their new places and were always restored instantly to their former places, so that their given position remained unchanged throughout an entire periodic month, the total variation of the inclination during the time of that month would be to $33''10'''33^{iv}$ as the sum of all the areas $HpMh$ which are generated during a revolution of the point p (these areas being summed according to their proper signs $+$ and $-$) multiplied by $AZ \times TZ \times \frac{Pp}{PG}$ is to $Mp \times AT^3$, that is, as the whole circle $QAqa$ multiplied by $AZ \times TZ \times \frac{Pp}{PG}$ is to $Mp \times AT^3$, that is, as the circumference $QAqa$ multiplied by $AZ \times TZ \times \frac{Pp}{PG}$ is to $2Mp \times AT^2$.

COROLLARY 3. Accordingly, in a given position of the nodes, the mean hourly variation, from which, continued uniformly for a month, that monthly variation could be generated, is to $33''10'''33^{iv}$ as $AZ \times TZ \times \frac{Pp}{PG}$ to $2AT^2$, or as $Pp \times \frac{AZ \times TZ}{\frac{1}{2}AT}$ to $PG \times 4AT$, that is (since Pp is to PG as the sine of the above-mentioned inclination to the radius, and $\frac{AZ \times TZ}{\frac{1}{2}AT}$ is to $4AT$ as the sine of twice the angle ATn to four times the radius), as the sine of that same inclination multiplied by the sine of twice the distance of the nodes from the sun to four times the square of the radius.

COROLLARY 4. Since the hourly variation of the inclination, when the nodes are in the quadratures, is (by this proposition) to the angle $33''10'''33^{iv}$ as $IT \times AZ \times TG \times \frac{Pp}{PG}$ to AT^3, that is, as $\frac{IT \times TG}{\frac{1}{2}AT} \times \frac{Pp}{PG}$ to $2AT$, that is, as the sine of twice the distance of the moon from the quadratures multiplied by $\frac{Pp}{PG}$ is to twice the radius, it follows that the sum of all the hourly variations, in the time in which the moon in this situation of the nodes passes from quadrature to syzygy (that is, in the space of $177\frac{1}{6}$ hours), will be to the sum of the same number of angles $33''10'''33^{iv}$, or $5,878''$, as the sum of all the sines of twice the distance of the moon from the quadratures multiplied by $\frac{Pp}{PG}$ is to the sum of the same number of diameters; that is, as the diameter multiplied by $\frac{Pp}{PG}$ is to the circumference; that is, if the inclination is $5°1'$, as $7 \times \frac{874}{10,000}$ to 22, or 278 to 10,000. And accordingly the total variation, composed of the sum of all the hourly variations in the aforesaid time, is 163″, or 2′43″.

Proposition 35
Problem 16

To find the inclination of the moon's orbit to the plane of the ecliptic at a given time.

Let AD be the sine of the greatest inclination, and AB the sine of the least inclination. Bisect BD in C, and with center C and radius BC describe a circle BGD. On AC take CE in the ratio to EB that EB has to 2BA. Now if, for the given time, the angle AEG is set equal to twice the distance of the nodes from the quadratures, and the perpendicular GH is dropped to AD, then AH will be the sine of the required inclination.

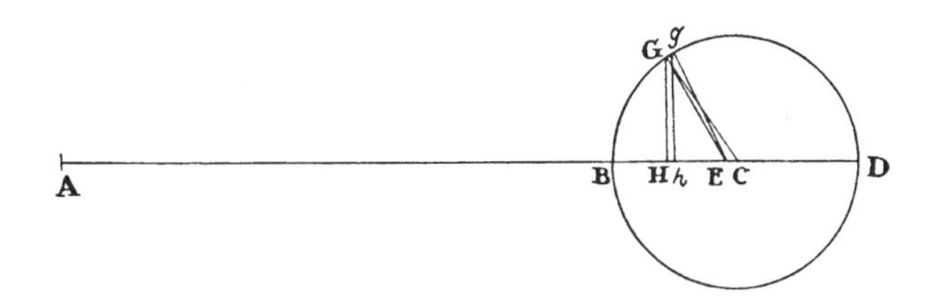

For GE^2 is equal to $GH^2 + HE^2 = BH \times HD + HE^2 = HB \times BD + HE^2 - BH^2 = HB \times BD + BE^2 - 2BH \times BE = BE^2 + 2EC \times BH = 2EC \times AB + 2EC \times BH = 2EC \times AH$. And thus, since 2EC is given, GE^2 is as AH. Now let AE*g* represent twice the distance of the nodes from the quadratures after some given moment of time has been completed, and the arc G*g* (because the angle GE*g* is given) will be as the distance GE. Moreover, H*h* is to G*g* as GH to GC, and therefore H*h* is as the solid [or product] GH × G*g*, or GH × GE; that is, as $\frac{GH}{GE} \times GE^2$ or $\frac{GH}{GE} \times AH$, that is, as AH and the sine of the angle AEG jointly. Therefore, if AH, in any given case, is the sine of the inclination, it will be increased by the same increments as the sine of the inclination, by corol. 3 of the preceding prop. 34, and therefore will always remain equal to that sine. But when the point G falls upon either point B or D, AH is equal to this sine and therefore remains always equal to it. Q.E.D.

In this demonstration, I have supposed that the angle BEG, which is twice the distance of the nodes from the quadratures, increases uniformly. For there is no time to consider all the minute details of inequalities. Now suppose that the angle BEG is a right angle and that in this case G*g* is the hourly increment of twice the distance of the nodes and sun from each other; then (by corol. 3 of prop. 34) the hourly variation of the inclination in the same case will be to $33''10'''33^{iv}$ as the solid [or product] of the sine AH of the inclination and the sine of the right angle BEG (which is twice the distance of the nodes from the sun) is to four times the square of the radius; that is, as the sine AH of the mean inclination to four times the radius; that is (since that mean inclination is about $5°8\frac{1}{2}'$), as its sine (896) to four times the radius (40,000), or as 224 to 10,000. And the total variation, corresponding to BD, the difference of the sines, is to that hourly variation as the diameter BD to the arc *Gg*; that is, as the diameter BD to the semicircumference BGD and the time of $2,079\frac{7}{10}$ hours (during which the node goes from the quadratures to the syzygies) to 1 hour jointly; that is, as 7 to 11 and $2,079\frac{7}{10}$ to 1. Therefore, if all the ratios are combined, the total variation BD will become to $33''10'''33^{iv}$ as $224 \times 7 \times 2,079\frac{7}{10}$ to 110,000, that is, as 29,645 to 1,000, and hence that variation BD will come out $16'23\frac{1}{2}''$.

This is the greatest variation of the inclination insofar as the place of the moon in its orbit is not considered. For if the nodes are in the syzygies, the inclination is not at all changed by the various positions of the moon. But if the nodes are in the quadratures, the inclination is less when the moon is in the syzygies than when it is in the quadratures, by a difference of $2'43''$, as we have indicated in corol. 4 of prop. 34. And the total mean variation BD, diminished when the moon is in its quadratures by $1'21\frac{1}{2}''$ (half of this excess), becomes $15'2''$; while in the syzygies it is increased by the same amount and becomes $17'45''$. Therefore, if the moon is in the syzygies, the total variation in the passage of the nodes from quadratures to syzygies will be $17'45''$; and so if the inclination, when the nodes are in the syzygies, is $5°17'20''$, it will be $4°59'35''$ when the nodes are in the quadratures and the moon in the syzygies. And that these things are so is confirmed by observations.

If now it is desired to find the inclination of the orbit when the moon is in the syzygies and the nodes are in any position whatever, let AB become to AD as the sine of $4°59'35''$ is to the sine of $5°17'20''$, and take the angle ABG equal to twice the distance of the nodes from the quadratures; then AH will be the sine of the required inclination. The inclination of the orbit is equal to this inclination when the moon is 90° distant from the nodes. In other positions of the moon, the monthly inequality that occurs in the variation of the inclination is compensated for in the calculation of the latitude of the moon (and, in a manner, canceled) by the monthly inequality in the motion of the nodes (as we have said above) and thus can be neglected in calculating that latitude.

Scholium

[a]I wished to show by these computations of the lunar motions that the lunar motions can be computed from their causes by the theory of gravity. By the same theory I found, in addition, that the annual equation of the mean motion of the moon arises from the varying dilatation [and contraction] of the orbit of the moon produced by the force of the sun, according to book 1, prop. 66, corol. 6. When the sun is in perigee, this force is greater and dilates the orbit of the moon; when the sun is in apogee, the force is smaller and permits the orbit to be contracted. The moon revolves more slowly in the dilated orbit, more swiftly in the contracted one; and the annual equation which compensates for this inequality vanishes in the apogee and perigee of the sun, rises to roughly 11′50″ in the mean distance of the sun from the earth, and in other places is proportional to the equation of the center of the sun; and it is added to the mean motion of the moon when the earth is going from its aphelion to its perihelion and is subtracted when the earth is in the opposite part of the orbit. Assuming the radius of the earth's orbit to be 1,000 and the eccentricity of the earth to be 16⅞, this equation, when it is greatest, came out 11′49″ by the theory of gravity. But the eccentricity of the earth seems to be a little greater; and if the eccentricity is increased, this equation should be increased in the same ratio. If the eccentricity is taken at $16\frac{11}{12}$, the greatest equation will be 11′51″.

aa. Ed. 1 has: "Up to now no consideration has been taken of the motions of the moon insofar as the eccentricity of the orbit is concerned. By similar computations, I found that the apogee, when it is in conjunction with or in opposition to the sun, moves forward 23′ each day with respect to the fixed stars but, when it is in the quadratures, regresses about 16⅓′ each day and that its mean annual motion is about 40°. By the astronomical tables which the distinguished Flamsteed adapted to the hypothesis of Horrocks, the apogee moves forward in its syzygies with a daily motion of 24′28″ but regresses in the quadratures with a daily motion of 20′12″ and is carried forward [or in consequentia] with a mean annual motion of 40°41′. The difference between the daily forward motion of the apogee in its syzygies and the daily regressive motion in its quadratures is 4′16″ by the tables but 6⅔′ by our computation, which we suspect ought to be attributed to a fault in the tables. But we do not think that our computation is exact enough either. For by means of a certain calculation the daily forward motion of the apogee in its syzygies and the daily regressive motion in its quadratures came out a little greater. But it seems preferable not to give the computations, since they are too complicated and encumbered by approximations and not exact enough."

I found also that the apogee and nodes of the moon move more swiftly in the perihelion of the earth (because of the greater force of the sun) than in its aphelion, [and this] inversely as the cube of the distance of the earth from the sun. And from this there arise annual equations of these motions proportional to the equation of the sun's center. Now, the motion of the sun is inversely as the square of the distance of the earth from the sun, and the greatest equation of the center that this inequality generates is $1°56'20''$, corresponding to the above-mentioned eccentricity of the sun of $16^{11}/_{12}$. But if the motion of the sun were inversely as the cube of the distance, this inequality would generate a greatest equation of $2°54'30''$. And therefore the greatest equations that the inequalities of the motions of the apogee and nodes of the moon generate are to $2°54'30''$ as the daily mean motion of the apogee and the daily mean motion of the nodes of the moon are to the daily mean motion of the sun. Accordingly, the greatest equation of the mean motion of the apogee comes out $19'43''$, and the greatest equation of the mean motion of the nodes $9'24''$. And the first of these equations is added and the second subtracted when the earth is going from its perihelion to its aphelion, and the opposite happens in the opposite part of the orbit.

By the theory of gravity it was also established that the action of the sun upon the moon is a little greater when the transverse diameter of the moon's orbit is passing through the sun than when this diameter is at right angles to the line joining the earth and the sun; and therefore the moon's orbit is a little greater in the first case than in the second. And hence arises another equation of the moon's mean motion, one that depends on the position of the apogee of the moon with respect to the sun; this equation is greatest when the apogee of the moon is in an octant with the sun, and vanishes when the apogee reaches the quadratures or syzygies, and is added to the mean motion in the passage of the apogee of the moon from quadrature of the sun to syzygy, and is subtracted in the passage of the apogee from syzygy to quadrature. This equation, which I shall call semiannual, rises in the octants of the apogee (when it is greatest) to roughly $3'45''$, as far as I could gather from phenomena. This is its quantity at the mean distance of the sun from the earth. But it is increased and decreased inversely as the cube of the distance from the sun, and so at the greatest distance of the sun is $3'34''$ and at the least distance $3'56''$—very nearly; and when the apogee of the moon is situated outside the octants, it becomes less, and is to the greatest equation as the sine of twice the distance of the moon's apogee from the nearest syzygy or quadrature is to the radius.

By the same theory of gravity, the action of the sun upon the moon is a little greater when a straight line drawn through the nodes of the moon passes through the sun than when that line is at right angles to the straight line joining the sun and earth. And hence arises another equation of the moon's mean motion, which I shall call the second semiannual and which is greatest when the nodes are in the octants of the sun and vanishes when they are in the syzygies or quadratures, and in other positions of the nodes is proportional to the sine of twice the distance of either node from the next syzygy or quadrature; and it is added to the mean motion of the moon if the sun is ahead of [in antecedentia] the node nearest to it, and subtracted if beyond [in consequentia]; and in the octants, where it is greatest, it rises to 47″ at the mean distance of the sun from the earth, as I conclude from the theory of gravity. At other distances of the sun, this equation (which is greatest in the octants of the nodes) is inversely as the cube of the distance of the sun from the earth, and so in the perigee of the sun rises to about 49″ and in its apogee to about 45″.

By the same theory of gravity the apogee of the moon advances as much as possible when it is either in conjunction with the sun or in opposition, and regresses when it is in quadrature with the sun. And the eccentricity becomes greatest in the first case and least in the second, by book 1, prop. 66, corols. 7, 8, and 9. And these inequalities, by the same corollaries, are very great and generate the principal equation of the apogee, which I shall call the semiannual. And the greatest semiannual equation is roughly 12°18′, as far as I could gather from observations. Our fellow countryman Horrocks was the first to propose that the moon revolves in an ellipse around the earth, which is set in its lower focus. Halley placed the center of the ellipse in an epicycle, whose center revolves uniformly around the earth. And from the motion in this epicycle there arise the inequalities (mentioned above) in the advance and retrogression of the apogee and in the magnitude of the eccentricity. Suppose the mean distance of the moon from the earth to be divided into 100,000 parts, and let T represent the earth and TC the mean eccentricity of the moon, of 5,505 parts. Let TC be produced to B, so that CB is the sine of the greatest semiannual equation (12°18′) to the radius TC; then the circle BDA, described with center C and radius CB, will be that epicycle in which the center of the moon's orbit is located and which revolves according to the order of the letters BDA. Let the angle BCD be taken equal to twice the annual argument, or twice the distance of the true place of the sun from the moon's apogee equated one time [i.e., corrected by the equation applied once], and CTD will be the semiannual equation of the moon's apogee and TD the eccentricity of its orbit, tending to the apogee equated a second time. And once the moon's mean motion and apogee and eccentricity have been found, as well as the orbit's major axis of 200,000 parts, then from these data the true place of the moon in its orbit and its distance from the earth will be found by very well known methods.

In the perihelion of the earth, because of the greater force of the sun, the center of the moon's orbit moves more swiftly around the center C than in its aphelion, and does so inversely as the cube of the distance of the earth from the sun. Because the equation of the center of the sun is comprehended in the annual argument, the center of the moon's orbit moves more swiftly in the epicycle BDA inversely as the square of the distance of the earth from the sun. In order for the center of the moon's orbit to move still more swiftly, inversely in the simple ratio of the distance, draw a straight line DE from the center D of the orbit toward the apogee of the moon, or parallel to the straight line TC, and take the angle EDF equal to the excess of the above-mentioned annual argument over the distance of the apogee of the moon from the perigee of the sun in a forward direction [or in consequentia]; or, which is the same, take the angle CDF equal to the complement of the true anomaly of the sun to 360°. And let DF be to DC jointly as twice the eccentricity of the earth's orbit is to the mean distance of the sun from the earth and as the daily mean motion of the sun from the apogee of the moon is to the daily mean motion of the sun from its own apogee, that is, as 33⅞ to 1,000 and 52′27″16‴ to 59′8″10‴ jointly, or as 3 to 100. And suppose that the center of the moon's orbit is located in point F and revolves in an epicycle whose center is D and whose radius is DF, while the point D advances in the circumference of the circle DABD. For in this manner the velocity with which the center of the moon's orbit will move in a certain

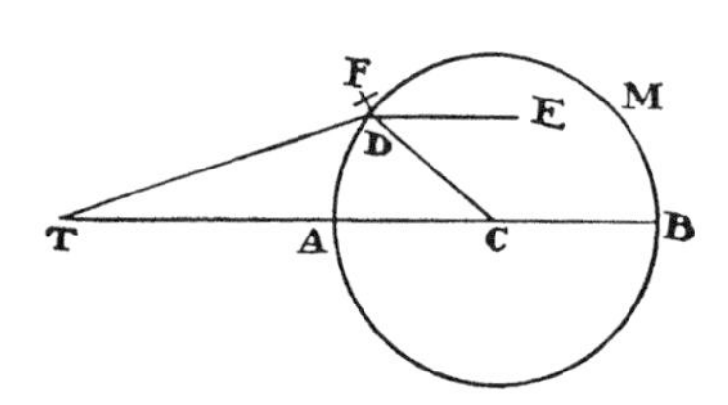

curved line described about the center C will be very nearly inversely as the cube of the distance of the sun from the earth, as it ought to be.

The computation of this motion is difficult, but it will be made easier by the following approximation. If the mean distance of the moon from the earth is 100,000 parts and the eccentricity TC is 5,505 parts as above, then the straight line CB or CD will be found to consist of 1,172¾ parts and the straight line DF of 35⅕ parts. And this straight line, at the distance TC, subtends at the earth the angle that the transfer of the center of the orbit from place D to place F generates in the motion of this center; and the same straight line DF doubled, in a position parallel to a line drawn from the earth to the upper focus of the moon's orbit, subtends the same angle, which of course that transfer generates in the motion of the focus; and at the distance of the moon from the earth it subtends the angle that the same transfer generates in the moon's motion and that therefore can be called the second equation of the center. And this equation, at the mean distance of the moon from the earth, is very nearly as the sine of the angle which that straight line DF contains with the straight line drawn from point F to the moon, and when it is greatest comes out 2′25″. And the angle which the straight line DF contains with the straight line drawn from point F to the moon is found either by subtracting the angle EDF from the mean anomaly of the moon or by adding the distance of the moon from the sun to the distance of the apogee of the moon from the apogee of the sun. And as the radius is to the sine of the angle thus found, so 2′25″ is to the second equation of the center, which should be added if that sum is less than a semicircle and subtracted if it is greater. In this way the longitude of the moon in the very syzygies of the luminaries will be found.

The atmosphere of the earth refracts the light of the sun up to a height of thirty-five or forty miles and, by refracting it, scatters it into the shadow of the earth, and by scattering the light at the edge of the shadow dilates the shadow. Hence, in lunar eclipses I add 1 minute, or 1⅓ minutes, to the diameter of the shadow as found from the parallax.

The theory of the moon, furthermore, should be examined and established by phenomena, first in the syzygies, then in the quadratures, and finally in the octants. And anyone who is going to undertake this task will not go wrong by using the following mean motions of the sun and moon at noon at the Royal Greenwich Observatory, on the last day of December 1700 (O.S.): namely, the mean motion of the sun ♑20°43′40″, and of its apogee ♋7°44′30″; and the mean motion of the moon ♒15°21′00″, and of its apogee ♓8°20′00″, and of its ascending node ♌27°24′20″; and the difference between the meridians of this observatory and the Royal Paris Observatory $0^{h}9^{m}20^{s}$; but the mean motions of the moon and of its apogee have not as yet been determined with sufficient exactness.[a]

Proposition 36
Problem 17

To find the force of the sun to move the sea.

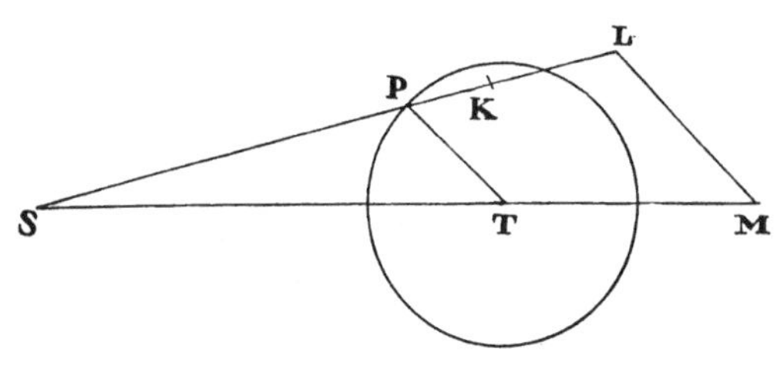

The sun's force ML or PT to perturb the motions of the moon, in the moon's quadratures, was (by prop. 25 of this book 3) to the force of gravity here on earth as 1 to 638,092.6. And the force TM−LN or 2PK in the moon's syzygies is twice as great. Now these forces, in the descent to the surface of the earth, are diminished in the ratio of the distances from the center of the earth, that is, in the ratio 60½ to 1; and so the first force on the surface of the earth is to the force of gravity as 1 to 38,604,600. By this force the sea is depressed in places that are 90 degrees distant from the sun. By the other force, which is twice as great, the sea is elevated both in the region directly under the sun and in the region opposite to the sun. The sum of these forces is to the force of gravity as 1 to 12,868,200. And since the same force arouses the same motion, whether it depresses the water in the regions that are 90 degrees distant from the sun or elevates the water in regions under the sun and opposite to the sun, this sum will be the total force of the sun to agitate the sea, and it will have the same effect as if all of it elevated the sea in regions under the sun and opposite it and had no action at all in regions that are 90 degrees distant from the sun.

This is the force of the sun to put the sea in motion in any given place when the sun is in the Zenith of the place as well as at its mean distance from the earth. In other positions of the sun, the force for raising the sea is directly as the versed sine of twice the altitude of the sun above the horizon of the place and inversely as the cube of the distance of the sun from the earth.

COROLLARY. The centrifugal force of the parts of the earth, arising from the daily motion of the earth (a force that is to the force of gravity as 1 to 289), causes the height of the water under the equator to exceed its height under the poles by a measure of 85,472 Paris feet (as was seen above in prop. 19); therefore, the solar force with which we have been dealing (since it is to the force of gravity as 1 to 12,868,200 and so to that centrifugal force as 289 to 12,868,200 or 1 to 44,527) will cause the height of the water in regions directly under the sun and directly opposite to the sun to exceed its height in places that are 90 degrees distant from the sun by a measure of only 1 Paris foot and $11\frac{1}{30}$ inches. For this measure is to the measure of 85,472 feet as 1 to 44,527.

Proposition 37
Problem 18

To find the force of the moon to move the sea.

[a]The force of the moon to move the sea is to be reckoned from its proportion to the force of the sun, and this proportion is to be determined from the proportion of the motions of the sea that arise from these forces. Before the mouth of the river Avon, at the third milestone below Bristol, in spring and autumn, the total ascent of the water in the conjunction and opposition of these two luminaries is (according to the observations of Samuel Sturmy) approximately 45 feet, but in the quadratures is only 25 feet. The first height arises from the sum of these two forces, the second from their difference. Therefore let the forces of the sun and the moon, when they are on the equator and at their mean distance from the earth, be S and L, and L + S will be to L − S as 45 to 25, or 9 to 5.

In Plymouth harbor, the tide of the sea (as observed by Samuel Colepress) is raised to approximately 16 feet in its mean height, and in spring and autumn the height of the tide in the syzygies can exceed its height in the quadratures by more than 7 or 8 feet. If the greatest difference of these heights is 9 feet, L + S will be to L − S as $20\frac{1}{2}$ to $11\frac{1}{2}$ or 41 to 23. And this proportion agrees well enough with the former one. Because of the magnitude of the tide in Bristol harbor, Sturmy's observations seem to be more trustworthy, and so, until something more certain is established, we shall use the proportion 9 to 5.

aa. In this proposition and its corollaries there are numerical differences in ed. 1 and sometimes also in ed. 2. The number 4.4815 at the end of the fifth paragraph, giving the ratio of the force of the sun to the force of the moon, was $6\frac{1}{3}$ in ed. 1.

But because of the reciprocating motions of the waters, the greatest tides do not occur at the syzygies of the luminaries but (as has been said earlier) are the third ones after the syzygies or follow next after the moon's third appulse to the meridian of the place after the syzygies, or rather (as is noted by Sturmy) are the third ones after the day of the new moon or full moon, or after approximately the twelfth hour from the new moon or full moon, and so occur at approximately the forty-third hour from the new moon or full moon. Now, in this harbor they occur at roughly the seventh hour from the appulse of the moon to the meridian of the place, and so they follow next after the appulse of the moon to the meridian, when the moon is approximately 18 or 19 degrees distant from the sun, or from the opposition of the sun, in a forward direction [or in consequentia]. The summer and winter reach their maximum, not in the solstices themselves, but when the sun has advanced through roughly a tenth of its whole circuit, or is approximately 36 or 37 degrees distant from the solstices. And similarly the greatest tide of the sea arises from the appulse of the moon to the meridian of the place, when the moon is distant from the sun by roughly a tenth part of its whole motion from one tide to the next. Let this distance be approximately 18½ degrees. Then the force of the sun at this distance of the moon from the syzygies and quadratures will be less effective to augment and to diminish that motion of the sea arising from the force of the moon than in the syzygies and quadratures themselves, in the ratio of the radius to the sine of the complement of twice this distance or the cosine of 37 degrees, that is, in the ratio of 10,000,000 to 7,986,355. And so in the above analogy, 0.7986355S ought to be written for S.

But additionally, the force of the moon must be diminished in the quadratures, because of the declination of the moon from the equator. For the moon in the quadratures, or rather at 18½ degrees beyond the quadratures, is in a declination of approximately $22°13'$. And the force of either luminary to move the sea is diminished when that luminary is declining from the equator, and diminished very nearly as the square of the cosine of the declination. And therefore the force of the moon in these quadratures is only 0.8570327L. Therefore $L + 0.7986355S$ is to $0.8570327L - 0.7986355S$ as 9 to 5.

Besides, the diameters of the orbit in which the moon would have to move (supposing no eccentricity) are to each other as 69 to 70; and thus the distance of the moon from the earth in the syzygies is to its distance in the quadratures as 69 to 70, other things being equal. And its distances when 18½° beyond the syzygies (where the greatest tide is generated) and then 18½° beyond the quadratures (where the least tide is generated) are to its mean distance as 69.098747 and 69.897345 to 69½. But the forces of the moon to move the sea are as the cubes of the distances inversely; and thus the forces at the greatest and least of these distances are to the force at the mean distance as 0.9830427 and 1.017522 to 1. Hence 1.017522L + 0.7986355S will be to 0.9830427 × 0.8570327L − 0.7986355S as 9 to 5; and S will be to L as 1 to 4.4815. Therefore, since the force of the sun is to the force of gravity as 1 to 12,868,200, the force of the moon will be to the force of gravity as 1 to 2,871,400.

COROLLARY 1. Since the water acted on by the force of the sun ascends to a height of 1 foot and 11⅟30 inches, by the force of the moon it will ascend to a height of 8 feet and 7⁵⁄22 inches, and by both forces to a height of 10½ feet, and when the moon is in its perigee the water will ascend to a height of 12½ feet and beyond, especially when the tide is made greater by winds. And so great a force is more than sufficient to give rise to all the motions of the sea and corresponds exactly to the quantity of the motions. For in seas that extend widely from east to west, as in the Pacific Ocean and the parts of the Atlantic Ocean and the Ethiopic [or South Atlantic] Sea, which are outside the tropics, the water is generally raised to a height of 6, 9, 12, or 15 feet. And in the Pacific Ocean, which is deeper and wider, the tides are said to be greater than in the Atlantic Ocean and the Ethiopic Sea. For, to have the tide be full, the width of the sea from east to west should be no less than 90 degrees. In the Ethiopic Sea the ascent of the water within the tropics is less than in the temperate zones, because of the narrowness of the sea between Africa and the southern part of America. In the middle of the sea the water cannot rise unless it simultaneously falls on both shores, both the eastern and the western; nevertheless, in our narrow seas, the water ought to rise alternately on the two shores, that is, rise on one shore while it falls on the other. For this reason the ebb and flow are generally very small in islands that are farthest from the shores. In certain harbors, where the water is compelled to flow in and flow out with great impetus through shallow places, so as to fill and empty bays alternately, the ebb and flow must be greater than usual, as at Plymouth and Chepstow Bridge in England, at Mont-Saint-Michel and the city of Avranches in Normandy, at Cambay and Pegu [b]in the East Indies.[b] In these places the sea, coming in and going back out with great velocity, at times inundates the shores and at other times leaves them dry for many miles. And the impetus of flowing in or going back out cannot be broken before the water is raised or depressed to 30, 40, or 50 feet and more. And the same is true of oblong and shallow straits, such as the Straits of Magellan and that channel by which England is surrounded [presumably, the channel and seas, but not the ocean, bordering England]. The tide in harbors and straits of this sort is increased beyond measure by the impetus of running in and back. But on shores that face the deep and open sea with a steep descent, where the water can be raised and can fall without the impetus of flowing out and coming back, the magnitude of the tide corresponds to the forces of the sun and moon.

COROLLARY 2. Since the force of the moon to move the sea is to the force of gravity as 1 to 2,871,400, it is evident that this force is far smaller than what can be perceived in experiments with pendulums or in any statical or hydrostatical experiments. It is only in the tides of the sea that this force produces a sensible effect.

COROLLARY 3.[c] Since the force of the moon to move the sea is to the similar force of the sun as 4.4815 to 1, and since those forces (by book 1, prop. 66, corol. 14) are as the densities of the bodies of the moon and sun and the cubes of their apparent diameters jointly, the density of the moon will be to the density of the sun as 4.4815 to 1 directly and as the cube of the diameter of the moon to the cube of the diameter of the sun inversely, that is (since the apparent mean diameters of the moon and the sun are $31'16\frac{1}{2}''$ and $32'12''$), as 4,891 to 1,000. Now, the density of the sun was to the density of the earth as 1,000 to 4,000, and therefore the density of the moon is to the density of the earth as 4,891 to 4,000, or 11 to 9. Therefore the body of the moon is denser and more earthy than our earth.

bb. The Latin here is "in *India* orientali," *lit.* "in east India." In Newton's day the terms "East India" and "East Indies" were collective names applied to the whole area consisting of India, Indochina, Malaya, and the Malay Archipelago (see *Oxford English Dictionary*, s.vv. "East India" and "East Indies"; *Webster's New Geographical Dictionary*, s.v. "East Indies"). Although that usage is now obsolete, modern English provides no alternative collective name for that area, and so we have chosen the rendering "in the East Indies" used by Motte. In modern geographical terms, Cambay is in western India, and Pegu in Burma.

c. For a gloss on this corollary see the Guide, §10.16.

COROLLARY 4. And since the true diameter of the moon, from astronomical observations, is to the true diameter of the earth as 100 to 365, the mass of the moon will be to the mass of the earth as 1 to 39.788.

COROLLARY 5. And the accelerative gravity on the surface of the moon will be about three times smaller than the accelerative gravity on the surface of the earth.

[d]COROLLARY 6. And the distance of the center of the moon from the center of the earth will be to the distance of the center of the moon from the common center of gravity of the earth and the moon as 40.788 to 39.788.

COROLLARY 7. And the mean distance of the center of the moon from the center of the earth (in the octants of the moon) will be nearly 60⅖ greatest semidiameters of the earth. For the greatest semidiameter of the earth was 19,658,600 Paris feet, and the mean distance between the centers of the earth and the moon, which consists of 60⅖ such semidiameters, is equal to 1,187,379,440 feet. And this distance (by the preceding corollary) is to the distance of the center of the moon from the common center of gravity of the earth and the moon as 40.788 to 39.788; and hence the latter distance is 1,158,268,534 feet. And since the moon revolves with respect to the fixed stars in $27^{d}7^{h}43\frac{4}{9}^{m}$, the versed sine of the angle that the moon describes in the time of one minute is 12,752,341, the radius being 1,000,000,000,000,000. And the radius is to this versed sine as 1,158,268,534 feet to 14.7706353 feet. The moon, therefore, falling toward the earth under the action of that force with which it is kept in its orbit, will in the time of one minute describe 14.7706353 feet. And by increasing this force in the ratio of $178\frac{29}{40}$ to $177\frac{29}{40}$, the total force of gravity in the orbit of the moon will be found by prop. 3, corol. [of this book 3]. And falling toward the earth under the action of this force, the moon will describe 14.8538067 feet in the time of one minute. And at $\frac{1}{60}$ of the distance of the moon from the center of the earth, that is, at a distance of 197,896,573 feet from the center of the earth, a heavy body—falling in the time of one second—will likewise describe 14.8538067 feet. [e]And so, at a distance of 19,615,800 feet (which is the mean semidiameter of the earth), a heavy body in falling will describe—in the time of one second—15.11175 feet, or 15 feet 1 inch and $4\frac{1}{11}$ lines. This will be the descent of bodies at a latitude of 45

dd. Ed. 1 lacks this.
ee. This is considerably different in ed. 2.

degrees. And by the foregoing table, presented in prop. 20, the descent will be a little greater at the latitude of Paris by about ⅔ of a line. Therefore, by this computation, heavy bodies falling in a vacuum at the latitude of Paris will—in the time of one second—describe approximately 15 Paris feet 1 inch and $4\frac{25}{33}$ lines. And if gravity is diminished by taking away the centrifugal force that arises from the daily motion of the earth at that latitude, heavy bodies falling there will—in the time of one second—describe 15 feet 1 inch and 1½ lines. And that heavy bodies do fall with this velocity at the latitude of Paris has been shown above in props. 4 and 19 [of this book 3].[e]

COROLLARY 8. [f]The mean distance between the centers of the earth and the moon in the syzygies of the moon is 60 greatest semidiameters of the earth, taking away roughly $\frac{1}{30}$ of a semidiameter. And in the moon's quadratures, the mean distance between these centers is $60\frac{5}{6}$ semidiameters of the earth. For these two distances are to the mean distance of the moon in the octants as 69 and 70 to 69½, by prop. 28.[f]

[g]COROLLARY 9. The mean distance between the centers of the earth and the moon in the syzygies of the moon is $60\frac{1}{10}$ mean semidiameters of the earth. And in the moon's quadratures, the mean distance of the same centers is 61 mean semidiameters of the earth, taking away $\frac{1}{30}$ of a semidiameter.

COROLLARY 10. In the moon's syzygies, its mean horizontal parallax at latitudes of 0°, 30°, 38°, 45°, 52°, 60°, and 90° is 57′20″, 57′16″, 57′14″, 57′12″, 57′10″, 57′8″, and 57′4″ respectively.[g]

In these computations I have not considered the magnetic attraction of the earth, the magnitude of which is very small anyway and is unknown. But if this attraction can ever be determined—and if the measures of degrees on the meridian, and the lengths of isochronous pendulums at various parallels of latitude, and the laws of the motions of the sea, and the moon's parallax, together with the apparent diameters of the sun and moon, are ever determined more accurately from phenomena—it will then be possible to undertake all this calculation over again with a higher degree of accuracy.[a d]

Proposition 38
Problem 19

To find the figure of the body of the moon.

If the body of the moon were fluid like our sea, the force of the earth to elevate that fluid in both the nearest and farthest parts would be to the force

ff. This is very different in ed. 2.
gg. Ed. 2 lacks this.

of the moon by which our sea is raised in the regions both under the moon and opposite to the moon as the accelerative gravity of the moon toward the earth is to the accelerative gravity of the earth toward the moon and as the diameter of the moon is to the diameter of the earth, jointly—that is, as 39.788 to 1 and 100 to 365 jointly, or as 1,081 to 100. Hence, since our sea is raised by the force of the moon to 8⅗ feet, the lunar fluid would have to be raised by the force of the earth to 93 feet. And for this reason the figure of the moon would be a spheroid, the greatest diameter of which, produced, would pass through the center of the earth and would exceed by 186 feet the diameters perpendicular to that one. Therefore, it is just such a figure that the moon has and must have had from the beginning. Q.E.I.

COROLLARY. And hence it happens that the same face of the moon is always turned toward the earth. For in any other position, the moon cannot remain at rest, but by a motion of oscillation will always return to this position. But those oscillations would nevertheless be extremely slow because the forces producing them are small in magnitude; so that the face of the moon that should always look toward the earth can (for the reason given in prop. 17) be turned toward the other focus of the moon's orbit and not at once be drawn back from there and turned toward the earth.

[a]Lemma 1

Let APE*p represent the earth, uniformly dense, with a center* C *and poles* P *and p and equator* AE, *and suppose a sphere* P*ape*[b] *to be described with center* C *and radius* CP. *Let* QR *be the plane on which a straight line drawn from the center of the sun to the center of the earth stands perpendicularly. Then, if the individual particles of the whole exterior earth* P*ap*APE*p*E, *which is higher than the sphere just described, endeavor to recede in both directions from the plane* QR, *and the endeavor of each particle is as its distance from the plane, I say, first of all, that the total force and efficacy of all the particles that lie in the circle of the equator* AE *(disposed uniformly outside the globe, in the manner of a ring completely encircling that globe) to rotate the earth around its center will be to the total force and efficacy of the same number of particles standing at point* A *of the equator (which is most distant from the plane* QR*) to move the earth with a similar circular*

aa. In ed. 1, with certain variants, lems. 1–3 and hyp. 2 are simply three lemmas, of which the first contains the statement of lem. 1 followed by the demonstration (which is, however, greatly altered in ed. 2 and ed. 3) of lem. 2, the second corresponds to lem. 3, and the third corresponds to hyp. 1.

b. Strictly speaking, a sphere is defined as a solid body all of whose points are equidistant from the center, but the context and the diagram leave no doubt that in lem. 1, Newton's "sphere P*ape*" ("sphaera P*ape*") is not truly spherical but of an ellipsoidal shape.

motion around its center as 1 is to 2. And that circular motion will be performed around an axis lying in the common section of the equator and the plane QR.

For let a semicircle INLK be described with center K and diameter IL. Suppose the semicircumference INL to be divided into innumerable equal parts, and from the individual parts N to the diameter IL drop the sines NM. Then the sum of the squares of all the sines NM will be equal to the sum of the squares of the sines KM, and both sums will be equal to the sum of the squares of the same number of semidiameters KN; and so the sum of the squares of all the sines NM will be one-half of the sum of the squares of the same number of semidiameters KN.

Now let the perimeter of the circle AE be divided into the same number of equal particles, and from each one of them F to the plane QR drop a perpendicular FG, as well as a perpendicular AH from the point A. Then the

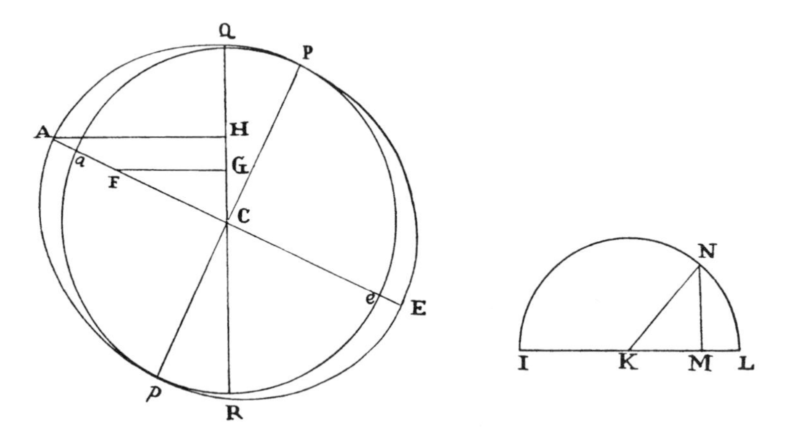

force by which the particle F recedes from the plane QR will (by hypothesis) be as that perpendicular FG, and this force multiplied by the distance CG will be the efficacy of the particle F to turn the earth around its center. And thus the efficacy of a particle in the place F will be to the efficacy of a particle in the place A as FG × GC to AH × HC, that is, as FC^2 to AC^2; and therefore the total efficacy of all the particles in their places F will be to the efficacy of the same number of particles in place A as the sum of all the FC^2 to the sum of the same number of AC^2, that is (by what has already been demonstrated), as 1 to 2. Q.E.D.

And since the particles act by receding perpendicularly from the plane QR, and do so equally from each side of this plane, they will turn the circumference of the circle of the equator, together with the earth adhering to it, around an axis lying in the plane QR as well as in the plane of the equator.

Lemma 2

Under the same conditions, I say, secondly, that the total force and efficacy of all the particles situated everywhere outside the globe to rotate the earth around the given axis is to the total force of the same number of particles, disposed uniformly throughout all of the circle of the equator AE *in the manner of a ring, to move the earth with a similar circular motion, as 2 is to 5.*

For let IK be any smaller circle parallel to the equator AE, and let L and *l* be any two equal particles situated in this circle outside the globe P*ape*.[c] And if perpendiculars LM and *lm* are dropped to the plane QR, which is perpendicular to a radius drawn to the sun, the total forces with which the particles recede from the plane QR will be proportional to the perpendiculars LM and *lm*. Now, let the straight line L*l* be parallel to the plane P*ape*; bisect L*l* at X; through the point X draw N*n* parallel to the plane QR and meeting the perpendiculars LM and *lm* at N and *n*; and drop a perpendicular XY to the plane QR. Then the contrary forces of the particles L and *l* to rotate the earth in opposite directions are as LM × MC and *lm* × *m*C, that is, as LN×MC+NM×MC and *ln*×*m*C−*nm*×*m*C, or LN×MC+NM×MC and LN × *m*C − NM × *m*C; and their difference LN × M*m* − NM × (MC + *m*C) is the force of both particles taken together to rotate the earth. The positive part of this difference, LN × M*m* or 2LN × NX, is to the force 2AH × HC of two particles of the same magnitude located at A as LX^2 to AC^2. And the negative part, NM × (MC + *m*C) or 2XY × CY is to the force 2AH × HC of the same particles located at A as CX^2 to AC^2. And accordingly the difference of the parts, that is, the force of the two particles L and *l* (taken together) to rotate the earth, is to the force of two particles equal to those and

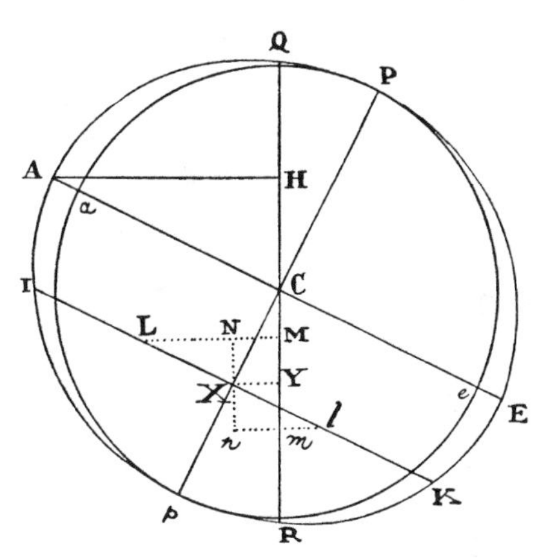

c. In moving ahead from lem. 1 to lem. 2, Newton has shifted his vocabulary from "sphere" to "globe." He now writes of a circle "outside the globe P*ape*" ("extra globum P*ape*"), where again the context and the diagram leave no doubt that the "globe P*ape*" is ellipsoidal.

standing in the place A, likewise to rotate the earth, as $LX^2 - CX^2$ to AC^2. But if the circumference IK of the circle IK is divided into innumerable equal particles L, all the LX^2 will be to as many IX^2 as 1 to 2 (by lem. 1), and to this same number of AC^2 as IX^2 to $2AC^2$; and just as many CX^2 will be to the same number of AC^2 as $2CX^2$ to $2AC^2$. Therefore the combined forces of all the particles in the circumference of the circle IK are to the combined forces of as many particles in the place A as $IX^2 - 2CX^2$ to $2AC^2$, and therefore (by lem. 1) to the combined forces of as many particles in the circumference of the circle AE as $IX^2 - 2CX^2$ to AC^2.

Now, if the diameter P*p* of the sphere[d] is divided into innumerable equal parts, on which the same number of circles IK stand, the matter in the perimeter of each circle IK will be as IX^2; and so the force of that matter to rotate the earth will be as IX^2 multiplied by $IX^2 - 2CX^2$. And the force of the same matter, if it stood in the perimeter of the circle AE, would be as IX^2 multiplied by AC^2. And therefore the force of all the particles of the total matter standing outside the globe in the perimeters of all the circles is to the force of as many particles standing in the perimeter of the greatest circle AE as all the IX^2 multiplied by $IX^2 - 2CX^2$ to as many IX^2 multiplied by AC^2, that is, as all the $AC^2 - CX^2$ multiplied by $AC^2 - 3CX^2$ to as many $AC^2 - CX^2$ multiplied by AC^2, that is, as all the $AC^4 - 4AC^2 \times CX^2 + 3CX^4$ to as many $AC^4 - AC^2 \times CX^2$, that is, as the total fluent quantity whose fluxion[e] is $AC^4 - 4AC^2 \times CX^2 + 3CX^4$ to the total fluent quantity whose fluxion is $AC^4 - AC^2 \times CX^2$; and accordingly, by the method of fluxions, as $AC^4 \times CX - \frac{4}{3}AC^2 \times CX^3 + \frac{3}{5}CX^5$ to $AC^4 \times CX - \frac{1}{3}AC^2 \times CX^3$, that is, if the whole of C*p* or AC is written in place of CX, as $\frac{4}{15}AC^5$ to $\frac{2}{3}AC^5$, or as 2 to 5. Q.E.D.

Lemma 3 *Under the same conditions, I say, thirdly, that the motion of the whole earth around the axis described above, a motion that is composed of the motions of all the particles, will be to the motion of the above-mentioned ring around the same axis in a ratio that is compounded of the ratio of the matter in the earth to the matter in the ring and the ratio of three times the square of the quadrantal arc of any circle to two times the square of the diameter—that is, in the ratio of the matter to the matter and of the number 925,275 to the number 1,000,000.*

d. Newton has here changed his terminology, reverting to "sphere."

e. Note that here Newton makes explicit use of the "method" of fluxions, that is, the calculus.

For the motion of a cylinder revolving around its axis at rest is to the motion of an inscribed sphere revolving together with it as any four equal squares to three of the circles inscribed in them; and the motion of the cylinder is to the motion of a very thin ring surrounding the sphere and cylinder at their common contact as twice the matter in the cylinder to three times the matter in the ring; and this motion of the ring, continued uniformly around the axis of the cylinder, is to the uniform motion of the ring about its own diameter (in the same periodic time) as the circumference of the circle to twice its diameter.

Hypothesis 2

If the ring discussed above were to be carried alone in the orbit of the earth about the sun with an annual motion (supposing that all the rest of the earth were removed from it), and if this ring revolved at the same time with a daily motion about its axis, inclined to the plane of the ecliptic at an angle of 23½ degrees, then the motion of the equinoctial points would be the same whether that ring were fluid or consisted of rigid and solid matter.[a]

Proposition 39
Problem 20

To find the precession of the equinoxes.

The mean hourly motion of the nodes of the moon in a circular orbit was, for the nodes in the quadratures, $16''35'''16^{iv}36^{v}$, and half of this, $8''17'''38^{iv}18^{v}$, is (for the reasons explained above [at the end of corol. 2 to prop. 30]) the mean hourly motion of the nodes in such an orbit; and in a whole sidereal year the mean motion adds up to $20°11'46''$ [see beginning of prop. 32]. Therefore, since in a year the nodes of the moon would, in such an orbit, move backward [or in antecedentia] through $20°11'46''$; and since, if there were more moons, the motion of the nodes of each (by book 1, prop. 66, corol. 16) would be as the periodic times; it follows that if the moon revolved near the surface of the earth in the space of a sidereal day, the annual motion of the nodes would be to $20°11'46''$ as a sidereal day of $23^{h}56^{m}$ is to the periodic time of the moon, $27^{d}7^{h}43^{m}$—that is, as 1,436 to 39,343. And the same is true of the nodes of a ring of moons surrounding the earth, whether those moons do not touch one another, or whether they become liquid and take the form of a continuous ring, or finally whether that ring becomes rigid and inflexible.

Let us imagine therefore that this ring, as to its quantity of matter, is equal to all of the earth *PapAPepE* that lies outside of the globe *Pape*[a] (as in the figure to lem. 2). This globe is to the earth that lies outside of it as aC^2 to $AC^2 - aC^2$, that is (since the earth's smaller semidiameter PC or *a*C is to its greater semidiameter AC as 229 to 230), as 52,441 to 459. Hence, if this ring girded the earth along the equator and both together revolved about the diameter of the ring, the motion of the ring would be to the motion of the interior globe (by lem. 3 of this third book) as 459 to 52,441 and 1,000,000 to 925,275 jointly, that is, as 4,590 to 485,223; and so the motion of the ring would be to the sum of the motions of the ring and globe as 4,590 to 489,813. Hence, if the ring adheres to the globe and communicates to the globe its own motion with which its nodes or equinoctial points regress, the motion that will remain in the ring will be to its former motion as 4,590 to 489,813, and therefore the motion of the equinoctial points will be diminished in the same ratio. Therefore the annual motion of the equinoctial points of a body composed of the ring and the globe will be to the motion $20°11'46''$ as 1,436 to 39,343 and 4,590 to 489,813 jointly, that is, as 100 to 292,369. But the forces by which the nodes of the moons [i.e., a ring of moons] regress (as I have explained above), and so by which the equinoctial points of the ring regress (that is, the forces 3IT in the figure to prop. 30), are—in the individual particles—as the distances of those particles from the plane QR, and it is with these forces that the particles recede from the plane; and therefore (by lem. 2), if the matter of the ring were scattered over the whole surface of the globe, as in the configuration *PapAPepE*, so as to constitute that exterior part of the earth, the total force and efficacy of all the particles to rotate the earth about any diameter of the equator, and thus to move the equinoctial points, would come out less than before in the ratio of 2 to 5. And hence the annual regression of the equinoxes would now be to $20°11'46''$ as 10 to 73,092, and accordingly would become $9''56'''50^{iv}$.

[b]But because of the inclination of the plane of the equator to the plane of the ecliptic, this motion must be diminished in the ratio of the sine 91,706

a. See the notes to lem. 1 and lem. 2.

bb. Ed. 1 has: "This is the precession of the equinoxes that arises from the force of the sun. Now the force of the moon to move the sea was to the force of the sun as 6⅓ to 1, and this force in proportion to its quantity will also increase the precession of the equinoxes. And therefore the precession arising from both causes will now become greater in the ratio of 7⅓ to 1 and thus will be $45^{ii}24^{iii}15^{iv}$. This is the motion of the equinoctial points arising from the actions of the sun and the moon on the parts of the

(which is the sine of the complement of 23½ degrees [or the cosine of 23½ degrees]) to the radius 100,000. And thus this motion will now become $9''7'''20^{iv}$. This is the annual precession of the equinoxes that arises from the force of the sun.

Now, the force of the moon to move the sea was to the force of the sun as roughly 4.4815 to 1. And the force of the moon to move the equinoxes is to the force of the sun in this same proportion. And so the annual precession of the equinoxes that arises from the force of the moon comes out $40''52'''52^{iv}$, and the total annual precession arising from both forces will be $50''00'''12^{iv}$. And this motion of precession agrees with the phenomena. For, from astronomical observations, the precession of the equinoxes is more or less 50 seconds annually.

If the height of the earth at the equator exceeds its height at the poles by more than 17⅙ miles, its matter will be rarer at the circumference than

---

earth that lie on the globe *Pape*. For the earth cannot be inclined in any direction by those actions exerted upon the globe itself. [On Newton's use of "globe," see the note to lem. 2.]

"Now let APE*p* represent the body of the earth, possessing an elliptical shape and consisting of uniform matter. And if this is divided into innumerable elliptical, concentric, and similar figures APE*p*, BQ*bq*, CR*cr*, DS*ds*, . . . , whose diameters are in a geometric progression, then, since the figures are similar, the forces of the sun and the moon under the action of which the equinoctial points regress would cause those same equinoctial points of the remaining figures regarded separately to regress with the same velocity. And the case is the same for the motion of the single orbs AQE*q*, BR*br*, CS*cs*, . . . , which are the differences between those figures. [Newton here uses the word "orb" ("orbits") for the solid we would call an ellipsoid of revolution.] The equinoctial points of each orb, if it were alone, would have to regress with the same velocity. And it does not matter whether any orb

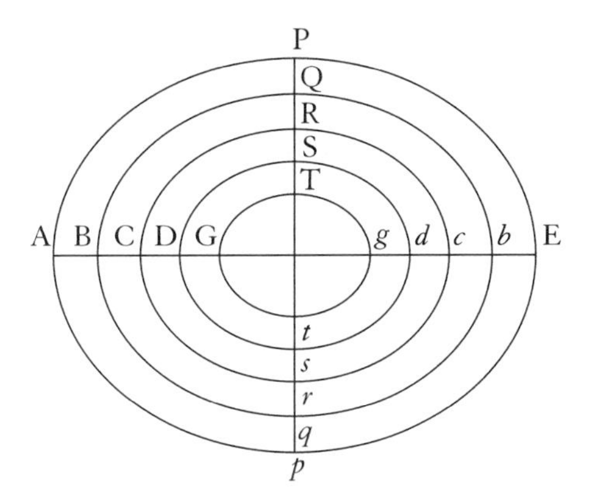

is denser or rarer, provided that it is made up of uniformly dense matter. Hence also if the orbs are denser at the center than at the circumference, the motion of the equinoxes of the whole earth will be the same as before, provided that each orb regarded separately consists of uniformly dense matter and that the shape of the orb is not changed. But if the shapes of the orbs are changed and if the earth now ascends higher than before at the equator AE because of the density of the matter at the center, the regression of the equinoxes will be increased as a result of the increase in the height, and will be increased in single separate orbs in the ratio of the greater height of the matter near the equator of that orb, and in the whole earth in the ratio of the greater height of the matter near the equator of an orb which is not the outermost AQE*q* and not the innermost G*g* but some mean orb CS*cs*. Moreover, we have implied above that the earth is denser at the center and therefore is higher at the equator than at the poles in a

at the center; and the precession of the equinoxes has to be increased because of that excess in height, and diminished because of the greater rarity.[b]

We have now described the system of the sun, the earth, the moon, and the planets; something must still be said about comets.

Lemma 4 *The comets are higher than the moon and move in the planetary regions.*

Just as the lack of diurnal parallax requires that comets be located beyond the sublunar regions, so the fact that comets have an annual parallax is convincing evidence that they descend into the regions of the planets. For comets which move forward according to the order of the signs are all, toward the end of their visibility, either slower than normal or retrograde if the earth is between them and the sun, but swifter than they should be if the earth is approaching opposition. And conversely those comets that move

---

greater ratio than 692 to 689. And the ratio of the greater height can be gathered approximately from the greater diminution of gravity at the equator than that which ought to follow from a ratio of 692 to 689. The deviations by which the length of a seconds pendulum oscillating on Gorée Island and on Cayenne exceeded the length of a pendulum oscillating at Paris in the same time were found by the French to be $\frac{1}{10}$ and $\frac{1}{8}$ of an inch, which, however, from the proportion of 692 to 689, came out $\frac{81}{1,000}$ and $\frac{89}{1,000}$. Therefore the length of a pendulum on Cayenne is greater than it should be in the ratio of $\frac{1}{8}$ to $\frac{89}{1,000}$, or 1,000 to 712, and on Gorée Island, in the ratio of $\frac{1}{10}$ to $\frac{81}{1,000}$, or 1,000 to 810. If we take a mean ratio of 1,000 to 760, the gravity of the earth will have to be diminished at the equator, and its height increased in the same place, in the ratio of 1,000 to 760 very nearly. Hence the motion of the equinoxes (as was said above), if increased in the ratio of the height of the earth, not at the outermost orb, not at the innermost, but at some intermediate orb, that is, not in the greatest ratio of 1,000 to 760, not in the least ratio of 1,000 to 1,000, but in some mean ratio, say 10 to $8\frac{1}{3}$ or 6 to 5, will come out to be $54^{ii}29^{iii}6^{iv}$ annually.

"Again, because of the inclination of the plane of the equator to the plane of the ecliptic, this motion must be diminished, and must be diminished in the ratio of the sine of the complement of the inclination to the radius. For the distance of each terrestrial particle from the plane QR, when the particle is farthest away from the plane of the ecliptic, being (so to speak) in its tropic, is diminished by the inclination of the planes of the ecliptic and the equator to each other, in the ratio of the sine of the complement of the inclination to the radius. And the force of the particle to move the equinoxes is also diminished in the ratio of that distance. The sum of the forces of that same particle is also diminished in the same ratio in places equally distant in both directions from the tropic, as could easily be shown from what has been demonstrated earlier, and therefore the whole force of that particle to move the equinoxes in an entire revolution, as well as the whole force of all the particles, and the motion of the equinoxes arising from that force, is diminished in the same ratio. Therefore since that inclination is $23\frac{1}{2}°$, the motion of $54^{ii}29^{iii}$ must be diminished in the ratio of the sine of 91,706 (which is the sine of the complement of $23\frac{1}{2}°$) to the radius 100,000. In this way that motion will now become $49^{ii}58^{iii}$. Therefore the points of the equinoxes regress with an annual motion (according to our calculation) of $49^{ii}58^{iii}$, nearly as the celestial phenomena require. For that annual regression, from the observations of astronomers, is 50^{ii}."

contrary to the order of the signs are swifter than they should be, at the end of their visibility when the earth is between them and the sun, and slower than they should be or retrograde if the earth is on the opposite side of the sun. This happens principally as a result of the motion of the earth in its different positions [with respect to the comets], just as is the case for the planets, which, according as the motion of the earth is either in the same direction or in an opposite one, are sometimes retrograde, and sometimes seem to advance more slowly and at other times more swiftly. If the earth goes in the same direction as the comet and by its angular motion is carried about the sun so much more swiftly that a straight line continually drawn through the earth and the comet converges toward the regions beyond the comet, then the comet as seen from the earth will appear to be retrograde because of its slower motion; but if the earth is going more slowly, the motion of the comet (taking away the motion of the earth) becomes at least slower. But if the earth goes in a direction opposite to the comet's motion, the motion of the comet will as a result appear speeded up. And from the acceleration or retardation or retrograde motion, the distance of the comet may be ascertained in the following way.

Let ♈QA, ♈QB, and ♈QC be three observed longitudes of a comet at the beginning of its [visible] motion, and let ♈QF be its last observed longitude, just as the comet ceases to be seen. Draw the straight line ABC,

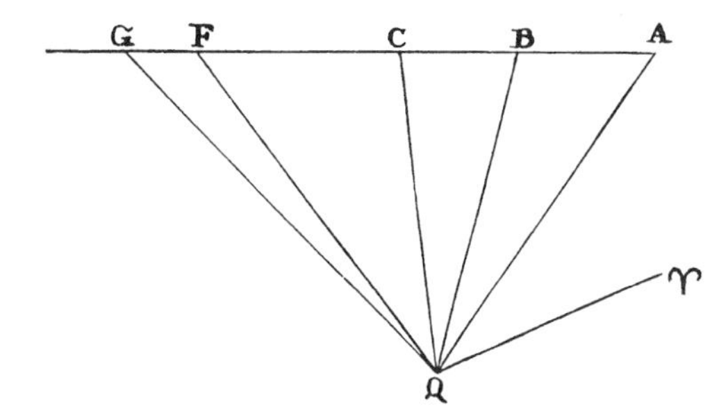

whose parts AB and BC placed between the straight lines QA and QB, and between the straight lines QB and QC, are to each other as the times between the first three observations. Let AC be produced to G, so that AG is to AB as the time between the first and the last observation is to the time between the first and the second observation, and let QG be joined. Then, if the comet moved uniformly in a straight line and the earth were either at rest or also moved forward in a straight line with uniform motion, the angle ♈QG would be the longitude of the comet at the time of the last observation. Therefore, the angle FQG, which is the difference between the longitudes, arises from the inequality of the motions of the comet and of the earth. And this angle, if the earth and comet move in opposite directions, is added to the angle ♈QG, and thus makes the apparent motion of the comet swifter; but if the comet is going in the same direction as the earth, this angle is subtracted from that same angle ♈QG and makes the motion of the comet either slower or possibly retrograde, as I have just explained. Therefore this angle arises chiefly from the motion of the earth and on that account is rightly regarded as the parallax of the comet, ignoring, of course, any increase or decrease in it which could arise from the nonuniform motion of the comet in its own orbit. And the distance of the comet may be ascertained from this parallax in the following manner.

Let S represent the sun, *ac*T the earth's orbit, *a* the place of the earth in the first observation, *c* the place of the earth in the third observation, T the place of the earth in the last observation, and let T♈ be a straight line drawn toward the beginning of Aries. Let angle ♈TV be taken equal to angle ♈QF, that is, equal to the longitude of the comet when the earth is in T. Let *ac* be drawn and produced to *g*, so that *ag* is to *ac* as AG to AC; then *g* will be the place which the earth would reach at the time of the last observation, with its motion uniformly continued in the straight line *ac*. And so if *g*♈ is drawn parallel to T♈, and the angle ♈*g*V is taken equal to the angle ♈QG, this angle ♈*g*V will be equal to the longitude of the comet as seen from place *g*, and the angle TV*g* will be the parallax that arises from the transfer of the earth from place *g* to place T; and accordingly V will be the place of the comet in the plane of the ecliptic. And this place V is ordinarily lower than the orbit of Jupiter.

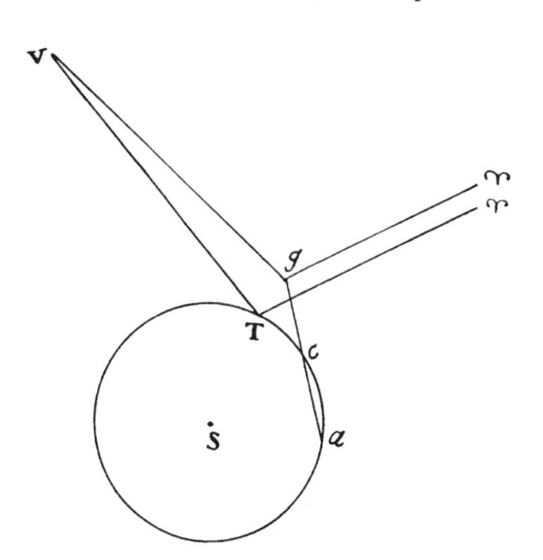

The same may be ascertained from the curvature of the path of comets. These bodies go almost in great circles as long as they move more swiftly, but at the end of their course, when that part of their apparent motion which arises from parallax has a greater proportion to the total apparent motion, they tend to deviate from such circles, and whenever the earth moves in one direction, they tend to go off in the opposite direction. Because this deviation corresponds to the motion of the earth, it arises chiefly from parallax, and its extraordinary quantity, according to my computation, has placed disappearing comets quite far below Jupiter. Hence it follows that when comets are closer to us, in their perigees and perihelions, they very often descend below the orbits of Mars and of the inferior planets.

The nearness of comets is confirmed also from the light of their heads. For the brightness of a heavenly body illuminated by the sun and going off into distant regions is diminished as the fourth power of the distance; that is, it is diminished as the square because of the increased distance of the body from the sun and diminished as the square again because of the diminished apparent diameter. Thus, if both the quantity of light [i.e., brightness] and the apparent diameter of the comet are given, its distance will be found by taking its distance to the distance of some planet directly in the ratio of diameter to diameter and inversely as the square root of the ratio of light to light. Thus, as observed by Flamsteed through a sixteen-foot telescope and measured with a micrometer, the least diameter of the coma[a] of the comet of the year 1682 equaled $2'0''$, while the nucleus or star in the middle of the head occupied scarcely a tenth of this width and therefore was only $11''$ or $12''$ wide. But in the light and brilliance of its head it surpassed the head of the comet of the year 1680 and rivaled stars of the first or second magnitude. Let us suppose that Saturn with its ring was about four times brighter; then, because the light of the ring almost equals the light of the globe within it, and the apparent diameter of the globe is about $21''$, so that the light of the globe and the ring together would equal the light of a globe whose diameter was $30''$, it follows that the distance of the comet will be to the distance of Saturn as 1 to $\sqrt{4}$ inversely and $12''$ to $30''$ directly, that

a. The Latin word "coma" means "head of hair" and is used today to designate the nebulous envelope surrounding the nucleus or head of a comet. Another Latin word for "head of hair" is "capillitium." In book 3, lem. 4, Newton uses "capillitium" for the "head of hair" of a comet, but in prop. 41 he uses "coma." We have translated both as "coma," the term commonly used in English.

is, as 24 to 30 or as 4 to 5. Again, on the authority of Hevelius, the comet of April 1665 surpassed in its brilliance almost all the fixed stars, and even Saturn itself (that is, by reason of its far more vivid color). Indeed, this comet was brighter than the one which had appeared at the end of the preceding year and was comparable to stars of the first magnitude. The width of the comet's coma was about 6′, but the nucleus, when compared with the planets by the aid of a telescope, was clearly smaller than Jupiter and was judged to be sometimes smaller than the central body of Saturn and sometimes equal to it. Further, since the diameter of the coma of comets rarely exceeds 8′ or 12′, and the diameter of the nucleus or central star is about a tenth or perhaps a fifteenth of the diameter of the coma, it is evident that such stars generally have the same apparent magnitude as the planets. Hence, since their light can often be compared to the light of Saturn and sometimes surpasses it, it is manifest that all the comets in their perihelions should be placed either below Saturn or not far above. Those who banish the comets almost to the region of the fixed stars are, therefore, entirely wrong; certainly in such a situation, they would not be illuminated by our sun any more than the planets in our solar system are illuminated by the fixed stars.

In treating these matters, we have not been considering the obscuring of comets by that very copious and thick smoke by which the head is surrounded, always gleaming dully as if through a cloud. For the darker the body is rendered by this smoke, the closer it must approach to the sun for the amount of light reflected from it to rival that of the planets. This makes it likely that the comets descend far below the sphere of Saturn, as we have proved from their parallax.

But this same result is, to the highest degree, confirmed from their tails. These arise either from reflection by the smoke scattered through the aether or from the light of the head. In the first case the distance of the comets must be diminished, since otherwise the smoke always arising from the head would be propagated through spaces far too great, with such a velocity and expansion as to be unbelievable. In the second case, all the light of both the tail and the coma must be ascribed to the nucleus of the head. Therefore, if we suppose that all this light is united and condensed within the disc of the nucleus, then certainly that nucleus, whenever it emits a very large and very bright tail, will far surpass in its brilliance even Jupiter itself. Therefore, if it has a smaller apparent diameter and is sending forth more light, it will be much more illuminated by the sun and thus will be much closer to the sun. By the same argument, furthermore, the heads ought to be located below the orbit of Venus, when they are hidden under the sun and emit tails both very great and very bright like fiery beams, as they do sometimes. For if all of that light were understood to be gathered together into a single star, it would sometimes surpass Venus itself, not to say several Venuses combined.

Finally, the same thing may be ascertained from the light of the heads, which increases as comets recede from the earth toward the sun and decreases as they recede from the sun toward the earth. Thus the latter comet of 1665 (according to the observations of Hevelius), from the time when it began to be seen, was always decreasing in its apparent motion and therefore had already passed its perigee; but the splendor of its head nevertheless increased from day to day until the comet, concealed by the sun's rays, ceased to be visible. The comet of 1683 (also according to the observations of Hevelius) at the end of July, when it was first sighted, was moving very slowly, advancing about 40′ or 45′ in its orbit each day. From that time its daily motion kept increasing continually until 4 September, when it came to about 5°. Therefore, in all this time the comet was approaching the earth. This is gathered also from the diameter of the head, as measured with a micrometer, since Hevelius found it to be on 6 August only 6′5″ including the coma, but on 2 September 9′7″. Therefore the head appeared far smaller at the beginning than at the end of the motion; yet at the beginning the head showed itself far brighter in the vicinity of the sun than toward the end of its motion, as Hevelius also reports. Accordingly, in all this time, because of its receding from the sun, it decreased with respect to its light, notwithstanding its approach to the earth.

The comet of 1618, about the middle of December, and that of 1680, about the end of the same month, were moving very swiftly and therefore were then in their perigees. Yet the greatest splendor of their heads occurred about two weeks earlier, when they had just emerged from the sun's rays, and the greatest splendor of their tails occurred a little before that, when they were even nearer to the sun. The head of the first of these comets, according to the observations of [Johann Baptist] Cysat, seemed on 1 December to be greater than stars of the first magnitude, and on 16 December (being now in its perigee) it had failed little in magnitude, but very much in the splendor or clarity of its light. On 7 January Kepler, being uncertain about its head, brought his observing to an end. On 12 December the head of the second of these comets was sighted, and was observed by Flamsteed at a distance of 9° from the sun, a thing which would scarcely have been possible in a star of the third magnitude. On 15 and 17 December it appeared as a star of the third magnitude, since it was diminished by the brightness of clouds near the setting sun. On 26 December, moving with the greatest speed and being almost in its perigee, it was less than the mouth of Pegasus, a star of the third magnitude. On 3 January it appeared as a star of the fourth magnitude, on 9 January as a star of the fifth magnitude, and on 13 January it disappeared from view, as a result of the splendor of the crescent moon. On 25 January it scarcely equaled stars of the seventh magnitude. If equal times are taken on both sides of the perigee (before and after), then the head, being placed at those times in distant regions, ought to have shone with equal brilliance because of its equal distances from the earth, but it appeared brightest in the region [on the side of the perigee] toward the sun and disappeared on the other side of the perigee. Therefore from the great difference of light in these two situations, it is concluded that there is a great nearness of the sun and the comet in the first of these situations. For the light of comets tends to be regular and be greatest when the heads move most swiftly, and accordingly are in their perigees, except insofar as this light becomes greater in the vicinity of the sun.

COROLLARY 1. Therefore comets shine by the sun's light reflected from them.

COROLLARY 2. From what has been said it will also be understood why comets appear so frequently in the region of the sun. If they were visible in the regions far beyond Saturn, they would have to appear more often in the parts of the sky opposite to the sun. For those that were in these parts would be nearer to the earth; and the sun, being in between, would obscure the others. Yet in running through the histories of comets, I have found that four or five times more have been detected in the hemisphere toward the sun than in the opposite hemisphere, besides without doubt not a few others which the sun's light hid from view. Certainly, in their descent to our regions comets neither emit tails nor are so brightly illuminated by the sun that they show themselves to the naked eye so as to be discovered before they are closer to us than Jupiter itself. But by far the greater part of the space described about the sun with so small a radius is situated on the side of the earth that faces the sun, and comets are generally more brightly illuminated in that greater part, since they are much closer to the sun.

COROLLARY 3. Hence also it is manifest that the heavens are lacking in resistance. For the comets, following paths that are oblique and sometimes contrary to the course of the planets, move in all directions very freely and preserve their motions for a very long time even when these are contrary to the course of the planets. Unless I am mistaken, comets are a kind of planet and revolve in their orbits with a continual motion. For there seems to be no foundation for the allegation of some writers, basing their argument on the continual changes of the heads, that comets are meteors. The heads of comets are encompassed with huge atmospheres, and the atmospheres must be denser as one goes lower. Therefore, it is in these clouds, and not in the very bodies of the comets, that those changes are seen. Thus, if the earth were viewed from the planets, it would doubtless shine with the light of its own clouds, and its solid body would be almost hidden beneath the clouds. Thus, the belts of Jupiter are formed in the clouds of that planet, since they change their situation relative to one another, and the solid body of Jupiter is seen with greater difficulty through those clouds. And the bodies of comets must be much more hidden beneath their atmospheres, which are both deeper and thicker.

Proposition 40
Theorem 20

Comets move in conics having their foci in the center of the sun, and by radii drawn to the sun, they describe areas proportional to the times.

This is evident by corol. 1 to prop. 13 of the first book compared with props. 8, 12, and 13 of the third book.

COROLLARY 1. Hence, if comets revolve in orbits, these orbits will be ellipses, and the periodic times will be to the periodic times of the planets as the $\frac{3}{2}$ powers of their principal axes. And therefore comets, for the most part being beyond the planets and on that account describing orbits with greater axes, will revolve more slowly. For example, if the axis of the orbit of a comet were four times greater than the axis of the orbit of Saturn, the time of a revolution of the comet would be to the time of a revolution of Saturn (that is, to 30 years) as $4\sqrt{4}$ (or 8) to 1, and accordingly would be 240 years.

COROLLARY 2. But these orbits will be so close to parabolas that parabolas can be substituted for them without sensible errors.

Corollary 3. And therefore (by book 1, prop. 16, corol. 7) the velocity of every comet will always be to the velocity of any planet, [considered to be] revolving in a circle about the sun, very nearly as the square root of twice the distance of the planet from the center of the sun to the distance of the comet from the center of the sun. Let us take the radius of the earth's orbit (or the greatest semidiameter of the ellipse in which the earth revolves) to be of 100,000,000 parts; then the earth will describe by its mean daily motion 1,720,212 of these parts, and by its hourly motion 71,675½ parts. And therefore the comet, at the same mean distance of the earth from the sun, and having a velocity that is to the velocity of the earth as $\sqrt{2}$ to 1, will describe by its daily motion 2,432,747 of these parts, and by its hourly motion 101,364½ parts. But at greater or smaller distances, both the daily and the hourly motion will be to this daily and hourly motion as the square root of the ratio of the distances inversely, and therefore is given.

Corollary 4. Hence, if the latus rectum of a parabola is four times greater than the radius of the earth's orbit, and if the square of that radius is taken to be 100,000,000 parts, the area that the comet describes each day by a radius drawn to the sun will be 1,216,373½ parts, and in each hour that area will be 50,682¼ parts. But if the latus rectum is greater or smaller in any ratio, then the daily and hourly area will be greater or smaller, as the square root of that ratio.

Lemma 5 *To find a parabolic curve that will pass through any number of given points.*

Let the points be A, B, C, D, E, F, . . . , and from them to any straight line HN, given in position, drop the perpendiculars AH, BI, CK, DL, EM, FN, ...

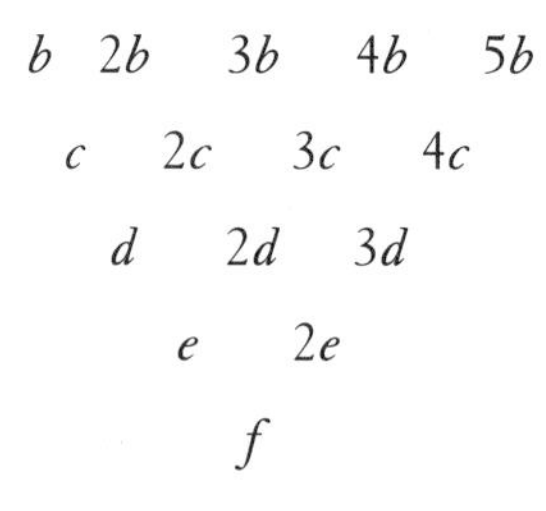

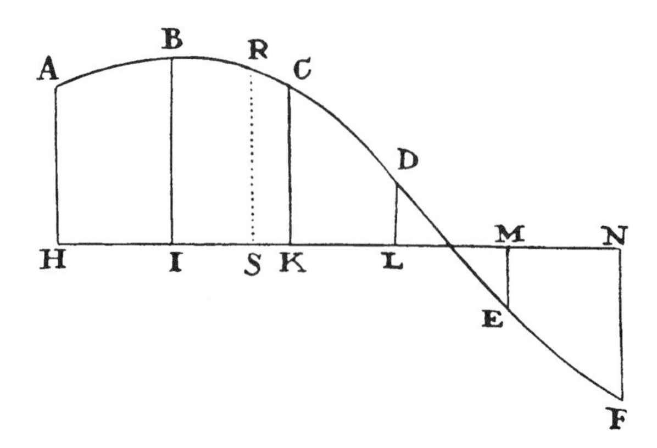

CASE 1. If the intervals HI, IK, KL, ... between the points H, I, K, L, M, N are equal, take the first differences b, b_2, b_3, b_4, b_5, ... of the perpendiculars AH, BI, CK, ...; the second differences c, c_2, c_3, c_4, ...; the third differences d, d_2, d_3, ...; that is, in such a way that $\text{AH} - \text{BI} = b$, $\text{BI} - \text{CK} = b_2$, $\text{CK} - \text{DL} = b_3$, $\text{DL} + \text{EM} = b_4$, $-\text{EM} + \text{FN} = b_5$, ..., then $b - b_2 = c$, ..., and go on in this way to the last difference, which here is f. Then, if any perpendicular RS is erected, which is to be an ordinate to the required curve, in order to find its length, suppose each of the intervals HI, IK, KL, LM, ... to be unity, and let AH be equal to a, $-\text{HS} = p$, $\frac{1}{2}p \times (-\text{IS}) = q$, $\frac{1}{3}q \times (+\text{SK}) = r$, $\frac{1}{4}r \times (+\text{SL}) = s$, $\frac{1}{5}s \times (+\text{SM}) = t$, proceeding, that is, up to the penultimate perpendicular ME, and prefixing negative signs to the terms HS, IS, ..., which lie on the same side of the point S as A, and positive signs to the terms SK, SL, ..., which lie on the other side of the point S. Then if the signs are observed exactly, RS will be $= a + bp + cq + dr + es + ft + \ldots$.

CASE 2. But if the intervals HI, IK, ... between the points H, I, K, L, ... are unequal, take b, b_2, b_3, b_4, b_5, ..., the first differences of the perpendiculars AH, BI, CK, ... divided by the intervals between the perpendiculars; take c, c_2, c_3, c_4, ..., the second differences divided by each two intervals; d, d_2, d_3, ..., the third differences divided by each three intervals; e, e_2, ..., the fourth differences divided by each four intervals, and so on—that is, in such a way that $b = \dfrac{\text{AH} - \text{BI}}{\text{HI}}$, $b_2 = \dfrac{\text{BI} - \text{CK}}{\text{IK}}$, $b_3 = \dfrac{\text{CK} - \text{DL}}{\text{KL}}$, ..., and then $c = \dfrac{b - b_2}{\text{HK}}$, $c_2 = \dfrac{b_2 - b_3}{\text{IL}}$, $c_3 = \dfrac{b_3 - b_4}{\text{KM}}$, ..., and afterward $d = \dfrac{c - c_2}{\text{HL}}$, $d_2 = \dfrac{c_2 - c_3}{\text{IM}}$ .... When these differences have been found, let AH be equal to a, $-\text{HS} = p$, $p \times (-\text{IS}) = q$, $q \times (+\text{SK}) = r$, $r \times (+\text{SL}) = s$, $s \times (+\text{SM}) = t$, proceeding, that is, up to the penultimate perpendicular ME; then the ordinate RS will be $= a + bp + cq + dr + es + ft + \ldots$.

COROLLARY. Hence the areas of all curves can be found very nearly. For if several points are found of any curve which is to be squared [i.e., any curve whose area is desired] and a parabola is understood to be drawn through them, the area of this parabola will be very nearly the same as the area of that curve which is to be squared. Moreover, a parabola can always be squared geometrically by methods which are very well known.

Lemma 6 *From several observed places of a comet, to find its place at any given intermediate time.*

Let HI, IK, KL, LM represent the times between the observations (in the figure to lem. 5), HA, IB, KC, LD, ME five observed longitudes of the comet, and HS the given time between the first observation and the required longitude. Then, if a regular curve ABCDE is understood to be drawn through the points A, B, C, D, E, and if the ordinate RS is found by the above lemma, RS will be the required longitude.

By the same method the latitude at a given time is found from five observed latitudes.

If the differences of the observed longitudes are small, say only 4 or 5 degrees, three or four observations would suffice for finding the new longitude and latitude. But if the differences are greater, say 10 or 20 degrees, five observations must be used.

Lemma 7 *To draw a straight line* BC *through a given point* P, *so that the parts* PB *and* PC *of that line, cut off by two straight lines* AB *and* AC, *given in position, have a given ratio to each other.*

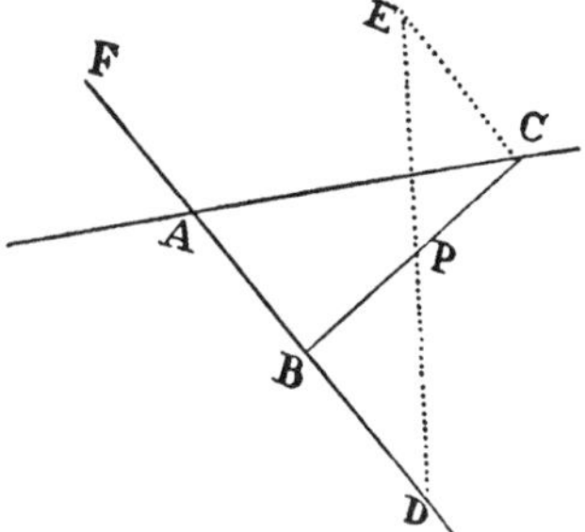

From that point P draw any straight line PD to either of the straight lines, say AB, and produce PD toward the other straight line AC as far as E, so that PE is to PD in the given ratio. Let EC be parallel to AD; and if CPB is drawn, PC will be to PB as PE to PD. Q.E.F.

Lemma 8 *Let* ABC *be a parabola with focus* S. *Let the segment* ABCI *be cut off by the chord* AC *(which is bisected at* I*), let its diameter be* Iμ, *and let its vertex be* μ. *On* Iμ *produced, take* μO *equal to half of* Iμ. *Join* OS *and produce it to* ξ, *so that* Sξ *is equal to* 2SO. *Then, if a comet* B *moves in the arc* CBA, *and if* ξB *is drawn cutting* AC *in* E, *I say that the point* E *will cut off from the chord* AC *the segment* AE *very nearly proportional to the time.*

For join EO, cutting the parabolic arc ABC in Y, and draw μX so as to touch the same arc in the vertex μ and meet EO in X; then the curvilinear area AEXμA will be to the curvilinear area ACYμA as AE to AC. And thus, since triangle ASE is in the same ratio to triangle ASC as the ratio of those curvilinear areas, the total area ASEXμA will be to the total area

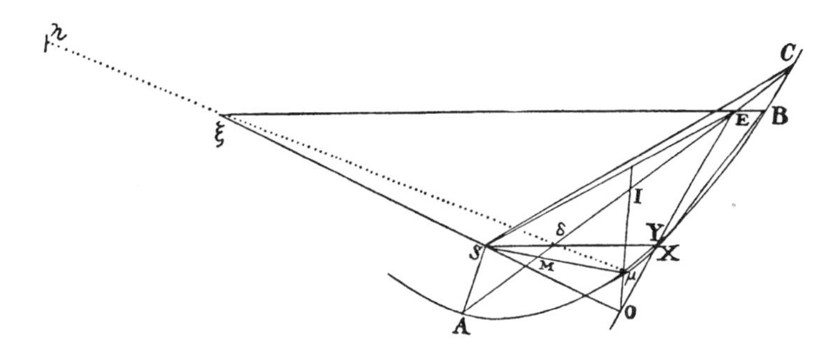

ASCYμA as AE to AC. Moreover, since ξO is to SO as 3 to 1, and EO is in the same ratio to XO, SX will be parallel to EB; and therefore, if BX is joined, the triangle SEB will be equal to the triangle XEB. Thus, if the triangle EXB is added to the area ASEXμA and from that sum the triangle SEB is taken away, there will remain the area ASBXμA equal to the area ASEXμA, and thus it will be to the area ASCYμA as AE to AC. But the area ASBYμA is very nearly equal to the area ASBXμA, and the area ASBYμA is to the area ASCYμA as the time in which the arc AB is described to the time of describing the total arc AC. And thus AE is to AC very nearly in the ratio of the times. Q.E.D.

COROLLARY. When point B falls upon the vertex μ of the parabola, AE is to AC exactly in the ratio of the times.

Scholium

If $\mu\xi$ is joined, cutting AC at δ, and if ξn, which is to μB as 27MI to 16Mμ, is taken in this line, then when Bn is drawn it will cut the chord AC more nearly in the ratio of the lines than before. But the point n is to be taken so as to lie beyond point ξ if point B is more distant than point μ from the principal vertex of the parabola; and contrariwise if B is less distant from that vertex.

Lemma 9

The straight lines Iμ *and* μM *and the length* $\frac{\text{AIC}}{4\text{S}\mu}$ *are equal to one another.*

For 4Sμ is the latus rectum of a parabola, extending to the vertex μ.

Lemma 10

Let Sμ *be produced to* N *and* P, *so that* μN *is one-third of* μI, *and so that* SP *is to* SN *as* SN *to* Sμ. *Then, in the time in which a comet describes the arc* AμC, *it would—if it moved forward always with the velocity that it has at a height equal to* SP*—describe a length equal to the chord* AC.

For if the comet were to move forward in the same time uniformly in the straight line that touches the parabola at μ, and with the velocity that it has in μ, then the area that it would describe by a radius drawn to point S would be equal to the parabolic area ASCμ. And hence the space determined by the length described along the tangent and the length Sμ would be to the space determined by the lengths AC and SM as the area ASCμ to the triangle ASC, that is, as SN to SM. Therefore, AC is to the length described along the tangent as Sμ to SN. But the velocity of the comet at the height SP is (by book 1, prop. 16, corol. 6) to its velocity at the height Sμ as the square root of the ratio of SP to Sμ inversely, that is, in the ratio of Sμ to SN; hence the length described in the same time with this velocity will be to the length described along the tangent as Sμ to SN. Therefore, since AC and the length described with this new velocity are in the same ratio to the length described along the tangent, they are equal to each other. Q.E.D.

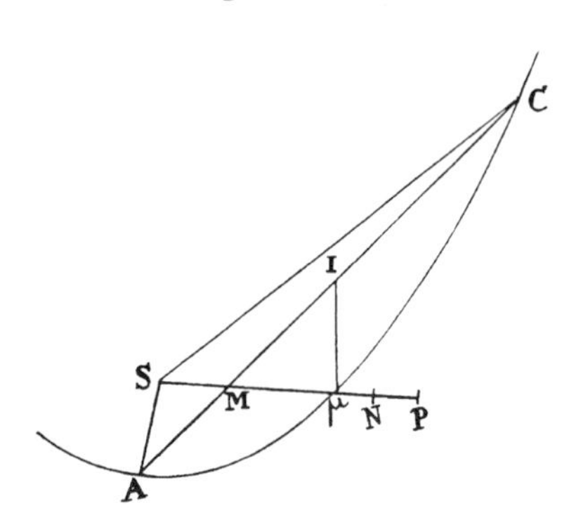

COROLLARY. Therefore, in that same time, the comet, with the velocity that it has at the height $S\mu + \frac{2}{3}I\mu$, would describe the chord AC very nearly.

Lemma 11 *Suppose a comet, deprived of all motion, to be let fall from the height* SN *or* $S\mu + \frac{1}{3}I\mu$, *so as to fall toward the sun, and suppose this comet to be urged toward the sun always by that force, uniformly continued, by which it is urged at the beginning. Then in half of the time in which the comet describes the arc* AC *in its orbit, it would—in this descent toward the sun—describe a space equal to the length* Iμ.

For by lem. 10, in the same time in which the comet describes the parabolic arc AC, it will—with the velocity that it has at the height SP—describe the chord AC; and hence (by book 1, prop. 16, corol. 7), revolving by the force of its own gravity, it would—in that same time, in a circle whose semidiameter was SP—describe an arc whose length would be to the chord AC of the parabolic arc in the ratio of 1 to $\sqrt{2}$. And therefore, falling from the height SP toward the sun with the weight that it has toward the sun at that height, it would in half that time (by book 1, prop. 4, corol. 9) describe a space equal to the square of half of that chord, divided by four times the height SP, that is, the space $\frac{AI^2}{4SP}$. Thus, since the weight of the comet toward the sun at the height SN is to its weight toward the sun at the height SP as SP to Sμ, the comet—falling toward the sun with the weight that it has at the height SN—will in the same time describe the space $\frac{AI^2}{4S\mu}$, that is, a space equal to the length Iμ or Mμ. Q.E.D.

Proposition 41
Problem 21

To determine the trajectory of a comet moving in a parabola, from three given observations.

Having tried many approaches to this exceedingly difficult problem, I devised certain problems [i.e., propositions] in book 1 which are intended for its solution. But later on, I conceived the following slightly simpler solution.

Let three observations be chosen, distant from one another by nearly equal intervals of time. But let that interval of time when the comet moves more slowly be a little greater than the other, that is, so that the difference of the times is to the sum of the times as the sum of the times to more or less six hundred days, or so that the point E (in the figure to lem. 8) falls very nearly on the point M and deviates from there toward I rather than toward A. If such observations are not at hand, a new place of the comet must be found by the method of lem. 6.

Let S represent the sun; T, t, and τ three places of the earth in its orbit; TA, tB, and τC three observed longitudes of the comet; V the time between the first observation and the second; W the time between the second and the third; X the length that the comet could describe in that total time [V + W] with the velocity that it has in the mean distance of the earth from the sun (which length is to be found by the method of book 3, prop. 40, corol. 3); and let tV be a perpendicular to the chord Tτ. In the mean observed longitude tB, let the point B be taken anywhere at all for the place of the comet in the plane of the ecliptic, and from there toward the sun S draw line BE so as to be to the sagitta tV as the content[a] of SB and St^2 to the cube of the hypotenuse of the right-angled triangle whose sides are SB and the tangent of the latitude of the comet in the second observation to the radius tB. And

a. Here "content" has the sense of Newton's "solid," that is, the result of multiplying SB by the square of St.

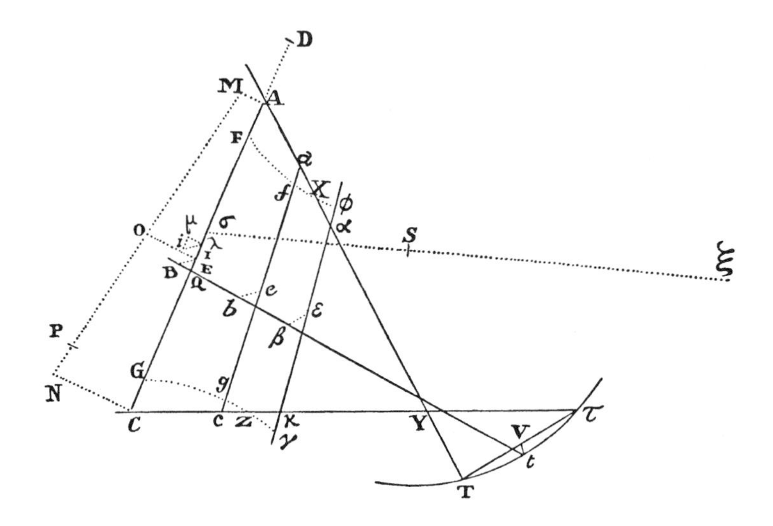

through point E (by lem. 7 of this third book) draw the straight line AEC so that its parts AE and EC, terminated in the straight lines TA and τC, are to each other as the times V and W. Then A and C will be the places of the comet in the plane of the ecliptic in the first and third observations very nearly, provided that B is its correctly assumed place in the second observation.

Upon AC, bisected in I, erect a perpendicular I*i*. Through point B let a line B*i* be imagined,[b] drawn parallel to AC. Let S*i* be a line imagined as cutting AC at λ, and complete the parallelogram $iI\lambda\mu$. Take Iσ equal to 3Iλ, and through the sun S draw the dotted line $\sigma\xi$ equal to $3S\sigma + 3i\lambda$.

b. In prop. 41, Newton refers to the line B*i* as "[lineam] occultam B*i*," directing that "Per punctum B age occultam B*i*," literally, "Through point B draw the occult line B*i*." In the next sentence, the same adjective, "occult," is applied to the line S*i*; in the following sentence, the line $\sigma\xi$ is said to be "occult." (In the first and second editions, there is another "occult" line OD.) In the final paragraph, there is a reference to the "occult" line AC.

In Newton's day, according to the *Oxford English Dictionary*, the adjective "occult" was used to denote "a line drawn in the construction of a figure, but not forming part of the finished figure," and also to denote a dotted line. The diagram for prop. 41 does not show any lines B*i*, S*i*, OD, or AC, but $\sigma\xi$ does appear as a dotted line. Accordingly, we have translated "occult" in the sense of "imagined" in the case of lines S*i*, B*i*, and AC, and as "dotted" in the case of line $\sigma\xi$.

Perhaps the reason why Newton has referred to all these lines (both invisible and dotted) as "occult" is that, in the original diagram that he drew for the cutter of the wood block, he did not show $\sigma\xi$ as a dotted line. In this case, all of these lines would have been "occult" or hidden from view, invisible and only imagined.

And after deleting the letters A, B, C, and I, let a new imagined line BE be drawn from the point B toward the point ξ so that it is to the former line BE as the square of the distance BS to the quantity $S\mu + \frac{1}{3}i\lambda$. And through point E again draw the straight line AEC according to the same rule as before, that is, so that its parts AE and EC are to each other as the times V and W between observations. Then A and C will be the places of the comet more exactly.

Upon AC, bisected in I, erect the perpendiculars AM, CN, and IO, so that, of these perpendiculars, AM and CN are the tangents of the latitudes[c] in the first and third observations (to the radii TA and τC). Join MN, cutting IO in O. Construct the rectangle $iI\lambda\mu$ as before. On IA produced, take ID equal to $S\mu + \frac{2}{3}i\lambda$. Then on MN, toward N, take MP so that it is to the length X found above as the square root of the ratio of the mean distance of the earth from the sun (or of the semidiameter of the earth's orbit) to the distance OD. If point P falls upon point N, then A, B, and C will be three places of the comet, through which its orbit is to be described in the plane of the ecliptic. But if point P does not fall upon point N, then on the straight line AC take CG equal to NP, in such a way that points G and P lie on the same side of the straight line NC.

Using the same method by which points E, A, C, and G were found from the assumed point B, find from other points b and β (assumed in any way whatever) the new points e, a, c, g, and ε, α, κ, γ. Then if the circumference of circle $Gg\gamma$ is drawn through G, g, and γ, cutting the straight line τC in Z, Z will be a place of the comet in the plane of the ecliptic. And if on AC, ac, and $\alpha\kappa$, there are taken AF, af, and $\alpha\varphi$, equal respectively to CG, cg, and $\kappa\gamma$, and if the circumference of a circle $Ff\varphi$ is drawn through points

c. Basically, here Newton is determining a comet's distance from its latitude and longitude as determined by a terrestrial observer. He guesses a position B of the comet in the plane of the ecliptic and then determines the altitude (or distance above the ecliptic), and so can construct a right triangle, of which one side is SB (a line drawn from the sun to the point B on the ecliptic) and other is tB times the tangent of the latitude of the comet (which Newton writes as: the tangent of the latitude of the comet in the second observation to the radius tB).

For an extensive gloss on prop. 41 and on Newton's theory of comets, see A. N. Kriloff, "On Sir Isaac Newton's Method of Determining the Parabolic Orbit of a Comet," *Monthly Notices of the Royal Astronomical Society* 85 (1925): 640–656; see also the enlightening discussion by S. Chandrasekhar, *Newton's "Principia" for the Common Reader* (Oxford: Clarendon Press, 1995), pp. 514–529, especially the diagram on p. 514.

F, f, and φ, cutting the straight line AT in X, then point X will be another place of the comet in the plane of the ecliptic. At the points X and Z, erect the tangents of the latitudes of the comet (to the radii TX and τZ), and two places of the comet in its orbit will be found. Finally (by book 1, prop. 19), let a parabola with focus S be described through those two places; this parabola will be the trajectory of the comet. Q.E.I.

The demonstration of this construction follows from the lemmas, since the straight line AC is cut in E in the ratio of the times, by lem. 7, as required by lem. 8; and since BE, by lem. 11, is that part of the straight line BS or Bξ which lies in the plane of the ecliptic between the arc ABC and the chord AEC; and since MP (by lem. 10, corol.) is the length of the chord of the arc that the comet must describe in its orbit between the first observation and the third, and therefore would be equal to MN, provided that B is a true place of the comet in the plane of the ecliptic.

But it is best not to choose the points B, b, and β any place whatever, but to take them as close to true as possible. If the angle AQt, at which the projection of the orbit described in the plane of the ecliptic cuts the straight line tB, is known approximately, imagine the straight line AC drawn at that angle so that it is to $\frac{4}{3}$Tτ as the square root of the ratio of SQ to St. And by drawing the straight line SEB, so that its part EB is equal to the length Vt, point B will be determined, which may be used the first time around. Then, after deleting the straight line AC and drawing AC anew according to the preceding construction, and after additionally finding the length MP, take point b on tB according to the rule that if TA and τC cut each other in Y, the distance Yb is to the distance YB in a ratio compounded of the ratio of MP to MN and the square root of the ratio of SB to Sb. And the third point β will have to be found by the same method, if it is desired to repeat the operation for the third time. But by this method two operations would, for the most part, be sufficient. For if the distance Bb happens to be very small, then after the points F, f and G, g have been found, the straight lines Ff and Gg (when drawn) will cut TA and τC in the required points X and Z.

EXAMPLE. Let the comet of 1680 be proposed as the example. The following table shows its motion as observed by Flamsteed and as calculated by him from these observations, and corrected by Halley on the basis of the same observations.

	Apparent time	*True time*	*Longitude of the sun*	*Longitude of the comet*	*North latitude of the comet*
	h m	h m s	° ′ ″	° ′ ″	° ′ ″
1680, Dec. 12	4 46	4 46 0	♑ 1 51 23	♑ 6 32 30	8 28 0
21	6 32½	6 36 59	11 6 44	♒ 5 8 12	21 42 13
24	6 12	6 17 52	14 9 26	18 49 23	25 23 5
26	5 14	5 20 44	16 9 22	28 24 13	27 0 52
29	7 55	8 3 2	19 19 43	♓ 13 10 41	28 9 58
30	8 2	8 10 26	20 21 9	17 38 20	28 11 53
1681, Jan. 5	5 51	6 1 38	26 22 18	♈ 8 48 53	26 15 7
9	6 49	7 0 53	♒ 0 29 2	18 44 4	24 11 56
10	5 54	6 6 10	1 27 43	20 40 50	23 43 52
13	6 56	7 8 55	4 33 20	25 59 48	22 17 28
25	7 44	7 58 42	16 45 36	♉ 9 35 0	17 56 30
30	8 7	8 21 53	21 49 58	13 19 51	16 42 18
Feb. 2	6 20	6 34 51	24 46 59	15 13 53	16 4 1
5	6 50	7 4 41	27 49 51	16 59 6	15 27 3

To these add certain observations of my own.

	Apparent time	*Longitude of the comet*	*North latitude of the comet*
	h m	° ′ ″	° ′ ″
1681, Feb. 25	8 30	♉ 26 18 35	12 46 46
27	8 15	27 4 30	12 36 12
Mar. 1	11 0	27 52 42	12 23 40
2	8 0	28 12 48	12 19 38
5	11 30	29 18 0	12 3 16
7	9 30	♊ 0 4 0	11 57 0
9	8 30	0 43 4	11 45 52

These observations were made with a seven-foot telescope, and a micrometer the threads of which were placed in the focus of the telescope; and with these instruments we determined both the positions of the fixed stars in relation to one another and the positions of the comet in relation to the fixed stars. Let A represent the star of the fourth magnitude in the left heel of Perseus (Bayer's o), B the following star of the third magnitude in the left foot (Bayer's ζ), C the star of the sixth magnitude in the heel of the same

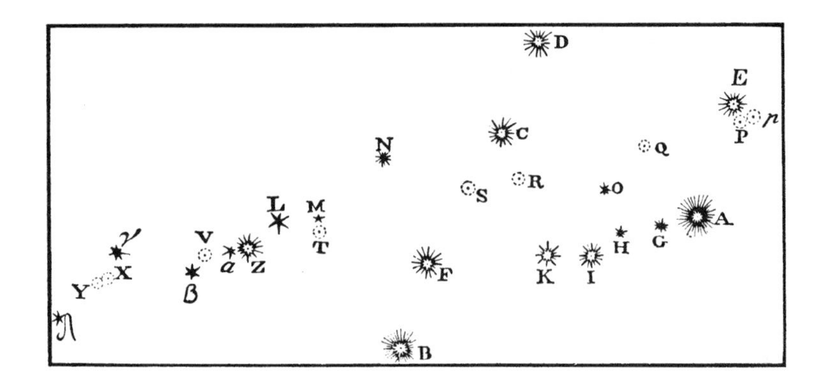

foot (Bayer's *n*), and D, E, F, G, H, I, K, L, M, N, O, Z, α, β, γ, and δ other smaller stars in the same foot. And let *p*, P, Q, R, S, T, V, and X be the places of the comet in the observations described above; and, the distance AB being reckoned at $80\frac{7}{12}$ parts, AC was $52\frac{1}{4}$ parts, BC $58\frac{5}{6}$, AD $57\frac{5}{12}$, BD $82\frac{6}{11}$, CD $23\frac{2}{3}$, AE $29\frac{4}{7}$, CE $57\frac{1}{2}$, DE $49\frac{11}{12}$, AI $27\frac{7}{12}$, BI $52\frac{1}{6}$, CI $36\frac{7}{12}$, DI $53\frac{5}{11}$, AK $38\frac{2}{3}$, BK 43, CK $31\frac{5}{9}$, FK 29, FB 23, FC $36\frac{1}{4}$, AH $18\frac{6}{7}$, DH $50\frac{7}{8}$, BN $46\frac{5}{12}$, CN $31\frac{1}{3}$, BL $45\frac{5}{12}$, NL $31\frac{5}{7}$. HO was to HI as 7 to 6 and, when produced, passed between stars D and E in such a way that the distance of star D from this straight line was $\frac{1}{6}$CD. LM was to LN as 2 to 9 and, when produced, passed through star H. Thus the positions of the fixed stars in relation to one another were determined.

Finally our fellow countryman Pound again observed the positions of these fixed stars in relation to one another and recorded their longitudes and latitudes, as in the following table.

The fixed stars	*Longitudes*	*Latitudes north*	*The fixed stars*	*Longitudes*	*Latitudes north*
	° ′ ″	° ′ ″		° ′ ″	° ′ ″
A	♉ 26 41 50	12 8 36	L	♉ 29 33 34	12 7 48
B	28 40 23	11 17 54	M	29 18 54	12 7 20
C	27 58 30	12 40 25	N	28 48 29	12 31 9
E	26 27 17	12 52 7	Z	29 44 48	11 57 13
F	28 28 37	11 52 22	α	29 52 3	11 55 48
G	26 56 8	12 4 58	β	♊ 0 8 23	11 48 56
H	27 11 45	12 2 1	γ	0 40 10	11 55 18
I	27 25 2	11 53 11	δ	1 3 20	11 30 42
K	27 42 7	11 53 26			

I observed the positions of the comet in relation to these fixed stars as follows.

On Friday, 25 February (O.S.), at 8^h30^m P.M., the distance of the comet, which was at *p*, from star E was less than $\frac{3}{13}$AE, and greater than $\frac{1}{5}$AE, and thus was approximately $\frac{3}{14}$AE; and the angle A*p*E was somewhat obtuse, but almost a right angle. For if a perpendicular were dropped from A to *p*E, the distance of the comet from that perpendicular was $\frac{1}{5}$*p*E.

On the same night at 9^h30^m, the distance of the comet (which was at P) from star E was greater than $\frac{1}{4\frac{1}{2}}$AE and less than $\frac{1}{5\frac{1}{4}}$AE, and thus was very nearly $\frac{1}{4\frac{7}{8}}$AE, or $\frac{8}{39}$AE. And the distance of the comet from a perpendicular dropped from star A to the straight line PE was $\frac{4}{5}$PE.

On Sunday, 27 February, at 8^h15^m P.M., the distance of the comet (which was at Q) from star O equaled the distance between stars O and H; and the straight line QO, produced, passed between stars K and B. Because of intervening clouds, I could not determine the position of this straight line more exactly.

On Tuesday, 1 March, at 11^h P.M., the comet (which was at R) lay exactly between stars K and C; and the part CR of the straight line CRK was a little greater than $\frac{1}{3}$CK and a little smaller than $\frac{1}{3}$CK + $\frac{1}{8}$CR, and thus was equal to $\frac{1}{3}$CK + $\frac{1}{16}$CR, or $\frac{16}{45}$CK.

On Wednesday, 2 March, at 8^h P.M., the distance of the comet (which was at S) from star C was very close to $\frac{4}{9}$FC. The distance of star F from the straight line CS, produced, was $\frac{1}{24}$FC, and the distance of star B from that same straight line was five times greater than the distance of star F. Also, the straight line NS, produced, passed between stars H and I and was five or six times nearer to star H than to star I.

On Saturday, 5 March, at 11^h30^m P.M. (when the comet was at T), the straight line MT was equal to $\frac{1}{2}$ML, and the straight line LT, produced, passed between B and F four or five times closer to F than to B, cutting off from BF a fifth or sixth part of it toward F. And MT, produced, passed outside the space BF on the side of star B and was four times closer to star B than to star F. M was a very small star that could scarcely be seen through the telescope, and L was a greater star, of about the eighth magnitude.

On Monday, 7 March, at 9^h30^m P.M. (when the comet was at V), the straight line Vα, produced, passed between B and F, cutting off $\frac{1}{10}$BF from BF on the side of F, and was to the straight line Vβ as 5 to 4. And the distance of the comet from the straight line $\alpha\beta$ was $\frac{1}{2}$Vβ.

On Wednesday, 9 March, at 8^h30^m P.M. (when the comet was at X), the straight line γX was equal to $\frac{1}{4}\gamma\delta$, and a perpendicular dropped from star δ to the straight line γX was $\frac{2}{5}\gamma\delta$.

On the same night at 12^h (when the comet was at Y), the straight line γY was equal to $\frac{1}{3}\gamma\delta$ or a little smaller, say $\frac{5}{16}\gamma\delta$, and a perpendicular dropped from star δ to the straight line γY was equal to about $\frac{1}{6}$ or $\frac{1}{7}\gamma\delta$. But the comet could scarcely be discerned because of its nearness to the horizon, nor could its place be determined so surely as in the preceding observations.

From observations of this sort, by constructions of diagrams, and by calculations, I found the longitudes and latitudes of the comet, and from the corrected places of the fixed stars our fellow countryman Pound corrected the places of the comet, and these corrected places are given above. I used a crudely made micrometer, but nevertheless the errors of longitudes and latitudes (insofar as they come from my observations) scarcely exceed one minute. Moreover, the comet (according to my observations) at the end of its motion began to decline noticeably toward the north from the parallel which it had occupied at the end of February.

Now, in order to determine the orbit of the comet, I selected—from the observations hitherto described—three that Flamsteed made, on 21 December, 5 January, and 25 January. From these observations I found S*t* to be of 9,842.1 parts, and V*t* to be of 455 parts (10,000 such parts being the semidiameter of the earth's orbit). Then for the first operation, assuming *t*B to be of 5,657 parts, I found SB to be of 9,747, BE the first time 412, Sμ 9,503, $i\lambda$ 413; BE the second time 421, OD 10,186, X 8,528.4, MP 8,450, MN 8,475, NP 25. Hence for the second operation I reckoned the distance *tb* to be 5,640. And by this operation I found at last the distance TX to be 4,775 and the distance τZ to be 11,322. In determining the orbit from these distances, I found the descending node in ♋1°53′ and the ascending node in ♑1°53′, and the inclination of its plane to the plane of the ecliptic to be 61°20⅓′. I found that its vertex (or the perihelion of the comet) was 8°38′ distant from the node and was in ♐27°43′ with a latitude 7°34′ S; and that its latus rectum was 236.8, and that the area described each day by a radius drawn to the sun was 93,585, supposing the square of the semidiameter of the earth's orbit to be 100,000,000; and I found that the comet had advanced in this orbit in the order of the signs, and was on 8 December 0^h4^m A.M. in the vertex of the orbit or the perihelion. I made all these determinations graphically by a scale of equal parts and by chords of angles, taken from the table of natural sines, constructing a fairly large diagram, that is, one in which the semidiameter of the earth's orbit (of 10,000 parts) was equal to 16⅓ inches of an English foot.

Finally, in order to establish whether the comet moved truly in the orbit thus found, I calculated—partly by arithmetical and partly by graphical operations—the places of the comet in this orbit at the times of certain observations, as can be seen in the following table.

	Distance of the comet from the sun	*Calculated longitude*	*Calculated latitude*	*Observed longitude*	*Observed latitude*	*Difference in longitude*	*Difference in latitude*
		° ′	° ′	° ′	° ′		
Dec. 12	2,792	♑ 6 32	8 18½	♑ 6 31⅓	8 26	+1	− 7½
29	8,403	♓ 13 13⅔	28 0	♓ 13 11¾	28 10¹⁄₁₂	+2	−10¹⁄₁₂
Feb. 5	16,669	♉ 17 0	15 29⅔	♉ 16 59⅞	15 27⅖	+0	+ 2¼
Mar. 5	21,737	29 19¾	12 4	29 20⁶⁄₇	12 3½	−1	+ ½

[d]Later, our fellow countryman Halley determined the orbit more exactly by an arithmetical calculation than could be done graphically [*lit.* by the descriptions of lines]; and while he kept the place of the nodes in ♋1°53′ and in ♑1°53′ and the inclination of the plane of the orbit to the ecliptic 61°20⅓′, and also the time of the perihelion of the comet 8 December 0^h4^m, he found the distance of the perihelion from the ascending node (measured in the orbit of the comet) to be 9°20′, and the latus rectum of the parabola to be 2,430 parts, the mean distance of the sun from the earth being 100,000 parts. And by making the same kind of arithmetical calculation exactly (using these data), he calculated the places of the comet at the times of the observations, as follows.

dd. The next eight paragraphs (and the included tables) are not present in ed. 1. They were first published in ed. 2 and considerably revised and expanded in ed. 3.

True time	*Distance of the comet from the sun*	*Calculated longitude*	*Calculated latitude*	*Errors in longitude*	*Errors in latitude*
d h m		° ′ ″	° ′ ″	′ ″	′ ″
Dec. 12 4 46	28,028	♑ 6 29 25	8 26 0 N	−3 5	−2 0
21 6 37	61,076	♒ 5 6 30	21 43 20	−1 42	+1 7
24 6 18	70,008	18 48 20	25 22 40	−1 3	−0 25
26 5 21	75,576	28 22 45	27 1 36	−1 28	+0 44
29 8 3	84,021	♓ 13 12 40	28 10 10	+1 59	+0 12
30 8 10	86,661	17 40 5	28 11 20	+1 45	−0 33
Jan. 5 6 1½	101,440	♈ 8 49 49	26 15 15	+0 56	+0 8
9 7 0	110,959	18 44 36	24 12 54	+0 32	+0 58
10 6 6	113,162	20 41 0	23 44 10	+0 10	+0 18
13 7 9	120,000	26 0 21	22 17 30	+0 33	+0 2
25 7 59	145,370	♉ 9 33 40	17 57 55	−1 20	+1 25
30 8 22	155,303	13 17 41	16 42 7	−2 10	−0 11
Feb. 2 6 35	160,951	15 11 11	16 4 15	−2 42	+0 14
5 7 4½	166,686	16 58 25	15 29 13	−0 41	+2 10
25 8 41	202,570	26 15 46	12 48 0	−2 49	+1 14
Mar. 5 11 39	216,205	29 18 35	12 5 40	+0 35	+2 24

This comet also appeared in the preceding November and was observed by Mr. Gottfried Kirch at Coburg in Saxony on the fourth, sixth, and eleventh days of this month (O.S.); and from its positions with respect to the nearest fixed stars (observed with sufficient accuracy, sometimes through a two-foot telescope and sometimes through a ten-foot telescope), from the difference of the longitudes of Coburg and London, eleven degrees, and from the places of the fixed stars observed by our fellow countryman Pound, our own Halley has determined the places of the comet as follows.

On 3 November 17^h2^m, apparent time at London, the comet was in ♌29°51′ with latitude 1°17′45″ N.

On 5 November 15^h58^m, the comet was in ♍3°23′ with latitude 1°6′ N.

On 10 November 16^h31^m, the comet was equally distant from the stars σ and τ (Bayer) of Leo; it had not yet reached the straight line joining these stars, but was not far from it. In Flamsteed's catalog of stars, σ then was in ♍14°15′ with latitude about 1°41′ N, while τ was in ♍17°3½′ with latitude 0°34′ S. And the midpoint between these stars was ♍15°39¼′ with latitude 0°33½′ N. Let the distance of the comet from that straight line be about 10′ or 12′; then the difference of the longitudes of the comet and that midpoint will be 7′, and the difference of the latitudes roughly 7½′. And thus the comet was in ♍15°32′ with roughly latitude 26′ N.

The first observation of the position of the comet in relation to certain small fixed stars was more than exact enough. The second also was exact enough. In the third observation, which was less exact, there could have been an error of six or seven minutes, but hardly a greater one. And the longitude of the comet in the first observation, which was more exact than the others, being computed in the parabolic orbit mentioned above, was ♌29°30′22″, its latitude 1°25′7″ N, and its distance from the sun 115,546.

Further, Halley noted that a remarkable comet had appeared four times at intervals of 575 years—namely, in September after the murder of Julius Caesar; in A.D. 531 in the consulship of Lampadius and Orestes; in February A.D. 1106; and toward the end of 1680—and that this comet had a long and remarkable tail (except that in the year of Caesar's death the tail was less visible because of the inconvenient position of the earth); and he set out to find an elliptical orbit whose major axis would be 1,382,957 parts, the mean distance of the earth from the sun being 10,000 parts, that is, an orbit in which a comet might revolve in 575 years. Then he computed the motion of the comet in this elliptical orbit with the following conditions: the ascending node in ♋2°2′, the inclination of the plane of the orbit to the plane of the ecliptic 61°6′48″, the perihelion of the comet in this plane in ♐22°44′25″, the equated time of the perihelion 7 December 23^h9^m, the distance of the perihelion from the ascending node in the plane of the ecliptic 9°17′35″, and the conjugate axis 18,481.2. The places of this comet, as deduced from observations as well as calculated for this orbit, are displayed in the following table [page 912].

The observations of this comet from beginning to end agree no less with the motion of a comet in the orbit just described than the motions of the planets generally agree with planetary theories, and this agreement provides proof that it was one and the same comet which appeared all this time and that its orbit has been correctly determined here.[d]

[e]In this table we have omitted the observations made on 16, 18, 20, and 23 November as being less exact. Yet the comet was observed at these times also.[e] In fact, [Giuseppe Dionigi] Ponteo and his associates, on 17 November

ee. These two sentences were added in ed. 3.

True time	Observed longitude	Observed north latitude	Calculated longitude	Calculated latitude	Errors in longitude	Errors in latitude
d h m	° ′ ″	° ′ ″	° ′ ″	° ′ ″	′ ″	′ ″
Nov. 3 16 47	♌ 29 51 0	1 17 45	♌ 29 51 22	1 17 32 N	+0 22	−0 13
5 15 37	♍ 3 23 0	1 6 0	♍ 3 24 32	1 6 9	+1 32	+0 9
10 16 18	15 32 0	0 27 0	15 33 2	0 25 7	+1 2	−1 53
16 17 0			♎ 8 16 45	0 53 7 S		
18 21 34			18 52 15	1 26 54		
20 17 0			28 10 36	1 53 35		
23 17 5			♏ 13 22 42	2 29 0		
Dec. 12 4 46	♑ 6 32 30	8 28 0	♑ 6 31 20	8 29 6 N	−1 10	+1 6
21 6 37	♒ 5 8 12	21 42 13	♒ 5 6 14	21 44 42	−1 58	+2 29
24 6 18	18 49 23	25 23 5	18 47 30	25 23 35	−1 53	+0 30
26 5 21	28 24 13	27 0 52	28 21 42	27 2 1	−2 31	+1 9
29 8 3	♓ 13 10 41	28 9 58	♓ 13 11 14	28 10 38	+0 33	+0 40
30 8 10	17 38 0	28 11 53	17 38 27	28 11 37	+0 7	−0 16
Jan. 5 6 1½	♈ 8 48 53	26 15 7	♈ 8 48 51	26 14 57	−0 2	−0 10
9 7 1	18 44 4	24 11 56	18 43 51	24 12 17	−0 13	+0 21
10 6 6	20 40 50	23 43 32	20 40 23	23 43 25	−0 27	−0 7
13 7 9	25 59 48	22 17 28	26 0 8	22 16 32	+0 20	−0 56
25 7 59	♉ 9 35 0	17 56 30	♉ 9 34 11	17 56 6	−0 49	−0 24
30 8 22	13 19 51	16 42 18	13 18 28	16 40 5	−1 23	−2 13
Feb. 2 6 35	15 13 53	16 4 1	15 11 59	16 2 7	−1 54	−1 54
5 7 4½	16 59 6	15 27 3	16 59 17	15 27 0	+0 11	−0 3
25 8 41	26 18 35	12 46 46	26 16 59	12 45 22	−1 36	−1 24
Mar. 1 11 10	27 52 42	12 23 40	27 51 47	12 22 28	−0 55	−1 12
5 11 39	29 18 0	12 3 16	29 20 11	12 2 50	+2 11	−0 26
9 8 38	♊ 0 43 4	11 45 52	♊ 0 42 43	11 45 35	−0 21	−0 17

(O.S.) at 6^h A.M. in Rome, that is, at 5^h10^m London time, using threads applied to the fixed stars, observed the comet in ♎8°30′ with latitude 0°40′ S. Their observations may be found in the treatise which Ponteo published about this comet. [Marco Antonio] Cellio, who was present and sent his own observations in a letter to Mr. Cassini, saw the comet at the same hour in ♎8°30′ with latitude 0°30′ S. At the same hour Gallet in Avignon (that is, at 5^h42^m A.M. London time) saw the comet in ♎8° with null latitude; at which time, according to the theory, the comet was in ♎8°16′45″ with latitude 0°53′7″ S.

On 18 November at 6^h30^m A.M. in Rome (that is, at 5^h40^m London time) Ponteo saw the comet in ♎13°30′ with latitude 1°20′ S; Cellio saw it in

♎13°30′ with latitude 1°00′ S. Moreover, Gallet at 5^h30^m A.M. in Avignon saw the comet in ♎13°00′ with latitude 1°00′ S. And the Reverend Father Ango at the College of La Flèche in France, at 5^h A.M. (that is, at 5^h9^m London time), saw the comet midway between two small stars, of which one is the middle star of three in a straight line in the southern hand of Virgo, Bayer's ψ, and the other is the outermost star of the wing, Bayer's ϑ. Thus the comet was then in ♎12°46′ with latitude 50′ S. On the same day at Boston in New England, at a latitude of 42½°, at 5^h A.M. (that is, 9^h44^m London time), the comet was seen near ♎14° with latitude 1°30′ S, as I was informed by the distinguished Halley.

On 19 November at 4^h30^m A.M. in Cambridge, the comet (according to the observation of a certain young man) was about 2 degrees distant from Spica Virginis toward the northwest. And Spica was in ♎19°23′47″ with latitude 2°1′59″ S. On the same day at 5^h A.M. at Boston in New England, the comet was 1 degree distant from Spica Virginis, the difference of latitudes being 40 minutes. On the same day on the island of Jamaica, the comet was about 1 degree distant from Spica. On the same day Mr. Arthur Storer, at the Patuxent River, near Hunting Creek in Maryland, which borders on Virginia, at latitude 38½°, at 5^h A.M. (that is, 10^h London time), saw the comet above Spica Virginis and almost conjoined with Spica, the distance between them being about ¾ of a degree. And comparing these observations with one another, I gather that at 9^h44^m in London the comet was in ♎18°50′ with latitude roughly 1°25′ S. And by the theory the comet was then in ♎18°52′15″ with latitude 1°26′54″ S.[f]

On 20 November, Mr. Geminiano Montanari, professor of astronomy in Padua, at 6^h A.M. in Venice (that is, 5^h10^m London time), saw the comet in ♎23° with latitude 1°30′ S. On the same day at Boston the comet was distant from Spica by 4 degrees of longitude eastward and so was approximately in ♎23°24′.

On 21 November, Ponteo and his associates at 7^h15^m A.M. observed the comet in ♎27°50′ with latitude 1°16′ S, Cellio in ♎28°, Ango at 5^h A.M. in ♎27°45′, Montanari in ♎27°51′. On the same day on the island of Jamaica the comet was seen near the beginning of Scorpio and had roughly the same

f. Newton referred to the star α Virginis in the constellation Virgo as "spica ♍" and simply as "spica." We have rendered these as "Spica Virginis" and "Spica."

latitude as Spica Virginis, that is, 2°2′. On the same day at 5^h A.M. at Balasore in the East Indies (that is, at 11^h20^m the preceding night, London time) the comet was distant 7°35′ eastward from Spica Virginis. It was in a straight line between Spica and the scale [or pan of the Balance] and so was in ♎26°58′ with latitude roughly 1°11′ S, and after 5 hours and 40 minutes (that is, at 5^h A.M. London time) was in ♎28°12′ with latitutde 1°16′ S. And by the theory the comet was then in ♎28°10′36″ with latitude 1°53′35″ S.

On 22 November, the comet was seen by Montanari in ♏2°33′, while at Boston in New England it appeared in approximately ♏3°, with about the same latitude as before, that is, 1°30′. On the same day at 5^h A.M. at Balasore the comet was observed in ♏1°50′, and so at 5^h A.M. in London the comet was approximately in ♏3°5′. On the same day at London at 6^h30^m A.M. our fellow countryman Hooke saw the comet in approximately ♏3°30′, on a straight line that passes between Spica Virginis and the heart of Leo, not exactly indeed, but deviating a little from that line toward the north. Montanari likewise noted that a line drawn from the comet through Spica passed, on this day and the following days, through the southern side of the heart of Leo, there being a very small interval between the heart of Leo and this line. The straight line passing through the heart of Leo and Spica Virginis cut the ecliptic in ♍3°46′, at an angle of 2°51′. And if the comet had been located in this line in ♏3°, its latitude would have been 2°26′. But since the comet, by the agreement of Hooke and Montanari, was at some distance from this line toward the north, its latitude was a little less. On the 20th, according to the observation of Montanari, its latitude almost equaled the latitude of Spica Virginis and was roughly 1°30′; and by the agreement of Hooke, Montanari, and Ango, the latitude was continually increasing and so now (on the 22d) was sensibly greater than 1°30′. And the mean latitude between the limits now established, 2°26′ and 1°30′, will be roughly 1°58′. The tail of the comet, by the agreement of Hooke and Montanari, was directed toward Spica Virginis, declining somewhat from that star—southward according to Hooke, northward according to Montanari; and so that declination was hardly perceptible, and the tail, being almost parallel to the equator, was deflected somewhat northward from the opposition of the sun.

On 23 November (O.S.) at 5^h A.M. at Nuremberg (that is, at 4^h30^m London time) Mr. [Johann Jacob] Zimmermann saw the comet in ♏8°8′ with latitude 2°31′ S, determining its distances from the fixed stars.

On 24 November before sunrise the comet was seen by Montanari in ♏12°52′ on the northern side of a straight line drawn through the heart of Leo and Spica Virginis, and so had a latitude a little less than 2°38′. This latitude (as we have said), according to the observations of Montanari, Ango, and Hooke, was continually increasing, and so it was now (on the 24th) a little greater than 1°58′, and at its mean magnitude can be taken as 2°18′ without perceptible error. Ponteo and Gallet would have the latitude decreased now, and Cellio and the observer in New England would have it retained at about the same magnitude, namely 1 or 1½ degrees. The observations of Ponteo and Cellio are rather crude, especially those that were made by taking azimuths and altitudes, and so are those of Gallet; better are the ones that were made by means of the positions of the comet in relation to fixed stars by Montanari, Hooke, Ango, and the observer in New England, and sometimes by Ponteo and Cellio. On the same day at 5^h A.M. at Balasore, the comet was observed in ♏11°45′, and so at 5^h A.M. at London it was nearly in ♏13°. And by the theory the comet was at that time in ♏13°22′42″.

On 25 November before sunrise Montanari observed the comet approximately in ♏17¾°. And Cellio observed at the same time that the comet was in a straight line between the bright star in the right thigh of Virgo and the southern scale of Libra, and this straight line cuts the path of the comet in ♏18°36′. And by the theory the comet was at that time approximately in ♏18⅓°.

Therefore these observations agree with the theory insofar as they agree with one another, and by such agreement they prove that it was one and the same comet that appeared in the whole time from the 4th of November to the 9th of March. The trajectory of this comet cut the plane of the ecliptic twice and therefore was not rectilinear. It cut the ecliptic not in opposite parts of the heavens, but at the end of Virgo and at the beginning of Capricorn, at points separated by an interval of about 98 degrees; and thus the course of the comet greatly deviated from a great circle. For in November its course declined by at least 3 degrees from the ecliptic toward the south, and afterward in December verged from the ecliptic 29 degrees toward the north: the two parts of its orbit, in which the comet tended toward the sun and returned from the sun, declining from each other by an apparent angle of more than 30 degrees, as Montanari observed. This comet moved through nine signs, namely from the last degree of Leo to the beginning of Gemini, besides [that part of] the sign of Leo through which it moved before it began to be seen; and there is no other theory according to which a comet may travel over so great a part of the heaven with a motion according to some rule. Its motion was extremely nonuniform. For about the 20th of November it described approximately 5 degrees per day; then, with a retarded motion between 26 November and 12 December, that is, during 15½ days, it described only 40 degrees; and afterward, with its motion accelerated again, it described about 5 degrees per day until its motion began to be retarded again. And the theory that corresponds exactly to so nonuniform a motion through the greatest part of the heavens, and that observes the same laws as the theory of the planets, and that agrees exactly with exact astronomical observations cannot fail to be true.

Furthermore, it seemed appropriate to show the trajectory that the comet described and the actual tail that it projected in different positions, as in the accompanying figure, in the plane of the trajectory; in this figure, ABC denotes the trajectory of the comet, D the sun, DE the axis of the trajectory,

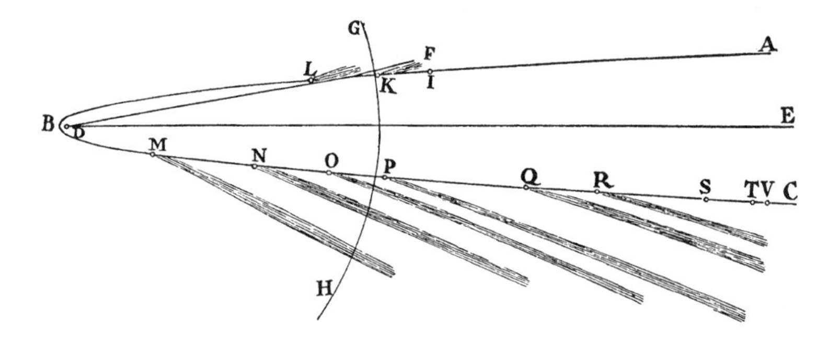

DF the line of nodes, GH the intersection of the sphere of the earth's orbit with the plane of the trajectory, I the place of the comet on 4 November 1680, K its place on 11 November, L its place on 19 November, M its place on 12 December, N its place on 21 December, O its place on 29 December, P its place on 5 January of the following year, Q its place on 25 January, R its place on 5 February, S its place on 25 February, T its place on 5 March, and V its place on 9 March. I used the following observations in determining the tail.

On 4 and 6 November the tail was not yet visible. On 11 November the tail, which had now begun to be seen, was observed through a ten-foot telescope to be no more than half a degree long. On 17 November the tail was observed by Ponteo to be more than 15 degrees long. On 18 November the tail was seen in New England to be 30 degrees long and directly opposite to the sun, and it was extended out to the star ♂ [i.e., the planet Mars], which was then in ♍9°54′. On 19 November, in Maryland, the tail was seen to be 15 or 20 degrees long. On 10 December the tail (according to the observations of Flamsteed) was passing through the middle of the distance between the tail of Serpens (the Serpent of Ophiuchus) and the star δ in the southern wing of Aquila and terminated near the stars A, ω, b in Bayer's tables. Therefore the end of the comet's tail was in ♑19½° with a latitude of about 34¼° N. On 11 December the tail was rising as far as the head of Sagitta (Bayer's α, β), terminating in ♑26°43′, with a latitude of 38°34′ N. On 12 December the tail was passing through the middle of Sagitta and did not extend very much further, terminating in ♒4°, with a latitude of about 42½° N.

These things are to be understood of the length of the brighter part of the tail. For when the light was fainter and the sky perhaps clearer, on 12 December at $5^{h}40^{m}$ in Rome, the tail was observed by Ponteo to extend to 10 degrees beyond the uropygium of Cygnus [i.e., the rump of the Swan], and its side toward the northwest terminated 45 minutes from this star. Moreover, in those days the tail was 3 degrees wide near its upper end, and so the middle of it was 2°15′ distant from that star toward the south, and its upper end was in ♓22° with a latitude of 61° N. And hence the tail was about 70 degrees long.

On 21 December the tail rose almost to Cassiopeia's Chair, being equally distant from β and Schedar [= α Cassiopeiae] and having a distance from each of them equal to their distance from each other, and so terminating in ♈24° with a latitude of 47½°. On 29 December the tail was touching Scheat, which was situated to the left of it, and exactly filled the space between the two stars in the northern foot of Andromeda; it was 54 degrees long; accordingly it terminated in ♉19° with a latitude of 35°. On 5 January the tail touched the star π in the breast of Andromeda on the right side and the star μ in the girdle on the left side, and (according to our observations) was 40 degrees long; but it was curved, and its convex side faced to the south. Near the head of the comet, the tail made an angle of 4 degrees with the circle passing through the sun and the head of the comet; but near the other end, it was inclined to that circle at an angle of 10 or 11 degrees, and the chord of the tail contained an angle of 8 degrees with that circle. On 13 January the tail was visible enough between Alamech and Algol [= β Persei], but it ended in a very faint light toward the star κ in Perseus's side. The distance of the end of the tail from the circle joining the sun and the comet was 3°50′, and the inclination of the chord of the tail to that circle was 8½ degrees. On 25 and 26 January the tail shone with a faint light to a length of 6 or 7 degrees; and, a night or so later, when the sky was extremely clear, it attained a length of 12 degrees and a little more, with a light that was very faint and scarcely to be perceived. But its axis was directed exactly toward the bright star in the eastern shoulder of Auriga, and accordingly declined from the opposition of the sun toward the north at an angle of 10 degrees. Finally on 10 February, my eyes armed [with a telescope], I saw the tail to be 2 degrees long. For the fainter light mentioned above was not visible through the glasses. But Ponteo writes that on 7 February he saw the tail with a length of 12 degrees. On 25 February and thereafter, the comet appeared without a tail.

Whoever considers the orbit just described and turns over in his mind the other phenomena of this comet will without difficulty agree that the bodies of comets are solid, compact, fixed, and durable, like the bodies of planets. For if comets were nothing other than vapors or exhalations of the earth, the sun, and the planets, this one ought to have been dissipated at once during its passage through the vicinity of the sun. For the heat of the sun is as the density of its rays, that is, inversely as the square of the distance of places from the sun. And thus, since the distance of the comet from the center of the sun on 8 December, when it was in its perihelion, was to the distance of the earth from the center of the sun as approximately 6 to 1,000, the heat of the sun on the comet at that time was to the heat of the summer sun here on earth as 1,000,000 to 36, or as 28,000 to 1. But the heat of boiling water is about three times greater than the heat that dry earth acquires in the summer sun, as I have found [by experiment]; and the heat of incandescent iron (if I conjecture correctly) is about three or four times greater than the heat of boiling water; and hence the heat that dry earth on the comet would have received from the sun's rays, when it was in its perihelion, would be about two thousand times greater than the heat of incandescent iron. But with so great a heat, vapors and exhalations, and all volatile matter, would have to have been consumed and dissipated at once.

Therefore the comet, in its perihelion, received an immense heat at [i.e., when near] the sun, and it can retain that heat for a very long time. For a globe of incandescent iron, one inch wide, standing in the air would scarcely lose all its heat in the space of one hour. But a larger globe would preserve its heat for a longer time in the ratio of its diameter, because its surface (which is the measure according to which it is cooled by contact with the surrounding air) is smaller in that ratio with respect to the quantity of hot matter it contains. And so a globe of incandescent iron equal to this earth of ours—that is, more or less 40,000,000 feet wide—would scarcely cool off in as many days, or about 50,000 years. Nevertheless, I suspect that the duration of heat is increased in a smaller ratio than that of the diameter because of some latent causes, and I wish that the true ratio might be investigated by experiments.

Further, it should be noted that in December, when the comet had just become hot at the sun, it was emitting a far larger and more splendid tail than it had done earlier in November, when it had not yet reached its perihelion. And, universally, the greatest and brightest tails all arise from comets immediately after their passage through the region of the sun. Therefore the heating up of the comet is conducive to a great size of its tail, and from this I believe it can be concluded that the tail is nothing other than extremely thin vapor that the head or nucleus of the comet emits by its heat.

There are indeed three opinions about the tails of comets: that the tails are the brightness of the sun's light propagated through the translucent heads of comets; that the tails arise from the refraction of light in its progress from the head of the comet to the earth; and finally that these tails are a cloud or vapor continually rising from the head of the comet and going off in a direction away from the sun. The first opinion is held by those who are not yet instructed in the science of optics. For beams of sunlight are not seen in a dark room except insofar as the light is reflected from particles of dust and smoke always flying about through the air, and for this reason in air darkened with thicker smoke the beams of sunlight appear brighter and strike the eye more strongly, while in clearer air these beams are fainter and are perceived with greater difficulty, but in the heavens, where there is no matter to reflect these beams of sunlight, they cannot be seen at all. Light is not seen insofar as it is in the beam, but only to the degree that it is reflected to our eyes; for vision results only from rays that impinge upon the eyes.

Therefore some reflecting matter must exist in the region of the tail, since otherwise the whole sky, illuminated by the light of the sun, would shine uniformly.

The second opinion is beset with many difficulties. The tails are never variegated in color, and yet colors are generally the inseparable concomitants of refractions. The light of the fixed stars and the planets which is transmitted to us is distinct [i.e., clearly defined]; this demonstrates that the celestial medium is empowered with no refractive force. It is said that the Egyptians sometimes saw the fixed stars surrounded by a head of hair, but this happens very rarely, and so it must be ascribed to some chance refraction by clouds. The radiation and scintillation of the fixed stars also should be referred to refractions both by the eyes and by the tremulous air, since they disappear when these stars are viewed through telescopes. By the tremor of the air and of the ascending vapors it happens that rays are easily turned aside alternately from the narrow space of the pupil of the eye but not at all from the wider aperture of the objective lens of a telescope. Thus it is that scintillation is generated in the former case while it ceases in the latter; and the cessation of scintillation in the latter case demonstrates the regular transmission of light through the heavens without any sensible refraction. And to counter the argument that tails are not generally seen in comets when their light is not strong enough, for the reason that the secondary rays do not then have enough force to affect the eyes, and that this is why the tails of the fixed stars are not seen, it should be pointed out that the light of the fixed stars can be increased more than a hundred times by means of telescopes, and yet no tails are seen. The planets also shine with more light, but they have no tails; and often comets have the greatest tails when the light of their heads is faint and exceedingly dull. For such was the case for the comet of 1680; in December, at a time when the light from its head scarcely equaled stars of the second magnitude, it was emitting a tail of notable splendor as great as 40, 50, 60, or 70 degrees in length and more. Afterward, on 27 and 28 January, the head appeared as a star of only the seventh magnitude, but the tail extended to 6 or 7 degrees in length with a very faint light that was sensible enough; and with a very dim light, which could scarcely be seen, it stretched out as far as 12 degrees or a little further, as was said above. But even on 9 and 10 February, when the head had ceased to be seen by the naked eye, the tail—when I viewed it through a telescope—was 2 degrees long. Further, if the tail arose from refraction by celestial matter, and if it deviated from the opposition of the sun in accordance with the form of the heavens, then, in the same regions of the heavens, that deviation ought always to take place in the same direction. But the comet of 1680, on 28 December at 8^h30^m P.M. London time, was in ♓8°41′ with a latitude of 28°6′ N, the sun being in ♑18°26′. And the comet of 1577, on 29 December, was in ♓8°41′ with a latitude of 28°40′ N, the sun again being in approximately ♑18°26′. In both cases the earth was in the same place and the comet appeared in the same part of the sky; yet in the former case the tail of the comet (according to my observations and those made by others) was declining by an angle of 4½ degrees from the opposition of the sun toward the north, but in the latter case (according to the observations of Tycho) the declination was 21 degrees toward the south. Therefore, since refraction by the heavens has been rejected, the remaining possibility is to derive the phenomena of comets' tails from some matter that reflects light.

Moreover, the laws which the tails of comets observe prove that these tails arise from the heads and ascend into regions turned away from the sun. For example, if the tails lie in planes of the comets' orbits which pass through the sun, they always deviate from being directly opposite the sun and point toward the region which the heads, advancing in those orbits, have left behind. Again, to a spectator placed in those planes, the tails appear in regions directly turned away from the sun; while for observers not in those planes, the deviation gradually begins to be perceived and appears greater from day to day. Furthermore, other things being equal, the deviation is less when the tail is more oblique to the orbit of the comet, and also when the head of the comet approaches closer to the sun, especially if the angle of deviation is taken near the head of the comet. And besides, the tails that do not deviate appear straight, while those that do deviate are curved. Again, this curvature is greater when the deviation is greater, and more sensible when the tail, other things being equal, is longer; for in shorter tails the curvature is scarcely noticed. Then, too, the angle of deviation is smaller near the head of the comet and larger near the other extremity of the tail; and thus the convex side of the tail faces the direction from which the deviation is made and which is along a straight line drawn from the sun through the head of the comet indefinitely. Finally, the tails that are more extended and wider and that shine with a more vigorous light are a little more resplendent on their convex sides and are terminated by a less indistinct limit than on their concave sides. For all these reasons, then, the phenomena of the tail depend on the motion of the head and not on the region of the sky in which the head is seen; and therefore these phenomena do not come about through refraction by the heavens, but arise from the head supplying the matter. For as in our air the smoke of any ignited body seeks to ascend and does so either perpendicularly (if the body is at rest) or obliquely (if the body is moving sideways), so in the heavens, where bodies gravitate toward the sun, smoke and vapors must ascend with respect to the sun (as has already been said) and move upward either directly, if the smoking body is at rest, or obliquely, if the body by advancing always leaves the places from which the higher parts of the vapor have previously ascended. And the swifter the ascent of the vapor, the less the obliquity, namely in the vicinity of the sun and near the smoking body. Moreover, as a result of this difference in obliquity, the column of vapor will be curved; and since the vapor on that side of the column in the direction of the comet's motion is a little more recent [i.e., more recently exhaled], so also the column will be somewhat more dense on that same side, and therefore will reflect light more abundantly and will be terminated by a less indistinct limit. I add nothing here concerning sudden and uncertain agitations of the tails, nor concerning their irregular shapes (which are sometimes described), because either these effects may arise from changes in our air and the motions of the clouds that may obscure those tails in one part or the other; or, perhaps, these effects may arise because some parts of the Milky Way may be confused with the tails as they pass by and may be considered as if they were parts of the tails.

Moreover, the rarity of our own air makes it understandable that vapors sufficient to fill such immense spaces can arise from the atmospheres of comets. For the air near the surface of the earth occupies a space about 850 times greater than water of the same weight, and thus a cylindrical column of air 850 feet high has the same weight as a foot-high column of water of the same width. Further, a column of air rising to the top of our atmosphere is equal in weight to a column of water about 33 feet high; and therefore if the lower part, 850 feet high, of the whole air column is taken away, the remaining upper part will be equal in weight to a column of water 32 feet high. And hence (by a rule confirmed by many experiments, that the compression of air is as the weight of the incumbent atmosphere and that gravity is inversely as the square of the distance of places from the center of the earth), by making a computation using the corollary of prop. 22, book 2, I found that air, at a height above the surface of the earth of one terrestrial semidiameter, is rarer than here on earth in a far greater ratio than that of all space below the orbit of Saturn to a globe described with a diameter of one inch. And thus a globe of our air one inch wide, with the rarity that it would have at the height of one terrestrial semidiameter, would fill all the regions of the planets as far out as the sphere of Saturn and far beyond. Accordingly, since still higher air becomes immensely rare and since the coma[g] or atmosphere of a comet is (as reckoned from the center) about ten times higher than the surface of the nucleus is, and the tail then ascends even higher, the tail will have to be exceedingly rare. And even if, because of the much thicker atmosphere of comets and the great gravitation of bodies toward the sun and the gravitation of the particles of air and vapors toward one another, it can happen that the air in the celestial spaces and in the tails of comets is not so greatly rarefied, it is nevertheless clear from this computation that a very slight quantity of air and vapors is abundantly sufficient to produce all those phenomena of the tails. For the extraordinary rarity of the tails is also evident from the fact that stars shine through them. The terrestrial atmosphere, shining with the light of the sun, by its thickness of only a few miles obscures and utterly extinguishes the light not only of all the stars but also of the moon itself; yet the smallest stars are known to shine, without any loss in their brightness, through the immense thickness of the tails, which are likewise illuminated by the light of the sun. Nor is the brightness of most cometary tails generally greater than that of our air reflecting the light of the sun in a beam, one or two inches wide, let into a dark room.

The space of time in which the vapor ascends from the head to the end of the tail can more or less be found by drawing a straight line from the end of the tail to the sun and noting the place where this straight line cuts the trajectory. For if the vapor has been ascending in a straight line away from the sun, then the vapor that is now in the end of the tail must have begun to ascend from the head at the time when the head was in that place of intersection. But the vapor does not ascend in a straight line away from the sun, but rather ascends obliquely, since the vapor retains the motion of

g. See note a on p. 891 above.

the comet which it had before its ascent and this motion is compounded with the motion of its own ascent. And therefore the solution of the problem will be nearer the true one if the straight line that cuts the orbit is drawn parallel to the length of the tail, or rather (because of the curvilinear motion of the comet) if it diverges from the line of the tail. In this way I found that the vapor that was in the end of the tail on 25 January had begun to ascend from the head before 11 December and thus had spent more than forty-five days in its total ascent. But all of the tail that appeared on 10 December had ascended in the space of those two days that had elapsed after the time of the perihelion of the comet. The vapor, therefore, rose most swiftly at the beginning of its ascent, in the vicinity of the sun, and afterward proceeded to ascend with a motion always retarded by the vapor's own gravity; and as the vapor ascended, it increased the length of the tail. The tail, however, as long as it was visible, consisted of almost all the vapor which had ascended from the comet's head since the time of the comet's perihelion; and that vapor which was the first to ascend, and which composed the end of the tail, did not disappear from view until its distance both from the sun which illuminated it and from our eyes became too great for it to be seen any longer. Hence it happens, also, that in other comets which have short tails, those tails do not rise up with a swift and continual motion from the heads of the comets and soon disappear, but are permanent columns of vapors and exhalations (propagated from the heads by a very slow motion that lasts many days) which, by sharing in the motion that the heads had at the beginning of the exhalations of the vapors, continue to move along through the heavens together with the heads. And hence again it may be concluded that the celestial spaces are lacking in any force of resisting, since in them not only the solid bodies of the planets and comets but also the rarest vapors of the tails move very freely and preserve their extremely swift motions for a very long time.

The ascent of the tails of comets from the atmospheres of the heads and the movement of the tails in directions away from the sun are ascribed by Kepler to the action of rays of light that carry the matter of the tail along with them. And it is not altogether unreasonable to suppose that in very free [or empty] spaces, the extremely thin upper air should yield to the action of the rays, despite the fact that gross substances in the very obstructed regions here on earth cannot be sensibly propelled by the rays of the sun. Someone else believes that there can be particles with the property of levity as well as gravity and that the matter of the tails levitates and through its levitation ascends away from the sun. But since the gravity of terrestrial bodies is as the quantity of matter in the bodies and thus, if the quantity of matter remains constant, cannot be intended and remitted [or increased and decreased], I suspect that this ascent arises rather from the rarefaction of the matter of the tails. Smoke ascends in a chimney by the impulse of the air in which it floats. This air, rarefied by heat, ascends because of its diminished specific gravity and carries along with it the entangled smoke. Why should the tail of a comet not ascend away from the sun in the same manner? For the sun's rays do not act on the mediums through which they pass except in reflection and refraction. The reflecting particles, warmed by this action, will warm the aethereal upper air in which they are entangled. This will become rarefied on account of the heat communicated to it; and because its specific gravity, with which it was formerly tending toward the sun, is diminished by this rarefaction, it will ascend and will carry with it the reflecting particles of which the tail is composed. This ascent of the vapors is also increased by the fact that they revolve about the sun and endeavor by this action to recede from the sun, while the atmosphere of the sun and the matter of the heavens are either completely at rest or revolve more slowly only by the motion that they have received from the rotation of the sun.

These are the causes of the ascent of tails of comets in the vicinity of the sun, where the orbits are more curved, and the comets are within the denser (and, on that account, heavier) atmosphere of the sun and soon emit extremely long tails. For the tails which arise at that point, by conserving their motion and meanwhile gravitating toward the sun, will move about the sun in ellipses as the heads of the comets do; and by that motion they will always accompany the heads and will very freely adhere to them. For the gravity of the vapors toward the sun will no more cause the tails to fall afterward from the heads toward the sun than the gravity of the heads can cause them to fall from the tails. By their common gravity they will either fall simultaneously and together toward the sun or will be simultaneously retarded in their ascent; and therefore this gravity does not hinder the tails and heads of comets from very easily acquiring (whether from the causes already described or any others whatsoever), and afterward very freely preserving, any position in relation to one another.

The tails that are formed when comets are in their perihelia will therefore go off into distant regions together with their heads, and either will return to us from there together with the heads after a long series of years or rather, having been rarefied there, will disappear by degrees. For afterward, in the descent of the heads toward the sun, new little tails should be propagated from the heads with a slow motion, and thereupon should be immeasurably increased in the perihelia of those comets which descend as far as the atmosphere of the sun. For vapor in those very free spaces becomes continually rarefied and dilated. For this reason it happens that every tail at its upper extremity is broader than near the head of the comet. Moreover, it seems reasonable that by this rarefaction the vapor—continually dilated—is finally diffused and scattered throughout the whole heavens, and then is by degrees attracted toward the planets by its gravity and mixed with their atmospheres. For just as the seas are absolutely necessary for the constitution of this earth, so that vapors may be abundantly enough aroused from them by the heat of the sun, which vapors either—being gathered into clouds—fall in rains and irrigate and nourish the whole earth for the propagation of vegetables, or—being condensed in the cold peaks of mountains (as some philosophize with good reason)—run down into springs and rivers; so for the conservation of the seas and fluids on the planets, comets seem to be required, so that from the condensation of their exhalations and vapors, there can be a continual supply and renewal of whatever liquid is consumed by vegetation and putrefaction and converted into dry earth. For all vegetables grow entirely from fluids and afterward, in great part, change into dry earth by putrefaction, and slime is continually deposited from putrefied liquids. Hence the bulk of dry earth is increased from day to day, and fluids—if they did not have an outside source of increase—would have to decrease continually and finally to fail. Further, I suspect that that spirit which is the smallest but most subtle and most excellent part of our air, and which is required for the life of all things, comes chiefly from comets.

In the descent of comets to the sun, their atmospheres are diminished by running out into tails and (certainly in that part which faces toward the sun) are made narrower; and, in turn, when comets are receding from the sun, and when they are now running out less into tails, they become enlarged, if Hevelius has correctly noted their phenomena. Moreover, these atmospheres appear smallest when the heads, after having been heated by the sun, have gone off into the largest and brightest tails, and the nuclei are surrounded in the lowest parts of their atmospheres by smoke possibly coarser and blacker. For all smoke produced by great heat is generally coarser and blacker. Thus, at equal distances from the sun and the earth, the head of the comet which we have been discussing appeared darker after its perihelion than before. For in December it was generally compared to stars of the third magnitude, but in November to stars of the first magnitude and the second magnitude. And those who saw both describe the earlier appearance as a greater comet. For a certain young man of Cambridge, who saw this comet on 19 November, found its light, however leaden and pale, to be equal to Spica Virginis and to shine more brightly than afterward. And on 20 November (O.S.) the comet appeared to Montanari greater than stars of the first magnitude, its tail being 2 degrees long. And Mr. Storer, in a letter that came into our hands, wrote that in December, at a time when the largest and brightest tail was being emitted, the head of the comet was small and in visible magnitude was far inferior to the comet which had appeared in November before sunrise. And he conjectured that the reason for this was that in the beginning the matter of the head was more copious and had been gradually consumed.

It seems to pertain to the same point that the heads of other comets that emitted very large and very bright tails appeared rather dull and very small. For on 5 March 1668 (N.S.) at 7^{h} P.M., the Reverend Father Valentin Stansel, in Brazil, saw a comet very close to the horizon toward the southwest with a very small head that was scarcely visible, but with a tail so shining beyond measure that those who were standing on the shore easily saw its appearance reflected from the sea. In fact it had the appearance of a brilliantly shining torch with a length of 23 degrees, verging from west to south and almost parallel to the horizon. But so great a splendor lasted only three days, decreasing noticeably immediately afterward; and meanwhile, as its splendor was decreasing, the tail was increasing in size. Thus in Portugal the tail is said to have occupied almost a quarter of the sky—that is, 45 degrees—stretched out from west to east with remarkable splendor, and yet not all of the tail was visible, since in those regions the head was always hidden below the horizon. From the increase of the size of the tail and the decrease of the splendor, it is manifest that the head was receding from the sun and had been nearest to the sun at the beginning of its visibility, as was the case for the comet of 1680. And in the *Anglo-Saxon Chronicle*, one reads about a similar comet of 1106, "of which the star was small and dim [h](as was that of 1680),[h] but the splendor that came out of it stretched out extremely bright and like a huge torch toward the northeast" as Hevelius also has it from Simeon the Monk of Durham. This comet appeared at the beginning of February, and thereafter was seen at about evening toward the southwest. And from this and from the position of the tail it is concluded that the head was near the sun. "Its distance from the sun," says Matthew of Paris, "was about one cubit, as from the third hour (more correctly, the sixth) until the ninth hour it emitted a long ray from itself." Such also was that fiery comet described by Aristotle (*Meteor.* 1.6), "whose head, on the first day, was not seen because it had set before the sun, or at least was hidden under the sun's rays; but on the following day, it was seen as much as it could be. For it was distant from the sun by the least possible distance, and soon set. Because of the excessive burning (of the tail, that is), the scattered fire of the head did not yet appear, but as time went on," says Aristotle, "since (the tail) was now flaming less, the comet's own face came back to (the head). And it extended its splendor as far as a third of the sky (that is, to 60 degrees). Moreover, it appeared in the winter (in the 4th year of the 101st Olympiad) and, ascending up to Orion's belt, vanished there."

The comet of 1618 which emerged out of the sun's rays with a very large tail seemed to equal stars of the first magnitude, or even to surpass them a little, but a number of greater comets have appeared which had shorter tails. Some of these are said to have equaled Jupiter, others Venus or even the moon.

We said that comets are a kind of planet revolving about the sun in very eccentric orbits. And just as among the primary planets (which have no tails) those which revolve in smaller orbits closer to the sun are generally smaller, so it seems reasonable also that the comets which approach closer to the sun in their perihelia are for the most part smaller, since otherwise they would act on the sun too much by their attraction. I leave the transverse diameters of the orbits and the periodic times of revolution of the comets to be determined by comparing comets that return in the same orbits after long

hh. The clause in parentheses was added by Newton, as were the following expressions within parentheses.

intervals of time. Meanwhile the following proposition may shed some light on this matter.

Proposition 42
Problem 22

To correct a comet's trajectory that has been found [by the method of prop. 41].

OPERATION 1. Assume the position of the plane of the trajectory, as found by prop. 41, and select three places of the comet which have been determined by very accurate observations and which are as greatly distant from one another as possible; let A be the time between the first and second observations, and B the time between the second and third. The comet should be in its perigee in one of these places, or at least not far from perigee. From these apparent places find, by trigonometric operations, three true places of the comet in that assumed plane of the trajectory. Then through those places thus found, describe a conic about the center of the sun as focus, by arithmetical operations made along the lines of prop. 21, book 1; and let D and E be areas of the conic which are bounded by radii drawn from the sun to those places—namely, D the area between the first and second observations, and E the area between the second and third. And let T be the total time in which the total area D + E should be described by the comet, with the velocity as found by prop. 16, book 1.

OPERATION 2. Let the longitude of the nodes of the plane of the trajectory be increased by adding 20 or 30 minutes (which can be called P) to that longitude; but keep constant the inclination of that plane to the plane of the ecliptic. Then from the three aforesaid observed places of the comet, let three true places of the comet be found in this new plane (as in oper. 1); and also the orbit passing through those places, two of its areas (which can be called d and e) described between observations, and the total time t in which the total area $d + e$ should be described.

OPERATION 3. Keep constant the longitude of the nodes in the first operation, and let the inclination of the plane of the trajectory to the plane of the ecliptic be increased by adding 20 or 30 minutes (which can be called Q) to that inclination. Then from the aforesaid three observed apparent places of the comet, let three true places be found in this new plane; and also the orbit passing through those places, two of its areas (which can be called δ and ϵ) described between observations, and the total time τ in which the total area $\delta + \epsilon$ should be described.

Now take C so as to be to 1 as A to B, and take G to 1 as D to E, and g to 1 as d to e, and γ to 1 as δ to ϵ, and let S be the true time between the first and third observations; and carefully observing the signs + and −, seek the numbers m and n, by the rule that $2G - 2C = mG - mg + nG - n\gamma$; and $2T - 2S = mT - mt + nT - n\tau$. And if, in the first operation, I designates the inclination of the plane of the trajectory to the plane of the ecliptic, and K the longitude of either node, $I + nQ$ will be the true inclination of the plane of the trajectory to the plane of the ecliptic, and $K + mP$ will be the true longitude of the node. And finally if in the first, second, and third operations, the quantities R, r, and ρ designate the latera recta of the trajectory, and the quantities $\frac{1}{L}$, $\frac{1}{l}$, $\frac{1}{\lambda}$ the transverse diameters [or latera transversa] respectively, $R + mr - mR + n\rho - nR$ will be the true latus rectum, and $\frac{1}{L + ml - mL + n\lambda - nL}$ will be the true transverse diameter of the trajectory that the comet describes. And given the transverse diameter, the periodic time of the comet is also given. Q.E.I.

But the periodic times of revolving comets, and the transverse diameters [latera transversa] of their orbits, will by no means be determined exactly enough except by the comparison with one another of comets that appear at diverse times. If several comets are found, after equal intervals of times, to have described the same orbit, it will have to be concluded that all these are one and the same comet revolving in the same orbit. And then finally from the times of their revolutions the transverse diameters of the orbits will be given, and from these diameters the elliptical orbits will be determined.

To this end, therefore, the trajectories of several comets should be calculated on the hypothesis that they are parabolic. For such trajectories will always agree very nearly with the phenomena. This is clear not only from the parabolic trajectory of the comet of 1680, which I compared above with the observations, but also from the trajectory of that remarkable comet which appeared in 1664 and 1665 and was observed by Hevelius. He calculated the longitudes and latitudes of this comet from his own observations, but not very accurately. From the same observations our own Halley calculated the places of this comet anew, and then finally he determined the trajectory of the comet from the places thus calculated. And he found its ascending node in ♊21°13′55″, the inclination of its orbit to the plane of the ecliptic 21°18′40″, the distance of its perihelion from the node in the orbit 49°27′30″. The perihelion in ♌8°40′30″ with heliocentric latitude 16°1′45″ S. The comet in its perihelion on 24 November, 11^h52^m P.M. mean time [*lit.* equated time] at London, or 13^h8^m (O.S.) at Gdansk, and the latus rectum of the parabola 410,286, the mean distance of the earth from the sun being 100,000. How exactly the calculated places of the comet in this orbit agree with the observations will be evident from the following table calculated by Halley [p. 932].

In February, in the beginning of 1665, the first star of Aries, which I shall from here on call γ, was in ♈28°30′15″ with latitude 7°8′58″ N. The second star of Aries was in ♈29°17′18″ with latitude 8°28′16″ N. And a certain other star of the seventh magnitude, which I shall call A, was in ♈28°24′45″ with latitude 8°28′33″ N. And on 7 February at 7^h30^m Paris time (that is, 7 February at 8^h30^m Gdansk time) (O.S.), the comet made a right triangle with those stars γ and A, with the right angle at γ. And the distance of the comet from the star γ was equal to the distance between the stars γ and A, that is, 1°19′46″ along a great circle, and therefore it was 1°20′26″ in the parallel of the latitude of the star γ. Therefore, if the longitude 1°20′26″ is taken away from the longitude of the star γ, there will remain the longitude of the comet ♈27°9′49″. Auzout, who had made this observation, put the comet in roughly ♈27°0′. And from the diagram with which Hooke delineated its motion, it was then in ♈26°59′24″. Taking the mean, I have put it in ♈27°4′46″. From the same observation, Auzout took the latitude of the comet at that time to be 7° and 4′ or 5′ toward the north. He would have put it more correctly at 7°3′29″, since the difference of the latitudes of the comet and of the star γ was equal to the difference of the longitudes of the stars γ and A.

On 22 February at 7^h30^m in London (that is, 22 February at 8^h46^m Gdansk time), the distance of the comet from the star A, according to Hooke's observation (which he himself delineated in a diagram) and also according to Auzout's observations (delineated in a diagram by Petit), was a fifth of the distance between the star A and the first star of Aries, or 15′57″. And the distance of the comet from the line joining the star A and the first star of Aries was a fourth of that same fifth part, that is, 4′. And hence the comet was in ♈28°29′46″ with latitude 8°12′36″ N.

On 1 March at 7^h0^m at London (that is, 1 March at 8^h16^m Gdansk time), the comet was observed near the second star of Aries, the distance between

Apparent time at Gdansk, O.S.	*Observed distances of the comet*		*Observed places*	*Calculated places in the orbit*
d h m		° ′ ″	° ′ ″	° ′ ″
December				
3 18 29½	from the heart of Leo from Spica Virginis	46 24 20 22 52 10	Long. ♎ 7 1 0 Lat. S. 21 39 0	♎ 7 1 29 21 38 50
4 18 1½	from the heart of Leo from Spica Virginis	46 2 45 23 52 40	Long. ♎ 16 15 0 Lat. S. 22 24 0	♎ 6 16 5 22 24 0
7 17 48	from the heart of Leo from Spica Virginis	44 48 0 27 56 40	Long. ♎ 3 6 0 Lat. S. 25 22 0	♎ 3 7 33 25 21 40
17 14 43	from the heart of Leo from the right shoulder of Orion	53 15 15 45 43 30	Long. ♌ 2 56 0 Lat. S. 49 25 0	♌ 2 56 0 49 25 0
19 9 25	from Procyon from the bright star in the jaw of Cetus	35 13 50 52 56 0	Long. ♊ 28 40 30 Lat. S. 45 48 0	♊ 28 43 0 45 46 0
20 9 53½	from Procyon from the bright star in the jaw of Cetus	40 49 0 40 4 0	Long. ♊ 13 3 0 Lat. S. 39 54 0	♊ 13 5 0 39 53 0
21 9 9½	from the right shoulder of Orion from the bright star in the jaw of Cetus	26 21 25 29 28 0	Long. ♊ 2 16 0 Lat. S. 33 41 0	♊ 2 18 30 33 39 40
22 9 0	from the right shoulder of Orion from the bright star in the jaw of Cetus	29 47 0 20 29 30	Long. ♉ 24 24 0 Lat. S. 27 45 0	♉ 24 27 0 27 46 0
26 7 58	from the bright star in Aries from Aldebaran	23 20 0 26 44 0	Long. ♉ 9 0 0 Lat. S. 12 36 0	♉ 9 2 28 12 34 13
27 6 45	from the bright star in Aries from Aldebaran	20 45 0 28 10 0	Long. ♉ 7 5 40 Lat. S. 10 23 0	♉ 7 8 45 10 23 13
28 7 39	from the bright star in Aries from the Hyades	18 29 0 29 37 0	Long. ♉ 5 24 45 Lat. S. 8 22 50	♉ 5 27 52 8 23 37
31 6 45	from the girdle of Andromeda from the Hyades	30 48 10 32 53 30	Long. ♉ 2 7 40 Lat. S. 4 13 0	♉ 2 8 20 4 16 25
Jan. 1665 7 7 37½	from the girdle of Andromeda from the Hyades	25 11 0 37 12 25	Long. ♈ 28 24 47 Lat. N. 0 54 0	♈ 28 24 0 0 53 0
13 7 0	from the head of Andromeda from the Hyades	28 7 10 38 55 20	Long. ♈ 27 6 54 Lat. N. 3 6 50	♈ 27 6 39 3 7 40
24 7 29	from the girdle of Andromeda from the Hyades	20 32 15 40 5 0	Long. ♈ 26 29 15 Lat. N. 5 25 50	♈ 26 28 50 5 26 0
February 7 8 37			Long. ♈ 27 4 46 Lat. N. 7 3 29	♈ 27 24 55 7 3 15
22 8 46			Long. ♈ 28 29 46 Lat. N. 8 12 36	♈ 28 29 58 8 10 25
March 1 8 16			Long. ♈ 29 18 15 Lat. N. 8 36 26	♈ 29 18 20 8 36 12
7 8 37			Long. ♉ 0 2 48 Lat. N. 8 56 30	♉ 0 2 42 8 56 56

them being to the distance between the first and second stars of Aries, that is, to 1°33′, as 4 to 45 according to Hooke, or as 2 to 23 according to [Gilles François] Gottigniez. Accordingly, the distance of the comet from the second star of Aries was 8°16″ according to Hooke, or 8′5″ according to Gottigniez; or, taking the mean, was 8′10″. And according to Gottigniez the comet had now just gone beyond the second star of Aries by about a space of a fourth or a fifth of the course completed in one day, that is, roughly 1′35″ (and Auzout agrees well enough with this), or a little less according to Hooke, say 1′. Therefore, if 1′ is added to the longitude of the first star of Aries, and 8′10″ to its latitude, the longitude of the comet will be found to be ♈29°18′, and its latitude 8°36′26″ N.

On 7 March at 7^h30^m in Paris (that is, 7 March at 8^h37^m Gdansk time), the distance of the comet from the second star of Aries, according to Auzout's observations, was equal to the distance of the second star of Aries from the star A, that is, 52′29″. And the difference between the longitudes of the comet and of the second star of Aries was 45′ or 46′ or, taking the mean, 45′30″. And therefore the comet was in ♉0°2′48″. From the diagram of Auzout's observations that Petit constructed, Hevelius determined the latitude of the comet to be 8°54′. But the engraver curved the path of the comet irregularly toward the end of its motion, and Hevelius corrected the irregular curving in a diagram of Auzout's observations drawn by Hevelius himself, and thus made the latitude of the comet 8°55′30″. And by correcting the irregularity a little more, the latitude can come out to be 8°56′, or 8°57′.

This comet was also seen on 9 March and then must have been located in ♉0°18′ with latitude roughly 9°3½′ N.

This comet was visible for three months in all, during which time it passed through about six signs, completing about 20 degrees in each day. Its path deviated considerably from a great circle, being curved northward; and toward the end, its motion changed from retrograde to direct. And notwithstanding so unusual a path, the theory agrees with the observations from beginning to end no less exactly than theories of the planets tend to agree with observations of them, as will be clear upon examination of the table. Nevertheless, roughly 2 minutes must be subtracted when the comet was swiftest, and this will result by taking away 12 seconds from the angle between the ascending node and the perihelion, or by making that angle 49°27′18″. The annual parallax of each of the two comets (both this one and the previous one) was quite pronounced, and as a result it gave proof of the annual motion of the earth in its orbit.

The theory is confirmed also by the motion of the comet that appeared in 1683. It had a retrograde motion in an orbit whose plane contained almost a right angle with the plane of the ecliptic. Its ascending node (by Halley's calculation) was in ♍23°23′; the inclination of its orbit to the ecliptic 83°11′; its perihelion in ♊25°29′30″; its perihelial distance from the sun 56,020, the radius of the earth's orbit being taken at 100,000, and the time of its perihelion 2 July 3^h50^m. And the places of the comet in this orbit, as calculated by Halley and compared with the places observed by Flamsteed, are displayed in the following table.

1683 Mean [lit. equated] time	*Place of the sun*	*Calculated longitude of the comet*	*Calculated latitude north*	*Observed longitude of the comet*	*Observed latitude north*	*Difference in longitude*	*Difference in latitude*
d h m	° ′ ″	° ′ ″	° ′ ″	° ′ ″	° ′ ″	′ ″	′ ″
July 13 12 55	♌ 1 2 30	♋13 5 42	29 28 13	♋13 6 42	29 28 20	+1 0	+0 7
15 11 15	2 53 12	11 37 48	29 34 0	11 39 43	29 34 50	+1 55	+0 50
17 10 20	4 45 45	10 7 6	29 33 30	10 8 40	29 34 0	+1 34	+0 30
23 13 40	10 38 21	5 10 27	28 51 42	5 11 30	28 50 28	+1 3	−1 14
25 14 5	12 35 28	3 27 53	24 24 47	3 27 0	28 23 40	−0 53	−1 7
31 9 42	18 9 22	♊27 55 3	26 22 52	♊27 54 24	26 22 25	−0 39	−0 27
31 14 55	18 21 53	27 41 7	26 16 57	27 41 8	26 14 50	+0 1	−2 7
Aug. 2 14 56	20 17 16	25 29 32	25 16 19	25 28 46	25 17 28	−0 46	+1 9
4 10 49	22 2 50	23 18 20	24 10 49	23 16 55	24 12 19	−1 25	+1 30
6 10 9	23 56 45	20 42 23	22 47 5	20 40 32	22 49 5	−1 51	+2 0
9 10 26	26 50 52	16 7 57	20 6 37	16 5 55	20 6 10	−2 2	−0 27
15 14 1	♍ 2 47 13	3 30 48	11 37 33	3 26 18	11 32 1	−4 30	−5 32
16 15 10	3 48 2	0 43 7	9 34 16	0 41 55	9 34 13	−1 12	−0 3
18 15 44	5 45 33	♉24 52 53	5 11 15	♉24 49 5	5 9 11	−3 48	−2 4
			South		South		
22 14 44	9 35 49	11 7 14	5 16 53	11 7 12	5 16 50	−0 2	−0 3
23 15 52	10 36 48	7 2 18	8 17 9	7 1 17	8 16 41	−1 1	−0 28
26 16 2	13 31 10	♈24 45 31	16 38 0	♈24 44 0	16 38 20	−1 31	+0 20

The theory is confirmed also by the motion of the retrograde comet that appeared in 1682. Its ascending node (by Halley's calculation) was in ♉21°16′30″. The inclination of the orbit to the plane of the ecliptic 17°56′0″. Its perihelion in ♒2°52′50″. Its perihelial distance from the sun 58,328, the radius of the earth's orbit being 100,000. And the perihelion 4 September

7^h39^m mean [*lit.* equated] time. And the places calculated from Flamsteed's observations and compared with the places calculated by the theory are shown in the following table.

1682 Apparent time	*Place of the sun*	*Calculated longitude of the comet*	*Calculated latitude north*	*Observed longitude of the comet*	*Observed latitude north*	*Difference in longitude*	*Difference in latitude*
d h m	° ′ ″	° ′ ″	° ′ ″	° ′ ″	° ′ ″	′ ″	′ ″
Aug. 19 16 38	♍ 7 0 7	♌ 18 14 28	25 50 7	♌ 18 14 40	25 49 55	−0 12	+0 12
20 15 38	7 55 52	24 46 23	26 14 42	24 46 22	26 12 52	+0 1	+1 50
21 8 21	8 36 14	29 37 15	26 20 3	29 38 2	26 17 37	−0 47	+2 26
22 8 8	9 33 55	♍ 6 29 53	26 8 42	♍ 6 30 3	26 7 12	−0 10	+1 30
29 8 20	16 22 40	♎ 12 37 54	18 37 47	♎ 12 37 49	18 34 5	+0 5	+3 42
30 7 45	17 19 41	15 36 1	17 26 43	15 35 18	17 27 17	+0 43	−0 34
Sept. 1 7 33	19 16 9	20 30 53	15 13 0	20 27 4	15 9 49	+3 49	+3 11
4 7 22	22 11 28	25 42 0	12 23 48	25 40 58	12 22 0	+1 2	+1 48
5 7 32	23 10 29	27 0 46	11 33 8	26 59 24	11 33 51	+1 22	−0 43
8 7 16	26 5 58	29 58 44	9 26 46	29 58 45	9 26 43	−0 1	+0 3
9 7 26	27 5 9	♏ 0 44 10	8 49 10	♏ 0 44 4	8 48 25	+0 6	+0 45

[a]The theory is confirmed also by the retrograde motion of the comet that appeared in 1723. Its ascending node (by the calculation of Mr. Bradley, Savilian professor of astronomy at Oxford) was in ♈ 14°16′. The inclination of the orbit to the plane of the ecliptic 49°59′. Its perihelion in ♉ 12°15′20″. Its perihelial distance from the sun 998,651, the radius of the earth's orbit being 1,000,000, and the mean [*lit.* equated] time of the perihelion being 16 September 16^h10^m. And the places of the comet in this orbit, as calculated by Bradley and compared with the places observed by himself and his uncle Mr. Pound, and by Mr. Halley, are shown in the following table.[a]

By these examples it is more than sufficiently evident that the motions of comets are no less exactly represented by the theory that we have set forth than the motions of planets are generally represented by planetary theories. And therefore the orbits of comets can be calculated by this theory, and the periodic time of a comet revolving in any orbit whatever can then be determined, and finally the transverse diameters [*lit.* latera transversa] of their elliptical orbits and their aphelian distances will become known.

The retrograde comet that appeared in 1607 described an orbit whose ascending node (according to Halley's calculation) was in ♉ 20°21′; the in-

aa. This paragraph and the accompanying table (on p. 936) appeared for the first time in ed. 3.

1723 Mean [lit. equated] time	Observed longitude of the comet	Observed latitude north	Calculated longitude of the comet	Calculated latitude north	Difference in longitude	Difference in latitude
d h m	° ′ ″	° ′ ″	° ′ ″	° ′ ″	″	″
Oct. 9 8 5	♒7 22 15	5 2 0	♒7 21 26	5 2 47	+49	−47
10 6 21	6 41 12	7 44 13	6 41 42	7 43 18	−50	+55
12 7 22	5 39 58	11 55 0	5 40 19	11 54 55	−21	+ 5
14 8 57	4 59 49	14 43 50	5 0 37	14 44 1	−48	−11
15 6 35	4 47 41	15 40 51	4 47 45	15 40 55	− 4	− 4
21 6 22	4 2 32	19 41 49	4 2 21	19 42 3	+11	−14
22 6 24	3 59 2	20 8 12	3 59 10	20 8 17	− 8	− 5
24 8 2	3 55 29	20 55 18	3 55 11	20 55 9	+18	+ 9
29 8 56	3 56 17	22 20 27	3 56 42	22 20 10	−25	+17
30 6 20	3 58 9	22 32 28	3 58 17	22 32 12	− 8	+16
Nov. 5 5 53	4 16 30	23 38 33	4 16 23	23 38 7	+ 7	+26
8 7 6	4 29 36	24 4 30	4 29 54	24 4 40	−18	−10
14 6 20	5 2 16	24 48 46	5 2 51	24 48 16	−35	+30
20 7 45	5 42 20	25 24 45	5 43 13	25 25 17	−53	−32
Dec. 7 6 45	8 4 13	26 54 18	8 3 55	26 53 42	+18	+36

clination of the plane of its orbit to the plane of the ecliptic was 17°2′; its perihelion was in ♒2°16′; and its perihelial distance from the sun was 58,680, the radius of the earth's orbit being 100,000. And the comet was in its perihelion on 16 October at 3^h50^m. This orbit agrees very closely with the orbit of the comet that was seen in 1682. If these two comets should be one and the same, this comet will revolve in a space of seventy-five years and the major axis of its orbit will be to the major axis of the earth's orbit as $\sqrt[3]{(75 \times 75)}$ to 1, or roughly 1,778 to 100. And the aphelial distance of this comet from the sun will be to the mean distance of the earth from the sun as roughly 35 to 1. And once these quantities are known, it will not be at all difficult to determine the elliptical orbit of this comet. What has just been said will be found to be true if the comet returns hereafter in this orbit in a space of seventy-five years. The other comets seem to revolve in a greater time and to ascend higher.

But because of the great number of comets, and the great distance of their aphelia from the sun, and the long time that they spend in their aphelia, they should be disturbed somewhat by their gravities toward one another, and hence their eccentricities and times of revolutions ought sometimes to be increased a little and sometimes decreased a little. Accordingly, it is not to be expected that the same comet will return exactly in the same orbit, and with the same periodic times. It is sufficient if no greater changes are found to occur than those that arise from the above-mentioned causes.

And hence a reason appears why comets are not restricted to the zodiac as planets are, but depart from there and are carried with various motions into all regions of the heavens—namely, for this purpose, that in their aphelia, when they move most slowly, they may be as far distant from one another as possible and may attract one another as little as possible. And this is the reason why comets that descend the lowest, and so move most slowly in their aphelia, should also ascend to the greatest heights.

The comet that appeared in 1680 was distant from the sun in its perihelion by less than a sixth of the sun's diameter; and because its velocity was greatest in that region and also because the atmosphere of the sun has some density, the comet must have encountered some resistance and must have been somewhat slowed down and must have approached closer to the sun; and by approaching closer to the sun in every revolution, it will at length fall into the body of the sun. But also, in its aphelion, when it moves most slowly, the comet can sometimes be slowed down by the attraction of other comets and as a result fall into the sun. So also fixed stars, which are exhausted bit by bit in the exhalation of light and vapors, can be renewed by comets falling into them and then, kindled by their new nourishment, can be taken for new stars. Of this sort are those fixed stars that appear all of a sudden, and that at first shine with maximum brilliance and subsequently disappear little by little. Of such sort was the star that Cornelius Gemma saw in Cassiopeia's Chair on 9 November 1572; it was shining brighter than all the fixed stars, scarcely inferior to Venus in its brilliance. But he did not see it at all on 8 November, when he was surveying that part of the sky on a clear night. Tycho Brahe saw this same star on the 11th of that month, when it shone with the greatest splendor; and he observed it decreasing little by little after that time, and he saw it disappearing after the space of sixteen months. In November, when it first appeared, it equaled Venus in brightness. In December, somewhat diminished, it equaled Jupiter. In January 1573 it was less than Jupiter and greater than Sirius, and it became equal to Sirius at the end of February and the beginning of March. In April and May it was equal to stars of the second magnitude; in June, July, and August, to stars of the third magnitude; in September, October, and November, to stars of the fourth magnitude; in December and in January 1574, to stars of the fifth magnitude; and in February, to stars of the sixth magnitude; and in March, it vanished from sight. Its color at the start was clear, whitish, and bright; afterward it became yellowish, and in March of 1573 reddish like Mars or the star Aldebaran, while in May it took on a livid whiteness such as we see in Saturn, and it maintained this color up to the end, yet all the while becoming fainter. Such also was the star in the right foot of Serpentarius, the beginning of whose visibility was observed by the pupils of Kepler in 1604, on 30 September (O.S.); they saw it exceeding Jupiter in its light, although it had not been visible at all on the preceding night. And from that time it decreased little by little and in the space of fifteen or sixteen months vanished from sight. It was when such a new star appeared shining beyond measure that Hipparchus is said to have been stimulated to observe the fixed stars and to put them into a catalog. But fixed stars that alternately appear and disappear, and increase little by little, and are hardly ever brighter than fixed stars of the third magnitude, seem to be of another kind and, in revolving, seem to show alternately a bright side and a dark side. And the vapors that arise from the sun and the fixed stars and the tails of comets can fall by their gravity into the atmospheres of the planets and there be condensed and converted into water and humid spirits, and then—by a slow heat—be transformed gradually into salts, sulphurs, tinctures, slime, mud, clay, sand, stones, corals, and other earthy substances.

GENERAL SCHOLIUM[a]

The hypothesis of vortices is beset with many difficulties. If, by a radius drawn to the sun, each and every planet is to describe areas proportional to the time, the periodic times of the parts of the vortex must be as the squares of the distances from the sun. If the periodic times of the planets are to be as the $\frac{3}{2}$ powers of the distances from the sun, the periodic times of the parts of the vortex must be as the $\frac{3}{2}$ powers of the distances. If the smaller vortices revolving about Saturn, Jupiter, and the other planets are to be preserved and are to float without agitation in the vortex of the sun, the periodic times of the parts of the solar vortex must be the same. The axial revolutions [i.e., rotations] of the sun and planets, [b]which would have to agree with the motions of their vortices,[b] differ from all these proportions. The motions of comets are extremely regular, observe the same laws as the motions of planets, and cannot be explained by vortices. Comets go with very eccentric motions into all parts of the heavens, which cannot happen unless vortices are eliminated.

The only resistance which projectiles encounter in our air is from the air. With the air removed, as it is in Boyle's vacuum, resistance ceases, since a tenuous feather and solid gold fall with equal velocity in such a vacuum. And the case is the same for the celestial spaces, which are above the atmosphere

a. Ed. 1 lacks the General Scholium, which includes Newton's famous discussions of God and of hypotheses. This scholium is first printed in ed. 2 but is documented further by its changing versions in five extant earlier holograph drafts and is treated also in contemporaneous correspondence between Newton and Roger Cotes, editor of ed. 2. For details see *Unpublished Scientific Papers of Isaac Newton*, ed. A. Rupert Hall and Marie Boas Hall (Cambridge: Cambridge University Press, 1962), pp. 348–364; I. Bernard Cohen, *Introduction to Newton's "Principia"* (Cambridge, Mass.: Harvard University Press; Cambridge: Cambridge University Press, 1971), pp. 240–245.

bb. Ed. 2 lacks this.

of the earth. All bodies must move very freely in these spaces, and therefore planets and comets must revolve continually in orbits given in kind and in position, according to the laws set forth above. They will indeed persevere in their orbits by the laws of gravity, but they certainly could not originally have acquired the regular position of the orbits by these laws.

The six primary planets revolve about the sun in circles concentric with the sun, with the same direction of motion, and very nearly in the same plane. Ten moons revolve about the earth, Jupiter, and Saturn in concentric circles, with the same direction of motion, very nearly in the planes of the orbits of the planets. And all these regular motions do not have their origin in mechanical causes, since comets go freely in very eccentric orbits and into all parts of the heavens. And with this kind of motion the comets pass very swiftly and very easily through the orbits of the planets; and in their aphelia, where they are slower and spend a longer time, they are at the greatest possible distance from one another, so as to attract one another as little as possible.

This most elegant system of the sun, planets, and comets could not have arisen without the design and dominion of an intelligent and powerful being. And if the fixed stars are the centers of similar systems, they will all be constructed according to a similar design and subject to the dominion of *One*, especially since the light of the fixed stars is of the same nature as the light of the sun, and all the systems send light into all the others. [c]And so that the systems of the fixed stars will not fall upon one another as a result of their gravity, he has placed them at immense distances from one another.[c]

He rules all things, not as the world soul but as the lord of all. And because of his dominion he is called Lord God *Pantokrator*[d]. For "god" is a relative word and has reference to servants, and godhood[e] is the lordship of God, not over his own body [f]as is supposed by those for whom God is the world soul[f], but over servants. The supreme God is an eternal, infinite, and absolutely perfect being; but a being, however perfect, without dominion is

cc. Ed. 2 lacks this.

d. Newton's note a: "That is, universal ruler."

e. Newton here uses the word "deitas," a nonclassical term which signifies the essential nature of the divinity or "god-ness." Although "Godhead" does fit, the term "godhood" (which is more abstract) may more accurately convey the sense of Newton's "deitas."

ff. Ed. 2 lacks this.

not the Lord God. For we do say my God, your God, the God of Israel, the God of Gods, and Lord of Lords, but we do not say my eternal one, your eternal one, the eternal one of Israel, the eternal one of the gods; we do not say my infinite one, or my perfect one. These designations [i.e., eternal, infinite, perfect] do not have reference to servants. The word "god" is used far and wide to mean "lord,"[g] but every lord is not a god. The lordship of a spiritual being constitutes a god, a true lordship constitutes a true god, a supreme lordship a supreme god, an imaginary lordship an imaginary god. And from true lordship it follows that the true God is living, intelligent, and powerful; from the other perfections, that he is supreme, or supremely perfect. He is eternal and infinite, omnipotent and omniscient, that is, he endures from eternity to eternity, and he is present from infinity to infinity; he rules all things, and he knows all things that happen or can happen. He is not eternity and infinity, but eternal and infinite; he is not duration and space, but he endures and is present. He endures always and is present everywhere, and by existing always and everywhere he constitutes duration and [h]space.[h] Since each and every particle of space is *always*, and each and every indivisible moment of duration is *everywhere*, certainly the maker and lord of all things will not be *never* or *nowhere*.

[i]Every sentient soul, at different times and in different organs of senses and motions, is the same indivisible person. There are parts that are successive in duration and coexistent in space, but neither of these exist in the person of man or in his thinking principle, and much less in the thinking substance of God. Every man, insofar as he is a thing that has senses, is one and the same man throughout his lifetime in each and every organ of his senses. God is one and the same God always and everywhere.[i] He is omnipresent not only *virtually* but also *substantially*; for action requires substance [*lit.* for active power [virtus] cannot subsist without substance]. In him all things are contained and move,[j] but he does not act on them nor they on him. God

g. Newton's note b, which ed. 2 lacks: "Our fellow countryman Pocock derives the word 'deus' from the Arabic word 'du' (and in the oblique case 'di'), which means lord. And in this sense princes are called gods, Psalms 82.6 and John 10.35. And Moses is called a god of his brother Aaron and a god of king Pharaoh (Exod. 4.16 and 7.1). And in the same sense the souls of dead princes were formerly called gods by the heathen, but wrongly because of their lack of dominion."

hh. Ed. 2 has "space, eternity, and infinity."

ii. Ed. 2 lacks this.

j. Newton's note c: "This opinion was held by the ancients: for example, by Pythagoras as cited in Cicero, *On the Nature of the Gods*, book 1; Thales; Anaxagoras; Virgil, *Georgics*, book 4, v. 221, and *Aeneid*,

experiences nothing from the motions of bodies; the bodies feel no resistance from God's omnipresence.

It is agreed that the supreme God necessarily exists, and by the same necessity he is *always* and *everywhere*. It follows that all of him is like himself: he is all eye, all ear, all brain, all arm, all force of sensing, of understanding, and of acting, but in a way not at all human, in a way not at all corporeal, in a way utterly unknown to us. As a blind man has no idea of colors, so we have no idea of the ways in which the most wise God senses and understands all things. He totally lacks any body and corporeal shape, and so he cannot be seen or heard or touched, nor ought he to be worshiped in the form of something corporeal. We have ideas of his attributes, but we certainly do not know what is the substance of any thing. We see only the shapes and colors of bodies, we hear only their sounds, we touch only their external surfaces, we smell only their odors, and we taste their flavors. But there is no direct sense and there are no indirect reflected actions by which we know innermost substances; much less do we have an idea of the substance of God. We know him only by his properties and attributes and by the wisest and best construction of things and their final causes, [k]and we admire him because of his perfections;[k] but we venerate and worship him because of his dominion. [l]For we worship him as servants, and a god[l] without dominion, providence, and final causes is nothing other than fate and nature. [m]No variation in things arises from blind metaphysical necessity, which must be the same always and everywhere. All the diversity of created things, each in its place and time, could only have arisen from the ideas and the will of a necessarily existing being. But God is said allegorically to see, hear, speak, laugh, love, hate, desire, give, receive, rejoice, be angry, fight, build, form, construct. For all discourse about God is derived through a certain similitude

---

book 6, v. 726; Philo, *Allegorical Interpretation*, book 1, near the beginning; Aratus in the *Phenomena*, near the beginning. Also by the sacred writers: for example, Paul in Acts 17.27, 28; John in his Gospel 14.2; Moses in Deuteronomy 4.39 and 10.14; David, Psalms 139.7, 8, 9; Solomon, 1 Kings 8.27; Job 22.12, 13, 14; Jeremiah 23.23, 24. Moreover idolators imagined that the sun, moon, and stars, the souls of men, and other parts of the world were parts of the supreme god and so were to be worshiped, but they were mistaken." In ed. 2 this note reads: "This opinion was held by the ancients: Aratus in the *Phenomena*, near the beginning; Paul in Acts 7.27, 28; Moses, Deuteronomy 4.39 and 10.14; David, Psalms 139.7, 8; Solomon, Kings 8.27; Job 22.12; the prophet Jeremiah, 23.23, 24."

kk. Ed. 2 lacks this.

ll. Ed. 2 has: "For a god."

mm. Ed. 2 lacks this.

from things human, which while not perfect is nevertheless a similitude of some kind.[m] This concludes the discussion of God, and to treat of God from phenomena is certainly a part of [n]natural[n] philosophy.

Thus far I have explained the phenomena of the heavens and of our sea by the force of gravity, but I have not yet assigned a cause to gravity. Indeed, this force arises from some cause that penetrates as far as the centers of the sun and planets without any diminution of its power to act, and that acts not in proportion to the quantity of the *surfaces* of the particles on which it acts (as mechanical causes are wont to do) but in proportion to the quantity of *solid* matter, and whose action is extended everywhere to immense distances, always decreasing as the squares of the distances. Gravity toward the sun is compounded of the gravities toward the individual particles of the sun, and at increasing distances from the sun decreases exactly as the squares of the distances as far out as the orbit of Saturn, as is manifest from the fact that the aphelia of the planets are at rest, and even as far as the farthest aphelia of the comets, provided that those aphelia are at rest. I have not as yet been able to deduce from phenomena the reason for these properties of gravity, and I do not [o]feign[o] hypotheses. For whatever is not deduced from the phenomena must be called a hypothesis; and hypotheses, whether metaphysical or physical, or based on occult qualities, or mechanical, have no place in experimental philosophy. In this experimental philosophy, propositions are deduced from the phenomena and are made general by induction. The impenetrability, mobility, and impetus of bodies, and the laws of motion and the law of gravity have been found by this method. And it is enough that gravity really exists and acts according to the laws that we have set forth and is sufficient to explain all the motions of the heavenly bodies and of our sea.

[p]A few things could now be added concerning a certain very subtle spirit pervading gross bodies and lying hidden in them; by its force and actions, the particles of bodies attract one another at very small distances and cohere when

nn. Ed. 2 has "experimental."

oo. The word "fingo" in Newton's famous declaration, "Hypotheses non fingo," appears to be the Latin equivalent of the English word "feign." Andrew Motte translated "fingo" by "frame," a verb which at that time could have a pejorative sense. For details see the Guide, §9.1.

pp. The final paragraph of the General Scholium has attracted much scholarly attention, notably in an effort to discern what Newton intended (in the opening and closing sentences) by a "spirit" which may

they become contiguous; and electrical [i.e., electrified] bodies act at greater distances, repelling as well as attracting neighboring corpuscles; and light is emitted, reflected, refracted, inflected, and heats bodies; and all sensation is excited, and the limbs of animals move at command of the will, namely, by the vibrations of this spirit being propagated through the solid fibers of the nerves from the external organs of the senses to the brain and from the brain into the muscles. But these things cannot be explained in a few words; furthermore, there is not a sufficient number of experiments to determine and demonstrate accurately the laws governing the actions of this spirit.[p]

---

be operative in various types of phenomena. It might even appear that Newton was here introducing a speculation—we dare not call it a hypothesis—although Newton's actual language indicates that for him there was no question about whether this spirit "really" exists, only about the laws according to which this spirit acts.

A puzzle relating to the interpretation of this "spirit" is the appearance of the qualifying adjectives "electric and elastic," introduced in the original Motte translation and retained in the Motte-Cajori version. Although these words are not found in either the second or the third Latin editions, they have a Newtonian provenience since they occur in Newton's personal interleaved copy of the second edition as one of the proposed emendations. Furthermore, thanks to the research of A. Rupert Hall and Marie Boas Hall, we know that the spirit in question is indeed "electrical." In particular, as Newton worked toward the second edition of the *Principia*, he composed various drafts of proposed conclusions which, together with other manuscripts, provide evidence for the importance of electrical phenomena in his thinking about gravity during the years 1711–1713. For details see the Guide to the present translation, §9.3.

One possible reason why Newton decided not to insert the qualifying phrase "electric and elastic" into the text of the third edition (1726) is that in his interleaved copy of the second edition he has finally drawn a line through the whole paragraph, showing his intention of deleting it in a third edition. The reason for this decision seems to be that some time after 1713 Newton lost his enthusiasm for electricity as a possible agent in gravitation.

We may readily understand why Newton omitted to carry out either the revision or the proposed cancellation of the final paragraph. By the time that the third edition was fully printed, in about February 1726, Newton and Pemberton had spent several years revising the text and reading the proofs and Newton was within a little more than a year of his death. When Newton reached the last paragraph he was probably so weary that he overlooked his proposed alteration of the conclusion.

---

p. The third edition concludes with an "Index Rerum Alphabeticus" (pp. 531–536) and an advertisement of books sold by William and John Innys (pp. 537–538).

Contents of the Principia

Halley's Ode to Newton 379
Newton's Preface to the First Edition 381
Newton's Preface to the Second Edition 384
Cotes's Preface to the Second Edition 385
Newton's Preface to the Third Edition 400

DEFINITIONS 403

Definition 1. Quantity of matter is a measure of matter that arises from its density and volume jointly. 403
Definition 2. Quantity of motion is a measure of motion that arises from the velocity and the quantity of matter jointly. 404
Definition 3. Inherent force of matter is the power of resisting by which every body, so far as it is able, perseveres in its state either of resting or of moving uniformly straight forward. 404
Definition 4. Impressed force is the action exerted on a body to change its state either of resting or of moving uniformly straight forward. 405
Definition 5. Centripetal force is the force by which bodies are drawn from all sides, are impelled, or in any way tend, toward some point as to a center. 405
Definition 6. The absolute quantity of centripetal force is the measure of this force that is greater or less in proportion to the efficacy of the cause propagating it from a center through the surrounding regions. 406
Definition 7. The accelerative quantity of centripetal force is the measure of this force that is proportional to the velocity which it generates in a given time. 407
Definition 8. The motive quantity of centripetal force is the measure of this force that is proportional to the motion which it generates in a given time. 407
Scholium 408

AXIOMS, OR THE LAWS OF MOTION 416

Law 1. Every body perseveres in its state of being at rest or of moving uniformly straight forward, except insofar as it is compelled to change its state by forces impressed. 416
Law 2. A change in motion is proportional to the motive force impressed and takes place along the straight line in which that force is impressed. 416

Law 3. To any action there is always an opposite and equal reaction; in other words, the actions of two bodies upon each other are always equal and always opposite in direction. 417

Corollary 1. A body acted on by [two] forces acting jointly describes the diagonal of a parallelogram in the same time in which it would describe the sides if the forces were acting separately. 417

Corollary 2. And hence the composition of a direct force AD out of any oblique forces AB and BD is evident, and conversely the resolution of any direct force AD into any oblique forces AB and BD. And this kind of composition and resolution is indeed abundantly confirmed from mechanics. 418

Corollary 3. The quantity of motion, which is determined by adding the motions made in one direction and subtracting the motions made in the opposite direction, is not changed by the action of bodies on one another. 420

Corollary 4. The common center of gravity of two or more bodies does not change its state whether of motion or of rest as a result of the actions of the bodies upon one another; and therefore the common center of gravity of all bodies acting upon one another (excluding external actions and impediments) either is at rest or moves uniformly straight forward. 421

Corollary 5. When bodies are enclosed in a given space, their motions in relation to one another are the same whether the space is at rest or whether it is moving uniformly straight forward without circular motion. 423

Corollary 6. If bodies are moving in any way whatsoever with respect to one another and are urged by equal accelerative forces along parallel lines, they will all continue to move with respect to one another in the same way as they would if they were not acted on by those forces. 423

Scholium 424

BOOK 1: THE MOTION OF BODIES 431

SECTION 1: The method of first and ultimate ratios, for use in demonstrating what follows 433

Lemma 1. Quantities, and also ratios of quantities, which in any finite time constantly tend to equality, and which before the end of that time approach so close to one another that their difference is less than any given quantity, become ultimately equal. 433

Lemma 2. If in any figure A*ac*E, comprehended by the straight lines A*a* and AE and the curve *ac*E, any number of parallelograms A*b*, B*c*, C*d*, . . . are inscribed upon equal bases AB, BC, CD, . . . and have sides B*b*, C*c*, D*d*, . . . parallel to the side A*a* of the figure; and if the parallelograms *a*K*bl*, *b*L*cm*, *c*M*dn*, . . . are completed; if then the width of these parallelograms is diminished and their number increased indefinitely, I say that the ultimate ratios which the inscribed figure AK*b*L*c*M*d*D, the circumscribed figure A*albmcndo*E, and the curvilinear figure A*abcd*E have to one another are ratios of equality. 433

Lemma 3. The same ultimate ratios are also ratios of equality when the widths AB, BC, CD, . . . of the parallelograms are unequal and are all diminished indefinitely. 433

COROLLARIES 1–4 434

Lemma 4. If in two figures A*ac*E and P*pr*T two series of parallelograms are inscribed (as above) and the number of parallelograms in both series is the same; and if, when their widths are diminished indefinitely, the ultimate ratios of the parallelograms in one figure to the corresponding parallelograms in the other are the same; then I say that the two figures A*ac*E and P*pr*T are to each other in that same ratio. 434

COROLLARY 435

Lemma 5. All the mutually corresponding sides—curvilinear as well as rectilinear—of similar figures are proportional, and the areas of such figures are as the squares of their sides. 435

Lemma 6. If any arc ACB, given in position, is subtended by the chord AB and at some point A, in the middle of the continuous curvature, is touched by the straight line AD, produced in both directions, and if then points A and B approach each other and come together, I say that the angle BAD contained by the chord and the tangent will be indefinitely diminished and will ultimately vanish. 435

Lemma 7. With the same suppositions, I say that the ultimate ratios of the arc, the chord, and the tangent to one another are ratios of equality. 436

COROLLARIES 1–3 436

Lemma 8. If the given straight lines AR and BR, together with the arc ACB, its chord AB, and the tangent AD, constitute three triangles RAB, RACB, and RAD, and if then points A and B approach each other, I say that the triangles as they vanish are similar in their ultimate form, and that their ultimate ratio is one of equality. 436

COROLLARY 437

Lemma 9. If the straight line AE and the curve ABC, both given in position, intersect each other at a given angle A, and if BD and CE are drawn as ordinates to the straight line AE at another given angle and meet the curve in B and C, and if then points B and C simultaneously approach point A, I say that the areas of the triangles ABD and ACE will ultimately be to each other as the squares of the sides. 437

Lemma 10. The spaces which a body describes when urged by any finite force, whether that force is determinate and immutable or is continually increased or continually decreased, are at the very beginning of the motion in the squared ratio of the times. 437

COROLLARIES 1–5 438

Scholium 438

Lemma 11. In all curves having a finite curvature at the point of contact, the vanishing subtense of the angle of contact is ultimately in the squared ratio of the subtense of the conterminous arc. 439

CASES 1–3 439

COROLLARIES 1–5 440

Scholium 440

SECTION 2: To find centripetal forces 444

Proposition 1, Theorem 1. The areas which bodies made to move in orbits describe by radii drawn to an unmoving center of forces lie in unmoving planes and are proportional to the times. 444

COROLLARIES 1–6 445

Proposition 2, Theorem 2. Every body that moves in some curved line described in a plane and, by a radius drawn to a point, either unmoving or moving uniformly forward with a rectilinear motion, describes areas around that point proportional to the times, is urged by a centripetal force tending toward that same point. 446

CASES 1–2 446

COROLLARIES 1–2 447

Scholium 447

Proposition 3, Theorem 3. Every body that, by a radius drawn to the center of a second body moving in any way whatever, describes about that center areas that are proportional to the times is urged by a force compounded of the centripetal force tending toward that second body and of the whole accelerative force by which that second body is urged. 448

COROLLARIES 1–4 448

Scholium 449

Proposition 4, Theorem 4. The centripetal forces of bodies that describe different circles with uniform motion tend toward the centers of those circles and are to one another as the squares of the arcs described in the same time divided by the radii of the circles. 449

COROLLARIES 1–9 450

Scholium 452

Proposition 5, Problem 1. Given, in any places, the velocity with which a body describes a given curve when acted on by forces tending toward some common center, to find that center. 453

Proposition 6, Theorem 5. If in a nonresisting space a body revolves in any orbit about an immobile center and describes any just-nascent arc in a minimally small time, and if the sagitta of the arc is understood to be drawn so as to bisect the chord and, when produced, to pass through the center of forces, the centripetal force in the middle of the arc will be as the sagitta directly and as the time twice [i.e., as the square of the time] inversely. 453

COROLLARIES 1–5 454

Proposition 7, Problem 2. Let a body revolve in the circumference of a circle; it is required to find the law of the centripetal force tending toward any given point. 455

Another solution 456

COROLLARIES 1–3 456

Proposition 8, Problem 3. Let a body move in the semicircle PQA; it is required to find the law of the centripetal force for this effect, when the centripetal force tends toward a point S so distant that all the lines PS and RS drawn to it can be considered parallel. 457

Scholium 458

Proposition 9, Problem 4. Let a body revolve in a spiral PQS intersecting all its radii SP, SQ, . . . , at a given angle; it is required to find the law of the centripetal force tending toward the center of the spiral. 458

Another solution 458

Lemma 12. All the parallelograms described about any conjugate diameters of a given ellipse or hyperbola are equal to one another. 458

Proposition 10, Problem 5. Let a body revolve in an ellipse; it is required to find the law of the centripetal force tending toward the center of the ellipse. 459

Another solution 460

COROLLARIES 1–2 460

Scholium 460

SECTION 3: The motion of bodies in eccentric conic sections 462

Proposition 11, Problem 6. Let a body revolve in an ellipse; it is required to find the law of the centripetal force tending toward a focus of the ellipse. 462

Another solution 463

Proposition 12, Problem 7. Let a body move in a hyperbola; it is required to find the law of the centripetal force tending toward the focus of the figure. 463

Another solution 465

Lemma 13. In a parabola the latus rectum belonging to any vertex is four times the distance of that vertex from the focus of the figure. 465

Lemma 14. A perpendicular dropped from the focus of a parabola to its tangent is a mean proportional between the distance of the focus from the point of contact and its distance from the principal vertex of the figure. 465

COROLLARIES 1–3 465

Proposition 13, Problem 8. Let a body move in the perimeter of a parabola; it is required to find the law of the centripetal force tending toward a focus of the figure. 466

COROLLARIES 1–2 467

Proposition 14, Theorem 6. If several bodies revolve about a common center and the centripetal force is inversely as the square of the distance of places from the center, I say that the principal latera recta of the orbits are as the squares of the areas which the bodies describe in the same time by radii drawn to the center. 467

COROLLARY 467

Proposition 15, Theorem 7. Under the same suppositions as in prop. 14, I say that the squares of the periodic times in ellipses are as the cubes of the major axes. 468

COROLLARY 468

Proposition 16, Theorem 8. Under the same suppositions as in prop. 15, if straight lines are drawn to the bodies in such a way as to touch the orbits in the places where the bodies are located, and if perpendiculars are dropped from the common focus to these tangents, I say that the velocities of the bodies are inversely as the perpendiculars and directly as the square roots of the principal latera recta. 468

COROLLARIES 1–9 468

Proposition 17, Problem 9. Supposing that the centripetal force is inversely proportional to the square of the distance of places from the center and that the absolute quantity of this force is known, it is required to find the line which a body describes when going forth from a given place with a given velocity along a given straight line. 470

COROLLARIES 1–4 471

Scholium 472

SECTION 4: To find elliptical, parabolic, and hyperbolic orbits, given a focus 473

Lemma 15. If from the two foci S and H of any ellipse or hyperbola two straight lines SV and HV are inclined to any third point V, one of the lines HV being equal to the principal axis of the figure, that is, to the axis on which the foci lie, and the other line SV being bisected in T by TR perpendicular to it, then the perpendicular TR will touch the conic at some point; and conversely, if TR touches the conic, HV will be equal to the principal axis of the figure. 473

Proposition 18, Problem 10. Given a focus and the principal axes, to describe elliptical and hyperbolic trajectories that will pass through given points and will touch straight lines given in position. 473

Proposition 19, Problem 11. To describe about a given focus a parabolic trajectory that will pass through given points and will touch straight lines given in position. 474

Proposition 20, Problem 12. To describe about a given focus any trajectory, given in species [i.e., of given eccentricity], that will pass through given points and will touch straight lines given in position. 474

CASES 1–4 474

Lemma 16. From three given points to draw three straight slanted lines to a fourth point, which is not given, when the differences between the lines either are given or are nil. 477

CASES 1–3 477

Proposition 21, Problem 13. To describe about a given focus a trajectory that will pass through given points and will touch straight lines given in position. 478

Scholium 479

SECTION 5: To find orbits when neither focus is given 480

Lemma 17. If four straight lines PQ, PR, PS, and PT are drawn at given angles from any point P of a given conic to the four indefinitely produced sides AB, CD, AC, and DB of some quadrilateral ABDC inscribed in the conic, one line being drawn to each side, the rectangle PQ × PR of the lines drawn to two opposite sides will be in a given ratio to the rectangle PS × PT of the lines drawn to the other two opposite sides. 480

CASES 1–3 480

Lemma 18. With the same suppositions as in lem. 17, if the rectangle PQ × PR of the lines drawn to two opposite sides of the quadrilateral is in a given ratio to the rectangle PS × PT of the lines drawn to the other two sides, the point P from which the lines are drawn will lie on a conic circumscribed about the quadrilateral. 481

COROLLARY 482

Scholium 482

Lemma 19. To find a point P such that if four straight lines PQ, PR, PS, and PT are drawn from it at given angles to four other straight lines AB, CD, AC, and BD given in position, one line being drawn from the point P to each of the four other straight lines, the rectangle PQ × PR of two of the lines drawn will be in a given ratio to the rectangle PS × PT of the other two. 483

COROLLARIES 1–2 484

Lemma 20. If any parallelogram ASPQ touches a conic in points A and P with two of its opposite angles A and P, and if the sides AQ and AS, indefinitely produced, of one of these angles meet the said conic in B and C, and if from the meeting points B and C two straight lines BD and CD are drawn to any fifth point D of the conic, meeting the other two indefinitely produced sides PS and PQ of the parallelogram in T and R; then PR and PT, the parts cut off from the sides, will always be to each other in a given ratio. And conversely, if the parts which are cut off are to each other in a given ratio, the point D will touch a conic passing through the four points A, B, C, and P. 485

CASES 1–2 485

COROLLARIES 1–3 486

Lemma 21. If two movable and infinite straight lines BM and CM, drawn through given points B and C as poles, describe by their meeting-point M a third straight line MN given in position, and if two other infinite straight lines BD and CD are drawn, making given angles MBD and MCD with the first two lines at those given points B and C; then I say that the point D, where these two lines BD and CD meet, will describe a conic passing through points B and C. And conversely, if the point D, where the straight lines BD and CD meet, describes a conic passing through the given points B, C, and A, and the angle DBM is always equal to the given angle ABC, and the angle DCM is always equal to the given angle ACB; then point M will lie in a straight line given in position. 486

Proposition 22, Problem 14. To describe a trajectory through five given points. 488

Another solution 489

COROLLARIES 1–2 490

Scholium 490

Proposition 23, Problem 15. To describe a trajectory that will pass through four given points and touch a straight line given in position. 490

CASE 1 490

Another solution 491

CASE 2 491

Proposition 24, Problem 16. To describe a trajectory that will pass through three given points and touch two straight lines given in position. 492

Lemma 22. To change figures into other figures of the same class. 493

Proposition 25, Problem 17. To describe a trajectory that will pass through two given points and touch three straight lines given in position. 495

Proposition 26, Problem 18. To describe a trajectory that will pass through a given point and touch four straight lines given in position. 496

Lemma 23. If two straight lines AC and BD, given in position, terminate at the given points A and B and have a given ratio to each other; and if the straight line CD, by which the indeterminate points C and D are joined, is cut in K in a given ratio; I say that point K will be located in a straight line given in position. 497

COROLLARY 498

Lemma 24. If three straight lines, two of which are parallel and given in position, touch any conic section, I say that the semidiameter of the section which is parallel to the two given parallel lines is a mean proportional between their segments that are intercepted between the points of contact and the third tangent. 498

COROLLARIES 1–2 498

Lemma 25. If the indefinitely produced four sides of a parallelogram touch any conic and are intercepted at any fifth tangent, and if the intercepts of any two conterminous sides are taken so as to be terminated at opposite corners of the parallelogram; I say that either intercept is to the side from which it is intercepted as the part of the other conterminous side between the point of contact and the third side is to the other intercept. 499

COROLLARIES 1–3 499

Proposition 27, Problem 19. To describe a trajectory that will touch five straight lines given in position. 500

Scholium 501

Lemma 26. To place the three corners of a triangle given in species and magnitude on three straight lines given in position and not all parallel, with one corner on each line. 502

COROLLARY 504

Proposition 28, Problem 20. To describe a trajectory given in species and magnitude, whose given parts will lie between three straight lines given in position. 504

Lemma 27. To describe a quadrilateral given in species, whose corners will lie on four straight lines, given in position, which are not all parallel and do not all converge to a common point—each corner lying on a separate line. 505

COROLLARY 506

Proposition 29, Problem 21. To describe a trajectory, given in species, which four straight lines given in position will cut into parts given in order, species, and proportion. 507

Scholium 508

SECTION 6: To find motions in given orbits 510

Proposition 30, Problem 22. If a body moves in a given parabolic trajectory, to find its position at an assigned time. 510

COROLLARIES 1–3 510

Lemma 28. No oval figure exists whose area, cut off by straight lines at will, can in general be found by means of equations finite in the number of their terms and dimensions. 511

COROLLARY 513

Proposition 31, Problem 23. If a body moves in a given elliptical trajectory, to find its position at an assigned time. 513

Scholium 514

SECTION 7: The rectilinear ascent and descent of bodies 518

Proposition 32, Problem 24. Given a centripetal force inversely proportional to the square of the distance of places from its center, to determine the spaces which a body in falling straight down describes in given times. 518

CASES 1–3 518

Proposition 33, Theorem 9. Supposing what has already been found, I say that the velocity of a falling body at any place C is to the velocity of a body describing a circle with center B and radius BC as the square root of the ratio of AC (the distance of the body from the further vertex A of the circle or rectangular hyperbola) to ½AB (the principal semidiameter of the figure). 519

COROLLARIES 1–2 520

Proposition 34, Theorem 10. If the figure BED is a parabola, I say that the velocity of a falling body at any place C is equal to the velocity with which a body can uniformly describe a circle with center B and a radius equal to one-half of BC. 521

Proposition 35, Theorem 11. Making the same suppositions, I say that the area of the figure DES described by the indefinite radius SD is equal to the area that a body revolving uniformly in orbit about the center S can describe in the same time by a radius equal to half of the latus rectum of the figure DES. 521

CASES 1–2 522

Proposition 36, Problem 25. To determine the times of descent of a body falling from a given place A. 523

Proposition 37, Problem 26. To define the times of the ascent or descent of a body projected upward or downward from a given place. 523

Proposition 38, Theorem 12. Supposing that the centripetal force is proportional to the height or distance of places from the center, I say that the times of falling bodies, their velocities, and the spaces described are proportional respectively to the arcs, the right sines, and the versed sines. 524

COROLLARIES 1–2 524

Proposition 39, Problem 27. Suppose a centripetal force of any kind, and grant the quadratures of curvilinear figures; it is required to find, for a body ascending straight up or descending straight down, the velocity in any of its positions and the time in which the body will reach any place; and conversely. 524

COROLLARIES 1–3 526

SECTION 8: To find the orbits in which bodies revolve when acted upon by any centripetal forces 528

Proposition 40, Theorem 13. If a body, under the action of any centripetal force, moves in any way whatever, and another body ascends straight up or descends straight down, and if their velocities are equal in some one instance in which their distances from the center are equal, their velocities will be equal at all equal distances from the center. 528

COROLLARIES 1–2 529

Proposition 41, Problem 28. Supposing a centripetal force of any kind and granting the quadratures of curvilinear figures, it is required to find the trajectories in which bodies will move and also the times of their motions in the trajectories so found. 529

COROLLARIES 1–3 531

Proposition 42, Problem 29. Let the law of centripetal force be given; it is required to find the motion of a body setting out from a given place with a given velocity along a given straight line. 532

SECTION 9: The motion of bodies in mobile orbits, and the motion of the apsides 534

Proposition 43, Problem 30. It is required to find the force that makes a body capable of moving in any trajectory that is revolving about the center of forces in the same way as another body in that same trajectory at rest. 534

Proposition 44, Theorem 14. The difference between the forces under the action of which two bodies are able to move equally—one in an orbit that is at rest and the other in an identical orbit that is revolving—is inversely as the cube of their common height. 535

COROLLARIES 1–6 536

Proposition 45, Problem 31. It is required to find the motions of the apsides of orbits that differ very little from circles. 539

EXAMPLES 1–3 540

COROLLARIES 1–2 543

SECTION 10: The motion of bodies on given surfaces and the oscillating motion of simple pendulums 546

Proposition 46, Problem 32. Suppose a centripetal force of any kind, and let there be given both the center of force and any plane in which a body revolves, and grant the quadratures of curvilinear figures; it is required to find the motion of a body setting out from a given place with a given velocity along a given straight line in that plane. 546

Proposition 47, Theorem 15. Suppose that a centripetal force is proportional to the distance of a body from a center; then all bodies revolving in any planes whatever will describe ellipses and will make their revolutions in equal times; and bodies that move in straight lines, by oscillating to and fro, will complete in equal times their respective periods of going and returning. 547

Scholium 547

Proposition 48, Theorem 16. If a wheel stands upon the outer surface of a globe at right angles to that surface and, rolling as wheels do, moves forward in a great circle [in the globe's surface], the length of the curvilinear path traced out by any given point in the perimeter [or rim] of the wheel from the time when that point touched the globe (a curve which may be called a cycloid or epicycloid) will be to twice the versed sine of half the arc [of the rim of the wheel] which during the time of rolling has been in contact with the globe's surface as the sum of the diameters of the globe and wheel is to the semidiameter of the globe. 548

Proposition 49, Theorem 17. If a wheel stands upon the inner surface of a hollow globe at right angles to that surface and, rolling as wheels do, moves forward in a great circle [in the globe's surface], the length of the curvilinear path traced out by any given point in the perimeter [or rim] of the wheel from the time when that point touched the globe will be to twice the versed sine of half the arc [of the rim of the wheel] which during the time of rolling has been in contact with the globe's surface as the difference of the diameters of the globe and wheel is to the semidiameter of the globe. 548

COROLLARIES 1–2 550

Proposition 50, Problem 33. To make a pendulum bob oscillate in a given cycloid. 550

COROLLARY 552

Proposition 51, Theorem 18. If a centripetal force tending from all directions to the center C of a globe is in each individual place as the distance of that place from the center; and if, under the action of this force alone, the body T oscillates (in the way just described) in the perimeter of the cycloid QRS; then I say that the times of the oscillations, however unequal the oscillations may be, will themselves be equal. 552

COROLLARY 553

Proposition 52, Problem 34. To determine both the velocities of pendulums in individual places and the times in which complete oscillations, as well as the separate parts of oscillations, are completed. 553

COROLLARIES 1–2 555

Proposition 53, Problem 35. Granting the quadratures of curvilinear figures, it is required to find the forces by whose action bodies moving in given curved lines will make oscillations that are always isochronous. 556

COROLLARIES 1–2 557

Proposition 54, Problem 36. Granting the quadratures of curvilinear figures, to find the times in which bodies under the action of any centripetal force will descend and ascend in any curved lines described in a plane passing through the center of forces. 557

Proposition 55, Theorem 19. If a body moves in any curved surface whose axis passes through a center of forces, and a perpendicular is dropped from the body to the axis, and a straight line parallel and equal to the perpendicular is drawn from any given point of the axis; I say that the parallel will describe an area proportional to the time. 558

COROLLARY 559

Proposition 56, Problem 37. Granting the quadratures of curvilinear figures, and given both the law of centripetal force tending toward a given center and a curved surface whose axis passes through that center, it is required to find the trajectory that a body will describe in that same surface when it has set out from a given place with a given velocity, in a given direction in that surface. 559

SECTION 11: The motion of bodies drawn to one another by centripetal forces 561

Proposition 57, Theorem 20. Two bodies that attract each other describe similar figures about their common center of gravity and also about each other. 561

Proposition 58, Theorem 21. If two bodies attract each other with any forces whatever and at the same time revolve about their common center of gravity, I say that by the action of the same forces there can be described around either body if unmoved a figure similar and equal to the figures that the bodies so moving describe around each other. 562

CASES 1–2 562

COROLLARIES 1–3 563

Proposition 59, Theorem 22. The periodic time of two bodies S and P revolving about their common center of gravity C is to the periodic time of one of the two bodies P, revolving in orbit about the other body S which is without motion, and describing a figure similar and equal to the figures that the bodies describe around each other, as the square root of the ratio of the mass of the second body S to the sum of the masses of the bodies S + P. 563

Proposition 60, Theorem 23. If two bodies S and P, attracting each other with forces inversely proportional to the square of the distance, revolve about a common center of gravity, I say that the principal axis of the ellipse which one of the bodies P describes by this motion about the other body S will be to the principal axis of the ellipse which the same body P would be able to describe in the same periodic time about the other body S at rest as the sum of the masses of the two bodies S + P is to the first of two mean proportionals between this sum and the mass of the other body S. 564

Proposition 61, Theorem 24. If two bodies, attracting each other with any kind of forces and not otherwise acted on or impeded, move in any way whatever, their motions will be the same as if they were not attracting each other but were each being attracted with the same forces by a third body set in their common center of gravity. And the law of the attracting forces will be the same with respect to the distance of the bodies from that common center and with respect to the total distance between the bodies. 564

Proposition 62, Problem 38. To determine the motions of two bodies that attract each other with forces inversely proportional to the square of the distance and are let go from given places. 565

Proposition 63, Problem 39. To determine the motions of two bodies that attract each other with forces inversely proportional to the square of the distance and that set out from given places with given velocities along given straight lines. 566

Proposition 64, Problem 40. If the forces with which bodies attract one another increase in the simple ratio of the distances from the centers, it is required to find the motions of more than two bodies in relation to one another. 566

Proposition 65, Theorem 25. More than two bodies whose forces decrease as the squares of the distances from their centers are able to move with respect to one another in ellipses and, by radii drawn to the foci, are able to describe areas proportional to the times very nearly. 567

CASES 1–2 568

COROLLARIES 1–3 569

Proposition 66, Theorem 26. Let three bodies—whose forces decrease as the squares of the distances—attract one another, and let the accelerative attractions of any two toward the third be to each other inversely as the squares of the distances, and let the two lesser ones revolve about the greatest. Then I say that if that greatest body is moved by these attractions, the inner body [of the two revolving bodies] will describe about the innermost and greatest body, by radii drawn to it, areas more nearly proportional to the times and a figure more closely approaching the shape of an ellipse (having its focus in the meeting point of the radii) than would be the case if that greatest body were not attracted by the smaller ones and were at rest, or if it were much less or much more attracted and were acted on either much less or much more. 570

CASES 1–2 570

COROLLARIES 1–22 573

Proposition 67, Theorem 27. With the same laws of attraction being supposed, I say that with respect to the common center of gravity O of the inner bodies P and T, the outer body S—by radii drawn to that center—describes areas more nearly proportional to the times, and an orbit more closely approaching the shape of an ellipse having its focus in that same center, than it can describe about the innermost and greatest body T by radii drawn to that body. 585

Proposition 68, Theorem 28. With the same laws of attraction being supposed, I say that with respect to the common center of gravity O of the inner bodies P and T, the outer body S—by radii drawn to that center—describes areas more nearly proportional to the times, and an orbit more closely approaching the shape of an ellipse having its focus in the same center, if the innermost and greatest body is acted on by these attractions just as the others are, than would be the case if it is either not attracted and is at rest or is much more or much less attracted or much more or much less moved. 585

COROLLARY 586

Proposition 69, Theorem 29. If, in a system of several bodies A, B, C, D, . . . , some body A attracts all the others, B, C, D, . . . , by accelerative forces that are inversely as the squares of the distances from the attracting body; and if another body B also attracts the rest of the bodies A, C, D, . . . , by forces that are inversely as the squares of the distances from the attracting body; then the absolute forces of the attracting bodies A and B will be to each other in the same ratio as the bodies [i.e., the masses] A and B themselves to which those forces belong. 587

COROLLARIES 1–3 587

Scholium 588

SECTION 12: The attractive forces of spherical bodies 590

Proposition 70, Theorem 30. If toward each of the separate points of a spherical surface there tend equal centripetal forces decreasing as the squares of the distances from the point, I say that a corpuscle placed inside the surface will not be attracted by these forces in any direction. 590

Proposition 71, Theorem 31. With the same conditions being supposed as in prop. 70, I say that a corpuscle placed outside the spherical surface is attracted to the center of the sphere by a force inversely proportional to the square of its distance from that same center. 590

Proposition 72, Theorem 32. If toward each of the separate points of any sphere there tend equal centripetal forces, decreasing in the squared ratio of the distances from those points, and there are given both the density of the sphere and the ratio of the diameter of the sphere to the distance of the corpuscle from the center of the sphere, I say that the force by which the corpuscle is attracted will be proportional to the semidiameter of the sphere. 592

COROLLARIES 1–3 592

Proposition 73, Theorem 33. If toward each of the separate points of any given sphere there tend equal centripetal forces decreasing in the squared ratio of the distances from those points, I say that a corpuscle placed inside the sphere is attracted by a force proportional to the distance of the corpuscle from the center of the sphere. 593

Scholium 593

Proposition 74, Theorem 34. With the same things being supposed as in prop. 73, I say that a corpuscle placed outside a sphere is attracted by a force inversely proportional to the square of the distance of the corpuscle from the center of the sphere. 593

COROLLARIES 1–3 594

Proposition 75, Theorem 35. If toward each of the points of a given sphere there tend equal centripetal forces decreasing in the squared ratio of the distances from the points, I say that this sphere will attract any other homogeneous sphere with a force inversely proportional to the square of the distance between the centers. 594

COROLLARIES 1–4 594

Proposition 76, Theorem 36. If spheres are in any way nonhomogeneous (as to the density of their matter and their attractive force) going from the center to the circumference, but are uniform throughout in every spherical shell at any given distance from the center, and the attractive force of each point decreases in the squared ratio of the distance of the attracted body, I say that the total force by which one sphere of this sort attracts another is inversely proportional to the square of the distance between their centers. 595

Corollaries 1–9 596

Proposition 77, Theorem 37. If toward each of the individual points of spheres there tend centripetal forces proportional to the distances of the points from attracted bodies, I say that the composite force by which two spheres will attract each other is as the distance between the centers of the spheres. 597

Cases 1–6 597

Proposition 78, Theorem 38. If spheres, on going from the center to the circumference, are in any way nonhomogeneous and nonuniform, but in every concentric spherical shell at any given distance from the center are homogeneous throughout; and the attracting force of each point is as the distance of the body attracted; then I say that the total force by which two spheres of this sort attract each other is proportional to the distance between the centers of the spheres. 599

Corollary 599

Scholium 599

Lemma 29. If any circle AEB is described with center S; and then two circles EF and *ef* are described with center P, cutting the first circle in E and *e*, and cutting the line PS in F and *f*; and if the perpendiculars ED and *ed* are dropped to PS; then I say that if the distance between the arcs EF and *ef* is supposed to be diminished indefinitely, the ultimate ratio of the evanescent line D*d* to the evanescent line F*f* is the same as that of line PE to line PS. 599

Proposition 79, Theorem 39. If the surface EF*fe*, just now vanishing because its width has been indefinitely diminished, describes by its revolution about the axis PS a concavo-convex spherical solid, toward each of whose individual equal particles there tend equal centripetal forces; then I say that the force by which that solid attracts an exterior corpuscle located in P is in a ratio compounded of the ratio of the solid [or product] $DE^2 \times Ff$ and the ratio of the force by which a given particle at the place F*f* would attract the same corpuscle. 600

Proposition 80, Theorem 40. If equal centripetal forces tend toward each of the individual equal particles of some sphere ABE, described about a center S; and if from each of the individual points to the axis AB of the sphere, in which some corpuscle P is located, there are erected the perpendiculars DE, meeting the sphere in the points E; and if, on these perpendiculars, the lengths DN are taken, which are jointly as the quantity $\frac{DE^2 \times PS}{PE}$ and as the force that a particle of the sphere, located on the axis, exerts at the distance PE upon the corpuscle P; then I say that the total force with which the corpuscle P is attracted toward the sphere is as the area ANB comprehended by the axis AB of the sphere and the curved line ANB, which the point N traces out. 601

Corollaries 1–4 602

Proposition 81, Problem 41. Under the same conditions as before, it is required to measure the area ANB. 603

Examples 1–3 603

Proposition 82, Theorem 41. If—in a sphere described about center S with radius SA—SI, SA, and SP are taken continually proportional [i.e., SI to SA as SA to SP], I say that the attraction of a corpuscle inside the sphere at any place I is to its attraction outside the sphere at place P in a ratio compounded of the square root of the ratio of the distances IS and PS from the center, and the square root of the ratio of the centripetal forces, tending at those places P and I toward the center. 606

Proposition 83, Problem 42. To find the force by which a corpuscle located in the center of a sphere is attracted toward any segment of it whatever. 607

Proposition 84, Problem 43. To find the force with which a corpuscle is attracted by a segment of a sphere when it is located on the axis of the segment beyond the center of the sphere. 608

Scholium 608

SECTION 13: The attractive forces of nonspherical bodies 610

Proposition 85, Theorem 42. If the attraction of an attracted body is far stronger when it is contiguous to the attracting body than when the bodies are separated from each other by even a very small distance, then the forces of the particles of the attracting body decrease, as the attracted body recedes, in a more than squared ratio of the distances from the particles. 610

Proposition 86, Theorem 43. If the forces of the particles composing an attracting body decrease, as an attracted body recedes, in the cubed or more than cubed ratio of the distances from the particles, the attraction will be far stronger in contact than when the attracting body and attracted body are separated from each other by even a very small distance. 610

Proposition 87, Theorem 44. If two bodies, similar to each other and consisting of equally attracting matter, separately attract corpuscles proportional to those bodies and similarly placed with respect to them, then the accelerative attractions of the corpuscles toward the whole bodies will be as the accelerative attractions of those corpuscles toward particles of those bodies proportional to the wholes and similarly situated in those whole bodies. 611

COROLLARIES 1–2 611

Proposition 88, Theorem 45. If the attracting forces of equal particles of any body are as the distances of places from the particles, the force of the whole body will tend toward its center of gravity, and will be the same as the force of a globe consisting of entirely similar and equal matter and having its center in that center of gravity. 612

COROLLARY 613

Proposition 89, Theorem 46. If there are several bodies consisting of equal particles whose forces are as the distances of places from each individual particle, the force—compounded of the forces of all these particles—by which any corpuscle is attracted will tend toward the common center of gravity of the attracting bodies and will be the same as if those attracting bodies, while maintaining their common center of gravity, were united together and were formed into a globe. 613

COROLLARY 613

Proposition 90, Problem 44. If equal centripetal forces, increasing or decreasing in any ratio of the distances, tend toward each of the individual points of any circle, it is required to find the force by which a corpuscle is attracted when placed anywhere on the straight line that stands perpendicularly upon the plane of the circle at its center. 613

COROLLARIES 1–3 614

Proposition 91, Problem 45. To find the attraction of a corpuscle placed in the axis of a round solid, to each of whose individual points there tend equal centripetal forces decreasing in any ratio of the distances. 615

COROLLARIES 1–3 615

Proposition 92, Problem 46. Given an attracting body, it is required to find the ratio by which the centripetal forces tending toward each of its individual points decrease [i.e., decrease as a function of distance]. 618

Proposition 93, Theorem 47. If a solid, plane on one side but infinitely extended on the other sides, consists of equal and equally attracting particles, whose forces—in receding from the solid—decrease in the ratio of any power of the distances that is more than the square; and if a corpuscle set on either side of the plane is attracted by the force of the whole solid; then I say that that force of attraction of the solid in receding from its plane surface will decrease in the ratio of the distance of the corpuscle from the plane raised to a power whose index is less by 3 units than that of the power of the distances in the law of attractive force [*lit.* will decrease in the ratio of the power whose base is the distance of the corpuscle from the plane and whose index is less by 3 than the index of the power of the distances]. 618

CASES 1–2 618

COROLLARIES 1–3 619

Scholium 620

SECTION 14: The motion of minimally small bodies that are acted on by centripetal forces tending toward each of the individual parts of some great body 622

Proposition 94, Theorem 48. If two homogeneous mediums are separated from each other by a space terminated on the two sides by parallel planes, and a body passing through this space is attracted or impelled perpendicularly toward either medium and is not acted on or impeded by any other force, and the attraction at equal distances from each plane (taken on the same side of that plane) is the same everywhere; then I say that the sine of the angle of incidence onto either plane will be to the sine of the angle of emergence from the other plane in a given ratio. 622

CASES 1–2 622

Proposition 95, Theorem 49. With the same suppositions as in prop. 94, I say that the velocity of the body before incidence is to its velocity after emergence as the sine of the angle of emergence to the sine of the angle of incidence. 623

Proposition 96, Theorem 50. With the same suppositions, and supposing also that the motion before incidence is faster than afterward, I say that as a result of changing the inclination of the line of incidence, the body will at last be reflected, and the angle of reflection will become equal to the angle of incidence. 624

Scholium 625

Proposition 97, Problem 47. Supposing that the sine of the angle of incidence upon some surface is to the sine of the angle of emergence in a given ratio, and that the inflection of the paths of bodies in close proximity to that surface takes place in a very short space, which can be considered to be a point; it is required to determine the surface that may make all the corpuscles emanating successively from a given place converge to another given place. 626

COROLLARIES 1–2 627

Proposition 98, Problem 48. The same conditions being supposed as in prop. 97, and supposing that there is described about the axis AB any attracting surface CD, regular or irregular, through which the bodies coming out from a given place A must pass; it is required to find a second attracting surface EF that will make the bodies converge to a given place B. 627

Scholium 628

BOOK 2: THE MOTION OF BODIES 631

SECTION 1: The motion of bodies that are resisted in proportion to their velocity 633

Proposition 1, Theorem 1. If a body is resisted in proportion to its velocity, the motion lost as a result of the resistance is as the space described in moving. 633

COROLLARY 633

Lemma 1. Quantities proportional to their differences are continually proportional. 633

Proposition 2, Theorem 2. If a body is resisted in proportion to its velocity and moves through a homogeneous medium by its inherent force alone and if the times are taken as equal, the velocities at the beginnings of the individual times are in a geometric progression, and the spaces described in the individual times are as the velocities. 633

CASES 1–2 633

COROLLARY 634

Proposition 3, Problem 1. To determine the motion of a body which, while moving straight up or down in a homogeneous medium, is resisted in proportion to the velocity, and which is acted on by uniform gravity. 634

COROLLARIES 1–4 636

Proposition 4, Problem 2. Supposing that the force of gravity in some homogeneous medium is uniform and tends perpendicularly toward the plane of the horizon, it is required to determine the motion of a projectile in that medium, while it is resisted in proportion to the velocity. 636

COROLLARIES 1–7 638

Scholium 641

SECTION 2: The motion of bodies that are resisted as the squares of the velocities 642

Proposition 5, Theorem 3. If the resistance of a body is proportional to the square of the velocity and if the body moves through a homogeneous medium by its inherent force alone and if the times are taken in a geometric progression going from the smaller to the greater terms, I say that the velocities at the beginning of each of the times are inversely in that same geometric progression and that the spaces described in each of the times are equal. 642

COROLLARIES 1–5 643

Proposition 6, Theorem 4. Equal homogeneous spherical bodies that are resisted in proportion to the square of the velocity, and are carried forward by their inherent forces alone, will, in times that are inversely as the initial velocities, always describe equal spaces, and lose parts of their velocities proportional to the wholes. 644

Proposition 7, Theorem 5. Spherical bodies that are resisted in proportion to the squares of the velocities will, in times that are as the first motions directly and the first resistances inversely, lose parts of the motions proportional to the wholes and will describe spaces proportional to those times and the first velocities jointly. 644

COROLLARIES 1–5 644

Lemma 2. The moment of a generated quantity is equal to the moments of each of the generating roots multiplied continually by the exponents of the powers of those roots and by their coefficients. 646

CASES 1–6 648

COROLLARIES 1–3 649

Scholium 649

Proposition 8, Theorem 6. If a body, acted on by gravity uniformly, goes straight up or down in a uniform medium, and the total space described is divided into equal parts, and the absolute forces at the beginnings of each of the parts are found (adding the resistance of the medium to the force of gravity when the body is ascending, or subtracting it when the body is descending), I say that those absolute forces are in a geometric progression. 650

COROLLARIES 1–3 651

Proposition 9, Theorem 7. Given what has already been proved, I say that if the tangents of the angles of a sector of a circle and of a hyperbola are taken proportional to the velocities, the radius being of the proper magnitude, the whole time of ascending to the highest place will be as the sector of the circle, and the whole time of descending from the highest place will be as the sector of the hyperbola. 651

CASES 1–2 652

COROLLARIES 1–7 653

Proposition 10, Problem 3. Let a uniform force of gravity tend straight toward the plane of the horizon, and let the resistance be as the density of the medium and the square of the velocity jointly; it is required to find, in each individual place, the density of the medium that makes the body move in any given curved line and also the velocity of the body and resistance of the medium. 655

COROLLARIES 1–2 659

EXAMPLES 1–4 659

Scholium 664

RULES 1–8 665

SECTION 3: The motion of bodies that are resisted partly in the ratio of the velocity and partly in the squared ratio of the velocity 670

Proposition 11, Theorem 8. If a body is resisted partly in the ratio of the velocity and partly in the squared ratio of the velocity and moves in a homogeneous medium by its inherent force alone, and if the times are taken in an arithmetic progression, then quantities inversely proportional to the velocities and increased by a certain given quantity will be in a geometric progression. 670

COROLLARIES 1–2 671

Proposition 12, Theorem 9. With the same suppositions, I say that if the spaces described are taken in an arithmetic progression, the velocities increased by a certain given quantity will be in a geometric progression. 671

COROLLARIES 1–3 672

Proposition 13, Theorem 10. Supposing that a body attracted downward by uniform gravity ascends straight up or descends straight down and is resisted partly in the ratio of the velocity and partly in the squared ratio of the velocity, I say that if straight lines parallel to the diameters of a circle and a hyperbola are drawn through the ends of the conjugate diameters and if the velocities are as certain segments of the parallels, drawn from a given point, then the times will be as the sectors of areas cut off by straight lines drawn from the center to the ends of the segments, and conversely. 672

CASES 1–3 672

COROLLARY 674

Scholium 674

Proposition 14, Theorem 11. With the same suppositions, I say that the space described in the ascent or descent is as the difference between the area which represents the time and a certain other area that increases or decreases in an arithmetic progression, if the forces compounded of the resistance and the gravity are taken in a geometric progression. 674

Cases 1–3 675

Corollary 676

Scholium 678

Section 4: The revolving motion of bodies in resisting mediums 680

Lemma 3. Let PQR be a spiral that cuts all the radii SP, SQ, SR, . . . in equal angles. Draw the straight line PT touching the spiral in any point P and cutting the radius SQ in T; erect PO and QO perpendicular to the spiral and meeting in O, and join SO. I say that if points P and Q approach each other and coincide, angle PSO will come out a right angle, and the ultimate ratio of rectangle TQ × 2PS to PQ^2 will be the ratio of equality. 680

Proposition 15, Theorem 12. If the density of a medium in every place is inversely as the distance of places from a motionless center and if the centripetal force is in the squared ratio of the density, I say that a body can revolve in a spiral that intersects in a given angle all the radii drawn from that center. 680

Corollaries 1–9 682

Proposition 16, Theorem 13. If the density of the medium in every place is inversely as the distance of places from a motionless center and if the centripetal force is inversely as any power of that distance, I say that a body can revolve in a spiral that intersects in a given angle all the radii drawn from that center. 685

Corollaries 1–3 685

Scholium 685

Proposition 17, Problem 4. To find both the centripetal force and the resistance of the medium by means of which a body can revolve in a given spiral, if the law of the velocity is given. 686

Proposition 18, Problem 5. Given the law of the centripetal force, it is required to find in every place the density of the medium with which a body will describe a given spiral. 686

Section 5: The density and compression of fluids, and hydrostatics 687

Definition of a Fluid. A fluid is any body whose parts yield to any force applied to it and yielding are moved easily with respect to one another. 687

Proposition 19, Theorem 14. All the parts of a homogeneous and motionless fluid that is enclosed in any motionless vessel and is compressed on all sides (apart from considerations of condensation, gravity, and all centripetal forces) are equally pressed on all sides and remain in their places without any motion arising from that pressure. 687

Cases 1–7 687

Corollary 688

Proposition 20, Theorem 15. If every part of a fluid that is spherical and homogeneous at equal distances from the center and rests upon a concentric spherical bottom gravitates toward the center of the whole, then the bottom will sustain the weight of a cylinder whose base is equal to the surface of the bottom and whose height is the same as that of the fluid resting upon it. 689

Corollaries 1–9 690

Proposition 21, Theorem 16. Let the density of a certain fluid be proportional to the compression, and let its parts be drawn downward by a centripetal force inversely proportional to their distances from the center; I say that if the distances are taken continually proportional, the densities of the fluid at these distances will also be continually proportional. 692

COROLLARY 693

Proposition 22, Theorem 17. Let the density of a certain fluid be proportional to the compression, and let its parts be drawn downward by a gravity inversely proportional to the squares of their distances from the center; I say that if the distances are taken in a harmonic progression, the densities of the fluid at these distances will be in a geometric progression. 694

COROLLARY 695

Scholium 696

Proposition 23, Theorem 18. If the density of a fluid composed of particles that are repelled from one another is as the compression, the centrifugal forces [or forces of repulsion] of the particles are inversely proportional to the distances between their centers. And conversely, particles that are repelled from one another by forces that are inversely proportional to the distances between their centers constitute an elastic fluid whose density is proportional to the compression. 697

Scholium 698

SECTION 6: Concerning the motion of simple pendulums and the resistance to them 700

Proposition 24, Theorem 19. In simple pendulums whose centers of oscillation are equally distant from the center of suspension, the quantities of matter are in a ratio compounded of the ratio of the weights and the squared ratio of the times of oscillation in a vacuum. 700

COROLLARIES 1–7 700

Proposition 25, Theorem 20. The bobs of simple pendulums that are resisted in any medium in the ratio of the moments of time, and those that move in a nonresisting medium of the same specific gravity, perform oscillations in a cycloid in the same time and describe proportional parts of arcs in the same time. 701

COROLLARY 703

Proposition 26, Theorem 21. If simple pendulums are resisted in the ratio of the velocities, their oscillations in a cycloid are isochronal. 703

Proposition 27, Theorem 22. If simple pendulums are resisted as the squares of the velocities, the differences between the times of the oscillations in a resisting medium and the times of the oscillations in a nonresisting medium of the same specific gravity will be very nearly proportional to the arcs described during the oscillations. 703

COROLLARIES 1–2 704

Proposition 28, Theorem 23. If a simple pendulum oscillating in a cycloid is resisted in the ratio of the moments of time, its resistance will be to the force of gravity as the excess of the arc described in the whole descent over the arc described in the subsequent ascent is to twice the length of the pendulum. 704

Proposition 29, Problem 6. Supposing that a body oscillating in a cycloid is resisted as the square of the velocity, it is required to find the resistance in each of the individual places. 705

COROLLARIES 1–3 708

Proposition 30, Theorem 24. If the straight line *a*B is equal to a cycloidal arc that is described by an oscillating body, and if at each of its individual points D perpendiculars DK are erected that are to the length of the pendulum as the resistance encountered by the body in corresponding points of the arc is to the force of gravity, then I say that the difference between the arc described in the whole descent and the arc described in the whole subsequent ascent multiplied by half the sum of those same arcs will be equal to the area BK*a* occupied by all the perpendiculars DK. 708

Corollary 710

Proposition 31, Theorem 25. If the resistance encountered by an oscillating body in each of the proportional parts of the arcs described is increased or decreased in a given ratio, the difference between the arc described in the descent and the arc described in the subsequent ascent will be increased or decreased in the same ratio. 711

Corollaries 1–5 712

General Scholium 713

Section 7: The motion of fluids and the resistance encountered by projectiles 724

Proposition 32, Theorem 26. Let two similar systems of bodies consist of an equal number of particles, and let each of the particles in one system be similar and proportional to the corresponding particle in the other system, and let the particles be similarly situated with respect to one another in the two systems and have a given ratio of density to one another. And let them begin to move similarly with respect to one another in proportional times (the particles that are in the one system with respect to the particles in that system, and the particles in the other with respect to those in the other). Then, if the particles that are in the same system do not touch one another except in instants of reflection and do not attract or repel one another except by accelerative forces that are inversely as the diameters of corresponding particles and directly as the squares of the velocities, I say that the particles of the systems will continue to move similarly with respect to one another in proportional times. 724

Corollaries 1–2 725

Proposition 33, Theorem 27. If the same suppositions are made, I say that the larger parts of the systems are resisted in a ratio compounded of the squared ratio of their velocities and the squared ratio of the diameters and the simple ratio of the density of the parts of the systems. 725

Corollaries 1–6 726

Proposition 34, Theorem 28. In a rare medium consisting of particles that are equal and arranged freely at equal distances from one another, let a sphere and a cylinder—described with equal diameters—move with equal velocity along the direction of the axis of the cylinder; then the resistance of the sphere will be half the resistance of the cylinder. 728

Scholium 729

Proposition 35, Problem 7. If a rare medium consists of minimally small equal particles that are at rest and arranged freely at equal distances from one another, it is required to find the resistance encountered by a sphere moving forward uniformly in this medium. 731

Cases 1–3 731

Corollaries 1–7 732

Scholium 733

Proposition 36, Problem 8. To determine the motion of water flowing out of a cylindrical vessel through a hole in the bottom. 733

Cases 1–6 734

Corollaries 1–10 738

Lemma 4. The resistance of a cylinder moving uniformly forward in the direction of its length is not changed by an increase or decrease in length and thus is the same as the resistance of a circle described with the same diameter and moving forward with the same velocity along a straight line perpendicular to its plane. 741

Proposition 37, Theorem 29. If a cylinder moves uniformly forward in a compressed, infinite, and nonelastic fluid in the direction of its own length, its resistance arising from the magnitude of its transverse section is to the force by which its whole motion can be either destroyed or generated, while it is describing four times its length, very nearly as the density of the medium is to the density of the cylinder. 742

COROLLARIES 1–3 744

Scholium 745

Lemma 5. If a cylinder, a sphere, and a spheroid, whose widths are equal, are placed successively in the middle of a cylindrical channel in such a way that their axes coincide with the axis of the channel, these bodies will equally impede the flow of water through the channel. 746

Lemma 6. With the same suppositions, these bodies are equally urged by the water flowing through the channel. 746

Lemma 7. If the water is at rest in the channel, and these bodies go through the channel with equal velocity in opposite directions, the resistances will be equal to one another. 746

Scholium 747

Proposition 38, Theorem 30. The resistance to a sphere moving uniformly forward in an infinite and nonelastic compressed fluid is to the force by which its whole motion could either be destroyed or generated, in the time in which it describes 8/3 of its diameter, very nearly as the density of the fluid to the density of the sphere. 747

COROLLARIES 1–4 748

Proposition 39, Theorem 31. The resistance to a sphere moving uniformly forward through a fluid enclosed and compressed in a cylindrical channel is to the force by which its whole motion could be either generated or destroyed, while it describes 8/3 of its diameter, in a ratio compounded of three ratios, very nearly: the ratio of the opening of the channel to the excess of this opening over half of a great circle of the sphere, the squared ratio of the opening of the channel to the excess of the opening over a great circle of the sphere, and the ratio of the density of the fluid to the density of the sphere. 748

Scholium 748

Proposition 40, Problem 9. To find from phenomena the resistance of a sphere moving forward in a compressed, very fluid medium. 749

Scholium 750

EXPERIMENTS 1–14 751

SECTION 8: Motion propagated through fluids 762

Proposition 41, Theorem 32. Pressure is not propagated through a fluid along straight lines, unless the particles of the fluid lie in a straight line. 762

COROLLARY 762

Proposition 42, Theorem 33. All motion propagated through a fluid diverges from a straight path into the motionless spaces. 763

CASES 1–3 763

Proposition 43, Theorem 34. Every vibrating body in an elastic medium will propagate the motion of the pulses straight ahead in every direction, but in a nonelastic medium will produce a circular motion. 765

CASES 1–2 765

COROLLARY 767

Proposition 44, Theorem 35. If water ascends and descends alternately in the vertical arms KL and MN of a tube, and if a pendulum is constructed whose length between the point of suspension and the center of oscillation is equal to half of the length of the water in the tube, then I say that the water will ascend and descend in the same times in which the pendulum oscillates. 767

COROLLARIES 1–3 768

Proposition 45, Theorem 36. The velocity of waves is as the square roots of the lengths. 768

Proposition 46, Problem 10. To find the velocity of waves. 768

COROLLARIES 1–2 769

Proposition 47, Theorem 37. If pulses are propagated through a fluid, the individual particles of the fluid, going and returning with a very short alternating motion, are always accelerated and retarded in accordance with the law of an oscillating pendulum. 769

COROLLARY 772

Proposition 48, Theorem 38. The velocities of pulses propagated in an elastic fluid are as the square root of the elastic force directly and the square root of the density inversely, provided that the elastic force of the fluid is proportional to its condensation. 772

CASES 1–3 772

Proposition 49, Problem 11. Given the density and elastic force of a medium, it is required to find the velocity of the pulses. 773

COROLLARIES 1–2 775

Proposition 50, Problem 12. To find the distances between pulses. 775

Scholium 776

SECTION 9: The circular motion of fluids 779

Hypothesis. The resistance that arises from the friction [*lit.* lack of lubricity or slipperiness] of the parts of a fluid is, other things being equal, proportional to the velocity with which the parts of the fluid are separated from one another. 779

Proposition 51, Theorem 39. If an infinitely long solid cylinder revolves with a uniform motion in a uniform and infinite fluid about an axis given in position, and if the fluid is made to revolve by only the impulse of the cylinder, and if each part of the fluid perseveres uniformly in its motion, then I say that the periodic times of the parts of the fluid are as their distances from the axis of the cylinder. 779

COROLLARIES 1–6 780

Proposition 52, Theorem 40. If a solid sphere revolves with a uniform motion in a uniform and infinite fluid about an axis given in position, and if the fluid is made to revolve by only the impulse of this sphere, and if each part of the fluid perseveres uniformly in its motion, then I say that the periodic times of the parts of the fluid will be as the squares of the distances from the center of the sphere. 781

CASES 1–3 781

COROLLARIES 1–11 783

Scholium 786

Proposition 53, Theorem 41. Bodies that are carried along in a vortex and return in the same orbit have the same density as the vortex and move according to the same law as the parts of the vortex with respect to velocity and direction. 788

COROLLARIES 1–2 789

Scholium 789

BOOK 3: THE SYSTEM OF THE WORLD 791

RULES FOR NATURAL PHILOSOPHY 794

Rule 1. No more causes of natural things should be admitted than are both true and sufficient to explain their phenomena. 794

Rule 2. Therefore, the causes assigned to natural effects of the same kind must be, so far as possible, the same. 795

Rule 3. Those qualities of bodies that cannot be intended and remitted [i.e., qualities that cannot be increased and diminished] and that belong to all bodies on which experiments can be made should be taken as qualities of all bodies universally. 795

Rule 4. In experimental philosophy, propositions gathered from phenomena by induction should be considered either exactly or very nearly true notwithstanding any contrary hypotheses, until yet other phenomena make such propositions either more exact or liable to exceptions. 796

PHENOMENA 797

Phenomenon 1. The circumjovial planets [or satellites of Jupiter], by radii drawn to the center of Jupiter, describe areas proportional to the times, and their periodic times—the fixed stars being at rest—are as the 3/2 powers of their distances from that center. 797

Phenomenon 2. The circumsaturnian planets [or satellites of Saturn], by radii drawn to the center of Saturn, describe areas proportional to the times, and their periodic times—the fixed stars being at rest—are as the 3/2 powers of their distances from that center. 798

Phenomenon 3. The orbits of the five primary planets—Mercury, Venus, Mars, Jupiter, and Saturn—encircle the sun. 799

Phenomenon 4. The periodic times of the five primary planets and of either the sun about the earth or the earth about the sun—the fixed stars being at rest—are as the 3/2 powers of their mean distances from the sun. 800

Phenomenon 5. The primary planets, by radii drawn to the earth, describe areas in no way proportional to the times but, by radii drawn to the sun, traverse areas proportional to the times. 801

Phenomenon 6. The moon, by a radius drawn to the center of the earth, describes areas proportional to the times. 801

PROPOSITIONS 802

Proposition 1, Theorem 1. The forces by which the circumjovial planets [or satellites of Jupiter] are continually drawn away from rectilinear motions and are maintained in their respective orbits are directed to the center of Jupiter and are inversely as the squares of the distances of their places from that center. 802

Proposition 2, Theorem 2. The forces by which the primary planets are continually drawn away from rectilinear motions and are maintained in their respective orbits are directed to the sun and are inversely as the squares of their distances from its center. 802

Proposition 3, Theorem 3. The force by which the moon is maintained in its orbit is directed toward the earth and is inversely as the square of the distance of its places from the center of the earth. 802

COROLLARY 803

Proposition 4, Theorem 4. The moon gravitates toward the earth and by the force of gravity is always drawn back from rectilinear motion and kept in its orbit. 803

Scholium 805

Proposition 5, Theorem 5. The circumjovial planets [or satellites of Jupiter] gravitate toward Jupiter, the circumsaturnian planets [or satellites of Saturn] gravitate toward Saturn, and the circumsolar [or primary] planets gravitate toward the sun, and by the force of their gravity they are always drawn back from rectilinear motions and kept in curvilinear orbits. 805

Corollaries 1–3 806

Scholium 806

Proposition 6, Theorem 6. All bodies gravitate toward each of the planets, and at any given distance from the center of any one planet the weight of any body whatever toward that planet is proportional to the quantity of matter which the body contains. 806

Corollaries 1–5 809

Proposition 7, Theorem 7. Gravity exists in all bodies universally and is proportional to the quantity of matter in each. 810

Corollaries 1–2 811

Proposition 8, Theorem 8. If two globes gravitate toward each other, and their matter is homogeneous on all sides in regions that are equally distant from their centers, then the weight of either globe toward the other will be inversely as the square of the distance between the centers. 811

Corollaries 1–4 812

Proposition 9, Theorem 9. In going inward from the surfaces of the planets, gravity decreases very nearly in the ratio of the distances from the center. 815

Proposition 10, Theorem 10. The motions of the planets can continue in the heavens for a very long time. 815

Hypothesis 1. The center of the system of the world is at rest. 816

Proposition 11, Theorem 11. The common center of gravity of the earth, the sun, and all the planets is at rest. 816

Proposition 12, Theorem 12. The sun is engaged in continual motion but never recedes far from the common center of gravity of all the planets. 816

Corollary 817

Proposition 13, Theorem 13. The planets move in ellipses that have a focus in the center of the sun, and by radii drawn to that center they describe areas proportional to the times. 817

Proposition 14, Theorem 14. The aphelia and nodes of the [planetary] orbits are at rest. 819

Corollaries 1–2 819

Scholium 819

Proposition 15, Problem 1. To find the principal diameters of the [planetary] orbits. 819

Proposition 16, Problem 2. To find the eccentricities and aphelia of the [planetary] orbits. 820

Proposition 17, Theorem 15. The daily motions of the planets are uniform, and the libration of the moon arises from its daily motion. 820

Proposition 18, Theorem 16. The axes of the planets are smaller than the diameters that are drawn perpendicularly to those axes. 821

Proposition 19, Problem 3. To find the proportion of a planet's axis to the diameters perpendicular to that axis. 821

Proposition 20, Problem 4. To find and compare with one another the weights of bodies in different regions of our earth. 826

Proposition 21, Theorem 17. The equinoctial points regress, and the earth's axis, by a nutation in every annual revolution, inclines twice toward the ecliptic and twice returns to its former position. 832
Proposition 22, Theorem 18. All the motions of the moon and all the inequalities in its motions follow from the principles that have been set forth. 832
Proposition 23, Problem 5. To derive the unequal motions [i.e., the inequalities in the motions] of the satellites of Jupiter and of Saturn from the motions of our moon. 834
Proposition 24, Theorem 19. The ebb and flow of the sea arise from the actions of the sun and moon. 835
Proposition 25, Problem 6. To find the forces of the sun that perturb the motions of the moon. 839
Proposition 26, Problem 7. To find the hourly increase of the area that the moon, by a radius drawn to the earth, describes in a circular orbit. 840
Proposition 27, Problem 8. From the hourly motion of the moon, to find its distance from the earth. 843
Corollaries 1–2 843
Proposition 28, Problem 9. To find the diameters of the orbit in which the moon would have to move, if there were no eccentricity. 844
Proposition 29, Problem 10. To find the variation of the moon. 846
Proposition 30, Problem 11. To find the hourly motion of the nodes of the moon in a circular orbit. 848
Corollaries 1–2 850
Proposition 31, Problem 12. To find the hourly motion of the nodes of the moon in an elliptical orbit. 852
Corollary 854
Proposition 32, Problem 13. To find the mean motion of the nodes of the moon. 856
Proposition 33, Problem 14. To find the true motion of the nodes of the moon. 859
Corollary 860
Scholium 860

[J. Machin] *On the Motion of the Nodes of the Moon 861*
Proposition 1 861
Proposition 2 863
Corollary 863
Scholium 864

Proposition 34, Problem 15. To find the hourly variation of the inclination of the lunar orbit to the plane of the ecliptic. 864
Corollaries 1–4 865
Proposition 35, Problem 16. To find the inclination of the moon's orbit to the plane of the ecliptic at a given time. 867
Scholium 869
Proposition 36, Problem 17. To find the force of the sun to move the sea. 874
Corollary 874
Proposition 37, Problem 18. To find the force of the moon to move the sea. 875
Corollaries 1–10 877
Proposition 38, Problem 19. To find the figure of the body of the moon. 880
Corollary 881

Lemma 1. Let APE*p* represent the earth, uniformly dense, with a center C and poles P and *p* and equator AE, and suppose a sphere P*ape* to be described with center C and radius CP. Let QR be the plane on which a straight line drawn from the center of the sun to the center of the earth stands perpendicularly. Then, if the individual particles of the whole exterior earth P*ap*AP*ep*E, which is higher than the sphere just described, endeavor to recede in both directions from the plane QR, and the endeavor of each particle is as its distance from the plane, I say, first of all, that the total force and efficacy of all the particles that lie in the circle of the equator AE (disposed uniformly outside the globe, in the manner of a ring completely encircling that globe) to rotate the earth around its center will be to the total force and efficacy of the same number of particles standing at point A of the equator (which is most distant from the plane QR) to move the earth with a similar circular motion around its center as 1 is to 2. And that circular motion will be performed around an axis lying in the common section of the equator and the plane QR. 881

Lemma 2. Under the same conditions, I say, secondly, that the total force and efficacy of all the particles situated everywhere outside the globe to rotate the earth around the given axis is to the total force of the same number of particles, disposed uniformly throughout all of the circle of the equator AE in the manner of a ring, to move the earth with a similar circular motion, as 2 is to 5. 883

Lemma 3. Under the same conditions, I say, thirdly, that the motion of the whole earth around the axis described above, a motion that is composed of the motions of all the particles, will be to the motion of the above-mentioned ring around the same axis in a ratio that is compounded of the ratio of the matter in the earth to the matter in the ring and the ratio of three times the square of the quadrantal arc of any circle to two times the square of the diameter—that is, in the ratio of the matter to the matter and of the number 925,275 to the number 1,000,000. 884

Hypothesis 2. If the ring discussed above were to be carried alone in the orbit of the earth about the sun with an annual motion (supposing that all the rest of the earth were removed from it), and if this ring revolved at the same time with a daily motion about its axis, inclined to the plane of the ecliptic at an angle of 23½ degrees, then the motion of the equinoctial points would be the same whether that ring were fluid or consisted of rigid and solid matter. 885

Proposition 39, Problem 20. To find the precession of the equinoxes. 885

Lemma 4. The comets are higher than the moon and move in the planetary regions. 888

Corollaries 1–3 894

Proposition 40, Theorem 20. Comets move in conics having their foci in the center of the sun, and by radii drawn to the sun, they describe areas proportional to the times. 895

Corollaries 1–4 895

Lemma 5. To find a parabolic curve that will pass through any number of given points. 896

Cases 1–2 897

Corollary 897

Lemma 6. From several observed places of a comet, to find its place at any given intermediate time. 898

Lemma 7. To draw a straight line BC through a given point P, so that the parts PB and PC of that line, cut off by two straight lines AB and AC, given in position, have a given ratio to each other. 898

Lemma 8. Let ABC be a parabola with focus S. Let the segment ABCI be cut off by the chord AC (which is bisected at I), let its diameter be Iμ, and let its vertex be μ. On Iμ

produced, take μO equal to half of Iμ. Join OS and produce it to ξ, so that Sξ is equal to 2SO. Then, if a comet B moves in the arc CBA, and if ξB is drawn cutting AC in E, I say that the point E will cut off from the chord AC the segment AE very nearly proportional to the time. 898

Corollary 899

Scholium 899

Lemma 9. The straight lines Iμ and μM and the length $\frac{\text{AIC}}{4\text{S}\mu}$ are equal to one another. 899

Lemma 10. Let Sμ be produced to N and P, so that μN is one-third of μI, and so that SP is to SN as SN to Sμ. Then, in the time in which a comet describes the arc AμC, it would—if it moved forward always with the velocity that it has at a height equal to SP—describe a length equal to the chord AC. 899

Corollary 900

Lemma 11. Suppose a comet, deprived of all motion, to be let fall from the height SN or Sμ + ⅓Iμ, so as to fall toward the sun, and suppose this comet to be urged toward the sun always by that force, uniformly continued, by which it is urged at the beginning. Then in half of the time in which the comet describes the arc AC in its orbit, it would—in this descent toward the sun—describe a space equal to the length Iμ. 900

Proposition 41, Problem 21. To determine the trajectory of a comet moving in a parabola, from three given observations. 901

Example 904

Proposition 42, Problem 22. To correct a comet's trajectory that has been found [by the method of prop. 41]. 929

Operations 1–3 929

General Scholium 939

Index of Names

The following index lists all references to names of persons in the translation of the *Principia* and in footnotes to the translation. Italic page references indicate names in tables. Readers are also reminded that a complete table of contents for the *Principia*, proposition by proposition, is given on pages 945–971. A detailed analytical table of contents of the Guide may be found on pages 3–7.

Aaron, 941n
Alfonso, King, 398
Anaxagoras, 941n
Ango, Reverend Father, 913, 914, 915
Apollonius, 478, 480
Aratus, 942n
Archimedes, 600
Aristotle, 385, 779n, 809, 928
Auzout, Adrien, 931, 933

Bayer, Johann, 905–906, 913, 917
Bentley, Richard, 398–399
Borelli, Giovanni Alfonso, *797*
Boulliau, Ismael, 800
Boyle, Robert, 387, 939
Bradley, James, 400, 935
Brougham, Henry Lord, 546n

Caesar, Julius, 911
Cajori, Florian, 441n, 647n, 944n
Cambridge, young man at, 913, 927
Cassini, Gian Domenico or Jean-Dominique, *797*, 798–799, 820, 821n, 822, 825, 832n, 912
Cassini, Jacques, 822
Cellio, Marco Antonio, 912–913, 915

Châtelet, Gabrielle du, 647n
Cicero, 941n
Colepress, Samuel, 875
Collins, John, 649
Copernicus, Nicolas, 570n, 789, 803
Cotes, Roger, 399, 484n, 513n, 722n, 939n
Couplet, Pierre, 829–830
Cysat, Johann Baptist, 893

David, 942n
Desaguliers, Jean-Théophile, 758
Descartes, René, 393, 484n, 625, 722n, 809
Des Hayes, Gérard Paul, 829, 830, 831

Euclid, 442, 466, 603

Feuillée, Father, 830
Flamsteed, John, 812n, 813n, 869n, 891, 894, 904, 908, 910, 917, 934, 935

Galileo Galilei, 394–395, 424, 460, 621, 622, 625, 657, 756
Gallet, Jean Charles, 912, 913, 915
Gemma, Cornelius, 937
Gottigniez, Gilles François, 933
Grimaldi, Francesco Maria, 625

Hall, A. Rupert, 944n
Hall, Marie Boas, 944n
Halley, Edmond, 380, 383, 400, 452, 829, 871, 904, 909, 910, 911, 913, 930, 931, 934, 935
Harris, John, 381n, 646–647n, 793n
Hevelius, Johannes, 892, 893, 926, 928, 930, 933
Hipparchus, 938
Hooke, Robert, 452, 914, 915, 931, 933
Horrocks, Jeremiah, 813n, 869n, 871
Hudde, Jan, 650
Huygens, Christiaan, 389, 424–425, 427, 452, 555, 799, 803, 804, 812n, 813, 822n

Innys, William and John, 944n

Jeremiah, 942n
John, 942n

Kepler, Johannes, 404n, 513n, 800, 805, 813n, 893–894, 924, 938
Kirk/Kirch, Gottfried, 400, 910

La Hire, Philippe de, 479, 830–831
Lampadius, 911
Leibniz, Gottfried Wilhelm, 649n

Machin, John, 860–864
Mariotte, Edme, 425
Matthew of Paris, 928
Mercator, Nicolaus, 820
Mersenne, Marin, 776–777n
Montanari, Geminiano, 913–914, 915, 927
Moses, 941n, 942n
Motte, Andrew, 441n, 647n, 878n, 943n, 944n

Norwood, Richard, 821–822, 832n

Orestes, 911

Pappus of Alexandria, 381, 484n
Paul, 942n
Pemberton, Henry, 400, 558n, 860, 944n
Petit, Pierre, 931, 933
Philo, 942n
Picard, Jean, 821n, 822, 830, 832n
Pocock, Edward, 941n
Ponteo, Giuseppe Dionigi, 911–914, 915, 917, 918
Pound, James, 400, 797–798, 825, 906, 908, 910, 935
Ptolemy, 803
Pythagoras, 941n

Riccioli, Giambattista, 813n
Richer, Jean, 829, 831
Roberval, Gilles Personne de, 776–777n
Routh, Edward John, 546n

Sauveur, Joseph, 778
Simeon the Monk of Durham, 928
Sluse, René-François de, 649
Snel, Willebrord, 625
Solomon, 942n
Stansel, Reverend Father Valentin, 927
Storer, Arthur, 913, 927
Street [Streete], Thomas, 803
Sturmy, Samuel, 875–876

Thales, 941n
Thorp, Robert, 441n
Townly [Towneley], Richard, 797
Tycho Brahe, 803, 937

Varin, 829
Vendelin, Bernard, 803, 813n
Viète, François, 478
Virgil, 941–942n

Wallis, John, 381n, 424–425
Whiteside, D. T., 559n
Wren, Sir Christopher, 424–425, 427, 452, 555

Zimmermann, Johann Jacob, 914

Designer: Barbara Jellow

Compositor: Technical Typesetting, Inc.

Text and Display: Granjon

Printer and Binder: Edwards Brothers, Inc.